MODERN INORGANIC SYNTHETIC CHEMISTRY

MODERN INORGANIC SYNTHETIC CHEMISTRY

Edited by

RUREN XU, WENQIN PANG, QISHENG HUO

ELSEVIER

AMSTERDAM • BOSTON • HEIDELBERG • LONDON • NEW YORK • OXFORD
PARIS • SAN DIEGO • SAN FRANCISCO • SINGAPORE • SYDNEY • TOKYO

Elsevier
Radarweg 29, PO Box 211, 1000 AE Amsterdam, The Netherlands
Linacre House, Jordan Hill, Oxford OX2 8DP, UK

Library of Congress Cataloging-in-Publication Data
A catalog record for this book is available from the Library of Congress

British Library Cataloguing-in-Publication Data
A catalogue record for this book is available from the British Library.

ISBN: 978-0-444-53599-3

For information on all Elsevier Publications
visit our Web site at elsevierdirect.com

Printed and bound in the United Kingdom

Transferred to Digital Print 2011

Contents

Foreword

I have great pleasure in writing the Foreword for this fine book on Modern Inorganic Synthetic Chemistry written and edited by Professors Xu, Pang, and Huo. The book consisting of 24 chapters covers a variety of aspects of present-day synthetic inorganic chemistry. Inorganic chemistry itself has undergone great change in the last two decades or so wherein it has absorbed various aspects of materials science as well as chemical biology. The uniqueness of this present book is that it covers most of the recent developments in inorganic synthetic chemistry.

It is refreshing to see that the book has chapters on high-temperature as well as low-temperature synthesis, hydrothermal and solvothermal synthesis, high-pressure synthesis, and photochemical synthesis. The book covers preparation of a large variety of inorganic materials under different conditions which include the use of microwaves. The book has a chapter on coordination compounds and coordination polymers as well as cluster compounds. Organometallic chemistry has not been ignored and is suitably covered along with inorganic polymers. Special mention must be made of the fine coverage on porous materials, host–guest materials, ceramics, and amorphous materials as well as nanomaterials and membranes. In addition, there is also a chapter on crystal growth. Inorganic synthesis related to biomimetic synthesis constitutes a chapter. The book concludes with a discussion of synthetic design. All in all, I consider this book to be a very valuable reference book for students, teachers, and practitioners of modern inorganic chemistry in the broadest sense, and materials chemistry in particular. I have great pleasure in recommending this book to the chemical community. I wish the book great success.

C.N.R. Rao
National Research Professor and Linus Pauling Research Professor, Jawaharlal Nehru Centre for Advanced Scientific Research, Bangalore, India

Preface

In 1998, Stephen J. Lippard of MIT made an insightful remark about the future development of chemistry over the next 25-years: "What is most important about chemistry is that we make new things. We don't just study the natural world; we make new molecules, new catalysts, and new compounds of uncommon reactivity. Part of our subject allows us to be creatively artistic through the synthesis of beautiful and symmetric molecules. Our ability to rearrange atoms in new ways allows us a tremendous opportunity for creation that other sciences don't have." (Chemistry's Golden Age in ***C&EN News***, Jan 12, 1998) Ryoji Noyori, a Nobel laureate in 2001, further emphasized in his feature article "Synthesizing our future" (***Nature Chemistry*** 2009, April) that "Chemistry has a central role in science, and synthesis has a central role in chemistry. To begin, the central place of synthesis in chemistry is emphasized and extended to chemistry place in science." Indeed, synthetic chemistry is undoubtedly the core of chemistry, and it provides the most powerful tool for chemists to shape the world and thus the future of our society. Chemists have not only discovered and synthesized a large number of substances that exist in nature but also created many new compounds, phases, and states of matter. According to the latest information from the CAS, over 50 million organic and inorganic compounds have been registered. Many of these have become indispensable in our daily life and in industrial productions, providing a continuous driving force for the further advances of science and technology. With the advent of the 21st century and the rapid development of new technologies, there is greater demand for synthetic chemists to create increasingly more compounds and materials with novel structures and functions. In the mean-time, greater attentions have been made to the basic studies in areas relevant to green synthetic routes, biomimetic syntheses, inorganic syntheses under extreme conditions, molecular engineering, and rational tectonics, gearing toward syntheses of new matters in more efficient, focused, and economical manners. These are important for the sustainable and rapid advancement in both science and technology in the new century.

Examples of the role played by synthetic chemistry in driving industrial revolution and development of new sciences are numerous. In the early 20th century, Fritz Haber used nitrogen from the air and hydrogen from water gas as raw materials to produce ammonia, the main ingredient of fertilizers, which greatly helped to accelerate the growth of food production to feed the world's rapidly increasing population. Since the mid-20th century, the successful synthesis of a large number of new drugs has significantly curbed infectious diseases and improved the health conditions of mankind. The creation of three major synthetic materials, namely the synthetic fiber, plastic, and synthetic rubber, has formed an important basis for modern industrial and agricultural developments around the world. These are all powerful evidences of the contribution of chemistry, especially synthetic chemistry, in improving the overall well-being of mankind. Advances in synthetic chemistry have led to continuous creation and development of new compounds and materials and provided the basis for studying their structures, functions, and reactions. This has become a major driving force for the progress of chemistry and related disciplines. In fact, what lies beneath the most significant technological advancements of the 20th century, such as information technology, nuclear technology, laser, nanotechnology, and aerospace technology, is the chemical synthesis of new and advanced materials. Without the efforts of generations of synthetic chemists and their achievements in synthesizing new matter and materials, these advancements would not have been possible. Thus, the related scientific areas such as semiconductor, superconductor, cluster, and nanoscience could not have come into being.

As a crucial part of synthetic chemistry, modern inorganic synthesis has become a key branch of inorganic chemistry. Its scope encompasses both traditional synthesis and the new science of material preparation and assembly. With the synthesis and preparation of a large number of new compounds, phases, and materials each year, inorganic synthesis has quickly become a key area in the recent development of inorganic chemistry and related disciplines. The vibrant growth of emerging sciences and technologies in the 21st century has placed increased demands on the development of new inorganic materials. In addition, these new materials have fueled the developments of new industries and scientific disciplines.

With the development of synthetic chemistry, special synthesis techniques, structural chemistry, and theoretical chemistry, as well as their interplays with related disciplines such as life science, materials science, and computer science, the science of inorganic synthesis has expanded from conventional synthesis to synthesis under extreme conditions, rationally designed synthesis, and biomimetic synthesis focused on inorganic materials with specifically designed structures and functions. In view of its broad scope of research and its close relationship with other disciplines, we believe that it is necessary to present to our readers the science of inorganic synthesis chemistry in the context of modern synthetic and preparative chemistry in a systematic manner. Currently, there are a number of books and monographs focused on this discipline, but generally speaking, they tend to be manuals or specialized works. This prompts us to have devoted considerable effort to the writing of this book by experts in relevant fields to provide both a broad coverage and an in-depth discussion of the synthesis and preparative inorganic chemistry. We name this book ***Modern Inorganic Synthetic Chemistry***.

The 24 chapters of the book are grouped into four parts. The first part consists of Chapters 2-8. The discussion centers on inorganic synthesis and preparation routes under specific conditions, dealing with inorganic synthesis and preparative chemistry under specific conditions such as high temperature, low temperature and cryogenic condition, hydrothermal and solvothermal condition, high and superhigh pressure, photochemical, microwave irradiation, and plasma conditions. The second part consists of Chapters 9-14, focused on the synthesis, preparation, and assembly of six important categories of compounds, with wide coverage of distinct synthetic chemistry systems, namely coordination compounds, coordination polymers, clusters, organometallic compounds, nonstoichiometric compounds, and inorganic polymers. The third part consists of Chapters 15-22. Seven important representative inorganic materials are selected for discussion of their chemistry of preparation and assembly, including porous materials, advanced ceramic, amorphous and nanomaterials, inorganic membrane, and assembly of two types of advanced functional materials. The last part, unique to this book, is composed of two chapters that bring the reader to the frontiers of inorganic synthesis and preparative chemistry, with Chapter 23 on biomimetic synthesis and Chapter 24 on rationally designed synthesis. With the inclusion of these two chapters, we aim to introduce to the reader the two emerging areas in synthetic chemistry. Also worth mentioning is that this book has a collection of more than 3000 references, the majority of which are from the last decade. It is hoped that these references provide a comprehensive list of the relevant literature on the recent development and new directions in the field of modern synthetic inorganic chemistry.

To provide a comprehensive and an in-depth coverage of this rapidly evolving field, particularly about the frontiers of related research, we invited a group of scientists, who work at the forefront of their respective research areas, to write chapters in their areas of expertise. In this book, we have tried to present to the reader both mature achievements and recent developments in the field when making the high-level design of the book. We do recognize that we might not necessarily have the best balance between the two even with our best efforts. We also recognize that there could be some overlaps among different chapters written by different authors. We tried our best to minimize this issue, but overlaps at some level might still exist in the book. We sincerely hope that the reader can understand and would be willing to provide us feedbacks and comments.

This book not only highlights the core science, frontiers, and possible directions of different branches of inorganic synthesis chemistry but also touches on structural chemistry and functional features of synthesized materials. Therefore, it can serve as a textbook as well as a key reference book for university senior undergraduate students and graduate students in chemistry, chemical engineering, and materials science. It can also be used as a specialized reference book for researchers and technicians who work in synthetic chemistry and related fields. We, the editors, would like to take this opportunity to thank all the authors whose hard work made possible the completion of this book in a timely fashion. We also want to thank Professor Raj Pal Sharma of Panjab University for his careful work in fixing grammar throughout this book. Special thanks go to colleagues and students in the State Key Laboratory of Inorganic Synthesis and Preparative Chemistry at Jilin University, who did almost all the preparation work for this book, including paperwork, figure design, reference checks, and much more.

Ruren Xu, Professor
State Key Laboratory of Inorganic Synthesis and Preparative Chemistry Jilin University, Changchun, China

Contributors

Julia K.C. Abbott Department of Chemistry, The University of Tennessee, Knoxville, Tennessee 37996, USA

Jiesheng Chen School of Chemistry and Chemical Engineering, Shanghai Jiao Tong University, 800 Dongchuan Road, Shanghai 200240, China

Xiao-Ming Chen MOE Key Laboratory of Bioinorganic and Synthetic Chemistry School of Chemistry & Chemical Engineering Sun Yat-Sen University Guangzhou 510275, China

Shun-Liu Deng Department of Chemistry, College of Chemistry and Chemical Engineering, Xiamen University, Xiamen 361005, China

Brenda A. Dougan Department of Chemistry, The University of Tennessee, Knoxville, Tennessee 37996, USA

Xue Duan State Key Laboratory of Chemical Resource Engineering, Box 98, Beijing University of Chemical Technology, Beijing 100029, P. R. China

David G. Evans State Key Laboratory of Chemical Resource Engineering, Box 98, Beijing University of Chemical Technology, Beijing 100029, P. R. China

Shouhua Feng State Key Laboratory of Inorganic Synthesis & Preparative Chemistry, College of Chemistry, Jilin University, 2699 Qianjin Street, Changchun 130012, China

Jingkun Guo State Key Laboratory of High Performance Ceramics and Superfine Microstructures, Shanghai Institute of Ceramics, Chinese Academy of Sciences, 1295 Dingxi Road, Shanghai 200050, China

Guangyan Hong State Key Laboratory of Rare Earth Resources Utilization, 5625 Renmin Street, Changchun 130022, China

Zhuang-Qi Hu Institute of Metal Research, Chinese Academy of Sciences, 72 Wenhua Road, Shenyang 110016, China

Xiaobin Huang School of Chemistry and Chemical Engineering, Shanghai Jiao Tong University, 800 Dongchuan Road, Shanghai 200240, China

Qisheng Huo State Key Laboratory of Inorganic Synthesis & Preparative Chemistry, College of Chemistry, Jilin University, 2699 Qianjin Street, Changchun 130012, China

Lei Jiang Research Center for Biomimetic Smart Science and Technology, College of Chemistry and Environment, Beijing University of Aeronautics and Astronautics, Beijing 100191, China, Beijing National Laboratory for Molecular Sciences (BNLMS), Key Laboratory of Organic Solids, Institute of Chemistry, Chinese Academy of Sciences, Beijing 100190, China

Huamin Kou State Key Laboratory of High Performance Ceramics and Superfine Microstructures, Shanghai Institute of Ceramics, Chinese Academy of Sciences, 1295 Dingxi Road, Shanghai 200050, China

Guanghua Li State Key Laboratory of Inorganic Synthesis & Preparative Chemistry, College of Chemistry, Jilin University, 2699 Qianjin Street, Changchun 130012, China

Jiang Li State Key Laboratory of High Performance Ceramics and Superfine Microstructures, Shanghai Institute of Ceramics, Chinese Academy of Sciences, 1295 Dingxi Road, Shanghai 200050, China

Jerry Y.S. Lin Chemical Engineering School for Engineering of Matter, Transport and Energy, Arizona State University, Tempe, AZ 85260, USA

Xiaoyang Liu State Key Laboratory of Inorganic Synthesis & Preparative Chemistry, College of Chemistry, Jilin University, 2699 Qianjin Street, Changchun 130012, China

Xinsheng Liu BASF Corporation, R&D Center, 25 Middlesex-Essex Turnpike, Iselin, NJ 08830, USA

Jian Liu ARC Centre of Excellence for Functional Nanomaterials, Australian Institute for Bioengineering and Nanotechnology, The University of Queensland, Brisbane, QLD 4072, Australia

Kesong Liu Research Center for Biomimetic Smart Science and Technology, College of Chemistry and Environment, Beijing University of Aeronautics and Astronautics, Beijing 100191, China

Jun Lu State Key Laboratory of Chemical Resource Engineering, Box 98, Beijing University of Chemical Technology, Beijing 100029, P. R. China

Changsheng Lu State Key Laboratory of Coordination Chemistry, School of Chemistry and Chemical Engineering, Nanjing University, 22 Hankou Road, Nanjing 210093, China

Qingjin Meng State Key Laboratory of Coordination Chemistry, School of Chemistry and Chemical Engineering, Nanjing University, 22 Hankou Road, Nanjing 210093, China

Wenqin Pang State Key Laboratory of Inorganic Synthesis & Preparative Chemistry, College of Chemistry, Jilin University, 2699 Qianjin Street, Changchun 130012, China

Shi Zhang Qiao ARC Centre of Excellence for Functional Nanomaterials, Australian Institute for Bioengineering and Nanotechnology, The University of Queensland, Brisbane, QLD 4072, Australia

Gao Qing (Max) Lu ARC Centre of Excellence for Functional Nanomaterials, Australian Institute for Bioengineering and Nanotechnology, The University of Queensland, Brisbane, QLD 4072, Australia

Shriya K. Seshadri Chemical Engineering School for Engineering of Matter, Transport and Energy, Arizona State University, Tempe, AZ 85260, USA

Qiang Su School of Chemistry and Chemical Engineering, Sun Yat-sen University, Guangzhou 510275, China

Yuan-Zhi Tan Department of Chemistry, College of Chemistry and Chemical Engineering, Xiamen University, Xiamen 361005, China

Xiaozhen Tang School of Chemistry and Chemical Engineering, Shanghai Jiao Tong University, 800 Dongchuan Road, Shanghai 200240, China

Ji-Tao Wang Department of Microelectronics, Fudan University, Shanghai 200433, China

Su-Yuan Xie Department of Chemistry, College of Chemistry and Chemical Engineering, Xiamen University, Xiamen 361005, China

Ruren Xu State Key Laboratory of Inorganic Synthesis & Preparative Chemistry, College of Chemistry, Jilin University, 2699 Qianjin Street, Changchun 130012, China

Zi-Ling Xue Department of Chemistry, The University of Tennessee, Knoxville, Tennessee 37996, USA

Wenfu Yan State Key Laboratory of Inorganic Synthesis & Preparative Chemistry, College of Chemistry, Jilin University, 2699 Qianjin Street, Changchun 130012, China

Guo-Yu Yang State Key Laboratory of Structural Chemistry, Fujian Institute of Research on the Structure of Matter, Chinese Academy of Sciences, 155 West Yangqiao Road, Fuzhou 350002, China

Jihong Yu State Key Laboratory of Inorganic Synthesis & Preparative Chemistry, College of Chemistry, Jilin University, 2699 Qianjin Street, Changchun 130012, China

Jilin Zhang State Key Laboratory of Rare Earth Resources Utilization, 5625 Renmin Street, Changchun 130022, China

Jingping Zhao State Key Laboratory of Inorganic Synthesis & Preparative Chemistry, College of Chemistry, Jilin University, 2699 Qianjin Street, Changchun 130012, China

CHAPTER

1

Introduction - Frontiers in Modern Inorganic Synthetic Chemistry

Ruren Xu

Jilin University, China

Synthetic chemistry is at the core of modern chemistry; it provides the most powerful means for chemists to create the material foundation for our envisioned world. Its main objective is to create a large variety of compounds, phases, materials, and ordered chemical systems needed by our rapidly advancing society, going considerably beyond just finding and synthesizing naturally existing compounds. According to recently published studies, over 50 million compounds, naturally existing or not, have been discovered or synthesized, some of which have become indispensable to our daily life. These compounds have provided the basis for many scientific and technological advances in the recent history. In turn, these advances have created rapidly increasing needs for new materials with specific structures and functions, posing challenges to, as well as creating opportunities for synthetic chemists. Specifically, we see increasing needs in this new century, for novel synthesis strategies and techniques, as well as for the related scientific understanding, gearing toward green synthesis, biomimetic synthesis, inorganic synthesis under extreme conditions, and molecular and tectonic engineering of inorganic materials, in efficient, rationally designed and economic manners. We believe that these are among the most essential key elements for the continuing and rapid advancement of science and technology in this new century [1,2].

In the past century, advances in synthetic chemistry have often been the key driving force for the industrial revolutions and birth of new science and technologies; examples of this sort have been numerous [2]. For instance, F. Haber, in the early twentieth century, invented a high-pressure technique to synthesize ammonia, the key ingredient of chemical fertilizers, from the abundantly available H_2 and N_2 using osmium as the catalyst. Twenty years later, C. Bosch improved the technique by using inexpensive iron instead of expensive osmium as the catalyst, which laid a solid foundation for the human society to maintain a continued increase in food production to keep up with the human population increase; a major challenge that we have been facing since the past century. Because of their profound contributions to science as well as to the human society, Haber and Bosch received Nobel prizes in chemistry in 1918 and 1931, respectively. Health industry is another area where synthetic chemistry has been playing pivotal roles. Outstanding examples since the mid-twentieth century include the successful syntheses of SAS drug, penicillin, a variety of antibiotics and other medicines, which have substantially improved and continue to improve our overall abilities in treating human diseases and fighting against them. Our ability in producing the three major classes of synthetic materials, namely synthetic fiber, synthetic plastic, and synthetic rubber, has paved the way for many of the recent industrial and agricultural advances. There is no doubt that chemistry, especially synthetic chemistry, has been making considerable contributions to improve the living conditions of the human society.

From a scientific perspective, a pool of very large number of new materials created by synthetic chemistry has provided plenty of samples for studying the structure—function (property) relationships of materials as well as their syntheses, facilitating scientists to study the fundamental chemistry of these materials, which has become a driving force in the recent developments of chemistry and related sciences. For example, the successful preparation of single crystalline silicon and

Modern Inorganic Synthetic Chemistry, DOI: 10.1016/B978-0-444-53599-3.10001-0

numerous semiconductive materials has fueled the emergence of information technology; the production and posttreatment of nuclear fuel of uranium and plutonium, the key to the nuclear technology and safe application, have all been built on chemical technologies with roots in synthetic chemistry. Similar can be said about other high technologies such as laser, nanotech, aviation, and space technology. Without a doubt, the so-called *six great technology inventions* in the twentieth century would have never materialized without the foundational work by generations of synthetic chemists in the past. The same is true about other technological breakthroughs and growth points in related sciences such as semiconductor, super conduction, cluster, and nanotechnology.

Modern inorganic synthetic chemistry, an important branch of synthetic chemistry, has evolved considerably from the traditional synthesis and preparation of inorganic compounds, which now includes the synthesis, assembly, and preparation of supramolecular and high-level ordered structures in its studies. In recent years, we have been witnessing that an increasingly large number of new inorganic compounds, phases, and complex materials are being synthesized and assembled, having made inorganic synthetic chemistry a key driver for many new scientific and technological developments and advancements. We anticipate that inorganic synthetic chemistry will continue to play equally or more important roles in science as well as in our upcoming life.

1.1. DEVELOPMENT OF NEW SYNTHETIC REACTIONS, SYNTHETIC ROUTES, TECHNOLOGIES AND ASSOCIATED BASIC SCIENTIFIC STUDIES

1.1.1. The Basic Inorganic Compounds

This basic class includes covalently bonded molecular compounds, coordination compounds, cluster compounds, metal organic compounds, nonstoichiometric compounds and inorganic polymer, among others.

1.1.2. Inorganics and Materials with Specific Structures

Study of inorganic compounds and phases with specific structures is becoming increasingly important as the need for materials with specific properties and functions continues to rise. It is well accepted that the properties and functions of materials are determined by their structures and compositions. More specifically, such properties and functions are often determined by the characteristics of high-level molecular structures such as those of molecular aggregates, ordered molecular assemblies, and structures in condensed states instead of single molecular structures. Take defects for example, the properties and functions of materials often result from various forms of structural defects in their component compounds or phases in condensed state. A key reason that many complex oxides are being used as popular substrates for functional materials is that they can form many types of structural defects in addition to their many adjustable component elements. Hence, it has become a major topic at the forefront of inorganic chemistry research to study the preparation of solid-state matters with specific structural defects and the associated principles as well as related detection techniques. In addition, the key research topics in today's inorganic chemistry also include preparation of surfaces and interfaces with specific structures and properties, stacking of layered compounds, preparation of specific polytypes and their intergrowths as well as intercalation structures and low-dimensional structures of inorganic compounds, synthesis and preparation of inorganic compounds with mixed valence complexes and clustered compounds with specific structures, as well as the rapidly emerging and increasingly useful porous compounds with specific channel structures such as microporous crystals, meso- and hierarchical porous materials. Also particularly interesting is the preparation of phases that tend to form distinct structures and are able to form large varieties of distinct structures under extreme synthetic conditions like high or ultrahigh pressures. While a few synthesis examples with the aforementioned characteristics have been reported in the literature, such studies have generally been done in rather ad hoc manners, often accomplished through utilizing the particularity of specific reactions or specific synthesis techniques rather than based on new understanding of a general class of synthesis problems and new synthesis technologies. The latter is clearly more important for the future development of synthetic chemistry.

1.1.3. Inorganics and Materials in Special Aggregate States

Another important class of materials are the compounds in special aggregate state, such as in nano state, ultrafine particles, clusters, noncrystalline state, glass state, ceramic, single crystal, and other matters with varying crystalline morphologies such as whisker and fiber. The rapid emergence of nanoscience and technology strongly suggests that different aggregate states of the same matter could exhibit different properties and have different functions. The understanding of this could have substantial implications to the future development of science as well as new functional materials.

1.1.4. Assembly of High-level Ordered Structures

There is an emerging class of functional inorganic materials, commonly characterized as being highly ordered supramolecular systems, formed via self-assembly among molecules or molecular aggregates through molecular recognition. The key interaction forces in the formation of such large molecular assemblies are intermolecular non- or weak-bond interactions (van der Waals and hydrogen bond). Examples of such materials include coordination polymers, inorganic polymers, and molecular systems with specific structural features such as nanosystems, capsula, ultrathin membrane (monolayer membrane, multilayer composited membrane), interfaces, two-dimensional layered structures, and three-dimensional biological systems; many of which have been widely used for fabricating high-tech microdevices. Self-assembly is increasingly becoming a key and practical technique in the synthesis and preparation of complex functional systems. It has even been suggested that the introduction of self-assembly-based synthesis techniques could fundamentally advance the chemical production processes that are being widely used in the current industries [2].

1.1.5. Composition, Assembly, and Hybridization of Inorganic Functional Materials

The following areas have received considerable attention in recent years: (1) multi-phase composition of materials including enhanced or reinforced fiber- (or whisker-)based materials, the second-phase particle dispersion materials, two- or multi-phase composite materials, inorganic and organic materials, inorganics and metals, and functional gradient materials as well as nanomaterials; and (2) composite material-related host–guest chemistry, which represents a highly interesting and a very challenging research area. The research focuses include, for example, the assembly of different types of chemical entities in hosts with microporous or mesoporous frameworks such as quantum dot or super lattice-forming semiconductive clusters, nonlinear optical molecules, molecular conductors made of linear conductive polymers and electron transfer chains as well as D–A transfer pairs. All these complex composites could be assembled through synthetic routes consisting of ion exchanges, CVDs, "ships in bottle" and microwave dispersion; (3) nanohybridization of inorganics and organics, which represents a rapidly emerging interdisciplinary field. It studies the formation of new hybrid materials through combining polymerization and sol–gel processes. These hybrid materials possess those properties which are generally absent in pure inorganics or pure organics, and are increasingly being used in fiber optics, wave propagation, and nonlinear materials. It is worth noting that the first survey about this emerging field was published in 1996 by P. Judeinstein [3].

As outlined above, a key task in today's inorganic synthetic chemistry is to develop novel synthetic reactions, synthetic routes, and associated techniques aiming to create new functional materials with specifically desired multilevel structures in condensed states. As per the past experience, the discovery of a novel and effective synthetic route or technique has typically led to the creation of a large class of new matters and materials. For example, the advent of sol–gel synthetic route has been a key reason for the development and emergence of nano-states and nanocomposite materials, glass states and glass composites, ceramic and ceramic-based composites, fibers and related composites, inorganic membranes and composite membranes, and hybrid materials. The core chemistry of this synthetic route is hydrolysis and polymerization of starting reactant molecules (or ions) in aqueous solution, i.e., from molecular → polymeric state → sol → gel → crystalline state (or noncrystalline state). This synthetic process could possibly be regulated differently at each individual reaction step so as to create solid-state compounds or materials with different structures or in different aggregate states. While highly promising, we are clearly not there yet due to the complexity as well as our limited understanding of polymerization processes of inorganic molecules in both theoretical and experimental executions. Thus, fundamental studies of these issues represent key areas of focus in today's inorganic synthetic chemistry.

In summary, the near and intermediate-term objectives for today's inorganic synthetic chemists are to develop novel and more effective synthetic technologies and to carry out related theoretical studies aiming to gain better understanding of the desired new synthesis capabilities which are both economical and environment-friendly.

1.2. BASIC RESEARCH IN SUPPORT OF GREEN SYNTHESIS

The vast majority of known synthetic reactions, especially those used in the preparation of a large variety of rare elements from their ores or raw materials, in the production of fine chemicals as well as in medical and pharmaceutical industries, produce large amounts of by-products, which, along with the used chemicals, solvents, additives, and catalysts, often add major pollutants to our environment and have created considerable environmental issues in the past. Thus, it has become absolutely essential to study ways to considerably lower or completely remove environmental pollution produced by the current chemical industry. While this has posed substantial challenges for synthetic chemists, it has also created new opportunities to further develop synthetic chemistry toward new and healthier directions. Green chemistry, clean technologies, and environment-friendly chemical processes have now become a common conviction of many chemists. *Ideal synthesis*, a concept proposed by Wender [4] in 1996, aims to "make complex molecules from simple starting materials in a manner that is operationally simple, fast, safe, environmentally acceptable and resource efficacious." This definition has essentially defined the general direction for realizing green syntheses. In 2009, Noyori [5] proposed that we should aim at synthesizing target compounds with a 100% yield and 100% selectivity and avoid the production of waste. This process must be economical, safe, resource-efficient, energy efficient, and environmentally benign. In this regard, the atom economy and the *E*-factor should be taken into account. The 3Rs (reduction, recycling, and reuse) of resources are particularly important. Such "Green Chemistry" is creative and brings about prosperity. The following research directions have received considerable attention in the recent years, from many synthetic chemists [6]: development and applications of green synthetic reactions with efficient atomic economy, environmental friendliness, and energy efficiency; development and application of environment-friendly source materials, reaction media and solvents, additives and catalysts and highly efficient and selective synthetic reactions as well as associated theoretical studies. These have become the major focuses at the forefront of synthetic chemistry research.

1.3. BASIC RESEARCH ON SYNTHETIC AND PREPARATIVE ROUTES UNDER EXTREME CONDITIONS

There have been many cases of successfully synthesizing materials under extreme conditions such as ultrahigh pressure, high temperature, high vacuum, ultralow temperature, strong magnetic and electric fields, laser and plasma which are not possible to be synthesized under normal experimental conditions. A large variety of new compounds, phases and materials as well as new synthetic routes and techniques have been synthesized and developed specifically for chemical syntheses. For example, ultrapure crystals with no dislocation defects can be synthesized in ultrahigh vacuum with zero-gravity. It has even been suggested that the Periodic Table of Elements may need to be significantly modified under ultrahigh pressure since the width of the forbidden band and the distance between the internal and the external electronic orbits for many matters may be changed under such conditions which can lead to significant differences in the stable valances of an element under the normal versus ultrahigh pressures. It has also been observed that changes in reactivity and reaction rules of reactants under ultrahigh pressure have led to the formation of a variety of new species and more interestingly, of new phases. It is also worth noticing that compounds with specific valence, configuration, and crystal morphology can be formed under hydrothermal conditions with medium temperature and pressure, which helps to overcome the issue caused by the lack of successful synthesis routes in solid-states chemistry for many inorganic functional materials under high temperature. Hence, further studies of the general rules and principles of chemical synthesis under extreme conditions have become one of the major research frontiers in synthetic chemistry.

1.4. BIOMIMETIC SYNTHESIS AND APPLICATIONS OF BIOTECHNOLOGY IN INORGANIC SYNTHESIS

Biomimetic synthesis typically refers to syntheses that mimic biological synthesis processed in living organisms. An ultimate goal is to develop synthesis techniques and processes that can lead to the creation of new materials with similar or better/improved properties of naturally existing biological materials or to synthesize new materials with specifically desired properties using naturally existing materials. As a rapidly emerging research field, biomimetic synthesis has attracted great interest of researchers from a number of fields and is being considered a new frontier in synthetic chemistry in the twenty-first century. An interesting observation has been that some of the highly complex synthesis processes using traditional approaches become easy and efficient through biomimetic synthesis. Here we use "biominerals" and "biomineralization" as examples to illustrate some basic ideas of biomimetic synthesis. Various biomineralized materials have been formed as parts of living organisms as a result of genetic mutations and selection by evolution such as bones, teeth, pearls, shells, diatoms, and spider silk. The formation of such special tissues, though by accidents, has given special advantages to the relevant organisms and hence has been kept (selected) during evolution. The inorganic components in these special tissues such as calcium carbonate, calcium phosphates, calcium oxalate, metal sulfates, amorphous silica, iron oxide, and iron sulfide are generally called biominerals.

Biomineralization refers to the formation process of biominerals inside living organisms. The process typically involves a sequence of chemical reactions leading to the formation of new tissues mostly made of inorganic phases. The fundamental difference between biomineralization and mineralization in general is that in biomineralization the precipitation of inorganic mineral phases is accomplished through interactions between bio-macromolecules and inorganic ions at the interface between cells and body fluids which are controlled at the molecular level. Because of the unique formation process, biominerals often have special multilevel structures distinct from inorganic structures existing outside living organisms. They tend to have specific characteristics of crystals with highly uniform sizes, clear structure and composition boundaries, highly ordered spatial arrangements, complex morphologies, and well-defined crystal orientations and tend to have clearly defined multilevel structures. In a nutshell, biomineralization is a controlled precipitation and deposition process of biominerals with highly ordered, regular and multilevel structures as the final products. The biomineralization process inside a living organism generally consists of four intertwined and interactive steps: supramolecular preorganization, interfacial molecular recognition, vectorial regulation, and cellular regulation and processing.

A key characteristic of biomineralization is the nucleation and growth of inorganic minerals around supramolecular templates in a highly regulated manner. During the biomineralization process, the morphology, size, orientation, and structure of the biominerals are controlled in a sophisticated manner by organic components such as bio-macromolecules involved in the process. Understanding the mechanisms of biomineralization can be useful to guide biomimetic syntheses of new functional materials at multi-scales ranging from the meso- to macro-scale. This is rapidly becoming one of the important research directions in material chemistry as well as in inorganic synthetic chemistry. The highly interesting and unique properties of biomineralized materials, such as (a) lotus leaves and insect wings with self-cleaning properties, (b) cameo shells with specially high strength, toughness, and abrasion resistance, (c) rat's tooth enamel, (d) spider silk with superb strength and elasticity, and (e) iron oxides located inside fish heads serving as natural compasses, are all results of different structural characteristics of the self-assembled biominerals at multiple scales. Fueled by these observations, a new branch of chemistry, *biomimetic material chemistry*, is being formed and is rapidly growing with the key aims of elucidating relationships between functions and coordination effects among the multilevel structures of biomineralized materials, to design desired multilevel structures, to apply learned mechanisms of biomineralization to the synthesis of inorganic materials, and to synthesize materials with specific multilevel structures and desired properties. In addition, more and more attention is being paid to the development of new techniques that directly mimic biochemical processes in inorganic syntheses, preparation, and assemblies. For example, a number of synthesis methods such as widely used enzymatic catalysis, microorganism-mediated (such as virus and bacteria) synthetic reactions, and template effect used in synthesis and assembly of inorganic functional materials are all inspired by biological synthesis processes.

Another example is the emergence of combinatorial synthesis technique, which is regarded as a major breakthrough in the recent history of synthesis techniques. By organizing a large number of polypeptides in an array as catalysts, combinatorial synthesis allows rapid syntheses of astonishingly many new compounds within a short period of time. Such techniques have significantly shortened the screening time, for example, for potentially new drugs and new pesticides. As a result, combinatorial syntheses are being extensively used in the preparation and hydrothermal synthesis of inorganic materials.

1.5. RATIONAL SYNTHESIS AND MOLECULAR ENGINEERING OF INORGANIC COMPOUNDS WITH SPECIFIC STRUCTURES AND FUNCTIONS

There have been some cases of new material synthesis through molecular design and engineering in recent years. Traditionally, creation of new compounds with desired properties typically involves syntheses of a large number of compounds and a selection process for the desired compounds from these synthesized compounds. Since 1950s, the number of synthesized compounds has increased from 2 million to more than 50 million, which has formed a large and highly useful compound library.

The emerging field of molecular engineering takes a rather different approach to chemical synthesis. The basic idea is that it starts with desired functionalities of a to-be-synthesized material, designs the possible structures of the material based on the specified functionalities, and then creates the material through rational synthesis. The biggest impact of the emergence of molecular engineering on chemistry is that it has greatly broadened our view about the relationships among the functionalities, structures, and synthesis processes, allowing us to better appreciate and understand the relationships between functions and high-level structures beyond single molecular structures.

While this field is still in its nascent stage, it is already believed that this is the future direction of synthetic chemistry. Researchers have already started synthesizing new materials based on the general principles of molecular engineering in a number of selected fields. Among these studies, molecular design and rational synthesis of microporous crystal systems represent one of the relatively mature research areas.

Microporous crystals generally have specific and regular channel structures [7]. The chemical interactions between the guest molecules and the channels and the framework tend to be considerably stronger than those of other porous materials, and hence the structural features and properties of the channels of such materials, such as the pore size, shape, dimension, orientation, composition of the channel walls, cavities, cages, and structural defects, generally have significantly stronger effects on the diffusion, adsorption, and desorption, the formation of intermediates and the selectivity of molecular reactions inside the channels than those for other porous materials. Thus, microporous crystals represent the most unique system, and could potentially become one of the largest classes of catalysts and adsorption−separation materials. Microporous crystals as well as other porous materials such as mesoporous, macroporous, and porous metal-organic framework (MOF) materials are being increasingly used in emerging high technologies, showing great potentials in the development of new materials in the future. As of now, 194 framework types of microporous crystals and much more types of inorganic open-framework materials, have been synthesized in laboratories. Over the years, extensive studies about these structures have been done on the structural features, the framework structures, and their effects on the movement and reactivity of the molecules inside their channels, the rules and regularities of the pore-creating reactions, crystallization, and modifications of the channels, windows, and internal surfaces. Therefore, it is reasonable to select the microporous crystals as a case study in molecular engineering. While substantial work has been done in this area, it should be noted that only a small number of true success stories have been reported as of now.

The design and rational synthesis of microporous zeolites need to be done based on thorough study of the relationships between the specified functions of the to-be-synthesized material and channel structures. Initial channel models of the desired crystal could be done with the help of computer programs. Subsequently ideal structural models will be selected according to the established relationship between properties and structures of the microporous crystals derived from known structures and functions of such crystals in relevant databases. Finally, a rational synthesis plan of these ideal structures will be made based on the relationship between the structures and the synthesis conditions. But it is generally not possible to achieve true rational syntheses like those done in organic synthesis via analyzing reaction paths and steps, because their formation mechanism remains elusive and relationship between the synthetic parameters and structural characters remains unclear. Despite the difficulties associated with the rational synthesis, considerable efforts have been made to establish ways toward the rational design and synthesis of target zeolitic materials. Our group has built up a ZEOBANK that includes a database of zeolite synthesis and a database of zeolite structures with the aim to explore a novel way to guide the synthesis of zeolitic materials through data mining.

Engineering the synthesis of new matters with desired structures and functions has attracted considerable attention in the areas of chemistry and material science. In Chapter 24 of this book, we will describe our efforts toward the rational design and synthesis of zeolitic inorganic open-framework materials.

Currently, our group as well as several other research groups has been actively carrying out studies in the following areas: (1) method development for rational design of structures, (2) development and update of the ZEOBANK synthesis and structure database and synthetic approach guided by data mining for microporous compounds, (3) in-depth study of the formation mechanisms of microporous compounds and the structure-directing effect via experiments and computational simulation, (4) derivation of potential synthesis mechanisms as well as empirical relationship between synthesis conditions and resulting structures derived based on known synthesis data and computer simulation results, which could be used to guide rational structure design and directed synthesis of desired materials, (5) performing combinatorial synthesis for microporous compounds with specific structures and properties, and (6) structural modification, fine-tuning of the chemical properties of channels, windows and internal surfaces, and rational addition of specific active sites such as ions, metal particles, oxides or salts, complex ions, and clusters into specific channels or onto the internal surfaces based on the desired functions and properties of the microporous material.

I would like to end this chapter with the words of Ryoji Noyori, the winner of Nobel Prize in 2001, in his feature article "Synthesizing our future" [5] that "Synthesis has a central role in chemistry; chemical synthesis has now reached an extraordinary level of sophistication, but there is vast room for improvement; and chemical synthesis must pursue 'practical elegance' that is, it must be logically elegant but must at the same time lead to practical application."

References

[1] M.B. Rudy, Chemistry's golden age, C&EN News (1998). Jan 12.
[2] Committee on Challenges for the Chemical Sciences in the 21st Century, Beyond the Molecular Frontier: Challenges for Chemistry and Chemical Engineering, National Academy of Sciences, 2003.
[3] P. Judeinstein, C. Sanchez, J. Mater. Chem. 6 (1996) 511–525.
[4] P.A. Wender, Chem. Rev. 96 (1996) 1–2.
[5] R. Noyori, Nat. Chem. 1 (2009) 5–6.
[6] J.H. Clark, Nat. Chem. 1 (2009) 12–13.
[7] R.R. Xu, W.Q. Pang, J.H. Yu, Q.S. Huo, J.S. Chen, Chemistry of Zeolites and Related Porous Materials: Synthesis and Structure, John Wiley & Sons, Singapore, 2007.

CHAPTER

2

High-temperature Synthesis

Ruren Xu, Qiang Su

Jilin University, China, Sun Yat-sen University, China

High temperature is an important means of inorganic synthesis, particularly of new high-temperature phases, compounds, and materials. Though it is not the case for all operations, yet some processes do require very high temperature, for instance, the preparation of high purity melts, arc casting and material densification, sintering of high-melting metal powders, ceramic specimens, and crystal growth of refractory compounds. In some cases, high-temperature facilities are capable of operation in dry inert, reducing atmosphere and in high vacuum. Various facilities and operation techniques are required to attain high temperature. High-temperature laboratory furnaces are suitable for a wide variety of laboratory and small-scale syntheses and fabrication applications. They are simple to use, easy to install, and reliable.

2.1. ATTAINMENT OF HIGH TEMPERATURE-LABORATORY FURNACES [1] AND RELATED TECHNIQUES [2–6]

2.1.1. Resistance Furnaces

Resistance furnaces are the most common heating furnaces in the laboratory. Their advantages include simple construction, convenience in operation, and accurate temperature control. Different resistance materials can achieve different temperature limits. The maximum working temperature, operating environment, and temperature sensors are given in Table 2.1.

Furnace Construction

Based on model, the furnace shell is either heavy walled, seamless extruded 6061 T6 aluminum with an anodized interior and exterior or stainless steel (water-jacketed). The bulkheads are hard-anodized aluminum or nickel-plated copper (based on model) and the doors are nickel-plated copper. The shell and bulkheads have integral water-cooling passages. External surface temperatures are maintained at 65°C or less. O-ring seals are made of Viton.

The furnace has four $1^1/_8 - 6$ radial and two axial threaded ports. One radial port is equipped with a 16-mm diameter viewing port with an anti-fog gas diffuser. All other ports are plugged but will allow the addition of optional sight windows, thermocouples, pyrometers, feedthroughs, and valved adapters.

Power connections are at one end of the furnace through radially mounted, water-cooled copper feedthroughs, providing unimpeded access to the bulkheads and doors. Electrical connection between the feedthroughs and the element is made by simple internal clamp connections.

The hot zone has straight-through access for convenient loading from either end. Sight hearths in both ends of the furnace serve as thermal baffles and in the vertical position will support the work. A graphite tray is optionally available for attachment to either hearth to support work with the furnace in a horizontal position.

2.1.2. Crystal Grower Equipment

The broad range of crystal grower equipment makes it suitable for all popular crystal growing methods. A partial list includes the following:

Czochralski Technique: Barium sodium niobate, bismuth germanium oxide, calcium molybdate, lithium fluoride, lithium germanium oxide, lithium niobate, lithium tantalite, magnesium fluoride, potassium chloride, strontium barium niobate, Nd: YAG, sapphire, and silicon.

Modern Inorganic Synthetic Chemistry, DOI: 10.1016/B978-0-444-53599-3.10002-2

TABLE 2.1 Features of Some Resistance Furnaces in Laboratory

Resistance element	Max. working temperature (°C)	Operating atmosphere	Temp. sensors	note
Nichrome wire	1100	Air	Thermocouple	
Globar	1400	Air	Thermocouple	
Platinum wire	1400	Air	Thermocouple	
Mo (mesh, sheet)	2000–2500	Vac., inert and reducing atmosphere*	W5% vs W26% thermocouple (2000°C) optical pyrometer (3000°C)	**
Pyrometer (3000°C) W, Mo wire	1700	Ibid	thermocouple (2000°C)	
$ThO_2$95%, $La_2O_3$5%	1950	Air	Thermocouple (2000°C)	
Ta wire	2000	Vacuum		
High-density graphite	2500–3000	Vac, inert and reducing atm.	Type C thermocouple < 2000°C, optical pyrometers, <3000°C	***
Carbon tube	2500	Reducing atm.		

* *Gas purification system (Thermal technology's) purifies inert gas with a closed-loop temperature controlled, Ti getter furnace. Nitrogen can be purified simply by replacing the Ti charge tube with an Mn charge tube (the level of purity is equivalent to a vacuum level of 10^{-9} Torr, Oxygen content < 1 ppb).*
** *Model 1100 (Thermal Technology's).*
*** *Model 1000 (Thermal Technology's).*

Bridgman Technique: Calcium fluoride, cadmium fluoride, cadmium telluride, gallium arsenide, magnesium fluoride, silver gallium selenide, and silver gallium sulfide.
Kyropoulos: Sapphire and titanium-doped sapphire.

The two most important factors for critical growth are stability and control. Both factors are necessary for achievement of consistency, repeatability, and uniformity. Stability provides a constant environment responsible for crystal growth. It assures uniform, tightly defined temperatures, and thermal gradients for consistent melts and zone refining. Well-controlled gaseous or vacuum environments are required to achieve stability. All must be controlled by the computer system.

2.1.3. Arc Melting Furnace

Arc Melting Furnace is effective for powder melting, arc casting, metallic and nonmetallic buttons, annealing, compound synthesis, and material densification. It is easy to operate, dependable, versatile, and affordable while offering instant high temperature and extremely pure melts. It accepts a variety of hearth configurations and can accommodate a titanium getter.

For the preparation of high purity melts: Because of its simplicity, ease of access into the ARC melting furnace and small volume, the furnace is easily purged. Therefore, work specimens are quickly prepared and a high degree of purity is maintained. Arc melting furnace consists of a water-cooled stainless steel bell jar hinged to a fixed base plate. The electrode (stinger) shaft penetrates down the top of the bell jar. The stinger is sealed at the top of the bell jar by means of a ball joint and stainless steel bellows. The hearth is water-cooled copper with interchangeable top surface. Various cavity configurations can be provided in the hearth top surface. Light and window ports ease inside observation. Clamps are provided for operation at a slightly positive pressure. Pressure compensation for "zero feel" is incorporated into the seal design. Provision is made for attachment to a vacuum pump for evacuation prior to back-filling with inert gas. Also included are ports and gauges for inert gas.

2.1.4. Spark Plasma Sintering

As a revolutionary high-speed powder consolidation technology, Spark Plasma Sintering (SPS) utilizes high amperage DC pulse current to rapidly and evenly disperse spark plasma energy between particles. An SPS system can process conductive, nonconductive, and composite materials to any level of density (including full density) with high homogeneity and especially strong bonds between particles.

The SPS plasma theory is based on the electrical spark discharge phenomenon wherein a high-energy pulse current momentarily generates spark plasma at high localized temperatures (up to 10,000°C) between particles. The spark energy vaporizes contaminants and oxidizes the surface of the particles prior to neck formation. Joule heat is concentrated on particle surfaces producing plastic deformation which aids in high-density applications. Spark Plasma Sintering (SPS) is a pressure-assisted process utilizing up to 300 tons of force. The atmosphere employed is vacuum with the ability to use inert gas. Heat feedback is provided via thermocouples or pyrometer.

2.2. TYPES OF HIGH-TEMPERATURE SYNTHETIC REACTIONS AND ROUTES

Some important types of high-temperature reactions and synthetic routes are outlined as follows:

1. High temperature solid-state synthetic reactions

   The most common routes for synthesizing a wide variety of inorganic multicomponent solid materials such as advanced functional complex oxides and nonoxide ceramics are through direct reactions among the solid components at high temperatures.

   This is because such solids are generally difficult to react with each other at a low temperature, except in limited cases such as solid-state coordination compounds, cluster compounds, and inorganic molecular solids [7–10]—even thermodynamics favors product formation. High temperatures are necessary to achieve appreciable reaction rates and production at industrial scales.

   The most important includes [11]:

   (1) Ceramic method
   (2) Carbothermal reduction
   (3) Sintering and materials densification

2. High-temperature solid–gas state synthetic reactions

   The most important of them includes:

   (1) Thermoreduction

   Almost all metals and some nonmetals are prepared by the reduction of their compounds (such as halides, oxides, and sulfides) with H_2, CO, etc. and alkali, alkali-earth metals vapor (metalepsis) as reductants at high temperatures.

   (2) Oxidation, Chlorination at high temperatures
   (3) Nitridation at high temperatures

   For example, high-temperature direct nitridation for the preparation of important ceramic Si_3N_4 powders:

   i. $$3Si(s) + 2N_2(g) \xrightarrow{1250-1350\ ^\circ C} Si_3N_4(s)$$

   ii. $$3SiO_2(s) + 6C(s) + 2N_2(g) \xrightarrow{1550\ ^\circ C} Si_3N_4(s) + 6CO(g)$$

   iii. $$3Si(s) + 4NH_3(N_2)(g) \xrightarrow{plasma} Si_3N_4(s) + 6H_2(g)$$

   (4) Intercalation at high temperature

   The insertion of guest species (gas or vapors) into a crystalline host lattice (three-dimensional with channel systems or low-dimensional, i.e., layered lattices or chain structures consisting of one-dimensional stacks) results in the modification of the properties and gives rise to materials with new functions.

3. High-temperature combustion synthesis reactions

   The combustion synthesis uses highly exothermic reactions. The reactions are initiated by an external source, and sufficient heat is released to render the reactions self-sustaining.

   The combustion synthesis can be conducted in two modes:

   (1) Self-propagation mode, referred to as the "self-propagating high-temperature synthesis (SHS)" approach;
   (2) Simultaneous combustion mode, also referred to as the "thermal explosion" approach;

4. High-temperature electro-reduction synthesis in molten system:

   (1) Molten salt electrolysis;
   (2) Electro-reduction synthesis of compounds of middle or unusually low valency;

5. High-temperature "phase-transformation" synthetic reactions;
6. High-temperature smelting and alloy preparation;
7. Glass formation and glass melting at high temperature;
8. Crystal growth and zone refining at high temperature; and
9. Ultra high-temperature synthetic reactions.

The first four and the ninth type of the above reactions will be discussed in detail in this chapter, whereas the other types will be dealt with in other chapters, due to particularities of the contents. High-temperature synthesis is closely related to chemical thermodynamics, especially high-temperature thermodynamic behavior and reaction kinetics, particularly that of high-temperature solid-state reactions. Therefore, the general aspects of solid-state reactions as well as some general principles along with relevant data are presented in the following discussions on synthetic reactions.

2.3. HIGH-TEMPERATURE SOLID-STATE REACTION [11–14]

This is an important type of synthetic reaction. Many inorganic functional materials and compounds of special properties and functions are obtained directly from high-temperature (generally 1000–1500°C) solid reactions. This type of reaction also has outstanding features. An example will be given on $MgO(s) + Al_2O_3(s) \rightarrow MgAl_2O_4$ (spinal type) to illustrate in detail the mechanism and characters of a solid-state synthetic reaction.

2.3.1. Mechanism and Characters of Solid-state Reaction

From the thermodynamics point of view, the reaction $MgO(s) + Al_2O_3 \rightarrow MgAl_2O_4(s)$ can go to completion. However, it can hardly proceed at temperatures <1200°C, and requires several days to proceed at 1500°C. High-temperature requirement for the reaction can be primarily explained with Fig. 2.1.

Under certain conditions of high temperature, the reaction can proceed at the crystal boundary of MgO and Al_2O_3 crystallites and form a product layer of spinal type $MgAl_2O_4$. The first stage of this reaction is to form nuclei of $MgAl_2O_4$ crystallites at the crystal lattice of the reactants, or adjacent to, their boundary. The nucleation reaction is not easy since the nuclei are from different reactants and are different in structure. Therefore, nucleation needs to go through structural rearrangement, including breaking the cation–anion bonds of reactant molecules, releasing, diffusing, and repositioning of Mg^{2+} and Al^{3+} ions in the crystal lattice. A high temperature favors all these processes and thus nucleus formation occurs. Similarly, the growth of nucleus is also quite difficult. Mg^{2+} and Al^{3+} ions in the reactants need to diffuse through two boundaries (Fig. 2.1) to grow on the nucleus to thicken the product layer. It is obvious that the controlling step of the reaction is the diffusion of Mg^{2+} and Al^{3+} ions, which is favored by an increase in temperature, and so is the reaction. On the other hand, the reaction rate decreases with the thickening of the product layer. The solid reaction forming another spinal type $NiAl_2O_4$ has confirmed that the diffusion of Ni^{2+} and Al^{3+} cations in the product layer is the controlling step of the reaction. According to the general principle, the following relationship exists:

$$\frac{dx}{dt} = kx^{-1} \tag{2.1}$$

$$x = (k't)^{1/2} \tag{2.2}$$

Where x is the thickness of $NiAl_2O_4$ product layer; t is the reaction time; and k and k' are the reaction rate constants.

Similarly in the case of $MgAl_2O_4$, a linear relationship between x^2 and t can be found as shown in Fig. 2.2. The reaction rate can be obtained from the slope of the line, and the activation energy of the reaction can be obtained by plotting the $lnk'-T^{-1}$ curve.

From the above analysis and existing experimental results, the mechanism of $MgAl_2O_4$ formation can be expressed with the following two equations (a) and (b) (Fig. 2.1(b)):

(a) at the $MgO/MgAl_2O_4$ boundary

$$2Al^{3+} - 3Mg^{2+} + 4MgO \rightarrow \ = MgAl_2O_4$$

(b) at the $MgAl_2O_4/Al_2O_3$ boundary

$$3Mg^{2+} - 2Al^{3+} + 4Al_2O_3 \rightarrow 3MgAl_2O_4$$

The overall reaction is:

$$4MgO + 4Al_2O_3 \rightarrow 4MgAl_2O_4$$

From the above reactions at boundaries, the amount of product from reaction (b) is three times that from reaction (a). As illustrated in Fig. 2.2, the thickening rate of the right layer is three times that of the left. This has been confirmed experimentally.

In summary, the following three factors influence the rate of this type of reaction: (a) the surface and

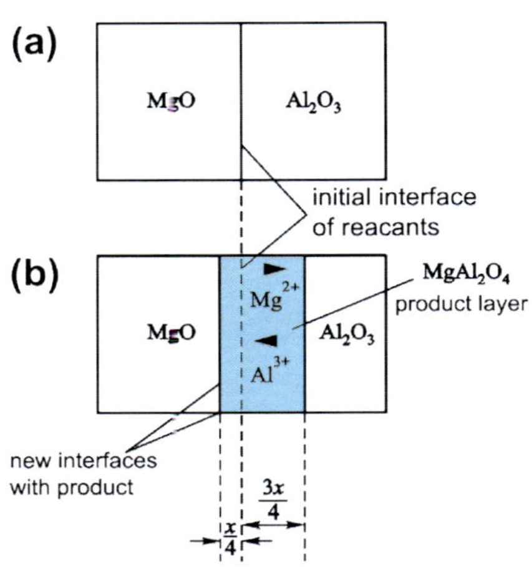

FIGURE 2.1 Schematic diagram of reaction mechanism. *Anthony R. West, Solid State Chemistry and Its Applications. Page 6. 1992. Copyright Wiley-VCH Verlag GmbH & Co. KGaA. Reproduced with permission.*

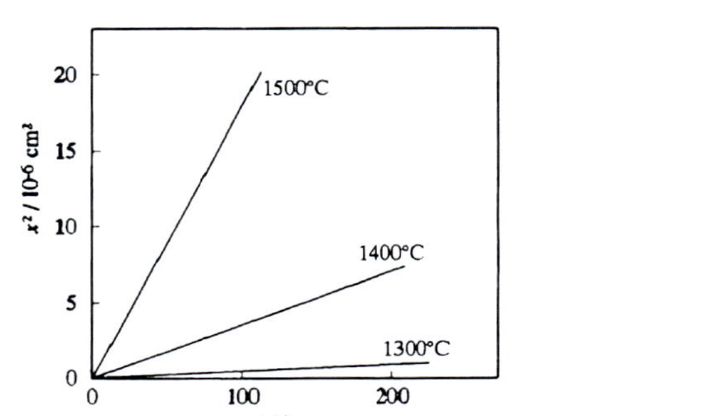

FIGURE 2.2 x^2–t Relationship for $MgAl_2O_4$ at various temperatures. *Anthony R. West, Solid State Chemistry and Its Applications. Page 6. 1992. Copyright Wiley-VCH Verlag GmbH & Co. KGaA. Reproduced with permission.*

contacting areas of the reactants; (b) the nucleation rate of the product; and (c) the ionic diffusion rate at the phase boundaries and particularly through the product layer.

From the above knowledge of the mechanism and characters of this type of solid reaction, it is clear that reaction between two solids may not occur even if thermodynamic considerations favor product formation, diffusion rates of ions through the various phases. An understanding of this will enable us to control and develop high-temperature solid synthetic reactions.

2.3.2. Some Aspects of Synthesis Via Solid-state Reaction

1. Surface area and contacting area of reactants

The surface area as well as the contacting area of reactants can be increased by prolonged crushing milling and pressing, or by various chemical routes to prepare fine particle reactants that have high specific surface area and high surface activity. The methods of enhancing solid reaction include making uniform contact by palletizing, hot pressing, or making reactant precursors through coprecipitation or other types of reactions. Some examples are given below.

Example 1: preparation of precursors for synthesizing spinal $ZnFe_2O_4$. The starting materials $Fe_2[(COO)_2]_3$ and $Zn(COO)_2$ in 1:1 ratio are dissolved in water and thoroughly agitated. The solution is heated to evaporate the water solvent, and $Fe_2[(COO)_2]_3$ and $Zn(COO)_2$ coprecipitate slowly to an oxalate mixture. This solid solution has almost uniform distribution of Fe^{3+} and Zn^{2+} ions. After filtering and calcining, a fine precursor for solid reaction is obtained. This precursor can be used to synthesize $ZnFe_2O_4$ at a much lower temperature (about 1000°C) lower than an ordinary reaction. The overall reaction is:

$$Fe_2[(COO)_2]_3 + Zn(COO)_2 \rightarrow ZnFe_2O_4 + 4CO\uparrow + 4CO_2\uparrow$$

2. Reactivity of solid precursors [15]

When the precursor and product are similar in structure, the nucleation is easier due to the convenience in structural rearrangement. For instance, in the above reaction, the oxygen ions in MgO and spinal $MgAl_2O_4$ are similar in structural arrangement. It is easy to form and grow $MgAl_2O_4$ nuclei in the lattice or in the vicinity of the MgO boundary via a topotactic or epitactic reaction. In addition, the precursor reactivity is also related closely to the material source, preparation history, state of matter, and structure particularly. Precursors are generally in the form of polycrystalline powders.

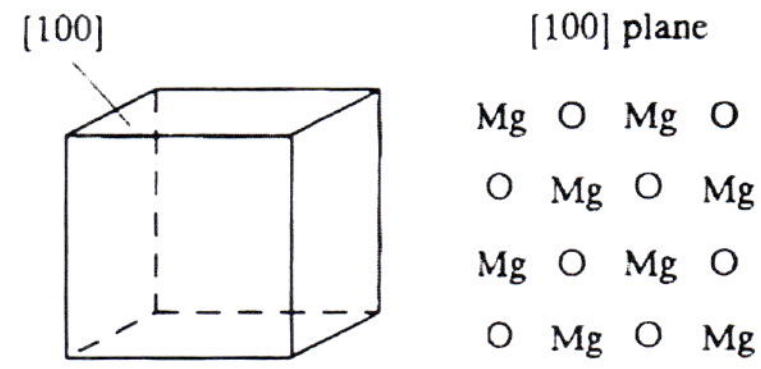

FIGURE 2.3 Idealized MgO crystal structure and arrangement of Mg^{2+} and O^{2-} ions at [100] plane. *Anthony R. West, Solid State Chemistry and Its Applications. Page 11. 1992. Copyright Wiley-VCH Verlag GmbH & Co. KGaA. Reproduced with permission.*

Due to the deficiency of crystal structure, reactivity at the surface of different parts can be different. For example, the structure of ideal MgO crystals belongs to the cubic NaCl type, where Mg^{2+} and O^{2-} ions are alternatively arranged at the [100] plane (Fig. 2.3). For a deficient polycrystalline MgO such as the form shown in Fig. 2.4, the crystal surface has [100] and [111] planes simultaneously. The [111] plane may consist of only Mg^{2+} or O^{2-} ions as illustrated in Fig. 2.4. It is obvious that different planes of the crystal are of different reactivity. The solid reactivity relating to the existence of any defects in crystals will not be discussed here in further detail due to limited space. Increasing the reactivity of a precursor through the preparation route, condition, and material source is beneficial and enhances a solid-state reaction. For example, high reactivity precursors with fine particle size or high specific surface area, of amorphous or metastable phase, can be prepared before, or at the same time with, a solid-state reaction. Newly formed precursors through precipitation, decomposition, oxidation–reduction, or phase transformation appear to have very high reactivity due to the instability of their structure. In the above solid reaction forming $MgAl_2O_4$, MgO, and α-Al_2O_3 can be replaced, respectively, with $MgCO_3$ and newly precipitated Al $(OH)_3$. These materials decompose and form in situ MgO and Al_2O_3 at 600–900°C, and the solid reaction is thus enhanced.

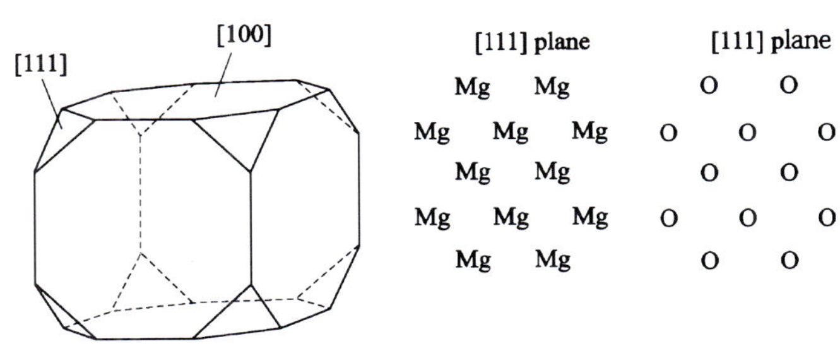

FIGURE 2.4 [111] Plane of MgO structure. *Anthony R. West, Solid State Chemistry and Its Applications. Page 11. 1992. Copyright Wiley-VCH Verlag GmbH & Co. KGaA. Reproduced with permission.*

3. Property of product from solid-state reaction

Being heterogeneous, a solid-state reaction advances mainly at the phase boundaries. The controlling step of the reaction—ionic diffusion through the phases—is influenced by many undefined factors. Therefore, the composition and structure of the product from this type of reaction are usually nonstoichiometric and heterogeneous. Taking the $MgO-Al_2O_3$ system again as an example, below approximately 1500°C, the reaction product is an $MgAl_2O_4-Mg_{0.75}Al_{2.18}O_4$ solid solution. In other words, the spinal product at this primary stage of solid reaction has a variable composition of a certain range. The spinal crystals formed at the $MgO/MgAl_2O_4$ boundaries are Mg-rich $MgAl_2O_4$; whereas at the $MgAl_2O_4/Al_2O_3$ boundaries are Mg poor-$Mg_{0.75}Al_{2.18}O_4$. This results in structural as well as compositional heterogeneity. If the reaction is allowed to continue, it is difficult to obtain a stoichiometry of 1:3 even after a prolonged period of time. This phenomenon exists in almost all high-temperature solid-state reactions.

In the above discussions, it has been mentioned many times that ionic diffusion in lattices and through phases is an important factor influencing a solid-state reaction, or even the controlling step for reaction of this type. However, it can be seen from the above discussions that it is very difficult to access, in further detail, the principle of ionic diffusion. This is due to the structural characters of reactants (usually unstable, or metastable polycrystalline, or amorphous solids) and products. Further studies in this area are thus required.

2.4. PREPARATION OF RARE EARTH-CONTAINING MATERIALS

Rare earths are a group of lithophile elements with the tendency to combine with oxygen. Therefore, rare earth compounds containing oxygen such as oxides and complex oxides are easily formed when they are synthesized in air. Rare earth halides, chalcogenides, pnictides, and their oxy-compounds are also presented in this section.

2.4.1. Oxides and Complex Oxides

The melting point of rare earth oxides RE_2O_3 is very high, ranging from 2593 K for La_2O_3 to 2713 K for Y_2O_3 and 2763 K for Lu_2O_3[16]. Their single crystals have to grow at very high temperature.

Most rare earth complex oxides are also prepared by high-temperature solid-state reaction methods. The melting point of pure rare earth oxides can be lowered by adding other components such as Al_2O_3 in the Y_2O_3 (or Lu_2O_3)$-Al_2O_3$ binary system. The congruent melting point of $Y_3Al_5O_{12}$ (YAG) is 2203 K [16] and it can be grown at this temperature by Czochralski pulling method (CZ) with iridium crucible and under high purity N_2 or Ar atmosphere in radio frequency (RF) induction furnace.

By using this method, Ce^{3+}-doped $Lu_3Al_5O_{12}$ or YAG single crystal (LuAG:Ce^{3+} and YAG:Ce^{3+}) has been grown as shown in Fig. 2.5. LuAG:Ce^{3+} is a scintillation crystal [17] used for positron emission tomography (PET) and YAG:Ce^{3+} thin crystal plate can be used as cover on top of InGaN chip to prepare white light emitting diode (WLED) for lighting [18].

For preparation of polycrystalline YAG:Ce^{3+} which is widely used recently as phosphor for WLED, the synthesis temperature can be lowered further. For example, if solid-state (SS) reaction method is used for preparation, Y_2O_3, CeO_2, and $Al(OH)_3$ are used as starting materials and CO as reducing atmosphere to transform Ce^{4+} to Ce^{3+}. The synthesis temperature is at about 1773 K [19]. This solid-state reaction method is to be compared below with several soft chemical methods such as (1) coprecipitation method with ammonium bicarbonate as precipitant (CP); (2) sol−gel (SG) method with citric acid; and (3) combustion (CB) method with urea as fuel. The synthesis temperature can be lowered still further to 1273 K.

Because organic reagents are used in the SG and CB methods, when the organic reagents are decomposed

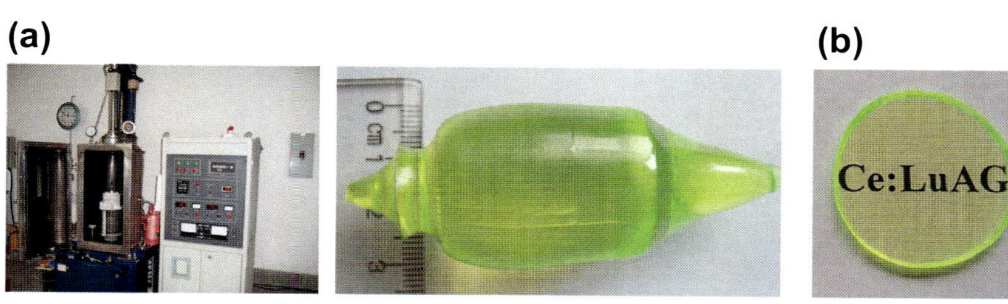

FIGURE 2.5 LuAG:Ce^{3+} single crystal grown along the <111> direction by the Cz method. (a) The as-grown crystal boule, (b) the polished crystal wafer with the size of 1.0 mm in thickness [17]. *Reprinted from Journal of Rare Earths, 25 (2007), Zhong Jiuping, Liang Hongbin, Su Qiang, Zhang Guobin, Dorenbos Pieter, Birowosuto Muhammad Danang, Effects of Annealing Treatments on Luminescence and Scintillation Properties of Ce:Lu$_3$ Al$_5$ O$_{12}$ Crystal Grown by Czochralski Method, page 568, Copyright (2007), with permission from Elsevier.*

at the synthesis temperature, reducing atmosphere is produced itself, so calcinations can be carried out in air by using these two methods. Among these methods, although the synthesis temperatures of SG and CB are the lowest, the emission intensity of Ce^{3+} prepared by the SS method at higher temperature is the strongest. Increasing the sintering temperature could promote the luminescence intensity, because it improves the crystallization of YAG particles.

2.4.1.1. *Preparation of Materials Containing Divalent Rare Earth Ions*

Eu^{2+} is a good activator for luminescent materials. Because the 5d electron is located in the outer orbit, its $4f^7 \rightarrow 4f^65d$ emission depends strongly on the environment of Eu^{2+}, its emission wavelength, and intensity changes with the composition of the host and the symmetry of Eu^{2+} ion. The emission wavelength of Eu^{2+} ion can be changed from blue to red. On the other hand, the ionic radius of divalent Eu^{2+} is similar to those of divalent alkaline earth ions (Ca^{2+}, Sr^{2+}, and Ba^{2+}), especially to that of divalent Sr^{2+} ion, when coordination number (CN) is 8, the ionic radii are 125 pm for Eu^{2+} and 126 pm for Sr^{2+}. Therefore, Sr^{2+} ion can be replaced by Eu^{2+} ion easily in the strontium compounds. Many alkaline earth compounds doped with Eu^{2+} ion are widely studied and used as luminescent materials, such as $SrAl_2O_4$:Eu^{2+}, Dy^{3+} for long-lasting phosphor [20], and Ba_2SiO_4:Eu^{2+} for light emitting diode (LED) [21].

In order to reduce the trivalent rare earth ions such as Eu^{3+}, Sm^{3+}, Tm^{3+}, and Yb^{3+} into divalent ions RE^{2+} in solid state, $H_2(10\%) + N_2(90\%)$ is usually used as reducing agent. For Eu^{3+}, CO can also be used as reducing agent, which is produced by graphite heated at high temperature (carbothermal method).

2.4.1.2. *Synthesis of Divalent Rare Earth Materials in Air with Defects Produced by Aliovalent Substitution*

It is found that some Sm^{3+}, Eu^{3+}, Tm^{3+}, and Yb^{3+} ions in the alkaline earth borates, phosphates or borophosphates containing tetrahedral BO_4, PO_4, and BPO_5 groups can be reduced into divalent state even in air at high temperature by the electron carried on defects produced by aliovalent substitution. The following requirements seem to be necessary for the reduction of RE^{3+} to RE^{2+} in air at high temperature [22]:

(1) No oxidizing ions are present in the host.
(2) The dopant trivalent RE^{3+} ions must replace the lower valence cations of the host (aliovalent substitution). There must be no other stable trivalent rare earth ions such as La^{3+} present as one of components of the host. In these cases, in order to keep charge neutrality, some vacancies and electron carried defects will emerge, which can be used as reductants.
(3) The substituted cations must have radii similar to those of the corresponding divalent RE^{2+} ions. For example, Sr^{2+} is the best substituted ion when Eu^{2+} is used as dopant.
(4) The host compound must have an appropriate structure. For example, structure containing stiff tetrahedron BO_4 is one of the best.

The reduction process may be expressed as follows [23]:

$$3Sr^{2+} + 2Eu^{3+} = V''_{sr} + 2Eu^{\cdot}_{sr}$$
$$V''_{sr} = V_{sr} + 2e$$
$$2Eu^{\cdot}_{sr} + 2e = 2Eu^{x}_{sr}$$

According to these requirements, some rare earth ions such as Eu^{3+}, Sm^{3+}, Yb^{3+}, and Tm^{3+} can be reduced into divalent state in air at high temperature for certain alkaline earth borates, phosphates, borophosphates, aluminates, and silicates containing tetrahedron anion groups as SrB_4O_7, SrB_6O_{10}, BaB_8O_{13}, $Sr_2B_5O_9Cl$, $MBPO_5$ (M = Ca, Sr, Ba), $Sr_4Al_{14}O_{25}$, $BaMgSiO_4$, etc. [24].

This provides a convenient and safe method for the synthesis of materials doped with divalent rare earth ions.

2.4.2. Halides

Rare earth halides are a group of important compounds. Anhydrous rare earth chlorides and fluorides are starting materials for the preparation of rare earth metals. Anhydrous $LaCl_3$:Ce^{3+} or $LaBr_3$:Ce^{3+} single crystal is a good scintillator used for detecting high-energy particles, but they are hygroscopic compounds.

The solubility of rare earth fluorides in aqueous solution is very low. Their hydrates can be obtained by precipitation method with HF as precipitant from rare earth chloride or nitrate solutions. In order to prepare anhydrous fluoride, they must be dehydrated with NH_4F at 473−873 K or under HF + N_2 atmosphere at 973 K using graphite as container. But the product obtained by this dry method contains trace amounts of oxyfluoride impurity.

Complex fluorides, for example, ABF_4 (A = Li, Na; B = trivalent rare earth), are important materials for application. The electronegativity of *F*(3.98) is high, so the chemical bonds in fluorides and complex fluorides are dominantly ionic. The energy gap of some fluorides such as $LiYF_4$ is large ($\sim$12.5 eV), so they are transparent in the vacuum ultraviolet (VUV) range and are a useful matrix for laser or VUV phosphor. $NaREF_4$ can be prepared by solid-state reaction under HF atmosphere at high temperature by using NaF and REF_3 as starting materials, but the product obtained contains traces of oxyfluoride impurity. They can also be synthesized by hydrothermal method under mild conditions. For instance, $NaHoF_4$ and $NaEuF_4$ with fluoride structure have been prepared by this method. The molar ratio of the initial reaction mixture for the synthesis of $NaHoF_4$ and $NaEuF_4$ is 6.0 NaOH:1.0 Ho_2O_3 (or Eu_2O_3):6.0 NH_4F:6.0 HF:100 H_2O. The reaction mixture is sealed with 80% fill in a Teflon-lined stainless steel autoclave for hydrothermal crystallization at 513 K. After the autoclave is cooled and depressed, the final powder product is washed with deionized water and dried in air at room temperature [25]. $NaYF_4$ co-doped with Yb^{3+}, Er^{3+} (or Tm^{3+}) can also be prepared by hydrothermal method using NaF, NH_4HF_2, and lanthanide nitrates as starting materials, Er^{3+} (or Tm^{3+}) as activator and Yb^{3+} as sensitizer. Over a wide range of reaction temperature and time, the hydrothermal route can successfully lead to an easily controlled synthesis of both cubic phase nanospheres and hexagonal phase microtubes of $NaYF_4$. It is a good upconversion phosphor and can convert the 980 nm infrared light to red, green, and blue visible light. The hexagonal structure $NaYF_4$:Yb^{3+}, Er^{3+} (or Tm^{3+}) microtubes display considerably stronger upconversion emission than the cubic $NaYF_4$:Yb^{3+}, Er^{3+} nanospheres [26].

Hydrated rare earth chloride or bromide can be prepared by simply dissolving the corresponding rare earth oxide in HCl or HBr acid and then evaporating to crystallize. Hydrous chloride or bromide with different crystalline water is obtained. But anhydrous rare earth halide is more difficult to prepare, because oxyhalide impurity is formed if trace of oxygen or moisture is present during the synthesis. For preparation of anhydrous $LaCl_3$ or $CeCl_3$, the starting material $LaCl_3 \cdot 7H_2O$ or $CeCl_3 \cdot 7H_2O$ can be dehydrated in dry HCl atmosphere. The single crystal of anhydrous $LaCl_3$ or $CeCl_3$ can be grown by the Bridgman technique using a static vertical silica ampoule and a moving furnace. To minimize the effect of oxygen and moisture, all handling for putting the anhydrous $LaCl_3$ or $CeCl_3$ in the ampoule is done in a nitrogen-purged glovebox. Gas in the glovebox is recirculated through a mixed bed of reduced copper catalyst and dry molecular sieve. The oxygen and moisture concentration in the glovebox can be lowered to less than 10 ppm [27].

2.4.3. Chalcogenides

In order to prepare chalcogenides such as sulfides or selenides containing rare earth, attention must be paid to prevent oxidation by oxygen in air. For sulfides, H_2S, CS_2, or S is usually used as sulfurizing reagents and rare earth oxides RE_2O_3 or chlorides $RECl_3$ as starting materials. If S is used as sulfurizing agent, 25% mass excess of S is thoroughly mixed with RE_2O_3 and put in the quartz ampoule, evacuated to 1×10^{-6} Torr and sealed, and then fired in furnace at the synthesis temperature and time. If rare earth metals RE and S are used as starting materials, the reaction is RE + S = RES. If H_2S or CS_2 is used as the sulfurizing agent, the rare earth oxides are put in the graphite crucible and inserted into a tube furnace. After flushing with Ar gas at room temperature, the furnace is heated to the synthesis temperature in Ar atmosphere and then H_2S or CS_2 gas is introduced. After necessary time, furnace is cooled down to room temperature in Ar atmosphere. The reactions are:

$$RE_2O_3 + 3H_2S = RE_2S_3 + 3H_2O;$$
$$2RECl_3 + 3H_2S = RE_2S_3 + 6HCl$$

or

$$2RE_2O_3 + 3CS_2 = 2RE_2S_3 + 3CO_2.$$

Because Eu^{3+} is easily reduced into Eu^{2+} in this condition, only EuS is formed instead of Eu_2S_3 by using this method.

There are a variety of rare earth sulfides with different compositions, for example, monosulfides RES, sesquisulfides RE_2S_3, disulfides RES_2, RE_3S_4, etc. Even for sesquisulfides RE_2S_3, there are many kinds of structures. A series of RE_2S_3 and EuS have been prepared. The starting materials, sulfurization conditions (1093−1523 K), structures, and colors of the products are discussed in more detail. It is interesting to note that the color of EuS (dark brown) and different structure types of RE_2S_3 are different such as α-RE_2S_3 (Pnma, Maroon), β-($I4_1/acd$, yellow to green), γ-(I43d, red to yellow), δ-($P2_1/m$, yellow to tan), and ε-(R3C, white to yellow) [28,29]. They can be used as environmentally friendly inorganic pigments, especially the γ-Ce_2S_3, which is a red pigment and can substitute for the toxic elements such as Cd, Cr, and Pb.

Rare earth selenides can be prepared by methods similar to sulfides, but H_2Se or Se is used instead of H_2S or S [16].

2.4.3.1. *Sulfides*

Because the ionic radii of rare earth ions are similar to those of alkaline earth ions (Ca^{2+}, Sr^{2+}, Ba^{2+}), the latter ions in the alkaline earth compounds can be substituted by the former ions. Especially the divalent ion Sr^{2+} can be easily replaced by the divalent ion Eu^{2+} and isostructural compounds are formed. For example, the complex sulphide $SrGa_2S_4$ is isostructural with $EuGa_2S_4$. They belong to the orthorhombic crystals. Their lattice parameters are $a = 20.840$, $b = 20.495$, and $c = 12.212$ Å for $SrGa_2S_4$ and $a = 20.716$, $b = 20.404$, and $c = 12.200$ Å for $EuGa_2S_4$ [30]. They can be prepared by solid-state reaction. The starting materials are $SrCO_3$, Ga_2O_3, and Eu_2O_3. Stoichiometric quantities are weighed and thoroughly mixed by grinding in mortar, then put in the corundum crucible and inserted into a tube furnace. After flushing with Ar gas, and fired at 1173–1273 K in H_2S for 2–3 h, the furnace is cooled down to room temperature in Ar atmosphere [30]. $EuGa_2S_4$ is a green light emitting phosphor. In order to avoid the concentration quenching of Eu^{2+}, its concentration has to be reduced by doping it in the isomorphous $SrGa_2S_4$. $SrGa_2S_4:Eu^{2+}$ is a good green phosphor, which can be excited by 400 or 460 nm light emitted from InGaN chip and used for LED [31,32] and other devices.

The Pauling electronegativities of S (2.58) and Se (2.55) are smaller than that of O (3.44), so the covalent character of sulfides and selenides is larger than that of oxides. This makes the band gap (eV) of rare earth sulfides and selenides smaller than that of oxides and the colors of the rare earth sulfides and selenides deeper than that of trivalent rare earth oxides [28,29]. Due to the greater covalent character of sulfides and selenides, the induced larger nephelauxetic effect also shifts the excitation and emission spectra of the doped rare earth ions to longer wavelength (Red Shift). Therefore, the excitation and emission wavelengths of the rare earth ions can be tuned this way.

Complex sulfides such as $CaAl_2S_4$ doped with Eu^{2+}, a green-emitting phosphor for LED with near-ultraviolet InGaN chip, can also be obtained by using evacuated quartz ampoules, in which the stoichiometric amounts of CaS, Al, EuS, and Na_2CO_3 with 50 mass% S are put and heated to 1323 K for 5 h. In this method [33], aluminum powder is used as it is a cheap and stable starting material for preparation of $CaAl_2S_4$ instead of expensive and weakly hygroscopic Al_2S_3.

2.4.3.2. *Oxysulfides*

$YO_2S:Eu^{3+}$ is a well-known oxysulfide, a good red-emitting phosphor widely used for color TV. A good blue-emitting phosphor of oxysulfide $CaLaGa_3S_6O:Ce^{3+}$ with CIE chromaticity coordinates $x = 0.147$, $y = 0.089$ was proposed for the application of LED excited by 400 nm InGaN chip [34]. Its pure phase can be prepared by a two-step method. The first step is to prepare the starting sulfide materials β-La_2S_3, Ga_2S_3, and γ-Ce_2S_3 at high temperature in horizontal tube furnace. β-La_2S_3 and γ-Ce_2S_3 are prepared from La_2O_3 and CeO_2 under CS_2 atmosphere at 1523 K for 3 h. Ga_2S_3 is prepared from Ga_2O_3 under flowing H_2S at 1223 K for 3 h. The second step is to prepare the final product $CaLaGa_3S_6O:Ce^{3+}$ with the starting sulfide materials by the following solid-state reaction:

$$CaO + (1-x)/2La_2S_3 + 3/2Ga_2S_3 + x/2Ce_2S_3 \rightarrow CaLa_{1-x}Ce_xGa_3S_6O$$

The stoichiometric amounts of materials such as CaO, β-La_2S_3, Ga_2S_3, and γ-Ce_2S_3 are thoroughly mixed by grinding, and then are sintered at 1223 K for 2 h under Ar atmosphere.

2.4.4. Pnictides and oxypnictides

2.4.4.1. *Nitrides*

Nitrogen (Pauling's electronegativity = 3.04) reacts with rare earth metals RE, whose electronegativity is about 1. Stable mononitrides REN are formed at high temperature, which are characterized by high melting points, around 2773 K, and are highly sensitive to hydrolysis. In consequence of the large triple bond energy in dinitrogen (941 kJ mol^{-1}), compared to oxygen (499 kJ mol^{-1}), the amount of nitrides is much less than oxides. Direct combination with nitrogen generally requires elevated temperatures, and total absence of oxygen is needed in order to prevent a preferential reaction with oxygen or an attack on the nitride by oxygen at high temperature as follows: nitride + O_2 → oxide + $N_2\uparrow$.

Complex nitrides such as the isostructural compounds $MYbSi_4N_7$ (M = Sr, Eu) and $Sr_{1-x}Eu_xYSi_4N_7$ ($0 \le x \le 1$) can be prepared as follows [35]. The starting materials are β-Si_3N_4, metallic Y, Eu, and Sr. SrNx ($x \sim 0.65$) and EuN are presynthesized by a nitriding reaction of Sr and Eu metals under nitrogen atmosphere at 1073–1123 K, and then grinding them into fine powders. The starting mixtures are thoroughly mixed and ground with an agate motar and pestle. All manipulations are carried out in a protecting atmosphere in a glovebox filled with dry nitrogen because of the great air sensitivity of most of the raw materials. Subsequently the well mixed starting powders are placed in a molybdenum crucible and fired at 1673 K and 1933 K for 12 and 16 h, respectively, under a flowing gas mixture 5% H_2–95% N_2 in horizontal tube furnaces with an intermediate grinding between the firing steps.

Recently, compounds containing nitrogen element, nitrides, and oxynitrides have attracted much attention for LED applications. The key point is that N^{3-} shows higher formal charge and larger nephelauxetic effect (covalence), both of which contribute to the strengthening crystal field splitting of the 5d states of rare earth (Ce^{3+}, Eu^{2+}) than in an analogous oxygen environment. Consequently, oxynitride and nitride phosphors doped with rare earth ions are anticipated to show longer excitation and emission wavelengths than their oxide counterparts. These properties meet the requirements for the fabrication of LED using InGaN chip, which emits 400 nm near-UV or 460 nm blue light for excitation.

Other isostructural nitrides $M_2Si_5N_8$ (M = Sr^{2+}, Eu^{2+}) can be prepared by carbothermal reduction and nitridation reaction (CRN) at 1773 K [36]:

$$Si_3N_4 + SrO + Eu_2O_3 + C + N_2 \rightarrow Sr_2Si_5N_8 : Eu + CO_2\uparrow$$

Some nitrides doped with rare earth ions as phosphors can also be prepared by Gas Pressure Sintering (GPS) method using a furnace with a graphite heating element and a nitrogen atmosphere under pressure in the range of 0.1–1.0 MPa. For example, $CaAlSiN_3:Eu^{2+}$ is formed by the reaction among metal nitride starting powders at 1873 K under 1.0 MPa N_2 atmosphere:

$$Si_3N_4 + AlN + Ca_3N_2 + EuN \rightarrow CaAlSiN_3 : Eu$$

2.4.4.2. Oxynitrides

If the $(AlO)^+$ in the alkaline earth aluminates MAl_2O_4 is replaced by $(SiN)^+$, oxynitrides $MAl_{2-x}Si_xO_{4-x}N_x$ can be prepared [37]. A $BaAl_{2-x}Si_xO_{4-x}N_x:Eu^{2+}$ phosphor is synthesized by using $BaCO_3$, Al_2O_3, α-Si_3N_4, and Eu_2O_3 as starting materials and sintering at 1473–1673 K for 4–8 h in an $H_2 + N_2$ reductive atmosphere. Yellowish green light emitting diode was fabricated by this phosphor and near-ultraviolet (~395 nm) GaN chip [38].

>By using Si_3N_4 as source of Si and N, $MSi_2O_2N_2$ (M = Ca, Sr) and their Eu^{2+}-doped solid solution $(Ca_{1-x-y}Sr_x)Si_2O_2N_2{:}yEu^{2+}$ can be prepared by solid-state reaction. The stoichiometric mixtures of the raw materials $CaCO_3$, $SrCO_3$, SiO_2, α-Si_3N_4, and Eu_2O_3 are thoroughly ground and then fired in alumina crucible at 873 K for 2 h under N_2 atmosphere. After being reground, they are sintered at 1773 K for 6 h in $90\%N_2/10H_2$ reducing atmosphere in horizontal tube furnace. Intense green LEDs were fabricated by this phosphor and near-ultraviolet (~395 nm) GaN chip [39].

Gas-reduction nitridation (GRN) method is another effective and cheap method for synthesizing $RE_aMO_yN_z$ oxynitrides and nitrides. In this approach, a corresponding RE–M–O ternary oxide precursor is synthesized by "Pechini" or amorphous citrate wet-chemical process using citric acid as a complexing agent and ammonia as precipitant. After evaporation under vacuum and calcination, an RE–M–O ternary oxide precursor is obtained. The reaction is generally performed in an alumina boat containing the oxide precursor powder loaded inside an alumina or quartz tube through which NH_3 or NH_3–CH_4 gas flows at appropriate rates at high temperatures (1173–1873 K). The NH_3 or NH_3–CH_4 gas acts as both reducing and nitriding agents. This reaction is able to create RE–N bonds from RE–O bonds, in accordance with the general equation [40]:

$$RE_aMO_x + zNH_3 \rightarrow RE_aMO_yN_z + (x-y)H_2O\uparrow$$

The above reaction is also called thermal ammonolysis method. M stands for Ta, Nb, Mo, W, Ti, etc. Quaternary RE–M–O–N oxynitrides, like RE–M–O oxides, do not show any hygroscopic character [40].

Oxynitrides, whether they are of simple or complex type, such as $LaTiO_2N$, $La(Mg,Ta)(O,N)_3$, and $Ba(Sc,Ta)(O,N)_3$, have band gaps of 1.7–2.4 eV. These values fall in the visible range of electromagnetic spectrum [41]. The color originates from electronic interband transitions between the occupied valence band and the vacant conduction band. Upon substituting the higher electronegativity of O (3.44) by the lower electronegativity of N (3.04), an increased covalent character induces a decrease in the width of the optical band gap and in a possible selective absorption of visible light.

Among $RE_aMO_yN_z$, $(Ca,La)Ta(ON)_3$ are inorganic yellow–red pigments without toxic metals [42]. Their performance is better than that of commercial cadmium sulfoselenides.

2.4.4.3. Oxyphosphides and Arsenides

Oxypnictides such as LaFePO and LaNiPO have been known to show superconductivity at low temperature (4–7 K). The publication of a superconducting transition at 26 K in the fluoride-doped rare earth iron arsenide oxide $LaFeAsO_{1-x}F_x$ [43] has sparked tremendous interest in rare earth pnictides and oxypnictides containing iron for superconductor, and higher critical temperatures (T_c) have emerged quickly. By changing La^{3+} with equivalent valence Sm^{3+} and substituted aliovalently for Sr^{2+}, the T_c increased from 26 K to 56 K in $Sr_{1-x}Sm_xFeAsF$ ($x = 0.5$) [44]. This superconductor is synthesized by the following solid-state reaction using SrF_2, SrAs, SmAs, and Fe_2As as starting materials:

$$0.5SrF2 + (0.5-x)SrAs + xSmAs + 0.5Fe_2As \rightarrow Sr_{1-x}Sm_xFeAsF.$$

SrAs is presynthesized by heating the mixture of Sr and As powders in an evacuated quartz tube at 873 K for 10 h. SmAs and Fe_2As are prepared with Sm, Fe, and As powder by a similar process at 1073 K for 10 h, respectively. All processes except for firing are carried out in glovebox (O_2, $H_2O < 1$ ppm) in which highly pure argon atmosphere is introduced.

2.5. SOL–GEL PROCESS AND PRECURSORS IN HIGH-TEMPERATURE SOLID SYNTHESIS [45–47]

The sol–gel synthesis is a recently developed method for preparing ceramics, glasses, and many other solid materials as an alternative to high-temperature solid synthesis. This technique offers the following advantages in comparison with the traditional high-temperature powder synthesis.

1. By mixing reactant solutions, it is very easy to obtain a desired homogeneous multiple component system.
2. The temperature required in this type of preparation can be lowered significantly. Therefore, ceramics, glasses, and other functional materials can be synthesized under moderate conditions [48].
3. The rheologic properties of sols favor the preparation of films, fibers, and deposit materials using techniques such as spraying, dip-coating, or impregnation.

Some special forms that usually require special conditions can now be prepared using this technique. This will be illustrated with the following example on $YBa_2Cu_3O_{7-\delta}$ superconducting oxide film [49]. Superconducting oxides can be obtained through a variety of routes such as the traditional high-temperature solid-state reaction, coprecipitation, electron beam deposition, sputtering, and laser vaporization. When the high-temperature synthesis is used, the processed material is repeatedly ground and sintered in order to obtain a homogeneous product. Otherwise, it generally requires special synthetic conditions. Compared with the above methods, the sol–gel technique is simple and has a relatively low cost. Furthermore, by employing the characteristic rheologic property of a sol, high-quality films can be made. Two different sol–gel routes can be used to prepare superconducting $YBa_2Cu_3O_{7-\delta}$ films. One is to use $Y(NO_3)_3 \cdot 5H_2O$, $Ba(NO_3)_2$, and $Cu(NO_3)_2 \cdot H_2O$ stoichiometric nitrates as the starting materials. The starting materials are dissolved in ethylene glycol to form a homogeneous solution. The solution is brought to a temperature (say 130–180°C) by refluxing and the solvent is evaporated. The gel obtained is then calcined at a high temperature (950°C) under an oxygen atmosphere to obtain pure orthorhombic $YBa_2Cu_3O_{7-\delta}$. The other route is to use stoichiometric metallorganic compounds ($Y(OC_3H_7)_3$, $Cu(O_2CCH_3)_2 \cdot H_2O$, and Ba $(OH)_2$) as the starting materials. They are heated and dissolved in ethylene glycol with vigorous agitation. After the solvent is evaporated, the gel can be calcined to superconducting $YBa_2Cu_3O_{7-\delta}$. Some $YBa_2Cu_3O_{7-\delta}$ films of uniform thickness (10–100 μm) with high superconductivity can also be obtained by applying, with a fine brush, the gel prepared from the above two routes to a certain substrate. This substrate can be the [110] surface of sapphire, the [100] surface of $SrTiO_3$ single crystal, or the [001] surface of ZrO_2 single crystal. (1) The applied film can be heated successively to 400°C (2°C min^{-1}) and 950°C (5°C min^{-1}) and then cooled to room temperature (3 °C min^{-1}) under O_2 atmosphere. This procedure is repeated 2–3 times. Finally the film is annealed at 800°C under O_2 for 12 h and cooled down to room temperature at a rate of 3°C min^{-1}. (2) The applied thin film is calcined at 950°C for 10 min under air atmosphere, then another fresh layer is applied. The calcinations and layer application procedure is repeated several times. The film is finally annealed under O_2 atmosphere at 550–950°C for 5–12 h.

2.5.1. The Chemistry of Sol–gel Synthesis [45–47,50]

The sol–gel technique has been used recently to prepare films of many unusual oxides such as V_2O_5, TiO_2, MoO_2, WO_3, ZrO_2, and Nb_2O_5.

Some compounds or materials of certain structure and aggregation state can be synthesized using the sol–gel technique based on the understanding and control of process chemistry. This is because the basic chemistry of this technique lies in the hydrolysis and condensation of molecules or ions of precursor materials, i.e., molecules → polymers → sol → gel → crystals (or noncrystals). The reactant precursors of a sol–gel process are usually aqueous solutions of metal salts or metal organic compounds. Therefore, the hydrolysis and condensation of these two groups of materials will be discussed.

1. Hydrolysis and condensation of inorganic salts

   When a cation M^{z+} dissolves in pure water, the following solvating reaction can occur:

$$M^{Z+} + :O\begin{matrix}H\\H\end{matrix} \rightarrow \left[M \leftarrow O\begin{matrix}H\\H\end{matrix}\right]^{Z+}$$

   In many cases (such as for transition metallic ions), the solvation results in partial covalent bonds. Water molecules become more acidic due to the partial transfer of an electric charge from the filled $3a_1$ bonding orbital of a water molecule to the empty d orbital of the transition metallic ion. Depending on the extent of electric charge transfer, the solvated molecules go through the following changes:

$$[M-OH_2]^{Z+} \rightleftharpoons [M-OH]^{(Z-1)+} + H^+ \rightleftharpoons [M{=}O]^{(Z-2)+} + 2H^+$$

   In an ordinary aqueous solution, there are possible ligands for a metallic ion. They are aquo ligand (OH_2), hydroxo ligand (–OH), and oxo ligand (=O). If N is the coordination number or the number of water molecules covalently bonded to the cation M^{z+}, an approximate formula can be written as: $[MO_NH_{2N-h}]^{(Z-h)+}$, where h is defined as the molar ratio of hydrolysis. When $h = 0$, the precursor is in the form of "aquo-ion" $[M(OH_2)_N]^{Z+}$, while for $h = 2N$, it is in the form of "oxo-ion" $[MO_N]^{(2N-Z)-}$. When $0 < h < 2N$, the precursor can be in the form of either oxo–hydroxo complex $[MO_x(OH)_{N-x}]^{(N+x-Z)-}$

($h > N$), hydroxo–aquo complex $[M(OH)_h(OH_2)_{N-h}]^{(Z-h)+}$ ($h < N$), or hydroxo complex $[M(OH)_N]^{(N-Z)-}$ ($h = N$). The hydrolysis product of a metallic ion (the precursor) can be roughly determined from the "Charge–pH" diagram.

Under different conditions, these complex ions can condense in different ways to form dimers or polymers, and some can further form framework structures. If the hydroxo bridge M–OH–M is formed via nucleophilic substitution (SN_1), the reaction of the hydroxo–aquo precursor $[M(OH)_x \cdot (OH_2)_{N-x}]^{(Z-x)+}$ ($x < N$) also follows the SN_1 mechanism. The charged precursor ($Z - h \geq 1$) cannot condense indefinitely to form a solid phase, mainly due to the changes in nucelophilic strength (with partial electric charge δ) of the hydroxyl groups during condensation. For instance, the dimerization of Cr(III) is:

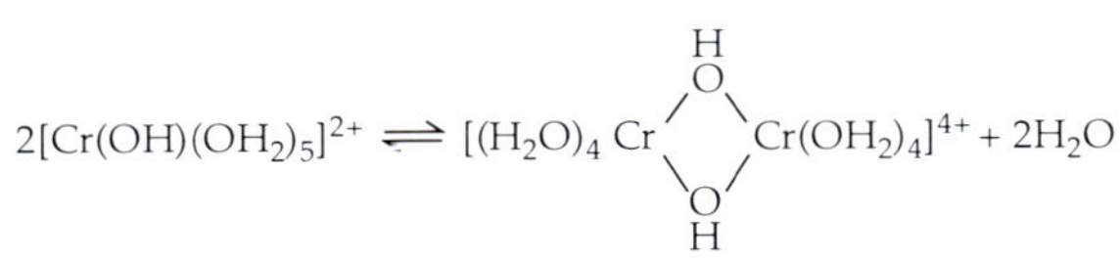

The partial electric charge of the OH group in a monomer is negative ($\delta(OH) = -0.02$). In a dimer, it becomes positive ($\delta(OH) = +0.01$), which indicates that OH in a dimer loses the ability of further condensation. Zero-charged precursor ($h = Z$) can condense to a solid via the hydroxo groups, and the final product is a hydroxide $M(OH)_Z$.

There are many factors involved in the preparation of gels from aquo–hydroxo inorganic complex precursor such as pH gradient, concentration, reactant feeding method, gelling rate, and temperature. Since nucleation and growth are diffusion-controlled processes via mainly the condensation of hydroxo bridges, all the above factors should be taken into account. To prepare pure materials, one needs to obtain stable gels. Some metals can form stable hydroxo bridge and then give rise to fine $M(OH)_Z$ of well-defined structure. This may not be the case for some other metals that form amorphous gelatinous precipitate $MO_{x/2}(OH)_{Z-x} \cdot YH_2O$. The amorphous precipitate of this kind finally forms $MO_{Z/2}$ via oxolation as a result of continuous water loss. For multivalent elements such as Mn, Fe, and Co, the situation is more complicated. This is because electron transfer can occur in the solution, in the solid phase, or even at the oxide–water interface.

Another type of condensation is through the reaction of oxo ligands, forming M–O–M oxo bridges. Such a condensation process requires that no aquo complex exists in the layer of metal complex such as an oxo–hydroxo precursor $[MO_x(OH)_{N-x}]^{(N+x-Z)-}$, $x < N$. If $[MO_3(OH)]^-$ (M = W, Mo) monomers form tetramers $[M_4O_{12}(OH)_4]^{4-}$ via a nucleophilic addition mechanism, the edge of $(_2(O)_2)$ or face of $(_2(O)_3)$ oxo bridges can form during the reaction. Another example is the reaction with a mechanism of addition–subtraction ($AN\beta E_1$ and $AN\beta E_2$) such as the dimerization of Cr(VI) ($h = 7$):

$$[HCrO_4]^- + [HCrO_4]^- \rightleftharpoons [Cr_2O_7]^{2-} + H_2O$$

and the condensation of vanadate:

$$[VO_3(OH)]^{2-} + [VO_2(OH)_2]^- \rightleftharpoons [V_2O_6(OH)]^{3-} + H_2O$$
$$[VO_3(OH)]^{2-} + [V_2O_4(OH)_3] \rightleftharpoons [V_3O_9]^{3-} + 2H_2O$$

2. Hydrolysis–condensation of metallorganic molecules

Metal alkoxides $M(OR)_n$ are a common molecular precursor in the sol–gel synthesis of metal oxides. Almost all metals (including lanthanides) can form this kind of compound. An $M(OR)_n$ reacts thoroughly with water and forms a hydroxide or hydrous oxide:

$$M(OR)_n + nH_2O \rightarrow M(OH)_n + nROH$$

In fact, the overall reaction process involving hydrolysis and condensation is very complicated. Hydrolysis generally occurs in aqueous or water and ethanol solutions and forms active M–OH. The reaction takes place in three stages:

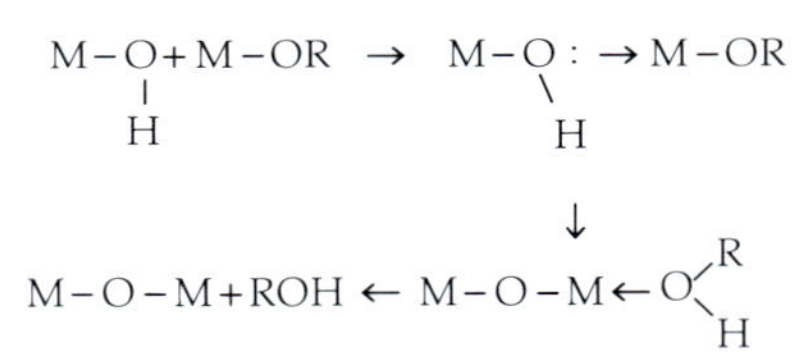

With the formation of hydroxo ligands, condensation further proceeds. There are three types of condensation corresponding to experimental conditions.

(1) Alkoxylation

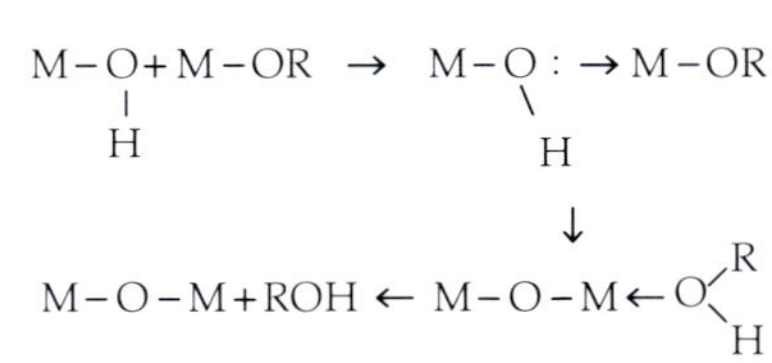

(2) Oxolation

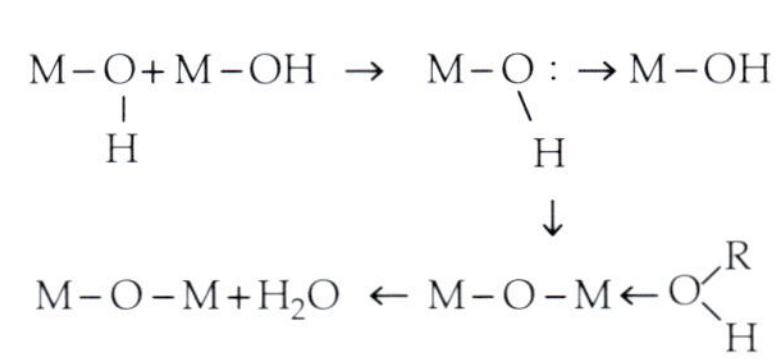

(3) Olation

$$M{-}OH + M \leftarrow O\begin{smallmatrix}R\\H\end{smallmatrix} \rightarrow M{-}\overset{H}{O}{-}M + ROH$$

$$M{-}OH + M \leftarrow O\begin{smallmatrix}H\\H\end{smallmatrix} \rightarrow M{-}\overset{H}{O}{-}M + H_2O$$

In addition to the compounds discussed above, alkoxyl chlorides and acetates can also be used as the organometallic molecular precursors.

3. Examples of synthesizing compounds of special structure via sol–gel processes [50]

(1) $V_2O_5 \cdot 1.6H_2O$ fiber

$V_2O_5 \cdot 1.6H_2O$ fiber is a proton and electron mixed conductor. It can be prepared using many methods such as: (1) acidification of $NaVO_3$ solutions (by HCl, HNO_3, or by proton strong-acid cation-exchange resins); (2) addition of V_2O_5 melts directly to cold water; and (3) hydrolysis of $VO(OR)_3$ ($R = Et, Pr^i, Pr, Bu$, and Am^t) with excessive water.

Vanadate becomes acidified at pH = 2, and forms a precursor $[VO(OH)_3]^0$ of $h = 5$. V is tetrahedrally coordinated and has a high electron affinity ($\delta(V) = +0.62$). Addition of any nucleophilic ligand can induce the transformation from tetrahedral to octahedral. In addition, since the precursor is acidic:

$$[VO(OH)_3]^0 \rightleftharpoons [VO_2(OH)2]^- + H^+$$
$$\delta(O) = -0.35 \quad \delta(O) = -0.44$$

Addition and condensation of such a tetrahedral precursor lead to the formation of decavanadate ($x \geq 4$):

$$(10 - x)[VO(OH)_3]^0 + x[VO_2(OH)_2]^- \rightleftharpoons [H_{6-x}V_{10}O_{28}]^{x-} + 12H_2O$$

When the nucleophile, water ($\delta(O) = -0.40$) is added, V_2O_5 fibers can be obtained. The possible reaction procedure is shown in Fig. 2.6. At first, two water molecules are introduced, which increases the C.N. from 4 to 6. Further olation forms $[VO(OH)_3(OH_2)]_n^0$ chains. Then the oxolation between chains can occur, fulfilling the transformation from unstable $_2(OH)_1$ to stable $_3(O)_1$ bridges. The condensation of these double chains forms the fiber-like structure.

(2) $MO_3 \cdot H_2O$ layer compounds (M = Mo, W)

As with vanadium acidification, the precursor $[VO_2(OH)_2]^0$ of $h = 6$ is also formed at pH = 2. The addition of a nucleophilic ligand leads to a C.N. transformation from 4 to 6 ($\delta(W) = +0.64$). When $h = 6$, the precursor is acidic:

$$[MO_2(OH)_2]^0 \rightleftharpoons [MO_3(OH)]^- + H^+$$
$$\delta(O) = -0.31 \quad \delta(O) = -0.42$$

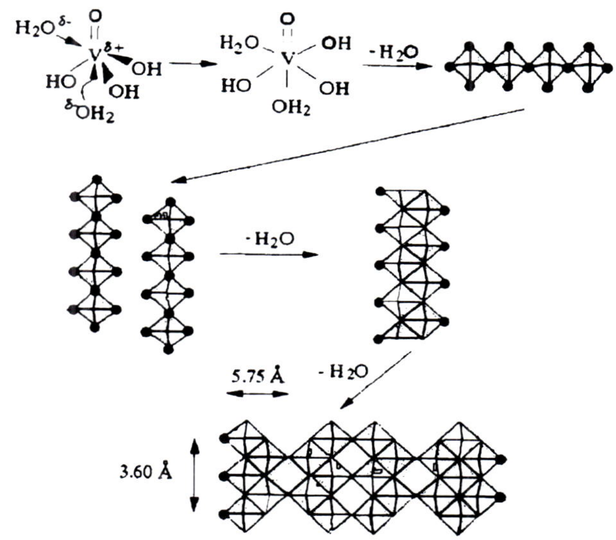

FIGURE 2.6 Mechanism of synthesizing V_2O_5 fiber through SN and ANßE processes from monomer precursor ($h = 5$). [50]. *Reprinted from Progress in Solid State Chemistry, 18 (1988), J. Livage, M. Henry, C. Sanchez, Sol-gel chemistry of transition metal oxides, page 259, Copyright (1988), with permission from Elsevier.*

Addition and condensation of these tetrahedral and hexahedral precursors form isopolyanions:

$$(10 - x)[WO_2(OH)_2]^0 + x[WO_3(OH)]^- \rightleftharpoons [H_{4-x}W_{10}O_{32}]^{x-} + 8H_2O$$
$$(6\text{-}x)[MO_2(OH)_2]^0 + x[MO_3(OH)]^- \rightleftharpoons [H_{2-x}M_6O_{19}]^{x-} + 5H_2O \quad M = Mo, W$$

If $x = 0$, water molecules can enter the coordination sphere. As the precursor of $h = 6$ has two oxo ligands, two water molecules can be added in para-positions. In this case, the functionality of the precursor is $f = 2$. Therefore, condensation can occur only through oxolation, forming linear or cyclic polymers:

$$n[MO_2(OH)_2(OH_2)_2]^0 \rightleftharpoons [MO_3(OH_2)_2]_n + nH_2O \ (M = Mo, W)$$

For *W*, due to the dissociation of one of the water molecules, the precursor $[WO(OH)_4(OH_2)]^0$ of $h = 6$ can be formed. The precursor condenses through olation to layered structure. The hydrogen bonding between the layers results in $WO_3 \cdot 2H_2O$ and $WO_3 \cdot H_2O$ layer compounds. The reaction process is shown in Fig. 2.7. The water dissociation process is very slow with Mo. Nevertheless, it can be speeded up by heating and leads to $MoO_3 \cdot 2H_2O$ and $MoO_3 \cdot H_2O$ analogously.

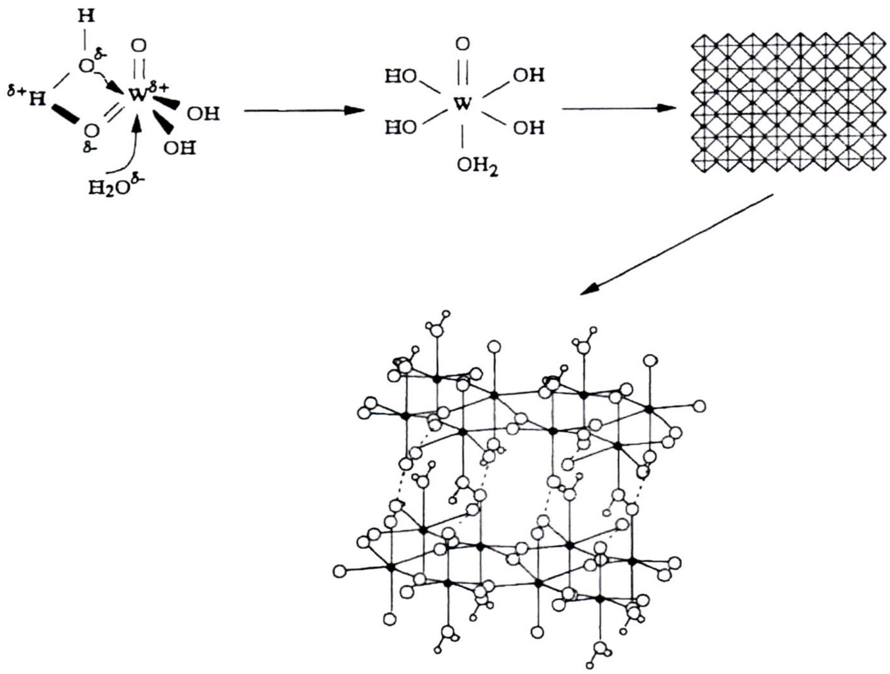

FIGURE 2.7 Formation mechanism of $WO_3 \cdot H_2O$ layer [50]. *Reprinted from Progress in Solid State Chemistry, 18 (1988), J. Livage, M. Henry, C. Sanchez, Sol-gel chemistry of transition metal oxides, Page 259, Copyright (1988), with permission from Elsevier.*

2.5.2. The Precursors in High-temperature Solid-state Synthesis [46]

In this section, the precursors used for the processes described in preceding sections are classified in groups. The precursors can be molecular or polymeric or a colloidal suspension of particles.

1. Inorganic salts
   Inorganic salts are often used as molecular precursors in high-temperature solid-state synthesis. Inorganic salts are ionic compounds; examples are listed in Table 2.2.
2. Metal organic compounds
   Metal organic compounds are covalent or inorganic coordinate compounds in which the metal center is bonded to the ligand via a noncarbon atom such as oxygen, sulfur, phosphorus, or nitrogen.
   In the literature, the organometallic compounds described in the next section are also referred to as metal organic compounds. Metal organic compounds are used as precursors. Examples of metal organic compounds are provided in Table 2.3.
3. Organometallic compounds
   Organometallic compounds are covalent or coordinate compounds in which the ligand is bonded to the metal center via a carbon atom. Like metal organic compounds, organometallic compounds are used as precursors for both wet chemical- and dry vapor-related processes. Commonly used organometallic precursors are listed in Table 2.4.
4. Polymeric precursors
   In some processes, such as MOD and sol–gel processing, polymeric precursors can be used as starting materials for producing glassy or ceramic compositions. These polymers are sometimes referred to as preceramic polymers. Selected examples of such polymers are given in Table 2.5.
5. Colloidal suspension
   Suspensions of molecular precursors or a suspension of oxide, hydroxide, sulfide, and other such powders in a given solvent can also be used as a starting material for preparing ceramic or glassy materials. A widely used silica colloidal suspension, Stober spheres, with monodispersed particles can produce a variety of glassy products.

TABLE 2.2 Inorganic Salt Precursors

Inorganic Salts	Examples
Metal halides	$MgCl_2$, LiF, KCl, $SiCl_4$, $TiCl_4$, $CuCl_2$, KBr, $ZrOCl_2$
Metal carbonates	$MgCO_3$, $CaCO_3$, Na_2CO_3, $SrCO_3$
Metal Sulfates	$MgSO_4$, $BaSO_4$, K_2SO_4, $PbSO_4$
Metal nitrates	$LiNO_3$, KNO_3, $Fe(NO_3)_2$
Metal hydroxides	$Ca(OH)_2$, $Mg(OH)_2$, $Al(OH)_3$, $Fe(OH)_3$, $Zr(OH)_4$
Salts with mixed ligands	$(CH_3)_3SnNO_3$, $(C_2H_5)_3SiCl$, $(CH_3)_2Si(OH)_2$

TABLE 2.3 Metal Organic Compounds

Metal Organic Compounds	General Formula	Selected Examples
Metal alkoxides	–M(–OR), where R is an alkyl	$Al(OC_3H_7)_3$, $Si(OCH_3)_4$, $Ti(OC_3H_7)_4$, $Zr(OC_4H_9)_4$
Metal carboxylates	–M(–OC(O)R)$_x$, where R is an alkyl	$Al(OC(O)CH_3)_3$, $Pb(OC(O)(CH_3)_2$, – acetates $Pb(OC(O)CH_2CH_3)_4$, – propionate $Al(OC(O)C_6H_5)_3$, – benzoate
Metal ketonates	–M(–OC(R)CH(R′)CO–)$_x$, where R is an alkyl or aryl	$Ca(OC(CH_3)CH(CH_3)CO)_2$, – pentanedionate $Al(OC(C(CH_3)_3)CH(C(CH_3)_3)CO)_2$ – heptanedionate
Metal amides (sometimes also referred to as amines)	—	$(CH_3)_2AlNH_2$, $(C_2H_5)_2AlN(CH_3)_2$, $(CH_3)BeN(CH_3)_2$, $(iC_3H_7)_3GeNH_2$, $(C_3H_7)_3PbN(C_2H_5)_2$
Metal thiolates	–M(–SR)$_x$, where R is an alkyl or an aryl	$(CH_3)_2Ge(SC_2H_5)$, $Hg(C_4H_3S)_2$, $(SCH_3)Ti(C_5H_5)_2$, $(CH_3)Zn(SC_6H_5)$
Metal azides	–MN_3	$(CH_3)_3SnN_3$, CH_3HgN_3
Metal thiocyanides	–M(–NCS)$_x$	$(C_2H_5)_3Sn(NCS)$
Metal organic compounds with mixed functional group	—	$(C_4H_9)Sn(OC(O)CH_3)_3$, $(C_5H_5)_2TiCl_2$, $(C_5H_5)Ti(OC(O)CH_3)_3$

TABLE 2.4 Organometallic Precursors

Organometallic Compounds	Selected Examples
Metal alkyls	$As(CH_3)_3$, $Ca(CH_3)_2$, $Sn(CH_3)_4$, – methyl
Metal aryls	$Ca(C_6H_5)_2$, – phenyl
Metal alkenyls	$Al(CH{=}CH_2)_3$, $Ca(CH{=}CHCH_3)_2$, – vinyl, propenyl
Metal alkynyls	$Al(CCH)_3$, $Ca(CCH)_2$, – acetylnyl
Metal carbonyls	$Co_2(CO)_8$, $Mn_2(CO)_{12}$, $W(CO)_6$, – carbonyl
Mixed organometallic ligands	$Ca(CCC_6H_5)_2$ – phenylacetylnyl $(C_5H_5)_3U(CCH)$ – cyclopentadienyl/ethynyl

TABLE 2.5 Polymeric Precursors

Polymer	Formula	Miscellaneous
Polycarbosilane	–[(RR′)Si–CH_2–]$_x$	Precursor to SiC in MOD and sol-gel-type processes where R is an active functional group such as an olefin, acetylene, H
Polysilazane	–[(RR′)Si–NR–]$_x$, where R is a organic unit or H	Precursor to Si_3N_4 or silicon carbonitride in a manner similar to polycarbosilanes
Polysiloxanes	1. –[Si(RR′)O–]$_x$: linear, where R is an alkyl or aryl 2. Sesquisiloxane: ladder 3. –[$Si(CH_3)_2OSi(CH_3)_2(C_6H_4)_m$–] siloxane-silarylene 4. Random and block copolymers of the above	Used in sol-gel processing and *in situ* multiphase systems used as precursors to SiO_2 or silicon oxycarbide
Polysilane	–[Si(RR′)–]$_n$, where R is an alkyl or aryl –[Si(CH)(CH)–] – Silalkylene	Precursors to SiC, as photoresist and photoinitiators
Borazines	–[BRNR′–]$_n$: cyclic or chain repeat units	Precursors to BN in CVD/MOCVD-type process or sol-gel-type process
Carboranes	Cage compounds of B and C	Precursors to B_4C in MOCVD- or MOD-type processes
Phophasphazenes	–[N = P(R_2)–]$_n$, where R is an organic, organometallic, or inorganic unit	Most common types of functional groups include alkoxy, aryloxy, arylamide, carboxylate, and halide
Polystannoxanes	–[Sn(R)$_2$–O–R′–O–Sn(R)$_2$–O–]$_n$: chain, where R is an organic unit. Drum- or ladder-type structures also possible	—
Polygermanes	–[Ge(RR′)–]$_n$	Can be used in microlithographic applications such as polysilanes

2.6. SELF-PROPAGATING HIGH-TEMPERATURE SYNTHESIS (SHS) [51–60]

2.6.1. General aspects

Development of efficient and energy-saving technologies is of great importance today. Self-propagating high-temperature synthesis (SHS) or combustion synthesis is a relatively novel and simple method for making certain advanced ceramic, composites, and intermetallic compounds (Fig. 2.8). This method has received considerable attention as an alternative to conventional furnace technology [61–65].

The SHS or combustion synthesis approach uses highly exothermic reactions. Such reactions typically have high activation energies and generate a substantial amount of heat. Once the reactions are initiated, to render self-sustaining reactions and to form a combustion wave, the temperature of the combustion can be very high (as 5000 K) and the rate of wave propagation can be very rapid (as 25 cm/s). This process offers an opportunity to investigate reactions in extreme thermal gradients (as 10^5 K/cm) conditions.

In a typical combustion synthesis, the reactants are usually fine powders, mixed and pressed into a pellet to increase an intimate contact between them. The reactant mixture is placed in a refractory container and ignited in vacuum or inert atmosphere. The products of the reaction are extremely porous, typically 50% of theoretical density (Fig. 2.9) .

Reactions between particulate materials are an alternative way to produce various types of materials considering the extreme simplicity of the process, relatively low energy requirement, high purity of the products obtained, the possibility to obtain metastable phases, and the possibility of simultaneous synthesis and densification. Higher purity of products is the consequence of high temperature associated with the combustion. Volatile impurities are expelled as the wave propagates through the sample. The possibility of forming metastable phases is based on high thermal gradients and rapid cooling rate associated with the reaction.

Two approaches are being used in SHS technology. The first approach is the production of intermediate products, which are then used as raw materials in further processing; the second one is based on direct production of finished articles, in which case synthesis, structuring, and shaping are carried out in one stage (simultaneous synthesis and densification).

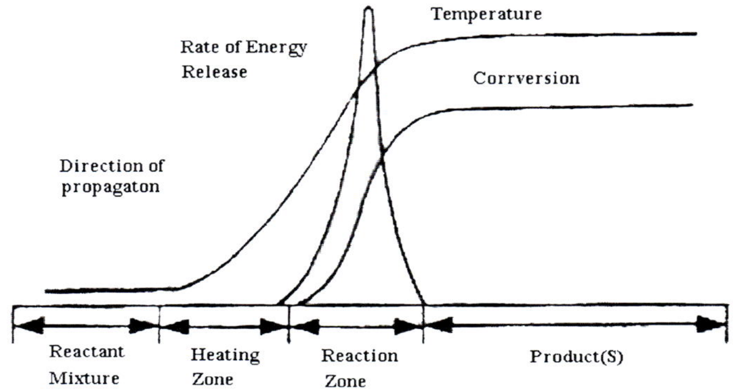

FIGURE 2.9 Structure of the combustion wave [66].

Actually over 500 compounds have been synthesized by the SHS method. Some of these materials are listed in Table 2.6.

2.6.2. Chemical Classes of SHS Reactions

The chemistry of SHS is versatile. The most important types of reactions are given below.

- Synthesis from the elements

$$Ti + C \rightarrow TiC$$
$$Ni + Al \rightarrow NiAl$$
$$3Si + 2N_2 \rightarrow Si_3N_4$$
$$Zr + H_2 \rightarrow ZrH_2$$

- Redox reactions

$$B_2O_3 + 3Mg + N_2 \rightarrow 2BN + 3MgO$$
$$B_2O_3 + TiO_2 + 5Mg \rightarrow TiB_2 + 5MgO$$
$$MoO_3 + B_2O_3 + 4Al \rightarrow MoB_2 + 2Al_2O_3$$
$$3TiO_2 + 3C + 4Al \rightarrow 3TiC + 2Al_2O_3$$
$$2TiCl_4 + 8Na + N_2 \rightarrow 2TiN + 8NaCl$$

- Oxidation of metals with complex oxides

$$3Cu + 2BaO_2 + 1/2Y_2O_3 + 0.5(1.5 - x)O_2 \rightarrow YBa_2Cu_3O_{7-x}$$
$$Nb + 2.5Li_2O_2 + 2Nb_2O_5 \rightarrow 5LiNbO_3$$
$$8Fe + SrO + 2Fe_2O_3 + 6O_2 \rightarrow SrFe_{12}O_{19}$$

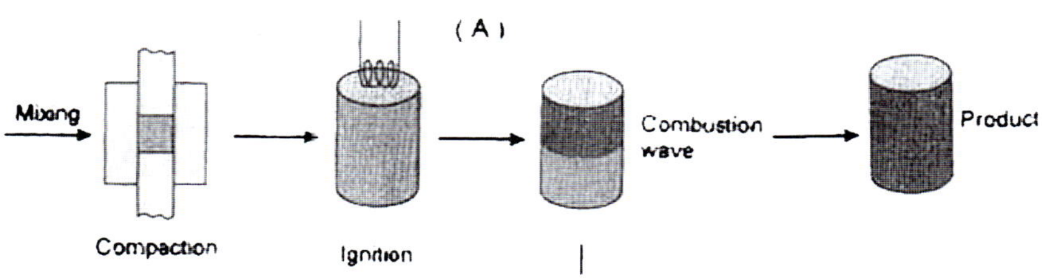

FIGURE 2.8 Schematic illustration of the SHS process.

TABLE 2.6 Some Materials Produced by SHS [53]

Borides	CrB, HfB, NbB, NbB_2, TaB_2, TiB, LaB_6, MoB, MoB_2, MoB_4, Mo_2B, WB, W_2B_5, WB_4, ZrB_2, VB, V_3B_2, VB_2
Carbides	TiC, ZrC, HfC, NbC, SiC, Cr_3C_2, B_4C, WC, TaC, Ta_2C, VC, Al_4C, Mo_2C
Nitrides	Mg_3N_2, BN, AlN, SiN, Si_3N_4, TiN, ZrN, HfN, VN, NbN, Ta_2N, TaN
Aluminides	NiAl, CoAl, $NbAl_3$
Silicides	TiS_3, Ti_5Si_3, ZrSi, Zr_5Si_3, $MoSi_2$, $TaSi_2$, Nb_5Si_3, $NbSi_2$, WSi_2, V_5Si_3
Hydrides	TiH_2, ZrH_2, NbH_2,CsH_2, PrH_2, IH_2
Intermetallics	NiAl, FeAl, NbGe, $NbGe_2$, TiNi, CoTi, CuAl
Carbonitrides	TiC–TiN, NbC–NbN, TaC–TaN, ZrC–ZrN
Cemented carbides	TiC–Ni, TiC–(Ni, Mo), WC–Co, Cr_3C–(Ni,Mo)
Chalcogenides	MgS, $NbSe_2$,$TaSe_2$, MoS_2, $MoSe_2$, WS_2, WSe_2
Binary compounds	TiB_2–MoB_2, TiB_2–CrB_2, ZrB_2–CrB_2, TiC–WC, TiN–ZrN, MoS_2–NbS_2, WS_2–NbS_2
Composites	TiB_2–Al_2O_3, TiC–Al_2O_3, B_4C–Al_2O_3, TiN–Al_2O_3, TiC–TiB_2, $MoSi_2$–Al_2O_3, MoB–Al_2O_3, Cr_2C_3–Al_2O_3, 6VN–5Al_2O_3, ZrO_2–Al_2O_3–2Nb

- Synthesis from compounds

$$PbO + WO_3 \rightarrow PbWO_4$$

- Reaction of the elements with decomposition product

$$2TiH_2 + N_2 \rightarrow 2TiN + 2H_2$$
$$4Al + NaN_3 + NH_4Cl \rightarrow 4AlN + NaCl + 2H_2$$

- Thermal decomposition

$$2BH_3N_2H_4 \rightarrow 2BN + N_2 + 7H_2$$

2.6.3. SHS Process and its Characterization

1. Combustion in SHS processes

For SHS systems the most popular are the following modes of combustion:

(1) Gasless combustion
(2) Solid–gas combustion
(3) Multiphase combustion

The combustion reaction is initiated locally by supplying a short intense burst of energy (e.g., electric spark, laser beam, resistively heated wire, chemical igniter, etc.). A combustion wave then propagates through the unreacted materials. A typical structure of the combustion wave is shown in Fig. 2.9. However, this structure might be more complicated when the melting of a product and/or the formation of new phases takes place. Combustion characteristics include the rapid propagation of a combustion wave (1×10^{-4}–0.15 m/s), generations of high temperatures (2073–3973 K; 1800–3700°C), and the rapid heating of the product (10^3–10^6 K/s). Due to relatively fast cooling rates, a solid product consisting of nonequilibrium phases might also be formed (Borovinskaya, 1974).

2. Characterization of SHS reactions

SHS reactions can be characterized by the adiabatic combustion temperature T_{ad}. This can be calculated by assuming that the enthalpy of the reaction heats up the products and that no energy is lost to the surrounding environment. Thus, T_{ad} is a measure of the exothermicity of the reaction and defines the upper limit for any combustion system. Some selected thermochemical data of binary ceramic compounds are shown in Table 2.7. As a rule-of-thumb, if $T_{ad} < 1200°C$, combustion does not occur, and if $T_{ad} > 2200°C$, self-propagating combustion occurs. In the range $1200°C < T_{ad} < 2200°C$, a combustion wave cannot propagate but can be made to do so by special techniques such as preheating of the reactants. For example, the reactions

$$Si + C \rightarrow SiC(T_{ad} = 1527\ ^\circ C,\ T_{ig} = 1300\ ^\circ C) \text{ or}$$
$$Ti + Al \rightarrow TiAl(T_{ad} = 1245\ ^\circ C,\ T_{ig} = 640\ ^\circ C)$$

are not self-sustaining when $T_{start} = 25°C$, while the reaction is self-sustaining.

$$Ti + C \rightarrow TiC(T_{ad} = 2937\ ^\circ C,\ T_{ig} = 1027\ ^\circ C) \text{ or}$$

The reaction Ti + Al → TiAl can be made self-sustaining by preheating above 100°C.

3. Fundamental SHS reaction parameters:

There are a number of reaction parameters in the SHS process such as particle size of reactants, stoichiometry of reactants, green density, gas pressure, size of pellets, and ignition mode.

All of these parameters have an effect on the completion of reaction, the temporal sequence of the reaction, the temperature profile of the combustion zone, the velocity of the combustion wave, and the morphology and macroscopic structure of combustion products.

2.6.4. Activation of SHS Processes [53,67–69]

The SHS process is very efficient only for reactions that are highly exothermic. If the reactions are not so

TABLE 2.7 Heat Capacity C_p, Melting Temperature T_m, Heat of Formation $\Delta_f H_{298}$, and Adiabatic Temperature Rise ΔT_{ad}, for Selected Carbides, Borides, Silicides, and Nitrides [66]

	$C_P = a + b10^{-3}T + C10^5T^{-2}$ (J/mol K)						
Compound	*a*	*b*	*c*	**Temperature internal (K)**	T_m **(K)**	$\Delta_f H_{298}$ **(Kj/mol)**	T_{ad} **(K)**
B_4C	96.190	22.594	44.852	298–1200	2620	39.0	702
TiC	49.497	3.347	14.949	298–3210	3210	183.1	2912
TaC	43.292	11.339	8.786	298–4270	4270	144.7	2402
SiC	41.714	7.615	15.230	298–3100	3100	72.5	1502
TiB_2	62.718	15.606	20.836	298–3190	3190	278.2	2892
MoB_2	38.493	48.869	—	298–1000	2520	96.2	1205
FeB	48.367	7.740	7.573	298–1810	1920	71.1	1402
Ti_5Si_3	243.592	24.016	110.709	298–1200	2390	580.3	2202
$TaSi_2$	22.845	37.321	−93.052	298–2100	2470	115.9	1502
$MoSi_2$	67.488	12.523	7.406	298–1200	2300	131.4	1602
W_5Si_3	111.587	75.843	−144.306	1200–2200	2590	194.6	902
TiN	49.831	3.933	12.385	298–3220	3220	334.4	4602
Si_3N_4	134.198	20.041	9.665	298–4000	2170	745.8	4002
BN	15.230	30.292	—	298–1200	3270	248.3	3402
AlN	22.886	32.635	—	298–900	2500	318.4	2602

exothermic, the front does not propagate. Such is the case for a large number of single-phase and composite materials, such as ceramic and intermetallic compounds, but sometimes these compounds are very interesting from a technological point of view. In addition, SHS reactions are limited by kinetic considerations. These include the rate of the reactions and the rate of heat transfer ahead of the reaction front. The latter is governed by the effective thermal conductivity of the product phases.

The need for activation stems from the thermodynamic and kinetic limitation of SHS.

Many researchers have developed a new SHS process coupling with other activation techniques that are able to sustain the reaction, furnishing supplementary energy for the propagation of the combustion wave.

1. Electric field-activated SHS (FACS) [53]

The effect of an electric field on reactions can be multifaceted; the field (in reality a current) can have a thermal effect (Joule heating), a mass transport effect (electro-migration), or a more complex effect arising from reported plasma creation in the powder compact. Modeling studies are based on the assumption that the main effect is Joule heating [69].

With the application of an electric field it is possible to synthesize very interesting composites, for example AlN–SiC. A common method for the synthesis of this solid solution is to heat a powder mixture of AlN and SiC at above 2100°C for several hours, using field-activation. The mixture was obtained by the following reaction:

$$Si_3N_4 + 4Al + 3C \rightarrow 4AlN + 3SiC$$

This reaction is highly exothermic, but no self-sustaining reaction can be established without field-activation. However, the SHS reaction doesn't occur unless the applied field is ≥8 V/cm. At the field between 8 and 25 V/cm, the reaction is complete with the product containing AlN and SiC.

2. Mechanically activated SHS (MRS and MASHS) [53]

Another way to increase the reactivity of the green powders is by combining simultaneously or successively ball milling (MA) and an SHS reaction, which can be classified as follows: (1) Mechanically induced Self-propagating Reactions (MRS) for which an SHS reaction occurs during milling inside the ball mill in the same activation time, when the powder reaches a well-defined critical state. (2) Mechanically Activated Self-propagating High-temperature Synthesis (MASHS) which consists of a short duration high-energy ball milling step followed by SHS.

3. SHS-coupled gas transport [53]

The activation through gas phase diffusion transport augments the solid-phase diffusion transport that is required to sustain the propagation of combustion wave. Even small quantities of gas can have a significant impact on SHS combustion, leading to pronounced changes in burning velocity, temperature, product elongation, and reactants-to-products conversion efficiency. The presence of gas transport agents is particularly important in the systems where the adiabatic flame temperature is below the melting point of reactants, products, and intermediate species. Considering the reaction:

$$Ta + C \rightarrow TaC$$

I_2 vapor and CO_2 were chosen as the gas transport agents. For example, TaI_5 is produced via

$$Ta(s) + 5\tfrac{1}{2} I_2(g) \rightarrow TaI_5(g)$$

At low temperature (<1000 K) and then activated Ta is condensed by reverse reaction at >1400 K, and carbon simply diffuses in the activated higher specific surface area Ta to form TaC.

The SHS process has a good control of the chemical composition, so a new application of materials product by SHS process can be Thermal Spray.

Another recent application of SHS technique is the synthesis of lithium nickelate ($Li_{0.88}Ni_{1.12}O_2$), lithium cobaltate ($LiCoO_2$), and lithium manganate ($LiMn_2O_4$); these compounds are three most popular 4 V cathode materials used in lithium-ion batteries. For this synthesis the stoichiometries of the reaction mixture were taken as follows:

$$LiNO_3 + Co(NO_3)_2 + 2.5NH_2 - CO - NH_2 \text{ (Urea)}$$
$$LiNO_3 + Ni(NO_3)_2 + 2.5NH_2 - CO - NH_2 \text{ (Urea)}$$
$$LiNO_3 + 2Mn(NO_3)_2 + 4.2NH_2 - CO - NH_2 \text{ (Urea)}$$

The nitrate is responsible for the oxidation of urea to yield a high temperature, which results in the oxidation of the transition metals.

Recently the production of composite thin films is becoming more important in the engineering of surfaces; the combustion synthesis provides an alternative way for the production of composite films. An example is the production of TiC–TiB_2 composite targets, which could be used in sputtering thin films. This target is produced with the application of uniaxial pressure during the reaction, and after it is possible to sputter it in order to produce a film. The film produced has high thermal stability.

2.7. HIGH-TEMPERATURE PREPARATION OF METAL VAPORS AND ACTIVE MOLECULES FOR USE IN CRYOSYNTHESIS [70–72]

In developing organometallic compounds, a unique synthetic route—cryosynthesis (cryogenic method or matrix-isolation)—has been exploited [70]. Accordingly, with the enrichment of synthetic chemistry, a new area named cryochemistry has been developed in the last 10 years. Cryochemistry mainly deals with the spontaneous and simple synthetic reactions of high-temperature (HT) species with other gaseous molecules or solutions at low temperatures (of liquid nitrogen or even lower). New compounds that are impossible or difficult to synthesize or separate by other routes are thus readily obtained. The so-called HT species in the reactions may be simple substances (mainly metals) or active molecules formed at high temperatures (from electron bombardments, laser, and so forth). Some typical examples are given below:

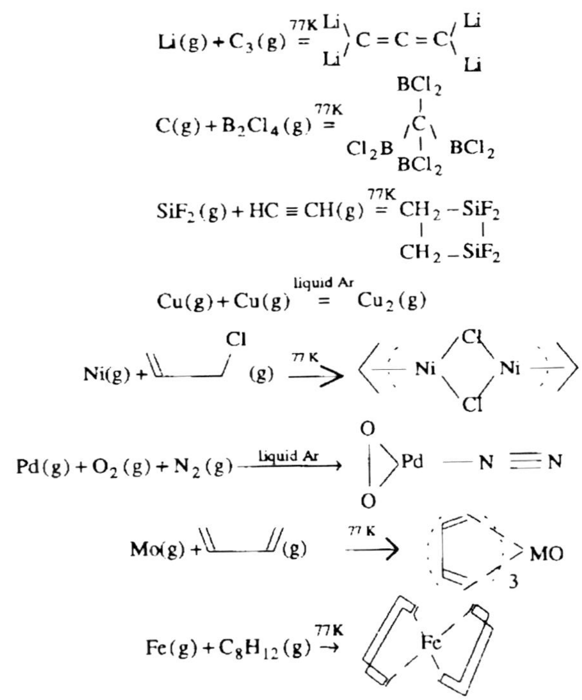

From the above reactions, the key step of this type of reaction is to use the temperature technique to form metals, other simple substances, or activated molecules. The following is the high-temperature reactions for preparing such high-temperature species.

1. Solid or liquid vaporization

$$CaF_2(s) \xrightarrow[\sim 1\ Pa]{1300\ ^\circ C} CaF_2(g)$$

$$C \xrightarrow[\sim 1\ Pa]{2500\ ^{\circ}C} C_1(g) + C_2(g) + C_3(g)$$

2. Solid–solid or solid–liquid reaction

$$Si(s) + SiO_2(s) \xrightarrow[\sim 1\ Pa]{1350\ ^{\circ}C} 2SiO(g)$$

$$2B(s) + 2B_2O_3(l) \xrightarrow[\sim 1\ Pa]{1300\ ^{\circ}C} 3(O = B = B = O)$$

3. Gas–solid reaction

$$2B(s) + BF_3(g) \xrightarrow[\sim 10^2\ Pa]{2000\ ^{\circ}C} 3BF(g)$$

$$2B(s) + 2H_2S(g) \xrightarrow[\sim 10^2\ Pa]{1200\ ^{\circ}C} 2HBS(g) + H_2(g)$$

4. Gas decomposition

$$P_2F_4(g) \xrightarrow[\sim 0.1\ Pa]{900\ ^{\circ}C} 2PF_2(g)$$

$$S_8(g) \xrightarrow[\sim 1\ Pa]{1300\ ^{\circ}C} 4S_2(g)$$

The above HT species are mostly unstable. This is also true for all the atom or molecular vapors of metals or nonmetals (C_2, C_3, C_4, P_2, As_2, Sb_2, S_2, and Se_2), lower chlorides (such as BX, CX_2, SiX_2 (X = F, Cl, Br), PI, PI_2, AlF, ScF, and LaCl) and lower oxides, sulfides, nitrides, and carbides (such as B_2O_2, CS, PN, and BC_2) in excited state or as free radicals. They are the necessary basis for the cryosynthetic reactions at low temperatures (cryosynthesis will be discussed in more detail in Chapter 3). The HT species mentioned above can be obtained using the following techniques.

2.7.1. Techniques of Metal Vapor Preparation

1. High-temperature vaporization using resistors

A metal can be vaporized or sublimated in a resistor-heated refractory oxide crucible or directly on a resistor under vacuum or in an inert gas atmosphere. An alumina crucible is commonly used as shown in Fig. 2.10.

Generally, a temperature of 1800°C can be achieved with molybdenum or tungsten wire resistor. Such a set-up is suitable for the vaporization or sublimation of chromium, manganese, iron, cobalt, nickel, palladium, copper, silver, gold, indium, and tin. The vaporization rate is usually 10–50 $g \cdot h^{-1}$, which is suitable for laboratory operations. Some metals require special oxide crucibles, e.g., ThO_2 crucibles for titanium, and BeO crucibles for silicon. For higher efficiency and less

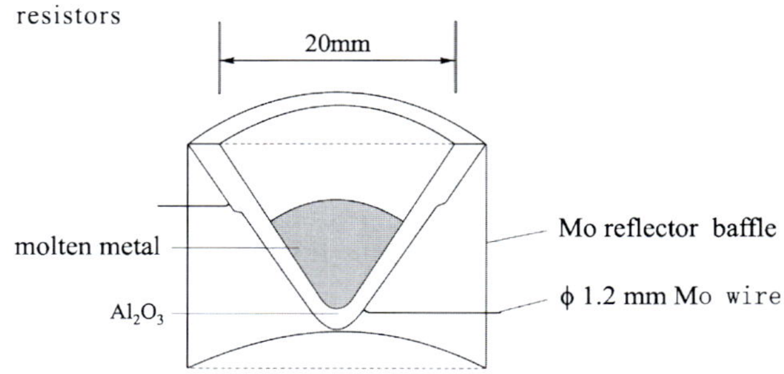

FIGURE 2.10 Alumina crucible for metal evaporation [71].

contamination, many metals such as calcium, magnesium, rate earths, silicon, uranium, vanadium, and titanium are evaporated or sublimated directly on refractory metal resistors. The metal resistors can be either wires, baskets, or boats of molybdenum or tungsten. The simplified diagrams of the experimental arrangement are shown in Figs 2.11 and 2.12.

2. Arc vaporization

It has high efficiency and can produce metal atoms of excited state for further reactions. The experimental arrangement is shown in Fig. 2.13.

3. Electron bombardment

An electric beam can be focused by passing through an electrostatic or magnetic field.

The energy density of the electric beam can reach a level of 10 W mn^{-2}. This suffices the rapid vaporization or sublimation of any substances. Figure 2.14 shows the typical electrostatic focusing equipment. Two points should be stressed when

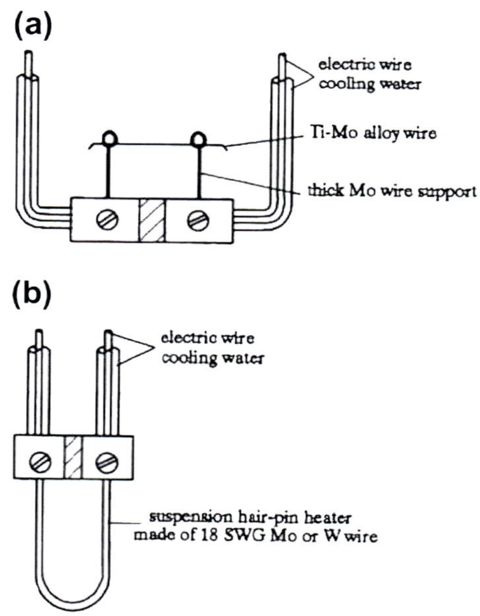

FIGURE 2.11 Metal evaporation on resistance wires [71].

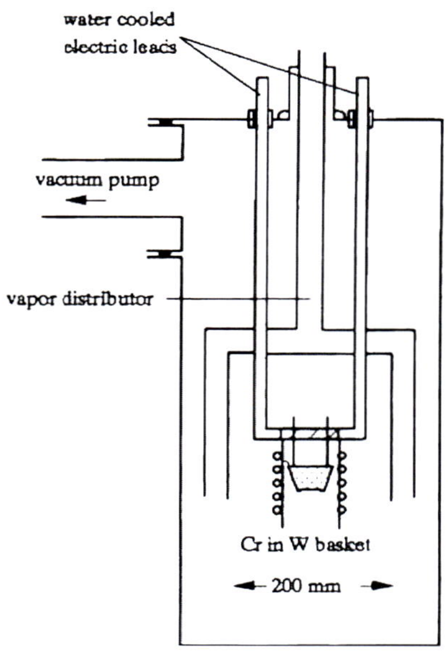

FIGURE 2.12 Metal evaporation using W wire basket [71].

using this technique in cryochemical synthesis. One is the possible by-reactions of some scattered electrons reflected from the heated metal with other reactant gases. The other is the stringent requirement for pressure in the reactor. A discharge occurs at $p > 0.1$ Pa, which can influence the metal vaporization. Nevertheless, the metal vaporization can be improved progressively by the installation of auxiliary instruments such as tungsten wires. Generally, metals are difficult to evaporate or sublimate but rare earths, molybdenum, niobium, tantalum, ruthenium, titanium, zirconium, and hafnium can be prepared using this method.

4. Laser beam vaporization

A ~200 W (niobium) yttrium–aluminum garnet laser unit of a miniature vaporizer is shown in Fig. 2.15. The focused laser beam has a high energy.

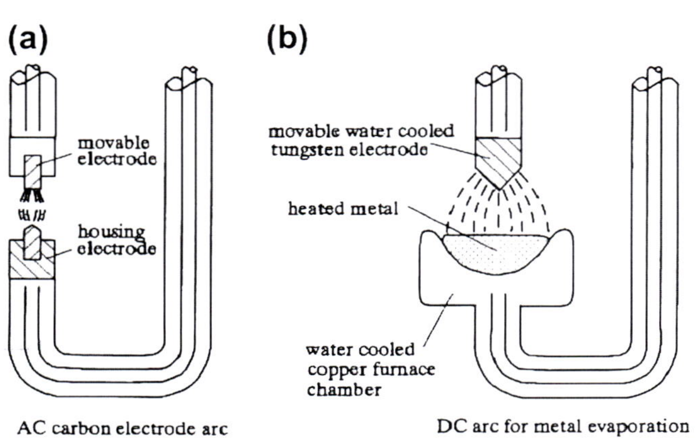

FIGURE 2.13 Arc evaporation [71].

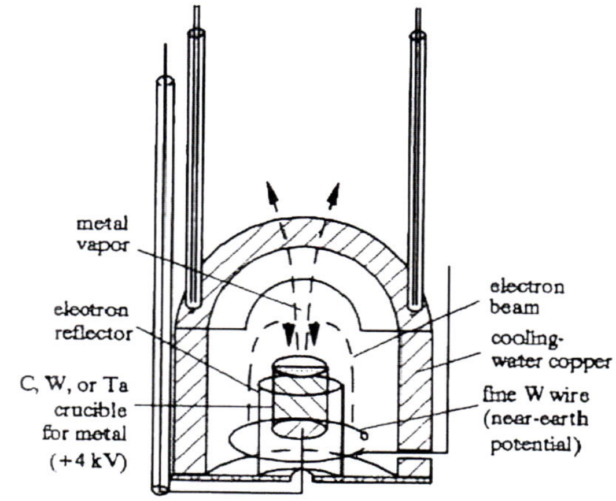

FIGURE 2.14 Electron beam evaporator—electrostatic focusing type [71].

This favors not only metal vaporization but also formation of excited atoms for an easier subsequent synthetic reaction.

2.7.2. Attainment of HT species

Many molecular species are also obtained from other high-temperature routes similar to those mentioned above, such as gas–solid reaction, vapor thermolysis, and discharge reaction. The attainment

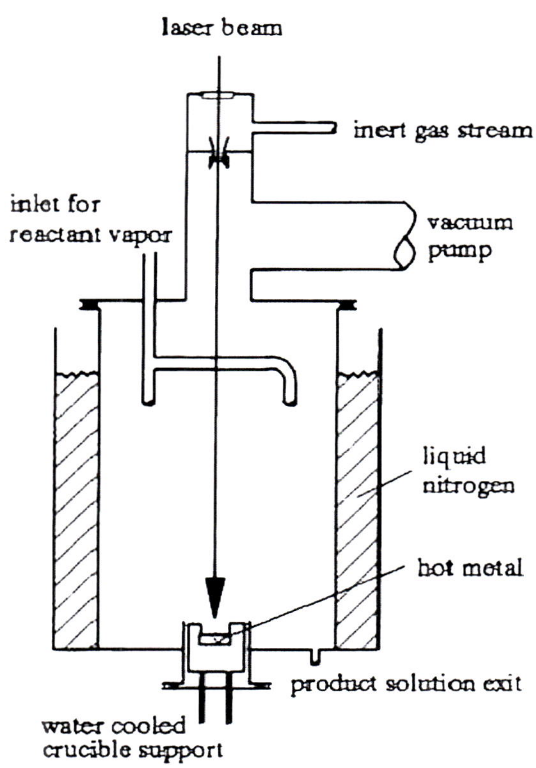

FIGURE 2.15 Laser evaporator [71].

of HT species will be illustrated in the following three examples.

1. High-temperature synthesis of BF via gas–solid reaction. BF is obtained through the following reaction:

$$2B(s) + BF_3(g) \xrightarrow[\text{Vacuum}]{1800\ ^\circ C} 3BF(g)$$

The apparatus is shown in Fig. 2.16. A graphite crucible holding boron particles of 4–10 μm is placed in the inducing coils. A temperature of 1600–1800°C can be achieved by applying a high frequency current. The boron particles react at low pressure (~10^2 Pa) BF_3 is introduced to the system and form BF. The production rate can be as high as 95%. This high-temperature gas–solid reaction apparatus is also suitable for synthesizing many lower chlorides of boron and silicon. In a liquid nitrogen reactor, the BF obtained can undergo a series of cryochemical synthetic reactions such as:

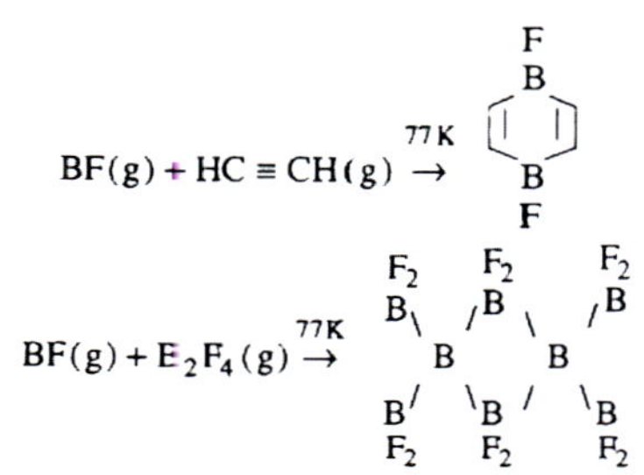

2. Preparation of BCl by flash thermolysis of B_2Cl_4

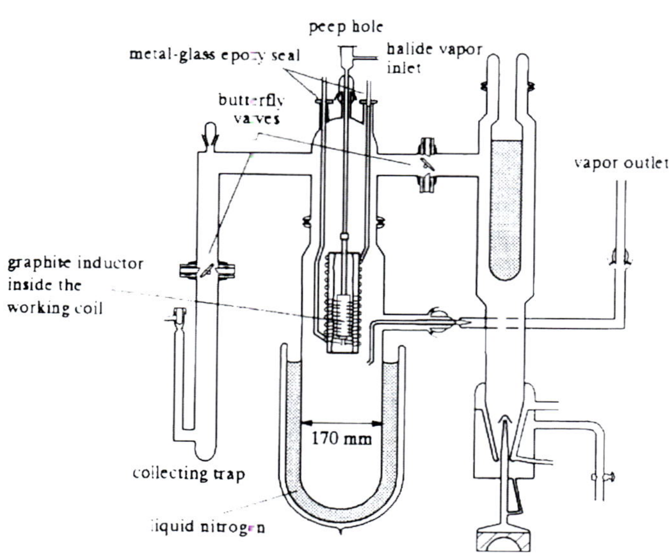

FIGURE 2.16 Apparatus for high-temperature gas-solid reaction [72].

There are generally two routes in B_2Cl_4 thermolysis:

$$B_2Cl_4(g) \rightarrow \begin{cases} \rightarrow B(s) + BCl_3(g) & (1) \\ \rightarrow BCl(g) + BCl_3(g) & (2) \end{cases}$$

Route (2) is rapid and referred to as flash thermolysis. Figure 2.17 shows an apparatus for flash thermolysis. At low pressure (feeding pressure 8×10^2 Pa) B_2Cl_4 gas is passed at a rate of 1–1.5 $mmol \cdot min^{-1}$ into a quartz tube of 20 mm long and 1 mm pore. The thermolysis zone ($T = 1100°C$) is heated with a molybdenum wire. The thermolysis product expands and enters the next stage of synthetic reaction in the cryochemical synthesizing zone conditioned with liquid nitrogen.

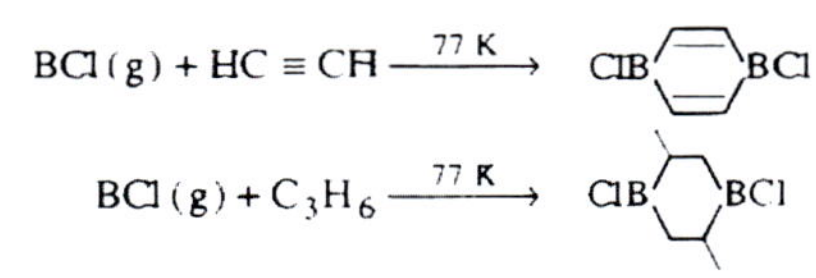

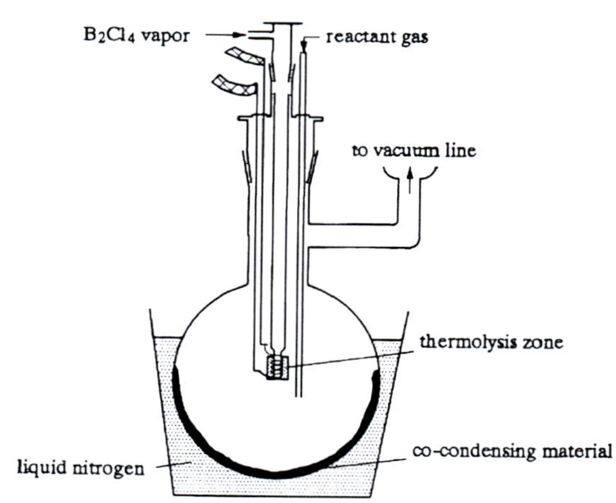

FIGURE 2.17 Apparatus for flash thermolysis of B_2Cl_4 [72].

3. Discharge synthesis of CS

CS is obtained using the discharge apparatus shown in Fig. 2.18 via the reaction: $CS_2(g) \xrightarrow{\text{AC discharge}} CS(g) + 1/8S_8(s)$. The main part of the apparatus is a Pyrex electric discharge tube of 400 mm long and 18 mm in diameter. The electrodes are welded inside the tube. The flow rate of CS_2 vapor entering the discharge tube is 40–150mmol·h^{-1} with a maximum pressure of 10^2 Pa. The discharge conditions are 12 kV and 10–30 mA. The production rate is 25%. The CS from the reaction can then react with HCl and HBr under the condition of liquid nitrogen and form a series of special compounds:

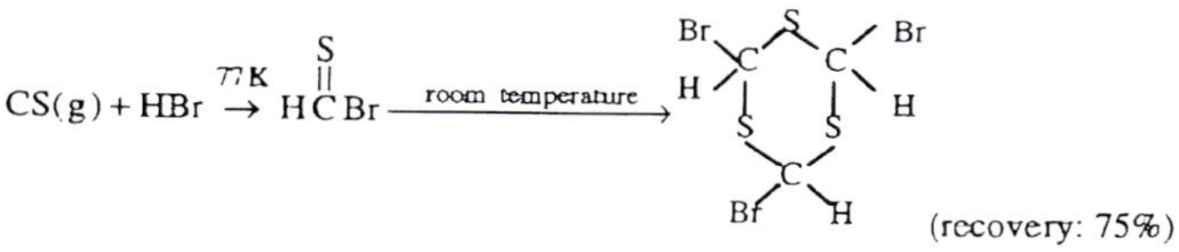

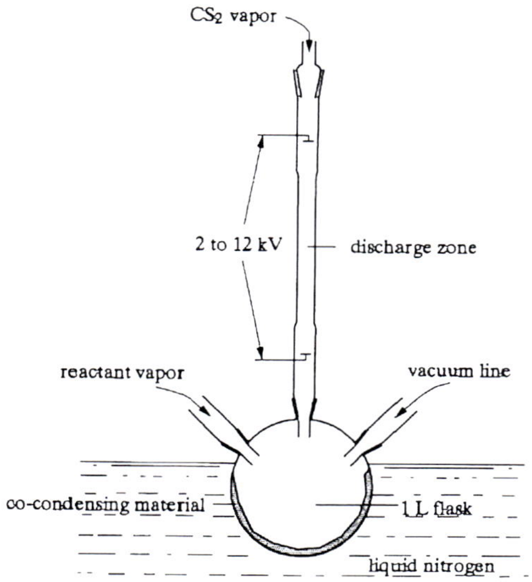

FIGURE 2.18 Apparatus for discharge synthesis of CS [72].

2.8. HIGH-TEMPERATURE ELECTROLYSIS IN MOLTEN SALT SYSTEM [73–79]

2.8.1. Basic of Molten Salt Electrolysis

Despite the wide application of molten salts, the research on their physical and chemical properties as ionic melts are far from adequate due to the high temperature and corrosivity involved. But generally, the properties of molten salts have theoretical and practical significance, such as specific gravity (Density), depression of freezing point, viscosity, vapor pressure, and surface tension, which significantly influences the Batch Composition, temperature, other electrolytic conditions and industrial practice of molten salt electrolysis [74–77].

2.8.2. Electrochemical Series of Molten Salts

The electrochemical series of molten salts can be obtained from the decomposition potentials calculated from the thermodynamic functions of the compounds, or measured directly from electrochemical experiments.

The relationship between the decomposition potential of molten salts and temperature can be deduced from the Gibbs–Helmholtz equation, whereas that between the mono-electrode potential and temperature can be obtained from the Nernst equation:

$$E = E^0 + \frac{RT}{nF} \ln a \tag{2.3}$$

With increasing temperature, the decomposition potential decreases, i.e., the temperature factor is negative.

The decomposition potential of a single molten salt can be determined theoretically from the change of the Gibbs energy of the primary cell during the electrochemical process, i.e.:

$$V_d = -\frac{\Delta G}{nF} \tag{2.4}$$

Where V_d is the decomposition potential; ΔG is the change of free energy; and n is the valence potential. Alternatively, these thermodynamic data can be obtained from the electrochemical measurements.

Using the graphical method on the $I-V$ curves drawn from experiment measurements, the decomposition potentials of single molten salts as well as salt mixtures can be determined.

2.8.3. Anode Effect

In some electrolytic processes of molten salts, an anode effect can sometimes occur. When it happens, the terminal voltage increases rapidly and the current density decreases sharply. Meanwhile, a phenomenon of poor wetting between the electrolyte and electrode appears. The electrolyte seems to be cut off from the anode by a layer of gas, and fine coronas due to spark discharge also appear around the anode.

The anode effect occurs only when the current density exceeds a "critical value". The critical current density of molten salts varies, and is related to the electrolytic conditions (e.g., temperature, electrolyte composition, and anode material). For example, the electrolysis of molten $Na_3AlF_6-Al_2O_3$ to Al is also carried out with a current density close to the critical current density, so the anode effect occurs frequently.

There are various opinions about the causes of the anode effect. It is well accepted that high interfacial tension between the electrolyte and anode results in poor wetting of the anode and induces the anode effect. When a large amount of Al_2O_3 (O^{2-}, AlO^+, and AlO_2^- ions) is contained in the cryolite melt, the interfacial tension between cryolite and carbon is low and the wettability is good. The gas is released at the anode surface in form of small gas bubbles as shown in Fig. 2.19. The contact between the electrolyte and anode is good, and the electrolysis can be carried out in a normal manner.

When the content of surface active Al_2O_3 is decreased in cryolite, the wettability of the electrolyte to the anode becomes deteriorated. The liberated anode gas is in the form of large gas bubbles, leaving convex bubble nuclei behind after the liberation as shown in Fig. 2.20. Eventually, the gas bubbles expel the electrolyte from most of the anode surface, forming an almost impermeable gaseous film, and the anode effect occurs.

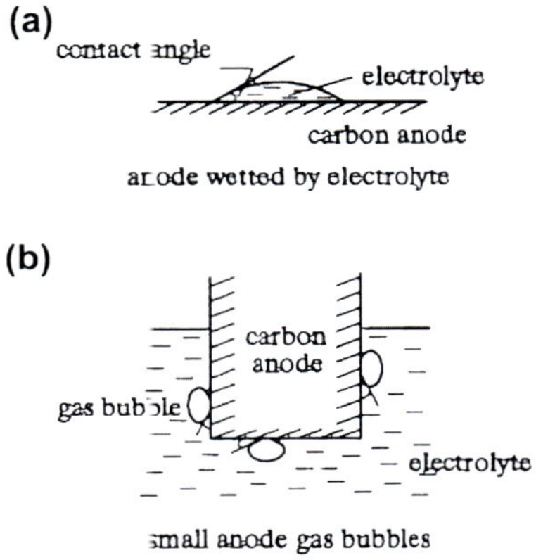

FIGURE 2.19 Normal electrolysis [77].

2.8.4. Examples of Molten Salt Electrolysis-Rare Earth Metal Preparation [76]

1. Introduction

There are two electrolyte systems for preparing rare earth metals from molten salt electrolysis: $RECl_3-MCl$ and $REF_3-MF-RE_2O_3$ (where M = alkali metal). The electrolytic preparation of mixed or single rare earth metals of melting points lower than 1000°C is usually carried out at a temperature higher than the melting points. The metals are in liquid form during the electrolysis, and form lump products after cooling. In the preparations of yttrium and heavy rare earth metal using molten salt electrolysis, low melting metals such as magnesium, zinc, and cadmium are usually used as liquid anodes to form alloys. The low melting metals are distilled afterward and rare earth metals are obtained. Liquid metals are also obtained directly from the electrolyses of oxide–fluoride melts.

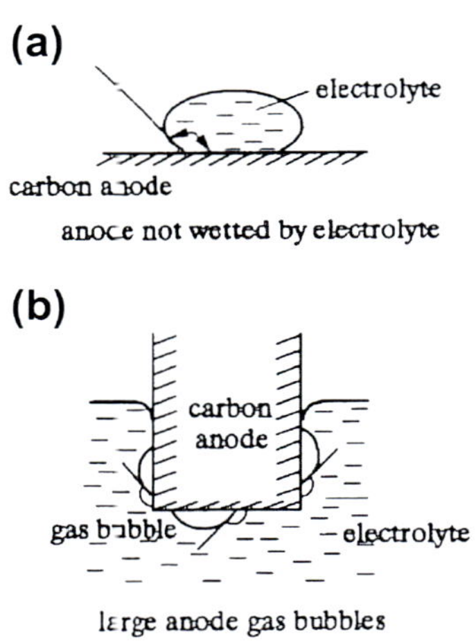

FIGURE 2.20 Electrolysis with anode effect [77].

2. Electrolytic process of rare earth chlorides

(a) Cathode process

In the electrolysis of the mixed chlorides of rare earth and alkali metals, the overall cathode process can be generally divided into three stages according to the studies on the polarography of molybdenum cathode. In this situation, the relationship between the current density of cathode and potential (based on chlorine reference electrode) is of particular concern.

1. Region of potential more positive than the equilibrium potential of rare earth metals. The cathode potential is between −1.0 and −2.6 V, at which impure cations having more positive potentials are discharged at the cathode. Meta-valent rare earth ions such as Sm^{3+} and Eu^{3+} also undergo incomplete discharge:

$$RE^{3+} + e^- \rightarrow RE^{2+}$$

2. Region of potential close to the equilibrium potential of rare earth metals. The cathode potential is around −3.0 V, and the cathode current density is from 0.1 to several $A \cdot cm^{2-}$ (determined by the $RECl_3$ content in the electrolyte and temperature). In this situation, rare earth metals are obtained directly from the electrochemical reduction:

$$RE^{3+} + 3e^- \rightarrow RE$$

From experiments, the liberation of rare earth metals is carried out at a potential close to the equilibrium potential (of corresponding concentration and temperature), and the overpotential is insignificant.

As mentioned previously, the rare earth metals liberated from electrolysis can re-dissolve in rare earth chlorides:

$$RE + 2RECl_3 \rightleftharpoons 3RECl_2$$

Or undergo exchange reactions with alkali chlorides:

$$RE + 3MCl \rightleftharpoons RECl_3 + 3M$$

The higher the electrolytic temperature, the more vigorous are these processes. Therefore, the temperature of the electrolysis should not be too high. In addition, the secondary reaction should be controlled by operation conditions, electrolyte composition, and cell construction (e.g., decreasing the contacting area between rare earth metals and molten salt).

It has been reported that at a potential between potential regions 1 and 2 mentioned above, the reduction of alkali ions to meta-ions also occurs:

$$2M^+ + e^- \rightarrow M_2^+$$

The alkali meta-ions further reduce RE^{3+} to metal particles that are then dispersed or dissolved in the electrolyte.

3. Region of potential more negative than the equilibrium potential of rare earth metals. The cathode potential is from −3.3 to −3.5 V, at which the reduction of alkali metals occurs:

$$M^+ + e^- \rightarrow M$$

This reaction proceeds in the following circumstances. The concentration of rare earth ions decreases gradually during the electrolysis. When the current is increased to a diffusion limit, the cathode polarization potential increases rapidly to a value at which alkali metals are liberated. To avoid this, the content of rare earth chlorides should be kept sufficiently high. In addition, the potential and current density of the cathode should be controlled within the range of discharging rare earth metals.

(b) Anode process

In a normal electrolytic process, chlorine is produced at the graphite anode. The main process is:

$$Cl^- - e^- \rightarrow Cl$$
$$2Cl \rightarrow Cl_2\uparrow$$
(Adsorption)

3. Current efficiency

Generally, the current efficiency of RE separation from molten salt electrolysis is low. This is because the secondary reaction such as the significant dissolution of rare earth metals in molten salts forms substances unfavorable to the electrolysis.

(1) Dissolution of rare earth metals in molten salts

It has long been known that a rare earth metal reacts with its own molten salt. This is believed to be a common phenomenon in molten salt electrolysis (a character distinguishable from the aqueous electrolysis) and a very important cause for the decrease in current efficiency. The solubility of a rare earth metal in its molten chloride is over 10–30 mol%.

The solubility of single light rare earth metal in its molten chlorides increases with atomic number. This is in correspondence with the decrease in current efficiency as shown in Table 2.8.

The significant difference in the solubility of rare earth metals in their molten chlorides can be explained with "contraction of lanthanum series". Neodymium and samarium have smaller atomic radii than lanthanum and cerium, and are easier to enter the voids of molten salt and thus have higher dissolution losses.

Because neodymium has higher solubility than lanthanum and cerium, more water-insoluble substances are formed from the dissolved neodymium under the effect of air in the same electrolysis period. In an Nd–$NdCl_3$ system, in addition to $NdCl_2$, $NdCl_{2.27}$, and $NdCl_{2.37}$, two other compounds have been found.

On one hand, rare earth metals of large atomic number have relatively high solubility in molten salts, and thus high losses from the secondary reaction; on the other hand, they also undergo incomplete electric liberation (e.g., $Sm^{3+} + e^- \rightarrow Sm^{2+}$) and result in relatively large idle electricity consumption. This is the reason for the decreased electric efficiency of light rare earth metals with increasing atomic number.

The dissolution loss of a metal in its salt has many factors. Generally, the loss increases with increasing temperature.

Some studies have been carried out to minimize the dissolution loss of metal in the

TABLE 2.8 Relationship Between the Solubility of Single Light Rare Earth Metal in its Molten Chloride and the Current Efficiency of Electrolysis

Metal	Molten salt	T / °C	Solubility mol. metal in salt	Current efficiency / %	Water insoluble substance after 1 hour of electrolysis / %
La	$LaCl_3$	1000	12	80	5.6
Ce	$CeCl_3$	900	9	77	6.8
Pr	$PrCl_3$	927	22	60	
Nd	$NdCl_3$	900	31	50.1	11.8
Sm	$SmCl_3$	>850	>30		
Mg	$MgCl_2$	800	0.2 ~ 0.3	>80	
Li	LiCl	640	0.5 ± 0.2		

molten salt. It has been found that the salt of a cation that has a more negative potential can decrease the solubility of a rare earth metal. For example the more the KCl added, the less the lanthanum loss.

Addition of NaCl can also decrease the dissolution loss of rare earth metals in their respective salts, but with a less significant effect.

(2) Effects of slag

In the electrolysis of rare earth molten salt, slag such as rare earth oxides, hydroxides, nitrides, and carbides are likely to form from the secondary reaction:

$$2RECl_3 + RE \rightarrow 3RECl_2$$
$$RECl_2 + O_2 + RE \rightarrow 2REOCl$$
$$2RE + N_2 \rightarrow 2REN$$
$$xRE + yC \rightarrow RE_xC_y$$
(such as RE_2C, RE_2C_3, and REC_2)
$$2RECl_3 + H_2O \rightarrow RE_2O_3 + 6HCl$$
$$RECL_3 + H_2O \rightarrow REOCl + 2HCl$$
$$2CeCL_3 + O_2 + 2H_2O \rightarrow CeO_2 + 4HCl + Cl_2$$

In open electrolytic cells at 580–600°C, an $RECl_3$–KCl melt becomes turbid due to the formation of RE_2OCl_4:

$$4RECl_3 + O_2 \rightarrow 2RE_2OCl_4 + 2Cl_2$$

RE_2OCl_4 encounters, and further reacts with, the dissolved rare earths, oxygen, and water in the gaseous environment:

$$2RE_2OCl_4 + RE \rightarrow 2REOCl + 3RECl_2$$
$$2RE_2OCl_4 + O_2 \rightarrow 4REOCl + 2Cl_2$$
$$RE_2OCl_4 + H_2O \rightarrow 2REOCl + 2HCl$$

Most of these products are insoluble in molten chloride and become the so-called slag. Although rare earth oxochlorides (REOCl and RE_2OCl_4) can dissolve in a chloride system with high solubility, they are influenced by the environment and change to precipitates, eventually becoming part of the slag. These components of slag are different from $RECl_3$ as they do not dissolve in water at room temperature. Therefore, they can be detected using water. This kind of slag is also known as "water-insoluble substance." The water-insoluble substance is a challenge in the RE electrolysis industry. It either deposits at the cell bottom and covers the cathode surface, preventing the accumulation of metal and thus normal electrolysis; or disperses in the melt as particles, increasing the electrolyte viscosity and the loss of metals via the formation of metal fogs. The main sources of slag are the oxo-RE compounds from water and oxygen. When the dehydration conditions for crystalline material are not appropriate (such as a low vacuum degree, a low NH_4Cl content, or a too high temperature), part of the RE chloride undergoes the following hydrolytic reaction:

$$RECl_3 \cdot H_2O \rightarrow REOCl + 2HCl$$

In addition to the water contained in the feed, the dehydrated feed absorbs water (especially in a rainy season) during storage and transportation if not sealed or thermally insulated. Therefore, slag is formed during electrolysis as described in the above reaction.

4. Preparation of RE chloride feed

There are two ways to prepare the anhydrous RE chlorides for molten salt electrolysis: the dehydration of crystalline material and chlorination by ammonia chloride.

In the first preparation method, the hydrated rare earth chloride $RECl_3 \cdot 6H_2O$ (crystallized material) is slowly heated to a high temperature in a vacuum dehydrator. Generally, it is heated at 100, 120, and 200°C and takes 6–10 h to dehydrate:

$$RECl_3 \cdot 6H_2O \xrightarrow{100\ ^\circ C} RECl_3 \cdot 3H_2O + 3H_2O$$

$$RECl_3 \cdot 3H_2O \xrightarrow{120\ ^\circ C} RECl_3 \cdot H_2O + 2H_2O$$

$$RECl_3 \cdot H_2O \xrightarrow{200\ ^\circ C} RECl_3 + H_2O$$

Hydrolysis is very likely when the last hydration water is lost:

$$RECl_3 \cdot H_2O \rightarrow REOCl + 2HCl$$

Two methods can be used to prevent hydrolysis: one is to increase the vacuum degree and thus reduce the water partial pressure; the other is to add sufficient amount of hydrogen chloride to the system and force the reaction to proceed in the direction of forming $RECl_3$:

$$REOCl + 2HCl \rightarrow RECl_3 + H_2O$$

Hydrogen chloride can be added at the final stage of dehydration, which can result in fairly pure anhydrous RE chlorides. Alternatively, if the subsequent operations are not affected, an appropriate amount of ammonia chloride (1.5–5%) can be added to the precondensed crystallized material while it is still warm. The advantage of the latter procedure is that the HCl resulting from the decomposition of NH_4Cl at an elevated temperature has a protection effect on $RECl_3$ from the hydrolysis:

$$NH_4Cl \rightarrow NH_3 + HCl$$

The second RE chloride preparation method is the chlorination using a dry ammonia chloride powder. The mechanism is to obtain anhydrous RE chlorides by mixing loose RE oxides or carbonates with solid ammonia chloride:

$$RE_2O_3 + 6NH_4Cl \overset{200-400^\circ C}{=} 2RECl_3 + 3H_2O + 6NH_3$$

$$RE_2(CO_3)_3 + 6NH_4Cl \overset{200-450^\circ C}{=} 2RECl_3 + 3H_2O + 6NH_3 + 3CO_2$$

The reaction container can be an open ceramic evaporator. An electric furnace can be used as the means of heating. The material can be stirred with a nickel spade to achieve uniform temperature and to diffuse the resultant water. A tube furnace can also be used. Care should be taken that the container should not be full, and a space should be reserved for the expansion of the reactant materials during the reaction. In addition, the gas outlet should be downward so that water can be discharged in time. In the preparation of RE chlorides, especially heavy RE chlorides containing little water-insoluble substances, it is feasible to exercise multiple chlorination cycles. The sample is withdrawn after primary chlorination and more ammonia chloride is mixed to carry out the secondary chlorination. Since most of the water has been expelled in the previous chlorination, the secondary chlorination can be very complete. From the differential thermal analysis, the reactions proceed stepwise. Chlorides and hydroxides react with excess ammonia chloride and form various compounds. Only at a temperature higher than 430°C, ammonia chloride volatilizes completely from the system. If the product obtained at 200–360°C is a hard mass, which is due to poor water discharge, the dehydration reaction should be continued at >350°C until the complete discharge of water. A loose packed product material can then be obtained.

5. Operation conditions of electrolysis

(1) Electrolyte composition

The electrolyte used in the electrolytic production of mixed or single rare earth metals is composed of rare earth chloride and sodium chloride or potassium chloride (usually $CaCl_2$ and $BaCl_2$ are also contained). The rare earth chloride accounts for 30–50%, or sometimes 20–35% of the electrolyte mass in industrial production, or 40–60% (equivalent of a 1:2–3 mol ratio of $RECl_3$:KCl) in the laboratory preparation. A too high or too low initial concentration of rare earth chloride may result in a decreased current efficiency. This is because K^+ (or Na^+) and RE^{3+} ions can be discharged together at a low $RECl_3$ content, and the electrolyte reacts with oxygen and becomes viscous at a too high $RECl_3$ content causing a high resistance and a high loss of the rare earth metal in the salt.

(2) Electrolytic temperature

This is an important factor. The common temperature ranges used are: 820–900°C (870°C) for mixed rare earth metals; 820–910°C (850–870°C) for cerium; 850–960°C (920–930°C) for lanthanum; and 920–960°C (950°C) for praseodymium. The effects of electrolytic temperature can be illustrated by the electrolyses of cerium and lanthanum. At too low temperature, the metal disperses in the melt and is unfavorable for accumulation; whereas at a too high temperature, the rare earth metal has an even higher reactivity, its loss thus increases, and so does the salt loss due to evaporation. At high temperature, interactions between the metal, electrolytic atmosphere, and construction materials become more vigorous, and the current efficiency, recovery, and metal quality decreases. Therefore, some studies of electrolysis at temperatures below the melting point of the metal have been carried out. For instance, spongy neodymium has been synthesized from the electrolysis of $NdCl_3$–NaCl–KCl at 750°C. The product is melted after the electrolysis.

(3) Current density

The current density J at the cathode is generally 3–6 $A \cdot cm^{-2}$. The current density is determined by the concentration of the rare earth ions, circulation of the electrolyte, and electrolytic temperature. An appropriate increase in current density can result in a more negative cathode potential, and favor the discharge of rare earth metals by reducing the relative loss of the metal and increasing the current efficiency. This is because the current density at cathode determines the relative speed of metal dissolution and discharge. However, if the current density at the cathode is too high, the possibility of alkali metal reduction increases, and the cathode zone or even the overall fused salt can be overheated.

6. Purity of rare earth metal products

The purity of industrial electrolytic products of mixed rare earth metals is generally around 98%, and that of single rare earth metals is around 99%. Purer metals can be obtained under laboratory electrolytic conditions. For instance, for lanthanum with a rare earth impurity level below the lower limit of spectroscopy, a maximum non-RE impurity of 0.03%

has been obtained. When preparing pure rare earth metals by electrolysis, in addition to the pure feed and high-quality construction materials (metal container, graphite, and sleeve tube), the electrolytic conditions such as a low temperature and a closed system should be used.

2.8.5. Other Applications of Synthesis by Molten Salt Electrolysis [73–75]

1. Preparation of alloys

In addition to the preparation of rare earth metals by molten salt electrolysis, the electrosynthesis of RE-CM alloys has also been developed. So far, binary alloys such as RE-Al, RE-Mg, RE-Fe, and RE-Sn and ternary composition alloys such as RE-Al–Sn and RE-Al–Si have been obtained from molten salt electrolysis. Furthermore, the characteristics of large alloying heat of Al-rich and Mg-rich zones have been used to produce Al-base and Mg-base rare earth alloys using the reduction method. This simple and convenient technique can be used in the smelting furnaces in plants, and is proved to be feasible in reducing the production cost of rare earth alloys.

2. Coating films on metals

Using the interaction between high-melting metals and molten salts, a high-melting metal can be plated on another metal at a temperature well below its melting point. For instance, a steel specimen and a piece of titanium (molybdenum or chromium) can be added to a molten salt of KCl–NaF at 800°C. A layer of titanium can be plated to the steel surface after a period of time, although steel is not in contact with titanium. An alloy layer is formed between the plated film and the substrate, therefore the plated film is dense and rigid. This is a phenomenon similar to the anode corrosion. When the substrate specimen and the high-melting metal are respectively connected to the cathode and anode while applying DC, the growth of the plated film can be controlled and a high-quality film can be obtained.

3. Synthesis of high-melting binary metal compounds of ceramic type

There have been many high-melting binary compounds such as those of metal with carbon, boron, silicon, phosphorus, sulfur, selenium, and tellurium. They all can be synthesized through the cathode discharge from the molten salt electrolysis. For example, an $MgO–MgF_2–2B_2O_3–½TiO_2$ system can be electrolyzed to TiB_2 at 1000°C, 20 A, and 7 V. An $Na_2O–NaF–2B_2O_3–(1/3–1/9)WO_3$ system can be electrolyzed to WB at 1000°C, 20 A, and 4 V. CaB_6, BaB_6, LaB_6, CeB_6, CrB, TiB_2, and ZrB_2 are all obtained using the same method. Other possible substances from the molten salt electrolysis are the carbides of calcium, molybdenum, silicon, tungsten, iron, niobium, and tantalum, the phosphides of iron, gallium, indium, manganese, molybdenum, and tungsten, $CaSi_2$, Li_6Si_2, and a variety of sulphides of other metals. Some binary plated films have also been prepared.

4. Synthesis of middle valence compounds

Some middle compounds that cannot be prepared by the electrolysis of aqueous solutions can usually be prepared by the electrochemical redox reaction of molten salt. For example, a series of middle oxides of high-melting metals such as Mo_2O_3, MoO_2, U_4O_9, VO, VO_2, NbO, and TaO, and middle chlorides such as ZrCl, $ZrCl_3$, $HfCl_2$, and $HfCl_3$ have been prepared from molten salt electrolysis. Russian scientists Kaliev and Baraboshkin reported many complexes of bronze compound series of vanadium, molybdenum, and tungsten crystallized through the electrolysis of molten salt [78,79]. Another example is to reduce the molten mixture of WO_3 and alkaline tungstate electrochemically in neutral atmospheres, according to the reaction [80]:

$$x/2Na_2WO_4 + (1-x)WO_3 \xrightarrow[\text{neutral atmosphere}]{800°C} \underset{\text{on the cathode}}{Na_xWO_3} + \underset{\text{on the anode}}{x/4O_2}$$

The electrodes can be graphite as anode and Pt or CrNi alloy as cathode. Well-formed single crystals can grow by applying a seed on the cathode and using about 20 mA cm^{-2} current density.

Figure 2.21 shows a double cell, which was constructed by Perloff and Wold for preparing tungsten oxide bronze single crystals by electrolysis from a melt.

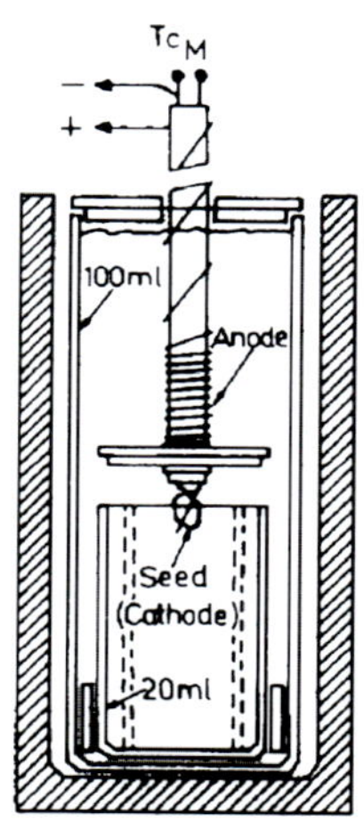

FIGURE 2.21 Double electrolytic cell [80]. *Reprinted from International Journal of Refractory Metals and Hard Materials, 13 (1995), L. Bartha, A. B. Kiss, T. Szalay, Chemistry of tungsten oxide bronzes, page 77, Copyright (1995), with permission from Elsevier.*

References

[1] http://www.thermalthechologies.com, Thermal Technologies, Inc.
[2] J.L. Margrave, R. Hauge, W.R. Bryaut (Eds.), High Temperature Technique in Techniques of Chemistry Chemical Experimentation under Extreme Conditions, John Wiley & Sons, New York, 1980.
[3] J.H. Lemelson, High Temperature Reaction Method, US, 1997.
[4] P. Bernard, F. Prudtion, High Temperature Reaction Apparatus, US, 1986.
[5] J.H. Lemelson, High Temperature Reaction Apparatus, US, 1996.
[6] T. Nishimaki, Y. Nakaminato and K. Fujimoto, Apparatus and Processes for Preparing High Temperature Reaction Products, US, 1984.
[7] X.Q. Xin, L.M. Zheng, Journal of Solid State Chemistry 106 (1993) 451.
[8] F. Toda, K. Tanaka, A. Sekikawa, Journal of the Chemical Society-Chemical Communications (1987) 279.
[9] J.G. Li, X.Q. Xin, Z.Y. Zhou, K.B. Yu, Journal of the Chemical Society-Chemical Communications (1991) 249.
[10] T.Y. Zhou, X.Q. Xin, Nanotechnology 15 (2004) 534.
[11] U. Schubert, N. Hüsing, Synthesis of Inorganic Material, Wiley-VCH, Weinheim, 2004.
[12] A.R. West, Solid State Chemistry and Application, John Wiley and Sons Inc., New York, 1992.
[13] P. Hagenmuller, Preparative Methods in Solid State Chemistry, Academic Press, New York, 1972.
[14] J.M. Honig, C.N.R. Rao, Preparation and Characterization of Materials, Academic Press, New York, 1981.
[15] V.V. Boldyrev, Reactivity of Solids: Past, Present and Future: A "Chemistry for the 21st Century Monograph", Blackwell Science Ltd, Chichester, 1996.
[16] Q. Su, "Chemistry of Rare Earths", (in Chinese) Publisher of Henan Science and Technology, 1993.
[17] J.P. Zhong, H.B. Liang, Q. Su, G.B. Zhang, P. Dorenbos, M.D. Birowosuto, Journal of Rare Earths 25 (2007) 568.
[18] Q. Su, H. Xie, F. Fang, J. Wang, J. Zhong, X. Li, H. Xu, N. Wu, L. Pan and J. Li, Chinese Patent ZL 2005 1 0102388.7, China, 2005.
[19] Y.X. Pan, M.M. Wu, Q. Su, Materials Science and Engineering B-Solid State Materials for Advanced Technology 106 (2004) 251.
[20] T.Z. Zhang, Q. Su, The Journal of the Society for Information Display 8 (2000) 27.
[21] M. Zhang, J. Wang, Q.H. Zhang, W.J. Ding, Q. Su, Materials Research Bulletin 42 (2007) 33.
[22] Z.W. Pei, Q. Su, J.Y. Zhang, Journal of Alloys and Compounds 198 (1993) 51.
[23] Q. Su, Q.H. Zeng, Z.H. Pei, Chinese Journal of Inorganic Chemistry 16 (2000) 293.
[24] Q. Su, H.B. Liang, T.D. Hu, Y. Tao, T. Liu, Journal of Alloys and Compounds 344 (2002) 132.
[25] X.M. Xun, S.H. Feng, J.Z. Wang, R.R. Xu, Chemistry of Materials 9 (1997) 2966.
[26] J.L. Zhuang, L.F. Liang, H.H.Y. Sung, X.F. Yang, M.M. Wu, I.D. Williams, S.H. Feng, Q. Su, Inorganic Chemistry 46 (2007) 5404.
[27] W.M. Higgins, A. Churilov, E. van Loef, J. Glodo, M. Squillante, K. Shah, Journal of Crystal Growth 310 (2008) 2085.
[28] J. Flahaut, in K.A. Gscheidner, Jr. and L. Eyring (Eds.), Handbook on the Physics and Chemistry of Rare-Earths, Vol. 4, North-Holland, Amsterdam, 1979.
[29] H.B. Yuan, J.H. Zhang, R.J. Yu, Q. Su, Journal of Rare Earths 27 (2009) 308.
[30] T.E. Peters, J.A. Baglio, Journal of the Electrochemical Society 119 (1972) 230.
[31] Q. Su, J. Xu, J. Zhang, Chinese Patent ZL 02 1 52035.6, China, 2004
[32] Q. Su, X. Zhang, J. Xu, J. Zhang, Chinese Patent ZL 200310111931.0, China, 2006
[33] R.J. Yu, J. Wang, J.H. Zhang, H.B. Yuan, Q. Su, Journal of Solid State Chemistry 181 (2008) 658.
[34] R.J. Yu, J. Wang, M. Zhang, J.H. Zhang, H.B. Yuan, Q. Su, Chemical Physics Letters 453 (2008) 197.
[35] Y.Q. Li, C.M. Fang, G. de With, H.T. Hintzen, Journal of Solid State Chemistry 177 (2004) 4687.
[36] R.J. Xie, N. Hirosaki, Science and Technology of Advanced Materials 8 (2007) 588.
[37] Y.Q. Li, G. de With, H.T. Hintzen, Journal of the Electrochemical Society 153 (2006) G278.
[38] M. Zhang, B.H. Li, J. Wang, Z.Y. Zhang, Q.H. Zhang, Q. Su, Journal of Materials Research 24 (2009) 2589.
[39] M. Zhang, J. Wang, Z. Zhang, Q. Zhang, Q. Su, Applied Physics B-Lasers and Optics 93 (2008) 829.
[40] F. Tessier, R. Marchand, Journal of Solid State Chemistry 171 (2003) 143.
[41] Y.I. Kim, P.M. Woodward, Journal of Solid State Chemistry 180 (2007) 3224.
[42] M. Jansen, H.P. Letschert, Nature 404 (2000) 980.
[43] Y. Kamihara, T. Watanabe, M. Hirano, H. Hosono, Journal of the American Chemical Society 130 (2008) 3296.
[44] G. Wu, Y.L. Xie, H. Chen, M. Zhong, R.H. Liu, B.C. Shi, Q.J. Li, X.F. Wang, T. Wu, Y.J. Yan, J.J. Ying, X.H. Chen, Journal of Physics-Condensed Matter 21 (2009) 142203.
[45] U. Schubert, N. Hüsing, Synthesis of Inorganic Materials, Wiley-VCH, 2005. Vol. 4.5, Sol-Gel Process.
[46] D. Sangeeta, R. LaGraff, Inorganic Material Chemistry Desk Reference, second ed. CRC Press, 2005.
[47] L.L. Hench, J.K. West, Chemical Reviews 90 (1990) 33.
[48] M. Galan, J. Llorens, J.M. Gutierrez, C. Gonzalez, C. Mans, Journal of Non-Crystalline Solids 147 (1992) 518.
[49] M. Nagano, M. Greenblatt, Solid State Communications 67 (1988) 595.
[50] J. Livage, M. Henry, C. Sanchez, Progress in Solid State Chemistry 18 (1988) 259.
[51] A.E. Sytschev, Self-Propagating High-Temperature Synthesis: Theory and Practice, Territorya, Chernogolovka, 2001 (in Russian).
[52] G.A. Merzhanov, Self-Propagating High-Temperature Synthesis: Twenty Years of Search and Findings, ISMAN, Chernogolovka, 1989.
[53] P. Mossino, Ceramics International 30 (2004) 311.
[54] M. Koizumi, Chemistry of Combustion Synthesis, Mir Publ., Moscow, 1998.
[55] S. Yin, Combustion Synthesis (in Chinese), Beijing, 1998.
[56] E.A. Levashov, A.S. Rogachev, V.I. Yukhvid, L.P. Borovinskaya, Physico-Chemical and Technological Foundations of Self-Propagation High-Temperature Synthesis, Binom, Moscow, 1999.
[57] Z.A. Munir, American Ceramic Society Bulletin 67 (1988) 342.
[58] A.G. Merzhanov, Ceramics International 21 (1995) 371.
[59] A.G. Merzhanov, Combustion Processes and Materials Synthesis, ISMAN, Chernogolovka, 1998.
[60] A.G. Merzhanov, Solid Flame Combustion, Izd. ISMAN, Chernogolovka, 2000 (in Russian).
[61] A.W. Weimer, Carbide, Nitride and Boride Materials Synthesis and Processing, Chapman & Hall, London-Weinheim-New York-Tokyo-Melbourne-Madras, 1997.
[62] S.Y. Sharivker, A.G. Merzhanov, SHS-Produced Powders and Their Processing, Izd. ISMAN, Chernogolovka, 2000.
[63] F. Bernard, E. Gaffet, International Journal of SHS 10 (2001) 109.

[64] H.C. Yi, J.J. Moore, Journal of Materials Science 25 (1990) 1159.
[65] L.L. Wang, Z.A. Munir, Y.M. Maximov, Journal of Materials Science 28 (1993) 3693.
[66] J.A. Pusszynski, Thermochemistry and kinetics, in: A.W. Weimer (Ed.), Carbide Nitride and Boride Material, Chapman and hall, London, 1997, pp. 183–228.
[67] I.A. Filimonov, N.I. Kidin, Combustion Explosion and Shock Waves 41 (2005) 639.
[68] S. Gedevanishvili, Z.A. Munir, Materials Science and Engineering A-Structural Materials Properties Microstructure and Processing 246 (1998) 81.
[69] R. Orra, G. Cao, Z.A. Mumir, Metallurgical and Materials Transaction A 30A (1999) 1101.
[70] B.I. Rharisov, A.D. Garnovskii, L.M. Blanco, et al., Journal of Coordination Chemistry 49 (1999) 113.
[71] M. Moskovits, G.A Ozin, Cryochemistry, John Wiley and Sons Inc., New York, 1976, pp. 75–81.
[72] M. Moskovits, G.A Ozin, Cryochemistry, John Wiley and Sons Inc., New York, 1976, pp. 99–113.
[73] G.E. Marcelle, Molten Salts: From Fundamental to Applications, Kluwer Academic Publisher, 2001.
[74] M.J. Zhang, S.W. Wang, Electrochemistry of Molten Salts: Fundamental and Applications, (in Chinese) Chemical Industry Press, 2006.
[75] D.X. Tang, Electrometallurgy of Molten Salts, (in Chinese) Beijing Iron and Steel Institute Press, Beijing, 1985.
[76] D.X. Tang, Rare Earth, (in Chinese) Metallurgical Industry Press, Beijing, 1995.
[77] S.Y. Shen, F.H. Hu, Fundamental of Electrochemistry of Molten Salts, (in Chinese) Chinese Industry Press, Beijing, 1965.
[78] K.A. Kaliev, A.N. Baraboshkin, S.M. Zakharyash, Soviet Electrochemistry 20 (1984) 307.
[79] A.N. Baraboshkin, K.A. Kaliev, A.G. Aksentev, Soviet Electrochemistry 17 (1981) 606.
[80] L. Bartha, A.B. Kiss, T. Szalay, International Journal of Refractory Metals Hard Materials 13 (1995) 77.

CHAPTER

3

Synthesis and Purification at Low Temperature

Wenqin Pang

Jilin University, China

3.1. ATTAINMENT AND MEASUREMENT OF LOW AND ULTRALOW TEMPERATURES

3.1.1. Attainment

1. Common cold baths

(1) Ice–salt eutectic systems

Thorough mixing of crushed ice with salt (use ice mill to mill it) can give low temperatures. For example, different temperatures can be reached using the following ice–salt mixtures:

3 parts ice + 1 part NaCl → −21°C

3 parts ice + 1 part $CaCl_2$ → −40°C

2 parts ice + 1 part con. HNO_3 → −56°C

(2) Dry ice bath

This low temperature bath is also commonly used. Its sublimation temperature is −78.5°C. When it is used, inert solvents such as acetone, alcohol, and chloroform should always be added to improve the thermal conduction.

2. Phase change cold baths

This type of low temperature bath can maintain constant temperature. For example, in CS_2 case −111.6°C can be obtained. This temperature is the solid–liquid equilibrium point (i.e., melting point) for carbon disulfide under standard atmospheric pressure. Commonly used constant-temperature cold baths (phase change) are listed in Table 3.1.

3. Laboratory Cryogenic Systems

This is an important route for synthesizing many compounds of rare-gas such as HXY molecules (X = Kr, Xe, Ar, Y = halides CN, SH) and many volatile inorganic compounds. Recently, a complete range of both cryogen-cooled and cryogen-free equipment for low and ultralow temperature synthetic research has been developed. Refer to Table 3.2 below for guidance in choosing the right system (for example, we introduce the products from Janis Research Co. Inc., USA) for laboratory research applications.

A full range of accessories and ancillary equipment for various cryogenic equipment needs are given below: Flexible Helium Transfer Lines; Liquid Helium Storage Dewars and Accessories; Liquid Nitrogen Storage Dewars; Mechanical Pumping Stations and High Vacuum Pumping Stations; Cryostat Accessories; and Temperature Controllers.

TABLE 3.1 Phase Change Temperature of Some Common Low Temperature Baths

Low temperature bath	Temperature (°C)
Ice + water	0.0
Carbon tetrachloride	−22.8
Liquid ammonia	−33.0 to −45.0
Chlorobenzene	−45.2
Chloroform	−63.5
Dry ice	−78.5
Ethyl acetate	−83.6
Toluene	−95.0
Carbon disulfide	−111.6
Methylcyclohexane	−126.3
n-Pentane	−130.0
iso-Pentane	−160.5
Liquid oxygen	−183.0
Liquid nitrogen	−196.0

Modern Inorganic Synthetic Chemistry, DOI: 10.1016/B978-0-444-53599-3.10003-4

TABLE 3.2 Laboratory Cryogenic Systems Guide

		ST	STVP	SVT	VT	Immersion	VNF	VPF	CCR	SCON
Useable cryogen	Liquid Nitrogen	●	○	○	●	●	●	●		
	Liquid Helium	●	●	●	●	●				●
Temperature range	<2 K to 77 K	●	●	●	●				*	●
	77 K to 325 K	●	●	●	●			●	●	●
	325 K to 500 K	●			○		●	●	●	
	500 K to 600+K	○						●	○	
Sample environment	Vacuum	●		○	●			●	●	
	Vapor/Liquid		●	●			●			●
	Exchange gas		○	○	●				●	
Cryogen usage liters/h	LHe at 4.2 K	0.5	1.1	0.3	0.3	0.15	N/A	N/A	N/A	0.3
	LHe at 100 K	0.1	0.2	0.2	0.4	N/A	N/A	N/A	N/A	0.3
	LN2 at 100 K	0.1	0.2	0.1	0.2	0.1	0.2	0.1	N/A	N/A
Cryostat hold time (hours)		N/A	N/A	8	6	8	6–8	5	N/A	12
Initial cool-down time (min.)		15	15	60	60	60	30	15	60	120

●, Standard model available; ○, available as an option; and *, not available <4 K.

Note: Above are approximate values for basic models and vary depending on the exact configuration.

Model types of cryogenic equipment, (product from Janis Research Company Inc.): VNF: Liquid Nitrogen, Variable Temperature Cryostats-sample in Vapor; VPF: Liquid Nitrogen, Variable Temperature Cryostats-sample in Vacuum; ST: Liquid Helium, Continuous Flow Cryostat Systems-sample in Vacuum; STVP: Liquid Helium, Continuous Flow Cryostat Systems-Sample in Vapor; SVT: Liquid Helium, Variable temperature (Reservoir) Cryostats-Flowing Vapor; VT: Liquid Nitrogen, Variable Temperature Cryostats-Vacuum or Exchange gas; SCON: Liquid Helium, Superconducting Magnet Systems; and CCR: Cryogen-free, Closed Cycle Refrigeration.

3.1.2. Thermometry

The most common thermometers in laboratory for measuring low and ultralow temperature are: (1) Thermal resistance thermometer, (2) Thermocouple, and (3) Thermistor.

Some types (from products of Lakeshore Co. Inc., USA) and related properties are given in Table 3.3.

3.2. VACUUM TECHNIQUE AND ITS APPLICATION IN INORGANIC SYNTHESIS

Vacuum technique is a very important experimental method in inorganic synthesis. The synthesis or separation of many volatile compounds and those sensitive to oxygen, air, and water vapor needs to be carried out under vacuum conditions. Some precision measurements such as LEED (Low-Energy Electron Diffraction), AES (Auger Electron Spectroscopy), and ISS (Ion Scattering Spectroscopy) also need to be operated under ultrahigh vacuum of 10^{-9} Pa. And many experiments in the synthetic reactions, purifications, and characterizations are apparently dependent on the vacuum conditions. Therefore, it is highly necessary for inorganic synthesists to understand and master vacuum technique.

Vacuum is not absolute emptiness, but is a state of gaseous space that has a pressure lower than that of the standard atmosphere. The scarceness level of gas in a vacuum state is usually called the vacuum degree, and expressed with a value of pressure (Pa). To use the international system of units (SI) suggested worldwide, the conversion of Torr and mbar to Pa is:

$$1\ \text{Torr} = 1\ \text{mmHg} = 133.322\ \text{Pa}$$

$$1\ \text{mbar} = 100\ \text{Pa}$$

For convenience in practice, vacuum degree is classified into the following groups. This is done by referring to the physical properties of gaseous space, the effective ranges of common vacuum pumps and gauges, and the application features of vacuum technique.

TABLE 3.3 Types of Thermometers in Laboratory

Type	Model	Temperature range (K)	Relative accuracy (20 K)/mk	Reproducibility (4.2 K)/mk
Silicon	DT-670-SD	1.4–500	±14	±10
Diode	DT-471-SD	10–500	±14	±10
GaAs Diode	TG-120-SD	1.4–500	±14	±10
Platinum RTD	PT-103	14–873	±10	±5
Cernox™ RTD	CX-1010-SD	0.1–325	±9	±3
	CX-1050-SD-HT	1.4–420	±9	±3
	CX-1080-SD-HT	20–420	±9	±3
Germanium RTD	GR-200A-30	0.05–4.2	±9	±0.5
	GR-200A-250	0.5–100	±90	±0.5
Carbon-Glass RTD	CGR-1-500	1.4–325	±8	±0.75
Rox-type	RX-102A-AA	0.05–40	±37	±15
Rhodium-Iron	RF-800-4	0.65–800	±14	±5
Capacitance	CS-501-GR	1.4–290	NA	±10
Thermocouple	Type K	3.2–1543	NA	NA
	Type E	3.2–953	NA	NA
Thermometer	Chromel-AuFe	1.2–600	NA	NA

Low vacuum	10^5–10^2 Pa
Medium vacuum	10^2–10^{-1} Pa
High vacuum	10^{-1} to 10^{-5} Pa
Ultrahigh vacuum	10^{-6} to 10^{-9} Pa
Extreme vacuum	$<10^{-9}$ Pa

3.2.1. Vacuum Attainment [1–3]

The process of achieving a vacuum is known as evacuating. The equipment used is called a vacuum pump. The pump can be water jet pump, mechanical pump, diffusion pump, condensation pump, getter-ion pump, or turbomolecular pump. Since covering a pressure range of 10^5–10^{-11} Pa—a total of 16 orders of magnitude, a vacuum is not attained with only one but a combination of many pumps. In the laboratory, mechanical, diffusion, and various condensation pumps are commonly used. Table 3.4 lists the various techniques of attaining vacuum and the achievable pressures.

The working characteristics of a vacuum pump are usually described with four parameters: (1) initial pressure, which is the highest permissible pressure for the pump to operate; (2) critical back pressure, which is the highest achievable back pressure at the exhaust of the pump; (3) ultimate pressure, which is the lowest achievable pressure by the vacuum pump after prolonged evacuating of a tight system; and (4) pumping speed, which is the throughput of the concerned pump at a given pressure and temperature. It is very important to understand these four parameters. For instance, a mechanical pump has an initial pressure of 101.3 Pa and a critical pressure of 0.1 Pa, whereas a diffusion

TABLE 3.4 Techniques of Attaining Vacuum and Achievable Pressure

Vacuum (Pa)	Main vacuum pumps
10^5–10^2	Water jet pumps, mechanical pumps, and various rough pumps
10^2–10	Mechanical pumps, oil or mechanical boosters, and condensation pumps
10–10^{-5}	Diffusion pumps and getter-ion pumps
10^{-5} to 10^{-11}	Trap diffusion pumps, turbomolecular pumps, and getter-ion pumps
$<10^{-11}$	Combination pumps (getter-ion pump, molecular pump, etc.)

pump has an initial pressure of 10 Pa. Therefore, prior to the use of diffusion pump, the mechanical pump should be used to evacuate the enclosure to a pressure lower than 10 Pa. Therefore, mechanical pumps are often referred to as fore pumps, whereas diffusion pumps as secondary pumps.

1. Rotary vane mechanical pump

   A rotary vane mechanical pump consists of a pump cavity, a rotor, vanes, an exhaust valve, and an inlet port. All these parts are immersed in mechanical oil contained in a pump casing. The pumping mechanism is based on volume displacement. The working principle of a single-stage rotary vane mechanical pump is illustrated in Fig. 3.1.

   Generally, a single-stage pump has an ultimate vacuum of approximately 1 Pa. Two single-stage pumps can be connected in a series to operate as a two-stage pump, achieving an ultimate vacuum as high as 10^{-2} Pa.

2. Oil diffusion pump

   An oil diffusion pump is the main equipment for attaining a high vacuum. The working principle is based on the diffusion of gas molecules into directional vapor streams. Figure 3.2 is a schematic illustration of a three-stage metal diffusion pump. The working medium is usually an oil of low vapor pressure.

   Generally, the critical backpressure of diffusion pumps is about 10 Pa, so they should be used in conjunction with mechanical pumps. The ultimate vacuum of diffusion pumps is generally as high as 10^{-4} Pa, though there are slight differences according to the oil used and the pump capacity. The appropriate working pressure range of a diffusion pump is 10^{-2} to 10^{-4} Pa.

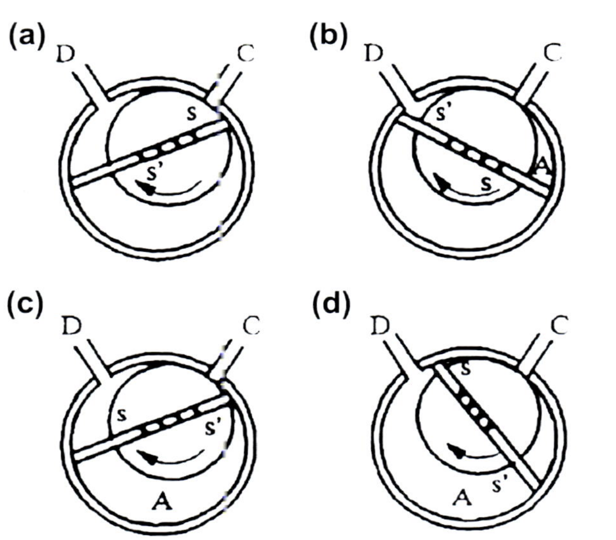

FIGURE 3.1 Working principle of rotary vane mechanical pump.

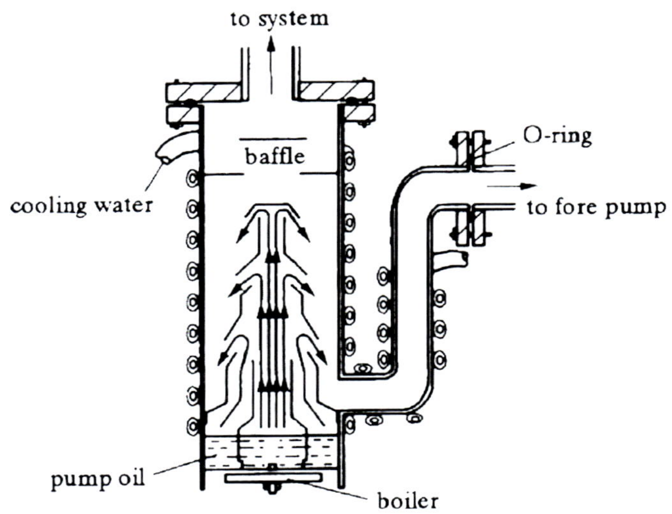

FIGURE 3.2 Three-stage metal diffusion pump.

3. Oil-free vacuum pump

   To obtain an ultrahigh vacuum, a pump that does not use oil as the working medium is necessary. There are four types of such pumps with mechanical movement, vapor stream, and adsorption as the evacuation method: Molecular pump (Compound- and Turbomolecular Pump); Getter-ion pumps; Molecular sieve sorb-pump; and Titanium sublimation pump.

3.2.2. Vacuum Measurement Principles and Typical Measurement Ranges [1]

Some vacuum applications require a negative pressure of only a few Pascals compared to atmospheric pressure, whereas others reach absolute pressures below 10^{-7}. This is a thousandth of a billionth of the normal atmospheric pressure and points out the complexity of vacuum measurement. The most sensitive vacuum gauges measure absolute pressures of less than 10^{-10} Pa. This is 15 decades smaller than atmospheric pressure. Due to physical constraints, there exists no measurement method that covers this huge range. Instead, for different pressure ranges we have different principles of measurement with different accuracies.

Gas-type-independent measuring principles utilize mechanical force interactions between gas particles and surfaces. This is referred to as a "direct" measurement. Pressure is defined as force per surface area (in SI units: $1\ N/m^2 = 1$ Pa). According to the kinetic gas theory, the force depends on the number of gas particles per volume, but is independent of the molar mass of the particles, i.e., the particular gas type does not affect the measurement. Typical gauges are Bourdon gauges and diaphragm gauges.

Gas-type-dependent (indirect) measuring principles do not utilize force interactions, but other physical properties of gases, which also depend on the gas density and therefore indirectly on the pressure. Two such properties are thermal conductivity and ionization probability. A problem is, however, that in addition to the density, the chemical composition of the gas also affects the measurement. To obtain comparability between different gas-type-dependent gauges, they are usually calibrated for nitrogen, since nitrogen is the main component (approx. 80%) of normal air. Typical gauges are heat loss gauges and ionization gauges (Fig. 3.3).

Vacuum measurement can be divided roughly into gas-type-independent and gas-type-dependent measuring principles. According to the function of vacuum gauges, some important types used for measuring high and ultrahigh vacuum are introduced as follows.

Ionization vacuum gauges. To measure high and ultrahigh vacuum, ionization gauges are the most commonly used devices. For this purpose the gauge ionizes the residual gas inside a defined gauge volume, collects the ions, and measures the resulting ion current, which is proportional to the pressure. The measurement is gas-type-dependent in principle, since each type of gas has a specific ionization probability. Therefore each particular gas or gas mixture demands a specific calibration or at least a correction factor. Ionization vacuum gauges are typically built in two versions: (1) Cold cathode ionization vacuum gauges, (2) Hot cathode ionization vacuum gauges, and (3) Combined gauges.

Many vacuum processes require the application of two or more measurement principles in order to meet the specified measurement range and/or accuracy. In some cases, it is reasonable to use combination gauges that combine two or more measurement methods in one single device. An example in this catalogue is the ATMION wide-range vacuum gauge, which integrates a Piranic sensor into a Bayard-Alpert ionization gauge and thus covers a pressure range over 13 decades from atmosphere to 10^{-8} Pa.

3.2.3. Common Vacuum Equipments in the Laboratory [2,4]

A vacuum unit for use in the laboratory usually consists of three parts: vacuum pumps, vacuum gauges, and lines or apparatus designed to meet the requirement of a specific experiment.

Valves (or stopcocks) are essential in a vacuum system. The selection and assembly of valves influence directly the vacuum degree of the system. A vacuum valve is a component used to adjust or cut off the gas stream in a vacuum system. There are various valves of different materials and structure designed for different purposes. Traps are often installed to reduce the influences of oil, water, and mercury vapors and other corrosive gases on the system. Sometimes, traps are also used to separate substances or simply increase the vacuum degree of the system.

Special vacuum lines or devices are mainly used for handling materials that are volatile or reactive with air or water vapor. These materials are common in inorganic synthesis.

1. Valves

Glass valves (or stopcocks) are the earliest and most convenient valves. They have advantages such

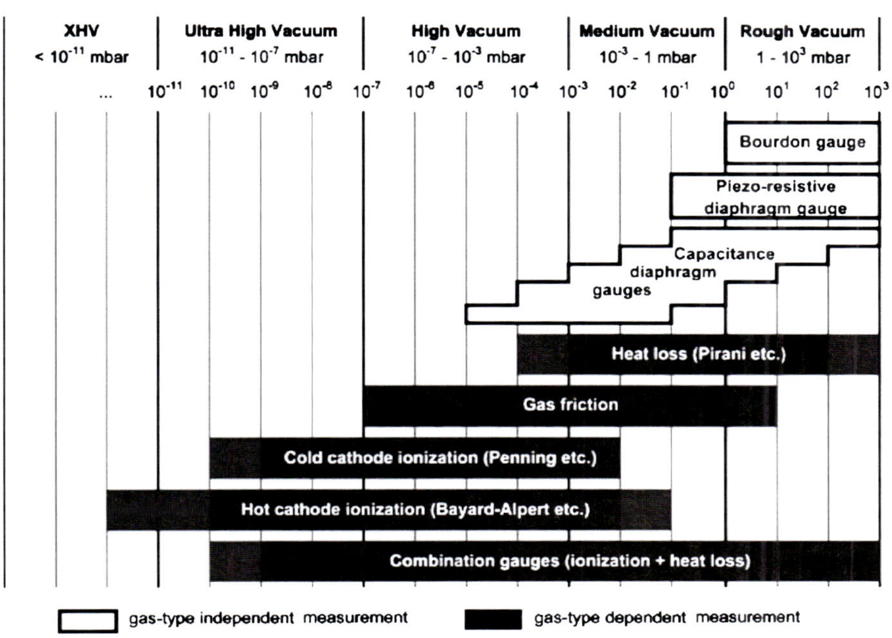

FIGURE 3.3 Vacuum degree and measurement [1].

as being easy to clean and make, chemically stable, highly insulating, and convenient for leakage checking. When using this type of ground stopcock, a uniform thin layer of seal grease should be applied at the interface of the two halves of the stopcock. The saturated vapor pressures of common vacuum greases are all lower than 10^{-4} Pa. The application temperature varies with the type of the grease. Many common seal greases are suitable to use at room temperature up to 130°C. Special greases can be used at −40 to 200°C. Therefore, an appropriate grease should be selected according to the experimental needs.

Figure 3.4 shows several two-way and three-way glass stopcocks. Figure 3.5 shows a "needle" glass valve. The coil-controlled needle is mobile inside the capillary. The diameters of the glass needle and capillary are determined by the volumetric throughput of gas and the intended degree of control. In addition, ultrahigh vacuum glass stopcocks such as ball-ground electromagnetic valves and oil-free glass stopcocks can be made to meet the requirement. These stopcocks are mostly opened or closed by means of magnet. There are two types of oil-free glass stopcocks: one uses the ground seal for opening and closing under vacuum conditions; and the other utilizes the property of a low melting Ga—In alloy that remains in a liquid state at room temperature to cut off the gas stream. Figure 3.6 shows these oil-free glass stopcocks. Tin metal can also be used to make liquid-metal stopcocks. The melting point of tin is 232°C, and its vapor pressure below 450°C is 10^{-11} Pa. At an elevated temperature, the cap of stopcock can be raised or lowered with a screw. This simple type of stopcock has always been highly remarkable.

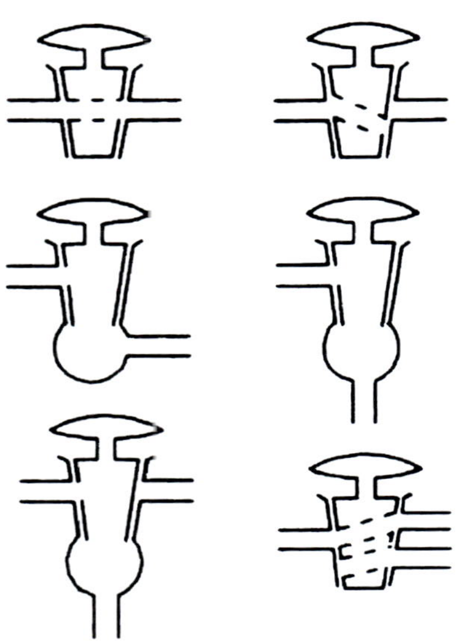

FIGURE 3.4 Glass stopcocks.

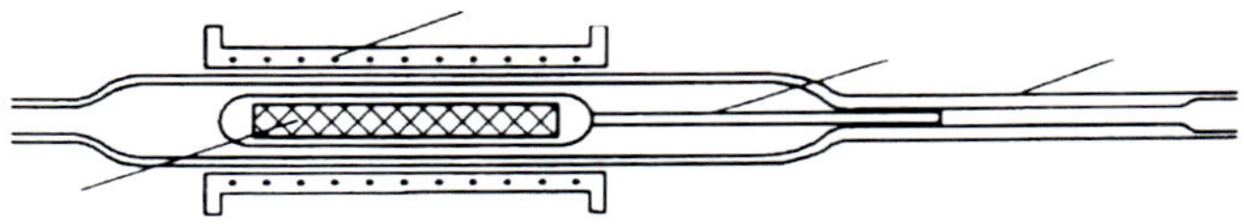

FIGURE 3.5 Glass needle valve.

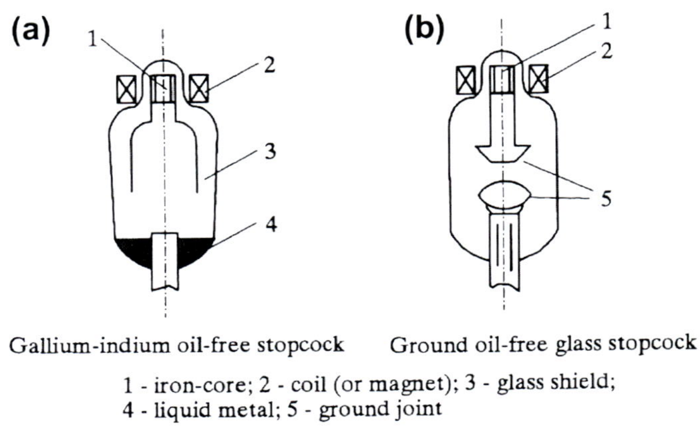

FIGURE 3.6 Two oil-free glass stopcocks.

In a chemical preparation process, when a joint needs to be open or close only once, a fusing seal (for closing) can be used in conjunction with a breaking valve (for opening). Since a tube with a large diameter is difficult to seal well using the fusing method after a vacuum has been achieved, its diameter should be reduced and the wall thickened at the position of seal. This can be done by heating the tube to soft and pressing axially to thicken the wall, then stretching it along the axial direction to reduce the diameter. When sealing such a tube, this part of the tube should be evenly heated and stretched along the axial direction, or using a glass rod to pull it apart.

Using a breaking valve (Fig. 3.7), a joint can be opened. The end of a glass tube is stretched to form a thin bent tip or a thin-wall bulb. The tube is then fusing sealed to a larger tube to form a breaking valve.

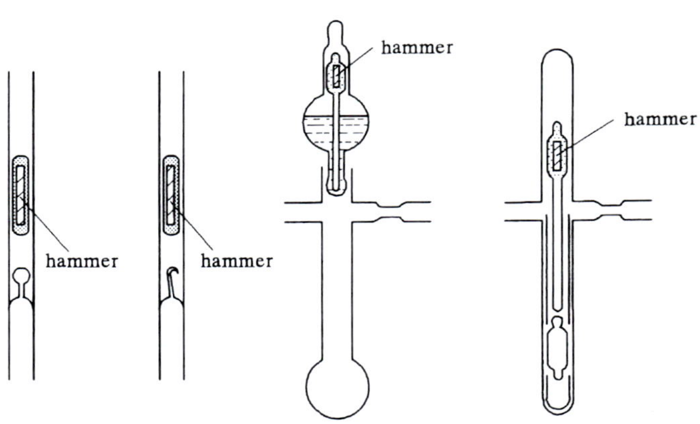

FIGURE 3.7 Breaking valves.

A hammer is made by enclosing a piece of steel wire into a glass tube. The breaking valve is arranged vertically. The hammer is slipped gently down in the glass tube onto the bulb or the tip using a strong magnet. Then the two ends of the valve are fusing connected to other parts of the system. To open the valve, the hammer is lifted up for several centimeters with an electromagnet and dropped down to break the bulb or tip to allow the gas to flow through. There are many other types of breaking valves in similar principles. Using this method, gaseous, liquid, or solid reagents can be introduced to the system for once.

2. Traps

The main types of traps are mechanical trap, cold trap, thermoelectric trap, ion trap, and sorption trap. A mechanical trap with an added condensation mechanism is used to prevent the back stream of oil vapor in a diffusion pump. It is also used to prevent the oil vapor from entering the fore pump. Generally, the oil vapor pressure of a diffusion pump at the freezing point is 10^{-6} Pa. Therefore it is essential to eliminate the oil vapor in order to obtain a vacuum degree of 10^{-8} Pa. In a cold trap, liquid nitrogen is usually used as the coolant to achieve a low temperature of $-196°C$. Using the cold trap, vapor pressures of various harmful gases in the system can be reduced considerably, and a relatively high ultimate vacuum can be obtained. According to experimental requirements, other coolants such as tap water, low temperature salt water, dry ice, Freon, and liquid helium can be used. Some common traps are shown schematically in Fig. 3.8.

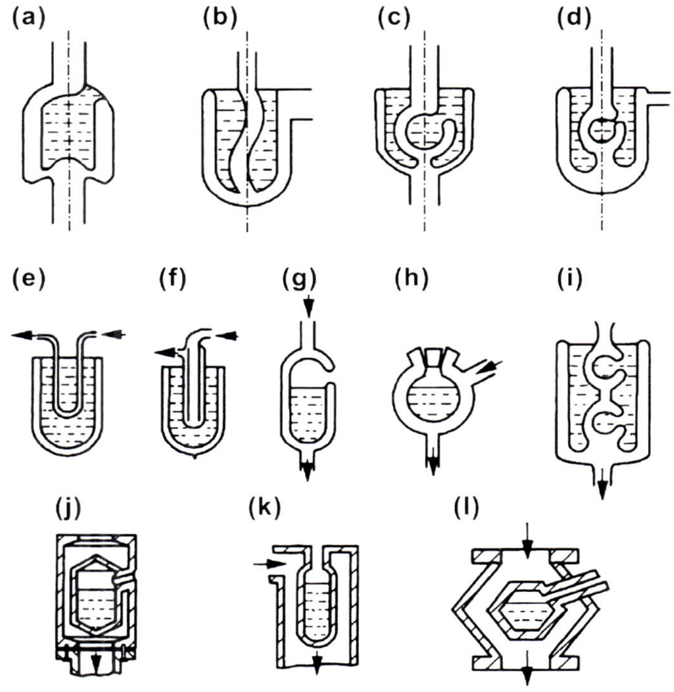

a to i - glass cold traps; j to l - metal cold traps

FIGURE 3.8 Cold traps.

Recently, there are also many metal valves, including metallic gate valve, used in the vacuum or ultra-vacuum lines.

In a hot trap, decomposition of a hydrocarbon compound on a heating board is used to form gaseous materials (hydrogen, carbon monoxide) and solid carbon. Latter is used to adsorb the vapors. The trap is placed between diffusion and a mechanical pump to prevent the back stream of oil vapor from low boiling oil of mechanical pump. Various sorption traps can be made using porous adsorptive materials. This type of trap is very effective for eliminating inert gases that cannot be adsorbed by ordinary cold traps, and cleans the system while decreasing the partial pressure. For instance, molecular sieve traps, active alumina traps, and active carbon traps are all used in attaining ultrahigh vacuum. Figure 3.9 shows two cold adsorption traps with folded copper foil. They can be used to achieve an ultrahigh vacuum of 10^{-8} Pa. The arrangement of a trap in an ultrahigh vacuum line is shown in Fig. 3.10.

In synthetic experiments under vacuum conditions, traps (usually cold traps) are commonly used to store substances that are volatile at room temperature, or to condense volatile components in the reactor. Cold traps are also used in the separation of volatile compounds, such as in fractional condensation.

3. Vacuum lines for handling materials sensitive to air and water vapor

Figure 3.11 shows a standard vacuum line for handling materials sensitive to air and water vapor. It comprises three parts: the liquid or solvent stock, the

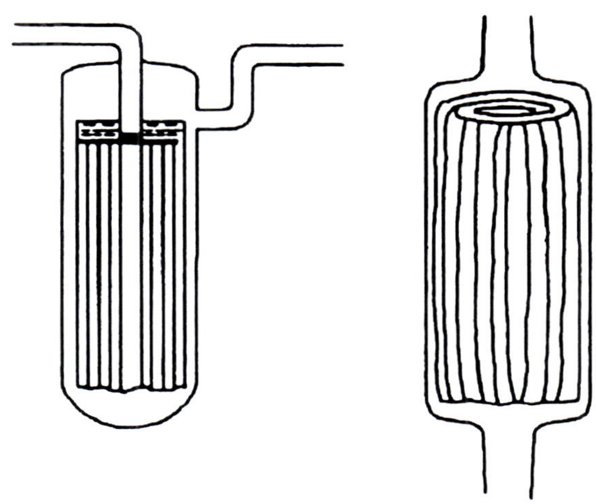

FIGURE 3.9 Two cold adsorption traps consisting of folded copper foils.

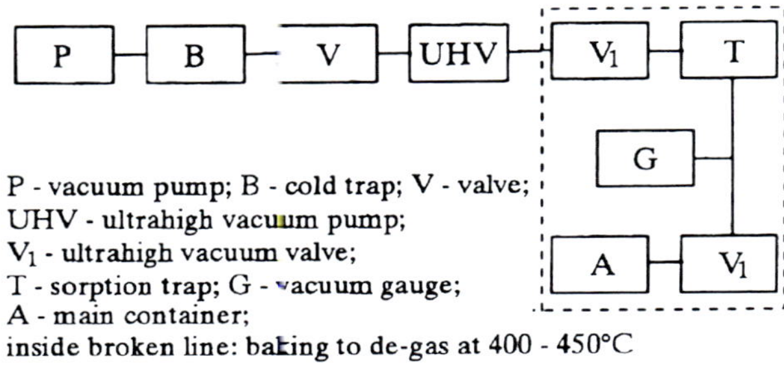

FIGURE 3.10 Block diagram of ultrahigh vacuum apparatus.

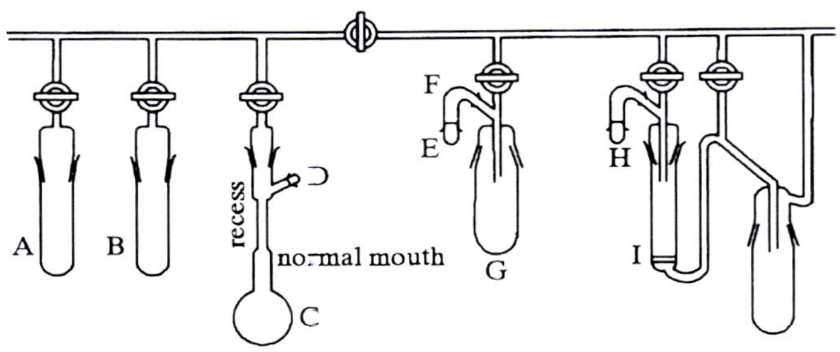

FIGURE 3.11 Standard vacuum line for materials sensitive to air and water vapor.

reactor, and the separation system. The overall vacuum line is carefully preevacuated to eliminate water vapor. A and B in the Fig. 3.11 are two stock tubes; C is a reaction flask; D is the inlet port for solid materials; and the rest are two filtering units.

First, the liquid material or solvent is put into stock tubes A and B. The solid material is stored in a ground tube under nitrogen in advance, and is then introduced to the reaction flask C at joint D against a nitrogen stream. The system is evacuated. The liquid or solvent is transferred to the reaction flask from the stock tubes by means of distillation, and the distillate is collected to the reaction flask using a refrigerant such as liquid nitrogen or dry ice—acetone. When the reactant and the solvent are in, the flask is sealed through fusing and disconnected from the vacuum system. The flask and its contents are then heated up to a temperature, and shaken to achieve a complete reaction. The flask is usually opened when cooled down under a nitrogen atmosphere. If the product is in a liquid state, the flask can be connected to the filtering section at joint E under nitrogen atmosphere. By evacuating, the solution is passed through filter board F to collector G. The intended product is collected through solvent evaporation. If the product is a solid, the flask can be connected to the filtering section at joint H. The product can be filtered through filter board I. The product can also be washed with a solvent, and trace solvent can be eliminated through continuous evacuation.

The reaction of metal halides with appropriate alkyl nitrile forming $MX_4 \cdot R.CN$ can be carried out in the above vacuum line. The synthesis of many complex compounds can also follow the above operation procedures. For instance, $[(C_2H_2)_5NH_2]_2TiCl_6$ can be synthesized from $TiCl_4 \cdot 2C_2H_5CN$ and diethylamine chloride using chloroform as a solvent.

3.3. PURIFICATION AND SEPARATION OF INORGANICS AT LOW TEMPERATURE

In general, the reactions between nonmetallic compounds and other gas types of inorganic molecules are not possible to proceed to completion due to the involvement of many side reactions. So the purification and separations of raw materials and by-products are very important and mainly depend on the following separation methods at low temperatures:

Low temperature fractional condensation under reduced pressure;
Low temperature fractional vaporization under reduced pressure;
Low temperature selective adsorption;
Low temperature fractional distillation; and
Low temperature chemical separation.

3.3.1. Low Temperature Fractional Condensation [3]

1 Description of method:

In low temperature fractional condensation, one allows a gaseous mixture to pass through a series of cold traps of different temperatures. The gases condense separately in the cold traps of different temperatures due to the different boiling points of gases, thus achieving the separation purpose.

However, there are some problems that need to be clarified. After the gas passes through the trap, how low should the vapor pressure be for the gas to be considered as completely condensed? Under what condition will it not be condensed? It is generally agreed that the compound is completely condensed and quantitatively collected in the cold trap when its vapor pressure is less than 1.3 Pa after it has passed through the cold trap. On the other hand, if the pressure is higher than 133.3 Pa after the gas passes through cold trap, it is considered as not condensable.

Frequently, for pressures in the region of 1.3 Pa, the temperature—vapor pressure data are unavailable for compounds of interest or are not readily calculable.

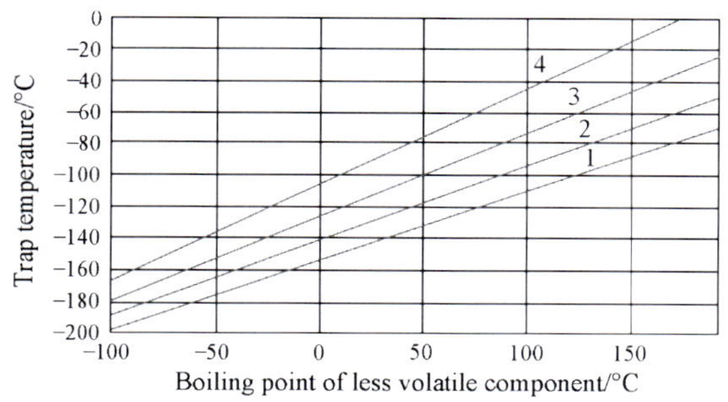

1- Very good condensation and trapping when Δb.p.>120°C
2- Good condensation and trapping when Δb.p.>90°C
3- Fair condensation and trapping when Δb.p.>60°C
4– Poor condensation and trapping when Δb.p.>40°C

FIGURE 3.12 Cold-trap temperatures suggested for separating volatile mixtures [3].

This poses a problem of selecting the cold traps. However, to separate the two compounds, we can choose a suitable trap temperature according to their boiling points or sublimation points at 0.1 MPa. Under certain conditions, Fig. 3.12 can also be used.

For example, if we want to separate diethyl ether (b.p. = 34.6°C) and stibine (b.p. = −18.4°C), we choose a cold trap that allows ether to condense quantitatively and let stibine pass through. How do we select the cold trap in order to achieve this purpose? By referring to the figure we note that the boiling point of ether is 34.6°C and reading up from here to reach line "3" (because the boiling point difference between ether and stibine is 53°C) we find that the trap close to −100°C should be used to condense the ether. From Table 3.1 we can see that a toluene slush bath (−95°C) is very suitable. If CS_2 bath is used, some stibine may be condensed out; it is only usable when the distillation is carried out very slowly.

One should be aware that mixed vapors should not pass through the trap too rapidly or else the efficiency of separation will be affected because of the following reasons:

(1) The less volatile component may not be completely condensed in the trap and may be carried away by the more volatile component. Therefore, the more volatile component may contain some less volatile component.

(2) Due to the moderately high pressure in the system, the more volatile component may partially be condensed in the trap and hence the less volatile component may be contaminated with the more volatile component.

The mixture also should not pass through the trap too slowly because due to slow motion, some of the less volatile component in the trap can evaporate and escape (even at the low temperature of the trap it has a finite vapor pressure). Therefore the more volatile component may contain the less volatile fraction.

Then what is considered as the appropriate speed of the mixed vapors passing through the trap? In general, when the rate is 1 mmol/min the effect of separation is good.

Furthermore, a mixture with the boiling points of the components differing by less than 40°C cannot be quantitatively separated by fractional condensation. However, they can be separated by repeated fractional condensation, which has a low recovery rate.

2. Application examples:

Under standard condition for temperature and pressure (STP) condition, allowing a mixture of 83.3 ml $B_3N_3H_6$ and 23.8 ml BCl_3 to react for 116 hours can give the following mixture of compounds: $B_3N_3H_4Cl_2$, $B_3N_3H_5Cl$, $B_3N_3H_6$, B_2H_6, and H_2.

		$B_3N_3H_4Cl_2$	(b.p. 151.9°C)
	room temp.	$B_3N_3H_5Cl$	(b.p. 109.5°C)
$B_3N_3H_6 + BCl_3$	$\longrightarrow$	$B_3N_3H_6$	(b.p. 50.6°C)
	116 hours	B_2H_6	(b.p. -86°C)
		H_2	(b.p. -253°C)

(Mixture of reactants and products)

How do we separate such a mixture of compounds? We can use the following diagram to illustrate:

Initially, suitable traps are chosen. The first trap selected is chlorobenzene, the second trap is dry ice, the third trap is carbon disulfide and the fourth trap is liquid nitrogen.

When the mixture passes through the first trap, $B_3N_3H_4Cl_2$ condenses. Its vapor pressure is related to temperature by the following equation:

$$\log p_1 = \frac{-1994}{T} + 7.572$$

When $T = -45.2°C$ (227.8 K) is substituted into the equation we obtain $p_1 = 8.8$ Pa, which shows that $B_3N_3H_4Cl_2$ is basically condensed out. For $B_3N_3H_5Cl$, the relation between the vapor pressure and temperature is:

$$\log p_2 = \frac{-1864}{T} + 7.703$$

When $T = 227.8$ K we obtain $p_2 = 53$ Pa, showing that $B_3N_3H_5Cl$ has basically escaped. In this way $B_3N_3H_4Cl_2$ is separated from the other mixture of compounds. Following this argument step-by-step, all will be finally separated (Fig. 3.13).

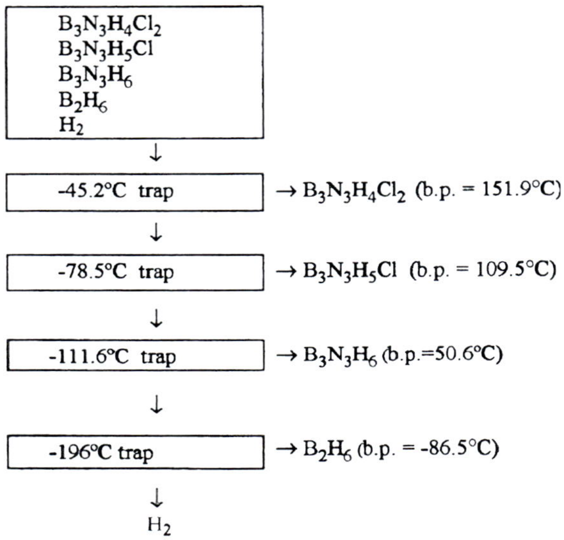

FIGURE 3.13 Separation by low temperature fractional condensation [3].

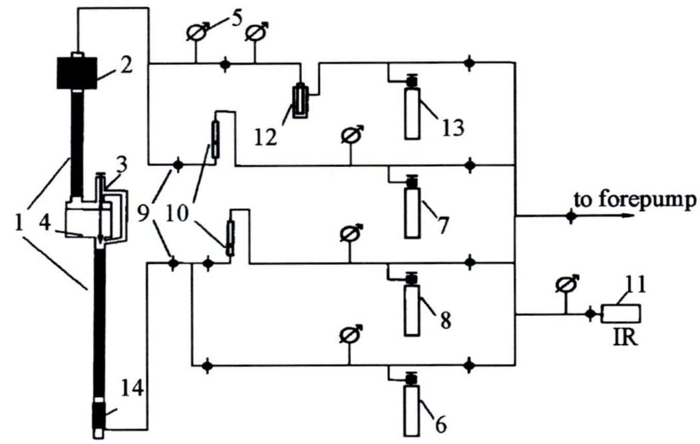

FIGURE 3.14 Schematic representation of the SiF_4 and $^{28}SiH_4$ distillation system: 1. Rectifying sections, 2. Condenser, 3. Flow-control valve, 4. Feed tank, 5. Pressure gage, 6–8. Cylinders, 9. Metering valves, 10. Flowmeters, 11. IR spectroscopic cell, 12. Petryanov filter, 13. Receiving cylinder, and 14. Reboiler [5]. *With kind permission from Springer Science+Business Media: Inorganic Materials, Preparation and Fine Purification of SiF_4 and $^{28}SiH_4$, 38, 2002, pages 283–287, A.D. Bulanov.*

3.3.2. Low Temperature Fractional Distillation

A good example, purification of SiF_4 and $^{28}SiH_4$ realized by fractional distillation at low temperature [5], is described in more detail.

Silicon tetrafluoride, owing to its high saturation vapor pressure and chemical stability and the fact that ^{19}F is the only fluorine isotope, is a convenient compound for separating silicon isotopes with the purpose of obtaining chemically pure monoisotopic silicon, a promising electronic material. SiF_4 was obtained by the thermal decomposition of pure-grade Na_2SiF_6 followed by purification by fraction distillation at low temperature. The setup used to purify SiF_4 is schematized in Fig. 3.14.

SiF_4 was distilled in a middle-fed stainless steel column with foamed-plastic insulation. The rectifying sections 1 were packed with 3 × 3 × 0.25 mm nichrome prismatic coils. First, the condenser 2 was cooled to set the temperature by liquid nitrogen fed at a controlled rate. Afterwards, up to 1.5 kg of SiF_4 was introduced into the column from cylinder 6 through the condenser. Liquid SiF_4 was fed from the tank 4 into the bottom section through a flow-control valve 3, initially at a high rate to wet the packing and then at 1.5 l/h. The SiF_4 pressure was measured with gauge 5 and controlled by varying the liquid nitrogen supply to the condenser. The reboiler 14 was maintained at a constant temperature with the use of a Proterm −100°C microprocessor-based temperature controller. The condenser temperature was −70°C, and the gauge pressure was 0.3 MPa. The column was operated in the closed-distillation mode for 4 h, and then the overhead and bottoms were simultaneously drawn off at a constant rate into cylinders 7 and 8, respectively. The withdrawal rate was controlled with metering valves 9 and measured with flowmeters 10. The withdrawal ratio was ≈0.002. The overhead, bottoms, and main product were sampled at intervals for IR spectroscopic analysis into cell 11 through a sampling line. Hydrogen fluoride, silane, trifluorosilane, and methane accumulated in the overhead; hexafluorodisiloxane, carbon dioxide, and silicon hydroxyfluoride, in the bottoms. The desired product stayed in the feed tank. The process was terminated after the Si_2OF_6 concentration remained constant. The purified product was transferred through Petryanov filter 12 to cylinder 13. Distillation reduced the Si_2OF_6 concentration in SiF_4 by one order of magnitude. The purified product was isotopically enriched by centrifugation at OAO GAZ. Four enriched SiF_4 specimens were obtained, in which the ^{28}Si abundance was 99.6, 99.82, 99.93, and 99.96%. Note that the concentration of Si_2OF_6 (which was ^{28}Si-enriched to the same extent) in the enriched SiF_4 specimens was 10^{-1} to 10^{-2} mol%.

Next, SiF_4 was ^{28}Si-enriched by centrifugation and converted to silane by calcium hydride reduction:

$$^{28}SiF_4(g) + 2CaH_2(s) \xrightarrow{200^\circ C} {}^{28}SiH_4(g) + 2CaF_2(S)$$

In the absence of a solvent the yield of silane was found to be very low. This may be attributed to the presence of impurities in commercially available CaH_2, such as calcium oxide and hydroxide, formed due to its reaction with atmospheric moisture. The reaction of these compounds with SiF_4 is thermodynamically plausible.

Therefore, to obtain a purer CaH_2, distilled calcium can be reacted with Pd-diffused hydrogen.

The silane resulting from the reaction was found to be isotopically diluted by less than 0.1%. The SiF_4 content, determined by IR spectroscopy, was below 0.01 mol%, but the total fluorine content, determined by ion-exchange chromatography, was 0.1 wt%.

The reaction yield was 90–98% in terms of SiF_4 and 6–9% in terms of CaH_2. The capacity of this apparatus was about 10 g/h. The basic disadvantages of this process are a low yield in terms of CaH_2 and the necessity of grinding the hydride in a very dry atmosphere. Small amounts of moisture may passivate CaH_2 and thus reduce the reaction yield in terms of SiF_4. To reduce the total concentration of fluorine, whose compounds are corrosive, the unreacted SiF_4 was sorbed by sodium fluoride or further hydrogenated with CaH_2. As a result, the total fluorine content was as low as 10^{-4} wt%. To remove the hydrocarbons, $^{28}SiH_4$ was distilled in a quartz or metallic column similar to the column in Fig. 3.14, but with a smaller diameter of the rectifying sections and a smaller feed tank (300 cm^3). The charge weight was 70–180 g. The condenser temperature was about −110°C, and the pressure was slightly above atmospheric. The reflux rate was 0.8 of the maximum and was ≈900 ml/h in the metallic column and 360 ml/h in the quartz column. The column was operated in the closed-distillation mode for 3 h, and then the overhead and bottoms were simultaneously drawn off at a constant rate. The withdrawal ratio was 0.001. Methane and hydrogen accumulated in the overhead, whereas, the other hydrocarbons and disilane, in the bottoms. The purified product was transferred through a Petryanov filter to a receiver.

3.3.3. Low Temperature Selective Adsorption

The amount of substance adsorbed decreases with increase in temperature because the process of physical adsorption is exothermic. However, when the size of the adsorbed gas molecule (like N_2, Ar, CO, etc.) is close to that of the pores of the adsorbent, the peculiar effects of temperature on the amount adsorbed will appear. Figure 3.15 shows the isopiestic adsorption curves of O_2, N_2, Ar, and CO on 4A molecular sieves [6]. The amount of O_2 adsorbed increases with decrease in temperature; at 0°C, only trace amount is adsorbed while at −196°C, the amount adsorbed can reach 130 ml/86 (18.6%). For N_2, Ar, and CO, between 0 and 80°C the amount adsorbed increases with decrease in temperature whereas between −80 and −196°C the amount adsorbed decreases with decrease in temperature. In other words, the amount adsorbed at about −80°C is maximum. This is because the sizes of N_2, Ar, and CO are very close to the pore diameter of the

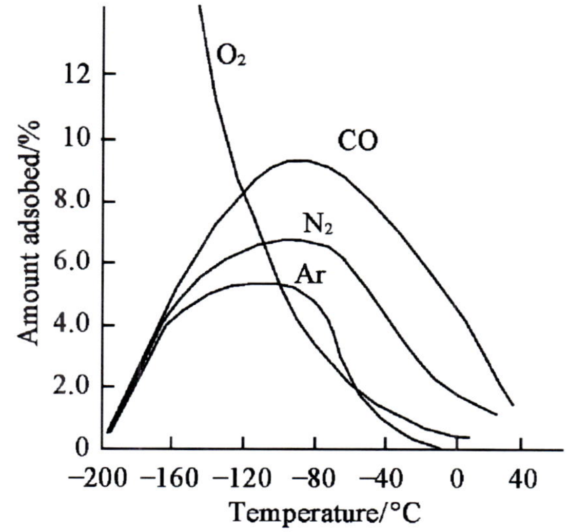

FIGURE 3.15 The isopiestic adsorption curves on 4A molecular sieves [6].

sieve. At very low temperatures, their activation energies are very low and the sieve pores contract, resulting in the more difficult diffusion of these molecules in the cavities. Therefore this would cause a decrease in the amount adsorbed as temperature decreases. From this discussion, low temperature seems to be better to separate O_2 from the other gases.

For the separation of helium and neon at low temperatures let us look at the isothermal adsorption curves of these two gases on the 5A and 13X molecular sieves (−196°C), which are shown in Fig. 3.16. If 13X molecular sieve is used as adsorbent, its separation coefficient, α, is 5.3 when the adsorption temperature is at −196°C; the isothermal line of neon is also linear. When separation

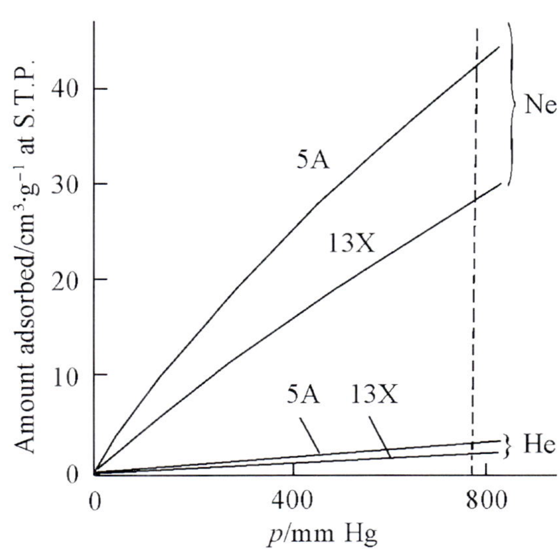

FIGURE 3.16 The isothermal adsorption curves of neon and helium on 5A and 13X molecular sieves (−196 °C) [6].

by selective adsorption is carried out at an appropriate pressure, 99.5% pure neon can be obtained with the recovery rate greater than 98%.

Pure CO is a raw material used for the synthesis of many important chemicals, such as acetic acid, acetic anhydride, oxalates, methyl carbonate, formic acid, *N*,*N*-dimethylformamide (DMF), phosgene, polyisocyanates (TDI and MDI), and metal carbonyls. Sources of CO, such as synthesis gas and by-product gases in steel plants, often contain N_2, H_2, CO_2, CH_4, H_2O, etc. Separation of CO from the gas mixtures is of great interest in industries. The conventional way of separating CO from gas mixture is the cryogenic process. This process requires high investment and high energy consumption, because it operates at low temperature and high pressure. In addition, the process cannot be used to separate CO from syngas with high N_2 content, because N_2 has a boiling point close to CO.

Xie and Tang [7,8] have developed a CuCl/zeolite adsorbent with high CO adsorption capacity and selectivity over H_2, N_2, CH_4, and CO_2. By using this adsorbent in a VPSA (Vacuum Pressure Swing Adsorption) process, many large-scale plants have been designed and built in China to produce high-purity CO from syngas.

Common zeolites cannot be used to separate CO from syngas, because they have low CO adsorption selectivity over N_2, CH_4, and CO_2. It is well known that Cu^+ ions can form π-complex with CO. If Cu^+ ions are incorporated onto the surface of a zeolite, the zeolite might become an adsorbent with high adsorption capacity and selectivity of CO. It has been found in Xie's lab that salts can disperse spontaneously onto the surface of supports to form a monolayer or submonolayer [7]. Based on this principle, a mixture of CuCl and a zeolite was heated at suitable temperature, and then CuCl was dispersed onto the surface of the zeolite to form a CuCl/zeolite adsorbent, which showed high adsorption capacity and selectivity for CO.

Adsorbents of CuCl/zeolites are prepared by mixing and grinding CuCl with a variety of zeolites, and then heating the mixtures at 673 K for 4 h under nitrogen atmosphere to make the CuCl disperse onto the surface of the zeolite. XRD shows that the loading of CuCl dispersed on zeolite has a threshold value. When CuCl loading is higher than the threshold, crystalline CuCl can be detected, while no crystalline CuCl can be found if the loading is lower than the threshold, indicating that all the CuCl has highly dispersed onto the surface of the zeolite. Adsorption measurement shows that a CuCl/zeolite adsorbent has a maximum CO capacity when its CuCl loading is equal or close to the threshold.

Figure 3.17 shows the CO adsorption isotherms of Cu^+Y, CuCl/NaA, CuCl/NaX, CuCl/NaY, and CuCl/Cu^+Y zeolite. Adsorbents with CuCl loading are close to the threshold of the zeolites respectively. All the

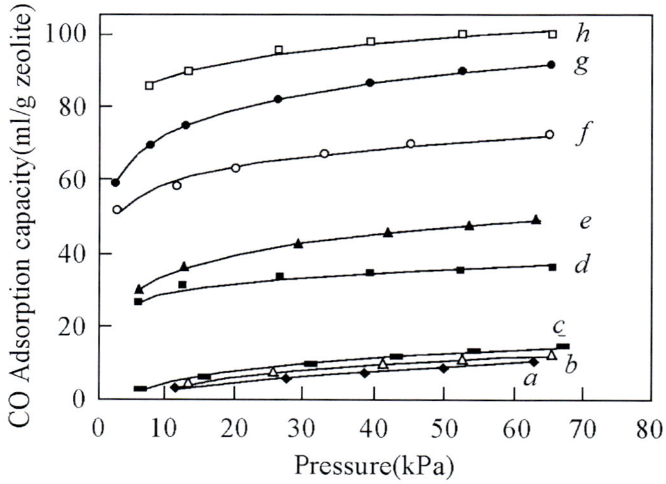

FIGURE 3.17 Adsorption isotherms of CO on various adsorbents at 293 K [7]. *Reprinted from Advances in Catalysis, 37 (1990), You-Chang Xie, You-Qi Tang, Spontaneous Monolayer Dispersion of Oxides and Salts onto Surfaces of Supports: Applications to Heterogeneous Catalysis, 1–42, Copyright (1990), with permission from Elsevier.*

adsorbents containing Cu^+ adsorb much more CO than the zeolites without copper. Adsorbent Cu^+Y, which is made by ion exchange of NaY with $CuCl_2$ and reduced with hydrogen, has a CO capacity lower than the CuCl/zeolite adsorbents, because the ion-exchangeable sites in NaY are limited. The CuCl/NaY adsorbs much more CO than CuCl/NaX and CuCl/NaA. The reason is that when CuCl is heated with zeolites, both dispersion of CuCl onto the surface of the zeolites and solid ion-exchange reaction between Cu^+ and Na^+ ions occur at the same time. NaX and NaA zeolites have more Na^+ ions than NaY, and then more NaCl would be formed after the solid ion-exchange reaction and remained in the caves of NaX and NaA to obstruct adsorption of CO. NaY zeolite has more cave space than NaX and NaA, therefore, CuCl/NaY has higher CO adsorption capacity. Peking University Pioneer Technology Company has used this technology to produce a highly efficient CO adsorbent with a commercial name PU-1. The adsorbent uses an optimum CuCl/zeolite recipe and some binder. Figure 3.18 gives the isotherms of CO, CO_2, CH_4 N_2, and H_2 of PU-1 adsorbent. The adsorbent adsorbs more CO than CO_2 and much more than CH_4, N_2, and H_2.

3.3.4. Low Temperature Chemical Separation [3]

Sometimes the two compounds that cannot be easily separated on the basis of the difference in their volatilities can be separated by adding an excess of the third compound that can form a relatively nonvolatile compound with one of the components. After removing

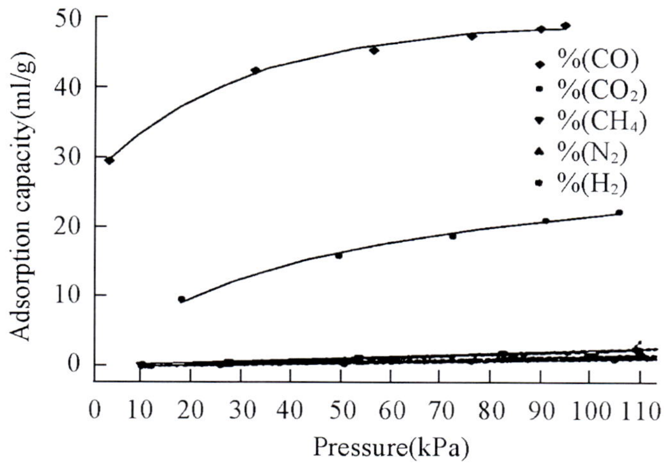

FIGURE 3.18 Adsorption isotherms of CO, CO_2, CH_4, N_2, H_2 on PU-1 adsorbent [7]. *Reprinted from Advances in Catalysis, 37 (1990), You-Chang Xie, You-Qi Tang, Spontaneous Monolayer Dispersion of Oxides and Salts onto Surfaces of Supports: Applications to Heterogeneous Catalysis, 1–42, Copyright (1990), with permission from Elsevier.*

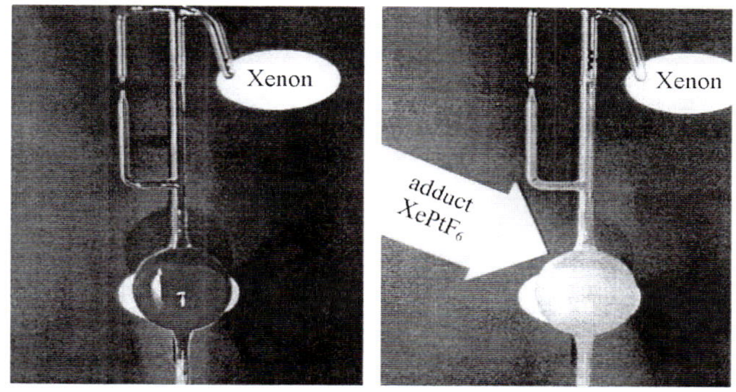

FIGURE 3.21 The Xe + PtF6 reaction: left, the gases separated by a glass break seal; right, the yellow reaction product formed on breaking the seal. The major product (yellow solid) which is not soluble in HF, is probably polymeric $(XeF^+)_n(PtF_5)_n^{n-}$ [18]. *Reprinted from Coordination Chemistry Reviews, Page 197 (2000), Lionell Graham, Oliver Graudejus, Narendra K. Jha, Neil Bartlett, Concerning the nature of $XePtF_6$, Page 321, Copyright (2000), with permission from Elsevier.*

the volatile component, an excess of the fourth compound is added to the nonvolatile product to form a nonvolatile compound with the third compound added earlier and to displace the original component of the mixture. Let us now refer to Fig. 3.19 for an example.

As shown above, the impurities in sulfur tetrafluoride are SF_6 and SOF_2. When a third compound BF_3 is added, an adduct $SF_4 \cdot BF_3$ of low volatility is formed only with SF_4. At this time, the whole system is cooled to $-78°C$ and the more volatile components BF_3, SF_6, and SOF_2 are pumped away with a vacuum pump leaving behind the less volatile complex $SF_4 \cdot BF_3$. To this complex, an excess of compound Et_2O is added, forming $Et_2O \cdot BF_3$ due to the greater extent of complex formation of Et_2O with BF_3 than that of SF_4 with BF_3. $Et_2O \cdot BF_3$ has low volatility and pumping at $-112°C$ pumps off SF_4 only, leaving $Et_2O \cdot BF_3$ behind.

For another example see Fig. 3.20. Here an excess of a third compound, HCl, is added to a mixture of GeH_4 and PH_3 that is to be separated. PH_3 and HCl form PH_4Cl of low volatility but GeH_4 does not react. Then evacuation is carried out at $-112°C$ pumping away GeH_4 and leaving behind PH_4Cl. Subsequently, they

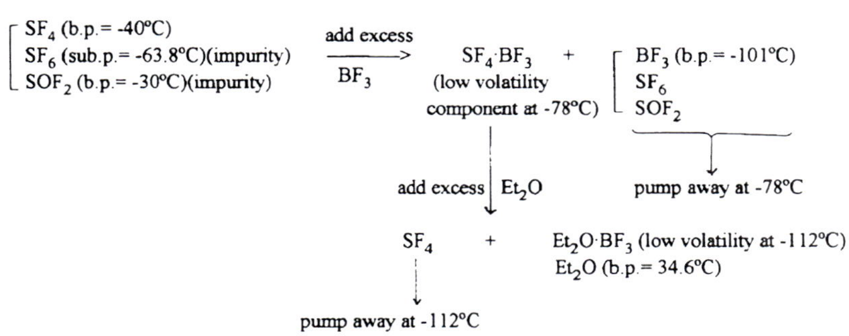

FIGURE 3.19 Purification of SF_4 at low temperatures [3].

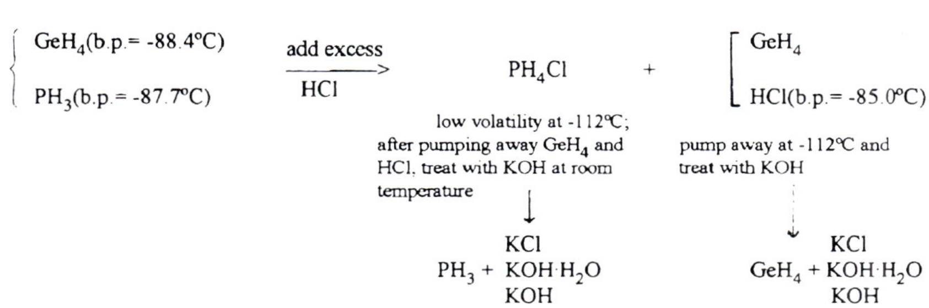

FIGURE 3.20 Low temperature separation of GeH_4 and PH_3 [3].

can separately be treated with KOH to give pure PH_3 and GeH_4.

3.4. THE SYNTHESIS OF VOLATILE INORGANIC COMPOUNDS AT LOW TEMPERATURE

The synthesis and purification of a large amount of volatile inorganic compounds, such as fluorine-containing compounds, volatile nonmetallic halides, chalcogenides, hydrides, carbon-containing inorganic compounds (CO, C_3O_2, COS, $(CN)_2$, HCN, CNCl, HCNO, $(SCN)_2$), and some carbonyl compounds, should be carried out in vacuum at low temperature. These conditions help in handling highly volatile compounds and complexity of by-products produced from side reactions.

3.4.1. Synthesis of High-purity AsF_5 [9–11]

AsF_5 has found extensive use as a fluoride acceptor and oxidant in the field of synthetic inorganic fluorine chemistry. AsF_5 (m.p. −79.8°C, b.p. −52.8°C) is a volatile gas at room temperature. It is usually prepared by fluorination of elemental As [12] or As_2O_3 [13] in the flow system. Other methods include the fluorination of AsF_3 at low temperature [9,11,14], or fluorination of As powder at −196°C [15] in a Ni can. Here, we present a laboratory-scale synthesis of high-purity HF-free AsF_5 from AsF_3 and F_2 at low temperature. The static conditions as well as the detailed synthesis of HF-free AsF_3 are described below.

Hydrogen fluoride-free AsF_5 was replaced by the direct fluorination of AsF_3

$$AsF_3 + F_2 \rightarrow AsF_5$$

Cylindrical Ni vessels were used for the synthesis and storage of AsF_5. All Ni vessels were statically tested to 70 atm with N_2 gas, dried under vacuum and passivated with 2–3 atm of F_2 at 200–300°C for a minimum of 12 h prior to use, all valves that had been degreased and lubricated with KCl-F wax and outfitted with a Teflon packing washer to render it suitable for fluorine service.

Purified AsF_3 (30 g, 0.227 mol) was distilled into a preweighed vacuum dried and F_2 passivated vessel, weighed, and then condensed into a 1 L Ni reaction vessel. HF and condensable impurities were removed from commercial F_2 (1.35 atm) into a 2 L Ni vessel followed by condensation into a 30 mL Ni vessel at −196°C.

The vessel was warmed to −183°C in a liquid oxygen bath and the F_2 was allowed to expand through a metal vacuum line manifold and was recondensed into the reaction vessel containing AsF_3 at −196°C. The F_2 was condensed, in three equal amounts, resulting in a 50 mol% excess of F_2 (0.34 mol). After each addition, the valve of the reaction vessel was closed and the assembly was allowed to warm to ambient temperature behind a blast shield. After the final addition, the reaction vessel was allowed to warm to room temperature. The reaction vessel was cooled to −196°C and excess F_2 was removed under vacuum by pumping it through a soda-lime column. The yield of AsF_5 was essentially quantitative. The product was subsequently stored in and dispensed from the reaction vessel. It is important that the amount of HF contaminant in AsF_5 be kept to a minimum as it promotes attack of the metal vessel and valve, forming metal hexafluoroarsenates that corrode the storage vessel and that may lead to valve blockage. Hydrogen fluoride contaminant in the product could also render it unsuitable for synthetic work in glass vessels.

In instances where low levels of HF contaminant can be tolerated, the above synthesis of AsF_5 may be simplified by leaving out the F_2 purification step. If commercial or unpurified F_2 is used without further purification, the NaF treatment must be repeated.

The purified AsF_5 was transferred to a 30 mL nickel vessel for storage. The degree to which AsF_5 loss occurs as a result of reaction with NaF to form [Na][AsF_6] is difficult to assess; however, 10–20% losses have been noted.

3.4.2. Syntheses of $R_3SiCo(CO)_4$ Type Compounds [16]

The syntheses of several compounds having the formulas $R_3SiCo(CO)_4$ (R = Cl, C_2H_5O, C_2H_5) and $H_3SiCo(CO)_4$ were reported in 1965. The synthetic route given here is representative of these synthetic methods, which has been used by other workers to prepare compounds such as $Cl_3SiMn(CO)_5$, $(C_6H_5)_3SiMn(CO)_5$, $Cl_3SiMo(CO)_3C_5H_5$. The general scheme for reaction consists of the interaction of a silicon hydride with a transition-metal carbonyl compound.

The complete synthetic reaction takes place by a two-step interaction at low temperature.

The first step results in the formation of $HCo(CO)_4$ and $R_3SiCo(CO)_4$:

$$R_3SiH + Co_2(CO)_8 \rightarrow R_3SiCo(CO)_4 + HCo(CO)_4$$

The second step results in the generation of hydrogen gas and the formation of another molecule of $R_3SiCo(CO)_4$:

$$R_3SiH + HCo(CO)_4 \rightarrow R_3SiCo(CO)_4 + H_2$$

Each synthesis is carried out in a glass pressure vessel [16]. This vessel has been found to be suitable for reaction mixtures having an autogenous pressure of up to 50 atm. The vessel under pressure is normally manipulated behind a safety shield with the operator wearing a leather glove. Volatile reactants are purified by the use of normal vacuum-line procedures and then condensed directly into the reactor. At the end of the reaction period, if the reactor is cooled to liquid nitrogen temperature and then opened to a manometer, noncondensable material (presumed to be H_2 and CO of decomposition) will be observed.

The silyl compounds described in the following preparations react with oxygen and water vapor. All nonvacuum-line manipulations of these materials require the use of inert atmospheres (deoxidized nitrogen or carbon monoxide) and an efficient fume hood. The toxicity of these compounds is unknown, but they are assumed to be extremely dangerous as are most transition-metal carbonyls.

Example: Synthesis of Tetracarbonyl(trichlorosilyl) cobalt

$$2Cl_3SiH + Co_2(CO)_8 \rightarrow 2Cl_3SiCo(CO)_4 + H_2$$

Commercially available trichlorosilane, $HSiCl_3$ (about 8 g), is placed in a glass tube which can be attached to the vacuum system through a stopcock. The $HSiCl_3$ is then cooled to liquid nitrogen temperature (−196°C), and then the tube is evacuated. The stopcock is then closed and the nitrogen bath removed. After the sample has melted, it is solidified at −196°C and the tube evacuated again. The freezing and melting processes are continued until the sample has been freed of noncondensable impurities (air). The contents of the tube are then warmed slowly to room temperature while distilling through a −132°C trap (*n*-pentane slush) into a −196°C trap. The material which stops in the −132°C trap ($HSiCl_3$) is cooled to −196°C and then again warmed slowly to room temperature and allowed to distill through a −132°C trap into a −196°C trap. This process is repeated until no HCl is observed in the −196°C trap. The process may be represented by:

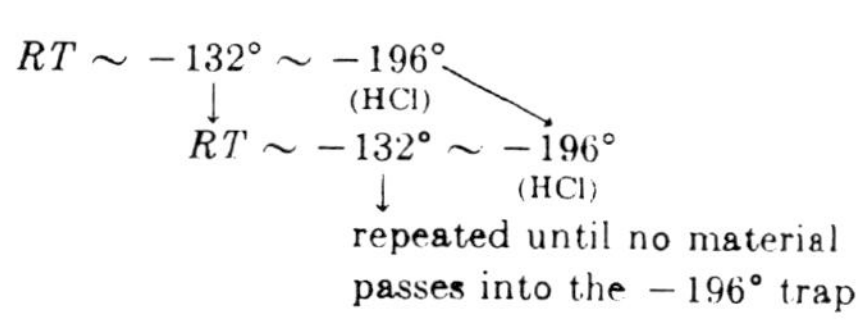

where RT means to warm the material slowly from −196°C to room temperature. The process is carried out without pumping unless noted. The material in the −196°C trap (HCl) is discarded. The trichlorosilane in the −132°C trap will have an infrared spectrum and a molecular weight consistent with the literature.

A 2.0 g sample of powdered dicobalt octacarbonyl is placed in a 25 mL glass reactor in a nitrogen-filled glove bag. The reactor is evacuated; and then 5 g of trichlorosilane is condensed from the vacuum system into the reactor which has been cooled to −196°C. The Teflon stopcock is closed, the reaction vessel allowed to warm to room temperature, and then the reactants are permitted to stand for 24 h. Cool the reactor to −42°C (diethyl ketone slush), open the Teflon stopcock, and remove the excess silane and noncondensable substances into the vacuum system with pumping. The remaining dry, solid material is then transferred in a nitrogen (or carbon monoxide) filled glove bag to a sublimation apparatus. The solid is then sublimed in vacuo twice at 30°C using a cold finder at −5°C to give tetracarbonyl(trichlorosilyl)cobalt. The product is then scraped from the cold finger in an inert atmosphere and placed in bottles with tight-fitting caps. The compound recovered in a 64% yield based on the cobalt carbonyl, $Co_2(CO)_8$, should be stored in a freezer (−10°C).

The tetracarbonyl(trichlorosilyl)cobalt obtained will be a light yellow solid (m.p. 44.5–45.0°C).

3.5. FORMATION OF RARE-GAS MOLECULES AT CRYOGENIC CONDITION

The rare gases have a particularly stable electronic configuration, comprising fully filled s and *p* valence orbitals. This makes these elements relatively nonreactive, and they exist at room temperature as monatomic gases. Pauling predicted in 1933 that heavier noble gases, whose valence electrons are screened by core electrons and thus less strongly bound, could form stable molecules. This prediction was verified in 1962 by the preparation of xenon hexafluoroplatinate, $XePtF_6$, the first compound to contain rare-gas atom [17,18] and in 1963 by KrF_2 [19], the first compound to contain Kr. Since then, a range of different compounds containing radon, xenon, and krypton have been theoretically anticipated and prepared [20–23]. Although the lighter noble gases helium, neon and argon are also expected to be reactive under suitable conditions [24,25], they remain the last three long-lived elements of the periodic table for which no stable compound is known before 1991.

3.5.1. Synthesis of $Xe[PtF_6]$ at Low Temperature [17,18]

PtF_6, diluted with SF_6 (1:6), was bubbled slowly through a length of 1/8 inch copper tubing into a Kel-F vessel while simultaneously flowing xenon into the same vessel through a separate length of tubing but at a faster flow rate. The flow rate of the reactant gases was regulated by passing them each through Monel Nupro SS-4MA metering valves. These valves were each coupled to the 1/8 inch lengths of copper tubing via a 1/4–1/8 o.d. reducing union. The lower portion of approximately 6 inch Kel-F reactor was maintained at 77 K. Thus the product formed by the gas–gas interaction was quickly quenched, along with excess xenon and sulfur hexafluoride diluent. This technique allowed for preparation of the approximately 1:1 product in gram amounts. The resultant, finely divided, mustard-yellow solid was retained on removal of sulfur hexafluoride and xenon at −80°C. Further warming to room temperature produced no other volatiles.

Many experiments were carried out at −20°C, which involved the simple mixing by diffusion of Xe and PtF_6 after a break seal separating them had been broken (see Fig. 3.21 for Xe + PtF_6 reaction). These reactions [26,27] were followed both tensimetrically, using a sensitive nickel diaphragm gauge as described by Cromer [28] to sense the pressure [29], and gravimetrically. When the quantity of PtF_6 greatly exceeded that of the Xe, the stoichiometry was closer to Xe $(PtF_6)_2$ and the product appeared deep red and more sticky. On the other hand, when Xe was more abundant, the deep yellow solid contained a composition closer to $XePtF_6$.

3.5.2. Photochemical Synthesis of KrF_2 at Low Temperature [19,30]

In 1962, Weeks et al. failed to obtain krypton difluoride when they used UV light to irradiate a mixture of krypton and fluorine at −60°C. In 1963, Streng mixed krypton and fluorine (or F_2O) in 1:1 ratio in a hard-glass container and exposed them to sunlight for 5 weeks at normal temperature and pressure. Initially, the product was believed as KrF_2, but the results were not reproduced and therefore, the claim was rejected. The photochemical synthetic approach to prepare KrF_2 was delayed until 1975, when Slivnik obtained KrF_2 (4.7 g) by lowering the reaction temperature to −196°C and irradiating the liquid krypton and fluorine mixture in a 100 mL hard-glass reaction flask with UV light. Experiment shows large influence of temperature on the reaction. A slight increase in temperature will not give KrF_2. For example, at −78°C, UV irradiation (200–470 nm) of a mixture of krypton and fluorine was unable to produce krypton fluoride. This explains the failures of the photochemical synthesis of krypton fluoride in the early years because the temperatures used were −60°C and room temperature, which are higher than −78°C.

In the photochemical synthesis of KrF_2, molecular fluorine get excited to dissociate into the fluorine atoms, which combine with krypton to give KrF radicals. Then

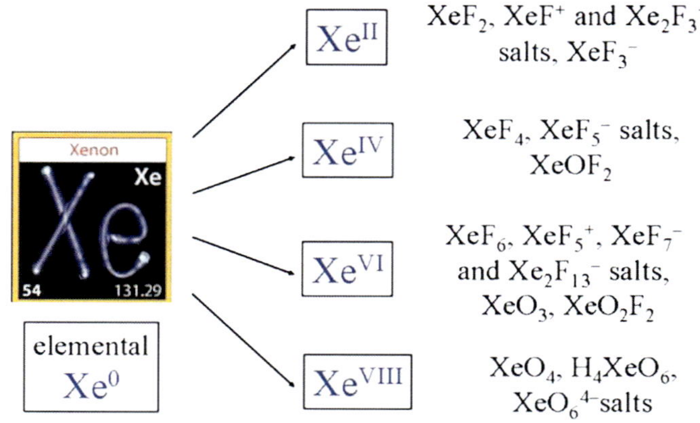

FIGURE 3.22 Classical chemistry of Xe involving connections to F and O ligands at various oxidation states of a rare gas [31]. *Reproduced by permission of The Royal Society of Chemistry.*

KrF collides with another KrF or F atom to give KrF_2. Therefore, the wavelength of light has a significant effect on the quantum yield (yield of KrF_2/W h).

For example, the quantum yield at 3100 Å is largest because the absorption band of fluorine molecule is in 2500–3000 Å region, which is conducive to the excitation of molecule and its dissociation to fluorine atom. The impurities in the system also have a very large effect on the quantum yield. For example, if oxygen content in Kr(s)–F_2(g) system is 10–15%, then quantum yield will decrease by 1–2 times. The effect of xenon is even larger and leads to the production of mainly XeF_2 instead of KrF_2. When BF_3 is added, the quantum yield doubles. This may be due to the reaction of BF_3 with KrF• radical to produce $KrBF_4$, which helps to stabilize KrF• and reduces its dissociation. On the other hand, BF_3 combines with the KrF_2 form $KrF_2 \cdot BF_3$, which prevents the photolysis of KrF_2.

About 500 compounds of heavier rare gas have been synthesized since 1962 [31]. He refers to compounds predicted by Pauling, etc. so long ago which contain fluoride, oxide, and related ligands and contain Xe at oxidation states +2, +4, +6, +8 (Fig. 3.22).

3.5.3. Synthesis of HXY Molecules by Matrix Photogeneration at Cryogenic Conditions

A fruitful route for preparing new rare gas species has been developed by Räsänen et al. [32]. They synthesized whole series of neutral rare-gas molecules of the type (X = Kr, Xe; Y = H, halogen, pseudohalogen) (Fig. 3.23) in solid rare-gas matrices. In short, the synthesis involves homolytic photodissociation of an HY precursor generating isolated H atoms and Y fragments, which are trapped in the matrix. The diffusion of hydrogen in the matrix is effectively controlled by temperature, and a selective mobilization of the hydrogens is achieved by careful annealing. The diffusing

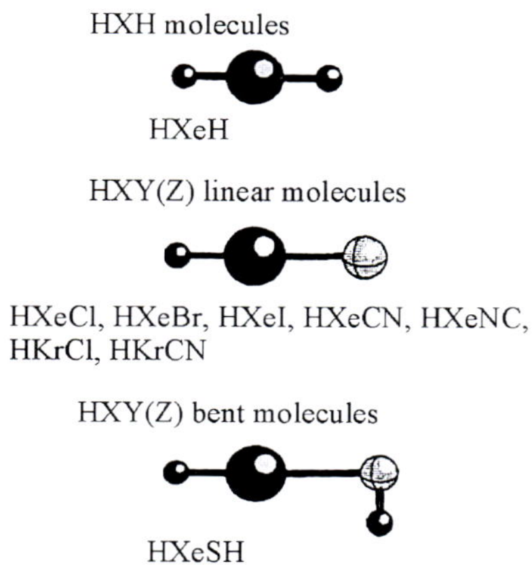

FIGURE 3.23 The HXH, HXY and HXY (Z) molecules [32].

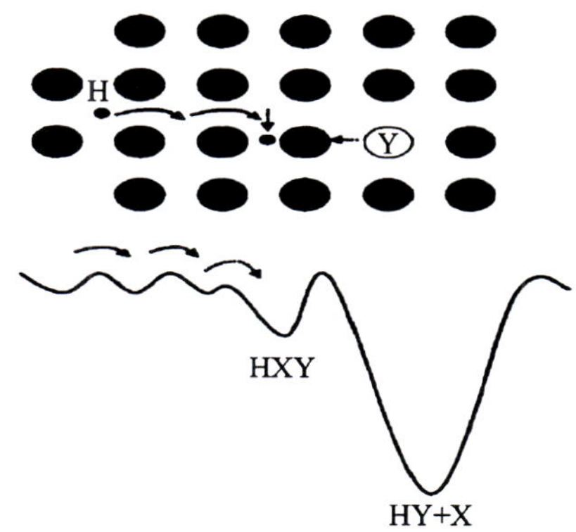

FIGURE 3.24 A schematic picture of the generation of the HXY (X = Kr, Xe; Y = Cl, Br, I, CN) molecules in solid rare gases; the lower curve describes the potential energy [32].

hydrogen can then approach a rare-gas atom with a Y fragment as neighbor, upon which the new HXY molecule is formed.

In the following we describe shortly the solid-state experiments for preparing the new rare-gas containing molecules. Solid rare gases (5–50 K) are used as the medium for synthesis and for spectroscopic identification of the formed species. The precursor molecules HY (hydrogen halides, HCN, H_2S etc.) are mixed with rare gases in the gas phase and isolated in a typical 1000-fold excess of the host atoms in a polycrystalline solid deposited onto an optically transparent substrate. The isolated species and their photoproducts can be monitored by measuring their infrared absorption spectra or by measuring the time-resolved luminescence spectra.

In order to promote the isolated precursor to photodissociative repulsive states, different light sources can be used. The dissociation energies (D_0) of the hydrogen halides HCl, HBr, and HI are 4.43, 3.75, and 3.05 eV, respectively [33], and various UV lamps work well for decomposing these precursors. For H_2S, photodissociation at 193 nm (ArF laser) is very efficient [34], since the absorption maximum of the precursor is near this wavelength. Also HCN can be photodissociated at 193 nm in solid Kr and Xe even though the efficiency is low. Additionally, water, which is always present as an impurity, produces H atoms into the solid host upon 193 nm excitation.

Upon photolysis, the IR absorptions of the precursor disappear and isolated H atoms and electronegative fragments Y are formed. This can be seen for instance from the appearance of the spin-orbit ($^2P_{1/2} \leftarrow {}^2P_{3/2}$) absorption at ca. 7600 and 3600 cm^{-1} for iodine and bromine atoms, respectively [35]. Both the H atoms and Y fragments are stable below the mobilization temperature of hydrogen atoms (30–50 K). In very diluted matrices (M/A ratio > 1000), it is possible to convert a major part of the precursor molecules into hydrogen atoms and Y fragments. In more concentrated matrices, other processes like clusterization and photo-induced reactions of hydrogen atoms produce more complicated products and hydrogen molecules [35].

The next step in the synthesis is the raising of the temperature to the point at which the photogenerated hydrogen atoms start diffusing in solid Xe. Several separate experiments have shown diffusion of hydrogen atoms at around 45 K [36–39]. Randomly distributed hydrogen atoms diffuse until they find an Xe atom which has a fragment Y as a nearest neighbor. At this stage, these three fragments H, Xe, and Y react and form an HXY molecule. This process is described in Fig. 3.24. Another possibility is a reaction of two hydrogen atoms with an Xe atom between them. This reaction produces xenon dihydride (HXeH) and it happens in any Xe matrix which contains hydrogen atoms. Most importantly, since HXY molecules are formed from H and Y which are too far from each other to be able to reform the precursor HY, species HXY and HXeH are the major trapping sites for H atoms and Y fragments in solid rare gases after the mobilization of hydrogen atoms [40]. Some successful results for synthesizing of HXY molecules (X = Kr, Xe; Y = X^-, CN, SH...) are reported in the late 1990s [41–45].

3.5.4. First Synthesis of a Stable Argon Compound HArF at Cryogenic Condition

In 2000, M. Räsänen was the first who reported the synthesis of a stable argon compound HarF [45] by matrix photogeneration route. Hydrogen fluoride was prepared in an argon matrix by passing argon over an HF–pyridine polymer (Fluka) at room temperature, and condensing the mixture onto a CsI substrate kept at 7.5 K. The argon used was either ^{40}Ar (100%; AGA) or ^{36}Ar (99.5% isotopically enriched; ICON Services).

By optimizing the experimental conditions, fairly monomeric samples of HF were obtained in the matrix (infrared bands corresponding to H−F stretching vibrations were found at $\upsilon_{HF} = 3962$ and $3954\ cm^{-1}$; $\upsilon_{DF} = 2895\ cm^{-1}$), as evidenced by comparison with previous HF/Ar matrix-isolation studies. A very high degree of deuteration (>90%) was achieved by passing the gaseous HF/Ar mixture through a volume containing small amounts of liquid deuterated sulfuric acid.

Photolysis of the HF was performed by illuminating the HF/Ar matrix (through an MgF_2 window) with a Kr vacuum-ultraviolet continuum discharge lamp (Opthos), which emitted in the spectral region of 127−160 nm wavelength. Typically, only a fraction (<30%) of HF dissociated when this light source was used, even after extended periods of photolysis; this most probably indicates self-limitation of the photolysis due to formation of absorbing species, possibly involving charge-transfer excitation of impurity oxygen atoms.

After photolysis of the HF, three previously unknown bands appeared in the infrared spectra at about 1969.5, 687.0, and $435.7\ cm^{-1}$ in an ^{40}Ar environment; the highest-frequency band displayed three components, assigned to different matrix sites. The intensities of these new absorptions increase proportionally on annealing the sample at about 18 K, and the absorber was found to be stable in these experiments at temperatures below 27 K. A similar set of new absorptions was obtained in photolyzed and annealed $DF/^{40}Ar$ matrices at 1466.3, 513.0, and $435.3\ cm^{-1}$. In $HF/^{36}Ar$ samples, three absorptions shifted to higher frequencies, these shifts being approximately +2, +2, and $+7\ cm^{-1}$ from the values given above for HF in ^{40}Ar matrix.

3.6. INORGANIC SYNTHESIS IN LIQUID AMMONIA

3.6.1. Reactions of Metals with Liquid Ammonia [46]

Some major reactions are summarized below:

1. Reaction of liquid ammonia with alkali metals

The solutions of alkali metals in liquid ammonia are metastable. The reactions are rather slow under most conditions, but in the presence of a catalyst the reaction is rapid, forming metallic amide and liberating H_2.

$$M + NH_3(l) = MNH_2 + \frac{1}{2}H_2\uparrow$$

The rate of this reaction increases with an increase in temperature and the atomic weights of the alkali metals.

Certain alkali metal compounds can also react with liquid ammonia:

$$MH + NH_3 = MNH_2 + H_2$$
$$M_2O + NH_3 = MNH_2 + MOH$$

2. Reactions between alkali earth metals and liquid ammonia

Beryllium and magnesium do not dissolve and react in liquid ammonia. However, in the presence of small amount of ammonium ions as catalyst, magnesium can react with liquid ammonia to form insoluble amide. The reactions are:

$$Mg + 2NH_4^+ = Mg^{2+} + 2NH_3 + H_2$$
$$Mg^{2+} + 4NH_3 = Mg(NH_2)_2 + 2NH_4^+$$

Other alkali earth metals, similar to the alkali metals, are also soluble in liquid ammonia, forming solutions that slowly decompose to form metallic amides.

The salts of alkali earth metals can also react with liquid ammonia to form the corresponding amides.

3. Reactions between rare earth metals and liquid ammonia

In 2006, Imamura et al. [47] reported that rare earth metals Eu and Yb are also soluble in liquid ammonia, which easily deposit to $Eu(NH_2)_2$, $Yb(NH_2)_2$, and $Yb(NH_2)_3$ at room temperature that decompose to form corresponding amides at 503 K and 470 K. All the products have good catalytic properties of hydrogenation. The reactions are:

$$Eu(NH_2)_2 \rightarrow EuN + \frac{1}{2}H_2 + NH_3$$
$$Yb(NH_2)_2 \rightarrow YbNH + NH_3 + H_2 + \frac{1}{2}N_2$$
$$Yb(NH_2)_3 \rightarrow YbN + 2NH_3$$

3.6.2. Reactions of Nonmetals with Liquid Ammonia

Among the nonmetals sulfur is most readily dissolved in liquid ammonia. The reaction in the solution may be represented by the following equation:

$$10S + 4NH_3 = S_4N_4 + 6H_2S$$

Ozone at −78°C reacts with liquid ammonia to give ammonium nitrate. The reactions are:

$$2NH_3 + 4O_3 = NH_4NO_3 + H_2O + 4O_2$$
$$2NH_3 + 3O_3 = NH_4NO_2 + H_2O + 3O_2$$

The percentage yield of ammonium nitrate is 98%, whereas that of ammonium nitrite is 2%.

3.6.3. Ammonolysis of Inorganic Compounds in Liquid Ammonia

Many compounds in liquid ammonia can undergo ammonolysis to give the corresponding amide compounds. For example:

$$BCl_3 + 6NH_3 = B(NH_2)_3 + 3NH_4Cl$$

If the amide $B(NH_2)_3$ is warmed to above 0°C, it decomposes to give imide:

$$2B(NH_2)_3 = B_2(NH)_3 + 3NH_3$$

Study shows that boron tri-iodide in liquid ammonia at −33°C can directly yield imide:

$$3BI_3 + 9NH_3 = B_2(NH)_3 + 6NH_4I$$

Compounds such as As_4S_6 and P_4S_3 can also undergo ammonolysis with liquid ammonia, for example: As_4S_6 in liquid ammonia gives a bright yellow ammonium salt, which on warming to 0°C yields a deep orange red imide of arsenic.

3.6.4. Substitution Reactions in Liquid Ammonia

−36°C

Example 1: $[Co(H_2O)_6]^{2+} + 6NH_3$
$= [Co(NH_3)_6]^{2+} + 6H_2O$

Example 2: $(\eta^5\text{-}C_5H_5)2TiCl + 4NH_3$
$= (\eta^5\text{-}C_5H_5)TiCl(NH_2)\cdot 3NH_3$
$+ C_5H_6(\eta^5\text{-}C_5H_5)TiCl(NH_2)\cdot 3NH_3$
$= (\eta^5\text{-}C_5H_5)TiCl(NH_2)3NH_3$

3.6.5. Synthesis of $MgCl_2\cdot 6NH_3$ in Liquid Ammonia [48]

Magnesium chloride hexammoniate ($MgCl_2\cdot 6NH_3$) has emerged as interesting functional material due to its new properties and applications. For example, due to the remarkable reversible ammonia storage property, $MgCl_2\cdot 6NH_3$ has potential application in solid-state ammonia storage/delivery system. Furthermore, this new ammonia storage material has a high volumetric ammonia density (up to 93% of liquid ammonia). It has also been considered for the separation of NH_3 from low-pressure ammonia plants and as sorption systems for refrigeration. In recent years, Danish researchers have proposed a new pathway to safe, high density, and convenient storage of hydrogen in the form of pellets of $MgCl_2\cdot 6NH_3$, which can hold more than 9% hydrogen by weight. This ammonia-mediated pathway based on $MgCl_2\cdot 6NH_3$ may be a new breakthrough in hydrogen storage technologies. Up to now, a variety of synthesis processes have been developed to fabricate $MgCl_2\cdot 6NH_3$.

Here we present a large-scale and low-cost synthesis route to $MgCl_2\cdot 6NH_3$. It can be prepared by the reaction between magnesium silicide (Mg_2Si) and ammonium chloride (NH_4Cl) in liquid ammonia, which was developed by Komatsu Electronic Metals Corp. to produce silane (SiH_4). The $MgCl_2\cdot 6NH_3$ product was characterized by X-ray diffraction (XRD), Fourier transform infrared spectroscopy (FTIR), and thermogravimetry (TGA). XRD study showed that the solid product prepared by the experiment using an Mg_2Si/NH_4Cl mass ratio of 1:2.8 was phase-pure and highly crystalline $MgCl_2\cdot 6NH_3$.

The reaction mechanism of this process is described as follows:

$$Mg_2Si + 4NH_4Cl + 8NH_3 \rightarrow 2MgCl_2\cdot 6NH_3 + SiH_4$$

The synthetic procedure is described as follows: $MgCl_2\cdot 6NH_3$ products were synthesized by the magnesium silicide method, with the experimental apparatus schematically shown in Fig. 3.25. The starting reagents used in this work were chemical purity-grade magnesium silicide (Mg_2Si), anhydrous ammonium chloride (NH_4Cl), and liquid ammonia (NH_3). Four kilograms of Mg_2Si powder and 11.2 kg of NH_4Cl powder were mixed and transferred into the stainless steel vessel. The vessel was then sealed and evacuated

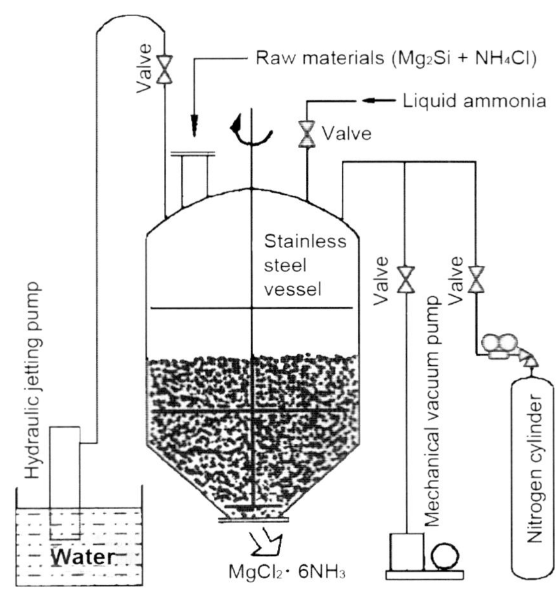

FIGURE 3.25 Schematic diagram of the experimental apparatus [48]. *Reprinted with permission from [48]. Copyright 2009 American Chemical Society.*

TABLE 3.5 Cocondensation and Warmup Reactions of Metal Atoms with Inorganic and Organic Coordinating Ligands [54]

Coordinating	Metal	Product
PF_3	Cr, Pd, Ni	$Cr(PF_3)_6$, $Pd(PF_3)_4$, $Ni(PF_3)_4$
	Fe	$Fe(PF_3)_5$, 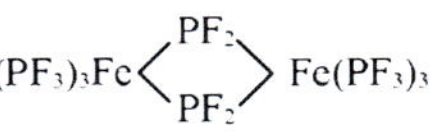
	Co	$Co_2(PF_3)_8$
PF_3/PH_3	Ni	$Ni(PF_3)_3(PH_3)$, $Ni(PF_3)_2(PH_3)_2$
PF_2Cl	Ni	$Ni(PF_2Cl)_4$
$NO/BF_3/PF_3$	Mn	$Mn(PF_3)(NO)_3$
C_6H_6	Cr	$Cr(C_6H_6)_2$
C_5H_6	Cr, Fe	$Cr(C_5H_5)_2$, $Fe(C_5H_5)_2$
	Co, Mo	$(C_5H_6)Co(C_5H_5)$, $(C_5H_5)_2MoH_2$
	Ti, V	$(C_5H_5)_2Ti$, $(C_5H_5)_2V$
	Ni	$(C_5H_7)Ni(C_5H_5)$, $(C_5H_5)_2Ni$
$1,3\text{-}C_4H_6/L$	Fe	$Fe(1,3\text{-}C_4H_6)LFe(1,3\text{-}C_4H_6)_2$
(L=CO, PF_3)	Cr	$Cr(1,3\text{-}C_4H_6)(CO)_4$
C_6H_6	Fe	C_6H_{10}, C_6H_6 and $Fe(C_6H_6)(C_6H_8)$
toluene/PF_3	Fe	Fe(toluene)$(PF_3)_2$
toluene/$1,3\text{-}C_4H_6$	Fe, Co, Ni, Cr	M(toluene)/$(1,3\text{-}C_4H_6)$
BCl_3	Cu	CuCl, B_2Cl_4
PCl_3	Cu, Ag	$CuCl(PCl_3)_x$, P_2Cl_4
RX(alkyl halide)	Mg	When RMgX is preheated, the unsolvated Grignard
	Ni	reagent produces$(CH_2{=}CH\text{-}CH_2NiBr)_2$
	Pt	$[PtCl(C_3H_5)]_4$

by the mechanical vacuum pump. High-purity nitrogen gas was then introduced into the vessel, and it was re-evacuated. This flushing and evacuation process was repeated three times. After complete evacuation, excess liquid ammonia was introduced into the vessel, and the reaction between Mg_2Si and NH_4Cl was carried out. At the same time, silane (SiH_4) gas was produced. After 20 min, the silane gas and residual ammonia were exhausted by the hydraulic jetting pump. Finally, a high purity of $MgCl_2 \cdot 6NH_3$ product was obtained by using an Mg_2Si/NH_4Cl mass ratio of 1:2.8 at low cost.

TABLE 3.6 Reactions of Si, SiO, B and BF Generated at High Temperatures with the Inorganic and Organic Molecules [54]

Monomeric unit	Active species	Product
B_2F_4		$BF \xrightarrow{-196\ °C} BF_2BFBF_2 \xrightarrow{-50\ °C} B_8F_{12}$
$CH_3C{\equiv}CCH_3$	BF	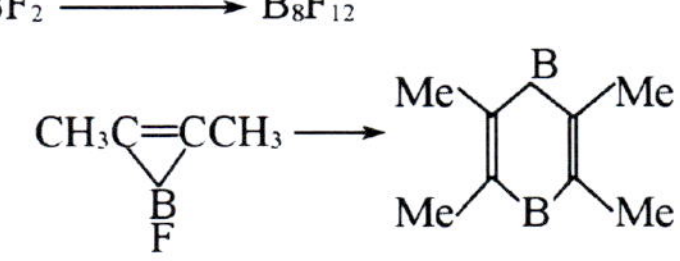
$CH{\equiv}CH$	BF	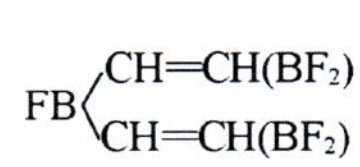
HCl	B	H_2, Cl_2, $HBCl_2$, BCl_3
$(CH_3)_3SiH$	Si	$(CH_3)_3Si\text{-}SiH + Me_3SiH \rightarrow CH_3)_3SiSiH_2Si(CH_3)_3$
alkene, 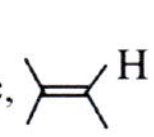	SiO	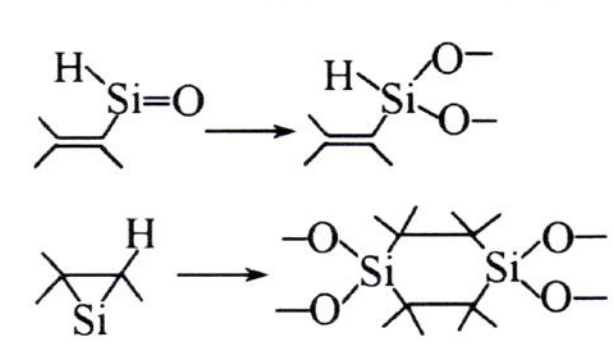

TABLE 3.7 Reactions of Silicon dihalide Radicals [54]

Studies on the radical reactions

$SiF_2 + BF_3 = Si_2BF_7$[a]

$SiF_2 + CO = SiF_2CO$

$SiF_2 + NO = N_2O +$ 

Synthetic reactions:

$SiF_2/BF_3 = SiF_2BF_3$

$SiF_2/BF_3 = Si_2BF_7,\ Si_3BF_9,\ Si_4BF_{11}$

$SiF_2/BCl_3 = Si_nF_{2n+1-z}Cl_zBF_{2-x}Cl_x (x=2, 1, 0;\ z=3, 2, 1, 0)$

$SiF_2/C_2H_2 = HC{\equiv}C(SiF_2)_2CH{=}CH_2$

$SiF_2/C_2H_4 =$ 

SiF_2/benzene= 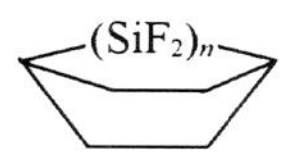 (n=2, 3)

$SiF_2/RC{\equiv}CR'$[b] 2:1 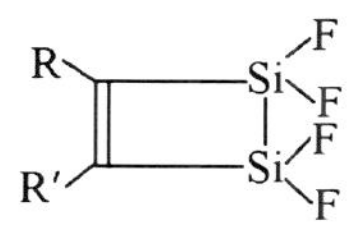

2:2 $RC{=}CR'{-}(SiF_2)_2{-}CR'{=}CR$

$SiCl_2/C_3H_6 =$ 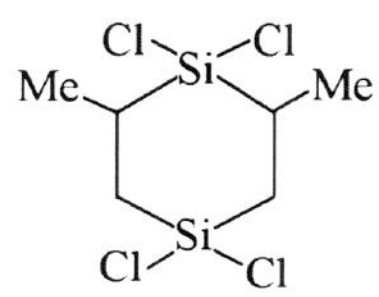

$SiCl_2/PCl_3 = SiCl_3PCl_2$

$SiCl_2/BCl_3 = SiCl_3BCl_2$

$SiCl_2/B_2Cl_4 = SiCl_3BCl_2$

$SiCl_2/SiCl_4 = Si_2Cl_6$[c]

$SiBr_2/SiBr_4 = Si_2Br_6$[c]

a: ESR study shows that in this reaction, the SiF_2 free radical exists in the form of the dimeric radical, Si_2F_4; b. may undergo cyclization or H-transfer from R to R′; c: SiF_2/SiF_4 cannot react to give Si_2F_6.

Phase identification of the $MgCl_2 \cdot 6NH_3$ products was carried out by an X-ray diffractometer.

Other than this some synthesis of binary metal chalcogenides and pnictides were carried out by metathesis synthetic reactions in liquid ammonia [49–51].

3.7. CRYOSYNTHESIS OF UNUSUAL INORGANIC COMPOUNDS [52–55]

Interaction of metal vapor or high-temperature (HT) inorganic chemical species with inorganic and organic ligands or molecules under cryogenic conditions led to the creation of some new families of coordination and unusual inorganic compounds which are difficult to prepare by traditional methods.

Typical examples of this class of reactions and the general methods of preparing high temperature species have been illustrated in detail in Chapter 2. In this part, some other types of cryosynthetic reactions and the related synthetic techniques will be specifically discussed in more detail.

3.7.1. Types of Cryosynthetic Reactions

In this type of cryosynthetic reactions, cocondensation is carried out in vacuum at temperature as low as that of liquid nitrogen or liquid hydrogen. It may also be carried out in two steps, viz., cocondensation and warm up, such as:

$$BF \xrightarrow{-196^\circ C} BF_2BFBF_2 \xrightarrow{-50^\circ C} B_8F_{12}$$

There are three types of reactions:

1. Reaction of the metal atom vapor with inorganic or organic molecule, sometimes known as the metal vapor synthesis (MVS).

TABLE 3.8 Cocondensation and Warmup Reactions of Carbon Vapor with Organic Groups [54] (C_1 in ^3P Ground State and ^1D, ^1S Metastable States)

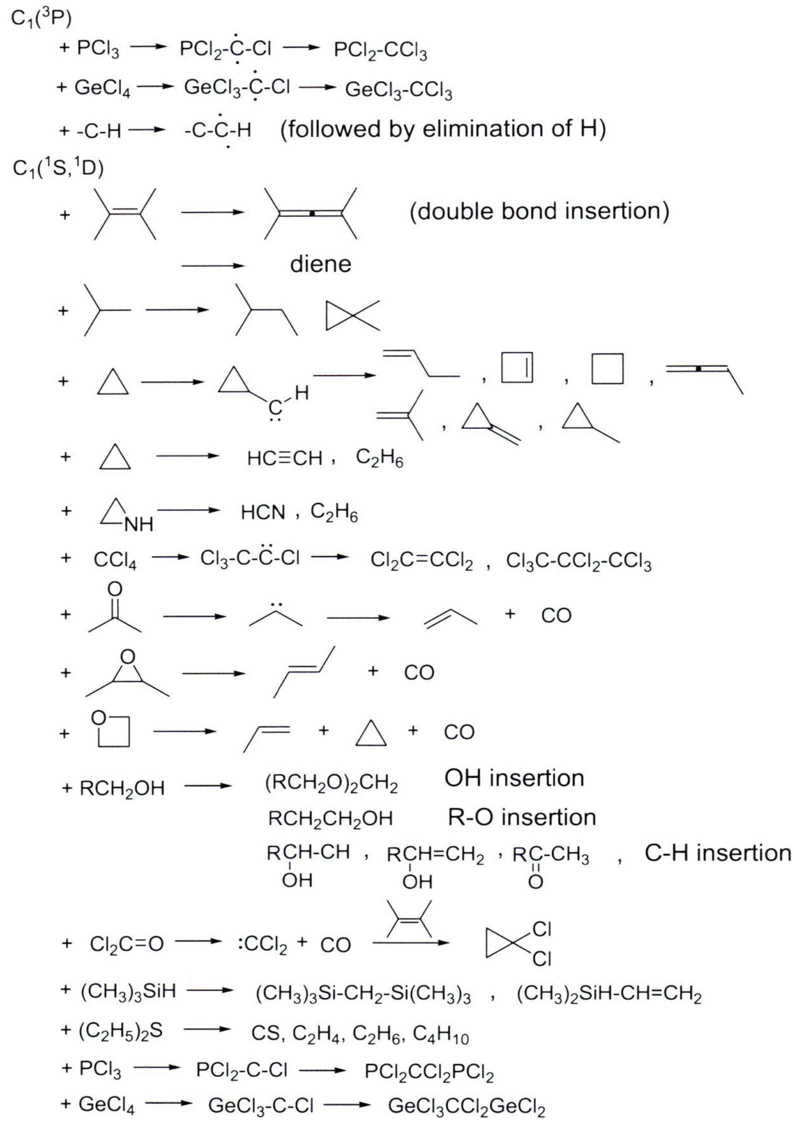

2. Reaction of nonmetallic high temperature molecular species or free radical with inorganic or organic molecule.
3. Reaction of carbon atom vapor (including ground and excited states) with inorganic or organic molecule.

Typical examples of the three types of reactions are listed in Tables 3.5–3.8, respectively.

3.7.2. Basic Apparatus for the Synthetic Reactions

In cryogenic chemical synthesis, the six types of basic apparatus for cocondenstation at liquid nitrogen of high temperature species and reactant (vapor or solution) are shown in Fig. 3.26(a–f).

The reaction apparatus (a) and (b) are generally not suitable, mainly because their condensation area is rather small. The choice of other apparatus is decided by the characteristics of the high temperature species and the requirements of the reaction. It is to be noted that in the apparatus (d) and (f), because of the reaction vessel, which is placed in liquid nitrogen, is rotating and the metal or the high temperature species vapor is mixed with the reactant through jetting, when the high temperature species reacts with the gaseous reactant or reactant solution, the reacting substances are uniformly

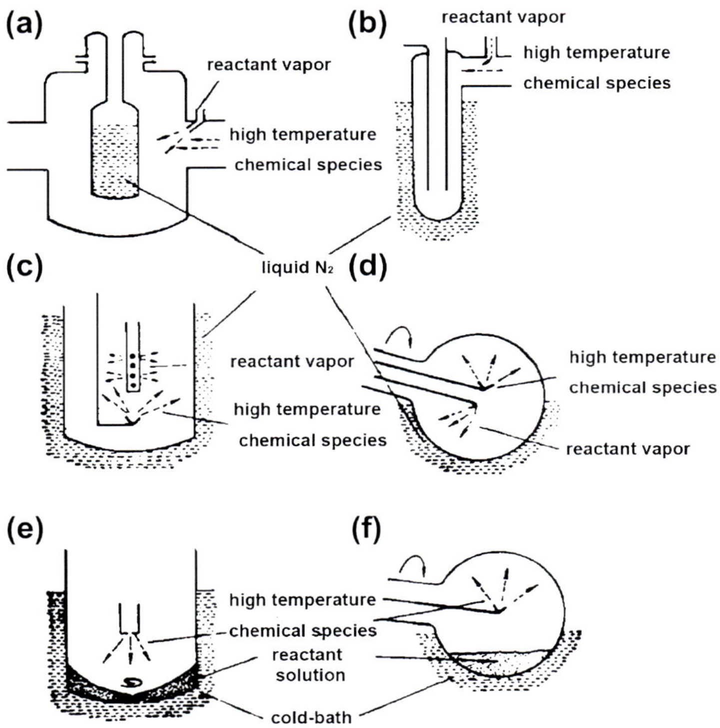

FIGURE 3.26 Six basic apparatus for cocondensation reaction [53,54].

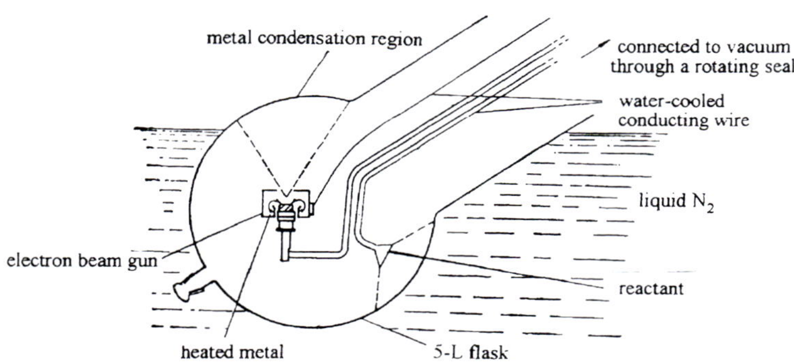

FIGURE 3.27 Electron-gun type apparatus for gas–liquid cocondensation reaction [53,54].

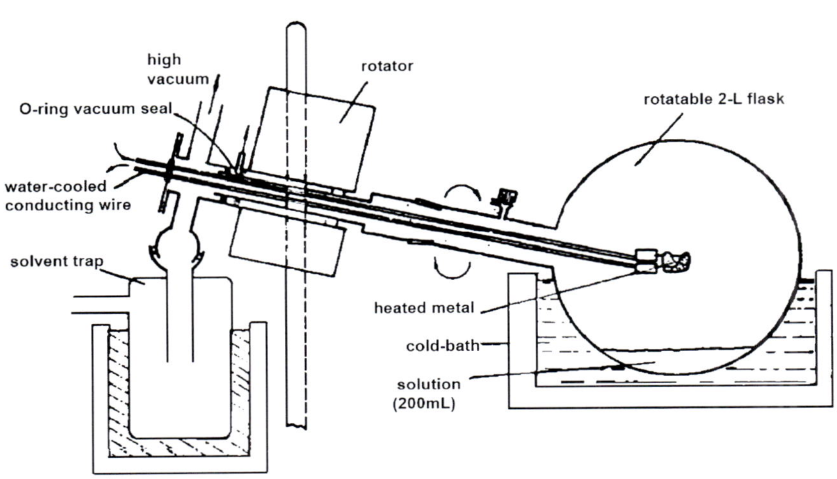

FIGURE 3.28 Apparatus for reacting metal atoms with ligand solution [53,54].

mixed and the condensation area becomes larger, resulting in better performance of the reaction. Given below (Figs. 3.27 and 3.28) are the more detailed diagrams of these two apparatus

References

[1] VACOM Vacuum, Inc. http://www.vacom-vacuum.com/
[2] H. Melville, B.G. Gowenlock, Experimental Methods in Gas Reactions, second ed., Macmillan & CO LTD, London, ST MARTIN's Press, 1964.
[3] W.L. Jolly, The Synthesis and Characterization of Inorganic Compounds, Prentice Hall Inc., Englewood Cliffs, NJ, 1970.
[4] G. Brauer, Handbuch der Präprativen Anorganischen Chemie III, in: E. Lindner, W. Wassing, R. Fawzi, M. Steimann (Eds.), Ferdinand Enke, Stuttgart, 1981.
[5] A.D. Bulanov, V.V. Balabanov, D.A. Pryakhin, O.Y. Troshin, Inorganic Materials 38 (2002) 283.
[6] Zeolites Research Group, Dalian Institute of Chemical Physics Molecular Sieve Zeolites (in Chinese), Scientific Press, Beijing, 1978.
[7] Y.C. Xie, Y.Q. Tang, Advances in Catalysis 37 (1990) 1.
[8] Y.C. Xie, Y.Q. Tang, US Patent 4, 917, 711, 1990.
[9] A.A.A. Emara, J.F. Lehmann, G.J. Schrobilgen, Journal of Fluorine Chemistry 126 (2005) 1373.
[10] Z. Mazej, B. Zemva, Journal of Fluorine Chemistry 126 (2005) 1432.
[11] D.R. Aris, C. Knapp, J. Passmore, X.P. Wang, Journal of Fluorine Chemistry 126 (2005) 1368.
[12] O. Ruff, W. Menzel, H. Plant, Zeitschrift Fur Anorganische Und Allgemeine Chemie 206 (1932) 59.
[13] F. Seel, O. Detmer, Zeitschrift Fur Anorganische Und Allgemeine Chemie 301 (1959) 113.
[14] H.P.A. Mercier, J.C.P. Sanders, G.J. Schrobilgen, S.S. Tsai, Inorganic Chemistry 32 (1993) 386.
[15] R. Faggiani, R.J. Gillespie, R. Kapoor, C.J.L. Lock, J.E. Vekris, Inorganic Chemistry 27 (1988) 4350.
[16] F.A. Cotton (Ed.), Inorganic Syntheses, Vol. XIII, McGRAW-HILL Book CO, 1972, pp. 65–73.
[17] N. Bartlett, Proceedings of the Chemical Society of London (1962) 218.
[18] L. Graham, O. Graudejus, N.K. Jha, N. Bartlett, Coordination Chemistry Reviews 197 (2000) 321.
[19] J.J. Turner, G.C. Pimentel, Science 140 (1963) 974.
[20] L.Y. Nelson, G.C. Pimentel, Inorganic Chemistry 6 (1967) 1758.
[21] V.E. Bondybey, Matrix Isolation Search for Transient Species, Univ. California, Berkeley, 1971.
[22] L. Stein, Nature 243 (1973) 30.
[23] J.H. Holloway, E.G. Hope, Advances in Inorganic Chemistry 46 (1999) 51.
[24] G. Frenking, D. Cremer, Structure and Bonding 73 (1990) 17.
[25] G. Frenking, W. Koch, D. Cremer, J. Gauss, J.F. Liebman, Journal of Physical Chemistry 93 (1989) 3410.
[26] N. Bartlett, N.K. Jha, in: H.H. Hyman Jr. (Ed.), Nobel Gas Compounds, University of Chicago Press, Chicago, IL, 1963.
[27] N.K. Jha, Ph.D. Thesis, University of British Columbia, 1965.
[28] S. Cromer, U.S.A.E. C Declassified Report, MDDC-803, 1947
[29] T.E. Mallouk, G.L. Rosenthal, G. Muller, R. Brusasco, N. Bartlett, Inorganic Chemistry 23 (1984) 3167.
[30] F. Schreiner, J.G. Malm, J.C. Hindman, Journal of the American Chemical Society 87 (1965) 25.
[31] W. Grochala, Chemical Society Reviews 36 (2007) 1632.
[32] M. Pettersson, J. Lundell, M. Rasanen, European Journal of Inorganic Chemistry (1999) 729.
[33] H. Okabe, Photochemistry of Small Molecules, Wiley, New York, 1978.
[34] L. Khriachtchev, M. Pettersson, E. Isoniemi, M. Rasanen, Journal of Chemical Physics 108 (1998) 5747.
[35] M. Pettersson, J. Nieminen, Chemical Physics Letters 283 (1997) 1.
[36] R. Schriever, M. Chergui, N. Schwentner, Journal of Physical Chemistry 95 (1991) 6124.
[37] D. Labrake, E. Weitz, Chemical Physics Letters 211 (1993) 430.
[38] J. Eberlein, M. Creuzburg, Journal of Chemical Physics 106 (1997) 2188.
[39] H. Muto, K. Nunome, M. Iwasaki, Journal of Physical Chemistry 84 (1980) 3402.
[40] M. Pettersson, Ph.D. Thesis, University of Helsinki, 1998.
[41] M. Pettersson, J. Lundell, M. Rasanen, Journal of Chemical Physics 102 (1995) 6423.
[42] M. Pettersson, J. Lundell, L. Khriachtchev, M. Rasanen, Journal of Chemical Physics 109 (1998) 618.
[43] M. Pettersson, J. Lundell, M. Rasanen, Journal of Chemical Physics 103 (1995) 205.
[44] M. Pettersson, J. Lundell, L. Khriachtchev, E. Isoniemi, M. Rasanen, Journal of the American Chemical Society 120 (1998) 7979.
[45] L. Khriachtchv, M. Pettersson, N. Runebag, J. Lundell, M. Räsänen, Nature 406 (2000) 874.
[46] D. Nichols, Inorganic Chemistry in Liquid Ammonia, Elsevier Scientific Publishing Company, 1979.
[47] H. Imamura, Y. Sakata, Y. Tsuruwaka, S. Mise, Journal of Alloys and Compounds 408 (2006) 1113.
[48] H.L. Zhu, X.Y. Gu, K.H. Yao, L.H. Gao, J.J. Chen, Industrial & Engineering Chemistry Research 48 (2009) 5317.
[49] G.A. Shaw, I.P. Parkin, Inorganic Chemistry 40 (2001) 6940.
[50] L. Sun, Z.J. Zhang, Z.H. Wang, Z.S. Wu, H.X. Dang, Materials Research Bulletin 40 (2005) 1024.
[51] H.M. Zhu, D.R. Sadoway, Journal of Materials Research 16 (2001) 2544.
[52] B.I. Kharisov, A.D. Garnovskii, L.M. Blanco, A.S. Burlov, A. Garcia-Luna, Journal of Coordination Chemistry 49 (1999) 113.
[53] G.A. MoskovitsMartinOzin, Cryochemistry, John Wiley & Sons Inc., 1976.
[54] G.A. Ozin, A.V. Voet, Cryogenic Inorganic Chemistry-Progress in Inorganic Chemistry, Vol. 19, John Wiley & Sons Inc., 1976, p. 105.
[55] P.L. Timms, Low temperature condensation of high temperature species as a synthetic method, in: , Advances in Inorganic Chemistry and Radiochemistry, Vol. 14, Academic Press, 1972, p. 121.

CHAPTER

4

Hydrothermal and Solvothermal Syntheses

Shouhua Feng, Guanghua Li

Jilin University, China

Hydrothermal and solvothermal syntheses are important branches of inorganic synthesis. Hydrothermal synthesis refers to the synthesis through chemical reactions in aqueous solution above boiling point of water and solvothermal synthesis in nonaqueous solution at relatively high temperatures. The development of hydrothermal and solvothermal techniques has a history of more than 100 years, beginning with the synthesis of minerals and the extraction of elements from minerals. Nowadays, these techniques have widely been applied in the syntheses of conventional and advanced materials, treatment of wastes, and mimicking geothermal and bio–hydrothermal processes. Zeolite and quartz industries further promote the basic studies of hydrothermal and solvothermal chemistry, and more and more scientists realize the importance of developing hydrothermal and solvothermal reactions, upon which effective syntheses would be established. Therefore, the investigation of hydrothermal or solvothermal reactions is the first step for the synthesis of novel materials. Moreover, one needs to consider the matching of the reactions and crystal growth if the crystallization is involved in the process of synthesis. Finally, we need to think about green synthesis. Developing trends of hydrothermal or solvothermal synthesis are going to the high-tech materials and biomolecules according to liquid nucleation mechanisms of hydrothermal or solvothermal process, different from the defusing mechanisms of solid-state reaction. In this chapter, we present the foundation of hydrothermal and solvothermal syntheses of materials, hydrothermal biochemistry, supercritical water oxidation process, technique and methods, and ionothermal synthesis.

4.1. FOUNDATION OF HYDROTHERMAL AND SOLVOTHERMAL SYNTHESES [1]

4.1.1. Features of Hydrothermal Synthetic Reactions

The hydrothermal technique is historically rooted in the geological sciences. In the mid-nineteenth century, the term "hydrothermal" was first used by a British Geologist, Sir Roderick Murchison (1792–1871), to describe the formation of minerals by hot water solutions rising from cooling magma. Since then, extensive study has been performed to study the synthesis of new materials, the development of new hydrothermal methods, and the understanding of reaction mechanism. Chemists today commonly refer to a hydrothermal or solvothermal synthesis as the synthesis by chemical reactions of substances in a sealed and heated aqueous solution or organic solvent at appropriate temperature (100–1000°C) and pressure (1–100 MPa). Normally, hydrothermal and solvothermal reactions are conducted in a specially sealed container or high-pressure autoclave under subcritical or supercritical conditions of solvent. The studies on hydrothermal and solvothermal syntheses have mainly focused on the reactivity of the reactants, regularities of synthetic reactions and conditions, and their relationship with the structures and properties of products.

The main difference between hydrothermal and solid-state reactions lies in "reactivity" which is reflected in their different reaction mechanisms. Solid-state reactions depend on the diffusion of the raw materials at interface, whereas in hydrothermal and solvothermal

Modern Inorganic Synthetic Chemistry, DOI: 10.1016/B978-0-444-53599-3.10004-6

reactions the reactant ions and/or molecules react in solution. Obviously, the difference of reaction mechanisms may lead to different final structures of products, even if the same reactants are used. Hydrothermal and solvothermal syntheses deal with the chemistry in preparation, synthesis, and assembly of special compounds or materials through solution route. More importantly, many compounds or materials with special structures and properties, which cannot be prepared from solid-state synthesis, may be obtained by hydrothermal and solvothermal reactions. In some cases, hydrothermal or solvothermal reaction offers an alternative and mild synthetic method for solid-state reactions by lowering reaction temperature.

The nonideal and nonequilibrium states are key features of the hydrothermal and solvothermal reaction systems [2]. Water or other solvents can be activated under high temperature and pressure conditions. A series of hydrothermal and solvothermal reactions have become increasingly important routes for the preparation of most inorganic functional materials, special compositions, structures, and condensed states and particular morphology, such as nano- and ultrafine powders, sol–gel, noncrystalline state, inorganic membrane, and single crystals.

The operability and tunableness of hydrothermal and solvothermal chemistry is another feature, which bridges the synthetic chemistry and physical properties of as-made materials. Some new hydrothermal reactions have been discovered. Compared to other synthetic techniques, hydrothermal and solvothermal techniques have their preponderance. Till now, a variety of materials and crystals used in many technology fields can be prepared by hydrothermal and solvothermal methods. The chemical and physical properties of the resulting materials have their own specificity and superiority. Based on hydrothermal and solvothermal chemistry, we may conduct unique synthetic reactions which cannot take place in solid-state reactions due to evaporation of reactants at high temperatures, prepare new materials with special valence states, metastable structure, condensed and aggregation states, produce metastable phases or materials with low melting point, high vapor pressure, and low thermal stability, grow perfect single crystals with thermodynamically equilibrium defect, controllable morphology and particle size, and make ion-doping directly in synthetic reaction.

4.1.2. Classification of Hydrothermal Reactions

(1) Synthesis: powders or single crystals are formed by chemical reaction of multicomponent reactants.

$$Nd_2O_3 + H_3PO_4 \rightarrow NdP_5O_{14}$$
$$CaO \cdot nAl_2O_3 + H_3PO_4 \rightarrow Ca_5(PO_4)_3OH + AlPO_4$$
$$La_2O_3 + Fe_2O_3 + SrCl_2 \rightarrow (La, Sr)FeO_3$$
$$FeTiO_3 + KOH \rightarrow K_2O \cdot nTiO_2 (n = 4, 6)$$

(2) Heated treatment: crystal materials could be hydrothermally treated to change or modify its properties.

Artificial fluorine asbestos → Artificial fluorine mica

(3) Transformation: thermodynamic or dynamic stability of crystalline materials driving the phase transformation.

Feldspar → Kaolinite
Peridot → Serpentine
NaA → NaS

(4) Ion exchange.
(5) Crystal growth.
(6) Dehydration: $Mg(OH)_2 + SiO_2 \rightarrow$ Chrysotile
(7) Decomposition:

$$FeTiO_3 \rightarrow FeO + TiO_2$$
$$ZrSiO_4 + NaOH \rightarrow ZrO_2 + Na_2SiO_3$$
$$FeTiO_3 + K_2O \rightarrow K_2O \cdot nTiO_2 + FeO\ (n = 4, 6)$$

(8) Extraction.
(9) Disproportionation.
(10) Redox:

$$Cr + H_2O \rightarrow Cr_2O_3 + H_2$$
$$Zr + H_2O \rightarrow Zr_2O_3 + H_2$$
$$Me + nL \rightarrow MeLn(Me = \text{metal},\ L = \text{organic ligand})$$

(11) Precipitation:

$$KF + MnCl_2 \rightarrow KMnF_3$$
$$KF + CoCl_2 \rightarrow KCoF_3$$

(12) Crystallization:

$$CeO_2 \cdot xH_2O \rightarrow CeO_2$$
$$ZrO_2 \cdot H_2O \rightarrow M - ZrO_2 + T - ZrO_2$$
Amorphous aluminosilicate (gel) → zeolite

(13) Hydrolyzation.
(14) Agglomeration.
(15) Hydrothermocompression.
(16) Solidification.

According to reaction temperature, hydrothermal and solvothermal syntheses can be classified into subcritical and supercritical synthesis reactions. The temperature is in the range of 100–240°C in subcritical synthesis reaction, which is applicable to industrial and laboratory operations. The hydrothermal and solvothermal syntheses of zeolites are typical subcritical syntheses. In the supercritical synthesis, the temperature could reach 1000°C and the pressure could reach 0.3 GPa. By using the special properties of solvent (water or organic solvent) and other reactants under supercritical temperature and pressure, various syntheses with specific features could be conducted, resulting in the formation of numerous crystalline materials with interesting structures. In addition, it should be pointed out that some crystal materials cannot be obtained by using other preparation approaches. Hydrothermal and solvothermal chemistry has been applied in the synthesis and preparation of complex inorganic materials. For instance, a large number of functional materials, such as nonlinear optical materials $NaZrP_3O_{12}$ and $AlPO_4$, acousto-optical crystal zinc lithium aluminates, multifunctional laser crystal $LiNbO_3$, and $LiTaO_3$ can be prepared under high temperature and pressure hydrothermal conditions. Hydrothermal and solvothermal syntheses have also been widely used in the preparation of ZnO_2, ZrO_2, GeO_2, and CrO_2, ferroelectric, magnetoelectric, and photoelectric solid materials such as $LaFeO_3$, $LiH_3(SeO_2)_2$, superconducting membrane $BaPb_{1-x}BiO_3$, modern artificial jewels and colored quartz.

4.1.3. Property of Reaction Medium [3]

Reaction medium is an important constituent for any hydrothermal or solvothermal system. It exhibits unique properties, especially under supercritical conditions. Water is environmentally safe material and cheaper than other solvents and thus remains the most widely used solvent in hydrothermal and solvothermal syntheses as compared to organic solvents. It can be seen that the solvent properties of water under high temperature and pressure hydrothermal condition can be significantly changed. For example, under high temperature and pressure hydrothermal condition, the density, surface tension, and viscosity of water will be lower, and the vapor pressure and ion product will be higher.

There are three features of hydrothermal system: first, accelerating the reaction rate among the complex ions; second, intensifying the hydrolyzation reaction; and third, significantly changing the redox potential of the reactants. Normally, there are two extreme types of basic chemical reactions. One is ionic reaction including metathesis of inorganic materials which could be instantly finished at environment temperature. Another is free-radical reaction including explosive reaction of organic compounds. According to the electronic theory, the reactions of organic compounds with polar bonds usually have some characteristics of ionic reaction. Therefore, when water is used as a medium and the reaction system is heated above its boiling point in a sealed reactor, the ionic reaction rate will be certainly accelerated, which is consistent with Arrhenium equation: $d\ln k/dT = E/RT^2$, e.g., the reaction rate constant k will exponentially increase with the increase of reaction temperature. Because of the increased ionization constant of water caused by the increased temperature, the ionic or hydrolysis reaction for those indissolvable inorganic materials at environment temperature or organic compounds can be promoted under the high temperature and pressure hydrothermal condition.

It is well-known that the chart of pVT of water with the temperature as high as 1000°C and the pressure as high as 1 GPa has been accurately measured with an error of less than 1% [4]. As shown in Fig. 4.1, within the studied range, the ion product of water rapidly increases with the increase in temperature and pressure. At temperature 1000°C and pressure 1.0 GPa, the $-\lg k_w$ is 7.85 ± 0.3; while under very high PT conditions (1000°C and 15–20 GPa), the density of water is about 1.7–1.9 g/cm^3. At this state, medium water is completely dissociated into H_3O^+ and OH^-, behaving like a molten salt.

The viscosity of water decreases with the increase in temperature. At temperature 500°C and pressure 0.1 GPa, the viscosity of water only amounts to 10% of its value under normal condition [5]. Hence, the mobility of molecules and ions in water under hydrothermal condition is much higher than under normal condition.

When using water as a solvent, the dielectric constant is one of the very important properties of water. As shown in Fig. 4.2, the dielectric constant will decrease with the increase of temperature and increase with the increase of pressure [6]. In fact the temperature factor plays a leading role in the dielectric constant of water. Knowledge of the temperature dependence of the relative dielectric constant is of great importance for understanding the hydration/dehydration phenomena and reactivity of a variety of solutes in supercritical water. According to Franck's results, the high dielectric

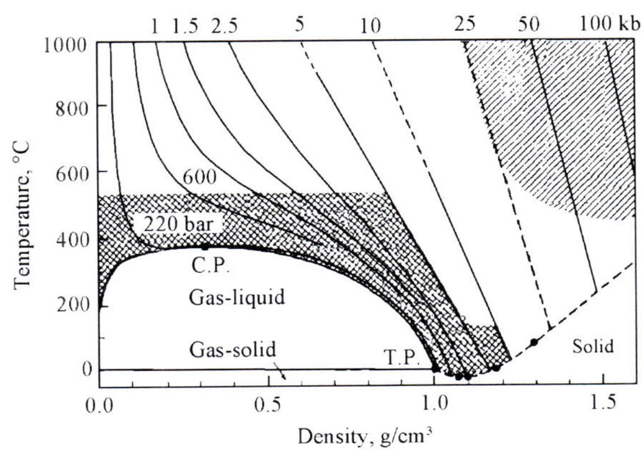

FIGURE 4.1 The plots of temperature–density of water with pressure as a parameter. *Images reproduced with permission [4].*

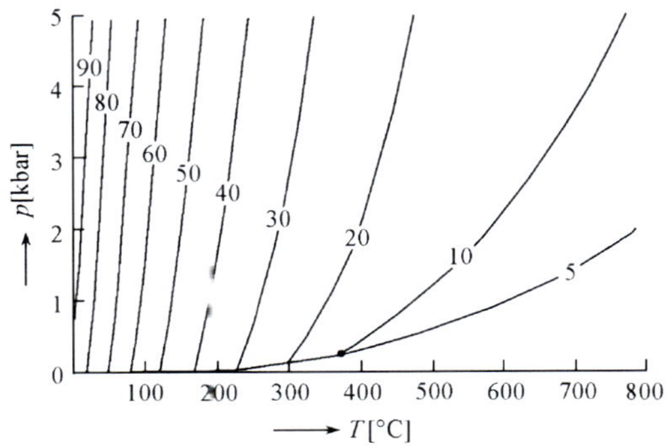

FIGURE 4.2 A plot of dielectric constant of water versus temperature and pressure [6]. *Reprinted from Phys. Chem. Earth, 13/14, T. M. Seward, Metal complex formation in aqueous solutions at elevated temperatures and pressures, p113, Copyright (1981), with permission from Elsevier.*

constant of water is confined to a small region of low temperatures and high pressures, while the dielectric constant of water is about 10–30 in the supercritical region, which is a large range [7]. Under normal conditions, electrolyte will be totally dissociated in water. However, the dissociated ions will tend to recombine together with rising temperature and for most of the substances this transition will occur at 200–500°C [8].

In order to better study hydrothermal synthesis, it is important to study the correlation of pressure and temperature for water. The pressure prevailing under hydrothermal conditions depends on the degree of fill of the reaction solution (as shown in Fig. 4.3) [9]. For a typical hydrothermal synthesis experiment, the degree of fill is usually between 50 and 80% and the pressure between 0.02 and 0.3 GPa.

In short, under high temperature and pressure hydrothermal condition, water may act as a solvent or a part of the reactants in some cases, change the chemical and physical properties of reactants and products, accelerate reaction, and transfer pressure.

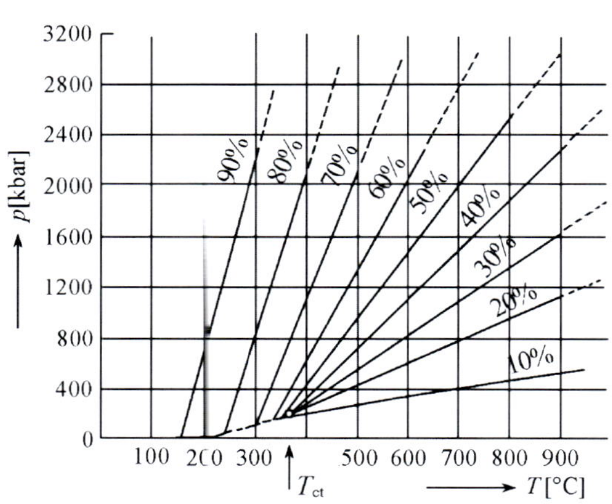

FIGURE 4.3 Correlation of pressure and temperature for water, with the filling factor of the sealed reactor as a parameter [3]. *Images reproduced with permission from Wiley-VCH.*

The reaction medium of hydrothermal synthesis is water, while the reaction mediums of solvothermal synthesis are various organic solvents. The variety of organic solvents with different properties offers more room for new materials. In solvothermal synthesis, the organic solvent not only supplies a reaction medium but also dissolves or partially dissolves the reactants to form solvent–reactant complex, which will affect the chemical reaction rate. In addition, the organic solvent can also affect the concentration and state of the active species of the reactants, which could finally change the reaction process. The first consideration in choosing an organic molecule as solvent is the role it will play in the synthesis. So far, several dozens of alcohols have been used in the solvothermal synthesis. The classification of organic solvent is usually based on the macroscopic and microscopic molecular parameters and empirical solvent polar parameters of compound, such as molecular weight (*Mr*), density (*d*), melting point (mp), boiling point (*bp*), molecular volume, heat of evaporation, dielectric constant (ε), dipole moment (μ), and solvent polarity (E_N^T). Among these parameters, solvent polarity, which is defined as the sum of the interaction of solvent and solute including Coulomb force, induction force, dispersion force, H-bond, and charge transport force, is the key one to describe the solvation property of solvent. Table 4.1 lists the solvent polarity and other parameters for commonly used solvents in solvothermal synthesis.

4.2. FUNCTIONAL MATERIALS FROM HYDROTHERMAL AND SOLVOTHERMAL SYSTEMS

Inorganic solids may be prepared by two synthetic methods, e.g., solid-phase reaction and solution reaction. The solid-state synthesis usually requires relatively high temperatures to overcome difficulties in transporting the reactants to the sites of the reaction. The high temperature of solid-phase reactions also tends to provide routes to the thermodynamically more favored phases in the systems of interest. This synthesis method is a traditional approach to prepare solid-state oxides. Transport in liquid phase is obviously much easier than in solids. So the synthetic reactions in solution need much lower temperature. Understanding the mechanism of hydrothermal reactions is particularly necessary for the suitable application of a method to a specific synthesis and the exploration of new materials with desired properties. Basically, the mechanism of hydrothermal reactions follows a liquid nucleation model. It is different from that of solid-state reactions, where the reaction mechanism mainly involves

TABLE 4.1 Physical Parameters of the Solvents used in Solvothermal Synthesis (Unit: Omitted)

Solvent	M_r	d	mp	bp	ε	μ	E^T_N
Water	18.01	1.000	0	100	80.4	1.94	1.000
Methyl alcohol	32.04	0.791	−98	65	32.6	1.70	0.762
Ethyl alcohol	46.07	0.785	−130	78	24.3	1.69	0.654
Propanol	60.10	0.804	−127	97	20.1	1.66	0.602
2-Propanol	60.10	0.785	−90	82	18.3	1.66	0.546
Butanol	74.12	0.810	−90	118	17.1	1.66	0.602
2-Butanol	74.12	0.807	−115	98	15.8		0.506
2-Methyl-1-propanol	74.12	0.802	−10	108	17.7	1.64	0.552
2-Methyl-2-propanol	74.12	0.786	25	83			0.389
Pentanol	88.15	0.811	−78	137	13.9	1.80	0.568
2-Pentanol	88.15	0.809		120	13.8	1.66	
3-Methyl-1-butanol	88.15	0.809	−11	130	14.7	1.82	0.565
2-Methyl-2-butanol	88.15	0.805	−12	102	7.0	1.70	0.321
Hexyl alcohol	102.18	0.814	−52	157	13.3		0.559
1-Heptanol	116.20	0.822	−36	176	12.1		0.549
2-Methyl-2-hexanol	116.20	0.8119		139.4			
Tetradecanol	214.39	0.823	39	289			
Cyclohexanol	100.16	0.963	21	160	15.0	1.90	0.500
Benzyl alcohol	108.14	1.045	−15	205	13.1	1.70	0.608
Ethylene glycol	62.07	1.109	−11	199	37.7	2.28	0.790
1,3-Propanediol	76.10	1.053	−27	214	35.0	2.50	0.747
1,2-Propanediol	76.10	1.036	−60	187	32.0	2.25	0.722
1,4-Butanediol	90.12	1.017	16	230	31.1	2.40	0.704
1,3-Butanediol	90.12	1.004	−50	207			0.682
Diethylene glycol	106.12	1.118	−10	245			0.713
Triethylene glycol	150.18	1.123	−7	287	23.7	5.58	0.704
Tetraethylene glycol	194.23	1.125	−6	314			0.664
Glycerol	92.09	1.261	20	180	42.5		0.812
Diglycerol	166.18	1.300					

diffusion of atoms or ions at the interface between reactants. However, in the supercritical region of water, less data are available at present, and only those for pure water and simple saltwater solutions are known. New methods and strategies play an important role in the investigation of hydrothermal synthesis. Various methods such as induced growth through crystal seeds, structure-directing agent technique, mineralizing, templating, complexing, nonaqueous solvothermal routes, high temperature and pressure technique, and redox environment control make hydrothermal reactions particularly available for new advanced solid materials.

Hydrothermal synthesis or solvothermal synthesis has been successful for the preparation of important solids such as zeolites, open-framework compounds, organic–inorganic hybrid materials, metal-organic framework materials, superionic conductors, chemical sensors, electronically conducting solids, complex oxide ceramics and fluorides, magnetic materials, and luminescence phosphors. It is also a route to unique condensed materials including nanometer particles, gels, thin films, equilibrium defect solids, distinguished helical and chiral structures, and particularly stacking-sequence materials. In addition to the synthesis of new materials, hydrothermal synthesis has been important in biology and environmental sciences, for example, in the origin of life and in the supercritical water oxidation process for decomposing organic wastes.

Hydrothermal or solvothermal synthesis often takes place in the supercritical regime at reaction temperature higher than 300°C. Until 1980s, supercritical hydrothermal or solvothermal synthesis was mainly used to produce large single crystals. The vast majority of hydrothermal or solvothermal synthesis preformed today in the synthesis of inorganic materials are conducted under much milder reaction condition (temperature < 300°C). In this section, except the production of large single crystals, we will only focus on the hydrothermal or solvothermal synthesis of functional materials under mild reaction conditions. The supercritical hydrothermal or solvothermal technique will be discussed in Section 4.5.

4.2.1. Single Crystals

One of the most important applications of hydrothermal technique is the synthesis and growth of bulk single crystals [10–12]. Hydrothermal technique is conducive to the growth of bigger, purer, and dislocation-free single crystals. Hydrothermal crystal growth differs from conventional liquid-solution crystal growth at room temperature. In a process of conventional liquid-solution crystal growth, reagents should have a high solubility otherwise the crystal growth will face a severe problem of diffusion. However, slow diffusion has little effect on crystal growth in a process of hydrothermal crystal growth because the viscosity of hydrothermal mediums is much lower than that of near-ambient solutions. Because diffusion is significantly faster at hydrothermal conditions, the growth rate of crystals is faster and reagents of low solubility can be used to grow crystals.

In general, the stages of crystal formation include small aggregations of chemical precursors that give unstable germ nuclei, some of these embryos become large enough to be stable nuclei and spontaneous deposition of more material on such nuclei results in larger crystallites. Numerous detailed studies have been performed on kinetics and thermodynamic aspects of hydrothermal crystal growth, and basic theoretical studies of this field have been more mature. For example, Barrer has systematically studied the nucleation, crystal growth, and reaction variables of zeolites [2]. Because of space limitations the basic theory of hydrothermal crystal growth will be not discussed here. The readers can get more information about this topic from related literature source.

Hydrothermal technique has widely been proved to be one of the most effective approaches for crystal growth and practically all inorganic crystals, both in nature and in laboratory, such as native elements, oxides, silicates, phosphates, chalcogenides, halides, germanates, carbonates, and so on, have been prepared and grown by this technique. Moreover, hydrothermal technique is being employed on a large scale to prepare electronic, magnetic, optic, ceramic, and a host of other materials, both as single crystals and polycrystalline materials. For example, α-quartz crystal, as a key piezoelectric material, is already grown commercially. In fact, hydrothermal technique is the only technique to prepare and grow some of the inorganic crystals. For example, because of solubility and diffusion limitations, α-quartz crystals cannot be grown by liquid-solution crystal growth approach. Because α-quartz phase is not stable at the melting point and because SiO_2 melts are so viscous that they form glasses rather than crystals when they are cooled, quartz cannot also be grown from a melt. Under hydrothermal conditions, however, quartz crystals grow at rates as high as 2 mm per day without severe faults or dendritic growth, even though solubility is only a few percent (Fig. 4.4).

In contrast to other crystal growth approaches, hydrothermal technique offers several advantages. As we mentioned above, hydrothermal technique can be used to synthesize and grow some crystals with special valence, which are usually difficult to obtain by other approach. It is an effective method for the preparation

FIGURE 4.4 Harvesting an early quartz growth run, AT&T Merrimac Valley Factory, Massachusetts [12]. *Reprinted from Journal of Crystal Growth, 264, R. S. Feigelson, 50 Years of Progress in Crystal Growth, p185, Copyright (2004), with permission from Elsevier.*

and growth of low temperature phases, such as α-quartz and α-berlinite. It is the only approach to synthesize some metastable materials.

In the fields of semiconductors and lasers, crystal materials are the core components of the devices and equipment. The crystal quality generally determines the quality of devices and equipment. So high-quality bulk crystals are required in order to develop devices and equipment with enhanced performance. Technological advances are important for growing high-quality crystals with fewer defects and dislocations. In recent years, numerous scientists are focusing their attention on developing new technologies under hydrothermal or solvothermal conditions to improve the quality of bulk crystals.

ZnO, a third-generation semiconductor, is a potential candidate for preparing high luminous efficiency UV/visible light emitting diodes and laser diodes owing to the characteristics of a wide direct band gap and large exciton binding energy at room temperature. ZnO single crystal can be prepared and grown by many methods. Hydrothermal or solvothermal technique, benefiting from a relatively low growth temperature and an approximate thermodynamic equilibrium growth condition, can produce ZnO crystals with fewer defects and dislocations.

Recently, chemists have developed new hydrothermal or solvothermal approach to grow high-quality and large-sized ZnO single crystals. In 2004, Ohshima et al. reported the growth of high purity and transparent ZnO single crystals with a large size of $50 \times 50 \times 15\ mm^3$ by the hydrothermal method with a platinum inner container (Fig. 4.5) [13]. In the experiment, mineralizers of crystal growth are LiOH and KOH. In 2005, Maeda et al. also produced 2-inch ZnO single crystals of high purity and high crystallinity by the hydrothermal method combined with a platinum inner container [14]. The surface quality of the ZnO wafer was improved by annealing and the step-and-terrace structure was formed. In 2006, Ehrentraut et al. reported the largest ZnO crystals of 3-inch in diameter which were grown by a high-pressure medium-temperature hydrothermal process employing alkaline-metal mineralizer for solubility enhancement [15]. In their work, almost 100 specimens 2 inch each in size can be grown on mainly (0001) and $(10\bar{1}0)$ oriented seeds from one growth run. All crystals developed excellent facets and the average crystal thickness was about 1 cm in the [0001] direction. In 2008, Dem'yanets and Lyntin reported ZnO single crystals with size of 3-inch were grown using commercial 500-L autoclave and protective vessels from Ti-based alloy [16]. In the same year, Zhang et al. produced ZnO crystals with dimensions of $30 \times 38 \times 8\ mm^3$ by the hydrothermal method using a mixed solution of KOH, LiOH, and H_2O_2 [17]. In 2009, Huang and coworkers designed and developed new mineralizers for the growth of ZnO crystals, and they produced ZnO crystals with dimensions of $30 \times 25 \times 5\ mm^3$ by the hydrothermal method [18]. ZnO crystal grown by the new mineralizers has a unique electrical property of higher carrier mobility with the presence of high carrier concentration.

Recently, in addition to ZnO single crystals, hydrothermal or solvothermal technique was used to grow some optical crystal materials, such as γ-$LiBO_2$ [19], $KTiOPO_4$ (KTP) [20], $KBe_2BO_3F_2$ (KBBF) [21], $RbBe_2BO_3F_2$ (RBBF) [21], $CsBe_2BO_3F_2$ (CBBF) [21], $TlBe_2BO_3F_2$ (TBBF) [21], YVO_4, Nd:YVO_4 [22], and so on. The photographs of some of the optical crystals are shown in Fig. 4.6.

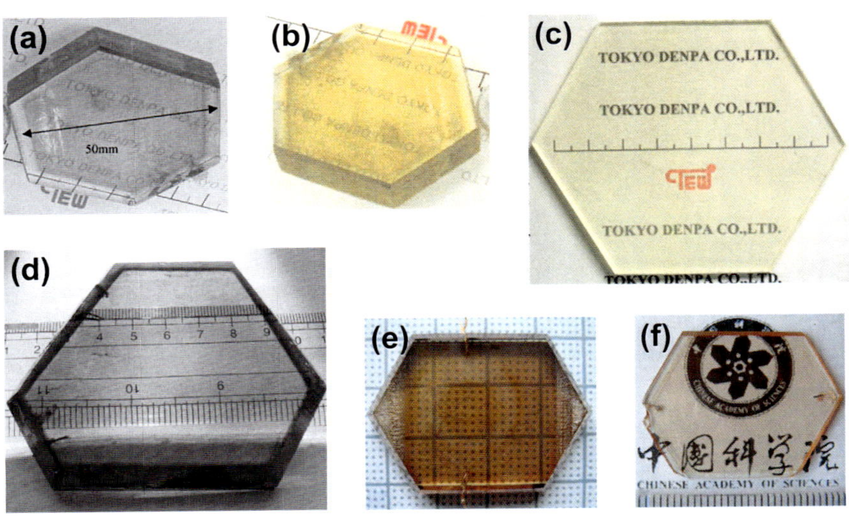

FIGURE 4.5 Photograph of ZnO single crystals grown by (a). *Ohshima et al.; Reprinted from Elsevier, Vol. 260, Eriko Ohshima, Hiraku Ogino, Ikuo Niikura, Katsumi Maeda, Mitsuru Sato, Masumi Ito, Tsuguo Fukuda, Growth of the 2-in-size bulk ZnO single crystals by the hydrothermal method, Page 166, Copyright (2004), with permission from Elsevier [13]. (b) Maeda et al.; Images reproduced with permission [14]. (c) Ehrentraut et al.; Reprinted from Elsevier, Vol. 52, Dirk Ehrentraut, Hideto Sato, Yuji Kagamitani, Hiroki Sato, Akira Yoshikawa, Tsuguo Fukuda, Solvothermal growth of ZnO, Page 280, Copyright (2006), with permission from Elsevier [15]. (d) Dem'yanets et al.; Reprinted from Elsevier, Vol. 310, L.N. Dem'yanets, V.I. Lyutin, Status of hydrothermal growth of bulk ZnO: Latest issues and advantages, Page 993, Copyright (2008), with permission from Elsevier [16]. (e) Zhang et al.; Reprinted from Elsevier, Vol. 310, Author(s), Hydrothermal growth and characterization of ZnO crystals, Page 1819, Copyright (2008), with permission from Elsevier [17]. (f) Huang et al.; Reprinted with permission from [18] Copyright 2009 American Chemical Society.*

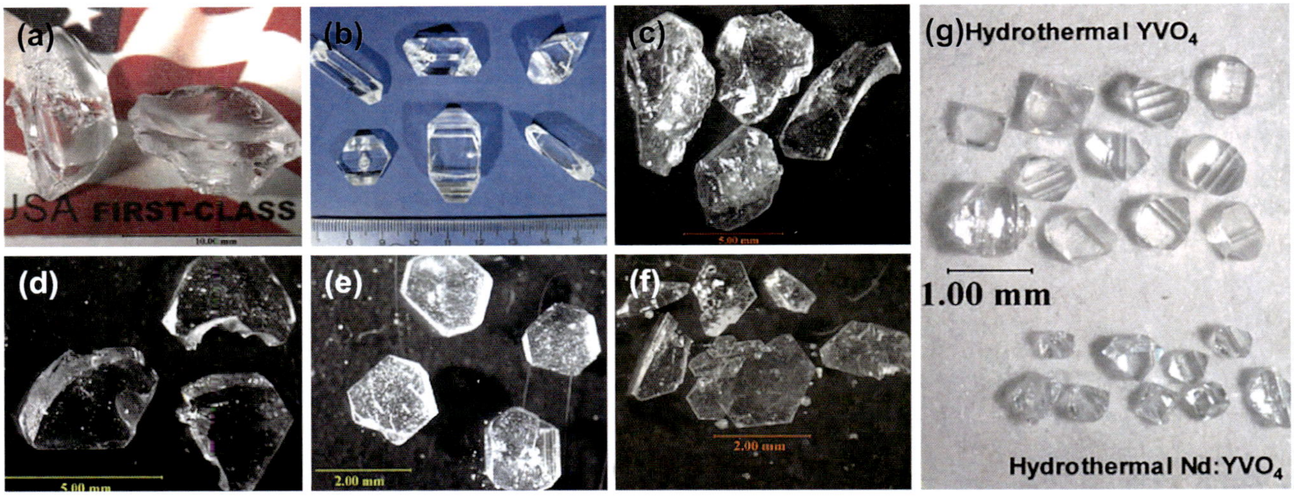

FIGURE 4.6 Photographs of optical crystals grown by hydrothermal methods. (a) γ-$LiBO_2$. *Reprinted from Elsevier, Vol. 310, Colin D. McMillen, Henry G. Giesber, Joseph W. Kolis, The hydrothermal synthesis, growth, and optical properties of* γ-$LiBO_2$*, Page 364, Copyright (2008), with permission from Elsevier.* (b) *KTP; Reprinted from Elsevier, Vol. 292, Chang-long Zhang, Ling-xiong Huang, Wei-ning Zhou, Ge Zhang, Han-de Hou, Qing-feng Ruan, Wei Lei, Shi-jie Qin, Fu-hua Lu, Yan-bin Zuo, Hong-yuan Shen, Guo-fu Wang, Growth of KTP crystals with high damage threshold by hydrothermal method, Page 364, Copyright (2006), with permission from Elsevier. (c) KBBF; (d) RBBF; (e) CBBF; and (f) TBBF; Reprinted from Elsevier, Vol. 310, Colin D. McMillen, Joseph W. Kolis, Hydrothermal crystal growth of* $ABe_2BO_3F_2$ *(A = K, Rb, Cs, Tl) NLO crystals, Page 2033, Copyright (2008), with permission from Elsevier. (g)* YVO_4 *and* $Nd:YVO_4$*; Reprinted from Elsevier, Vol. 310, Alfred R. Forbes, Colin D. McMillen, Henry G. Giesber, Joseph W. Kolis, The hydrothermal synthesis, solubility and crystal growth of* YVO_4 *and* $Nd:YVO_4$*, Page 4472, Copyright (2008), with permission from Elsevier.*

4.2.2. Zeolites and Related Materials [1,23]

Zeolites are basically prepared from hydrothermal systems. In 1940s, zeolites were first hydrothermally synthesized by R.M. Barrer and his coworkers through the simulation of the geothermal formation of natural zeolites. In the beginning of 1960s, scientists at Mobil Corp. began to use organic amines and quaternary alkyl-ammonium cations as templates in the hydrothermal synthesis of high-silica zeolites, which is considered a milestone in the development of zeolite synthesis.

In 1982, Wilson et al. (UCC (Union Carbide Corporation)) hydrothermally synthesized a novel family of microporous aluminophosphates $AlPO_4$-n [24]. Ever since then the microporous aluminophosphates and closely related materials with various structures, known as open-framework materials, have been successfully prepared [25–41] (see Chapter 16). The structures of the open-framework materials with extra-large channels are shown in Fig. 4.7.

In recent decades, the use of new synthetic techniques, such as the solvothermal synthetic route, has opened up many new possibilities in the preparation of new compounds with previously unseen compositional and structural diversity. Recently, many low-dimensional open-framework materials were observed in hydrothermal or solvothermal synthesis. These low-dimensional materials, which constitute an important area of materials chemistry, are rare in nature, but in nonaqueous systems they show reasonable stability in either one-dimensional chains or two-dimensional layers. Among low-dimensional open-framework materials, a number of members of a novel class of low-dimensional aluminophosphates (AlPOs) provide good examples, for they were well-defined by both theoretical network modeling and experimental synthesis [42,43]. Apparently, the organic solvent plays an important role in decreasing the dimensionality in AlPOs. Part of the reason for this may be the coverage and space effects of organic solvents on certain reactive groups of aluminum and phosphor species. Indeed, in their structures various basic units of Al-centered polyhedra such as AlO_4, AlO_5, $AlO_4(OH_2)_2$, and terminal P–OH and/or P=O groups exist, leading to their Al/P ratios less than unity; the common Al/P ratios are 1/2, 2/3, 3/4, 3/5, 4/5, and 13/18, and as a result, the networks are negatively charged with the empirical formula $Al_nP_{n+1}O_{4(n+1)}{}^{3-}$ (n =1, 2, 3, and 4). For instance, a variety of low-dimensional structures with different Al and P stoichiometries are noteworthy in open-framework AlPOs. 2D layered AlPOs are found as $AlPO_4(OH)_2{}^-$, $AlP_2O_8{}^{3-}$, $Al_2P_3O_{12}{}^{3-}$, $Al_3P_4O_{16}{}^{3-}$, $Al_4P_5O_{20}{}^{3-}$, $Al_{13}P_{18}O_{72}{}^{15-}$, etc., among which the stoichiometry of $Al_3P_4O_{16}{}^{3-}$ is known in a large number of 2D layered compounds. A number of AlPOs also occur as 1D infinite chain with stoichiometry of $AlP_2O_8{}^{3-}$ and $Al_3P_5O_{20}{}^{6-}$. These 2D layers and 1D chains are stabilized by protonated organic amines (templates) through H-bonding interaction with certain regularity. The templating ability of various organic

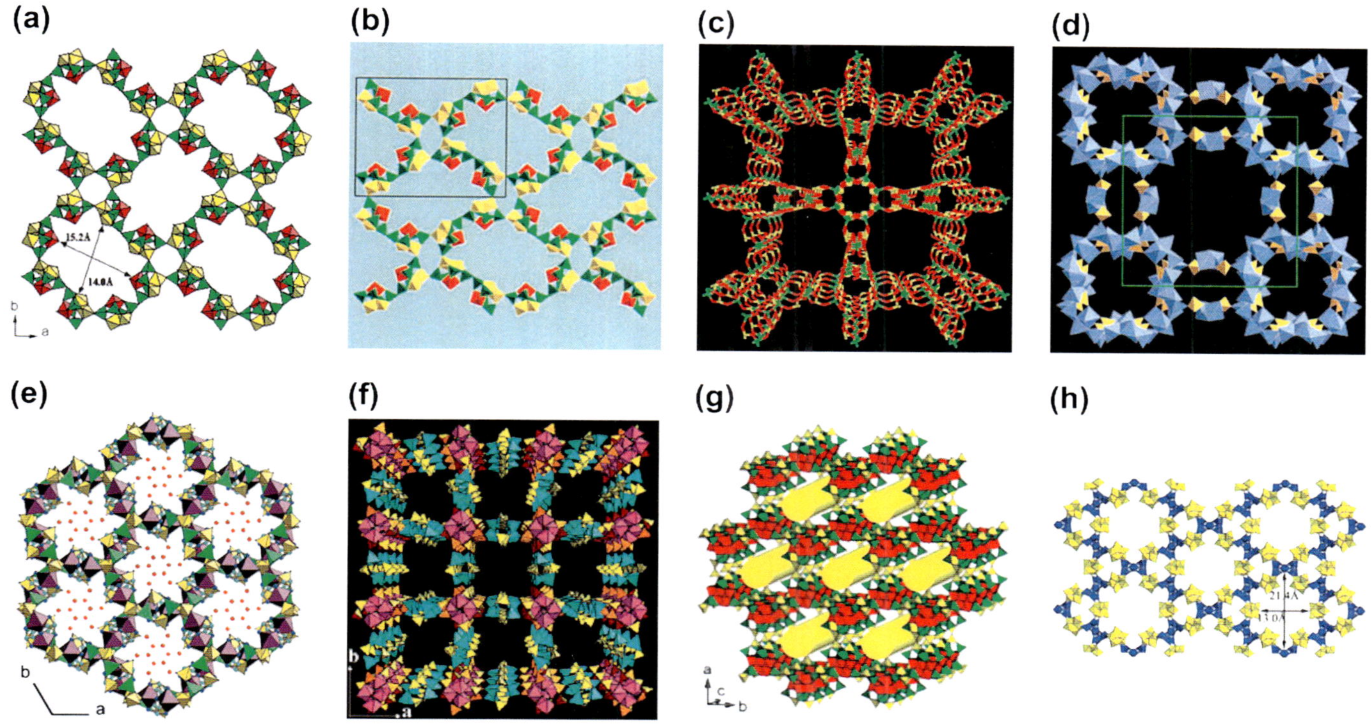

FIGURE 4.7 The structures of open frameworks with large pore sizes. (a) SU-12. *Reprinted with permission from [27], Copyright 2005 American Chemical Society (b) ASU-16; Reprinted with permission from [30], Copyright 2001 American Chemical Society. (c) ZnHPO-CJ1; J. Liang, J. Y. Li, J.H. Yu, P. Chen, Q. R. Fang, F. X. Sun, R. R. Xu, Angew. Chem. Int. Ed. 45 (2006) 2546, Copyright Wiley-VCH Verlag GmbH & Co. KGaA. Reproduced with permission. (d) VPO; Reprinted with permission from [34], Copyright 1996 American Chemical Society. (e) NTHU-1; Reprinted with permission from [37], Copyright 2001 American Chemical Society. (f) NTHU-5; Reprinted with permission from [38], Copyright 2007 American Chemical Society. (g) SU-61; Reprinted with permission from [39], Copyright 2008 American Chemical Society. (h) JLG-12; Reprinted with permission from [40], Copyright 2009 American Chemical Society.*

amines for the inorganic layers or chains depends on the energies of the template–layer interactions.

4.2.3. Organic–inorganic Hybrid Materials

The field of organic–inorganic hybrid materials has been widely recognized as one of the most promising and rapidly emerging research areas in material chemistry. In the past decade, a tremendous number of papers about organic–inorganic hybrid materials have been published. Organic–inorganic hybrid materials are generally produced using various methodologies of "soft" inorganic chemistry in liquid or sol–gel medium.

Hydrothermal and solvothermal reactions have been extensively applied to the synthesis of organic–inorganic hybrid materials, such as coordination polymers and clusters. In contrast to other synthesis methods, the hydrothermal and solvothermal synthetic methods have the ability to increase the solubility of reactants, enhance the reactivity of reactants, grow perfect crystal materials, and prepare important metastable phases which are difficult to prepare or cannot be prepared by traditional reactions.

Over the last decades, crystalline organic–inorganic hybrid materials have attracted much attention due to their enormous variety of interesting structures and wide potential applications as functional materials in many fields. We developed our strategy based on the hydrothermal synthesis of novel coordination polymers and clusters by using organic N and O-based ligands, Férey and coworkers designed and prepared the phosphate-free hybrid solids by the use of linear diphosphonates and dicarboxylates with hydrothermal method, Zubieta described the VO and MOXI-*n* hybrid family in hydrothermal system, Rao and Cheetham et al. discovered the porous oxalatophosphates, Yaghi paved the way for metal-organic frameworks, and a lot of chemists with either inorganic, coordination chemistry, or even organic cultures came to the field of organic–inorganic hybrid materials prepared by hydrothermal or solvothermal technique [44]. This leads to an explosion of results and a tremendous number of papers and it is rather difficult to classify the results of such "boiling" activity. The coverage of this section is limited to some examples of the recent hydrothermal or solvothermal syntheses as many works in the field have been well reviewed.

Recent efforts in this field are focused on the design and hydrothermal synthesis of novel coordination polymers and clusters by using neutral donor ligands (i.e., 4,4′-bipyridine, pyrimidine, pyrazine), strictly anionic ligands (i.e., carboxylate), and their combination. For example, the hydrothermal reactions of V_2O_5, $H_2C_2O_4$, $Ni(NO_3)_2$, ethylenediamine (en) or 1,2-diaminopropane (enMe), and water yield two layered vanadium oxides with interlayer metal coordination complexes, $[Ni(en)_2]_{0.5}[V_3O_7]$ and $[Ni(enMe)_2]_{0.5}[H_2enMe]_{0.5}[V_6O_{14}]$ [45]. Both compounds consist of 3D mixed-valence vanadium oxides with interlayers of $\{Ni(en)_2\}^{2+}$, or $\{Ni(enMe)\}^{2+}$ and H_2enMe^{2+}. A family of inorganic–organic hybrid compounds with formula $CuL(VO_2)(PO_4)$ (L = 4,4′-bipyridine (4,4′-bipy), 1,10-phenanthroline (1,10-phen), and 2,2′-bipyridine (2,2′-bipy)) were hydrothermally synthesized at 160°C for 120 h [46]. The use of different bidentate organodiamine ligands in the initial reaction systems gave raise to a variety of structures of the products with the same inorganic composition.

Metal-organic framework (MOF, also known as porous coordination polymer) materials are currently the hottest materials in the field of hybrid materials and have attracted wide scientific attention. Such a high interest is caused not only by the enormous variety of interesting molecular topologies but also due to their excellent properties, with promising applications such as the storage of gases, molecular separation from the gaseous and liquid mixtures, catalysis, sometimes showing the enantioselectivity, and sensors for special classes of molecules. They can also be designed as multifunctional materials with excellent physical properties like magnetism, luminescence, and optoelectronics. A wide variety of synthesis methods to obtain metal-organic framework materials have been explored, such as solvent evaporation method, diffusion method, hydrothermal or solvothermal method, microwave reaction, and ultrasonic methods. Hydrothermal or solvothermal method is an important approach to prepare MOFs and this method exploits the self-assembly of products from soluble or insoluble precursors. The operational temperature range is 80–260°C inside a closed space (autoclave) under autogenous pressure. A tremendous number of MOFs have been prepared by hydrothermal or solvothermal method. It is very difficult to discuss all MOFs one by one. Here, we only give two typical cases of hydrothermal or solvothermal synthesis of metal-organic framework materials.

In 2005, Férey and coworkers reported the hydrothermal reactions of H_2BDC (terephthalic acid), $Cr(NO_3)_3 \cdot 9H_2O$, fluorhydric acid, and water produced a highly crystallized green powder of the chromium terephthalate with formula $Cr_3F(H_2O)_2O[(O_2C)C_6H_4(CO_2)]_3 \cdot nH_2O$ (MIL-101) [47]. MIL-101 has an augmented MTN zeotype architecture and the structure is constructed by BDC and trimeric inorganic building blocks formed by the assembly of three octahedra sharing a μ_3-O common vertex (as shown in Fig. 4.8). MIL-101 shows very large pore sizes and surface area. Its zeotype cubic structure has a giant cell volume (about 702,000 Å^3), a hierarchy of extra-large pore sizes (about 30–34 Å), and a Langmuir surface area for N_2 of about 5900 ± 300 $m^2 g^{-1}$. Besides the usual properties of porous compounds, this solid has potential as a nanomold for monodisperse nanomaterials, as illustrated by the incorporation of Keggin polyanions within the cages.

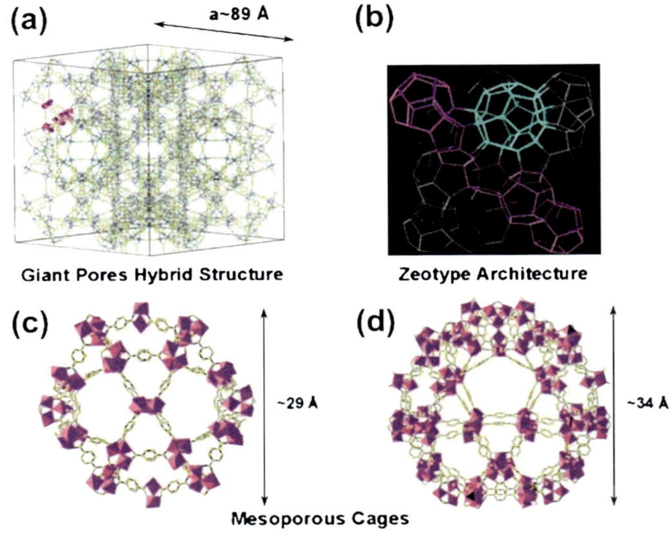

FIGURE 4.8 The structure of MIL-101. (a) Ball-and-stick representation of one unit cell; (b) Schematic 3D representation of the MTN zeotype architecture; (c and d) Ball-and-stick view of the two cages. *From [47]. Reprinted with permission from AAAS.*

In 2008, Banerjee et al. prepared 25 different zeolitic imidazolate frameworks (ZIFs) by a high-throughput approach under solvothermal conditions (Fig. 4.9) [48]. All ZIFs were synthesized by using the fully integrated high-throughput synthetic apparatus utilizing a 96-well glass plate as the reaction vessel. Their experimental procedure for the synthesis of ZIFs, in general, is described as follows: a solution of imidazole(s) in DMF/DEF and a solution of a hydrated metal salt (usually nitrate) in DMF/DEF are used as stock solutions. To an imidazole(s) stock solution, a metal salt stock solution is added. After the 96-well glass plate is loaded with mixtures of stock solutions dispensed by a programmed liquid handler, it is covered with a polytetrafluroethylene (PTFE) sheet, sealed by fastening the sheet with a metal clamp, then heated in an oven and allowed to react solvothermally for 24–96 h. All of the ZIF structures have tetrahedral frameworks: 10 of which have two different links (heterolinks), 16 of which are

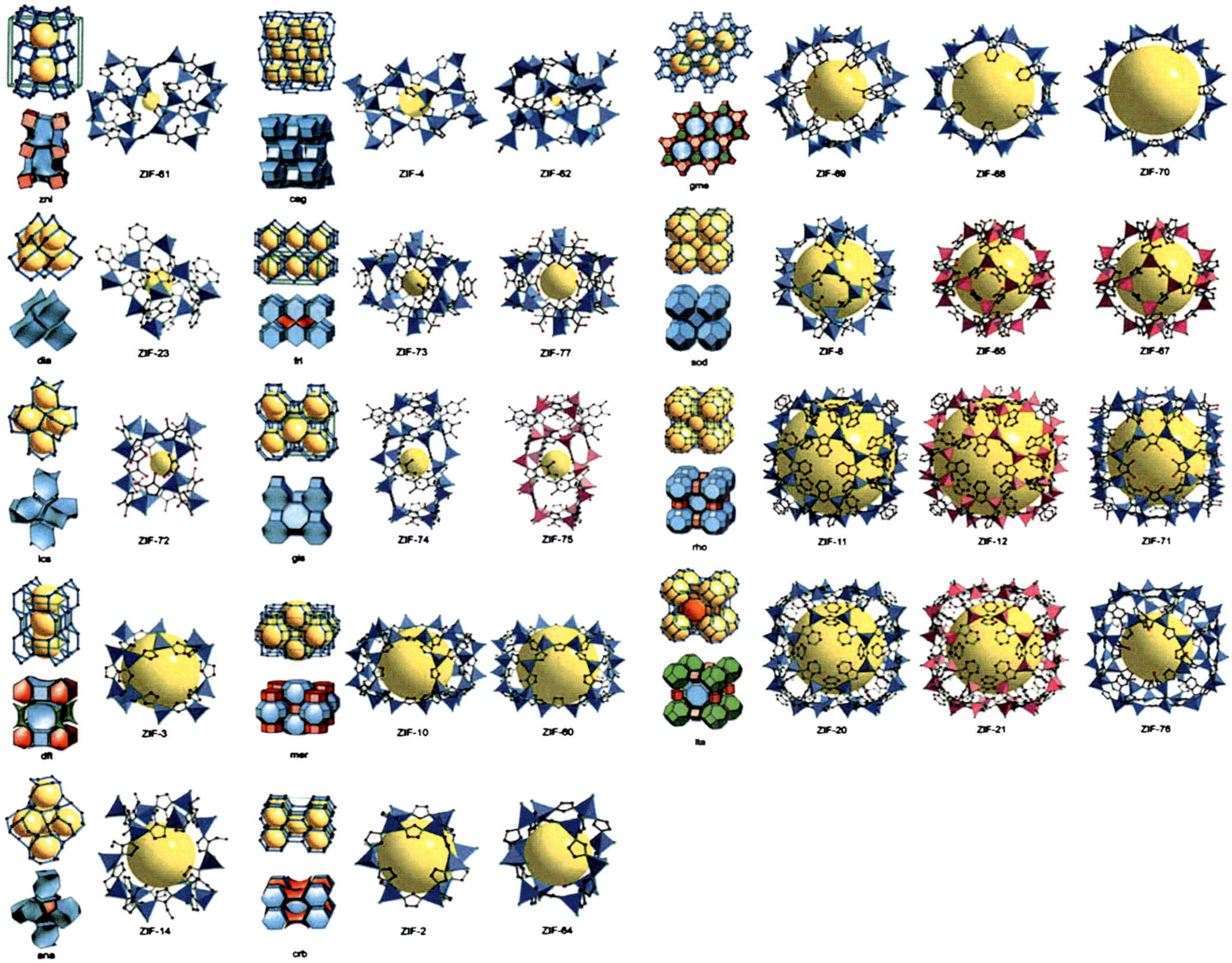

FIGURE 4.9 The crystal structures of ZIFs prepared by Yaghi et al. *From [48]. Reprinted with permission from AAAS.*

previously unobserved compositions and structures, and five of which have topologies as yet unobserved in zeolites. Members of a selection of these ZIFs (termed ZIF-68, ZIF-69, and ZIF-70) have high thermal stability (up to 390°C) and chemical stability in refluxing organic and aqueous media. Their frameworks have high porosity (with surface areas up to 1970 $m^2 g^{-1}$), and they exhibit unusual selectivity for CO_2 capture from CO_2/CO mixtures and extraordinary capacity for storing CO_2: 1 L of ZIF-69 can hold about 83 L of CO_2 at 273 K under ambient pressure.

Recently, an attractive advance in the synthetic chemistry of MOFs is to synthesize new MOFs by click reaction. Devic et al. employed the click reaction pre-MOF synthesis to produce a novel flexible ligand in 2007 [49]. In 2008, Goto et al. performed the cycloaddition on previously assembled frameworks and synthesized an azide-containing MOF material that may be modified directly [50]. In 2009, Nguyen and coworkers prepared a Zn-based metal-organic framework material bearing silyl-protected acetylenes and the MOF was postsynthetically modified using "click" chemistry [51]. Using a solvent-based, selective deprotection strategy, two different organic azides were "clicked" onto the MOF crystals, resulting in a porous material whose internal and external surfaces are differently functionalized.

Another attractive advance in the synthetic chemistry of MOFs is to synthesize new MOFs by hydrothermal or solvothermal in situ metal/ligand reactions. As a nonconventional method, solvothermal in situ ligand reactions were first reported by Li et al. in 1998 on the rearrangement of 2,2′-dipyridylamine into dipyrido-[1,2-α:2,′3′-d]imidazole in the synthesis of one-dimensional coordination metal halides [52]. Till date, a number of hydrothermal or solvothermal in situ metal/ligand reactions have been reported. For example, in 2008, we prepared two luminescent copper halide coordination polymers, $(CuCl)_3C_6H_7N_3$ and $(CuI)_2C_6H_7N_3$, by in situ metal/ligand reaction under solvothermal conditions [53]. In the process of solvothermal reaction, nitrogen heterocycle ligand 2,3-dihydroimidazo-[1,2-α]-pyrimidine was formed in situ as illustrated in Fig. 4.10. The mechanism for the formation of 2,3-dihydroimidazo[1,2-α]-pyrimidine in our system involves four steps. The first step is the change of some ethanol to iodoethane, which is an easy reaction because there are some I- anions in our system. The second step is the alkylation of 2-amino-pyrimidine, in which iodoethane attacks the N atom on the ring of 2-amino-pyrimidine, making aromaticity diverted. The third step, which is the most important one, is HIO_4 oxidation of $-CH_3$ to $-CHO$. The fourth step is the formation of five-membered ring. Each $-CHO$ and each $-NH_2$ dehydrate, generating 2,3-dihydroimidazo [1,2-α]-pyrimidine. Similar to other in situ reactions, pressure has also been proved to be effective for the dehydrogenation of methyl when suitable transition metal salt is present. We speculate that the process of Cu^{2+} to Cu^{+} is very important because when we changed Cu to other metals such as Zn, Mn, or Co, we could not obtain this reaction. Some important solvothermal (including hydrothermal) in situ metal/ligand reactions and their mechanisms, including dehydrogenative carbon–carbon coupling, hydroxylation of aromatic rings, cycloaddition of organic nitriles with azide and ammonia, transformation of inorganic and organic sulfur, as well as the Cu^{II} to Cu^{I} reduction, are outlined by Chen and Tong [54]. Here we will not discuss them one by one.

4.2.4. Ionic and Electronic Conductors [55]

Ionic conductors are useful materials in many fields such as energy conversion, chemical sensors, combustion control, high-temperature membrane reactors, and chemical processing. Traditionally, ionic conductors are

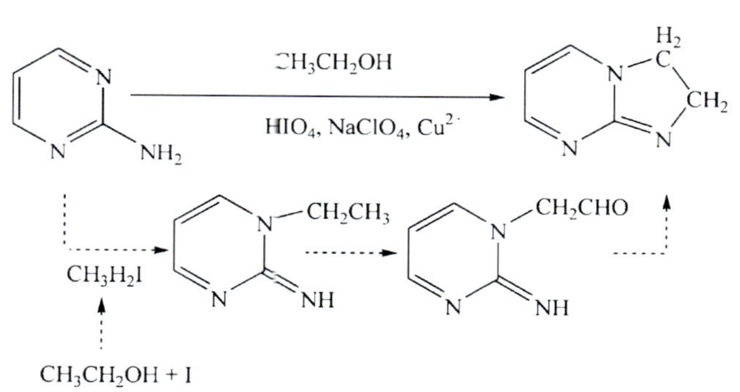

FIGURE 4.10 The in situ reaction of 2-amino-pyrimidine and EtOH [53].

produced from high-temperature solid-state reactions with the resulting aggregates requiring a subsequent milling process. Since fabrication of advanced and complex ionic conductors always requires homogeneous and pure powders with fine and uniform particle sizes, an effective synthesis procedure is needed. Mild hydrothermal synthesis below 240°C and under autogenous pressure has been a promising route for preparing novel ionic conductors. Compared with traditional synthesis methods, the hydrothermal synthesis method has some advantages, for example, mild synthesis temperature, reaction homogeneity, high purity, and controlled size and morphology which tend to improve or dramatically change the physical properties of the final products.

Co-doped nanocrystalline ceria-based compositions, $Ce_{1-x}M_xBi_{0.4}O_{2.6-x}$ (M = Ca, Sr, and Ba, $x = 0.01-0.15$), serve as good examples for describing the hydrothermal synthesis and doping for ionic conductors [56,57]. CeO_2-based materials doped with aliovalent cations show promising ionic transport properties in application to solid oxide fuel cells and oxygen pumps. Pure phase $CeO_{2-\delta}$ with the cubic fluorite structure shows mixed electronic and ionic conduction, and the introduction of lower-valence metal ions into the lattice of CeO_2 greatly improves their ionic conduction due to the quantitative formation of oxygen vacancies and in some cases eliminates its electronic conduction.

$Ce_{1-x}M_xBi_{0.4}O_{2.6-x}$ was prepared by a hydrothermal method. The particle sizes of all the products were in the nanometer range due to the use of a hydrothermal synthesis technique. The ionic conductivity has been greatly enhanced in these co-doped systems. With mild hydrothermal synthesis method, various types of ionic conductors, such as $HZr_2P_3O_{12}$, $Bi_{17}V_3O_{33}$, $K_8Sb_8P_2O_{29}\cdot 8H_2O$, $M_3HGe_7O_{16}\cdot xH_2O$ (M = NH_4^+, Li^+, Na^+, K^+, Rb^+, and Cs^+), can also be prepared [58–63]. All the materials mentioned above exhibit excellent ionically conducting properties.

The hydrothermal technique is a promising method for preparing advanced complex oxides and fluorides. Hydrothermal synthesis is advantageous due to the relatively mild conditions required, one-step synthetic procedure, and controllable particle size distribution. With the mild hydrothermal synthetic method, we have prepared a series of functional complex oxides with fluorite (AO_2), perovskites (ABO_3), scheelite (ABO_4), spinel (AB_2O_4), pyroligneous ($A_2B_2O_7$), and garnet ($A_3B_2(OH)_3$) structures (A, B = chemical elements of the main block and the transition series).

Hydrothermal synthesis favors the control of morphology, particle size, elemental valence of synthetic materials, and unique equilibrium defect solids. Perovskite oxides have provided magical structural models for superconducting and colossal magnetoresistance, and the search for nanoscale and/or atomic-scale devices with particular property by specific preparations in the same systems has been extensively conducted.

With an aim toward stabilizing particularly mixed valences of metal in crystals, Feng et al. carried out the hydrothermal synthesis of normal manganese perovskite oxides through the disproportionation reactions of MnO_2 [64]. They obtained a family of manganese perovskite oxides, $La_{1-x-y}Ca_xK_yMnO_3$ ($x = 0.74-0.18$, $y = 0.01-0.14$) through the partial substitution of Ca^{2+} and K^+ for La^{3+} [64]. We found the three-oxidation states of manganese (Mn^{3+}, Mn^{4+}, Mn^{5+}) in the perovskite oxide, $La_{0.66}Ca_{0.29}K_{0.05}MnO_3$ (structures and physical properties are shown in Fig. 4.11). This solid shows rectifying effect as atomic-scale $p-n$ junctions of single crystals and films. Mn average oxidation states in $La_{0.66}Ca_{0.29}K_{0.05}MnO_3$ were first measured by oxidation–reduction titration. The identification of Mn^{5+} was made by its characteristic near-infrared (IR) absorption spectroscopy, Mn *K*-edge X-ray absorption near-edge spectroscopy (XANES), and laser-induced (1064 nm) luminescence spectroscopy. Structurally, the charges of three-oxidation state Mn ions at B-sites were compensated by their three corresponding A-site ions of La^{3+}, Ca^{2+}, and K^+ to maintain the charge neutrality.

Complex fluorides show various interesting structures and physical properties such as piezoelectric characteristics, photoluminescence behavior, ionic conductivity, and nonmagnetic insulating behavior. The preparation of complex fluorides by high-temperature solid-state reactions was previously reported, but requirements for the synthetic apparatus due to the corrosive nature of fluorides limited the study by solid-state synthesis. Mild hydrothermal synthesis can enhance the reactivity of reactants in solution and can be used to prepare important crystalline solids which are difficult to prepare or cannot be prepared by solid-state reactions. Comparing with conventional high-temperature solid-state reactions, the hydrothermal synthetic apparatus has the advantage of preventing the corrosive nature of fluorides. Till date, a series of functional complex fluorides with fluorite (AF_2), perovskites (ABF_3), scheelite, and wolframite (ABF_4) structures (A, B = chemical elements of the main block and the transition series) and complex fluorides doped with

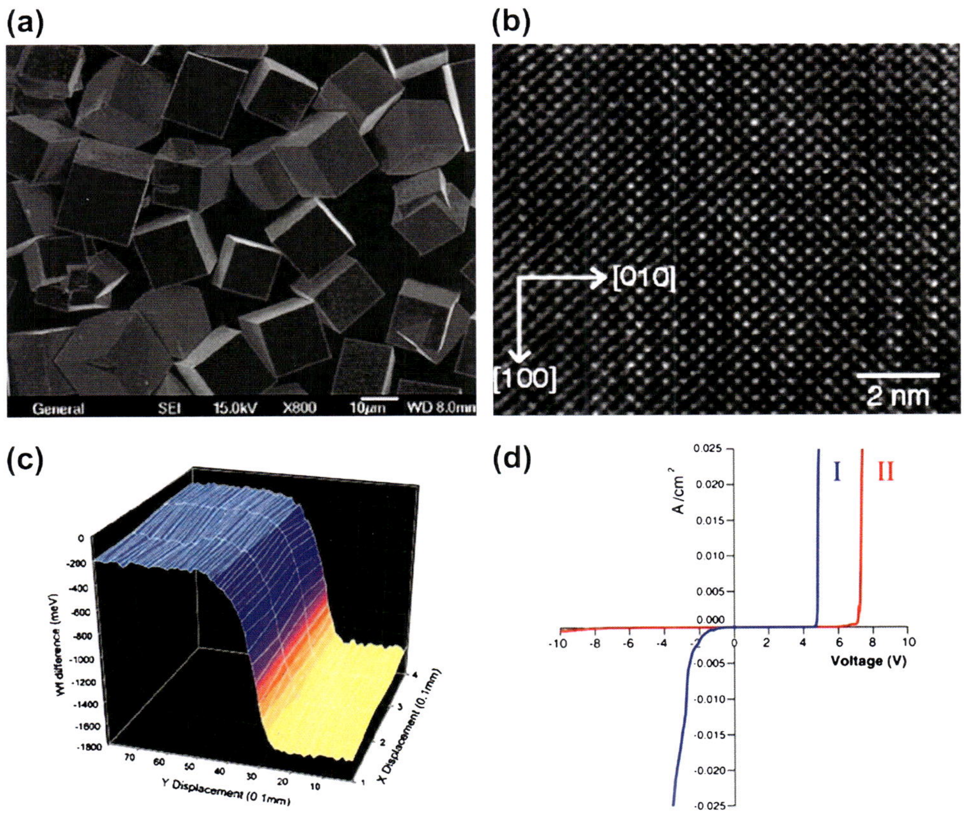

FIGURE 4.11 The structure and properties of $La_{0.66}Ca_{0.29}K_{0.05}MnO_3$. (a) SEM photo; (b) HR-TEM image; (c) the work function; (d) $I-V$ characteristics of the atomic-scale $p-n$ junctions. *With kind permission from Springer Science+Business Media: Journal of Materials Science, Three oxidation states and atomic-scale* p–n *junctions in manganese perovskite oxide from hydrothermal systems, volume 43, 2008, page 2131, Shouhua Feng, Figure 4. [64].*

photosensitive rare earth elements have been successfully prepared in hydrothermal system. For instance, hydrothermal syntheses of $LiBaF_3$ and $KMgF_3$ with perovskite structures can be carried out at 120–240°C in the presence of a mineralizer, HF or NH_4HF_2 [65]. Crystalline $KMgF_3$ can be obtained by mixing KF and MgF_2 with deionized water, adding hydrofluoric acid, and heating at 120–240°C under autogenous pressure for several days. $LiBaF_3$ can be synthesized in the same way either from a reaction mixture of LiF, BaF_2, and water, or from LiOH and $Ba(OH)_2$ solutions to which NH_4HF_2 was added. The complex fluorides, ABF_4 (A = Li, Na, K; B = Y, Eu, Ho), can be synthesized at 140–240°C, and doping of Eu^{3+} ions in $LiYF_4$ and KYF_4 can be carried out by directly adding Eu_2O_3 into the reaction mixtures [66–68].

4.2.5. Nanomaterials

The hydrothermal or solvothermal technique has become one of the most important methods for the fabrication of nanostructural materials. In recent years, a tremendous number of nanomaterials have been processed by hydrothermal or solvothermal method and thousands of science papers about the hydrothermal/solvothermal synthesis of nanomaterials have been published. From the perspective of morphology of nanomaterials, the hydrothermal technique has been used to process nanomaterials with a variety of morphological features, such as nanoparticle, nanosphere, nanotube, nanorod, nanowire, nanobelt, nanoplate, and so on. From the perspective of composition of nanomaterials, the hydrothermal technique can be used to process almost all types of advanced materials like metal, alloy, oxides, semiconductors, silicates, sulphides, hydroxides, tungstates, titanates, carbon, zeolites, ceramics, and a variety of composites. The hydrothermal or solvothermal technique is not only used to process simplest nanomaterials, but also acts as one of the most attractive techniques for processing nanohybrid and nanocomposite materials. In short words, the great advantages of hydrothermal technology for nanomaterials processing are the production of particles that are monodispersed with total control over their shape and size in addition to their chemical homogeneity with the highest dispersibility. The latest progress in nanomaterials prepared with hydrothermal or solvothermal technique was reviewed by Byrappa and Adschiri [69]. Readers can get more information about this topic from this literature source. It is not possible to discuss the processing of all these nanomaterials using hydrothermal or solvothermal technology. Instead, the coverage of this section is limited to some representative nanomaterials synthesized by hydrothermal or solvothermal methods in last two years.

A family of layered-lanthanum crystalline nanowires with hierarchical pores was synthesized by mild hydrothermal reactions (Fig. 4.12) [70]. The diameter of the pores ranged from 2 to 50 nm, covering both the micropore scale and the mesopore scale. Luminescence was introduced by doping nanowires with Eu^{3+} ion. By combining the merits of hierarchical porous nanowires and layered hydroxides, these as-synthesized products have shown a unique bioengineering application of capturing and releasing short DNA fragments rapidly in dilute solution, and environmental engineering applications due to their remarkable capability to remove organic dye from water.

Yang et al. demonstrated a new solvothermal synthetic route for morphology-controlled preparation of high-quality anatase TiO_2 single crystals using 2-propanol as a synergistic capping agent and reaction medium together with HF [71]. The roles of 2-propanol and HF in facilitating the growth of anatase single-crystal nanosheets were explored by first-principle theoretical calculations, revealing that the presence of 2-propanol can enhance the stabilization effect associated with fluorine adsorption over the (001) surface. Thus, they construe that a synergistic functionality of chemisorbed F to lower the (001) surface energy and 2-propanol to enhance this stabilization and act as protective capping agent leads to the formation of anatase TiO_2 single-crystal nanosheets, which is a thermodynamically favored morphology under these conditions. Liu et al. developed a facile hydrothermal method to grow oriented single-crystalline rutile TiO_2 nanorod films on transparent conductive substrates [72]. Their results show the growth parameters including the growth time, the growth temperature, the initial reactant concentration, acidity, and types of additives that could be selectively chosen to prepare TiO_2 nanorod film with desired lengths and densities. Jia et al. presented a novel approach for synthesizing single-crystal α-Fe_2O_3 nanorings, employing a double anion-assisted hydrothermal method [73]. Their results show the cooperative action of the phosphate and sulfate ions, involving adsorption and coordination, as a crucial factor in bringing about the formation of the nanorings. By varying the ratios of phosphate and sulfate ions to ferric ions, they produced a series of hollow nanostructures with tunable size, morphology, and surface architecture. These hollow α-Fe_2O_3 nanostructures were converted via a reduction or reduction–oxidation process to single-crystal Fe_3O_4 or γ-Fe_2O_3, respectively, having exactly the same morphology. Yu et al. reported a straightforward solvothermal route to synthesize stable bracelet-like hydrophilic Ni–Co magnetic alloy flux-closure nanorings in high yield [74]. The Ni–Co nanorings were solvothermally prepared by a synchronous reduction of cobalt(II) acetylacetonate ($Co(acac)_2$) and nickel

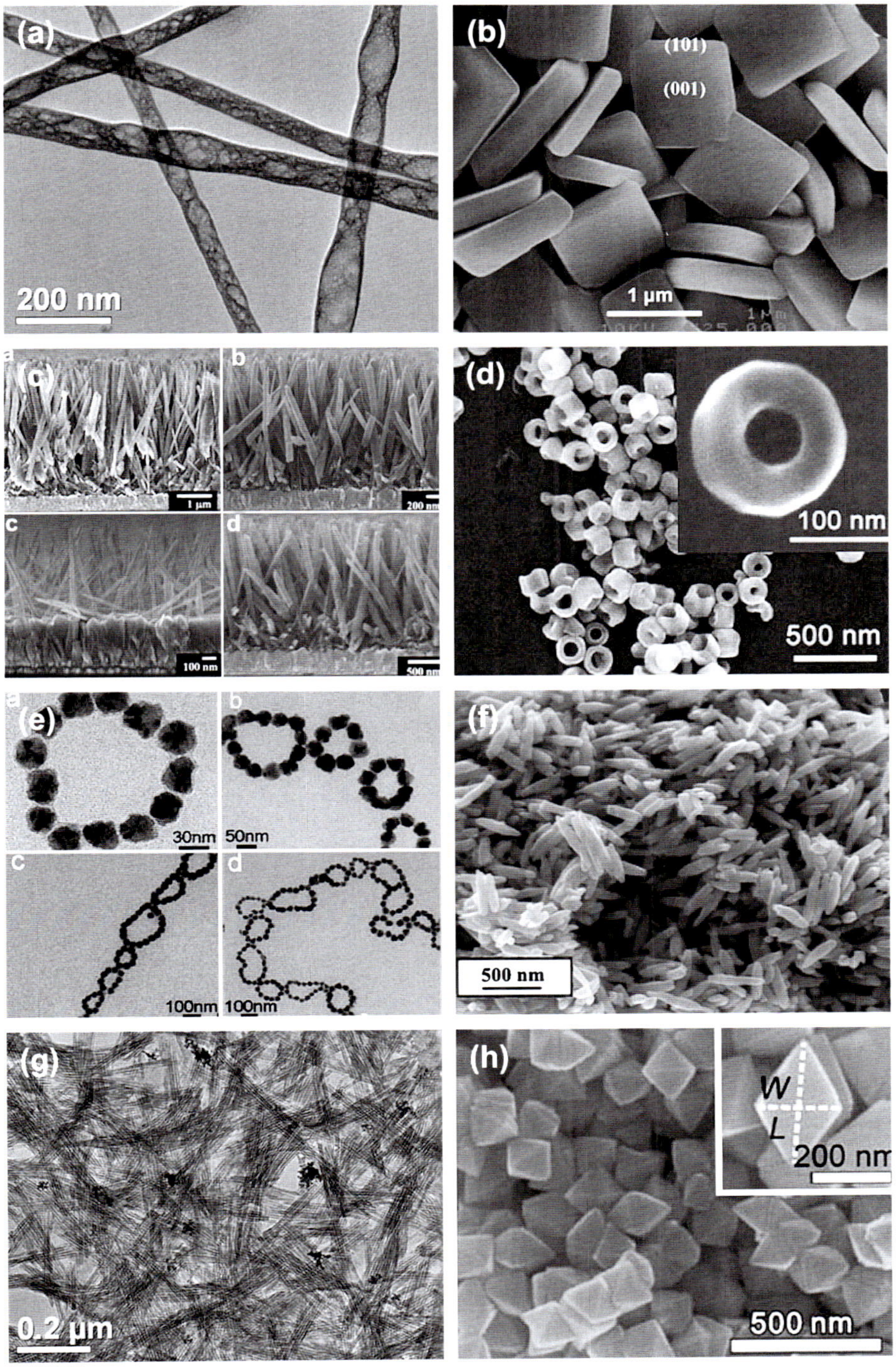

FIGURE 4.12 SEM and TEM images of nanomaterials discussed in this section. (a) Layered-lanthanum crystalline nanowires. *Reprinted with permission from [70]. Copyright 2009 American Chemical Society. (b) Anatase TiO_2 nanosheets; Reprinted with permission from [71], Copyright 2009 American Chemical Society. (c) Rutile TiO_2 nanorods; Reprinted with permission from [72], Copyright 2009 American Chemical Society. (d) α-Fe_2O_3 nanorings; Reprinted with permission from [73], Copyright 2008 American Chemical Society. (e) Ni–Co alloy nanorings; Reprinted with permission from [74], Copyright 2008 American Chemical Society (f) $VO_{1.52}(OH)_{0.77}$ nanorods; Reprinted with permission from [75], Copyright 2008 American Chemical Society. (g) MoO_3 single-walled nanotubes; Reprinted with permission from [76], Copyright 2008 American Chemical Society. (h) SnO_2 nanoparticles. X. G. Han, M. S. Jin, S. F. Xie, Q. Kuang, Z. Y. Jiang, Y. Q. Jiang, Z. X. Xie, L. S. Zheng, Angew. Chem. Int. Ed. 48 (2009) 9180, Copyright Wiley-VCH Verlag GmbH & Co. KGaA. Reproduced with permission.*

(II) acetylacetonate (Ni(acac)$_2$) at 240°C for 3 h with a capacity of 22 mL using triethylene glycol (TREG) as solvent in the presence of poly(vinylpyrrolidone) (PVP). Their results proved high nickel content in the alloy and high PVP concentration that favor the fabrication of nanorings. Djerdj et al. reported the solvothermal synthesis of ellipsoidal nanorods by using $VOCl_3$ and benzyl alcohol as reagents. The ellipsoidal nanorod has a novel hollandite-type structure and the formula is $VO_{1.52}(OH)_{0.77}$ [75]. Wang et al. synthesized uniform MoO_3 single-walled nanotubes with a simple hydrothermal method based on a thiol-assisted mechanism [76]. Han et al. reported that octahedral SnO_2 particles with exposed high-energy {221} facets were synthesized by a simple hydrothermal route with the assistance of HCl and PVP [77]. The morphologies of the SnO_2 particles can be controlled by changing the amount of HCl. The octahedral SnO_2 particles show excellent gas-sensing performance due to the high chemical activity of the exposed {221} facets. Their results demonstrate that it is feasible to improve gas-sensing properties of SnO_2-based sensors by the surface engineering strategy.

4.3. HYDROTHERMAL BIOCHEMISTRY

In addition to new materials, currently attractive studies on the origin of life and environmental issues are related to hydrothermal synthesis or hydrothermal chemistry [55]. It was hypothesized that rich chemical reactions occurred in the warm sea and all microorganisms have high-temperature ancestors. It is found that most microorganisms living at very high temperatures are archaea on the molecular biological tree. More and more evidence supports the model of the hydrothermal origin of life. However, the most interesting research, such as that into nonenzyme hydrothermal synthesis from inorganic to organic or biological species, needs to be explored further for both deeper understanding of life evolution and possible industrial applications. In contrast to the mild hydrothermal synthesis, the supercritical water oxidation process will effectively decompose almost all organic wastes into smaller molecules and CO_2, serving as an environmental friendly tool. Accordingly, hydrothermal synthesis (chemistry) provides new materials, helps to clean up our environment, and helps to understand the origin of life.

Current research of hydrothermal biochemistry provides evidences of chemical synthesis, microbiology, molecular phylogenetic tree, and exploration of the sea, such as simulated hydrothermal conditions, the abiotic synthesis of H_2, NH_3, CH_4, CH_3COOH, cytosine, uracil, peptide, and computer simulation of thermodynamic of amino acids synthesis, molecular simulation of amino acids into peptides on zeolite as well.

4.3.1. Warm pond: Hydrothermal Seafloor

In order to test Oparin's hypothesis [78], early Earth with reducing atmosphere, Miller, a student of University of Chicago did a well-known experiment [79]. An apparatus was built to circulate CH_4, NH_3, H_2O, and H_2 with an electric discharge. "Hot springs," the hydrothermal-vent origin-of-life hypothesis was built up. The hypothesis is that life originated from the Earth surface, inorganic molecules, getting energy from light and lightning, became organic molecules, and they were enriched in the surface of water. Furthermore, the organic molecules formed macromolecules and at last macromolecules with self-organization function were evolved into cell with replication and membrane. In 1977, during an expedition led by Robert Ballard and sponsored by the National Oceanic and Atmospheric Administration (NOAA), Alvin discovered and documented the existence of black smokers around the Galapagos Islands [80]. Existing at a depth of more than 2000 m, black smokers emitted a strong flow of black, smoky water, superheated to over 400°C (750°F). Corliss first proposed the origin-of-life hypothesis [81]. This theory was improved by increasing geological evidences, homology analyses, and fieldworks. Two views argued what was the source of the energy. One point agreed that energy was from sun and another agreed that it was from geothermal and chemical energy. Now more and more evidences of cosmology [82], geology [83], molecular phylogenetic tree [84,85], and the simulation of chemical evolution support the last view [86–91]. But Fischer–Tropsch type reaction from CO and H_2 to organic molecules did not succeed in solution, and catalyst poisoning happened at high concentration of H_2S [92]. When temperature goes up to 250°C, racemization rate of many of amino acids is faster than decomposition rate. It is not conducive to the formation of peptide if a large amount of water exists. Even though peptide was formed, it decomposed quickly. In water (pH = 7, T = 350°C), the half-life period of RNA is only 4 s. High-energy phosphate bonds would be destroyed when temperature is above 250°C, glycosyl would decompose at the temperature. Of course, it is just to discuss the thermal stability of organic matter isolating, regardless of salt effect, pressure, mineral which can stabilize organic molecules.

4.3.2. Evolutionary Tree and Time Evidence

Only at the molecular level can we understand evolution. Molecular sequence can explain the essence better than the classical standard form and molecular function to reveal evolutionary relevance. Ribosome which makes proteins out of amino acids is at the core. It is easy to separate ribosome RNA which is stable and whose sequence changes relatively slowly [93]. A small prokaryotic ribosomal subunit, the 16S rRNA, is used to classify all kinds of life by examining the conserved sequence and the degree of difference, and the molecular phylogenetic trees were issued [93–95]. The molecular phylogenetic tree (Fig. 4.13) classified life into three categories: archaea, bacteria, and eukaryota [96]. Archaea was close to the roots, thermophiles and extreme thermophiles were more close to the roots. So life came from high temperature was put forward.

By researching geographical and chemical conditions of early Earth, it is found that Earth originated from ca. 4600 million years ago. The primitive Earth had a high-temperature surface. Maher and Stevenson found that probiotic chemical synthesis had commenced in hydrothermal system at the bottom of ocean before 4–4.2 billion years, but 3.7–4.0 billion years on land [97]. Some evidences proved that photosynthetic bacteria lived before 3.5 billion years. ^{13}C isotope analysis proved life started before 3.85 billion years. Prebiotic chemical synthesis started on early Earth which has weak reducing atmosphere (abundant CO_2 and N_2) and hot surface and was often impacted by other planets. Possibly the bottom of ocean was the cradle of life. In 1977, Alvin found a large quantity of chemical autotrophic bacteria at a depth of 2.5 km, in the Eastern Equatorial Pacific Ocean, and the concentration of bacteria was about 10^8–10^9/mL. In 1991, Alvin observed volcanic activity at the Eastern Pacific ridge about 2500 m below sea level. In March of 1992 Alvin revisited the area, and this time Alvin found bacterial colonies at the vents, and a 30-cm long *Tevnia jerichonana* – a kind of tube worm was found. In December of 1993, Alvin revisited here again, and found a 1.5-m long tube worm [98]. This proved that biology could grow quickly under hydrothermal condition. Geochemical research for the hydrothermal vents at the bottom of ocean shows that the highest temperature was about 380°C, pressure was 20–30 MPa, and the concentration of reducing gas, especially H_2S was high in seafloor vent [99].

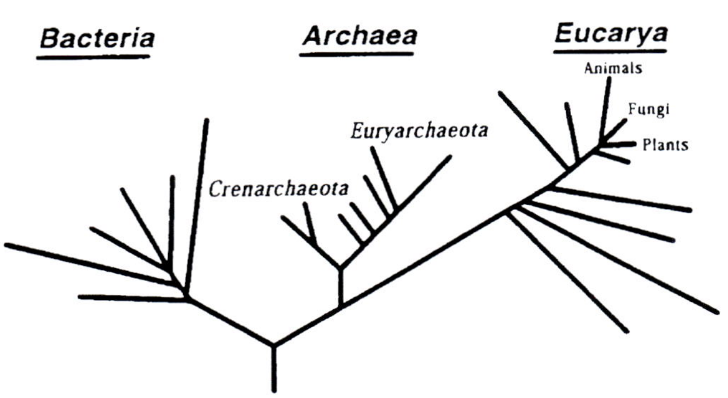

FIGURE 4.13 The phylogenetic tree drawn by C.R. Woese [96]. *Copyright (2000) National Academy of Sciences, U.S.A.*

4.3.3. Chemical Ladder: Synthesis and Evolution

The oxidation of some inorganic compounds with strong reducibility can release a large amount of energy which is the driving force for the evolution of life. In the seafloor, high concentrations of H_2S at the hydrothermal vent on seafloor enable people to explore their role. In anaerobic conditions and the closed system of 100°C, Drobner et al. obtained FeS_2 which is very stable under hydrothermal conditions [100]. H_2 was produced during the reaction, FeS was the important starting material, FeS_2 was the usual product and the intermediate HS group was the lively reaction group: $FeS + H_2S \rightarrow Fe(HS)_2$ and $Fe(HS)_2 \rightarrow FeS_2 + H_2$.

Mantle environment was simulated and NH_3 was prepared from N_2 in solution of 300–800°C and 10.1–40 TPa: $3(1-x)\quad Fe + N_2 + 3H_2O \rightarrow 3Fe_{(1-x)}O + 2NH_3$ [101]. NH_3 was prepared at pH = 4 and 100°C through the reaction among FeS, H_2S ,and NO_3^- [102] and it would be possible for further synthesis of amino acids. In laboratory, HCO_3^- and H_2 can transform to CH_4 with Ni–Fe catalysis: $HCO_3^- + 4H_2 \rightarrow CH_4 + OH^- + 2H_2O$ [86]. Because of hydrothermal activity, H_2S, CO_2, CO, and trace of CH_3SH would be released. On the surface of NiS and FeS, the reaction of CH_3SH and CO achieved the fixation of carbon. Wächtershäuster found that it had a high yield (pH = 6.5, 100°C). The mechanism of this reaction is schemed in Fig. 4.14 [87].

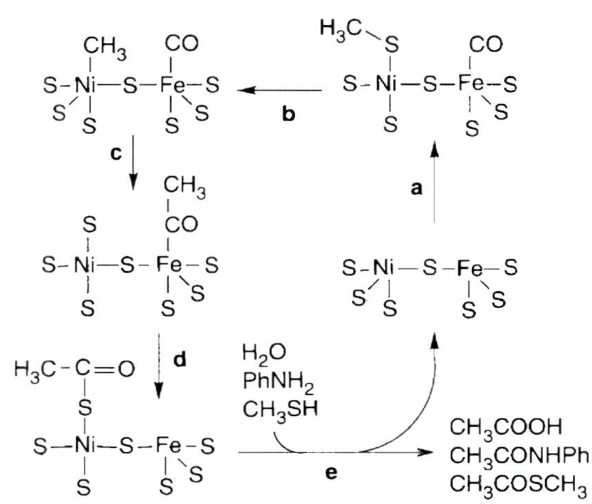

FIGURE 4.14 Notional representation of a hypothetical mechanism of acetic acid formation from CO and CH_3SH on NiS–FeS. Step a: uptake of CO by an Fe center and of CH_3SH by an Ni center. Step b: formation of a methyl-Ni center. Step c: migration of methyl to a carbonyl group, forming an Fe-bonded (or Ni-bonded) acetyl group. Step d: migration of acetyl to a sulfido (or sulfhydryl) ligand, forming a thioacetate ligand of Ni (or Fe). Step e: hydrolytic formation of acetic acid. The free valences of the sulfur ligands are either bonded to another metal center or to H (or CH_3). Alternatively, in step d the acetyl group may migrate to a CH_3–S-ligand to form the methylthioester CH_3–CO–SCH_3, which subsequently detaches. *From [87]. Reprinted with permission from AAAS.*

It was reported that the reaction of CO_2 with Fe_3O_4 in the presence of different amounts of water and obtained acetaldehyde, acetic acid and ethanol: $CO_2 + H_2O + Fe_3O_4 \rightarrow CH_3CHO + CH_3CH_2OH + CH_3COOH + \text{Others}$ [103]. Feng and coworkers conducted the hydrothermal synthesis of organic molecules from $NaHCO_3$ and CO_2, and examined the effect of catalysts [104,105]. Since sodium hydrogen carbonate can be translated into CO_2, which simulates the ocean environments of the primordial Earth in a certain geological time interval. On the basis of the observation of the final product phenol and intermittent formic acid and formaldehyde in the hydrothermal reactions, they proposed a possible reaction mechanism for the formation of phenol (Fig. 4.15). The processes seem complicated but basically involve two simple types of reactions: the oxidative coupling reactions (denoted as 1) and rearrangement reactions (denoted as 2).

$$Fe + H_2O \rightarrow Fe^{2+} + H_2 \quad HCO_3^- + H_2O \rightarrow CO_2 + H_2O$$

They carried out the reactions of formaldehyde with water in the presence of metal iron powder under hydrothermal conditions, and obtained some products (formic acid, acetic acid, propionic acid, methyl acetate, propyl propionate, and propyl isobutyrate), which were important to the origin of life.

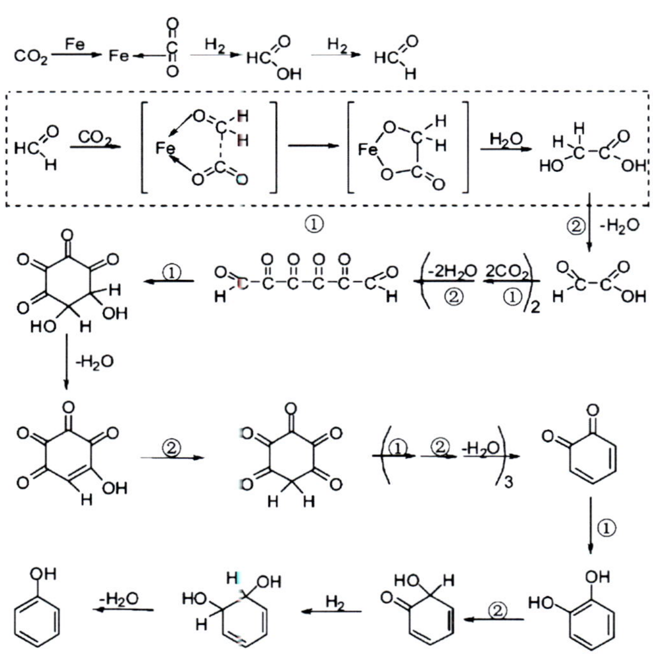

① Refers to the oxidative coupling reactions. ② Refers to the rearrangement reactions.

FIGURE 4.15 Proposed Mechanism of Phenol Formation. *Reprinted with permission from [104]. Copyright 2007 American Chemical Society.*

$$CH_3COOH + NH_3 \xrightarrow[\text{Hydrothermal conditions}]{\text{Metal Oxides}} \underset{NH_2}{\underset{|}{CH_2}}COOH$$

$$CH_3CH_2COOH + NH_3 \xrightarrow[\text{Hydrothermal conditions}]{\text{Metal Oxides}} (D,L)\ CH_3\underset{NH_2}{\underset{|}{C}}HCOOH$$

They continued their experiments with addition of ammonia into the reaction system. During the reaction of ammonia with acetic acid or propionic acid, they witnessed the formation of two amino acids, glycine and racemic D, L-alanine in the presence of Fe_2O_3. Furthermore, they carried out hydrothermal synthesis of peptide from the racemic amino acids without any catalyst. The yield of DD- or LL-dialanine is higher than that of DL- or LD-dialanine. The racemization occurred during the formation of oligopeptides in hydrothermal systems and this process was dynamic. HO−CH=CH−CN and urea were used to form cytosine and uracil [89].

$$CH_3COOH + NH_3 \xrightarrow[\text{Hydrothermal conditions}]{\text{Metal Oxides}} \underset{NH_2}{\underset{|}{CH_2}}COOH$$

$$CH_3CH_2COOH + NH_3 \xrightarrow[\text{Hydrothermal conditions}]{\text{Metal Oxides}} (D,L)\ CH_3\underset{NH_2}{\underset{|}{C}}HCOOH$$

High temperature was favorable to form peptides and low temperature was favorable to stabilize peptides. So alternating hot and cold could prolong oligopeptides. To prove these views, Imai et al. simulated submarine hydrothermal system [91]. In this experiment, when fluid containing glycine was repeatedly circulated through the hot and cold regions in a reactor, oligopeptides were made from glycine. When $CuCl_2$ was added under acidic conditions, oligoglycine was elongated up to tetraglycine and hexaglycine. This observation suggests that prebiotic monomers could be oligomerized in the vicinity of submarine hydrothermal vents on primitive Earth.

4.3.4. Expectation

The possibility of life originating under hydrothermal condition has been understood deeply and the research will be continued. The reaction from inorganics to organics occurred certainly under reductive hydrothermal condition without sunlight, oxygen, high temperature, and high pressure. Life can grow and reproduce under these conditions. The geothermal power and chemical energy supplied energy for the original life which was chemoautotrophy. It is a beneficial study for carbon dioxide fixation, greenhouse gas emissions reduction, and new energy sources development through hydrothermal origin of life's theory.

4.4. SUPERCRITICAL WATER—A NOVEL REACTION SYSTEM [106–110]

Supercritical water (SCW) is attractive medium for chemical reactions because of its unique properties. Supercritical water is defined as water above its critical parameters (i.e., above its critical temperature and pressure). In practice, however, the term "supercritical" is usually restricted to water close to its critical point (374.1°C and 22 MPa), hence close to their critical densities. At 25°C and 0.1013 MPa, liquid water has a density of 0.997 g cm^{-3} and water vapor has a density of 2×10^{-5} g cm^{-3}. With increasing temperature, the density of the liquid decreases and that of the vapor increases. At the critical point both phases become identical and the dividing meniscus disappears. The critical temperature, pressure, and density are 374.1°C, 22.1 MPa, and 0.322 g cm^{-3}. Supercritical water is distinguished from the critical fluids such as CO_2, N_2 and propane. It is a clean, powerful, and stable solvent for organic molecules even though its molecules show a similar polarity to the ambient water.

Most early industrial applications of the supercritical water technology were limited to the unit operations of separation and extraction. However, destruction of hazardous wastes in supercritical water has recently been commercialized after many years of process development. New areas of research in reactions in supercritical water include biomass conversion and gasification, synthesis of organic chemicals and organometallic catalysts, and material processing. Fundamental understanding of the unique properties of supercritical water has improved through both experimental investigations and molecular simulations.

4.4.1. Properties of Supercritical Water [106]

The physicochemical properties of water are highly dependent on the intensive variables of temperature and pressure. At the critical point (374.2°C and 22 MPa), the vapor and liquid phases of water become indistinguishable. At or near this state, water has low density, high diffusivity, low viscosity, high compressibility and a break down of hydrogen bonding. Above the critical point, a single homogenous fluid phase exists with properties intermediate between the gas phase and liquid phase. The density, compressibility, viscosity, and diffusivity of water in this region are extremely sensitive to changes in temperature and pressure.

(1) Ion products and dielectric constants

Two of the most important properties of water are its ion product and dielectric constant; both have been experimentally measured over a wide range of conditions. The dissociation constant K_w increases with increasing temperature and density. The density effect is greater than temperature effect. The dissociation constant K_w of liquid water (25°C and 0.1013 MPa) is 10^{-14}. At 450°C and 25 MPa, K_w of water declines to 10^{-22}, which is at least eight orders of magnitude lower than that of 25°C. At constant pressure, the dielectric constant of water exponentially decreases with increasing the temperature. The dielectric constant of water at room temperature and high pressures is about 80, reflecting its highly polar nature. The high dielectric constant of normal water allows dissolution of polar species and salvation of ions. In the near-critical region 250–350°C, the dielectric constant of water assumes a value between 30 and 20, equivalent to common organic solvents such as acetone and methanol. In this region, dramatic increases in the solubility of organic molecules are expected. However, the dielectric constant remains large enough to support ionic and polar materials. Near the critical temperature and pressure, dramatic variation in the dielectric constant is observed, with the dielectric constant assuming values similar to that of steam. In the temperature range of 400–500°C, extreme pressures (>60 MPa) maintain the dielectric constant above 4. As a consequence of this low dielectric constant in the supercritical region, inorganic salts, such as sodium chloride, become insoluble in water.

(2) Structural transformation

The variations in dissociation constant and the low dielectric constant suggest the structural evolution of supercritical water. At the supercritical state, the hydrogen bonding formed between water molecules breaks, which has been proved by the neutron diffraction data and the simulation of molecular dynamics. With the experimental data of IR and X-ray, and Raman spectroscopy, scientists have found the water molecular dimmers in the supercritical water. The isotropic Raman spectra data show that more than 80% hydrogen bonding is dissociated in supercritical water with density 0.8 g cm^{-3} at 400°C. The water molecular dimmers were present as bifurcated hydrogen bonding. The calculations of simulated molecule dynamics also indicated that no water rings were present in the supercritical water. The results were actually contrary to the explanation of water-cage effect in the supercritical water oxidation of methane. However, the research of the molecule dynamics (MD) simulation showed that the presence of hydrogen bonding was detected at every supercritical state; two hydrogen-bonded molecules, which allowed cavities, substituted the tetrahedral structure, typical of liquid water at room temperature. Therefore, the cage effect in SCW might actually be caused by cavities of two hydrogen-bonded molecules.

(3) Thermal conductivity, viscosity, and diffusion coefficient

The structural evolution of supercritical water leads to significant changes in the properties of water, such as thermal conductivity, diffusion coefficient, and viscosity. Under normal circumstances, the thermal conductivity of supercritical water is slightly lower than that of liquid water. Liquid water has a thermal conductivity of 0.598 $W\,m^{-1}\,K^{-1}$, and the thermal conductivity of supercritical water is 0.418 $W\,m^{-1}\,K^{-1}$. The thermal-conducting behavior of subcritical and supercritical water (>300°C) is similar to that of dissociating gases, the dissociating reaction equilibriums of which are strongly shifted by temperature changes.

The viscosity of supercritical water at 400°C and 50 MPa is about 15 times lower than that of liquid water. At normal conditions, the viscosities of gases and liquids differ by about two orders of magnitude. With increasing temperature at constant density, viscosity increases slightly at gas-like densities and decreases markedly at liquid-like densities. In the low-density regime, momentum transport is dominated by translational transfer, which has been well described by the kinetic theory of gases. In the high-density regime, collisional transfer prevails, which is not yet understood in detail. In the scope of 400–800°C and 0.2–0.8 $g\,cm^{-3}$, momentum transport is composed of translational transfer and collisional transfer, and the viscosity weakly depends on temperature and density. In this range, viscosity of water amounts only to about one-tenth of that of liquid water. The low viscosity causes high mobility of both water and solute molecules. As the consequences of high mobility, solute diffusion coefficients, self-diffusion coefficients, and ionic mobility are much larger in supercritical water than in liquid water, which help to form homogeneity in mixtures of supercritical water and solutes.

The diffusion coefficient of supercritical water is much larger than that of liquid water. For instance, the diffusion coefficient of liquid water at 25°C and 0.1013 MPa is $7.8 \times 10^{-6}\,cm^2\,s^{-1}$, and that of supercritical water at 450°C and 27.0 MPa is $7.7 \times 10^{-4}\,cm^2\,s^{-1}$.

4.4.2. Chemical Applications of Supercritical Water

Supercritical water has received attention as a medium for the synthesis of various specialty and commodity chemicals. Water has an advantage over most supercritical fluids as it is safe, nontoxic, readily available, inexpensive, and environmentally benign.

Mn-doped zinc silicate particles, $Zn_2SiO_4{:}Mn^{2+}$ (ZSM), serve as a good example for describing the fabrication of inorganic materials in supercritical water system [111]. ZSM was produced by Smith et al. at supercritical conditions with batch reactors. Water density in the batch reactor was 0.3 $g\,cm^{-3}$ to give a calculated pressure of 29 MPa at the reaction temperature of 400°C. The batch reactors were heated in a molten salt bath at the given temperature (400°C) and time (1–4320 min) and quenched in water at room temperature. The most stable shape and phase of Mn-doped zinc silicate in supercritical water at 400°C and 29 MPa was needle-like-shaped α-ZSM that was formed via two routes. One route was homogenous nucleation from dissolved raw materials in supercritical water producing needle-like-shaped α-ZSM particles in one step. The other route accompanied the transformation of β-ZSM into α-ZSM through solid-state diffusion and the alteration of shape through dissolution in supercritical water and recrystallization for long reaction time. This research contributes to future work in the field of inorganic material science and the possibility of using supercritical solvents to control the physical form and possible chemical composition of a material.

In recent years there has been an increasing interest in the fabrication of nanomaterials in subcritical or supercritical water system. An attractive advantage of supercritical hydrothermal synthesis in the field of nanomaterials synthesis is short reaction time. Under the special reaction conditions of supercritical water, some nanomaterials with different morphologies have been produced. In 2007, Adschiri et al. reported a simple, rapid and green chemistry approach to control the shape of ceria nanocrystals by tuning the interaction of organic molecules with various crystallographic planes of fluorite cubic ceria, using supercritical water as a reaction medium [112]. In their experiment, the hydrothermal reaction was performed in the reactor at 400°C for 10 min. With similar approach, they also produced high-quality transparent cobalt blue nano pigment in 2007 [113]. Hydrothermal reaction was also performed at 400°C and 38 MPa pressure for 10 min. In 2008, Takami et al. synthesized cubic nanoassemblies of octahedral CeO_2 nanocrystals by simply heating an aqueous solution of cerium nitrate in the presence of hexanedioic acid in a lab-scale plug-flow reactor [114]. In their experiment, they mixed heated water with flow rate of 16.0 $mL\,min^{-1}$ and reagent solution with flow rate of 4.0 $mL\,min^{-1}$ to conduct the hydrothermal reaction. The temperatures of the heated water and the reactants flow were set to about 300 and 250°C, respectively. The flows of the reactants and heated water were mixed at a junction of the plug-flow reactor and the residential time in the isothermal zone was only 1.9 s.

The use of supercritical water as a reaction medium for organic chemistry is a vast topic and a multitude of publications has been devoted to this research, far too many to be adequately covered here. Almost any type of organic reaction, such as hydrogenation/dehydrogenation, rearrangements, hydration/dehydration, elimination, hydrolysis, partial oxidation, H–D exchange, can be performed in supercritical water. For a more comprehensive look at reactions in supercritical water, Savage provided an excellent review [115], and almost an entire issue of Chemical Reviews (Chem. Rev. 99 (1999) 353–634) was devoted to the subject as well.

Supercritical water is a unique medium for organic chemical reactions. The role of supercritical water is an important factor in organic chemical reactions. Akia and Savage's studies have provided unambiguous evidence that water has remarkable effects (as shown in Table 4.2) for organic chemical reactions in supercritical water [107].

4.4.3. Technological Applications of Supercritical Water [109,110,116]

Supercritical water oxidation (SCWO) is a novel and innovative treatment process for the destruction of hazardous waste, which is made possible due to special properties of supercritical water. Among all of the research with applications in supercritical water, supercritical water oxidation has received the most attention because SCWO has showed extremely promising potential as a waste treatment technology. In principle, SCWO can be simply described as pressurized streams of aqueous waste and oxidant, such as oxygen or hydrogen peroxide, mixed and heated to supercritical conditions. Typical SCWO process is conducted at temperatures and pressures of 500–650°C and 25–35 MPa, respectively. When water achieves critical temperature and critical pressure (374.1°C and 22 MPa), it can act as a nonpolar solvent because of greatly reduced hydrogen bonding and polarity. Additionally, the dissociation constant of water decreases from 10^{-14} at normal conditions to about 10^{-23} at typical SCWO conditions. Organic substances exhibit high to complete solubility and gases are completely miscible in supercritical water. On the contrary, inorganic salts exhibit greatly reduced solubility in supercritical water. Numerous studies have proved that a homogeneous single-phase mixture exists at typical SCWO temperature and pressure. Additionally, at typical SCWO condition, viscosity is approximately 1/40th of its value at normal conditions, resulting in high diffusivities. Thus, the reactants, including water, organic wastes, and the oxidant, are brought together in a homogeneous single-phase where oxidation proceeds rapidly and should be limited primarily by chemical kinetics and not by transport processes. Therefore, the residence time required for complete oxidation in the SCWO is generally less than 1 m, with efficiencies often greater than 99.99%. The

TABLE 4.2 Summary of Effects of Water for Organic Chemical Reactions in High-temperature Water (HTW)

Role of water	Applicable conditions	Affected reactions
Reactant/product	Any, but importance increases at higher water densities	Reaction in which water is a reactant (e.g., hydrolysis, hydration, hydrogen abstraction)
Catalyst	Any	Proton-transfer reactions in which water interacts with reactant (s), typically via hydrogen bonding
Acid/base catalyst precursor	More important at higher temperatures and liquid-like densities	Acid/base-catalyzed reactions
Preferential solvation/ desolvation of transition state	Any, but probably less important at gas-like densities	Reactions with change in solute–solvent interactions (e.g., electrostatic, hydrogen bonding, etc.) between reactant(s) and transition state; examples include reactions with a change in polarity along the reaction coordinate
Hydrophobic effect	More important at near-ambient conditions than in HTW	Condensation reactions with immiscible organic reactants
solvent dynamics	more important at liquid-like densities	Very fast reactions (faster than solvent reorganization)
Density inhomogeneities	Supercritical conditions	Any reaction can be affected by the local composition differing from the bulk composition
Energy transfer	More important at gas-like densities; at liquid-like densities, energy transfer is not likely to be rate limiting	Nominally unimolecular elementary reactions
Cage effects	more important at liquid-like densities	Bimolecular reactions (in either forward or reverse direction)

Reprinted with permission from [107]Copyright 2002 American Chemical Society.

oxidation of organic wastes will only result in products such as CO_2, molecular nitrogen (N_2 or N_2O), and water. Heteroatoms such as chlorine, sulfur, and phosphorus can be transformed to their mineral acids and consequently be neutralized using a suitable base. Typical undesired and noxious byproducts SO_2 or NO_x of combustion processes will not be formed because their oxidation pathways are not favored by the lower temperatures of SCWO comparing to incineration which is the most commonly used waste treatment technology. SCWO has also been found successful in treating different chemical warfare agents.

There are several common and alternative waste treatment processes besides SCWO which include wet-air oxidation (WAO) process, incineration, and biological treatment. Although they have shown different advantages when treating organic wastes, some intrinsic drawbacks have prohibited them from being used at several circumstances. Wet-air oxidation (WAO) is a thermal oxidation process most commonly used for the treatment of sludge. WAO is typically operated at a temperature range of 175–315°C and at pressures up to 20 MPa (below the critical point). One major drawback of WAO is the long residence time (1–2 h) and another one is that the oxidation is generally incomplete. The major shortcoming of incineration process is that oxidation is incomplete and the oxidation products may contain some environmentally harmful substances such as CO and NO_x The biggest problem for biological treatment process is that it can be used only if the organic compounds are biodegradable. Compared to these alternative waste treatment processes, the merits of SCWO are obvious. SCWO can be used for wastes with concentrations of chemical compounds from 0 to 100%. SCWO is efficient (99.99% conversion rate) and quick (less than 1 min residence time). Oxidation products of SCWO are complete and environmentally benign.

The following are the three reasons why SCWO has not yet become a current waste treatment process.

(1) Severe reactor corrosion caused by acids, which are formed during the waste treatment process.
(2) Serious plugging of the reactors caused by precipitating salts at supercritical temperatures and low densities.
(3) Due to lack of experimental data, cost evaluations, especially for the scale-up of SCWO plants to an industrial scale, are unreliable.

4.5. TECHNIQUES AND METHODS

Hydrothermal and solvothermal synthesis technique involves reaction containers (e.g., autoclaves), reaction control system, hydrothermal and solvothermal synthesis process, characterization techniques (in situ), and intermediate species.

4.5.1. Reaction Containers [11]

High-pressure vessel, popularly known as an autoclave, is the basic equipment of hydrothermal and solvothermal synthetic experiment. Advances in hydrothermal and solvothermal synthesis of research depend largely on the equipment available. Crystal growth or materials processing under hydrothermal and solvothermal conditions requires high-pressure vessels to have outstanding capability of containing highly corrosive reagents, resisting high temperature and high pressure. Experimental studies under hydrothermal and solvothermal conditions require facilities that must operate routinely and reliably under extreme conditions.

For different objectives and tolerances of hydrothermal synthetic experiment, there is a great variety of high-pressure vessels used for hydrothermal technology. Most of the earlier workers had dealt with this aspect in their own way. Because of the limit of space, we will discuss only the commonly used autoclaves or hydrothermal reactors in the hydrothermal synthesis; they were all designed and fabricated during the twentieth century. Generally speaking, an ideal hydrothermal autoclave should have the main characteristics as follows:

(1) Should have high mechanical strength in order to bear high pressure and temperature experiments for long duration.
(2) Should have excellent acid, alkali, and oxidant resistance.
(3) Should have a simple mechanical structure and easy to operate and maintain.
(4) Should have good sealing performance in order to obtain the required temperature and pressure.
(5) Should have a suitable size and shape in order to obtain a desired temperature gradient.

The top-priority parameter to be considered in selecting a suitable autoclave is the experimental pressure and temperature conditions and the corrosion resistance in a hydrothermal solution. In accordance with the above, the materials, thick glass cylinders, thick quartz cylinders, and high strength refractory alloys are usually used to fabricate hydrothermal autoclaves. Some hydrothermal reactions, in which the reagents or solvents are noncorrosive, can take place directly in the hydrothermal autoclaves. In contrast, in the majority of hydrothermal reactions, the reagents or solvents used are highly corrosive and they can attack the autoclaves in the high-pressure temperature range. It requires the autoclaves to contain inert linings, liners, or cans. As far as we know, some materials used in the manufacturing of the linings, liners, and cans include Teflon, Pyrex, quartz, graphite, Armco iron, titanium, silver, platinum, tantalum, copper, nickel, and gold. Figure 4.16 shows the most popular Teflon-lined, stainless steel autoclave, in which the mild hydrothermal and solvothermal reactions are commonly performed. Such autoclaves can be used at up to 270°C, and depending on the engineering specification of the steel walls, pressures of ~150 MPa can be withheld.

Up to now, there is no uniform standard for the classification of hydrothermal autoclaves in the world. In accordance with different classification standard, there is more than one name for a hydrothermal autoclave. The information of classification of hydrothermal autoclaves is introduced as follows:

(a) Classification by sealing methods: self-tightened seal autoclave and external-tightened seal autoclave.
(b) Classification by mechanical structure: flange-disc type autoclave; internal-plug-screw type autoclave; big-screw-cap type autoclave; and lever-press type autoclave.
(c) Classification by pressure formation methods: internally pressurized type autoclave and externally pressurized type autoclave.
(d) Classification by the name of designers: Morey autoclave; Smyth autoclave; Tuttle autoclave; and Barnes rocking reactors.
(e) Classification by heating methods: internally heated autoclave and externally heated autoclave.
(f) Classification by experimental system: high-pressure autoclave and flow reactors and diffusion reactors.

(1) Externally heated and internally pressurized autoclave

This type of autoclave is popularly known as a Morey autoclave since Morey first designed this simple, gasketed, sealed steel autoclave of 25–100 mL volume in 1913 [117]. The cross-section of a typical Morey autoclave is shown in Fig. 4.17.

FIGURE 4.16 General purpose autoclave popularly used for hydrothermal or solvothermal synthesis.

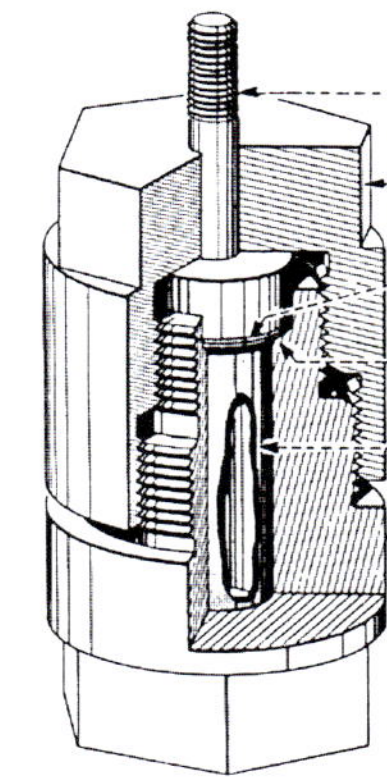

FIGURE 4.17 Cross-section of a typical Morey autoclave. *Reprinted with permission from [117]. Copyright 1913 American Chemical Society.*

The usual dimensions of a Morey autoclave are 10–20 cm length and 2.5 cm inner diameter. The autoclave generates an autogeneous pressure depending on the degree of filling, the fluid, and the temperature. In the initial design stages, the autoclave is sealed by an internally pressurized seal gasket. For a long time reaction, the autoclave can be used with high temperature of 600°C and high pressure of 0.04 GPa. For a short time reaction, the autoclave is able to withstand a high operating temperature of up to 700°C and a high working pressure of up to 0.07 GPa. If the working pressure is very high, the autoclave sealed by an internally pressurized seal gasket will have two drawbacks. First, the autoclave is prone to leakage; second, it is difficult to open the autoclave. In a later improved version of Morey autoclave, the sealing way is redesigned as a self-tightened seal. For a long time reaction, the improved autoclave can be used with high temperature of 600°C and high pressure of 0.2 GPa. Today most autoclaves popularly used for hydrothermal or solvothermal synthesis in laboratory are almost the improved versions of Morey autoclave.

In addition to the Morey-type autoclave, there are also several important externally heated and internally pressurized autoclaves. One is self-tightened cold-seal autoclave. There are two main differences between this type of autoclave and the Morey-type autoclave. First, the closure of the autoclave is a self-tightened closure; second, the plug and seal of the autoclave are outside the furnace. Another important externally heated and internally pressurized autoclave is cone-seal autoclave. Unlike the Morey-type autoclave, this type of autoclave has a cone closure. Moreover, the plug and seal of the autoclave are also outside the furnace.

(2) Externally heated and externally pressurized autoclave

This type of autoclave is popularly known as a Tuttle autoclave (as shown in Fig. 4.18) since Tuttle first designed this ingenious cone-seal steel autoclave with very small volume in 1948 [118]. The working chamber of Tuttle autoclave is closed by a cone-in-cone seal, so another name of this autoclave is cold-cone-seal autoclave. In the initial design stages, the Tuttle autoclave has a very small volume and the entire autoclave is kept inside the furnace. Although this apparatus can be used under very high temperature, and the pressure can be well controlled, it has not been widely used for hydrothermal studies. In a later improved version of Tuttle autoclave redesigned by Tuttle, the length of working chamber is increased and the plug and seal are outside the furnace. Pressure is transmitted to the sample, which is contained in a sealed working chamber, through a hole in the closure. This apparatus is able to withstand a high operating temperature of up to 750°C and a high working pressure of up to 1.2 GPa. When pressure is up to 0.7 GPa, Argon is used as the pressure medium in all experiments. The reason is that water will freeze at room temperature when pressure is up to 0.7 GPa. This apparatus, with features of simple structure, convenient operation, and low cost, is widely used for hydrothermal synthesis and materials processing.

Barnes rocking reactor, which was first designed by Barnes in 1963, is also an externally heated and internally pressurized autoclave [119]. The reaction vessel is of Type 310 stainless steel, heavily chrome-plated on interior surfaces and with a capacity near 1100 mL. The closure is of the same materials and uses a simple compression seal with an annealed, Type 304 stainless steel ring for the gasket. The entire valve and furnace assembly is agitated to speed equilibration by rocking at 36 rpm through an arc of 30° about a horizontal axis. Tubing connections to the reaction vessel, therefore, are made using flexible capillary tubing or spiral coils of tubing. All tubing and valves are stainless steel. This apparatus is widely used for the studies about solubility of minerals, mineral synthesis, $P-V-T$ relations of gases and liquids, and conductivities of liquids. This apparatus is able to withstand a high working pressure of up to 0.05 GPa at 25°C or 0.03 GPa at 400°C.

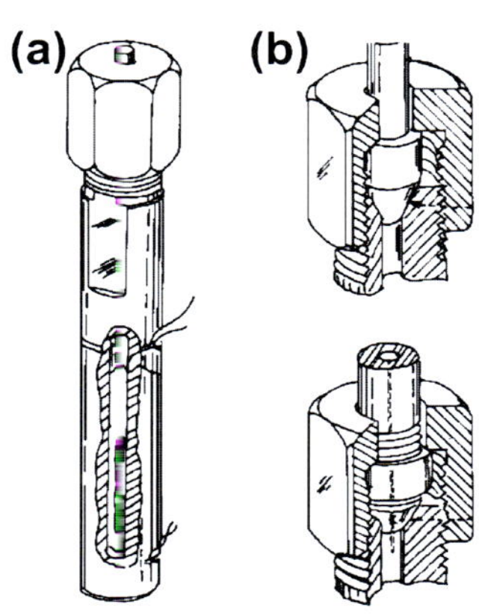

FIGURE 4.18 Tuttle autoclave (a) overall appearance; (b) cross-section of cap [118].

(3) Internally heated and externally pressurized autoclave

All the autoclaves discussed above are externally heated apparatuses. In contrast to these apparatuses, another class of apparatuses in which the pressure vessel is cooled and the sample is heated internally has been designed by several workers. For an internally heated and externally pressurized autoclave, the heating system is installed inside the equipment inside and experimental pressure is supplied by an external high-pressure system (Fig. 4.19). Although, all internally heated apparatuses are significant, we will only discuss several representative examples.

(a) Smyth and Adams autoclave [120]. This apparatus was firstly designed by Smyth and Adams in 1923, and was modified by Goranson in 1931 [121]. This type of autoclave is able to withstand a high operating temperature of up to 1400°C and a high working pressure of up to 1.0 GPa.
(b) Yoder autoclave [122]. This apparatus was designed by Yoder in 1950, and it was an improved version of Smyth and Adams autoclave. This type of autoclave is able to withstand a high operating temperature of up to 1600°C and a high working pressure of up to 1.0 GPa.
(c) Burnham, Holloway, and Davis autoclave [123]. This apparatus was designed by Burnham et al. in 1969. This type of autoclave is able to withstand a high operating temperature of up to 1500°C.

Till date, the number of hydrothermal autoclaves has exceeded one hundred, including some novel designs for specific studies such as controlled growth of crystal, kinetics, thermodynamics, hydrothermal hot pressing, microwave-enhanced hydrothermal synthesis, and so on. Moreover, many of them have been commercially produced by the United States, Germany, Japan, China, and other countries. Figure 4.20 shows the photographs of commercial hydrothermal autoclaves produced by Parr Instrument Company (Moline, Illinois) [124]. It would be very difficult to discuss all autoclaves in this book. Byrappa and Yoshimura (2001) have reviewed

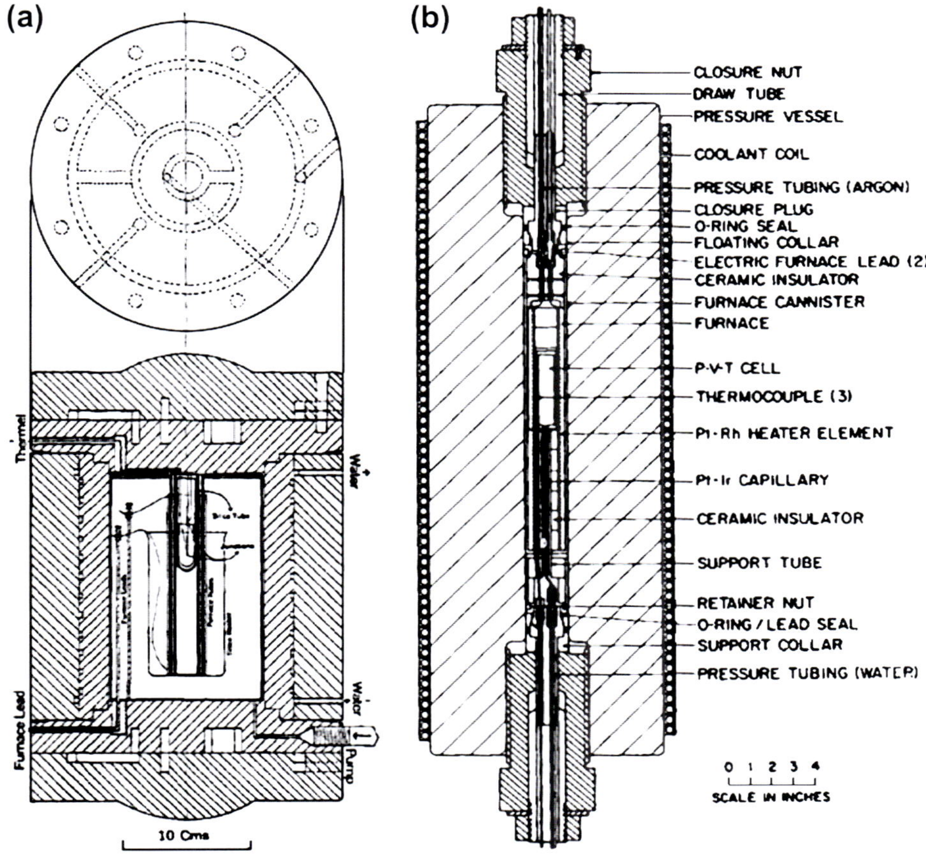

FIGURE 4.19 Internally7 heated autoclave (a) Smyth and Adams autoclave; Reprinted with permission from [120]. Copyright 1923 American Chemical Society. (b) Burnham, Holloway and Davis autoclave [123].

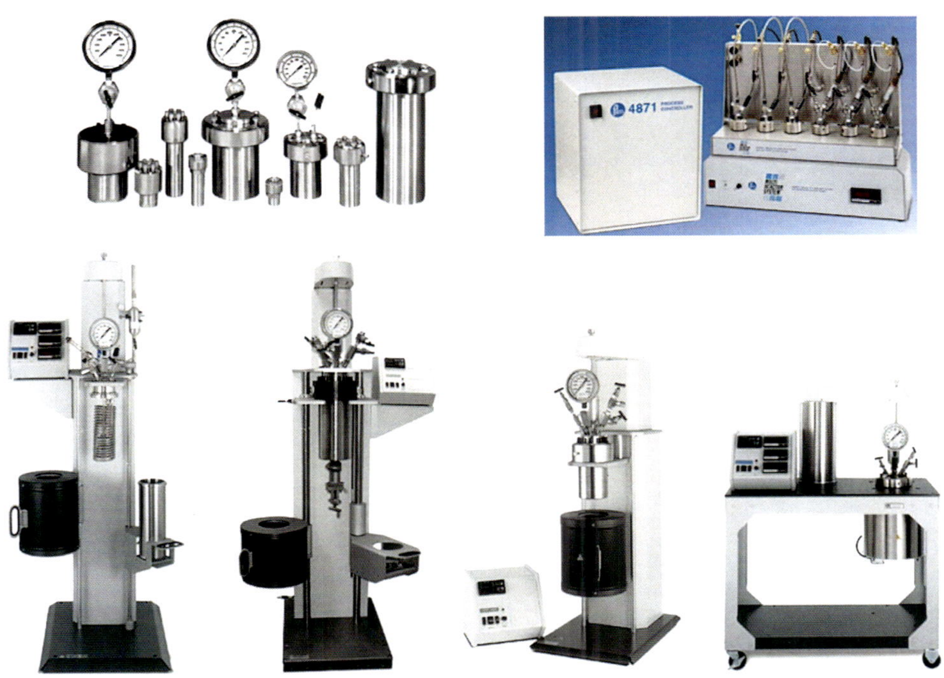

FIGURE 4.20 Parr Instrument Company Pressure Vessels and Reactors [124].

various autoclave designs in detail [11], and readers can get more information about this topic from this literature source.

4.5.2. Reaction Control Systems

Safety is the utmost important issue when working in chemical laboratory. In hydrothermal or solvothermal research, we should attach great importance to safety of hydrothermal reactors/equipments. In general, a hydrothermal reactor/equipment should have several safety control system components including temperature control system, pressure control system, and closed system control.

Temperature has important effects not only on the structure of the materials used for a hydrothermal reactor but also on the speed and result of a hydrothermal reaction. Therefore, we should determine a suitable temperature for a hydrothermal reaction. In general, the temperature control component of a hydrothermal reactor should include high-precision temperature controllers and high sensitive temperature sensors.

Pressure is crucial to the safety of hydrothermal experiments. For an internally pressurized autoclave, the pressure of a hydrothermal reaction depends on the degree of fill and the properties of reaction mediums. The value of pressure can be calculated from the Clapeyron equation ($PV = nRT$). In general, for safety reason, the degree of fill is usually between 50 and 80% for a typical hydrothermal synthesis experiment. For an externally pressurized autoclave, experimental pressure is supplied by an external high-pressure system. The externally pressurized autoclave works based on the principle of internal and external pressure balance. The internal pressure in a sealed working chamber depends on the degree of fill and the properties of reaction mediums, hence the values can be calculated from the Clapeyron equation. The value of external pressure is manually set according to the internal pressure. During the hydrothermal reaction process, the value of external pressure must be adjusted at any time to make it close to the internal pressure. Otherwise a large pressure difference between external and internal pressure will lead to the burst of sealed working chamber.

4.5.3. General Experimental Procedure

Hydrothermal or solvothermal technology is constantly developing to keep in step with the ever-evolving research needs of scientists. Hydrothermal synthesis or solvothermal synthesis has been successful in the preparation of important solids such as zeolites, open-framework compounds, organic–inorganic hybrid materials, superionic conductors, chemical sensors, electronically conducting solids, complex oxide ceramics and fluorides, magnetic materials, and luminescence phosphors. Whatever is the purpose of hydrothermal or solvothermal experiment, an ideal experimental procedure should be built on the basis of understanding of reaction mechanisms and the accumulation of chemical experience. Experimental procedure of hydrothermal or solvothermal synthesis depends on the purpose of the research. Here, we only present a general experimental procedure of hydrothermal synthesis.

(1) To choose suitable reagents;
(2) To determine the mole ratio of reagents;
(3) To explore addition order of reagents and mix the reagents;
(4) To put the mixture of reagents into an autoclave and seal the autoclave;
(5) To choose suitable reaction temperature, reaction time, and reaction state (dynamic or static crystallization);
(6) To take out the autoclave from oven and cool the autoclave to room temperature;
(7) To open the autoclave and take out the products from the autoclave;
(8) To process the products (such as wash, filtrate, and dry);
(9) To observe the appearance of products by using an optical microscope; and
(10) To characterize the products by using suitable research Instruments.

4.5.4. In Situ Characterization Techniques

In the early mechanism studies of hydrothermal or solvothermal reactions, chemists studied the changes of hydrothermal/solvothermal reaction system and the structures of the products by optical/X-ray or other physical techniques after they rapidly broke off the hydrothermal or solvothermal reaction. This is an effective and commonly used method by earlier chemists. However, one of the major drawbacks of this method, which cannot be ignored, is that reaction intermediates can only be speculated and cannot be observed. Along with the development of scientific and technical means, spectroscopic techniques were used for the mechanism studies of hydrothermal or solvothermal reactions. Because vibrational spectrum is an important and effective means for determining the types of reaction intermediates, the structures of final products, and reaction rates, UV/visible spectroscopy is first applied for studying hydrothermal or solvothermal reactions. For a common steel autoclave, the hydrothermal or solvothermal reaction process cannot be monitored by "looking." Moreover, it is not very reliable that reaction intermediate states are inferred by the final products. For real-time observation or detection of intermediate states, scientists designed several hydrothermal autoclaves with quartz and sapphire windows. In recent years, along with the improvement of hydrothermal autoclaves, scientists have developed many in situ characterization techniques for studying hydrothermal or solvothermal reactions [125–130]. Thus, the hydrothermal autoclave is no longer a "black box."

4.6. IONOTHERMAL SYNTHESIS

Recently, a novel synthesis method for preparing inorganic functional materials, known as ionothermal synthesis, has prompted a significant amount of research. The term "ionothermal" was used to describe reactions that are conducted in ionic liquids (ILs) at high temperature. Ionothermal reactions are analogs of hydrothermal or solvothermal approaches using ionic liquids as solvents.

Ionic liquids, as the reaction medium of ionothermal, are now being defined as salts composed solely of ions with melting points below 100°C, although this strict definition is often given some latitude as more and more applications and insight arise from the increased scrutiny of heterogeneous liquids. The first ILs, or organic molten salts, were discovered in 1914, but the discovery of air- and water-resistant species has motivated a resurgence of interest during the past decade. In contrast with inorganic molten salts, ILs are composed solely of ions and have a high cohesive energy density, but in contrast their melting points are often less than 100°C. In particular, the reasons for the low melting point of ionic liquids are not clear, although an explanation is usually given: both cations and anions of ILs are relatively complicated ions where in most cases the charge is delocalized or shielded by side groups. A great advantage of ILs is the variation of the ion species: 10^{12}–10^{18} different liquids, binary and ternary mixtures with versatile properties have been predicted. The cationic parts of most common ILs are organic-based moieties such as imidazolium, pyridinium, quaternary ammonium, quaternary phosphonium cations, or nitrogen-rich alkyl-substituted heterocyclic cations. The anionic part can be organic or inorganic and include such entities as some halides, nitrate, acetate, hexafluorophosphate (PF_6), tetrafluoroborate (BF_4), trifluoromethylsulfonate, and bis(trifluoromethanesulfonyl)imide.

Ionic liquids exhibit unique physicochemical properties which are influenced both by their cationic and their anionic moieties. In contrast with traditional solvents, such as water or mostly volatile organic compounds, the common properties of most ILs include low melting temperature, tunability, nonvolatility, nonflammability, low viscosity, high thermal and chemical stability, zero toxicity, ionic conductivity, and heating behavior under microwave radiation. ILs have low to negligible vapor pressures (1–10 Pa) which enables experiments at temperature up to 300°C. The extremely low vapor pressures also allow experiments under vacuum conditions; thus the surface of the liquid can be investigated by surface sensitive techniques. ILs exist as liquids over a rather wide temperature range as large as 200°C, maintaining at the same time a very low vapor pressure. This fact is related to their high cohesive energy density, which is mainly given by the bare Coulombic interactions between the constituents.

ILs may be the most complex of all solvents because they are capable of virtually all possible types of interactions with solutes. ILs can solubilize a variety of organic and inorganic compounds. They can be designed to be immiscible or miscible with water and a number of organic solvents. ILs have attracted considerable attention of chemists looking for potentially environmentally benign alternatives to the traditional solvents due to their unique physicochemical properties. Today, the applications of ionic liquids can be numerous, from synthesis, separations, catalysis, electrochemistry, photoelectrochemistry, phase transfer, and magnetic or nanostructured materials. In a special issue of Accounts of Chemical Research (Acc. Chem. Res. 40(11) (2007)), the latest progress in ILs was reviewed by some of the leading IL research groups. Readers can get more information about this topic from this literature source.

It should be noted there is a special class of ILs named as deep eutectic solvent (DES). Unlike ILs, DES is a type of mixed solvent composed of two or more compounds which form a eutectic. A DES consists of ionic compounds and molecular compounds. The melting point of DES is much lower than either of the individual components. As ionic liquids, DESs share many characteristics of conventional ILs. DES exhibits unusual solvent properties and is able to dissolve metal salts, organic aromatic acid, amino acids, and some metal oxides. Many DESs are nontoxic and biodegradable. Compared with ILs, DESs are inert in water, and easily prepared in a pure state. In addition, DESs based on relatively available components such as urea and choline chloride are cheaper to make than ILs, which makes DESs particularly desirable for applications in the large-scale synthesis of new functional materials.

ILs have widely been used in organic and polymer synthesis, separations, catalysis, electrochemistry, and photoelectrochemistry. The synthesis of inorganic materials in ILs is a rather new development and has attracted increasing interest in recent years. In 2004, ionothermal synthesis method was reported by Cooer et al. for preparing molecular sieves [131]. By using this method, Morris et al. have prepared several aluminophosphate and cobalt aluminophosphate compounds in an imidazolium-based ionic liquid (1-ethyl-3-methylimidazolium bromide ([emim]Br)). In ionothermal synthesis, inorganic materials are prepared in ionic liquids, which is different from traditional hydrothermal and solvothermal synthesis which employs molecular liquids. Ionothermal synthesis could in principle be used in any situation where hydro(solvo)thermal synthesis has been successfully implemented. Compared with the traditional hydro(solvo)thermal synthesis, ionothermal synthesis has interesting features and potential advantages (as described below).

(1) ILs have many unique physicochemical properties and the coexistence of ionic and organic groups and the large temperature windows of ILs make them ideal media for the reactions between metal ions and organic ligands. With these attributes, ionothermal synthesis is considered to be more environment-friendly and safer.

(2) In the synthesis of inorganic materials, the IL, the reaction medium of ionothermal synthesis, can participate as solvent, structure-directing agent, or template agent, structure-inducting agent, charge-compensating group, or ionic liquid (crystal) precursor.

(3) The IL solvent systems offer a novel chemical environment that can uniquely influence the course of chemical reactions. Therefore, ionothermal synthesis can result in inorganic materials with special structure or properties, which are difficult or impossible to obtain by using other traditional synthesis routs.

(4) The ionothermal synthesis can be performed at ambient pressure, which is due to the negligible vapor pressures of almost all ILs. As a result, ionothermal synthetic reactions avoid the high pressures of traditional hydro(solvo)thermal reactions and therefore eliminate safety concerns associated with high pressures.

(5) ILs have been proven to be good medium for absorbing microwaves. Therefore, microwave technique can be safely used in the ionothermal synthesis reactions.

As a new synthesis method, ionothermal synthesis has been studied to prepare many types of solid materials such as molecular sieves and molecular-sieve analogous, organic–inorganic hybrid materials, metal-organic frameworks (MOFs), semiconductors, inorganic solid materials, nanomaterials, and so on. The latest progress of ionothermal synthesis has recently been reviewed by Morris and coworkers [132,133].

The most successful cases of current ionothermal synthesis are the preparation of molecular sieves, molecular-sieve analogous, and related porous materials, and a number of molecular sieves and related materials with both known and previously unknown structure types have been prepared [134–144]. In general, molecular sieves are typically prepared by hydrothermal or solvothermal methods in aqueous or nonaqueous conditions by using organic amines or quaternary ammonium ions as the templates or structure-directing agents (SDAs). Distinguishing from traditional hydro(solvo)thermal syntheses, ILs in ionothermal synthesis can not only act as reaction solvents but also provide the template ions during the reaction process. One most important feature of this strategy is the removal of competition between the template–framework and solvent–framework interactions, present in hydrothermal or solvothermal synthesis (Fig. 4.21). In principle this may lead to improved templating of the growing molecular sieve crystal structure. The first work in this area, published in 2004 [131], used 1-ethyl-3-methylimidazolium bromide and urea/choline chloride deep eutectic solvents to prepare several different materials (named as SIZ-1, SIZ-2, SIZ-3, SIZ-4, SIZ-5, and AlPO-CJ2) depending on the conditions.

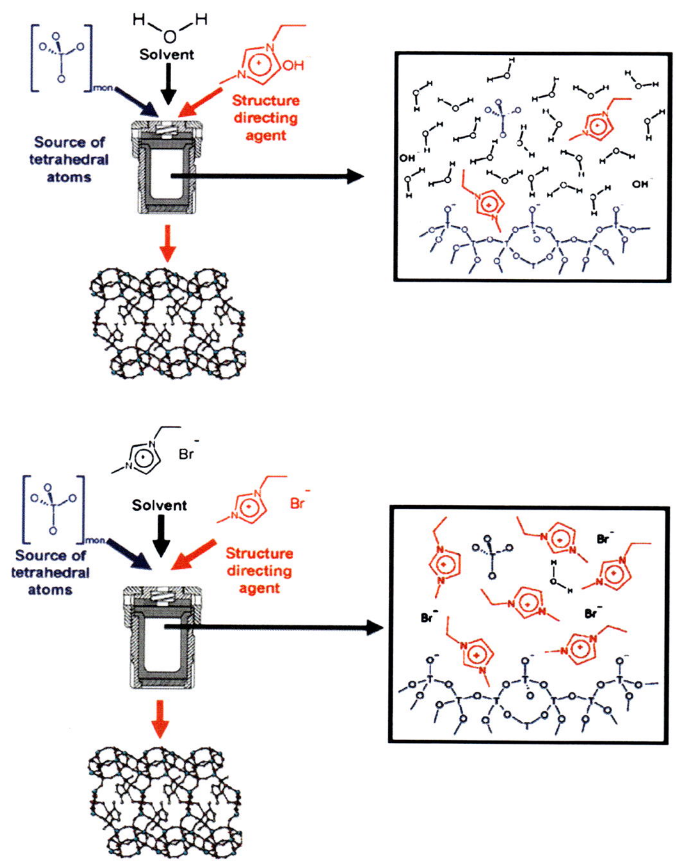

FIGURE 4.21 Schematic representations of the synthesis of a tetrahedral (zeotype) framework under hydrothermal (Left) and ionothermal (Right) conditions. *Reproduced by permission of The Royal Society of Chemistry [133].*

Templating or structure-directing effect is an important issue for the synthesis of molecular sieves. If an additional template agent, such as an organic amine, is introduced into ionothermal synthesis system, the crystallization process of molecular sieves will be altered. The addition of organic amines into ILs will result in better phase selectivity. Recently, Yu et al. of our lab reported that organic amines and ILs can act as co-template agents to prepare molecular sieve (JIS-1) [145]. The structure of JIS-1 consists of an anionic open-framework $[Al_6P_7O_{28}H]^{2-}$. Protonated methylimidazolium cations along with 1-ethyl-3-methylimidazolium cations act as the templates to coexist in the intersection of the tridirectional channels in the structure of JIS-1 (Fig. 4.22). The role of organic amines in ionothermal synthesis has been studied by Bao and coworkers [146]. Their study result indicates that a hybrid of ILs with organic amine cations can be formed through a hydrogen bond during the crystallization of molecular sieves. This hybrid that is formed in the gel stage could act as a cooperative structure-directing agent for the specific structure in the final solids. Their results may suggest a synthesis mechanism of molecular sieves in ILs in which the IL–organic-amine hybrid is required in the nucleation step, whereas the crystal growth occurs through the occlusion of ILs.

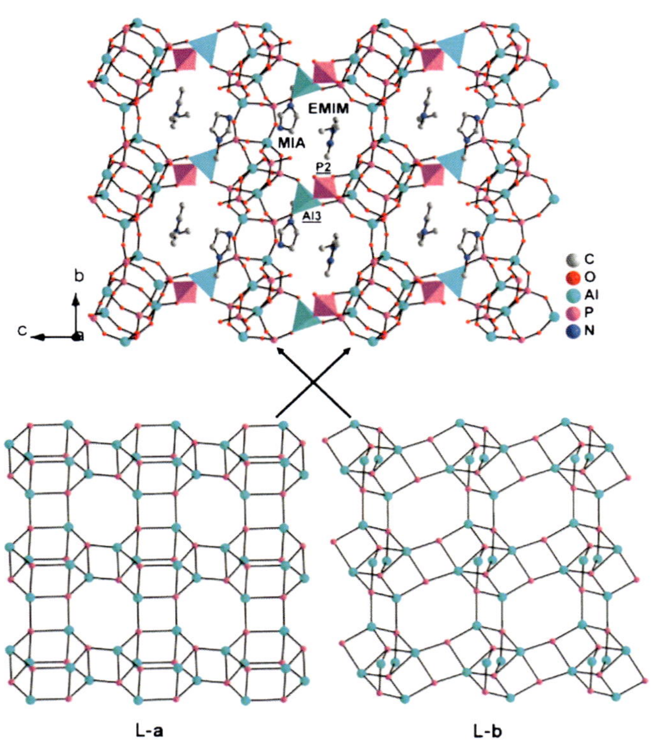

FIGURE 4.22 View of JIS-1 structure along [100] direction with co-templates of the $MIAH^+$ and $EMIM^+$ cations located in the channels (Top) and the two component layers (Bottom), L-a and L-b. *Reprinted with permission from [145]. Copyright 2008 American Chemical Society.*

In ionothermal synthesis, the role of water is an important factor, and the quantity of water in reaction system can influence phase selection. Some studies have provided unambiguous evidence that water has a remarkable effect on the ionothermal synthesis of molecular sieves [147,148]. Addition of reagent quantities of water can enhance the crystallization kinetics of molecular sieves greatly, and the main product phase is molecular sieve. If the quantity of water is equimolar in concentration with ILs, it will lead to the formation of dense phases. Too much water in the ionothermal synthesis system is detrimental to the formation of molecular sieves.

It is worth noting that microwave-enhanced ionothermal synthesis is an effective strategy for the preparation of molecular sieves and other inorganic solid materials [149,150]. Microwave dielectric heating synthesis is an effective method developed for the synthesis of molecular sieves. Compared with conventional heating, microwave-enhanced hydrothermal synthesis has obvious advantages, such as fast crystal growth and high selectivity obtained. However, the drawbacks of microwave-enhanced hydrothermal synthesis are that experiment is unsafe and organic templates used in the synthesis will decompose under microwave irradiation at high temperatures. ILs have been proven to be good medium for absorbing microwaves, and the negligible vapor pressure of ILs make them be heated to relatively high temperature without the production of autogenous pressure. Therefore, microwave-enhanced ionothermal synthesis is expected to be a promising approach to the synthesis of inorganic materials. Cai et al. have prepared extremely well-oriented zeolite coatings on copper-containing aluminum alloys by microwave-enhanced ionothermal synthesis method [150]. In their experiment, they prepared two different types of zeolite coating with AEL structure topology. One is pure aluminophosphate (AlPO) and another is a silicoaluminophosphate (SAPO). Figure 4.23 shows the morphology and microstructure of SAPO coating.

In addition to the synthesis of molecular sieves, another focus of ionothermal synthesis is the preparation of metal-organic frameworks (MOFs). Up to now, a number of MOFs and related materials with both known and previously unknown structure types have been prepared [151–159]. Similar to the synthesis of molecular sieves, ILs in ionothermal synthesis of MOFs can act as reaction solvents and template reagents. Besides this, ILs can also act as reactants or structure-inducting agents. In some cases of ionothermal synthesis, ILs, as reactants, can participate in the constructing of the frameworks of final products. For instance, under specific conditions, some ILs base on dialkylated imidazolium cations can break down to

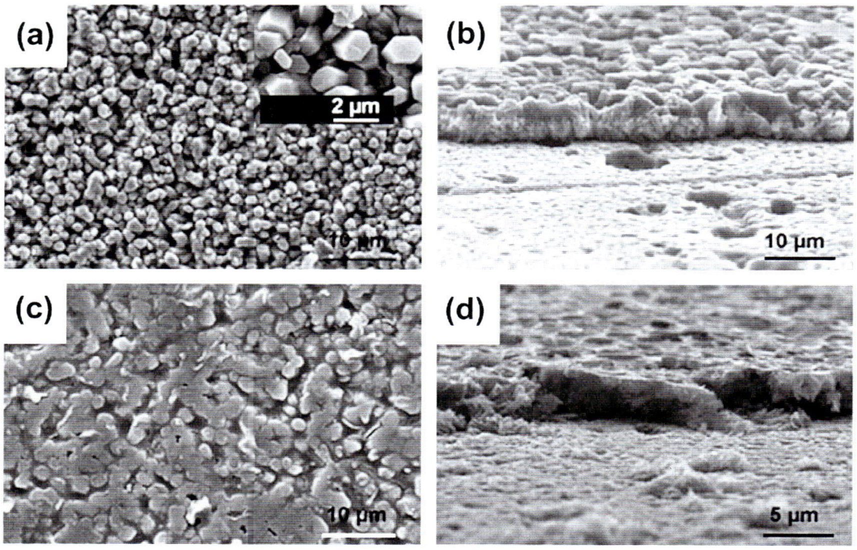

FIGURE 4.23 SEM images of different as-synthesized AEL coatings on AA 2024-T3. (a) SAPO-11 (surface, inset is higher magnification); (b) SAPO-11 (cross-section, mildly polished surface); (c) SAPO-11 with spin-on BTSM-MEL (surface); and (d) SAPO-11 with spin-on BTSM-MEL (cross-section). *R. Cai, M.W. Sun, Z.W. Chen, R. Munoz, C. O'Neill, D.E. Beving, Y.S. Yan, Angew. Chem. Int. Ed. 47 (2008) 525. Copyright Wiley-VCH Verlag GmbH & Co. KGaA. Reproduced with permission [150].*

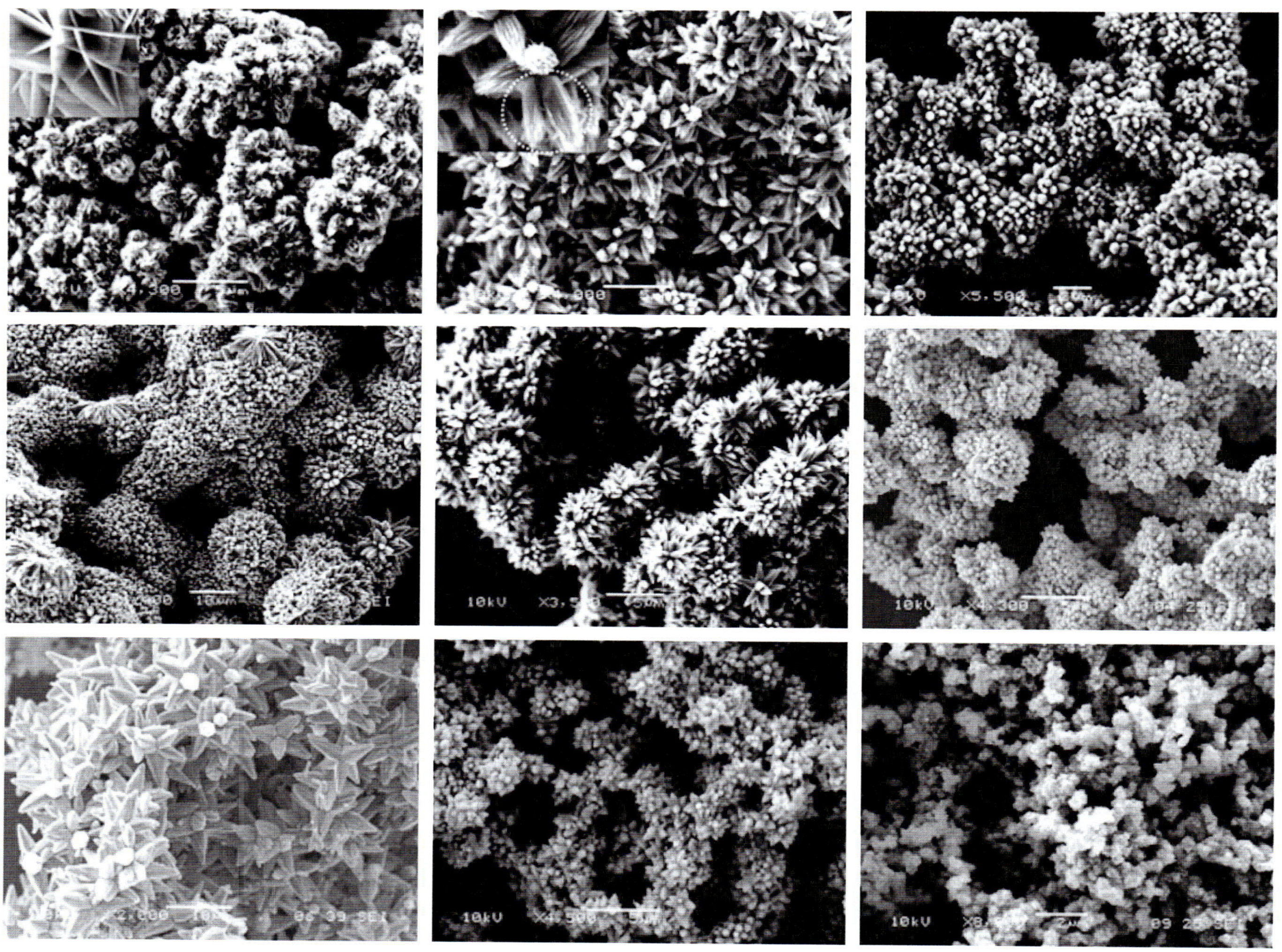

FIGURE 4.24 SEM images of hierarchical ZnO nanostructures. *Reprinted with permission from [165]. Copyright 2002 American Chemical Society.*

leave the monoalkylated imidazole species that can coordinate to metals.

ILs are composed solely of cations and anions, and the nature of the anions plays important parts in controlling the nature of the ILs. It is noteworthy that anions of ILs in ionothermal synthesis of MOFs have extremely important roles. The first important role of anions is that they can be occluded into the structure of MOFs as template reagents in certain circumstances. In most cases, the template role of anions combines with that of IL cations. The second important role of anions is structure-inducting effect. In some case, the anions themselves are not occluded into the structure of MOFs, but the nature of anions determines the structure constructing of final products.

In addition to molecular sieves and MOFs, ionothermal synthesis method has been used to prepare many other types of inorganic materials [160–165]. For instance, a number of nanomaterials, such as aluminophosphate nanorolls, $NaYF_4$:Yb^{3+}, $Er^3 + /Tm^{3+}$ nanoparticles, IL/$Al(OH)_3$ hybrid nanorods/nanofibres, and hierarchical ZnO nanostructures, were successfully synthesized in ILs. Figure 4.24 shows the morphology and microstructure of some reported nanomaterials. In general, the ILs can, in most cases, participate as ionic liquid (crystal) precursors. The detailed information of nanomaterials prepared by ILs is given in literature sources [166–175].

References

[1] R.R. Xu, W.Q. Pang, J.H. Yu, Q.S. Huo, J.S. Chen, Chemistry of Zeolites and Related Porous Materials, John Wiley, 2007.
[2] R.M. Barrer, Hydrothermal Chemistry of Zeolites, Academic Press, New York, 1982.
[3] A. Rabenau, Angew. Chem. Int. Ed. Engl. 24 (1985) 1026.
[4] E.U. Franck, Pure Appl. Chem. 24 (1970) 13.
[5] E.U. Frank, Int. Corros. Conf. Ser. (1973) 109.
[6] T.M. Seward, Phys. Chem. Earth 13/14 (1981) 113.
[7] E.U. Franck, Phys. Chem. Earth 13/14 (1981) 65.
[8] H.P. Eugster, Phys. Chem. Earth 13/14 (1981) 461.
[9] A. Rabenau, H. Rau, Philips Tech. Rundsch. 30 (1969/70) 53
[10] K. Byrappa, T. Ohachi, Crystal Growth Technology, William Andrew, New York, 2003.
[11] K. Byrappa, M. Yoshimura, Handbook of Hydrothermal Technology: a Technology for Crystal Growth and Materials Processing, William Andrew, New York, 2001.
[12] R.A. Laudise, in: R.S. Feigelson (Ed.), 50 Years Progress in Crystal Growth: a Reprint Collection, Elsevier, Amsterdam, 2004, p. 185.
[13] E. Ohshima, H. Ogino, I. Niikura, K. Maeda, M. Sato, M. Ito, et al., J. Cryst. Growth 260 (2004) 166.
[14] K. Maeda, M. Sato, I. Niikura, T. Fukuda, Semicond. Sci. Technol. 20 (2005) S49.
[15] D. Ehrentraut, H. Sato, Y. Kagamitani, H. Sato, A. Yoshikawa, T. Fukuda, Prog. Cryst. Growth Charact. Mater. 52 (2006) 280.
[16] L.N. Dem'yanets, V.I. Lyntin, J. Cryst. Growth 310 (2008) 993.
[17] C.L. Zhang, W.N. Zhou, Y. Hang, Z. Lü, H.D. Hou, Y.B. Zuo, et al., J. Cryst. Growth 310 (2008) 1819.
[18] W.W. Lin, D.G. Chen, J.Y. Zhang, Z. Lin, J.K. Huang, W. Li, et al., Cryst. Growth Des. 9 (2009) 4378.
[19] C.D. McMillen, H.G. Giesber, J.W. Kolis, J. Cryst. Growth 310 (2008) 299.
[20] C.L. Zhang, L.X. Huang, W.N. Zhou, G. Zhang, H.D. Hou, Q.F. Ruan, et al., J. Cryst. Growth 292 (2006) 364.
[21] C.D. McMillen, J.W. Kolis, J. Cryst. Growth 310 (2008) 2033.
[22] A.R. Forbes, C.D. McMillen, H.G. Giesber, J.W. Kolis, J. Cryst. Growth 310 (2008) 4472.
[23] A.K. Cheetham, G. Férey, T. Loiseau, Angew. Chem. Int. Ed. 38 (1999) 3268.
[24] S.T. Wilson, B.M. Lok, R.M. Flanigen, US Patent 4,310,440 (1982).
[25] M.E. Davis, C. Saldarriaga, C. Montes, J.M. Garces, C. Crowder, Nature 331 (1988) 698.
[26] Q.S. Huo, R.R. xu, S.G. Li, Z.G. Ma, J.M. Thomas, R.H. Jones, et al., J. Chem. Soc. Chem. Commun. (1992) 875.
[27] L.Q. Tang, M.S. Dadachov, X.D. Zou, Chem. Mater. 17 (2005) 2530.
[28] Y.M. Zhou, H.G. Zhu, Z.X. Chen, M.Q. Chen, Y. Xu, H.Y. Zhang, et al., Angew. Chem. Int. Ed. 40 (2001) 2166.
[29] Z.E. Lin, J. Zhang, J.T. Zhao, S.T. Zheng, C.Y. Pan, G.M. Wang, et al., Angew. Chem. Int. Ed. 44 (2005) 6881.
[30] J. Plévert, T.M. Gentz, A. Laine, H.L. Li, V.G. Young, O.M. Yaghi, et al., J. Am. Chem. Soc. 123 (2001) 12706.
[31] N. Guillou, Q.M. Gao, M. Noguès, R.E. Morris, M. Hervieu, G. Férey, et al., C.R. Acad. Sci. Paris Ser. IIc 2 (1999) 387.
[32] N. Guillou, Q.M. Gao, P.M. Forster, J.S. Chang, M. Noguès, S.E. Park, et al., Angew. Chem. Int. Ed. 40 (2001) 2831.
[33] J. Liang, J.Y. Li, J.H. Yu, P. Chen, Q.R. Fang, F.X. Sun, et al., Angew. Chem. Int. Ed. 45 (2006) 2546.
[34] M.I. Khan, L.M. Meyer, R.C. Haushalter, A.L. Schweitzer, J. Zubieta, J.L. Dye, Chem. Mater. 8 (1996) 43.
[35] G.Y. Yang, S.C. Sevov, J. Am. Chem. Soc. 121 (1999) 8389.
[36] J. Zhu, X.H. Bu, P.Y. Feng, G.D. Stucky, J. Am. Chem. Soc. 122 (2000) 11563.
[37] C.H. Lin, S.L. Wang, K.H. Lii, J. Am. Chem. Soc. 123 (2001) 4649.
[38] Y.L. Lai, K.H. Lii, S.L. Wang, J. Am. Chem. Soc. 129 (2007) 5350.
[39] K.E. Christensen, C. Bonneau, M. Gustafsson, L. Shi, J.L. Sun, J. Grins, et al., J. Am. Chem. Soc. 130 (2008) 3758.
[40] X.Y. Ren, Y. Li, Q.H. Pan, J.H. Yu, R.R. Xu, Y. Xu, J. Am. Chem. Soc. 131 (2009) 14128.
[41] X.D. Zou, T. Conradsson, M. Klingstedt, M.S. Dadachov, M. O'Keeffe, Nature 437 (2005) 716.
[42] J.H. Yu, R.R. Xu, Acc. Chem. Res. 36 (2003) 481.
[43] J.H. Yu, R.R. Xu, Chem. Soc. Rev. 35 (2006) 593.
[44] G. Férey, Chem. Mater. 13 (2001) 3084.
[45] Z. Shi, L.R. Zhang, G.S. Zhu, G.Y. Yang, J. Hua, H. Ding, et al., Chem. Mater. 11 (1999) 3565.
[46] Z. Shi, S.H. Feng, L.R. Zhang, G.Y. Yang, J. Hua, Chem. Mater. 12 (2000) 2930.
[47] G. Férey, C. Mellot-Draznieks, C. Serre, F. Millange, J. Dutour, S. Surblé, et al., Science 309 (2005) 2040.
[48] R. Banerjee, A. Phan, B. Wang, C. Knobler, H. Furukawa, M. O'Keeffe, et al., Science 319 (2008) 939.
[49] T. Devic, O. David, M. Walls, J. Marrot, F. Couty, G. Férey, J. Am. Chem. Soc. 129 (2007) 12614.
[50] Y. Goto, H. Sato, S. Shinkai, K. Sada, J. Am. Chem. Soc. 130 (2008) 14354.
[51] T. Gadzikwa, O.K. Farha, C.D. Malliakas, M.G. Kanatzidis, J.T. Hupp, S.T. Nguyen, J. Am. Chem. Soc. 131 (2007) 13613.
[52] J.Y. Lu, B.R. Cabrera, R.J. Wang, J. Li, Inorg. Chem. 37 (1998) 4480.

[53] M.H. Bi, G.H. Li, J. Hua, Y.J. Lin, J.J. Cao, Z. Shi, et al., Eur. J. Inorg. Chem. (2008) 1035.
[54] X.M. Chen, M.L. Tong, Acc. Chem. Res. 40 (2007) 162.
[55] S.H. Feng, R.R. Xu, Acc. Chem. Res. 34 (2001) 239.
[56] H. Zhao, S.H. Feng, Chem. Mater. 11 (1999) 958.
[57] G.S. Li, Y.C. Mao, L.P. Li, S.H. Feng, M.Q. Wang, X. Yao, Chem. Mater. 11 (1999) 1259.
[58] S.H. Feng, M. Greenblatt, Chem. Mater. 5 (1993) 1277.
[59] G.S. Pang, S.H. Feng, Y.C. Tang, C.H. Tan, R.R. Xu, Chem. Mater. 10 (1998) 2446.
[60] Y.L. An, S.H. Feng, Y.H. Xu, R.R. Xu, Chem. Mater. 8 (1996) 356.
[61] S.H. Feng, M. Tsai, M. Greenblatt, Chem. Mater 4 (1992) 388.
[62] S.H. Feng, M. Greenblatt, Chem. Mater. 4 (1992) 462.
[63] S.H. Feng, M. Tsai, S.P. Szu, M. Greenblatt, Chem. Mater. 4 (1992) 468.
[64] S.H. Feng, H.M. Yuan, Z. Shi, Y. Chen, Y.W. Wang, K.K. Huang, et al., J. Mater. Sci. 43 (2008) 2131.
[65] C.Y. Zhao, S.H. Feng, Z.C. Chao, R.R. Xu, C.S. Shi, J.Z. Ni, Chem. Commun. (1996) 1641.
[66] C.Y. Zhao, S.H. Feng, R.R. Xu, C.S. Shi, J.Z. Ni, Chem. Commun. (1997) 945.
[67] X.M. Xun, S.H. Feng, J.Z. Wang, R.R. Xu, Chem. Mater. 9 (1997) 2966.
[68] X.M. Xun, S.H. Feng, R.R. Xu, Mater. Res. Bull. 33 (1998) 369.
[69] K. Byrappa, T. Adschiri, Prog. Cryst. Growth Charact. Mater. 53 (2007) 117.
[70] P.P. Wang, B. Bai, S. Hu, J. Zhuang, X. Wang, J. Am, Chem. Soc. 131 (2009) 16953.
[71] H.G. Yang, G. Liu, S.Z. Qiao, C.H. Sun, Y.G. Jin, S.C. Smith, et al., J. Am. Chem. Soc. 131 (2009) 4078.
[72] B. Liu, E.S. Aydil, J. Am. Chem. Soc. 131 (2009) 3985.
[73] C.J. Jia, L.D. Sun, F. Luo, X.D. Han, L.J. Heyderman, Z.G. Yan, et al., J. Am. Chem. Soc. 130 (2008) 16968.
[74] M.J. Hu, Y. Lu, S. Zhang, S.R. Guo, B. Lin, M. Zhang, et al., J. Am. Chem. Soc. 130 (2008) 11606.
[75] I. Djerdj, D. Sheptyakov, F. Gozzo, D. Arčon, R. Nesper, M. Niederberger, J. Am. Chem. Soc. 130 (2008) 11364.
[76] S. Hu, X. Wang, J. Am. Chem. Soc. 130 (2008) 8126.
[77] X.G. Han, M.S. Jin, S.F. Xie, Q. Kuang, Z.Y. Jiang, Y.Q. Jiang, et al., Angew. Chem. Int. Ed. 48 (2009) 9180.
[78] A.I. Oparin, The Origin of Life, Moscow Worker Publisher, Moscow, 1924; English translation: A.I. Oparin, The Origin and Development of Life (NASA TTF-488), D.C.L GPO., Washington, 1968.
[79] S.L. Miller, Science 117 (1953) 528.
[80] J.M. Edmond, K.L.V. Damm, R.E. McDuff, C.I. Measures, Nature 297 (1982) 187.
[81] J.B. Corliss, J.A. Baross, S.E. Hoffman, Oceanol. Acta 4 (1981) 59.
[82] J. Maddox, Nature 367 (1994) 409.
[83] J.F. Kasting, Science 259 (1993) 920.
[84] N.R. Pace, Cell 65 (1991) 531.
[85] J.B. Corliss, Nature 347 (1990) 624.
[86] J. Horita, M.E. Berndt, Science 285 (1999) 1055.
[87] C. Huber, G. Wächtershäuster, Science 276 (1997) 245.
[88] R.H. Crabtree, Science 276 (1997) 222.
[89] M.P. Robertson, S.L. Miller, Nature 375 (1995) 772.
[90] H. Huber, G. Wächtershäuster, Science 281 (1998) 670.
[91] E. Imai, H. Honda, K. Hatori, A. Brack, K. Matsuno, Science 283 (1999) 831.
[92] S.L. Miller, J.L. Baca, Nature 334 (1988) 609.
[93] G.E. Fox, E. Stackebrandt, R.B. Hespell, J. Gibson, J. Maniloff, T.A. Dyer, R.S. Wolfe, et al., Science 209 (1980) 457.
[94] M.L. Sogin, G. Hinkle, D. Lelpe, Nature 362 (1993) 795.
[95] N.R. Pace, Science 276 (1997) 734.
[96] C.R. Woese, Proc. Natl. Acad. Sci. U.S.A. 97 (2000) 8392.
[97] K. Maher, D.J. Stevenson, Nature 331 (1988) 612.
[98] R.A. Lutz, T.M. Shank, D.J. Fonari, R.M. Haymon, M.D. Lilley, K.L. Von Damm, et al., Nature 371 (1994) 663.
[99] H.W. Jannasch, M.J. Mottl, Science 229 (1985) 717.
[100] E. Drobner, H. Huber, G. Wächtershäuster, D. Rose, K.O. Stetter, Nature 346 (1990) 742.
[101] J.A. Brandes, N.Z. Boctor, G.D. Cody, B.A. Cooper, R.M. Hazen, H.S. Yoder, Nature 395 (1998) 365.
[102] E. Blöchl, M. Keller, G. Wächtershäuster, K.O. Stetter, Proc. Natl. Acad. Sci. U.S.A. 89 (1992) 8117.
[103] Q.W. Chen, D.W. Bahnemann, J. Am. Chem. Soc. 122 (2000) 970.
[104] G. Tian, H.M. Yuan, Y. Mu, C. He, S.H. Feng, Org. Lett. 9 (2007) 2019.
[105] S.H. Feng, G. Tian, C. He, H.M. Yuan, Y. Mu, Y.W. Wang, et al., J. Mater. Sci. 43 (2008) 2418.
[106] H. Weingärtner, E.U. Franck, Angew. Chem. Int. Ed. 44 (2005) 2672.
[107] N. Akiya, P.E. Savage, Chem. Rev. 102 (2002) 2725.
[108] G. Brunner, J. Supercrit, Fluids 47 (2009) 373.
[109] G. Brunner, J. Supercrit, Fluids 47 (2009) 382.
[110] P.E. Savage, J. Supercrit, Fluids 47 (2009) 407.
[111] M. Takesue, A. Suino, Y. Hakuta, H. Hayashi, R.L. Smith, J. Solid State Chem. 181 (2008) 1307.
[112] J. Zhang, S. Ohara, M. Umetsu, T. Naka, Y. Hatakeyama, T. Adschiri, Adv. Mater. 19 (2007) 203.
[113] D. Rangappa, T. Naka, A. Kondo, M. Ishii, T. Kobayashi, T. Adschiri, J. Am. Chem. Soc. 129 (2007) 11061.
[114] S. Takami, S. Ohara, T. Adschiri, Y. Wakayama, T. Chikyow, Dalton Trans. (2008) 5442.
[115] P.E. Savage, Chem. Rev. 99 (1999) 603.
[116] S. Koda, J. Supercrit, Fluids 47 (2009) 400.
[117] G.W. Morey, P. Niggli, J. Am. Chem. Soc. 35 (1913) 1086.
[118] O.F. Tuttle, Am. J. Sci. 246 (1948) 628.
[119] H.L. Barnes, Econ. Geol. 58 (1963) 1054.
[120] F.H. Smyth, L.H. Adams, J. Am. Chem. Soc. 45 (1923) 1167.
[121] R.W. Goranson, Am. J. Sci. 22 (1931) 481.
[122] H.S. Yoder, Trans. Am. Geophys. Union 31 (1950) 827.
[123] C.W. Burnham, J.R. Holloway, N.F. Davis, Am. J. Sci. 267 (1969) 70.
[124] www.parrinst.com
[125] A.M. Beale, A.M.J. van der Eerden, D. Grandjean, A.V. Petukhov, A.D. Smith, B.M. Weckhuysen, Chem. Commun. (2006) 4410.
[126] F.T. Fan, Z.C. Feng, G.N. Li, K.J. Sun, P.L. Ying, C. Li, Chem. Eur. J. 14 (2008) 5125.
[127] M. Bremholm, M. Felicissimo, B.B. Iversen, Angew. Chem. Int. Ed. 48 (2009) 4788.
[128] F.T. Fan, Z.C. Feng, K.J. Sun, M.L. Guo, Q. Guo, Y. Song, et al., Angew. Chem. Int. Ed. 48 (2009) 8743.
[129] C. Kongmark, V. Martis, A. Rubbens, C. Pirovano, A. Löfberg, G. Sankar, et al., Chem. Commun. (2009) 4850.
[130] K. Simmance, G. Sankar, R.G. Bell, C. Prestipino, W. van Beek, Phys. Chem. Chem. Phys. 12 (2010) 559.
[131] E.R. Cooer, C.D. Andrews, P.S. Wheatley, P.B. Webb, P. Wormald, R.E. Morris, Nature 430 (2004) 1012.
[132] E.R. Parnham, R.E. Morris, Acc. Chem. Res. 40 (2007) 1005.
[133] R.E. Morris, Chem. Commun. (2009) 2990.
[134] L. Liu, D.S. Wragg, H.Y. Zhang, Y. Kong, P.J. Byrne, T.J. Prior, et al., Dalton Trans. (2009) 6715.
[135] P.J. Byrne, D.S. Wragg, J.E. Warren, R.E. Morris, Dalton Trans. (2009) 795.
[136] P.A. Wright, R.E. Morris, P.S. Wheatley, Dalton Trans. (2007) 5359.

[137] E.A. Drylie, D.S. Wragg, E.R. Parnham, P.S. Wheatley, A.M.Z. Slawin, J.E. Warren, et al., Angew. Chem. Int. Ed. 46 (2007) 7839.
[138] E.R. Parnham, R.E. Morris, Chem. Mater. 18 (2006) 4882.
[139] E.R. Parnham, R.E. Morris, J. Am. Chem. Soc. 128 (2006) 2204.
[140] E.R. Parnham, P.S. Wheatley, R.E. Morris, Chem. Commun. (2006) 380.
[141] Z.J. Lin, D.S. Wragg, P. Lightfoot, R.E. Morris, Dalton Trans. (2009) 5287.
[142] L. Liu, X.P. Li, H. Xu, J.P. Li, Z. Lin, J.X. Dong, Dalton Trans. (2009) 10418.
[143] E.P. Ng, S.S. Sekhon, S. Mintova, Chem. Commun. (2009) 1661.
[144] L. Wang, Y.P. Xu, B.C. Wang, S.J. Wang, J.Y. Yu, Z.J. Tian, et al., Chem. Eur. J. 14 (2008) 10551.
[145] H.Z. Xing, J.Y. Li, W.F. Yan, P. Chen, Z. Jin, J.H. Yu, et al., Chem. Mater. 20 (2008) 4179.
[146] R.S. Xu, W.P. Zhang, J. Guan, Y.P. Xu, L. Wang, H.J. Ma, et al., Chem. Eur. J. 15 (2009) 5348.
[147] D.S. Wragg, A.M.Z. Slawin, R.E. Morris, Solid State Sci. 11 (2009) 411.
[148] H.J. Ma, Z.J. Tian, R.S. Xu, B.C. Wang, Y. Wei, L. Wang, et al., J. Am. Chem. Soc. 130 (2008) 8120.
[149] Y.P. Xu, Z.J. Tian, S.J. Wang, Y. Hu, L. Wang, B.C. Wang, et al., Angew. Chem. Int. Ed. 45 (2006) 3965.
[150] R. Cai, M.W. Sun, Z.W. Chen, R. Munoz, C. O'Neill, D.E. Beving, et al., Angew. Chem. Int. Ed. 47 (2008) 525.
[151] Z.J. Lin, D.S. Wragg, R.E. Morris, Chem. Commun. (2006) 2021.
[152] Z.J. Lin, A.M.Z. Slawin, R.E. Morris, J. Am. Chem. Soc. 129 (2007) 4880.
[153] Z.J. Lin, D.S. Wragg, J.E. Warren, R.E. Morris, J. Am. Chem. Soc. 129 (2007) 10334.
[154] J. Zhang, T. Wu, S.M. Chen, P.Y. Feng, X.H. Bu, Angew. Chem. Int. Ed. 48 (2009) 3486.
[155] J. Zhang, S.M. Chen, X.H. Bu, Angew. Chem. Int. Ed. 47 (2008) 5434.
[156] L. Xu, S.H. Yan, E.Y. Choi, J.Y. Lee, Y.U. Kwon, Chem. Commun. (2009) 3431.
[157] S.M. Chen, J. Zhang, X.H. Bu, Inorg. Chem. 47 (2008) 5567.
[158] W.X. Chen, Y.P. Ren, L.S. Long, R.B. Huang, L.S. Zheng, CrystEngComm 11 (2009) 1522.
[159] T. Hogben, R.E. Douthwaite, L.J. Gillie, A.C. Whitwood, CrystEngComm 8 (2006) 866.
[160] X.K. Guo, Q.L. Ma, X.F. Guo, W.P. Ding, Y. Chen, Chem. Commun. (2009) 3443.
[161] X.M. Liu, J.W. Zhao, Y.J. Sun, K. Song, Y. Yu, C.A. Du, et al., Chem. Commun. (2009) 6628.
[162] H.S. Park, Y.S. Choi, Y.M. Jung, W.H. Hong, J. Am. Chem. Soc. 130 (2008) 845.
[163] A. Taubert, Z. Li, Dalton Trans. (2007) 723.
[164] H. Park, Y.S. Choi, Y. Kim, W.H. Hong, H. Song, Adv. Funct. Mater. 17 (2007) 2411.
[165] H.G. Zhu, J.F. Huang, Z.W. Pan, S. Dai, Chem. Mater 18 (2006) 4473.
[166] L.L. Li, W.M. Zhang, Q. Yuan, Z.X. Li, C.J. Fang, L.D. Sun, et al., Cryst. Growth Des. 8 (2008) 4165.
[167] Z.H. Li, P. Rabu, P. Strauch, A. Mantion, A. Taubert, Chem. Eur. J. 14 (2008) 8409.
[168] Y. Qin, Y. Song, N.J. Sun, N. Zhao, M.X. Li, L.M. Qi, Chem. Mater. 20 (2008) 3965.
[169] A. Taubert, A. Uhlmann, A. Hedderich, K. Kirchhoff, Cryst. Growth Des. 8 (2008) 1640.
[170] P. Kuhn, A. Forget, J. Hartmann, A. Thomas, M. Antonietti, Adv. Mater. 21 (2009) 897.
[171] J.S. Lee, X.Q. Wang, H.M. Luo, G.A. Baker, S. Dai, J. Am. Chem. Soc. 131 (2009) 4596.
[172] Z.H. Li, Y.X. Luan, Q.Z. Wang, G.S. Zhuang, Y.X. Qi, Y. Wang, et al., Chem. Commun. (2009) 6273.
[173] X.M. Liu, J.W. Zhao, Y.J. Sun, K. Song, Y. Yu, C.A. Du, X.G. Kong, H. Zhang, Chem. Commun. (2009) 6628.
[174] E.P. Ng, S.S. Sekhon, S. Mintova, Chem. Commun. (2009) 1661.
[175] H. Park, Y.C. Lee, B.G. Choi, Y.S. Choi, J.W. Yang, W.H. Hong, Chem. Commun. (2009) 4058.

CHAPTER

5

High Pressure Synthesis and Preparation of Inorganic Materials

Xiaoyang Liu
Jilin University, China

Ultrahigh pressure inorganic synthesis is a very effective method for synthesizing new inorganic compounds which cannot be obtained under normal conditions. Conditions for ultrahigh pressure synthesis is that the pressure should be greater than 1.01×10^9 Pa at temperature ranging from ambient to 3000°C. The inorganic compounds synthesized under these high pressure conditions can have special valences, unique structures, and unprecedented physical and chemical properties [1].

As compared to the systematic study of chemistry in the past two centuries, research can now be done at atomic level with assistance from the latest development in diffraction and spectroscopic techniques. However, most studied compounds in the laboratory were obtained under ambient pressure (1.01×10^5 Pa) or pressure very close to 1.01×10^5 Pa [2]. In the infinite universe, at least 90% substances exist under ultrahigh pressure more than 1.01×10^{10} Pa and this means that only a small amount of substances in the universe have been studied by scientists [3]. In order to expand chemical research, pressure is a very important normal thermodynamic variant to explore. Due to the complicacy, difficulty in data acquisition, and great expense of high-pressure experiments, the research in high pressure science has improved very slowly and laboriously for a long time. Recently, due to the rapid development of ultrahigh pressure technique, it is possible that its application in basic research allows for more and more discovery in the scientific and technical research. In fact, ultrahigh pressure research was listed by *Science* magazine as one of the five most important tools in scientific and technique discovery, highlighting its growing importance as an experimental technique in modern science [4].

Ultrahigh pressure is a very important research method for physics, chemistry, and material synthesis because it can efficiently change the atomic distances and electron states of the substances. It can therefore be used to tune atomic distances, to probe for information and to aid in other special areas of the most frontier research; even the periodic table of elements is considered to be rewritten under ultrahigh pressure conditions. Pressure, as a thermodynamic parameter, has extreme influence on substances. Pressure is one part of the Gibbs energy equation, which along with temperature, controls chemical reaction and phase transition. Ultrahigh pressure develops another dimension of reaction parameter and increases the possibility of the chemical reaction. With the advances of high pressure technique, ultrahigh pressure is not only an experimental method or extreme condition, but also the third physical parameter along with temperature and chemical stoichiometry. When ultrahigh pressure technique is applied to inorganic synthesis, there are some amazing physical changes (electrical conductivity, optical absorption, and magnetic properties in addition to the densification of the solid). The atoms in the inorganic compounds synthesized under high pressure have more of first or the second coordinates and unusual oxidation state. In modern inorganic synthesis, ultrahigh pressure is widely applied for synthesis that cannot be carried out under normal conditions and many new compounds, new phases, new physical states, and new synthetic routes are obtained. The obtained results show that there are at least five new phases of the substances that are stable under normal pressure at 1.01×10^{11} Pa. That is to say, new substance whose amount is many times that of current substance will be obtained if a "pressure" parameter is applied.

Modern Inorganic Synthetic Chemistry, DOI: 10.1016/B978-0-444-53599-3.10005-8

High pressure synthesis means that external high pressure is applied to make the substances transform or react with each other to form new phases, new compounds, or new materials. The structure and property changes under high pressure are normally reversible and revert back when high pressure is unloaded. Therefore, high pressure and high temperature are used together in high pressure synthesis, which is called high pressure–high temperature synthesis and the special structure and properties found under high pressure and high temperature can remain after the pressure is unloaded and the temperature is decreased [1].

P.W. Bridgman developed high pressure technique over his lifetime and opened a new area of phase transition and physical properties under high pressure. After he won the Nobel Prize in 1946, high pressure synthesis of new substances and new materials began to receive more attention. After F.P. Bundy synthesized the first artificial diamond under high pressure conditions [5], synthesis of new substance under high pressures became a hot area of research. The next important material synthesized under high pressure conditions was cubic boron nitride (c-BN) [6], which cannot be found in nature, but it is isoelectronic to diamond and its hardness is next to that of diamond. Since then, the research on high pressure synthesis received more and more attention.

5.1. EXPERIMENTAL METHODS OF INORGANIC SYNTHESIS UNDER HIGH PRESSURE

Since the invention of the high pressure device by Bridgeman in the 1940s [7], a lot of developments have been made in high pressure technology such as pressure and temperature measurement technology and improvements in the hydrostatic pressure. In recent years, there has been rapid advancement in both instrumentation and techniques, e.g., the application of "Walker type" large volume apparatus (LVP) which has greatly improved experimental pressure and temperature. Therefore, high pressure technology has been extensively used in inorganic synthesis and preparation.

5.1.1. High Pressure Apparatus

In the field of high pressure research, the pressure scale has a range of 0.1–500 GPa or greater. Generating the wide range of high pressure conditions for materials exploration and synthesis requires a selection of dynamic and static compression techniques, which is distinguished from the time duration of pressing. Dynamic extreme high pressure can be generated through a high speed object bump on the sample or the dynamic shock wave of an explosion to compress the sample; therefore, it can also be called shock wave high pressure and can generate a maximum pressure of 10^{12} Pa. But the equipment of shock wave is too expensive and the press time duration is in the microsecond range, so it is of restricted use in inorganic chemistry research. There are many physical phenomena connected with the generation of press, which can also apply in high pressure techniques, such as thermal expansion and phase transition. Actually, the most effective method is to press the object through various techniques. The volume of object is reduced when bearing the press and high pressure is produced inside the object as long as the compression is large enough. Using this kind of technique, we can keep the press maintained for a long time, so it is called static high pressure technique.

In order to compress the sample, two items are necessary: (i) one or more removable assembly to load the press (called pressure head or anvil) and (ii) a compression resistance chamber (called pressure chamber) in which the sample is placed. The spacing between anvil and chamber is prerequisite for the moveable anvil. The sample will flow into the spacing when it is compressed, so the spacing must be sealed to obtain high pressures. A pressure-transmitting media is also needed between the anvils and chamber to obtain hydrostatic pressure. The pressure-transmitting media has a significant effect on the pressure conditions within the sample chamber; and the ideal pressure-transmitting media should remain fluid at high pressure.

Among the high-pressure apparatuses, diamond anvil cells (DACs) are currently the most widely used. The DAC gives laboratory access to very high static pressures in a relatively easy and safe manner, allowing experimental measurements at pressure above 100 GPa. In a DAC the sample is placed between the flat faces (culets) of two brilliant-cut diamonds. The sample placed into DAC pressure chamber is no more than 100 μm in diameter. This sample size is in contrast with other types of high pressure apparatus which use larger sized sample (no less than 2 mm), called large volume apparatus (LVP).

The simplest high pressure device is piston-cylinder apparatus, as shown in Fig. 5.1 [8]. As a type of LVP, the removable piston compresses the pressure media placed in the cylinder to obtain high pressure. Simultaneously, the use of a webbing cylinder, compounding cylinder, and self-tight cylinder technique can greatly improve the pressure limit of this type in high pressure apparatus. At present, the maximum pressure and temperature obtained for these kinds of devices are 4 GPa and 1700°C, respectively.

Bridgman invented the high pressure anvil in 1940s. Figure 5.2 shows the schematic sketch of typical Bridgman anvil [8]. Two opposite flat anvils press the

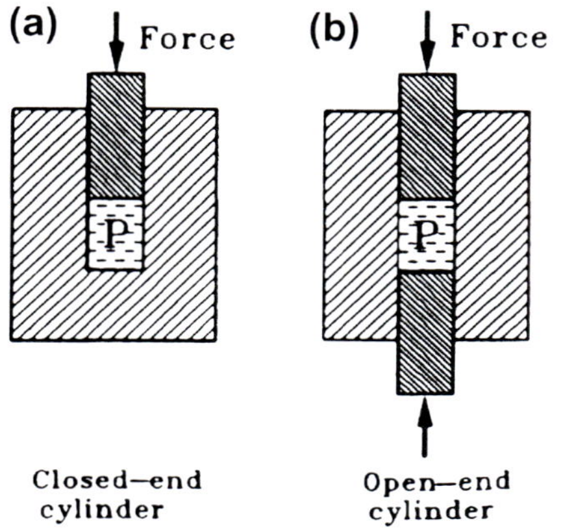

FIGURE 5.1 Piston-cylinder cell (schematic). *With permission from Oxford University Press.*

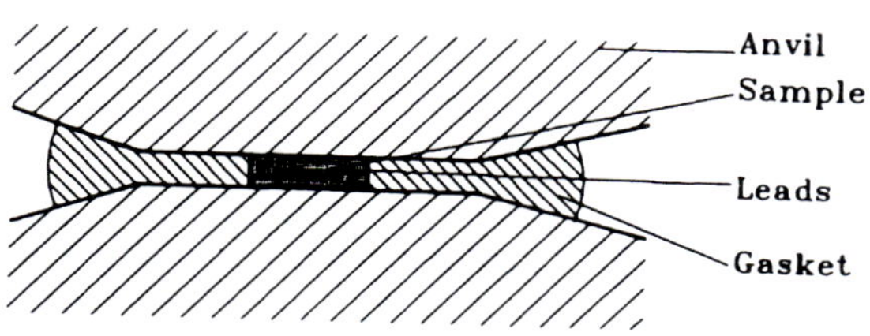

FIGURE 5.2 Schematic diagram of a Bridgman anvil. *With permission from Oxford University Press.*

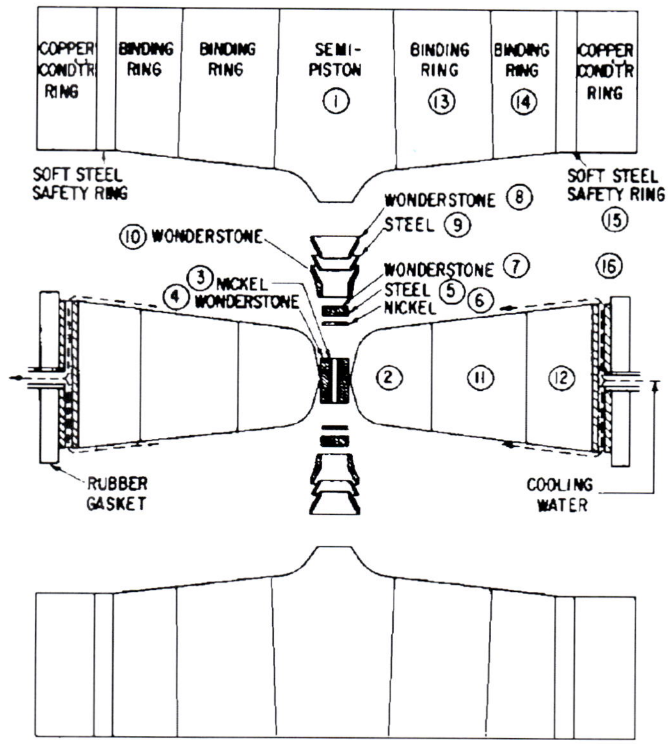

FIGURE 5.3 The "belt": high-temperature, high-pressure apparatus. "Exploded" assembly. ② Carboloy chamber, ⑪⑫ Hardened steel binding ring ⑯ soft steel safety ring.

compressible gasket in which the sample is placed. Pressures up to 10 GPa can be achieved in the Bridgman anvil apparatus. The electrical properties of samples at high pressure are extensively studied using this kind of device.

The above two types of LVP designs have their disadvantages when they are used in laboratories. For example, the piston-cylinder device can press large samples but cannot generate very high pressure, while Bridgman anvil device can produce high pressure but with smaller sample size. Alternatively, Hall has invented the Belt type device in 1960s to solve these problems. The schematic sketch of Belt high pressure device is shown in Fig. 5.3 [9]. High pressure (up to 10 GPa) and high temperature (up to 2000°C) can be generated with the Belt designs. Actually, the Belt device is an improvement on piston-cylinder designs with two "frustum pyramid" tungsten carbide (WC) anvils and a large sample chamber with homologous cone angle. Because the anvils are supported by toroidal side and the chamber is protected by multilayer cylinder, the Belt device can generate high pressure.

Beyond the limit of piston-cylinder (maximum 4 GPa) and Belt apparatus (maximum 10 GPa), multianvil devices can be used for reaching pressure exceeding 25 GPa. The pressure attained depends on the geometrical design of the devices, the frictional strength of the gasket, and the compression strength of the anvil. Developments and improvements in this area led to several designs of multi-anvil devices. Huppertz et al. developed the three-dimensional tetrahedral anvil devices in 1960s, which served as a basis for the hexahedral and octahedral anvils [10]. These devices are all called multi-anvil apparatus. Figure 5.4 shows the schematic sketch of tetrahedral and hexahedral anvil devices [11]. In a multi-anvil system, a sample volume is compressed between variable numbers of identical shaped anvils in two stages, which move each other forward. The multi-anvil apparatus simply consists of a hydraulic press. The applied uniaxial load can vary and in the standard experiments load can vary between 600 and 1000 tons.

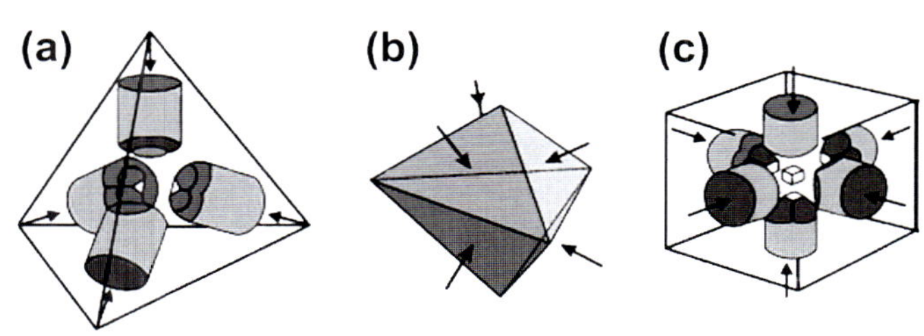

FIGURE 5.4 Tetrahedral (a) trigonal–bipyramidal (b) and cubic (c) multianvil geometries. *Reprinted with permission from [11]. Copyright 1969, American Institute of Physics.*

FIGURE 5.5 Inside the DIA apparatus, uniaxial forces on steel guide blocks apply horizontal forces on steel bolsters and attached anvils onto cell assembly. *Images reproduced with permission from Mineralogical Society of America.*

Hydraulic machine is commonly used in multi-anvil module to produce pressure by equally applying load in all directions at the same time. The cubic module also uses DIA set (DIA is the abbreviation for Diamond, as the earliest diamond was synthesized in this apparatus in 1953) where six anvils are driven into the sample via uniaxial load to produce the pressure [10], as shown in Fig. 5.5. In this module, six sets of anvil and wedge drivers are placed in the octahedron cavity, driven by both of the module plates, i.e., the top and bottom module plates. The 45° angle of the wedge face ensures that the same load is applied and moving distance of six anvils gives a uniform pressure in the module.

In 1990, Walker et al. developed a similar module design called the Walker module high-pressure apparatus [12], which attained the high pressure in all of the large-volume press under the basic principle of two-stage press, as shown in Fig. 5.6 [10]. Eight truncated cubes, usually made from tungsten carbide (WC), form a large cube with an octahedral hole, where the pressure-transmitting media is placed. The large cube is placed

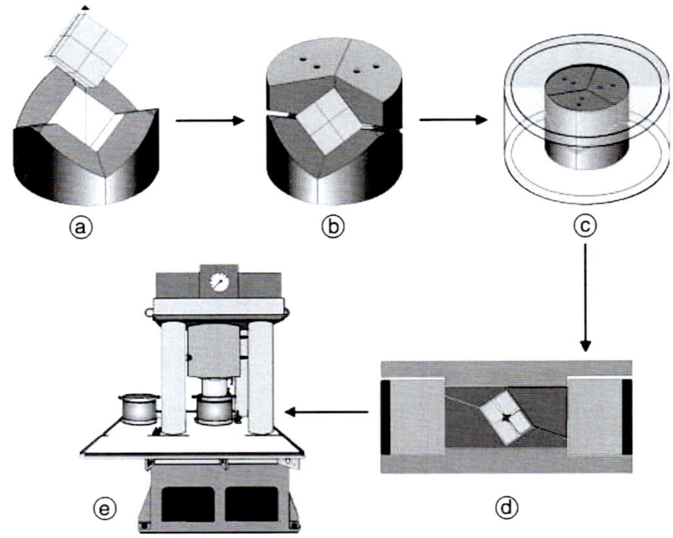

FIGURE 5.6 Schematic drawing of the assemblage of the Walker-type module.

in a cubic gap formed by six wedged block made of tool steel, as shown in Figure 5.6 (a and b). The cube and wedged blocks are then placed into a steel loop, as shown in Figure 5.6 (c). Finally, the formed piston-cylinder apparatus is placed into the hydraulic machine to be pressed, shown in Figure 5.6 (d). The pressure is first applied to the wedge, then to the cube. Pressure of 30 GPa and temperature of 2300°C can be attained if the cubes are made from tungsten carbide. Pressure of approximately 90 GPa may be attained by using the anvils coated by diamond. By using the cube with a hole and the appropriate pressure-transmitting media, such as Boron, etc., the apparatus, combined with the synchrotron radiation technology, can be used for in situ high-pressure research on the structure and physical property of the materials.

The anvils in the high-pressure apparatus are usually made from materials with high hardness such as high-strength steel or tungsten carbide. Because diamond is the hardest material in nature, the apparatus using diamond as anvils is called Diamond Anvil Cell (DAC) and it will produce higher pressure. As natural diamond is tiny, the module with the diamond anvils is a micro high-pressure apparatus. In 1959, the pressure of 10 GPa was achieved by Wier et al. by means of the diamond anvils [13]. Due to the development of the mechanical processing technology, the DAC could reach pressure as high as several million bars, by using the gasket-packaged pressure media, Bevelled diamond and the perfect center-aligning technique [14]. A typical DAC apparatus is the Bridgman apparatus with diamond as the anvils, which is shown in Fig. 5.7 [15]. The gasket with indentation and punch, which is usually made from stainless steel, is placed between

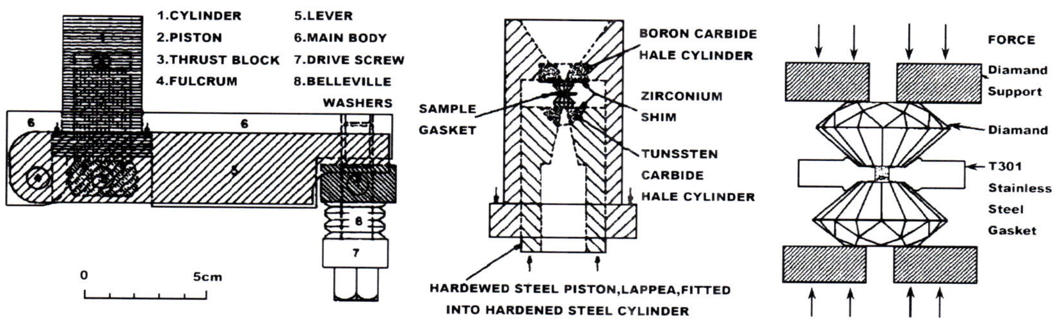

FIGURE 5.7 Schematic drawing for Mao-Bell type of Diamond anvil cell. *Reprinted with permission from A. Jayaraman, J. Rev. Sci. Instrum, 57, 1013. Copyright 1986, American Institute of Physics.*

the two diamond anvils. The media and sample are placed in a hole with the diameter of 100 microns. High pressure is obtained by applying load on the diamond. The classical Mao-Bell device, which is shown in Fig. 5.7, achieves its high pressure by following the lever principle and uses the piston-cylinder to center it, while others use the slider. As diamond is transparent, heat conductive, and insulated, the DAC is usually combined with synchrotron radiation X-ray diffraction and Raman diffraction, etc., to do research on phase transition, high pressure synthesis, and physical properties in situ under high pressure.

5.1.2. Choice of Pressure Transmitting Media

When the sample is pressurized, the high-pressure is directly loaded on the pressure transmitting media, instead of the sample. Without doubt, liquid or gas media are ideal pressure media due to the homogeneous pressure distributions around the sample chamber. However, for most of the pressure generating ways except for DAC, gas or liquid are very difficult to be encapsulated, so the most popular choice of the pressure media is solid. Some requirements for the properties of solid media are electrical insulation, thermal insulation, softness (with the tiny shear strength), and chemical stability. The gaskets squeezed in the multianvil apparatus should not only seal the sample chamber but also support the anvil faces, the solid media requires additional properties: a high coefficient of internal friction and high hardness. Two different types of pressure media are employed to ensure that the pressure generated around the samples is as much hydrostatic as possible. The outside media seals the sample chamber and the internal one needs to be soft enough to maintain a quasi-hydrostatic pressure environment around the sample [16].

For the large-volume presses, pyrophyllite is the most common choice of pressure media; other frequently used media are mica, lithium hydride, carbon, sulfide, hexagonal boron nitride, and Teflon. Additionally, AgCl and CsCl also are commonly used as pressure media because of their low melting points and high flowability. Other media used for high pressure research include $CdBr_2$, CdI_2, $CoCl_2$, $CuBr_2$, HgI_2, Ag_2S, metallic lead and indium, hard paraffin, polyethylene (PE), nylon, etc. In several special cases, such as under high temperature conditions, semi-sintered ceramics with a certain amount of porosity are employed, where the porosity can be regulated by changing the sintering temperature. MgO, BeO, ZrO_2, AlP, Fe_2O_3, and Mg_2SiO_4 are employed as semi-sintered ceramics [17].

Pressure media can not only transmit pressure but are also used for encapsulating materials. In order to increase the efficiency of pressure generated, the compressibility of pressure media must be as small as possible. Meanwhile, to obtain nearly hydrostatic pressure, the shear strength of the media materials should be minimized. Therefore, the choice of pressure media must be made carefully, with consideration to both low compressibility and minimal shear strength.

When considering pressure in relationship to the compressibility and shear strength of the media, AgCl is the most ideal choice as a pressure media [16]. Further consideration is needed for the large volume presses because the pressure media is also used to seal the samples in this case. Additionally, the pressure media requires a high friction coefficient and hardness, and the ideal candidate for above requirements is pyrophyllite. However, under the conditions of high temperatures and high pressures up to 8–10 GPa, pyrophyllite will decompose into stishovite and alumina, which will be very hard and unsuitable as a media anymore. In these cases, semi-sintered ceramics are the proper choice, and Fig. 5.8 [18] shows the relationship between pressure and applied load when using semi-sintered ceramics as the pressure media, which indicates that the efficiency of pressure generation is much higher when employing semi-sintered ceramics instead of pyrophyllite as the pressure media. Therefore, semi-sintered ceramics are the most ideal media under these special cases. Moreover, when taking into account the

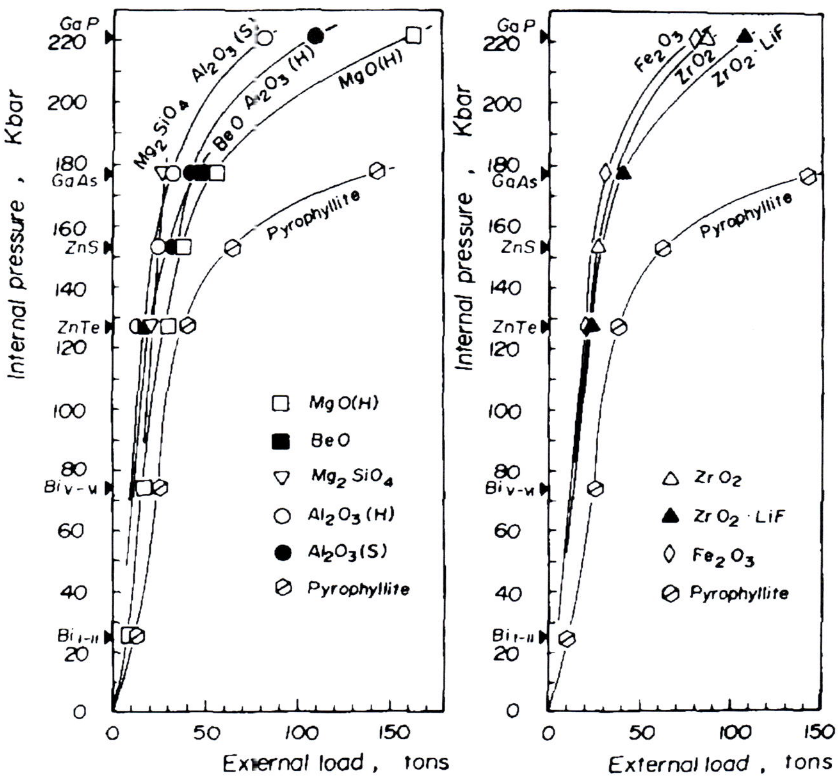

FIGURE 5.8 Pressure-vs-applied-external-load relations for various oxide media. *Reprinted with permission from A. Onodera, K. Suito and N. Kawai, Journal of Applied Physics, 51, 315. Copyright 1980, American Institute of Physics.*

need for materials with low X-ray absorption, it can be concluded that boron is a suitable pressure media for in situ high pressure measurements carried out on synchrotron radiation facilities.

Pressure media used for DAC equipment are different from that applied in the large volume presses, because DAC's anvils are supported by the gaskets, instead of the pressure media in a large volume press. Therefore, the choice of pressure media for DAC is mainly dependent on the liquidity of materials in order to generate hydrostatic pressure. According to the specific needs, a gas, liquid, or solid can be chosen as the media. For the liquid pressure media, a 4:1 (volume ratio) methanol–ethanol mixture has been used to ensure the hydrostatic conditions up to 10 GPa. Above 10 GPa, this mixture will transform into a glass state. There are several other available liquid media for higher pressure experiments. Some examples of these mixtures in volume ratio and their corresponding highest hydrostatic pressures are as follows: 16:3:1 mixture of methanol:ethanol:water, 14.5 GPa; methanol, 8.6 GPa; 1:1 mixture of *n*-pentane:isopentane, 7.4 GPa; 2-isopropoxyethanol, 4.3 GPa; petroleum ether, 6 GPa; and silicone oil, 15 GPa. For practical applications, the substances chosen as the liquid media need to be chemically inert, pollution-free for the diamond anvils, transparent, free from fluorescence, etc. Based on these desired characteristics, liquid helium is the best liquid pressure media in theory, but it is very hard to operate in practice. Considering the values of shear strength, the best solid media is NaCl, because in a DAC NaCl will become transparent under lower pressure. Other examples of solid materials used for pressure transmitting media are In, Pb, AgCl, KI, KCl, KBr, Teflon, mica, and pyrophyllite. The most suitable substances used for the gas pressure media are helium and hydrogen. Up to 70 GPa, R lines of ruby fluorescence spectra are not broadened, which means the pressure is hydrostatic. If hydrogen, argon, and neon are chosen for the pressure transmitting media, they will generate the hydrostatic pressure up to 177 GPa, 9 GPa, and 16 GPa, respectively [8].

5.1.3. Pressure Calibration

The Definition of High Pressure

In the ideal system, which is symmetrical, hydrostatic, and independent of time, pressure is force per unit, e.g., pressure equals the applied force (F) divided by the contact area (s), or $p = F/s$. Furthermore, this equation exists only for low viscosity liquid, not suitable for those conditions, such as solid and high viscosity liquid.

It should be noted that hydrostatic pressure here means every point is hydrostatic, but the pressure of each point does not have to be symmetrical in all spaces. Due to the effect of factors such as gravity and electric field in space, there are slight differences in pressure at each point. Therefore, hydrostatic pressure is a constant in space instead of symmetrical pressure.

Based on the definition of pressure above, we can see that pressure has temporal distribution besides spatial distribution. Temporal distribution means that when pressure changes, the whole system will transform from one thermodynamic equilibrium to another. Meanwhile, it always needs some time to reach thermodynamic equilibrium, and this time is named as the relaxation time τ. The relaxation process is characterized as hysteresis of depressurization during the pressure measurement. There are mainly two factors which will induce the influences:

(1) Influences relate to pressure transmitting media, such as excessively high viscosity coefficient and slow release of pressure. For solids with higher viscosity coefficients, the influences appear more obviously, which results in a system that is not hydrostatic.
(2) Influences relate to high pressure devices. For example, in piston-cylinder devices, the influence is led by the slow release of pressure inside cylinder during depressurization.

Pressure Unit

In the International System of Units, the pressure unit (force per unit area) is Pascal (Pa, $1\ \mathrm{Pa} = 1\ \mathrm{N/m^2}$). On testing apparatus, the unit of bar is normally used, where $1\ \mathrm{bar} = 10^5\ \mathrm{Pa}$. However, depending on the experimental conditions or literature sources, units of pressure can vary further. MPa is commonly taken as unit of pressure in current chemical papers: 0.1 MPa equals approximately one standard atmospheric pressure, or 1000 bar = 1 kbar = 100 MPa. For chemical research under extreme high pressure, GPa is the pressure unit, where 1 GPa = 10 kbar.

Pressure Measurement

There are many approaches to measuring pressure which may be classified into several scales, such as primary scale, thermodynamic absolute pressure scale, fixed pressure points, secondary pressure scale, equation of states, and ruby scale [8].

Primary scale uses the fundamental definition of pressure, $p = F/s$, and is limited to piston-cylinder apparatus. Relative error of pressure measurement is up to 0.24%, with pressures of 2.6 GPa. Measurements at higher pressures, up to 8 GPa, exhibit a greater relative error of approximately 4%.

Fixed pressure point scale uses fixed phase transition pressure of some materials to fit the pressure dependence of an applied load, which can be used to measure pressure. This scale is not so important for DAC, because the pressure inside of DAC can be mainly measured using the ruby scale. However, the fixed pressure point scale is still used for extremely high pressure over 200 GPa, and it also remains practical for large volume apparatus for calibration of applied loads and measurements of internal pressure. Construction of the scale relies on the phase transition of some materials, where generally there is a change in the resistance, melting point, and volume of the material. For practical reasons there are some requirements for phase transition of the materials selected:

(1) The measured effect should be large;
(2) Hysteresis of the phase transition induced by pressure should be small, because it is related to the error of measurement; and
(3) The temperature dependence of the phase transition must be small or nonexistent.

The recommended phase transitions for fixed pressure points are listed in Table 5.1. The pressure dependence of Bi's resistance is shown in Fig. 5.9 [19]. It is valuable to mention that the connecting methods for resistance measurements, horizontal or perpendicular, should be particularly noted, which is related to temperature gradient of the pressure transmitting media. A good pressure transmitting media, such as AgCl, should be chosen near the measurement point. It is also available to use this approach as secondary pressure scale in other direct pressure measurement methods.

Pressure can also be calibrated with the aid of an equation of state (EOS) through precise X-ray diffraction and neutron diffraction. For thermodynamic equation of state:

$$p = -\frac{\delta E_c}{\delta V} + \gamma\frac{E_{1T}}{V} + \gamma_e\frac{E_e}{V} \tag{5.1}$$

where γ is the Grüneisen parameter, E_c (the energy of the static lattice), and E_e (the energy of the thermal excitation electrons) can be calculated using the Anderson linear muffin-tin orbitals (LMTO) method, and E_{lT} (the energy of lattice vibrations) can be evaluated from the Debye

TABLE 5.1 Recommended Phase Transitions for Fixed Pressure Points [8]

Phase transition	Pressure (GPa)	Remarks
Bi_{I-II}	2.55 ± 0.006	Midpoint of hysteresis
Tl_{I-II}	3.67 ± 0.03	Midpoint of hysteresis
Ba_{I-II}	5.5 ± 0.2	Midpoint of hysteresis
Bi_{III-IV}	7.7 ± 0.3	Resistance change to its value in the high-pressure phase

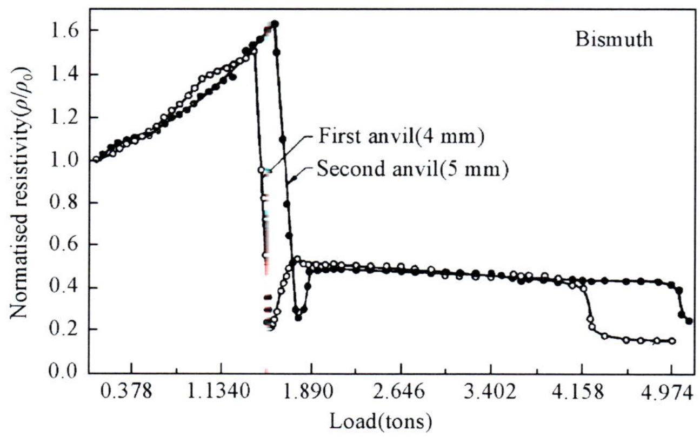

FIGURE 5.9 Phase transitions in bismuth under high pressure. *With kind permission from Springer Science+Business Media: Bulletin of Materials Science, Dual anvil high pressure cell, volume 2, 1980, A K Bandyopadhyay, Figure 1. [19].*

Model, but typical accuracy is not high, at 5–10%. In practice, a semiempirical equation of state is usually used, such as the Murnaghan equation:

$$p = \frac{K_0}{K_0'}\left[\left(\frac{V_0}{V}\right)^{K_0'} - 1\right] \tag{5.2}$$

where the parameters K_0 and K_0' can be obtained at low pressure.

It should be noted that there are many forms of EOSs used under different pressure ranges. The EOS of Al can be applied to pressure calibration for pressures up to 220 GPa [20]. The EOS of NaCl can also be used, however, there is a drawback of a B1–B2 phase transition near 29 ± 1 GPa.

The ruby-luminescence scale is normally used for DAC apparatus, and Barnett et al. [21] discovered that its R_1 line shifted linearly with pressure when they calibrated a ruby with the EOS of NaCl; so ruby is now often used as a pressure calibration material. Mao et al. calibrated the ruby scale up to 100 GPa under nonhydrostatic pressure through EOSs of some metals obtained from shock wave data; Bell et al. continued it up to 180 GPa based on using Au and Cu as internal standards; Mao et al. also calibrated the ruby scale in argon media up to 110 GPa under hydrostatic conditions [22]. Two scales can be shown in one formula [23]:

$$p = \frac{1904}{B}\left[\left(1 + \frac{\delta\lambda}{694.24}\right)^{B} - 1\right] \tag{5.3}$$

where p is pressure in GPa and $\delta\lambda$ is wavelength shift of R_1 line in nanometers. Parameter B is equal to 7.665 for quasi-hydrostatic conditions and 5 for nonhydrostatic conditions. The accuracy of pressure determination is ~0.03 GPa, if R_1 and R_2 peaks are well resolved, with highest accuracy up to 0.005 GPa at low temperatures. Furthermore, for measurements with the greatest accuracy, ruby needs to be annealed to relieve the internal stress before packed into the DAC. Temperature effect should also be taken into account for the measurement; the ruby luminescence lines shift with temperature, with a slope of 0.0068 nm/K. This means that a ΔT of ~0.5 K produces the same shift as the application of 0.1 GPa of pressure. For the most precise determination of pressure, temperature should be maintained at 20°C under atmospheric pressure.

5.1.4. Generation of High-temperature

The high temperature and high pressure apparatus allows both of the thermodynamic variables, temperature and pressure, to be changed, and it will lead to many applications in different fields: it can be used to study synthesis of new materials, phase transition, and geophysics. In a large volume anvil apparatus, the temperature is usually generated by means of placing the heater into the assembly. But for DAC, there are two heating methods: one is direct resistance heating, and another, laser heating.

The large volume anvil apparatus has the advantage of a larger volume, but its disadvantage is that it is very difficult to realize in situ measurements. As compared to DAC, it is also very difficult to determine the internal experimental conditions. For the large volume anvil apparatus, the temperature is produced by inserting the heater into the assembly, the actual temperature is greatly related to the sample assembling methods. The typical assembly for the piston-cylinder apparatus is shown in Fig. 5.10 [24].

Usually, graphite and a metal such as Ta, Mo, or Pt are used as heater. If the assembly contains Fe, Co, and Ni transition metals, these metals could catalyze the transformation of graphite to diamond or another intermediate, and result in heating termination where these metals contact with graphite; however, NaCl or ZrO_2 is needed to protect the graphite. At high temperature, when the pressure is higher than 14 GPa, the transition of graphite to diamond or semi-diamond will occur. In this case, Pt, Mo, or $LaCrO_4$ will be chosen as a heater. It is also possible to obtain pressure and temperature greater than 36 GPa and 1900 K, respectively, if the mixture of MgO and TiC or diamond is used as a heater. Tungsten carbide (WC) anvil is an excellent heat conductor, so it is necessary to minimize the contact area between the anvils and the heater, which will reduce the temperature gradient. For large volume press, the pyrophyllite is usually used as the pressure transmitting media. At high temperatures and pressures above 8–10 GPa, the semi-sintered ceramics, such as MgO, BeO, ZrO_2, are chosen as pressure-transmitting media.

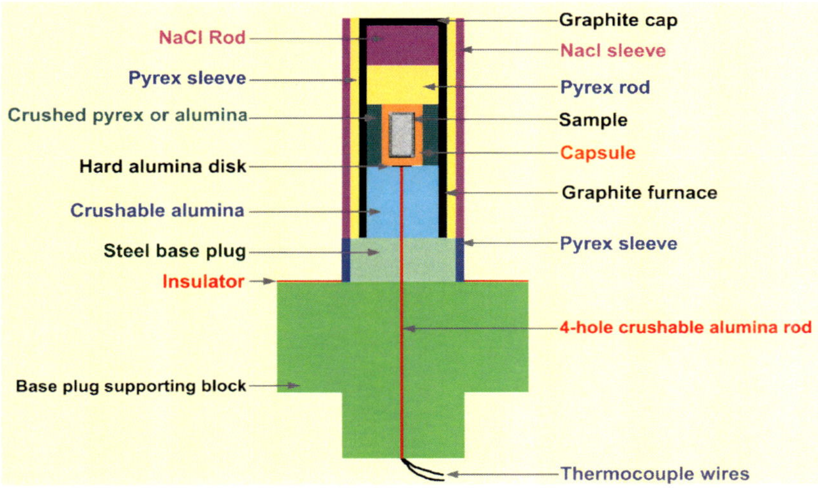

FIGURE 5.10 Experimental configurations for nonend loaded piston-cylinder.

A DAC can be heated internally or externally using resistance wire heating. Internal heating can be achieved by placing the resistance wire, for example, Fe wire, into the sample. Manghnani et al. obtained experimental conditions of 36 GPa and 2000 K using this method [25]. Generally, the sample can be heated to the temperature of 2000 K if the heating power is 7 W, and the temperature can reach 3700 K if the heating power is 8 W (W wire). However, this method has the temperature gradient. External heating involves placing a small furnace around the DAC, which will lead to even temperature distribution, but there are temperature differences between sample and outside of diamond. This heating method could result in the oxidation of DAC assembly parts. As a general example, diamond, Fe, etc. is oxidized around 800°C, but in a vacuum diamond can be kept stable for several hours at temperatures above 1500°C. So when operating above 800°C, a special DAC and gas protection device are needed.

Presently, laser heating is widely used in DAC. This method has the characteristic of high temperature, but the temperature distribution is extremely uneven as a Gaussian distribution along the gasket and anvil, with a temperature gradient around 1000 K/μm. Laser heating can be greatly improved by improving the design; for example, surrounding the sample with a heat-absorbing substance such as Pt would form an environment similar to the furnace and consequently reduce the temperature gradient. Even without improvement, laser heating of DAC can achieve pressures and temperatures of 200 GPa and 1500–5000 K, respectively [8].

Generally, an Nd:YAG laser can be used for heating, with a wavelength of 1.064 μm and the laser beam power of 20 W. At this frequency, many samples are transparent, meaning that they don't absorb the laser. In this case, the samples can be indirectly heated by addition of platinum powder to the samples, or heated by a carbon dioxide laser. Whichever heating method is used, the stability of the laser is important during the heating process, as the fluctuation of power will result in a deviation of 200 K with temperature up to 2000 K, although the fluctuation is normally 2 or 3%.

Two methods can be used to measure temperature for laser heating. One method is radiation spectroscopy, which utilizes the Plank function $r(\lambda,T) = r_0(\lambda,T) \cdot \varepsilon(\lambda,T)$. The method utilizes the melting point of metal, such as Fe (1808 K), Zr (2125 K), Mo (2890 K), Ta (3269 K), and W (3680 K), to fit the curve of radiation spectroscopy, and then the temperature can be calibrated through the curve. Another method to measure temperature is to compare the relative intensity ratio of the Stokes lines with anti-Stokes lines in Raman spectrum.

Due to big effect of temperature on pressure while the sample is heated by laser in DAC, care should be taken when measuring the pressure. Because the temperature in DAC presents as Gauss distribution and temperature gradient is large, the temperature in the edge of gasket is usually very low, the ruby should be placed at the edge of the gasket hole to calibrate pressure. The deviation of pressure is normally about 2–3 GPa in high temperature conditions. Pressure can also be calibrated through the fluorescent property of other materials, such as Sr_4BO_7:Sm^{3+}. In addition, when heating is carried out by laser, argon is usually used as pressure transmitting media in order to insulate the diamond anvil with the sample. Other solid materials can also be chosen as pressure transmitting media, such as Al_2O_3, MgO, BN, and NaCl.

High temperature and high pressure apparatuses have achieved considerable progress, with attained pressures greater than 450 GPa and temperatures up to 5000 K. In the synthesis and preparation experiments, the large volume press is always used, however, different apparatuses have strengths and weaknesses in their use. Piston-cylinder press with a small volume can be used to synthesize a large sample. The cubic module can also be used to synthesize large samples, but the apparatus is very large and finds greater use to produce diamond in industry. The concave anvil cell, which evolved from Bridgman anvil press, can be used to synthesize large sample in a shorter time. Even with this advantage, the apparatus is titanic. The Belt-type press with a moderate volume can be used to synthesize large samples. Because Belt-type press can generate higher pressure than that of piston-cylinder press, Belt-type press is a popular choice at present. Nowadays, the most advanced press is a 1000-ton super high pressure apparatus designed by the American Rockland Research. The two module types available are the Cubic module and Walker module. The former can be used to synthesize large samples, while the latter can generate varied pressure according to different truncated edge length (TEL) with the smaller chamber, the higher the pressure that can be generated. The sinter diamond used as anvil can achieve 90 GPa, which is the highest pressure obtained by a large volume press. Only in the DAC, can we achieve pressures more than 100 GPa in inorganic synthesis. In conclusion, inorganic synthesis at extra high pressures requires different types of high pressure apparatus, according to the size of sample, the pressure and the laboratory conditions.

5.2. EFFECTS OF HIGH PRESSURE ON BASIC STATES OF MATTERS

As a thermodynamic parameter, pressure has a tremendous impact on material. With the continuous advance in high-pressure technology, high pressure is no longer just an experimental means and extreme condition, but has become a third independent physical parameter that is beyond the temperature and chemical composition. Under high pressure and high temperature, almost every kind of material will experience a change in physical and chemical properties. For instance, even the simplest substances, such as hydrogen and water, will unexpectedly undergo a series of novel changes under high pressure.

5.2.1. Gas Under High Pressure

With the increase of pressure, gas density will gradually increase to form a molecular solid, as shown in Fig. 5.11. If the pressure continues to increase, molecular solids will be converted into atomic or ionic solids. Even the noble gases which exhibit strong chemical inertness at normal pressure conditions undergo change at high pressures; their chemical reactivity can become greatly enhanced. In fact, not only can the inert gas undergo oxidation with oxygen to form inert gas oxides such as ArO_3, combination reactions may also occur among these inert gases to form new compounds, such as $NeHe_2$ [26].

Hydrogen is the simplest chemical element. At normal temperature and pressure ranges, hydrogen atoms exist in molecular form, or as hydrogen gas (H_2). When the temperature is below 20.4 K, hydrogen gas changes into liquid; when the temperature is further reduced to 14 K, the liquid hydrogen is transformed into a solid. In 1935, Wigner and Huntington predicted by theoretical calculations that hydrogen is likely to exist in the form of metal under high-pressure conditions [27]. In 1988, Hemley et al. from Carnegie Institute in the United States found that under the conditions of 150 GPa and 77 K, the hydrogen stretching vibration oscillator frequency showed discontinuous changes, which indicated that a hydrogen phase transition might have occurred [28]. This was later confirmed by the experiments of Silver's group from Harvard [29]. In 1989, Mao et al. raised the pressure up to 250 GPa and found that the hydrogen molecular state still existed under this pressure, while the sample transformed into a black opaque material [30], and the reasons leading to this phenomenon are unclear yet. In 1990, Mao and Hemley found in their experiments that the infrared reflection coefficient of hydrogen significantly increased under the condition of 150 GPa and 295 K, so that they firmly believed that free-electron behavior must have occurred in hydrogen [31]; however, this view was later proved false by many other experiments [32–34]. Recent experiments show that even under 300 GPa, solid hydrogen remains in the molecular state, while there are only some very weak interactions between these

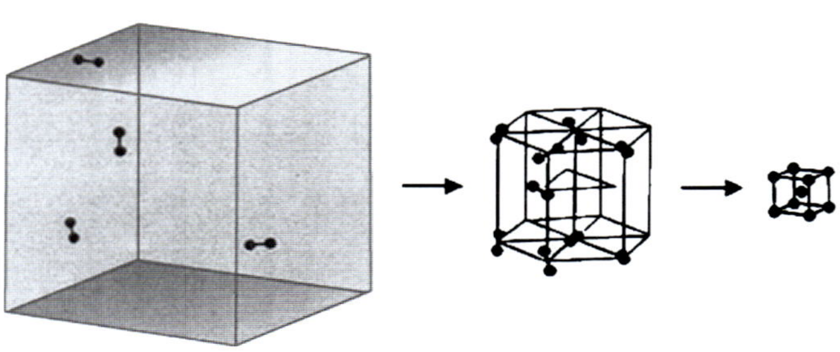

FIGURE 5.11 The compression of a molecular fluid or gas to form a molecular solid; with further increasing in pressure, the intermolecular and intramolecular distances become comparable and typically induce phase transformations, including molecular dissociation, shown here to form a monatomic solid.

hydrogen molecules. Recently, Loubeyre and Pickard et al. predicted that hydrogen's metallization would occur under the pressure of 450 GPa [35,36].

Carbon dioxide is a kind of gas closely related to daily life. One major issue of study on carbon dioxide under high pressure is whether carbon dioxide would become a polymer like silica if carbon dioxide is shifted from three-coordination to four-coordination under high pressure. In 1999, Yoo et al. found in their experiments that in the high-pressure conditions of 10–20 GPa, carbon dioxide molecules will aggregate to form a three-dimensional network structure similar to Tridymite (a high-temperature form of quartz found as thin hexagonal crystals in some igneous rocks and stony meteorites) [37–39]. As shown in Fig. 5.12 [38], carbon atoms form CO_4 tetrahedrons coordinated by sp^3 hybridization with the adjacent oxygen atoms. In 2006, Santoro et al. showed that "carbonia glass" could be made when a high-density form of solid CO_2 was melted and cooled under high-pressure conditions [40]. This phenomenon further unveils that carbon has properties similar to silicon and germanium from the same family of elements.

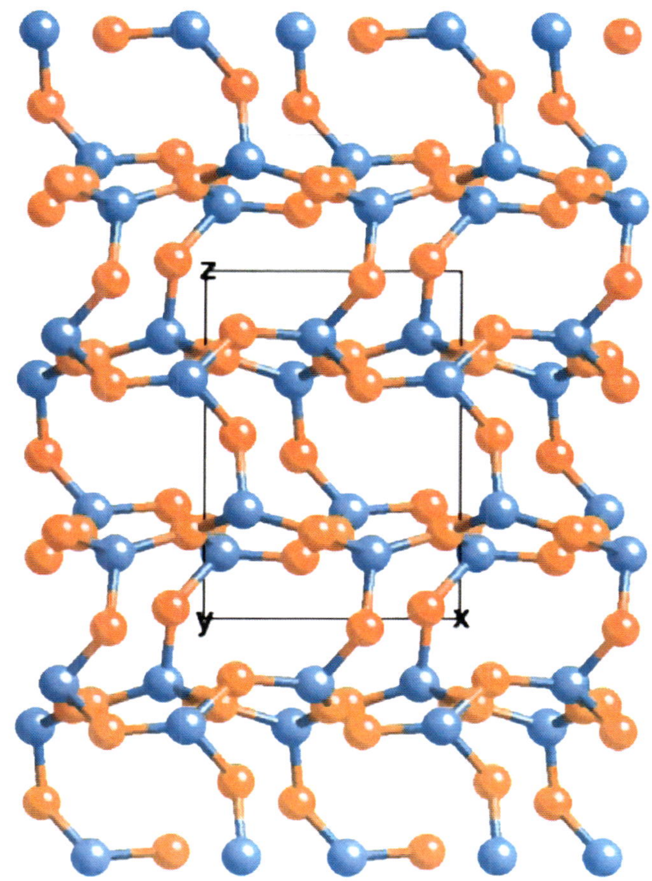

FIGURE 5.12 An "ideal" crystal structure of CO_2-V in $P2_12_12_1$ with eight molecules per unit cells. Carbon atoms (blue) are tetrahedrally bonded to four oxygen atoms (yellow) [38]. *Reprinted figure with permission from C.S. Yoo, H. Cynn, F. Gygi, G. Galli, V. Iota, M. Nicol, S. Carlson, D. Hausermann and C. Mailhiot, Physical Review Letters, 83, 5527, 1999. Copyright (1999) by the American Physical Society.*

5.2.2. Solids Under High Pressure

It is well known that solids under high pressure conditions undergo dramatic changes in their physical and chemical properties. The main reason for this change is that high pressure can alter the atomic structure of the outermost electron shell, and the outermost electrons will determine the various physical and chemical properties of the atoms. For example, calcium metal under high pressure exists as hexagonal close-packed structures, but at pressure above 20 GPa, the metals calcium are transformed into body-centered structure, and the density reduced. This phenomenon can be explained by the orbital hybridization of the 4*s* and 3*d* electron that occurs when calcium metal is subjected to high pressures, where Ca exhibits properties similar to the transition metal elements. Another example is that while the alkali metal potassium will not react with transition metals at atmospheric pressure, but under high pressure conditions, it will form K–Ni, K–Ag alloy compounds and other solid solution materials [41].

Under high pressure, metals exhibit increasingly shorter interatomic distance, widen the valence and conduction bands, and pronounce a free-electron-like behavior. Core electrons overlapped by compression can now be achieved experimentally. Such as lithium and sodium, the simple free-electron metals, the electronic properties have been changed by the overlap and they will transform under high pressure into insulating states. The electrical resistance measurements on lithium in a diamond anvil cell are up to a pressure of 105 GPa, as shown in Fig. 5.13 [42]. The data thus provide unambiguous experimental evidence for a pressure-induced metal-to-semiconductor transition in a "simple" metallic element. And as shown in Fig. 5.14 [43], sodium transforms into an optically transparent phase at ~200 GPa, experimental and computational data identify the new phase as a wide band-gap dielectric.

High pressure is beneficial for the formation and stability of high oxidation state of transition metals [44]. For example, oxidation state of Cu is 3+ in the compounds $La_2CuO_{4+\delta}$ and $La_{2-x}Nd_xCuO_{4+\delta}$, which are synthesized under the conditions of high oxygen pressure. Under the conditions of high fluorine pressure, the synthesized compound Cs_2AgF_6, has both trivalent and pentavalent Ag. Table 5.2 shows how high-pressure conditions can be used to further stabilize the high oxidation states of some transition metal

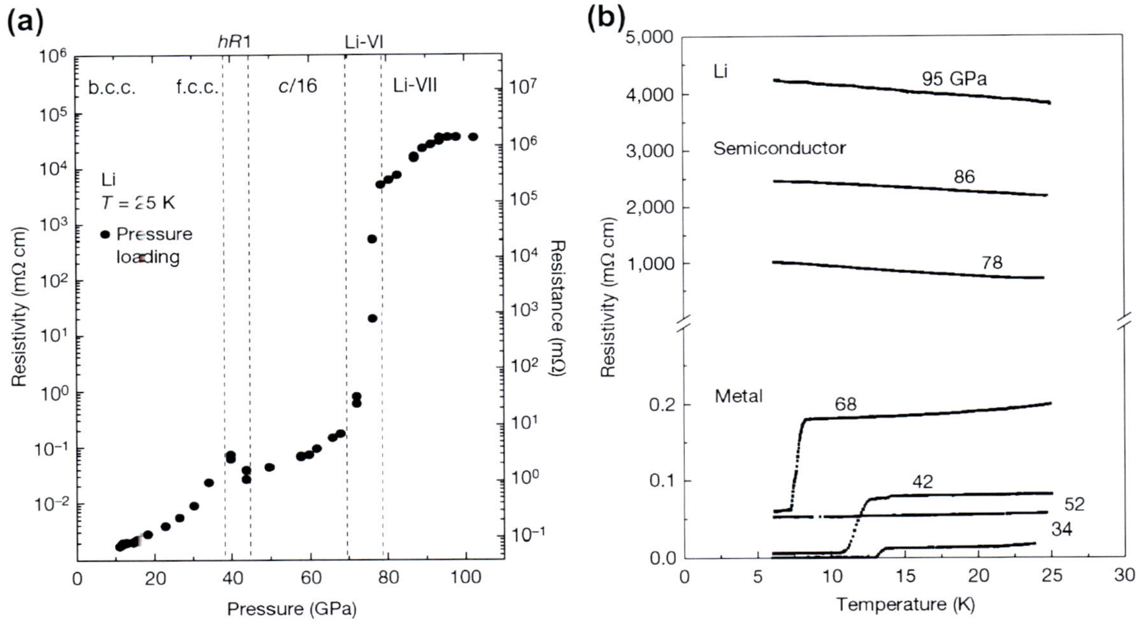

FIGURE 5.13 Measured resistance values (right vertical axis) and derived electrical resistivity values (left vertical axis) at 25 K as a function of applied pressure. *Reprinted by permission from Macmillan Publishers Ltd: Nature [42], Copyright (2009).*

compounds. In addition, some high pressure experiments found in transition metal oxides and fluorides, the $M^{n+}-X$ (where M is the transition metal ions, X is either O^{2-} or F^{-}) bond energy is rapidly increased with an increased n.

High pressure is also conducive in increasing the coordination number of cations. For example, the basic structural unit of SiO_2 phase is SiO_4 tetrahedron, but in the high-pressure phase of SiO_2, such as the ultraquartz crystal (formed by SiO_2 under pressure greater than 9 GPa), the silicon coordination numbers change from 4 to 6. The positive quadrivalent cations with the coordination number of 6 also exist in Si_3N, Ge_3N_4, Sn_3N_4, and other compounds with the spinel structure [45].

5.2.3. Water Under High Pressure

Pressure and temperature have an effect on both the structural form and the bonding of water. At 0°C or so, with lower pressure such as several megapascals, water molecules can be aggregated to form ice with a structure of a plane [46]. Under low-temperature and high-pressure conditions, water and natural gas can form natural gas hydrates. The combination of water molecules is through strong hydrogen bonds, while the acting force between the gas molecules and water molecules arises from van der Waals forces. The natural gas hydrate found has three kinds of structure: type I, type II, and type H. Cages formed by water molecules can capture gases such as CH_4 and CO_2, whose structure are shown in Fig. 5.15 [47]. These natural gas hydrates formed by water and CH_4 are abundant in nature. Around the world, marine and continental strata have proven reserves that have more than twice the energy of the world's traditional fossil fuels (coal, oil, natural gas, oil shale, etc.) [48]. The nature of water can capture CO_2 to serve as a CO_2 storage media to achieve manual

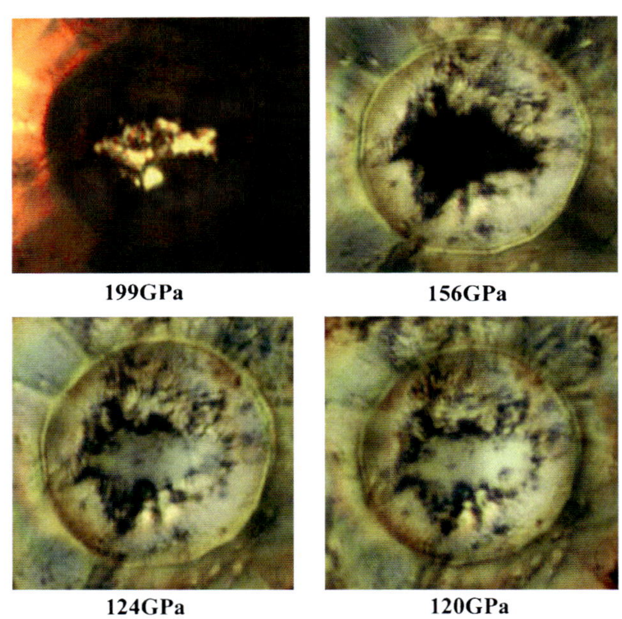

FIGURE 5.14 Photographs of the Na sample taken under combined transmitted and reflected illumination. *Reprinted with permission from Macmillan Publishers Ltd:* Nature *[43], Copyright (2009).*

TABLE 5.2 Stability of High Oxidation State of Transition Metals [44]

Element	Oxidation state	Electron configuration	Compounds
Cr	Cr(V)	$t_{2g}^2 e_g^0$	$La_2LiV_{1-x}Cr_xO_6$
Fe	Fe(IV)	$t_{2g}^3 d_{z^2}^1 d_{x^2-y^2}^0$	$A_{0.5}La_{1.5}Li_{0.5}Fe_{0.5}O_4$ (A = Ca, Sr, Ba) (SrLa) $M_{0.5}Fe_{0.5}O_4$ (M = Mg, Zn)
Fe	Fe(V)	$t_{2g}^3 e_g^0$	La_2LiFeO_6
Co	Co(III)	LS → HS	$LnCoO_3$ (Ln = rare earth element) $SrLnCoO_4$
Co	Co(IV)	$t_{2g}^6 e_g^0$	$(Sr_{0.5}La_{1.5})Li_{0.5}Co_{0.5}O_4$
Ni	Ni(III)	$t_{2g}^6 \sigma^{*1}$	$LnNiO_3$(Ln = rare earth element) $SrLnNiO_4$
Cu	Cu(III)	$t_{2g}^6 d_{z^2}^2 d_{x^2-y^2}^0 t_{2g}^6 \sigma^{*2}$	$La_2Li_{0.5}Cu_{0.5}O_4$ $LaCuO_3$, $SrLaCuO_4$
Ag	Ag(III,V)	$t_{2g}^6 d_{z^2}^2 d_{x^2-y^2}^0 t_{2g}^4 \sigma^{*2}$	Cs_2AgF_6

control of CO_2 content in the air, thereby reducing the carbon dioxide green house effect.

The structure of water can vary widely under high-pressure conditions, with 15 kinds of structure currently known, as shown in Fig. 5.16 [49]. The phase I_h is the normal hexagonal phase, which accounts for almost all of the ice of the Earth's entire biosphere; phase II is the trigonal crystal phase, where the proton is a highly ordered structure. When the phase I_h is pressurized at 190−210 K, it can change into the phase II; phase III is crystal phase in tetragonal system, and can be formed at about 0.3 GPa and 250 K. Phase IV is a metastable trigonal crystal phase, and can be only formed under the conditions of existing nucleation media; phase V is monoclinic crystal phase, and can be formed at the pressure of about 0.5 GPa when the temperature is approximately 250 K. This phase is the most complex of all the known water phases; phase VI is a tetragonal crystal phase, and can be formed at pressures of about 1 GPa, with temperatures around 270 K; phase VII is ice in cubic form where the hydrogen atom position in its structure is unordered; phase VIII is a proton ordered phase that can be obtained by cooling phase VII. This phase is also antiferroelectric. Phase IX is a metastable tetragonal crystal phase that can be formed by cooling phase III. Phase X is proton-ordered symmetry phase and formed at pressures around 70 GPa. The phase XI is the orthorhombic crystal phase of phase I_h proton rearrangement at low temperatures and is ferroelectric. The phase XII is metastable high-density ice of tetragonal crystal phase which can be obtained under 810 MPa by heating high-density amorphous ice [26].

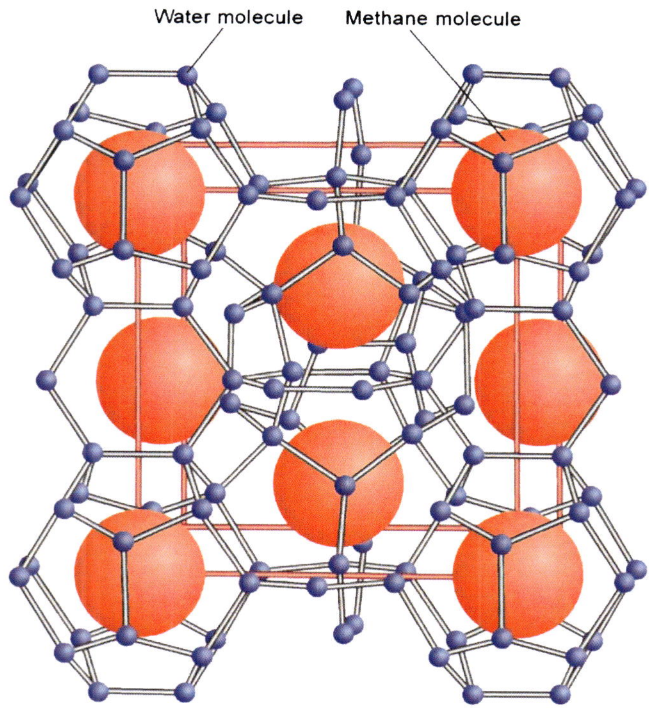

FIGURE 5.15 Structure I clathrate lattice with entrapped methane molecules [47].

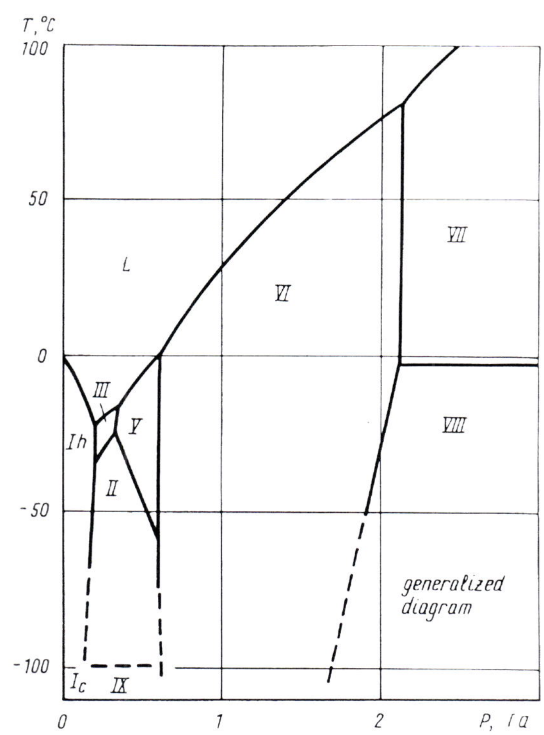

FIGURE 5.16 The phase diagram of H_2O [49].

5.3. EFFECTS OF HIGH PRESSURE ON INORGANIC CHEMICAL REACTIONS

For chemists, the synthesis of ammonia through Haber process and the transformation of all allotropes of carbon are two of the first examples highlighting the effect of pressure on inorganic reactions. The more comprehensive understanding of the effect of pressure came from many aspects such as the various gas laws, pressure as a variable in the application of the Principle of Le Chatelier, the osmotic pressure of solution, hydrostatic pressure as it relates to the marine environment, and pressure as a parameter of many thermodynamic equations. Haber synthesis and synthesis of diamond are gas and solid phase reactions, respectively. In this section, most of the reactions discussed are liquid-phase reactions. We are not only concerned about the physical changes caused by pressure, but also the effect of pressure on chemical reactions.

In general, the pressure variable can be used in many ways, such as the synthesis of inorganic compounds, the study of chemical kinetics, and the description of reaction mechanism. The approach to use pressure is not merely a matter of randomly choosing a reaction and subjecting it to pressure. Based on a predictive analysis of the system in question, the consequence of applying pressure should ideally result either in an improved yield of product or in an optimized yield of a specifically desired product in a preparative procedure, or it can be expected to provide mechanistic insight. Basic inorganic reaction is the foundation for studying inorganic synthesis. Knowing and understanding the effect of pressure on simple inorganic reactions will help further advancements in the research of high pressure inorganic synthesis [50].

5.3.1. Influence of Pressure on Thermodynamics and Dynamics of Inorganic Reaction

In the thermodynamic equilibrium of an inorganic reaction system, increasing pressure raises the temperature of the solution while decreasing pressure reduces the temperature. When elevating the pressure of a system, many characteristics may change, such as the solubility of the solute, the freezing point, or the viscosity of the solvent. Under high pressure conditions, instead of activation energy, the diffusion effect will control the reaction rate. It is the consideration of these aforementioned characteristics that will improve the experimental design and thus enhance the possibility of success.

Only thermodynamics and its application of thermodynamic principle to the absolute reaction rate theory can be used by chemists to explain and predict the effect of pressure on reactions, either a synthesis reaction or a mechanism reaction [51,52]. What we are interested in it is how pressure can affect the equilibrium constant (k), free energy ($\Delta G^{0\prime}$), or the free energy of activation ($\Delta G^{\neq}$) of the reaction. The yields of the reaction and the reaction rate will increase if pressure causes an increase in k or a decrease in $\Delta G^{\neq}$. Knowing this, we can determine if pressure affects equilibrium by altering the chemical potential of the reactant or if pressure affects the kinetics of the reaction through the transition state. Most people tend to consider only the changes in molar volume of the initial and transition states, or if possible, would consider the changes in molar volume of the products. For a reaction, if the volume of the transition state is less than the volume of initial state, pressure will speed up the reaction. To go one step further, if the volume of the products is different from the volume of initial state, pressure will affect the reaction rate. Strictly speaking, it should be first stated that the volume of a chemical in solution is different from the volume of the pure chemical, and it is easy to determine the molar volume of materials in aqueous solution by density.

5.3.2. The Impact of Pressure on Inorganic Reaction

Synthetic Reaction

Common chemical reactions, especially inorganic reactions, have specific relationship with pressure. Pressure can speed up or slow down the rate of synthetic reactions, or regulate the selectivity of specific product. This knowledge has been applied to many inorganic and organometallic systems, such as the reaction of gallium with nitrogen or hydrogen.

As the metal surface directly interacts with N_2 or H_2 under high pressure, more attention is paid to the synthesis of metallic nitride and hydride [53]. Heating pure gallium can produce gallium nitride in a nitrogen atmosphere [54], and this technology has been used in studying the growth of gallium nitride crystal as well [55]. If the sample attains its equilibrium pressure at a certain temperature, we can also study the decomposition reaction of the gallium nitride. Using this method, the generation and decomposition reactions of gallium nitride, as well as its dynamical decomposition reaction [56], can be investigated at temperatures up to 1700°C and pressures up to 2×10^3 GPa. In addition information on thermodynamics and dynamics, which is relevant to the temperature and pressure, can be obtained.

Addition–elimination Reactions

A common reaction is oxidative addition, where a four-coordinated metal complex is oxidized to a six-coordinated compound, which usually has a larger and more negative value of $\Delta V^{\neq}$, resulting from bond formation and the flex of transition state enhanced by solvent power which is caused by change in charge. When adding H_2, CH_3I, or HCl to Vaska compound ([*trans*-$IrCl(CO)(PPh_3)_2$]) or an analogous compound, it develops a negative $\triangle V^{\neq}$ and its behavior is dependent on solvent. An array of $\triangle V^{\neq}$ is related to the function q_p of solvent polarizability and consequently, we can deduce that the range of $\triangle V^{\neq}$ is about 17–18 $cm^3\ mol^{-1}$. Two Ir–H bonds are formed simultaneously, resulting in a *cis*-dihydro complex by the addition of H_2, while the addition of CH_3I causes a linear transition state [$I-CH_3-IrL_4$] [57].

Electron Transfer Reaction

It is very useful to study the electron transfer reactions under high pressure conditions. A multitude of research involves the comparative study of the experimental value and the calculated value predicted from the Marcus–Hush theory [58] in the outer electron transfer reaction. For the reduction reactions, the class of the inner region or the exterior region domain has been completed for many years. The initial activation volume value is determined as a standard and is used in reactions to distinguish between the two-metal-center electron transfer reactions generated in the preformed region of ligand coordination or in the region of coordination to the metal.

It has been paid more attention to the outer electron transfer reactions, especially the symmetric self-exchange reactions. They are more suitable for one theoretical study due to a simplified zero-reaction volume. An example is the following chemical reaction equation:

$$MnO_4^- + MnO_4^{2-} \rightarrow MnO_4^{2-} + MnO_4^-$$

This reaction is monitored by nuclear magnetic resonance (NMR) [59], and has the value of $\Delta V^{\neq}$, $-21\ cm^3\ mol^{-1}$. This reaction can be catalyzed by some independent ions, such as sodium and potassium, which also shows significantly different values of $-4\ cm^3\ mol^{-1}$ and $-1\ cm^3\ mol^{-1}$.

The Radiation-induced Process

There are three possible pathways for the thermodynamic ligand substitution reaction, combination (A), dissociation (B), or exchange (I), which all display unique relationship with pressure. The question caused by this is whether the mechanisms still exist in the process of optical excitation or radiation. It can be concluded that in the electronic excited states of either inorganic molecules or metal organic molecules, both chemical and physical processes exhibit characteristics that are related to pressure [60]. The relationship between pressure and products or pressure and lifetime of excited state, which are observed in light-induced substitution reaction, clarifies the nature of excitation of the substitution reaction in accordance to three types of mechanisms as mentioned above. The success of this approach led to a series of ammonia complexes of Rhodium(III) and Cadmium(III). The equation below shows the reaction where a light-induced (313 nm) M $(CO)_6$ complex is generated. This is accompanied by a significant positive $\Delta V^{\neq}$ value, which shows that reaction mechanism is dissociation mechanism.

$$M(CO)_6 + L \rightarrow M(CO)_5L + CO$$

where M = cadmium, molybdenum, or tungsten; and L = piperdine, pyridine, or pyruvonitrile. The reaction mechanism for the reaction the light-excited CO replacing the complex of $M(CO)_4$(phen) shows that this reaction is controlled by the property of excitation state, which may indicate a mechanism through disassociating into LF excitation or associating into MLCT excitation. The significant differences of these related mechanisms can be solved by studying the relationship between pressure and the light substitution reactions induced by radio waves. The results clearly revealed that activation volume of LF excited state is positive, while the activation volume of MLCT excited state is negative. Similar studies have been used to determine the mechanism of the light substitution reactions of $(CO)_5ReMn(CO)_3$ and $CpFe(CO)_2(COCH_3)$.

The above-mentioned examples demonstrated the effects of pressure on inorganic reactions. With the development of solution reaction kinetics, some valid data will be revised. Study on reaction kinetics will help to verify the validity and accuracy of inorganic reaction mechanisms previously studied. In many experiments with pressure less than 150 MPa, the relationship between pressure and activation volume has not been found. The Marcus–Hush theory is more suitable for interpreting the high-pressure reaction kinetics in the electron transfer reactions, because the electron transfer reactions predict a function of the activating volume and pressure in self-exchanged oxidation–reduction reactions. These studies will have important contributions to chemical research. It can be predicted that further research on inorganic and classical metal organic chemical reactions will offer more complex and useful information by using variable high pressure technology.

5.4. EFFECTS OF HIGH PRESSURE ON CRYSTAL AND ELECTRONIC STRUCTURES OF INORGANIC COMPOUNDS

5.4.1. Changes in Crystal Structure at High Pressure

During the research on the synthesis methods of inorganic compounds, high pressure methods have been commonly used for realizing the crystal structure transitions of inorganic compounds and the syntheses of inorganic compounds with new crystal structures. Due to strong interaction between atoms in the crystal structure of an inorganic compound, newly obtained inorganic compounds with novel crystal structures can be easily maintained at normal pressure, despite the difficulties in synthesis [61–63].

In 1940s, Bridgman from Harvard University designed a high pressure device which could generate the pressure up to 40 GPa [7]. He pioneered the field of high pressure science and technology and also indicated that high-quality diamond could be synthesized by combining electric heating and high pressure. Commonly, the starting material used to synthesize diamond is graphite. The crystal structure of graphite has a hexagonal cell with each unit cell consisting of four atoms, as seen in Fig. 5.17. In the same lattice plane, carbon atoms are linked by a σ-bond and π-bond stacking, with a 1.42×10^{-10} m C–C bond length. The atom layers are linked by van der Waals interactions, with a layer distance of 3.35×10^{-10} m. Diamond consists of carbon atoms that are arranged in a tetrahedral configuration. Figure 5.18 shows a unit cell of diamond with a face-centered cubic (fcc) structure. The lattice constant of diamond is $a = 0.35667$ nm, and the space group is $Fd3m$. The difference between the structure of graphite and that of diamond is due to the different chemical bonds in the atoms. The structure of graphite is built on the interaction of three sp^2 hybridized orbitals and one $2p_z$ orbital from the valence electrons of the carbon atoms. The plane layer, with a hexagonal net structure, comprises a covalent bond and a metallic bond. These layers are linked by van der Waals interactions. In contrast, the structure of diamond is built on the tetrahedral configuration formed from the interaction of the valence electrons of carbon atoms along the four sp^3 hybridized orbitals, meaning the atoms of diamond are linked by pure covalent bonds.

Up to now, the mechanism of the structure transformation of graphite to diamond under the conditions of high temperature and pressure is still not confirmed. Gou believed that [55] carbon atoms in these layers approach each other perpendicularly along the c-axis under high pressure, which means that the layer distance of 335 pm is compressed. The vibration of carbon atoms becomes intense at high temperature. Because the carbon atoms in the layers are staggered while the vibrational directions of vicinal atoms in the layer are opposing, the corresponding atoms between layers frequently approach and attract each other, which makes their distances shorten. Half of the carbon atoms previously at the hexagonal grid crunode are perpendicularly displaced upward and the remaining half is perpendicularly displaced downward. Due to these displacements, plane hexagonal grids are generally contorted into tortile hexagonal grids. The atoms above and below attract each other at the same time, meaning free $2p_z$ electrons are centralized between the two atoms and eventually form covalent bonds perpendicular to the layers. As a result, free $2p_z$ electrons, which formed metallic bonds in the hexagonal grids, form covalent bonds perpendicular to the layers and link the vicinal atoms above and below. Only covalent bonds exist in the tortile hexagonal grids, so each carbon atom is covalently bonded to four vicinal atoms. This illustrates the

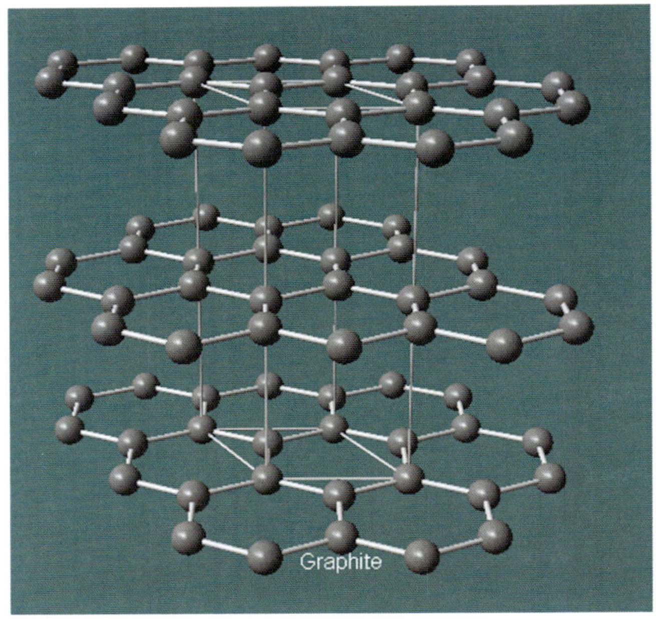

FIGURE 5.17 The crystal structure of hexagonal graphite.

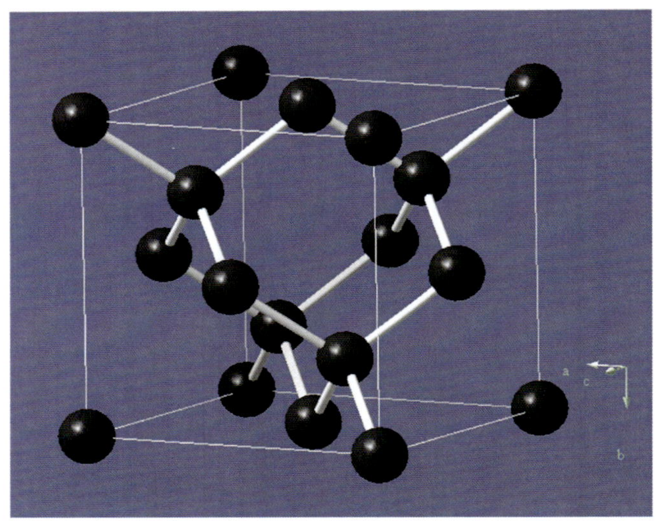

FIGURE 5.18 The cell of diamond.

transformation of graphite into diamond at high temperature and high pressure. Graphite can transform into diamond at lower pressure (5–6 GPa) and temperature (1300–2000 K) by using graphite starting material containing metallic catalyst, which is a typical example of catalytic synthesis method at high static pressure and high temperature.

Crystals with perovskite ($CaTiO_3$) structure is the important category of crystalline material, whose formula is usually represented as ABX_3. The ideal cubic $CaTiO_3$ cell is shown in Fig. 5.19. The cornered A site represents the cations with large diameters, such as La^{3+}, Nd^{3+}, Pr^{3+}, Sm^{3+}, Y^{3+}, Ca^{2+}, Sr^{2+}, Ba^{2+}, and Pb^{2+}, while the body-centered B site represents the cations with small diameters, such as Mn^{3+}, Mn^{4+}, Cr^{3+}, Fe^{3+}, Ti^{4+}, and Co^{3+}. In addition, the face-centered O site represents the anions of O^{2-}, S^{2-}, Cl^{-}, etc. As a result, O^{2-} and A site cations fall in the cubic close-packed arrangement and the small B site cations with sixfold coordinates reside in octahedral voids packed by O^{2-}, forming corner-sharing BO_6 octahedra. Meanwhile, the large A site cations with 12-fold coordinates reside in the voids formed by eight corner-sharing BO_6 octahedra. Because of the variety of the relative tilting between the BO_6 octahedra (23 types) [56,64] as well as the distortion of the BO_6 octahedra itself, the perovskite-related materials possess many excellent optical, electronic, or magnetic properties. Furthermore, the lower mantle of the Earth mainly consists of orthorhombic perovskite Mg(Fe, Al)SiO_3, meaning the structure evolution of perovskite under high temperature and pressure plays a crucial role in exploring deep earth structure. Consequently, perovskite-related materials have been intensely studied as an important subject in chemistry, material, and Earth sciences.

At ambient pressure, the low symmetry perovskite transforms into the highly symmetric prototype cubic perovskite with increasing temperature; this occurs through a series of intermediates by decreasing the tilting angle of BO_6 octahedra to zero [65]. In contrast, under high pressure, early experimental studies found that there are two structure evolution routes for the orthorhombic perovskite [66,67]: one is that the tilting angle of the BO_6 octahedra, as well as the distortion of AO_{12} polyhedra, increases with pressure, i.e., crystal structure tends to transform into a low symmetry form under high pressure, and another is that the tilting angle of the octahedra as well as the distortion of AO_{12} polyhedra decrease with pressure [68,69], i.e., crystal structure tends to transform into a high symmetry form under high pressure. The structure evolution of perovskite under high pressure has close relationship with the relative compressibility between BO_6 octahedra and AO_{12} polyhedra. If BO_6 octahedra are more compressible than AO_{12} polyhedra, the tilting angle of BO_6 octahedra decreases with pressure. On the contrary, if BO_6 octahedra are less compressible than AO_{12} polyhedra, the tilting angle of BO_6 octahedra increases with pressure [70]. Till date, the factors which determine the relative compressibility between BO_6 octahedra and AO_{12} polyhedra are not clear. For many years, scientists have attempted to answer this question, because it provides a possibility to predict high pressure compressibility of perovskite compounds. Recently, Zhao et al. [71] proposed a bond-valence matching relationship for the structure evolution of perovskite under high temperature and high pressure based on the bond-valence model. Specifically, a relationship between relative compressibility of BO_6 octahedra and AO_{12} polyhedra in orthorhombic perovskite was established [70]. Based on the bond-valence model, valence V_i of cation i is shared with several nearest anion j, and the individual valence S_{ij} of single bond is $S_{ij} = \exp[(R_o - R_{ij})/B]$, in which R_o is bond-valence parameter of the element [72] (each different element corresponds a to different R_o); B is commonly taken to be a universal constant equal to 0.37 Å; and R_{ij} is the bond length between cation i and anion j. Therefore, the sum of all the individual valences S_{ij}, that is, the bond valence sum V_i equals to the valence V_0 of cation i. Therefore, the bond-valence sum rule can be written as $V_i = \sum S_{ij} = V_0$ [72]. Under high pressure, the distance between atoms decreases, or the bond length becomes shorter. According to bond-valence sum rule, the ion valence increases with pressure, leading to strains at site A and site B. If the strains at both A and B sites are not balanced, the crystal structure will be unstable. Therefore, minimum strain energy is obtained only if the valence variations of these sites are the same [73]. On the other hand, the relationship between compressibility β_i and position parameter M_i of BO_6 and AO_{12} polyhedra is $\beta_B/\beta_A = M_A/M_B$. Position parameter M_i is defined as $M_i = R_i N_i / B \exp[(R_0 - R_i)/B]$, in which N_i is the coordinate of cation i, and R_i the average bond length. M_i represents the

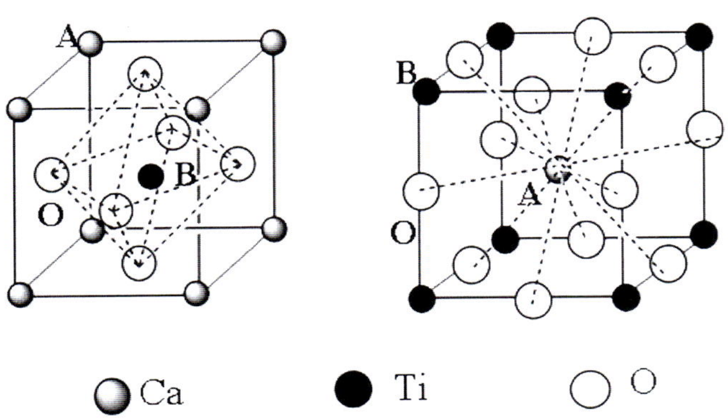

FIGURE 5.19 The cell of the perfect Perovskite structure.

variation of bond valence sum of the centered cation i of the polyhedra induced by the variation of the average bond length of the cation. Based on the crystal structure data at ambient pressure and bond-valence parameter R_0, M_i can be calculated. The relative compressibility of the polyhedra can then be attained. For BO_6 and AO_{12} polyhedra, the bond-valence-sum variation M_i induced by per average bond length is usually different. Therefore, according to the bond valence relationship, BO_6 and AO_{12} polyhedra have different compressibility. If the ratio of $M_A/M_B < 1$, then $\beta_B < \beta_A$, the tilting of BO_6 octahedra as well as the distortion of BO_6 and AO_{12} polyhedra will increase with pressure, such as the 2:4 type perovskites ($A^{2+}B^{4+}O^{2-}{}_3$): $CaSnO_3$ [74] and $MgSiO_3$ [76]. If the ratio of $M_A/M_B > 1$, then $\beta_B \backslash > \beta_A$, the tilting of BO_6 octahedra as well as the distortion of AO_{12} will decrease with pressure, such as the 3:3 type perovskites ($A^{3+}B^{3+}O^{2-}{}_3$): $YAlO_3$, $GdAlO_3$, and $GdFeO_3$ [69]. It has been demonstrated by high pressure single crystal X-ray diffraction that the bond valence matching relationship is consistent with the experimental results [70]. Therefore, the ratio of M_A/M_B determines the high pressure compressible behavior of perovskite. The greater the M_A/M_B deviation, the greater the compressibility difference between BO_6 and AO_{12} polyhedra, and the more unstable BO_6 octahedra become. This will further increase the changes in the tilting angle of BO_6 with increasing pressure, making BO_6 easier to compress.

ZnS belongs to IIB−VIA compound semiconductor materials and is a wide band gap semiconductor ($E_g = 3.7$ eV), which is considered as one of the most important materials for preparation of short-wavelength light-emitting devices [75]. As shown in Fig. 5.20 (a, b), ZnS typically has two kinds of structures at ambient pressure; namely, wurtzite and sphalerite. Sphalerite ZnS will change into the rock salt cubic structure [Fig. 5.20 (c)] under the pressure between 15 and 21 GPa, and it is a reversible phase transition [76−78]. Under pressure, wurtzite ZnS will first transform into sphalerite structure and eventually change into the rock salt cubic structure with increasing pressure. In a decreasing pressure process, the rock salt structure of ZnS can only be changed back to sphalerite structure, but will not return to wurtzite structure [79]. Sphalerite ZnS is a semiconductor with the direct band gap [80]. In previous electronic measurements under high pressure, there is a significant decline in ZnS resistance at the phase transition point where sphalerite changes into the cubic phase. The sample also changed from the initial transparent state to opaque and thus ZnS was considered to be in a metallic phase transition at the corresponding pressure [81].

Another example related to this topic is the study on the new types of super-hard materials. Superhard materials can be divided into two categories: one is the strong covalent compounds formed by light elements (B, C, N, and O), such as cubic BN and diamond, and the other one is compounds formed by precious metals (Pt, Ir, Os, Ru, etc.) and light elements that have some covalent and some ionic bond characters. For example, precious metal Pt commonly used for high pressure experiments can usually form halides, oxides, and sulfides, but the formation of nitrides and borides is difficult. In 2004, Gregoryanz et al. [82] first synthesized platinum nitride (PtN_x) using a laser-heat diamond anvil cell (LH-DAC) under the pressure of 40−50 GPa and above 2000 K. Platinum nitride has attracted great attention due to its high elastic modulus ($B = 392$ GPa) and special electronic structure. At present, the theoretical and experimental research of precious metal compounds (Pt, Ir, Os, etc.) is becoming one of the important frontier topics involving high pressure synthesis and materials science.

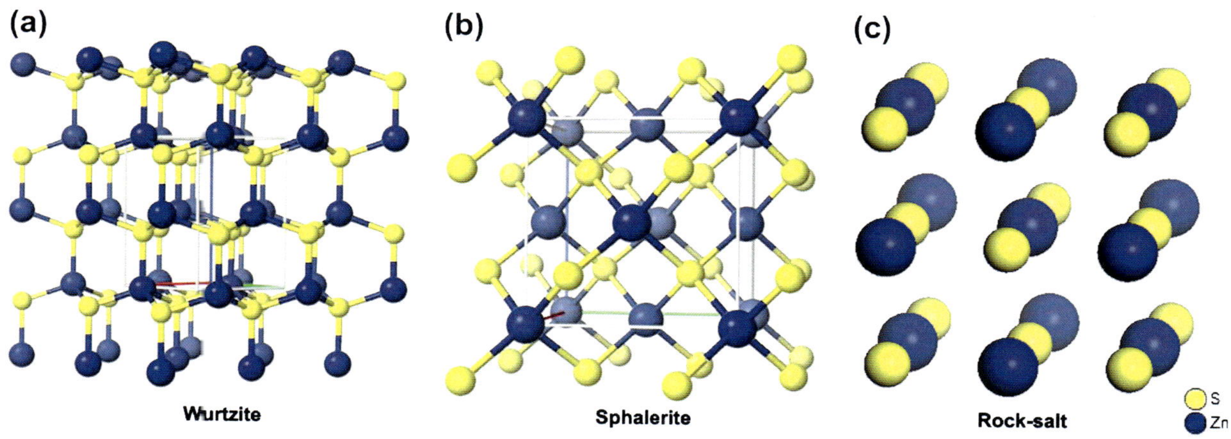

FIGURE 5.20 The crystal structures of ZnS. Sulfur and zinc atoms are denoted as yellow and blue balls, respectively.

5.4.2. Electronic Structure Changes Under High Pressures

A thermodynamic variable like temperature, pressure has a significant effect on the properties of materials. Pressure reduces the volume of materials under same temperature; at the atomic level, pressure shortens the distance between molecules or atoms. As the physical and chemical properties of materials depend on the interaction between outer electron orbits, when the distance between atoms or molecules changes under the external pressure, the interaction will also change, which will lead to the influences in the energy levels to varying degrees, and affect the nature of the material. Properties such as the band gap, light absorption coefficient, elastic coefficient, and bulk modulus may be altered under the condition of high pressure.

Pressure normally affects the electronic structure by increasing the electronic orbital overlap of the adjacent molecules or atoms, and thus affects the nature of the electronic orbitals. The way the pressure affects the electronic energy levels can be used to characterize the electronic state, verify the basic theory, observe electron transfer to form a new ground state (such as an insulator to metal transition), and so on. For example, lanthanide and transition metal oxide $LaCoO_3$ with perovskite structure is nonmagnetic semiconductor under normal pressure and low temperature. The ground state of Co^{3+} (t_{2g}^{6}, s = 0) is in low-spin state. However, when the temperature reaches 100 K, the magnetic susceptibility of the oxides changes significantly, becoming paramagnetic; the oxides begin to transform from semiconductor to metal at 425 K. Under ultrahigh pressure, the original cell volume and lattice constants of $LaCoO_3$(s) crystals are changed significantly, as shown in Fig. 5.21. Through neutron diffraction experiments we can further determine the changes of Co–O bond length and Co–O–Co bond angle of the $LaCoO_3$(s) crystals, as shown in Fig. 5.22. At room temperature, as the pressure increases, Co–O bond length decreases linearly, while the Co–O–Co bond angle increases [83].

Raman spectroscopy results of $LaCoO_3$(s) crystals at high pressure are shown in Fig. 5.23. The results reveal that the vibration frequency of A_{2g} breathing vibration mode and E_g vibration mode (Jahn–Teller distortion) increases as pressure increases. When the temperature reaches 550 K at 2.5 GPa, the intensities of the A_{2g} mode and the E_g mode are significantly reduced, which signifies that $LaCoO_3$(s) crystals change from semiconductor to metallic state, which also indicates that the transition temperature of $LaCoO_3$(s) crystals shifted is from 425 K at ambient pressure to 550 K under high pressure.

Magnetite is an important compound found in the deep earth, and was the first magnetic material discovered by human beings. The molecular formula of magnetite is represented as $Fe^{3+}{}_{(Td)}(Fe^{2+}\ Fe^{3+})_{(Oh)}O_4$, where T_d stands for tetrahedral position and O_h stands for octahedral position. At normal temperatures, magnetite has an anti-spinel type structure with half of the Fe^{3+} occupying the tetrahedral voids, while Fe^{2+} and the remaining half of Fe^{3+} occupy the octahedral positions. At pressures above 25 GPa, the phase transition of magnetite will take place. Mao et al. demonstrated that the structure of high-pressure phase magnetite has monoclinic symmetry where all of the

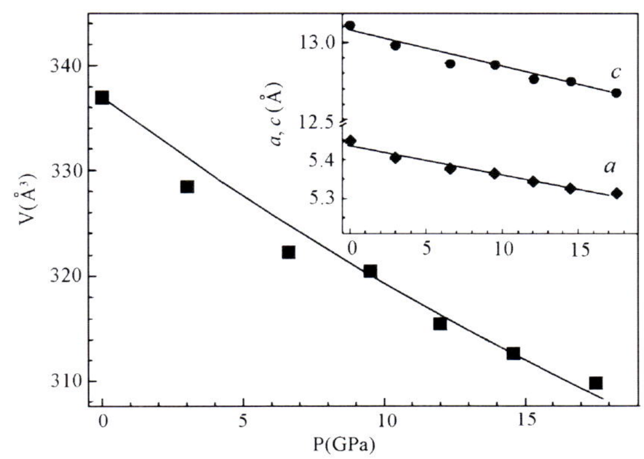

FIGURE 5.21 Pressure dependences of the unit-cell volume fitted by the Birch–Murnaghan equation of state and lattice parameters (in the hexagonal setting, inset) of $LaCoO_3$ at ambient temperature fitted by linear functions [83]. *Reprinted figure with permission from D.P. Kozlenko, N.O. Golosova, Z. Jirak, L.S. Dubrovinsky, B.N. Savenko, M.G. Tucker, Y. Le Godec and V.P. Glazkov, Physical Review B, 75, 064422, 2007. Copyright (1999) by the American Physical Society.*

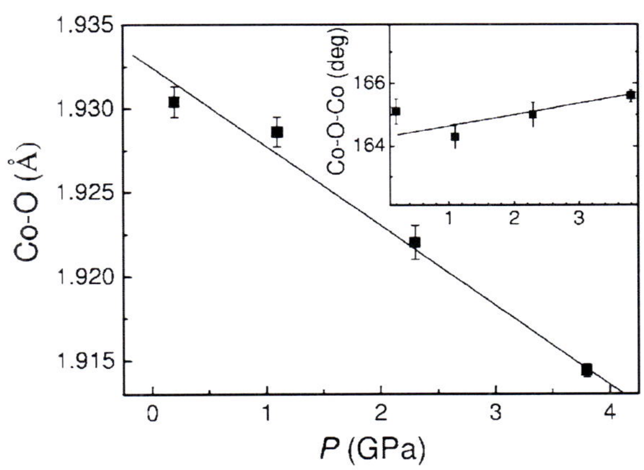

FIGURE 5.22 Pressure dependences of the Co–O bond length and Co–O–Co angle (inset) at ambient temperature [83]. *Reprinted figure with permission from D.P. Kozlenko, N.O. Golosova, Z. Jirak, L.S. Dubrovinsky, B.N. Savenko, M.G. Tucker, Y. Le Godec and V.P. Glazkov, Physical Review B, 75, 064422, 2007. Copyright (1999) by the American Physical Society.*

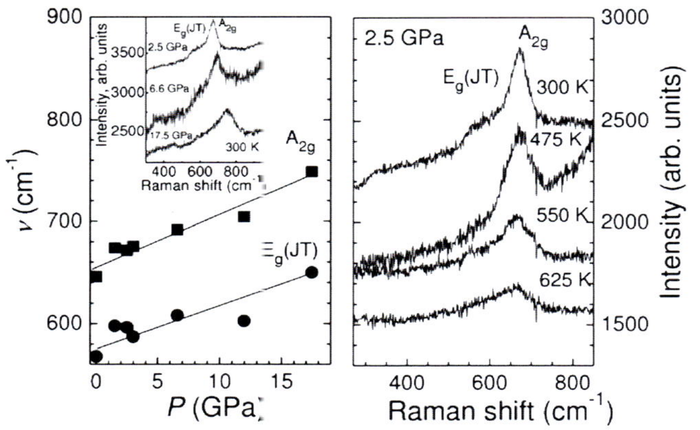

FIGURE 5.23 Left panel: pressure dependences of A_{2g} breathing and quadrupole (JT) E_g mode frequencies of $LaCoO_3$ and Raman spectra of $LaCoO_3$ at different pressures and ambient temperature (inset). Right panel: temperature dependencies of Raman spectra of $LaCoO_3$ at P = 2.5 GPa [33]. *Reprinted figure with permission from D.P. Kozlenko, N.O. Golosova, Z. Jirak, L.S. Dubrovinsky, B.N. Savenko, M.G. Tucker, Y. Le Godec and V.P. Glazkov, Physical Review B, 75, 064422, 2007. Copyright (1999) by the American Physical Society.*

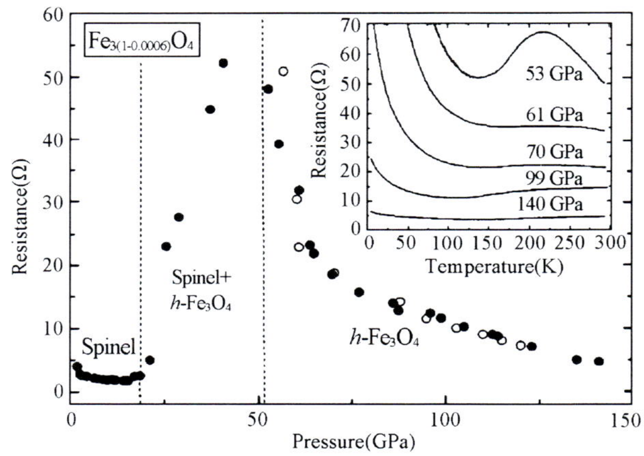

FIGURE 5.24 The pressure dependence of the resistance of *h*-Fe_3O_4 recorded at 295 K. The open (○) and solid (●) circles are data points obtained during the cycles of compression and decompression, respectively, of the highly stoichiometric $Fe_{3(1-\delta)}O_4(\delta = 0.0006)$ sample [85]. *Reprinted figure with permission from W.M. Xu, G.Y. Machavariani, G.K. Rozenberg and M.P. Pasternak, Physical Review B, 70, 174106, 2004. Copyright (2004) by the American Physical Society.*

Fe atoms occupy octahedral sites, named *h*-Fe_3O_4 [84]. As shown in Fig. 5.24 [85], the electrical property of magnetite also changes accordingly under high pressure. A significant increase in the resistance of magnetite was observed at pressures from 20 to 40 GPa, followed by the phase transition from an insulator to a metal at approximately 50 GPa. Ding et al. used X-ray magnetic circular dichroism (XMCD) to study the magnetic changes of magnetite under high pressure [86]. The results show that the magnetic dichroism decreased by about 50% at pressure range from 12 to 16 GPa, which could be attributed to the change from high-spin state (HS) to intermediate-spin state (IS) of the octahedral sited Fe^{2+}.

Under high pressure environment of the Earth's deep mantle, high pressure may change the spin state of iron, which is one of the most important element in the Earth, from a high-spin to a low-spin state, which can cause the valence change of Fe^{2+} in the iron-bearing silicates or oxides [87]. The configurations of electrons in the 3*d* orbitals of Fe^{2+} with high-spin and low-spin state (under high pressure) are shown in Fig. 5.25. According to Hund's rule, the energy of 3*d* electrons with antiparallel spins (low-spin state) is higher than that with the parallel spin (high-spin state). Under low pressure, the 3*d* electrons of Fe^{2+} will occupy the e_g and t_{2g} orbitals, which cause the high spin state. When high pressure is applied, the energy of the electron orbitals increases with increasing pressure, and the energy of e_g orbitals increases more than that of t_{2g} orbitals due to more overlaps with the nearest neighbor oxygen orbitals. Therefore, six 3*d* electrons preferentially occupy the t_{2g} orbitals with lower energy (low spin state), and Fe^{2+} doesn't have spin magnetic moment. The Speziale et al.'s results of in situ high-pressure Mossbauer spectroscopy on magnesiowüstite [(Mg_x Fe_{1-x})O] confirmed the spin change of Fe^{2+} [87].

Under high pressure, the superconducting transition temperature (T_C) of pure metals could change a lot. For example, at ambient pressure, the T_C of lithium metal is ~0.4 mK [88], but T_C was increased to 5 K at high pressure of 20 GPa, or 14–20 K at high pressure of 30 GPa [89,90]. The calcium metal does not have the superconducting change under ambient

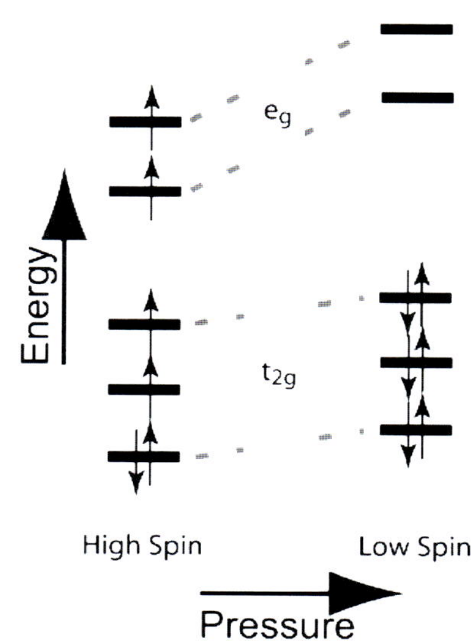

FIGURE 5.25 The spin diagram of electrons among 3*d* orbitals for sixfold coordinated Fe^{2+} with the varieties under pressure [87].

pressure, but it changes to the superconducting state at the temperature of 1.2 K [91] and high pressure of 50 GPa, its corresponding T_C rises to 25 K when the pressure is 160 GPa [92], which is the T_C record of pure metals.

Since 1980, the research on the high temperature superconductor becomes more and more important, and one of its roles is to find the new high temperature superconducting material. It has been found that high-pressure technique is a very effective method for exploring new superconductors. For example, Jin et al. synthesized new chlorine-containing high-T_C superconductor $Sr_{3-x}\ Ca_x\ Cu_2\ O_{4+\delta}\ Cl_{2-y}$ under the pressure of 5 GPa and at the temperature of 1000°C [93], which has the T_C of 80 K. The doping mechanism of this superconductor is neither the interlayer ion, nor the cation substitution, but a new doping, i.e., apical oxygen doping. The partial substitution of the monovalent Cl^- with divalent O^{2-} introduces the hole-like carriers. This superconductor is the first *p*-type high temperature superconductor obtained by anion doping.

A characteristic feature of all high-T_C superconductors is their layered structure containing Cu—O layers, which are essential for the occurrence of superconductivity in these materials and their T_C is determined by the concentration of the carrier. The use of doping and high-pressure experiments has confirmed that superconducting transition temperature is closely related to their carrier concentration [94,95], so they are the important methods in the search for new high temperature superconductor and the understanding of the mechanism of their superconductivity. High pressure can change the crystal structure parameters, carrier concentration, Fermi level, etc., which is more direct and simple than the doping. Therefore, studying the change of the parameters of high temperature superconductors via high pressure has attracted more and more attention.

5.5. MAJOR ROLES OF HIGH PRESSURE METHOD IN INORGANIC SYNTHESIS

High-temperature and high-pressure synthetic methods can lead to the formation of many new compounds, which are very difficult to synthesize under conventional conditions. This can be attributed to the unusual physical and chemical phenomena often observed when these thermodynamic parameters are used together and consequently result in a number of new inorganic compounds with unique structures. Moreover, there are many special functions associated with the use of high temperature and high pressure experimental methods that conventional techniques cannot achieve [44].

5.5.1. Preventing the Decomposition of Thermally Unstable Starting Reagents

As the decomposition temperature for the compounds that are unstable thermodynamically will increase under high-pressure, these compounds can be used directly as starting materials in high-temperature and high-pressure syntheses for synthesizing some special inorganic compounds. For example, under high-temperature and normal pressure, compounds such as HgO, PbO_2, and MnO_2 will decompose into metal oxides with lower chemical valence, making them unsuitable as starting materials. While under high pressure, the decomposition temperature of these compounds will be much higher, so that they can be used as reactants in inorganic synthesis, which leads to the new route for divalent Hg synthetic chemistry.

As early as 1968, HgO and its corresponding oxides were used to synthesize $Hg_2M_2O_7$ (M = V, Nb, Ta, Sb) complex oxides with pyrochlore structure under high temperature and high pressure [96], and the mercury cuprate family superconductor $HgBa_2Ca_2Cu_3O_{8+\delta}$ (Hg-1223) with high transition temperature T_C of 133 K was synthesized under high-pressure by Morikawa et al. using HgO as one of the starting materials [97,98]. In other studies, $Ba_2Ca_3Cu_4O_{10+\delta}$ (Hg-1234 and Hg-1223) type superconductors were also synthesized by using the same method, at pressure of 2–4 GPa [99].

The complex oxide $Ln_2M_2O_7$ (M = Pb, Mn; Ln = rare earth elements) with pyrochlore structure can also be directly synthesized under high pressure and high temperature, using PbO_2 or MnO_2 and the corresponding rare earth element oxides as starting materials. Some more examples include rare earth plumbite that was obtained under the conditions of 300 MPa and 700°C [100], or $Y_2Mn_2O_7$ synthesized at even higher pressure [101]. It is worth noting that PbO_2 will decompose under atmospheric pressure at a temperature less than 300°C, but is stable at higher temperatures when under high-pressure conditions.

5.5.2. Improving the Densification Effect and Consequently to Help the Preparation of Compact Structures

The chemical reaction of A and B to C is presented in the following equation,

$$\begin{array}{ccc} A & + B \rightarrow & C \\ V/Z & V/Z & V/Z \end{array}$$

where V/Z is the molar volume of each compound. When $(V_A/Z_A + V_B/Z_B) > V_C/Z_C$, the use of pressure will be conducive in the synthesis of compound C.

The compounds shown in Table 5.3 were synthesized by means of the high-pressure densification method,

TABLE 5.3 Compounds Synthesized Under High Pressure.

Compounds	Synthetic conditions	Crystal structures
MV_2O_6 (M = Ni, Mg, Co, Zn, Mn, Cd)	5–8 GPa; 800–1200°C	Orthogonal Columbite [113]
$Pb_2(B, B')O_6$ (B = W, Mo; B′ = Zn, Fe, Co, Mn)	1~3 GPa; 850°C	Perovskite [114, 115]
$PbSnO_3$	6~7 GPa; 500°C	
YbO	4 GPa; 1200°C	Rock Salt [102]
$Ln_2Ge_2O_7$ (Ln = Gd, Dy, Ho, Et, Tm, Yb, Lu, Tl, Y, Sc, In)	5–6 GPa; 900–1250°C	Pyrochlore [106]
$LaCo_2$	1–6.5 GPa; 1050–1350°C	$MgCu_2$ [111,112]
$LnFe_2$ (Ln = Pt, Nd, Yb)	2–3 GPa; 1000°C	

which resulted in products with more compact structures than those under atmospheric condition. For example, MV_2O_6 (M = Ni, Mg, Co, Zn, Mn, Cd) have the structure of brannerite under normal pressure, but will have a more closed packed structure of orthogonal columbite when synthesized under high pressure.

Under high pressure and high temperature, the monoxides of rare earth elements such as YbO are obtained through the following chemical reaction [102],

$$Yb_2O_3 + Yb \rightarrow 3YbO$$

In general, such chemical reactions will not take place under the conditions of ambient pressure and high temperature, however, application of high pressure will make the divalent ytterbium stable. Moreover, some novel materials can be synthesized under high pressure using the different compressing ratios of the elements which consist of the target compounds. The compressing ratio of cation or anion is decided by their electronic structures or the original sizes. Because of the differential compressible ratio, the crystal structures of the corresponding inorganic compounds may be different, resulting in a variety of structure types. For example, the compound $A_2B(IV)_2O_7$ has a pyrochlore structure, and the maximum r_A/r_B ratio for its stable structure is less than 1.55 [103–105]; however, under the conditions of high pressure and high temperature (6.5 GPa, 900°C < T < 1250°C), the same structure type can be kept until the element of Ge(IV) ($r \approx 0.053$ nm), and the corresponding r_A/r_B ratio is about 1.8 [106]. This reaction takes advantage of the different compressing ratios between cation and oxygen anion, and the smaller cation was pressed into the crystal lattice to yield a compound that is impossible to form under normal pressure. Other inorganic compounds obtained using same methods include those having structures similar to rutile, corundum, or perovskite, such as SiO_2, $MgGeO_3$, and $SrGeO_3$, respectively [107–110]. Meanwhile, compounds bearing the structure of $MgCu_2$ such as $LaCo_2$ and $LnFe_2$ (Ln = Pt, Nd, Yb) have also been synthesized under high pressure according to the characteristics of the different element compressing ratios [111,112].

5.5.3. Decreasing Reaction Temperature and Shortening Reaction Time

High-pressure can increase the chemical reaction rate and the conversion rate of the product, therefore greatly reducing the synthesis temperature and the reaction time. Since high pressure can make the tolerance factor small, under high-temperatures and high-pressures it is easy to synthesize compounds which are very difficult to obtain at high temperatures and ambient pressure, for example, the synthesis of compounds in the mix systems of O–F, N–F, O–N, and double rare earth oxides $LnLn'O_3$ (Ln = rare earth elements). In the mixed systems of O–F, N–F, and O–N, if the corresponding compounds were synthesized using conventional methods, they often have low thermal stability because of the differences in their reactivity and physical and chemical properties; however, under high pressure we can synthesize a lot of stable compounds in the system of O–F, such as $KTiO_2F$ and Tl (I)Tl(III)O_2F with a perovskite structure [116]. As shown in Table 5.4, the experimental results demonstrate the significant difference between ambient pressure and high pressure reaction conditions, in which the synthesis of $LaLnO_3$ (Ln = Ho, Er, Yb, Lu) under ambient pressure and at a very high temperature (1950 K) for a few hundreds hours resulted in only the C-type solid solution and a small amount of $LaLnO_3$ [117] product, but pure $LaLnO_3$ was obtained in less than 30 min when the reaction was conducted at high pressure (2.9 GPa) and lower temperature

TABLE 5.4 Synthetic Conditions for the Synthesis of $LaLnO_3$.

Compounds	Pressure and temperature	Duration
$LaErO_3$ (Type C solid solution and small amount of $LaErO_3$)	0.1 MPa;1950 K	8 days [117]
$LaLnO_3$ (Ln = Ho, Er, Yb, Lu)	2.9 GPa; < 1550 K	Less than 30 min [118]

(<1550 K) [118]. This also demonstrates the difficulty in synthesizing double rare-earth oxides under ambient pressure, which can be overcome easily by using high temperature and high pressure reaction conditions, thus making it possible to synthesize a series of oxides $LaLnO_3$ [119].

5.5.4. Stabilizing the Highest Oxidation States of Transition Metals Through the Development of High Oxygen or Fluorine Pressure

The transition element-containing compounds synthesized under ambient pressure usually have the normal oxidation states of transition elements; in contrast, if they are synthesized under high pressure, they will have higher oxidation states of transition elements. For example, under high pressure and high temperature conditions, the transition element-containing compounds with Fe^{4+} and Cu^{3+} have been obtained, which is the unique character and the advantage of high-temperature and high-pressure synthesis method.

In order to explore the relationship between the oxidation states of the chemical bond of M^{n+}—X and the physical properties of the final reaction products, the experiments are employed under high oxygen and fluorine pressure, respectively. As the oxidation states of M increases, the strength of chemical bond M^{n+}—X will promptly get stronger [119,120]. Under high oxygen pressure, a series of new transition element-containing compounds with new oxidation states are obtained, and they are shown in Table 5.5 in which the relationship between the oxidation states and the binding energy is listed.

In the study of high T_C superconductor materials, high oxygen pressure (up to 100 MPa) was used to remove the impurities and further change the T_C of the cuprate superconductor [122]. Karpinski et al. reported the synthesis of the low-dimensional compound $Sr_{0.73}CuO_2$ with infinite chain structure using high oxygen pressure (220 MPa). This low-dimension compound is not a superconductor, but an insulator that exhibits antiferromagnetism at 135 K [123]. In the magnetic field, there will be an obvious change when the temperature is below 10 K, which is the superconductive transition temperature for the analogous compound $Sr_{0.4}Ca_{13.6}Cu_{24}O_{41.84}$. For the compounds of $La_2CuO_{4+\delta}$ or $La_{2-x}Nd_xCuO_{4+\delta}$ [122,124], high oxygen pressure can cause the mixing valence state of Cu(II) and Cu(III), which will induce their superconductivity.

TABLE 5.5 Comparison of Oxidation State $n+$ with Binding Energy Δ (eV) for the Compounds with Perovskie Structures [121].

Compounds	Oxidation state, $n+$	Binding energy Δ (eV)
La_2MgIrO_6	Ir(IV)	3.18
La_2LiIrO_6	Ir(V)	3.67
Ba_2CaIrO_6	Ir(VI)	3.89
Sr_2CaIrO_6	Ir(VII)	3.90

Reprinted from Journal of Alloys and Compounds, 262-263, G. Demazeau, D.Y. Jung, A. Largeteau, C. Cros and J.H. Choy, High oxygen pressures and the stabilization of the highest oxidation states of transition elements, p191, Copyright (1997), with permission from Elsevier.

Under high fluorine pressure conditions, the compounds with unusual oxidation states such as Na_3MF_6 [M = Co(III), Ni(III), Cu(III)] and Cs_2AgF_6 [Ag (III), Ag(V)] have been synthesized [125].

5.5.5. Hindering the Disproportionation of Intermediate Oxidation States

High pressure synthesis conditions can also stabilize the compounds with intermediate oxidation states, which will easily undergo a disproportionation reaction. For example, due to the instability of Cr(IV) in the tetrahedral position, a disproportionation reaction will take place: 3Cr(IV) → 2Cr(III) + Cr(VI). High pressure can shift the above reaction toward the left, and produce the pure Cr(IV). Pd(III) also easily undergoes a disproportionation reaction: 2Pd(III) → Pd(II) + Pd(IV), but high pressure can inhibit this reaction, and thus make it possible to synthesize the fluoride of Pd(III) [44].

5.5.6. The Synthesis of Light Element Compounds which have Special Physical Properties

Diamond and diamond-like compounds made up of B, C, N, O, and other light elements exhibit excellent physical and chemical properties, in particular, the unparalleled hardness. This series of compounds are therefore known as superhard compounds, or because they comprised light elements, as light element compounds.

In 1965, Hall and Compton [126] used boron and boron oxides as reagents to synthesize the boron suboxides B_2O, which is isoelectronic specie of the carbon, under conditions of high temperature and pressure (1200–1800°C, 5–7 GPa). B_2O was red-brown in color and hexagonal in structure. The compounds, which consist of the elements in groups I, II, III, ..., are called the "periodic compounds" [127]. These compounds may be symmetrical, such as BN, BeO, and LiF; or asymmetrical, such as BeN_2, LiN_3, B_2O, B_6O, B_3F, the cubic phase C_3N_4, and boron oxides. High-temperature and high pressure methods play an important role in the synthesis of these covalent compounds, which have constantly attracted scientists' attention [128–130].

5.5.7. Preparation of Important Amorphous and Quasicrystalline Materials using High Pressure Methods

High pressure can be used to prepare a number of important amorphous and quasicrystalline materials, or induce crystal into an amorphous phase. This is because the quenching rate under conditions of high temperature and high pressure is at least 1000 times faster than that under atmospheric pressure. For example, under conditions of high temperature and high pressure (≈1000°C, 0.5–1.5 GPa), in the $Li_3N-Ca_3N_2-P_3N_5$ system, a pure glass phase, in which only N^{3-} is as the skeleton anion, can be synthesized. Another example for the synthesis of amorphous phase under high pressure is amorphous ZrW_2O_8 by Perottoni et al., which have special property of thermal anti-expansion [131]. This type of material can be widely used in various fields of modern science because it has a lot of unique properties, such as transmittance, hardness, high-refraction angle, and high softening temperature. [132]. High pressure is also used to induce $AlPO_4$ partially amorphous under the conditions of nonhydrostatic pressure, which is used to improve the surface wear resistance of 9Cr18 steel [133]. High pressure quenching technology is also widely used to synthesize a large number of quasicrystal materials.

The development of high pressure technology promotes the synthesis of new materials in the field of materials science. The compact structure and the increasing of the strength of chemical bond caused by pressure lead to a series of special physical and chemical properties which have great value in many aspects in the field of science and technology. The syntheses of the nitride glass and covalent materials are typical examples. Furthermore, the synthesis of high oxidation state transition metal oxides can be achieved under conditions of high oxygen or high fluorine pressure. Finally, high pressure results in the formation of an enhanced M–X covalent bond, which can be used to further study the relationship between the physicochemical property of materials and the chemical bindings. With the development of solid-state science, the contribution of pressure parameters to the synthetic routes of materials will become increasingly important.

5.6. SOME IMPORTANT INORGANIC COMPOUNDS SYNTHESIZED UNDER HIGH PRESSURE

High-pressure methods have many advantages that the conventional synthesis methods cannot match. In the chemical reaction process, high pressure has many functions, including inhibiting the diffusion of atom, accelerating the action rate, improving the conversion efficiency, reducing the synthesis temperature, increasing the ligancy and symmetry of atom, shortening the bond length, improving the speed of the material condensing, intercepting a variety of metastable phases, and even improving the single-phase nature and crystallinity of products. For decades, high pressure methods have led to the syntheses of a large number of inorganic materials that have important applications and significant economic value.

5.6.1. Synthesis of Diamond Under High Pressure

Diamond and graphite are two inorganic materials composed of pure carbon. However, the different structures of these two compounds lead to great differences in their nature. For example, diamond is the hardest material in nature, while graphite is a good solid lubricant; diamond is a broadband semiconductor, while graphite is a real conductor; diamond is transparent, however graphite is opaque. The crystal structure of diamond is shown in Fig. 5.18, where each sp^3 hybridized carbon atom combines with four adjacent carbon atoms by covalent bond to form a stable tetrahedral structure, with a bond length of 0.15 nm and an angle of 109° 28′ between the bonds. The crystal structure of graphite is shown in Fig. 5.17; each carbon atom (sp^2 hybridization) combines with three adjacent carbon atoms by covalent bond to form the carbon atom sheet. Graphite layers are connected by van der Waals force, so they are easy to slip.

The production of diamond in nature is very rare. Usually, the production of finished diamond is only a billionth of what is mined, so it is difficult to meet the growing industrial demand, leading to a need for synthetic diamond. In 1796, British scientist S. Tennant's precise combustion test first revealed that diamond is composed of pure carbon. Since then, mankind has begun a long exploration of synthetic diamond. In 1953, scientists in Sweden General Motor-Brown Boveri United Company (ASEA) synthesized the world's first man-made diamond at the pressure of about 8.4 GPa [134], but this discovery was kept a secret at the time, thus they lost the laurel of the inventor of the first synthetic diamond. One year later, under high temperature and pressure conditions, Hall et al. in the US General Electric (GE) Company used metal as a catalyst to repeat the feat of synthetic diamond. They published their experimental result in the journal *Nature* [5], and after many years, the road to a synthetic diamond had finally been successful.

After it was proven that diamond is a mineral composed of pure carbon and formed in the Earth's mantle under high temperature and high pressure, people have always wanted to synthesize diamond under the condition of high temperature and high pressure. It was not until the mid-twentieth century, after Simon and Berman obtained the phase boundary equation of the graphite–diamond [135] (Eq. (5.4)) through experiments and speculation, the study of synthetic diamond developed rapidly.

$$p/(108\ \text{Pa}) = 12.0 + 0.0301\ T/^{\circ}\text{C} \qquad (5.4)$$

The ground-state energy of the graphite and diamond is very close from the thermodynamic point of view, making it seem feasible to synthesize diamond with graphite as the starting material at lower temperature and pressure conditions; however, the transformation rate of graphite into diamond at lower pressure is almost nonexistent due to the enormous energy potential barrier between diamond and graphite. Further experimental studies have shown that the transformation of graphite into diamond at a high rate requires very high temperature and pressure conditions of 3300 K and 1.3×10^4 GPa [136]. Fortunately, scientists have found that the use of metal catalysts (Fe, Ni, Co, etc.) can reduce the energy potential barrier between diamond and graphite, allowing diamond to be synthesized at a relatively low temperature and pressure (1300–1500 K, 6–8 GPa) [5]. In 1958, the synthesis of diamond under high temperature and pressure conditions became commercial. Since then, the production of synthetic diamond has gradually overcome the production of natural diamond. Currently, more than 20 countries have the capacity to synthesize diamond: the United States, Britain, China, South Korea, Japan, France, Russia, Germany, etc.

In recent years, many scientists in the field of high pressure have tried to use nongraphitic carbon compounds, such as amorphous carbon, glassy carbon, carbon nanotubes, and C_{60}, as the starting material for diamond synthesis [137–144]. Because C_{60} has a special structure and shorter C–C bond lengths, it shows promise of making up the high-quality synthetic diamond under mild conditions. The crystal structure of C_{60} is shown in Fig. 5.26 [145]. The compound looks like a hollow sphere composed of C atoms; its surface is composed of six- and five-membered rings connecting to each other. Six-membered rings are also the structural motif of the graphite layer. In graphite, the C–C bond length is shorter than that in diamond structure (0.142 nm vs 0.154 nm), and the rigidity of C_{60} is even greater. After the high temperature and high pressure treatment of C_{60}, it was found that the ratio of sp^3 hybridization in production increases with the increase of reaction temperature. At pressure of 12.5 GPa and temperature of 450°C, the Vickers hardness of C_{60} polymer was 70 GPa. When the pressure exceeded 13 GPa and the temperature exceeded 800 K, the hardness of the C_{60} polymer obtained appeared to exceed the hardness of the diamond indenter used for hardness testing [142,143]; however, subsequent experiments determined that this was not the case [146–148].

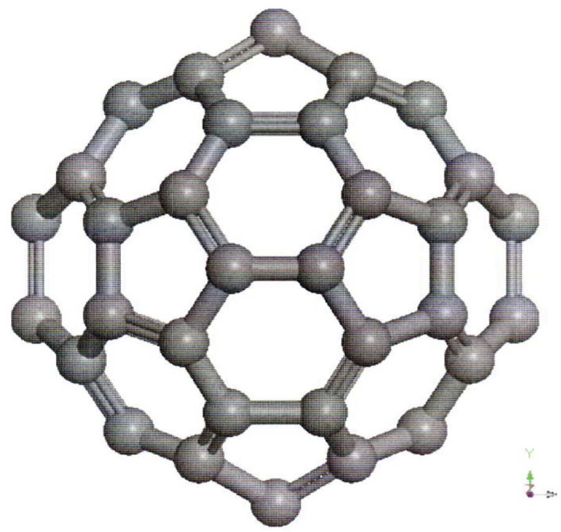

FIGURE 5.26 The crystal structure of C_{60} *Reprinted with permission from Macmillan Publishers Ltd:* Nature *[145], Copyright (1991).*

In 2003, Professor Irifune at Ehime University and his research team synthesized the nano-polycrystalline diamond with graphite as the starting material [149]. The sample was prepared under 12–25 GPa pressure and 2300–2500°C temperature conditions, using the Kawai-type multianvil high-temperature high-pressure devices. Transmission electron microscope (TEM) photographs showed that the nano-polycrystalline diamond is composed of a large number of diamond particles with 10–20 nm in size, the arrangement of particles is nondirectional. The Knoop hardness of this kind of diamond is up to 120–140 GPa and it is significantly better than single crystal diamond (Knoop hardness 60–120 GPa) due to its nondirectional particle arrangement [150]. After several years of effort, the same research team could synthesize nano-polycrystalline diamond with diameters of about 5 nm and thickness of several millimeters; there are no cracks and impurities in glomerocrysts [151].

In summary, although the production of synthetic diamond far exceeds the production of natural diamond, the current size of synthetic diamond can only be from a few microns to several millimeters due to the limitations of high temperature and high pressure technology, which severely restricted the application of synthetic diamond in large scale. Therefore, it will take a long time to replace the natural diamond with the synthetic diamond completely.

5.6.2. Synthesis of Cubic Boron Nitride (c-BN) Under High-pressure

The hexagonal boron nitride (*h*-BN) is a kind of pure man-made material composed of B and N. B, C, and N are adjacent elements in Periodic Table of Elements, which are similar in many aspects. For example, their atomic radii are approximately equal and they are also isoelectric. The hexagonal boron nitride (*h*-BN) and graphite have very similar crystal structures (Fig. 5.27). Because graphite can be converted into diamond under high temperature and pressure condition, it is believed that *h*-BN could also be converted into another compound whose structure is similar to diamond, that is, cubic boron nitride (*c*-BN). In 1957, Wentorf et al. in the US General Electric synthesized *c*-BN from *h*-BN at high temperature and pressure [6]. The *c*-BN has a sphalerite structure, with an sp^3 hybridized bond between B and N. The structure of *c*-BN is similar to that of diamond, which is nested by two face-centered cubic lattices along the cube diagonal stagger a quarter length (Fig. 5.28) [152]. However, there are some differences between them. In diamond structure, the two face-centered cubic lattices consist of the same atoms, but in *c*-BN structure, they are different. The atoms in *c*-BN are close packing of spheres which are formed by B atom layers and N atom layers arranged alternately with each other.

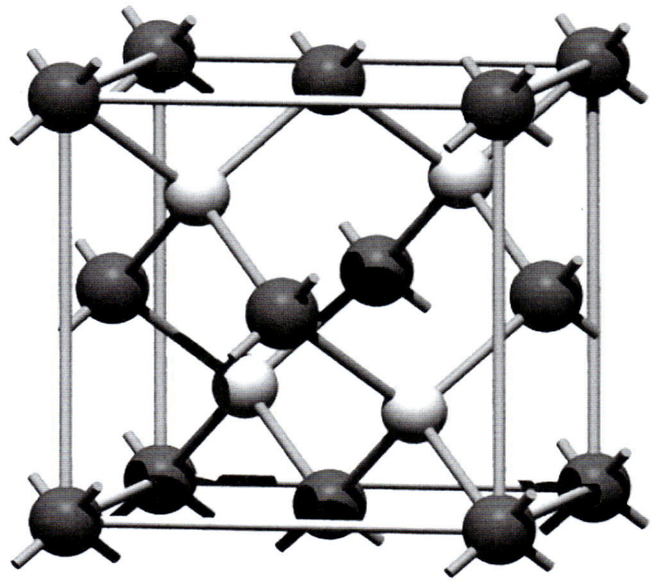

FIGURE 5.28 Crystal structure of cubic BN. Light spheres represent B, dark spheres represent N [152].

The *c*-BN has many characteristics such as high hardness, wide band gap, high stability, and high impedance, which lead to a wide range of applications. Although the hardness of *c*-BN is less than that of diamond, its chemical inertness and the thermal stability are superior to diamond. This gives *c*-BN an obvious advantage in the mechanical processing of high speed steel containing high vanadium, high tantalum, high cobalt, and other alloy structural steel, tool steel, mold steel, etc. The *c*-BN is greatly complementary with the diamond in this aspect.

During earlier periods, people thought the *h*-BN/*c*-BN system was analogous to the graphite/ diamond system. It was believed that *h*-BN could be converted to *c*-BN at high temperature and pressure because the *h*-BN was the thermodynamically stable phase at room temperature and ambient pressure, while the *c*-BN is the thermodynamically stable phase only at high temperature and pressure [153,154]. Recent experiments and calculations have shown that *c*-BN is also thermodynamically stable phase in the ambient temperature and

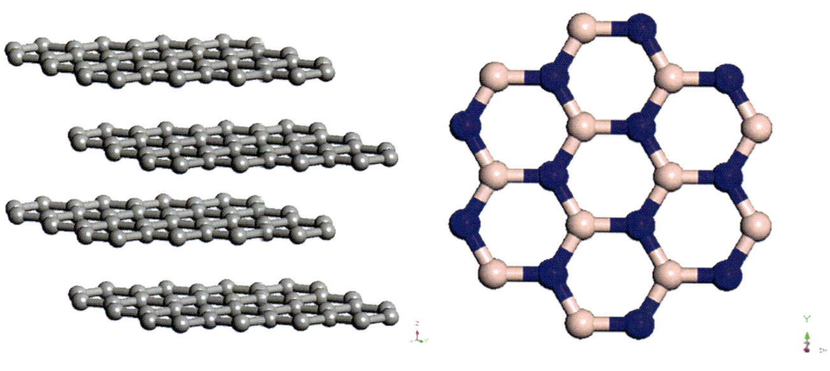

FIGURE 5.27 The crystal structures of graphite (left) and *h*-BN (right).

pressure. But this doesn't mean that *c*-BN can be formed directly from *h*-BN under ambient temperature and pressure conditions. This is because there is an enormous energy potential barrier between *h*-BN and *c*-BN. From a thermodynamics point of view, although *h*-BN could be converted into the *c*-BN at ambient temperature and pressure conditions, from a kinetic point of view, there need higher temperature and pressure for the phase transition from *h*-BN to *c*-BN. In addition, it was also found that adding a catalyst in the *h*-BN could greatly reduce the energy barrier and thus decrease the temperature and pressure required for the synthesis of the *c*-BN [155].

Wentorf first synthesized *c*-BN in 1957 under high temperature (1400°C) and high pressure (6 GPa) condition, using Mg as catalyst and *h*-BN as a raw material. Since then, the synthesis of *c*-BN has been rapidly developed. Many kinds of catalysts have been discovered and used for the synthesis of the *c*-BN, for example, alkali metals, alkaline earth metals and their nitrides, borides, boron nitride, water, ammonium salt, etc.

Usually when *h*-BN is heated to 1400−1600°C at a pressure of 4.0−6.0 GPa in the presence of a catalyst, *c*-BN is formed if given enough time. This has become the main method to produce *c*-BN in industry after many continuous improvements and developments. To increase the pressure of hydraulic unit, enlargement of the reaction chamber, search for high efficient catalyst, and more reasonable technology are the research objectives for the synthesis of *c*-BN single crystals with excellent quality, large sizes, and high hardness

5.6.3. High Pressure Synthesis of C_3N_4

The diamond had been regarded as the hardest substance for a long time, therefore the hardness of other substances were frequently measured using diamond as reference. At the end of 1980s, this conventional idea was first challenged by Liu and Cohen, the Physics Professors of California University at Berkeley. Starting with the crystal structure of β-Si_3N_4, they replaced the silicon element with carbon and applied the First-Principles with Pseudo-potentials method [156,157] to predict a theoretical new compound β-C_3N_4 that did not exist in nature. The calculated results indicated that β-C_3N_4 combined with covalent bonding, and the C−N bond length (0.147 nm) was shorter than C−C bond length (0.154 nm) in diamond. In the early 1990s, Liu confirmed that the elastic modulus of β-C_3N_4 exceeds that of diamond on the basis of ab initio calculations. Thus, it is thought that the hardness of β-C_3N_4 may surpass that of diamond. It must be pointed out, however, that the shear modulus of β-C_3N_4 represents the hardness of the material rather than the elastic modulus. That is to say, the value of bulk elastic modulus is not always consistent with the value of hardness. Even so, it was the first time for human to predict theoretically a new material with the hardness close to that of the diamond, and people paid more attention to the synthesis of C_3N_4.

Based on the importance of new C_3N_4 materials both in theory and application, many researchers began to investigate the materials by theoretical calculation after the structure of β-C_3N_4 was depicted by Cohen. So far, five different types of structures of C_3N_4 were obtained by calculation: such as α phase (α-C_3N_4), β phase (β-C_3N_4), cubic phase (*c*-C_3N_4), pseudo-cubic phase (pc-C_3N_4), and graphitic phase (g-C_3N_4). Except for g-C_3N_4, the theoretical hardness of the four C_3N_4 polymorphs is comparable with the diamond.

Many reports have claimed that superhard material C_3N_4 has been synthesized, but there has been no convincing evidence. For example, in 1998 Nguyen et al. reported that C_3N_4 materials were synthesized under high temperature and pressure [128]. But in their papers, only several peak positions of the relatively sharp diffraction peaks of the corresponding powder XRD pattern in their products are consistent with theoretical values, and there was no evidence to support a synthetic compound which simultaneously consisted of silicon and carbon elements. In the same year, Nesting et al. also reported two kinds of the C−N materials that were synthesized under high temperature and pressure [158]. Energy loss spectroscopy verified that the first C−N materials indeed consisted of C and N elements, but the elemental analysis showed that the ratio of C:N was 3:2. The second C−N materials crystallized in hexagonal system with $a = 0.7815$ nm, $c = 0.6219$ nm by indexing analysis. It is not generally in good agreement with any theoretical structures of C_3N_4. So far, no theoretical predication about the structure of C_3N_4 has been verified.

According to the discovery of g-C_3N_4, it is thought that the synthesis of superhard material C_3N_4 has a bright prospect. In 2006, a new C_3N_4 phase [159] was synthesized by Ming et al. using g-C_3N_4 as the starting material by means of diamond anvil cell (DAC), under conditions of 3−23 GPa and 1300−3000 K. X-ray diffraction data revealed that the structure of the new C_3N_4 phase was cubic, which is inconsistent with the theoretical calculation. In 2009, the shock-wave technique was used to synthesize the superhard material C_3N_4 under high pressure (16 GPa) and dicyandiamide ($C_2H_4N_4$) was used as the precursor [160]. The results show that the XRD pattern was in accordance with the theoretical value. For example, α-C_3N_4, β-C_3N_4, and g-C_3N_4 are recognizable in the products. The mass ratio of carbon to nitrogen is about 1.00:2.98 and nitrogen is mostly bonded with carbon.

5.6.4. The Synthesis of Superconductors Under High Pressure

In 1911, the Dutch physicist, Heike Kamerlingh Onnes discovered that the resistance of pure mercury suddenly vanished at 4.2 K [161] and he stated that mercury had passed into a new state, which on account of its extraordinary electrical properties may be called the superconductive state. There was no complete microscopic theory on superconductivity until 1957, when the BCS theory was finally proposed by three American physicists: John Bardeen, Leon Cooper, and Robert Schrieffer [162]. At that time, all known critical transition temperatures (T_C) of discovered superconductors were very low—until 1985, the highest T_C was only 23.2 K [163]. In 1986, the Sweden physicists, Bednorz and Mueller discovered superconductivity in a lanthanum-based cuprate perovskite material, which had a T_C value of 36 K and the first of the high temperature superconductors [164]. Subsequently, high temperature superconductor researches have been developed rapidly and the T_C scores are constantly updated.

High pressure plays an indispensable role in the synthesis of high T_C superconductors. In general, several kinds of pressure effects should be taken into account:

1. Increasing the density of material;
2. Improving the coordination number of ions;
3. Propitious to the process of substitution;
4. Reducing evaporation of volatile components;
5. Changing of solubility of individual components; and
6. Providing the atmosphere for reduction or oxidation.

Thus, high pressure technology plays a crucial role and cannot be replaced in the synthesis of new superconductors. Soon after the new discovery of superconductivity at T_C as high as 36 K in the La–Ba–Cu–O (LBCO) cuprates [164], Chu et al. reported that the corresponding T_C value was raised to 52 K under high pressure [165]. An unprecedented large positive pressure coefficient of T_C of LBCO led the researchers to realize that high pressure could lead to an even higher transition temperature. They then tried to use another means—chemical pressure to replace physical pressure to improve the superconducting transition temperature. They simulated the effects of external pressure by replacing La ion with the smaller isovalent Y ion. This led to the next giant leap in raising T_C above the temperature of liquid nitrogen, as first demonstrated in the Y–Ba–Cu–O (YBCO) cuprate [166]. Later, by in situ high-pressure study, it was found that pressure may increase the T_C value of p-type superconductors to a large extent. However, pressure had little impact on the transition temperature of n-type superconductors without vertex oxygen in the Cu–O plane, revealing the importance of the vertex oxygen. In addition, compared with the other cuprate superconductors, spin-ladder type compound $Ca_{13.6}Sr_{0.4}Cu_{24}O_{41}$ had no Cu–O plane and it had no superconductivity at ambient pressure, but became superconducting at high pressure with $T_C = 12$ K [167]. Therefore, the unique structure and superconductivity of this compound have become a focus of attention.

The crystal structures of the known high-temperature superconductor materials are of the perovskite-type close packed structure, which makes the high-temperature and high-pressure synthesis method play a unique role in the research of high T_C superconductors. In the recent years, almost all the new cuprate superconductors have been synthesized under high temperature and high pressure, including the following several series: infinite layer, Cu-based, halogen-based, ferrous pnictide compounds, and Ba–Cu–O compounds. It is now common practice to use high oxygen pressures (up to several hundred or several thousand bars) during the synthesis to control the oxygen content and hence vary T_C of cuprate superconductors, in addition by suppressing the appearance of unwanted impurity phases [122]. In 1997, Moriwaki et al. [98] demonstrated a high pressure synthesis route of these compounds directly from a mixture of the component oxides. In the same year, Lokshin et al. [99] reported a series of Hg-1234 ($HgBa_2Ca_3Cu_4O_{10+\delta}$) and Hg-1223 ($HgBa_2Ca_2Cu_3O_{8+\delta}$) synthesized at the pressure of 2–4 GPa. After carefully studying the relationship between the products purity and the reaction temperature and pressure, the ratio of the starting materials, they found that the amount of Hg-1234 in the products was related to the oxygen content in the starting materials. In addition, they also found that there is a strong correlation between the value of T_C and the unit cell parameter. For example, Hg-1223 with $a = 0.3852$ nm exhibits superconductivity at the maximum transition temperature of 135 K. Attifield et al. had investigated a series of A_2CuO_4 (A = La, Nd, Ca, Sr, Ba) and found that T_C reached the maximum value of 39 K at a mean-A site cation radius of 0.122 nm [168]. Both of these works indicated great sensitivity of superconducting critical transition temperature to crystal lattice strains. Locquet et al. have utilized this dependency to double the T_C value (from 25 K to 49 K) in $La_{0.9}Sr_{0.1}CuO_4$ thin films grown on the $SrLaAlO_4$ substrate [169]. This work could provide a powerful route for turning the properties of the superconductor thin film.

Another interesting family of superconductors synthesized under high pressure is the rare earth-containing superconductors, particularly those containing Pr^{3+} ions. But the parent substance itself of these superconductors including Pr^{3+} ions, $PrBaCuO_2$ does not manifest superconductivity. The reason for this phenomenon is still not clear. In 1997, Chen and his coworkers synthesized a series of compounds $(R_{1-x}Pr_x)_2$ $Ba_4Cu_7O_{14-\delta}$ (R = Nd, Sm, Eu, Gd, Ho, and Tm) and found that the superconductivity was suppressed with increasing Pr content [170,171]. Yao et al. have synthesized nearly single phase orthorhombic Ca-doped Pr-123 in a highly oxidizing environment (using $KClO_4$ as oxygen source) [172]. This material was superconducting with $T_C = 52$ K. By changing the environment, a high tetragonal material was formed and its T_C value was increased to 97 K. High pressure synthetic methods have also been used to explore new families of ceramic superconductors. In 1997, Iyo et al. reported the synthesis of (M, C) $(Ba, Sr)_2Ca_2Cu_3O_9$ (M = Al and Ga) superconductors [173]. The partial substitution of C^{4+} for Al^{3+} or Ga^{3+} was found to enhance T_C.

In recent years, an iron-based family of high temperature superconductors obtained via high pressure synthesis has been receiving attention. This discovery soared research into the iron-based "pnictide" family of superconductors. In 2008, Takahashi and his colleagues found lanthanum oxygen fluorine iron arsenide ($LaO_{1-x}F_xFeAs$), an oxypnictide that superconducted below 26 K [174]. Meanwhile, research from other groups, mainly from China, demonstrated that replacing the lanthanum in $LaO_{1-x}F_xFeAs$ with other rare earth elements such as cerium [175], samarium [176], neodymium [177], and praseodymium [178] led to superconductors that work at 52 K. It is believed that the discovery of iron-based pnictide family will help researcher to better understand the mechanisms of high temperature superconductivity. High pressure method offers more opportunities to expand the search for higher T_C and better properties of superconductor materials.

5.6.5. The High Pressure Synthesis and Properties of the Compounds Containing Special Oxidation State Transition Metals

High pressure helps in the formation and subsequent stability of high oxidation state transition metals [36]. In 1972, Demazeau et al. synthesized the compound $LaCuO_3$ containing Cu^{3+} by using La_2CuO_4 and CuO as raw materials, $KClO_3$ as the oxygen source at 6 GPa and 950°C [179]. In 1995, $La_{1-x}Sr_xCuO_3$, including Cu^{3+} and Cu^{4+}, was synthesized by Darracq et al. under high oxygen pressure conditions at 7 GPa and 1000°C [180].

In recent years, inorganic synthesis under ultrahigh pressure has made great and rapid progress resulting in a large number of novel and important research results as well as the construction of ultrahigh pressure inorganic synthesis laboratories. So far, scientists have made remarkable research results using high pressure inorganic synthesis technology in the research frontiers of nanomaterials, superconducting materials, superhard materials, silicate with special structure, pressure-induced phase-change inorganic materials, etc. [181–204].

In the near future, the research topics in the field of inorganic synthesis under ultrahigh pressure should include:

(1) Reaction of inorganic compounds and mechanism of chemical combination, summarization of the synthetic regularities, synthesis of inorganic compounds that help to deeply understand the new physical phenomena and have important application backgrounds;
(2) Design of the various precursors, nanomaterials, and synthetic products at the atomic and molecular levels, research of the boundary chemical reaction in solid nanomaterials, and ultrahigh pressure synthesis of compounds which are difficult to synthesize under normal conditions;
(3) Synthesis and reaction mechanism of the single crystals of new inorganic compounds under ultrahigh pressure;
(4) High pressure capture of metastable phases with new structures, new physical properties, and new applications through controlling conditions and research on kinetic theory under high pressure;
(5) Synthesis of new compounds that does not exist in thermodynamic equilibrium phase diagram, combining ultrahigh pressure synthesis technology with nonequilibrium synthesis technology;
(6) Research on the inorganic synthesis reaction mechanism under the conditions of in situ ultrahigh pressure;
(7) Study of the application of the compounds which have been synthesized under ultrahigh pressure;
(8) Research on ultrahigh pressure devices with large volume chamber used for inorganic synthesis; and
(9) Search of new synthetic routes in order to lower the pressure and temperature for the synthesis of the known important materials already made up under high pressure.

References

[1] X.Y. Liu, Progress in Chemistry 21 (2009) 1373.
[2] P.F. McMillan, Chemical Society Reviews 35 (2006) 855.
[3] G. Demazeau, Comptes Rendus Chimie 12 (2009) 933.
[4] P.H. Abelson, Science 283 (1999) 1263.
[5] F.P. Bundy, H.T. Hall, H.M. Strong, J. Nature 176 (1955) 51.
[6] R.H. Wentorf, Journal of Chemical Physics 26 (1957) 956.
[7] P.W. Bridgman, J. Phys. Rev. 57 (1940) 237.

[8] M.I. Erements, High pressure experimental methods, Oxford University Press, Oxford, 1996.
[9] H.T. Hall, Review of Scientific Instruments 31 (1960) 125.
[10] H. Huppertz, Zeitschrift Fur Kristallographie 219 (2004) 330.
[11] H.T. Hall, Review of Scientific Instruments 29 (1969) 267.
[12] D. Walker, M.A. Carpenter, C.M. Hitch, American Mineralogist 75 (1990) 1020.
[13] C.E. Weir, E.R. Lippincott, A. Vanvalkenburg, E.N. Bunting, Journal of Research of the National Bureau of Standards Section a-Physics and Chemistry 63 (1959) 55.
[14] D.J. Dunstan, I.L. Spain, Journal of Physics E-Scientific Instruments 22 (1989) 913.
[15] A. Jayaraman, J. Rev. Sci. Instrum 57 (1986) 1013.
[16] J.H. King, Journal of Scientific Instruments 42 (1965) 374.
[17] A. Onodera, J. High Temp. High Press, 19 (1987) 579.
[18] A. Onodera, K. Suito, N. Kawai, Journal of Applied Physics 51 (1980) 315.
[19] A.K. Bandyopadhyay, B.S. Sivaram, S.V. Subramanyam, J. Bull. Mater. Sci. 2 (1980) 121.
[20] R.G. Greene, H. Luo, A.L. Ruoff, Physical Review Letters 73 (1994) 2075.
[21] J.D. Barnett, S. Block, Gj Piermari., Review of Scientific Instruments 44 (1973) 1.
[22] H.K. Mao, R.J. Hemley, M. Hanfland, Physical Review B 45 (1992) 8108.
[23] H.K. Mao, J. Xu, P.M. Bell, Journal of Geophysical Research-Solid Earth and Planets 91 (1986) 4673.
[24] C. Wang, High pressure and high temperature synthesis and characterization of silicates with special structures, Jilin University, Changchun, P R China, 2006.
[25] M.H. Manghnani, S.Y., High-pressure research in mineral physics, Terra Scientific Publishing Co, Tokyo, 1987.
[26] R.J. Hemley, Annual Review of Physical Chemistry 51 (2000) 763.
[27] E. Winger, B. Huntington, J. Chem. Phys. 3 (1935) 764.
[28] R.J. Hemley, H.K. Mao, Physical Review Letters 61 (1988) 857.
[29] H.E. Lorenzana, I.F. Silvera, K.A. Goettel, Physical Review Letters 63 (1989) 2080.
[30] H.K. Mao, R.J. Hemley, Science 244 (1989) 1462.
[31] H.K. Mao, R.J. Hemley, M. Hanfland, Physical Review Letters 65 (1990) 484.
[32] J.H. Eggert, F. Moshary, W.J. Evans, H.E. Lorenzana, K.A. Goettel, I.F. Silvera, et al., Physical Review Letters 66 (1991) 193.
[33] N.H. Chen, E. Sterer, I.F. Silvera, Physical Review Letters 76 (1996) 1663.
[34] R.J. Hemley, H.K. Mao, A. Goncharov, M. Hanfland, V. Struzhkin, Physical Review Letters 76 (1996) 1667.
[35] P. Loubeyre, F. Occelli, R. LeToullec, Nature 416 (2002) 613.
[36] C.J. Pickard, R.J. Needs, Nature Physics 3 (2007) 473.
[37] V. Iota, C.S. Yoo, H. Cynn, Science 283 (1999) 1510.
[38] C.S. Yoo, H. Cynn, F. Gygi, G. Galli, V. Iota, M. Nicol, et al., Physical Review Letters 83 (1999) 5527.
[39] S. Serra, C. Cavazzoni, G.L. Chiarotti, S. Scandolo, E. Tosatti, Science 284 (1999) 788.
[40] M. Santoro, F.A. Gorelli, R. Bini, G. Ruocco, S. Scandolo, W.A. Crichton, Nature 441 (2006) 857.
[41] P.F. McMillan, High Pressure Research 23 (2003) 7.
[42] T. Matsuoka, K. Shimizu, Nature 458 (2009) 186.
[43] Y.M. Ma, M. Eremets, A.R. Oganov, Y. Xie, I. Trojan, S. Medvedev, et al., Nature 458 (2009) 182.
[44] G. Demazeau, European Journal of Solid State and Inorganic Chemistry 34 (1997) 759.
[45] P.F. McMillan, Chemical Communications (2003) 919.
[46] D. Londono, W.F. Kuhs, J.L. Finney, Nature 332 (1988) 141.
[47] A. Hardage, H. Roberts, J. The Leading Edge (2006) 566.
[48] K.A. Kvenvolden, J. Chem, Geo. 71 (1996) 41.
[49] E.Y. Tonkov, High pressure phase transformation, Vol. 1.2, Gordon and Breach Science Publishers, Philadelphia, 1992.
[50] R. van Eldik, C.D. Hubbard, Chemistry under extreme or non-classical conditions, Wiley-Spektrum, New York, 1997.
[51] P.W. Atkins, Physical Chemistry, Oxford University Press, Oxford, 1994.
[52] J.W. Moore, R.G. Pearson, Kinetics and mechanism, John Wiley, New York, 1981.
[53] Y. Yamamoto, K. Maruyama, K. Matsumoto, Organometallics 3 (1984) 1583.
[54] J. Karpinski, J. Jun, S. Porowski, Journal of Crystal Growth 66 (1984) 1.
[55] Q.Q. Gou, Solid state physics, Remin Education Press, Beijing, 1978.
[56] A.M. Glazer, Acta Crystallographica Section B-Structural Science, B 28 (1972) 3384.
[57] R. Schmidt, M. Geis, H. Kelm, Zeitschrift Fur Physikalische Chemie-Frankfurt 92 (1974) 223.
[58] R.A. Marcus, Journal of Chemical Physics 24 (1956) 966.
[59] N.S. Hush, Transactions of the Faraday Society 57 (1961) 557.
[60] P.C. Ford, D.R. Crane, Coordination Chemistry Reviews 111 (1991) 153.
[61] H. Fukuoka, S. Yamanaka, Chemistry of Materials 22 (2010) 47.
[62] C. Wang, X.Y. Liu, M.E. Fleet, J.X. Li, S.H. Feng, R.R. Xu, et al., CrystEngComm 12 (2010) 1617.
[63] M.E. Fleet, X.Y. Liu, American Mineralogist 89 (2004) 396.
[64] A.M. Glazer, Acta Crystallographica Section A 31 (1975) 756.
[65] C.J. Howard, H.T. Stokes, Acta Crystallographica Section A 61 (2005) 93.
[66] D. Andrault, J.P. Poirier, Physics and Chemistry of Minerals 18 (1991) 91.
[67] T. Yagi, H.K. Mao, P.M. Bell, Physics and Chemistry of Minerals 3 (1978) 97.
[68] N.L. Ross, J. Zhao, R.J. Angel, Journal of Solid State Chemistry 177 (2004) 1276.
[69] N.L. Ross, J. Zhao, R.J. Angel, Journal of Solid State Chemistry 177 (2004) 3768.
[70] J. Zhao, N.L. Ross, R.J. Angel, Acta Crystallographica Section B-Structural Science 60 (2004) 263.
[71] J. Zhao, N.L. Ross, R.J. Angel, J. Physics 35 (2006) 461.
[72] I.D. Brown, D. Altermatt, Acta Crystallographica Section B-Structural Science 41 (1985) 244.
[73] M.W. Lufaso, P.M. Woodward, Acta Crystallographica Section B-Structural Science 57 (2001) 725.
[74] J. Zhao, N.L. Ross, R.J. Angel, Physics and Chemistry of Minerals 31 (2004) 299.
[75] E. Kato, H. Noguchi, M. Nagai, H. Okuyama, S. Kijima, A. Ishibashi, Electronics Letters 34 (1998) 282.
[76] H. Drickamer, Review of Scientific Instruments 41 (1970) 1667.
[77] F. Birch, Journal of Geophysical Research-Solid Earth and Planets 91 (1986) 4949.
[78] M.J. Walter, Y. Thibault, K. Wei, R.W. Luth, Canadian Journal of Physics 73 (1995) 273.
[79] S.B. Qadri, E.F. Skelton, A.D. Dinsmore, J.Z. Hu, W.J. Kim, C. Nelson, et al., Journal of Applied Physics 89 (2001) 115.
[80] A. Nazzal, A. Qteish, Physical Review B 53 (1996) 8262.
[81] J.Z. Jiang, L. Gerward, D. Frost, R. Secco, J. Peyronneau, J.S. Olsen, Journal of Applied Physics 86 (1999) 6608.
[82] E. Gregoryanz, C. Sanloup, M. Somayazulu, J. Badro, G. Fiquet, H.K. Mao, et al., Nature Materials 3 (2004) 294.
[83] D.P. Kozlenko, N.O. Golosova, Z. Jirak, L.S. Dubrovinsky, B.N. Savenko, M.G. Tucker, et al., Physical Review B 75 (2007) 064422.

[84] H.K. Mao, T. Takahash, W.A. Bassett, G.L. Kinsland, L. Merrill, Journal of Geophysical Research 79 (1974) 1165.
[85] W.M. Xu, G.Y. Machavariani, G.K. Rozenberg, M.P. Pasternak, Physical Review B 70 (2004) 174106.
[86] Y. Ding, D. Haskel, S.G. Ovchinnikov, Y.C. Tseng, Y.S. Orlov, J.C. Lang, et al., Physical Review Letters 100 (2008) 045508.
[87] S. Speziale, A. Milner, V.E. Lee, S.M. Clark, M.P. Pasternak, R. Jeanloz, Proceedings of the National Academy of Sciences of the United States of America 102 (2005) 17918.
[88] J. Tuoriniemi, K. Juntunen-Nurmilaukas, J. Uusvuori, E. Pentti, A. Salmela, A. Sebedash, Nature 447 (2007) 187.
[89] K. Shimizu, H. Ishikawa, D. Takao, T. Yagi, K. Amaya, Nature 419 (2002) 597.
[90] V.V. Struzhkin, M.I. Eremets, W. Gan, H.K. Mao, R.J. Hemley, Science 298 (2002) 1213.
[91] S. Okada, K. Shimizu, T.C. Kobayashi, K. Amaya, S. Endo, Journal of the Physical Society of Japan 65 (1996) 1924.
[92] T. Yabuuchi, T. Matsuoka, Y. Nakamoto, K. Shimizu, Journal of the Physical Society of Japan 75 (2006) 083703.
[93] C.Q. Jin, X.J. Wu, P. Laffez, T. Tatsuki, T. Tamura, S. Adachi, et al., Nature 375 (1995) 301.
[94] D.T. Jover, R.J. Wijngaarden, R.S. Liu, J.L. Tallon, R. Griessen, Physica C-Superconductivity and Its Applications 218 (1993) 24.
[95] D.T. Jover, R.J. Wijngaarden, A. Schilling, H.R. Ott, R. Griessen, Physica C 235 (1994) 893.
[96] A.W. Sleight, Inorganic Chemistry 7 (1968) 1704.
[97] A. Schilling, M. Cantoni, J.D. Guo, J. Nature (1998) 886.
[98] Y. Moriwaki, T. Tatsuki, A. TokiwaYamamoto, T. Tamura, S. Adachi, K. Tanabe, Journal of Materials Science Letters 16 (1997) 115.
[99] K.A. Lokshin, I.G. Kuzemskaya, L.F. Kulikova, E.V. Antipov, E.S. Itskevich, Physica C 279 (1997) 11.
[100] A.W. Sleight, Inorganic Chemistry 8 (1969) 1807.
[101] H. Fujinaka, N. Kinomura, Y. Miyamoto, in: B. Vodar, P. Marteau (Eds.), High pressure science and technology, Vol. 1, Pergamon Press, New York, 1980, p. 556.
[102] J.M. Leger, J. Maurion, L. Albert, in: B. Vodar., P.h. Marteau (Eds.), High pressure science and technology, Vol. 1, Pergamon Press, New York, 1980, p. 558.
[103] J.M. Haschke, Inorganic Chemistry 9 (1970) 851.
[104] G. Brauer, Zeitschrift Fur Anorganische Und Allgemeine Chemie 356 (1967) 46.
[105] O. Knop, F. Brisse, Canadian Journal of Chemistry 47 (1969) 971.
[106] R.D. Shannon, A.W. Sleight, Inorg. Chem. 7 (1968) 1649.
[107] R.D. Shannon, Solid State Communications 4 (1966) 629.
[108] R.D. Shannon, Inorganic Chemistry 6 (1967) 1474.
[109] A.E. Ringwood, M. Seabrook, Journal of Geophysical Research 67 (1962) 1690.
[110] A. Grzechnik, H. Hubert, P. McMillan, W. Petuskey, Integrated Ferroelectrics 15 (1997) 191.
[111] D.L. Robertson, J.F. Cannon, H.T. Hall, Materials Research Bulletin 7 (1972) 977.
[112] J.F. Cannon, H.T. Hall, Dl Robertso., Materials Research Bulletin 7 (1972) 5.
[113] M. Gondrand, A. Collomb, J.C. Joubert, R.D. Shannon, Journal of Solid State Chemistry 11 (1974) 1.
[114] T. Fujita, O. Fukunaga, T. Nakagawa, S. Nomura, Materials Research Bulletin 5 (1970) 759.
[115] F. Sugawara, Y. Syono, S. Akimoto, Materials Research Bulletin 3 (1968) 529.
[116] B.L. Chamderland, Materials Research Bulletin 6 (1971) 311.
[117] U.J. Berndt, J. Solid State Chem 13 (1975) 131.
[118] W.H. Su, D.M. Wu, X.F. Ma, Z.G. Qian, Y.F. Wang, X.Y. Li, et al., Physica B & C 139 (1986) 661.
[119] W.H. Su, X.X. Liu, M.Z. Jin, W.M. Xu, D.M. Wu, M.L. Liu, Physical Review B 37 (1988) 35.
[120] G. Demazeau, M. Pouchard, B. Buffat, P. Hagenmuller, Journal De Physique 45 (1984) 345.
[121] G. Demazeau, D.Y. Jung, A. Largeteau, C. Cros, J.H. Choy, Journal of Alloys and Compounds 262-263 (1997) 191.
[122] J.P. Attfield, A.L. Kharlanov, J.A. McAllister, Nature 394 (1998) 157.
[123] J. Karpinski, H. Schwer, G.I. Meijer, K. Conder, E.M. Kopnin, C. Rossel, Physica C 274 (1997) 99.
[124] F. Arrouy, A. Wattiaux, C. Cros, in: A.I. Singh (Ed.), High pressure research (proceedings XIIIth AIRAPT conference, Bangalore, India) "Recent trends in high pressure research", Oxford & IBH publishing, New-Delhi, 1991, p. 562.
[125] A.C. Dash, R.K. Nanda, Journal of Inorganic & Nuclear Chemistry 38 (1976) 119.
[126] H.T. Hall, L.A. Compton, Inorganic Chemistry 4 (1965) 1213.
[127] H.T. Hall, J. Science 148 (1963) 1331.
[128] J.H. Nguyen, W.A. Caldwell, L.R. Benedetti, J. Mater. Res. Symp 499 (1998) 303.
[129] X.Y. Liu, X.D. Zhao, W.M. Hou, W.H. Su, Journal of Alloys and Compounds 223 (1995). L7.
[130] H. Hubert, L.A.J. Garvie, P.R. Buseck, W.T. Petuskey, P.F. McMillan, Journal of Solid State Chemistry 133 (1997) 356.
[131] C.A. Perottoni, J.A.H. da Jornada, Science 280 (1998) 886.
[132] T. Grande, S. Jacob, J.R. Holloway, P.F. McMillan, C.A. Angell, Journal of Non-Crystalline Solids 184 (1995) 151.
[133] P. Gillet, J. Badro, B. Varrel, P.F. McMillan, Physical Review B 51 (1995) 11262.
[134] H. Liander, J. ASEA Journal (1955) 97.
[135] R. Berman and F. Simon, J.Z. Electrochem, 59 (1955) 333.
[136] F.P. Bundy, H.M. Strong, R.H. Wentorf, Chemistry and physics of carbon, Marcel Dekker, New York, 1973.
[137] A. Onodera, K. Higashi, Y. Irie, Journal of Materials Science 23 (1988) 422.
[138] H. Yusa, Diamond and Related Materials 11 (2002) 87.
[139] N. Dubrovinskaia, L. Dubrovinsky, F. Langenhorst, S. Jacobsen, C. Liebske, Diamond and Related Materials 14 (2005) 16.
[140] V. Brazhkin, G. Lyapin, J. Rev. High Pressure Sci. Technol. 7 (1998) 817.
[141] A.G. Lyapin, V.V. Brazhkin, S.G. Lyapin, J. Rev. High Pressure Sci. Technol. 7 (1998) 881.
[142] V.D. Blank, S.G. Buga, G.A. Dubitsky, N.R. Serebryanaya, M.Y. Popov, B. Sundqvist, Carbon 36 (1998) 319.
[143] P.D. Ownby, X. Yang, J. Liu, Journal of the American Ceramic Society 75 (1992) 1876.
[144] C.Z. Wang, K.M. Ho, J. Phys. Rev. Lett. 71 (1993) 1184.
[145] W.I.F. David, R.M. Ibberson, J.C. Matthewman, K. Prassides, T.J.S. Dennis, J.P. Hare, et al., Nature 353 (1991) 147.
[146] S. Prawer, K.W. Nugent, D.N. Jamieson, J.O. Orwa, L.A. Bursill, J.L. Peng, Chemical Physics Letters 332 (2000) 93.
[147] M. Popov, Y. Koga, S. Fujiwara, B.N. Mavrin, V.D. Blank, New Diamond and Frontier Carbon Technology 12 (2002) 229.
[148] A.V. Talyzin, L.S. Dubrovinsky, M. Oden, U. Jansson, Diamond and Related Materials 10 (2001) 2044.
[149] T. Irifune, A. Kurio, S. Sakamoto, T. Inoue, H. Sumiya, Nature 421 (2003) 599.
[150] T. Irifune, A. Kurio, S. Sakamoto, T. Inoue, H. Sumiya, K. Funakoshi, Physics of the Earth and Planetary Interiors 143 (2004) 593.
[151] H. Sumiya, T. Irifune, J.SEI Tech. Rev. 66 (2008) 85.
[152] A. Zerr, R. Riedel, T. Sekine, J.E. Lowther, W.Y. Ching, I. Tanaka, Advanced Materials 18 (2006) 2933.
[153] V.L. Solozhenko, V.Z. Turkevich, W.B. Holzapfel, Journal of Physical Chemistry B 103 (1999) 2903.

[154] V.L. Solozhenko, V.Z. Turkevich, N.V. Novikov, J.P. Petitet, Physical Chemistry Chemical Physics 6 (2004) 3900.
[155] B.P. Singh, V.L. Solozhenko, G. Will, Diamond and Related Materials 4 (1995) 1193.
[156] A.Y. Liu, M.L. Cohen, Science 245 (1989) 841.
[157] A.Y. Liu, M.L. Cohen, Physical Review B 41 (1990) 10727.
[158] D.C. Nesting, J. Kouvetakis, J.V. Badding, National institute for research in inorganic materials Tsukuba (1998) 1.
[159] L.C. Ming, P. Zinin, Y. Meng, X.R. Liu, S.M. Hong, Y. Xie, Journal of Applied Physics 99 (2006) 033520.
[160] Y.W. Yu, Y.C. Liu, H.L. Zhang, C.C. Chen, Journal of Inorganic Materials 24 (2009) 627.
[161] K. Onnes, J. Comm. Phys. Lab. Univ. Leiden (1911) 122.
[162] J. Bardeen, L.N. Cooper, J.R. Schrieffer, Physical Review 108 (1957) 1175.
[163] J.R. Gavaler, M.A. Janocko, C.K. Jones, Journal of Applied Physics 45 (1974) 3009.
[164] J.G. Bednorz, K.A. Muller, Zeitschrift Fur Physik B-Condensed Matter 64 (1986) 189.
[165] C.W. Chu, P.H. Hor, R.L. Meng, L. Gao, Z.J. Huang, Science 235 (1987) 567.
[166] M.K. Wu, J.R. Ashburn, C.J. Torng, P.H. Hor, R.L. Meng, L. Gao, et al., Physical Review Letters 58 (1987) 908.
[167] H. Takahashi, N. Mori, T. Nakanishi, T. Nagata, M. Uehara, J. Akimitsu, et al., Physica B 237 (1997) 112.
[168] A.J. Williams, B.M. Sobotka, J.P. Attfield, Journal of Solid State Chemistry 173 (2003) 456.
[169] J.P. Locquet, J. Perret, J. Fompeyrine, E. Machler, J.W. Seo, G. Van Tendeloo, Nature 394 (1998) 453.
[170] T.M. Chen, F.S. Kao, Journal of Solid State Chemistry 132 (1997) 73.
[171] T.M. Chen, Y.L. Lai, F.S. Kao, Physica C 282 (1997) 789.
[172] Y.S. Yao, Y.F. Xiong, D. Jin, J.W. Li, F. Wu, J.L. Luo, et al., Physica C 282 (1997) 49.
[173] A. Iyo, T. Saya, K. Tokiwa, Y. Tanaka, N. Terada, M. Tokumoto, et al., Physica C 282 (1997) 509.
[174] H. Takahashi, K. Igawa, K. Arii, Y. Kamihara, M. Hirano, H. Hosono, Nature 453 (2008) 376.
[175] G.F. Chen, Z. Li, D. Wu, G. Li, W.Z. Hu, J. Dong, et al., Physical Review Letters 100 (2008) 247002.
[176] X.H. Chen, T. Wu, G. Wu, R.H. Liu, H. Chen, D.F. Fang, Nature 453 (2008) 761.
[177] Z.A. Ren, J. Yang, W. Lu, W. Yi, X.L. Shen, Z.C. Li, et al., Europhysics Letters 82 (2008) 57002.
[178] Z.A. Ren, J. Yang, W. Lu, W. Yi, G.C. Che, X.L. Dong, et al., Materials Research Innovations 12 (2008) 105.
[179] G. Demazeau, A. Baranov, G. Heymann, H. Huppertz, A. Sobolev, I. Presniakov, Solid State Sciences 9 (2007) 376.
[180] S. Darracq, S.G. Kang, J.H. Choy, G. Demazeau, Journal of Solid State Chemistry 114 (1995) 88.
[181] M.E. Fleet, X.Y. Liu, Acta Crystallographica Section B-Structural Science 56 (2000) 940.
[182] A. San-Miguel, Chemical Society Reviews 35 (2006) 876.
[183] Y.H. Liu, G.C. Che, K.Q. Li, Z.X. Zhao, Z.Q. Kou, N.L. Di, et al., Physical Review B 71 (2005) 104503.
[184] J.C. Crowhurst, A.F. Goncharov, B. Sadigh, C.L. Evans, P.G. Morrall, J.L. Ferreira, et al., Science 311 (2006) 1275.
[185] M.E. Fleet, X.Y. Liu, Biomaterials 26 (2005) 7548.
[186] I. Goncharenko, M.I. Eremets, M. Hanfland, J.S. Tse, M. Amboage, Y. Yao, et al., Physical Review Letters 100 (2008) 045504.
[187] X.Y. Liu, M.E. Fleet, Journal of Physics-Condensed Matter 14 (2002) 11223.
[188] J.J. Wu, X.Y. Liu, S.H. Tolbert, Journal of Physical Chemistry B 104 (2000) 11837.
[189] A. Yoshiasa, K. Sugiyama, S. Sakai, H. Isobe, D. Sakamoto, K. Ota, et al., Journal of Crystal Growth 311 (2009) 974.
[190] F. Mezzadri, M. Calicchio, E. Gilioli, R. Cabassi, F. Bolzoni, G. Calestani, et al., Physical Review B 79 (2009) 014420.
[191] T. Nishino, N. Kitamura, K. Murotani, Journal of Polymer Science Part a-Polymer Chemistry 47 (2009) 754.
[192] L. Lei, D.W. He, Crystal Growth & Design 9 (2009) 1264.
[193] A. Haberer, H. Huppertz, Journal of Solid State Chemistry 182 (2009) 888.
[194] K. Miyazawa, K. Kihou, M. Ishikado, P.M. Shirage, C.H. Lee, N. Takeshita, et al., New Journal of Physics 11 (2009) 045002.
[195] J.G. Zhao, L.X. Yang, Y. Yu, F.Y. Li, R.C. Yu, C.Q. Jin, Inorganic Chemistry 48 (2009) 4290.
[196] A.V. Ponomareva, A.V. Ruban, N. Dubrovinskaia, L. Dubrovinsky, I.A. Abrikosov, Applied Physics Letters 94 (2009) 181912.
[197] L. George, V. Drozd, A. Durygin, J.H. Chen, S.K. Saxena, International Journal of Hydrogen Energy 34 (2009) 3410.
[198] A. Merlen, P. Toulemonde, S. Le Floch, G. Montagnac, T. Hammouda, O. Marty, et al., Carbon 47 (2009) 1643.
[199] H. Huppertz, S.A. Hering, C.E. Zvoriste, S. Lauterbach, O. Oeckler, R. Riedel, et al., Chemistry of Materials 21 (2009) 2101.
[200] T.C. Su, X.P. Jia, H.A. Ma, F.R. Yu, Y.J. Tian, G.H. Zuo, et al., Journal of Applied Physics 105 (2009) 073713.
[201] A. Zerr, G. Miehe, J.W. Li, D.A. Dzivenko, V.K. Bulatov, H. Hofer, et al., Advanced Functional Materials 19 (2009) 2282.
[202] D. Moser, D.J. Bull, T. Sato, D. Noreus, D. Kyoi, T. Sakai, et al., Journal of Materials Chemistry 19 (2009) 8150.
[203] S. Glisic, I. Lukic, D. Skala, Bioresource Technology 100 (2009) 6347.
[204] H.Y. Chen, L.P. Wang, J.M. Bai, J.C. Hanson, J.B. Warren, J.T. Muckerman, et al., Journal of Physical Chemistry C 114 (2010) 1809.

CHAPTER

6

Inorganic Photochemical Synthesis

Xinsheng Liu

BASF Corporation, USA

Inorganic photochemical synthesis with light irradiation (both ultraviolet and visible) has been one of the most fruitful synthesis methods for obtaining new and unique structured inorganic compounds and materials. The early studies of inorganic photochemical synthesis were mainly focusing on organometallic coordination compounds due to interest in their unique photophysical properties [1], electron transfer mechanism [2,3], and their possible use in solving energy crisis [4,5]. Later, the study on photochemical synthesis was greatly expanded to the synthesis of organometallic coordination compounds with unique structures, and specific photochromic, catalytic, and electronic properties. Over the last several decades, inorganic photochemical synthesis made significant progress along with the rapid establishment of numerous photophysical investigation techniques [6] and electron transfer theory [7]. In addition, great progress in our understanding of photophysical and photochemical processes occurred during the study of photochemical reactions. Investigations of the harvest of sunlight have resulted in the development of photochemistry of compounds ranging from small organometallic compounds to supramolecular systems [8,9]. The studies on photochemical, photo-assisted reactions, and photolysis have led to the establishment of many new facile synthetic routes for syntheses of metal, metal oxide and semiconductor films, and nano-sized particles [10,11]. The studies on photochemical reactions and photocatalysis have manifested that photochemical synthesis of clean fuels such as H_2 may become an ultimate solution for our energy crisis and global warming problems [12–14].

Photochemical synthesis is different from other synthesis methods because it is selective. This selectivity occurs because light absorption features of reactants determine their reaction products. In a general way, the synthesis processes can be photochemical, photo-assisted/photo-induced, and even photolytic, depending on the reaction systems. The species generated by photon excitation or photolysis can be primary species for final products or precursors that react subsequently to form new species and then to final products.

In this chapter, we will first briefly introduce the basic concepts and experimental methods for inorganic photochemical reactions and then outline the photochemical synthesis of organometallic compounds, inorganic compounds, films, nanomaterials, and photocatalytic generation of H_2 via water splitting. We do not intend to give a comprehensive review of the field but rather focus on the basic ideas of the syntheses.

6.1. THE BASIC CONCEPTS

Photochemical reaction is a chemical reaction of excited electronic states of substances. The electrons in the excited electronic states undergo various photophysical and chemical processes to lower their energy and carry out electron transfer reactions [15].

For organometallic complexes, electron transitions can take place within the transition metal ion, within a ligand, between different ligands, from metal to ligand, and from ligand to metal. When these different transitions occur upon photo-excitation, different photochemical reactions may be involved. For transition metal complex containing more than one metal species, electron transfer between the different metal species can also occur. Light with different wavelengths can selectively excite these transitions, and thereby direct the photochemical reaction pathways of the metal complex toward desired products.

Inorganic solid crystal materials upon photo-excitation undergo different photophysical processes compared to molecules due to their band structures. Light with energy larger than the band gap of the solids excite the electrons in the valence band to the conduction band. A thereafter electron migration occurs toward

Modern Inorganic Synthetic Chemistry, DOI: 10.1016/B978-0-444-53599-3.10006-X the solid surface and on the way of migration the electrons can recombine with the positively charged holes, and be trapped by defects and imperfections in the solid. The number of defect sites and imperfections in the crystal structure affect the migration of the excited electrons and reactions occurring on the solid surface.

Photon energy. The photon energy of light (normally with 200–800 nm wavelengths) in photochemical reactions is inversely proportional to the wavelength:

$$E = h\upsilon = hc/\lambda, \quad (6.1)$$

where E is energy (J), h is Planck constant, 6.62×10^{-34} J s, ν is the frequency of the light (s^{-1}), c is the light velocity under vacuum, 2.998×10^8 m s^{-1}, λ is the wavelength of the light (nm).

The molar absorbed photon energy is described by:

$$E = N_A h\upsilon = N_A hc/\lambda, \quad (6.2)$$

where N_A is Avogadro constant, 6.023×10^{23} mol^{-1}. Therefore, the energy obtained by a substance absorbing one mole of photons (6.023×10^{23} photons or one Einstein) is:

$$E = (6.62 \times 10^{-34} \times 2.998 \times 10^8 \times 6.023 \times 10^{23})/10^{-9} \times \lambda \times 10^3 = 1.197 \times 10^5/\lambda (\text{kJ mol}^{-1}). \quad (6.3)$$

Using this equation, if the light wavelength is known, the light energy can be easily calculated using the above equation.

Light absorption. A photochemical reaction takes place under light irradiation. During this process, light is absorbed and the efficiency of light absorption depends on the electronic structures of the substance and the match of the bond or band energy with the light energy. For organic compounds, the common transitions are $n \rightarrow \pi^*$, $\pi \rightarrow \pi^*$, $n \rightarrow \sigma^*$, and $\sigma \rightarrow \sigma^*$. Here, n represents a nonbonding orbital and * stands for an antibonding orbital. These transitions, in general, have high extinction coefficients, and the value can be as high as 10^5 L mol^{-1} cm^{-1}. Organic ligands in organometallic complexes generally maintain their noncomplexed absorption characteristics but differences can be observed due to interactions of the ligands with their central transition metal ions.

The *d–d* transitions of a transition metal ion in a complex have a much lower absorption extinction coefficient compared to their organic ligands due to their (octahedron) symmetry forbidden property. The d–d transition occurs in the visible spectral region with extinction coefficients only about 10^1–10^2 L mol^{-1} cm^{-1}. Transitions from central metal ion to ligand and from ligand to central transition metal ion generally occur in the UV and short wavelength visible spectral region. Their absorption extinction coefficient is relatively large and can be in the order of 10^4 L mol^{-1} cm^{-1}.

Light absorption by inorganic solids such as semiconductors is determined by their band structures. Their absorption spectra generally have a critical absorption wavelength below which the light with higher energy (short wavelength) is absorbed. The critical absorption wavelength depends on the band gap of the semiconductor. The larger the band gap, the shorter the critical absorption wavelength. The band gap can vary when particle size of a semiconductor reaches its nano-size regime (quantum size effect [16]). The quantum size effect enlarges the band gap of the semiconductor, resulting in an absorption onset shift toward shorter wavelength (higher energy).

Light absorption by a dilute solution of a compound without photochemical reaction upon light irradiation can be described by Beer–Lambert law:

$$I = I_0 e^{-acl}, \quad (6.4)$$

where I_0 is the incident light intensity, I is the transmitted light intensity, c is the concentration of the compound (mol L^{-1}), l is the light pathlength, i.e., the thickness of the solution cell (cm), and α is the absorption extinction coefficient. In practical applications, the equation has the form:

$$\log(I_0/I) = \varepsilon cl. \quad (6.5)$$

Here, ε is the molar extinction coefficient of the light-absorbing compound, $\varepsilon = \alpha/2.303$.

For solid powder materials, due to scattering of light by particles of the materials, the Beer–Lambert law cannot be used. In this case, the Kubelka–Munk equation [17] is applied under diffuse reflectance conditions:

$$f(R_\infty) = (1 - R_\infty)^2/2R = k/s, \quad (6.6)$$

where R is the absolute reflectivity of the solid layer, s is the scattering coefficient, k is molar absorption coefficient, and ∞ means that the solid layer should be thick enough with no transmitted light. From the above equation, it can be seen that when s is a constant, the Kubelka–Munk equation can be expressed as:

$$f(R_\infty) = (1 - R_\infty)^2/2R = c/k'. \quad (6.7)$$

Here, c is the sample concentration, k' is the parameter associated with particle size and molar absorptivity, $k' = s/2.303\varepsilon$. Under diffuse reflectance conditions, this equation can be regarded as the "Beer–Lambert law" for solid powder materials.

Quantum yield. The Start–Einstein law states that a chemical reaction can take place only after a molecule absorbs one quanta of photon energy. This law is no doubt proper for the initial reaction. However, in some cases, a molecule, after it absorbs one photon, can undergo chain reactions to form more than one molecule or can continuously absorb two photons in order to form a molecule. Under these conditions, the Start–Einstein law is no longer applicable. To overcome this difficulty in accounting for the photon utilization efficiency, the concept that describes the photochemical reaction yield—quantum yield, Φ—is introduced and it is defined as:

$$\Phi = \text{molecules produced or molecules consumed} / \text{photons absorbed}$$

Following this equation, the quantum yield can be obtained if the number of molecules produced or consumed and the number of photons absorbed are known. From the quantum yield, one can clearly understand the efficiency of a photochemical reaction Thus, the concept of quantum yield is more meaningful than the Stark–Einstein law in practice.

6.2. EXPERIMENTAL TECHNIQUES

Light sources. Nowadays, most popularly used light sources are mercury lamps. Other light sources like halogen lamps, xenon arc lamps, and phosphor-coated lamps are also used. Light emitting diodes (LED) may also be used but their narrow wavelength range may limit their use. The mercury lamps are convenient and easy to use and provide light with wavelengths ranging from ultraviolet to visible (200–750 nm). Based on the mercury vapor pressure in the lamps, they can be categorized into low pressure (0.6665–13.33 Pa), medium pressure (1.013×10^4–1.013×10^5 Pa) and high pressure (2.026×10^7 Pa) mercury lamps. The low pressure mercury lamp mainly emits 253.7 and 184.9 nm light at ambient temperature. The high energy 184.9 nm light is normally of low intensity due to absorption of the glass material of the lamp. If high-quality quartz is used for the lamp, this wavelength of light can be intense. For the low pressure mercury lamps, other wavelength light is also emitted but the intensity is low. Medium pressure mercury lamps operate at a relatively higher temperature and a few minutes warm-up is normally needed. The light emitted from this type of lamp has wavelengths of 265, 310, and 365 nm with relatively high intensities. This type of lamp has a wider wavelength range and is suitable for excitation of organic compounds. High pressure mercury lamps have even more characteristic emission lines compared to the low and medium pressure mercury lamps and for certain wavelength regions, the emission becomes continuous.

For some photochemical reactions, single monochromic light is needed. In these cases, laser light sources can be used. If more than one single wavelength light is needed, a combination of different lasers with different wavelengths can be assembled. For a laser unit, sometimes different wavelengths lasers can be obtained by changing the type of gases or mixing different gases with different ratios in the laser resonance cavity. The current tunable laser instrument can provide radiation with a range of wavelengths. The most commonly used lasers include argon and argon-pumped lasers, metal vapor-pumped dye lasers, and solid-state lasers.

If lasers are not available due to cost issues, the narrow wavelength window light can be obtained by using window or cutoff glass filters. Selection of these filters depends on the details of the photochemical experiments where selective excitation of some chromophores or removal of some unwanted light from the light sources to avoid side reactions may be needed.

Reactors. Reactors for photochemical synthesis vary, depending on the purposes. Effective use of the light source is normally the main concern. The Pyrex vessels used for photochemical reactions are transparent to the light with wavelengths > 300 nm, while quartz vessels allow light of wavelengths > 200 nm to pass through. The light source can be located in reactor center surrounded by photochemical reactants or can be set around a reaction vessel with reflection mirrors to reflect all the light to the reaction vessel. Sometimes, optical lenses may be used to focus light into reaction zones or areas. This is often encountered in film and nanoparticle making processes.

Photon actinometer. For measurement of quantum yield, the number of photons from the light source and the number of photons absorbed by a substance are needed. To obtain such information, actinometers are used. Two types of actinometers are commonly employed in photochemical reactions: a solution chemical actinometer and an electronic actinometer [18]. For solution photochemistry with light wavelengths of 250–450 nm, the potassium ferrioxalate solution actinometer is the most widely accepted standard actinometer [19]. This actinometer uses a solution of potassium ferrioxalate solution with a concentration of about 3×10^{-3} mol L^{-1}. Under photo-irradiation of the solution without de-gasing, ferrioxalate anions absorb photons to form nonphoton-absorbing ferrooxalate anions with one-photon-to-two-ferrooxalate-anions correspondence. The photochemical reactions are described as follows:

$$\left[Fe^{III}(C_2O_4)_3\right]^{3-} \xrightarrow{h\nu} \left[Fe^{II}(C_2O_4)_2\right]^{2-} + \cdot C_2O_4^{-} \quad (A)$$

$$\cdot C_2O_4^{-} + \left[Fe^{III}(C_2O_4)_3\right]^{3-} \rightarrow C_2O_4^{2-} + \cdot\left[Fe^{III}(C_2O_4)_3\right]^{2-} \quad (B)$$

$$\cdot\left[Fe^{III}(C_2O_4)_3\right]^{2-} \rightarrow \left[Fe^{II}(C_2O_4)_2\right]^{2-} + 2CO_2 \quad (C)$$

The ferrooxalate anion can be measured by monitoring the 510 nm absorption of its red colored complex with 1,10-pyrrolidone. In the wavelength range of 254–360 nm, the quantity of the ferrooxalate anion generated photochemically is almost constant, having an average quantum yield of 1.24. With this known quantum yield and the measured concentration of ferrooxalate anion, the number of photons absorbed by the ferrioxalate solution is obtained.

Besides this most popular solution chemical actinometer, many other solution chemical actinometers have also been developed. A compilation of them can be found in the IUPAC technical report given by Kuhn et al. [18]. Depending on the systems of interest, different chemical actinometers can be chosen, but before using an actinometer, details of the method of use need to be known and the original references need to be consulted to obtain a full understanding of the methods.

An electronic actinometer [18] is an integrated system with an Si photodiode or a photomultiplier that compares the transmitted light with the reflected light from a beam splitter. This type of actinometer is more convenient to use, but it is easily damaged by intense light and self-deterioration may lead to large measurement errors. Therefore, frequent calibration is needed.

6.3. PHOTOCHEMICAL SYNTHESIS OF ORGANOMETALLIC COMPLEXES

Organometallic complexes are a group of compounds with wide applications. Study of their photochemical properties and reactions has led to the photochemical syntheses of these materials. Almost all transition metals in the periodic table [20] are involved. The most studied systems are those with versatile photochemical reactions [21], specific photophysical properties [22], and catalytic applications [23]. Photochemical synthesis has been a route for synthesizing or "molecular engineering" new and unusual organometallic complexes and a huge number of new compounds with novel structures and properties have been obtained [24,25]. The uniqueness of photochemical synthesis is that it allows the synthesis of organometallic complexes which cannot be easily made or cannot be made at all by conventional thermochemical synthesis methods. In addition, photochemical synthesis often has processing and cost advantages for synthesis of compounds when thermochemical synthesis is possible. From a practical point of view, simple routes of synthesis are always preferred.

Photochemical syntheses via photochemical reactions involve many reactions such as photosubstitution, photo-isomerization, photosensitized metal–metal bond cleavage, photosensitized electron transfer, and photooxidation and reduction. However, no matter which reaction is referred, the key step of these photochemical reactions is to activate the metal centers for reactivity. In the early days, the field was very active due to interest in photosynthesis occurring in nature and in photochemistry of transition metal coordination compounds [1,20,24]. Nowadays the photochemical reaction studies have been greatly expanded to syntheses of organometallic compounds with more practical applications such as supramolecular assemblies [26], photochemical molecular devices and machines [8,9], and photocatalysis [23]. In the following sections we briefly discuss these photochemical reactions.

6.3.1. Photosubstitution

Photosubstitution is the most commonly used synthesis method for making organometallic complexes [20,21,27]. The initial photochemical step in photosubstitution is to photochemically detach a ligand of the organometallic compound to create a vacant coordination site on the metal ion center. To achieve this, matching the wavelength of light with the optical absorption feature of the bond of the ligand targeted to be detached is important. After the ligand is photochemically detached from the metal ion center, a coordinatively unsaturated intermediate is formed. The intermediate is chemically very reactive and has the tendency to bond to any species nearby to saturate the coordination [28]. The detached ligand can come back to the metal ion center or other species can occupy the vacant site. Competition exists. If the other species gets coordinated to the ion center, photosubstitution is achieved. For synthesis purpose, one needs to consider all the factors involved including property of the solvent, property of the ligands, concentrations of the ligands, and chemical nature of the ligands, to create a condition which favors the desired product. Competing processes of ligands other than the desired ones are always present and our task is to minimize the unwanted processes. Overall, the type of photosubstitution reactions and the degree of substitution depend on the properties of central metal ions, the ligand field, the type of excited states, and the reaction conditions such as temperature, pressure, and solvent. For photochromic applications using photosubstitution reactions, ligand configuration also needs to be considered [29].

A lot of studies on photosubstitution in the past were focused on transition metal complexes which are not thermally reactive. These complexes are mainly those of transition metal complexes having d^3, low spin d^5 and d^6 six-coordination, and d^8 square-planar configurations [20,30–32]. Many photosubstitution reactions take place via one-step reaction of the complex excited state with a substituting ligand as follows:

$$\left[[ML_x]^{n+}\right]^* + L' \rightarrow \left[ML_{x-1}L'\right]^{n+} + L$$

where L is the ligand, M is the central metal ion, L' is a substituting ligand, and * represents an excited state. This type of ligand photosubstitution is very common for d^3 and d^6 six-coordinated transition metal complexes [24].

If the metal coordination compound has more than one type of ligand, detaching which ligand depends on the properties of the coordination compounds and the wavelength of light used. Under visible light irradiation, photo-aquation of Cr(III) amine/chlorine coordination compounds takes place mainly on the NH_3 ligand not on the Cl^- ligand. This is in contrast to the thermal chemical reaction where Cl^- substitution is the main reaction. The difference between the photochemical and thermochemical substitutions lies in the different reactivity of the ground state and the excited state of the complex. Using photo-aquation reactions, different ligand-substituted transition metal coordination compounds have been synthesized and details have been summarized in [24] with the types of products and their quantum yields.

For metal coordination compounds with a chain ligand such as $(C_5H_4R)Mn(CO)_3$, $R{=}COCH_3$, $COCH_2SCH_3$, $CO(CH_2)_2SCH_3$, $COCH_2OCH_3$, $(CH_2)_2CO_2CH_3$, and $CH_2CO_2CH_3$ after detaching a CO ligand by photolysis, the side chain could effectively trap the coordinatively unsaturated metal center, forming a ring through the coordination of S and O on the side chain to the metal center [33]. Due to the favorable conformation of the side chain on the metal center, ring closure following photolysis is extremely fast and the quantum yield for the side chain substitution can reach unity. Comparing complexes with different side chains, the quantum yields can vary to some extent, probably due to the different ring strain and different rate for replacement of solvent coordinated intermediates [33].

Photosubstitution does not just occur on one ligand. In some cases, one ligand can substitute two carbonyl ligands on the complex to form new coordination compounds [34]:

$$\eta^5\text{-}C_5H_5V(CO)_4 + RC{\equiv}CH \xrightarrow[C_6H_6]{h\nu} \eta^5\text{-}C_5H_5V(CO)_2RCCH + 2CO$$

where $R = H$, $n\text{-}C_6H_5$, $n\text{-}C_4H_9$, CMe_3, and in the absence of other ligands, photosubstitution led to dimerization [35], and thus multimetal-containing coordination compounds are obtained:

$$2\eta^5\text{-}C_5H_5V(CO)_4 \xrightarrow[THF]{h\nu} (\eta^5\text{-}C_5H_5)_2V_2(CO)_5 + 3CO$$

For the reaction above, the yield of the two-metal-containing compound can reach 90% [36].

For highly symmetric transition metal carbonyl, $M(CO)_6$ with inert d^6 configuration, photosubstitution results in release of CO and forms coordination compounds with different ligands:

$$M(CO)_6 + L \xrightarrow{h\nu} M(CO)_5L + CO$$

The eighth subgroup transition metal-containing coordination compounds, $M(CO)_5$, can be photo-substituted similarly [20].

For coordination compounds containing two metals, the two metals can either form a metal–metal bond or be linked by a ligand bridge. The former, upon light irradiation, undergoes metal–metal bond hemolytic cleavage and forms two active metal carbonyl radicals. The two active metal carbonyl radicals then undertake thermal substitution reactions [37], for example,

$$\left[Mn(CO)_5\right]_2 + 2pph_3 \xrightarrow{h\nu} \left[Mn(CO)_4(PPh_3)\right]_2 + 2CO$$

Further reaction under light irradiation leads to removal of CO ligand and formation of a metal–metal bond,

$$\eta^5\text{-}C_5H_5(CO)_2Mn(AsMe_2)M(CO)_n \xrightarrow[THF]{h\nu} \eta^5\text{-}C_5H_5(CO)_2\overline{Mn(AsMe_2)M}(CO)_{n-1} + CO$$

where $M = Mn$, Re, $n = 5$; $M = Co$, and $n = 4$ [38].

Similar to two-metal-containing compounds, photoreaction of tri-metal carbonyl coordination compounds mainly leads to metal–metal bond cleavage. However, the bond-cleaved intermediates can further react to form either segments of the coordination compounds or ligand-substituted complex, for example [39],

$$OS_3(CO)_{12} + PPh_3 \xrightarrow[Toluene]{h\nu} OS_3(CO)_{11}PPh_3 + CO$$
$$OS_3(CO)_{11} + PPh_3 \xrightarrow[Toluene]{h\nu} OS_3(CO)_{10}(PPh_3)_2 + CO$$
$$OS_3(CO)_{10} + PPh_3 \xrightarrow[Toluene]{h\nu} OS_3(CO)_9(PPh_3)_3 + CO$$

Similarly, photolysis of some four-metal coordination compounds leads to compound segments, but for some compounds, photosubstitution can be the main reaction leading to multimetal multiligands coordination compounds, particularly, for coordination compounds containing second and third row transition metal ions [40,41].

Besides the metal carbonyl coordination compounds, other coordination compounds with ligands such as hydride, isocyanide, olefin, aromatics, and alkyls can also undergo photosubstitution to form corresponding ligand-substituted compounds [24,42]. Metal hydride coordination compounds form dehydrogenated coordination compounds upon photolysis, which cannot be obtained thermally. The photo-generated dehydrogenated compounds are active reactants and can be used for further synthesis of many other organometallic complexes [43].

Photolysis of metal–isocyanide coordination compounds leads to detachment of ligands and therefore generates active species for substitution reactions. The Cr, Mo, and W six-phenyl-isocyanide coordination compounds experience photolysis in very pure pyridine solvent, forming pyridine-substituted products:

$$M(CNR)_6 \xrightarrow[Py]{h\nu} M(CNR)_5(py) + CNR$$

Reference [44] summarizes the quantum yields of these reaction products upon irradiation of light with different wavelengths.

For aromatic-ligand transition metal coordination compounds, photolysis can lead to partial or complete ligand substitution, forming new complexes. For example [45]:

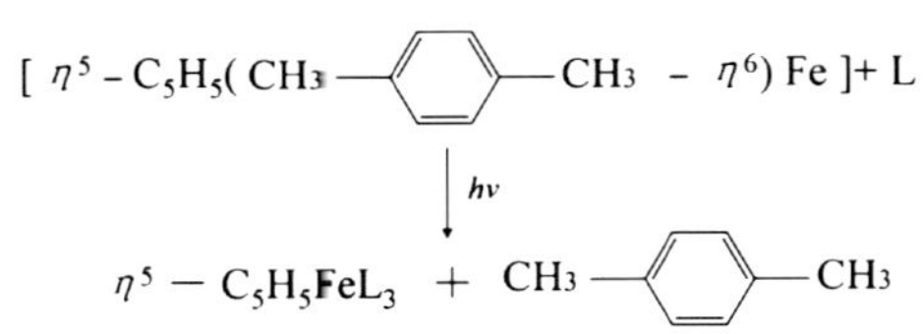

Here L_3 can be monodentate ligands such as $(CNC_6H_4Me)_3$, $(CO)_3$, and $(P(OPh)_3)_3$ or tridentate ligands such as h^6-C_6Me_6 and $PhP(CH_2PPh_2)_2$. High quantum yields were obtained for this reaction, for example, when $L_3 = PhP(CH_2PPh_2)_2$, the photosubstitution quantum yield for *p*-xylene at 436 nm is 0.57 [45].

In some cases, photosubstitution can also take place between the coordinating aromatic ligand and the aromatic solvent molecules:

$$(\eta^6 - \text{arene}_1)RuCl_2(PR_3) + \text{arene}_2 \xrightarrow{h\nu} (\eta^6 - \text{arene}_2)RuCl_2(PR_3) + \text{arene}_1$$

This substitution reaction, in some cases, is quite efficient, for example, with cumene as solvent, the substitution efficiency can reach 100% [46].

Recently, it is found that photosubstitution of polypyridine ruthenium(II) complexes is an efficient route to obtain new ruthenium(II) compounds, although thermal synthesis of this type of organometallic coordination compounds is common. The reason for extensive study of this type of organometallic coordination compounds is their potential use in solar energy conversion processes [47,48] and as components of luminescent sensors [49–51], light emitting diodes [52], and photoswitchable molecular devices [53–55]. Photosubstitution of polypyridine ruthenium(II) complexes takes place after their metal-to-ligand charge-transfer (MLCT) band is photo-excited. Thermal population of their low-lying *d–d* states from their MLCT excited state leads to expulsion of their weaker ligands and subsequent substitution by other ligands [56,57]. Starting from $[Ru(terpyridine)(phen)Cl]^+$ complex, a series of photo-substituted ruthenium $[Ru(terpyridine)(phen)L]^{n+}$ complexes, such as $L = CH_3CN$, CN^-, pyridine, isoquinoline, dmap (4-dimethylaminopyridine), ptz (phenothiazine), and mqt (4-(4′-methylpyridinium)pyridine), were synthesized by irradiating their corresponding degassed acetone solutions with white light from a slide projector for 3–8 h at ambient temperature [58]. The product yields range from 78% to 96%. The photochemical synthesis is shown to be a very efficient route for obtaining these new polypyridine ruthenium complexes.

Photosubstitution can also be used to synthesize piano-stool organo-iron complexes $[CpFeL_1L_2L_3]^+$ bearing a variety of ligands ($Cp = \eta^5$-C_5H_5, L = neutral 2-electron ligand) via visible light photolysis of any member of the family of $[FeCp(arene)][PF_6]$ sandwich complexes (arene = η^6-arene) including those in which the arene is mono-, bis or tri-substituted [59]. A more facile access to the family of piano-stool family with the Cp* ligand involves photolysis/photosubstitution of the isolable tricarbonyl Fe complex using visible light [60,61].

6.3.2. Photo-isomerization

Here the photo-isomerization reaction is the stereoisomerization of organometallic compounds. The purpose of the study of this type of reaction is to synthesize, under light irradiation, those stereoisomers which either cannot be synthesized by other synthetic means or can be synthesized but with undesirably low reaction rate. Many photo-isomerization reactions are reversible and the reaction conditions determine the direction of the reactions. This sets up the basis for the photochromic applications [53] and molecular switch. However, these applications mainly involve organic compounds and are much less common for organometallic compounds [62].

(1) Reversible cis- to trans-isomerization.

$Ru(II)(bipyridine)_2$ coordination compound undergoes cis- and trans-photo-isomerization [63]:

$$cis - [Ru(bipy)_2(H_2O)_2]^{2+} + \overset{h\nu}{\rightleftharpoons} trans - [Ru(biby)_2 \times (H_2O)_2]^{2+}$$

This reaction is reversible but the forward reaction has a higher quantum yield than the reverse reaction

(0.043 versus 0.025). d^6 ion coordination compounds such as $Co(NH_3)_3(H_2O)$ also undergo similar photo-isomerization reactions [64]. The *trans*-(n-Pr_3P)$PdCl_2$ coordination compound in chloroform converts to the light-stable cis-isomer under UV light irradiation [65,66]:

$$trans-(n-Pr_3P)_2PdCl_2 \rightleftharpoons cis-(n-Pr_3P)_2PdCl_2$$

Due to its very long decay time (10^3 h), characterization of the cis-isomer can be carried out with normal optical spectrometry. Cis- to trans-photo-isomerization reaction of Pt planner coordination compounds is also reversible, but the relative concentration of the cis-/trans-isomers changes with solvent. For *trans*-$PtCl_2(Et_3P)_2$ to its cis-isomer reaction, the quantum yield is about 10^{-2} when the ligand field band is photo-irradiated ($\lambda > 300$ nm) [67].

(2) Isomerization of ligands in organometallic coordination compounds. Some organometallic coordination compounds undergo isomerization and form ligand isomerized compounds upon light irradiation. For example, solid *trans*-$[Ru(NH_3)_4Cl(SO_2)]Cl$, when irradiated with the 365 nm light at low temperature, isomerizes [68]:

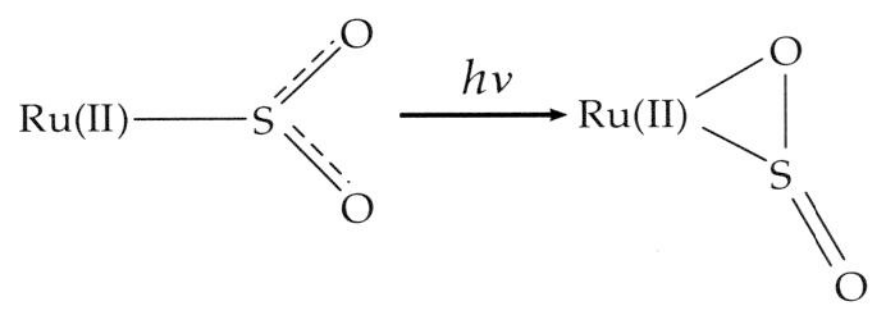

The reaction is reversible at room temperature. Similarly, Co coordination compounds can also undergo such reaction [69]:

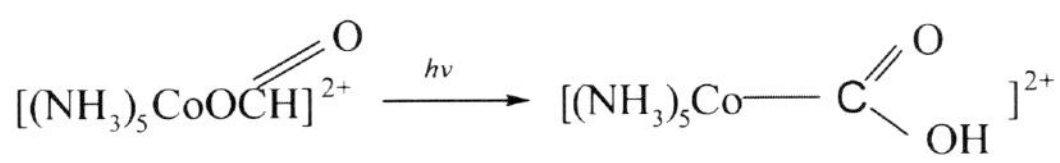

However, the above isomer product is unstable to temperature and it either slowly returns back to the original reactant or decomposes to Co ion.

Because of the instability of the ligand isomerized products, prolonged light irradiation can cause the ligand isomerized products themselves to further react with solvent molecules or with other ligands [70]:

$$[Pd(MeE+4dien)NO_2]^+ \xrightarrow{hv} [Pd(MeE+4dien)(ONO)]^+ \xrightarrow{hv} [Pd(MeE+4dien)S]^2 + [NO_2]^-$$

where S represents solvent molecule, MeE+4dien = 4-methyl-1,1,7,7,-tetraethyl diethylenetriamine.

Recently, studies on photochromic compounds of ruthenium-polypyridine-dmso compounds such as *trans*- and *cis*-$[Ru(tpy)(Me\text{-}pic)(dmso)](OSO_2CF_3)$ and *cis*-[Ru (tpy)(Br-pic)(dmso)]-(PF_6) (tpy = 2,2′: 6′,2″-terpyridine; Me-pic = 6-methyl-2-pyridinecarboxylate; dmso = dimethyl sulfoxide, Br-pic = 6-bromo-2-pyridinecarboxylate) have shown photo-triggered S-to-O isomerization following metal-to-ligand charge-transfer (MLCT) excitation [71–76]. The S-to-O photo-isomerization is controlled by the rotation vibration of the dmso ligand about the Ru–S bond. The quantum yield of the isomersization of these compounds therefore depends on the interaction of other ligands with the dmso ligand.

6.3.3. Photosensitized Metal–metal Bond Cleavage Reaction

Photosensitized metal–metal bond cleavage reaction is one of the most important photochemical reactions for synthesis [24]. Here the coordination compounds are either binuclear or multinuclear, as seen from the examples given above in the photosubstitution reactions. Photosensitized metal–metal bond cleavage is not specific on the kind of metals. As far as the reaction product is concerned, the reactions proceed with fragmentation followed by formal oxidation and substitution on the metal centers.

(1) *Metal–metal bond cleavage reaction of binuclear metal coordination compounds*. For metal–metal bond cleavage reactions of binuclear metal coordination compounds, the most studied are the binuclear metal carbonyl coordination compounds, $M_2(CO)_{10}$, and their derivatives (M = group VI, VII, and VIII transition metals) [77–80]. In the metal–metal bond cleavage reactions, solvent determines whether the metal–metal bond is broken, by a homolytic or heterolytic process. The cleavage pathway determines the final products formed by either substitution or insertion. For synthesis purposes, it is extremely important to choose the proper solvent.

Metal Mo and W carbonyl coordination compounds $(h^5\text{-}C_5H_5)_2M_2(CO)_6$ exhibit different homolytic cleavage rates in the presence of different halogenated hydrocarbons [81]:

$$(\eta^5-C_5H_5)_2M_2(CO)_6 + 2PhCH_2Cl \xrightarrow[C_6H_6]{hv} 2\eta^5-C_5H_5M(CO)_3Cl + PhCH_2CH_2Ph$$

The reactivity order follows the sequence $CCl_4 > CHCl_3 > PhCH_2Cl > CH_2Cl_2$. If donor solvents are used, these binuclear coordination compounds undergo heterolytic cleavage-substitution reactions [82,83],

$$(\eta^5-C_5H_5)_2Mo_2(CO_6) + X \xrightarrow{hv} \eta^5-C_5H_5Mo(CO)_3X + [\eta^5-C_5H_5Mo(CO)_3]^-$$

here X = Cl, Br, and SCN.

(2) *Metal–metal bond cleavage reaction of multinuclear metal coordination compounds.* Study of metal–metal bond cleavage reactions of multinuclear metal coordination compounds has also focused on metal carbonyl coordination compounds and their derivatives. $M_3(CO)_{12}$ (M = Fe, Ru, Os) coordination compounds mainly carry out photofragmentation. For example [84,85],

$$Ru_3(CO)_9(PPh_3)_3 + 3L \xrightarrow{hv} 3Ru(CO)_3PPh_3L$$

$$[Fe_3(CO)_{11}]^{2-} - 4PPh_3 \xrightarrow{hv} 2Fe(CO)_3(PPh3) + [Fe(CO)_4]^{2-} + CO$$

here L = CO and PPh_3.

6.3.4. Photo-induced Electron Transfer and Redox Reactions

One of the most important consequences of electron excitation is the increase of electron affinity and the decrease of ionization potential of the molecule. Long-lived excited state molecules easily transfer their electrons to other molecules. The early inorganic photochemists' attention mainly focused on intramolecular photoredox and ligand photosubstitution reactions [24], and nowadays, studies are more focused on conversion of solar energy to chemical energy using coordination compounds and new coordination compounds with unusual oxidation states and chemical properties [20,31].

The electron-excited states in the electron transfer reactions vary. Based on the molecular orbital involved in the electron transitions, the excited states can be divided into the following types:

(1) Metal centered (MC) or ligand-field excited states;
(2) Inner ligand or ligand centered (LC) excited states; and
(3) Electron transfer excited states.

The charge transfer can occur from metal to ligand (MLCT) or from ligand to metal (LMCT). In addition, charge transfer to solvent (CTTS) as well as between metals in the multinuclear coordination compounds can also occur. However, it needs to be mentioned that the assignment for the various excited states is arbitrary and when the states cannot be described with localized molecular orbital configurations, the assignment becomes unimportant. In addition, the sequence of orbital energy levels can change, which, upon excitation, leads to different excited states. The details of orbital energy levels depend on the type of ligand, the nature of metal, and the metal oxidation state. For example, the coordination compounds with the same Ir^{3+} ion, such as $Ir(phen)Cl_4^-$, $Ir(phen)_2Cl_2^+$, and $Ir(phen)_3^{3+}$, have different lowest excited states, MC, MLCT, and LC, respectively. The same type of coordination compounds with different metal ions such as $Rh(bipy)_2Cl_2^+$ and $Ir(bipy)_2Cl_2^+$ have lowest excited states MC and MLCT, respectively.

Via photo-redox reactions, transition metal coordination compounds with low valence states can be prepared [86]:

$$Fe(\eta^5 - C_5H_5)_2 \xrightarrow[CCl_4]{hv} [Fe(\eta^5 - C_5H_5)_2]^{\cdot}Cl^-$$

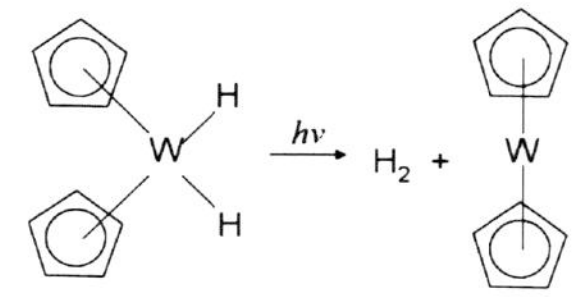

The first reaction above is a photo-redox reaction which generates a univalent iron coordination compound and the second reaction is a photochemical reduction which removes H_2.

Some square planar d^8 coordination compounds also undertake such reaction [87]:

$$M(C_2O_4)L_2 \xrightarrow{hv} 2CO_2 + ML_2$$

Here M = Ni, Pd, and Pt. An interesting $16e^-$ square planar coordination compound can be converted into inorganic analogue—carbon-olefin $14e^-$ metal coordination compound upon light irradiation. By this reaction, coordination compounds with low valence metals such as Pt^0 and Rh^+ can be synthesized [88].

Photochemical reactions of organometallic compounds normally need to avoid the presence of oxygen because the excited states and intermediates react with oxygen to form metal oxides. However, under certain conditions, this oxidation may be used to synthesize metal oxides with organometallic ligands [89–92]. The synthesis of $(CpV)_mO_n$ clusters by irradiation of $CpV(CO)_4$ in the presence of controlled amounts of oxygen [92], and the synthesis of Cp^*MoO_2Cl from $Cp^*Mo(CO)_3Cl$ [91] are examples of this kind and give novel partially oxidized organometallic compounds.

6.4. PHOTOCHEMICAL SYNTHESIS OF INORGANIC COMPOUNDS

6.4.1. Photochemical Synthesis Via Photosensitization

Photosensitized reactions are photochemical reactions in the presence of sensitizers. The function of a sensitizer is to transfer energy or to participate in photochemical reactions to form a radical which then reacts with reactants and converts back to the sensitizer. In many inorganic photochemical reactions, mercury is a common sensitizer. Upon light irradiation, mercury is excited and becomes excited mercury. The excited mercury collides with reactant molecules and transfers the energy to the reactant molecules to make them more reactive.

a. Hydrides of main group elements

The hydrides of the main group elements such as B, Si, and Ge, P, As, and Se, polymerize via mercury photosensitized reactions [93–98]:

$$2B_2H_6 + Hg(^3P_1) \rightarrow B_4H_{10} + H_2$$
$$2SiH_4 + Hg(^3P_1) \rightarrow Si_2H_6 + H_2$$
$$2GeH_4 + Hg(^3P_1)) \rightarrow Ge_2H_6 + H_2$$
$$2PH_3 + Hg(^3P_1)) \rightarrow P_2H_6$$
$$2AsH_3 + Hg(^3P_1)) \rightarrow As_2H_6$$

These reactions occur because the M–H (M = B, Si, Ge, P, As, Se) bond energies are within the range of 294–378 kJ mol^{-1} and absorb light with energy matching the released energy of excited mercury. Many nonmetal halides such as BCl_3, $SiCl_4$, CCl_4, and $AsCl_5$ also have similar bond energies and can be mercury-sensitized to form corresponding compounds [93,101].

Polymerization of silicon hydride and ethylene or acetylene to form a Si-containing alkane has been achieved via mercury photosensitization reaction and the resulting Si-containing alkane products have very high purity and are easily separated and purified [99]. Cyclic C_3F_6 compound can also be synthesized using mercury photosensitization reaction [100].

b. Metal carbonyl complexes

In addition to the examples given above, mercury photosensitization can also be used to synthesize metal carbonyl and their ligand-substituted complexes. This is also due to the energy match of transition metal-carbonyl bond energies with the energy released from the excited mercury, as were observed for the hydrides of the main group IV, V, and VI elements. The photosensitization reaction is also one of the main synthesis routes for these complexes [102].

6.4.2. Photochemical Synthesis of Boranes and their Derivatives

Although photochemical synthesis of boron compounds was not studied as extensively as their thermochemical synthesis due to the lack of visible light absorption of boranes, the study of photochemistry of boranes using short wavelength UV light (<280 nm) provided synthetic strategies for the formation of a variety of new boron compounds [93]. The synthesis can be carried out via mercury sensitization as mentioned in the above section or via direct photolysis. The latter generates radicals and the subsequent reactions between the radicals form the products. By this method, tetraborane, decaborane, and icosaborane were photochemically synthesized [93,103–105]:

$$B_2H_6(h\nu = 184.9\ \text{nm}) \rightarrow B_4H_{10} + B_5H_{11} + H_2 + (B_2H_2)_n$$
$$B_5H_9(h\nu = 193\ \text{nm}) \rightarrow B_{10}H_{16} + B_{10}H_{14} + H_2$$
$$2B_{10}H_{14}(h\nu = 272\ \text{nm}) \rightarrow B_{20}H_{26} + H_2(10\%\ \text{yield})$$

For icosaborane, $B_{20}H_{26}$, isomers are present in the final product [93,104,105].

6.4.3. Synthesis of Chlorine Gas Via Photolysis

Industrial chlorine gas is obtained via either electrolysis of aqueous NaCl solution or oxidation of HCl on a Cu catalyst [106]. The former demands a lot of electricity while the latter necessitates handling of corrosion at 450°C and catalyst deactivation. Photochemical synthesis of chlorine gas can overcome these problems and may be a useful approach to provide chlorine gas for wastewater treatment for poor and developing countries [32]. The photolysis is based on the flowing reactions:

$$[Fe(OH_2)_5Cl]^{2+} + H_2O \xrightarrow{h\nu} [Fe(OH_2)_6]^{2+} + Cl^{\cdot}$$
$$[Fe(OH_2)_4Cl_2]^{+} + H_2O \xrightarrow{h\nu} [Fe(OH_2)_5Cl] + +Cl^{\cdot}$$
$$Cl_2^{\cdot -} + Cl_2^{\cdot -} \rightarrow Cl_3^{\cdot -} + Cl^-$$
$$Cl_3^- \rightarrow Cl_2 + Cl^-$$

In these reactions, photolysis of $[Fe(OH_2)_5Cl]^{2+}$ and $[Fe(OH_2)_4Cl_2]^+$ with 270–400 nm light generates $Cl^{\cdot}$ radicals. The $Cl^{\cdot}$ radical and Cl^- react to form $Cl_2^{\cdot -}$, and subsequently, $Cl_2^{\cdot -}$ disproportionates to Cl_2 and Cl^-. For these reactions, the concentrations (need to be high) of $[Fe(OH_2)_5Cl]^{2+}$, $[Fe(OH_2)_4Cl_2]^+$ and Cl^- and pH (need to be low) are important factors to control. The quantum yield of the process is in the range of 0.09–0.13 [32]. The kinetic spectroscopic method gives a quantum yield in the range of 0.46–0.57 [107].

6.4.4. Photochemical Synthesis of Fluorides

XeF_2 is one of the inorganic compounds synthesized when the era of noble gas chemistry just began [108,109]. It has been widely used as a fluorinating and oxidizing agent to react with metals or metal oxides to form metal fluorides or oxyfluorides, as a medium strong Lewis base to react with metal penta- and tetrafluorides and with the compounds that are weak fluoride ion acceptors to form ionic or covalent adduct compounds, and as a ligand to metal ions to form coordination compounds [110]. XeF_2 can be prepared from gaseous mixtures of xenon and fluorine thermally or photochemically. The thermal preparation method uses a large excess of xenon over fluorine in the presence of catalyst in order to minimize formation of tetra- or higher fluorides [111]. In contrast, photochemical synthesis with the same mixture is simple and gives rise to pure XeF_2 product. The reaction can be performed in Pyrex flasks with sunlight or UV light from a mercury lamp but the pressure within the reaction vessels needs to be carefully monitored in order to avoid XeF_4 formation (a compound that forms highly explosive XeO_3 when it is hydrolyzed). Irradiation of a 2:1 molar ratio Xe/F_2 gaseous mixture with UV light and about 1 mol% of HF as catalyst can give up to 1 kg pure XeF_2 in a 10 L reactor [112].

Similarly, KrF_2, O_2F_2, SF_6, and ClF_5 can be prepared via irradiation of corresponding gas mixtures at liquid nitrogen temperature [113–115]. Other compounds such as S_2F_{10} and $CF_2(NO)_2$ can also be synthesized via photodecomposition of SF_5Br and photolysis of a mixture of CF_2I_2 and NO in the presence of mercury [116,117]. Ninety-nine percent yield can be achieved for S_2F_{10} [118] while for $CF_2(NO)_2$, the yield is about 10% [119].

6.5. SYNTHESIS OF INORGANIC THIN FILMS VIA PHOTOCHEMICAL REACTIONS

6.5.1. Laser-assisted Thin Film Formation of Inorganic Materials

Inorganic metal and metal oxide thin and ultrathin films are technologically important materials due to their applications in integrated circuits and related optoelectronic components [120,121], as gating materials for information processing, data storage and high density charge capacitance [122–130], as electrolytes for electrochemical energy conversion and storage [131–134], as electron conductors and transparent electrodes in a variety of photochemical technologies such as solar cells, membranes for water splitting and gas separation [135–139], and as passive oxide films for corrosion resistance of metal surfaces [140]. Chemical vapor deposition (CVD) and atomic layer deposition (ALD) are well-known processes for making such inorganic thin films on substrates. In recent years, laser-assisted processes have become useful routes for making them [141,142]. In the laser-assisted processes, laser photolysis of precursors generates reactive species to form thin films on substrates and for oxide ultrathin films formed via oxidation of a surface of metal or nonmetal substrate, photon irradiation of gas phase oxygen generates active oxygen species such as atomic oxygen radicals and ozone molecules which then oxidize the substrates surface forming oxide films. In term of the precursor formation mechanism, laser-assisted film formation can be a thermal process, similar to what was seen in the chemical vapor deposition (CVD), but as far as photochemical synthesis is concerned, laser-assisted film formation should be a process where photochemical (photolysis) process is the main step. Gas or adsorbed reactants absorb laser light and form dissociated or ionized precursors which then form films on the surface of substrates or substrates absorb laser light forming electron–hole pairs on the substrate surface, and then reactants react with the electron–hole pairs, depositing a film on the substrate surface. The latter process is the main reaction step in laser lithography. Compared to the conventional CVD and ALD film-making processes, laser-assisted film deposition and laser-assisted substrate surface oxidation for ultrathin oxide films provide technically easily controlled processes and high quality films. For some materials, laser-assisted process is even the only process for making films.

When carrying out laser-assisted film formation procedures, the following factors need to be considered:

Reactant: The reactants should contain the atoms needed for the film formation. Their absorption spectra, volatility, and possible routes of cleavage should be clearly understood. The gas phase reactants normally used are hydrides, halogenated compounds, alkyl, carbonyl, and acetylacetone compounds of corresponding metals, semiconductors, or insulators. For oxide films where the metal ions are provided by the substrates, laser generated atomic oxygen or ozone molecules in the gas phase are the reactants. For laser-assisted film deposition, ideal precursors are compounds which can be photolyzed with low intensity of light in the range of the irradiative laser wavelengths and form thin film with no impurities. The newly formed species and their volatile products should be relatively inert and the newly formed species should have enough vapor pressure to favor a rapid deposition onto the substrate surface. Therefore, the choice of proper precursors is important and determines the quality of the thin films and whether the specification of the films can be achieved for actual applications. For the ultrathin oxide films formed via reaction between photo-generated active atomic oxygen or ozone and the substrate elements, the photochemistry of oxygen molecules and the chemical properties of the formed oxide films need to be understood.

Substrate: The features of absorption spectra and the thermal conductivity of the substrates should be known for film depositing on them. For the ultrathin oxide films formed via oxidation of the substrates, the band gap of the oxides and their corresponding optical properties for light absorption need to be clearly understood.

Laser and its geometric shape: The wavelength of laser should satisfy the need for cleavage of the reactants while for generating atomic oxygen and ozone molecules for ultrathin oxide films, a short wavelength UV light matching the oxygen and ozone molecule absorption bands needs to be employed. A widely used light source is the low pressure (<0.1 Torr) mercury lamp (185 and 254 nm). In some cases, other light sources such as 222 nm KrCl* laser and high pressure Hg lamp with filter (restricted to 280–420 nm) have also been used [143,144]. The selection of the geometric shape of the laser depends on the real applications. In the process of laser-assisted film formation, there are two geometric categories for laser irradiations: the focused type and the unfocused or weakly focused type. In the focused type, the light beam is focused via lenses onto the substrate surface covered with reactant vapors. This type of laser arrangement is often called "laser direct writing". Photolysis of molecules near or adsorbed on the substrate surface causes newly generated species to deposit onto the substrate surface. If the substrate is laterally moving or fixed, thin films with different shape (line or point) are formed on the substrate surface. Taking advantage of this kind of laser geometric shape, sub-micron-sized films can be deposited on the substrate surface. In the unfocused or weakly focused type, an intense laser pulse is employed. With this type of irradiation shape, large area of thin films can be prepared. Another form of unfocused or weakly focused irradiation is to focus the laser into a line shape with cylindrical lenses and allow it to pass just above the substrate surface. In this case, the laser initiates the chemical reactions in the gas phase but not on the substrate surface, thus avoiding heating the substrate surface by the laser light. The third form of the unfocused or weakly focused irradiation is projection printing. This is also a large area irradiation but the lucifuge pattern needs to be placed in the light path. With this irradiation shape, deposited thin films with different designed patterns can be obtained. The above-mentioned three types of irradiation shapes with no focus or weak focus can be used individually or in combination.

Deposition and film formation models: Many photolysis deposition models have been proposed to illustrate and predict the local photolysis film growth rate of gas phase and surface adsorbed reactants. These film growth models only take into consideration the initial photolysis process, and not the chemical reactions of the photolysis products and their recombination. An early simple model [145] predicts that if the gas phase photolysis was the main process, the thin film growth rate at the deposition center is positively proportional to $1/\omega$. Here ω is the radius of the laser beam. If the surface photolysis is the main process, the deposition rate is positively proportional to $1/\omega^2$. A recent rather detailed model for laser gas phase deposition with a Gauss distribution laser beam is described as [146]:

$$k_{\text{photolysis}} = \frac{fn\phi_t\sigma_d(\lambda)}{2[\pi(\omega^2+r^2)]^{1/2}}\,\frac{8w^4+r^4}{8w^4+\pi^{1/2}r^4}.$$

Here, n is the gas density, Φ_t is the photon flux, σ_d is the photolysis cross-section, r is the radial coordinate, ω is the laser half-width at $1/e$ of the laser intensity, $f=\alpha_p/2$, and α_p is the sticking coefficient of the surface-deposited atoms. The above equation predicts that, for $r/\omega << 1$, the growth rate decreases with $1-(r/\omega)^2/2$. For $r/\omega >> 1$, the growth rate decreases with $1/r$. Therefore, due to the slow decrease of the growth rate when r/ω is large, it may be expected that the photolysis deposition area is more spread out than that of a laser with a Gaussian distribution.

For the ultrathin oxide films formed via reactions of gas phase photo-generated atomic oxygen and the surface of substrate, various film formation and growth models have been proposed [142]. Different from the deposition models discussed above, these oxide film formation models only consider the initial stage of oxidation of the substrate surface and transportation of electronic and ionic species through the newly formed oxide film. The function of photon irradiation is to increase the concentration of active oxygen species and to create an electric field. The latter originates from the space charge formed within the oxide film by photon irradiation and accelerates the diffusion of the charged species in the oxide film. A detailed model proposed by Kazor and Boyd [147] has the following form:

$$X^2 = \alpha/\beta[1-\exp(\beta t)]$$

Here, X is the thickness of the oxide film, $\alpha = \Omega D_{\text{eff}}\Delta n$ and $\beta = [(e\Omega D_{\text{eff}}\sigma)/2\varepsilon kT](n_1/6+n_2/3)$, t is time. $\alpha/\beta = X^2_{\text{lim}}$ represents the square of the limiting oxide thickness. Ω is the volume of the oxide per oxidant unit. D_{eff} is the effective diffusion coefficient of the ionic oxidant. $\Delta n = n_2 - n_1$, n_1 and n_2 are the oxidant concentrations at the surface and interface of an oxide layer of thickness at time t, $X(t)$, respectively. σ is the effective planar charge density, ε is the dielectric constant of the oxide, k is the Boltzmann constant, and T is the substrate temperature. This kinetic rate model has been used to explain the low temperature photo-enhanced oxide growth on Si [147–150].

After considering these factors for laser-assisted film formation, we now give a brief summary of inorganic thin films made with this technique. The laser-assisted thin film studies mainly focused on fabricating integration circuits, production of solar energy cell, infrared sensor manufacture, and exploration of the corresponding materials. In recent years, the field of photolysis of solid surface modulated adsorbed molecules has developed rapidly.

(1) Metal films

The precursors most widely used for making metal films are those of alkylates of metals such as Al, Zn, Cd, and Hg; carbonyl compounds of metals such as Cr, Mo, W, Fe, and Ni; halogenated compounds of metals such as Ti and W; and acetylacetonate compounds of metals such as Cu, Pd, Pb, and Au [141]. Depending on the nature of the precursors, the lasers with wavelengths of 193 nm (ArF laser), 248 nm (KrF laser), 257 nm (Ar^+ ion laser), 308 nm (XeCl laser), etc. have been selected. All these lasers are within the UV range and provide light which matches the absorption bands of metal to ligand electronic transitions. This match is important because the laser photolysis of the metal coordination compounds is actually a metal–ligand bond cleavage process. Another aspect is that metal precursors generally have multiple ligands, and photolysis of them normally needs more than one photon to complete the last metal–ligand bond cleavage to generate the species for film formation. The detail of the specific process depends on the kind of metal complexes and the number of ligands. Due to the existence of organic ligands in the metal precursors, the most easily occurring contamination during the laser photolysis film formation process is carbon. To avoid this, precursors with no carbon such as hydrides and fluorophosphides or with organic fragments that are stable during photolysis can be used. Carbon contamination can also be avoided by controlling the gas phase by adding some compounds with high volatility.

(2) Semiconductor films

The most studied systems are those of the group IV element semiconductors such as Si, Ge, Ge–Si alloy, C, and SiC; the group III–V element semiconductors such as GaAs, InSb, and InP; and the group II–VI element semiconductors such as HgTe, HgCdTe, and CdTe [141]. For Si films, the most studied compounds are monomer and polymeric silanes. Chlorosilane had also been used recently [141]. For group III–V compound semiconductors, films have been obtained via photolysis of methyl and ethyl compounds or hydrides of the III and V group elements. For group II–VI compound semiconductor thin films, formation of an abrupt interface structure is important owing to their narrow band gaps. Films with abrupt interface could be obtained via laser photolysis of precursors at low temperatures in order to avoid diffusion of atoms. For such cases, photolysis of gas phase compounds needs lens-focused laser beams.

(3) Insulator films

Using suitable semiconductor or metal precursors and N_2O as an oxygen source and NH_3 as a nitrogen source, oxide and nitride insulator films such as SiO_2, Si_3N_4, Al_2O_3, AlN, and $GeO_2–SiO_2$ have been synthesized [141]. Other insulator thin films such as Cr_2O_3/CrO_2 and TiC have also been made. In the formation processes of these insulator films, one reactant strongly absorbs the light while the other weakly absorbs the light. The strongly light-absorbing reactant decomposes during the photolysis and dominates the whole reaction process. The excitation of the reactants can be achieved directly or via mercury sensitization.

(4) Ultrathin oxide films via photo-assisted oxidation of substrate surfaces

Tsuchiya et al. [142] have reviewed this field and found that band gap is one of the most important parameters for the photon-assisted film formation. They discussed the semiconductor substrates such as Si, Ge, SiGe, GaAs, InAs, InP, and GaSb and oxide film materials and classified the latter into four groups based on their band gaps: (a) highly transparent oxides (>8 eV) such as Al_2O_3 and SiO_2; (b) intermediate bandgap oxides (5–6 eV) such as ZrO_2, Y-doped ZrO_2, HfO_2, and Y_2O_3; (c) low bandgap oxides (3–4 eV) such as CeO_2, TiO_2, $SrTiO_3$, $Ba_{1-x}Sr_xTiO_3$ (BST), ZnO, Ta_2O_5, $InSnO_2$, $Ti_{1-x}Co_xO_{2-\delta}$, and Fe-doped $SrTiO_3$; and (d) strongly correlated electron compounds such as VO_2 and $YBi_2Cu_3O_{7-\delta}$. For large bandgap oxide films such as Al_2O_3 and SiO_2, the self-limiting thickness is ~1.5 nm for SiO_2 [151] and ~1 nm for Al_2O_3 [152] at near room temperature due to transparency of the oxide film materials to UV light and very slow diffusion of oxygen or aluminum through the film. For intermediate bandgap oxide films, the primary photon energy (for 185 nm) is close to the band gap of the materials and the penetration of the 185 nm light is generally sufficient to go through the entire nanoscale film. For the low bandgap oxide films, due to strong absorption of the 254 and 185 nm light, the electron concentration is significantly perturbed by the photon irradiation and the light penetration is strictly limited to near the surface. For the last strongly correlated electron compound films, the photon irradiation may create a difference in defect concentration throughout the film thickness if UV light is used. For the semiconductor materials given above, ultrathin oxide film formation under UV irradiation can be carried out at room temperature and lead to good quality and dense conformational oxide films with good electrical characteristics. For the III–V compounds, ultrathin films containing segregated oxides can form with linear, logarithmic, or parabolic growth rates, depending on the systems ([142] and references therein). Overall, compared to the conventional film synthesis methods such as CVD and ALD, the synthesis of films of these oxides with different band gaps by the oxidation method under UV light irradiation gives films with less oxygen vacancies, better crystallinity, sharp abrupt interface, less contamination, lower leakage current, and better electric properties. In addition, the film syntheses under UV light irradiation can be carried out at much lower temperatures compared to those for conventional syntheses. In addition, the thickness of the oxide films obtained under UV light irradiation is self-limited and determined by the diffusion of oxygen species through the oxide films. This is different from films from conventional synthesis methods. The superior quality of the films formed under UV light irradiation results from better control of the oxygen vacancies and stoichiometry in the films.

It should be pointed out that although many studies have been carried out on the laser-assisted thin film formation of metals, semiconductors, and insulators, the details of the reaction mechanisms are still unclear. The initial reaction involves laser photolysis, but the film formation involves a lot of intermediate species. Which species determines film formation and how the film is formed still need to be clarified. Two possibilities are from atomic species or from the molecules with subsequent decomposition. For UV-light-assisted ultrathin oxide films via oxidation of substrate surfaces, a detailed understanding is still needed including the role of species migration in oxide films and the kinetics of film formation.

6.5.2. Film Formation Via Solution Photochemical Deposition

In the preceding section, we introduced film formation via photolysis of gases on solid substrate surfaces and photo-assisted oxidation of substrate surfaces. In recent years, film formation via photochemical deposition on solid surfaces in liquid solutions has been widely studied [153–163]. This photochemical deposition proceeds on a substrate plate surface in the solution which contains the species to be deposited. Upon photochemical reactions, the newly formed species react with those ions that already existed in the solution and form insoluble compound. Then these compounds deposit on the surface of the substrate plate being photo-irradiated and a film forms. With this method, numerous semiconductor thin films such as sulfides, CdS, ZnS, PbS, CuS, In_2S_3, SnS, Bi_2S_3, $Cd_{1-x}Zn_xS$, $CuInS_2$, and $(Bi, Sb)_2S_3$; selenides, CdSe and ZnSe; and oxides, ZnO and SnO_2, have been fabricated. Because the films can only be formed in the photo-irradiated areas, this method can be used to fabricate patterned thin films with the help of a masking template. The photochemical deposition method is easy to operate, scale up, and control. The light source can be just mercury lamps. Depending on the nature of the compounds to be photochemically activated, different wavelengths from light sources can be selected.

For metal sulfide thin film formation, the solutions need to have not only the metal ions but also photo-generated electrons and sulfide anions such as thiosulfate or sulfite. The latter two are responsible for providing electrons and controlling redox reactions for S formation. Detailed reactions are as follows:

$$S_2O_3^{=} + h\nu \rightarrow S + SO_3^{=}$$
$$2S_2O_3^{=} + h\nu \rightarrow S_4O_6^{=} + 2e^-$$
$$SO_3^{=} + S_2O_3^{=} + h\nu \rightarrow S_3O_6^{=} + 2e^-$$

Then metal ions such as Cd^{2+} react with S and electrons and form a CdS thin film,

$$Cd^{2+} + S + 2e^- \rightarrow CdS.$$

For the reactions to proceed smoothly, the solution acidity needs to be controlled. For a good quality of thin films, sometimes, post-thermal annealing is a necessary step. When the same method is used for preparing multiple sulfide thin films such as $Cd_{1-x}Zn_xS$, $CuInS_2$, and $(Bi, Sb)_2S_3$, the reactions need to proceed in two steps. Thermal annealing is also needed after the film formation. The concentration and pH in each step need to be controlled.

A similar method can also be used for making telluride thin films using sodium tellurite as a telluride source in the presence of sulfite anions. The latter provides photo-generated solvated electrons upon light irradiation to reduce tellurite. For telluride thin film formation, only UV light can initiate the reaction. Oxide films such as ZnO and SnO_2 can also be made using the same method. Here photo-generation of solvated electrons is an important step, but the reaction mechanism is still unclear.

Photochemical synthesis of Prussian blue (PB) and its analogues films from an acidic ferricyanide solution forms another class of film materials with promising applications in various fields such as magnetic materials [164–167], molecular sieves [168], catalysis [169,170], solid-state batteries [171–173], electrochromic devices [174,175], biosensors [176–178], photo-image formation [179,180], and ion-selective detection [181–183]. In the photochemical film formation reaction with 365 nm UV light irradiation [184], the rate-determining step is the photochemical reduction of $Fe(CN)_6^{3-}$ ions to $Fe(CN)_6^{4-}$ ions in the aqueous solution of 1 mM $Fe(CN)_6^{3-}$ and 0.2 M KCl. The $Fe(CN)_6^{4-}$ ions subsequently coordinate with free Fe^{3+} ions dissociated from $Fe(CN)_6^{3-}$ in acidic medium (pH = 1.6) to form Prussian blue film on the illuminated gold CDtrode (Kodak, Mexico) electrode surface. The reactions are described below:

$$Fe\,(CN)_6^{3-} + H_2O \xrightarrow{hv} Fe(CN)_6^{4-} + H^+ + OH^{\bullet}$$
$$K^+ + Fe^{3+} + Fe(CN)_6^{4-} \rightarrow KFeFe(CN)_6$$

In the first photochemical reaction, water is an electron donor. The deposition rate of the blue film was found to be determined by ferricyanide concentration, solution pH, and light intensity [185,186].

Another photochemical method for making thin films is the thin film-photolysis method [187–193]. The general procedure for this method involves several steps. First, one needs to choose suitable precursors for organometallic coordination compounds in solution; second, one needs to coat the solutions onto substrate plates using spin-coating techniques; the third step is irradiation of the spin-coated films with light of proper wavelengths to decompose the organometallic coordination compounds to form new metal or metal oxide films; and the last step is to remove the organic debris from the films using proper solvents. Because the reactions take place only in the light irradiated area, this method can be used to make thin films with designed patterns by using masked templates. Using this method, thin metal films such as Co, Cu, Ni, Pd, and Pt; thin metal oxide films such as Cr_2O_3, CuO, NiO, δ-UO_3, $NiCoO_X$, ZrO_2, HfO_2, and ReO_3; and thin complex oxides films such as $SrBi_2Ta_2O_9$ (SBT), $PbZr_xTi_{1-\&x}O_3$ (PZT), $Pb_{0.67}La_{0.22}(Zr_{0.2}Ti_{0.8})O_3$ (PLZT), and $Pb_{0.988}(Zr_{0.52}Ti_{0.48})_{0.976}Nb_{0.024}O_3$ (PZTN) have been fabricated. Compared to the conventional sol–gel and chemical vapor deposition methods, the thin-film-photolysis method has advantages of simplicity, convenience, easy of control and lack of corrosion damage, etc.

6.6. PHOTOCHEMICAL SYNTHESIS OF NANOMATERIALS

Nano-sized materials exhibit many unique physical and chemical properties. Syntheses of them have been very active in recent years. Among the numerous synthesis methods for nano-sized materials, photochemical synthesis plays a very important role [194–211]. The synthesis can lead to nano-sized metal, semiconductor, binary or tertiary metal and semiconductor composite materials, and even 2D particles. Distinct from the other synthesis methods, photochemical syntheses of nano-sized materials have to be carried out in an oxygen-free environment because oxygen if present in the solution competes for the photo-generated reductants. Surfactants and polymers are also used in the photochemical syntheses of metal nanoparticles for stabilization of the newly formed nano-sized particles. In recent years, other compounds such as dendrimers, ionic liquids, and DNA macromolecules have also been used for this purpose ([212] and references therein).

Depending on the chemical nature of the precursors, photochemical syntheses can proceed via either direct photolysis or photo-induced and photosensitized reactions. If direct photolysis is used to make nano-sized metal particles, the metals in the corresponding metal precursors must exist in their zero valence state. After breaking the metal–ligand bonds by photolysis, nano-sized metal particles are then formed. With this method, uniform and narrow-distributed Pd metal particles with sizes of 0.22, 5 and 10 nm have been synthesized [197]. The amount of surfactant used in the preparation seems very important, the higher the concentration of the surfactant, the smaller the metal particles.

If the photo-induced reductions are used to make nano-sized metal particles, the metal ion precursors are reduced either by solvent molecules or their ligands upon direct photo-excitation of the precursors [212]. Via this method, Ag, Au, Pt, Co, Ni, and Fe metal nanoparticles have been synthesized from their corresponding inorganic and organic precursors and UV light of wavelengths 254, 308, and 353 nm [213–241]. In a very recent study, tin nanoparticles were synthesized by decomposition of tin(II) dimethylamine compound with UV light [242]. Inspired by this work, a photochemical polythiol process [243] has been developed for making monodispersed bismuth, copper, lead, and antimony metal nanoparticles. In this process, thiol first reacts with the metal salts forming metal thiolates and then visible light leads to ligand-to-metal charge-transfer (LMCT) in the metal thiolates to reduce the metal cations. Via this method, very high yields of metal nanoparticles have been obtained by using very small amounts of solvent.

If photosensitized reductions are used to make nano-sized metal particles, organic or inorganic sensitizers are added. Photo-excitation of the sensitizers generates intermediates which then reduce the co-existing metal ions to M^0. Compared to the photo-induced reductions, the photosensitized reductions have advantages of fast and efficient nanoparticle formation and flexibility of light excitation wavelength because the excitation is on the sensitizers and not on the metal ion precursors. Via this method, Ag, Cu, Au, AuAg alloy, and Ag/Cu composite nanoparticles have been synthesized [194,244–251].

Photophysical properties of semiconductor nanoparticles can also be used to help making metal and metal–semiconductor nano-sized composite particles [200–211]. The basic idea of this method is that, photoexcitation of semiconductor nanoparticles in a suspension generates electrons in its conduction band and holes in its valence band. The holes in the valence band react with the reducing agents such as ethanol present in the suspension forming $CH_3CH \cdot OH$ radicals and the electrons in the conduction band migrate out and are trapped on the semiconductor nanoparticle surface. The $CH_3CH \cdot OH$ radicals further react with the semiconductor nanoparticles to give up their electrons. Then the metal ions in the solution react with the electrons accumulated on the semiconductor nanoparticle surface and are reduced to form metal nanoparticles which deposit on the surface of the semiconductor nanoparticles. After completion of all these reactions, a metal–semiconductor nano-sized composite material is formed. By changing the photochemical reaction conditions, various composition-controlled metal–semiconductor nanostructures, multicomponent alloys, and core–shell structures are synthesized. In recent years, using semiconductor nanoparticles such as TiO_2, ZnO, SnO_2, CdS, and ZnS, metal–semiconductor nano-sized composite materials having photochemical reduced metals such as Ag, Au, Cu, Pt, Pd, and Cd have been prepared [200–211].

Similarly, metal alloys can also be introduced to form metal–semiconductor nano-sized composite materials. But it is worth mentioning that if the two metals in a tertiary two-metal alloy composite have different reducibility, the easily reducible metal after reduction can facilitate the reduction of the less easily reducible metal to form an alloy. For example, in the composites of ZnO/Ag/Cu, ZnO/Ag/Cd, and ZnO/Ag/Zn, the easily reducible Ag made the Cu, Cd, and Zn reduction much easier [200].

Another phenomenon, which was observed when photophysical properties of semiconductor nanoparticles were used to make metal–semiconductor nano-sized composite particles, is the photocorrosion. Photocorrosion depends on the particle size of the semiconductors and is determined by the reduction potential of the semiconductor nanoparticles which changes with particle size. If the reduction potential of the semiconductor nanoparticles is increased with the increase of the particle size to the value that reduction of the metal ions of the semiconductors cannot easily occur, the photocorrosion of the semiconductor is stopped. Therefore, controlling the particle size of the semiconductor nanoparticles is important to avoid photocorrosion during the process.

The examples given above in this section all involve syntheses using photoreduction reactions of photo-generated electrons. Actually, photo-generated positively charged holes can also be used in photochemical synthesis. An example is given below to demonstrate this strategy which involves synthesis of a specific composite material simultaneously using photo-generated electrons and holes—photochemical synthesis of Pt particle-containing nano-sized hollow carbon sphere composite material [252]. In this synthesis, a solution was prepared which contains Pt(IV) precursor and benzophenol as well as highly dispersed nano-sized TiO_2 particles as a photocatalyst. Upon photon excitation of the TiO_2 particles, conduction band electrons and valence band holes were generated. The electrons reduced the Pt(IV) precursor to Pt nanoparticles and the holes oxidized the benzophenol which then polymerized on the surface of the TiO_2 nanoparticles. A subsequent 700°C-vacuum treatment carbonized the polymers to form a carbon shell on the TiO_2 nanoparticles. The TiO_2 nanoparticles were then removed by HF treatment resulting in a specific composite material of hollow carbon spheres with a wall thickness of about 4 nm. Each hollow sphere contained a few Pt nanoparticles.

6.7. PRODUCTION OF H_2 VIA PHOTODECOMPOSITION OF WATER

H_2 can be produced via various photochemical approaches [12,13,253]: photocatalytic dehydrogenation of hydrocarbons and alcohols, photocatalytic reduction of H^+ from acidic solutions, photo-assisted water–gas shift reactions, photobiological synthesis, photodecomposition of H_2S, and photocatalytic water splitting. Among these approaches, photocatalytic water splitting for H_2 production is one of the most widely studied subjects. Since Fujishima and Honda in Japan in the early 1970s [12,13] discovered that TiO_2 could split water to form H_2 and O_2 upon light irradiation, H_2 production via photo-splitting of water has become a popular hope for solving the energy crisis. The study of this subject has continued for a number of decades due to concerns about the energy crisis, global warming, and the search for new clean energy sources. In laboratory studies, man-made light sources are normally used while in the actual applications sunlight should be used. Therefore, using sunlight to produce H_2 via photo-splitting of water has become the ultimate goal of current research of this field. Indeed, if this can be achieved with sufficient efficiency, the world will no longer have the energy crisis and global warming problems.

The systems for using visible light to photocatalytically split water to produce H_2 can be categorized into photochemical, semiconductor, photobiological, and hybrid systems. In the redox reactions of photocatalytical water splitting for producing H_2 and O_2, the main step is to assemble a system which absorbs visible light. The absorbed light is then effectively transformed into oxidation and reduction energies (charge separation), which are adequate and sufficient for driving multi-electron transfers to split water into H_2 and O_2. Thus, the transformation from light energy to chemical energy (H_2 formation) is completed. When trying to build up a system like this, spectroscopic properties, photophysical processes, thermodynamics, and dynamic processes need to be considered.

The main reactions of photocatalytic water splitting to produce H_2 and O_2 can be described as follows:

$$S + A \overset{hv}{\rightleftharpoons} S' + A^-$$
$$4S' + 2H_2O \overset{cat_1}{\rightarrow} 4S + 4H^+ + O_2$$
$$2A^- + 2H_2O \overset{cat_2}{\rightarrow} 2A + 2OH + H_2$$

Here S^+ can be transition metal coordination compound ions, metal ions, or photo-generated positive charged holes of semiconductors, A^- can be transition metal coordination compound ions or photo-generated electrons of semiconductors. The first reaction is mainly a photo-generated charge separation process for generating strong oxidizing and reducing agents. The second and third reactions are the subsequent catalytic oxidation and reduction of water in the presence of catalysts. The standard oxidation and reduction potentials of water, E_0 (pH = 7) are +0.82 V and −0.41 V, respectively, therefore, for transferring one electron, we need 1.23 eV energy. In order to achieve oxidation and reduction of water, the potentials of S^+ and A^- must satisfy the requirement for the oxidation and reduction reactions.

Homogeneous and heterogeneous visible light photochemical systems for producing H_2 were widely studied in the middle of 1970s and very recently, study of this system has become active again [254–257]. The most widely studied system for visible light photochemical hydrogen production is the system containing transition metal complex ion $[Ru(bpy)_3]^{2+}$, bpy = bipyridine, as a photosensitizer, methyl viologen molecule as an electron carrier, EDTA molecule reacting with PS^+ and subsequently being decomposed, colloidal Pt as a catalyst, and visible light of wavelengths 400–600 nm. In a study with $[Ru(bpy)_3]^{2+} = 5.65 \times 10^{-5}$ M, $MV^{2+} = 3 \times 10^{-3}$ M, EDTA = 0.1 M and colloidal Pt = 1.92×10^{-5} M, pH = 5, the quantum yield reaches 0.17 [254]. However, the actual technical assessment of the system gives the quantum yield no more than 0.1 [256]. In recent years, investigation of these systems has continued. The main problems for practical application involve the stability of the photosensitizers, side reactions of hydrogenation of MV^{2+}, and subsequent treatment of the oxidized EDTA products.

Semiconductor photochemical systems for H_2 production started in the early 1970s when Fujishima and Honda [12,13] discovered water splitting in the presence of TiO_2 upon UV light irradiation. These investigations continued for a number of decades. The original discovery involved using a TiO_2 crystal as an anode and Pt metal as a cathode in a photo-electrochemical cell containing an aqueous solution of electrolyte. When the TiO_2 crystal electrode was irradiated with UV light, O_2 formed on the TiO_2 crystal electrode, and H_2 was formed on the Pt electrode. Later, Bard [258,259] in the United States proposed that the same process that occurred in the normal photo-electrochemical cell can be achieved in a microphoto-electrochemical cell with nano-sized Pt particles deposited on TiO_2 crystal powder. Because TiO_2 crystals have a band gap of 3.0 eV and can only absorb UV light with low quantum yield, much of the work in this field focused on finding stable semiconductor materials that adsorb visible light. The main reactions of water photo-splitting to H_2 and O_2 on semiconductors (taking TiO_2 as an example) can be described as follows:

$$TiO_2 \overset{hv}{\rightarrow} TiO_2(e^-_{cb} + h^+)$$
$$2e^-_{cb} + 2H^1 \overset{P_t}{\rightarrow} H_2$$
$$4h^+ + 2H_2O \rightarrow O_2 + 4H^+$$

where cb means conduction band. The reduction potential of photo-generated conduction band electrons and the oxidation potential of the valence band positively charged holes should satisfy the oxidation and reduction potentials required for water oxidation and reduction reactions, i.e., e^-_{cb} should have more negative potential than that for reduction of water to H_2, and valence band positively charged holes should have more positive potential than that for oxidation of water to O_2.

Although photophysical properties of many semiconductors satisfy or partially satisfy the requirement for water photo-splitting, they are not stable under water photo-splitting conditions. Until now, more than 130 semiconductor materials with different chemical compositions such as oxides and complex oxides, and semiconductors materials with tunnel and layer structures have been studied [260–262], but most of these materials require UV light and H_2 yields are only between several tens and a few hundred micromoles (μmol h^{-1} g^{-1}) [14,263]. Complex oxides such as $CuFeO_2$, $CuMnO_2$, $CuCrO_2$, $CuAlO_2$, which have a delafossite structure and absorb visible light, have been reported [264–266]. These complex oxides have E_g about 1.25–1.34 eV and when they are combined with Cu_2O and also present in a basic solution containing

S^{2-}, SO_3^{2-}, and $S_2O_3^{2-}$ reducing agents, H_2 is produced but with very low rates. The material with the current quantum yield record is Cr/Rh-modified GaN/ZnO which gives a quantum yield of 0.025 for photo-splitting pure water with visible light at wavelengths of 420–440 nm [267–269].

For hydrogen production using semiconductor photochemical systems, adding NaOH and Na_2CO_3 facilitates hydrogen generation, and increases the rate of hydrogen formation [14]. The particle size of semiconductors and the perfection of crystals have effects on charge separation. Gas phase water vapor and water thin films are easier to be photo-split compared to aqueous solutions [14]. Searching for stable semiconductors with visible light sensitivity and high quantum yield is still the target of the research in this field.

For practical application of a visible light water splitting system, the efficiency of H_2 generation has to be greater than the energy consumption for operating the system. Recently, Feng et al. estimated the critical conversion efficiency of light energy to be ~15% using a hypothetical system with 20,000 tons annual hydrogen production with Na_2S/Na_2SO_3 as sacrificial reagent [270]. To achieve an economically viable system, the efficiency still has a long way to go.

Hybrid systems appeared to be interesting alternatives to produce H_2 in recent years [271–275]. Combining ultrasonic and photocatalysis process to produce H_2 and O_2 is one example of this kind. Normally, visible light cannot directly split water into hydrogen and oxygen because water does not absorb the light. However, ultrasonic can directly split water molecules into H• and OH• radicals. These radicals can react with each other forming H_2 and H_2O_2,

Ultrasonic splits water into $H^{\cdot}$ and $OH^{\cdot}$ radicals	$H_2O \rightarrow H^{\cdot} + OH^{\cdot}$
Radical reactions	$H^{\cdot} + H^{\cdot} \rightarrow H_2$
	$OH^{\cdot} + OH^{\cdot} \rightarrow H_2O_2$
Photocatalysis decomposes H_2O_2 into O_2 and H_2O	$H_2O_2 \rightarrow 1/2O_2 + H_2O$

Thus, combining ultrasonic wave chemistry and photocatalysis can achieve direct water splitting into H_2 and O_2. This method is simple and can be realized, as long as an aqueous solution that contains photocatalyst nanoparticles is simultaneously irradiated with ultrasonic and UV or visible light. However, to effectively separate H_2 and O_2, ultrasonic wave irradiation and light irradiation need to be provided alternately, ultrasonic wave being first, and followed by light irradiation. In this way, ultrasonic wave irradiation generates H_2 and light irradiation generates O_2. In laboratories, the ultrasonic source can be such as frequency 200 kHz and power 200 W. The light source can be a 500 W xenon lamp. The photocatalysts can be either TiO_2 or $BiVO_4$. For TiO_2, UV light is needed while for $BiVO_4$, visible light can be employed. The reaction needs to be operated under inert gas. In an experiment with 200 mg TiO_2 photocatalyst, ultrasonic wave irradiation for 1 h generated about a few tens of micromoles of H_2 and for the same period of light irradiation, about half of the quantity of O_2 was obtained. It has been found experimentally that this method is suitable for producing H_2 and O_2 with seawater [271].

Another example is solar energy cell electrolysis of aqueous solution for H_2 production [272–275]. The idea of this method is simple: electrolysis of aqueous solution needs electricity and the electricity is provided by a solar energy cell. Electrolysis of aqueous solution for production of H_2 is therefore achieved by using a solar cell photovoltaic panel for the electrolysis cell. In a recent study [275], a system with a commercially available amorphous silicon photovoltaic panel (peak voltage is 8–10 V, current is 0.95 A, area is 1280 cm^2, solar energy to electric energy efficiency is 4.5%), BiO_x and Bi-doped TiO_2 mixture as a anode, stainless steel as a cathode, NaCl and Na_2SO4 aqueous solution as electrolyte, can produce H_2 with a rate of 0.18 mmol/min and O_2 with a rate of 0.028 mmol/min, H_2/O_2 = ~7. If organic reducing agents such as benzophenol and organic acid were added to the aqueous solution, the H_2 production rate can be increased by 20–30%. The energy usage efficiency for total H_2 production can reach 30–60% which is comparable to the energy usage of the industrial scale electrolysis using alkaline electrolyte or proton exchange film.

Hydrogen production in large scale using solar energy via photolysis of water with sufficient efficiency is our ultimate goal and still has a long way to go. As mentioned above, the practical application of a sunlight-driven water splitting system needs the efficiency well over the critical conversion efficiency of light energy (compensation for the energy needed for the system to operate). The discovery and development of new types of photocatalyst materials which can fulfill the requirement will be the targets for this field.

6.8. SUMMARY

Inorganic photochemical synthesis is one of the most important methods for synthesis of inorganic compounds, multifunctional inorganic materials, new types of inorganic nanoparticles, and clean fuels such as H_2 for solving the energy crisis and global warming problems. Many of the compounds and materials synthesized via photochemical methods exhibit unique structural features and physicochemical properties for which the materials via conventional synthetic methods are not able to provide. This uniqueness of photochemical synthesis arises from the difference between excited state chemistry and ground state chemistry, as exemplified by many examples of the syntheses of organometallic coordination compounds. For other inorganic materials such as film and nanoparticles, the uniqueness is from the better control of the synthesis and of the photo-generated electrons and electronic holes. In recent years, combination of photochemical synthesis with other synthesis methods opened up new and simple synthesis routes for materials science and engineering. Research on water splitting to produce H_2 using visible light (sunlight) has made the photochemical synthesis method even more attractive. If the massive production of clean fuel H_2 using solar energy could be realized in the future, we would not have any energy crisis and the world would also become much cleaner than ever before. At that time, inorganic photochemical synthesis would become a part of our daily life. If solar energy was used to produce ammonia from photolysis of N_2 and the synthesis of fertilizer via photochemical method became a method to provide enough fertilizer for crops, inorganic photochemical synthesis would become a solid basis for human life and growth. It is certain that inorganic photochemical synthesis will become more and more important in the future.

References

[1] V. Balzani, V. Carassiti, Photochemistry of Coordination Compounds, Academic Press, London, 1970.
[2] D.M. Klassen, W.G. Perkins, G.A. Crosby, J. Chem. Phys. 43 (1965) 1498–1503.
[3] G.A. Crosby, R.J. Watts, D.H.W. Carstens, Science 170 (1970) 1195–1196.
[4] G. Ciamician, Science 36 (1912) 385–394.
[5] N. Armaroli, V. Balzani, Angew. Chem. Int. Ed. 46 (2007) 52–66.
[6] N.V. Tkachenko, Optical Spectroscopy, Methods and Instrumentations, Elsevier 2006.
[7] R.A. Marcus, Angew. Chem. Int. Ed. Engl. 32 (1993) 111–1121.
[8] V. Balzani, G. Bergamini, P. Ceroni, Coord. Chem. Rev. 252 (2008) 2456–2469.
[9] V. Balzani, G. Bergamini, S. Campagna, F. Puntoriero, Top. Curr. Chem. 280 (2007) 1–36.
[10] I.P. Herman, Chem. Rev. 89 (1989) 1323–1357.
[11] R.F. Khairutdinov, Russ. Chem. Rev. 67 (1998) 109–122.
[12] A. Fujishima, K. Honda, Bull. Chem. Soc. Jpn. 44 (1971) 1148–1150.
[13] A. Fujishima, K. Honda, Nature (London) 238 (1972) 37–38.
[14] M. Kaneko, I. Okura (Eds.), Photocatalysis Science and Technology, Kodansha and Springer, 2002.
[15] K. Kalyanasundaram, Photochemistry in Microheterogeneous Systems, Academic Press, New York, 1987.
[16] Y. Kayanuma, Phys. Rev. B38 (1988) 9797–9805.
[17] P. Kubelka, F. Munk, Zeit. Für Tekn. Physik. 12 (1931) 593–601.
[18] H.J. Kuhn, S.E. Braslavsky, R. Schmidt, Pure Appl. Chem. 76 (2004) 2105–2146.
[19] C.G. Hatchard, C.A. Parker, New sensitive chemical actinometer II, potassium ferrioxalate as a standard chemical actinometer, Proc. Roy. Soc. Lond. 1235 (1956) 518–536. Part I: C.A. Parker. Proc. R. Soc. London A220, (1953) 104–116.
[20] M. Wrighton, Chem. Rev. 74 (1974) 401–430.
[21] T.E. Bitterwolf, Coord. Chem. Rev. 250 (2006) 388–413.
[22] G.A. Crosby, Acc. Chem. Res. 8 (1975) 231–238.
[23] F. Chanon, M. Chanon, in: N. Serpone, E. Pelizzetti (Eds.), Photocatalysis: Fundamentals and Applications, Wiley, New York, 1989, pp. 489–540.
[24] J.J. Zuckerman, Inorganic Reactions and Methods, vol. 15, VCH Publishers Inc., Deerfield Beach, Florida, 1986.
[25] T.E. Bitterwolf, J. Organomet. Chem. 689 (2004) 3939–3952.
[26] L. Ronconi, P.J. Sadler, Coord. Chem. Rev. 251 (2007) 1633–1648.
[27] E. Tfouni, Coord. Chem. Rev. 196 (2000) 281–305.
[28] R.N. Perutz, Chem. Soc. Rev. 22 (1993) 361–369.
[29] T.T. To, E.J. Heilweil, C.B. Duke III, K.R. Ruddick, C.E. Webster, T.J. Burkey, J. Phys. Chem. A113 (2009) 2666–2676.
[30] N.E. Leadbeater, J. Photochem. Photobiol. A Chem. 106 (1997) 67–74.
[31] K. Kalyanasundaram, M. Gratzel, E. Pelizzett, Coord. Chem. Rev. 69 (1986) 57–125.
[32] M. Lim, K. Chiang, R. Amal, J. Photochem. Photobiol. A Chem. 183 (2006) 126–132.
[33] T. Jiao, Z. Pang, T.J. Burkey, R.F. Johnston, T.A. Heimer, V.D. Kleiman, E.J. Heilweil, J. Am. Chem. Soc. 121 (1999) 4618–4624.
[34] R. Tsumura, N. Hagihara, Bull. Chem. Soc. Jpn. 38 (1965) 1901–1903.
[35] R.J. Kinney, W.D. Jones, R.G. Bergmann, J. Am. Chem. Soc. 100 (1978) 7902–7915.
[36] W.A. Herrmann, J. Plank, Chem. Ber. 112 (1979) 392–393.
[37] M.S. Wrighton, D.S. Ginley, J. Am. Chem. Soc. 97 (1975) 2065–2072.
[38] R. Mueller, H. Vahrenkamp, Chem. Ber. 110 (1977) 3910–3919.
[39] D.R. Tyler, M. Altobelli, H.B. Gray, J. Am. Chem. Soc. 102 (1980) 3022–3024.
[40] P.F. Heveldt, B.F.G. Johnson, J. Lewis, P.R. Raithby, G.M. Sheldrick, J. Chem. Soc., Chem. Commun. (1978) 340–341.
[41] S. Bhaduri, B.F.G. Johnson, J.W. Kelland, J. Lewis, P.R. Raithby, S. Rehani, G.M. Sheldrick, K. Wong, M. McPartlin, J. Chem. Soc., Dalton Trans. (1979) 562–568.
[42] J.J. Zuckermann, Z. Stasicka, E. Wasielewska, Coord. Chem. Rev. 159 (1997) 271–294.
[43] G.L. Geoffroy, M.S. Wrighton, Organometallic Photochemistry, Academic Press, New York, 1979.
[44] K.R. Mann, H.B. Gray, G.S. Hammond, J. Am. Chem. Soc. 99 (1977) 306–307.
[45] T.P. Gill, K.R. Mann, Inorg. Chem. 19 (1980) 3007–3010.
[46] M.A. Bennett, A.K. Smith, J. Chem. Soc., Dalton Trans. (1974) 233–241.
[47] A. Juris, V. Balzani, F. Barigelletti, S. Campagna, P. Belser, A. Von Zelewsky, Coord. Chem. Rev. 84 (1988) 85–277.
[48] J.P. Sauvage, J.P. Collin, J.C. Chambron, S. Guillerez, C. Coudret, V. Balzani, F. Barigelletti, L. De Cola, L. Flamigni, Chem. Rev. 94 (1994) 993–1019.
[49] A.P. De Silva, H.Q.N. Gunaratne, T. Guunlaugsson, A.J.M. Huxley, C.P. McCoy, J.T. Rademacher, T.E. Rice, Chem. Rev. 97 (1997) 1515–1566.
[50] M.H. Keefe, K.D. Benkstein, J.T. Hupp, Coord. Chem. Rev. 205 (2000) 201–228.
[51] C.W. Rogers, Y. Zhang, B.O. Patrick, W.E. Jones, M.O. Wolf, Inorg. Chem. 41 (2002) 1162–1169.

[52] S. Welter, K. Brunner, J.W. Hofstraat, L. De Cola, Nature 421 (2003) 54–57.
[53] P. Belser, S. Bernhard, C. Blum, A. Beyeler, L. De Cola, V. Balzani, Coord. Chem. Rev. 191-192 (1999) 155–169.
[54] J.P. Collin, A.C. Laemmel, J.P. Sauvage, New J. Chem. 25 (2001) 22–24.
[55] A.C. Laemmel, J.P. Collin, J.P. Sauvage, C.R. Acad. Sci. Paris 3 (2000) 43–49.
[56] C.R. Hecker, P.E. Fanwick, D.R. MacMillin, Inorg. Chem. 30 (1991) 659–666.
[57] H.F. Suen, S.W. Wilson, M. Pomerantz, J.L. Walsh, Inorg. Chem. 28 (1989) 786–791.
[58] S. Bonnet, J.P. Collin, N. Gruber, J.P. Sauvage, E.R. Schofield, Dalton Trans. (2003) 4654–4662.
[59] J. Ruiz Aranzaes, D. Astruc, Inorg. Chim. Acta 361 (2008) 1–4.
[60] D. Catherine, D. Astruc, J. Organomet, Chem. 248 (1983) C9–C12.
[61] D. Catherine, D. Astruc, J. Organomet. Chem. 272 (1984) 417–426.
[62] A.A. Rachford, J.L. Petersen, J.J. Rack, Inorg. Chem. 45 (2006) 5953–5960.
[63] B. Durham, S.R. Wilson, D.J. Hodgson, T.J. Meyer, J. Am. Chem. Soc. 102 (1980) 600–606.
[64] L.G. Vanquickenborne, A. Ceulemans, Inorg. Chem. 17 (1978) 2730–2736.
[65] M. Cusamano, G. Gugliemo, V. Ricevuto, P. Traverso, T.J. Kemp, J. Chem. Soc. Chem. Commun. (1979) 775–776.
[66] M. Cusamano, G. Gugliemo, V. Ricevuto, S. Sostero, O. Traverso, T.J. Kemp, J. Chem. Soc., Dalton Trans. (1981) 302–305.
[67] P. Haake, T.A. Hylton, J. Am. Chem. Soc. 84 (1962) 3774–3775.
[68] D.A. Johnson, V.C. Dew, Inorg. Chem. 18 (1979) 3273–3274.
[69] A.W. Zanella, A.F. Fucaloro, Inorg. Nucl. Chem. Lett. 16 (1980) 515–519.
[70] L. Costanzo, A. Guiffrida, G. Guglielmo, V. Ricevuto, Inorg. Chim. Acta 33 (1979) 29–33.
[71] J.J. Rack, J.R. Winkler, H.B. Gray, J. Am. Chem. Soc. 123 (2001) 2432–2433.
[72] J.J. Rack, N.V. Mockus, Inorg. Chem. 42 (2003) 5792–5794.
[73] J.J. Rack, A.A. Rachford, A.M. Shelker, Inorg. Chem. 42 (2003) 7357–7359.
[74] A.A. Rachford, J.L. Petersen, J.J. Rack, Inorg. Chem. 44 (2005) 8065–8075.
[75] N.V. Mockus, J.L. Petersen, J.J. Rack, Inorg. Chem. 45 (2006) 8–10.
[76] A.A. Rachford, J.L. Petersen, J.J. Rack, Inorg. Chem. 45 (2006) 5953–5960.
[77] S.A. Hallock, A. Wojcicki, J. Organomet, Chem. 54 (1973) C27–C29.
[78] M.S. Wrighton, D.S. Ginley, J. Am, Chem. Soc. 97 (1975) 4246–4251.
[79] M.S. Wrighton, D.S. Ginley, J. Am. Chem. Soc. 97 (1975) 2065–2072.
[80] D.S. Ginley, M.S. Wrighton, J. Am. Chem. Soc. 97 (1975) 4908–4911.
[81] M.S. Wrighton, D.S. Ginley, J. Am. Chem. Soc. 97 (1975) 4246–4251.
[82] D.M. Allen, A. Cox, T.J. Kemp, Q. Sultana, R.B. Pitts, J. Chem. Soc., Dalton Trans. (1976) 1189–1193.
[83] A.R. Burkett, T.J. Meyer, D.G. Written, J. Organomet. Chem. 67 (1974) 67–73.
[84] J.L. Graff, R.D. Sanner, M.S. Wrighton, J. Am. Chem. Soc. 101 (1979) 273–275.
[85] D.R. Tyler, H.B. Gray, J. Am. Chem. Soc. 103 (1981) 1683–1686.
[86] J.C. Green, J.N. Harvey, R. Poli, J. Chem. Soc., Dalton Trans. (2002) 1861–1866.
[87] R.S. Paonessa, W.C. Trogler, Organometal 1 (1982) 768–770.
[88] A.W. Addison, R.D. Gillard, P.S. Sheridan, L.R.H. Tipping, J. Chem. Soc., Dalton Trans. (1974) 709–716.
[89] M.J. Almond, R.W. Atkins, J. Mol. Struct. 449 (1998) 99–109.
[90] M.J. Almond, R.W. Atkins, R.H. Orrin, J. Chem. Soc., Dalton Trans. (1994) 311–313.
[91] M.K. Trost, R.G. Bergman, Organometal 10 (1991) 1172–1178.
[92] K. Hino, Y. Inokuchi, K. Kosugi, H. Sekiya, Y. Hosokoshi, K. Inoue, N. Nishi, J. Phys. Chem. B 106 (2002) 1290–1293.
[93] A. Pelter, R.T. Pardasani, P. Pardasani, Tetrahedron 56 (2000) 7339–7369.
[94] M.V. Piserchio, F.W. Lampe, J. Phorochem, Photobiol. A Chem. 60 (1991) 11–26.
[95] E.R. Austin, F.W. Lampe, J. Phys. Chem. 81 (1977) 1134–1138.
[96] T.L. Pollock, H.S. Sandhu, A. Jodhan, O.P. Strausz, J. Am. Chem. Soc. 95 (1973) 1017–1024.
[97] P. John, J.H. Purnell, J. Chem. Soc. Faraday Trans. I, 69 (1973) 1455–1461.
[98] G.A. Gibbon, Y. Rousseau, C.H. Van Dyke, G.J. Mains, Inorg. Chem. 5 (1966) 114–117.
[99] D.G. White, E.G. Rochow, J. Am. Chem. Soc. 76 (1954) 3897–3902.
[100] D. Saunders, J. Heicklen, J. Am. Chem. Soc. 87 (1965) 2088–2092.
[101] T. Hirata, J.E. Gunning, J. Chem. Phys. 27 (1957) 477.
[102] R.B. King, in: J.J. Eisch, R.B. King (Eds.), Organometallic Synthesis, vol. 1, Academic Press, New York, 1965, p. 93.
[103] W.C. Kreye, R.A. Marcus, J. Chem. Phys. 37 (1962) 419–427.
[104] N.N. Greenwood, J.D. Kennedy, T.R. Spalding, D. Taylorson, J. Chem. Soc., Dalton Trans. (1979) 840–846.
[105] S.K. Boocock, N.N. Greenwood, J.D. Kennedy, W.S. McDonald, J. Staves, J. Chem. Soc., Dalton Trans. (1980) 790–796.
[106] G.C. White, Handbook of Chlorination, Van Nostrand Reinhold Company, New York, 1972.
[107] V.A. Nadtochenko, J. Kiwi, Inorg. Chem. 37 (1998) 5233–5238.
[108] R. Hoppe, W. Dähne, H. Mattauch, K.M. Rödder, Angew. Chem. 74 (1962) 903.
[109] C.L. Chernick, H.H. Claassen, P.R. Fields, H.H. Hyman, J.G. Malm, W.M. Manning, M.S. Matheson, L.A. Quarterman, F. Schreiner, H.H. Selig, I. Sheft, S. Siegel, E.N. Sloth, L. Stein, M.H. Studier, J.L. Weeks, M.H. Zirin, Science 138 (1962) 136–138.
[110] M. Tramšek, B. Žemva, Acta Chim. Slov. 53 (2006) 105–116.
[111] B. Žemva, Croat. Chim. Acta 61 (1988) 163–187.
[112] A. Šmalc, K. Lutar, in: R.N. Grimes (Ed.), Inorganic Syntheses, vol. 29, Wiley-Interscience, New York, USA, 1992, pp. 1–4.
[113] A. Smalc, K. Lutar, B. Zemva, Inorganic Syntheses vol. 29 (1992) 11–15.
[114] S.A. Kinkead, J.R. FitzPatrick, J. Foropoulos Jr., R.J. Kissane, J.D. Purson, in: J.S. Thrasher, S.H. Strauss (Eds.), Inorganic Fluorine Chemistry toward the 21st Century, ACS Symposium Series 555, American Chemical Society, Washington, DC, 1994, pp. 40–55. Chapter 3.
[115] K. Lutar, A. Smalc, Eur. J. Solid State Inorg. Chem. 28 (1991) 631.
[116] R. Winter, P.G. Nixon, G.L. Gard, J. Fluorine Chem. 87 (1998) 85–86.
[117] E.O. John, R.L. Kirchmeier, J.M. Shreeve, Inorg. Chem. 31 (1992) 329–331.
[118] R. Winter, P.G. Nixon, G.L. Gard, J. Fluorine Chem. 87 (1998) 85–86.
[119] E.O. John, R.L. Kirchmeier, J.M. Shreeve, Inorg. Chem. 31 (1992) 329–331.

[120] I.P. Herman, in: D. Biuerle (Ed.), Laser Processing and Diagnostics, Chemical Physics, vol. 39, Springer, New York, 1984, p. 396.
[121] K.G. Ibbs, R.M. Osgood, in: K.G. Ibbs, R.M. Osgood (Eds.), Laser Chemical Processing for Microelectronics, Cambridge University Press, Cambridge, 1989, p. 1.
[122] M.T. Bohr, IEEE Trans. Nanotechnol. 1 (2002) 56–62.
[123] R. Chau, S. Datta, M. Doczy, B. Doyle, J. Kavalieros, M. Metz, IEEE Electron. Dev. Lett. 25 (2004) 408–410.
[124] C.J. Forst, C.R. Ashman, K. Schwarz, P.E. Blochl, Nature 427 (2004) 53–56.
[125] D.A. Muller, T. Sorsch, S. Moccio, F.H. Baumann, K. Evans-Lutterodt, G. Timp, Nature 399 (1999) 758–761.
[126] J. Robertson, Rep. Prog. Phys. 69 (2006) 327–396.
[127] G.D. Wilk, R.M. Wallace, J.M. Anthony, J. Appl. Phys. 89 (2001) 5243–5275.
[128] D.G. Schlom, J.H. Haeni, MRS Bull. 27 (2002) 198–204.
[129] M. Ritala, K. Kukli, A. Rahtu, P.I. Raisanen, M. Leskela, T. Sajavaara, et al., Science 288 (2000) 319–321.
[130] C.O. Chui, S. Ramanathan, B.B. Triplett, P.C. McIntyre, K.C. Saraswat, IEEE Electron. Dev. Lett. 23 (2002) 473–475.
[131] A. Weber, E. Ivers-Tiffee, J. Power Sources 127 (2004) 273–283.
[132] S.C. Singhal, Solid State Ionics 135 (2000) 305–313.
[133] M.C. Williams, J.P. Strakey, S.C. Singhal, J. Power Sources 131 (2004) 79–85.
[134] S.C. Singhal, Solid State Ionics 152 (2002) 405–410.
[135] Z.B. Zhang, C.C. Wang, R. Zakaria, J.Y. Ying, J. Phys. Chem. B 102 (1998) 10871–10878.
[136] O. Carp, C.L. Huisman, A. Reller, Prog. Solid State Chem. 32 (2004) 33–177.
[137] D.A. Tryk, A. Fujishima, K. Honda, Electrochim. Acta 45 (2000) 2363–2376.
[138] B. Oregan, M. Gratzel, Nature 353 (1991) 737–740.
[139] A. Hagfeldt, M. Gratzel, Chem. Rev. 95 (1995) 49–68.
[140] M.T.A. Saif, S. Zhang, A. Haque, K.J. Hsia, Acta Mater. 50 (2002) 2779–2786.
[141] I.P. Herman, Chem. Rev. 89 (1989) 1323–1357.
[142] M. Tsuchiya, S.K.R.S. Sankaranarayanan, S. Ramanathan, Prog. Mater. Sci. 54 (2009) 981–1057.
[143] N. Kaliwoh, J.Y. Zhang, I.W. Boyd, Appl. Surf. Sci. 168 (2000) 288–291.
[144] R. Merkle, R.A. De Souza, J. Maier, Angew. Chem. Int. Ed. 40 (2001) 2126–2129.
[145] T.H. Wood, J.C. White, B.A. Thacker, Appl. Phys. Lett. 42 (1983) 408–410.
[146] C.J. Chen, J. Vac. Sci. Technol. A 5 (1987) 3386–3398.
[147] A. Kazor, I.W. Boyd, J. Appl. Phys. 75 (1994) 227–231.
[148] S.-i. Kimura, E. Murakami, K. Miyake, T. Warabisako, H. Sunami, T. Tokuyama, J Electrochem. Soc. 132 (1985) 1460–1466.
[149] S.K. Sharma, B.C. Chakravarty, S.N. Singh, B.K. Das, J. Mater. Sci. Lett. 9 (1990) 982–984.
[150] Y. Yasuda, S. Zaima, T. Kaida, Y. Koide, J. Appl. Phys. 67 (1990) 2603–2607.
[151] D. Chi, P.C. McIntyre, Appl. Phys. Lett. 85 (2004) 4699–4701.
[152] L.P.H. Jeurgens, W.G. Sloof, F.D. Tichelaar, E.J. Mittemeijer, J. Appl. Phys. 92 (2002) 1649–1656.
[153] M. Ichimura, F. Goto, Y. Ono, et al., J. Crystal. Growth 198/199 (1999) 308–312.
[154] T. Miyawaki, M. Ichimura, Mater. Lett. 61 (2007) 4683–4686.
[155] R. Kumaresan, M. Ichimura, N. Sato, et al., Mater. Sci. Eng. B96 (2002) 37–42.
[156] J. Podder, R. Kobayashi, M. Ichimura, Thin Solid Films 472 (2005) 71–75.
[157] J. Podder, T. Miyawaki, M. Ichimura, J. Crystal Growth 275 (2005) e937–e942.
[158] K.M. Santheep, N.P. Rajesh, M. Ichimura, et al., Mater. Lett. 62 (2008) 591–593.
[159] R. Kumaresan, M. Ichimura, E. Arai, Thin Solid Films 414 (2002) 25–30.
[160] H. Sasaki, K. Shibayama, M. Ichimura, et al., J. Crystal Growth 237–239 (2002) 2125–2129.
[161] M. Gunasekaran, M. Ichimura, Sol. Energy Mater. Sol. Cells. 91 (2007) 774–778.
[162] M. Ichimura, K. Shibayama, K. Masui, Thin Solid Films 466 (2004) 34–36.
[163] M. Ichimura, T. Narita, K. Masui, Mater. Sci. Eng. B96 (2002) 296–299.
[164] M. Verdaguer, A. Bleuzen, V. Marvaud, J. Vaissermann, M. Seuleiman, C. Desplanches, A. Scuiller, C. Train, R. Garde, G. Gelly, C. Lomenech, I. Rosenman, P. Veillet, C. Cartier, F. Villain, Coord. Chem. Rev. 190–192 (1999) 1023–1047.
[165] F.H. Koehler, R. Lescoueezec, Angew. Chem., In. Ed. 43 (2004) 2571–2573.
[166] P.H. Zhou, D.S. Xue, H.Q. Luo, X.G. Chen, Nano Lett. 2 (2002) 845–849.
[167] D.S. Xue, F.S. Li, Hyperfine Interact. 156/157 (2004) 31–40.
[168] M. Pyrasch, A. Toutianoush, W.Q. Jin, J. Schnepf, B. Tieke, Chem. Mater. 15 (2003) 245–254.
[169] A.S. Kumar, J.M. Zen, Chem. Phys. Chem. 5 (2004) 1227–1231.
[170] K. Itaya, N. Shoji, I. Uchida, J. Am. Chem. Soc. 106 (1984) 3423–3429.
[171] A. Eftekhari, J. Power Sources 126 (2004) 221–228.
[172] K. Kuwabara, J. Nunome, K. Sugiyama, Solid State Ionics 48 (1991) 303–308.
[173] M. Jayalakshmi, F. Scholz, J. Power Sources 91 (2000) 217–223.
[174] D.M. Delongchamp, P.T. Hammond, Chem. Mater 16 (2004) 4799–4805.
[175] Y.H. Huang, L.C. Chen, K.C. Ho, Solid State Ionics 165 (2003) 269–277.
[176] D. Zhang, K. Zhang, Y.L. Yao, X.H. Xia, H.Y. Chen, Langmuir 20 (2004) 7303–7307.
[177] A.A. Karyakin, E.A. Puganova, I.A. Budashov, I.N. Kurochkin, E.E. Karyakina, V.A. Levchenko, V.N. Matveyenko, S.D. Varfolomeyev, Anal. Chem. 76 (2004) 474–478.
[178] Y.J. Zhang, Y. Wen, Y. Liu, D. Li, J.H. Li, Electrochem. Commun. 6 (2004) 1180–1184.
[179] M. Nishizawa, S. Kuwabata, H. Yoneyama, J. Electrochem. Soc. 143 (1996) 3462–3465.
[180] Y. Wu, B.W. Pfennig, E.P. Vicenzi, A.B. Bocarsly, Inorg. Chem. 34 (1995) 4262–4267.
[181] M. Hermes, F. Scholz, J. Solid State Electrochem. 1 (1997) 215–220.
[182] L.F. Schneemeyer, S.E. Spengler, D.W. Murphy, Inorg. Chem. 24 (1985) 3044–3046.
[183] K.M. Jeerage, W.A. Steen, D.T. Schwartz, Chem. Mater 14 (2002) 530–535.
[184] Y.L, J.H. Hu, W. Yuan, K. Chen, X.H. WangXia, Electrochem. Commun 7 (2005) 1252–1256.
[185] D. Zhang, K. Wang, D.C. Sun, X.H. Xia, H.Y. Chen, Chem. Mater 15 (2003) 4163–4165.
[186] D. Zhang, K. Wang, D.C. Sun, X.H. Xia, H.Y. Chen, J. Solid State Electrochem. 7 (2003) 561–566.
[187] G.E. Buono-Core, M. Tejos, G. Cabello, N. Guzman, R.H. Hill, Mater. Chem. Phys. 96 (2006) 98–102.
[188] G.E. Buono-Core, M. Tejos, J. Lara, F. Aros, R.H. Hill Mater, Res. Bull. 34 (1999) 2333–2340.
[189] G.E. Buono-Core, M. Tejos, G. Alveal, R.H. Hill, J. Mater, Sci. 35 (2000) 4873–4877.
[190] G.E. Buono-Core, M. Tejos, G. Cabello, F. Aros, R.H. Hill, Bol. Soc. Chil. Quim. 47 (2002) 495–500.

[191] G.E. Buono-Core, M. Tejos, A.H. Klahn, R. Schrebler, R.H. Hill, J. Chil. Chem. Soc. 49 (2004) 223–226.
[192] H.-H. Park, H.-H. Park, T.S. Kim, et al., Sensors and Actuators B130 (2008) 696–700.
[193] H.-H. Park, S. Yoon, H.-H. Park, Thin Solid Films 447–448 (2004) 669–673.
[194] S. Kapoor, T. Mukherjee, Chem. Phys. Lett. 370 (2003) 83–87.
[195] L. Wang, G. Wei, C.L. Guo, et al., Colloids Surf. A Physicochem. Eng. Aspects 312 (2008) 148–153.
[196] K. Mallick, Z.L. Wang, T. Pal, J. Photochem. Photobiol. A. Chem. 140 (2001) 75–80.
[197] H. Tan, T. Zhan, W.Y. Fan, Chem. Phys. Lett. 428 (2006) 352–355.
[198] S. Giuffrida, G.G. Condorelli, L.L. Costanzo, et al., Chem. Mater. 16 (2004) 1260–1266.
[199] P.D. Cozoli, R. Comparelli, E. Fanizza, J. Am. Chem. Soc. 126 (2004) 3868–3879.
[200] A.L. Stroyuk, V.V. Shvalagin, Y.S. Kuchmii, J. Photochem. Photobiol. A Chem. 173 (2005) 185–194.
[201] A. Wood, M. Giersig, P. Mulvaney, J. Phys. Chem. B 105 (2001) 8810–8815.
[202] R.F. Khairutdinov, Russ. Chem. Rev. 67 (1998) 125–139.
[203] L.C. Chen, F.-R. Tsai, C.-M. Huang, J. Photochem. Photobiol. A Chem. 170 (2005) 7–14.
[204] J.-K. Yang, S.-K. Lee, J. Colloid Interface Sci. 282 (2005) 5–10.
[205] E. Szabo-Bardos, H. Czili, A. Horvath, J. Photochem, Photobiol. A Chem. 154 (2003) 195–201.
[206] H.M. Sung-Suh, J.R. Choi, H.J. Hah, S.M. Koo, Y.C. Bae J, Photochem. Photobiol. A Chem. 163 (2004) 37–44.
[207] C. He, Y. Xiong, J. Chen, C. Zha, X. Zhu, J. Photochem. Photobiol. A Chem. 157 (2003) 71–79.
[208] S. Yamazaki, N. Takemura, Y. Yoshinaga, A. Yoshida, J. Photochem. Photobiol. A Chem. 161 (2003) 57–60.
[209] A. Yamakata, T. Ishibashi, H. Onishi, J. Photochem. Photobiol. A Chem. 160 (2003) 33–36.
[210] V. Vamathevan, R. Amal, D. Beydoun, G. Low, S. McEvoy, J. Photochem. Photobiol. A Chem. 148 (2002) 233–245.
[211] N. Toshima, T. Yonezawa, New J. Chem. 22 (1998) 1179–1201.
[212] M. Sakamoto, M. Fujistuka, T. Majima, J. Photochem. Photobio. C Photochem. Rev. 10 (2009) 33–56. and references therein.
[213] H. Hada, Y. Yonezawa, A. Yoshida, A. Kurakake, J. Phys. Chem. 80 (1976) 2728–2731.
[214] Y. Yonezawa, T. Sato, M. Ohono, H. Hada, J. Chem. Soc., Faraday Trans. 83 (1987) 1559–1567.
[215] K. Kurihara, J. Kizling, P. Stenius, J.H. Fendler, J. Am. Chem. Soc. 105 (1983) 2574–2579.
[216] L. Bronstein, D. Chernyshov, P. Valetsky, N. Tkachenko, H. Lemmetyinen, J. Hartmann, S. Förster, Langmuir 15 (1999) 83–91.
[217] M.Y. Han, C.H. Quek, Langmuir 16 (2000) 362–367.
[218] K. Malone, S. Weaver, D. Taylor, H. Cheng, K.P. Sarathy, G. Mills, J. Phys. Chem. B 106 (2002) 7422–7431.
[219] E. Leontidis, K. Kleitou, T. Kyprianidou-Leodidou, V. Bekiari, P. Lianos, Langmuir 18 (2002) 3659–3668.
[220] S. Eustis, H.Y. Hsu, M.A. El-Sayed, J. Phys. Chem. B 109 (2005) 4811–4815.
[221] S. Eustis, M.A. El-Sayed, J. Phys. Chem. B 110 (2006) 14014–14019.
[222] L. Longenberger, G. Mills, J. Phys. Chem. 99 (1995) 475–478.
[223] Y. Yonezawa, I. Kawabata, T. Sato, Ber. Bunsen Ges. Phys. Chem. 100 (1996) 39–45.
[224] E. Gachard, H. Remita, J. Khatouri, B. Keita, L. Nadjo, J. Belloni, New J. Chem. 22 (1998) 1257–1265.
[225] A. Henglein, Langmuir 15 (1999) 6738–6744.
[226] R.E. Cameron, A.B. Bocarsly, J. Am. Chem. Soc. 107 (1985) 6116–6117.
[227] R.E. Cameron, A.B. Bocarsly, Inorg. Chem. 25 (1986) 2910–2913.
[228] V.P. Grivin, I.V. Khmelinski, V.F. Plyusnin, I.I. Blinov, K.P. Balashev, J. Photochem. Photobiol. A Chem. 51 (1990) 167–178.
[229] V.P. Grivin, I.V. Khmelinski, V.F. Plyusnin, J. Photochem, Photobiol. A: Chem. 51 (1990) 379–389.
[230] V.P. Grivin, I.V. Khmelinski, V.F. Plyusnin, J. Photochem. Photobiol. A Chem. 59 (1991) 153–161.
[231] K. Kurihara, J.H. Fendler, J. Am. Chem. Soc. 105 (1983) 6152–6153.
[232] H. Einaga, M. Harada, Langmuir 21 (2005) 2578–2584.
[233] M. Harada, H. Einaga, Langmuir 22 (2006) 2371–2377.
[234] B. Marciniak, G.E. Buono-Core, J. Photochem. Photobiol. A Chem. 52 (1990) 1–25.
[235] G.G. Condorelli, L.L. Costanzo, I.L. Fragalà, S. Giuffrida, G. Ventimiglia, J. Mater. Chem. 13 (2003) 2409–2411.
[236] S. Giuffrida, G.G. Condorelli, L.L. Costanzo, I.L. Fragalà, G. Ventimiglia, G. Vecchio, Chem. Mater. 16 (2004) 1260–1266.
[237] R. Krasnansky, S. Yamamura, J.K. Thomas, R. Dellaguardia, Langmuir 7 (1991) 2881–2886.
[238] K. Judai, J. Nishijo, C. Okabe, O. Ohishi, H. Sawa, N. Nishi, Synth. Met. 155 (2005) 352–356.
[239] J. Nishijo, C. Okabe, J. Bushiri, K. Kosugi, N. Nishi, H. Sawa, Eur. Phys. J. D 34 (2005) 219–222.
[240] K. Kosugi, M.J. Bushiri, N. Nishi, Appl. Phys. Lett. 84 (2004) 1753–1755.
[241] J. Nishijo, O. Oishi, K. Judai, N. Nishi, Chem. Mater 19 (2007) 4627–4629.
[242] K. Soulantica, A. Maisonnat, M.-C. Fromen, M.-J. Casanove, B. Chaudret, Angew. Chem. Int. Ed. 42 (2003) 1945–1949.
[243] S.C. Warren, A.C. Jackson, Z.D. Cater-Cyker, F.J. DiSalvo, U. Wiesner, J. Am. Chem. Soc. 129 (2007) 10072–10073.
[244] M. Sakamoto, T. Tachikawa, M. Fujitsuka, T. Majima, Chem. Phys. Lett. 420 (2006) 90–94.
[245] M. Sakamoto, T. Tachikawa, M. Fujitsuka, T. Majima. Erratum, Chem. Phys. Lett. 442 (2007) 170.
[246] M. Sakamoto, T. Tachikawa, M. Fujitsuka, T. Majima, Adv. Funct. Mater 17 (2007) 857–862.
[247] M. Sakamoto, T. Tachikawa, S.S. Kim, M. Fujitsuka, T. Majima, ChemPhysChem. 8 (2007) 1701–1706.
[248] M. Sakamoto, T. Tachikawa, M. Fujitsuka, T. Majima, Chem. Mater 20 (2008) 2060–2062.
[249] T. Itakura, K. Torigoe, K. Esumi, Langmuir 11 (1995) 4129–4134.
[250] K. Esumi, T. Matsumoto, Y. Seto, T. Yoshimura, J. Colloid Interface Sci. 284 (2005) 199–203.
[251] N. Kometani, Y. Kohara, Y. Yonezawa, Colloids Surf. A 313–314 (2008) 43–46.
[252] Y.H. Ng, S. Ikeda, T. Harashi, et al., Chem. Mater. 20 (2008) 1154–1160.
[253] A.J. Esswein, D.G. Nocera, Chem. Rev. 107 (2007) 4022–4047.
[254] E. Amouyal, Sol. Energy Mater. Sol. Cell 38 (1995) 249–276.
[255] C. Konigstein, R. Bauers, Int. J. Hydrogen Energy 22 (1997) 471–474.
[256] J.R. Bolton, Sol. Energy 57 (1996) 37–50.
[257] K. Sakai, H. Ozawa, Coord. Chem. Rev. 251 (2007) 2753–2766.
[258] A.J. Bard, J. Photochem. 10 (1979) 59–75.
[259] A.J. Bard, Science 207 (1980) 139–144.
[260] F.E. Osterloh, Chem. Mater. 20 (2008) 35–54.
[261] A. Kudo, Int. J. of Hydrogen Energy 32 (2007) 2673–2678.
[262] L.J. Guo, L. Zhao, D.W. Jing, Y.J. Lu, H.H. Yang, B.F. Bai, X.M. Zhang, L.J. Ma, X.M. Wu, Energy 34 (2009) 1073–1090.
[263] R.M. Navarro, F. del Valle, J.A. Villoria de la Mano, M.C. Alvarez-Galvan, J.L.G. Fierro, Adv. Chem. Eng. 36 (2009) 111–143. and refernces therein.

[264] S. Saadi, A. Bouguelia, M. Trari, Solar Energy 80 (2006) 272–280.
[265] Y. Bessekhouad, M. Trari, J.P. Doumerc, Int. J. Hydrogen Energy 28 (2003) 43–48.
[266] M. Younsi, A. Aider, A. Bouguelia, M. Trari, Sol. Energy 78 (2005) 574–580.
[267] K. Maeda, K. Teramura, D.L. Lu, T. Takata, N. Saito, Y. Inoue, K. Domen, J. Phys. Chem. B 110 (2006) 13753–13758.
[268] K. Maeda, K. Teramura, D.L. Lu, T. Takata, N. Saito, Y. Inoue, K. Domen, Nature 440 (2006). 295–295.
[269] K. Maeda, K.K. Teramura, D.L. Lu, N. Saito, Y. Inoue, K. Domen, Angew. Chem. Int. Ed. 45 (2006) 7806–7809.
[270] X. Feng, W. Mao, W. Yan, Int. J. Hydrogen Energy 33 (2008) 3644–3650.
[271] M. Kaneko, I. Okura (Eds.), Photocatalysis Science and Technology, Kodansha and Springer, 2002, p. 203.
[272] R. Friberg, Int. J. Hydrogen Energy 18 (1993) 853–882.
[273] G.E. Ahmad, E.T. El Shenawy, Renew. Energy 31 (2006) 1043–1054.
[274] P.A. Lehman, C.E. Chamberlin, G. Pauletto, M.A. Rocheleau, Int. J. Hydrogen Energy 22 (1997) 465–470.
[275] H. Park, C.D. Vecitis, W. Choi, O. Weres, M.R. Hoffmann, J. Phys. Chem. C 112 (2008) 885–889.

CHAPTER

7

CVD and its Related Theories in Inorganic Synthesis and Materials Preparations

Ji-Tao Wang

Fudan University, China

7.1. BRIEF HISTORY OF CHEMICAL VAPOR DEPOSITION

Chemical vapor deposition (CVD) is a technology used for forming solid powders in the gas phase or depositing films on substrates from precursors in the gas phase. Many precursors are liquids or solids under ordinary conditions, and they are first vaporized for use in CVD. The name of CVD comes from the book "Vapor Deposition" written by Dr J. M. Blocher and others at the beginning of the 1960s [1]. Dr Blocher was respected as "Sir CVD" for his promotion of CVD international academic activities. During 1960s this technology had another term of "vapor plating," but later the term of "vapor deposition" was generally accepted. Based on the process mainly being physical or chemical, vapor deposition has been divided into two kinds, namely physical vapor deposition (PVD) and chemical vapor deposition. For instance, vacuum vaporization, spattering, ion plating, etc. usually belong to PVD; and vapor depositions based on chemical reactions or plasma-enhanced chemical reactions are called CVD or plasma-enhanced chemical vapor deposition (PECVD or PCVD). In fact, due to the development of science and technology, there are some cases that crossly belong to both parts, such as reaction spattering and reaction ion plating, in which metals are vaporized by spattering or ion bombardment and reacted to produce oxides or nitrides, that combines both physical and chemical processes.

The ancient original form of chemical vapor deposition might be cast back to the cave-dweller's time. Cave dwellers burnt wood for warming or cooking, and deposited black carbon layers on the cave wall or on the rock. These carbon deposits were formed from decomposition of organic gases during the burning of wood or cooking food. Therefore, the site of cave dwellers' barbecue discovered by archaeologists was also the most ancient site of original chemical vapor deposition. However, these were unconscious carry-over of cave dwellers. Their purpose was only warming, protecting from wild animals, or cooking food at that time. With the progress of human beings, the technology of chemical vapor deposition had also been developed consciously. Especially in the ancient China, necromancers or magicians like the alchemists in ancient Europe looked for elixir to become "celestial" or "immortal". They were using a technology of "sublimation and refinement", which was the early stage of chemical vapor deposition. As written by a famous Chinese scientist in the Preface of the book, "*Crystal Growth*" [2]: "We should pay attention on the manufacture of silvered red. Silvered red was synthetic cinnabar. Li Shi-Zhen cited a book called '*Secrets of Elixir*' written by Hu Yan: 'For sublimation and refinement of mercury sulfide powders (silvered red), put 1 kg sulfur crystals into a new frying pan. After sulfur is melt, add 0.5 kg mercury in it. Stir-fry the mixture till it becomes sandlike materials. After no liquid mercury particles could be found, grind and put the material into a pot. Close the pot with a flat stone cover, and truss up the pot with cover by iron lines. Seal the pot, and calcine it with fire. After cooling, open the pot. Mercury sulfide powders adhere to the pot wall, and cinnabar crystals close to the cover.' Here, the synthesis of cinnabar from mercury and sulfur by chemical vapor deposition has been described, and the process was called 'sublimation and refinement' at that time. During the transformation in chemical vapor deposition, different sizes of forming crystals are located at different places. The smaller ones are called silvered

Modern Inorganic Synthetic Chemistry, DOI: 10.1016/B978-0-444-53599-3.10007-1

red, and the bigger ores are called cinnabar. The growth of optoelectronic crystals like gallium arsenide is still mainly based on the technology of 'sublimation and refinement'. Such a technology was generally used in China in the ancient time of elixir." Up to now, the synthesis of cinnabar from mercury and sulfur cited by Li Shi-Zhen (1518−1593, Ming Dynasty of China) from "*Secrets of Elixir*" written by Hu Yan might be the most ancient written record about chemical vapor deposition in the history of human beings. Dr Blocher in 1989 at the beginning of the seventh European CVD Conference made an introduction about that to scientists and experts from the other parts of the world [3]. During 1950s the technology of CVD attained its modern development stage. At that time the technology of CVD was mainly applied to coatings on cutting tools. This was due to a strong necessity of the mechanical industry in Europe. If tungsten carbide base hard alloy cutting tools are coated with CVD complex layers of Al_2O_3, TiC, and TiN, the cutting performance will be much better. Coated cutting tools have good toughness with long cutting life, and their economic benefits are very remarkable. Therefore, the technology has been spread and applied practically. Usually the golden yellow TiN layer is the last layer on cutting tools, so complex coated tools have a bright gold color. German scientist Dr W. Ruppert was one of the leading scientists of CVD coating technology in Europe. Since 1960s and 1970s due to the researches and production of semiconductors and integrated circuits (IC), CVD technology has been developed more quickly and more generally. The CVD technology of ultrapure silicon, called polycrystalline silicon or simply poly-silicon, was the unique way to produce semiconductor grade of silicon material and epitaxial silicon in industry. The productions of epitaxial single-crystal gallium arsenide and other III−V or II−VI semiconductors are also based on CVD technology. Many film depositions, such as various doped epitaxial single-crystal semiconductor films, polycrystal films, semi-insulated oxygen-doped poly-silicon films, insulated silicon dioxide films, silicon nitride films, phosphosilicate glass, borosilicate glass, and metal tungsten films, were widely deposited using CVD technology for producing integrated circuits. During manufacturing various special semiconductor devises, the CVD technology of gallium phosphoarsenide and gallium nitride epitaxial layers, silicon−geranium alloy epitaxial layers and silicon carbide epitaxial layers, and so on have been very important in the preparation of light emission devices (LED). Japan made outstanding progress in the key process of gallium nitride epitaxial growth and first realized the industrial production of blue light emission devices in the world. With the increase of the integration of integrated circuits and the minimization of semiconductor devices the new technology of atomic layer chemical vapor deposition (ALCVD or ALD) became one of the basic processes in fabrication of integrated circuits. In about 1970, former Soviet Union scientists Deryagin, Spitsyn, Fedoseev, etc. created a new technology of the activated low-pressure CVD diamond synthesis by introducing atomic hydrogen. Research on the new diamond technology became a world-wide upsurge after 1986. Especially, at the beginning of the twenty-first century the growth of about 10-carat-size gem-quality single-crystal diamonds has been reached by American scientists Chih-shiue Yan, R.J. Hemley, Ho-kwang Mao, etc. This is an important breakthrough in diamond syntheses. Some fundamental creative progresses in researches on CVD have been reached in China. In 1987, Dr Blocher as Chairman of the 10th international CVD conference highly commended the simulation model of low-pressure chemical vapor deposition (LPCVD) in his invitation letter to J.-T. Wang: "On behalf of the Program Planning Committee of the Tenth International Conference on Chemical Vapor Deposition, I am pleasure to ask your participation in the Conference as an invited speaker, to present your work on the modeling of LPCVD reactors. Such modeling has not only improved the fundamental understanding of the technology in a scientific sense, but has led to very signification increases in the efficiency of processing of silicon wafers for use in microelectronic applications." In 1990 J.-T. Wang and J.-O. Carlsson proposed a thermodynamic coupling model for the activated low-pressure CVD diamond growth. In 2002, J.-T. Wang further created a new field of nonequilibrium nondissipative thermodynamics, which provided a theoretical base for the calculation of nonequilibrium phase diagrams, and demonstrated experimentally nonspontaneous reaction(s) taking place with energy provided from other simultaneous spontaneous reaction(s) through thermodynamic coupling. The transformation from graphite to diamond under low pressure is a typical nonspontaneous reaction, the realization of which is driven by the association of atomic hydrogen. There are really a lot of nonspontaneous reactions in the living bodies. These nonspontaneous reactions in living bodies may be discussed by comparing with the activated low-pressure CVD diamond growth (by providing external energy). Due to these researches, modern thermodynamics together with the whole discipline of thermodynamics has drastically been developed.

7.2. TECHNICAL FUNDAMENTALS OF CVD

CVD is a kind of technology for deposition of solid materials from their gas or vapor precursor(s) through chemical reaction(s). Therefore, the CVD technology in inorganic synthesis and materials preparations has the following characters.

- Step Coverage and Shape Preservation. Usually, the depositing reactions take place on the gas–solid surface of solid substrates, so thin films are covered on substrates with nice step coverage and shape preserving characters. Therefore, this is very nice for the application of CVD to coating tools and to the fabrications of IC and other semiconductors. Now it is very important in filling up deep trenches (more than several hundred nanometers deep and about 65 nm wide) in the fabrications of IC. This is also the reason for CVDs being used more generally than PVD in the IC industry. We can also know that the CVD technology is a very fine process in inorganic synthesis and materials preparations.
- Purification of Materials. CVD technology can also be used for purification of some solid materials. For instance, silicon can be changed into silane (silicon tetrahydride, SiH_4), and then after pyrolysis of silane in CVD process, bulk or ingot ultrapurified polycrystalline semiconductor silicon (or simply called poly-Si) can be obtained. For purification of silicon, silane trichloride ($SiHCl_3$) can also be reduced by hydrogen in CVD process, and bulk or ingot ultrapurified semiconductor silicon can be obtained similarly. The bulk or ingot poly-Si can be fused for growing single-crystalline silicon ingot for fabrication of integrated circuits.
- Free-Standing Films. Free-standing films can also be grown on some substrates, which are easy to be separated from the grown films. For instance, free-standing diamond films, deposited using DC Arc Plasma Jet CVD method onto graphite substrates with titanium interlayers, have been reached. The Ti interlayers were deposited by arc ion plating equipment. The titanium carbide (TiC) was detected on both sides of the interlayers, which play an important role with respect to reasonable adhesion with film and diamond nucleation. The semi-translucent or even the optical grade free-standing diamond films are obtained. The shape of the free-standing diamond films only depends on the shape of the substrate.
- Crystals or Powders Formation. In CVD technology crystals or powders can also be produced. For instance, cinnabar crystals or mercury sulfide powders can be produced by CVD processes. During the CVD process, the deposition reaction can take place inside the gas phase but not on the surface of substrate. The synthetic inorganic material can be formed as very fine powders, or even as nanometer-sized particles, called nanometer-sized ultrafine powders and nanowires. This is a new technology with nice prospects. Nanowires and nano-powders are usually of some new characters of excellence. For instance, nano-sized silicon dioxide called silica hydrated white carbon black may be used as a replacement for carbon black in rubber with better quality, nanometer titanium dioxide is a kind of photocatalysis, carbon nanotubes (CNTs) are of ultra-strength character, and so on.

There are some requirements of CVD technology to precursors, products, and reaction types. For instance,

- The raw materials should be gases, or liquids and solids, which are easy to be vaporized.
- The corresponding reactions should produce solid deposits and keep other byproducts in the gas phase.
- The reactions should be controlled.

CVD reactions can usually be divided into the following types [4].

7.2.1. Simple Pyrolysis Reactions

Usually, hydrides of low period elements in groups of IVB, IIIB, and IIB are gaseous compounds, such as CH_4, SiH_4, GeH_4, B_2H_6, PH_3, AsH_3, and so on. They are easy to decompose and release the corresponding elements. All of them are very nice to be used as precursors in CVD technology. Among them CH_4 and SiH_4 can be decomposed to directly form solid films. GeH_4 can be mixed with SiH_4, after pyrolysis Si–Ge alloy film can be obtained. Some typical reactions are:

$$CH_4 \xrightarrow{600-1200^\circ C} C + 2\,H_2$$

$$SiH_4 \xrightarrow{600-800^\circ C} Si + 2\,H_2$$

$$0.95\,SiH_4 + 0.05\,GeH_4 \xrightarrow{550-800^\circ C} Ge_{0.05}Si_{0.95}\ \text{(Si-Ge alloy)} + 2\,H_2$$

Some organic alkoxide compounds of elements are unstable under high temperature. After pyrolysis the corresponding oxide of the element will be produced, such as:

$$2Al(OC_3H_7)_3 \xrightarrow{\sim 420^\circ C} Al_2O_3 + 6C_3H_6 + 3H_2O$$

$$Si(OC_2H_5)_4 \xrightarrow{750-800^\circ C} SiO_2 + 4C_2H_4 + 2H_2O$$

The instabilities of hydrides or organic alkyl compounds can be used to produce other solid deposits through gas phase reactions with other gas compounds, such as:

$$Ga(CH_3)_3 + AsH_3 \xrightarrow{630-675^\circ C} GaAS + 3CH_4$$

$$Cd(CH_3)_2 + H_2S \xrightarrow{475^\circ C} CdS + 2CH_4$$

Besides, some carbonyl compounds of metals are gases and easy to be vaporized. After pyrolysis, the metal film is formed with release of CO, so these carbonyl compounds are suitable to be used in CVD technology. For instance,

$$Ni(CO)_4 \xrightarrow{140-240^\circ C} Ni + 4CO$$

$$Pt(CO)_2Cl_2 \xrightarrow{600^\circ C} Pt + 2CO + Cl_2$$

Note that metal compounds are usually inorganic salts and are very difficult to be changed into gases or vapors and to be used in CVD technology. However, their organic alkyl compounds are of much higher volatility. So these organic alkyl metal compounds are used as precursors in CVD technology for deposition of these metal films. Therefore, such special CVD processes are usually called metal-organic chemical vapor deposition (or MOCVD). Sometimes, other organic alkoxides, such as tri-isopropoxy-aluminum [$Al(OC_3H_7)_3$] and some metal ethyl acetoacetate coordination compounds (or β-diketone coordination compounds), do not include C−M bond (carbon−metal bond), so these compounds are not really metal organic compounds, but metal-containing compounds or metal coordination compounds. These compounds are usually volatile, often used in CVD technology, and are included in the MOCVD technology.

7.2.2. Reduction−oxidation Depositions

Hydrides or organic alkyl compounds of some elements are usually gases or liquids and solids with very high volatility. So they are suitable for use in CVD technology as precursors. When they are injected into the reactors together with oxygen, then the corresponding oxide films are deposited. For instance,

$$SiH_4 + 2O_2 \xrightarrow{325-475^\circ C} SiO_2 + 2H_2O$$

$$2SiH_4 + 2B_2H_6 + 10O_2 \xrightarrow{300-500^\circ C} 2B_2O_3 \cdot SiO_2 + 10H_2O$$

$$Al_2(CH_3)_6 + 12O_2 \xrightarrow{450^\circ C} Al_2O_3 + 9H_2O + 6CO_2$$

Halogen is of one negative valence. Many halides are often gases or volatile matters. They are also suitable for use as precursors in CVD technology. The corresponding element films are formed through the reduction of hydrogen. For instance,

$$WF_6 + 3H_2 \xrightarrow{\sim 300^\circ C} W + 6HF$$

$$SiCl_4 + 2H_2 \xrightarrow{1150-1200^\circ C} Si + 4HCl$$

Hydrogen reduction of trichlorosilane is the main process for the industrial production of semiconductor ultrapure silicon with the purity of >99.9999999%, or called 9 N (Nine) purity.

$$SiHCl_3 + H_2 \xrightarrow{1100-1150^\circ C} Si + 3HCl$$

7.2.3. Deposition Through Synthetic Reactions

In most of the CVD technology, the inorganic film or special form depositions take place through synthetic reactions from two or more precursors. For instance,

$$3SiH_4 + 4NH_3 \xrightarrow{750^\circ C} Si_3N_4 + 12H_2$$

$$3SiCl_4 + 4NH_3 \xrightarrow{850-900^\circ C} Si_3N_4 + 12HCl$$

$$2TiCl_4 + N_2 + 4H_2 \xrightarrow{1200-1250^\circ C} 2TiN + 8HCl$$

7.2.4. Deposition Through Chemical Mass Transportation

Some materials are very easy to decompose at higher temperature, after transportation to colder places the film or crystal deposition together with the synthetic reaction takes place, such as the above mentioned CVD HgS. The detailed reaction equation can be written as:

$$2HgS(s) \underset{T_1}{\overset{T_2}{\rightleftharpoons}} 2Hg(g) + S_2(g)$$

Sometimes the original materials are not easy to decompose, so some kinds of other materials (called transport reagent or "transporter") should be added to promote the gaseous intermediate formation. For instance,

$$2ZnS(s) + 2I_2(g) \underset{T_1}{\overset{T_2}{\rightleftharpoons}} 2ZnI_2(g) + S_2(g)$$

Usually, in these chemical mass transportations $T_2 > T_1$, that is, the formation of gaseous compounds is at the higher temperature T_2, and the reverse deposition is at the lower temperature T_1. However, sometimes the situations are different, and the reverse deposition is at higher temperature. For instance, inside the tube of iodine tungsten lamp (or bromine tungsten lamp) the chemical mass transport is from the lower temperature to the higher temperature places. For approaching the sunlight the working temperature of tungsten wires should be increased. After the increase of the tungsten wire working temperature (2800−3000°C) the volatility of tungsten increases very much. In this way the volatilized tungsten is condensed on the lower temperature (~1400°C) inside wall of the quartz tube, the quartz tube is darkened and the lifetimes of the tungsten wire and the lamp are shortened. If a little of iodine (or bromine) is sealed inside the lamp quartz tube, when the lamp is working, the gaseous iodine (or bromine) will react with the deposited tungsten film on the inside wall of the quartz tube to form tungsten tetraiodide (or tetrabromide). Tungsten tetraiodide (or tetrabromide) at that time will be a gas and will be transported inside the tube. The gaseous tungsten tetraiodide (or tetrabromide) will be decomposed and deposited on the thinner white heated tungsten wire, make the thinner part of the tungsten wire recover into its original size, and iodine (or bromine) is released into the gas phase for the continuous cyclical process. This is a very clever way to greatly increase the tungsten wire temperature and the lifetime of halogen tungsten lamps, by utilization of the chemical mass transportation CVD process.

$$W(s) + 3I_{2\,(g)} \underset{\sim 3000°C}{\overset{1400°C}{\rightleftharpoons}} WI_{6\,(g)}$$

7.2.5. Plasma-enhanced Chemical Vapor Deposition (PECVD or PCVD)

Under the low vacuum conditions, DC, AC, radio frequency (RF), microwave (MW), or electron cyclotron resonance (ECR) can be used for generation of the glow discharge to produce plasma inside the deposition reactor. Due to collisions between positive ions, electrons, and neutral reaction molecules the deposition temperature may be greatly decreased. For instance, the reaction between silane and ammonium takes place usually at about 850°C for the deposition of silicon nitride, but under plasma-enhanced reaction conditions, the formation of silicon nitride will be only at about 350°C. In this way, the CVD technology can be extensively applied to the final passivation process of IC chips. Under the high temperature as 800°C, IC chips will be destroyed. In PECVD process at 350°C, the deposition of silicon nitride films will not destroy the IC chips, but will improve the stability of devices by the passivative layers. These low-temperature deposited films are usually hydrogen-containing films, and their formulas are somewhat alterable, so they are usually written as SiO_x (or SiO_xH_y). Some examples of PECVD reactions are:

$$SiH_4 + xN_2O \xrightarrow{\sim 350°C} SiO_x(\text{or } SiO_xH_y) + \ldots$$

$$SiH_4 + xNH_3 \xrightarrow{\sim 350°C} SiN_x(\text{or } SiN_xH_y) + \ldots$$

$$SiH_4 \xrightarrow{\sim 350°C} \alpha - Si(H) + 2H_2$$

The last reaction equation of pyrolysis of silane may be used for manufacture of amorphous solar cells.

7.2.6. Atomic Layer Chemical Vapor Deposition (ALCVD or ALD)

This technology had been found many years ago, but it has been developed for a practical important technology only after a long time. Only one atomic layer is deposited in each deposition step. The ALD process of aluminum oxide (Al_2O_3) films may be taken as an example. Tri-methyl-aluminum (TMA) is taken as a precursor. TMA reacts with hydroxyl groups on the surface of substrate to form one atomic layer of aluminum, and at the same time methane is released until the original surface hydroxyl groups are exhausted. Only one atomic layer of aluminum is deposited on the substrate. The aluminum layer is covered by methyl groups, so it is impossible to deposit more atomic layers of aluminum. This is the self-limited character of ALD. In the next step water is the precursor for the oxygen atom deposition, the remainder methyl groups on the aluminum atoms are hydrolyzed by water, and released as methane molecules at the same time. After such an alternate deposition period of aluminum atoms and oxygen atoms, the substrate surface is recovered by hydroxyl groups, and is ready for the next period of atomic later deposition. Each period of deposition takes about a few seconds, and after several periods of deposition, or longer, the required film thickness can be reached. The controlling precision in ALD technology is much better than that in ordinary CVD technology, So ALD is especially suitable for filling deep trenches with excellent coverage. Therefore, ALD is an important technology for IC fabrication at the present time. ALD technology can be used for deposition of insulated films, semiconductor films, metallic films, and so on.

7.2.7. Enhanced Deposition by Other Energy Resources

With the development of new technology, laser enhanced chemical vapor deposition has also been used, such as:

$$W(CO)_6 \xrightarrow{\text{laser beam}} W + 6CO$$

Usually the reaction takes place at about 300°C on the surface of substrate without laser beam. If the laser beam is parallel to the surface of substrate, and the distance between the laser beam and the substrate surface is about 1 mm, then the bright tungsten film will be deposited on the substrate surface at the room temperature.

Other energy resources, such as flame combustion or hot filament, can also be used for the enhancement of deposition reactions. The introduction of the external energy may also be used for changing the quality or crystal structure of the deposit. For instance, methane or other hydrocarbons can be pyrolyzed at high temperature for production of carbon black. Carbon black consists of amorphous carbon and tiny graphite particles.

$$CH_4 \xrightarrow{800-1000°C} C\ (\text{carbon black}) + 2H_2$$

The pyrolysis of 1% methane in hydrogen at high temperature and low pressure will produce graphite and amorphous carbon as well. However, if the hot filament or plasma is introduced for the formation of atomic hydrogen from molecular hydrogen, then diamond but not graphite will be deposited under the pressure of about 0.1 MPa or lower.

$$CH_4 \xrightarrow[\quad]{\text{hot-filament or plasma}\ 800-1000°C} C\ (\text{diamond}) + 2H_2$$

Under such conditions graphite can be etched away together with the deposition of diamond. This realizes the transformation from graphite to diamond under low pressure, which has been regarded as thermodynamically impossible before.

$$C\ (\text{graphite}) + H_2 \xrightarrow{\text{plasma}\ 800-1000°C} CH_4C_2H_2 + K$$

$$+ \xrightarrow{\text{plasma}\ 800-1000°C} C\ (\text{diamond}) + 2H_2$$

7.3. EQUIPMENT OF CHEMICAL VAPOR DEPOSITION

CVD equipment usually consists of: (i) gas controlling part, (ii) deposition reactor, (iii) deposition temperature controller, (iv) pressure controlling part, and (v) vacuum pump. For plasma-enhanced or other energy resource-enhanced CVD the energy power controlling part is also needed. The structure and working principle of the CVD reaction chamber is the most important part, which is usually designed specially for different kinds of processes. Usually CVD equipment can be divided into the following types [4,5].

7.3.1. Equipment for Manufacture of Ultrapure Semiconductor Poly-Si

Figure 7.1 shows a reaction camber for the poly-Si deposition under normal pressure. In the camber there is heating bridge, consisting of three silicon rods and connected with electric power to keep it at about 1150°C. At the center of the bottom the gas mixture of trichlorosilane and hydrogen comes into the camber through a gas inlet tube. Ultrapure silicon deposits on the heating rods continuously and the rods become bigger and bigger. Finally, ultrapure poly-Si can be used for manufacture of semiconductor single-crystalline silicon.

$$SiHCl_3 + H_2 \xrightarrow{1100-1150°C} Si + 3HCl$$

7.3.2. Equipment for Silicon Epitaxy and Poly-Si Film Deposition

Figure 7.2 shows equipment for silicon epitaxy or poly-Si film deposition under atmospheric pressure. Figure 7.2(a) shows the simplest horizontal reactor and Fig. 7.2(b) shows a vertical reactor. Due to the ultrapure requirement the camber is usually made of ultrapure quartz and the susceptor is made of ultrapure graphite, which is easy to be heated by RF or infra-red (IR). These reactors are mainly for silicon epitaxy by hydrogen reduction of $SiCl_4$ to grow a few micrometers of silicon epitaxial layers on silicon wafers. Epitaxy is the method of depositing a monocrystalline film on a monocrystalline substrate with the same frame of crystal. Homoepitaxy is a kind of epitaxy performed with only one material. In homoepitaxy, a crystalline film is grown on a substrate of the same material. Heteroepitaxy is a kind of epitaxy performed with materials that are different from each other. For instance, silicon epitaxial layers may grow on sapphire (Al_2O_3) or spinel ($MgAl_2O_4$). That is heteroepitaxy, in which a crystalline film grows on a crystalline substrate of a different material. They are often used in semiconductor industry. The equipment, as shown in Fig. 7.2, may also be used for epitaxial growth of GaAs, GaPAs, Ge–Si alloy, SiC, and so on. They can also be used for deposition of silicon oxide, silicon nitride, poly-Si, or metal films. From Fig. 7.2(a) to (b), the throughput in each run changes from 3–4 wafers/run to 6–18 wafers/runs, or even more.

7.3.3. Equipment of Hot Wall LPCVD

Figure 7.3 shows equipment of hot wall low-pressure chemical vapor deposition (LPCVD), which has been regarded as an important breakthrough of processes in the IC industry at the end of 1970s [6]. Referring to the ordinary CVD processes at atmosphere pressure, as shown in Fig. 7.2, the excellences of LPCVD are: (1) the throughput in each run has been increased from several or ten wafers per run to 100–200 wafers/run, (2) the thickness within-wafer uniformity has been

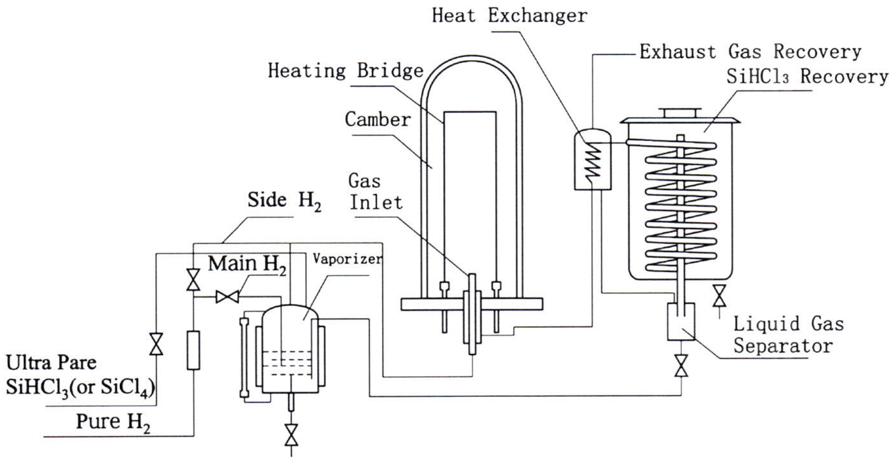

FIGURE. 7.1 Industrial equipment for manufacture of semiconductor ultrapure silicon (poly-Si) by hydrogen reduction of trichlorosilane.

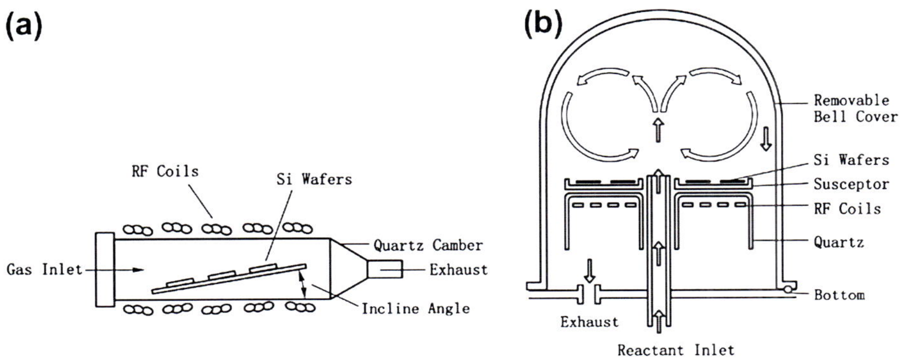

FIGURE. 7.2 Equipment for epitaxial growth or for film deposition.

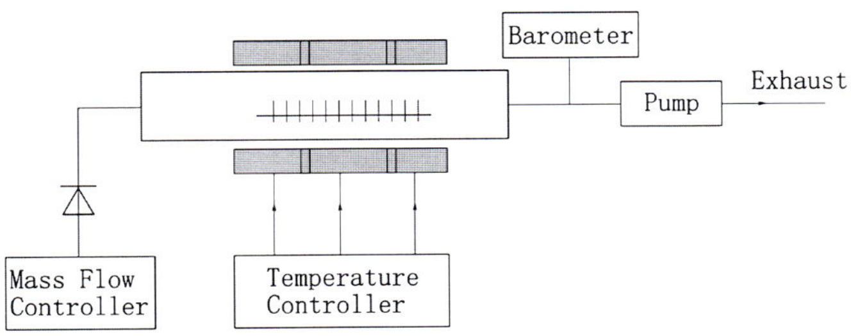

FIGURE. 7.3 Equipment of hot wall LPCVD.

improved from ±(10–20%) to ±(1–3%), and (3) the production cost has been decreased about one-tenth. LPCVD process in IC industry has widely been used even up to now. With the wafer diameter increases (from 3–4 inches in 1970s to 12 inches now), the furnace has been rotated from the horizontal (as shown in Fig. 7.3) into vertical structures, but the principles are still the same. The key factor in this process is how to keep the thickness uniformity of deposited films on wafers at different places (at the different wafer positions inside the furnace as shown in Fig. 7.3). A simulation model for the film thickness distribution in the LPCVD process was proposed by Ji-Tao Wang in 1980 [5,7–12]. Theoretical calculated results agree with experimental data. The model can be used as a theoretical guide for operation and automatic control. The details of this model will be discussed later in this chapter.

7.3.4. Equipment of Plasma-enhanced CVD (PECVD)

The CVD deposition temperature may be decreased down to several hundred degrees even to the room temperature through the plasma enhancement. Figure 7.4 shows two kinds of PECVD equipment. Figure 7.4(a) shows the simplest PECVD equipment in laboratory with plasma generated by inductance coupling. Figure 7.4 (b) shows a kind of PECVD equipment with two parallel planar electrodes for plasma generation. Substrates are loaded on the lower planar electrode with temperature controller. Pressure keeps usually at about 133 Pa. Plasma is generated by capacitance coupling discharge. Due to collisions between activated particles (electrons, neutral reactant molecules, species, and ions) in the gas phase, the substrate keeps usually at about 350°C, and deposited SiO_x or SiN_x films are used as the final passivation layers of integrated circuits to increase their reliability.

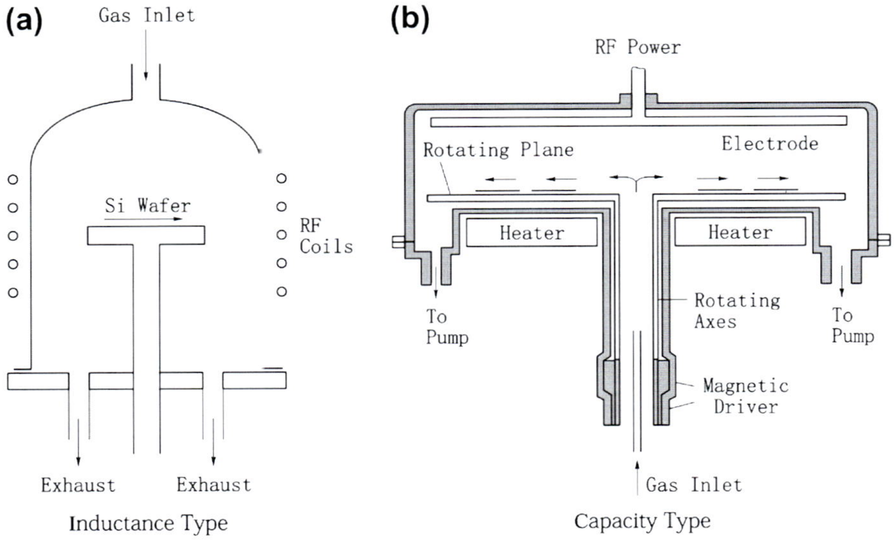

FIGURE. 7.4 Equipment of plasma-enhanced CVD (PECVD).

7.3.5. Equipment of Atomic Layer Deposition (ALD)

Figure 7.5 shows the principle and equipment of atomic layer deposition (ALD). Figure 7.5(a) shows steps in a period for ALD of Al_2O_3. (1) Inject the water vapor carried by nitrogen to cover the surface of substrate. (2) Purge the byproduct of methane with nitrogen and keep a monolayer of hydroxyl groups left on the substrate. (3) Inject tri-methyl-aluminum (TMA) carried by nitrogen onto the substrate covered by monolayer of hydroxyl groups. (4) Purge the byproduct of methane with nitrogen and keep a monolayer of aluminum with methyl groups left on the substrate. In this way ALD process is going on cycle by cycle to reach the Al_2O_3 thickness as required. The technology of ALD fulfills the accurate requirement in the development of IC industry. Figure 7.5(b) shows a kind of ALD equipment.

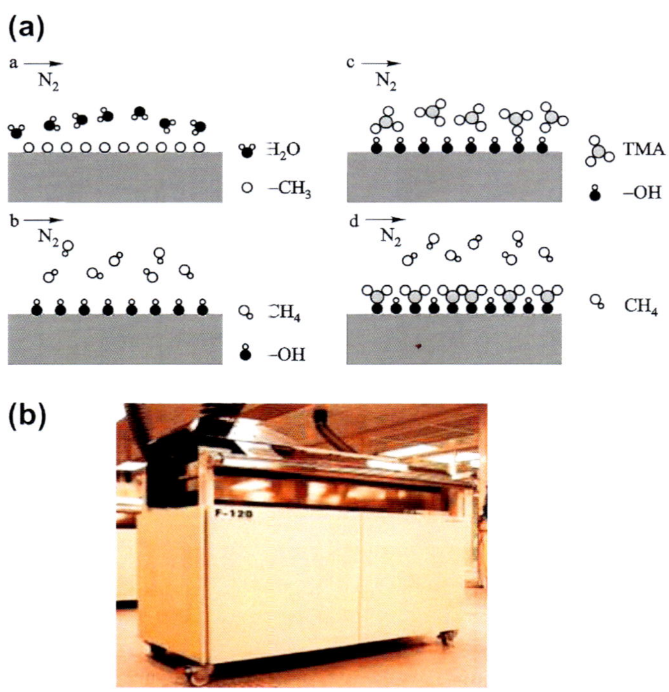

FIGURE. 7.5 (a) Principle and (b) equipment of atomic layer deposition (ALD).

7.3.6. Equipment of Tracklayer APCVD

For the industrial fabrications, equipment of tracklayer atmosphere pressure chemical vapor deposition (APCVD) was designed to continuously produce phosphosilicate glass films ($SiO_2 \cdot xP_2O_5$) from silane (SiH_4), phorsphine (PH_3) and oxygen at 400°C, as shown in Fig. 7.6. Equipment can also be used for low-temperature silicon oxide or other film deposition.

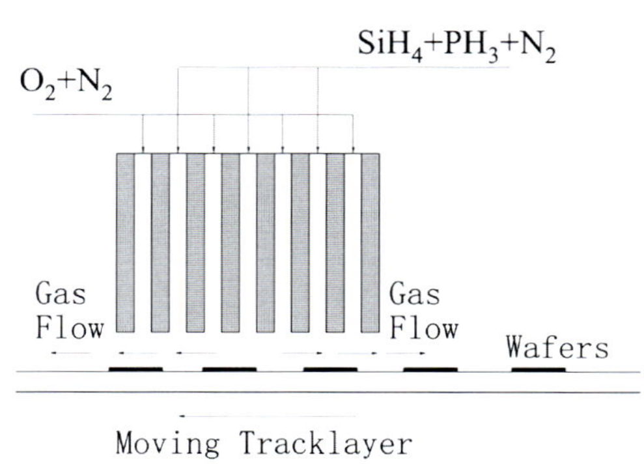

FIGURE. 7.6 Equipment of tracklayer APCVD.

7.3.7. Equipment of Barrel CVD Reactor

For the coatings on hard alloy cutting tools, equipment of barrel CVD reactor, as shown in Fig. 7.7, is often used in industry. The advantages are of very high throughput and independent of tool shapes. Several thousands of coated cutting tools can be produced in each run.

7.3.8. Equipment of Gallium Arsenide (GaAs) Epitaxial Growth

As mentioned above, structures of CVD reactors may be quite different, and should be based on the special requirement. For instance, equipment of gallium arsenide (GaAs) epitaxial growth should be designed with both the gas source inlet and the solid source and with special temperature distribution, as shown in Fig. 7.8. At the beginning, the reaction is mainly between arsenic trichloride ($AsCl_3$) and liquid gallium. After a solid film of GaAs is covered on the liquid gallium, the further reactions of GaAs epitaxial growth take place between $AsCl_3$, byproduct HCl, and other intermediates together with chemical transformations. Inside the reactor, reactions are rather complex. Some reactions close to the gallium source are as follows.

$$4AsCl_3 + 6H_2 \rightarrow As_4 + 12HCl$$

$$As_4 + 4Ga \rightarrow 4GaAs(s)$$

$$4GaAs(s) + 4HCl \rightarrow 4GaCl + 2H_2 + AS_4$$

$$AS_4 \rightarrow 2AS_2$$

$$GaCl + 2HCl \rightarrow CaCl_3 + H_2$$

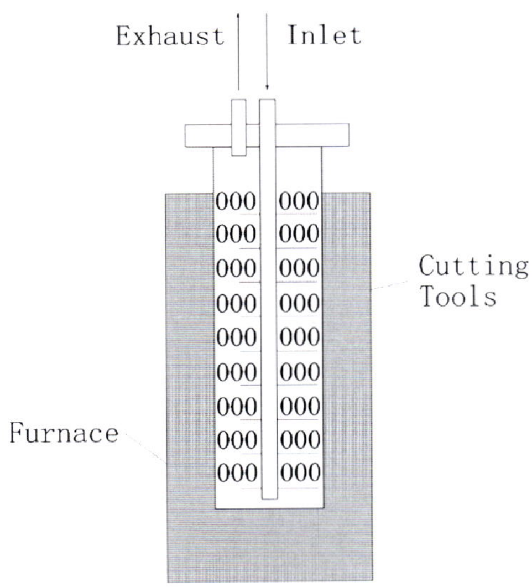

FIGURE. 7.7 Barrel CVD reactor.

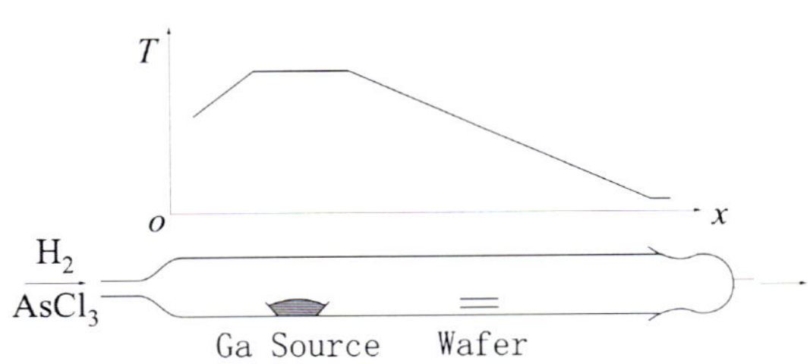

FIGURE. 7.8 Equipment of GaAs epitaxial growth.

7.4. SOME THEORETICAL MODELS OF CVD TECHNOLOGY

7.4.1. The Simple CVD Kinetic Model

There were some general kinetic descriptions for silicon epitaxial growth and film deposition, but the simple CVD kinetic model called the CVD growth rate model proposed by Grove [5] will first be discussed here. If the reaction related with mass transport is a first-order reaction, then this model provides a simple explanation for the difference between the mass transport-limited regime (or called diffusion-limited regime) and surface reaction-limited regime (or called kinetically limited regime). Theoretical predictions agree with experimental observations. Therefore, this model is often cited in textbooks.

Figure 7.9 shows schematically the model. Supposing that the inlet gas flow is perpendicular to the paper, C_G is the $SiCl_4$ concentration in the main flow, C_S is the $SiCl_4$ concentration on the growing surface, J_1 is the flux of $SiCl_4$ due to the mass transport from the main flow to the growing surface, and J_2 is the flux of $SiCl_4$ consumed by the chemical reaction for the epitaxial growth or for the film deposition, (atoms $cm^{-2} s^{-1}$). Therefore, J_1 can approximately be expressed by a linear equation.

$$J_1 = h_G(C_G - C_S) \tag{7.1}$$

Here, h_G is the mass transport coefficient of $SiCl_4$ in the gas phase ($cm\ s^{-1}$). We know that there is a nearly immobile layer on the solid substrate in the gas flow, which is

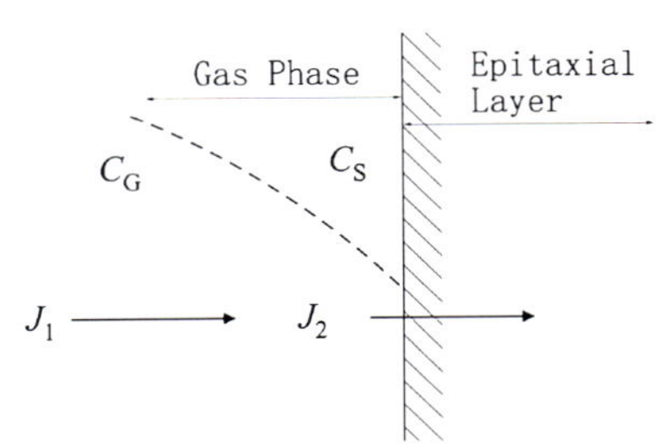

FIGURE. 7.9 The simple kinetic model of CVD.

called a stagnant layer. Supposing that the thickness of stagnant layer is δ and the concentration difference between two sides of the stagnant layer is $(C_G - C_S)$, then the concentration gradient in the stagnant layer is approximately $(C_G - C_S)/\delta$. Based on Fick's law, the flux of $SiCl_4$ molecules across the stagnant layer is:

$$J_1 = D_G \frac{(C_G - C_S)}{\delta} \quad (7.2)$$

Here, D_G is the $SiCl_4$ diffusivity or diffusion coefficient in H_2 ($cm^2 s^{-1}$). Comparing Eqn (7.1) with Eqn (7.2), the following equation is obtained.

$$h_G = \frac{D_G}{\delta} \quad (7.3)$$

On the other hand, the hydrogen reduction of $SiCl_4$ for the Si epitaxial growth is usually regarded as a first-order reaction, so

$$J_2 = k_r C_S \quad (7.4)$$

The proportional coefficient k_r is called chemical surface-reaction rate constant ($cm\ s^{-1}$). The relationship between chemical reaction rate and temperature T can usually be expressed by Arrhenius equation, i.e., reaction rate is proportional to $e^{-E/k_B T}$. Here, E is the activated energy of chemical reaction. k_B is the Boltzmann constant. Therefore, the value of k_r will decrease quickly with the decrease of temperature T.

Under stationary growth conditions, $J_1 = J_2 = J$ relation is established, so $h_G(C_G - C_S) = k_r C_S$. The following equation can be obtained:

$$C_S = \frac{h_G C_G}{k_r + h_G} \quad (7.5)$$

From Eqn (7.5), it is known that if $k_r >> h_G$, $C_S \approx 0$. It means that if the reaction rate on the growing surface is much higher than the mass transport rate of $SiCl_4$ in the gas phase, then the concentration of $SiCl_4$ approaches zero. The process belongs to the diffusion (or mass transport) limited regime. If $k_r << h_G$, $C_S \approx C_G$. It means that if the reaction rate on the growing surface is much slower than the mass transport rate of $SiCl_4$ in the gas phase, then the concentration of $SiCl_4$ approaches its concentration in the main flow. The process belongs to the surface reaction (or kinetically) limited regime. From the flux J under stationary growth conditions, it is very easy to get the epitaxial growth rate or the film deposition rate r. That is:

$$r = \frac{J}{N} = \frac{k_r C_S}{N} = \frac{k_r \frac{h_G C_G}{k_r + h_G}}{N} \quad (7.6)$$

After rearrangement,

$$r = \frac{k_r h_G}{k_r + h_G} \frac{C_S}{N} \quad (7.7)$$

Here N is the number of silicon atoms incorporated into a unit volume of the film (5.02×10^{22} atoms cm^{-3}). If the total number of molecules per cubic centimeter in the gas phase is C_T, and molar fraction of $SiCl_4$ is x, then $C_G = C_T x$. That is:

$$r = \frac{k_r h_G}{k_r + h_G} \frac{C_T}{N} x \quad (7.8)$$

From Eqn (7.8), the following conclusion can be obtained:

- The epitaxial growth rate or the film deposition rate is proportional to the molar fraction of $SiCl_4$ in the inlet gas. This agrees with experimental observations, if the concentration of $SiCl_4$ is sufficiently low.
- If $k_r << h_G$, then $r = k_r \frac{C_T}{N} x$. Based on Arrhenius relation between k_r and T, it is easy to conclude that the growth rate is proportional to the negative exponential of the reciprocal of temperature. That is, the growth rate increases exponentially with the temperature, which agrees with experimental data as shown in the low temperature region of Fig. 7.10.
- If $k_r >> h_G$ then $r = h_G \frac{C_T}{N} x$. Based on theoretical analysis of gas, the mass transport of gas molecules is not sensitively related to temperature. Under such conditions the epitaxial growth rate or the film deposition rate is also not sensitively related to temperature. This agrees with experimental data as shown in the high temperature region of Fig. 7.10.

Some factors such as the substrate etching of byproducts and the influences of gas flow on the growth rate are out of consideration in this simple model, so it is difficult to accurately describe the change of the growth rate in practical reaction systems. However, the general descriptions of some kinetic characters are nice for the epitaxial growth and the film deposition.

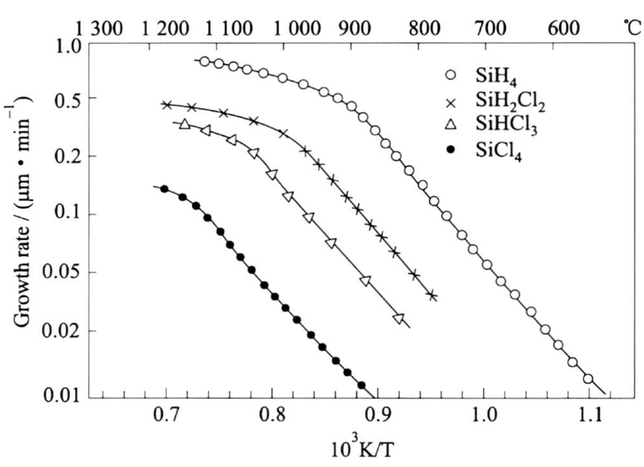

FIGURE. 7.10 Relationship between epitaxial growth rate or poly-Si film deposition rate and temperature.

7.4.2. LPCVD Simulation Model

Hot wall low-pressure chemical vapor deposition (LPCVD) technology has been extensively applied to manufacture integrated circuits. In each run of this process 100–200 wafers can be loaded, the throughput has been greatly increased, the production cost is lowered, and that is an important breakthrough in the IC fabrication technology [6]. In 1976 this technology was first utilized in production by Motorola and Fujitsu, and then it was quickly extended to other factories in the late 1970s. The within-wafer uniformity of each wafer is excellent. So the key effect is to ensure the wafer-to-wafer uniformity of this process during the large-scale fabrication. In 1980, a simulation model of the LPCVD process was proposed by Ji-Tao Wang [5,7–12] to mainly control such a wafer-to-wafer uniformity [13–17]. Based on the practical operation in LPCVD process, Figure 7.11 can be used as a physical model for the mathematic treatment. j is the wafer number or called the wafer position. The total number of wafers is L. $\eta(j)$ is the conversion efficiency of silane in the gas flow at the position j. p is the pressure at the gas inlet end. Δp is the pressure difference between the inlet end and the exhaust end. In this physical model an assumption has been included. That is, the radial concentration gradient of the reactant is zero in the reactor. Because the within-wafer uniformity of each wafer is excellent in LPCVD technology (thickness deviation is usually $<\pm 1\%$), the assumption is well accepted. Based on the assumption, the conversion efficiency η of silane in the reactor can be written as a one-dimensional function of the distance z from the inlet end of the deposit region. If j is used for indicating the position of the silicon wafer, then $\eta = \eta(j)$ can be obtained.

For simplicity, the poly-Si LPCVD process will first be discussed. Under such a low pressure, the factor of absorption can be negligible, and the pyrolysis of silane can approximately be regarded as a first-order reaction. That is, the deposition rate is proportional to the partial pressure of silane.

$$r = k_r p C \tag{7.9}$$

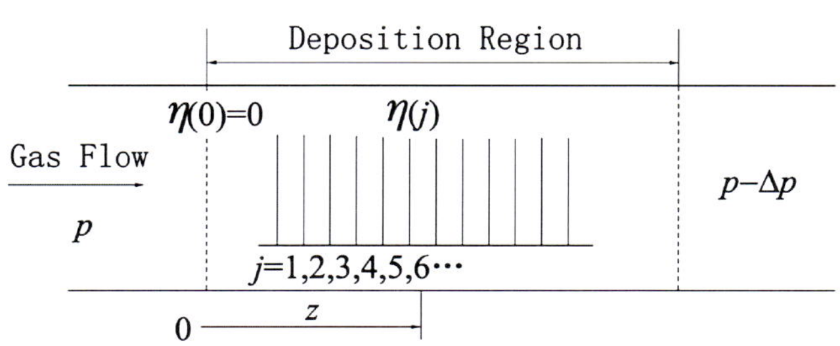

FIGURE. 7.11 Simulation model for poly-Si LPCVD process.

where k_r is the reaction rate constant and C is the concentration of silane.

During the pyrolysis of one silane molecule, two hydrogen molecules are produced.

$$SiH_4 \xrightarrow{600-650^\circ C} Si + 2H_2$$

Therefore, the relationship between C and η should not directly be written as $C_0(1-\eta)$. A correctional term of $1/(1+\eta C_0)$ for the total amount of gas should be introduced.

$$C = C_0(1-\eta)/(1+\eta C_0) \tag{7.10}$$

Put Eqn (7.10) into Eqn (7.9), and consider all of r, k_r, p, and η as a one-dimensional function of the axial position j, the following equation can be obtained.

$$r(j) = k_r p(j) C_0[1-\eta(j)]/[1+\eta(j)C_0] \tag{7.11}$$

Based on Arrhenius equation, it is known that:

$$k_r(j) = Ae^{-E/RT(j)} \tag{7.12}$$

where E is the activated energy, R is the gas constant, A is the factor before the exponent term, and $T(j)$ is the temperature at the wafer position j. From Fig. 7.11 the value of $p(j)$ is:

$$p(j) = p - j\Delta p/L \tag{7.13}$$

In the LPCVD system, usually $\Delta p << p$. For simplicity, p may be used instead of $p(j)$.

According to the gradient consumption of silane inside the deposition region, it is known that silane in the gas flow at the position $(j+1)$ has been consumed along its way from the position 0 to the position j. Therefore, $\eta(j+1)$, the conversion efficiency of silane at the position $(j+1)$ equals:

$$\eta(j+1) = \sum_{i=0}^{j}[S(i)r(i)]\omega/FC_0 \tag{7.14}$$

Here, $\omega = d_{Si}\tilde{V}/M$. $r(0)$ is the average deposition rate from the beginning of the deposition region to the position 1. $r(i)$ is the deposition rate at the position i. $S(0)$ is the deposition area before the position 1. $S(i)$ is the deposition area of wafer, tube wall, and boat at the position i. F is the total gas flow at the inlet end. ω is the ratio of the consumed gas volume of silane and the formed deposit volume of the solid thin films. d_{Si} is the density of silicon. $\tilde{V}$ is the volume of 1 mol gas under the standard conditions. M is the weight of 1 mol Si (g).

Combining the equation of $r(j)$ with the equation of $\eta(j+1)$ together with the initial condition (the conversion efficiency of silane at the inlet end equals zero), a set of cyclical step-up simulation formulas for LPCVD poly-Si process will be obtained as follows:

$$
\begin{aligned}
r(0) &= 0 \\
&\cdots\ \cdots \\
r(j) &= Ae^{-E/RT(j)}p(j)C_0[1-\eta(j)]/[1+\eta(j)C_0] \qquad (7.15)\\
\eta(j+1) &= \sum_{i=0}^{j}[S(i)r(i)]\omega/FC_0 \\
&\cdots\ \cdots
\end{aligned}
$$

This is a set of cyclical step-up algebraic equations, so it is very easy to write a software and to calculate by a computer. Figure 7.12(a) and (b) shows the comparison between theoretical curves and LPCVD poly-Si experimental data reported in the literature. The agreement between theory and experiments confirms that the theoretical simulation model is acceptable, and could be used as guide for the LPCVD process. During the theoretical derivation mentioned above, the kinetics of the silane pyrolysis has been regarded as a first-order reaction without the consideration of absorption effect. For the practical use, the absorption and some other effects should be under consideration, if necessary.

The theoretical simulation model is not only suitable for the LPCVD poly-Si process but also suitable for other LPCVD processes, such as for the LPCVD silicon nitride process [18]. In LPCVD poly-Si process there is only one reactant of silane, but in LPCVD silicon nitride process two reactants – dichlorosilane (DCS, SiH_2Cl_2) and ammonia (NH_3) – are needed. Therefore, the physical model of LPCVD silicon nitride for the mathematic treatment is shown in Fig. 7.13.

There are two main reactions in the LPCVD silicon nitride process:

$$SiH_2Cl_2(g) + \frac{4}{3}NH_3(g) = \frac{1}{3}Si_3N_4(s) + 2HCl(g) + 2H_2(g)$$

$$NH_3(g) = \frac{1}{2}N_2(g) + \frac{3}{2}H_2(g)$$

Their kinetic equations are as follows, respectively.

$$r_1 = K_s \cdot p_{SiH_2Cl_2} \cdot p_{NH_3}/[1 + K_1 \cdot p_{SiH_2Cl_2} + K_2 \cdot p_{NH_3}] \qquad (7.16)$$

$$r_2 = K_d \cdot p_{NH_3} \qquad (7.17)$$

where r_1 and r_2 are the deposition rate of silicon nitride and the pyrolysis rate of ammonium. K_s, K_1, K_2, and K_d are kinetic constants. p_i is the partial pressure of species i. If the molar ratio of precursors (NH_3/SiH_2Cl_2) is N, then initial concentrations of SiH_2Cl_2 and NH_3 are as follows, respectively.

$$X_1 = 1/(N+1) \qquad (7.18)$$

$$N/(N+1) \qquad (7.19)$$

The initial concentrations of NH_3 could be divided into two parts:

- Stoichiometric NH_3 part is equal to:

$$4/3(N+1) \qquad (7.20)$$

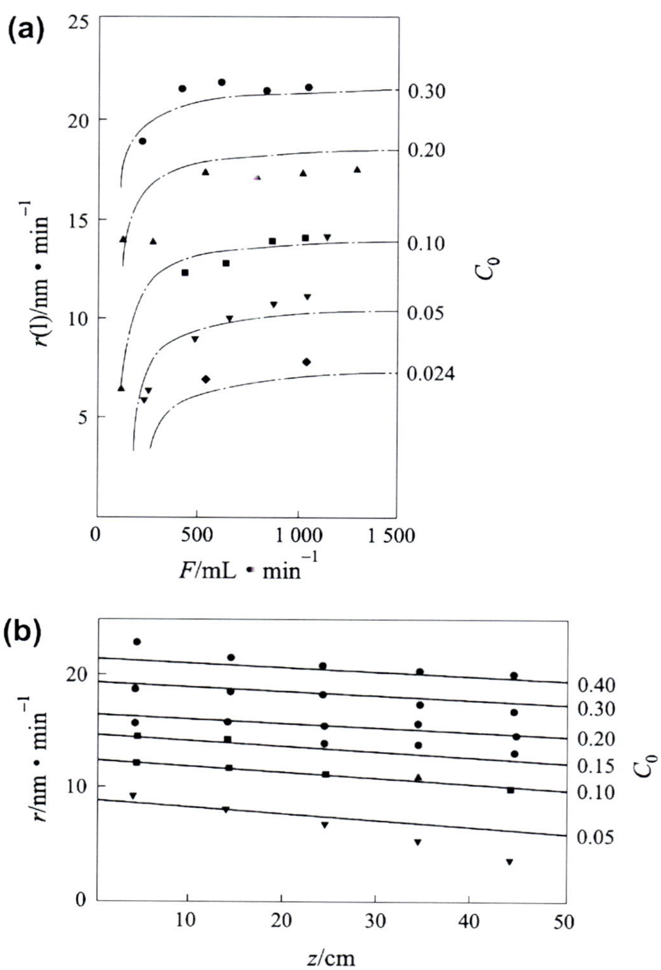

FIGURE. 7.12 Comparison between theoretical curves and LPCVD poly-Si experimental data reported in the literature [13].

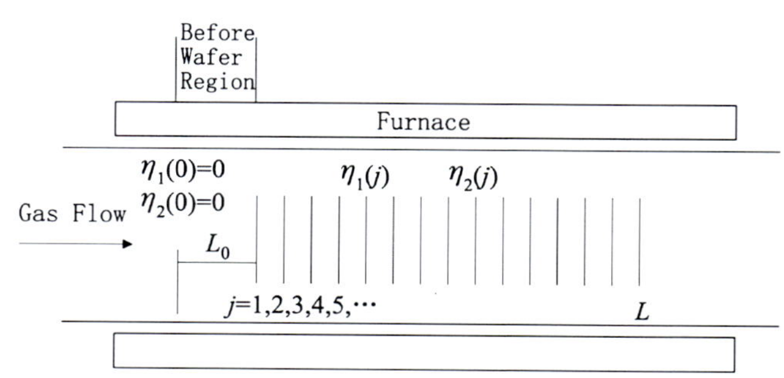

FIGURE. 7.13 Simulation model of LPCVD silicon nitride "Images reproduced with permission."

- Excess NH_3 part is equal to:

$$X_2 = (N - 4/3)/(N + 1) \quad (7.21)$$

When the conversion efficiencies of SiH_2Cl_2 and the conversion efficiencies of excess NH_3 are represented by η_1 and η_2, respectively, then after both the deposition of silicon nitride and the pyrolysis of ammonium the total gas volume will be a function of η_1 and η_2.

$$V = 1 + \frac{5}{3}X_1 \cdot \eta_1 + X_2 \cdot \eta_2 \quad (7.22)$$

Therefore, concentrations of SiH_2Cl_2 and NH_3 should be:

$$C_1 = X_1(1 - \eta_1)/V \quad (7.23)$$

$$C_2 = [4X_1(1 - \eta_1)/3 + X_2(1 - \eta_2)]/V \quad (7.24)$$

Their kinetic equations could be rewritten as:

$$r_1 = K_s \cdot C_1 \cdot C_2 \cdot p^2/[1 + K_1 \cdot C_1 \cdot p + K_2 \cdot C_2 \cdot p] \quad (7.25)$$

$$r_2 = K_d \cdot C_2 \cdot p \quad (7.26)$$

where p_i is the total pressure. A set of cyclical step-up simulation formulas for LPCVD silicon nitride process will similarly be obtained like that for LPCVD poly-Si process.

$$\eta_1(0) = 0$$
$$\eta_2(0) = 0$$
$$\cdots \cdots$$
$$\cdots \cdots$$
$$r_1(j) = K_s \cdot C_1(j) \cdot C_2(j) \cdot p^2/[1 + K_1 \cdot C_1(j) \cdot p + K_2 \cdot C_2(j) \cdot p]$$
$$r_2(j) = K_d \cdot C_2(j) \cdot p$$
$$\eta_1(j+1) = \sum_{i=-L_0+1}^{j} [S(i) r_1(i)]\omega'/FC_1^0$$
$$\eta_2(j+1) = \sum_{i=-L_0+1}^{j} [S(i) \cdot r_2(i) + \tfrac{4}{3}S(i) \cdot r_1(i) \cdot \omega']/FC_2^0$$
$$\cdots \cdots$$
$$\cdots \cdots$$

Here, L_0 is the length (the unit is a wafer number) of the deposition region before the wafer region, ω' is the volume ratio of the consumed SiH_2Cl_2 gas volume and the volume of the solid deposit. F is the total gas flow at the inlet end. The unit of gas flow is cubic centimeters of the gas volume per minute under the standard conditions (sccm).

Figure 7.14(a)–(e) shows a series of comparisons between the theoretical simulation calculations and experimental data reported in the literature [16]. The agreement between theory and experiments is excellent. Based on this simulation model and the corresponding software a series of virtual experiments can be done by computer. It will be very nice to master the rules of LPCVD process, and to use them for controlling the practical operation in fabrication. Table 7.1 shows a set of apparent kinetic parameters for LPCVD silicon nitride process. In fact, all of these kinetic data and kinetic equations (Eqns (7.16) and (7.17)) are obtained with the help of this theoretical simulation model.

7.4.3. Thermodynamic Coupling Model of Activated Low-pressure CVD Diamond Synthesis

Diamond is the hardest material in all known materials. Moreover, diamond has many other outstanding characteristics. The heat conductivity of diamond is better than that of silver and copper. Diamond is of a high refractive index and excellent transparency. Diamond is very valuable in industry and is also regarded as jewelry and symbol of treasure. In the nineteenth century and at the beginning of the twentieth century the diamond synthesis has been claimed many times, but all of those were not true. The phase diagram of carbon had been calculated based on the classical thermodynamics in about 1920s and 1930s. It was known that under low pressures the stable phase of carbon is graphite, and diamond is a metastable phase of carbon. Scientists even calculated that a very high pressure of about 15,000 times of atmosphere is needed to realize the transformation from graphite to diamond. Under such a theoretical guide, the high-pressure diamond synthesis had been reached in 1954. This is an important breakthrough in diamond synthesis, but it has also incorrectly been concluded that it is thermodynamically impossible to realize the diamond synthesis under low pressures. In about 1970, the former Soviet Union scientists Deryagin, Spitsyn, etc. realized the diamond synthesis under low pressures via the gas phase from methane or graphite with the activation of super-equilibrium concentration of atomic hydrogen (super-equilibrium atomic hydrogen, SAH). In 1976, they published the new progress together with beautiful photos of the diamond crystals grown on non-diamond substrates, as shown in Fig. 7.15 [19]. However, nearly no scientist in the West believed their achievement. Later, the new technology of the activated low-pressure diamond synthesis was confirmed by Japanese scientists Setaka and his colleagues in about 1980 [20], and by American scientist Roy in 1986 [21]. Since then the activated low-pressure diamond synthesis, or called the activated CVD diamond growth, has become a major focus of research throughout the world in 1990s.

The activated CVD diamond syntheses can be reached by different activation methods, such as hot filament process, plasma process, combustion process, and so on, as shown in Fig. 7.16. The low-pressure CVD diamond growth can also be realized from graphite via the vapor phase, as shown in Fig. 7.16(c). From these figures, it can be found that all of these processes are closely connected with the generation of super-equilibrium atomic

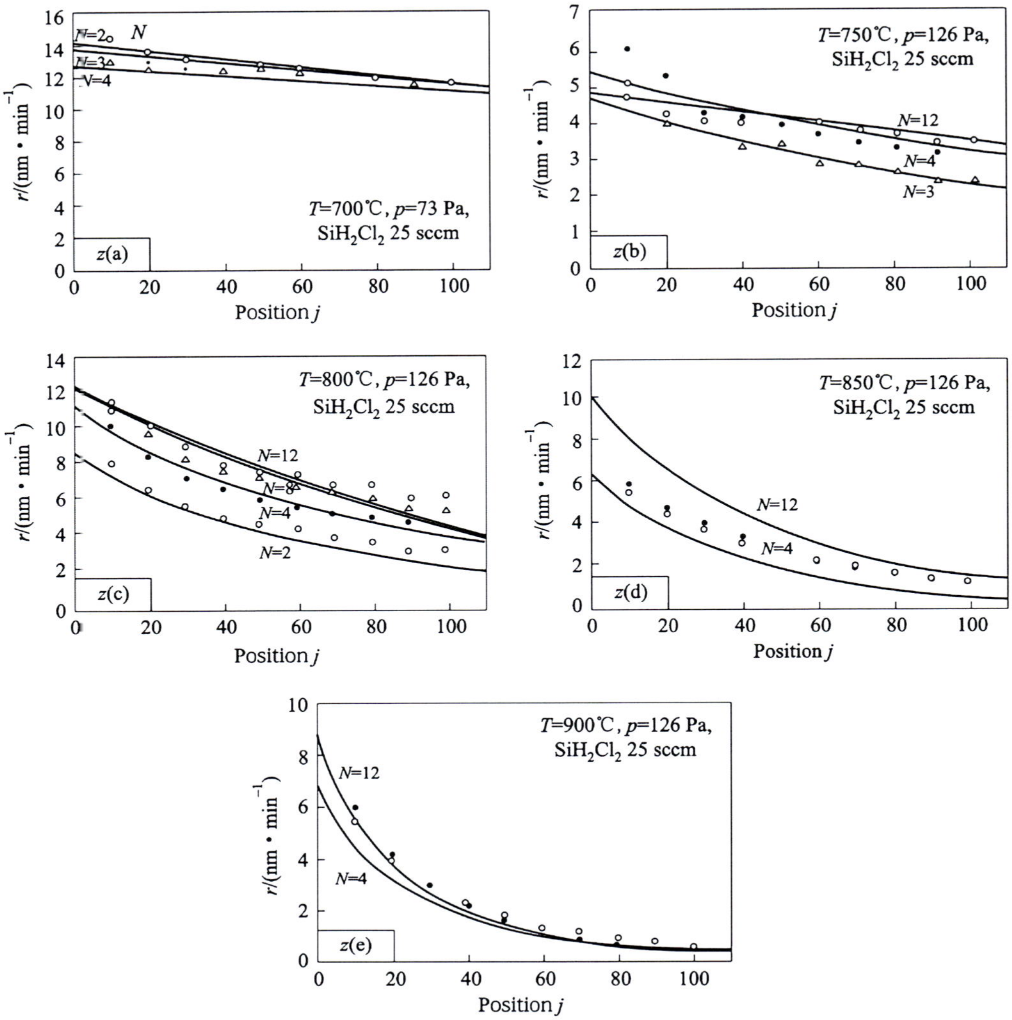

8-15

FIGURE. 7.14 Comparison between theoretical simulation curves (dashed lines) and experimental data of LPCVD silicon nitride "Images reproduced with permission."

TABLE 7.1 Apparent Kinetic Parameters for LPCVD Silicon Nitride Process and L_0 Values (For Simulation Calculations of Fig. 7.14)

T (°C)	K_S (nm min^{-1} Pa^{-1})	K_1 (Pa^{-1})	K_2 (Pa^{-1})	K_d (sccm Pa^{-1} cm^{-2})	L_0 (wafer numbers)
700	5.6×10^2	0.94	0.23	4.5×10^{-6}	5
750	6.8×10^2	0.75	0.026	1.5×10^{-5}	10
800	1.1×10^3	0.49	0.0045	3.8×10^{-5}	20
850	1.3×10^3	0.26	0.0030	1.5×10^{-3}	40
900	2.3×10^3	0.19	0.00025	2.9×10^{-3}	40

hydrogen (SAH). In recent years, about 10-carat-size single-crystal gem-quality diamonds have been obtained by Chih-shiue Yan, R.J. Hemley, and Ho-kwang Mao in Geophysical Laboratory, Carnegie Institution of Washington, as shown in Fig. 7.17 [22]. This is a more important breakthrough in low-pressure diamond syntheses and in diamond syntheses. Up to now, the carat-size single-crystal gem-quality diamonds are still very difficult to be obtained by high-pressure processes.

There were at least three theoretical models proposed in 1989 for explanation of the reason why diamond can

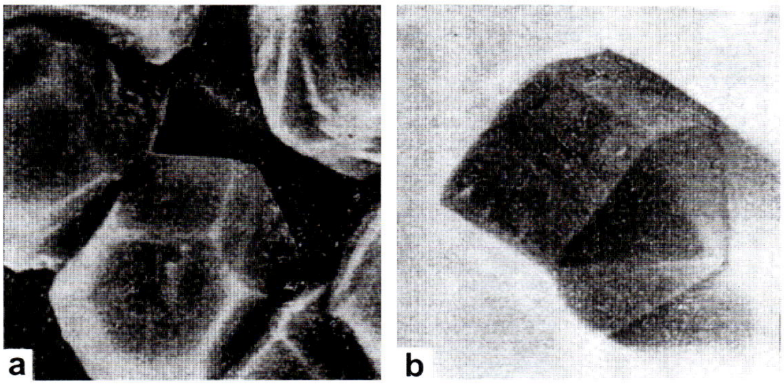

FIGURE. 7.15 Beautiful photos of the diamond crystals grown on non-diamond substrates were published by Deryagin, Spitsyn, etc. in 1976.

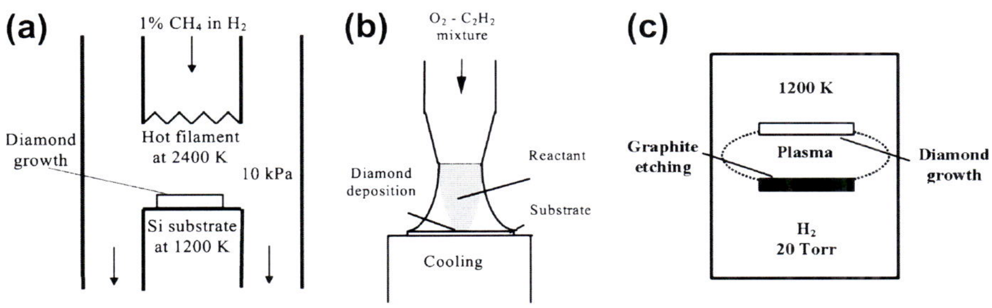

FIGURE. 7.16 Equipment for activated CVD diamond growth: (a) hot filament process, (b) combustion process, and (c) microwave plasma process.

grow from the vapor phase under low pressures. For instance, Sommer's quasi-equilibrium model [23], Yarbough's surface reaction model [24], and Bar-Yam's defects stabilization model [25] all of them were proposed in 1989. Among them, the defects stabilization model was published in the journal *Nature*. However, due to the influence of classical thermodynamics, no satisfying model has been found among them. In the February 1990 issue of the journal *Science*, a review paper "*Current Issue and Problems in the Chemical Vapor Deposition of Diamond*" was published. An objective comment on these models was that "All of these formulations suffer from being inconsistent with at least some experimental results." [26] Especially, the phenomenon of the diamond growth together with the simultaneous graphite etching takes place under low pressures, as show in Fig. 7.16(c), which puzzled a lot of scientists. In 1990s or even in 2005 it was said that the low-pressure CVD diamond growth was still regarded as a "thermodynamic paradox" perhaps "violating the second law of thermodynamics"; or explained by an incorrect so-called "nano-thermodynamics" [27–33].

FIGURE. 7.17 Synthetic diamond obtained in Carnegie Institution of Washington. (left): A brilliant cut single-crystal diamond produced from activated low-pressure chemical vapor deposition process (placed on mirror). (right): Five-carat single-crystal CVD diamond (without seed) 12 mm high and 6.7 mm in diameter cut from a 10-carat ($8 \times 8 \times 12\ mm^3$) block.

In April 1990, a "chemical pump" model was proposed by Ji-Tao Wang and Jan-Otto Carlsson at the Diamond Symposium of the Eighth International Conference on Thin Films of the Eighth International Conference on Thin Films at San Diego of United States [34]. It has been pointed out that the diamond stable growth under low pressures with the existence of SAH is a normal phenomenon. The transfer of carbon atoms from the stable graphite phase in the lower energy level to the metastable diamond phase is driven by the "chemical pump" due to the association of SAH. Such an idea includes the most fundamental conception of modern thermodynamics – thermodynamics coupling (or called reaction coupling in chemistry). That is, a non-spontaneous reaction may take place with compensation (or coupling) by another simultaneous spontaneous reaction. However, in quantitative calculations there were some obvious shortcomings at that time. In 1995, thermodynamic data of SAH and activated graphite was quantitatively derived. In this way the chemical pump model was improved into a thermodynamic coupling model for the activated low-pressure diamond growth [35–45]. In 2002, a new field of nondissipative thermodynamics, which is the theoretical base of calculations for nonequilibrium phase diagrams, was established [46,47]. In nonequilibrium phase diagrams of carbon, diamond may be more stable than graphite with the existence of SAH. Such a conclusion is directly against the traditional conclusion coming from classical thermodynamics.

Carnot theorem proposed by Carnot in 1824 told us that reversible heat engines are of the highest efficiency of energy conversion from heat to work. However, now it is known that Carnot theorem should be extended into the extended Carnot theorem. That is, nondissipation (not reversibility) is the necessary and sufficient condition for the highest efficiency of any energy conversion (including heat-work conversion by heat engines). For simplicity, the assumption of cyclical process, such as Carnot cycle, was introduced into classical thermodynamics as early as in 1865 by Clausius. Therefore, classical thermodynamics is only suitable for simple systems without thermodynamic coupling (or called uncoupling systems, i.e., only spontaneous processes under consideration) This is the reason for classical thermodynamics to be limited into equilibrium phase diagram calculations. In modern thermodynamics, complex systems with both nonspontaneous processes and spontaneous processes (or called coupling systems) are under consideration, so under nonequilibrium conditions nonequilibrium phase diagrams can also be calculated. In equilibrium conditions it is impossible to grow diamond from graphite under low pressures, but in nonequilibrium conditions it is possible to stably grow diamond from graphite under low pressures. Both of these conclusions agree with the second law of thermodynamics. A lot of experimental data for the activated low-pressure diamond growth reported in the literature agree excellently with theoretical calculations of nonequilibrium phase diagrams.

7.5. THERMODYNAMIC COUPLING DURING THE LOW-PRESSURE CVD DIAMOND GROWTH

7.5.1. Key Points of Thermodynamic Coupling Model in Modern Thermodynamics

We know that: for the low-pressure diamond growth from the vapor phase a large amount of superequilibrium atomic hydrogen (SAH) generated by hot filament, microwave, or RF-plasma is needed. So there are two main reactions inside the system.

Reaction 1:

$$C\ (\text{gra}) = C\ (\text{dia}),$$
$$\Delta G_1 > 0\ (T, p \leq 10^5\ \text{Pa})$$

Under isothermal isobaric conditions, the change of Gibbs free energy $\Delta G_1 > 0$ means that the transformation from graphite (gra) to diamond (dia) (i.e., Reaction 1, denoted by subscript 1) cannot take place spontaneously. T is the temperature of substrate. P is the pressure.

Reaction 2:

$$\text{H}* = 0.5\text{H}_2,$$
$$\Delta G_2 << 0\ (T_{\text{activation}} >> T,\ p \leq 10^5\ \text{Pa})$$

$\Delta G_2 << 0$ means that the association of SAH (H$*$) (i.e., Reaction 2, denoted by subscript 2) can take place spontaneously releasing a large amount of energy.

Whole System or Whole Reaction = (Reaction 1) + χ (Reaction 2):

$$C\ (\text{gra}) + \chi\text{H}* = 0.5\chi\text{H}_2 + C\ (\text{dia}),$$
$$\Delta G = \Delta G_1 + \chi\Delta G_2\ (T_{\text{activation}} >> T,\ p \leq 10^5\ \text{Pa})$$

If the reaction rate ratio $\chi = r_2/r_1$ is not too small ($\chi >> |\Delta G_1/\Delta G_2|$, i.e., with sufficient concentration of SAH), it must be $\Delta G < 0$. That is, the reaction between graphite and SAH to produce diamond under low pressures (i.e., the Gibbs free energy change of the whole system $\Delta G < 0$, without subscript here) is thermodynamically reasonable. From theoretical calculations it is known that $|\Delta G_1/\Delta G_2| \approx 0.05$, but it was known from hot filament process: $\chi = 0.28 >> 0.05$. Therefore, it has been demonstrated that the whole reaction toward the right side agrees with the second law of thermodynamics. In this way, the second law of thermodynamics for isothermal isobaric complex systems in modern thermodynamics should be written as ($\Delta G_1 > 0$, $\Delta G_2 < 0$, and $\Delta G \leq 0$). This model provides not only an explanation for the possibility to grow diamond under the activated low-pressure conditions, but also an explanation for the important role of SAH during the stable diamond growth from the vapor phase. This does not only agree with experiments, but also provides a theoretical basis for establishment of a new field, nondissipative thermodynamics, i.e., ($\Delta G_1 > 0$, $\Delta G_2 < 0$, and $\Delta G = 0$) under isothermal isobaric conditions. This is also the theoretical basis for calculations of nonequilibrium phase diagrams.

7.5.2. Mechanism of Thermodynamic Coupling

As mentioned in the last paragraph, the key for the low-pressure diamond growth is the introduction of SAH generated by plasma or hot filament. If atomic hydrogen in Reaction 2 is of the equilibrium

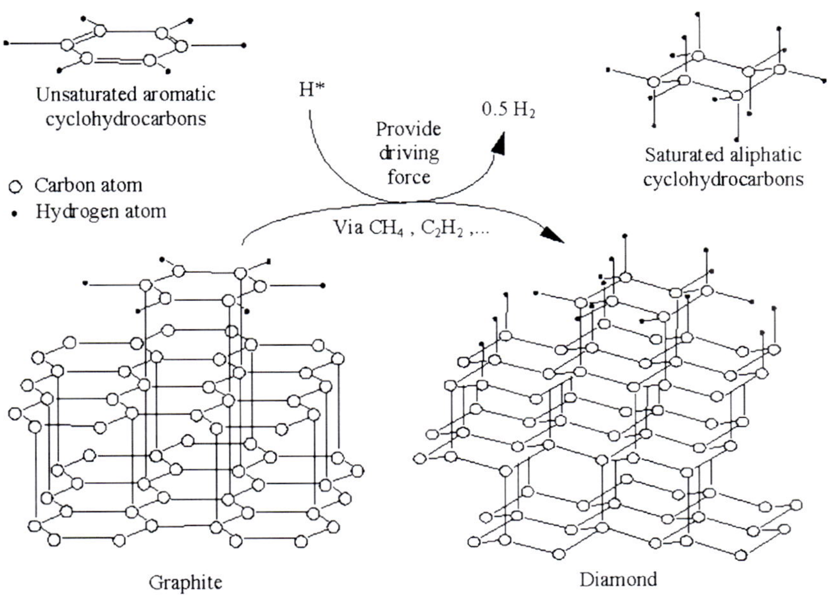

FIGURE. 7.18 Mechanism for the activated low-pressure CVD diamond growth.

concentration, then ΔG_2 equals zero. There is no driving force or thermodynamic coupling, so there is no possibility to realize the stable diamond growth under low pressures. Moreover, thermodynamics provides information for the possibility of the reaction, but for the practical realization a reasonable kinetic mechanism or path is still needed.

Under the activated low-pressure condition, the mechanism for the diamond (but not graphite) growth from a mixture of methane or hydrocarbons and hydrogen is very attractive, as shown in Fig. 7.18. This mechanism can also provide an explanation for the diamond growth from graphite via the gas phase. Figure 7.18 shows only one of the idealized possible mechanisms. For brevity, only the simplest six-member ring clusters on surfaces are discussed. The simplest six-member aromatic hydrocarbon is benzene, and the simplest six-member aliphatic hydrocarbon is cyclohexane. A single six-member cluster on a graphite surface in hydrogen can then be treated as a benzene molecule, which is physically adsorbed on the graphite surface. However, a single six-member cluster on a (111) diamond surface in hydrogen can be considered as a cyclohexane molecule, which has adsorbed chemically on the surface, reacted with the diamond hydrogenated surface and released three hydrogen molecules. The graphite clusters are unsaturated. On acceptance of SAH more saturated hydrocarbons (cyclohexane, cyclohexene, decahydronaphthalene, etc.) are formed. The atomic hydrogen provides sufficient energy for decomposition of the cluster into smaller fragments (CH_4, C_2H_2, C_2H_4, etc.). This results in etching of the graphite cluster as well as in etching of graphite itself. However, hydrogen shows only weak etching effect or no etching effect on diamond. SAH promotes the diamond growth with the simultaneous graphite etching.

The overall effect of SAH from an energy point of view is to increase the energy of graphite and to change the energy of diamond slightly. SAH plays the role of changing the relative chemical potential energy and the original stable graphite phase to a metastable phase.

Up to now, it is very clear that the activated low-pressure diamond growth is reasonable in both thermodynamics and kinetics. The shortcoming of kinetics is mainly to provide a qualitative explanation. For quantitative calculations for the influences of temperature, pressure, composition, and so on should be on the basis of thermodynamics, especially modern thermodynamics.

7.6. NONEQUILIBRIUM PHASE DIAGRAMS FOR THE LOW-PRESSURE CVD DIAMOND GROWTH

Phase diagram has its clear definition in thermodynamics. Phase diagrams for equilibrium systems are called equilibrium phase diagrams, while phase diagrams for nonequilibrium systems are called nonequilibrium phase diagrams. There is a special case in nonequilibrium systems, which is independent of time, called stationary nonequilibrium systems. The corresponding phase diagrams are called stationary nonequilibrium phase diagrams. Stationary nonequilibrium phase diagrams are independent of time, so they can be drawn on paper for discussion. Only stationary nonequilibrium phase diagrams are discussed here, but the word "stationary" is often omitted. The activated low-pressure diamond growth system is a typical example close to the idealized stationary nonequilibrium system. The diamond growth rate is very slow, and all other parameters nearly keep constant for several hours, several days, or even longer. Calculation of phase diagrams (CALPHAD) on the basis of nondissipative thermodynamics, i.e. ($\Delta G_1 > 0$, $\Delta G_2 < 0$, and $\Delta G = 0$) under isothermal isobaric conditions, is really a creative research work. According to the whole reaction equation mentioned above, diamond is more stable than graphite with the existence of SAH. Such a basic phenomenon should also be included in nonequilibrium phase diagrams. Experiments show that the graphite etching rate increases with the concentration of SAH, but the diamond etching rate is close to zero and nearly independent of the concentration of SAH. The etching rates are closely connected with their stability, so the whole reaction can be rewritten as follows:

$$C\ (\text{gra}) + \chi(H^* - 0.5\chi H_2) = C\ (\text{dia}),$$
$$or\ C\ (\text{gra}^*) = C\ (\text{dia}),$$
$$\Delta G' = \Delta G\ (T_{\text{activation}} >> T,\ p \le 10^5\ \text{Pa})$$

Here, $C(\text{gra}*) = [C(\text{gra}) + \chi(H* - 0.5H_2)]$, and is called activated graphite. Supposing that the concentration of SAH is in equilibrium concentration of atomic hydrogen at the activated temperature, then all thermodynamic data of SAH (such as entropy S, enthalpy H, and heat capacity at constant pressure C_p) can be quantitatively calculated on the basis of fundamental thermodynamic principles. Put these data of SAH together with thermodynamic data of graphite and molecular hydrogen into the expression of activated graphite, such as entropy of activated graphite $S_{\text{gra}*} = S_{\text{gra}} + \chi(S_{H*} - 0.5\ S_{H2})$, so all thermodynamic data of activated graphite can be obtained. Using these data of activated graphite instead of ordinary graphite data, and the ordinary CALPHAD software, which is based on the minimization of Gibbs free energy, a series of new nonequilibrium phase diagrams for the activated low-pressure diamond growth can be obtained.

Figures 7.19–23 are some typical nonequilibrium phase diagrams of C–H, C–O, and C–H–O systems for the activated low-pressure diamond growth. In all of them there is a diamond stable region for the diamond growth. Inside the diamond stable or growth region the diamond growth together with the graphite etching may take place simultaneously. That is, in ordinary equilibrium phase diagram diamond is a metastable phase and graphite is a stable phase under low pressures, but in nonequilibrium phase diagrams here diamond is a stable phase and graphite is a metastable phase during the existence of SAH even under low pressures. All of these nonequilibrium phase diagrams agree with the world-wide reported reliable experiments. Especially, in Fig. 7.23 theoretical calculated results

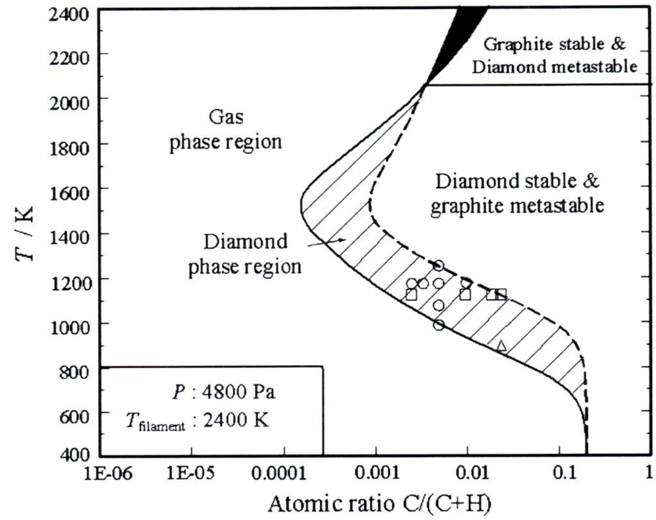

FIGURE. 7.19 $T-X$ nonequilibrium phase diagrams of C–H system. Diagonal region is the diamond growth region, and experimental data for the diamond growth reported in the literature (□, △ & ○) are just dropped in the diamond growth region. (The abscissa shows the composition of the system by atomic ratio, similar cases thereinafter will not be repeated.)

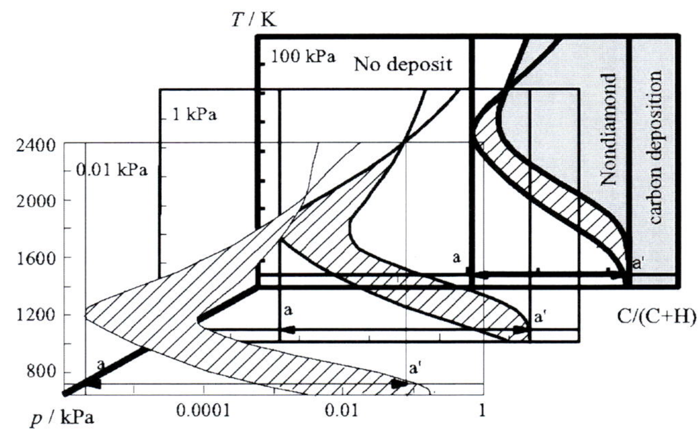

FIGURE. 7.20 $T-p-X$ nonequilibrium phase diagrams of C–H system. (Diagonal region is the diamond growth region.)

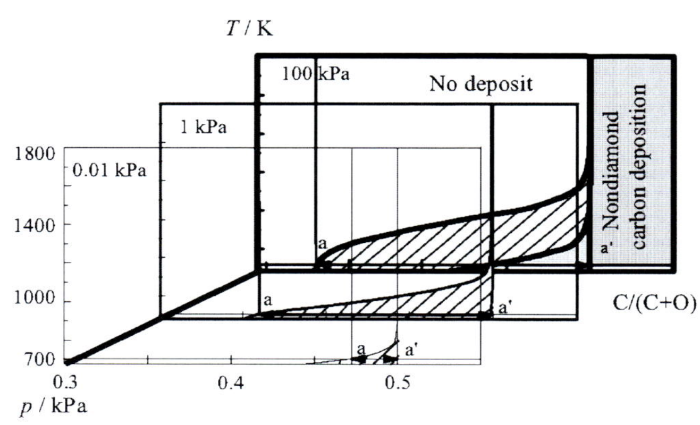

FIGURE. 7.21 $T-p--X$ nonequilibrium phase diagrams of C–O system. (Diagonal region is the diamond growth region.)

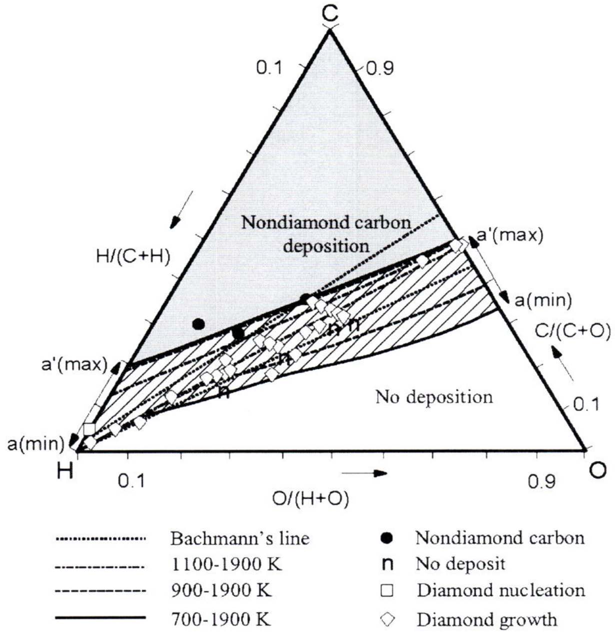

FIGURE. 7.22 Ternary composition projective nonequilibrium phase diagram of C–H–O system (0.01–100 kPa, Diagonal region is the diamond growth region between 700 and 1900 K).

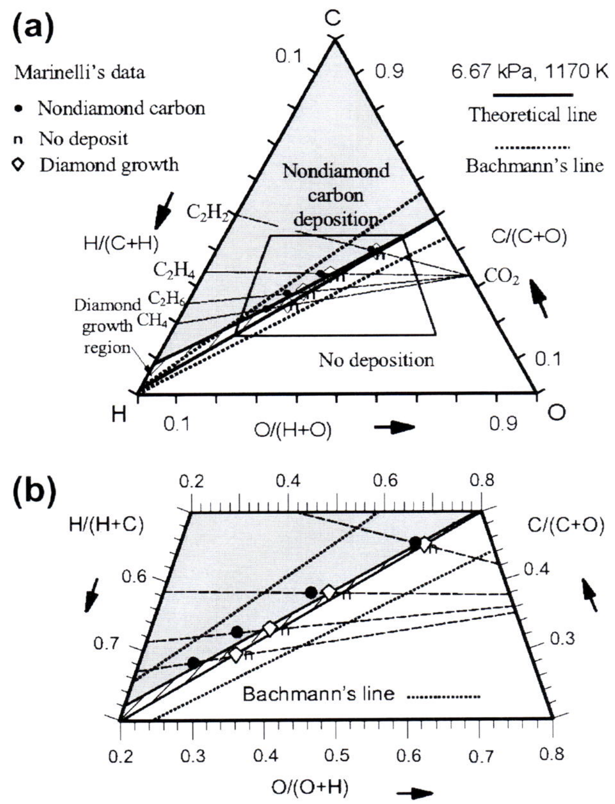

FIGURE. 7.23 Comparison of the theoretical ternary nonequilibrium phase diagram at fixed temperature and pressure with Marimelli's critical experiments. (b) An enlarged part. Diagonal region is the diamond growth region.

agree quantitatively with Marinelli's critical experimental data, measured by spectroscopic analysis. It well confirmed correctness of thermodynamic coupling model and theory of nonequilibrium phase diagrams in modern thermodynamics.

The conception of thermodynamic coupling (or called reaction coupling in chemistry) had already been referred to 1865 Clausius' famous paper, in which thermodynamic coupling was called "compensation" between nonspontaneous process and spontaneous process, but Clausius did not write out its mathematic expression. In the same system, two reactions with $\Delta G_1 > 0$ and $\Delta G_2 < 0$ may take place simultaneously, if Gibbs free energy of the system $\Delta G = (\Delta G_1 + \chi\ \Delta G_2) \leq 0 (\chi = r_2/r_1)$. Based on the conception of reaction coupling, a lot of phenomena and processes in living systems may be explained. However, no quantitative example, especially in living system, had been found. Therefore, in the last 60 years or longer, arguments on reaction coupling in chemistry lasted. Now, the success of the activated low-pressure diamond growth and its thermodynamic coupling model provides a powerful and quantitative demonstration. For instance, under the typically activated low-pressure CVD diamond growth conditions (substrate temperature $T = 1200$ K, total pressure $p = 10^4$ Pa and hot filament temperature $T_{\text{activation}} = 2400$ K), the transformation from graphite to diamond ($\Delta G_1 = 6.96$ kJ) does not take place spontaneously. The association of SAH ($\Delta G_2 = -112.72$ kJ) is a spontaneous reaction. Based on the experimental data of the reaction rate ratio $\chi = 0.28$, then the change of Gibbs free energy for the whole system or for the whole coupled reaction $\Delta G = 6.96 - 0.28 \times 112.72 = -24.60$ kJ < 0. It means that the reaction from graphite and SAH to produce diamond and molecular hydrogen agrees with the second law of thermodynamics. This creates a new way for inorganic syntheses or material preparations on the basis of thermodynamic coupling in the twenty-first century. In fact, thermodynamic coupling has already been used for the low-pressure cubic boron nitride synthesis and could also be used for the preparation of new materials, such as CN_x. A lot of phenomena in living systems also show that thermodynamic coupling has been well utilized. This also opens a nice prospect for bionic syntheses.

Due to the promotion of thermodynamic coupling model for the activated low-pressure diamond growth, Carnot theorem has already been extended into the extended Carnot theorem. Reversibility is a sufficient but not a necessary condition for the highest energy conversion efficiency, while nondissipation is a necessary and sufficient condition for the highest energy conversion efficiency. Based on the extended Carnot theorem, without any out-of-thermodynamic assumption, a complete classification of thermodynamics

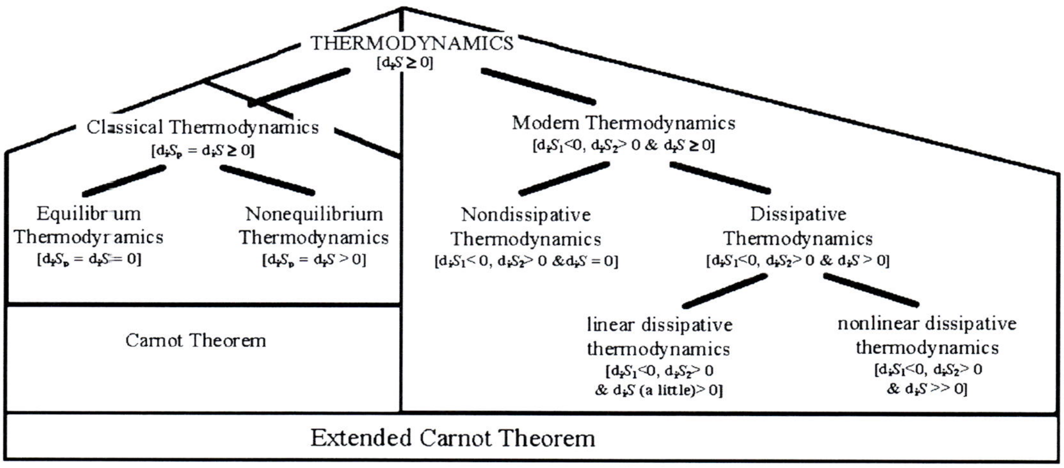

FIGURE. 7.24 A building image of the complete thermodynamics discipline on the basis of the extended Carnot theorem (Mathematic expressions of the second law of thermodynamics are listed in square brackets for each fields). Under isothermal isobaric conditions, $dG = -Td_iS$.

discipline can be constructed, as shown in Fig. 7.24 [48]. More about modern thermodynamics are out of the scope of this book. For readers who are interested in them, related books or references written by Ji-Tao Wang [49–55] may be referred.

References

[1] C.F. Powell, J.H. Oxley, J.M. Blocher, Vapor Deposition, Wiley, New York, 1966.

[2] X.S. Lu, Preface, in: K.C. Zhang, L.H. Zhang (Eds.), Crystal Growth, Science Press, Beijing, 1981.

[3] J.T. Wang, Proceedings of the 13th International Conference on Chemical Vapor Deposition, Los Angeles, 1996, T.M. Besmann, The Electrochem Soc, p. 651.

[4] G.Y. Meng, Chemical Vapor Deposition and New Inorganic Materials, Science Press, Beijing, 1984.

[5] J.T. Wang, M.D. Liu, Semiconductor Materials, High Education Press, Beijing, 1990.

[6] R.S. Rosler, Solid State Technol. 20 (1977) 63.

[7] J.T. Wang, J. Semiconductors 1 (1980) 6.

[8] J.T. Wang, Chin. Phys. 1 (1981) 461.

[9] J.T. Wang, Sci. Sin. (Series A), English edn 26 (1983) 283.

[10] J.T. Wang, Selected Works on CVD, 1, in: J.T. Wang (Ed.), Fudan University Press, Shanghai, 1985.

[11] J.T. Wang, S.L. Zhang, Y.F. Wang, Solid-State Electron. 29 (1986) 999.

[12] W. Kaplan, S.L. Zhang and J.T. Wang, Seattle, 1990, G.W. Cullen and K.E. Spear PV90-12, The Electrochem. Soc. Inc., p. 381.

[13] A.E.T. Kuiper, C.H.J. van den Brekel, J. de Groot, et al., J. Electrochem. Soc. 129 (1982) 2288.

[14] K.F. Jensen, D.B. Graves, J. Electrochem. Soc. 130 (1983) 1950.

[15] K.F. Roenigk, K.F. Jensen, Sweden, Uppsala, June 17–20 1985, p. 207.

[16] K.F. Roenigk, K.F. Jensen, J. Electrochem. Soc. 134 (1987) 1777.

[17] M.G. Joshi, J. Electrochem. Soc. 134 (1987) 3118.

[18] J.T. Wang, S.L. Zhang, Y.F.e. Wang, J. Phys. Colloques 5 (1989) 67.

[19] B.V. Deryagin, B.V. Spitsyn, B.V. Builov, et al., Dokl. Akad. Nauk. 231 (1976) 333.

[20] S. Matsumoto, Y. Sato, M. Tsutsumi, et al., J. Mater. Sci. 17 (1982) 3106.

[21] K.E. Spear, Earth Miner. Sci. 56 (1987) 53.

[22] C.S. Yan, Y.K. Vohra, H.K. Mao, R.J. Hemley, Proc. Natl. Acad. Sci. U.S.A. 99 (2002) 12523.

[23] M. Sommer, K. Mui, F.W. Smith, Solid State Commun. 69 (1989) 775.

[24] W.A. Yarbrough, MRS Fall Meeting, Boston, Nov 28 1989, Paper F1.3.

[25] Y. Bar-Yam, T.D. Moustakas, Nature 342 (1989) 786.

[26] W.A. Yarbrough, R. Messier, Science 247 (1990) 688.

[27] Y.F. Zhang, F.Q. Zhang, G.H. Chen, J. Mater. Res. 9 (1994) 2845.

[28] N.M. Hwang, J.H. Hahn, D.Y. Yoon, J. Cryst. Growth 160 (1996) 87.

[29] E. Marklova, A. Fojtaskova, N.M. Hwang, et al., J. Cryst. Growth 162 (1996) 55.

[30] N.M. Hwang, J. Cryst. Growth 204 (1999) 85.

[31] W. Piekarczyk, J. Mater. Sci. 33 (1998) 3443.

[32] W. Piekarczyk, Cryst. Res. Technol. 34 (1999) 553.

[33] C.X. Wang, G.W. Yang, Mater. Sci. Eng. R 49 (2005) 157.

[34] J.T. Wang, J.O. Carlsson, Surf. Coat. Technol. 43/44 (1990) 1.

[35] J.T. Wang, C.B. Cao, P.J. Zheng, J. Electrochem. Soc. 141 (1994) 278.

[36] J.T. Wang, P.J. Zheng, Chin. Sci. Bull Eng. Ed. 40 (1995) 1141.

[37] J.T. Wang, Z.Q. Huang, Y.Z. Wan, et al., J. Mater. Res. 12 (1997) 1530.

[38] J.T. Wang, Y.Z. Wan, D.W. Zhang, et al., J. Mater. Res. 12 (1997) 3250.

[39] D.W. Zhang, Y.Z. Wan, J.T. Wang, J. Cryst. Growth 177 (1997) 171.

[40] D.W. Zhang, Y.Z. Wan, Z.J. Liu, et al., J. Mater. Sci. Lett. 16 (1997) 1349.

[41] J.T. Wang, Y.Z. Wan, Z.J. Liu, et al., Mater. Lett. 33 (1998) 311.

[42] D.W. Zhang, Z.J. Liu, Y.Z. Wan, et al., Appl. Phys. 66 (1998) 49.

[43] Y.Z. Wan, D.W. Zhang, Z.J. Liu, et al., Appl. Phys. A 67 (1998) 225.

[44] J.T. Wang, Stationary Nonequilibrium Phase Diagrams – Thermodynamics of Low-pressure Diamond Syntheses from the Vapor Phase, Science Press, Beijing, 2000.

[45] J.T. Wang, D.W. Zhang, Z.J. Liu, Thermodynamic Coupling Model for Low-pressure Diamond Growth from the Vapor Phase, Science Press, Beijing, 1998.
[46] J.T. Wang, Nonequilibrium Nondissipative Thermodynamics – with Application to Low-pressure Diamond Synthesis, Springer, Heidelberg, 2002.
[47] J.T. Wang, Modern Thermodynamics – and a View of Whole Thermodynamics, Fudan University Press, Shanghai, 2005.
[48] J.T. Wang, Modern Thermodynamics – Based on the Extended Carnot Theorem, Science Press (English Ed), 2009. Fudan Univ. Press (Chinese Ed), Beijing. Shanghai, 2010.
[49] Y. Gogotsi, S. Welz, D.A. Ersoy, M.J. McNallan, Nature 411 (2001) 283.
[50] B. Heimann, V. Raiko, V. Buck, Int. J. Refract. Metals Hard Mater. 19 (2001) 169.
[51] M. Pons, D. Ballutaud, Chem. Vapor Deposition 10 (2004) 123.
[52] J.D. Felske, Appl. Mech. Rev. 57 (2004) B16.
[53] Consultation Group on Nonferrous Metallic Materials (Ed.), Consultation Report on Nonferrous Metallic Materials, Vol. 229, Engineering Academy of China, Academy of China, Status Quo of Materials Development of China and Countermeasure for Going to the New Century, Consultation Program, Shanxi, 2001.
[54] D.H. Dai, K.S. Zhou, Metallurgy Industry Publisher, Beijing, 2001, p. 191.
[55] D.H. Dai, K.S. Zhou, Z.H. Yuan, Metallurgy Industry Publisher, Beijing, 2004, p. 372.

CHAPTER

8

Microwave-assisted Inorganic Syntheses

Jingping Zhao, Wenfu Yan

Jilin University, China

The use of microwave as an energy source for chemical reactions and processes has been extensively investigated over a wide range of subjects since the last century [1–8]. Large numbers of reactions in organic and inorganic synthesis have been studied. Their applications include cyclo-addition reactions [9], synthesis of radio-isotopes [10], fullerene chemistry [11], polymers [12], carbohydrates [13], homogeneous [14]/heterogeneous catalysis [14,15], medicine [16,17], green chemistry [18], SiC [19], GaN [20], $CuFe_2O_4$ [21], and La_2CuO_4 [21]. Many review papers have covered almost every aspect from accelerated synthesis speed, microwave effects to microwave equipment [22–27].

8.1. BASIC PRINCIPLE OF MICROWAVE RADIATION, MICROWAVE HEATING, AND MICROWAVE EQUIPMENT

8.1.1. Basic Principle of Microwave Radiation and Microwave Heating

Microwave heating has been widely used in chemical synthesis. If microwave heating is regarded as an alternative to conventional oven heating, then its interest and potential in synthesis chemistry are limited. It is very important to understand the principle of microwave radiation and microwave heating because many chemical reactions show significant differences under microwave heating conditions in aspects of the reaction rate and product selectivity, as compared to conventional heating.

The basic principle of microwaves can be described by the following equation:

$$\lambda_0 = \frac{c}{f} \tag{8.1}$$

The wavelength λ_0 of microwave is related to the frequency f according to Eq. (8.1). The frequency indicates the numbers of oscillations of the electric or magnetic field in 1 s. Microwaves lie in the electromagnetic spectrum between infrared waves and radio waves. Usually, the wavelength is from 0.01 to 1 m and the frequency range is between 0.3 and 30 GHz. A typical electromagnetic spectrum is shown in Fig. 8.1 [28].

Microwave radiations in general are of a low energy as compared to other radiations (Table 8.1 [28]). The quantum energy of microwave is only about 0.0016 ev at 2450 MHz and it is too weak to break any chemical bond even hydrogen bond. However, microwave energy will affect many chemical reactions in different ways.

When a material is exposed to a microwave radiation, three different sets of conditions may occur depending on the nature of the material. If the material is

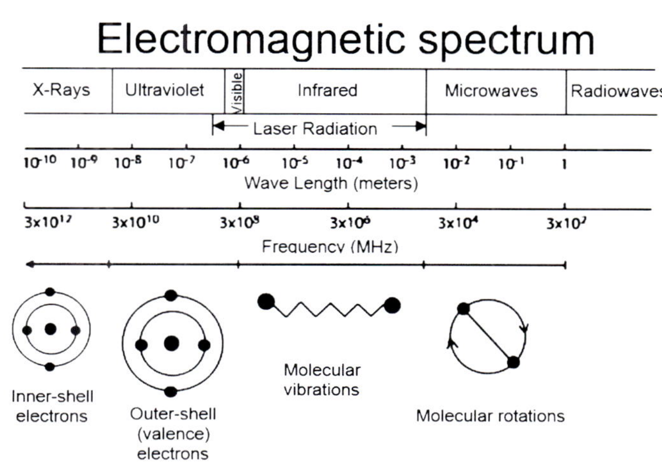

FIGURE 8.1 A typical spectrum of common electromagnetic radiations [28].

TABLE 8.1 Typical Energy of Common Radiation and Chemical Bonds [28]

Radiation type	Typical frequency (MHz)	Quantum energy (eV)
Gamma rays	3.0×10^{14}	1.24×10^{6}
X-rays	3.0×10^{13}	1.24×10^{5}
Ultraviolet	1.0×10^{9}	4.1
Visible light	6.0×10^{8}	2.5
Infrared light	3.0×10^{6}	0.012
Microwaves	2450	0.0016
Radio	1	4×10^{-9}

Chemical bond type	Chemical bond energy (eV)
H–OH	5.2
$H–CH_3$	4.5
$H–NHCH_3$	4.0
$H_3C–CH_3$	3.8
$PhCH_2–COOH$	2.4
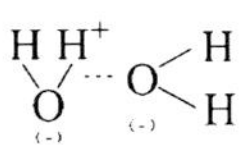	0.21

a conductor such as metals, microwave will be reflected from these materials and will not generate heat. If the material is insulator such as china, the material is transparent to microwave energy and again no heat will be generated. If the material is dielectric such as water, microwave energy will be absorbed by the material and heat will be generated. The last case is the most interesting aspect to chemists.

In general, microwave energy absorption has two main mechanisms. One is dipole rotation and another is ionic conduction.

Dipole rotation is the mechanism normally referred to as dielectric heating. Molecular rotation occurs in materials containing polar molecules having an electrical dipole moment, which will completely or partially align themselves continuously in an electromagnetic field. As the electromagnetic field alternates, the molecules change their direction. The rotating molecules push, pull, and collide with other molecules distributing the energy to adjacent molecules and atoms in the material. Molecules can rotate with electromagnetic field up to 10^6 Hz in liquid or glass systems. However, the rotation cannot follow the field changes all the time as it depends on the size and the dielectric coefficient of the molecules. In such cases, the field energy is converted to kinetic and thermal energy. Temperature is the average kinetic energy of the atoms or molecules in a material. Thus, dipole rotation is a mechanism by which energy in the form of electromagnetic radiation is converted to heat energy.

Molecular friction is often used as an easy understanding term for describing this behavior. The mobility and ability to orient molecules or the "molecular friction" decide the nature of absorption and conversion of the energy.

Ionic conduction is a mechanism normally referred as ions movement with the electronic field. When the electric field is changed, ions with the same charge migrate away from the field. Obviously, the ion movement will affect the chemical reaction in a different way as compared to the reaction under conventional heating.

Based on the above principle, microwave radiation affects chemical reactions mainly by dipole rotation and ionic conduction of the reactants. If a simple chemical reaction can be shown as:

$$A + B \rightarrow C \tag{8.2}$$

The reaction rate of the reaction can be described as:

$$k = Ae^{-Ea/RT} \tag{8.3}$$

The equation gives the dependence of the rate constant k of chemical reactions on the temperature T and activation energy E_a, where A is the pre-exponential factor or simply the *pre-factor* and R is the gas constant. The A factor can be understood as a description of the probability of reaction molecules to align in right orientation for band formation. Given the small temperature range in which kinetic studies are carried, it is reasonable to approximate the activation energy as being independent of the temperature. The most that people agree is that the activation energy E_a is most likely not to be significantly affected by microwave radiation because of the nature of microwave radiation. The temperature T is the physical measurement temperature which is the combination of Bulk temperature T_b and Instantaneous temperature T_i. The bulk temperature is achieved with conduction. T_b is equal to T_i in most of conventional heating reactions. However, in a microwave heating reaction system, the transfer of microwave energy is rapid and direct with any absorbing materials. The microwave energy can be directly delivered to the molecules. The rapid energy transfer creates nonequilibrium condition resulting in high instantaneous temperature T_i. The high T_i activates a higher percentage of molecules above the required activation energy. Due to rapid energy transfer under microwave heating condition, T_i is much larger than T_b. Thus, reaction kinetics is controlled by T_i under microwave radiation condition. In general, a reaction can be accelerated from 10 to 1000 times under microwave heating. According to some researchers, when a reaction is at 150°C

($T_b = 150$), and E_a is 50 kcal/mol, the reaction rate will increase 10 times if T_i is 167°C. If T_i increases to 185°C, the reaction speed is about 100 times faster. If T_i rises to 206°C, the reaction will speed up 1000 times. The instantaneous temperature is associated with the "superheating" or "hot spots" [29,30] which are generated by rapid transfer of microwave energy on local area of particular dielectric substances. However, it should be indicated that "superheating" is a local molecule-level overheating and is different from the overheating caused by delay of boiling under a conventional heating condition. The "superheating" and "hot spots" are the characteristics of microwave-heated reactions.

Based on the basic understanding for microwave heating, rapid and direct energy delivery is a unique feature affecting the reactions. Due to the rapid energy transferring under microwave heating, a microwave-heated reaction system may reduce reaction dramatically up to 10–1000 times, increase yield and product selectivity. It may also allow those reactions to occur which are otherwise not possible. The microwave accelerated-processes may be greener than a conventional process.

8.1.2. Microwave Equipment

Although the microwave frequencies cover from 0.3 to 30 GHz, there are only very limited frequency bands that can be used in microwave heating equipment. The frequencies used in microwave ovens were chosen on the basis of two strict constraints: (i) there should be one of the industrial, scientific, and medical (ISM) frequency bands set aside for noncommunication purposes; and (ii) besides 2.45 GHz, three additional ISM bands exist in the microwave frequencies range, but are not used for microwave heating. Bands at 5.8 GHz and 24.125 GHz are not used for microwave heating because of the very high cost of power generation at these frequencies. The third, centered at 433.92 MHz, is a narrow band which requires expensive equipment to generate sufficient power without creating interference outside the band and is only available in some countries. For household purposes, 2.45 GHz has the advantage over 915 MHz in that 915 MHz is only an ISM band in the ITU Region 2 while 2.45 GHz is available worldwide. Therefore, almost the entire commercial microwave heating equipment either for scientific or for domestic purposes has a fixed frequency at 2.45 GHz. Consequently, any results and conclusions discussed here are based on the microwave heating with fixed frequency at 2.45 GHz. If the microwave frequency is different from the fixed 2.45 GHz, many of these results and conclusions may not be reliable.

Microwave accelerated reactions are carried out under domestic household microwave oven and laboratory microwave heating equipment. Most domestic microwave ovens allow users to choose between several power levels from 500 to 1500 W. In most ovens, however, there is no change in the intensity of the microwave radiation. The temperature is controlled by turning on and off the magnetron in which the microwaves are generated. The temperature inside microwave oven is not monitored. Therefore, the reaction temperature in such oven can vary to a large/great extent depending on the microwave absorption of the reactants, reaction medium, and quantity of the reaction materials. Under such condition, the reaction temperature is not controlled. This type of microwave oven is only used on some of the studies in 1970s–1980s of last century. Due to the lack of temperature control of this type of microwave oven, the reactions carried out in this oven are usually without a reliable temperature. The results are almost of no comparison value.

With the development of microwave heating equipment, more sophisticated laboratory microwave ovens have become available. One of such microwave equipment is developed by CEM (Fig. 8.2). A laboratory microwave oven usually has sophisticated temperature control system which is able to monitor the temperature of single or batch of reactors directly or indirectly. Microwave power can be controlled continually. Using such microwave oven, the reaction temperature can be accurately controlled and results are much more comparable. Based on the principle of microwave heating, we know that absorption of microwave for a given substance also depends on the wavelength of the microwave radiation. Therefore, the microwave heating

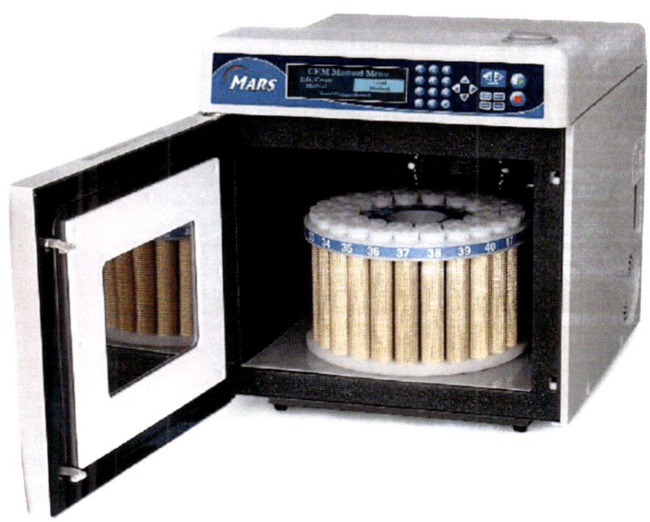

FIGURE 8.2 CEM Laboratory Microwave Oven [28]. *Courtesy of CEM (www.cem.com).*

equipment with variable frequency and continuous variable power output is ideal equipment for comprehensive studies on microwave associated synthesis chemistry. However, under current regulation and technique restrictions, majority of the research associated with microwave heating is carried out at a fixed microwave frequency. If the associated microwave frequency is changed, the results and conclusions are needed to be verified.

8.2. SYNTHESIS OF INORGANIC MATERIALS UNDER MICROWAVE HEATING

As described above, microwave heating can be a promising alternative to conventional heating. The following sections will discuss the syntheses of the various inorganic materials.

8.2.1. Reactions in Liquid Media

8.2.1.1. Microwave-assisted Hydrothermal Synthesis: porous Materials

A large number of compounds or materials have been synthesized under hydrothermal conditions: porous materials, superionic conductors, chemical sensors, electronically conducting solids, complex oxides and ceramics, fluorides, magnetic materials, luminescence phosphors, etc. [31].

Porous materials like zeolites are usually synthesized under hydrothermal conditions. Synthesis of zeolites under microwave heating condition has demonstrated some special effects comparing to the synthesis carried out under conventional hydrothermal condition. Much has been written about special "microwave effects", particularly where it has been claimed that reactions proceeded faster in microwave heating environment than under conventional conditions at the same temperature.

The early application of microwave heating in zeolite synthesis is NaA and ZSM-5 which was patented by Mobil in 1988. Since then, the microwave heating has been applied in many microporous and mesoporous materials synthesis. Some of the results are reported and summarized [23]

A typical explanation for accelerated zeolite synthesis under microwave heating conditions includes the following aspects.

1. Microwave heating is able to transfer microwave energy rapidly to the reaction system resulting in the rise in the temperature of the reaction mixture and hence accelerating the reaction (crystallization) speed [32].
2. Microwave heating may lead to more uniform temperature distribution in the reaction mixture [33] as it can deliver the microwave energy directly to the molecules of the reactants rather by the conventional heat conduction.
3. Microwave radiation can generate "superheating" in the reaction mixture [34]. Under atmospheric pressure, the boiling point of a liquid under microwave heating may rise temperature by 10–30°C to that of the conventional boiling point.
4. In heterogeneous systems, there may be significant different heating effects due to the heterogeneity of the dielectric species. In these cases, it is very possible for some regions to be hotter than others resulting in regional hot spots in macro or microscale. The superheating, hot spots as well as the ionic conduction my affect the orientation of the reactant molecules or crystallization route [29,30].
5. Microwave heating changes the association between reaction species in the reaction mixture [35,36].
6. Microwave energy enhances the dissolution of the reaction gel [35,37,38].

8.2.1.1.1. MICROWAVE-ASSISTED SYNTHESIS OF ALUMINOUS-SILICATE ZEOLITES

Following the early studies on the microwave-assisted syntheses of zeolites, many laboratories reported the microwave-assisted syntheses of aluminous-silicate zeolites. A typical research on the $Na_2O–SiO_2–A_2O_3–H_2O$ system shows that under microwave heating, the crystallization fields for hydroxysodalite of NaA and NaX were essentially same as those obtained under conventional heating. However, the crystallization time of the NaA under conventional heating is 10 times longer than that under microwave heating. The crystal size distribution obtained from the microwave-assisted synthesis is slightly narrower (0.5–1.4 μm) than that obtained from the conventional hydrothermal synthesis (0.2–1.3 μm).

For the same synthesis system, when the compositions of the mixture were $x Na_2O–1.0Al_2O_3–y SiO_2–240H_2O$, where $x = 6.2$ and $y = 10.0$, respectively, a well-crystallized NaY could be synthesized in 1.5 h at 110°C under microwave heating without seeding. Using the same reaction mixture, the synthesis takes 9–10 h at the same temperature under conventional heating condition. Thus, the synthesis is speeded up for about six times under microwave heating.

It is well known that pure NaY is usually difficult to be synthesized at a higher temperature than 150°C under conventional heating condition because P-type zeolite is an unavoidable impurity phase frequently obtained at higher temperature [39]. Therefore, most of the industrial processes for NaY production usually

TABLE 8.2 Aluminosilicate Zeolites Prepared by Microwave Synthesis

Zeolite type	Microwave hydrothermal treatment—composition, microwave oven, power, time, temperature	Synthesis characteristics
Sodalite (SOD)	1.9:1:2.66:128 ($SiO_2/Al_2O_3/Na_2O/H_2O$), CEM MARS X microwave, 1200 W max, 7 min at 160°C no gel aging	Spherical particles ≈ 2 μ size
NaY (FAU)	10:1:14:800 ($SiO_2/Al_2O_3/Na_2O/H_2O$), Snowtex-N Colloidal Silica source, CEM MARS-5, XP-1500 oven 300 W max, 100°C for 0.5–4 h, 20 mL in a 50 mL vessel heating rate of 20–200°C min^{-1}	2 h at 100°C produced 180–280 crystal size, 96% yield, particle size increased with reaction time, heating rate did not strongly effect the formation of NaY
NaY (EMT)	2.4:0.5:0.85–1.0:10:0.5–0.8:140 ($Na_2O/NaF/Al_2O_3/SiO_2/$[18-crown-6]$/H_2O$), fumed silica (BDH), CEM MDS-2100 oven, 14 h at 170°C, or 20 min at 150°C, high silica NaY in 1.5 h at 150°C	≈6× reduction in synthesis time over conventional, synthesis can be carried out at higher temperatures without formation of zeolite P
NaY (FAU)	$SiO_2/Al_2O_3 = 25$, $Na_2O/SiO_2 = 0.88$, $H_2O/Na_2O = 45–90$, domestic oven Matsushita Electric Co. NE-S30, 120 s to 100–120°C at 600 W, cooled, then repeated using 200 W power	10 min at 100°C fully crystalline NaY
NaY(FAU)	8–10:1:3–4:100–135($SiO_2/Al_2O_3/Na_2O/H_2O$) Aerosil 200 silica source, domestic Sharp R-10R-50, 200 mL Teflon autoclave in which two 30 mL vessels are placed, 30 s, 120°C then 10 min at 100°C	Impurities such as Zeolite P observed, fast heating avoids the formation of secondary phases, decrease in crystallinity observed with increase in Si/Al
NaY(FAU)	16:1:3:400 ($SiO_2/Al_2O_3/Na_2O/H_2O$), Aerosil 200 silica, Panasonic NN 7856, 900 W max pulsed, 900 W 120°C in 30 s, 18–25 min @ 120°C, solution aged for 24 h, PTEE vessels with 5 cm inside diameter	Used pulsed power to control the reaction temperature
NaY(FAU)	1:10:6:400 ($SiO_2/Al_2O_3/Na_2O/H_2O$), 10–60 min refluxing	<100 nm crystal sizes formed, uniform crystal size produced
NaY(FAU)	10:1:6:240 ($SiO_2/Al_2O_3/Na_2O/H_2O$), Ludox HS-40 silica, Custom continuous microwave reactor, Sharp r-10R50 oven, 140°C, 3.5 mL min^{-1}, 12 min residence	Average crystal size of 0.4 μm, $SiO_2/Al_2O_3 = 4$, conventional synthesis with rapid heating produced similar reaction results as microwave synthesis

Tompsett, G. A.; Conner, W. C.; Yngvesson, K. S.: Microwave synthesis of nanoporous materials. ChemPhysChem. 2006. Vol.7, Page 296. Copyright Wiley-VCH Verlag GmbH & Co. KGaA. Reproduced with permission from [3].

use a low synthesis temperature (usually ca. 90–100°C). At this temperature, the NaY crystallization usually takes about 10–30 h. However, under microwave heating, the FAU structure of NaY could be detected by XRD after about 5 min at 150°C, and a complete crystallization took only further 10–15 min [5]. The crystalline product had an average crystal size of 0.5–1.0 μm without any XRD detectable crystalline impurity. Even after extending the microwave heating time to a total of 120 min, there was no other XRD detectable crystalline phase formed. This suggests that the microwave energy has made an effective and apparently selective contribution to the formation of FAU structure. It appears that use of microwave energy may have a particularly advantage in encouraging the nucleation of single phase under favorable conditions rather than exerting any unique influence upon crystal growth.

Highly siliceous zeolite Y is widely used as catalysts in FCC process to increase the yield and selectivity. Usually, a high silica zeolite Y needs a longer crystallization time and many cases require templates and seeding or posttreatment of dealumination. However, a fairly siliceous and pure NaY can be synthesized in less than 1.5 h at 150°C using microwave heating with seeding. A typical product had an SiO_2/Al_2O_3 ratio of 4.7 as calculated from unit cell parameters. Unfortunately, the true comparison of microwave heating with conventional heating at higher temperature cannot be made because pure NaY cannot be obtained from conventional heating under these conditions.

The acceleration effect of the microwave heating on the crystallization of Faujasite-type zeolite is summarized in Table 8.2 [3]. It demonstrates that microwave heating can reduce about 2/3rd of the crystallization time at 110°C compared to conventional heating. A much higher crystallization temperature of up to 150°C can also be used under microwave heating without generating impure crystalline phases.

8.2.1.1.2. MICROWAVE-ASSISTED SYNTHESIS OF [Al]ZSM-5

[Al]ZSM-5 can be synthesized from a reaction mixture having a typical composition of $5.0Na_2O–0.2A_2O_3–60SiO_2–4.0TPABr–900H_2O$. The very well-crystallized

[Al]ZSM-5 can be obtained from this gel under microwave heating at 170°C without seeding in less than 3 h. The XRD patterns of the products show that the nucleation takes less than 1.65 h and the crystal growth needs about 1 h.

The above result shows that the [Al]ZSM-5 can be formed very rapidly under microwave heating. The particle size of the [Al]ZSM-5 synthesized under microwave heating is about 3–4 μm and no other crystalline phase is observed. Compared to conventional oven heating, the time taken for complete crystallization of the gel under microwave heating is only few hours, which speeds up the reaction (twice or more than twice). With further studies, 5% (by weight, based on the total amount of silica) of colloidal ZSM-5 was added into the typical reaction mixture as seeds. Using the seeded mixture, the crystallization of [Al]ZSM-5 takes about 2–3 h at 175°C under conventional heating. However, very well-crystallized [Al]ZSM-5 can be obtained from the same reaction mixture in 5 min at 175°C under microwave heating [30]. The particle size is about 0.3–0.5 μm.

A similar study on [Ti]ZSM-5 is also carried out at similar conditions using a reaction mixture having composition of $1.0NH_4F{-}0.03TiCl_3{-}1.0SiO_2{-}0.5TPABr{-}30H_2O$. The crystallization is similar to the aluminum-based fluoride system of $2.0NH_4F{-}0.14AlCl_3{-}1.0SiO_2{-}0.5TPABr{-}30H_2O$. The nucleation takes about 7 h and the crystal growth needs only about 1.5 h which is similar to that in an alkaline system. This shows that microwave heating has no unique influence on the crystal growth in either fluoride or alkaline synthesis systems. It suggests that nucleation is the major factor which controls the overall crystallization time in the fluoride-based MFI synthesis system. Under conventional heating, the crystallization will take about 20 h at 175°C. All of these studies discussed in this section suggest that the microwave heating can usually speed up the crystallization of zeolite with MFI structure 2–3 times.

For MFI type of zeolites, microwave heating can cut at least half of the crystallization time as compared to the conventional heating without seeding. But under seeding condition, microwave heating can accelerate the crystallization by more than 20 times.

Some synthesis details of MFI and MEL zeolites synthesized under microwave heating are summarized in [3].

8.2.1.1.3. MICROWAVE-ASSISTED SYNTHESIS OF MICROPOROUS ALUMINOPHOSPHATES

Microporous aluminophosphates $AlPO_4$-n have neutral frameworks, without any associated cations and significant catalytic activity. However, the substitution of heteroelements into these frameworks may introduce possibilities for many types of applications as catalysts and catalyst supporting materials. When silicon or metal substitution happened, the frameworks will produce a negative charge and resulting potential acid sites, whilst the presence of the framework transition metal may generate redox activity.

An early study reported [35] that $AlPO_4$-5 would be synthesized in 60 s at 180°C under microwave heating. A large series of synthetic parameters such as compositions, temperature, and heating time were investigated in order to determine their effects on the synthesis. An interesting process occurred during the use of two-stage synthesis in which the mother liquor was decanted after 60 s heating and heated again at 180°C for further 60 s. Although the particle size of the products at the first stage was fairly big and contained amorphous materials, the product at the second stage had a very narrow size distribution of the single crystals and was free from impurity. It was also found that a small amount of HF would increase the crystal size. Du et al. [40] found that a broader range of composition of the reaction mixture can result in the formation of $AlPO_4$-5 crystals under microwave heating condition, which is different from that formed under conventional heating. So far, many porous phosphates have been successfully synthesized under microwave heating, such as SAPO, MAPO, VAPSO, CoAPO, and MnAPO. Some of the results are summarized in [3]. For this type of porous materials, the synthesis is speeded up in most of the cases by microwave heating.

8.2.1.1.4. MICROWAVE-ASSISTED SYNTHESIS OF MESOPOROUS MATERIALS

Mesoporous materials with uniform porous dimensions were discovered by Mobil in 1992 [41]. The surface area of mesoporous materials is usually high (over $1000\ m^2\ g^{-1}$) with a narrow pore size distribution. A vast number of mesoporous materials containing pore size from 2 to 100 nm were produced [42] and modified by incorporation of metal elements. Although this kind of mesoporous materials exhibits unique catalytic, adsorption, and separation properties, their synthesis often required several days. In 1996, mesoporous silica (MCM-41) was synthesized by Wu [43] under microwave heating. Recent studies found that many of the mesoporous materials could be prepared under microwave heating condition and the synthesis could be significantly accelerated. Some of these results are summarized in Table 8.3 [3]. The particle size of the mesoporous materials prepared under microwave heating is usually smaller than those prepared under conventional heating, which suggests that the nucleation is fast and homogeneous under microwave heating conditions. Kim et al. [44] reported that the additive of ethylene glycol (EG) could improve the crystallinity and morphology of MCM-41 due to the higher dissipation factor of EG than that of water leading to fast energy transformation from microwave energy to thermal energy.

TABLE 8.3 Mesoporous Silica-based Materials Prepared Under Microwave Heating

Mesoporous material	Composition and synthesis	Product characteristics
MCM-41	$NaAlO_2$ (0.083 g), $C_{16}H_{33}NMe_3Cl/OH$ (9 mL) + TMA silicate (2 g; 10%) + silica (1 g), Questron microwave, 160°C for 1 min then 150°C for 80 min	Hexagonal phase forms after 20 min at 150°C, small crystallites
MCM-41	$NaAlO_2$, $C_nH_{2n+1}NMe_3X$ ($X \equiv$ Cl, Br, $n \equiv 12, 16$), mesitylene, aerosil or TEOS, 650 W, 1–30 min	
MCM-41	1:0.167:0.5:40.5:0–8.1 (SiO_2/MTAB/NaOH/H_2O/EG), CEM MDS-2000 630 W max, 100–150°C in 1–30 min, then 100°C 30 min (60 W)	Uniform fine particles, rodlike
MCM-41	1:0.167:0.5:40.5:0–8.1 (SiO_2/MTAB/NaOH/H_2O/EG), CEM MDS-2000 630 W max, 100–120°C for 40 min	Ethylene glycol improves crystallinity and formation of uniform, round ≈0.2 mm MCM-41 crystals. EG has a higher dissipation factor
Ti-MCM-41	1:0.02:0.2:0.25:31 (SiO_2/TiO_2/TEAOH/CTMACI/H_2O), CEM MDS-2100 950 W max, 100–120°C 10 min–5 h	Accelerated rate, 3 nm pores, narrow distribution, morphology disordered and wormlike, no long-rage order
B-MCM-41	1:1:0.6:0.2:55 (HDTMABr/SiO_2/EN/H_3BO_3/H_2O), CEM MDS 200, 95°C for 4 h	
MCM-48		Prepared microwave radiation-induced dispersion of magnesia or calcium oxide, impregnation with magnesium acetate solution, and ion exchange with cesium nitrate solution
MCM-48	1:0.15:0.5:80 (TEOS/CTABr/NaOH/H_2O), 100°C for 1 or 2 h, CEM MDS-2000 microwave oven	Spherical 1rodlikeμm crystals formed
SBA-15	SiO_2 (0.01 mol), $EO_{20}BO_{70}EO_{20}$(1 g), NaCl (7.7 mol), HCl (0.06 mol), ethanol (0.44 mol), H_2O(1.94 mol)., CEM MARS-5 microwave 1200 W max, 15–120 min at 100°C	
Ti-SBA-15	SiO_2/TiO_2 (0.02 mol), $EO_{20}BO_{70}EO_{20}$(2 g), HCl (0.12 mol), H_2O(3.92 mol)., CEM MARS-5 microwave 1200 W max, 120 min at 100°C	
Zr-SBA-15	SiO_2/TiO_2 (0.02 mol), $EO_{20}BO_{70}EO_{20}$(2 g), HCl (0.12 mol), H_2O(3.92 mol)., CEM MARS-5 microwave 1200 W max, 20 min at 100°C	Si/Zr ratios of 10, 20, 40, and 80 successfully synthesized
SBA-16	1:0.000317:6.68:137.9:1.4–8.4 (SiO_2/F127/HCl/H_2O/EG), CEM MARS-5 microwave 1200 W max, 60 min at 100°C	dodecahedral particles ≈ 2μm size.
FDU-1	1:0.00735:6.155 (TEOS/$EO_{39}BO_{47}EO$/HCl/H_2O), 15–180 min at 100 °C, microwave not specified	best sample obtained after 60 min, uniform nanopores, increased synthesis times decreased the uniformity of pore entrance sizes
PSU-1	1:0.12:2.6:100–280 (SiO_2/CTACI/HCl/H_2O), CEM MARS-5 1200 W max, static 100°C 2 h	Cage-like framework produced, similar to FDU-1 and SBA-16, Not obtained using conventional hydrothermal synthesis
micro–meso disordered silica	SiO_2 (0.01 mol):$EO_{20}BO_{70}EO_{20}$ (0/034/0.069/0.10/0.14/0.17): C10TAB (3.56.2.85/2.14/1.42/0.71) mmol, or C12TAB (3.34/2.59/1.94/1.29/0.65) mmol, or C16TAB (2.74/2.19/1.64/1.09/0.55) mmol, HCl (0.06 mol): H_2O (1.94 mol), CEM MARS-5 1200 W max, 2 h at 100°C	

Tompsett, G. A.; Conner, W. C.; Yngvesson, K. S.: Microwave synthesis of nanoporous materials. ChemPhysChem. 2006. Vol.7, Page 296. Copyright Wiley-VCH Verlag GmbH & Co. KGaA. Reproduced with permission from [3].

8.2.1.1.5. MICROWAVE-ASSISTED PREPARATION OF MEMBRANES OF POROUS MATERIALS

The preparation of nanoporous membranes is a very attractive aspect because the membranes often provide enhanced separations, catalytic activity and selectivity. However, the preparation of high-quality membranes is not easy. Under microwave heating condition, the formation of particles with smaller and more uniform size make it possible to prepare thinner, dense, better orientated, and more aligned membranes.

(a) $AlPO_4$-5 and SAPO-5 membranes

In 1995, the preparation of $AlPO_4$-5 membrane from very small orientated crystals under microwave heating condition was reported [45]. The temperature, microwave heating time, power, and aging time affected the thickness of the membrane and the orientation of the $AlPO_4$-5 crystals. The $AlPO_4$-5 crystals in the membranes prepared under conventional heating are not usually oriented. However, they may be oriented under microwave heating condition. Tsai et al. [46] prepared a well-aligned SAPO-5 membrane under microwave heating. Some studies [47] suggested that microwave heating limits the nonisothermal process, resulting in homogeneous heating in the reaction solution. Therefore, a large number of nuclei would form in the reaction solution in a very short time, leading to the very small and uniform crystals which help to form dense and well-orientated membranes.

(b) LTA membranes

LTA membranes have been extensively studied because of the convenient and facile synthesis of LTA-type zeolite. Some of the reported results are summarized in [26]. According to these results, comparison of a proposed mechanism of microwave heating and conventional heating effects on membrane formation is illustrated in Fig. 8.3 [26]. It is generally agreed that microwave heating significantly enhanced the formation of small and uniform crystals, resulting in more controllable, better orientated, less gradient membranes. It is found that microwave heating can affect the morphology of LTA-type zeolite during membrane formation, resulting in less perfect, aggregated, and dense membrane.

Usually it is a problem to use metallic materials in microwave environment because of the arcing. However, some laboratories have studied zeolite A membrane preparation on copper and stainless steel substrates under microwave heating [48]. According to their results, zeolite A membranes with 0.3–1 μm thickness could be rapidly formed on the copper but not on stainless steel base. Their results suggested that the nucleation of zeolites on the metallic substrates depended on the active sites of the surface, heating method, and the aging procedure of the reaction solution.

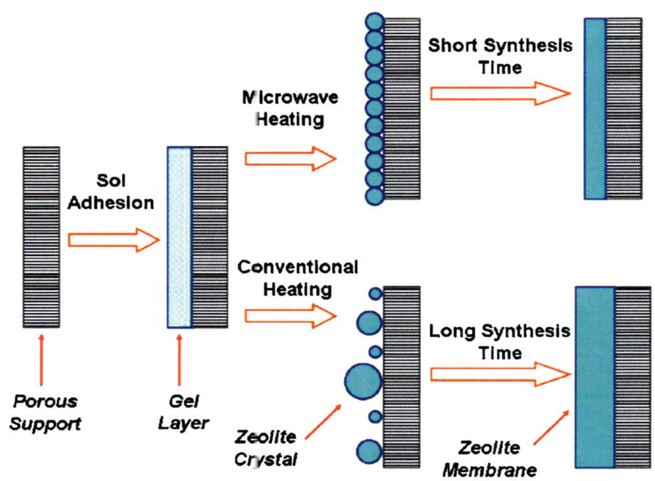

FIGURE 8.3 Comparison of synthetic model of zeolite membrane under different heating environments. *Reprinted from Journal of Membrane Science, Vol. 316, Yanshuo Li, Weishen Yang, Microwave synthesis of zeolite membranes: A review, Page 3, Copyright (2008), with permission from Elsevier [26].*

(c) Silicalite-1 (MFI structure) membranes

Silicalite-1 (MFI structure) membranes were prepared on silicon wafer support under microwave heating [49,50]. It was found that microwave heating provided an efficient preparation process for such membranes.

(d) Other zeolite-type membranes

Other zeolite-type membranes such as Sodalite, Faujasite, and ETS-4 have also been successfully prepared under microwave heating [26]. Some of the results are summarized in [26].

A well-crystallized Sodalite membrane supported on α-Al_2O_3 was synthesized in 45 min under microwave heating [51]. It is eight times faster than the synthesis conducted under conventional heating condition. The microwave heating also shows the preference to the Sodalite formation, resulting in the pure Sodalite membrane rather than the NaX and NaA co-existed membrane which is often obtained by conventional heating. A Faujasite membrane supported on γ-Al_2O_3 was prepared by two-stage synthesis under microwave heating [52]. Titanosilicate (ETS-4) membrane was also synthesized in less than 1 h by a microwave heating associated preparation procedure [53], which is significantly shorter than the preparation time of 36–48 h under the conventional heating. It shows the significant acceleration effects of microwave heating again in membrane preparation.

8.2.1.2. Microwave-assisted Sol–gel Synthesis

Sol–gel processing refers to the hydrolysis and condensation of alkoxide-based precursors such as $Si(OEt)_4$ (tetraethyl orthosilicate). Disregarding the nature of the precursors, the sol–gel process can be characterized by two steps: (1) Formation of stable solutions of the alkoxide or solvated metal precursor (the formation of the sol). (2) Gelation resulting from the formation of an oxide- or alcohol-bridged network (the formation of the gel) by a polycondensation reaction, which usually results in a dramatic increase in the viscosity of the solution [54]. Microwave-assisted sol–gel synthesis has been applied to prepare cathode material [55], positive electrode materials [56], carbon ceramic electrode [57], titanium dioxide, and perovskite thin films [58,59], phosphors [60], organic–inorganic hybrid [61], aluminum fluoride [62], oxide [63], etc. The advantages of the microwave heating on the preparation of materials via sol–gel process will be illustrated with the help of microwave-assisted sol–gel synthesis of $LaPO_4{:}Eu^{3+}$, Li^+ nanophosphors as an example [60].

In phosphors, small and spherical particles are desirable because it is easy to process them into devices with high resolution and high screen coverage, intense emission, and long service life. The conventional synthesis method via solid-state reaction conducted at high temperature for $LaPO_4$ normally resulted in irregular particles shape and size. Therefore, over the last few decades various solution-mediated routes, including sol–gel, precipitation, water–oil microemulsion, polyol-mediated process, and hydrothermal method, have been tried to reduce the reaction temperature and obtain high-quality $LaPO_4$-based nanoparticles. However, the success of the efforts on the simple and mass fabrication of $LaPO_4$ nanocrystals with narrow particle size distribution and uniform morphology is still very limited. Among these methods, the sol–gel process is superior to other preparation methods because the well-mixed reactants ensure the homogeneity of the final product. However, the cost and the toxicity of the conventional sol–gel process are high because of the utilization of alkoxide precursors. In the microwave-assisted sol–gel approach that combines the merits of sol–gel and the special heating effect of the microwave irradiation, the fast and mass preparation of uniform, spherical Li^+ and Eu^{3+}, co-doped $LaPO_4$ nanoparticles are achieved [60]. In this method, the use of ethylenediamine tetraacetic acid (EDTA) and citric acid (CA) as double chelating agents ensure the formation of homogeneous, transparent, metal–citrate–EDTA gels, while that of microwave irradiation as a heating source promotes the prompt thermal decomposition of the gels into uniform spherical $LaPO_4$-based nanoparticles. In addition, the general inorganic salts instead of alkoxides are used as the main precursors.

The detailed synthetic procedure can be found in Ref. [60]. Figure 8.4 shows the thermogravimetric/differential scanning calorimetry (TG/DSC) study of the thermal decomposition procedure of the La–P–citrate–EDTA gel precursors [60]. There are three weight loss regions from ambient temperature to 950°C. The first region occurred at 25–200°C (about 4.8%), the second one occurred at 200–400°C (about 29.5%), and the third one occurred at 400–800°C (about 47.6%). Correspondingly, three discrete, phase transformation regions can be observed in the DSC curve centered at around 100, 287, and 691°C. The first weight loss was assigned to the elimination of water. The second weight loss region can be ascribed to the pyrolysis of NO_3^- and organic phases (CA and EDTA) to give an amorphous inorganic phase, which is confirmed by the XRD analysis. Subsequent elimination of the remaining organic materials (carbon and organic compounds) occurs in the temperature ranging from 400 to 500°C and crystallized $LaPO_4$ inorganic phase is simultaneously formed.

The XRD studies of the as-obtained $Li_{0.1}La_{0.9}PO_4$ nanoparticles as a function of microwave heating temperature show that the materials heated at below 500°C are almost amorphous. The main planes of [111], [120], [−112]/[012], and [103]/[−311], referring to the monoclinic structured $LaPO_4$ appears, which indicates the formation of inorganic phase Li–La–P–O at a temperature ranging from 300 to 400°C. For the sample heated at 500°C, all diffraction peaks can be readily indexed to phase-pure $LaPO_4$. However, the wide and weak diffraction peaks suggest the poor crystallinity as well as fine particle size. When the heating temperature was further increased from 600 to 1000°C the intensity of

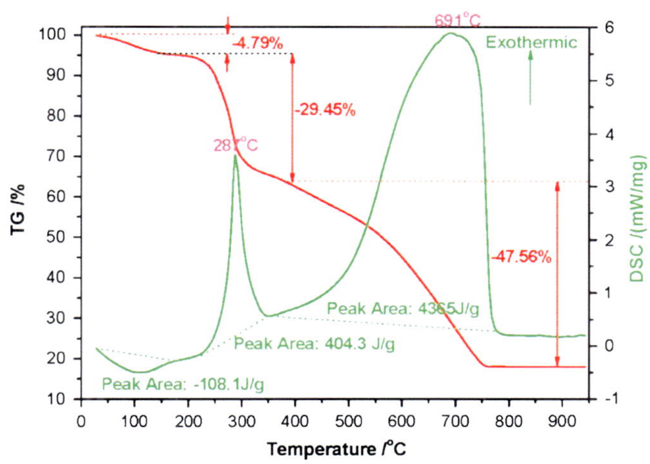

FIGURE 8.4 TG–DSC curves of the as-synthesized metal–citrate–EDTA precursors. *Reprinted with permission from [60]. Copyright 2008 American Chemical Society.*

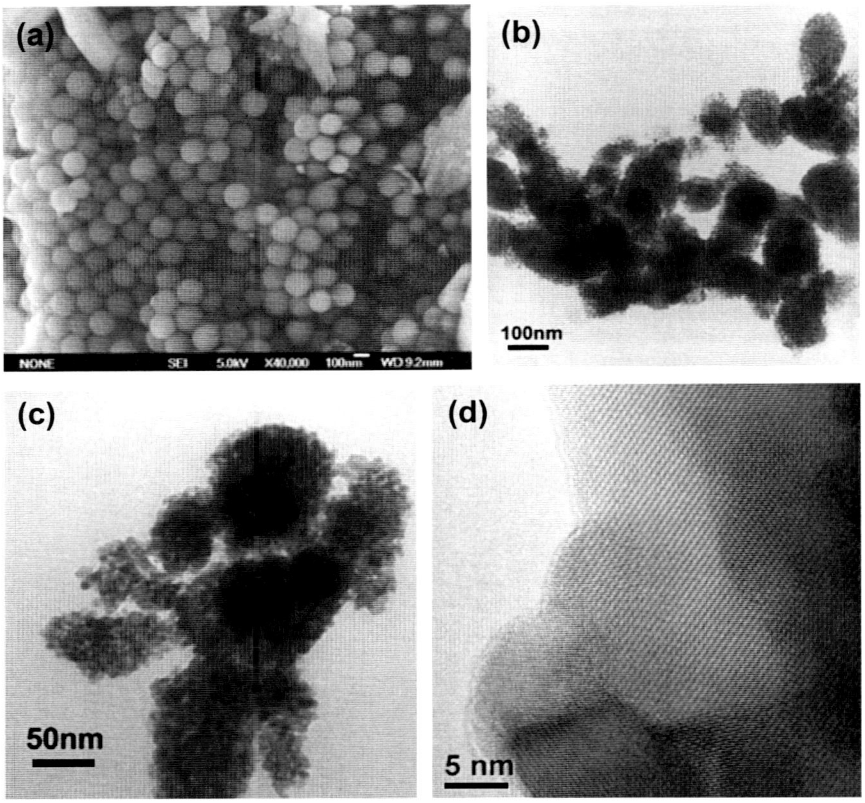

FIGURE 8.5 (a) SEM and TEM images of the as-obtained $Li_{0.05}Eu_{0.05}La_{0.9}PO_4$ nanoparticles under (b) low and (c) high magnification. (d) HRTEM pattern of this sample. *Reprinted with permission from [60]. Copyright 2008 American Chemical Society.*

all diffraction peaks gradually increased and the peak width gradually narrowed.

Figure 8.5(a) shows an scanning electron microscopy (SEM) image of the $Li_{0.05}Eu_{0.05}La_{0.9}PO_4$ sample heated at 800°C for 20 min. The sample is composed of uniform and spherical nanoparticles with a size of about 100 nm. The transmission electron microscopy (TEM) images of this sample are shown in Fig. 8.5(b and c), respectively, which clearly demonstrates that the obtained $Li_{0.05}Eu_{0.05}La_{0.9}PO_4$ particles were spherical and aggregated from smaller nanoparticles. Figure 8.5(d) is an high resolution transmission electron microscopy (HRTEM) image of this sample, which indicates that the obtained $Li_{0.05}Eu_{0.05}La_{0.9}PO_4$ primary nanoparticles are well crystallized and ordered in crystallography.

8.2.1.3. MICROWAVE-ASSISTED CRYSTALLIZATION

Due to the special heating mechanism and the high heating efficiency, the microwave irradiation can significantly affect the crystallization behavior of the materials during the synthesis. Compared to the synthesis conducted under conventional heating, the microwave heating can result in narrower particle size distribution, more uniform shape, and novel morphology [64]. More importantly, the crystallization behavior can be finely tuned by adjusting the synthetic parameters such as the reaction temperature and time and the composition of the synthetic system [65–68]. The influence of the microwave heating on the crystallization of the materials will be discussed with the morphology control of silicalite-1 crystals [66,68] as examples.

Xu and coworkers systematically studied the influence of the properties of solvent on the crystallization behavior of the silicalite-1 (Si-MFI) crystals [65,66,68]. Usually, silicalite-1 (Si-MFI) crystals can be synthesized from the reaction system TEOS–TPAOH-H_2O under microwave heating with a typical coffin-like shape. In their studies, the properties of the solvent were finely tuned by introducing second or third alcohol as cosolvent. First, a series of alcohols including ethylene glycol, methanol, ethanol, 1-propanol, isopropanol, *n*-butanol, and hexanol were used and the molar ratio of the components in the final gel was $1.0SiO_2$:0.357TPAOH:$x$alcohol:4.0EtOH:$1H_2O$, where x is 2, 4, or 8 for each batch. The microwave synthesis was typically conducted in a two-step model. For each step, the synthetic conditions (temperature, duration, maximum power) were defined as (T_1, t_1, P_1) and (T_2, t_2, P_2), respectively. The detailed information on the synthetic parameters and resulting crystal morphologies are summarized in Table 8.4 [68] and the SEM images of the resulting products are shown in Fig. 8.6 [68].

The results show that the polarity of alcohols used as the cosolvent significantly affects the morphology of the resulting Si-MFI crystals. The alcohols with relatively

TABLE 8.4 The Summary of the Synthetic Parameters and Morphologies of the Resulting Si-MFI Crystals

Sample no.	Cosolvent	Dielectric constant	Morphology of product
1	Ethylene	37.0	Isolated single crystals
2	Methanol	32.6	Isolated single crystals
3	Ethanol	24.3	Isolated single crystals
4	1-Propanol	20.1	Stacked + isolated
5	Isopropanol	18.3	Stacked crystals
6	*n*-Butanol	17.8	Stacked crystals
7	Hexanol	13.3	Stacked crystals

Aging time: 24 h; reaction composition (in molar ratio): TPAOH: SiO_2:Alcohol:EtOH:H_2O = 0.357:1:8:4:21; Microwave condition: $t_1 = 90$ min $T_1 = 80°C$, $P_1 = 250$ W; $t_2 = 60$ min, $T_2 = 180°C$, $P_2 = 400$ W.
Reprinted from Microporous and Mesoporous Materials, Vol. 104, Xiaoxin Chen, Wenfu Yan, Wanling Shen, Jihong Yu, Xuejing Cao, Ruren Xu, Morphology control of self-stacked silicalite-1 crystals using microwave-assisted solvothermal synthesis, Page 296, Copyright (2007), with permission from Elsevier [68].

high polarity (dielectric constant) resulted in isolated single crystals of Si-MFI, while alcohols with relative low polarity led to the self-stacked Si-MFI crystals. As shown in Fig. 8.6(a–c), when methanol, ethanol, and ethylene glycol with relative high dielectric constants were used as the cosolvent, the crystal shape is well defined. Further decreasing the dielectric constants of the cosolvent resulted in the loss of well-defined shape of the Si-MFI crystals, as well as the stacking of individual crystals (Fig. 8.6(d–g)). The crystals are stacked on top of one another along their *b* direction to form "fiber-like" self-stacked morphology. The "fiber" cannot be destroyed even by long time and strong ultrasonication which indicates that strong chemical bonds might exist between the individual crystals. In contrast to the microwave-assisted synthesis, the synthesis of all the above-described batches under conventional heating gave only single crystals with hexagonal plate shape. Considering the fact that the self-stacking phenomenon of crystals is related to the condensation of surface Si–OH bonds among individual crystals at the early stage of the synthesis, it is suggested that low polarity (dielectric constant) alcohol solvents might favor the formation of abundant Si–OH groups in the precursor species and on the surface of the nanocrystals formed at the early stage of the microwave-assisted crystallization. The Si–OH groups of individual crystals might undergo further condensation to form self-stacked crystals. Therefore, it is believed that the self-stacking of the Si-MFI crystals under microwave conditions is related to the rapid condensation of abundant surface T–OH groups among individual crystals.

While the diols were used as the cosolvent under microwave heating condition, the reaction system of TEOS–TPAOH–H_2O–Diol with the molar ratio of SiO_2:TPAOH:EtOH:Diol:H_2O = 1.0:0.357:4.0:x:y, where $x = 6–9$, $y = 21.55–45.55$ can produce the Si-MFI crystals with tunable sizes, shapes, and aspect ratios [66]. The diols included the ethylene glycol (EG), diethylene glycol (DEG), triethylene glycol (TEG), and tetraethylene glycol (tEG). The average length (L), width (W), thickness (T), and the aspect ratio (L/W and $L/W/T$) of the Si-MFI crystals are summarized in Table 8.5 [66] and the SEM images of the Si-MFI crystals are shown in Fig. 8.7 [66].

The results in Table 8.5 and Fig. 8.7 show that the well-defined Si-MFI crystals become longer, narrower, and thinner when the ratio of the number of carbon and the hydroxyl of the diols (C/OH), a parameter roughly reflecting the hydrophobic and hydrophilic properties of the molecule of the cosolvent, is gradually increased. In contrast to the microwave-assisted synthesis, the synthesis of all above-described batches under conventional heating gave only typical Si-MFI crystals with hexagonal plate shape and with almost same size and aspect ratios of L/W and $L/W/T$. This indicates that the special heating mechanism of the microwave affects the growth rates of the Si-MFI crystals along different planes. The combination of the microwave heating and the utilization of the diols significantly changed the growth kinetics of the Si-MFI crystals.

8.2.1.4. MICROWAVE-ASSISTED SYNTHESIS OF NANOMATERIALS

Nano is a hot topic in the past decades because nanomaterials, notable for their extremely small feature size, have the potential for wide-ranging industrial, biomedical, and electronic applications [69–78]. Many methods have been developed to synthesize and prepare the nanomaterials. Among these methods, microwave-assisted synthetic method has been proved to be highly efficient on the morphology control and shape selectivity. Nanosized oxide [79], metal [80], sulfide [81], carbide [82], carbon [83], fluoride [84], semiconductor [85,86], and phosphate [87] have been successfully synthesized and prepared by this method. The "specific microwave effect" on the control of nanomaterial growth will be illustrated with the microwave-assisted synthesis of CdSe and CdTe nanocrystals in microwave-nonabsorbing alkanes [86]. The controlled growth of nanomaterials can be achieved by selective heating the chalcogenide precursor. The selectivity offered by microwave heating shows the "specific microwave effect", which is defined as the ability to selectively heat molecular precursors that are highly polarizable in the presence of molecules that are less polarizable. Here, the chalcogenide precursor selectively absorbs the microwave energy, which appears to result in the instantaneous nucleation and growth upon microwave irradiation. The nanocrystals can be

FIGURE 8.6 SEM images of Si-MFI crystals crystallized by using different alcohol cosolvents under microwave radiation conditions: (a) ethylene glycol, (b) methanol, (c), ethanol (d) 1-propanol, (e) isopropanol, (f) *n*-butanol and (g) hexanol. *Reprinted from Microporous and Mesoporous Materials, Vol. 104, Xiaoxin Chen, Wenfu Yan, Wanling Shen, Jihong Yu, Xuejing Cao, Ruren Xu, Morphology control of self-stacked silicalite-1 crystals using microwave-assisted solvothermal synthesis, Page 296, Copyright (2007), with permission from Elsevier [68].*

TABLE 8.5 Influence of the Type of Diols on the L, W, T, L/W Ratio and L/W/T Ratio of Si-MFI Crystals

Diols	EG	DEG	TEG	tEG
Ratio of the carbon and hydroxyl of the diols (C/OH)	1	2	3	4
Length (μm)	1.05	1.23	1.20	1.19
Width (μm)	0.56	0.33	0.27	0.26
Thickness (μm)	0.25	0.16	0.13	0.11
L/W	1.88	3.74	4.43	4.56
L/W/T	7.52	23.38	34.08	41.45

Gel composition (in molar ratio):
SiO_2:TPAOH:EtOH:Diols:H_2O = 1:0.357:4.0:7:21.55.
Reprinted from Microporous and Mesoporous Materials, Vol. 119, Xiaoxin Chen, Wenfu Yan, Xuejing Cao, Jihong Yu, Ruren Xu, Fabrication of silicalite-1 crystals with tunable aspect ratios by microwave-assisted solvothermal synthesis, Page 6, Copyright (2009), with permission from Elsevier [66].

grown rapidly and controlled by a combination of reactant concentration and power, while size is indicated by the reaction temperature. The detailed preparation information of nanocrystalline CdSe and CdTe can be found in Ref. [86].

Since the precursor is only activated under microwave irradiation (the cleavage of the TOP-Se or TOP-Te bond), the accessibility of the chalcogenide monomer to growing nanocrystal is controlled by the microwave. This results in a remarkable controllability over the reaction path which reveals that the precursor is the dominant absorber in the reaction and the solvent acts merely as a reaction moderator to control the explosive growth. Figure 8.8 shows the ability to synthesize CdSe and CdTe nanocrystals in nonpolar, low boiling alkane solvents (decane, octane, heptanes) under microwave heating condition [86].

The isolated nanocrystals are elliptical (aspect ratio 1.2) and exhibit narrow size dispersities (6% rms in CdSe, 12% rms in CdTe) based on TEM analysis (Fig. 8.9). The CdSe crystal motif is wurtzite while the CdTe is like zinc blend as confirmed by their powder XRD patterns.

8.2.2. Solid-state Reactions

Solid-state reactions conducted under convention heating usually involved high temperature and long reaction time due to their special reaction mechanism. However, this kind of reaction can be significantly accelerated under microwave irradiation heating because of its rapid heating, selective coupling, and enhanced reaction kinetics. The key requirement is that one or more of the major constituents of the batch must be a microwave absorber and couple strongly to the microwave field at room temperature. So far, the microwave energy has been used to process many types of materials, including organics, ceramics, polymers, glasses, sol–gels, metals,

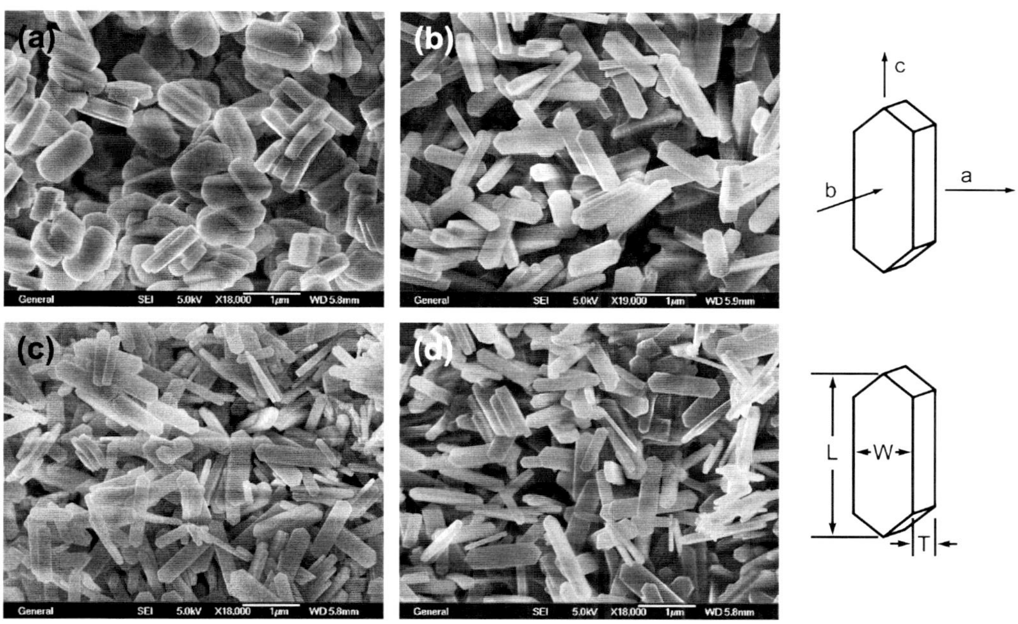

FIGURE 8.7 SEM images of the Si-MFI crystals crystallized from the microwave-assisted solvothermal synthesis system in the presence of the diols: (a) EG, (b) DEG, (c) TEG, and (d) tEG. A schematic identifying the crystal faces is shown on the right part of the figure. *Reprinted from Microporous and Mesoporous Materials, Vol. 119, Xiaoxin Chen, Wenfu Yan, Xuejing Cao, Jihong Yu, Ruren Xu, Fabrication of silicalite-1 crystals with tunable aspect ratios by microwave-assisted solvothermal synthesis, Page 217, Copyright (2009), with permission from Elsevier [66].*

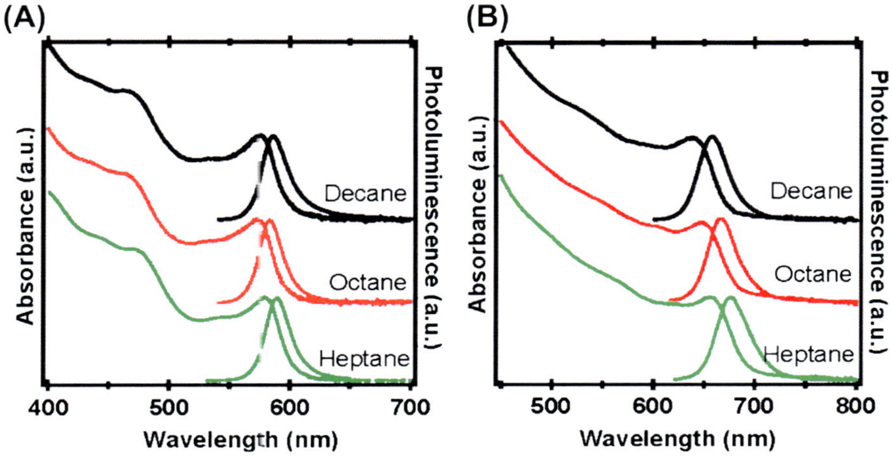

FIGURE 8.8 Optical absorption and photoluminescence of (A) CdSe and (B) CdTe using solvents of different alkyl chain lengths. CdSe/CdTe was grown at 240 °C/220 °C for 30 s/5 s with the power set to 300 W. *Reprinted with permission from [86]. Copyright 2008 American Chemical Society.*

and composites [88]. In the following sections, the solid-state reactions conducted under microwave heating will be illustrated under various categories, viz.: (1) microwave-assisted solid-state synthesis, (2) microwave-assisted processing of ceramics, (3) microwave-assisted processing of glass forming, (4) microwave-assisted SHS synthesis of materials, and (5) microwave-assisted solid–gas synthesis.

8.2.2.1. *Microwave-assisted Solid-state Synthesis*

The simplest, very illustrative, and technologically important microwave-assisted solid-state synthesis is that of β-silicon carbide (β-SiC) [89] which is so far the most widely used nonoxide ceramics for many industrial applications because of its attractive properties at high temperature such as high strength, high hardness, high wear, and thermal shock resistance. The carbothermal reduction of quartz by coke is the main method of β-SiC production. However, this method is an energy-intensive process and involves many steps in synthesizing a powder of pure phase β-SiC. In addition, the particle size of the powder is relatively coarse. Under microwave heating condition, formation of phase-pure β-SiC can be achieved by reacting Si and amorphous C powder mixture in a tubular microwave reactor at 1300°C in less than 5 min, resulting in sub-micron-sized particles. The detailed synthesis process can be found in Ref. [89]. The powder X-ray diffraction (XRD) patterns of the end product of Si and C reaction carried out at different temperatures under a microwave field are compared in Fig. 8.10 [89]. From Fig. 8.10, it can be noted that at 900°C, β-SiC is the major phase coexisting with unreacted Si. With the increase in reaction temperature, the amount of unreacted Si decreases until it completely disappears at 1200°C resulting in phase-pure β-SiC. The powder XRD patterns carried out on product synthesized at 1300°C for different holding times are shown in Fig. 8.11 [89]. At 1200°C, phase-pure β-SiC forms only after 15 min whereas at 1300°C, the phase-pure β-SiC forms in just 5 min.

8.2.2.2. *Microwave-assisted Processing of Ceramics*

Microwave heating can provide cheap and time effective processing routes for ceramic systems.

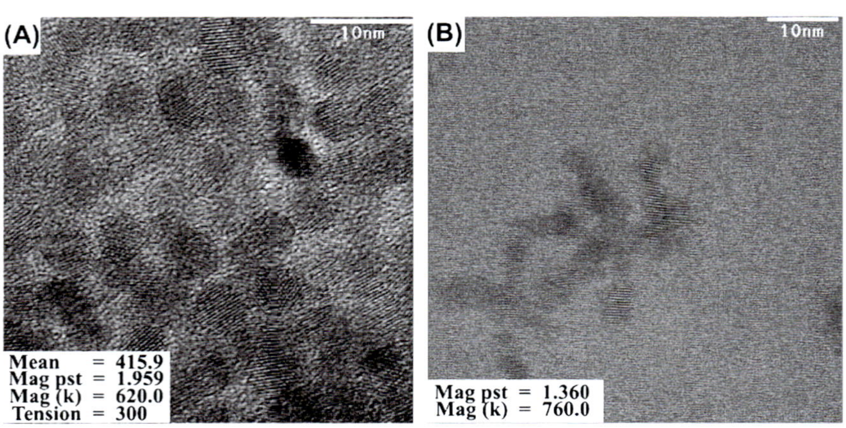

FIGURE 8.9 TEM images of (A) 5.8 nm CdSe and (B) 4.3 nm CdTe nanocrystals. Both images are taken at high magnification on a high resolution (Tietz Tem-Cam F415 slow scan CCD). *Reprinted with permission from [86]. Copyright 2008 American Chemical Society.*

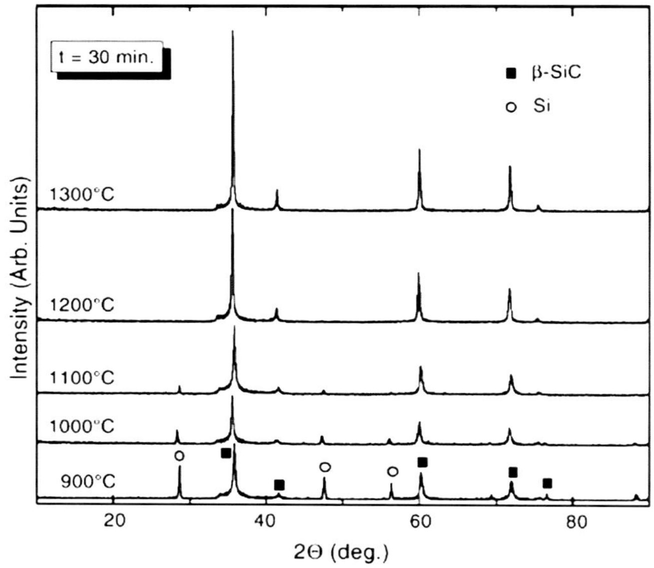

FIGURE 8.10 Powder X-ray diffraction patterns of powders which were synthesized in a microwave furnace at different temperatures and at a constant holding time of 30 min. *Reprinted from Materials Research Bulletin, Vol. 40, L.N. Satapathy, P.D. Ramesh, Dinesh Agrawal, Rustum Roy, Microwave synthesis of phase-pure, fine silicon carbide powder, Page 1871, Copyright (2005), with permission from Elsevier [89].*

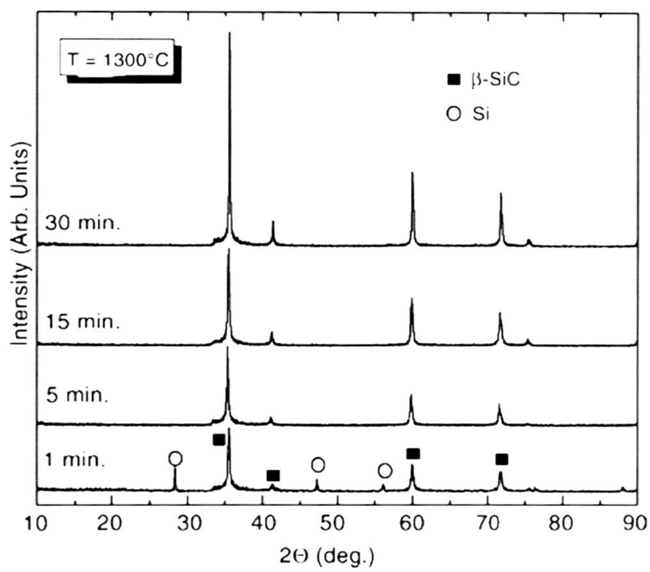

FIGURE 8.11 X-ray diffraction patterns of powders, which were synthesized in a microwave furnace at different holding times and at a constant temperature of 1573 K. *Reprinted from Materials Research Bulletin, Vol. 40, L.N. Satapathy, P.D. Ramesh, Dinesh Agrawal, Rustum Roy, Microwave synthesis of phase-pure, fine silicon carbide powder, Page 1871, Copyright (2005), with permission from Elsevier [89].*

Microwave-assisted processing of ceramics is fast emerging new field in ceramic processing and material synthesis. Under microwave heating, the synthesis of tungsten carbide (WC)-based ceramic composites, fabrication of transparent ceramics, sintering of powdered metals, and the design of a continuous microwave system have been achieved [90]. WC-based ceramics are commonly used for cutting tools and drilling instruments due to their unique combination of properties like hardness, toughness, and strength. Conventional methods for the synthesis of WC with Co as a binder phase involve high temperatures and lengthy synthesis cycles of the order of one day. In a conventional synthesis method, the carbide specimen is subjected to high temperatures (up to 1500°C) for long periods in order to achieve a high degree of sintering. Such conditions unfortunately favor undesirable WC grain growth in the presence of Co liquids. Consequently, the mechanical strength and hardness of the tool are diminished. It is a well-known fact that finer microstructures can provide superior mechanical properties and a longer lifetime of the product. After introducing the microwave heating, it was observed that microwave processed WC/Co bodies exhibited better mechanical properties than the conventionally processed ones. A fine and uniform microstructure (~1 micron size grains) with very little grain growth and nearly full density was achieved without adding any grain-growth inhibitors when sintered at 1250–1320°C for only 10–30 min. Table 8.6 provides a comparison between microwave and conventionally processed WC/Co cements [91].

TABLE 8.6 A Comparison of Microwave and Conventional Processes for the Synthesis of WC-Co Composites

	Microwave	Conventional
Sintering temperature (°C)	1300	1450
Total cycle time	90 min	12–24 h
Sintering time (minutes)	10	60
Density (% TD)	99.8	99.7
Average grain size (μm)	0.6	2
Bending strength (MPa)	1800	1700
Hardness (Rockwell A)	93	91

TD, theoretical density.
Reprinted from Current Opinion in Solid State & Materials Science, Vol. 3, Dinesh K Agrawal, Microwave processing of ceramics, Page 480, Copyright (1998), with permission from Elsevier [91].

In addition to the synthesis of WC/Co ceramic, microwave energy has been used to synthesize important ferroelectric materials and other ceramic powders by adopting the concept of the pre-reduction of phases, such as TiO_2 and Ta_2O_5, to give a highly microwave absorptive precursor material and thereby enhancing the reaction kinetics dramatically. The main idea was to create a defect structure to make microwave coupling more efficient at room temperature. It was reported that

TABLE 8.7 Summary of Main Applications of Microwave Irradiation in Glass Fabrication and Improvement

Process/study	Description/observations
PHYSICAL TREATMENT OF GLASS	
Bending of glass to obtain curved glass	Two MW sources were used; sometimes, conventional heating was used before MW treatment
Lamination of class sealing of glass	A film covered the glass and was subjected to further MW heating
Sealing of glass	This was used, in particular, to seal glasses to aluminum or hard plastics in vehicles. "Ceramic−glass−metal" and "ceramic−glass−ceramic" MW techniques were reported.
Vitrification of hazard wastes, in particular radioactive waste	A composite of a borophosphate or borosilicate glass with wastes was MW-formed
MW-heating studies for various types of glasses	High concentrations of alkali metals contributed to MW-heating glasses. Preheating, films on glasses, and MW melting of glasses were studied. Uniform MW heating was developed.
Glass repair	The glass was covered with a protective sol−gel precursor and MW-heated
Glass layers deposition	Glass layers were MW-deposited on a performed glass
Modification of glass surface	Interchange of K by Na in glasses with the use of microwaves
Studies of color changes	MW devitrification of $Li_2O-ZnO-Al_2O_3-SiO_2$ glass at 700−800 °C led to color changes from yellow to violet
Luminescence studies in glasses	Luminescence in MW-obtained glasses based on SiO_2, doped with Er^{3+} and containing additives of Ge, Al, K, N, F, was studied.
Elimination of bubbles and humidity	MW irradiation allowed their elimination
Superconductivity/ionic conduction	Local superconductivity in MW-obtained glasses was studied. Ionically conducting glass $2AgI-Ag_2O\text{-}2(0.95B_2O_3 \cdot 0.05SiO_2)$ was MW-synthesized.
SINTERIZATION OF VARIOUS TYPES OF GLASSES	
Common alumosilicate and phosphate glasses	A series of glasses (for instance, $15BaO-10PbO-5SrO-70P_2O_5$,$PbO-B_2O_3-ZnO-TiO2$) were MW-drying and polishing techniques for optical glasses were proposed
PREPARATION OF METAL-REINFORCED GLASSES	
Metal-reinforced glasses	borosilicate glasses, reinforced with 10% Mo, W, Al, Ti, Ni, and Fe (2−50 μm in size),were MW-fabricated
FILM DEPOSITION ON GLASS SURFACES	
Metals and nanodiamonds	Elemental Cu, Ag, Au, Sn/C, nanodiamonds were MW-deposited on glasses
Oxides	Various oxides, such as TiO_2,$MgO-Al_2O_3-TiO_2$, ZnO, Si−Al−O−N, and InO, were MW-deposited on glasses
Paints	MW-deposited paints for use in vehicle and architectural glass
SINTERIZATION OF GLASS−CERAMICS	
Obtaining glass−ceramics	MW heating was used to prepare a host of glass−ceramics, based on Li−Al−Si, $SrO-Fe_2O_3-B_2O_3$, $CaO-ZrO_2-SiO_2$, $R_2O-Al_2O_3-B_2O_3-SiO_2$ hydroxyapatite, perlite/glass, Ca−Mg−alumosilicate with SiC

Reprinted with permission from [88]. Copyright 2010 American Chemical Society.

by using pre-reduced TiO_2 and Ta_2O_5 oxides precursor, $BaTiO_3$, PZT ($Pb(Zr_{0.52}Ti_{0.48})O_3$), and BMT ($Ba(Mg_{0.33}Ta_{0.67})O_3$) could be synthesized at astonishingly low temperatures, i.e., between 300°C and 900°C in 5−12 min. Conventional methods for the synthesis of these phases require temperatures in the range of 900−1400°C and several hours. Pure stoichiometric metal oxides, such as Ta_2O_5 and TiO_2, do not couple with microwave energy efficiently unless heated to temperatures where they become dielectrically loss (>1000°C). By partially reducing these phases to oxygen defective states such as Ta_2O_{5-x} and TiO_{2-x}, their ability to absorb microwave energy at lower temperatures is radically enhanced.

8.2.2.3. *Microwave-assisted Processing of Glass Forming*

Microwave heating is a much-faster process (requiring minutes rather than hours), yielding good product quality, in comparison with the prolonged conventional thermal treatment of glass precursors and also provides an extremely easy and automatically temperature-controlled route to change/improve the properties of glasses already fabricated by conventional methods [92]. The only requirement is that at least one component of the charge used for making glass is microwave-active. Preheating the precursors, applying the microwave adsorbing coating, or hydration can significantly enhance the microwave adsorption. In the processing of glass forming, the principal advantages of microwave heating are rapid heating and a capability to affect hot areas of the glass selectively while leaving the cooler areas unaffected. Therefore, control over uniformity and rate of the heating is very difficult. It is still difficult to apply microwave heating widely in the glass industry because glasses will crack if theyare not uniformly heated. To obtain a uniform heating of glass, the uniform irradiation of electromagnetic wave is required. Usually, the glass-forming precursors can be melted in a domestic microwave oven, typically within 5 min and quenched into glasses [93]. Under microwave heating, silica glasses, phosphate glasses, optical glasses, glasses reinforced with metals, glasses containing other elements, coating glasses with inorganic and other films, and glass–ceramics and composites can be synthesized. The main applications of microwave heating in glass production and its improvement are given in Table 8.7 [88].

8.2.2.4. *Microwave-assisted Self-propagating High-temperature Synthesis (SHS)*

Self-propagating high-temperature synthesis (SHS) is a method that can produce materials via exothermic reaction. This combustion process is ignited by point-heating a small part of the prepared sample. The heat should be enough for initial burning of surrounding material which in turn, generates heat that burns the following part of the mixture. With this method it is possible to obtain various products of both inorganic and organic nature with unusual properties, for example, powders, metallic alloys, ceramics with high purity, corrosion-resistance at high-temperature or super-hardnessity. In addition, with the simultaneous application of external pressure, it is possible to make products in highly dense forms as has been demonstrated by investigations with TiC, TiB_2, and SiC [94]. Recently, microwave radiation has been used to ignite the SHS reactions [95]. Initiating combustion synthesis with microwaves typically results in ignition in the center of the body and the combustion wavefront propagating radially outward, which usually produces materials morphology completely different to that resulting from surface ignition, and may lead to a more complete conversion of reactants. This method will be illustrated by the synthesis of barium ferrite (BaF), $BaFe_{12}O_{19}$.

The starting materials are citric acid ($C_6H_8O_7 \cdot H_2O$), barium nitrate, $Ba(NO_3)_2$, and iron nitrate nonanhydrate, $Fe(NO_3)_3 \cdot 9H_2O$. The atomic ratio of Fe/Ba is strictly controlled to be 12 in order to obtain a single-phase $BaFe_{12}O_{19}$. First, the iron nitrates are dissolved into deionized water to get a clear solution, then an aqueous solution of citric acid (2 mol/L) is mixed with the nitrate solution with the ratio of metal ions/citric acid to be 1. The resulting solution is heated at 60°C in a water bath with vigorous stirring followed by the dropwise addition of $NH_3 \cdot H_2O$ until the pH value is stabilized at 7. A viscous gel is formed when water is evaporated and then the gel is treated at 300°C to get brown ashes. After grinded in an agate mortar, the brown ashes are calcined at 950°C in a microwave oven to get the pure barium ferrite (BaF), $BaFe_{12}O_{19}$.

8.2.2.5. *Microwave-assisted Solid–gas Synthesis*

Microwave-assisted reactions between solid (metal powders) and gases have been reported in the synthesis of metal halides, oxyhalides, and nitrides [96,97]. The types of apparatus developed for these syntheses are illustrated in Fig. 8.12 [96]. The fluidized-bed apparatus

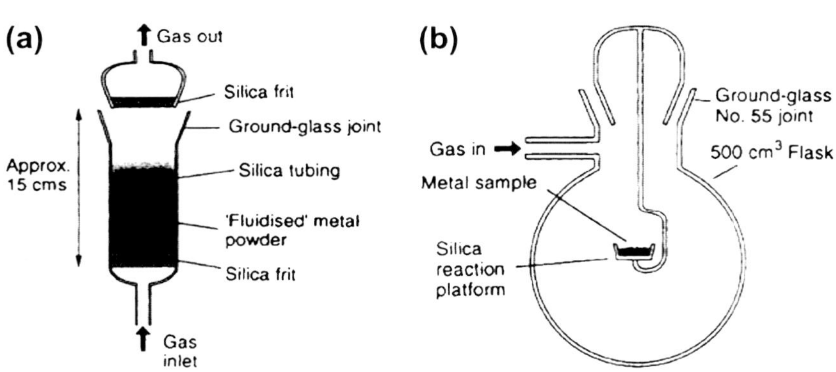

FIGURE 8.12 (a) Fluidized-bed apparatus and (b) suspended-sample apparatus. *Images reproduced with permission from [96].*

TABLE 8.8 Details of the Microwave-assisted Solid–gas Reactions

Compound	Mass/g	Size/mesh	Gas	Method	Characterization
$CrCl_3$	2.08	−100	Cl_2	*a*	XRPD
WCl_6	12.0	−16	Cl_2	*b* + *c*	IR, m.p. *c*, MS
$InCl_3$	3.21	−14	Cl_2	*a*	XRPD, *d*
$TeCl_4$	10.0	−100	Cl_2	*a*	XRPD, *d*
$WOCl_4$[e]	12.0	−16	$Cl_2 + O_2$	*b* + *c*	XRPD, *d*
WO_2Cl_2[e]	12.0	−16	$Cl_2 + O_2$	*b* + *c*	XRPD, *d*
TiN	50.0	−100	NH_3/N_2	*f*	XRPD
Cr_2N	20.0	−100	NH_3	*bf*	XRPD
$TiBr_4$	5.0	−100	Br_2 + Ar	*b* + *c*, *f*	XRPD, IR, MS, *d*
CuBr	8.0	Turnings	Br_2 + Ar	*b*	XRPD, *d*

[a] Suspended sample arrangement as described in the text. [b] A porcelain boat containing the metal is placed in a large Schlenk, containing the gas. Microwave power is applied, and further gas admitted as required or as in the case of Cr_2N, a flow system is created in a su table silica vessel. [c] The air/moisture-sensitive product(s) is then purified by sublimation onto a cold finger in the same Schlenk. [d] Elemental analysis. [e] $WOCl_4$ and WO_2Cl_2 were formed in the same reaction and separated by sublimation. [f] Fluidized-bed reactor

Images reproduced with permission from [96].

(a) is made wholly of silica. The metal powder is placed above the silica frit, and the reactive gas allowed to flow in a vertical direction. In the suspended-sample apparatus (b) the metal sample is placed on the reaction platform and the flask evacuated before filling with the reactive gas. The types of reaction conducted with this method, the quantities of metal used, the metal mesh size, and the techniques used to characterize the products are summarized in Table 8.8 [96].

8.3. SYNTHESIS OF INORGANIC MATERIALS ASSISTED WITH DIFFERENT MICROWAVE FREQUENCIES

Microwave heating is affected by mainly two factors. One is permittivity (dielectric constant) (ε') which is the extent to which the electric field is able to produce a polarized response in a molecule or assembly of molecules. The other is dielectric loss (ε'') which is indicative of the ability of a medium to convert dielectric energy to heat. The dependence of the heating rate is given by $\varepsilon''/\varepsilon'$, defined as loss tangent ($\tan\delta$). Dielectric loss tangent ($\tan\delta$) depends on the temperature, composition and physical state of the reactants, and the frequency of the electromagnetic waves. In the variable frequency microwave heating, a selected bandwidth is swept around a central frequency in a specified time which keeps the microwave energy focused at any given location for not more than a fraction of a second. Therefore, variable frequency microwave heating results in time-averaged heating. In the presence of standing waves of electric fields during fixed frequency microwave operation, arcing occurs from a charge build-up in conductive materials. Arcing problems and localized heating are eliminated with variable frequency microwave heating. Thus, the application of different microwave frequencies during the synthesis process could affect the properties of materials.

Nyutu et al. [98] have done a lot of research on the variable frequency microwave-assisted synthesis of nanocrystalline tetragonal barium titanate and cryptomelane-type manganese oxide [99]. The results of study, on the synthesis of nanocrystalline tetragonal barium titanate, are given in Table 8.9 [98] and the SEM images of samples prepared at different microwave frequencies for 2 h are shown in Fig. 8.13 [98]. Their results show that the particle sizes, morphologies, and surface areas of the products are influenced by the microwave frequency and bandwidth sweep time. High microwave frequency (5.5 GHz) and variable frequency (3–5.5 GHz to 1 s) led to spherical particles with narrow and more uniform particle size distributions. $BaTiO_3$ were prepared using the standard 2.45 GHz yielded particles with a cubic microstructure. They speculated that the dependence of crystallization behavior of barium titanate on microwave frequency could be due to different transverse magnetic modes at different frequencies.

8.4. PLASMA-ASSISTED SYNTHESIS OF INORGANIC MATERIALS

Plasma is a gaseous state of matter containing electrically charged particles where the sum of these charges is zero. The charged particles can be electrons, ionized atoms or molecules, and nanoparticles. Plasmas show electric conductivity because the charge carriers are mobile. In the plasma process, the degree of ionization is important and the ratio of charged particles over the neutral ones may be quite small. A plasma-containing particle is called "dusty plasma." The plasma can be classified as equilibrium and nonequilibrium plasmas. In equilibrium plasma, the energy of all charged particles, often called temperature, is equal; whereas in nonequilibrium plasma, electrons have the highest, the heavier ions a lower, and electrically neutral particles have the least energy.

There is a huge variety of plasma processes for the synthesis of inorganic materials, especially nanosized powders [100]. Plasma processes are a special variety of gas phase processes with a bundle of essential advantages in the synthesis of nanosized powders such as the high efficiency with respect to energy consumption and extreme narrow particle size distributions because particle formation in plasma synthesis is, in contrast to

TABLE 8.9 Summary of the Experimental Conditions and the Results[a]

Sample	Microwave frequency	Time (h)	Phases detected by XRD	Phases detected by Raman
1	2.45	2	C, BC	Td
2	4.00	2	C, BC	Td
3	5.50	2	C	Td
4	3–5.5, sweep 1 s	2	C	Td
5	2.45	10	C	Td
6	4.00	10	C, BC	Td
7	5.50	10	C	Td
8	3–5.5, sweep 1 s	10	C	Td
9	2.45	20	C	Td
10	4.00	20	C, BC	Td, Hex
11	5.50	20	C	Td
12	3–5.5, sweep 1 s	20	C	Td
13	3–5.5, sweep 5 s	20	C, BC	Td, Hex
14	CH	20	C	Td
15	5.50	40	C, BC	Td, Hex
16	3–5.5, sweep 1 s	40	C, BC	Td
17	CH	40	C, BC	Td

[a]*C, cubic-$BaTiO_3$* (Pm3m *symmetry); Td, tetragonal $BaTiO_3$* (P4mm *symmetry); Hex, hexagonal BaTiO3* (P6₃mmc *symmetry); BC, trace barium carbonate. All experiments were performed at 170 °C and total volume, 40 cm^3 in a sealed autoclave of the same size and volume.*
Reprinted with permission from [98]. Copyright 2008 American Chemical Society.

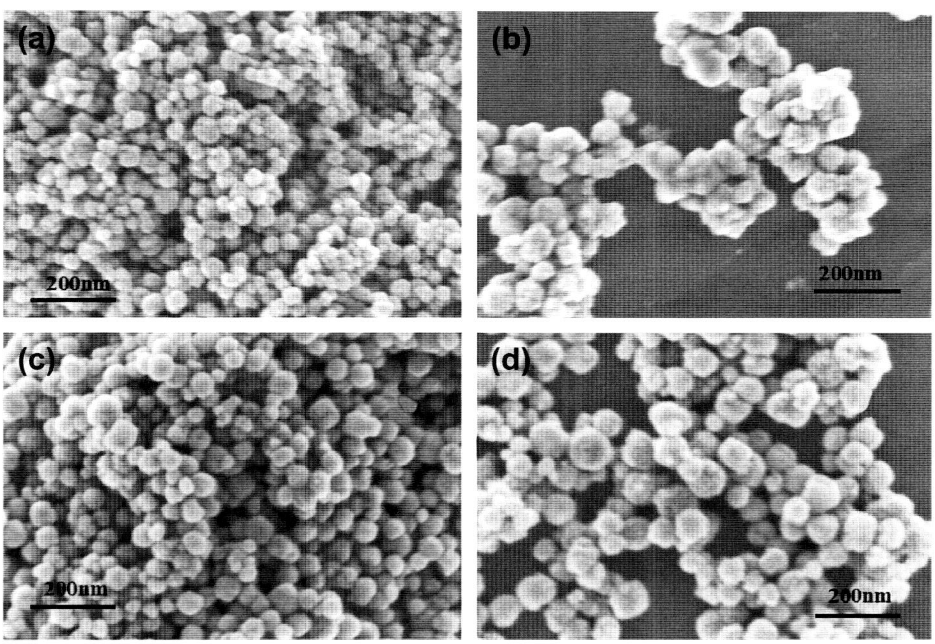

FIGURE 8.13 SEM images of samples prepared at different microwave frequencies for 2 h: (a) 2.45 GHz; (b) 4.0 GHz; (c) 5.5 GHz; and (d) 3–5.5 GHz, 1 s sweep time. *Reprinted with permission from [98]. Copyright 2008 American Chemical Society.*

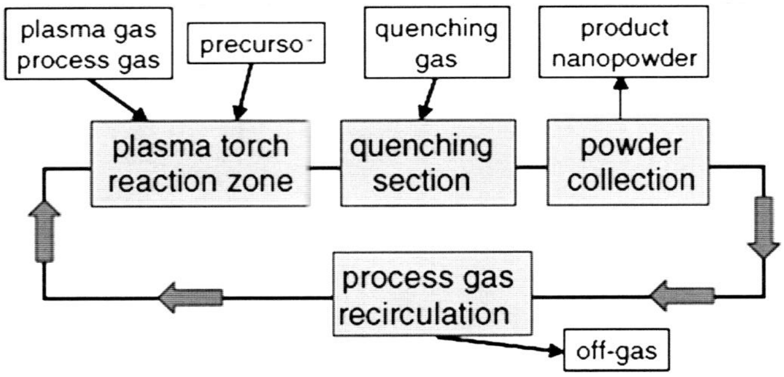

FIGURE 8.14 General layout of a powder production system based on a plasma system. With kind permission from Springer Science+Business Media: Journal of Nanoparticle Research, Plasma synthesis of nanopowders, volume 10, 2008, page 39, Dieter Vollath, figure number 1 [100].

conventional gas phase synthesis, not necessarily a random process.

Figure 8.14 shows the general layout of a system for plasma synthesis [100]. The devices or parameters for plasma-assisted synthesis include the temperature, gas pressure, frequency, or the existence of electrodes or not. Plasma-assisted synthesis process can be divided into two processes: (i) high-temperature processes, where the temperatures are significantly above 1000 K and (ii) low temperature processes, where the temperatures are below 1000 K. In most cases, high-temperature processes are related to higher gas pressures as compared to low temperature processes working at reduced gas pressure. Except for a few very special designs, the temperature processes do not use electrodes.

Figure 8.15 shows the basic designs of plasma burners for the synthesis of nanosized powder using electrodes [100]. The designs differ in the supply for the precursor. Both designs allow the use of liquids or powders as precursor. In both designs, plasma is burning between two co-axial electrodes. In the concentric space between the electrodes, process and carrier gases are supplied to the system. The gas stream blows the plasma out of the electrode system forming a plasma torch and cools the electrodes. The precursor is administered to the system either axially in a center bore of the inner electrode or outside of the electrode system perpendicularly to the plasma torch. The precursor may be solid, liquid, or gaseous. Liquid precursors are either suspensions of aqueous or organic solutions.

In the synthesis of nanosized powders, particles move randomly in the plasma. In addition, they have a drift movement in the direction of the gas stream. Therefore, there is a high probability for collision of the particles in the plasma which causes the formation of clusters leading to broad particle size distributions. This adverse effect can be reduced by a well-designed

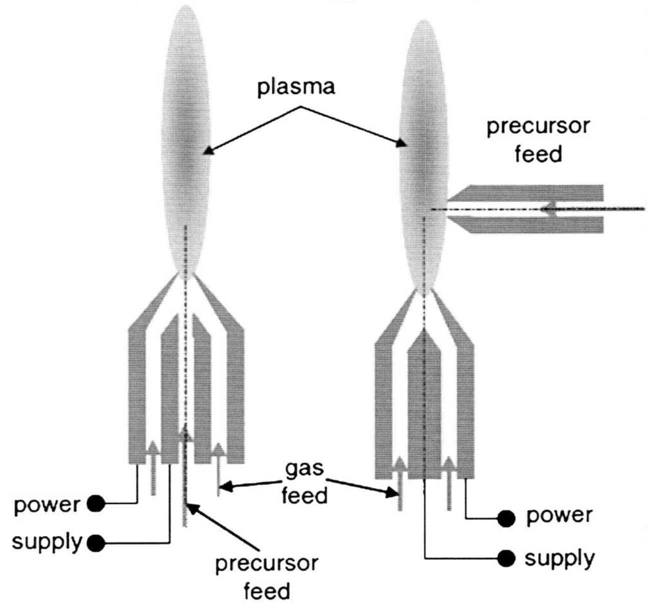

FIGURE 8.15 Basic designs of plasma burners for the synthesis of nanosized particles using electrodes. With kind permission from Springer Science+Business Media: Journal of Nanoparticle Research, Plasma synthesis of nanopowders, volume 10, 2008, page 39, Dieter Vollath, figure number 1 [100].

efficient quenching step directly after the plasma zone. Quenching gas is introduced either radially or axially against the flow direction into the system. However, it has been pointed out that quenching is necessary only in high-temperature processes; low temperature processes do not need quenching.

Figure 8.16 shows the GeO powder synthesized in a Radio Frequency plasma torch [100]. The SEM images show that the width of the particle size distribution can be significantly reduced by increasing the flow of the quenching gas (from 30–500 nm to 30–100 nm).

8.5. SOME OF THE BASIC CONCLUSIONS AND OUTLOOKS ABOUT MICROWAVE RADIATION ASSOCIATED CHEMISTRY

Effects on reaction rate

Microwave heating can be of great utility in the synthesis of inorganic materials. The most apparent advantages are the significant shortening of the reaction time which is found in virtually every case and the rapid heating rate and/or "superheating" which may change the reaction mechanism. Some of the concluding assessment is necessary on the topic of microwave effects in microwave-assisted syntheses. Some of earlier works on the synthesis of inorganic materials from liquid media may be considered ambiguous because of uncertainties in reaction

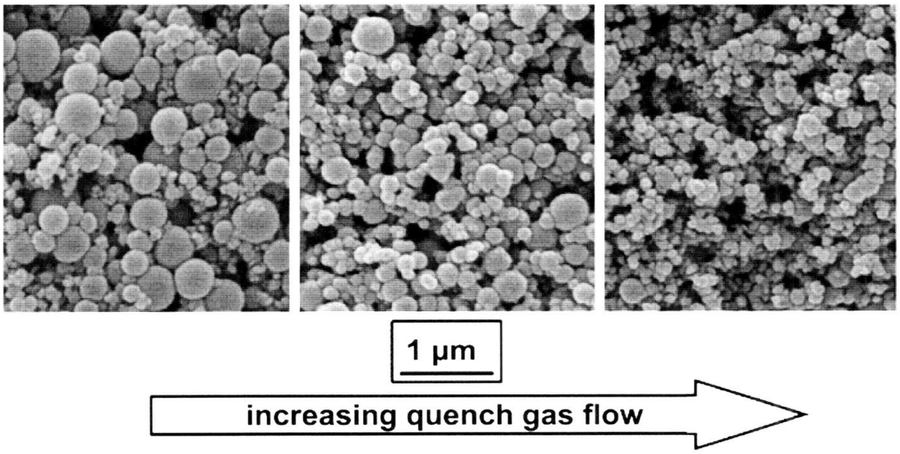

FIGURE 8.16 GeO powder synthesized in a Radio Frequency plasma torch. With kind permission from Springer Science+Business Media: Journal of Nanoparticle Research, Plasma synthesis of nanopowders, volume 10, year of publication 2008, page 39, Dieter Vollath, figure number 1 [100].

temperature (estimated or not measured at all) and in possible presence of temperature gradients (e.g., bulk vs. surface). In such cases where there have been no kinetic measurements possible, the actual reaction times quoted may be misleading. However, a major additional factor to be considered is that some of the inorganic material synthesis reactions are never homogeneous. The reaction mixture is a co-existed medium of microscopic gel, colloidal materials, and crystals. Based on such reaction system, one of the common acceptable reaction modes is that the materials format through thermal lag, the induction period before crystalline product is detected and the crystal growth is initiated. Thermal lag is always present in conventional syntheses although its extent depends on the reactor design. The effect increases with the size of the reactors under conventional heating environment and it is difficult to estimate if there is no internal measurement of temperature. Under microwave heating environment, this effect is reduced because of the extremely rapid rise to the operating temperature. Leaving aside all consideration of crystal growth at this stage, the induction period can be made up of two stages: (A) digestion and equilibration of the reactants and (B) the nucleation period. It is often a relatively long period in an "unseeded" synthesis system. According to many of experimental evidence, it is very likely that the reactant digestion step is much faster under microwave than under conventional heating condition. However, the reduction in overall reaction rate between comparable thermal and microwave-heated zeolite syntheses can be attributed to differences in the thermal lag of the system, the observed induction period, and pattern of crystal nucleation and growth.

In addition to the synthesis of inorganic materials from the liquid media, the microwave heating can significantly accelerate the solid−solid and solid−gas reactions known as solid-state synthesis, processing of ceramics and glass forming, and self-propagating high-temperature synthesis (SHS) of materials.

Selectivity or preference

Many research evidences suggest that microwave irradiation not only accelerate many of the reactions but also show the microwave special effects on product preferences and selectivity because of the different sensitivities to microwave energy of the reaction components. The differences of polarity and dielectric properties of reaction components, the nature of interfaces (particular heterogeneous reactions) and the presence of ions or conducting particles may fundamentally affect the final products resulting in higher yield, preferable product selectivity and even new products and morphologies which cannot be obtained under conventional heating condition.

Continuous processing

There is no doubt that the technique of microwave heating is becoming increasingly important not only in synthesis chemistry but also in chemical processes and environment. The very rapid synthesis rate may make many batch production processes to continue product process leading to evolutionary changes for some of the chemical productions. The examples of laboratory process in the synthesis of microporous crystalline materials using a continuous tube reactor under microwave heating can be found in Ref. [101−103].

"E"s processing

It is well accepted that the "sustainability" is the ultimate common goal for the chemical industry. Therefore, the chemical manufacturing processes should be *E*nvironmentally benign (clean), *E*conomic (minimum cost), *E*fficient (fast), and *E*nergy saving (low carbon) which are denoted as "*E*"s processing. Microwave-assisted synthesis is generally much faster (efficient and low carbon), cleaner (special phase selectivity or preference), and more economical than those conducted with conventional methods which provide us a powerful synthetic route in designing the future chemical processes.

References

[1] R. Gedye, F. Smith, K. Westaway, H. Ali, L. Baldisera, L. Laberge, et al., Tetrahedron Lett. 27 (1986) 279.
[2] H.M. Kingston, S.J. Haswell, Microwave-Enhanced Chemistry, ACS, Washington, DC, 1997.
[3] G.A. Tompsett, W.C. Conner, K.S. Yngvesson, ChemPhysChem 7 (2006) 296.
[4] A. Tavolaro, E. Drioli, Adv. Mater. 11 (1999) 975.
[5] J.P. Zhao, C. Cundy, J. Dwyer, in: H. Chon, S.K. Ihm, Y.S. Uh (Eds.), Progress in Zeolite and Microporous Materials, PTS A-C, vol. 105, Elsevier Science Publ B V, Amsterdam, 1997, p. 181.
[6] S. Caddick, R. Fitzmaurice, Tetrahedron 65 (2009) 3325.
[7] D. Dallinger, C.O. Kappe, Chem. Rev. 107 (2007) 2563.
[8] V. Santagada, F. Frecentese, E. Perissutti, F. Fiorino, B. Severino, G. Caliendo, Mini-Rev. Med. Chem. 9 (2009) 340.
[9] A. de la Hoz, A. Diaz-Ortis, A. Moreno, F. Langa, Eur. J. Org. Chem. (2000) 3659.
[10] N. Elander, J.R. Jones, S.Y. Lu, S. Stone-Elander, Chem. Soc. Rev. 29 (2000) 239.
[11] F. Langa, P. de la Cruz, E. Espildora, J.J. Garcia, M.C. Perez, A. de la Hoz, Carbon 38 (2000) 1641.
[12] L. Zong, S. Zhou, N. Sgriccia, M.C. Hawley, L.C. Kempel, J. Microw. Power Electromagn. Energ. 38 (2003) 49.
[13] S.K. Das, Synlett (2004) 915.
[14] M. Larhed, C. Moberg, A. Hallberg, Acc. Chem. Res. 35 (2002) 717.
[15] H. Will, P. Scholz, B. Ondruschka, CHEM-ING-TECH 74 (2002) 1057.
[16] C.O. Kappe, Curr. Opin. Chem. Biol. 6 (2002) 314.
[17] H.E. Blackwell, Org. Biomol. Chem. 1 (2003) 1251.
[18] R.S. Varma, Green Chem. 1 (1999) 43.
[19] P.D. Ramesh, B. Vaidhyanathan, M. Ganguli, K.J. Rao, J. Mater. Res. 9 (1994) 3025.
[20] B. Vaidhyanathan, K.J. Rao, Chem. Mat. 9 (1997) 1196.
[21] D.R. Baghurst, A.M. Chippindale, D.M.P. Mingos, Nature 332 (1988) 311.
[22] J.A. Gerbec, D. Magana, A. Washington, G.F. Strouse, J. Am. Chem. Soc. 127 (2005) 15791.
[23] W.C. Conner, G. Tompsett, K.H. Lee, K.S. Yngvesson, J. Phys. Chem. B 108 (2004) 13913.
[24] B. Panzarella, G. Tompsett, W.C. Conner, K. Jones, ChemPhysChem 8 (2007) 357.
[25] K.J. Rao, B. Vaidhyanathan, M. Ganguli, P.A. Ramakrishnan, Chem. Mat. 11 (1999) 882.
[26] Y.S. Li, W.S. Yang, J. Membr. Sci. 316 (2008) 3.
[27] W.C. Conner, G.A. Tompsett, J. Phys. Chem. B 112 (2008) 2110.
[28] CEM, Basic Microwave Theory – User Guide.
[29] P.M. Slangen, J.C. Jansen, H. vanBekkum, Microporous Mater. 9 (1997) 259.
[30] J.P. Zhao, C.S. Cundy, R.J. Plaisted, J. Dwyer, Proc. 12th Int. Zeolite Conf. (1999) 1591.
[31] S.H. Feng, R.R. Xu, Acc. Chem. Res. 34 (2001) 239.
[32] H. Katsuki, S. Furuta, S. Komarneni, J. Porous Mat. 8 (2001) 5.
[33] C. Stenzel, M. Brinkmann, J. Muller, R. Schertlen, Y. Venot, W. Wiesbeck, J. Microw. Power Electromagn. Energy 36 (2001) 155.
[34] O.H.R. Scharf, W. Schwieger, Deutsche Zeolith-Tagung (2002) 6.
[35] I. Girnus, K. Jancke, R. Vetter, J. Richtermendau, J. Caro, Zeolites 15 (1995) 33.
[36] M.A. Uguina, D.P. Serrano, R. Sanz, E. Castillo, Proc. 12th Int. Zeolite Conf. (1999) 1917.
[37] X.H. Xu, W.H. Yang, J. Liu, L.W. Lin, Sep. Purif. Technol. 25 (2001) 241.
[38] J.C. Jansen, A. Arafat, H. Vanbekkum, Abstr. Pap. Am. Chem. Soc. 202 (1991) 80.
[39] R. von Ballmoos, J.B. Higgins, Zeolites 10 (1990) 1.
[40] H.B. Du, M. Fang, W.G. Xu, X.P. Meng, W.Q. Pang, J. Mater. Chem. 7 (1997) 551.
[41] C.T. Kresge, M.E. Leonowicz, W.J. Roth, J.C. Vartuli, US Patent 5,098,684 (1992).
[42] A.R. Badiei, L. Bonneviot, Inorg. Chem. 37 (1998) 4142.
[43] C.G. Wu, T. Bein, Chem. Commun. (1996) 925.
[44] D.S. Kim, J.S. Chang, W.Y. Kim, H.Y. Kim, S.E. Park, Bull. Kor. Chem. Soc. 20 (1999) 408.
[45] I. Girnus, M.M. Pohl, J. Richtermendau, M. Schneider, M. Noack, D. Venzke, et al., Adv. Mater. 7 (1995) 711.
[46] T.G. Tsai, H.C. Shih, S.J. Liao, K.J. Chao, Microporous Mesoporous Mat. 22 (1998) 333.
[47] Y. Han, H. Ma, S.L. Qiu, F.S. Xiao, Microporous Mesoporous Mat. 30 (1999) 321.
[48] D. Baek, U.Y. Hwang, K.S. Lee, Y. Shul, K.K. Koo, J. Ind. Eng. Chem. 7 (2001) 241.
[49] L.M. Huang, Z.B. Wang, H.T. Wang, J.Y. Sun, Q.H. Li, D.Y. Zhao, et al., Microporous Mesoporous Mat. 48 (2001) 73.
[50] A. Julbe, J. Motuzas, F. Cazevielle, G. Volle, C. Guizard, Sep. Purif. Technol. 32 (2003) 139.
[51] X.C. Xu, Y. Bao, C.S. Song, W.S. Yang, J. Liu, L.W. Lin, Microporous Mesoporous Mat. 75 (2004) 173.
[52] K. Weh, M. Noack, I. Sieber, J. Caro, Microporous Mesoporous Mat. 54 (2002) 27.
[53] D. Coutinho, J.A. Losilla, K.J. Balkus, Microporous Mesoporous Mat. 90 (2006) 229.
[54] C. Jeffrey Brinker, G.W. Scherer, Sol-Gel Science, Academic Press, Inc., New York, 1990.
[55] F. Wu, L. Wang, C. Wu, Y. Bai, F. Wang, Mater. Chem. Phys. 115 (2009) 707.
[56] K. Suryakala, K.R. Marikkannu, G.P. Kalaignan, T. Vasudevan, J. Solid State Electrochem. 11 (2007) 1671.
[57] A. Abbaspour, A. Ghaffarinejad, Anal. Chem. 81 (2009) 3660.
[58] H.M. Yang, X.C. Zhang, Q.F. Tao, A.D. Tang, J. Optoelectron. Adv. Mater. 9 (2007) 2493.
[59] A.R. Phani, M. Passacantando, S. Santucci, J. Phys. Chem. Solids 68 (2007) 317.
[60] W. Li, J. Lee, J. Phys. Chem. C 112 (2008) 11679.
[61] K. Kuraoka, A. Hashimoto, J. Ceram. Soc. Jpn. 116 (2008) 832.
[62] D. Dambournet, G. Eltanamy, A. Vimont, J.C. Lavalley, J.M. Goupil, A. Demourgues, et al., Chem. Eur. J. 14 (2008) 6205.
[63] G. Fetter, P. Bosch, T. Lopez, J. Sol-Gel Sci. Technol. 23 (2002) 199.
[64] A.S. Vanetsev, Y.D. Tretyakov, Russ. Chem. Rev. 76 (2007) 397.
[65] X.X. Chen, W.F. Yan, X.J. Cao, R.R. Xu, Microporous Mesoporous Mat. 131 (2010) 45.
[66] X.X. Chen, W.F. Yan, X.J. Cao, J.H. Yu, R.R. Xu, Microporous Mesoporous Mat. 119 (2009) 217.
[67] X.L. Hu, J.C. Yu, Adv. Funct. Mater. 18 (2008) 880.
[68] X.X. Chen, W.F. Yan, W.L. Shen, J.H. Yu, X.J. Cao, R.R. Xu, Microporous Mesoporous Mat. 104 (2007) 296.

[69] Y. Xia, Y.J. Xiong, B. Lim, S.E. Skrabalak, Angew. Chem. Int. Ed. 48 (2009) 60.
[70] J.H. Gao, H.W. Gu, B. Xu, Acc. Chem. Res. 42 (2009) 1097.
[71] S. Laurent, D. Forge, M. Port, A. Roch, C. Robic, L.V. Elst, et al., Chem. Rev. 108 (2008) 2064.
[72] S.G. Kwon, T. Hyeon, Acc. Chem. Res. 41 (2008) 1696.
[73] J. Park, J. Joo, S.G. Kwon, Y. Jang, T. Hyeon, Angew. Chem. Int. Ed. 46 (2007) 4630.
[74] X. Chen, S.S. Mao, Chem. Rev. 107 (2007) 2891.
[75] J.P. Wilcoxon, B.L. Abrams, Chem. Soc. Rev. 35 (2006) 1162.
[76] P.D. Cozzoli, T. Pellegrino, L. Manna, Chem. Soc. Rev. 35 (2006) 1195.
[77] Y. Yin, A.P. Alivisatos, Nature 437 (2005) 664.
[78] E. Katz, I. Willner, Angew. Chem. Int. Ed. 43 (2004) 6042.
[79] H.F. Zhou, R. Yi, J.H. Li, Y. Su, X.H. Liu, Solid State Sci. 12 (2010) 99.
[80] Y.C. Yu, Y.X. Zhao, T. Huang, H.F. Liu, Mater. Res. Bull. 45 (2010) 159.
[81] T. Thongtem, A. Phuruangrat, S. Thongtem, Mater. Lett. 64 (2010) 136.
[82] K. Essaki, E.J. Rees, G.T. Burstein, J. Am. Ceram. Soc. 93 (2010) 692.
[83] H. Zhu, X.L. Wang, Y.L. Li, Z.J. Wang, F. Yang, X.R. Yang, Chem. Commun. (2009) 5118.
[84] C. Xu, H. Luo, W.D. Liu, T.K. Ying, Ceram. Int. 35 (2009) 917.
[85] G.D. Wei, W.P. Qin, W. Han, W.Y. Yang, F.M. Gao, G.Z. Jing, et al., J. Phys. Chem. C 113 (2009) 19432.
[86] A.L. Washington, G.F. Strouse, J. Am. Chem. Soc. 130 (2008) 8916.
[87] E.P. Ng, L. Delmotte, S. Mintova, Green Chem. 10 (2008) 1043.
[88] O.V. Kharissova, B.I. Kharisov, J.J.R. Valdes, Ind. Eng. Chem. Res. 49 (2010) 1457.
[89] L.N. Satapathy, P.D. Ramesh, D. Agrawala, R. Roy, Mater. Res. Bull. 40 (2005) 1871.
[90] H.S. Shulman, M.L. Fall, P. Strickland, Am. Ceram. Soc. Bull. 87 (2008) 34.
[91] D.K. Agrawal, Curr. Opin. Solid State Mat. Sci. 3 (1998) 480.
[92] D.V. Louzguine-Luzgin, G.Q. Xie, S. Li, A. Inoue, N. Yoshikawa, K. Mashiko, et al., J. Alloy. Compd. 483 (2009) 78.
[93] B. Vaidhyanathan, M. Ganguli, K.J. Rao, J. Solid State Chem. 113 (1994) 448.
[94] J.H. Peng, J. Binner, S. Bradshaw, Mater. Sci. Technol. 18 (2002) 1419.
[95] R.C. Dalton, I. Ahmad, D.E. Clark, Ceram. Eng. Sci. Proc. 11 (1990) 1729.
[96] A.G. Whittaker, D.M.P. Mingos, J. Chem. Soc. Dalton Trans. (1993) 2541.
[97] P.D. Ramesh, K.J. Rao, Adv. Mater. 7 (1995) 177.
[98] E.K. Nyutu, C.H. Chen, P.K. Dutta, S.L. Suib, J. Phys. Chem. C 112 (2008) 9659.
[99] K.A. Malinger, Y.S. Ding, S. Sithambaram, L. Espinal, S. Gomez, S.L. Suib, J. Catal. 239 (2006) 290.
[100] D. Vollath, J. Nanopart. Res. 10 (2008) 39.
[101] I. Braun, G. Schulz-Ekloff, D. Wohrle, W. Lautenschlager, Microporous Mesoporous Mat. 23 (1998) 79.
[102] S.E. Park, D.S. Kim, J.S. Chang, J.M. Kim, U.S. Pat Appl. Publ., US#2001054549 (2001).
[103] D.S. Kim, J.M. Kim, J.S. Chang, S.E. Park, Stud. Surf. Sci. Catal. 135 (2001) 573.

CHAPTER

9

Syntheses of Coordination Compounds

Changsheng Lu, Qingjin Meng
Nanjing University, China

Based on the classical Werner's theory of coordination compounds, various kinds of novel coordination compounds have been prepared and synthesized, such as metal clusters, macrocyclic compounds, π-coordination compounds, molecular nitrogen coordination compounds, organometallic compounds, metallocenes, and their superstructures. With the progress in host–guest chemistry and supramolecular chemistry, coordination compounds are becoming more and more comprehensive. It also covers the area of those compounds which are formed via intermolecular weak interactions, for instance second-sphere compounds, inclusion compounds, and clathrate compounds which consist of transition-metal-compound cores. The aqueous solution chemistry was boosted with the developments of classical coordination compounds. Although these compounds (usually contain ligands like water, amine, hydroxyl, fluorine, and chlorine) have been well studied, the novel ones that are based on nonaqueous solution chemistry are still as attractive as those prepared by the solid-state chemistry methodology.

This chapter intends to present and discuss some synthetic strategies of coordination compounds, covering the following sections as direct synthesis, components exchange method, redox interaction method, solid-state synthesis, inclusion complexation, and macrocyclic template method. It doesn't include some most recent methodologies as they have been organized elsewhere, such as hydrothermal or solvent–thermal syntheses, and metal-organic frameworks (MOFs). The reader can easily catch up with these latest progresses by referencing the special chapters in this book.

9.1. DIRECT SYNTHESIS METHOD

Direct synthesis is defined as one-pot reaction of the ligands and the metals or metal compounds in situ. It's divided into several branches, for example direct coordination reaction, precursor-induced complexation, metal vapor synthesis, and matrix isolation method [1].

9.1.1. Direct Coordination Reaction

In direct synthesis, the most frequently used starting metal compounds are oxides and hydroxides halides, acetates, sulfates, and so on. The ideal compound should readily react with ligands and consequently be easy to remove from the products. In the meantime, the solvent also plays a very important role in the synthesis. Therefore, it should not only dissolve the starting material but also avoid decomposition by hydrolysis or alcoholysis [2–8].

Water is a very important solvent. A lot of coordination reactions are carried out in aqueous solution where the ligands vary from halides and amines to acetonitrile and acetyl acetone. For example, potassium bioxalate copper(II) was synthesized in aqueous solution by simple mixing of potassium oxalate with copper sulfate [5].

$$CuSO_4 + 2K_2C_2O_4 \rightarrow [K_2Cu(C_2O_4)_2] + K_2SO_4$$

The acidity of the solution plays a very vital role in the isolation of final products and their yields. For example in the synthesis of $[Cr(C_5H_7O_2)_3]$, the final product is hard to be isolated from the starting material (chromic chloride and acetyl acetone) as it is water soluble. If the acidity of the mixture is finely controlled by ammonia, which could be released slowly upon the hydrolysis of urea in water, $[Cr(C_5H_7O_2)_3]$ will quantitatively yield in crystalline product [9].

$$CO(NH_2)_2 + H_2O \rightarrow 2NH_3 + CO_2$$
$$CrCl_3 + 3C_5H_8O_2 + 3NH_3 \rightarrow [Cr(C_5H_7O_2)_3] + 3NH_4Cl$$

Modern Inorganic Synthetic Chemistry, DOI: 10.1016/B978-0-444-53599-3.10009-5

The most often used organic solvents are alcohols, ethers, benzene, acetone, carbon tetrachloride, acetonitrile, and acylamines [10–14]. They work well for the syntheses where the ligands as halides, arsenic compounds, phosphates, phosphines, amines, and β-diketone are involved. For example, the diketone chelate of zirconium was prepared by dropwise addition of $CF_3COCH_2COCF_3$ into the turbid solution of $ZrCl_4$ in CCl_4, followed by refluxing until no hydrochloride was released from the mixture [15].

$$ZrCl_4 + 4CF_3COCH_2COCF_3 \xrightarrow[\text{Reflux}]{CCl_4} [Zr(CF_3COCHCOCF_3)_4] + 4HCl$$

Some ligands function as solvents during their reactions, such as acetonitrile, acetaldehyde, pyridine, and ethylene diamine.

$$Cu_2O + 2HPF_6 + 8CH_3CN \rightarrow 2[Cu(CH_3CN)_4]PF_6 + H_2O$$

As a matter of fact, mixed organic solvents are applied in some direct syntheses of coordination compounds. Taking advantage of the intermolecular interactions, direct synthesis can give rise to some unusual coordination compounds. For example, Raj Pal Sharma group prepared the bulky anion $[Hg_2(SCN)_7]^{3-}$ unexpectedly in the presence of hexaamminecobalt(III) when in the synthesis of $[Co(NH_3)_6]Cl[Hg(SCN)_4]$ [16,17].

9.1.2. Precursor-induced Complexation

The precursor-induced complexation features the self-assemblies of starting components under a well-defined order and in a proper ratio [2,18–21] as shown below.

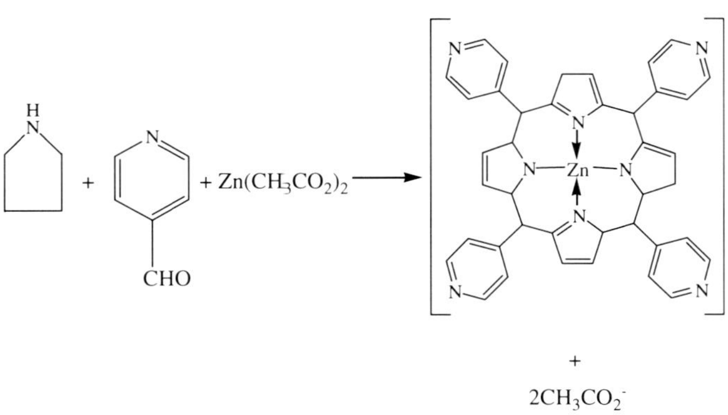

The saturated solution of zinc acetate in pyridine was dehydrated on molecular sieve, and then mixed with anhydrous pyrrole and pyridine-4-aldehyde in a sealed reactor. The reactants were degassed, followed by heating on oil bath at temperature between 130°C and 150°C for 48 h. After cooling down to room temperature, the precipitate was filtered from the mixture and washed with anhydrous ethanol to yield purple crystals of macrocyclic chelate of zinc. This method is specific for the syntheses of unstable coordination compounds as it excludes the separation and purification of the intermediates.

9.1.3. Metal Vapor Synthesis and Matrix Isolation Method

The use of metal vapors in the gas phase opened a new field of synthetic coordination chemistry, which is now known as cryosynthesis. Metal vapor synthesis (MVS) is used to prepare low-valent metal single-nuclear compounds or metal clusters, as well as the matrix isolation method which was later developed on the basis of MVS [22–24]. Actually, the cryosynthetic method has made a great contribution to the coordination chemistry of metal σ- and π-complexes [25].

Coordination compounds could be obtained by direct interactions of vaporized atomic metals and ligands in the gas phase. The apparatus of MVS vary from case to case, depending on many factors such as the chemical properties, the melting point and the vapor pressure of metals used for heating. Basically, it consists of a heat resistant device for evaporating metals, a stainless steel vacuum chamber for the reactions to take place, and a liquid-nitrogen-cooled-wall to condense metal atoms together with precursors to form final products. Metals vaporized in the heating device are highly reactive and react with the ligands (molecules or atomic clusters) and condense on the liquid-nitrogen-cooled-wall where the decomposition of final products is efficiently avoided.

The simplified apparatus for direct synthesis of $Co_2(PF_3)_8$ is illustrated in Fig. 9.1. The vapor inlet was filled with PF_3, while the metallic cobalt was collected in the resistive crucible. The whole system was then evacuated, after which the reaction vessel was cooled by liquid nitrogen and the crucible was heated to 1300°C. Further heating the crucible to 1600°C to evaporate metallic cobalt, while keeping the inlet of PF_3 at 10 mmol per minute, resulted in the condensation of the liquid-nitrogen-cooled-wall. The crucible was cooled down after the reaction was completed, and the vessel was filled with helium. Once the reactor was recovered to room temperature, the excessive PF_3 could be evacuated out while the final product $Co_2(PF_3)_8$ was retained on the liquid-nitrogen-cooled-wall for its less volatility [24].

The synthesis of $Fe(P(CH_3)_3)_4$ was also carried out in the similar apparatus, as well as those of $Ni(PF_2Cl)_4$, $Mn(PF_3)(NO)_3$, $Cr(PF_3)_6$, $Ni(PF_3)_3(PH_3)$, and other organometallic compounds.

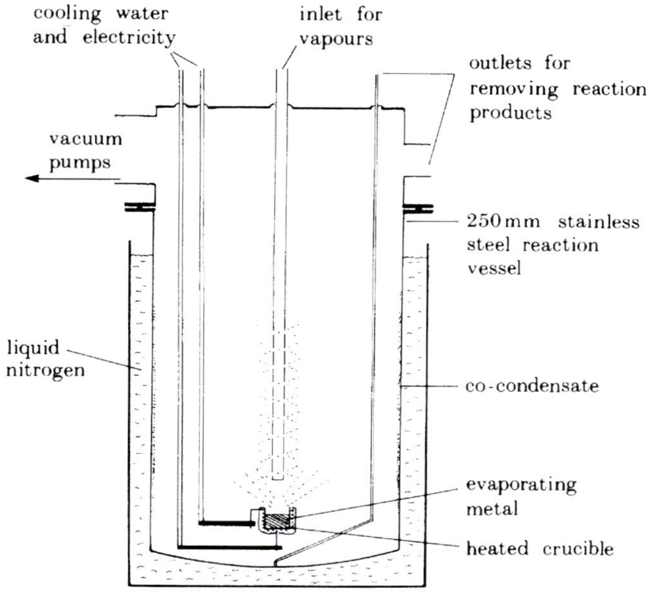

FIGURE 9.1 The reaction vessel for $Co_2(PF_3)_8$ based on MVS. *Reprinted from Proceedings of the Royal Society A: Mathematical, Physical and Engineering Sciences, Vol. 396, P.L. Timms, Page 1., Copyright (1984), with permission from Royal Society [24].*

However, such syntheses are not possible on the gram scale with ligands that are noncondensable at 77 K, for example N_2, O_2, H_2, CO, CH_4, NO, C_2H_4, etc. At 77 K, competitive metal atom diffusion/agglomeration processes in these volatile ligands overwhelm the desired metal–ligand complexation reaction, so that only the colloidal metal compositions rather than well-defined coordination compounds were prepared. But, if the temperature is further lowered below about one-third of the melting point of the ligand (the Tamman temperature, below which metal atom diffusion in the solid matrix is minimized), metal–ligand complexation predominates and yields desired products. A case in point is the temperature dependence of the nickel atom–nitrogen reactions:

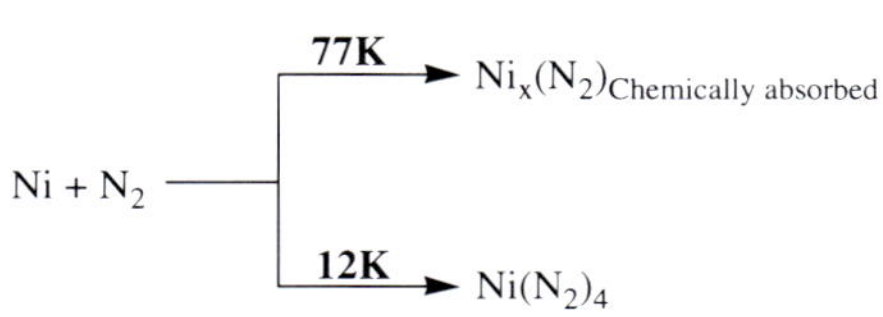

which requires low temperature for complex formation even though the decomposition temperature of Ni $(N_2)_4$ has been estimated to be about 80–100 K. With a judicious choice of metal atom concentration, ligand, deposition rate, and temperature, the yield of a desired M_xL_y compound can be optimized [26].

Matrix isolation is quite similar to MVS. The reaction vessel of matrix isolation from Ozin's group consists of an electron gun and a fairly high-capacity closed-cycle helium refrigerator (10 K) as the reaction chamber. The cocondensation reaction typically involved deposition of 10–100 mg of metal vapor with 10–100 g of CO onto a copper reaction shield maintained below 30 K. After deposition, the shield is slowly warmed to remove unreacted CO. Subsequently, the crude product is dissolved in a suitable solvent (pentane or toluene) and later purified by standard steps [26].

Various organic ligands have been used in cryosynthesis as well as inorganic ligands, which have been shown above as dinitrogen, dioxygen, carbon and nitrogen oxides, phosphorus trifluoride, etc. Till date, a lot of metal σ- and π-complexes have been synthesized efficiently by cryosynthesis methodology, where the ligands vary from unsaturated hydrocarbons (such as mono-olefins, diolefins, cyclic dienes, trienes, acetylenes, benzene and its derivatives, and so on) [27–34] to hetero-aromatic ligands (including thiophene, pyridine, and its derivatives) [35,36], oxygen-containing organic derivatives (for example, ethers, ketones, diketones, organic acids, anhydrides, acyl halides) [29,31], and even polymers [37–39]. Cryosynthesis of metal-containing polymers was carried out by cocondensation of metal vapors and polymers at 240–270 K according to the scheme showed below (where $x = 2–5$). This transformation opens the way to synthesize coordination metal polymers and metal–graphites.

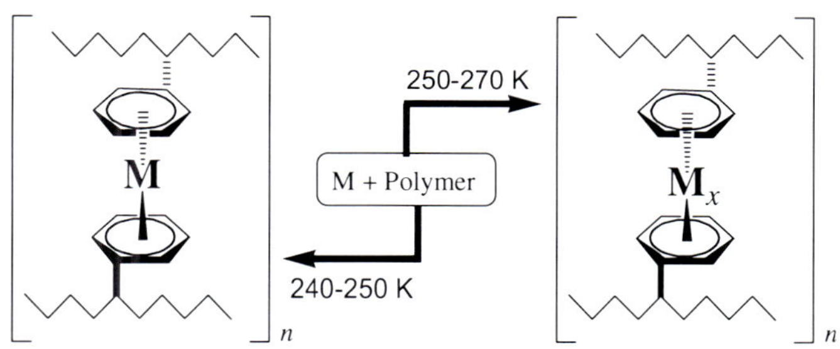

Another important application of metal vapor synthesis is the formation of intrazeolite metal clusters, where small metal clusters were deposited in the supercages of zeolites [38].

9.2. COMPONENT EXCHANGE METHOD [2,3,15]

9.2.1. Metal Exchange

Transition metal coordination compounds could interact with other metal ions and have themselves replaced by the foreign metals to form new compounds.

$$ML_x + M'^{n+} \rightarrow M'L_k + M^{m+} + (x - k)L$$

As described above, M might be some transition metals or hydrogen and M′ is transition metals. The metal exchange process will result in new coordination compounds, which are either thermodynamically or kinetically more stable. A proper example is shown below.

$$2Ln(NO_3)_3 + 3Ba(tfacam)_2 \rightarrow 2Ln(tfacam)_3 + 3Ba^{2+} + 6NO_3^-$$
$$tfacam = d\text{-3-trifluoroacetylcamphor}$$

The stabilities of coordination compounds, under such circumstance, depend on the coordinating ligands. For example, the metal exchange order was Cu > Ni > Zn > Mg in the case where the ligand was *N,N′*-disalicylidene ethylenediamine. Therefore, it's a quite simple method to get a series of new coordination compounds when starting from a single compound.

9.2.2. Ligand Substitution [2,3,40–44]

The ligand moieties of coordination compounds could be substituted by new ligands as well to result in new compounds under certain conditions, where the trans effect of ligands might dominate. A case in point is shown below.

$$Ni(CO)_4 + 4PCl_3 \rightarrow Ni(PCl_3)_4 + 4CO\uparrow$$

Anhydrous phosphorus trichloride was added into vigorously stirred tetracarbonylnickel under CO atmosphere. After reaction was completed, the mixture was filtered and dried to give the new substituted compound as final product. If partly substituted, the original compound would generate new coordination compound with mixed ligands, for example in the case of reaction below.

$$K_2PtCl_4 + 2(C_2H_5)_2S \rightarrow cis\text{-}\{Pt[(C_2H_5)_2S]_2Cl_2\} + 2KCl$$

The aqueous solution of potassium tetrachloroplatinate (II) was mixed with diethyl sulfide in a flask. After the reaction was completed, the mixture was evaporated to dryness and extracted by benzene. The extraction was then chilled with an ice-bath to give the yellow product.

Ligand substitution methods are also known as auxiliary ligand methods [45–47]. From this point of view, the auxiliary ligands may participate in two sequences. Under the first circumstance, they are mixed with so-called main ligands prior to the addition of transition metals. Otherwise, they are used to form precoordination compounds with transition metals and sequentially replaced by the main ligands to give the target products as well. Some most often used auxiliary ligands include pyridine, bi-pyridine, 1,10-Phenanthroline, hexamethyl phosphoryl triamide, triphenylphosphine, and triphenylphosphine oxide. The purpose of auxiliary ligands participation is supposed to increase the stability as well as the solubility of coordination compounds in solvents.

9.2.3. Subcomponent Exchange [2,48,49]

The subcomponents of ligands might as well react with other chemicals to form new coordination compounds.

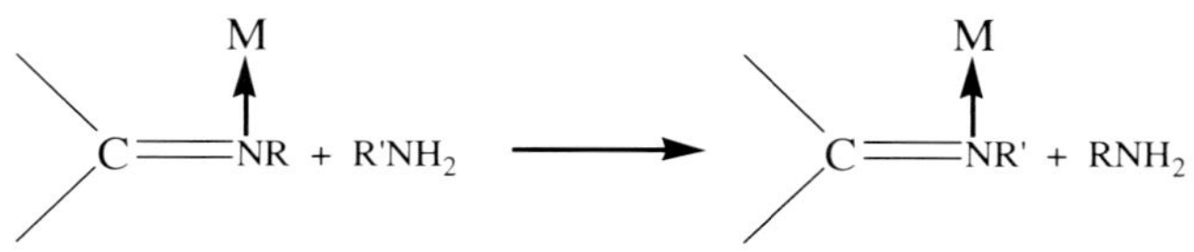

The imino group of Schiff base, in the reaction above, was substituted with primary amine to produce the expected compound. In addition to Schiff base compounds, some chemically active groups from parent acetyl acetone or azo compounds could also experience such sub-component exchange to result in new coordination compounds. For example, the original coordination compound was modified when *N*-bromo-succinimide was added into the stirred chloroform solution of acetone chromium(III), in the synthesis of tri(3-bromo-2,4-pentanedione) chromium (III). The mixture was heated and evaporated after the reaction was completed. The residue was then filtered and re-crystallized from benzene–heptane to give the final product.

9.3. REDOX INTERACTION METHOD [3]

9.3.1. Oxidation of Element Metals

It's been widely known and established that some metal salt hydrates can be obtained when the corresponding element metals are dissolved in acidic solution [50]. For example, element gallium was dissolved in perchloric acid (72%) and the mixture was boiled. On cooling, the expected product $[Ga(H_2O)_6](ClO_4)_3$ precipitated as crystals.

$$Ga + 3HClO_4 + 6H_2O \rightarrow [Ga(H_2O)_6](ClO_4)_3 + \frac{3}{2}H_2\uparrow$$

This methodology is applied in nonaqueous solution as well. For instance, the ether solution of iron and H(fod) was refluxed under ammonia atmosphere, and gave the final product as $Fe(fod)_3$ [51].

$$Fe + 3H(fod) \rightarrow Fe(fod)_3 + \frac{3}{2}H_2\uparrow$$
$$fod = CF_3CF_2CF_2C(OH) = CHCOC(CH_3)_3$$

Some coordination compounds of rare earth metals were synthesized by Mashima's group using this method [52]. The successful synthesis of [Ln($SC_5H_4N)_2$(hmpa)$_3$]I (Ln = Sm, Yb, hmpa = hexamethylphosphoric triamide) is illustrated as below, where the reaction could not be completed in the absence of iodine. And the similar coordination compounds, such as [{Yb(hmpa)$_3$}$_2$(μ-SPh)$_3$] [SPh] and Ln(SPh)$_3$(hmpa)$_3$ (Ln = samarium and ytterbium, hmpa = hexamethylphosphoric triamide), were also reported by the same group [45].

$$\mathrm{Ln} + \frac{1}{2}\mathrm{I_2} + \mathrm{PySSPy} \xrightarrow[\mathrm{THF}]{\mathrm{hmpa}} [\mathrm{Ln(SPy)_2(hmpa)}]\mathrm{I}$$

9.3.2. Oxidation of Low-valent Metals

High-valent transition metal coordination compounds can be prepared via oxidation of the corresponding low-valent compounds. A well-known case in point is the preparation of trivalent cobalt compound via oxidation of its bivalent counterpart [53].

$$2\mathrm{CoCl_2} + 2\mathrm{NH_4Cl} + 8\mathrm{NH_3} + \mathrm{H_2O_2} \rightarrow 2[\mathrm{Co(NH_3)_5Cl}]\mathrm{Cl_2} + 2\mathrm{H_2O}$$

To the solution of ammonia chloride in concentrated ammoniacal liquor was added $COCl_2 \cdot 6H_2O$, and the mixture was stirred and dropwise added with 30% hydroperoxide until no bubbles were evident. Hydrochloride was then added into the mixture, and the final product would precipitate as purple crystals.

Common oxidants include H_2O_2, air, halogens, $KMnO_4$, and PbO_2. And electrochemical oxidation method works as well. For instance, cerium(III) nitrate could be oxidized by air to give cerium(IV) nitrate. Chlorine is able to oxidize platinum(II) to platinum(IV) compound as shown below.

$$cis\text{-}[\mathrm{Pt(NH_3)Cl_2}] + \mathrm{Cl_2} \rightarrow cis\text{-}[\mathrm{pt(NH_3)_2CI_4}]$$

9.3.3. Reduction of High-valent Metals

On the other way around, low-valent transition metal coordination compounds can also be prepared via reduction of the high-valent counterparts. Reducing agents could be hydrogen, element potassium or sodium (including potassium amalgam and sodium amalgam), element zinc, hydrazine, and some organic reductants. Electrochemical reduction method works too. For example, low-valent rhodium(II) compound Rh[P($C_6H_5)_3]_2$(COO)Cl was obtained by the reduction of rhodium(III) chloride in the presence of triphenylphosphine and formaldehyde in anhydrous ethanol. The ligand could be used as reductant as well in the reaction below.

$$2\mathrm{Cu(NO_3)_2 \cdot 3H_2O} + 5\mathrm{P(C_6H_5)_3} \rightarrow 2\{\mathrm{Cu[P(C_6H_5)_3]_2(NO_3)}\} + \mathrm{OP(C_6H_5)_3} + 2\mathrm{HNO_3} + 5\mathrm{H_2O}$$

9.3.4. Middle-valent Metals

Some middle-valent transition metal coordination compounds were prepared by the reaction of their higher valent species with the lower valent ones. For instance, the methanol solution of arsenic triphenyl was added with copper nitrate and element copper (powder) under helium atmosphere, and the mixture was heated to reflux. After the reaction was over, the mixture was cooled down and filtered to give the final product as white crystals, as shown below.

$$\mathrm{Cu(NO_3)_2 \cdot 3H_2O} + \mathrm{Cu} + 6\mathrm{As(C_6H_5)_3} \rightarrow 2\{\mathrm{Cu[As(C_6H_5)_3]_3}(NO_3)\} + 3\mathrm{H_2O}$$

9.3.5. Electrochemical Synthesis [54–59]

Electrochemical preparation of coordination compounds displays one of the simplest and most direct methods of carrying out oxidation or reduction reactions, since the removal or addition of electrons can be achieved without the addition of redox reagents. This technique works in aqueous solution as well as in nonaqueous solution. The working electrodes can be inert (for instance, platinum) or sacrificial which will later participate in the redox reaction.

The apparatus for electrochemical synthesis of potassium enneachlorotungstate(III) is illustrated in Fig. 9.2. The electrolysis was carried out in a three-necked bottle, which was filled with prechilled concentrated hydrochloride and potassium tungstate. Gaseous hydrochloride was added through inlet 1, and water was filled through inlet 4 into the electrolysis tank. Once the solution around anode 3 turned red, the mixture was allowed to warm up to 45°C and the electrolysis was carried on until dark-green precipitate sedimented. The deposit was then isolated by filtration, and redissolved in a minimum volume of water. This solution was filtered again and the filtrate was precipitated by 90% ethanol to give the final product as $K_3[W_2C1_9]$.

The electrochemical synthesis technique is well studied in the case of halides and organic acids. For example, to the mixture of water and methanol was added acetyl acetone and chlorides, and the sequential electrolysis was carried out on a sacrificial iron anode (where the metal electrode was starting material) to yield pale-brown crystals, [Fe($C_5H_7O_2)_2$]. After electrolysis, the inlet of air flow or oxygen could oxidize the intermediate to produce [Fe($C_5H_7O_2)_3$] (Fig. 9.3).

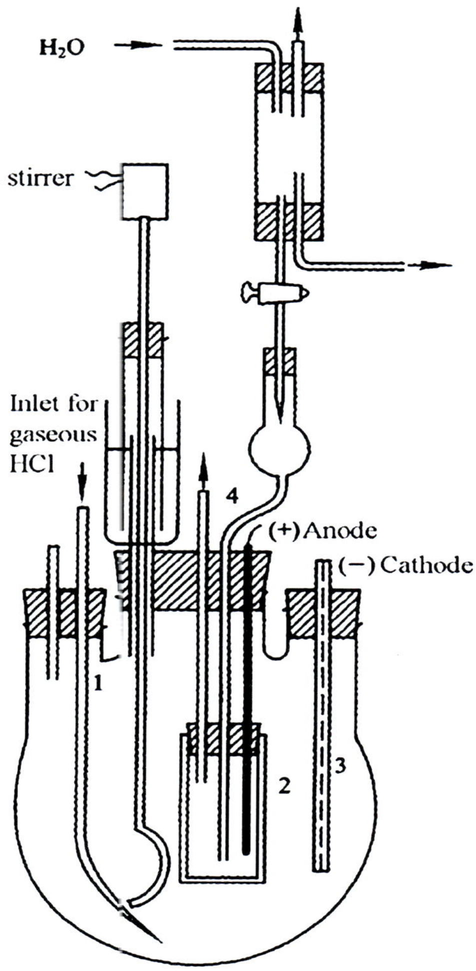

FIGURE 9.2 The apparatus of synthesis of $K_3[W_2Cl_9]$. *Reprinted with permission from [58]. Copyright 1957 American Chemical Society.*

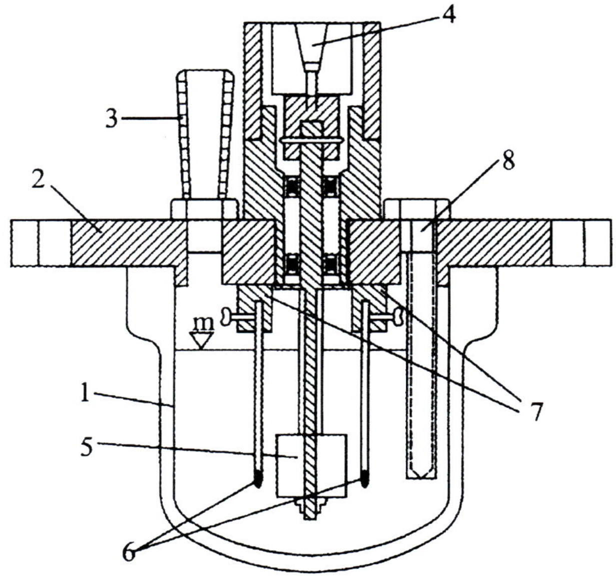

FIGURE 9.3 (1) Reaction vessel. (2) Sealed cap. (3) Outlet joint. (4 and 5) Electric stirrer. (6) Electrodes. (7) Electrode holders. (8) Thermocouple. *Lehmkuh, L.; Eisenback, W.: Justus Liebigs Annalen der Chemie. 1975. Volume 1975, Page 672, Copyright Wiley-VCH Verlag GmbH & Co. KGaA. Reproduced with permission from [56].*

In nonaqueous solution, the electrochemical synthesis technique has been well documented, especially in the preparation of some coordination compounds which are not readily accessible because of their strong hydrolysis. For instance, the electrolysis took place in $(Et_4N)_2[CoBr_4]$ with platinum as cathode and cobalt as anode.

9.3.6. Redox Under High Pressure

Some carbonyl compounds of transition metals were prepared under high pressure via the reduction of their oxides counterparts in the presence of CO. The reaction process is described below.

$$MoO_3 + 9CO \rightarrow Mo(CO)_6 + 3CO_2$$

9.4. Solid-state Synthesis Method

The solid-state reaction of coordination compounds is of crucial importance, but not readily accessible as it's very difficult to establish the reaction apparatus, i.e., to control the reaction conditions, and to monitor or detect the final products. However, it's possible to overcome these handicaps as the recent progresses have been achieved in product isolation and analysis methodologies. Therefore, most of the research works in this field still concentrate on simple cases instead of exploring this technique as preparative methodology.

9.4.1. Complexation of Ligands with Metal Precursors [60–62]

The solid-state synthesis technique has been applied in the complexation of ligands with metal precursor compounds. Under the reaction conditions, the ligands would turn into melts since they are most likely of low melting point. Therefore, solid-state synthesis usually takes place in the multiphase system. For example, the yellow coordination compound $Pd[P(C_6H_5)_3]_2Cl_2$ was obtained on heating of the mixture of triphenylphosphine with palladium chloride. The excessive ligand was removed by extraction. For the simplicity of this technique, many coordination compounds of transition metals (Co, Cu, Ni, Pd, and Pt) with ligands containing phosphines or arsenic derivatives have been reported.

9.4.2. Decomposition of Metal Precursors

In addition to the direct preparation, the new coordination compounds can also be obtained by the decomposition of their precursors under solid-state conditions [63,64]. For example, platinum tetra(triethylphosphine) was heated to 50–60°C under reduced pressure to produce $Pt(PEt_3)_3$ as red oil shown below.

$$Pt(PEt_3)_4 \xrightarrow[50-60\ ^{\circ}C]{vacuum} Pt(PEt_3)_3 + PEt_3$$

During the solid-state synthesis process, new metal–metal bond might come into being to form novel compound [65,66]. For instance, heating of $[K_2[Ni(CN)_4]$ in helium atmosphere gave rise to the formation of the nickel–nickel bonding in final product $K_4[Ni_2(CN)_6]$. The ligand substitution also plays a very important role in the preparation of new compounds [67]. The coordination compound, $[Co(NH_3)_5(H_2O)](ReO_4)_3 \cdot 2H_2O$, was dehydrated on an oil bath at 50°C for 2 h and then kept at 115–120°C for another 4 h to give the final product $[Co(NH_3)_5(OReO_3)](ReO_4)_2$.

9.5. INCLUSION COMPLEXATION METHOD

9.5.1. Layered Inclusion Compounds

Layered inclusion complexes of coordination compounds (such as graphite-like inclusion compounds and layered salts of selenic acid) have been prepared by the above-mentioned methodologies (for example, direct synthesis or component exchange). In the cases of tetravalent Zr, Ti, Sn, Ce, and Th, their layered coordination compounds were prepared via assembling the cations with XO_4 or RXO_3 counter ions (where X = P, As, and R = –H, –OH, $-CH_3$, and $-C_6H_5$). Basically, their structures are built up by the packing of polymeric macroanions consisting of cationic metal atoms lying in a plane and bridged through phosphate groups placed alternatively above and below this plane. Three oxygen atoms of each phosphate group are bonded to three metal atoms of the plane, and the fourth one is bonded to the hydrogen or substituted groups. The arrangement of the phosphate groups forms semicavities on both of the layer faces, so that the packing of the layers creates zeolite-like cavities in which the water molecule is filled. Thus, these intercalatable host compounds can be accessed by solvent molecules and, layer by layer, to form novel guest molecule-substituted compounds. They have potential applications in many fields such as electrochemical memory, electrochemical sensor, protonic conductor, and so on [68,69].

Among these compounds, the α-layered zirconium hydrogen phosphate (where zirconium is featured as being coordinated by oxygen atoms in an octahedron configuration) has been well studied to be able to include guest molecules to form intercalated compounds as

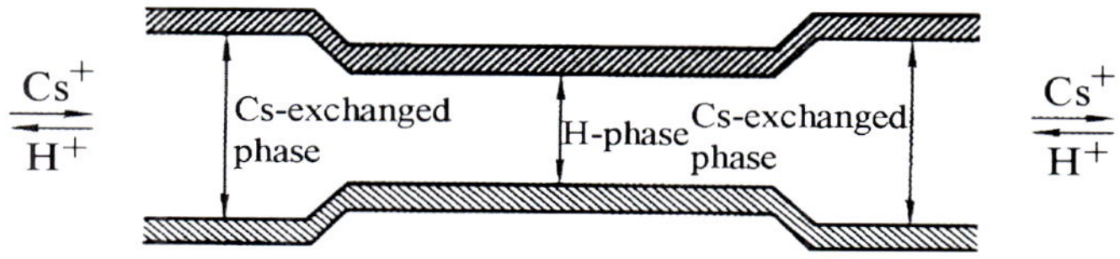

FIGURE 9.4 The ion-exchange in layered ZrP.

$M(RXO_3)_2 \cdot nS$ (where M = Zr, R = OH, X = P, S = metal cations, water, and solvent molecules).

For example, α-$Zr(HPO_4)_2 \cdot H_2O$ (ZrP) was reported to exhibit H^+/Cs^+ ion-exchange when its crystalline salt was dissolved in concentrated cesium solution. Although it's known that the foreign cation (Cs^+) has larger size than the maximum dimension of the windows connecting the cavities of ZrP and thus is not exchangeable, the activation energy of the H^+/Cs^+ exchange process could be lowered if a Cs-exchanged phase, having an interlayer distance larger than that of ZrP, is initially formed in the external part of the crystals of the exchanger (Fig. 9.4).

The enlargement of the external part of the crystals lowers the activation energy for the exchange of large cations. Thus, once started, the exchange can take place at an increasingly higher rate. This interface model is applicable in the field of layered inclusion compounds [70,71].

9.5.2. Multinuclear Compound and Metal Cluster-induced Inclusion Complexation

Nanometer-scaled particles, for example some multinuclear transition metal compounds or metal clusters, can be embedded in certain coordinating compounds to produce various novel nanocompounds via intermolecular recognition or inclusion complexation. Till date, two general methods are used in the preparation, which are defined as supramolecular assembly and nanoparticle inclusion complexation [72,73]. The former one is featured as the highly ordered intermolecular recognition between individually embedded nanoparticles, which are pre-prepared by the assemblies of embedding nanocrystals and embedded matrix (receptor).

In the preparation of nano-scaled titanium dioxide, the fresh nano-TiO_2 that was produced via hydrolysis of tetraisopropyl titanium(IV) turned into coated particles in the presence of hexadecyl ammonium bromide and pyridine derivatives (as receptors). Due to the weak interactions between these coated nanoparticles, a highly ordered or high-level ordering process took place and certain well-defined supramolecular assembly was obtained in the end (Fig. 9.5).

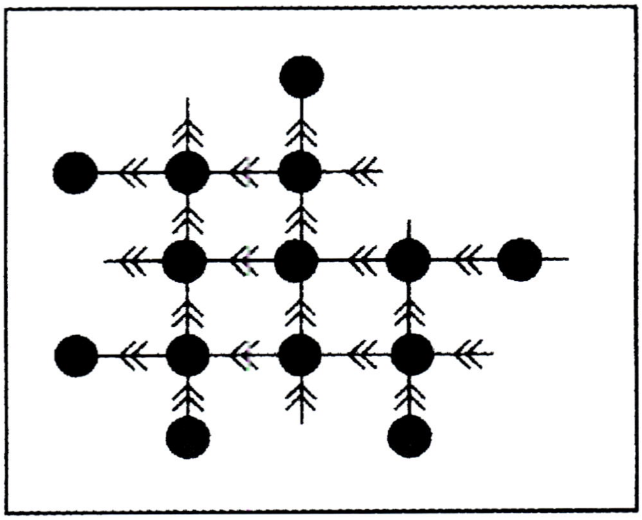

FIGURE 9.5 The preparation of nanomaterial via supramolecular assembly of TiO_2 with matrix (receptors).

Nanoparticle inclusion complexation is featured as the layer-by-layer encapsulation model [74]. For instance, ferromagnetic transition metal particles have to be encapsulated at the desire to preserve their performance from being decreased by oxidation. Therefore, the superfine magnetite particles with an average diameter of 10 nm were prepared prior to being coated with sodium oleate ($C_{17}H_{33}COONa$) by coprecipitation technique. Then as shown in Fig. 9.6, the surfactant-coated particles were placed 3 cm below the tungsten electrodes in an arc-discharge chamber. When the arc-discharge was performed, the helium atoms were ionized and supposed to bombard the surfactant and reduce it to carbon coating, which finally enveloped the core magnetite particles.

In addition to assembling coordination compounds on the surface of metal oxides, layer-by-layer technique has also enabled constructions of coordination compounds on metal surface, nanotubes, and even graphite. The possibility of building coordination compounds on solid surfaces has made possible the preparation of nano-scaled devices with new properties [75].

9.6. Macrocyclic Template Method [76–78]

In the preparation of macrocyclic coordination compounds with well-defined stereo-structure, metal ions could play a very important role as templates with proper sizes. Due to the strong interaction or high selectivity between metals templates and macrocyclic ligands, various multicyclic coordination compounds with high stabilities have been synthesized where the core metals varied from alkali metals to transition metals.

The anhydrous ethanol solution of 1,2-diaminobenzene was added dropwise into copper(II) acetate to precipitate deposits, which were filtered, washed with absolute ethanol, and finally resuspended. The anhydrous ethanol solution of 1,2-diaminoethane was then added at 10°C and the mixture was kept in dark with vigorous stirring for 24 h. The reaction mixture was later chilled to 2°C and was added with 2-bromo-1,3-propylene dialdehyde, which was kept for another 7 days to produce the final product as copper(II) 3,10-dibromo-1,5,8,12-tetraaza-benzocyclotetradecene.

In summary, several most often used methodologies in the preparation of coordination compounds have been introduced. Most likely, some combined synthetic techniques could work better than individual ones in real cases. With the progress in this field, more and more coordination compounds with novel functions and potential applications will be discovered, and vice versa.

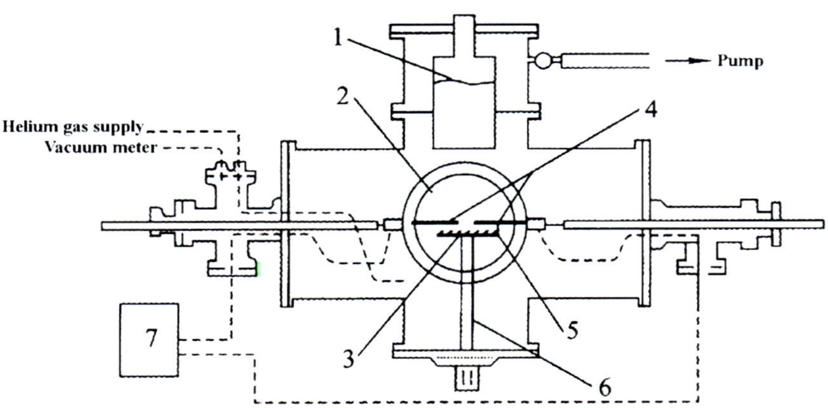

FIGURE 9.6 Illustration of the arc-discharge chamber. (1) Trap. (2) Observation window. (3) Surfactant-coated magnetite. (4) Tungsten rod. (5) Sample holder. (6) Stage height regulator. (7) Power supply. *Reprinted from Materials Science and Engineering: A, Vol. 217–218, B. Jeyadevan, Y. Suzuki, K. Tohji, I. Matsuoka, Encapsulation of nano particles by surfactant reduction, Copyright (1996), with permission from Elsevier [74].*

References

[1] A.D. Garnovskii, B.I. Khahsov, Direct Synthesis of Coordination and Organometallic Compounds, Elsevier, 1999.
[2] W.L. Jolly, Preparative Inorganic Reactions, Interscience Publishers, 1964.
[3] W.C. Fernelius, B.E. Bryant, in: T. Moeller (Ed.), The Inorganic Syntheses Series, vol. 5, McGraw-Hill Book Company, New York, 1957, p. 105.
[4] G.B. Kauffman, L.Y. Fang, in: L. Smith, J. Holt (Eds.), The Inorganic Syntheses Series, vol. 22, Wiley Interscience Publication, New York, 1983.
[5] S. Kirschner, in: E.G. Rochow (Ed.), The Inorganic Syntheses Series, vol. 6, McGraw-Hill Book Company, New York, 1960, p. 1.
[6] M. Wriedt, I. Jess, C. Nather, European Journal of Inorganic Chemistry (2009) 363.
[7] L. Wang, W. You, W. Huang, C. Wang, X.Z. You, Inorganic Chemistry 48 (2009) 4295.
[8] C. von Hanisch, O. Hampe, F. Weigend, S. Stahl, Angewandte Chemie International Edition 46 (2007) 4775.
[9] W.C. Fernelius, J.E. Blanch, in: T. Moeller (Ed.), The Inorganic Syntheses Series, vol. 5, McGraw-Hill Book Company, New York, 1957, p. 130.
[10] A.D. Garnovskii, A.P. Sadimenko, M.I. Sadimenko, D.A. Garnovskii, Coordination Chemistry Reviews 173 (1998) 31.
[11] J. Pons, J. Garcia-Anton, M. Font-Bardia, T. Calvet, J. Ros, Inorganica Chimica Acta 362 (2009) 2698.
[12] F. Marchetti, G. Pampaloni, S. Zacchini, Dalton Transactions (2009) 8096.
[13] E. Lopez-Torres, J.R. Dilworth, Chemistry – A European Journal 15 (2009) 3012.
[14] J. Lloret, F. Estevan, P. Lahuerta, P. Hirva, J. Perez-Prieto, M. Sanau, Chemistry – A European Journal 15 (2009) 7706.
[15] K.C. Joshi, V.N. Pathak, Coordination Chemistry Reviews 22 (1977) 37.
[16] R. Bala, R.P. Sharma, R. Sharma, B.M. Kariuki, Inorganic Chemistry Communications 9 (2006) 852.
[17] R. Sharma, R.P. Sharma, P. Brandao, V. Felix, Journal of Molecular Structure 886 (2008) 17.
[18] L.J. Hunez, G.L. Eichhom, Journal of the American Chemical Society (1962) 901.
[19] H.J. Butcher, E.B. Fleischer, in: R.W. Parry (Ed.), The Inorganic Syntheses Series, vol. 12, McGraw-Hill Book Company, New York, 1970, p. 256.
[20] N. Liu, Q. Yue, Y.Q. Wang, A.L. Cheng, E.Q. Gao, Dalton Transactions (2008) 4621.
[21] G.A. van Albada, S.A. Komaei, H. Kooijman, A.L. Spek, J. Reedijk, Inorganica Chimica Acta 287 (1999) 226.
[22] M. Moskovits, G.A. Ozin, Cryochemistry, Wiley Interscience Publication, New York, 1976.
[23] S. Taro, K.N. Ryoiki, 29 (1975) 59.
[24] P.L. Timms, Proceedings of the Royal Society of London Series A – Mathematical Physical and Engineering Sciences 396 (1984) 1.
[25] B.I. Kharisov, A.D. Garnovskii, L.M. Blanco, A.S. Burlov, A. Garcia-Luna, Journal of Coordination Chemistry 49 (1999) 113.
[26] J. Godber, H.X. Huber, G.A. Ozin, Inorganic Chemistry 25 (1986) 2909.
[27] K.J. Klabunde, Accounts of Chemical Research 8 (1975) 393.
[28] G.A. Domrachev, V.D. Zinovev, Uspekhi Khimii 47 (1978) 679.
[29] J.R. Blackborow, D. Young, Metal Vapor Synthesis in Organometallic Chemistry, Springer-Verlag Chemie, Berlin, Heidelberg, New York, 1979.
[30] J.J. Carroll, J.C. Weisshaar, Journal of the American Chemical Society 115 (1993) 800.
[31] M.A. Blitz, S.A. Mitchell, P.A. Hackett, Journal of Physical Chemistry 95 (1991) 8719.
[32] D.W. Ball, R.G.S. Pong, Z.H. Kafafi, Journal of the American Chemical Society 115 (1993) 2864.
[33] E.P. Kundig, P.L. Timms, B.A. Kelly, P. Woodward, Journal of the Chemical Society: Dalton Transactions (1983) 901.
[34] J.G. Brennan, F.G.N. Cloke, A.A. Sameh, A. Zalkin, Journal of the Chemical Society-Chemical Communications (1987) 1668.
[35] L.H. Simons, P.E. Riley, R.E. Davis, J.J. Lagowski, Journal of the American Chemical Society 98 (1976) 1044.
[36] T.G. Groshens, B. Henne, D. Bartak, K.J. Klabunde, Inorganic Chemistry 20 (1981) 3629.
[37] C.G. Francis, H. Huber, G.A. Ozin, Journal of the American Chemical Society 101 (1979) 6250.
[38] C.G. Francis, H. Huber, G.A. Ozin, Inorganic Chemistry 19 (1980) 219.
[39] C.G. Francis, H.X. Huber, G.A. Ozin, Angewandte Chemie International Edition in English 19 (1980) 402.
[40] W. Petz, Chemical Reviews 86 (1986) 1019.
[41] M. Draganjac, T.B. Rauchfuss, Angewandte Chemie International Edition in English 24 (1985) 742.
[42] X.L. Yu, Y. Sun, W.B. Hou, S.J. Zhang, Z.X. Cheng, Journal of Crystal Growth 182 (1997) 428.
[43] M. Rubio, A. Suarez, D. del Rio, A. Galindo, E. Alvarez, A. Pizzano, Organometallics 28 (2009) 547.
[44] A. Madalan, H. Roesky, M. Andruh, M. Noltemeyer, N. Stanica, Journal of the Chemical Society, Chemical Communications 15 (2002) 1638.
[45] K. Mashima, Y. Nakayama, T. Shibahara, H. Fukumoto, A. Nakamura, Inorganic Chemistry 35 (1996) 93.
[46] L.X. Zhang, X.G. Zhou, Z.E. Huang, R.F. Cai, L.B. Zhang, Q.J. Wu, Chinese Journal of Structural Chemistry 20 (2001) 40.
[47] A.U. Malik, F.R. Rahmani, Journal of Inorganic and Nuclear Chemistry 37 (1975) 1552.
[48] K.B. Takvorian, R.H. Barker, in: R.W. Parry (Ed.), The Inorganic Syntheses Series, vol. 12, McGraw-Hill Book Company, New York, 1970, p. 85.
[49] J.P. Collman, in: J. Kleinberg (Ed.), The Inorganic Syntheses Series, vol. 7, McGraw-Hill Book Company, New York, 1963, p. 134.
[50] L.S. Foster, in: W.C. Fernelius (Ed.), The Inorganic Syntheses Series, vol. 2, McGraw-Hill Book Company, New York, 1946, p. 26.
[51] R.E. Sievers, L.W. Connolly, in: R.W. Parry (Ed.), The Inorganic Syntheses Series, vol. 12, McGraw-Hill Book Company, New York, 1970, p. 72.
[52] K. Mashima, T. Shibahara, Y. Nakayama, A. Nakamura, Journal of Organometallic Chemistry 501 (1995) 263.
[53] G.G. Schlessinger, in: J.S. Young Tyree (Ed.), The Inorganic Syntheses Series, vol. 9, McGraw-Hill Book Company, New York, 1967, p. 160.
[54] J.H. Jacob, G.T. Dennis, Journal of the Chemical Society, Chemical Communication (1975) 808.
[55] J.H. Jacob, N. Lynn, G.T. Dennis, Canadian Journal of Chemistry 55 (1977) 2631.
[56] L. Lehmkuh, W. Eisenback, Liebigs Annalen der Chemie (1975) 672.
[57] D.G. Duck, Pure and Applied Chemistry 51 (1979) 2005.

[58] H.B. Jonassen, A.R. Tarsey, in: T. Moeller (Ed.), The Inorganic Syntheses Series, vol. 5, McGraw-Hill Book Company, New York, 1957, p. 139.
[59] M. Hesari, D. Nematollahi, L. Fotouhi, Journal of Coordination Chemistry 61 (2008) 1744.
[60] H.A. Tayim, A. Bouldouk, F. Awad, Journal of Inorganic & Nuclear Chemistry 32 (1970) 3799.
[61] J. Chen, D. He, Y. Di, Y. Kong, W. Yang, W. Dan, et al., Chinese Journal of Chemistry 27 (2009) 1675.
[62] Y. Di, Y. Kong, Q. Wang, W. Yang, Z. Tan, Chinese Journal of Chemistry 26 (2008) 1764.
[63] J.H. Yuan, in doctor degree dissertation: The redox of coordination compounds, Department of Chemistry, Nanjing University, Nanjing, 1980.
[64] T. Yoshida, T. Matsuda, S. Otsuka, in: R.J. Angelici (Ed.), The Inorganic Syntheses Series, vol. 28, Wiley Interscience Publication, New York, 1990, p. 119.
[65] W.M. Xie, S.R. Huang, X.Q. Xin, Journal of Nanjing Univrsity 25 (1989) 68.
[66] J.S. Yoo, E. Griswold, J. Kleinber, Inorganic Chemistry 4 (1965) 365.
[67] E. Lenz, R.K. Murmann, in: R.W. Parry (Ed.), The Inorganic Syntheses Series, vol. 12, McGraw-Hill Book Company, New York, 1970, p. 214.
[68] M. Casciola, U. Costantino, F. Marmottini, Solid State Ionics 35 (1989) 67.
[69] G. Alberiti, R. Palombare, Solid State Ionics (1989) 153.
[70] G. Alberiti, M.G. Bemasconi, M. Casciola, U. Costantino, Journal of Inorganic and Nuclear Chemistry 42 (1980) 1631.
[71] M. Abe, Denki Kagaku 48 (1980) 344.
[72] J.R. Heath, Science 270 (1995) 1315.
[73] A.P. Bisson, F.J. Carver, C.A. Hunter, J.P. Waltho, Journal of the American Chemical Society 116 (1994) 10292.
[74] B. Jeyadevan, Y. Suzuki, K. Tohji, I. Matsuoka, Materials Science and Engineering A: Structural Materials Properties Microstructure and Processing 217 (1996) 54.
[75] R. Cao, A.M. Diaz-Garcia, Coordination Chemistry Reviews 253 (2009) 1262.
[76] G.L. Honeyboume, in L. Smith, J. Holt (Eds), The Inorganic Syntheses Series, vol. 22, New York, 1983.
[77] N.E. Dixon, W.G. Jackson, G.A. Lawrance, in: L. Smith, J. Holt (Eds.), The Inorganic Syntheses Series, vol. 22, Wiley Interscience Publication, New York, 1983.
[78] S.C. Jackels, in: L. Smith, J. Holt (Eds.), The Inorganic Syntheses Series, vol. 22, Wiley Interscience Publication, New York, 1983.

CHAPTER

10

Assembly Chemistry of Coordination Polymers

Xiao-Ming Chen
Sun Yat-sen University, China

Although the term coordination polymer first appeared in 1960s, and can be traced up to the first man-made coordination polymer or Prussian Blue by an accident in early eighteenth century, the current extensive interest in coordination polymers was triggered after the reports [1] of Robson and Hoskins [2]. They proposed that the new materials with interesting properties such as porosity and catalysis could be deliberately engineered through describing crystal structures in terms of nets that was proposed in 1960s by Wells [3]. This new research field now attracts a large number of chemists around the world.

A coordination polymer contains metal ions (or metal clusters) linked by coordinated organic ligands into an infinite array. This infinite net must be defined by coordination bonds. Therefore, a structure linked by coordination bonds in one direction and supramolecular interactions (such as hydrogen bonding and $\pi-\pi$ stacking) in two other directions is a one-dimensional (1D) coordination polymer, a structure linked by coordination bonds in two directions and supramolecular interactions (such as hydrogen bonding and $\pi-\pi$ stacking) in another direction is a two-dimensional (2D) coordination polymer, while that linked by coordination bonds in three directions is, of course, a three-dimensional (3D) coordination polymer.

As being well known that metal ions (or metal clusters) and organic bridging ligands are diversified in coordination behavior and geometry, the combinations via coordination bonds are naturally diversified. Actually, a huge number of structurally different or similar coordination polymers with different functionalities, such as porous frameworks for gas sorption and separation, catalysis, molecular and ion-exchange, molecular magnets, luminescence, and so on, have been established.

Nets of Coordination Polymers: Usually, coordination polymers are assembled by metal ions (or metal clusters) and bridging organic ligands, and sometimes with additional supramolecular interactions, into highly ordered 3D structures. Such regular 3D structures can be described in terms of topologic nets. Topologically, a polymeric metal-organic net can be reduced into a series of nodes in a certain symmetry (triangular, tetrahedral, or octahedral, etc.) and linkers that connect the nodes into infinite 1D–3D periodic net. Topological description not only is useful for understanding the net structure, but also can be used for the design of new coordination polymers. The net-based approach has been called reticular chemistry [4]. Usually, metal ions or metal clusters can be regarded as the nodes, while organic ligands as the linkers. In some cases, higher multitopic (e.g., tri-topic) organic ligands can also be regarded as the nodes, especially when the metal ions serve as low (e.g., 2 or 3) connectors in the net. In principle, by the use of appropriate nodes and linkers, one can assemble a specific net.

A large number of different nets have been documented so far. Among them, typical and simple nets are zigzag or helical, ladder, railway chains for 1D nets, square or rectangular, honeycomb and brick-wall structures for 2D nets, diamondoid, and cubic for 3D nets, as illustrated in Fig. 10.1. Of course, a lot of more complicated and intriguing nets have also been documented in the literature [2].

Molecular Assembly: Coordination bonds are usually stronger and have better directionality than hydrogen-bonding and other weak supramolecular interactions. Therefore, the coordination preference of metal ions (or clusters) and the structures of organic bridging ligands play a critical role in the formation of coordination

Modern Inorganic Synthetic Chemistry, DOI: 10.1016/B978-0-444-53599-3.10010-1

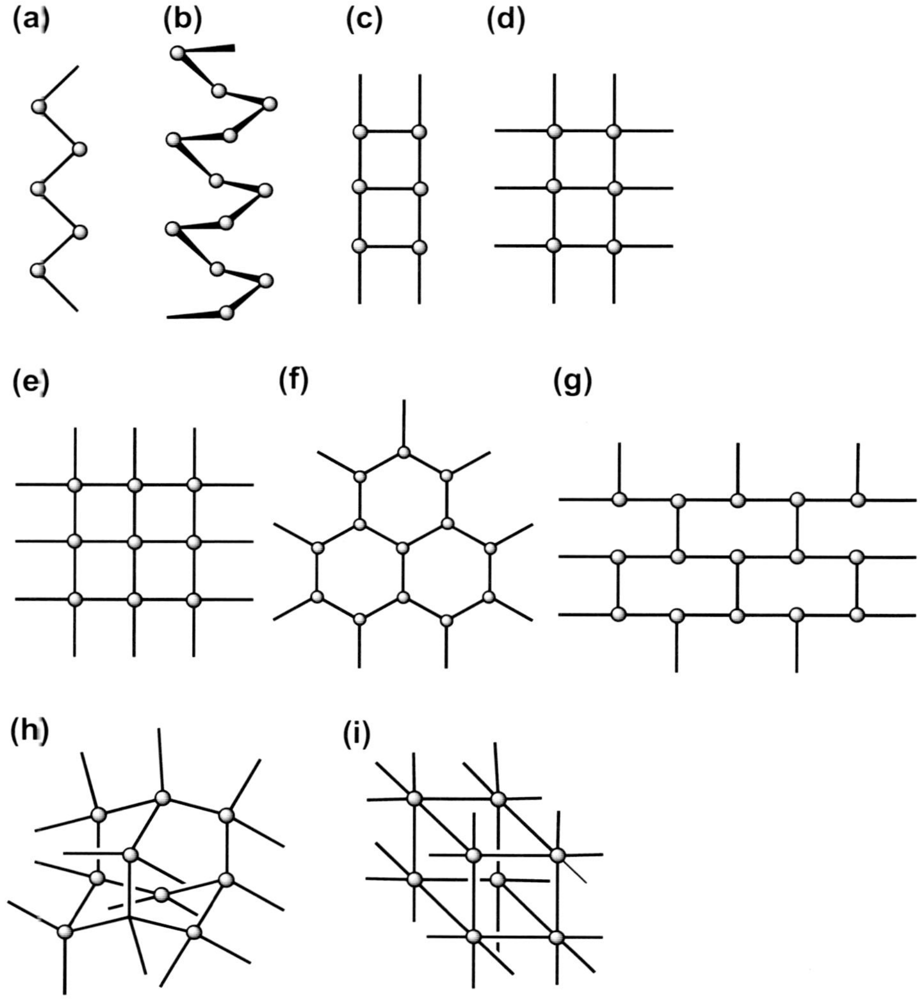

FIGURE 10.1 Some simple coordination polymer nets. (a) A zigzag chain, (b) a helical chain, (c) a ladder chain, (d) a railway chain, (e) a square net, (f) a honeycomb net, (g) a brick-wall net, (h) diamondoid net, and (i) a cubic net.

polymeric nets. Consequently, structural prediction of coordination polymers should be, in principle, easier than that for a molecular crystal.

However, coordination geometry of metal ions is usually diversified and even variable, and the structures of organic ligands are also diversified. The structures of coordination nets formed by the assemblies of the diversified metal ions (or clusters) and organic bridging ligands are almost infinite. Moreover, as a kind of weaker interaction than typical covalent bonding, coordination bonds are usually more flexible. Other weaker intermolecular interactions may also present in the self-assembly system Consequently, structural uncertainty is an intrinsic characteristic of a self-assembly system consisting of metal ions, organic ligands, counter anions, and solvent molecules. Many reaction parameters, such as the reaction/crystallization temperature, pH, template/additive, solvent, and counter ions, can play an important role in the formation of the coordination nets [5]. Thus, it would be difficult to predict the outcome of a complicated self-assembly system simply by molecular design (the choice of metal ions, metal clusters, and ligand structure). Actually, for a given set of metal and ligand at a specific molar ratio, different coordination polymers can be produced by different reaction conditions. In short, structural control of the coordination nets is rather challenging.

In a system for producing a particular coordination polymer, there are different components and solvent, even with other species such as template and additive. Under the direction of coordination bonding, and possibly in combination of other supramolecular interactions, the components can be combined into some tiny intermediate species (Fig. 10.2) before nucleation or crystallization. Such intermediate species can be combined into one or more kinds of large molecular

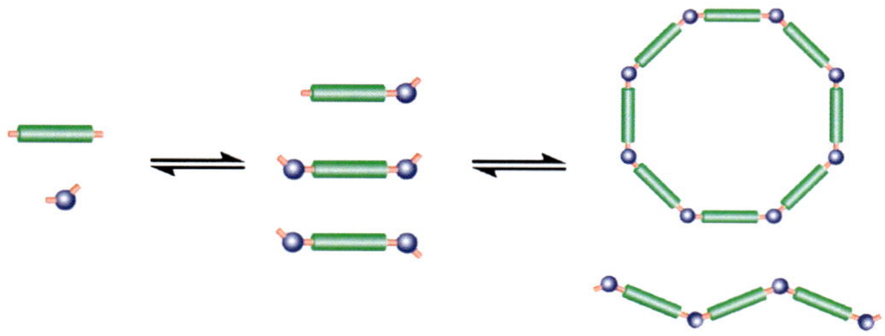

FIGURE 10.2 Illustration of two possible products assembled by metal ions and di-topic organic ligands.

assemblies, leading to the formation of crystal seeds and consequently crystals having the same or different superstructures. The structural diversity includes supramolecular isomerism, which is frequently observed for coordination polymers [5,6]. The different products may be either kinetically or thermodynamically favored. When different products can be generated from a reaction system, the energy barrier for a thermodynamically favored product is usually larger than that of a kinetically favored one (Fig. 10.3). Since thermal energy is proportional to the reaction temperature, the thermodynamically favored isomer associated with a high activation energy can be obtained at a higher reaction temperature, whilst lower temperature favors the kinetic product. Therefore, by control of the reaction temperature, and/or other reaction parameters, it is possible to alter the formation of products. Usually, high concentration and low temperature is favored for the kinetically favored product, while low concentration and high temperature is favored for the thermodynamically favored product. Practically, other reaction parameters, such as pH, template/additive, solvent, and counter ions, can also have significant impact on the

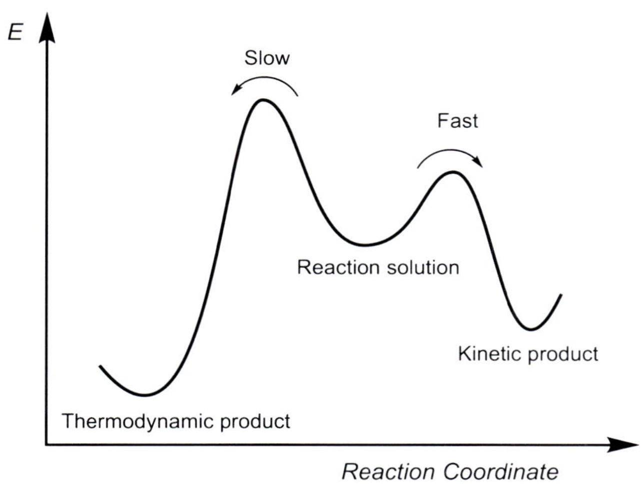

FIGURE 10.3 Thermodynamic and kinetic crystallization of products. *Ref. [5] – Reproduced with permission from The Royal Society of Chemistry.*

kinetic process, leading to the formation of different products.

10.1. MOLECULAR DESIGN OF COORDINATION POLYMERS

The coordination behaviors of both metal ions (and clusters) and organic ligands play a critical role in the self-assembly of coordination polymers, since the coordination bonding is a kind of directional and strong interaction in the process. Therefore, judicious choice of metal ions (and clusters) and organic ligands is critically important for a targeted structure. Three common strategies, or the utilizations of single metal ions, clusters as the nodes, and pillared-layered nets, are usually adopted.

10.1.1. Single-metal Noded Nets

Different metal ions have different electron configurations and ionic radii, hence exhibiting different coordination behaviors. Low coordination numbers, 2 or 3, are commonly found for Ag(I) and Cu(I) ions, tetrahedral and octahedral coordinations are frequently found for Zn(II) and Co(II) ions, while higher coordination numbers are usually found for lanthanide ions. The coordination geometry can be induced by different coordination environments to be idealized or distorted. Nevertheless, the directionality of such coordination geometry is rather marked, and thus can be used as single metal nodes for constructing coordination polymers.

Generally, transition metal ions have been more popular, due in part to the more predictable nature of their coordination geometries. Some simple coordination geometries of transition metal ions are shown in Fig. 10.4. On the other hand, lanthanides have also attracted increased attention recently, with their higher connectivity leading to interesting topologies, in addition to other inherent properties of interest (e.g., luminescence).

In principle, by judicious choice of metal ions and multitopic organic ligands with appropriate coordination geometry, as nodes and linkers, respectively, a particular framework can be assembled under certain reaction condition.

As illustrated in Fig. 10.4, organic ligands for coordination polymers should usually be ditopic or multitopic linkers, which can bridge single metal ions or metal clusters as nodes into infinite polymeric structures. Based on the different coordination geometries of both the nodes and linkers, the resulting polymeric nets are hence diversified.

The most simple polymeric structure is a linear chain, which can be interconnected by a linear, ditopic

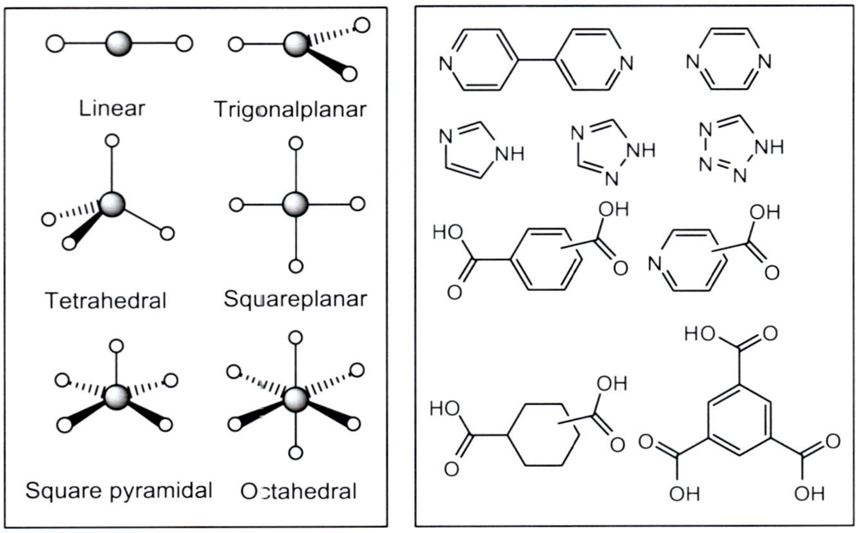

FIGURE 10.4 Typical coordination geometries of *d*-block metal ions (left) and multi-topic ligands (right).

organic linker, such as 4,4′-bipyridine (4,4′-bpy), and linear two-coordinate metal ions, such as Ag^{I}. For example, the cationic linear chain of [Ag(4,4′-bpy)] (NO_2) is shown in Fig. 10.5 [7].

When curved ditopic ligands are used, zigzag and/or helical chains can be generated. A typical example is the hydrothermal synthesis of copper(I) 2-methylimidazolate (Fig. 10.6), where Cu(I) is in situ generated from Cu(II) under weakly basic condition [8]. Due to the usually very poor solubility and rapid deposition, X-ray quality single crystals of such 1D polymers are usually difficult to grow under ambient condition. In contrast, hydrothermal (and solvothermal) synthesis is very promising in producing good quality single crystals, and thus is now widely used in the synthesis of different kinds of coordination polymers.

Square and rectangular nets are a kind of simple 2D nets, which can be simply assembled by metal ions preferring square-planar geometry. For example, assembly of Cu^{II} ions with linear, ditopic 4,4′-bpy or pyrazine ligands can yield a square net, as illustrated in Fig. 10.7. Meanwhile, octahedral metal ions, such as Zn^{II}, Cd^{II}, Fe^{II}, and Ni^{II}, can also be used to produce similar nets, when the axial positions of the metal ions are occupied by monodentate terminal ligands, such as water molecules. When mixed ditopic ligands of different lengths are used, the 2D nets become rectangular. For instance, by using 4,4′-bpy and pyrazine as bridging ligands, a rectangular net of $[Cu(4,4'\text{-bpy})(\text{pyrazine})]^{2+}$ can be generated in the presence of perchlorate ions as the counter ions [9]. Here, the reaction condition should be carefully selected to avoid the more easy formations of square nets that are interlinked uniquely by one kind of the ditopic ligands.

Combination of linear, ditopic ligands with tetrahedral metal ions, such as Zn^{II}, can lead to the formation of a 3D net. Noteworthy is that the coordination geometry of the metal ion and the structure of the ligand are critical in directing the 3D structure.

For example, a diamondoid net of $[Zn(ina)_2]$ (ina =isonicotinate), which was generated by in situ hydrolysis of 4-cyanopyridine with zinc salt (see Section 10.3.2), whereas the utilization of a curved ditopic ligand could furnish a more complicated and interesting net. For instance, the two nitrogen atoms in an imidazole have a bridging angle in the range of 135–145°, which are analogous to that of Si–O–Si bond angle (ca. 145°) in natural zeolites. Actually, using building blocks composed of bivalent tetrahedral metal ions and imidazolates is an effective strategy to form neutral coordination polymers with the same or similar topologies of natural zeolites. For example, a sodalite (SOD) net,

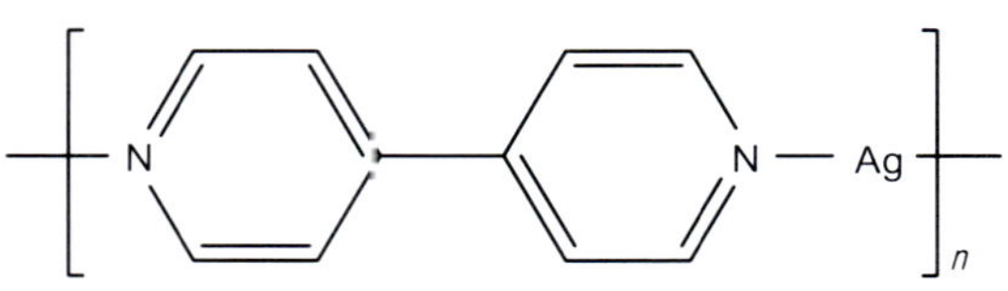

FIGURE 10.5 A linear chain interconnected by 4,4′-bpy and Ag^{I} ions.

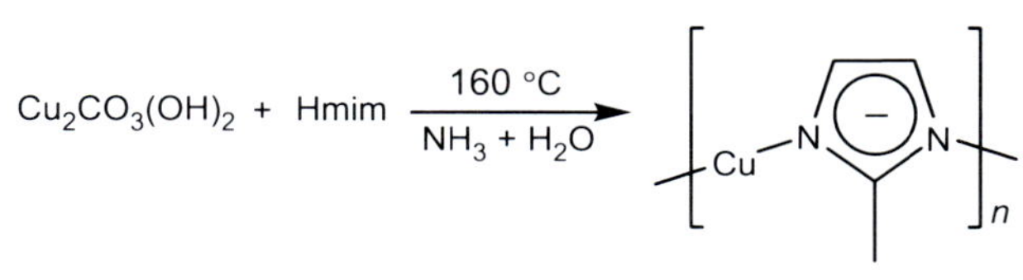

FIGURE 10.6 The hydrothermal reaction of Cu(II) and 2-methylimidazole.

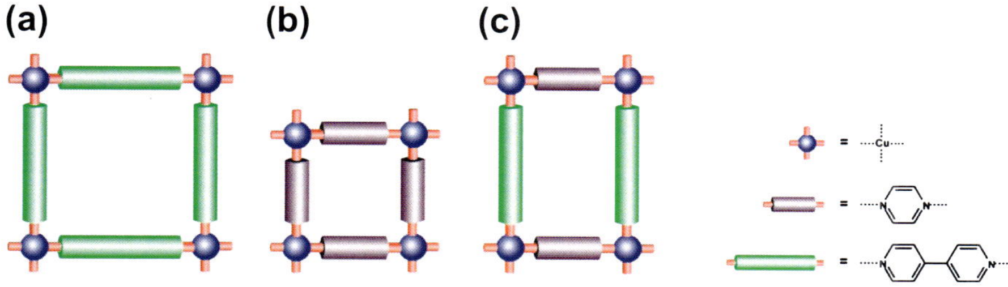

FIGURE 10.7 Structural units of square nets formed by 4,4′-bpy (a) and pyrazine (b), and a rectangular net (c) formed by mixed 4,4′-bpy and pyrazine ligands.

$[Zn(bim)_2]\cdot 1.67H_2O$, was constructed by benzimidazolate (bim) and zinc(II) (Fig. 10.8) [10]. A large number of metal imidazolates have been reported to exhibit different zeolite topologies [11]. These kinds of materials usually have large cages (typically 1.0–2.0 nm) and small windows (typically 0.22–0.74 nm) between the cages, and have very good gas sorption capacity and thermal stability.

10.1.2. Metal-cluster Noded Nets

Metal-clusters have been widely utilized in construction of coordination polymers, which may exhibit interesting properties, such as magnetism, photoluminescence, and sorption. This kind of metal clusters can be regarded as secondary building blocks (SBUs), which serve as nodes in their coordination polymers. The most commonly used metal-cluster nodes are paddle-wheel $[M_2(RCOO)_4]$, trimetallic $[M_3(\mu_3{-}O)(\mu{-}RCOO)_6]$ and $[M_3(\mu_3{-}OH)(\mu{-}RCOO)_6]$, as well as tetrametallic $[M_4(\mu_4{-}O)(\mu{-}RCOO)_6]$ SBUs (Fig. 10.9). In combination with different multitopic organic ligands, they can furnish quite a huge number of interesting coordination polymers. More importantly, the cluster SBUs can play a critical role in controlling the structures of coordination polymers, as they are usually rigid and have well-defined coordination geometries. In principle, a targeted coordination polymer can be designed through a combination of appropriate cluster SBUs and multitopic organic ligands.

In the paddle-wheel SBUs (Fig. 10.9a), the metal ions can be Cu^{II}, Zn^{II}, Co^{II}, Fe^{II}, Cd^{II}, etc. In this SBU, the two metal ions are bridged by four carboxylate ends into a square-planar node, which can be extended into layers by linear organic dicarboxylates. Since such kind of coordination polymers usually have very low solubility, they are usually synthesized by diffusion or hydro (solvo)thermal reactions.

Simply by diffusion of triethylamine and toluene into an aqua solution of zinc nitrate and 1,4-benzenedicarboxylic acid (1,4-bdcH$_2$), single crystals of $[Zn(1,4\text{-}bdc)(H_2O)]\cdot DMF$, as an early example, was obtained

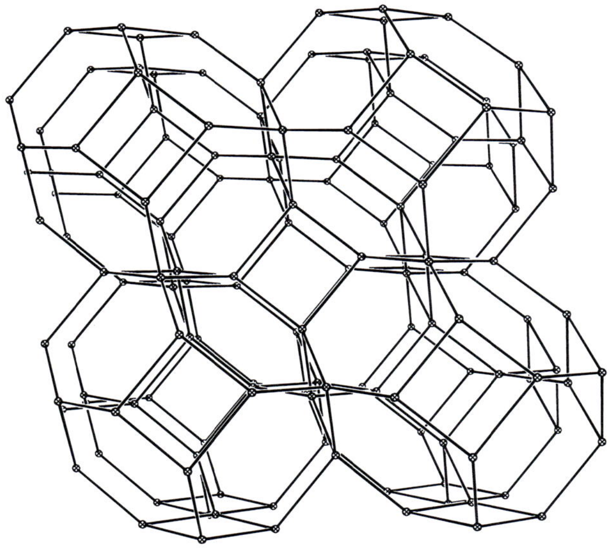

FIGURE 10.8 A sodalite (SOD) net constructed by tetrahedral metal ions (spheres) and imidazolates (bars). *Ref. [10] With kind permission from Springer Science+Business Media: Chinese Science Bulletin, $[Zn(bim)_2]\cdot(H_2O)_{1.67}$: A metal-organic open-framework with sodalite topology, Vol. 48, No. 15, August 2003, 1531, Xiaochun Huang, Jiepeng Zhang and Xiaoming Chen.*

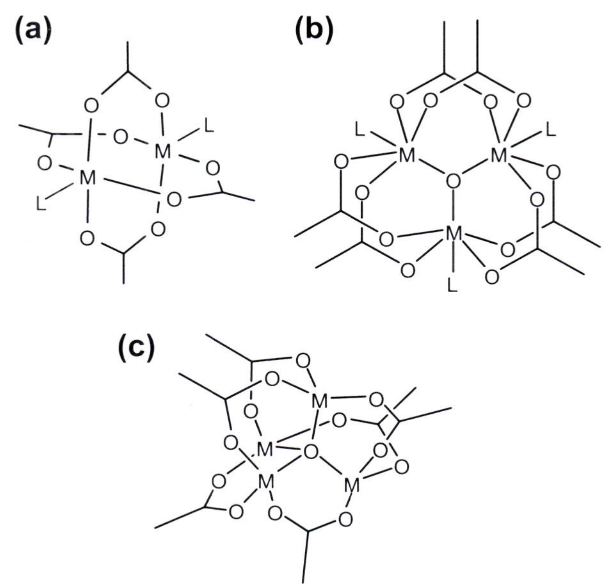

FIGURE 10.9 Three typical metal-cluster SBUs.

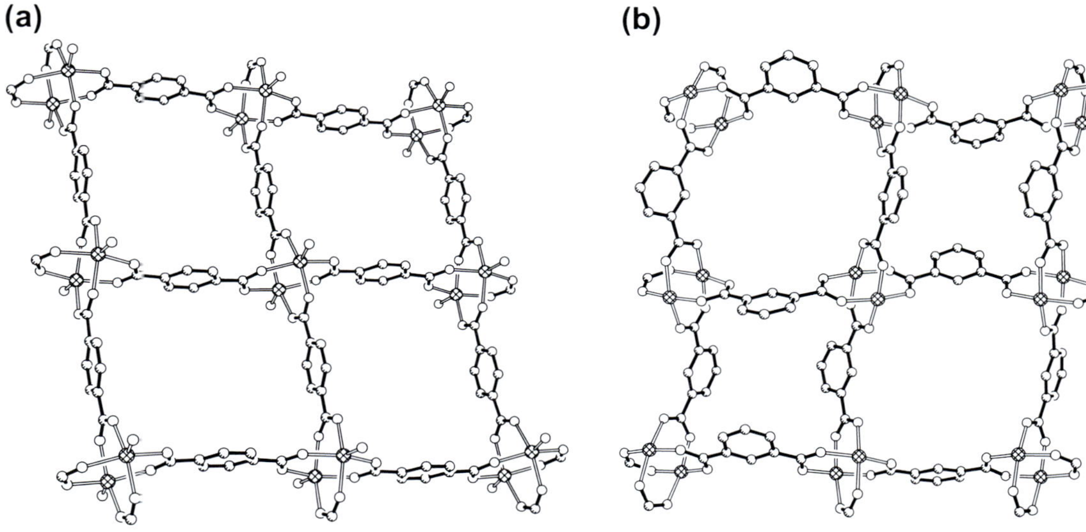

FIGURE 10.10 The 2D nets of $[Zn(1,4\text{-bdc})(H_2O)]$ (a) and $[Cu_2(1,3\text{-bdc})_2]$ (b).

(Fig. 10.10a) [12]. Owing to the hydrogen-bonding interactions between the terminal water molecules and carboxylate oxygen atoms from adjacent layers, the 2D nets of $[Zn(1,4\text{-bdc})(H_2O)]$ were stacked into a porous structure with 1D channels (size ca. 5 Å), which exhibits N_2 and CO_2 sorption behavior. It is also noteworthy that in some analogous microporous copper(II) dicarboxylates, such as $[Cu_2(1,3\text{-bdc})_2(DMF)]\text{-}(DMF)\cdot H_2O\cdot(C_2H_5OH)_{0.5}$, (1,3-bdc = 1,3-benzenedicarboxylate), which was solvothermally synthesized, the terminal solvent ligand at the axial positions of the dinuclear core is labile and can be removed via heating in vacuum to provide vacant sites for ligating potential ligands (e.g., ethanol) in the adsorption (Fig. 10.10b). Thus this compound can serve as a selective gas-sorption material based on a pore size effect and active sites [13].

When tritopic benzene-1,3,5-tricarboxylic acid ($btcH_3$) is used, the paddle-wheel SBUs can be extended into more interesting nets. For example, solvothermal reaction of cupric nitrate with $btcH_3$ in H_2O:EtOH (1:1) at 180°C gave turquoise crystals of $[Cu_3(btc)_2(H_2O)_3]$ (HKUST-1) [14]. HKUST-1 exhibits 3D square-shaped pores (9×9 Å^2) (Fig. 10.11), which are occupied by solvent molecules.

In the $[M_3O(COO)_6L_3]$ SBUs (Fig. 10.9b), the three metal ions are bridged by one μ_3-oxo group and six carboxylate ends into a triangular motif. They can be extended into a 2D three-connected net by ditopic ligands at the three L sites, a 3D six-connected net by dicarboxylates at the carboxylate sites, and even a 3D nine-connected net by mixed ditopic ligands at the six carboxylate sites and the three L sites, as illustrated in Fig. 10.12. Therefore, the $[M_3O(COO)_6]$ SBUs can be used as three-, six-, and nine-connected nodes to construct different coordination nets.

It seems that it is difficult to replace three terminal ligands in the SBUs to generate a three-connected net. So far only one example is reported in an iron(III) compound. By carefully layering an acetonitrile solution of $[Fe_3O(OAc)_6(H_2O)_3]Cl$ on an aqueous solution of $FeCl_3$ and NaOAc, red plate-like crystals of $[Fe_3(\mu_3\text{-}O)(\mu\text{-}OAc)_6(H_2O)_3][Fe_3(\mu_3\text{-}O)(\mu\text{-}OAc)_{7.5}]_2\cdot 7H_2O$ were obtained after months [15]. In this compound, $[Fe_3(\mu_3\text{-}O)(\mu\text{-}OAc)_{7.5}]^{1/2-}$ anions are connected by *anti-anti* μ-acetate

FIGURE 10.11 The structure of $[Cu_3(btc)_2(H_2O)_3]_n$ viewed down the *a*-axis, the terminal H_2O molecules are omitted for clarity. *From Ref. [14]. Reprinted with permission from AAAS.*

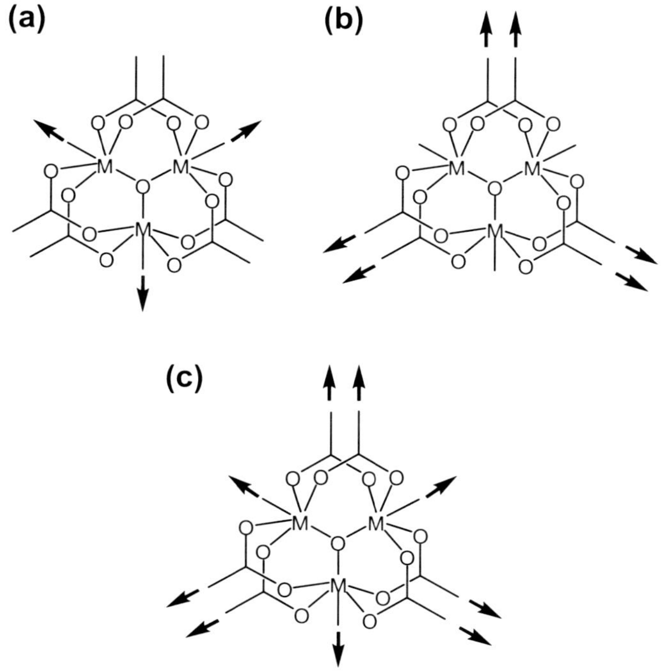

FIGURE 10.12 Three possible connecting fashions of [M_3O (carboxylate)$_6$].

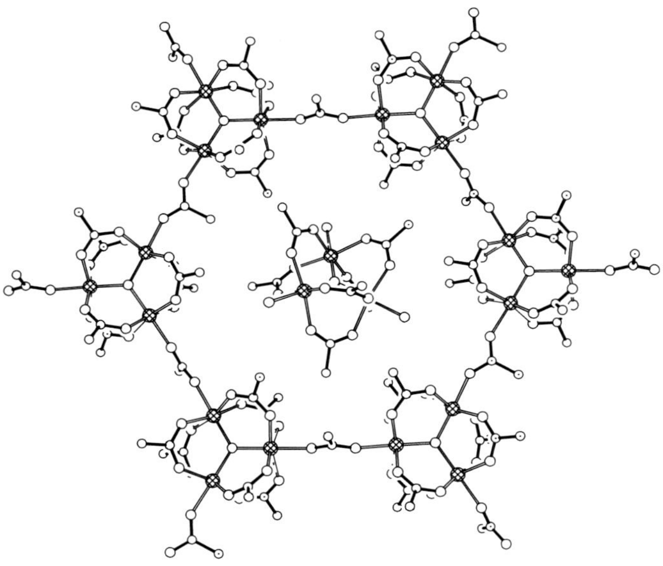

FIGURE 10.13 A structural unit of the "star" anionic framework.

bridges into three-connected 2D net with large voids, which host isolated [$Fe_3(\mu_3\text{-}O)(\mu\text{-}OAc)_6(H_2O)_3]^+$ cations (Fig. 10.13). Since the mean intra- and intercluster Fe···Fe distances are 3.307 and 6.130 Å, respectively, the anionic layer belongs to a distorted "star" lattice. Consequently, this compound exhibits the co-existence of spin-frustration and long-range magnetic order at low temperature.

When the SBUs are connected by dicarboxylates at the carboxylate sites, the SBUs become six-connected nodes, furnishing a 3D six-connected net. For example, hydrothermal reaction of equivalents of $Cr(NO_3)_3$, 1,4-bdcH$_2$ and HF at 220°C for 8 h yielded crystalline powder of $[Cr_3OF(H_2O)_2(1,4\text{-bdc})_3]\cdot nH_2O$ ($n \approx 25$) (MIL-101). In this material, as shown in Fig. 10.14, each $[M_3O(COO)_6]$ node is interlinked by six 1,4-bdc into 3D net of MTN zeolite topology with two kinds of large cages ($d = 2.9$ nm and 3.4 nm). The Langmuir surface is very large (5900 $m^2\,g^{-1}$) [16].

The SBUs can also act as nine-connected nodes by replacement of both the terminal ligands and carboxylate ends by mixed ditopic ligands into a 3D nine-connected net. By using naphthalene-2,6-dicarboxylate (ndc) and 4-(pyridin-4-yl)benzoate (pba) as mixed linkers, a robust porous material $[Ni^{II}_2Ni^{III}(\mu_3\text{-}OH)(pba)_3(ndc)_{1.5}]\cdot 9.5DMA\cdot 8.5H_2O$ (MCF-19, DMA = N,N'-di-methylacetylamide) with **ncb** topology was obtained (Fig. 10.15), which exhibits not only unique uninodal nine-connected net, but also interesting biporous intersecting channel system [17].

The tetrametallic $[M_4(\mu_4\text{-}O)(\mu\text{-}RCOO)_6]$ SBUs (Fig. 10.9c) are very useful for constructing 3D cubic nets with linear organic dicarboxylates. This kind of porous materials all have the general formula of $[M_4(\mu_4\text{-}O)L_3]$ (L = dicarboxylate) such that each SBU is surrounded by six carboxylate groups of the dicarboxylates. The materials can be easily prepared by the following approach. Solvothermal treatments of an N,N'-diethylformamide (DEF) solution mixture of $Zn(NO_3)_2\cdot 4H_2O$ and bdcH$_2$ at 85–105°C give crystalline $[M_4(\mu_4\text{-}O)(1,4\text{-bdc})_3]\cdot$guest in very high yield [18]. The framework of this material was termed to MOF-5 (Fig. 10.16a), which was the original one of this series of materials. MOF-5 exhibits 3D intersecting channels of size ca. 0.38 nm, and good sorption properties. More interestingly, the ditopic 1,4-bdc can be replaced by several analogues, two of which are shown in Fig. 10.16b, leading to the change of the channel size from 0.38 nm to 2.88 nm. The largest solvent accessible area of this series is up to 91.1%. It is noteworthy that 3,3′,5,5′-tetramethyl-4,4′-bipyrazolate can be used in place of the dicarboxylate to generate the SUB $[M_4(\mu_4\text{-}O)]$ and consequently the 3D nets similar to MOF-5 [19].

10.1.3. Pillared-layer Nets

Conventional reactions of $Cu(ClO_4)_2\cdot 6H_2O$ with Na$_2$pzdc (pzdc = pyrazine-2,3-dicarboxylate) and different pillar ligands, e.g., pyrazine, 4,4′-bpy and N-(4-pyridyl)-isonicotinamide (pia), in water or EtOH/H$_2$O (1/1), respectively, lead to three porous pillared-layer coordination polymers. In their crystal structures, Cu(pzdc) layers are interlinked by the ditopic N-ligands (L) to

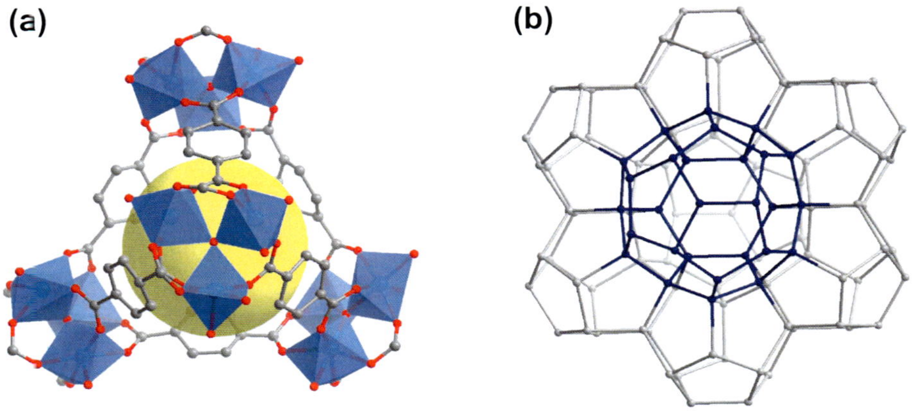

FIGURE 10.14 A supertetrahedron (ST) unit (a) constructed with 1,4-bdc and trinuclear SBUs and the 3D MTN topology (b) in $[Cr_3OF(H_2O)_2(1,4\text{-bdc})_3]$ framework (the vertices represent the centers of each ST).

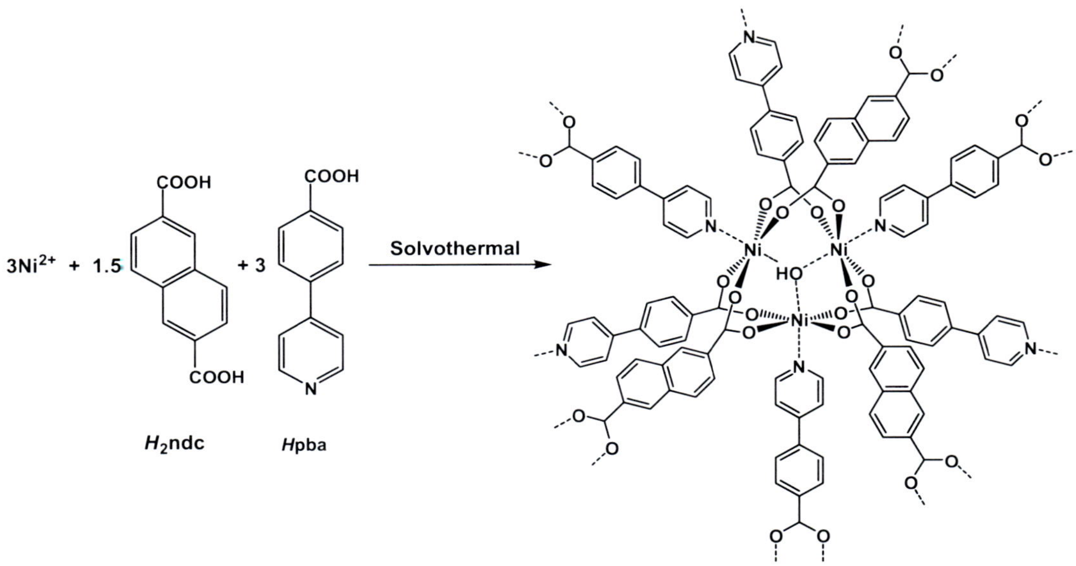

FIGURE 10.15 Construction of a structural unit of the uninodal, nine-connected net of MCF-19 by mixed linear organic ligands and the $Ni^{II}_2Ni^{III}(\mu_3\text{-OH})$ clusters.

exhibit 1D channels with different sizes of 4×6 Å^2, 9×6 Å^2 and 10×6 Å^2, respectively, depending on the lengths of ditopic N-ligands (Fig. 10.17) [20]. In other words, the channel size of a pillared-layer structure can be easily tuned by using similar ditopic ligands with different lengths.

This pillared-layer strategy can be applied to other coordination polymer systems. In the paddle-wheel SBUs of $[Zn_2(\text{dicarboxylate})_2(\text{solvent})_2]$, the apical solvent ligands (such as H_2O, DMF, MeOH, or EtOH) are labile, which can be replaced by other stronger ditopic N-ligands under solvothermal conditions to generate pillared-layer 3D porous structures [21]. Different dicarboxylates (with different *L* linkers) and neutral ditopic N-ligands can lead to different sizes of channels (see Fig. 10.18).

10.2. STRUCTURAL MODULATION BY REACTION CONDITIONS

As polymeric metal-organic compounds, coordination polymers are usually insoluble, so that recrystallization is usually not an appropriate technique to generate single crystals. Thus one of the challenges in the research is to grow single crystals that are suitable for detailed

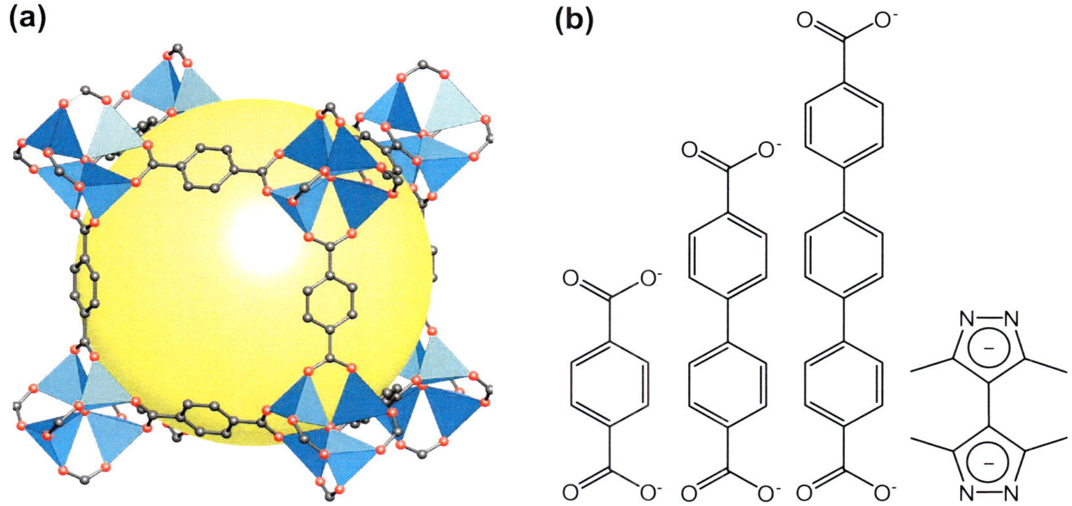

FIGURE 10.16 A structural unit of MOF-5 (a), three examples of the ditopic carboxylates and 3,3′,5,5′-tetramethyl-4,4′-bipyrazolate that can be used to form similar nets (b). From [16]. *Reprinted with permission from AAAS for (a).*

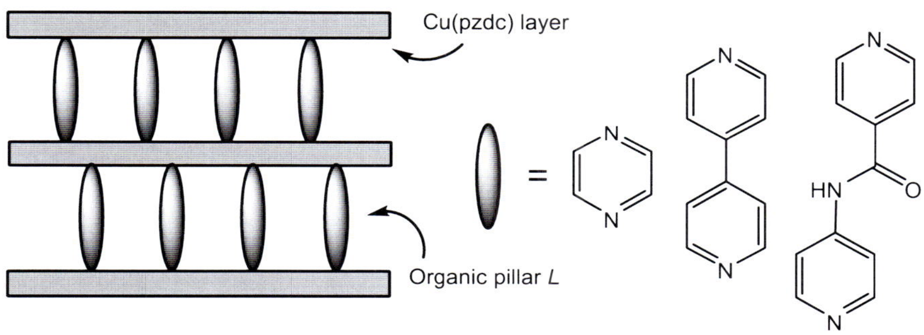

FIGURE 10.17 Schematic representation of the pillared-layer structures of $[Cu_2(pzdc)_2(L)]\cdot xH_2O$. *From Ref. [20]. Mitsuru Kondo, Takashi Okubo, Akiko Asami, Shin-ichiro Noro, Tomomichi Yoshitomi, Susumu Kitagawa, Tomohiko Ishii, Hiroyuki Matsuzaka, and Kenji Seki, Rational Synthesis of Stable Channel-Like Cavities with Methane Gas Adsorption Properties:*$[\{Cu_2(pzdc)_2(L)\}n]$*(pzdc=pyrazine-2,3-dicarboxylate; L=a Pillar Ligand), Angew. Chem. Int. Ed. 1999, 38, No. 1/2, Copyright Wiley-VCH Verlag GmbH & Co. KGaA. Reproduced with permission.*

crystallographic analysis by single-crystal diffraction. So far only very few of the crystal structures of coordination polymers have been established by powder diffraction.

Although coordination bonds are usually stronger and have better directionality than hydrogen bonding and other weak supramolecular interactions, structural prediction of coordination polymers should be, in principle, easier than that for a molecular crystal. However, structural diversity and uncertainty is an intrinsic problem for crystallization and construction of coordination polymers, due to the coordination flexibility of the metal ions and organic ligands, as well as the presence of other chemical variables such as solvent molecules and counter ions. Therefore, understanding the pivotal factors of individual variables that are involved in the crystallization processes should be very helpful for molecular design and controlling the structures. Among a number of factors that may contribute to the structures of coordination polymers, temperature, pH, solvent, counter ion, and template (or additive) play an important role in the formation of products. A number of examples will be given here to illustrate their significance in modulation of the structure.

10.2.1. Temperature Effect

Temperature plays an important role in the formation of either kinetic or thermodynamic product. The thermodynamically favored product can usually be obtained at higher reaction temperature, whilst low temperature favors the kinetic product. The effect of

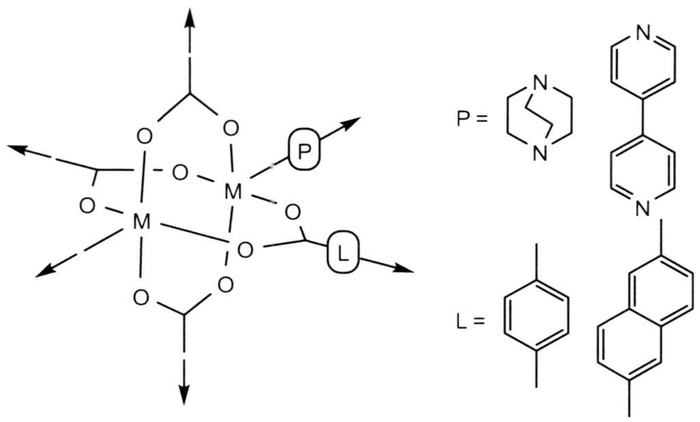

FIGURE 10.18 Linkers (L) and pillars (P) in paddle-wheel-based pillared-layer nets. *From Ref. [21]. Hyungphil Chun, Danil N. Dybtsev, Hyunuk Kim, and Kimoon Kim, Synthesis, X-ray Crystal Structures, and Gas Sorption Properties of Pillared Square Grid Nets Based on Paddle-Wheel Motifs: Implications for Hydrogen Storage in Porous Materials, Chem. Eur. J. 2005, 11, 3521–3529, Copyright Wiley-VCH Verlag GmbH & Co. KGaA. Reproduced with permission.*

temperature can be even more remarkable in hydro (solvo)thermal reactions, in which metastable kinetic phases can be more frequently isolated.

For a clear relationship between the reaction temperature and structure, all but the temperature should be set unchanged. A straightforward example was found in the hydrothermal reaction of $NiCl_2$, 3-(3-pyridyl) acrylic acid (pyaraH), and NaOH. At 180°C for 24 h, the reaction gives crystals of α-[Ni(pyara)$_2$(H_2O)$_2$], which has an interesting 3D (8,4) net. When the reaction is repeated at only 150°C, a totally different phase, or β-phase, is obtained, which consists of simple 1D chains (Fig. 10.19) [22].

Another nice example is the solvothermal reaction of 4,4′-bpy, 1,4-bdcH_2, and $Cd(NO_3)_2 \cdot 4H_2O$ in a 1:1:1 ratio in DMF/DEF (2:1 v/v) with different concentrations of starting materials (0.0125–0.10 M) at temperature range of 85–125°C, the reaction afforded two compounds, [Cd (4,4′-bpy)(1,4-bdc)]·3DMF·H_2O (α-phase) and [Cd(4,4′-bpy)(1,4-bdc)] (β-phase) (Table 10.1) [23,24]. Both phases bear the same (4,4) [Cd(4,4′-bpy)] layers based on the same dinuclear SBUs of {$Cd_2N_4O_8$}, which are further pillared into a neutral 3D framework. The most remarkable difference is that twofold interpenetration is found in β-phase to eliminate the porous space, while no interpenetration occurs in α-phase and the porous space is occupied by the solvent molecules (Fig. 10.20).

The results indicate that β-phase, as a thermodynamically stable, denser interpenetrated crystal form, is favored at higher temperature, while α-phase, as a kinetically stable, with an open and non-interpenetrated structure, is favored at lower temperature. On the other hand, when the temperature remains fixed and the concentration is changed, high concentration tends to favor β-phase over α-phase.

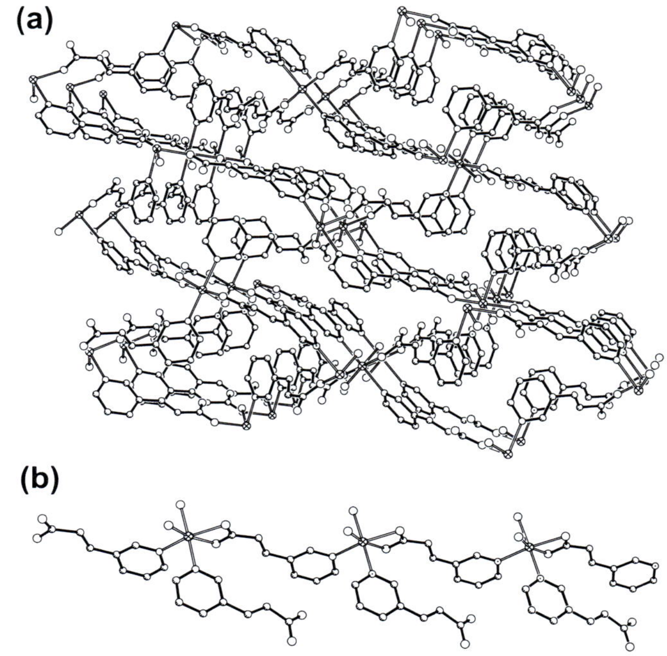

FIGURE 10.19 The 3D net of α-[Ni(pyara)$_2$(H_2O)$_2$] (a) and 1D chain of β-[Ni(pyara)$_2$(H_2O)$_2$] (b).

The third example is the reaction of cobalt(II) hydroxide, succinic acid, and water in a fixed stoichiometry ($Co(OH)_2$: succinic acid: water ≈ 1:1:28) with temperature as the only independent variable. At different temperatures (60–250°C), this reaction can produce five different coordination polymers which vary in the metal-to-ligand ratios, coordination geometries of the Co(II) and succinate, number of coordinated water and hydroxide, dimensionalities of the coordination frameworks, and contents of guest molecules [25].

As shown in Table 10.2, at 60°C and 100°C, the products are 1D chains with mononuclear Co(II) ions and edge-sharing trimers as the SBUs, respectively. In phases formed above 100°C, increasing temperatures in this series lead to fewer coordinated water molecules per Co(II) atom, increased edge-sharing connectivity, higher coordination numbers for carboxylate groups, and

TABLE 10.1 The Products Isolated at Different Temperatures and Concentrations

	0.1 M	0.05 M	0.025 M	0.0125 M
85 °C	α	α	α	α
95 °C	α	α	α	α
105 °C	$\alpha + \beta$	$\alpha + \beta$	α	α
115 °C	$\alpha\beta$	$\alpha\beta$	$\alpha + \beta$	α
125 °C	$\alpha\beta$	$\alpha\beta$	$\alpha + \beta$	unknown phase

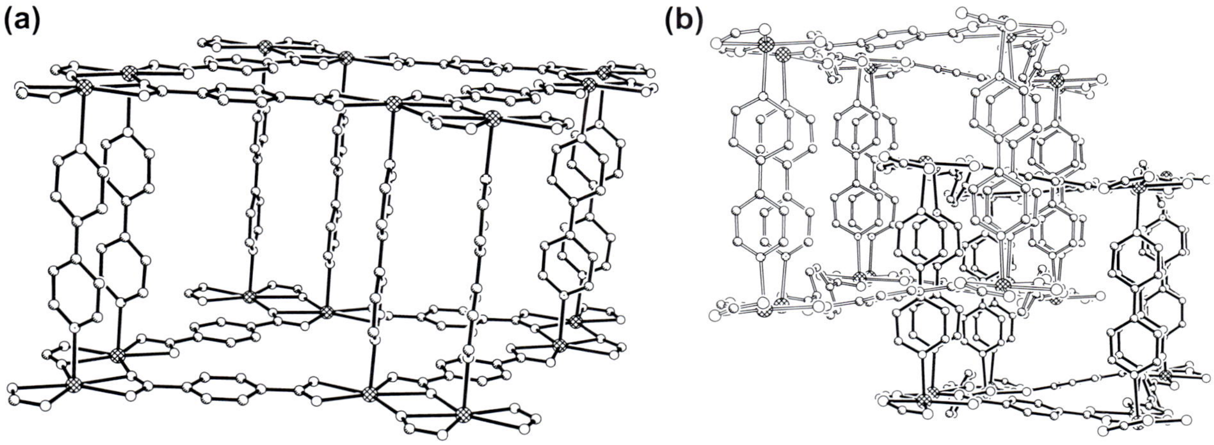

FIGURE 10.20 Pillared noninterpenetrated framework of [Cd(4,4′-bpy)(1,4-bdc)]·3DMF·H_2O (α-phase) (a) [Reprinted with permission from Ref. [23]. Copyright 2009 American Chemical Society.] and pillared interpenetrated framework of [Cd(4,4′-bpy)(1,4-bdc)] (β-phase) (b).[24] *Reproduced by permission of The Royal Society of Chemistry.*

incorporation of hydroxide. The amount of hydroxide incorporated into the structure does not appear to follow a clear progression, suggesting that structural considerations are more significant than temperature in determining the overall stoichiometry. However, it is very clear that the amount of water (total water or water ligands) decreases with increasing temperature. This trend can also be found in other reaction systems [26]. Meanwhile, the overall dimensionality of the structures also increases, from 1D chains up to 100°C, 2D sheets at 150°C, to 3D frameworks at 190°C and above. The M—O—M dimensionality increases as well, beginning with isolated Co(II) atoms or clusters through 100°C, and continuing to 2D Co—O—Co sheets at 150°C and above. The crystal densities are slightly increased at 150°C and above.

The above discussion strongly implies that the reaction temperature may be critical in both conventional and solvo(hydro)thermal reactions. However, it is very difficult to estimate an appropriate temperature for a targeted structure, as the assembly process is usually unknown and complex. Nevertheless, the above-mentioned trend should be considered and relevant literature should be consulted.

It should also be noted that, in hydro(solvo)thermal reactions, the reaction temperature can also have effect on the valence of some metal ions. The most frequently encountered metal ion of valence change is Cu(II), which can be easily changed into Cu(I) in the presence of organic heterocyclic compounds under hydro(solvo) thermal conditions. Such in situ reactions are usually dependent on several parameters, particularly temperature and pH. It has been widely observed that higher temperature and pH are favored for the reduction of Cu(II). For example, at different temperatures and pH, reaction of Cu(II) salt with imidazole yielded four coordination polymers, including 3D $[Cu^{II}(im)_2]$ of the SOD topology [27], $[Cu^{I}Cu^{II}(im)_3]$ of self-entangled uninodal four-connected 3D net, $[Cu^{I}_2Cu^{II}(im)_4]$ of simple four-connected 2D net [28], and $[Cu^{I}(im)]$ of 1D chain [29],

TABLE 10.2 Data for the Five Cobalt Succinates Synthesized at Different Temperatures

Synthesis temperature	Phase	H_2O/Co^{2+a}	Density g/cm³	Co^{2+}/CO_2	Dimensionality[b]
60 °C	$[Co(H_2O)_4(C_4H_4O_4)_2]$	4(4)	1.945	1	1(0)
100 °C	$[Co(H_2O)_2(C_4H_4O_4)_2]$	2(2)	1.926	2	1(0)
150 °C	$[Co_4(H_2O)_2(OH)_2(C_4H_4O_4)_3]\cdot 2H_2O$	1(1/2)	2.085	2.67	2(2)
190 °C	$[Co_6(OH)_2(C_4H_4O_4)_5]\cdot 2H_2O$	1/3(0)	2.197	2.8	3(2)
250 °C	$[Co_5(OH)_2(C_4H_4O_4)_4]$	0(0)	2.337	3	3(2)

[a]*The first number in the H_2O/Co^{2+} column refers to total water content, the second in parenthesis refers to water ligands.*
[b]*The first number in the Dimensionality column refers to the total dimensionality, the second to the M—O—M dimensionality.*

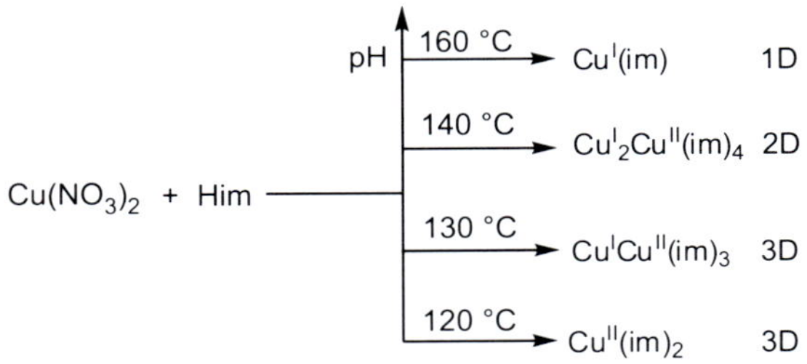

FIGURE 10.21 Temperature- and pH-dependent in situ Cu^{II} reduction in the formation of copper imidazolates.

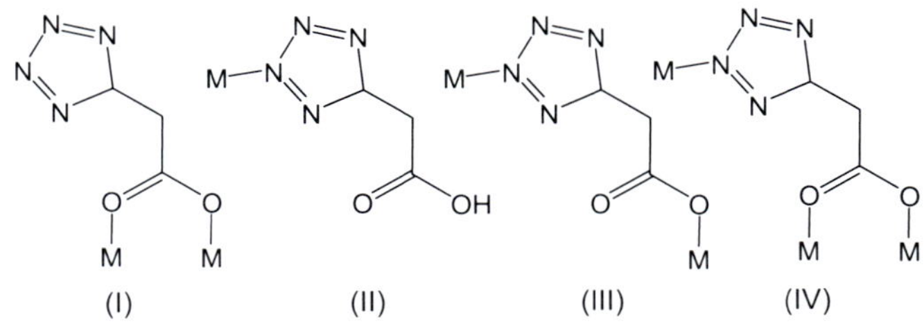

FIGURE 10.22 pH-dependent coordination modes of tetrazole-1-acetic acid (H_2tza). *Reprinted with permission from Ref. [32]. Copyright 2008 American Chemical Society.*

at 120–160°C (Fig. 10.21). Obviously, the difference in the valence of copper results in the difference in stoichiometry, and contributes to the structure.

10.2.2. pH Effect

Acid–base reactions are usually involved in the assembly of coordination polymers. Deprotonation of the donor groups of organic ligands for coordination to metal ions or clusters is then critical. Moreover, the reaction kinetics can be influenced by altering pH conditions in some cases [30].

Basicity is also very important to control the structures of coordination polymers of base/acid ligands. For example, when the Mn(II)/*p*-$H_2O_3PCH_2C_6H_4CO_2H$ molar ratio was fixed, the reaction of Mn(II) ions with phosphonocarboxylic acid *p*-$H_2O_3PCH_2C_6H_4CO_2H$ by increasing the pH led to the formations of three compounds Mn(*p*-$HO_3PCH_2C_6H_4CO_2H)_2 \cdot 2H_2O$, Mn(*p*-$O_3PCH_2C_6H_4CO_2H) \cdot H_2O$, and $Mn_2(OH)$(*p*-$O_3PCH_2C_6H_4CO_2) \cdot 2H_2O$ [31], which shows the stepwise deprotonation of phosphonate and carboxylate groups. Similarly, three copper(II) tetrazole-1-acetic acid (H_2tza) coordination polymers, $[Cu(tza)_2(Htza)_2] \cdot 2H_2O$, $[Cu(tza)_2]$, and $[Cu_4(tza)_6(\mu_3\text{-}OH)_2] \cdot 4H_2O$ can be prepared in the same molar ratio of the metal and ligand at the same temperature of 80°C by adjusting the pH of the reaction system to be 1.5, 2.5, and 5.0, respectively [32]. In this series, the H_2tza ligand exhibits the coordination modes I and II for $[Cu(tza)_2(Htza)_2] \cdot 2H_2O$, III for $[Cu(tza)_2]$, and I and IV for $[Cu_4(tza)_6(\mu_3\text{-}OH)_2] \cdot 4H_2O$ (Fig. 10.22).

It should also be mentioned that appropriate basicity is very important for constructing the cluster-based coordination polymers. For instance, in the reaction of Co(II) hydroxide and succinic acid (see Section 10.2.1), if other condition is kept the same and the temperature is fixed at 100°C, while the cobalt(II) hydroxide and succinic acid molar ratio is changed, the pH of the system will be altered. When the Co(II) hydroxide and succinic acid molar ratio is less than 1:1, the system is acidic, giving rise to a nonhydroxy product of $[Co(H_2O)_2(C_4H_4O_4)_2]$. When the Co(II) hydroxide and succinic acid molar ratio is larger than 1:1, the system is basic and the product is $[Co_7(H_2O)_3(OH)_6(C_4H_4O_4)_4] \cdot 7H_2O$ bearing hydroxy groups [26].

For many in situ metal/ligand reactions, pH is an important factor [33]. For example, decarboxylation and hydroxylation of aromatic ligand, as well as in situ metal/ligand redox reactions of Cu(II) are highly pH-dependent, and higher pH is usually favored for these reactions.

The hydrothermal reaction of Cu(II) salt with 1,2,3-benzenetricarboxylic acid (1,2,3-btcH_3) in the presence of 4,4′-bpy can be stepwise. Without addition of NaOH, no decarboxylation and no hydroxylation of 1,2,3-btc occurred (Fig. 10.23), leading to the formation of 3D porous framework $[Cu_2(1,2,3\text{-btc})(4,4'\text{-bpy})(H_2O)_2](NO_3)$ [34]. When 2 equiv. of NaOH was added, the 2-carboxylate was removed, giving rise to a mixed valence Cu(I,II) coordination polymer $[Cu_2(1,3\text{-bdc})(1,3\text{-bdcH})(4,4'\text{-bpy})_{1.5}]$. When 1,2,3-btc$H_3$ was reacted with 4 equiv. of NaOH, the 2-carboxylate was replaced by a hydroxy group, furnishing a mixed valence Cu(I,II) coordination polymer $[Cu_2(1,3\text{-bdcO})(4,4'\text{-bpy})]_n$ (1,3-bdcO = 2-hydroxy-1,3-benzenedicarboxylate). Actually, the conversion from 1,2,3-btc to 1,3-bdcO can be considered to be stepwise via the formation of 1,3-bdc as the intermediate, since 1,3-bdc can be hydroxylated into 1,3-bdcO under similar condition [35]. Meanwhile, other metal ions can also mediate similar in situ reactions of 1,2,3-btcH_3, which further confirms the importance of basicity [36].

The above observations indicate that: (1) the deprotonation of the donor group(s) is dependent on the pH, while at higher pH, more donor atoms of a multidonor ligand will be involved in coordination to the metal ions; and (2) inorganic and organic hydroxy groups may be formed at higher pH. Consequently, different pHs may lead to different in situ ligand reactions and crystal structures owing to different coordination modes of the ligand. It should also be noted that the reaction basicity is critical to reduction of Cu^{II} into Cu^{I} under hydrothermal condition, higher basicity is favored for the generation of Cu^{I} (see Fig. 10.21).

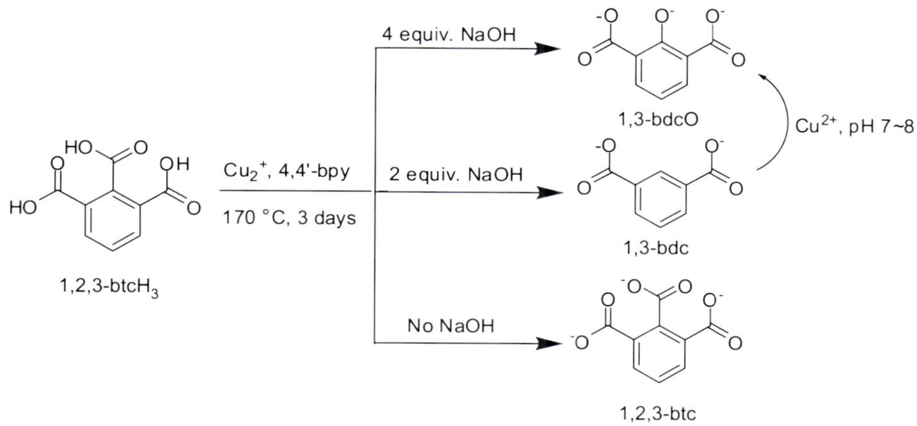

FIGURE 10.23 In situ reactions of 1,2,3-benzenetricarboxylic acid (1,2,3-btcH_3) and 1,3-bdcH_2 under different basicity. *From Ref. [34]. Reproduced with permission from The Royal Society of Chemistry.*

10.2.3. Template and Additive Effect

Templating strategy has been widely used in the synthesis of metal coordination complexes and inorganic porous materials. From the following examples, we can find it is also an effective approach for assembling coordination polymers, especially those with porous structures.

The isomeric coinage metal imidazolates serve as a representative example for template-induced isomerism. The chemical compositions and local coordination geometries of these binary metal imidazolates are readily controlled. The angular bridging mode of imidazolate satisfies the simplest model of supramolecular isomerism (ring, chain, and helix). For example, silver(I) 2-methylimidazolates can be prepared via liquid diffusions between the solution of Hmim (2-methylimidazole) in organic solvents and the solution of $AgNO_3$ in aqueous ammonia [37]. When different organic solvents were employed, e.g., methanol/benzene, *p*-xylene, and methanol, three supramolecular isomers of silver(I) 2-methylimidazolates, namely $[Ag_4(mim)_4(C_6H_6)]$,$[Ag_4(mim)_4(C_8H_{10})]$, and [Ag(mim)], could be obtained (Fig. 10.24). As a result of the template effect, the benzene solvated isomer features pseudo 8_1 helical chain with $Ag_8(mim)_8$ subunits and a helical pitch of 10.4307(4) Å, in which all the mim ligands are orientated in the *syn* fashion, and the *p*-xylene solvated isomer features an *S*-shaped chain with arc-shaped $Ag_4(mim)_4$ fragments, in which the mim ligands are orientated in both *syn* and *anti* fashions, whereas the unsolvated isomer exhibits simple zigzag chains with all the mim ligands in an *anti* fashion.

It is worth to mention that many porous coordination polymers are assembled by using template intentionally or unintentionally. With the similar effect mentioned above, the templates are usually very important in inducing the formation of some kinds of holes or cavities, via their impacts on the space-filling the holes or cavities. Indeed, nature abhorred a vacuum! If no appropriate space-filling molecules exist, and the holes or cavities in the coordination polymers are rather large, interpenetration of the nets can easily occur to decrease the volume of porosity [38]. Actually, when no intentional template is added into the assembly system, the solvent molecules may play a role as template in the formation of many porous coordination polymers.

Being different from templates that serve as space-filling components in the pores or cavities of the products, it has recently been found that some kinds of

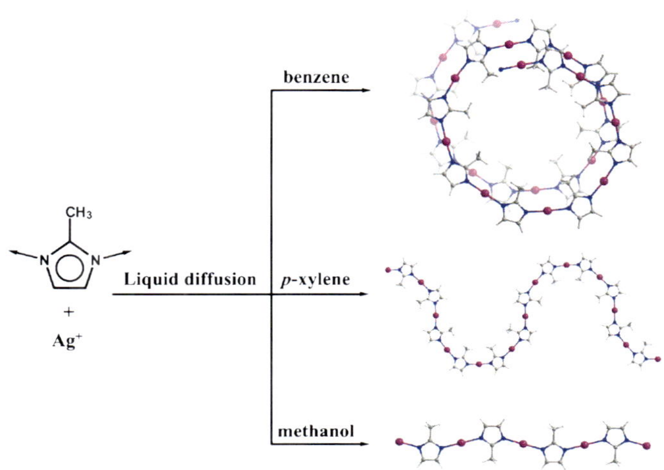

FIGURE 10.24 Solvent-induced supramolecular isomerism in silver(I) 2-methylimidazolate. *From Ref. [34]. Reproduced with permission from The Royal Society of Chemistry.*

molecules and ions in the reaction systems may also induce the formations of different structures but do not present in the products [39]. These molecules and ions can be termed as buffering additives. This structural directing approach may be used to fine-tune the local coordination structures for the formation of coordination supramolecular isomers including coordination polymers, although more investigations are needed to understand the intrinsic mechanism.

For example, a mixture of Cu(II) salt (2.0 mmol), aqueous ammonia (25%, 2.0 mL) and acetonitrile (5.0 mL) was sealed in a 15-mL Teflon-lined reactor and heated at 120−160°C for 3 days and slowly cooled to room temperature, giving rise to the in situ formation of copper(I) 3,5-dimethyl-1,2,4-triazolate (mtz), or [Cu(mtz)] [40]. The superstructures of [Cu(mtz)] can be tuned into three different phases, α, β, and γ-[Cu(mtz)], or three true supramolecular isomers (Fig. 10.25), by use of different Cu(II) sources and the presence or absence of 4,4′-bpy. Among them, the α-phase featuring a twofold interpenetration of the 4.8.10 nets was prepared by use of $Cu(OH)_2$, $Cu_2(OH)_2CO_3$, or $CuSO_4$ as the Cu(II) source, the β-phase featuring non-interpenatrating $8^2 10$-a topology was prepared by use of $Cu(OH)_2$ as the Cu(II) source and 4,4′-bpy (1-2 mmol) as an additive, while the (β-phase featuring a non-interpenatrating 6.10^2 net was prepared by use of $Cu(NO_3)_2$ as the Cu(II) source. In this case, the use of different Cu(II) sources and 4,4′-bpy led to the presence of different neutral and/or ionic additives in the system, which may give rise in different coordinatively buffering effect in the formation and crystallization of the products. The above phenomenon and other related examples indicate that the use of additives having different ligation abilities may have significant effect in tuning the superstructures, thus represents a new strategy for controlling of supramolecular isomerism of coordination polymers [41].

10.2.4. Solvent Effect

Solvent molecules have direct effects on solubilities of both reactants and products. Besides the role in dissolving the reactants and products, they can also impact on the superstructures of the intermediates and final products through supramolecular interactions between the products and solvents. Therefore, solvent molecules can have effect not only on the crystallization process, but also on the configurations of the certain groups in the products, leading to different superstructures. They play a role of template and occupy the voids of the products, or they only serve as the crystallization environment and do not exist in the products. As the template effect of solvent has been described in Section 10.2.3, only the role of crystallization environment is discussed below.

A very straightforward example was reaction of copper(II) salt with 2-ethylimidazole (Heim) [42]. When reacting $Cu_2CO_3(OH)_2$ (1.0 mmol) and Heim (1.0 mmol) in aqueous ammonia (25%, 5 mL) and water (2 mL) at 160°C for 80 h, 1D structured [Cu(eim)] (α-phase) was yielded. The α-phase crystallizes as a triple-strand helix, in which the hydrophobic ethyl groups are crowded inside the nanotubular cuprous imidazolate backbone, being similar to the case of micelles formation of amphiphilic molecules in aqueous solution. Since no other supramolecular interaction is found within the triple-strand helix, it should be mainly held by the hydrophobic interactions of the ethyl groups. When the solvent polarity is reduced (by use 3 mL of cyclohexane in place of water), [Cu(eim)] crystallizes as a simple zigzag chain isomer (β-phase) with

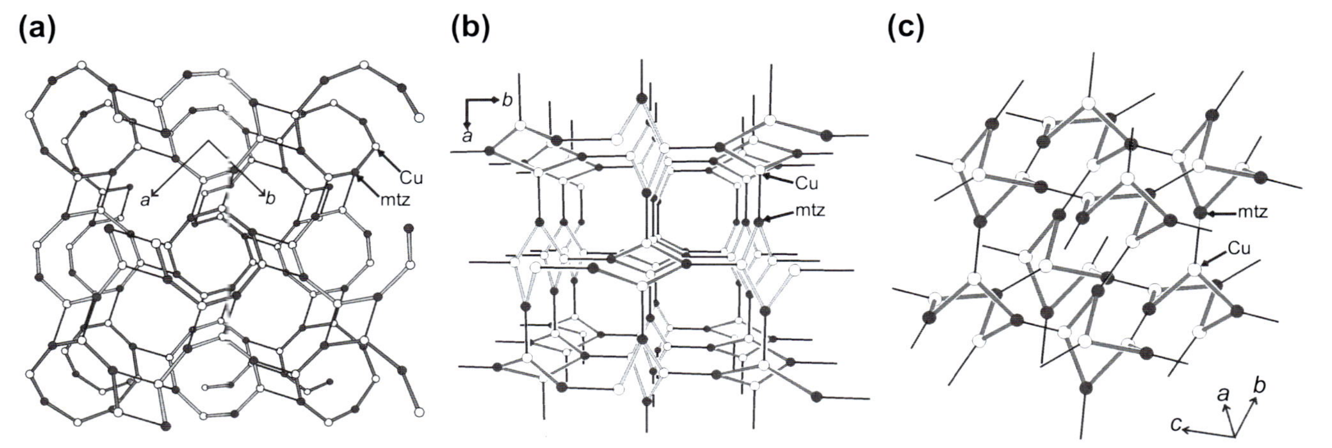

FIGURE 10.25 Topologic illustrations of α-Cu(mtz) (a), β-Cu(mtz) (b), and (γ-Cu(mtz) (c). *From Ref. [40]. Reproduced with permission from The Royal Society of Chemistry.*

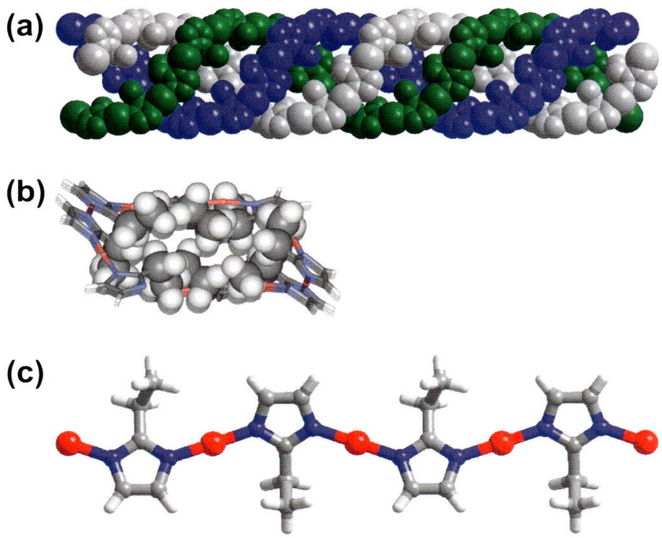

FIGURE 10.26 Side-view (a) and top-view (b) of the triple-strand helix of α-[Cu(eim)], and side-view (c) of the zigzag chain of β-[Cu (eim)]. *From Ref. [42]. Reproduced with permission from The Royal Society of Chemistry.*

alternately distributed ethyl groups (Fig. 10.26). Obviously, in a more polar solvent, the hydrophobic ethyl groups are forced to aggregate together and avoid strong interaction with the more hydrophilic solvent environment. This unique solvent-induced structural tuning demonstrates the significance of weak supramolecular interactions in directing self-assembled superstructures.

Besides that, many supramolecular isomerisms of coordination polymers induced directly by solvents were reported [5]. These findings demonstrate that the solvent effect is critical for assembling coordination polymers, leading to different supramolecular structure, via either templating or reaction environmental effects.

10.2.5. Counter-ion Effect

Counter-ions usually are the anions not involved in the coordination. In a general way, counter-ions can affect the crystallization of coordination polymer products, leading to different coordination structures. In some cases, especially in the assembly of silver(I) coordination polymers [43], the counter-ion can play a key role in the formation of final supramolecular structure. It is mainly because the different counter-ions have different coordination interactions with metal ions, especially silver(I) ions, the different H-bonding interaction with the other components, and then different influence to the crystallization speed for different supramolecular products.

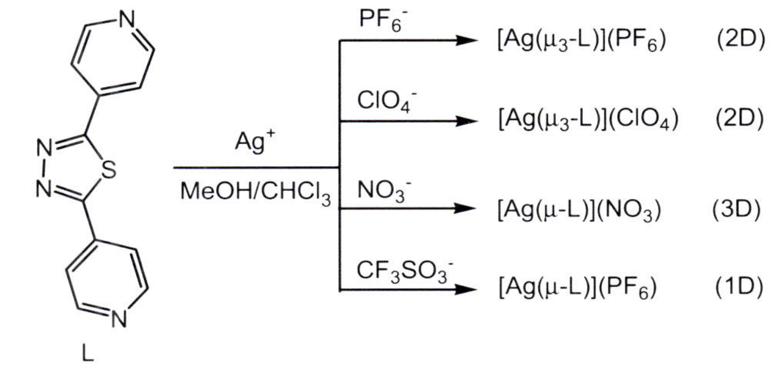

FIGURE 10.27 The influence of different counter-ion to the assembled products of Ag^{I} and organic ligand 2,5-bis(4-pyridyl)-1,3,4-thiadiazole.

For example, when various inorganic silver(I) salts were dissolved in methanol, and carefully layered on the solution of organic ligand 2,5-bis(4-pyridyl)-1,3,4-thiadiazole (L) in chloroform, a series of different products can be obtained after two weeks (Fig. 10.27) [44]. Due to the larger size and the weaker coordination ability of PF_6^- and ClO_4^- ions (shortest $Ag\cdots F = 0.2910$ nm, shortest $Ag\cdots O = 0.2701$ nm), they only play a role in the charge balance. Therefore, each L-ligand bridges three adjacent Ag^{I} centers in a tridentate coordination fashion to generate a non-interpenetrating 2D net. Whereas NO_3^- ion has a smaller size and a stronger coordination ability than PF_6^- and ClO_4^- ions, it can weakly bridge the Ag(μ-L) chains to be a 3D net ($Ag{-}O = 0.2583{-}0.2742$ nm). On the contrary, $CF_3SO_3^-$ ion, with larger size than PF_6^- and ClO_4^- ions, can decrease the bridge ability of ligand L, which only links two Ag(I) ions in a bidentate bridging mode with two pyridyl N donors to form an infinite chain.

The molecular rectangle assembled by 4,4′-bpy and pyrazine (Fig. 10.7c), which is mentioned in Section 10.2.1, is also a good example to demonstrate the importance of counter-ion on assembling such special crystal product. Since the ligands 4,4′-bpy and pyrazine have similar structure and property, only if the appropriate counter-ion, e.g., PF_6^- ion, was employed, they can assemble the rectangular grid rather than the square grid assembled by 4,4′-bpy or pyrazine, respectively [9].

10.3. IN SITU METAL/LIGAND REACTIONS

As a new synthesis method, in situ metal/ligand reaction has been widely employed in the synthesis of coordination polymers over the past decade. In general, the coordination polymers have poor solubility, thus solvothermal (including hydrothermal) reaction is a good approach to prepare coordination polymers. As higher temperature and pressure are employed, this method not only yields more stable coordination polymer product, but also gives rise to achieve some reactions inaccessible under general reaction condition. Up to date, many in situ metal/ligand reactions have been reported, including hydrolysis of carboxylate esters, organic nitriles, and aldehydes into the corresponding carboxylates, cleavage of acetonitrile/ethylene carbon–carbon bonds and 1,3,4-oxadiazole carbon–nitrogen/carbon–oxygen bonds, cleavage and formation of disulfide bonds, substitution of aromatic groups, as well as decarboxylation of aromatic carboxylates, hydroxylation of aromatic rings, cycloaddition of organic nitriles with azide and ammonia, and so on [33].

Herein, some coordination polymers with specific structures assembled by in situ metal/ligand reactions are outlined.

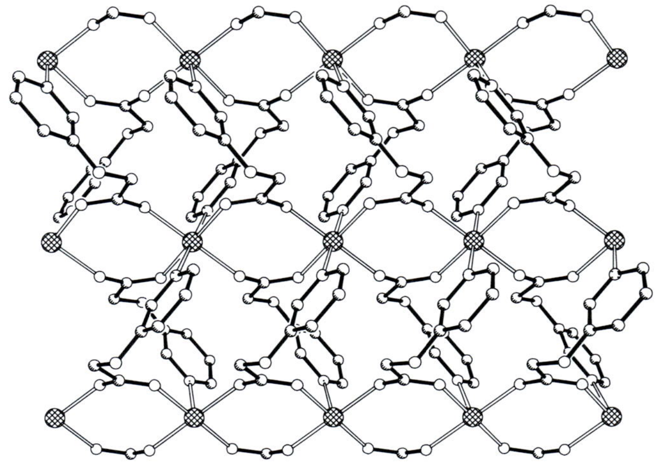

FIGURE 10.28 The 2D layer of $[Fe(pyoa)_2]$ prepared by the hydrothermal reaction of iron powder and pyoaH. *Reprinted with permission from Ref. [45]. Copyright 2008 American Chemical Society.*

10.3.1. In Situ Metal Redox Reactions

It is well-known that some metal ions have several valence-states in the solution or solid, and that the different valence-states could lead to different coordination surroundings, different aggregation structures, and hence different physical and chemical properties. Therefore, the valence-state control of metal ions in the product is a key point for assembly of the coordination polymers. Generally, the metal ions with targeted valence in product could be got from the reactant directly, but sometimes the reactant cannot provide the metal ions with specific valence requested in product. For instance, some special valence-state metal ions are unstable in solution, hence it is very difficult to assemble the corresponding product directly by using a reactant consisting of the specific valence metal ions. However, some controllable metal redox reactions, e.g., the oxidation of copper and iron, were found under the solvo (hydro)thermal reaction condition, giving rise to the functional coordination polymers with metal ions in the specific valence-state.

For example, under the solvo(hydro)thermal reaction condition, Fe(II) will be easily oxidized to Fe(III) by the oxygen in the air. Consequently, Fe(II) coordination polymers are hardly obtained. However, if iron powder is employed to react with the organic ligand, such as carboxylic acid, the solid, stable Fe(II) coordination polymers could be directly obtained via oxidation. For instance, a 2D coordination polymer $[Fe(pyoa)_2]$ [pyoa = 2-(pyridin-3-yloxy)acetate] was prepared by the direct reaction between iron powder and pyoaH in the degassed water at 140°C [45]. This compound consists of Fe(II)-carboxylate chains (Fig. 10.28), and reveals an unusual phase transition from the field-induced spin-canting antiferromagnet to single-chain-magnet, and it could not be prepared directly from a Fe(II) salt.

More importantly, the in situ metal redox reaction under the solvo(hydro)thermal reaction condition could be used to prepare the mixed-valent Fe(II,III) or Cu(I,II) coordination polymers. Practically, most of the mixed-valent metal coordination polymers were synthesized by the in situ metal redox reaction. For instance, a 3D mixed-valent Fe(II,III) *trans*-1,4-cyclohexanedicarboxylate (1,4-chdc) coordination polymer, $[Fe^{II}Fe^{III}(\mu_4\text{-}O)(1,4\text{-}chdc)_{1.5}]_\infty$, was hydrothermally synthesized by mixing iron powder and 1,4-chdcH_2, but it is hardly prepared by Fe(II) and/or Fe(III) salt directly [46]. This compound consists of ferrimagnetic chains with spin-frustration arising from competing exchange interactions between the Fe(II) and Fe(III) ions (Fig. 10.29). Besides that, some mixed-valent Cu(I,II) coordination polymers mentioned in Sections 10.2.1 and 10.2.2 are also good examples of mixed-valent coordination polymers synthesized by in situ metal redox reactions.

10.3.2. In Situ Ligand Formation

It is not a long history that in situ organic ligand reactions were employed in the synthesis of coordination polymers, but many kinds of in situ metal/ligand reactions have been uncovered to date [33]. Herein, some coordination polymers with specific structure assembled by in situ generated carboxylate ligands are outlined. These coordination polymers are hard to prepare by mixing the related ligand and metal ions directly.

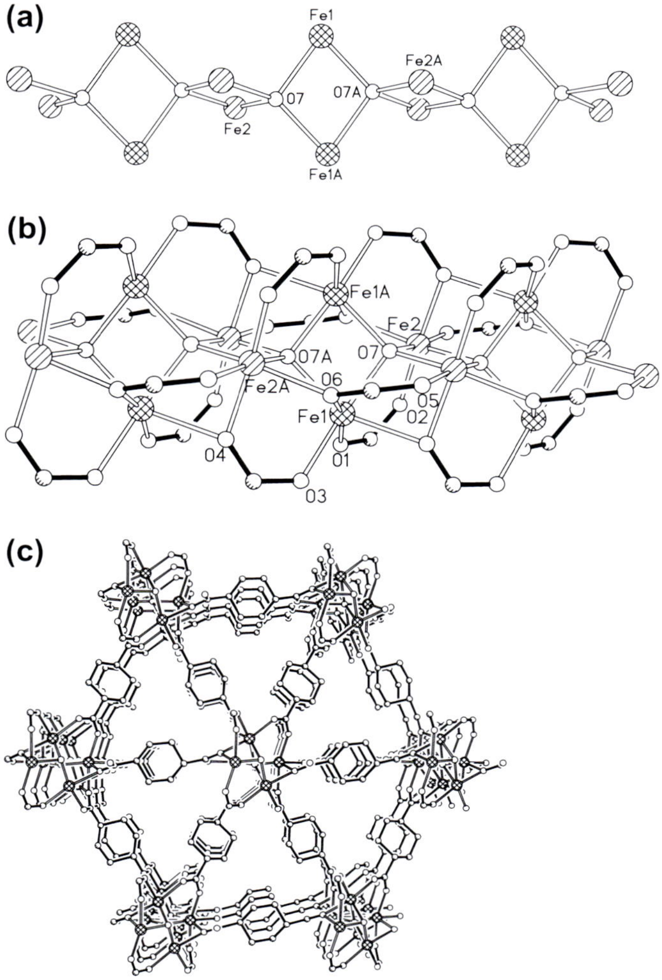

FIGURE 10.29 The carboxylate-supported tetrahedral chain (a, b), and the 3D framework (c) viewed along the *a*-axis in $[Fe^{II}Fe^{III}(\mu_4\text{-}O)(1,4\text{-chdc})_{1.5}]_\infty$. *Reprinted with permission from Ref. [46]. Copyright 2009 American Chemical Society.*

Carboxylic acids are the most widely used ligands to construct the coordination polymers including the acentric infinite coordination nets for use as NLO (second-order nonlinear optical) materials. To construct the ideal NLO material, a good electron donor and acceptor connected through a conjugated bridge are typically needed in NLO chromophore. However, the electronically asymmetric and highly dipolar NLO chromophores required in the efficient NLO materials tend to adopt centrosymmetric arrangements rather than the metastable acentric arrangements, as a result of the dominance of centrosymmetric dipole–dipole repulsions [47]. Therefore, the conventional reaction of carboxylate ligand and metal ions usually does not yield acentric or chiral coordination polymers. However, generating carboxylate groups, e.g., by hydrolysis of nitrile or ester groups in situ under solvo(hydro)thermal reaction condition, is an effective alternative synthesis method to obtain the NLO-active coordination polymers [47].

For instance, a coordination polymer $[Zn(ina)_2]$ (ina = isonicotinate), which crystallizes in chiral space group $P2_12_12_1$, was synthesized hydro(solvo)thermally by reaction of $Zn(ClO_4)_2 \cdot 6H_2O$ and 4-cyanopyridine in the ethanol:water (4:1) mixed-solvent at 130°C [48]. In other words, the final ligand ina is generated by in situ hydrolysis of 4-cyanopyridine. Each Zn(II) center in this compound is coordinated to two pyridyl N atoms of two isonicotinate groups, and to two carboxylate groups of two other isonicotinate groups in a monodentate fashion, resulting in a single 3D diamondoid net with a large void (Fig. 10.30). And the void space is erased by the formation of a threefold diamondoid structure, in which three independent diamondoid nets mutually interpenetrate. Some similar acentric or chiral coordination polymers were also prepared by this approach [47].

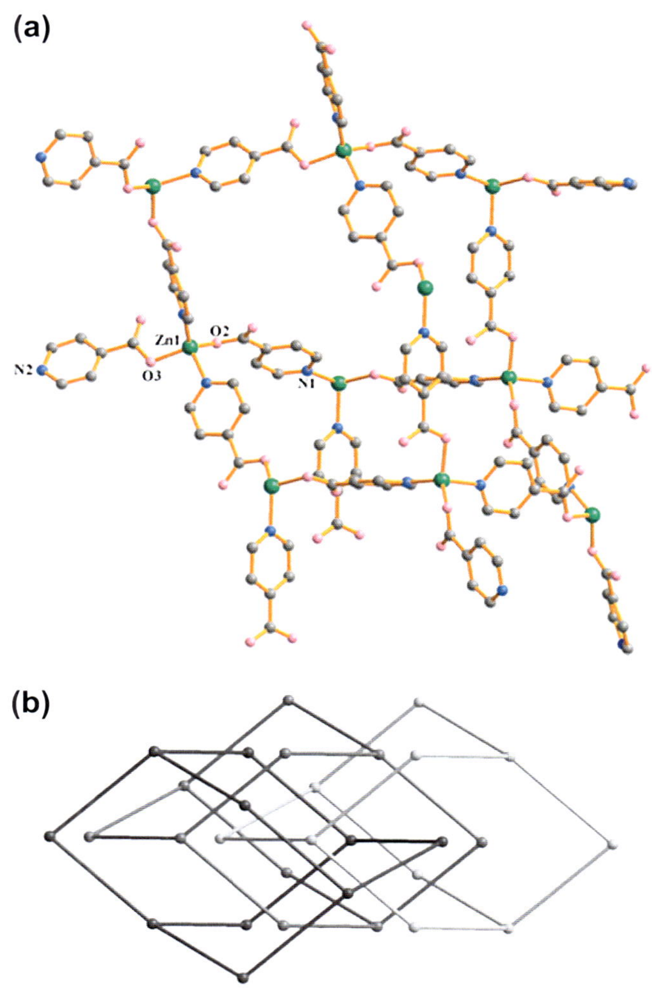

FIGURE 10.30 The single 3D diamondoid net (a) and the triple interpenetrated diamondoid nets (b) in $[Zn(ina)_2]$.

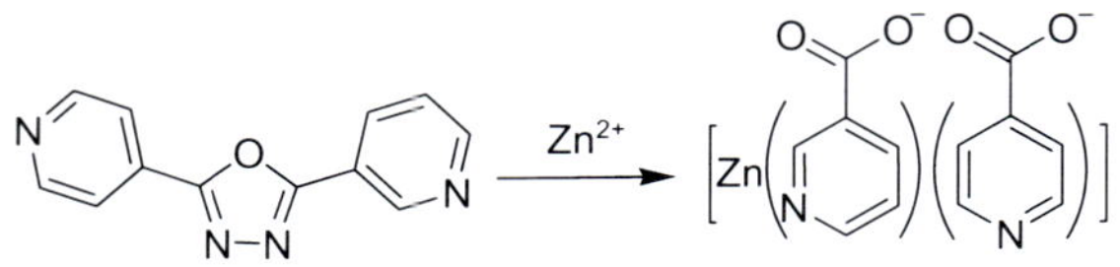

FIGURE 10.31 In situ hydrolysis of the asymmetrical oxadiazole ligand into the 1:1 mixture of Hna and Hina in [Zn(na)(ina)]. *Reprinted with permission from Ref. [49]. Copyright 2005 American Chemical Society.*

The in situ generated carboxylate ligands via the hydrolysis/oxidation of heterocycle and C=C groups, could also be used to construct metal-carboxylate coordination polymers. And it may result in some unique compounds, which cannot be synthesized by direct reaction of carboxylate ligands and metal ions, even the acentric or chiral coordination polymers. For example, an NLO-active mixed-ligand coordination polymer [Zn(na)(ina)] (na = nicotinate), which crystallizes in space group $Pna2_1$, was obtained by the solvothermal reaction of asymmetrical oxadiazole ligand 2-(2-pyridyl)-5-(4-pyridyl)-1,3,4-oxadiazole and Zn(II) salt (Fig. 10.31) [49]. This compound cannot be prepared by direct reactions of the Zn(II) salts with a 1:1 mixture of Hna and Hina under very similar reaction condition. Namely, the in situ hydrolysis of oxadiazole ligand plays a critical role in the formation of this chiral mixed-ligand coordination polymer.

References

[1] (a) B.F. Hoskins, R. Robson, J. Am. Chem. Soc. 111 (1989) 5962–5964.
(b) B.F. Hoskins, R. Robson, J. Am. Chem. Soc. 112 (1990) 1546–1554.
[2] S.R. Batten, S.M. Neville, D.R. Turner, Coordination Polymers: Design, Analysis and Application, the Royal Society of Chemistry, Cambridge, UK, 2009.
[3] (a) A.F. Wells, Three-Dimensional Nets and Polyhedra, Wiley-Interscience, New York, 1977.
(b) A.F. Wells, Further Studies of Three-dimensional Nets, ACA Monograph No. 8, American Crystallographic Association, Knoxville, TN, 1979.
(c) A.F. Wells, Structural Inorganic Chemistry, fifteenth ed., Oxford University Press, Oxford, 1984.
[4] O.M. Yaghi, M. O'Keeffe, N.W. Ockwig, H.K. Chae, M. Eddaoudi, J. Kim, Nature 423 (2003) 705–714.
[5] J.P. Zhang, X.C. Huang, X.M. Chen, Chem. Soc. Rev. 38 (2009) 2385–2396.
[6] B. Moulton, M.J. Zaworotko, Chem. Rev. 101 (2001) 1629–1658.
[7] A.J. Blake, N.R. Champness, M. Crew, P. Simon, New J. Chem. 23 (1999) 13–15.
[8] X.C. Huang, J.P. Zhang, X.M. Chen, J. Am. Chem. Soc. 126 (2004) 13218–13219.
[9] M.L. Tong, X.M. Chen, X.L. Yu, T.C.W. Mak, J. Chem. Soc., Dalton Trans. (1998) 5–6.
[10] X.C. Huang, J.P. Zhang, X.M. Chen, Chin. Sci. Bull. 48 (2003) 1531–1534.
[11] (a) X.C. Huang, Y.Y. Lin, J.P. Zhang, X.M. Chen, Angew. Chem. Int. Ed. 45 (2006) 1557–1559.
(b) R. Banerjee, A. Phan, B. Wang, C. Knobler, H. Furukawa, M. O'Keeffe, et al., Science 319 (2008) 939–943.
[12] H. Li, M. Eddaoudi, T.L. Groy, O.M. Yaghi, J. Am. Chem. Soc. 120 (1998) 8571–8572.
[13] D.X. Xue, Y.Y. Lin, X.N. Cheng, X.M. Chen, Cryst. Growth Des. 7 (2007) 1332–1336.
[14] S.S.Y. Chui, S.M.F. Lo, J.P.H. Charmant, A.G. Orpen, I.D. Williams, Science 283 (1999) 1148–1150.
[15] Y.Z. Zheng, M.L. Tong, W. Xue, W.X. Zhang, X.M. Chen, F. Grandjean, et al., Angew. Chem. Int. Ed. 46 (2007) 6076–6080.
[16] G. Férey, C. Mellot-Draznieks, C. Serre, F. Millange, J. Dutour, S. Surblé, et al., Science 309 (2005) 2040–2042.
[17] Y.B. Zhang, W.X. Zhang, F.Y. Feng, J.P. Zhang, X.M. Chen, Angew. Chem. Int. Ed. 48 (2009) 5287–5290.
[18] M. Eddaoudi, J. Kim, N. Rosi, D. Vodak, J. Wachter, M. O'Keeffe, et al., Science 295 (2002) 469–472.
[19] L. Hou, Y.Y. Lin, X.M. Chen, Inorg. Chem. 47 (2008) 1346–1351.
[20] M. Kondo, T. Okubo, A. Asami, S. Noro, T. Yoshitomi, S. Kitagawa, et al., Angew. Chem. Int. Ed. 38 (1999) 143–147.
[21] H. Chun, D.N. Dybtsev, H. Kim, K. Kim, Chem. Eur. J. 11 (2005) 3521–3529.
[22] M.L. Tong, X.M. Chen, S.R. Batten, J. Am. Chem. Soc. 125 (2003) 16170–16171.
[23] J.J. Zhang, L. Wojtas, R.W. Larsen, M. Eddaoudi, M.J. Zaworotko, J. Am. Chem. Soc. 131 (2009) 17040–17041.
[24] J. Tao, M.L. Tong, X.M. Chen, J. Chem. Soc., Dalton Trans. (2000) 3669–3674.
[25] P.M. Forster, A.R. Burbank, C. Livage, G. Férey, A.K. Cheetham, Chem. Commun. (2004) 368–369.
[26] P.M. Forster, N. Stock, A.K. Cheetham, Angew. Chem. Int. Ed. 44 (2005) 7608–7611.
[27] N. Masciocchi, S. Bruni, E. Cariati, F. Cariati, S. Galli, A. Sironi, Inorg. Chem. 40 (2001) 5897–5905.
[28] X.C. Huang, J.P. Zhang, Y.Y. Lin, X.L. Yu, X.M. Chen, Chem. Commun. (2004) 1100–1101.
[29] Y.Q. Tian, H.J. Xu, L.H. Weng, Z.X. Chen, D.Y. Zhao, X.Z. You, Eur. J. Inorg. Chem. (2004) 1813–1816.
[30] L.S. Long, CrystEngComm 12 (2010) 1354–1365.
[31] N. Stock, T. Bein, J. Mater. Chem. 15 (2005) 1384–1391.
[32] Q. Yu, X.Q. Zhang, H.D. Bian, H. Liang, B. Zhao, S.P. Yan, et al., Cryst. Growth Des. 8 (2008) 1140–1146.
[33] (a) X.M. Chen, M.L. Tong, Acc. Chem. Res. 40 (2007) 162–170.
(b) J.P. Zhang, X.M. Chen, in: M.C. Hong, L. Chen (Eds.), Design and Construction of Coordination Polymers, John-Wiley & Son Ltd, 2009, pp. 63–86 (Chapter 3).
[34] Y.Z. Zheng, M.L. Tong, X.M. Chen, New J. Chem. 28 (2004) 1412–1415.
[35] J. Tao, Y. Zhang, M.L. Tong, X.M. Chen, T. Yuen, C.L. Lin, et al., Chem. Commun. (2002) 1342–1343.
[36] Y.Z. Zheng, Y.B. Zhang, M.L. Tong, W. Xue, X.M. Chen, Dalton Trans (2009) 1396–1406.
[37] X.C. Huang, D. Li, X.M. Chen, CrystEngComm 8 (2006) 351–355.
[38] S.R. Batten, R. Robson, Angew. Chem. Int. Ed. 37 (1998) 1460–1494.
[39] J.P. Zhang, Y.Y. Lin, X.C. Huang, X.M. Chen, Chem. Commun. (2005) 1258–1260.
[40] J.P. Zhang, Y.Y. Lin, X.C. Huang, X.M. Chen, Dalton Trans. (2005) 3681–3685.
[41] J.P. Zhang, X.M. Chen, Chem. Commun. (2006) 1689–1699.
[42] X.C. Huang, J.P. Zhang, Y.Y. Lin, X.M. Chen, Chem. Commun. (2005) 2232–2234.

[43] M.A. Withersby, A.J. Blake, N.R. Champness, P. Hubberstey, W.S. Li, M. Schröder, Angew. Chem. Int. Ed. 36 (1997) 2327–2329.
[44] Z. Huang, M. Du, H.B. Song, X.H. Bu, Cryst. Growth Des. 4 (2004) 71–78.
[45] Y.Z. Zheng, W. Xue, M.L. Tong, X.M. Chen, F. Grandjean, G.J. Long, Inorg. Chem. 47 (2008) 4077–4087.
[46] Y.Z. Zheng, W. Xue, W.X. Zhang, M.L. Tong, X.M. Chen, F. Grandjean, et al., Inorg. Chem. 48 (2009) 2028–2042.
[47] O.R. Evans, W.B. Lin, Acc. Chem. Res. 35 (2002) 511–522.
[48] O.R. Evans, R.G. Xiong, Z.Y. Wang, G.K. Wong, W.B. Lin, Angew. Chem. Int. Ed. 38 (1999) 536–538.
[49] Y.T. Wang, H.H. Fan, H.Z. Wang, X.M. Chen, Inorg. Chem. 44 (2005) 4148–4149.

CHAPTER

11

Synthetic Chemistry of Cluster Compounds

Guo-Yu Yang

Fujian Institute of Research on the Structure of Matter, Chinese Academy of Sciences, China

11.1. DESCRIPTION OF THE CLUSTERS

11.1.1. Definition of a Cluster and the Clusters

Clusters can be viewed as solids at the nano-scale. A cluster is a group of the same or similar elements gathered or occurring closely together. In chemistry, a "cluster" is an ensemble of bound atoms intermediate in size between a molecule and a bulk solid. The clusters exist in diverse stoichiometries and nuclearities. For example, carbon and boron atoms form fullerene and borane or mixed carborane clusters, respectively. Transition metals, lanthanides, and main group elements form especially robust clusters [1].

The above-mentioned cluster is obviously different from neither the phrase "cluster" containing metal–metal bonds coined by F.A. Cotton in the early 1960s nor another cluster definition, i.e., a cluster contains a group of two or more metal atoms in which direct and substantial metal–metal bonding is present [2]. Therefore, the concept of the clusters has been greatly extended in modern chemistry.

11.1.2. Classification of the Clusters

According to the attribute of the central atoms, the clusters in chemistry can generally be divided into metal clusters and nonmetal clusters. Both clusters contain two main types in the light of coordination environments of the central atoms: one type is the clusters with ligands, and another is the ligand-free clusters. To the ligand-free clusters. The ligand-free clusters are also called the naked. clusters without stabilizing ligands, in which the Zintl clusters are very big and important family that are mostly formed by reduction of heavy main group metals and semimetals with alkali metals under a solid-state reaction in anhydrous and vacuum conditions or a solution in anhydrous and anaerobic liquid ethylenediamine or ammonia. In addition some unstable naked clusters such as certain aluminum clusters [3] and gold clusters [4,5], as well as fullerenes are often produced by laser-induced evaporation/ablation and observed in gas-phase by means of the mass spectrometry. To the clusters stabilized by the ligands, the typical ligands mainly include O-/N-/S-containing ligands, carbon monoxide, halides, isocyanides, alkenes, and hydrides, resulting in oxo clusters, metal chalcogenide clusters, transition metal (TM) carbonyl clusters, TM-halide clusters, boranes, carboranes, etc. Here, we only introduce some advances in hydro(solvo)thermal synthetic chemistry of the oxo and chalcogenide clusters.

11.1.3. The Oxo Clusters and Chalcogenide Clusters

In cluster chemistry, the oxo clusters are usually generated by combination of central atoms and oxygen ligands in which the oxygen ligands contain three forms: the O^{2-} anion, OH^- group, and O-donor of ligands, while the central atoms contain nonmetals and metals. Therefore, the oxo clusters can be divided into two types, oxo nonmetal and oxo metal clusters. The metal oxygen clusters are robust and generally contain oxo TM (TM–O), oxo main-group-metal (MGM–O), and oxo lanthanide (Ln–O) clusters. So far, some novel oxo mixed-metal clusters, such as early and late TM, Ln-/MGM-TM, and MGM-/Ln-MGM clusters, as well as some oxo nonmetal–metal clusters have been made.

Modern Inorganic Synthetic Chemistry, DOI: 10.1016/B978-0-444-53599-3.10011-3

In this chapter, some useful synthetic strategies are successfully applied to make novel oxo clusters. As to chalcogenide clusters, they are formed by assembly of central atoms and chalcogen-containing ligands such as sulfur, selenium, and tellurium atom as well as their derivatives.

11.2. SYNTHESIS OF THE OXO TM CLUSTERS UNDER HYDROTHERMAL CONDITIONS

11.2.1. Substituted Synthesis on Polyoxometalate (POM) Cages

1. Substituted synthesis on $\{V_{18}O_{42}\}$ cage by arsenic atoms

In 1978, the $V^{IV}_{18}O_{42}{}^{12-}$ cluster (Fig. 11.1) was isolated [6]. Until 1988, the arsenic substituted oxo vanadium (V—O) cluster, i.e., As—V—O cluster, $As_6V_{15}O_{42}^{6-}$ with As^{3+} and V^{4+} ions, was made at 85°C in aqueous solution [7]. In oxo anion $As_6V_{15}O_{42}{}^{6-}$, three V = O units of $V_{18}O_{42}{}^{12-}$ cluster were substituted by three As_2O groups. In 1991, a new $As_8V_{14}O_{42}{}^{4-}$ (denoted α-type) cluster was obtained by hydrothermal method [8], in which four V = O units of $V_{18}O_{42}{}^{12-}$ cluster were replaced by four As_2O groups. Subsequently, the related clusters $As_{2n}V_{18-n}O_{42}(X)^{m-}$ ($X = SO_3$, SO_4, H_2O, $n = 3,4$) [9] and $As_8V_{12}O_{40}{}^{8-}$ [10] have also been made. As compared to $\{As_8V_{14}O_{42}\}$, the $\{As_8V_{12}O_{40}\}$ is formed by losing two V = O units and rearranging two As_2O groups and one V-based trimer. Lately, other clusters based on $\{As_6V_{15}O_{42}\}$ and $\{As_8V_{14}O_{42}\}$ units have been made [11]. However, though many As—V—O clusters have been reported during the past 15 years, no new member or isomer of $\{As_{2n}V_{18-n}O_{42}\}$ series with $n = 1,2,5$ has been discovered. In 2005, the first isomer β-$\{As_8V_{14}O_{42}\}$, different from α-$\{As_8V_{14}O_{42}\}$ cluster, was hydrothermally synthesized [12]. The former is derived from α-$\{As_8V_{14}O_{42}\}$ by a 90° rotation of one-half of α-$\{As_8V_{14}O_{42}\}$ around its S_4 axis. In 2008, the first $\{As_4V_{16}O_{42}\}$ cluster (Fig. 11.1), derived from β-$\{As_8V_{14}O_{42}\}$ type, as the third member of the $\{As_{2n}V_{18-n}O_{42}\}$ series with $n = 2$ has been harvested [13] which filled the gap of the configurations of the $\{As_{2n}V_{18-n}O_{42}\}$ cluster. In addition, antimony can also substitute the V = O units of $\{V_{18}O_{42}\}$ shell to form Sb—V—O clusters with formula $\{Sb_{2n}V_{18-2n}O_{42}\}$ ($n = 2$—4) [14].

2. Substituted synthesis on $\{As_8V_{14}O_{42}\}$ cage by Zn, Ni, Cd atoms

In investigations on the secondary TM complexes link $\{As_{2n}V_{18-2n}O_{42}(X)\}$ ($n = 2$—4) clusters to form aggregate and extended structures, it was found that TMs, such as Zn^{2+}, Ni^{2+} and Cd^{2+}, can substitute one or two V = O groups of $\{As_8V_{14}O_{42}\}$ cluster to form TM-substituted As—V—O clusters [11,15—18]. In $[\{Zn(enMe)_2\}_2(enMe)_2\{Zn_2As_8V_{12}O_{40}(H_2O)\}]\cdot 4H_2O$ (enMe = 1,2-diaminopropane), two Zn^{2+} cations replace two V = O groups of α-$\{As_8V_{14}O_{42}\}$ cluster to form a new di-Zn-substituted cluster $\{Zn_2As_8V_{12}O_{40}\}$, while the *trans*-enMe bridge $\{Zn_2As_8V_{12}O_{40}\}$ cluster and $Zn(enMe)_2$ group (Fig. 11.2a, [15]). In chain-like polymer of $\{[As_8V_{13}NiClO_{41}][Ni(en)_2(H_2O)][Ni(en)_2]\{Ni(en)_2(H_2O)_2\}_{0.5}\}\cdot 4H_2O$ (en = ethylenediamine),

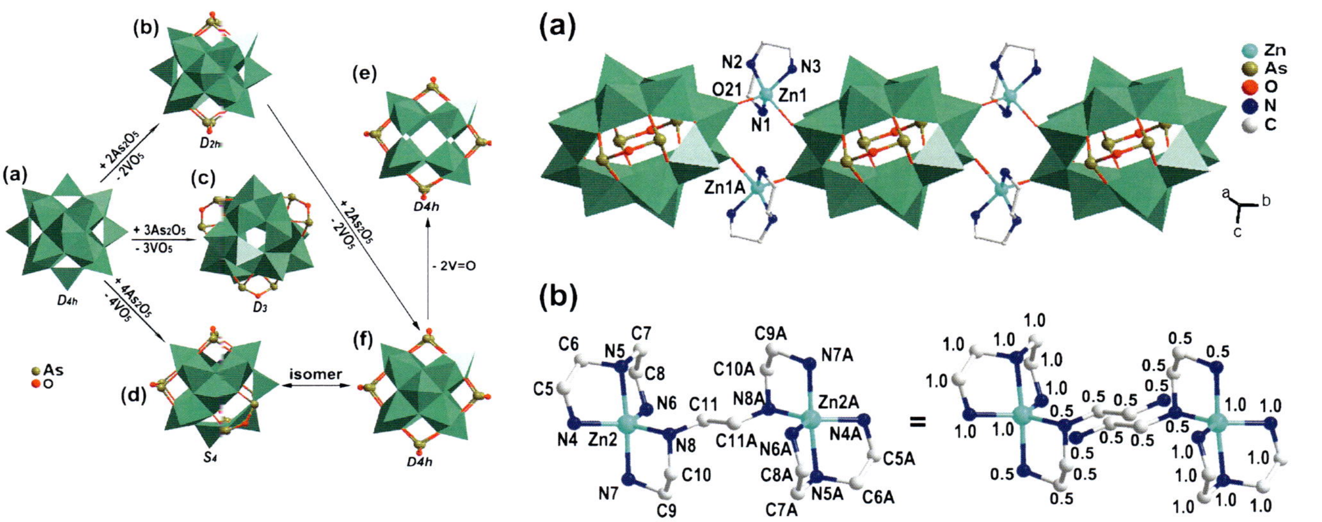

FIGURE 11.1 (Left) View of the relation between the related structures. (a) $\{V_{18}O_{42}\}$, (b) $\{As_4V_{16}O_{42}\}$, (c) $\{As_6V_{15}O_{42}\}$, d) α-$\{As_8V_{14}O_{42}\}$, (e) $\{As_8V_{12}O_{40}\}$, (f) β-$\{As_8V_{14}O_{42}\}$. (Right) (a) View of 1D chain in $[Zn_2(dien)_3]\{[Zn(dien)]_2As_4V_{16}O_{42}(H_2O)\}\cdot 3H_2O$. (b) View of isolated dinuclear $[Zn_2(dien)_3]^{4+}$ complex with disordered model. *Reproduced by permission of The Royal Society of Chemistry.*

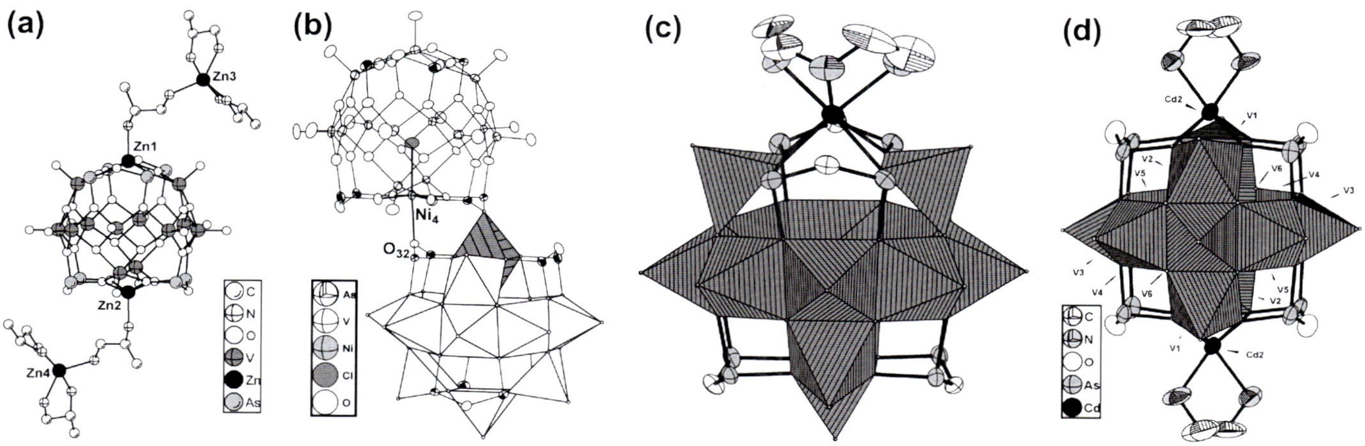

FIGURE 11.2 (a) View of di-Zn-substituted As–V–O cluster. (b) View of the dimeric As–V–O unit. Reprinted with permission from [16] Copyright (2004) American Chemical Society. (c) View of mono-Cd-substituted As–V–O cluster. (d) View of the di-Cd-substituted As–V–O unit. *Reprinted with permission from [17] Copyright (2004) American Chemical Society.*

an Ni^{2+} cation replace a V = O group to form $\{As_8V_{13}NiClO_{41}\}$, and then link an O atom of $\{As_8V_{13}NiClO_{41}\}$, resulting in a dimeric unit (Fig. 11.2b, [16]). When Zn^{2+} and Ni^{2+} are replaced by Cd^{2+} ion, $\{CdAs_8V_{13}O_{41}\}$ (Fig. 11.2c) and $\{Cd_2As_8V_{12}O_{40}\}$ (Fig. 11.2d) are obtained [17]. In the former, one Cd^{2+} replaces one V = O group of α-$\{As_8V_{14}O_{42}\}$, while two Cd^{2+} substitute two V = O groups of β-$\{As_8V_{14}O_{42}\}$ in the latter.

In 2007, two cluster dimers based on Zn-substituted As–V–O units linked by two kinds of bridges, organic ligand, and TM complex have been made [11]. The dimer $(ppz)\{\{Zn\text{-}(tepa)\}_2ZnAs_8V_{13}O_{41}\}_2$ (Fig. 11.3a) consists of two $\{ZnAs_8V_{13}O_{41}\}$ units and one ppz bridge (ppz = piperazine, tepa = tetraethylenepentamine). Interestingly, the formation of the dimer involves two new in situ ligand reactions under hydrothermal conditions: intermolecular and intramolecular deamination coupling reactions (DCRs) of dien, i.e., the dien molecules as a starting amine transform not only to ppz via in situ intermolecular DCR but also to tepa via in situ intramolecular DCR. Furthermore, other intermolecular DCRs based on different amines have also been found in our lab (Scheme 1). Such a kind of in situ ligand reaction has never been observed before in the literature. Notably, the same starting materials as dimer (ppz) $\{\{Zn(tepa)\}_2ZnAs_8V_{13}O_{41}\}_2$ but with different sampling sequences lead to another dimer $\{[Zn(dien)]_2(dien)_2[Zn_2As_8V_{12}O_{40}]\}_2$ (Fig. 11.3b) without involving in situ DCRs of dien. In 2008, the first di-Cd-substituted cluster $\{(enMe)_2Cd_2As_8V_{12}O_{40}\}$ (Fig. 11.3c) derived from α-$\{As_8V_{14}O_{42}\}$, and 1D cluster built by $\{(enMe)_2Cd_2As_8V_{12}O_{40}\}$ derived from β-$\{As_8V_{14}O_{42}\}$ and double $Cd(enMe)_2$ bridges have been made in the presence of enMe [18], respectively.

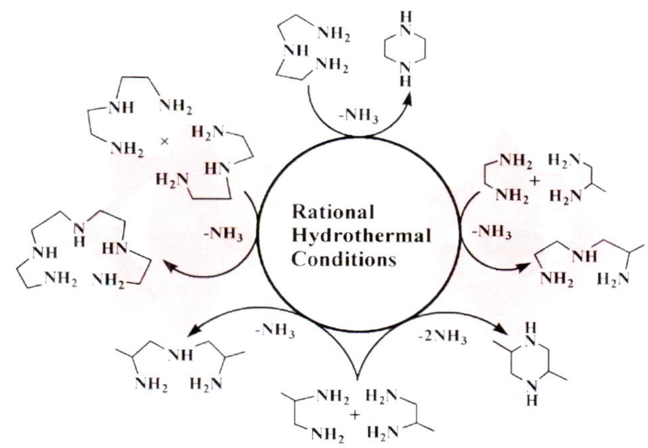

SCHEME 1 View of the modes of the intermolecular and intramolecular DCRs based on different amines under hydrothermal conditions.

11.2.2. Lacunary Directing Synthesis Via the Lacunary Sites of POM Fragments

Lacunary Keggin or Dawson POMs have attracted much attention because they can incorporate paramagnetic metals to form high-nuclear TM-substituted POMs (TMSPs), resulting in unique magnetic properties. Generally, TMSPs were made by conventional solution syntheses at atmospheric pressure and relatively low temperature (below 100°C). Before 2007, the investigation on syntheses of lacunary POM precursors combined with metal clusters under hydrothermal conditions remained largely undeveloped. Therefore, lacunary POM fragments combined with hydrothermal techniques may open a new avenue in making novel metal-substituted POMs. Here, the metal includes TM, Ln, MGM, etc. In 2007, we extended the synthesis of

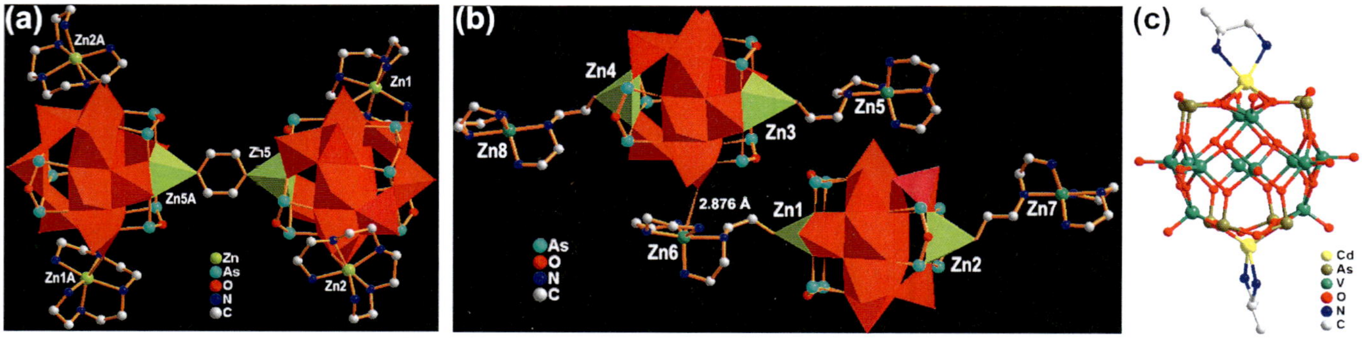

FIGURE 11.3 3 (a) View of ... by ppz bridge. *Reprinted with permission from [11]. Copyright 2007... Society.* (b) View of the neutral ... Zn-complex. *Reprinted with permission from [11]. Copyright ... Society.* (c) View of di-Cd-substituted ... cluster. *Reprinted with permission from [18]. Copyright 2005 Elsevier.*

TMSPs from conventional aqueous solution to hydrothermal techniques and explored new reaction systems containing lacunary POM fragments, metal cations, and organoamine based on the following ideas: (1) the trilacunary of $XW_9O_{34}/X_2W_{15}O_{56}$ or multilacunary sites of $P_2W_{12}O_{48}$ may act as the structure-directing agents (SDAs) and induce the metal ions to congregate large cluster aggregation under rational hydrothermal conditions, which is different from the invisible SDA of lone pair electron in Se(IV) and the visible SDAs of porous materials, such as inorganic/organic/complex cations and liquid crystal/surfactant. (2) The trilacunary or multilacunary fragments and the large cluster aggregates formed in situ may further act as the structural building units (SBUs) to combine subtly with each other, resulting in novel isolated or extended high-nuclear metal-substituted POMs (Fig. 11.4, [19]). Under the guidance of the above ideas, a series of high-nuclear TMSPs that cannot be made by conventional aqueous solution methods have been made [19–23].

1. Directed substitution by Ni_6/Ni_7 clusters on the lacunary sites of $XW_9O_{34}{}^{n-}$ (XW_9, X = Si/Ge/P) fragments

   In hexa-Ni-substituted POMs, the $\{Ni_6(\mu_3\text{-}OH)_3(H_2O)_6L_3\}$ (Ni_6, L = amine) cores capped to the top of the XW_9 (X = Si/P) fragments (Fig. 11.4) have different coordination environments owing to different orientations of water molecules or amines, resulting in structural isomers of hybrid Ni_6 units. These water ligands might be good centers for further structural derivation, i.e., intramolecular or

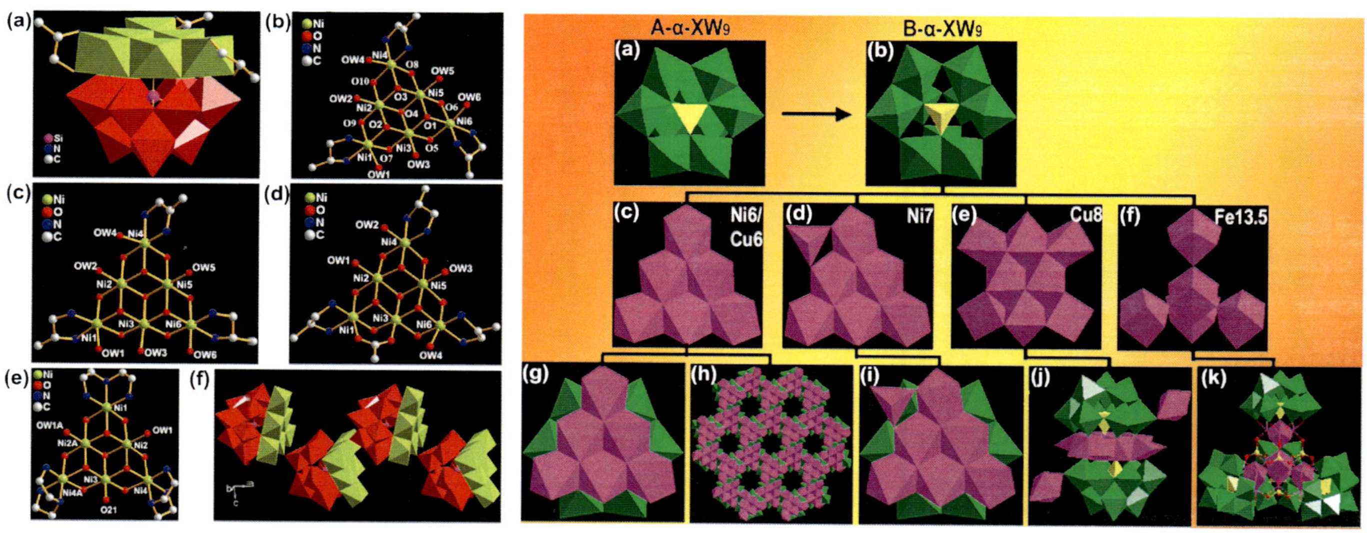

FIGURE 11.4 (Left) (a) View of cluster $[Ni_6(\mu_3\text{-}OH)_3(H_2O)_6(enMe)_3(B\text{-}\alpha\text{-}SiW_9O_{34})]$. (b)–(e) Views of the Ni_6 hybrid units in $[Ni(H_2O)_6][Ni_6(\mu_3\text{-}OH)_3(H_2O)_6(enMe)_3(B\text{-}\alpha\text{-}SiW_9O_{34})]_2\cdot 14H_2O$, $[Ni(enMe)_2][Ni_6(\mu_3\text{-}OH)_3(H_2O)_6(enMe)_3(B\text{-}\alpha\text{-}SiW_9O_{34})]_2\cdot 10H_2O$, $[Ni(enMe)_2(H_2O)_2][Ni_6(\mu_3\text{-}OH)_3(H_2O)_4(CH_3COO)(enMe)_3(B\text{-}\alpha\text{-}PW_9O_{34})]_2\cdot 10H_2O$ and $[Ni_6(\mu_3\text{-}OH)_3(H_2O)_2(dien)_3(B\text{-}\alpha\text{-}PW_9O_{34})]\cdot 4H_2O$, respectively. (f) View of the 1D structure of $[Ni_6(\mu_3\text{-}OH)_3(H_2O)_2(dien)_3(B\text{-}\alpha\text{-}PW_9O_{34})]$. (Right) View of the isomerization of $A\text{-}\alpha\text{-}XW_9 \rightarrow B\text{-}\alpha\text{-}XW_9$ and the assembly process of Ni_6/Cu_6, Ni_7, Cu_8, or $Fe_{13.5}$ and $B\text{-}\alpha\text{-}XW_9$ SBUs (X = P/Si).

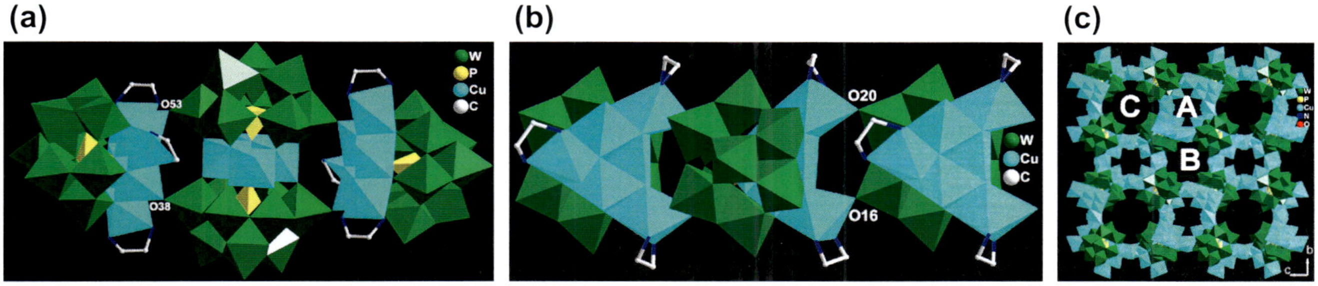

FIGURE 11.5 (a) The double sandwiched cluster of $[Cu_6(\mu_3\text{-}OH)_3(en)_3(H_2O)_2(B\text{-}\alpha\text{-}PW_9O_{34})]_2[Cu_4(H_2O)_2(B\text{-}\alpha\text{-}PW_9O_{34})_2]$; (b) the 1D chain of $[Cu_6(\mu_3\text{-}OH)_3(en)_3(H_2O)_3(B\text{-}\alpha\text{-}PW_9O_{34})]\cdot 4H_2O$; (c) the 3D openframework of $[Cu^{II}{}_6Cu^{I}{}_4(\mu_6\text{-}O)(OH)_{18}][Cu_6(\mu_3\text{-}OH)_3(enMe)_3(H_2O)_3(B\text{-}\alpha\text{-}PW_9O_{34})]_4\cdot 4H_3O\cdot 2H_2O$. *Reprinted with permission from [24]. Copyright 2009 American Chemical Society.*

intermolecular linkage via multidentate ligands, as well as intermolecular concentration, leading to novel or extended structures. In addition, a hepta-Ni cluster caps a PW_9 fragment and forms a novel Ni_7PW_9 cluster [20].

2. Directed substitution by Cu_6 clusters on the lacunary sites of XW_9 fragments

In isolated Cu_6XW_9 cluster, double sandwiched tetramer, 1D chain and 3D frameworks (Fig. 11.5), the Cu-polyhedral Jahn–Teller/*pseudo*-Jahn–Teller distortion plays a crucial role in forming these complex clusters [20,24]. In double sandwiched tetramers (Fig. 11.5a), the reason for substitution may be attributed to two major factors: (1) the steric hindrance of two penta-coordinated Cu-polyhedra is much less than that of two six-coordinated Cu-octahedra, (2) Jahn–Teller effect results in the axial elongation of Cu-octahedra, which can further reduce large steric hindrance when big $Cu_4(PW_9)_2$ unit attacks two Cu_6PW_9 units. This phenomenon can also be viewed as a "synergistic effect" between the distorted polyhedra (trigonal bipyramid and square pyramid) and Jahn–Teller effect of the axial elongate octahedra. It is noteworthy that such substitution has not been observed in Ni^{II}-octahedra of the Ni_6 core [19–23,25,26]. Similarly, both Jahn–Teller effect of the axial elongate Cu-octahedra and *pseudo*-Jahn–Teller effect of Cu-based square pyramid in 1D chain can reduce large steric hindrance when big Cu_6PW_9 units are further polymerized together (Fig. 11.5b). Therefore, the "synergistic effect" between *pseudo*-Jahn–Teller and Jahn–Teller effect results in a double-bridging chain, which differs from single-bridging chain in Ni_6-based TMSP [19], i.e., the formation of 1D chain is driven by Jahn–Teller/*pseudo*-Jahn–Teller effect of the Cu^{II}-polyhedra, while no such effect exists in the Ni^{II}-polyhedra. As to 3D framework (Fig. 11.5c), Jahn–Teller effect of the axial elongated Cu-octahedra reduces the steric hindrance and benefits the occurrence of the substitution reaction. So far, no such elongation was observed in Ni-octahedra.

11.2.3. Synergistic Directing Synthesis Via Two or More Lacunary XW_9 Fragments

In the above examples, we introduced the structure-directing viewpoint into the synthetic chemistry of POMs in which the structure-directing role only comes from the lacunary sites of the single POM fragment. In fact, the lacunary sites of two or more POM fragments in the structure may act as "synergistic directing agents" and play the synergistic directing roles in inducing the metal ions to assemble cluster, resulting in novel oligomers/poly(POM)s that cannot be induced by single POM fragment. Therefore, we call it as "synergistic directing synthesis". Now, some such examples with dimers and tetramers are shown as follows.

1. Tetra-metal sandwich-type cluster dimers

Up to now, numerous sandwich M_4-substituted POMs $[M_4(H_2O)_2(XW_9O_{34})_2]^{n-}$ ($X = Si/P/As$; $M = Mn/Fe/Co/Ni/Cu/Zn$) have been prepared by conventional solution syntheses at atmospheric pressure and relatively low temperature (below 100°C). However, in fact, almost all of reported M_4-sandwiched POMs are discrete structures. In 2007, attempts to make extended solids by using trilacunary Keggin XW_9 ($X = Si/Ge/P$) fragments as the precursors under hydrothermal conditions were based on the following considerations: (1) the hydrothermal techniques have been demonstrated to be a powerful tool in making extended POM-based solids; (2) our recent work has proved that the trilacunary Keggin XW_9 precursors can exist under hydrothermal conditions, but only undergo the isomerizations of $\{A\text{-}\alpha\text{-}XW_9\} \rightarrow \{B\text{-}\alpha\text{-}XW_9\}$ during the course of the reactions [19]; (3) under suitable conditions, the trilacunary Keggin XW_9 precursors have a strong tendency to form dimeric sandwich M_4-substituted POM anions that further act as SBUs for making the extended structures. As expected, a hybrid 2D network $[Cu(dien)(H_2O)]_2\{[Cu(dien)(H_2O)]_2[Cu(dien)(H_2O)_2]_2[Cu_4(SiW_9O_{34})_2]\}\cdot 5H_2O$ built by $Cu_4(SiW_9)_2$ dimers and Cu-complex bridges

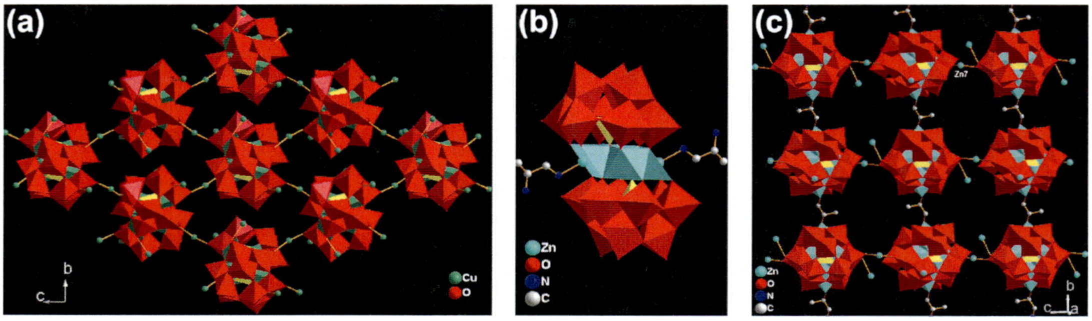

FIGURE 11.6 (a) View of the 2D network built by $Cu_4(PW_9)_2$ dimers and Cu-complex bridges. (b) View of $(HenMe)_2Zn_4(PW_9)_2$ dimer. (c) View of the 2D network linked by enMe and Zn-complex bridges.

has been made (Fig. 11.6a, [27]), in which the absent sites of the defective Cu_4 units may be good centers for further structural derivation, e.g., intramolecular decoration or intermolecular linkage via organic ligands to form hybrid structures. Fortunately, two such clusters based on Zn_4-substituted sandwich clusters were made [27]. To $[Zn(enMe)_2(H_2O)]_2$ $\{[Zn(enMe)_2]_2[Zn_4(HenMe)_2(PW_9O_{34})_2]\}\cdot 8H_2O$ built by $Zn_4(HenMe)_2(PW_9)_2$ dimers and Zn-complexes (Fig. 11.6b), two water molecules of the $Zn_4(H_2O)_2$ core were substituted by two mono-protonated enMe ligands. Such substitution reaction on the M_4 core has not been observed previously. The uncoordinated $-NH_2$ groups of enMe ligands suggest that the condensation of $\{Zn_4(HenMe)_2(PW_9)_2\}$ into extended solids via the linkage of enMe ligand in an end-to-end fashion could be feasible. A decrease in the initial quantity of enMe resulted in new solid $[Zn(enMe)_2(H_2O)]_4[Zn(enMe)_2]_2\{(enMe)_2\{[Zn(enMe)_2]_2[Zn_4(HSiW_9O_{34})_2]\}$ $\{[Zn(enMe)_2(H_2O)]_2[Zn_4(HSiW_9O_{34})_2]\}\}\cdot 13H_2O$ based on $\{[Zn(enMe)_2]_2[Zn_4(HSiW_9)_2]\}$ and $\{[Zn(enMe)_2(H_2O)]_2[Zn_4(HSiW_9)_2]\}$ dimers, linked by linear enMe ligands to form a 1D chain, which is extended into a 2D network by $[Zn(enMe)_2]$ bridges (Fig. 11.6c).

2. Hexa-metal sandwich-type cluster dimers

In 2007, a novel sandwich-type POM, $[Cu(enMe)_2]_2\{[Cu(enMe)_2(H_2O)]_2[Cu_6(enMe)_2(B\text{-}\alpha\text{-}SiW_9O_{34})_2]\}\cdot 4H_2O$ [28], had been made via synergistic directing roles of two SiW_9 units, in which two SiW_9 units sandwiched a belt-like Cu_6 core (Fig. 11.7a). In 2009, the first $CdSO_4$-type 3D framework, $[Cu(en)_2]_2[Cu(deta)(H_2O)]_2[Cu_6(en)_2(H_2O)_2(B\text{-}\alpha\text{-}GeW_9O_{34})_2]\cdot 6H_2O$, (deta = diethylenetriamine), built by $Cu_6(GeW_9)_2$ SBUs and Cu-complex bridges had also been made (Fig. 11.7b,c, [29]).

3. Octa-metal sandwich-type cluster dimers

The synergistic directing roles of two trilacunary fragments have induced secondary metal ions not only form tetra-M sandwich-type dimers but also from hexa-M sandwich-type dimers under hydrothermal conditions. Whether or not can the synergistic directing roles further induce the larger cluster? In our lab, some octa-Cu sandwiched dimers have also been made [20,30]. In the presence of aliphatic amines, four octa-Cu sandwiched dimers,

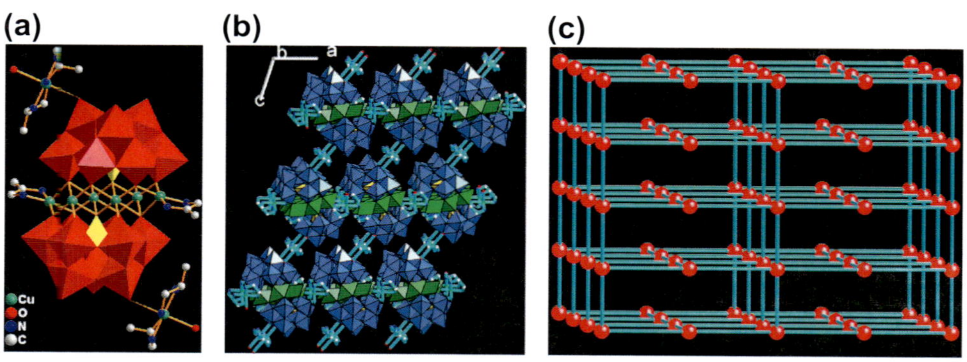

FIGURE 11.7 (a) Structure of $Cu_6(SiW_9)_2$ dimer. (b) View of The 3D framework. (c) The $6^5\cdot 8$ $CdSO_4$-type topology. *Reproduced by permission of The Royal Society of Chemistry.*

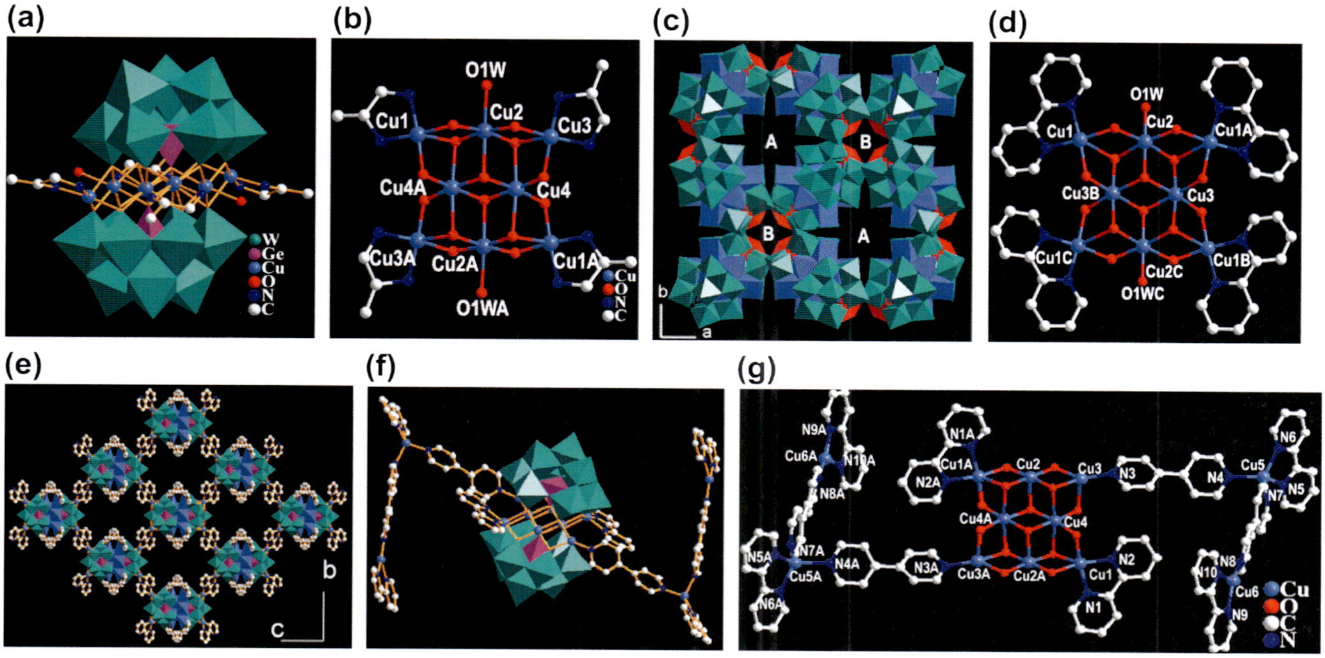

FIGURE 11.8 (a) View of dimer $Cu_8(dap)_4(GeW_9)_2$. (b) View of the $\{[Cu(dap)]_4Cu_4O_{14}(H_2O)_2\}$ cluster in dimer. (c) The 3D framework. d) The motif of the $\{[Cu^{II}(2,2'\text{-bpy})]_4Cu^{II}{}_4O_{14}(H_2O)_2\}$ cluster. (e) The 2D layer. (f) View of the sandwiched dimer hanged two Cu-complexes. (g) The dodeca-Cu cluster built by octa-Cu cluster $\{[Cu^{I}{}_2Cu^{II}{}_2(2,2'\text{-bpy})_2(4,4'\text{-bpy})_2]Cu^{II}{}_4O_{14}\}$ and two di-Cu^{I} cations $[Cu^{I}{}_2(2,2'\text{-bpy})_2(4,4'\text{-bpy})]^{2+}$.

$[Cu(dap)(H_2O)_3]_2[\{Cu_8(dap)_4(H_2O)_2\}(B\text{-}\alpha\text{-}SiW_9O_{34})_2]\cdot 6H_2O$ ([20]), $H_4[Cu_8(dap)_4(H_2O)_2(B\text{-}\alpha\text{-}GeW_9O_{34})_2]\cdot 13H_2O$, and $(H_2en)_2[Cu_8(en)_4(H_2O)_2(B\text{-}\alpha\text{-}XW_9O_{34})_2]\cdot nH_2O$ ($X = Ge/Si$, $n = 5/8$, Fig. 11.8a,b; [30]), as well as two 3D framework $[Cu(H_2O)_2]H_2[Cu_8(L)_4(H_2O)_2(B\text{-}\alpha\text{-}XW_9O_{34})_2]$ ($L = en/dap$, $X = Si/Ge$, Fig. 11.8c) have been obtained [31].

To dimer $Cu_8(L)_4(XW_9)_2$ ($L = en/dap$, $X = Si/Ge$), it not only is a novel Cu_8-substituted cluster but also sandwiches the maximum 3d TM cations in the reported trivacant Keggin or Dawson POM dimers to date (Fig. 11.8a,b). (3,6)-connected frameworks is built by $Cu_8(L)_4(HXW_9)_2$ ($L = en/dap$, $X = Si/Ge$) SBUs and $Cu(H_2O)_2$ bridges (Fig. 11.8c). However, on substitution of aliphatic amines by aromatic amines, two new high-nuclear substituted POMs, $[Cu_2(H_2O)_2(2,2'\text{-bpy})_2]\{[Cu(bdyl)]_2[Cu_8(2,2'\text{-bpy})_4(H_2O)_2(B\text{-}\alpha\text{-}GeW_9O_{34})_2]\}\cdot 4H_2O$ (bdyl = 2,2′-bipyridinyl, Fig. 11.8d,e) and $[Cu^{I}(2,2'\text{-bpy})(4,4'\text{-bpy})]_2\{[Cu^{I}{}_2(2,2'\text{-bpy})_2\ (4,4'\text{-bpy})]_2[Cu^{I}{}_2Cu^{II}{}_6(2,2'\text{-bpy})_2(4,4'\text{-bpy})_2\ (B\text{-}\alpha\text{-}GeW_9O_{34})_2]\}\cdot 2H_2O$ (Fig. 11.8f,g), have been made [30]. As shown in Fig. 11.8d, four 2,2′-bpy ligands chelated four Cu atoms to form a new hybrid $Cu_8(2,2'\text{-bpy})_4$ cluster that sandwiches between two GeW_9 fragments to form a $Cu_8(2,2'\text{-bpy})_4(GeW_9)_2$ dimer, which further links by four $[Cu_2(bdyl)]^{2+}$ bridges to produce 2D network (Fig. 11.8e). Notably, the bdyl ligand in the $[Cu_2(bdyl)]^{2+}$ bridges is derived from 2,2′-bpy via eliminating two protons, and adopts a unique *trans*-four-coordination mode to bond two Cu^{II} atoms. Such rollover metalation of 2,2′-bpy [32] is first observed in the system containing low-priced Cu^{II} cations under hydrothermal conditions. Interestingly, two 2,2′-bpy and two 4,4′-bpy ligands coordinate two Cu^{II} and two Cu^{I} atoms to form another novel mixed-valent octa-Cu sandwiched cluster and then further link two di-Cu^{I} complexes by 4,4′-bpy, forming a dodeca-Cu cluster (Fig. 11.8f,g).

4. High-nuclear iron incorporated cluster tetramers

When Fe ions were introduced, two high-nuclear Fe-substituted poly(POM)s, $(enH_2)_3H_{15}[\{Fe^{II}{}_{1.5}Fe^{III}{}_{12}(\mu_3\text{-}OH)_{12}(\mu_4\text{-}PO_4)_4\}(B\text{-}\alpha\text{-}PW_9O_{34})_4]\cdot ca130H_2O$ (Fig. 11.9a; $Fe_{13.5}(PW_9)_4$) and $(enH_2)_{3.5}H_{15}[\{Fe^{II}Fe^{III}{}_{12}(\mu_3OH)_{12}(\mu_4\text{-}PO_4)_4\}(B\text{-}\alpha\text{-}PW_9O_{34})_4]\cdot 83H_2O$ (Fig. 11.9c; $Fe_{13}(PW_9)_4$), were made [20,21]. The $Fe_{13.5}(PW_9)_4$ tetramer is built by four tri-Fe^{III}-substituted Fe_3PW_9 units linked by a central $Fe^{II}{}_4O_4$ cubane core and four $\mu_4\text{-}PO_4$ bridges. Each $\mu_4\text{-}PO_4$ unit links four $Fe^{II}Fe^{III}{}_3O(OH)_3$ cubanes; four Fe_3 trimers of Fe_3PW_9 units combine the central $Fe^{II}{}_4O_4$ cubane to form a $Fe_{13.5}$ cluster (Fig. 11.9b). Differently, the $Fe_{13}(PW_9)_4$ contains four tri-Fe^{III}-substituted Fe_3PW_9 units fused by an $Fe^{II}{}_2O_2$ unit and four $\mu_4\text{-}PO_4$ bridges, resulting in an $Fe^{II}Fe^{III}{}_{12}$ core, in which only two $Fe^{III}{}_3Fe^{II}O(OH)_3$ cubanes are joined (Fig. 9d).

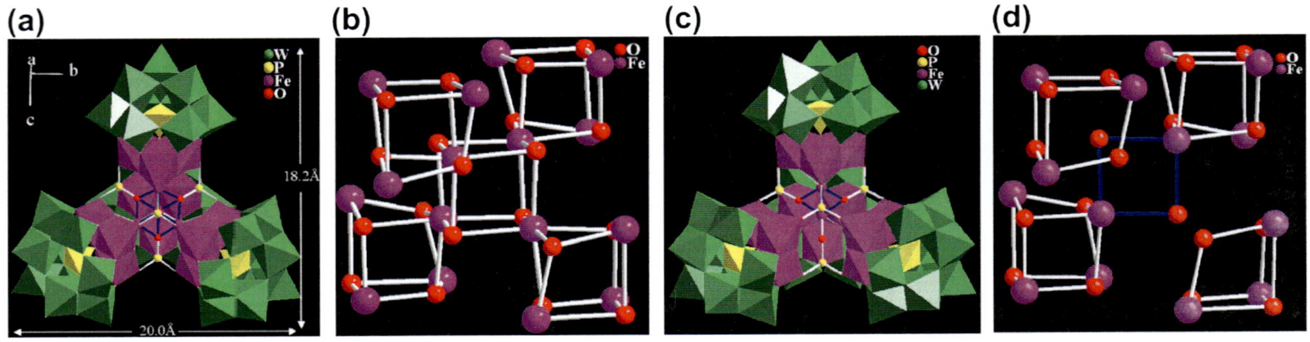

FIGURE 11.9 (a) Views of the tetrameric poly(POM) $Fe_{13.5}(PW_9)_4$. (b) Four Fe_3 trimers linking the central $Fe^{II}_4O_4$ cubane to form a $Fe_{13.5}$ cluster. (c) View of the tetramer $Fe_{13}(PW_9)_4$. (d) Distribution fashion of four Fe_3 trimers and the central $Fe^{II}{}_2O_2$ unit in Fe_{13} cluster. *Reprinted with permission from [21]. Copyright 2007 American Chemical Society.*

11.2.4. DESIGNED SYNTHESIS VIA THE PERIPHERAL SUBSTITUTION OF NI_6PW_9 SBUs

POM clusters with different shapes, sizes, and composition may provide a variety of SBUs for making novel POM–organic frameworks which belong to the cluster-organic frameworks (COFs). On the other hand, POM clusters are attractive inorganic SBUs owing to their nanosize and tunable acid/base, redox, magnetic, catalytic, and photochemical properties. Therefore, the designed synthesis of COFs will open up a new avenue for the creation of a variety of novel functional materials. In particular, the combination of POM clusters and organic ligands is more interesting because of their inherently different natures and possible synergetic effects in making COF materials. However, although many metal–organic frameworks (MOFs) built by metal-carboxylate SBUs have been made, only few examples of POM-based MOFs are reported [33]. So far, most of the reported POM-based materials with extended structures are based on the linkages of POM clusters and metal complexes bridged by oxygen atoms, which belong to the metal–oxygen frameworks. The linking of POMs with rigid carboxylates into COFs remains largely unexplored. This is because the POM clusters, usually having high negative charges and oxygen-rich compositions, prefer to bond metal cations rather than carboxylate anions. Therefore, the search for suitable POM clusters for making COFs is one of the most challenging issues in synthetic chemistry and material science. The following parts are two representative strategies for making COFs based on the metal cluster-substituted POM SBUs under hydrothermal conditions.

1. Directed assembly via multi-carboxylate ligands $[Ni_6(\mu_3\text{-}OH)_3(H_2O)_6(L)_3(B\text{-}\alpha\text{-}PW_9O_{34})]$ ($Ni_6PW_9(H_2O)_6$, L = en/enMe, Fig. 11.10a) as SBUs was chosen because of: (1) We have obtained well-defined conditions for in situ preparation of $Ni_6PW_9(H_2O)_6$ SBUs [19]; (2) Each SBU contains six

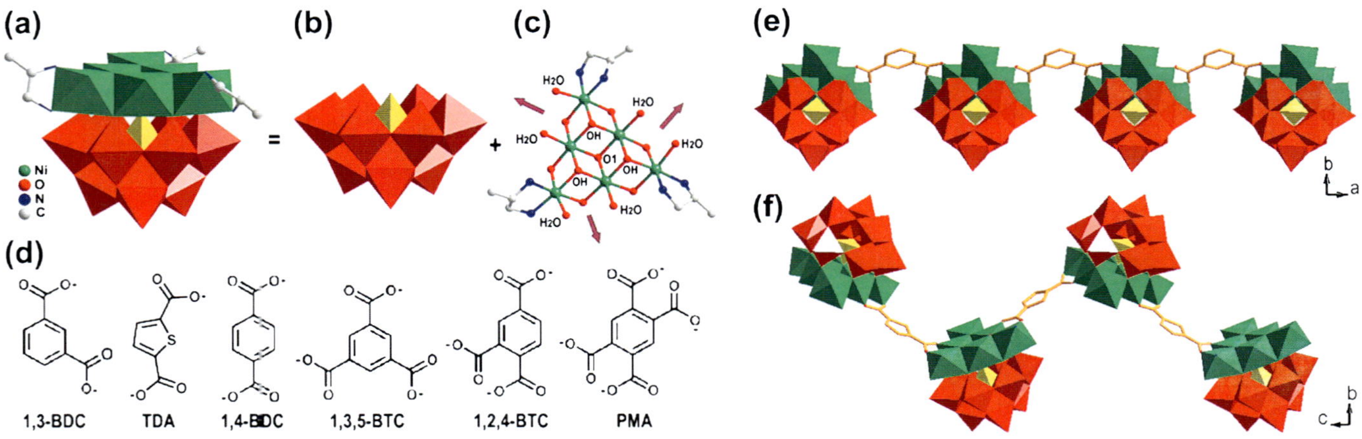

FIGURE 11.10 (a–c) Structure of $Ni_6PW_9(H_2O)_6$, PW_9 and Ni_6 unit. (d) Rigid carboxylate linkers. (e–f) View of 1D chain structures in Ni_6PW_9-1,3-bdc and Ni_6PW_9-tda.

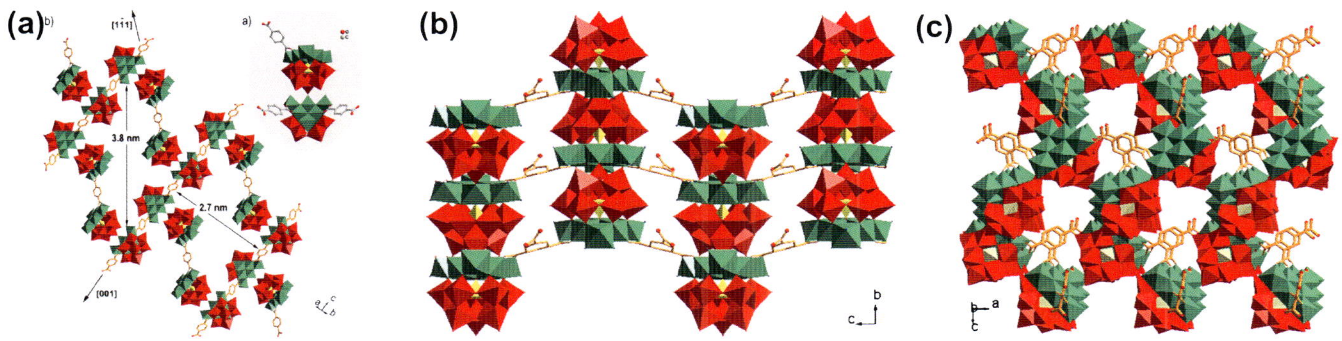

FIGURE 11.11 (a) View of 2-D layer in Ni_6PW_9-1,4-bdc. (b) View of 2-D layer in Ni_6PW_9-1,3,5-btc. (e) View of 3-D framework in Ni_6PW_9-1,2,4-btc.

terminal water ligands that offer the possibility for design of COFs by replacing the water with rigid carboxylates. Notably, the POM-based SBUs can be isolated and are stable in the absence of the carboxylate ligands [19], which differ from those metal-carboxylate SBUs that are only stable in the presence of the carboxylate ligands. Accordingly, we made a series of COFs based on Ni_6PW_9 SBUs via the directed assembly of the carboxylate linkers: {[$Ni_6(OH)_3(H_2O)_2(enMe)_3(PW_9O_{34})$](1,3-bdc)}[$Ni(enMe)_2$]·$4H_2O$ (Fig. 11.10e, Ni_6PW_9-1,3-bdc), {[$Ni_6(OH)_3(H_2O)(en)_4(PW_9O_{34})$](Htda)}·$H_3O$·$4H_2O$ (Fig. 11.10f, Ni_6PW_9-tda), {[$Ni_6(OH)_3(H_2O)(en)_3(PW_9O_{34})$][$Ni_6(OH)_3(H_2O)_4(en)_3(PW_9O_{34})$]$(1,4\text{-bdc})_{1.5}$}[$Ni(en)(H_2O)_4$]·$H_3O$ (Fig. 11.11a, Ni_6PW_9-1,4-bdc), {[$Ni_6(OH)_3(en)_3(PW_9O_{34})$](1,3,5-Hbtc)}[$Ni(en)(H_2O)_3$]·$2H_2O$ (Fig. 11.11b, Ni_6PW_9-1,3,5-btc), {[$Ni_6(OH)_3(H_2O)_5(PW_9O_{34})$](1,2,4-Hbtc)}·$H_2$enMe·$5H_2O$ (Fig. 11.11c, Ni_6PW_9-1,2,4-btc).

As shown in Fig. 11.10c, two terminal water ligands on each side of Ni_6 core can be substituted by a variety of rigid carboxylate ligands (Fig. 11.10d). Thus the Ni_6 core offers the opportunity to link $Ni_6PW_9(H_2O)_n$ ($n < 6$) units into COFs along three directions. On the other hand, owing to the ability of PW_9 unit sharing terminal oxo atoms with metal cations, the sizes of SBUs can be varied incrementally from discrete $Ni_6PW_9(H_2O)_n$ unit to aggregates and polymers via W = O–Ni linkages, which provide potential possibility for making tailor-made COFs. So the combination of Ni_6 and PW_9 makes $Ni_6PW_9(H_2O)_6$ SBU act as excellent candidate for making novel COFs. As expected, the addition of different rigid carboxylate ligands to the well-defined $Ni_6PW_9(H_2O)_6$-preparing solutions results in forming novel COFs with 1-, 2- and 3D structures [25]. The 1,3-bdc and tda ligands bridge Ni_6PW_9 SBUs to form 1D straight and zigzag chains (Fig. 11.10e,f), while 1,4-bdc ligands link the dimeric $(Ni_6PW_9)_2$ SBUs to produce 2D network (Fig. 11.11a). Furthermore, the Ni_6PW_9 units substitute each other to form zigzag and helical chain-like $\{Ni_6PW_9\}_\infty$ SBUs that are further bridged by the 1,3,5-btc and 1,2,4-btc ligands, resulting in 2D layer (Fig. 11.11b) and 3D framework (Fig. 11.11c).

2. Directed assembly via the spatial effects of the linkers

In above examples, the directed-combination of Ni_6PW_9 units with rigid carboxylate linkers yields a series of extended COFs. In fact, the spatial effects of the linkers play the crucial roles in forming the extended structures. Here, two examples, isolated nanoclusters and 1D nanowires, show the influence of the spatial effects of the linkers on the structure (Fig. 11.12, [26]). Poly(POM) of [$Ni(enMe)_2$]$_3$[$H_6Ni_{20}P_4W_{34}(OH)_4O_{136}(enMe)_8(H_2O)_6$]·$12H_2O$ ($Ni_{20}P_4W_{34}$-enMe) contains a 20-Ni-substituted $[H_6Ni_{20}P_4W_{34}(OH)_4O_{136}(enMe)_8(H_2O)_6]^{6-}$ (Fig. 11.12a,$Ni_{20}P_4W_{34}$) with two moieties of

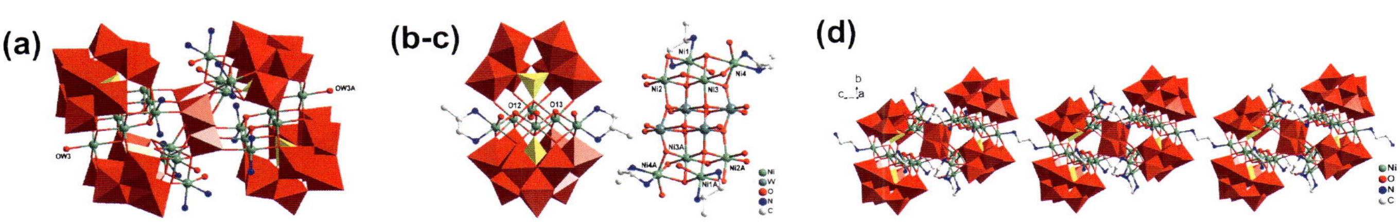

FIGURE 11.12 (a-d) Structures of $Ni_{20}P_4W_{34}$, $Ni_6P_2W_{15}$, Ni_8W_4, and 1D chain, respectively.

$\{Ni_6O_3(OH)_2(H_2O)(enMe)_2(PW_6O_{26})(PW_9O_{34})\}$ (Fig. 11.12b, $Ni_6P_2W_{15}$) linked by a central symmetric belt-like $\{Ni_8W_4O_{26}(H_2O)_4(enMe)_4\}$ (Fig. 11.12c, Ni_8W_4). In $Ni_{20}P_4W_{34}$, two terminal water ligands (OW3/OW3A) on the Ni_6 units point out of poly-(POT) unit. Our previous work confirmed that such terminal water ligands were good centers for further structural derivation, e.g., intramolecular decoration or intermolecular linkage by replacing terminal water with various organic ligands [25]. Therefore, the $Ni_{20}P_4W_{34}$ would be an effective precursor for making extended COFs. With these considerations in mind, continuous efforts were devoted to make such structures. However, we failed to obtain such complex by changing the reaction conditions of $Ni_{20}P_4W_{34}$-enMe, perhaps due to steric hindrance of the methyl group of enMe. In order to reduce the steric hindrance, we replaced enMe by en. As expected, a cluster-organic chain $[Ni(en)_2(H_2O)]_2[H_8Ni_{20}P_4W_{34}(OH)_4O_{136}(en)_9(H_2O)_4]\cdot 16H_2O$ (Fig. 11.12d, $Ni_{20}P_4W_{34}$-en) was isolated, in which en ligands not only replaced the terminal water ligands, but also acted as linear bridges to link adjacent SBUs in an end-to-end fashion.

11.3. SYNTHESIS OF THE OXO LANTHANIDE CLUSTERS UNDER HYDROTHERMAL CONDITIONS

Although the chemistry of the oxo TM clusters is well established, the chemistry of the oxo Ln clusters is less developed because the synthesis of high-nuclearity clusters is still a big challenge. In general, a common synthetic strategy for the oxo Ln clusters is to control hydrolysis of Ln salts in the presence of supporting ligands. So far, most of high-nuclearity oxo Ln clusters are discrete because the presence of hydrophobic groups in the periphery of oxo Ln cluster core prevents further aggregation. In order to investigate magnetic exchange interactions between 3d and 4f metals in the solid state, some Ln–TM clusters [34] and coordination polymers built by oxo Ln cluster units [35–36] were made. However, no systematic investigation on oxo Ln clusters and COFs built by oxo Ln clusters and TM clusters via organic linkers has been carried out under hydrothermal conditions. In our group, we focused our attention on exploring new reaction system to design and make novel COFs via the inductive roles of the functional organic ligands, and have made some great progress [37–40]. Compared with the oxo Ln clusters obtained by hydrolysis of Ln salts in aqueous solution, the Ln oxides, in our lab, were directly applied to make the COFs at high pH value under hydrothermal conditions.

11.3.1. INDUCED SYNTHESIS VIA THE LIGANDS

In our group, two types of the ligands, 4,5-imidazoledicarboxylic acid (H_3ImDC) and isonicotinic acid (HIN), were applied to make COFs under hydrothermal conditions.

1. Helical tubes induced by H_3ImDC ligands

   In various organic ligands, the H_3ImDC was chosen as the multidentate ligand because of the following considerations: (1) the flexible, multifunctional coordination sites provide a high likelihood for generation of structure with high dimensions. (2) It can be deprotonated to generate H_2ImDC^-, $HImDC^{2-}$, and $ImDC^{3-}$, which allow various, acidity-dependant coordination modes. (3) The skew coordination orientation of the carboxyl groups is favorable for making the helical structure. As expected, a series of COFs, $[Ln(ImDC)_2(H_2O)_3]\cdot nH_2O$ (Ln = Eu, Dy, Y, Pr, Nd, Sm, Gd, Tb, Er), were made by the hydrothermal reaction of Ln_2O_3 and H_3ImDC in water in the presence of HNO_3 at pH 2 and 170°C [37]. The structures display 1D achiral channels and left-/right-handed helical tubes arranged alternately, in which the helical tubes not only built by the $Ln_4(COO)_2$ cluster SBUs but also constructed from interweaving triple-helical and double-helical chains (Fig. 11.13a). Furthermore, other Ln-organic helical tubes have also been made by the use of H_3ImDC [41].

2. The nanosize Ln_{14} cluster induced by HIN ligands

   The HIN is a well-known ligand and has been widely used to make TM- or Ln-based coordination polymers. The HIN was chosen as the multifunctional ligand based on the following considerations: (1) it is a rigid ligand with O and N donors on opposite sides, enabling it to act as a linear bridge; and (2) the carboxyl group may induce the oxyphilic Ln ions to form oxo Ln cluster, the N atoms can coordinate to TM ions, and thus extended solids containing oxo Ln clusters and TM ions/clusters might be obtained. Accordingly, a series of COFs built by oxo Ln clusters and TM ions/clusters have been made by using Ln_2O_3, rather than Ln salts, as the Ln source in water at pH 2 under hydrothermal conditions: $[Ln_{14}(\mu_6\text{-}O)(\mu_3\text{-}OH)_{20}(IN)_{22}Cu_6X_4(H_2O)_8]\cdot 6H_2O$ (X = Cl, Ln = Y, Gd, Dy, Tb, Er; X = Br, Ln = Er, Gd, Y) [38], in which the high-nuclearity oxo Ln cluster $[Ln_{14}(\mu_6\text{-}O)(\mu_3\text{-}OH)_{20}(H_2O)_8]^{20+}$ (Ln_{14}) acts as SBUs to combine with copper ions/clusters via linear IN ligands, forming 3D frameworks (Fig. 11.13b). Three different centers, Ln_{14} cores, Cu_2Cl_2 dimers, and Cu atoms are joined by IN ligands with four coordination modes. These results provide an opening into a promising

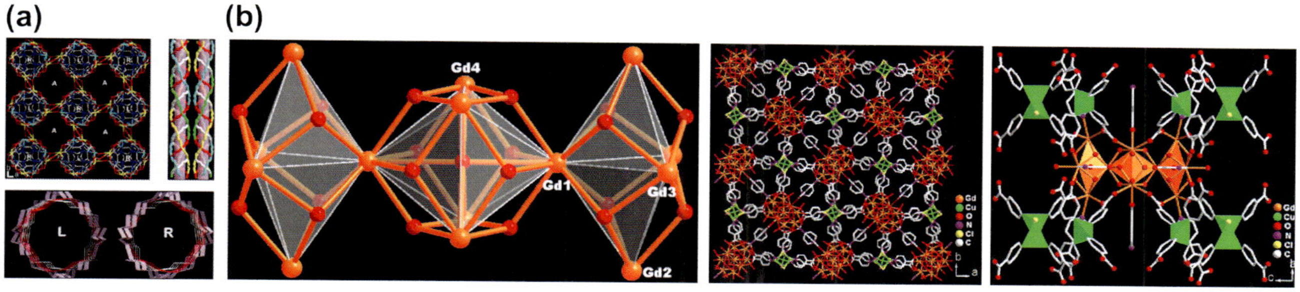

FIGURE 11.13 (a) The framework with two types of helical tubular and achiral channels. (b) View of the structure of $[Gd_{14}(\mu_6\text{-}O)(\mu_3\text{-}OH)_{20}]^{20+}$ (Gd_{14}) core containing one octahedral $[Gd_6(\mu_6\text{-}O)(\mu_3\text{-}OH)_8]^{8+}$ (Gd_6) unit and two trigonal bipyramidal $[Gd_5(\mu_3\text{-}OH)_6]^{4+}$ (Gd_5) units (left); the 3D structure formed three different centers, the Gd_{14} cores, Cu_2Cl_2 dimers and Cu ions, linked by IN ligands (middle), the linkages of the IN bridges and three different centers (right).

new field of 4f–3d mixed-metal functional materials and confirm the potential for developing new structural families of solid-state materials using nanosized oxo Ln cluster as SBUs.

11.3.2. Synergistic coordination Between the First and Secondary Ligands

The above COFs are constructed from nanosized Ln_{14} clusters, copper ions/clusters and IN linkers, their framework is very condensed because of four coordination modes of the IN ligands. How to make further open framework by reducing the coordination modes of the IN ligands is a great challenge. Here two approaches have been successfully applied to make such frameworks.

1. Synergistic coordination between two types of organic ligands

   In general, ligands in the periphery of metal ions play an important role in forming and stabilizing metal clusters or wheels. To 3D COFs built by Ln_{14} cores and two types of Cu centers [38], if the second ligand is introduced, the synergistic coordination or cooperativity of both ligands may lead to a new open framework that contains oxo Ln clusters different from the Ln_{14} core, which would be helpful for understanding the formation mechanism of the open frameworks. Now, we have extended this work and introduced a second ligand, 1,2-benzenedicarboxylic acid (H_2bdc) [39].

   The H_2bdc was selected as the second ligand based on the following considerations: (1) it has versatile coordination modes; (2) the chelating coordination of bdc may stabilize new oxo Ln cluster formed in the crystallization process via the synergistic coordination with the first ligands, IN. As expected, two novel COFs, $[Er_7(\mu_3\text{-}O)(\mu_3\text{-}OH)_6(bdc)_3](IN)_9[Cu_3X_4]$ (X = Cl/Br, denoted by FJ-2a/FJ-2b) have been successfully made under hydrothermal conditions, in which the wheel-shaped SBUs of $[Er_{36}(\mu_3\text{-}OH)_{30}(\mu_3\text{-}O)_6(bdc)_6]^{54+}$ (Er_{36}, Fig. 11.14a) with 18-ring is currently the largest oxo Ln wheel cluster [39]. Although the bdc ligand has more than 15 coordination modes, it adopts single μ_5-bdc mode found only in FJ-2; the number of coordination modes of IN is also reduced from four in Ln_{14} [38] to two in FJ-2, indicating the high cooperativity or synergistic coordination between IN and bdc ligands in the formation of FJ-2; the unique μ_5-bdc mode plays a key role in the forming and stabilizing of the Er_{36} wheel. Each Er_{36} cluster is linked to surrounding clusters, thus forming a highly ordered layered cluster network (Fig. 11.14b). Furthermore, the linkages between 2D hybrid cluster networks and Cu_3Cl_4 clusters by IN ligands give rise to 3D sandwich framework (Fig. 11.14c). These examples indicate that the unique cooperativity between IN and bdc ligands under hydrothermal conditions suggest that the synergistic coordination as a useful synthetic strategy has

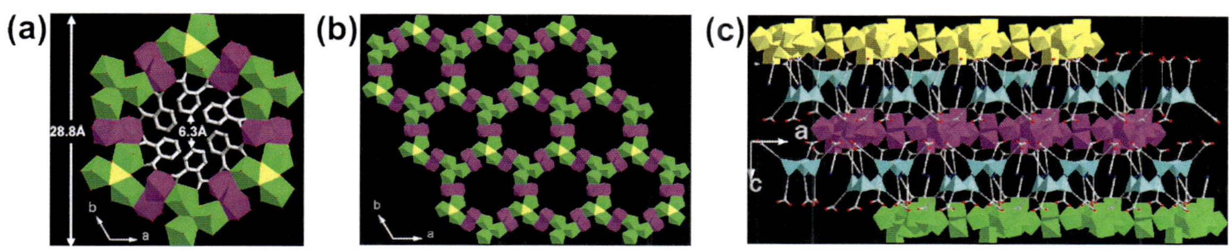

FIGURE 11.14 (a) View of the Er_{36} wheel trapped by the bdc ligands. (b) View of layered cluster network with 18-ring. (c) View of sandwich framework based on linkages of 2-D cluster layers and Cu_3Cl_4 cluster pillars by IN ligands.

potential application in design and synthesis of molecular materials.

2. Synergistic coordination between inorganic and organic ligands

As a continuation of our search for new COFs built from oxo Ln and TM clusters, we have introduced a larger halogen, iodine, based on the following considerations: the I^- ion has larger ionic radius than that of Cl^- and Br^-, and may favor higher coordination numbers and versatile coordination modes, resulting in larger TM-halide cluster than Cu_2X_2 [38] and Cu_2X_4 [39] cores. As expected, seven novel wheel-cluster organic frameworks (WCOFs): $[Ln_6(\mu_3\text{-}O)_2](IN)_{18}[Cu_8(\mu_4\text{-}I)_2(\mu_2\text{-}I)_3]\cdot H_3O$ (FJ-4, Ln = Y, Nd, Dy, Gd, Sm, Eu, Tb) have been made, showing the first example of sandwich WCOFs based on linkages of two distinct layered networks of nanosized wheels with 12 rings, $[Ln_{18}(\mu_3\text{-}O)_6(CO_2)_{48}]^{6-}$, Ln_{18}, and $[Cu_{24}(\mu_4\text{-}I)_6(\mu_2\text{-}I)_{12}]^{6+}$, Cu_{24}, formed by using IN ligands (Fig. 11.15a, [40]). So far, no systematic investigation on WCOFs, in which the distinct networks of metal wheels are linked by organic ligands, has been carried out.

Compared with the number of coordination modes of IN ligands, four in Ln_{14} [38] and two in FJ-2 [39], all IN ligands in FJ-4 only show one coordination mode when the I^- ions were introduced, which further indicates the high cooperativity or synergistic coordination between organic IN and inorganic I^- ligands in forming FJ-4. In FJ-4, six $[Ln_3(\mu_3\text{-}O)(CO_2)_6](Ln_3)$ cluster are linked by IN ligands to form a nanosized Ln_{18} wheel with 12-ring and the diameter of 26.7 Å, which differs from the Er_{36} [39] wheel in linkage modes. Adjacent Ln_{18} wheels are linked to each other to form a highly ordered layered Ln_{18} wheel-cluster network (Fig. 11.15b); six tetrameric I-centered tetrahedral $[Cu_4(\mu_4\text{-}I)]$ (Cu_4) cluster cores are linked to form another nanosized Cu_{24} wheel with 12 rings with the diameter of 26.4 Å. Adjacent Cu_{24} wheels are further linked to each other, forming another layered network of Cu_{24} wheels (Fig. 11.15c). Generally, it is not possible to form a large 3D metal wheel with only one type of bridging interaction. Unlike I-centered tetrahedra in Cu_{24} wheel network, tetrahedrally coordinated O atoms are not usually encountered in zeolite and oxide frameworks due to the small size of the O atom. Therefore, the larger size of I^- ions is critical for the formation of the Cu_{24} wheel networks of FJ-4. Like the Er_{36} wheel-cluster networks formed in FJ-2 due to the synergistic coordination between two different organic ligands [39], the Cu_{24}/Ln_{18} wheel-cluster networks in FJ-4 can also be described as the generation under the synergistic coordination between two kinds of ligands, inorganic I^-/O^{2-} and organic IN. Notice that the unusual trinuclear $Ln_3(\mu_3\text{-}O)$ and tetranuclear $Cu_4(\mu_4\text{-}I)$ cores are used as SBUs to make from large oligomers of Ln_{18} and Cu_{24} to 2D Ln_{18} and Cu_{24} networks, suggesting that the oligomerization of low-nuclearity SBUs may make larger wheels or WCOFs. Further work is in progress for making other new oxo Ln and TM wheels and WCOFs by using either larger oligomers of tri-/tetra-nuclear units, or oligomers of higher nuclearity SBUs, simultaneously we are trying to make larger wheels and WCOFs, especially mixed Ln and TM wheels, by selecting subtly another type of ligands or introducing shrewdly the second ligand.

11.4. SYNTHESIS OF THE OXO MAIN GROUP CLUSTERS UNDER HYDROTHERMAL CONDITIONS

11.4.1. Templated Synthesis of Borates

Borates have attracted much research interest due to their rich structural chemistry and diverse applications. Boron atom coordinates with oxygen not only in threefold (triangular, BO_3) but also in fourfold coordination (tetrahedral, BO_4). The BO_3 and BO_4 can further

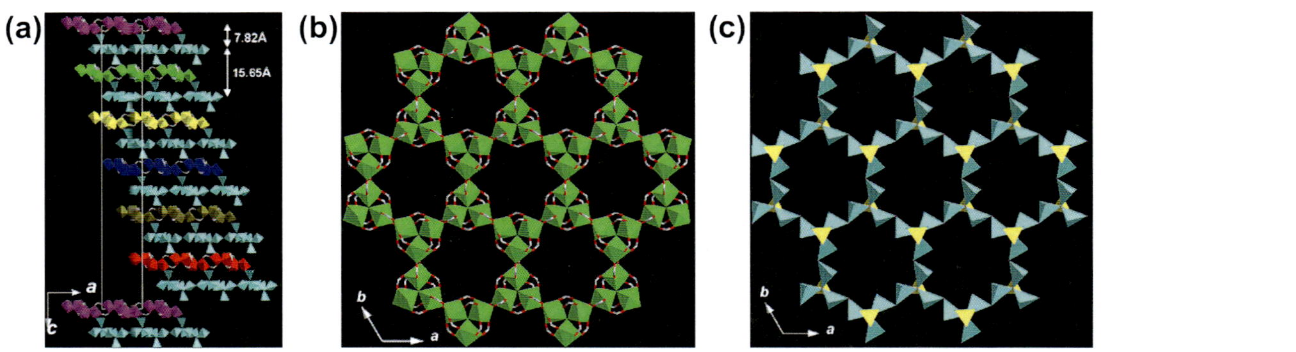

FIGURE 11.15 (a) View of sandwich framework based on linkages of the layered networks of Dy_{18} and Cu_{24} wheels stacked in parallel with – ABCDEF – alternations along the c-axis. (b) View of layered network of $\{Dy_{18}\}$ wheels. (c) View of layered network of $\{Cu_{24}\}$ wheels.

polymerize to all kinds of large oxo boron clusters through sharing vertices. Such oxo boron clusters can be considered as the fundamental building blocks (FBBs) to take part in constructing novel borates. Generally, most metal borates were made by high-temperature solid-state methods or boric acid flux methods. Therefore, borates have mainly formed the following systems: alkali and alkaline earth metal borates, main group and TM borates, as well as rare earth borates in the past seven decades. The most important application of borates is based on their optical properties, for example, nonlinear optical (NLO) property. The most well-known example is the discovery of β-BaB_2O_4 (BBO) [42], LiB_3O_5 (LBO) [43], CsB_3O_5 (CBO) [44], and $Sr_2Be_2B_2O_7$ (SBBO) [45]. However, these NLO borate materials are made by solid-state reactions, no systematic investigation on NLO borates has been carried out under hydro(solvo) thermal conditions. Since 2003, we have made considerable efforts to study hydro(solvo)thermal synthesis of borates and have made some advances in the systems of inorganic [46–49], organic [50–54] and TMC templated borates [55–57]. Importantly, some NLO borates have been successfully made under hydro(solvo) thermal conditions.

1. Inorganic templated borates
   In the exploration of novel NLO metal borates in M–O–B system, a new potassium templated acentric manganese borate, $K_7[(BO_3)Mn\{B_{12}O_{18}(OH)_6\}]\cdot H_2O$ (KMnBO) with NLO properties has been made in 2004 [46], in which KMnBO differs structurally from all known M–O–B in a remarkable way: for the first time we observed that the B and Mn atoms can link together via both vertex-sharing and edge-sharing (Fig. 11.16a).
2. Organic templated borates
   Since 2004, some organic templated borates [50,53] and borates linked by organic species [54] were reported. In the presence of *trans*-1,4-diaminocyclohexane acting as an SDA, a 2D borate, $[H_3N(C_6H_{10})NH_3][B_5O_8(OH)]$, was first made (Fig. 11.16b, [50]). In 2007, the first 1D borate $[NH_3CH_2CHCH_3NH_3][B_8O_{11}(OH)_4]\cdot H_2O$ templated by H_2enMe has been made [51]. Its structure consists of infinite open-branched borate chains built by $[B_3O_6(OH)]$ units, onto which the $[B_5O_7(OH)_3]$ clusters are grafted (Fig. 11.16c). In addition, two new borates $[H_2dab][B_7O_9(OH)_5]\cdot 2H_2O$ and $[H_2dab][B_7O_{10}(OH)_3]$ (dab = 1,4-diaminobutane) have also been obtained in the presence of dab [52]. The former contains $[B_7O_9(OH)_5]$ unit (Fig. 11.16d), which is the first example of organic templated hepta-borate, the latter consists of $[B_{14}O_{20}(OH)_6]$ unit (Fig. 11.16e) built by the dehydration of the $[B_7O_9(OH)_5]$ unit. In 2007, a novel hybrid borate $B_3O_4(OH)\cdot 0.5C_4H_{10}N_2$ was first made under milder hydro(solvo)thermal conditions (Fig. 11.16f, [54]).
3. TMC templated borates
   In 2004, the TMCs have first been introduced into the structure of the borates [55]. Two pentaborates,

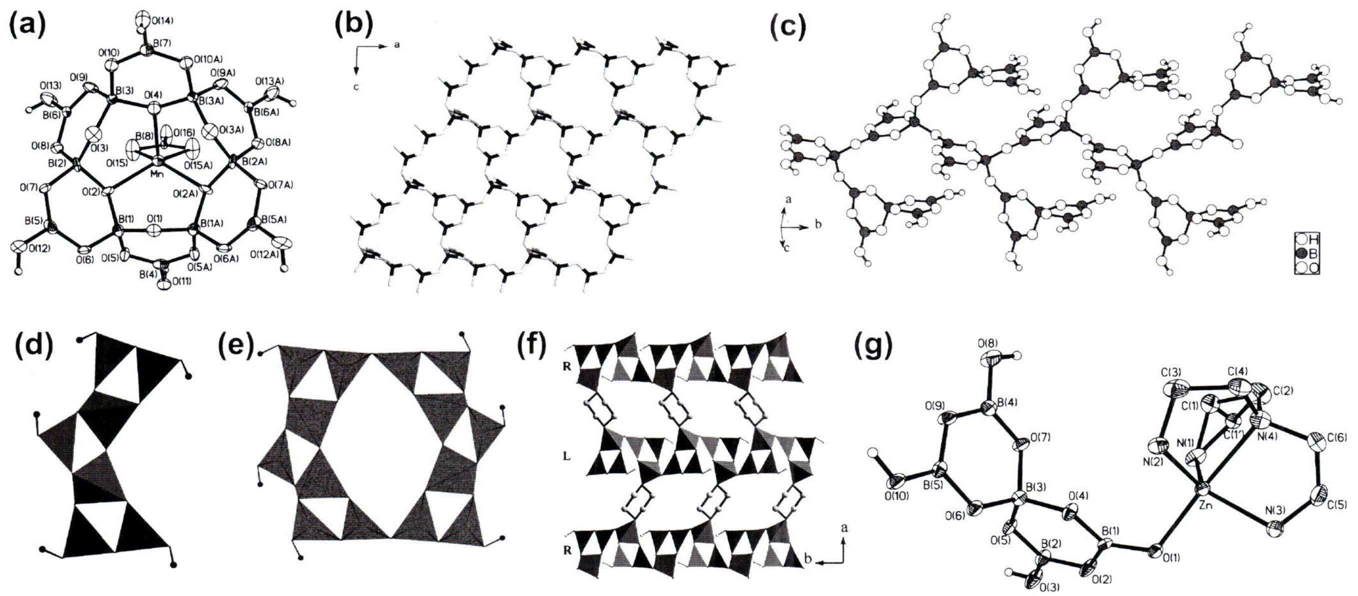

FIGURE 11.16 View of the … anion. *Reprinted with permission from [46], Copyright 2004 Elsevier.* (b) View of the layer built by B6O8(OH) FBBs. *Reprinted with permission from [50], Copyright 2004 Elsevier.* (c) View of … groups. *Reprinted with permission from [51], Copyright 2007 Elsevier.* … (f) View of ….cluster. *Reprinted with permission from [54], Copyright 2004 Elsevier.* (g) View of ….Zn-complex. *Reprinted with permission from [56], Copyright 2007 Elsevier.*

$[Zn(dien)_2][B_5O_6(OH)_4]_2$ and $[B_5O_7(OH)_3Zn(tren)]$ (tren = tris(2-ami-noethyl)amine), have been characterized in 2005 [56]. The former consists of isolated $[B_5O_6(OH)_4]^-$ anion and $[Zn(dien)_2]^{2+}$ cation, the latter is built by two distinct motifs, a $[B_5O_7(OH)_3]^{2-}$ cluster and a supporting [Zn $(tren)]^{2+}$ complex (Fig. 11.16g), showing the first example of the combination of B–O cluster with TMCs. In 2006, three new cobalt borates, $[Co(dien)_2]$ $[B_5O_6(OH)_4]_2$, $[B_5O_7(OH)_3Co(tren)]$, and $[Co_2(teta)_3]$ $[B_5O_6(OH)_4]_4$ (teta = triethylenetetramine), have been reported [57].

4. Templated synthesis of aluminoborates

Being in the same group as B, Al can exist in AlO_4, AlO_5, and AlO_6 forms. Being used in making zeolites, Al was introduced into borate system in 1973 giving porous aluminoborates (ABOs) analogous to zeolites [58]. After that, several ABO phases were revealed [59–65]. But their structures remain unknown because of the synthetic difficulties of crystals suitable for structure determination. Recently, Lin et al. made three open-framework ABOs of PKU-1, PKU-5, and PKU-6 [66–68] by using boric acid flux method. However, no systematic investigation was made on ABOs. In order to get open-framework borates, element Ge was introduced into borate frameworks [69–73]. As part of our ongoing work, we further extended our interest to make ABOs under hydro(solvo)thermal conditions, which is different from the traditional high-temperature solid-state or boric acid flux synthesis methods. Meanwhile, we chose $Al(i\text{-}PrO)_3$ (aluminum isopropoxide), not traditional Al_2O_3, $AlCl_3$, or Al $(NO_3)_3$, as an Al source based on the following considerations: (1) the $Al(i\text{-}PrO)_3$ is easily dissolved in organic solvents, and the chiral Al center may form from three-coordinated $Al(i\text{-}PrO)_3$ transforming to four-coordinated AlO_4 group via the hydrolysis of Al $(i\text{-}PrO)_3$ in the crystallization process; (2) the characteristic of acentric B–O clusters formed in situ via self-polymerization of H_3BO_3 can be transferred into the inorganic frameworks; and (3) the synergistic combination between chiral AlO_4 groups and acentric B–O clusters will not only greatly increase the likelihood of producing new acentric ABOs, but also offer us the opportunity for choosing efficient NLO materials from these products. Accordingly, we successfully make a series of open-framework ABOs: $Al[B_5O_{10}]\cdot H_2dab\cdot 2H_2O$, $[Al(B_4O_9)\text{-}(BO)]\cdot H_2en$, [Al $(B_4O_9)(BO)]\cdot H_2dap$ (dap = 1,3-diaminopropane), $K_2Al[B_5O_{10}]\cdot 4H_2O$, and $(NH_4)_{2-}$ $Al[B_5O_{10}]\cdot 4H_2O$. The former is centrosymmetric, whereas latter four are acentric and show good NLO activities [74]. Both $[Al(B_4O_9)(BO)]\cdot H_2en$ and $[Al(B_4O_9)(BO)]\cdot H_2dap$ are built by AlB_5O_{13} units and exhibit 10-ring channels (Fig. 11.17a,b). Another three ABOs are built up from AlO_4 tetrahedra and B_5O_{10} units to form 3D frameworks with 12-/11-/8-ring intersecting channels for $Al[B_5O_{10}]\cdot H_2dab\cdot 2H_2O$ (Fig. 11.17c), and odd 11-ring channels for two isomorphous, K_2Al $[B_5O_{10}]\cdot 4H_2O$ (Fig. 11.17d) and $(NH_4)_2Al[B_5O_{10}]\cdot$ $4H_2O$.

Furthermore, we have also extended inorganic and organic templates to TMCs in ABO systems, resulting in a series of TMC templated ABOs with 2-D layer and 3-D frameworks. In acentric $[Zn(dien)_2][\{Al(OH)\}$ $\{B_5O_9F\}]$ templated by $[Zn(dien)_2]$ ions, the $AlO_3(OH)$ tetrahedra and B_5O_9F clusters interconnect to form a layer of $[\{Al(OH)\}\{B_5O_9F\}]^{2-}$ (Fig. 11.18a, [75]). Series of open-framework ABOs, $[M(dien)_2]$ $[AlB_6O_{11}(OH)]$ (M = Co, Ni, Cd, Zn; type I) and [M $(en)_3][AlB_7O_{12}(OH)_2]\cdot(H_2O)_{0.25}$ (M = Co, Ni; type II), were classified as two structural types [76]: type I contains AlO_4 tetrahedra and $B_6O_{11}(OH)$ clusters, which link to form a 3D framework with 7-/9-ring helical channels and large 11-/13-ring channels (Fig. 11.18b), while type II is composed of AlO_4 tetrahedra, chain-like $B_4O_6(OH)_2$ tetramer and crab-like B_6O_{12} clusters, which interconnect to form another new 3D framework with 8-ring helical channels, rare 16-ring double-helical channels and larger odd 15-ring channels (Fig. 11.18c,d).

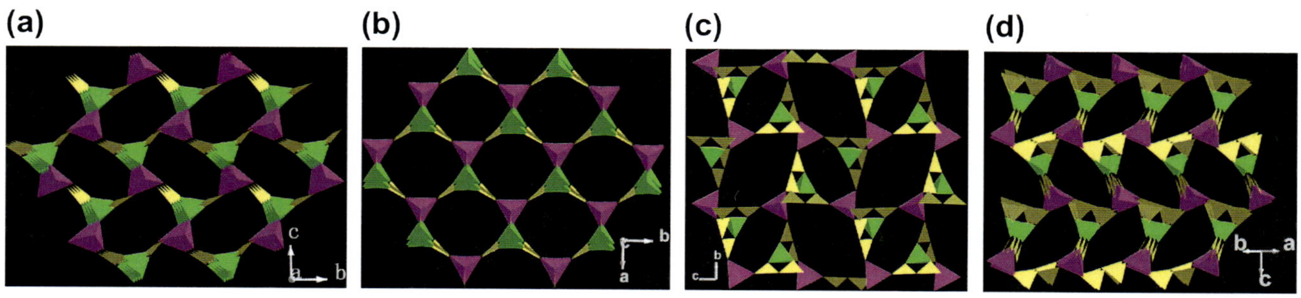

FIGURE 11.17 View of (a) the 10-ring channels in $[Al(B_4O_9)(BO)]\cdot H_2en$, (b) the 10-ring channels in $[Al(B_4O_9)(BO)]\cdot H_2dap$, (c) the 12-/8-ring channels in $Al[B_5O_{10}]\cdot H_2dab\cdot 2H_2O$, and (d) the 11-ring channels in $K_2Al[B_5O_{10}]\cdot 4H_2O$. *Reprinted with permission from [74]. Copyright 2009 American Chemical Society.*

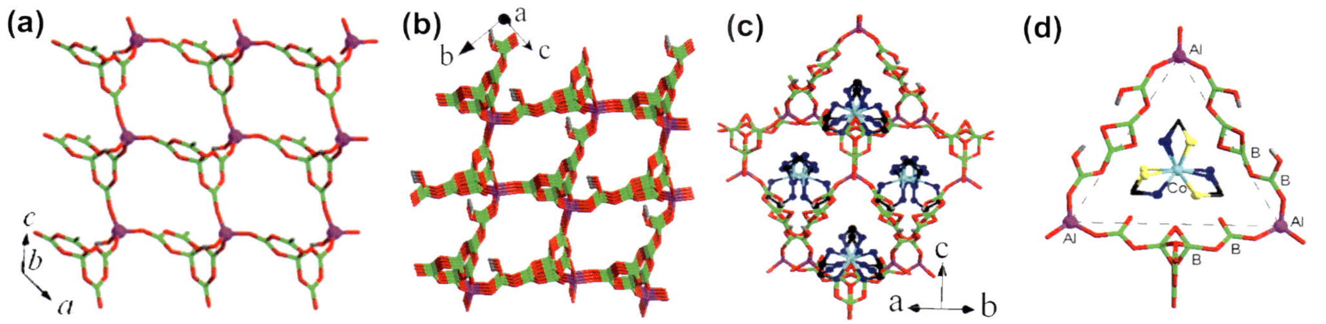

FIGURE 11.18 (a) View of 2D layer $[\{Al(OH)\}\{B_5O_9F\}]_n^{2n-}$. *Reproduced by permission of The Royal Society of Chemistry. (b) View of the 3-D framework in type I showing 13-ring channels. (c,d) $[M(en)_3]^{2+}$ cations located in 15-ring channels in type II.*

11.4.2. Templated Synthesis of Germanates and Borogermanates

1. Templated synthesis of germanates

Since the discovery of the first germanate with openframework in 1991 [77,78], a number of germanates with 2-, 3D, or isolated structure have been made by inorganic or organic templates. However, TMCs were not applied to make germanates, though TMCs as the templates have been widely used in making metal phosphates and borophosphates because the chirality of TMCs can be transferred into the inorganic frameworks. To understand the role of the chiral TMCs in determining inorganic frameworks of germanates, we introduced TMCs into germanates in 2003, and made a new germanate $[Ge_7O_{13}(OH)_2F_3]^{3-}\cdot Cl^-\cdot 2[Ni(dien)_2]^{2+}$ (FJ-6), which not only fills the gap left by the absence of 1D structures in germanate family but also is the first chiral germanate (Fig. 11.19a, [79]). Interestingly, the chirality of the *mer*-$Ni(dien)_2^{2+}$ ions transfer into adjacent germanate chains. As a continuation of our work, by using a racemic mixture of an $Ni(L)_3Cl_2$ complex (L = en/enMe) as template, we also made two novel germanate frameworks $Ni@Ge_{14}O_{24}(OH)_3\cdot 2Ni(L)_3$ (FJ-1a/1b), in which chiral $Ni@Ge_{14}O_{24}(OH)_3$ ($Ni@Ge_{14}$, Fig. 11.19b) clusters transferred from chiral TMCs containing Ni–Ge bonds and are further linked to form a 3D framework with 24-ring windows (Fig. 11.19c,d, [80]). FJ-1 is the first example of porous materials having T–M–T linkages (where T–M–T = Ge–Ni–Ge). In addition, FJ-1 exhibits stereospecificity and chiral molecular recognition between chiral TMC template and inorganic structural motif; the chirality and symmetry of the chiral TMC template can induce the chirality and symmetry of the inorganic motif/framework, while the symmetry of the achiral template can transfer to the inorganic motif. These phenomena originate from H-bonding between the template and the inorganic motif/framework. In 2007, a 2D germanate, $Ge_7O_{14}F_3\cdot 0.5$ In $(dien)_2\cdot 0.5H_3dien\cdot 2H_2O$, templated by indium complexes was first made [81], in which the unique germanate layers with uniform 10-ring apertures were built by $Ge_7O_{14}F_3$ clusters.

2. Templated synthesis of borogermanates

We focus on the investigation of the templated synthesis of germanates and are interested in introducing the element B into the germanate frameworks for the following reasons: (1) the flexible coordination geometries for Ge and B atoms and their ability to form Ge–O and B–O clusters as well as their unique 3-ring structures, the B–Ge–O system will

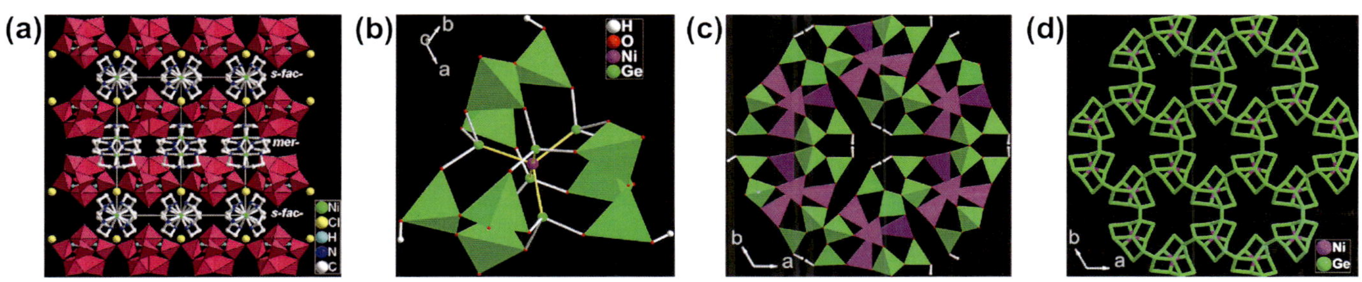

FIGURE 11.19 (a) A view of the ABAB... stacking sequence of chains and the relationship between the TMCs and the chains in **FJ-6**. *Reprinted with permission from [79]. Copyright 2003 American Chemical Society. (b) View of the $Ni@Ge_{14}$ unit. (c) View of the 24-ring channel built by three pairs of enantiomers of adjacent $Ni@Ge_{14}$ motifs. (d) The topological framework of FJ-1a.*

produce a new class of frameworks built by B—O and Ge—O clusters or mixed B—Ge—O units; (2) the characteristic of acentric B—O clusters formed in situ can be transferred into the inorganic frameworks to form acentric borogermanates, resulting in NLO properties of the products. Accordingly, the first chiral zeotype borogermanate $KBGe_2O_6$ (FJ-9, Fig. 11.20a, [69]) with 7-ring channels has been made. A 3D borogermanate $K_2[Ge(B_4O_9)]\cdot 2H_2O$ (FJ-16) with two pairs of interweaving double-helical 10-ring channels built by alternating linkages of B_4O_9 clusters and GeO_4 units (Fig. 11.20b, [70]), exhibits good NLO properties because of acentric structure. Different from FJ-9 and FJ-16 templated by K^+ ions, the first organically templated 3D borogermanate $(C_4N_3H_{15})[(BO_2)_2(GeO_2)_4]$ (FJ-17) was also made [72]. Its structure is built by two types of cluster units, the cyclic Ge_8O_{24} clusters with 8-ring and B_2O_7 dimers. The Ge_8O_{24} unit connects four same ones to form a 2D layer with 8,12-nets, which are further pillared by B_2O_7 clusters, resulting in a 3D framework with 12-/8-ring channels (Fig. 11.20c). Simultaneously, another two organically templated 2D borogermanates, $[H_2DACH][Ge_3B_2O_9(OH)_2]$ and $[H_2BAPPZ][Ge_3B_2O_9(OH)_2]\cdot 1.5H_2O$, have been obtained [71]. In 2008, a new organically templated borogermante $GeB_4O_9H_2en$ (FJ-18) was reported (Fig. 11.20d, [73]). Though its SBUs are the same as FJ-16, their structures are distinct because of different templates: large 12-ring channels in FJ-18, while 10-ring channels in FJ-16. Here, the concept of host—guest symmetry and charge matching [82] has further been extended from tetrahedral and tetrahedral—octahedral frameworks to tetrahedral—triangular frameworks.

Compared to the lighter group 13 elements B, Al, and Ga incorporated into the germanates, indium exhibits more flexible coordination geometries (tetrahedra, octahedra, and trigonal-bipyramids). To make novel indate—germanate frameworks, we are interested in introducing indium into germanates. In 2007, we first reported a novel organically templated indate—germanate $In_2Ge_6O_{15}(H_2dien)$ (FJ-3) [83], which was built by Ge—O layers and In—O chains, forming a 3D framework with 12-ring channels (Fig. 11.21a). Later, a new 2D hybrid indate—germanate $InGe_2O_5F(dap)\cdot 2H_2O$ has been made [84], in which two types of SBUs, $Ge_4O_{10}F_2$ clusters and $In_2O_6(dap)_2$ dimers, are interconnected to form 4- and 6-ring layers (Fig. 11.21b).

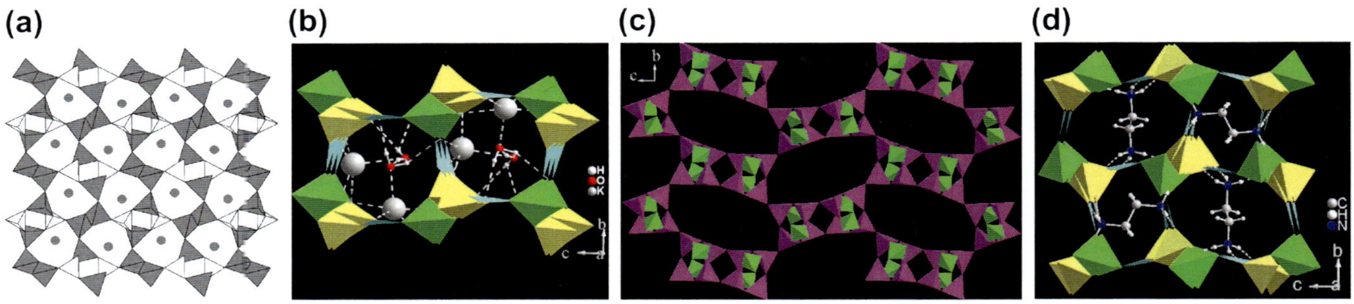

FIGURE 11.20 (a) View of the structure made by Ge_2O_7 and mixed $B_2Ge_2O_{12}$ cluster units in FJ-9. Reprinted with permission from [69]. Copyright 2003 American Chemical Society. (b) View of the 10-MR channels in FJ-16. View of the 12-ring channels of FJ-16. *Reprinted with permission from [70]. Copyright (2004) American Chemical Society. (c) View of the 8-/12-ring channels in FJ-17. Reprinted with permission from [72]. Copyright 2009 American Chemical Society. (d) View of the 8-/12-ring channels in FJ-18.*

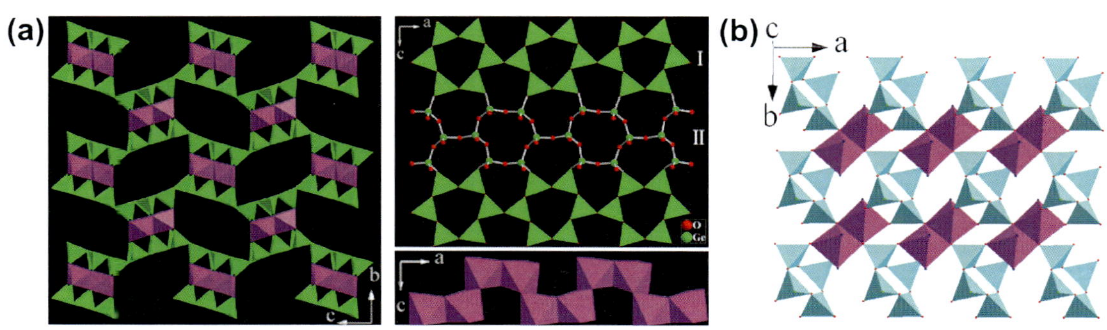

FIGURE 11.21 (a) The framework structure of FJ-3 built by the Ge-O layers and In—O chains, showing the 12-ring channels. (b) The 2-D layer built by $[Ge_4O_{10}F_2]$ clusters and $\{In_2O_6(dap)_2\}$ dimers. *Reprinted with permission from [84]. Copyright 2009 American Chemical Society.*

11.4.3. Self-polymerization and Induced Congregation of Lanthanide Germanate Clusters

So far, some main group elements and TMs have been introduced into germanate backbones under hydro (solvo)thermal conditions. However, except for several Ln germanates obtained by both flux-growth method and high-temperature, high-pressure hydrothermal method in supercritical water [85,86], no systematic investigation on Ln germanates has been carried out under mild hydrothermal method. This is because of the fact that the conditions for solubility of germanium dioxide (GeO_2, a traditional germanium source) and Ln oxide (Ln_2O_3) are incompatible in solution: acidic and basic solutions are favorable reaction systems for Ln_2O_3 and GeO_2, respectively. Recently, we are extending our interest from B−Ge−O to Ln−Ge−O system for making novel Ln germanate frameworks built from Ge−O and Ln−O or mixed Ln−Ge−O cluster units under mild hydrothermal conditions. Because of the flexible coordination geometries for Ge and Ln atoms, their ability to form Ge−O and Ln−O frameworks under rational conditions, and their unique cluster structural units, the Ln−Ge−O system is likely to provide novel structures with new topological frameworks. To make open-framework Ln germanates under mild hydrothermal conditions, three approaches have been tried in Ln−Ge−O system in our lab: (1) replacing Ln_2O_3 with Ln salts in basic solution; (2) replacing GeO_2 with soluble germanium alkoxides in acidic solution; and (3) using Ln salts and germanium alkoxide in acidic solution. However, the first approach failed because the Ln salts (oxide salts or alkoxides) were all hydrolyzed to form nondissolving Ln_2O_3 in basic solution, which could not further react with GeO_2; the latter two methods were also fruitless because of the very poor crystallization of the products, although germanium alkoxides and Ln_2O_3 (or Ln salts) have good solubility in acidic solution.

In previous works, we have made a series of Ln−O COFs, in which the Ln−O cluster unit is induced by carboxyl groups of the ligands. Accordingly, another type of germanium source containing carboxyls, bis (carboxyethylgermanium) sesquioxide $(HOOCCH_2CH_2)_2Ge_2O_3$ ($H_2E_2Ge_2O_3$, $E = -CH_2CH_2COO^-$), was chosen as the multifunctional ligand for the following reasons: (1) $H_2E_2Ge_2O_3$ with hydrophilic tails has good solubility in water, forming a highly acidic aqueous solution (pH 4) that may dissolve Ln_2O_3; (2) $H_2E_2Ge_2O_3$ contains two carboxyls and a Ge_2O_3 (O=Ge−O−Ge=O) core with two Ge=O bonds and thus not only offer the possibility for structural derivatives but also are effective precursors for making extended frameworks; and (3) the carboxyls may induce Ln ions to aggregate into Ln−O cluster, and the Ge_2O_3 core can further polymerize to form a Ge−O cluster when the Ge=O double bonds are opened; thus, the frameworks containing Ge−O and Ln−O clusters might be obtained. As expected, a series of Ln germanate frameworks have been made under mild hydrothermal conditions, $[Nd_8Ge_{12}(\mu_3\text{-}O)_{24}E_{12}(H_2O)_7]\cdot 13H_2O$ (FJ-19) and $[Ln_{11}Ge_{12}(\mu_3\text{-}O)_{24}E_{12}(pa)_6(H_2O)_{10}]\cdot(Cl,2OH)\cdot nH_2O$ (Hpa = 2-picolinic acid, FJ-20, Ln = Pr/Nd/Eu/Gd, FJ-20a–d; $n = 19$ for Pr/Nd; $n = 15/14$ for Eu/Gd) [87]. They contain high-nuclearity cage-shaped Ln germanate clusters (LGCs), $[Nd_8Ge_{12}(\mu_3\text{-}O)_{24}(H_2O)_7]^{24+}$ (Nd_8Ge_{12}) for FJ-19 and $[Ln_{11}Ge_{12}(\mu_3\text{-}O)_{24}(H_2O)_{10}]^{33+}$ ($Ln_{11}Ge_{12}$) for FJ-20, which act as the building blocks to further link each other via E groups, forming novel 3D Ln germanate cluster organic frameworks (LGCOFs), respectively. FJ-19 is a twofold interpenetrating enantiomorphic pair nets. In order to avoid interpenetration, a rigid chelating ligand, Hpa, was introduced into the reaction system because Hpa possess considerable steric hindrance and occupy enough space not only to avoid the formation of the interpenetrating networks but also to stabilize new LGC formed in situ through the chelating coordination modes. Thus, a series of LGCOFs, FJ-20, with openframeworks containing the $Ln_{11}Ge_{12}$ cluster units have been made (Fig. 11.22). These novel 3D LGCOFs made under mild hydrothermal conditions using an organogermanium compound as germanium source indicate that this strategy not only offers an effective and feasible way of making novel LGCOFs but also opens a new avenue for making metal germanate hybrid materials. Work in progress includes making novel functional LGCOFs by using larger rigid N-containing carboxylate ligands and metal germanate hybrid materials by using various metals salts under hydrothermal conditions. It is reasonable to believe that the present work will be important in expanding the study of germanates with open frameworks.

11.5. SYNTHESIS OF THE CHALCOGENIDE CLUSTERS UNDER HYDRO(SOLVO)THERMAL CONDITIONS

11.5.1. Tetrahedral Clusters

Since late 1980, the chalcogenide clusters have been studied to explore under hydro(solvo)thermal conditions. This is because the chalcogenide clusters not only provide a valuable opportunity for exploring the synthetic and structural chemistry but also become an important bridge between fundamental sciences and technological applications [88]. The tetrahedral clusters were not usually encountered in the oxo clusters, whereas they are common in chalcogenide clusters

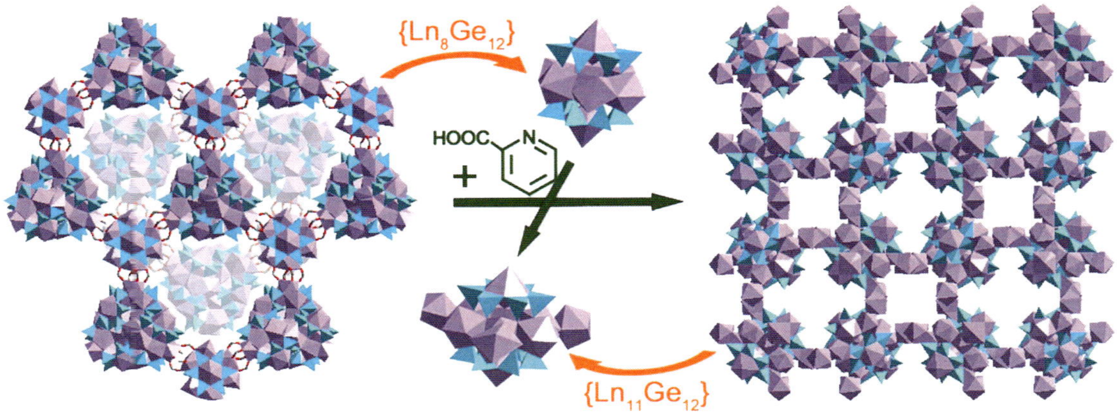

FIGURE 11.22 View of the transformation from the twofold interpenetrating nets of FJ-19 built by Nd_8Ge_{12} cluster units to the non-interpenetrating networks of FJ-20 constructed from $Ln_{11}Ge_{12}$ cluster units through introducing the second ligand, 2-picolinic acid. *Reprinted with permission from [87]. Copyright 2009 American Chemical Society.*

because both cations and anions in chalcogenide clusters may form tetrahedral coordination. One important reason for the difference between the oxo and chalcogenide clusters is the large size between oxygen and sulfur (selenium and tellurium) atoms, resulting in the coordination number of sulfur (selenium and tellurium) atoms higher than that of oxygen atoms. In case of S atom, its coordination number can easily go up to four even with large Cd^{2+} and In^{3+} cations. However, tetrahedrally coordinated O atoms are rare in oxo clusters due to the small size of the O atom (usually adopt bi- or tri-coordination in oxo clusters). In the chalcogenide clusters, three are mainly three series of tetrahedral clusters (Table 11.1): supertetrahedral clusters (denoted as T*n*) [88–90], pentasupertetrahedral clusters (denoted as P*n*) [90], and capped supertetrahedral clusters (denoted as C*n*) [89].

11.5.2. T*n* Clusters

T*n* clusters are the simplest tetrahedral clusters in supertetrahedral clusters (Fig. 11.23a) and have regular tetrahedrally shaped fragments of the cubic ZnS-type lattice. As shown in Fig. 11.23a, the number of anions in a T*n* cluster is equal to the number of cations in a $T(n+1)$ cluster. Here, n is the number of metal layers in each cluster. The number of cations (or anions) within each layer follows a simple series: 1, 3, 6, ... , $n(n+1)/2$. Therefore, T1 as MX_4 (M = metal cation and X = S^{2-}, Se^{2-}, Te^{2-}) tetrahedron is a special case in T*n* clusters. Now four types of the isolated T*n* clusters have been found, namely T2, T3, T4, and T5 with the formula of M_4X_{10}, $M_{10}X_{20}$, $M_{20}X_{35}$, and $M_{35}X_{56}$, respectively (Table 11.1). In a T*n* cluster (here $n \geq 2$), the numbers of the M and X ions are generally $[n(n+1)(n+2)]/6$ and $[(n+1)(n+2)(n+3)]/6$. In case of T2–T5 clusters, they all have four vertices that may share with other supertetrahedral cluster units to form the extended frameworks. So far, the largest supertetrahedral cluster with isolated form is T3 [91], whereas the largest supertetrahedral cluster unit in open-framework chalcogenides is T5 cluster [92,93].

TABLE 11.1 A Summary of Supertetrahedral Clusters

	Stoichiometry	Examples
T*n* series of supertetrahedral clusters		
T2	M_4X_{10}	$Ge_4X_{10}^{4-}$ (E = S, Se), $Sn_4E_{10}^{4-}$ (E = Se, Te), $In_4X_{10}^{8-}$ (X = S, Se), $M_4(SPh)_{10}^{2-}$ (M = Fe, Co, Cd; X = S, Se)
T3	$M_{10}X_{20}$	$M_{10}X_{20}^{10-}$ (M = Ga, In; X = S, Se), $M_{10}X_4(SPh)_{16}^{4-}$ (M = Zn, Cd; X = S, Se), $Ga_{10}S_{16}L_4^{2-}$ (L = 3,5-dimethylpyridine)
T4	$M_{20}X_{35}$	$M_4In_{16}S_{35}^{14-}$ (M = Mn, Fe, Co, Zn, Cd), $Zn_4Ga_{16}Se_{35}^{14-}$
T5	$M_{35}X_{56}$	$Cu_5In_{30}S_{56}^{17-}$, $Zn_{13}In_{22}S_{56}^{20-}$
≥T6	M_pX_q	None, here $p = [n(n+1)(n+2)]/6$; $q = [(n+1)(n+2)(n+3)]/6$
P*n* series of supertetrahedral clusters		
P1	M_8X_{17}	$M_4Sn_4S_{17}^{10-}$ (M = Mn, Fe, Co, Zn), $M_4Sn_4Se_{17}^{10-}$ (M = Mn, Zn), $ECd_8(E'Ph)_{16}^{2-}$ [E,E' = S, Se, Te], $In_8S_{16}(SH)^{9-}$
P2	$M_{26}X_{44}$	$Li_4In_{22}S_{44}^{18-}$, $Cu_{11}In_{15}Se_{16}(SePh)_{24}(PPh_3)_4$
≥P3	M_pX_q	None, $p = [4n(n+1)(n+2)] + [(n+1)(n+2)(n+3)]/6$; $q = [4(n+1)(n+2)(n+3)]/6 + [n(n+1)(n+2)]/6$
C*n* series of supertetrahedral clusters		
C1	$M_{17}X_{324}$	$Cd_{17}S_4(SPh)_{28}^{2-}$, $Cd_{17}S_4(SC_6H_4Me\text{-}4)_{28}^{2-}$
C2	$M_{32}X_5$	$Cd_{32}Se_{14}(SePh)_{36}(PPh_3)_4$, $Cd_{32}S_{14}(SPh)_{40}^{4-}$
C3	$M_{54}X_{84}$	$Cd_{54}X_{32}(SPh)_{48}(H_2O)_4^{4-}$ (X = S,Se)
≥C4	M_pX_q	None, $p = [n(n+1)(n+2)]/6 + [4(n+1)(n+2)]/2 + 4$; $q = [(n+1)(n+2)(n+3)]/6 + [4n(n+2)(n+3)]/2 + 4$

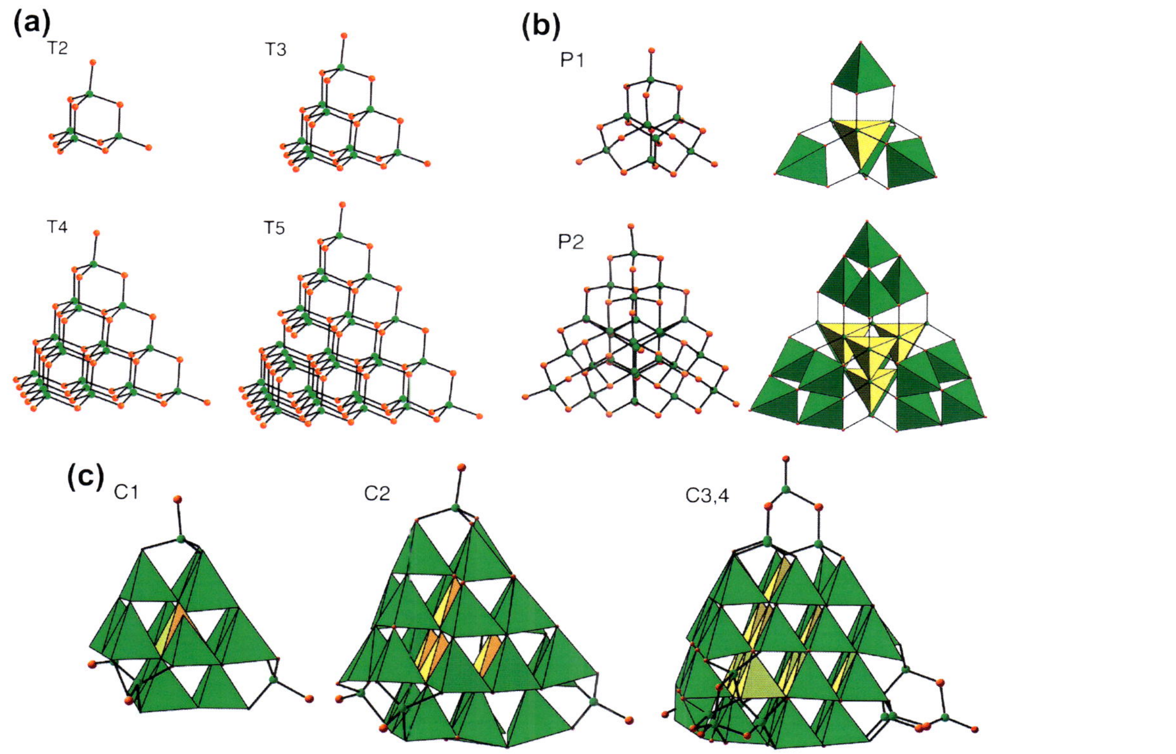

FIGURE 11.23 (a) Diagrams of T2, T3, T4, and T5 supertetrahedral clusters. Reprinted with permission from [88]. Copyright 2005 American Chemical Society. (b) View of P1 and P2 clusters. (c) The structural diagrams of C1, C2, and C3,4 capped supertetrahedral clusters. MX groups at the four corners of C*n* clusters and the rotated barrelanoid cages in C*m*,*n* cluster are shown in the ball-and-stick form [98].

11.5.3. P*n* Clusters

P*n* series of tetrahedral cluster can be considered as an assembly of four T*n* clusters tetrahedrally distributed onto four faces of one anti-T*n* cluster (Table 11.1, Fig. 11.23b). Here, an anti-T*n* cluster is defined as a T*n* cluster with interconverted cationic and anionic sites. Therefore, P1 cluster contains one anti-T1 tetrahedron (XM_4) located in the core and four T1 tetrahedra (MX_4) located in the corners, forming $(MX_4)_4(XM_4)$ [94]. Similarly, a P2 cluster has the composition of $(M_4X_{10})_4(X_4M_{10})$, such as isolated $Cu_{11}In_{15}Se_{16}(SePh)_{24}(PPh_3)_4$ [95], and 3D framework built by $[Li_4In_{22}S_{44}]^{18-}$ units [96]. In the P2 cluster, four T2 clusters bond to one anti-T2 cluster to form a P2 cluster (Fig. 11.23b). So far, P2 cluster is the largest size in P*n* series of supertetrahedral clusters.

11.5.4. C*n* Clusters

C*n* cluster consists of a regular T*n* cluster at the core covered with a shell of atoms the stoichiometry of which is also related to the T*n* cluster (Table 11.1), in which each face of the T*n* core unit is covered with a single sheet of atoms called the T($n+1$) sheet and each corner of this cluster is covered with an MX group. The T($n+1$) sheet is defined as the bottom atomic sheet of a T($n+1$) cluster [88]. In addition, the C*n*,*m* ($m = 1$–4) cluster is derived from the C*n* series and can be obtained by rotating the corner M_4X_5 group in the C*n* cluster by 60°. The structures of the C1, C2, and C3,4 clusters are shown in Fig. 11.23c, in which the largest C*n* cluster is C3,4 cluster with formula of $Cd_{54}X_{32}(SPh)_{48}(H_2O)_4$ (X = S, Se) [97,98].

11.5.5. Other Types of Surpertetrahedral Clusters

In addition to the above-mentioned clusters, other new clusters can also be made by adding atoms and removing atoms from the regular tetrahedral clusters [99]. In fact, the tetrahedral site in a T*q* cluster can be replaced by a T*p* cluster, resulting in novel and more complex super-surpertetrahedral cluster, T*p*,*q* [88,100,101].

11.5.6. Open-framework Chalcogenides made from Different Tetrahedral Clusters

Owing to the variety of chalcogenide tetrahedral clusters in sizes, shapes, and structures, they can act as the SBUs for making the extended structures. So far, some novel open-framework chalcogenides are built not only by the T*n*, P*n*, and C*n* cluster units but also by the mixed supertetrahedral cluster. Now, the frameworks constructed with different-size clusters from the same series of tetrahedral clusters (e.g., T3 and T4) or from different series (e.g., P1 and C1 or P1 and T2) have been successfully obtained [102–107]. For example, UCR-19 is made of alternating T3 ($Ga_{10}S_{20}{}^{10-}$) and T4 ($Zn_4Ga_{16}S_{35}{}^{14-}$) clusters [103]; UCR-15 is built from alternating T3 and coreless T5 clusters; when both T3 and pseudo-T5 clusters are replaced with T4 clusters, CdInS-44 and UCR-5MInS (M = Zn, Mn, Co) are obtained [104,105]; CMF-3 is built by P1 $[Cd_8S(SC_6H_4Me\text{-}3)_{16}]^{2-}$ and C1 $[Cd_{17}S_4(SC_6H_4Me\text{-}3)_{28}]^{2-}$ clusters [106], while HCF is built from P1 $[In_8S_{16}(SH)]^{9-}$ and T2 $[In_4S_9(SH)]^{7-}$ cluster units [107].

11.5.7. Open-framework Chalcogenides Based on Tetrahedral Clusters and Organic Linkers

Open-framework chalcogenides are usually linked together to form extended structures not only through corner-sharing chalcogens but also through organic ligands as the bridges. Compared with chalcogen atoms, organic ligands possess a wide variety of shapes and rich coordination chemistry, which may allow the creation of diverse chalcogenides with open frameworks. For example, the C*n* cluster (e.g., $Cd_{17}S_4(SPh)_{28}^{2-}$ and $Cd_{32}S_{14}(SPh)_{40}^{4-}$) units can be linked by neutral organic ligands (e.g., 4,4-bipyridine, 1,3-trimethylenedipyridine, or 1,2-bis(4-pyridyl)ethane) via substitution of four terminal corners in C*n* clusters, forming a series of cadmium chalcogenide clusters with 1- and 2D structures [108,109].

In brief, the chalcogenide superlattices are one of the most exciting research fields in cluster chemistry and porous materials chemistry. By combining synthetic and structural concepts in these two areas, many crystalline chalcogenide cluster solids not only be continuously created but also will exhibit varied properties, such as fast ion conductivity, photoluminescence, catalysis, solid-state ionics, and electrochemistry.

References

[1] J.E. Huheey, Inorganic Chemistry, Harper and Row, New York, 1983.
[2] D.M.P. Mingos, D.J Wales, Introduction to Cluster Chemistry (1990).
[3] J.U. Reveles, S.N. Khanna, P.J. Roach, A.W. Castleman, Proceedings of the National Academy of Sciences of the United States of America 103 (2006) 18405.
[4] S. Bulusu, X. Li, L.S. Wang, X.C. Zeng, Proceedings of the National Academy of Sciences of the United States of America 103 (2006) 8326.
[5] P. Gruene, D.M. Rayner, B. Redlich, A.F.G. van der Meer, J.T. Lyon, G. Meijer et al., Science 321 (2008) 674.
[6] G.K. Johnson, E.O. Schlemper, Journal of the American Chemical Society 100 (1978) 3645.
[7] A. Müller, J. Doring, Angewandte Chemie-International Edition in English 27 (1988) 1721.
[8] G. Huan, M.A. Greaney, A.J. Jacobson, Journal of the Chemical Society: Chemical Communications (1991) 260.
[9] A. Müller, J. Doring, Zeitschrift Fur Anorganische Und Allgemeine Chemie 595 (1991) 251.
[10] A. Müller, J. Doring, H. Bogge, Journal of the Chemical Society-Chemical Communications (1991) 273.
[11] S.T. Zheng, M.H. Wang, G.Y. Yang, Inorganic Chemistry 46 (2007) 9503.
[12] S.T. Zheng, J. Zhang, G.Y. Yang, Zeitschrift Fur Anorganische Und Allgemeine Chemie 631 (2005) 170.
[13] S.T. Zheng, J. Zhang, B. Li, G.Y. Yang, Dalton Transactions (2008) 5584.
[14] R. Kiebach, C. Nather, P. Kogerlerb, W.G. Bensch, Dalton Transactions (2007) 3221.
[15] S.T. Zheng, J. Zhang, G.Y. Yang, European Journal of Inorganic Chemistry (2004) 2004.
[16] X.B. Cui, J.Q. Xu, H. Meng, S.T. Zheng, G.Y. Yang, Inorganic Chemistry 43 (2004) 8005.
[17] S.T. Zheng, J. Zhang, G.Y. Yang, Inorganic Chemistry 44 (2005) 2426.
[18] D. Zhao, S.T. Zheng, G.Y. Yang, Journal of Solid State Chemistry 181 (2008) 3071.
[19] S.T. Zheng, D.Q. Yuan, H.P. Jia, J. Zhang, G.Y. Yang, Chemical Communications (2007) 1858.
[20] J.W. Zhao, H.P. Jia, J. Zhang, S.T. Zheng, G.Y. Yang, Chemistry – A European Journal 13 (2007) 10030.
[21] J.W. Zhao, J. Zhang, S.T. Zheng, G.Y. Yang, Inorganic Chemistry 46 (2007) 10944.
[22] J.W. Zhao, B. Li, S.T. Zheng, G.Y. Yang, Crystal Growth & Design 7 (2007) 2658.
[23] J.W. Zhao, J. Zhang, Y. Song, S.T. Zheng, G.Y. Yang, European Journal of Inorganic Chemistry (2008) 3809.
[24] B. Li, J.W. Zhao, S.T. Zheng, G.Y. Yang, Inorganic Chemistry 48 (2009) 8294.
[25] S.T. Zheng, J. Zhang, G.Y. Yang, Angewandte Chemie-International Edition 47 (2008) 3909.
[26] S.T. Zheng, J. Zhang, J.M. Clemente-Juan, D.Q. Yuan, G.Y. Yang, Angewandte Chemie-International Edition 48 (2009) 7176.
[27] S.T. Zheng, M.H. Wang, G.Y. Yang, Chemistry-an Asian Journal 2 (2007) 1380.
[28] S.T. Zheng, D.Q. Yuan, J. Zhang, G.Y. Yang, Inorganic Chemistry 46 (2007) 4569.
[29] J.W. Zhao, S.T. Zheng, Z.H. Li, G.Y. Yang, Dalton Transactions (2009) 1300.
[30] J.W. Zhao, C.M. Wang, J. Zhang, S.T. Zheng, G.Y. Yang, Chemistry – A European Journal 14 (2008) 9223.
[31] J.W. Zhao, J. Zhang, S.T. Zheng, G.Y. Yang, Chemical Communications (2008) 570.
[32] A.C. Skapski, V.F. Sutcliffe, G.B. Young, Journal of the Chemical Society – Chemical Communications (1985) 609.
[33] C. Streb, C. Ritchie, D.L. Long, P. Kogerler, L. Cronin, Angewandte Chemie-International Edition 46 (2007) 7579.
[34] X.M. Chen, S.M.J. Aubin, Y.L. Wu, Y.S. Yang, T.C.W. Mak, D.N. Hendrickson, Journal of the American Chemical Society 117 (1995) 9600.
[35] B.Q. Ma, D.S. Zhang, S. Gao, T.Z. Jin, C.H. Yan, G.X. Xu, Angewandte Chemie-International Edition 39 (2000) 3644.
[36] X.J. Zheng, L.P. Jin, S. Gao, Inorganic Chemistry 43 (2004) 1600.
[37] Y.Q. Sun, J. Zhang, Y.M. Chen, G.Y. Yang, Angewandte Chemie-International Edition 44 (2005) 5814.

[38] M.B. Zhang, J. Zhang, S.T. Zheng, G.Y. Yang, Angewandte Chemie-International Edition 44 (2005) 1385.
[39] J.W. Cheng, J. Zhang, S.T. Zheng, M.B. Zhang, G.Y. Yang, Angewandte Chemie-International Edition 45 (2006) 73.
[40] J.W. Cheng, J. Zhang, S.T. Zheng, G.Y. Yang, Chemistry – A European Journal 14 (2008) 88.
[41] Y.Q. Sun, G.Y. Yang, Dalton Transactions (2007) 3771.
[42] C.T. Chen, B.C. Wu, A.D. Jiang, G.M. You, Scientia Sinica Series B – Chemical Biological Agricultural Medical & Earth Sciences 28 (1985) 235.
[43] C.T. Chen, Y.C. Wu, A.D. Jiang, B.C. Wu, G.M. You, R.K. Li, Journal of the Optical Society of America B – Optical Physics 6 (1989) 616.
[44] Y.C. Wu, T. Sasaki, S. Nakai, A. Yokotani, H.G. Tang, C.T. Chen, Applied Physics Letters 62 (1993) 2614.
[45] C.T. Chen, Y.B. Wang, B.C. Wu, K.C. Wu, W.L. Zeng, L.H. Yu, Nature 373 (1995) 322.
[46] H.X. Zhang, J. Zhang, S.T. Zheng, G.Y. Yang, Inorganic Chemistry Communications 7 (2004) 781.
[47] G.M. Wang, Y.Q. Sun, G.Y. Yang, Journal of Solid State Chemistry 179 (2006) 398.
[48] G.M. Wang, Y.Q. Sun, S.T. Zheng, G.Y. Yang, Zeitschrift Fur Anorganische Und Allgemeine Chemie 632 (2006) 1586.
[49] H.X. Zhang, J. Zhang, S.T. Zheng, G.Y. Yang, Crystal Growth & Design 5 (2005) 157.
[50] G.M. Wang, Y.Q. Sun, G.Y. Yang, Journal of Solid State Chemistry 177 (2004) 4648.
[51] C.Y. Pan, G.M. Wang, S.T. Zheng, G.Y. Yang, Journal of Solid State Chemistry 180 (2007) 1553.
[52] C.Y. Pan, G.M. Wang, S.T. Zheng, G.Y. Yang, Zeitschrift Fur Anorganische und Allgemeine Chemie 633 (2007) 336.
[53] C.Y. Pan, G.M. Wang, S.T. Zheng, G.Y. Yang, Acta Crystallographica Section E-Structure Reports Online 63 (2007) o1207.
[54] G.Z. Liu, S.T. Zheng, G.Y. Yang, Inorganic Chemistry Communications 10 (2007) 84.
[55] H.X. Zhang, S.T. Zheng, G.Y. Yang, Acta Crystallographica Section C-Crystal Structure Communications 60 (2004) M241.
[56] G.M. Wang, Y.Q. Sun, G.Y. Yang, Journal of Solid State Chemistry 178 (2005) 729.
[57] G.M. Wang, Y.Q. Sun, G.Y. Yang, Journal of Solid State Chemistry 179 (2006) 1545.
[58] H.A. Lehmann, K. Teske, Zeitschrift Fur Anorganische und Allgemeine Chemie 400 (1973) 169.
[59] X.S. Liu, R.R. Xu, Journal of the Chemical Society – Chemical Communications (1989) 1837.
[60] J.H. Wang, S.H. Feng, R.R. Xu, Journal of the Chemical Society – Chemical Communications (1989) 265.
[61] J.H. Yu, R.R. Xu, Q.B. Kan, Y.H. Xu, B.Q. Xu, Journal of Materials Chemistry 3 (1993) 77.
[62] J.H. Yu, J.S. Chen, R.R. Xu, Y.H. Xu, Y. Yue, Polyhedron 15 (1996) 4127.
[63] J.H. Yu, R.R. Xu, J.S. Chen, Y. Yue, Journal of Materials Chemistry 6 (1996) 465.
[64] J.H. Yu, R.R. Xu, Y.H. Xu, Y. Yue, Journal of Solid State Chemistry 122 (1996) 200.
[65] S. Ayyappan, C.N.R. Rao, Chemical Communications (1997) 575.
[66] J. Ju, J.H. Lin, G.B. Li, T. Yang, H.M. Li, F.H. Liao, Angewandte Chemie International Edition 42 (2003) 5607.
[67] J. Ju, T. Yang, G. Li, F. Liao, Y. Wang, L. You, Chemistry 10 (2004) 3901.
[68] T. Yang, J. Ju, G.B. Li, F.H. Liao, X.D. Zou, F. Deng, Inorganic Chemistry 46 (2007) 4772.
[69] Z.E. Lin, J. Zhang, G.Y. Yang, Inorganic Chemistry 42 (2003) 1797.
[70] H.X. Zhang, J. Zhang, S.T. Zheng, G.M. Wang, G.Y. Yang, Inorganic Chemistry 43 (2004) 6148.
[71] G.M. Wang, Y.Q. Sun, G.Y. Yang, Crystal Growth & Design 5 (2005) 313.
[72] H.X. Zhang, J. Zhang, S.T. Zheng, G.Y. Yang, Inorganic Chemistry 44 (2005) 1166.
[73] C.Y. Pan, G.Z. Liu, S.T. Zheng, G.Y. Yang, Chemistry – A European Journal 14 (2008) 5057.
[74] C. Rong, Z.W. Yu, Q. Wang, S.T. Zheng, C.Y. Pan, F. Deng, Inorganic Chemistry 48 (2009) 3650.
[75] J. Zhou, S.T. Zheng, M.Y. Zhang, G.Z. Liu, G.Y. Yang, CrystEngComm 11 (2009) 2597.
[76] J. Zhou, W.H. Fang, C. Rong, G.Y. Yang, Chemistry – A European Journal 16 (2010) 4852.
[77] J. Cheng, R.R. Xu, G.D. Yang, Journal of the Chemical Society – Dalton Transactions (1991) 1537.
[78] J. Cheng, R.R. Xu, Journal of the Chemical Society – Chemical Communications (1991) 483.
[79] H.X. Zhang, J. Zhang, S.T. Zheng, G.Y. Yang, Inorganic Chemistry 42 (2003) 6595.
[80] Z.E. Lin, J. Zhang, J.T. Zhao, S.T. Zheng, C.Y. Pan, G.M. Wang, Angewandte Chemie International Edition 44 (2005) 6881.
[81] G.Z. Liu, H.X. Zhang, Z.E. Lin, S.T. Zheng, J. Zhang, J.T. Zhao, Chemistry – An Asian Journal 2 (2007) 1230.
[82] X.H. Bu, P.Y. Feng, G.D. Stucky, Science 278 (1997) 2080.
[83] G.Z. Liu, S.T. Zheng, G.Y. Yang, Angewandte Chemie International Edition 46 (2007) 2827.
[84] J. Zhou, Z.W. Yu, W.H. Fang, F. Deng, G.Y. Yang, Inorganic Chemistry 48 (2009) 10895.
[85] P.L. Chen, P.Y. Chiang, H.C. Yeh, B.C. Chang, K.H. Lii, Dalton Transactions (2008) 1721.
[86] M. Emirdag-Eanes, M. Krawiec, J.W. Kolis, Journal of Chemical Crystallography 31 (2002) 281.
[87] H. He, G.J. Cao, S.T. Zheng, G.Y. Yang, Journal of the American Chemical Society 131 (2009) 15588.
[88] P.Y. Feng, X.H. Bu, N.F. Zheng, Accounts of Chemical Research 38 (2005) 293.
[89] H.L. Li, A. Laine, M. O'Keeffe, O.M. Yaghi, Science 283 (1999) 1145.
[90] X.H. Bu, N.F. Zheng, P.Y. Feng, Chemistry – A European Journal 10 (2004) 3356.
[91] P. Vaqueiro, M.L. Romero, Chemical Communications (2007) 3282.
[92] N.F. Zheng, X.H. Bu, P.Y. Feng, Nature 426 (2003) 428.
[93] X.H. Bu, N.F. Zheng, Y.Q. Li, P.Y. Feng, Journal of the American Chemical Society 124 (2002) 12646.
[94] G.S.H. Lee, K.J. Fisher, D.C. Craig, M.L. Scudder, I.G. Dance, Journal of the American Chemical Society 112 (1990) 6435.
[95] A. Eichhofer, D. Fenske, Journal of the Chemical Society – Dalton Transactions (2000) 941.
[96] N.F. Zheng, X.H. Bu, P.Y. Feng, Angewandte Chemie-International Edition 43 (2004) 4753.
[97] N.F. Zheng, X.H. Bu, H.W. Lu, Q.C. Zhang, P.Y. Feng, Journal of the American Chemical Society 127 (2005) 11963.
[98] Z. Lin, P. Feng, X. Bu, Chapter Nano-/microporous materials: crystalline metal-chalcogenide superlattices, Nanomaterials: Inorganic and Bioinorganic Perspectives, edited by Charles M. Lukehart and Robert A. Scott, John Wiley & Sons, Ltd., 2008.
[99] E. Ruzin, A. Fuchs, S. Dehnen, Chemical Communications (2006) 4796.
[100] N.F. Zheng, X.G. Bu, B. Wang, P.Y. Feng, Science 298 (2002) 2366.
[101] H.L. Li, J. Kim, M. O'Keeffe, O.M. Yaghi, Angewandte Chemie-International Edition 42 (2003) 1819.

[102] C. Wang, X.H. Bu, N.F. Zheng, P.Y. Feng, Journal of the American Chemical Society 124 (2002) 10268.
[103] N.F. Zheng, X.H. Bu, P.Y. Feng, Journal of the American Chemical Society 125 (2003) 1138.
[104] C. Wang, Y.Q. Li, X.H. Bu, N.F. Zheng, O. Zivkovic, C.S. Yang, et al., Journal of the American Chemical Society 123 (2001) 11506.
[105] H.L. Li, J. Kim, T.L. Groy, M. O'Keeffe, O.M. Yaghi, Journal of the American Chemical Society 123 (2001) 4867.
[106] Q.C. Zhang, X.H. Bu, J. Zhang, T. Wu, P.Y. Feng, Journal of the American Chemical Society 129 (2007) 8412.
[107] Q.C. Zhang, X.H. Bu, L. Han, P.Y. Feng, Inorganic Chemistry 45 (2006) 6684.
[108] N.F. Zheng, X.H. Bu, J. Lauda, P.Y. Feng, Chemistry of Materials 18 (2006) 4307.
[109] N.F. Zheng, X.H. Bu, H.W. Lu, L. Chen, P.Y. Feng, Journal of the American Chemical Society 127 (2005) 14990.

CHAPTER

12

Synthetic Chemistry of Fullerenes

Shun-Liu Deng, Yuan-Zhi Tan, Su-Yuan Xie
Xiamen University, China

Fullerenes are a family of all-carbon clusters that are typically composed of a number of hexagons and exactly 12 pentagons. Soccer-ball shaped "Buckminsterfullerene" C_{60} having I_h symmetry is the most famous member discovered in 1985 [1]. The breakthrough experiment for the production of C_{60}, originally designed for studying the chemistry in interstellar space and circumstellar shells, was conducted by Kroto, Smalley, and Curl through vaporization of graphite using a pulsed laser and detection of carbon cluster by in situ time-of-flight mass spectrometry. Since the discovery of fullerenes, considerable efforts have been made to explore macroscopic synthesis and properties of fullerenes.

12.1. SYNTHESIS

12.1.1. Resistive Heating and Arc-Discharge of Graphite

In 1990, Kratschmer et al. [2] first reported the technique capable of producing milligram quantities of C_{60} by resistive heating of graphite. In a conventional glass bell jar evaporator filled with an inert gas, two graphite rods were kept in mechanical contact and evaporated by a suitably high current. Infrared spectral studies of the resulting black carbon soot scraped from the chamber showed four prominent sharp absorptions, close to those predicted for buckminsterfullerene [3]. Separated from the soot by extraction with benzene or sublimation in a vacuum (or in an inert atmosphere), about 100 mg per day of the purified fullerene materials could be obtained. The yield of fullerenes (C_{60}, C_{70}, etc.) in soluble products can be improved to 14% [4].

After some time for graphite evaporation, however, the contact between the graphite electrodes would be lost, leading to unsteady fluctuation of arc burning. Koch et al. [5] modified this arc technique with a gravity-feed idea and developed a fullerene bench top reactor. The apparatus consisted of a thin "consumable" graphite rod (3 mm diameter) and a thick graphite rod (12 mm diameter) which is essentially not degraded. When a direct current (DC) or alternating current (AC) was supplied, the thin rod evaporated and then slip downward to keep the constant electric contact. The fullerene C_{60} with a yield of about 0.25 g per day could be produced in this gravity-feed generator.

Haufler et al. [6] reported an efficient preparation of C_{60} by "contact-arc" vaporization of graphite rod. As compared with resistive heating technique, the evaporation of graphite electrodes can be maintained without mechanical contact. Figure 12.1 presents a schematic diagram of the C_{60} generator, in which a 6 mm o.d. graphite rod is kept in contact with a graphite disk by spring pressure and the fullerenes-containing soot is collected in a water-cooled cylinder. During vaporization, the spring tension is adjusted to maintain the arc between these two nearly contacting electrodes so that the bulk of power is dissipated in the arc, other than in the Ohmic heating of the rods. In such a "contact-arc" process, the content of toluene-soluble fullerenes in the produced soot is about $10 \pm 2\%$.

C_{60} and C_{70} are always predominantly produced in the graphite resistive heating or arc-discharge approaches. Depending on the doping of graphite reactant, however, the relative contents of fullerenes in the products can be varied to some extent. With a 1% B-doped graphite rod, the absolute yield of C_{60} was reduced from 8.85% to 2.75%, while the yield of higher fullerenes such as C_{78} and C_{84} were increased by a factor of 2 [7]. Ahmad et al. [8,9] reported the synthesis of higher fullerene C_{84} in DC arc discharge using Cu as a catalyst.

12.1.2. Laser Vaporization

Laser vaporization of graphite was studied in 1984 [10] which ultimately led to the discovery of Buckminsterfullerene C_{60} in 1985 [1]. However, no isolable amount of fullerenes was formed in the condensed soot.

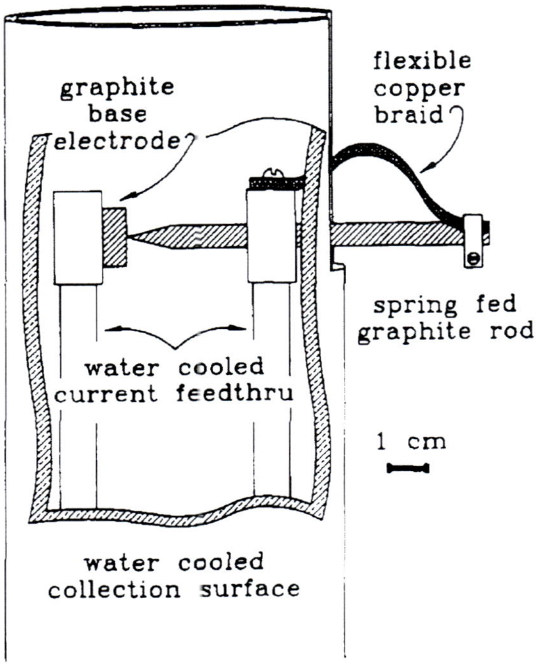

FIGURE 12.1 Schematic diagram of graphite rod contact-arc C_{60} generator [6]. *Reprinted with permission from [6]. Copyright (1990) American Chemical Society.*

A simple modification of the laser vaporization method [11], done in a "laser furnace" at the temperature around 1200°C, demonstrated that fullerene yields of over 40% of the vaporized carbon could be achieved [12]. Endohedral fullerenes such as La@C_{60} were readily produced as well [11,13]. Compared with arc method, the laser furnace is much more convenient, in which ablation laser power, buffer gas temperature and pressure can be controlled accurately. However, this process is not suitable for large-scale production of fullerenes simply because laser apparatus is too expensive.

C_{60} can also be synthesized by laser ablation of perchloroacenaphthylene ($C_{12}Cl_8$) [14] which is composed of two hexagons and a pentagon, the typical intermediate in C_{60} formation as predicted by Pentagon Road [15]. After 10 h of ablation by a pulsed laser beam in a vacuum, the products were collected and dissolved in toluene by an ultrasonic bath. Fullerene C_{60} together with a wide range of perchlorinated carbon clusters were observed on the basis of high-performance liquid chromatography combined with ultraviolet spectrometry and mass spectrometry (HPLC–UV–MS) analysis.

12.1.3. Solar Generation

To detour the photochemical destruction of C_{60} by UV radiation which is unavoidable in the traditional arc-discharge process, various research groups [12,16,17] have shown the possibility of fullerenes production in rarefied inert gas atmosphere by direct vaporization of graphite in focus sunlight. The apparatus "Solar 1" used at Rice University [12] is shown in Fig. 12.2. To produce soot, sunlight was collected by the parabolic mirror and focused onto the tip of a 0.4 mm diameter graphite rod which was mounted inside a Pyrex tube. In order to minimize the extent of conductive heat loss and to help anneal the carbon clusters as they grew from the vapor, the graphite rod was enclosed by a helical tungsten preheater. In 50 Torr argon atmosphere, the sunlight was focused directly on the tip of the graphite target to afford about 5 mg toluene-soluble carbon clusters within 3 h. A scale-up approach of solar reactor for fullerenes production was proposed by Flamant et al. [18], showing that a cylindrical target of 22 cm o.d. can be converted to 80–150 g/h carbon soot for an effective power of 350 kW.

12.1.4. Combustion Synthesis

The existence of positive and negative ions of fullerenes such as C_{60} and C_{70} in low-pressure premixed hydrocarbon sooting flames was first revealed by Homann and coworkers [19] in 1987. After that, numerous efforts have been focused on synthesis of fullerenes in macroscopic quantities from the soot made in laminar, premixed, low-pressure flame of hydrocarbon and oxygen with or without inert diluent gas. Howard and coworkers [20] demonstrated that both C_{60} and C_{70} fullerenes were produced in varying amounts from premixed laminar benzene–oxygen–argon flames under a range of conditions. HPLC and spectroscopic studies revealed that the mass of fullerenes ($C_{60}+C_{70}$) produced was around 0.003–9.2% of the soot. In addition to benzene [20,21], experiments have also been conducted using the fuels such as toluene [22], naphthalene [23,24], methane [25], and butadiene [26]. Richter et al. [27] demonstrated that the addition

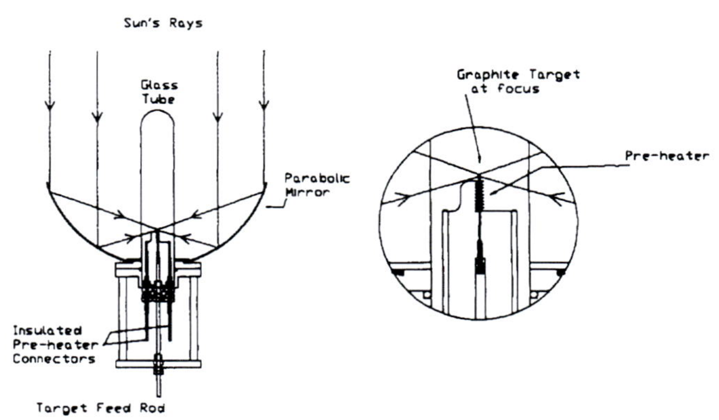

FIGURE 12.2 Schematic diagram of "Solar 1" fullerene generator [12]. *Reprinted with permission from [12]. Copyright (1993) American Chemical Society*

of 25% of Cl_2 led to a dramatic increase in the production of fullerene. Since the combustion method is a continuous and easily scalable process similar to those employed for commercial carbon black production, it can produce fullerenes with even greater efficiency than the vaporization of graphite methods. Based on this technology, large-scale production of fullerenes (tons per year) has been designed by the Frontier Carbon Corporation [28].

12.1.5. Thermal Pyrolysis

Polycyclic aromatic hydrocarbons (PAHs) have been suspected for a long time to be the precursors of fullerenes [29]. In 1993, Taylor et al. [30] first described the production of C_{60} and C_{70} from naphthalene pyrolysis at 1000°C in an argon stream. The yield of fullerenes was estimated to be less than 0.5% on the base of consumed naphthalene. A mechanism involving the "patch-assembly" of C_{10} units through various routes was proposed. Osterodt et al. [31] showed that thermal pyrolysis of various hydrocarbons and cyclopentadienide–metal complexes could produce a small amount of fullerene. In the pyrolytic process, bowl-shaped corannulene and benzo[*k*]fluoranthene can serve as precursors leading to fullerene C_{60} as well [32,33]. To improve the yield of C_{60}, efforts have been carried out on the pyrolysis of 1-chloro and 1-bromo derivatives of naphthalene [33,34]. 1-Bromonaphthalene was considered as a more suitable precursor than naphthalene, and the possibility of using perhaloarenes or oligomer of naphthalene as fullerene precursor in the pyrolysis method was also demonstrated [35,36].

12.1.6. Plasma Synthesis

Xie et al. [37,38] reported a continuous production of fullerene C_{60} and C_{70} via microwave plasma from chloroform at low-pressure argon atmosphere. The fullerenes soot was produced in a quartz column reactor as shown in Fig. 12.3 [37]. Yields of C_{60} (0.3–1.3%) and C_{70} (0.1–0.3%) and their ratio depended on the temperature gradient and the collision probability. The plasma synthesis of fullerenes can also be achieved in a glow plasma reaction of chloroform [39]. Fullerene yield obtained by this method is relatively low (less than 1%). Albeit the relatively low yield of fullerenes, the plasma method has lots of potential for the economical production of fullerene because of the continuous and controllable operation conditions. Furthermore, various fully chlorinated fullerene fragments, such as $C_{10}Cl_8$, $C_{12}Cl_8$, $C_{16}Cl_{10}$, and $C_{20}Cl_{10}$ formed together with C_{60} and C_{70} under the same reaction conditions, provide evidence shedding light on the formation mechanism of fullerenes.

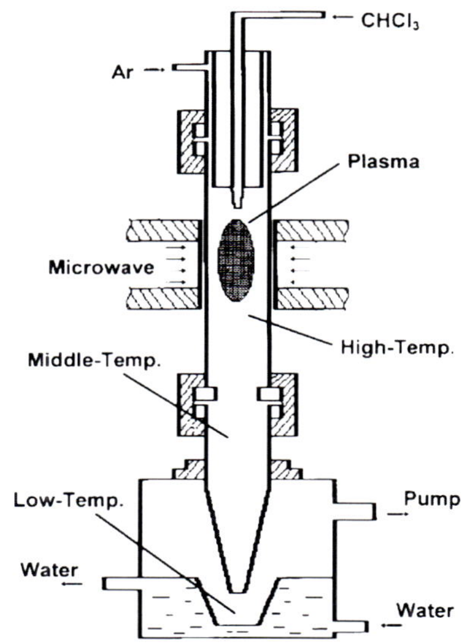

FIGURE 12.3 Schematic diagram of microwave synthesis system [37]. *Reprinted with permission from [37]. Copyright (1999), American Institute of Physics.*

12.1.7. Chemical Synthesis

Chemical synthesis of C_{60} involving related polyyne precursors was studied by several groups [40–43]. Recorded by mass spectrometry, C_{60} was produced because of zipping up of the macrocyclic polyalkynes subject of laser desorption/ionization. For example, the reactive cage polyyne $C_{60}H_6$ (**1**), supposed as a possible precursor to C_{60}, would collapse to I_h-symmetric cage structure with the loss of hydrogen atoms in gas phase. The initial studies of Rubin [43] demonstrated the possibility of complete dehydrogenation of $C_{60}H_6$. The polyyne cyclization mechanism for fullerene formation was also suggested by Tobe et al. [41,42] from the observation of C_{60} ions in mass spectra of the stable hydrocarbon $C_{60}H_6(Ind)_6$ (Ind = indane) (**2**) and its perchlorinated derivative $C_{60}Cl_6(Ind)_6$ (**3**). Diazafullerene anion $C_{58}N_2^-$ was also detected in the laser desorption mass spectrum from diazafullerene precursors with pyridine instead of benzene moieties [44].

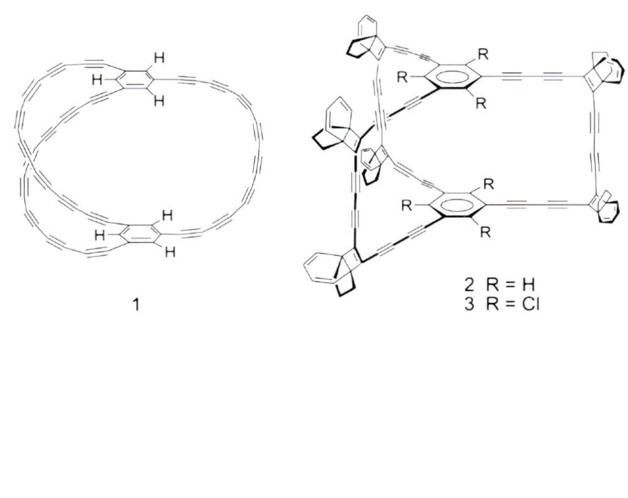

1

2 R = H
3 R = Cl

Fullerene C_{60} can also be produced with a high intensity UV laser irradiation of polycyclic aromatic hydrocarbon $C_{60}H_{30}$ (**4**) (Scheme 12.1), which was synthesized in nine steps by conventional laboratory methods [45,46]. According to the synthesis of bowl-shaped polycyclic aromatic hydrocarbon [47–51], incorporation of halogen atoms at strategic sites could facilitate the formation of molecular curvature [52]. Based on this guideline, polycyclic aromatic precursor $C_{60}H_{27}Cl_3$ (**5**), bearing chlorine substituents at key positions, was synthesized in 12 steps from commercially available starting materials [53]. Sufficient quantities of C_{60} were generated by flash-vacuum pyrolysis (FVP) of **5** at 1100°C (Scheme 12.1). No C_{70} or any other fullerenes were detected in the synthetic products. Higher fullerene precursors, such as $C_{84}H_{42}$ [54] and $C_{78}H_{48}$ [55] have been synthesized as well. The highly selective formation of C_{84} and C_{78} confirms the potentiality of using FVP technique for the direct synthesis of higher fullerenes from polycyclic aromatic hydrocarbon precursors. The formation of C_{60} and the triazafullerene $C_{57}N_3$ from corresponding aromatic precursors in cyclodehydrogenation process was also achieved on a platinum (111) surface at a temperature of 750 K [56].

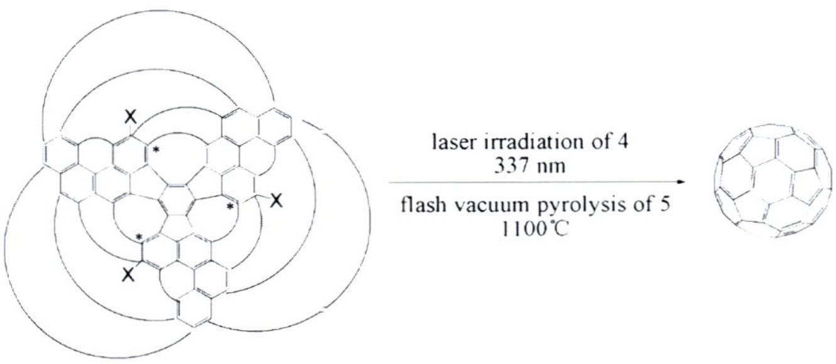

4 X = H
5 X = Cl

SCHEME 12.1 Final step in the chemical synthesis of C_{60} [46,53]. *From [53]. Reprinted with permission from AAAS.*

12.2. CHARACTERIZATION OF FULLERENES

In the original gas-phase experiment conducted in the laser vaporization conditions, there are plenty of fullerenes C_n (n is the even number larger than 20 but with the exception of 22) in the gas phase (Fig. 12.4) [1,10,57]. However, only a few bare hollow fullerenes survive the ambient conditions and have been isolated in the forms of stable hollow fullerenes. The most famous members of the stable hollow fullerenes are I_h-C_{60} and D_{5h}-C_{70} isolated in early 1990. (Note that the nomenclature is specified by symmetry to differentiate their isomers.)

The instability of most of the other fullerenes is due to their violation of the so-called Isolated Pentagon Rule

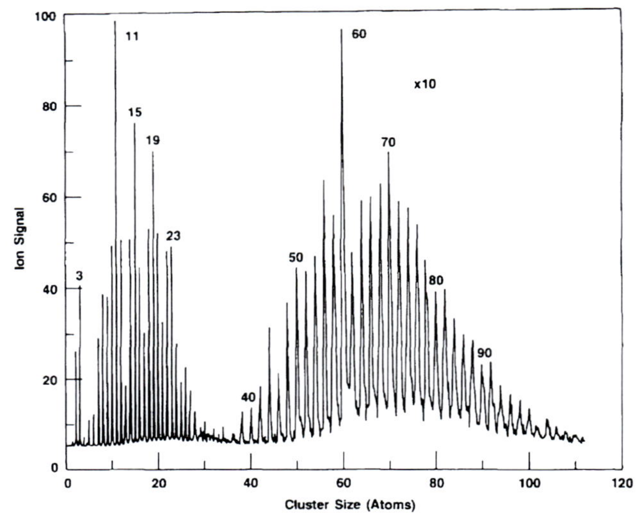

FIGURE 12.4 The mass spectra of fullerene family. *Reprinted with permission from [10]. Copyright (1984), American Institute of Physics.*

(IPR) proposed by Kroto [58]. This rule states that a stable hollow fullerene is the sort of carbon cage having each of its pentagons surrounded by five hexagons, i.e., all of the pentagons are separated from each other. Figure 12.5 shows the basic fused-pentagon configurations (with 2–3 pentagon subunits) in non-IPR fullerenes. Topologically, most of the fullerenes are the cages having at least a pair of their pentagons fused. Table 12.1 lists the isomer numbers of IPR-satisfying and non-IPR types of selected fullerenes [59]. As shown, the number of non-IPR fullerene isomers is awesomely larger than that of IPR-satisfying ones for a given cluster size. For example, there are 1811 non-IPR isomers and only one IPR-satisfying C_{60}, and all the fullerenes smaller than C_{60} are IPR-defying fullerenes.

Facing the huge number of non-IPR fullerenes that are therefore unstable, chemists face challenges to bring them into reality. Over the last 10 years, significant advances have been achieved in stabilizing and isolating the IPR-defying fullerenes by exohedral derivatization

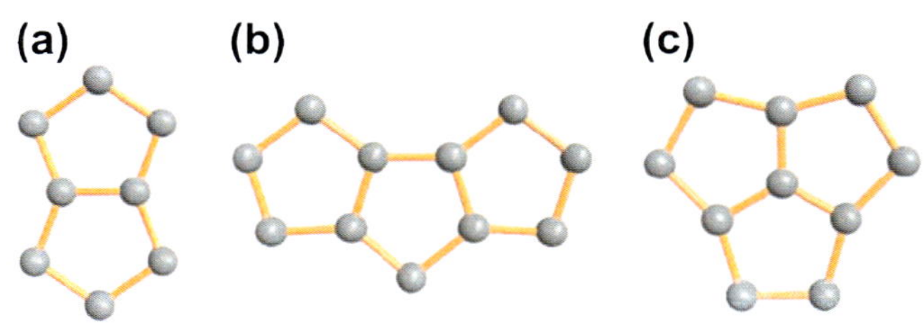

FIGURE 12.5 Basic fused-pentagon configurations (with 2–3 pentagon subunits) in non-IPR fullerenes. a, doubly-fused pentagons. b, triple sequentially fused pentagons. c, triple directly fused pentagons. Note that this set can be extended to multiple configurations with 4 or more pentagon subunits.

TABLE 12.1 The Numbers of IPR-Satisfying or Non-IPR Fullerene Isomers C_n Ranging from C_{20} to C_{100}

n	Non-IPR	IPR	*n*	Non-IPR	IPR
20	1	0	62	2385	0
24	1	0	64	3465	0
26	1	0	66	4478	0
28	2	0	68	6332	0
30	3	0	70	8148	1
32	6	0	72	11,189	1
34	6	0	74	14,245	1
36	15	0	76	19,149	2
38	17	0	78	24,104	5
40	40	0	80	31,917	7
42	45	0	82	39,710	9
44	89	0	84	51,568	24
46	116	0	86	63,742	19
48	199	0	88	81,703	35
50	271	0	90	99,872	46
52	437	0	92	126,323	86
54	580	0	94	153,359	134
56	924	0	96	191,652	187
58	1,205	0	98	230,758	259
60	1811	1	100	285,463	450

(chlorofullerenes, hydrofullerenes, or stabilized using other addends), by endohedral encapsulation of electron-donating metal atoms/clusters (such as metal(s), trimetal nitrides, metal carbides, or others), or by combinations thereof. The exohedral stabilization relies on means of relieving the molecular strain associated with fused pentagons and achieving local aromaticity of the carbon fragments, while the endohedral stabilization depends mainly on the matchable electron transformation from the encapsulated metal atom/cluster to the fullerene cage [60].

Krätschmer–Huffman arc-discharge method [2] is still the dominative approach to synthesize carbon cages in the forms of exohedral, endohedral as well as bare fullerenes, though combustion method has been used for macroscopic production of fullerenes including IPR-satisfying and non-IPR species [61]. The yield of target fullerenes is normally low and is sensitive to reaction conditions in the carbon arc process. Normally, the products from the carbon arc are complex mixtures. Separation and purification, usually by HPLC techniques, are necessary to obtain pure materials.

Traditional chemical methods have been used for the synthesis of I_h-C_{60} and dodecahedrane ($C_{20}H_{20}$) [62]. By chemical modification of bare fullerenes such as I_h-C_{60}, fullerene family has been expanded to include carbon cages incorporating four- and seven-membered rings [63–65].

All the classical characterization methods, including MS, FTIR, NMR, X-ray crystallography, can be used for identification of the structures of fullerenes. ^{13}C NMR spectrum is more important than ^{1}H NMR for structure determination since fullerene derivatives are carbon-rich molecules, but the solubility of fullerenes or their derivatives in common organic solvents is often not enough for acquirement of high-quality ^{13}C NMR spectra. Moreover, due to numerous structural isomers and poor symmetry of fullerenes, the ^{13}C NMR spectra are usually very complex and authentic structures are difficult to assign even in combination with theoretical computation. At present, X-ray crystallography has become the most important method for characterizing structures of fullerenes with high accuracy. However, single crystal growth is usually a serendipitous event. Moreover, bare spherical structures for fullerenes usually result in unavoidable cage disorder in the crystal. Exohedral derivatization by chemical groups or intermolecular interactions with, for example, M(OEP)s (M = metal, OEP = 2,3,7,8,12,13,17,18-octaethylporphinate) is thus necessary to order fullerene cages in the crystal. In the absence of X-ray crystallographic data, DFT computations coupled with spectroscopic data (such as optical band gap and vibrational spectra) may be considered as an alternative approach for the characterization of most stable endofullerene isomers.

Listed in Table 12.2 are the fullerenes synthesized/stabilized and characterized so far, including hollow fullerenes, endofullerenes, and exohedral fullerenes. In addition, heterofullerenes with nitrogen atom on the fullerene cage surface, such as $C_{59}N$ [66], $C_{69}N$ [67,68], and $M_2@C_{79}N$ (M = Y, Tb) [69], have been synthesized and identified as well. Note that any fullerenes which have not been fully characterized are excluded here.

12.3. CHEMICAL REACTIONS

Since the macroscopic synthesis of fullerenes (C_{60} and others) in 1990, fullerene reactivity has been an attractive area for chemists. In the first two decades of fullerene reactivity research, the studies are overwhelmingly predominated by chemical reactions involving buckminsterfullerene C_{60}. In contrast, investigations on the reactivity of other fullerenes such as endofullerenes are apparently less until recent years. In this section, the

TABLE 12.2 A List of Fullerenes Synthesized/Stabilized and Characterized so Far

Carbon atoms	Molecule stabilized	Pentagon configuration	Cage symmetry	References
20	$C_{20}H_{20}$, $C_{20}Cl_{16}$	1 duodenary directly fused	I_h	[70–73]
36	C_{36}, $C_{36}H_4$, $C_{36}H_6$ or $C_{36}H_6O$	1 sextuple sequentially fused	D_{6d}	[74–76]
50	$C_{50}Cl_{10}$	5 double fused	D_{5h}	[77]
54	$C_{54}Cl_8$	2 triple sequentially fused	C_2	[78]
56	$C_{56}Cl_{10}$	4 double fused	C_{2v}	[79]
	$C_{56}Cl_{12}$	2 double fused and 1 triple sequentially fused	C_1	[78]
58	$C_{58}F_{18}$ and $C_{58}F_{17}CF_3$	2 triple sequentially fused 1 heptagon[a]	C_2	[64]
60	C_{60}	No fused	I_h	[1,2]
	$C_{60}Cl_8$	2 double fused	C_{2v}	[80]
	$C_{60}Cl_{12}$	3 double fused	C_s	[80]
62	$C_{62}X_2$ (X = H, 4-MeC$_6$H$_4$, 2-Py, 3,5-(MeO)$_2$C$_6$H$_3$)	1 quadrangle[a]	C_2	[63,65]
64	$C_{64}H_4$ and $C_{64}Cl_4$	1 triple directly fused	C_3	[81,82]
66	$C_{66}Cl_6$ and $C_{66}Cl_{10}$	1 triple sequentially fused	C_2	[78]
	$Sc_2@C_{66}$	2 double fused	C_{2v}	[83]
68	$Sc_3N@C_{68}$	3 double fused	D_3	[84,85]
	$Sc_2C_2@C_{68}$	2 double fused	C_{2v}	[86]
70	C_{70}	No fused	D_{5h}	[1,2]
	$Sc_3N@C_{70}$	3 double fused	C_{2v}	[87]
72	$La_2@C_{72}$	2 double fused	D_2	[88–90]
	$La@C_{72}(C_6H_3Cl_2)$	1 double fused	C_2	[91]
74	C_{74}	No fused	D_{3h}	[92]
76	C_{76}	No fused	D_2	[93]
	$C_{76}(CF_3)_{12}$	No fused	T_d	[94]
	$C_{76}Cl_{24}$	5 double fused	C_2	[95]
	$DySc_2N@C_{76}$	2 double fused	C_s	[96]
78	C_{78}	No fused	D_3	[97,98]
	C_{78}	No fused	C_{2v}	[97,98]
	$C_{78}(CF_3)_{12}$	No fused	D_{3h}	[94,99]
	$C_{78}(OOCH_2C_6H_5)Cl_7$	1 double fused	C_1	[100]
	$M_3N@C_{78}$ (M = Dy, Tm, Gd)	2 double fused	C_2	[101,102]
80	C_{80}	No fused	D_{5d}	[103]
	$C_{80}(CF_3)_{12}$	No fused	C_{2v}	[94]
	$Sc_3N@C_{80}$, $Sc_3C_2@C_{80}$	No fused	I_h	[104,105]
	$Tm_3N@C_{80}$	No fused	D_{5h}	[106]
	$La@C_{80}(C_6H_3Cl_2)$	No fused	C_{2v}	[107]
82	$C_{82}(CF_3)_{12}$	No fused	C_2	[94]

(Continued)

TABLE 12.2 A List of Fullerenes Synthesized/Stabilized and Characterized so Far—cont'd

Carbon atoms	Molecule stabilized	Pentagon configuration	Cage symmetry	References
	$La@C_{82}$	No fused	C_{2v}	[108,109]
	$La@C_{82}$	No fused	C_s	[110]
	$Er_2@C_{82}$	No fused	C_s	[111]
	$Er_2@C_{82}$	No fused	C_{3v}	[112]
	$Gd_3N@C_{82}$	1 double fused	C_s	[113]
84	C_{84}	No fused	D_{6h}	[114]
	C_{84}	No fused	D_{3d}	[114]
	$Sc_2C_2@C_{84}$, $C_{84}(C_2F_5)_{12}$	No fused	D_{2d}	[115,116]
	$C_{84}(C_2F_5)_{12}$	No fused	C_{2v}	[115]
	$C_{84}(C_2F_5)_{12}$	No fused	C_2	
	$C_{84}(C_2F_5)_{12}$	No fused	C_s	
	$C_{84}(CF_3)_{12}$	No fused	D_{2d}	
	C_{84}, $C_{84}(CF_3)_{16}$, $C_{84}(C_2F_5)_{12}$	No fused	D_2	[115,117]
	$C_{84}Cl_{32}$	No fused, 1 heptagon[a]	C_s	[118]
	$Gd_3N@C_{84}$	1 double fused	C_s	[119]
86	C_{86}	No fused	C_2	[120]
	C_{86}	No fused	C_s	[120]
	$Tb_3N@C_{86}$	No fused	D_3	[122]
88	$C_{88}(CF_3)_{18}$	No fused	C_2	[121]
	$Tb_3N@C_{88}$	No fused	D_2	[122]
90	C_{90}	No fused	D_{5h}	[123]
	$C_{90}(CF_3)_{12}$	No fused	C_1	[124]
	$C_{90}Cl_{32}$	No fused	C_{2v}	[125]
	$C_{90}Cl_{32}$	No fused	C_s	[125]
92	C_{92}	No fused	C_2	[126]
	$C_{92}(CF_3)_{16}$	No fused	D_2	[121]
	$Gd_2C_2@C_{92}$	No fused	D_3	[127]
94	$Tm@C_{94}$, $Ca@C_{94}$	No fused	C_{3v}	[128]
	$C_{94}(CF_3)_{20}$	No fused	C_2	[129]
96	$C_{96}(C_2F_5)_{12}$	No fused	C_1	[129]
104	$Sm_2@C_{104}$	No fused	D_{3d}	[130]

[a]*The nonclassical fullerene incorporating heptagon or quadrangle.*

reactivities of buckminsterfullerene and endofullerenes are discussed.

12.3.1. Buckminsterfullerene C_{60}

The chemical reactions involving buckminsterfullerene C_{60} include reduction, cycloaddition, nucleophilic addition, radical addition, oxidation, hydrogenation, and halogenation reactions.

12.3.1.1. Reduction

Electrochemical reduction

Theoretical calculation predicted that the LUMO of C_{60} is able to accept up to six electrons to form diamagnetic $C_{60}{}^{6-}$. This predication was confirmed by cyclic voltammetry investigations [131]. Cyclic voltammetry studies showed that electrons can be added stepwise and reversibly to fullerenes. By the use of mixed solvent system (toluene and acetonitrile) at low temperature, the potential window was expanded down to −3.3 V vs Fc/Fc^+ allowing the observation of the generation of C_{60}^{6-}, C_{70}^{5-}, and C_{70}^{6-} even at very slow scan rate (100 mV/s).

Reduction by metal

The possibility of producing C_{60} polyanions by reaction with electropositive metals has been investigated. For example, reduction of C_{60} and C_{70} was carried out using Li metal in THF with the aid of ultrasound [132]. Although the fullerenes C_{60} and C_{70} are only slightly soluble in THF, the reduced fullerenes C_{60}^{n-} and C_{70}^{n-} are highly soluble and generate a deep red-brown solution after sonication. Alkylation of the C_{60} and C_{70} polyanion mixture with excess methyl iodide yields polymethylated fullerenes with up to 24 methyl groups. Reduction with metal is also achieved by solid-phase technique, in which alkali metals react directly with C_{60}. The discovery of superconductivity in alkali metal doped C_{60}[133] was achieved by this method.

Reduction by organic donors

Reaction with organic or metal organic donor molecules gives rise to fullerene-based charge-transfer (CT) complex. Due to the solubility of CT complexes in polar solvents, e.g., THF or benzonitrile, they can be obtained straightforwardly by precipitation with addition of nonpolar solvent. For example, anaerobic treatment of a crimson solution of Cr^{III}(TPP) in THF reacts with a magenta solution of C_{60} in toluene to produce a charge-transfer (CT) complex $Cr^{III}(TPP)^+C_{60}^-$, which exhibits remarkable electronic and magnetic properties [134]. Further studies have shown that strong electronic metallocene donors, such as cobaltocene ($CoCp_2$) [135], $[Fe^I(C_5H_5)(C_6Me_6)]$ [136], or decamethyl nickelocene [137] can be used to produce fullerene-based CT complexes.

12.3.1.2. *Cycloaddition*

[4 + 2] cycloaddition

In cycloaddition reactions the [6,6] double bonds of C_{60} exhibit a dienophilic character, which enables the molecule to undergo various Diels–Alder [4+2] type cycloadditions. With equimolar amounts of cyclopentadiene and C_{60}, monoadduct (Scheme 12.2) was achieved in high yield [138], while up to six addends can attach to the cage with an excess of cyclopentadiene [139]. The [4+2] cycloaddition of anthracene as well as its derivatives to C_{60} (across the 9-, 10-position) has been proved to be thermally reversible [140,141]. Selective hydrogenation and

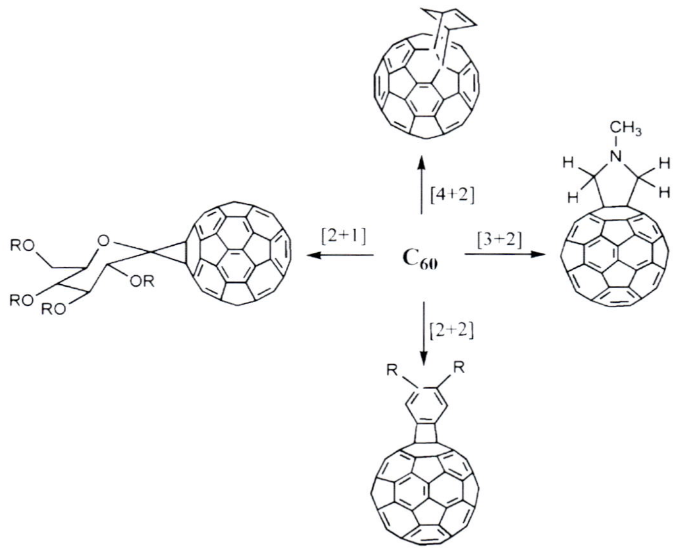

SCHEME 12.2 Examples of cyclo-additon reactions of C_{60}.

bromination of the addend or the introduction of aromaticity can overcome this problem and account for stable cycloadducts [139,142].

[3 + 2] cycloaddition

Maggini et al. [143] first reported the very general fullerene functionalization, based on the 1,3-dipolar cycloaddition of azomethine ylides to C_{60}. Azomethine ylides can be generated in situ from various starting materials [144]. The decarboxylation of immonium salts derived from the condensation of α-amino acids with aldehydes or ketones has been proved a successful approach. For example, fulleropyrrolidines were readily prepared by reaction of C_{60} with azomethine ylide which was generated by heating *N*-methylglycine and paraformaldehyde at reflux in toluene (Scheme 12.2).

[2 + 2] cycoladdition

Benzyne, generated in situ from anthranilic acid or 4,5-dimethoxyanthranilic acid, reacts with C_{60} to form a series of [2+2] cycloadducts [145,146]. The photochemical [2+2] cycloadditions of a variety of cyclic enones to C_{60} were studied using a high-pressure mercury lamp or with a XeCl excimer laser [147,148]. Cycloaddition of electron-rich alkyne (*N*,*N*-diethylpropynylamine) to C_{60} upon irradiation (>530 nm) of oxygen-free toluene at room temperature for 20 min results in a single adduct (C_{60} fused cyclobutenamine) in good yield [149].

[2 + 1] cycloaddition

Reactions between C_{60} and carbene, nitrene, and silylene lead to the formation of [2+1] cycloadducts. For example, a fullerene sugar (Scheme 12.2) can be produced in the reaction of C_{60} with carbene precursors O-benzyl- and O-pivaloyl-protected diazirine [150]. Addition of nitrenes, generated from photolysis of aroyl azides [151] or aryl azide [152] leads predominantly to the closed [6,6] bridged isomer. Isolable [2+1] cycloadduct is also formed by reaction of C_{60} and bis(2,6-diisopropyphen yl)silylene [153].

12.3.1.3. *Nucleophilic Addition*

Owning to the high electrophilic reactivity of fullerenes, nucleophilic addition reactions can take place readily with organolithium and Grignard compounds [154]. The reaction of C_{60} and *tert*-butyllithium in benzene gives $(t\text{-Bu})_nC_{60}{}^{n-}$ anions. Their protonation with acetic acid in THF yields the corresponding hydrofullerene derivatives $(t\text{-Bu})_nC_{60}H_n$ [154]. Due to the stability and low nucleophilicity, metal acetylides exhibit much less reactivity toward C_{60}. The nucleophilic addition can only occur under forced conditions [155]. For example, the addition of excess of [(trimethylsilyl)ethynyl]lithium in a refluxing toluene solution of C_{60} results in the formation of ethynylated fullerene anion (Scheme 12.3). The final products were obtained by quenching with acid. Desilylation of this (trimethylsilyl)ethynyl derivative followed by coupling reaction with CuCl leads to the formation of buckdumbbell [156].

Under certain conditions, organocopper compounds, prepared by transmetalation between Grignard reagent and $CuBr{\cdot}SMe_2$, can regioselectively react with C_{60} to produce penta-arylated, alkenylated, methylated, and silyl-methylated compounds ($C_{60}R_5H$) in good yields [160].

12.3.1.4. *Radical Addition*

C_{60} behaves like a radical sponge. A variety of photochemically or thermally in situ generated reactive radicals can be efficiently added to yield persistent radical adducts of C_{60}. The stability of radical adducts strongly depends on the nature of addition groups. A rather persistent radical $t\text{-BuC}_{60}$, with sterically demanding group, has been generated by irradiation of a saturated benzene solution of C_{60} in the presence of *tert*-butyl bromide [161]. The electron spin resonance (ESR) spectrum shows a 10-line hyperfine manifold having the binomial intensity distribution appropriate for nine equivalent protons interacting with a single unpaired electron. Along with *t*-Bu, a series of alkyl and aryl radicals can be similarly attached to C_{60} [162–165].

Since perfluorinated carbon-centered radicals are usually more stable than ordinary alkyl analogues, a large variety of $C_{60}(CF_3)_n$ derivatives, with even number of addends ($n = 2$–18) have been synthesized by reaction of C_{60} with any compounds that facilitate CF_3 radicals release on heating [166–170]. One of the available trifluoromethylation methods for the synthesis of $C_{60}(CF_3)_{16}$ (three isomers) and $C_{60}(CF_3)_{18}$ from the reaction between C_{60} and CF_3I was carried out in a glass ampoule around 400°C [167]. High-temperature reaction involving C_{60} and silver (I) trifluoroacetate was also carried out for the synthesis of $C_{60}(CF_3)_n$ ($n = 2$, 4, 6, 8, 10) [170]. Similarly, the high-temperature reactions of C_{60} with C_2F_5I [168] and $i\text{-}C_3F_7I$ [169] gave corresponding poly(perfluoroethyl)fullerenes $C_{60}(C_2F_5)_n$ ($n = 6$, 8) and poly(perfluoroisopropyl)fullerenes $C_{60}(i\text{-}C_3F_7)_n$ ($n = 6$, 8).

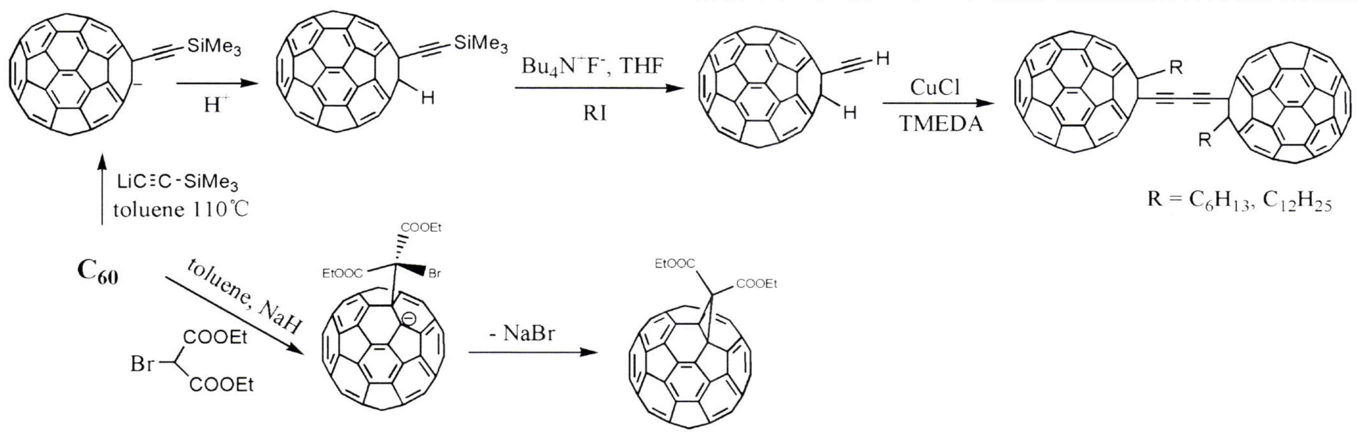

SCHEME 12.3 Examples of nucleophilic additions of C_{60}. *Adapted with permission from [153]. Copyright (1994) American Chemical Society.*

Besides protonation, alkylation, or benzylation, the intermediate $R_nC_{60}{}^{n-}$ anion generated in nucleophilic addition can also be stabilized by intramolecular nucleophilic substitutions. As shown by Bingel–Hirsch reaction [157], carbon nucleophiles generated by deprotonation of α-halo esters or α-halo ketones react with C_{60} resulting in the cyclopropanation of fullerenes (Scheme 12.3). Monoaddition product $C_{61}(COOEt)_2$ along with seven stable regioisomeric biadducts and chiral triadducts can be isolated [158]. Further cyclopropanation of C_3-symmetric triadduct in a stepwise ultimately gave T_h-symmetric hexakisadduct [159].

12.3.1.5. Oxidation

Oxygenation of C_{60} and C_{70} was detected in the fullerene soot generated by the vaporization of graphite [171,172] due to the small amount of oxygen present in the fullerene reactor. Photooxidation of C_{60} in benzene solution at room temperature leads to a single monoxide $C_{60}O$ in 7% yield [173]. Extensive oxidation and fragmentation of C_{60} can be achieved by UV irradiation in hexane or heating in the presence of oxygen [173,174].

Osmium tetraoxide is a powerful selective oxidant. Inspired by the pyridine-accelerated osmylation of polycyclic aromatic hydrocarbons [175], Hawkins reported the osmylation of C_{60} and provided the fully characterized C_{60} derivatives [176], thereby giving the first definitive proof of the buckminsterfullerene structure.

12.3.1.6. Hydrogenation

Birch reduction

The first successful hydrogenation of C_{60} was achieved by Smalley and coworkers using a Birch reduction process (Li, liquid NH_3, and *t*-BuOH) [6]. In experiment, the dark-colored C_{60} was converted to a light cream to off-white substance. Spectral analysis indicated the existence of isomers of $C_{60}H_{36}$ and $C_{60}H_{18}$. Further treated with 2,3-dichloro-5,6-dicyanobenzoquinone (DDQ) in reflux toluene, the hydrofullerenes can be converted back to C_{60}.

Zn−Cu couple reduction

Hydrogenation of fullerenes can also be achieved by the use of Zn (Cu) reduction in the presence of proton source. The pattern and the number of addend in hydrogenated fullerenes are controllable depending on reaction time, efficiency of stirring, and the ratio of metal to C_{60} [177,178]. For example, the simplest covalent derivative of C_{60}, $C_{60}H_2$, can be isolated in 66% yield after 1 h reaction, while extensive reaction times lead to the formation of $C_{60}H_4$ (2 h) and $C_{60}H_6$ (4 h).

Zn/HCl reduction

A more convenient and faster approach involves the use of Zn/HCl in toluene under nitrogen atmosphere. Hydrogenation of C_{60} with Zn and HCl results in the formation of $C_{60}H_2$, $C_{60}H_4$, $C_{60}H_6$, and $C_{60}H_{36}$ [179,180], while hydrogenation of C_{70} under the similar conditions gives the major products of $C_{70}H_{36}$ and $C_{70}H_{38}$ [180].

Hydrogenation via hydrozirconation and hydroboration

Upon hydrozirconation of C_{60} using $(\eta^5\text{-}C_5H_5)_2Zr(H)Cl$, the soluble intermediate $(\eta^5\text{-}C_5H_5)_2ZrClC_{60}H$ is hydrolyzed to give a dihydrofullerene $C_{60}H_2$. Further hydrozirconation and hydrolysis of this intermediate led to higher adducts $C_{60}H_4$ and $C_{60}H_6$ [181]. The reaction of fullerenes with BH_3 in toluene followed by hydrolysis has been used for the preparation of $C_{60}H_2$ and $C_{70}H_2$ [182].

Transfer hydrogenation

9,10-Dihydroanthracene is a typical transfer hydrogenation agent which can readily lose hydrogen to become aromatic [183,184]. The treatment of C_{60} in a sealed glass tube in a melt of 9,10-dihydroanthracene at 350°C for 30 min gives $C_{60}H_{36}$. Extending the reaction time to 24 h resulted in the formation of $C_{60}H_{18}$.

Hydrogenation with molecular hydrogen

Direct hydrogenation is also achieved under pressure in the presence of various catalysts. The level of hydrogenation can be controlled by hydrogen gas pressure and the reaction temperature. High degree of hydrogenation (up to $C_{60}H_{50}$) has been carried out by catalytic hydrogenation of C_{60} on activated carbon with Ru as catalyst in refluxing toluene [185]. The activity and selectivity of different catalysts have been systematically investigated: Reactions involving Ru, Rh, and Ir as catalyst gave mainly $C_{60}H_{18}$ while Pd, Pt, Co, and Ni predominantly gave $C_{60}H_{36}$. Au and Fe have very little activity for fullerene hydrogenation [186].

12.3.1.7. Halogenation

FLUORINATION

Among the halogenated fullerenes, fluorofullerenes represent a particularly interesting group of the compounds. First, due to the small size and strong electron withdrawing character of fluorine atom, fluorofullerenes behave like powerful electrophiles and are very susceptible to nucleophilic attack which makes them versatile fullerene building blocks for further derivatization. Second, significantly different retention times due to the high polarity and different fluorination level of fluorinated derivatives facilitate their separations and purifications by HPLC.

Initial fluorination of fullerene was carried out by reaction between F_2 gas and solid C_{60} [187,188]. The stoichiometry of the products was found ranging from $C_{60}F_{30}$ to $C_{60}F_{52}$ with $C_{60}F_{36}$ being particularly stable. Further studies of reaction between fullerenes and fluorine gas led to the conclusion that regardless of the fluorination conditions, a mixture of $C_{60}F_n$ with a wide range of compositions was formed ($n = 2-60$). However, under severe fluorinating conditions, higher degree of fluorination is accessible. For example, $C_{60}F_{48}$, the first isolated fluorofullerene, was favored only at very high temperature (315−350°C). An alternative process for the substantial formation of $C_{60}F_{48}$ was a two-stage reaction which involved direct fluorination of C_{60} at 250°C for 20 h and a subsequent fluorination of the isolated and purified intermediate product at 275°C for 30 h [189].

As compared to gaseous fluorine, metal fluorides are considered milder fluorination agents and widely used in selective fluorination of fullerenes. Previous experiments have been carried out with various binary metal fluorides (MnF_3, CoF_3, CeF_4, AgF, AgF_2) [190–193]. The reaction involving C_{60} with MnF_3 produced $C_{60}F_{36}$ selectively [190], while fluorination with AgF (or AgF_2) gave $C_{60}F_{18}$ (or $C_{60}F_{44}$) [193]. The smallest fluorofullerene $C_{60}F_2$ was generated using the very mild fluorination agents CuF_2 and FeF_3 [194]. Besides binary metal fluorides, ternary fluorides have also been used for preparing many fluorofullerenes under vacuum. For example, the $C_{60}F_{18}$ was isolated as a major product of fluorination with $KPtF_6$ at 450–520°C [195]. Reactions of C_{60} with ternary lead fluorides Cs_3PbF_7 and Cs_2PbF_6 resulted in high selective formation of $C_{60}F_{36}$ and $C_{60}F_{18}$, respectively [196].

Fluorination of C_{60} was also carried out either with XeF_2 in treating dichloromethane solutions [197] or with KrF_2 in anhydrous HF at room temperature [198]. Halogen fluorides such as ClF_3 and BrF_5 are known as more powerful fluorination agents than fluorine gas [199]. However, fluorination with rare gas fluoride or halogen fluoride generates inseparable mixture of products.

CHLORINATION

The chlorination of fullerenes has been intensively studied since early 1990s. For example, chlorination of pure C_{60} and C_{60}/C_{70} mixture was achieved in a hot glass tube with a slow stream of chlorine gas at the temperature of 250–400°C [200]. An average of 24 chlorine atoms was attached to C_{60} based on this chlorination method. By treatment with liquid chlorine around −35°C, C_{60} was quantitatively converted to multiply chlorinated products [201]. UV irradiation of a solution of C_{60} in CCl_4 saturated with Cl_2 resulted in $C_{60}Cl_{24}$ [202]. A family of chlorofullerenes, $C_{60}Cl_6$, $C_{60}Cl_8$, $C_{60}Cl_{10}$, $C_{60}Cl_{12}$, $C_{60}Cl_{14}$, and $C_{60}Cl_{26}$, has been synthesized as complex mixture based on the reaction involving C_{60} with ICl or ICl_3 [203]. Birkett et al. [204] first reported the synthesis and ^{13}C NMR characterization of chlorofullerene $C_{60}Cl_6$, and an improved synthetic procedure was developed by Kuvychko et al. [205] in which 90% C_s-$C_{60}Cl_6$ was synthesized in 7 min from C_{60} and iodine monochloride.

The highly chlorinated fullerene D_{3d}-$C_{60}Cl_{30}$ with a drum-shaped carbon cage was synthesized by high temperature (220–250°C) chlorination of C_{60} using ICl and $SbCl_5$ as reagents [206]. Chlorination of C_{60} using VCl_4 or $C_{60}Br_{24}$ using $SbCl_5$ [207] at 140–160°C allowed selective synthesis of highly symmetrical chlorofullerene, T_h-$C_{60}Cl_{24}$. Increase in the reaction time up to 2 months led to the formation of second isomer of $C_{60}Cl_{30}$ (point groups C_2) [208]. Chlorination of C_{60} with ICl at 120°C for 30 days resulted in the formation of C_1–$C_{60}Cl_{28}$ [208].

BROMINATION

Bromofullerenes were synthesized from bromination of C_{60} using concentrated solution of bromine or neat bromine. For example, the $C_{60}Br_{24}$ with T_h-symmetry, the highest symmetry available for a bromofullerene, was obtained by the reaction of C_{60} and liquid bromine [209]. The bromination of C_{60} in CS_2 as well as in chloroform resulted in the formation of $C_{60}Br_8$ [210]. Another bromide, $C_{60}Br_6$, was synthesized in either benzene or tetrachloromethane [210].

IODINATION

Iodination of fullerenes has not been observed to date, perhaps due to the weakness of C–I bond and the bulk size of iodine atom.

12.3.1.8. Organometallic Complex

A variety of organometallic fullerene derivatives have been synthesized through the reactions involving fullerenes and transition metal complexes. X-ray crystallographic investigations have shown that fullerenes act as electron-deficient rather than aromatic alkenes, thus η^2-derivatives are usually obtained instead of η^6-derivatives. For example, equimolar amounts of Vaska's compound, $Ir(CO)Cl(PPh_3)_2$, with C_{60} leads to the formation of $(\eta^2\text{-}C_{60})Ir(CO)Cl(PPh_3)_2$ (Fig. 12.6) [211]. Single-crystal X-ray structure analysis clearly demonstrated the η^2-binding of iridium to a [6,6] bond of C_{60}. Interestingly, replacement of one phenyl group by $-CH_2C_6H_4OCH_2Ph$ in Vaska's compound produces a supramolecular architecture in which C_{60} is surrounded by the arms of two phosphine ligands via π–π interaction [212].

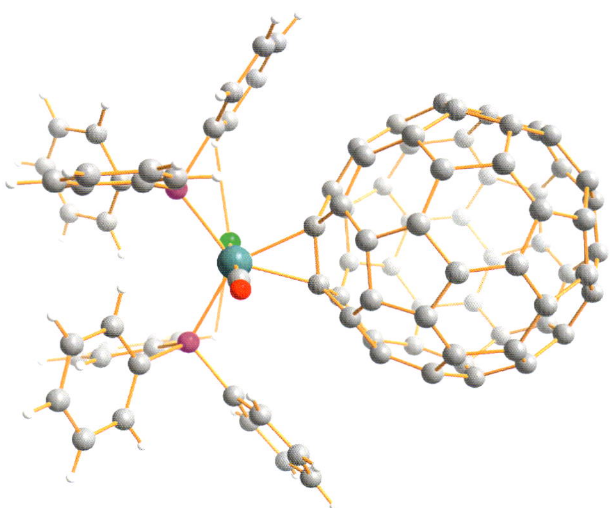

FIGURE 12.6 The molecular structure of $(\eta^2\text{-}C_{60})Ir(CO)Cl(PPh_3)_2$ [211]. *Reprinted with permission from [211]. Copyright 1991 American Chemical Society.*

Using smaller MEt_3 ligand (M = Pt, Pd), up to six metallic groups can be coordinated to C_{60} [213]. However, larger size of C_{70} allows attachment of four large $Pt(PPh_3)_2$ addends [214]. As for Vaska's complex, multiple addition products of C_{60} or C_{70} were synthesized when the phenyl groups on phosphorus were replaced with electron-donating substituents [215].

12.3.2. *Endofullerenes*

The chemical modifications on endofullerenes have been increasingly studied in addition to the synthesis of various pristine endofullerenes [216–220], because endofullerene derivatives are considered as promising materials with multiple potential applications in areas such as optics, molecular electronics, and medicine. The first derivatization reaction of endofullerenes was the silylation reaction of $La@C_{82}(C_{2v})$ reported by Akasaka et al. in 1995 [221]. Similar to the reactions performed with hollow fullerenes, the derivatization with endofullerenes includes cycloaddition, carbene, Bingel–Hirsch, hydroxylation, radical, and some other addition reactions. However, the derivatizations on endofullerenes seem more complex than their hollow analogues.

12.3.2.1. *Cycloaddition*

[4+2] CYCLOADDITION: DIELS–ALDER REACTION

The first [4+2] cycloaddition derivative of endofullerene was synthesized by refluxing for 24 h a solution of $Sc_3N@C_{80}(I_h)$ and excess 6,7-dimethoxyisochroman-3-one (99% ^{13}C-labeled) in 1,2,4-trichlorobenzene (TCB) (Scheme 12.4) [222]. NMR spectroscopic analysis suggested that the addition had mainly taken place at the [5,6] ring junction of the I_h C_{80} cage, which was confirmed by X-ray crystallography later on [223]. Some kinds of bisadducts of $Gd_3N@C_{80}$ were also synthesized by using the same approach but no regioselectivity was observed [224].

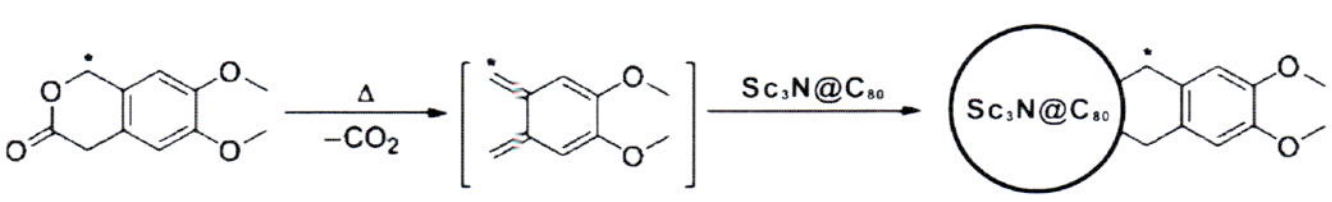

SCHEME 12.4 [4+2] Diels–Alder reaction of $Sc_3N@C_{80}(I_h)$ [222]. *Adapted with permission from [222]. Copyright (2002) American Chemical Society.*

$La@C_{82}(C_{2v})$ reacted with excess cyclopentadiene (Cp) in a degassed toluene over 2 h in the dark, produced monoadducts $La@C_{82}(C_{2v})(Cp)$ in 44% yield (Scheme 12.5). The retro-Diels–Alder reaction was obtained at 298 K in toluene, with the activation energy lower than that of the retro reaction of $C_{60}Cp$ [225]. Recently, a computational study of the Diels–Alder reaction involving 1,3-butadiene with $Y_3N@C_{78}(D_{3h})$ suggested that the addition was predominantly at its [6,6] junction with implied regioselective functionalization of $Y_3N@C_{78}(D_{3h})$ [226].

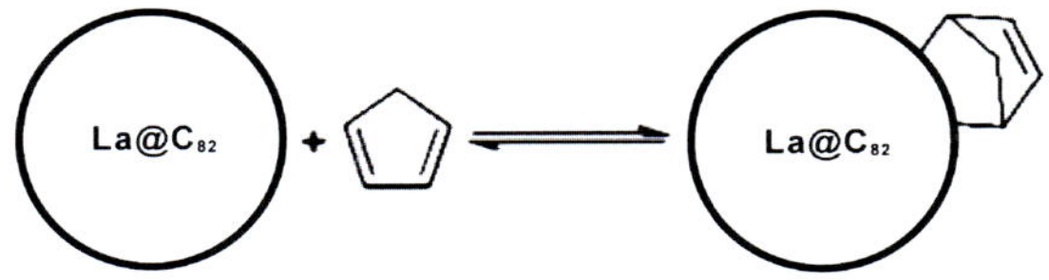

SCHEME 12.5 [4+2] Diels–Alder reaction of $La@C_{82}(C_{2v})$ with cyclopentadiene [225]. *Adapted with permission from [225]. Copyright (2005) American Chemical Society.*

[3+2] CYCLOADDITION: 1,3-DIPOLAR CYCLOADDITION In 1995, Akasaka et al. [221] reported the first functionalization reaction of endofullerene, $La@C_{82}(C_{2v})$, with 1,1,2,2-tetrakis(2,4,6-trimethylphenyl)-1,2-disilirane. When a degassed solution of $La@C_{82}(C_{2v})$ and disilirane in a sealed tube was irradiated with a tungsten-halogen lamp, a 1:1 mixture of monoadduct resulted (Scheme 12.6). Under the same condition, an analogous compound $Gd@C_{82}(C_{2v})(Mes_2Si)_2CH_2$ was obtained [227]. The same monoadducts were obtained under thermal conditions also. The similar reaction of $La@C_{82}(C_{2v})$ and 1,1,2,2-tetrakis(2,6-diethylphenyl)-1,2-digermirane occurred under photochemical and thermal conditions as well [228]. Up to date, the bissilylated monoadducts of $Pr@C_{82}$, $Ce@C_{82}$, $Y@C_{82}$, $Sc_3N@C_{80}$, and $Ce_2@C_{78}$ have been synthesized [229–233].

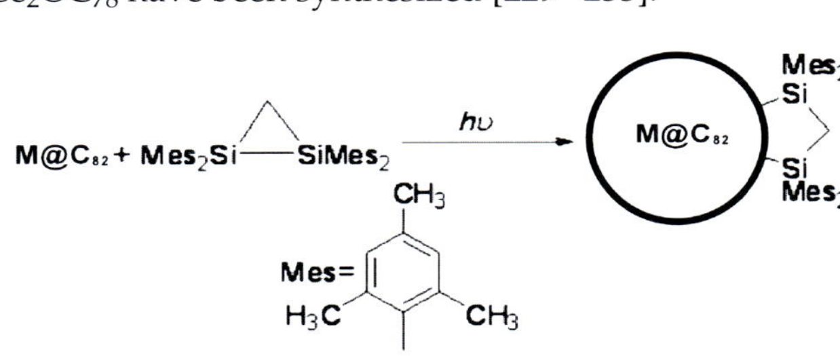

SCHEME 12.6 Silylation reaction of $M@C_{82}(C_{2v})$ (M = La, Gd) with disilirane [221].

The synthesis of a pyrrolidine adduct of $La@C_{82}(C_{2v})$ was reported by Cao et al. in 2004 [234]. A solution of $La@C_{82}(C_{2v})$ mixed with excess *N*-methylglycine and paraformaldehyde in toluene was heated at 100°C for 30 min in a sealed EPR quartz tube. Three bisadducts and two monoadducts were obtained. By HPLC isolation one mono- and one bisadduct formed which were isolated in pure form and further characterized by EPR spectra. The similar reaction was also carried out by Gu and coworkers with $M@C_{82}$ (M = Gd, Y) [235,236]. Utilizing ^{13}C-labeled formaldehyde, Echegoyen and coworkers [237] identified the addition pattern of a [5,6] bonded mono-pyrrolidine adduct of $Sc_3N@C_{80}(I_h)$ (Scheme 12.7) which was subsequently confirmed by Dorn and coworkers [238]. Starting from $Y_3N@C_{80}(I_h)$, the [6,6] addition of cycloaddition product was obtained

[239]. Further experiment [240] of thermal isomerization suggested that the [6,6] monoadduct was the kinetic product and the [5,6] monoadduct was the thermodynamic product. Such a thermal isomerization between two monoadducts of $Sc_3N@C_{80}(I_h)$ was also reported. The crystal structure of the [5,6] pyrrolidine derivative of $Y_3N@C_{80}$ was reported later on [241]. Moreover, the 1,3-dipolar retro-cycloaddition of the *N*-ethylpyrrolidino-$Sc_3N@C_{80}(I_h)$ derivative was achieved by heating in *o*-dichlorobenzene for 20–24 h in the presence of maleic anhydride [242].

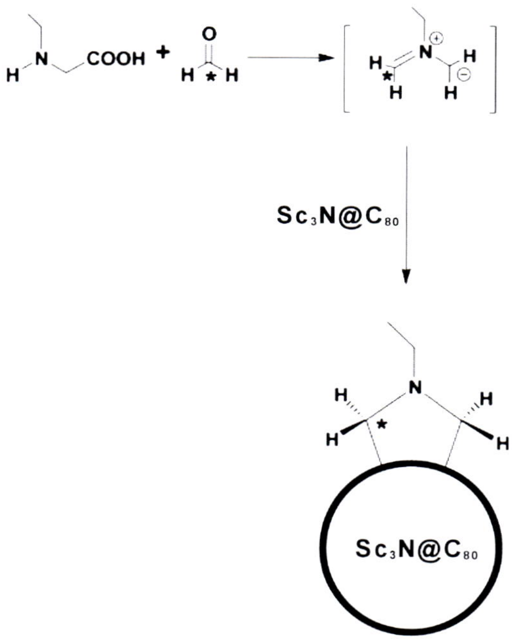

SCHEME 12.7 [1,3]-Dipolar cycloaddition of $Sc_3N@C_{80}(I_h)$ [237]. *Reprinted with permission from [237]. Copyright (2005) American Chemical Society.*

Using 3-triphenylmethyl-5-oxazolidinone as the reagent, both [5,6] and [6,6] pyrrolidine derivatives of $La_2@C_{80}$ were synthesized and the [6,6] adduct was identified by X-ray crystallography (Scheme 12.8) [243]. Two monoadducts ([5,6] and [6,6]) of $Sc_3N@C_{80}(I_h)$ were also synthesized and it was revealed that the [6,6] adduct was thermally more stable than the [5,6] adduct [244]. A similar 1,3-dipolar cycloaddition reaction was applicable with $Sc_3N@C_{80}(D_{5h})$ [245] and $Sc_3N@C_{78}(D_{3h})$ as well [246].

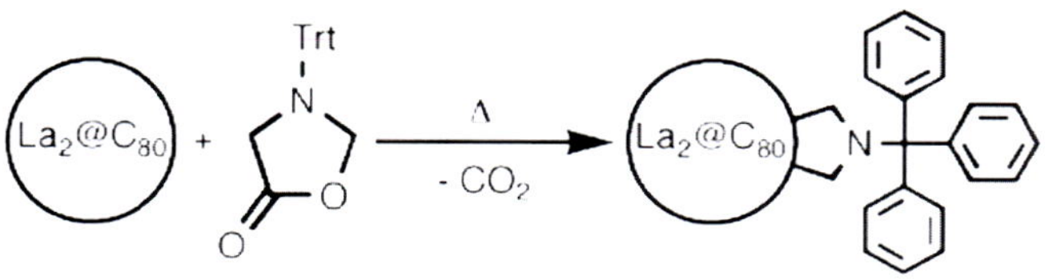

SCHEME 12.8 [1,3]-Dipolar cycloaddition of $La_2@C_{80}$[243]. *Reprinted with permission from [243]. Copyright (2006) American Chemical Society.*

12.3.2.2. CARBENE REACTION

The carbene reaction of 2-adamantane-2,3-[3*H*]-diazirine (Ad) with $La@C_{82}(C_{2v})$ was reported by Akasaka and coworkers (Scheme 12.9) [247]. The TCB/toluene solution of $La@C_{82}(C_{2v})$ and an excess amount of Ad in a seal tube was irradiated with a high-pressure mercury lamp, producing the monoadduct in 80% yield. An opened cyclopropane structure of this derivative was determined by X-ray diffraction. Under similar conditions, a monoadduct of Sc_3C_{82} was synthesized and purified. Its structure was unambiguously determined by X-ray crystallography, revealing surprising structure of metal carbide endofullerene $Sc_3C_2@C_{80}$ instead of Sc_3C_{82} [105]. To date, this reaction has been carried out with $Sc_2C_2@C_{82}$ [248], $La@C_{82}$ [249], $Gd@C_{82}$ [250], $La_2@C_{78}$ [251], and non-IPR $La_2@C_{72}$ [89,90], most derivatives of which have been characterized by X-ray crystallography. Especially in the case of non-IPR $La_2@C_{72}$, the derivatization reaction showed remarkable regioselectivity which was attributed to both the fused pentagons and encapsulated metal atoms [89,90].

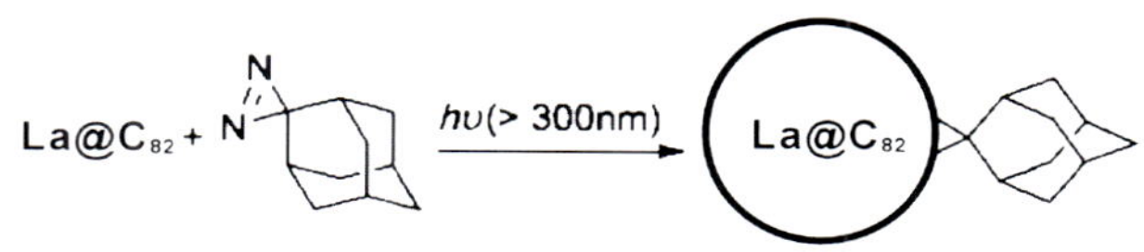

SCHEME 12.9 Carbene addition reaction of $La@C_{82}(C_{2v})$ [247]. *Reprinted with permission from [247]. Copyright (2004) American Chemical Society.*

12.3.2.3. BINGEL–HIRSCH REACTION

Akasaka and coworkers [252] synthesized a singly bonded monoadduct $La@C_{82}(C_{2v})CBr(COOC_2H_5)_2$ using Bingel–Hirsch reaction in which $La@C_{82}(C_{2v})$ and diethyl bromomalonate were mixed in the presence of 1,8-diazabicyclo[5.4.0]-undec-7-ene (DBU) in dry toluene over 2 h in inert atmosphere of Ar. One monoadduct was isolated by preparative HPLC in a yield of 40%. Its structure has been fully determined by NMR spectroscopic and X-ray crystallographic analyses. Different from the conventional Bingel adducts of empty fullerenes, the addition group bonds to the endofullerene with single bond in this derivative (Fig. 12.7). By increasing the reaction temperature to 60°C, a crystallographically identified bisadduct of $La@C_{82}(C_{2v})$ resulted with a high regioselectivity in a good yield [109]. The exohedral addition groups of this bisadduct were also bonded to the carbon cage with a single bond. However, the Bingel–Hirsch reaction of $Y_3N@C_{80}$ [239], to form a cyclopropane between $Y_3N@C_{80}$ and the added group is similar to those of empty fullerenes (Scheme 12.10). The monoadduct was characterized by NMR spectroscopy. Its subsequent

X-ray crystal structure determination revealed an open cyclopropane structure rather than a closed one [253]. Bingel–Hirsch reaction was also applied to other trimetal nitride endofullerenes such as $Er_3N@C_{80}$ [240], Gd_3N-endofullerenes (C_{80}, C_{84}, and C_{88}) [254], and $Sc_3N@C_{78}$ [255], all of which yielded cyclopropanated adducts.

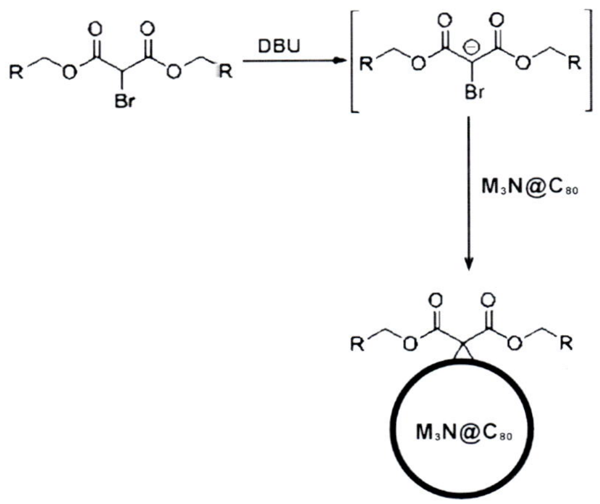

SCHEME 12.10 Bingel–Hirsch reaction of $M_3N@C_{80}(I_h)$ [239]. *Reprinted with permission from [239]. Copyright (2005) American Chemical Society.*

12.3.2.4. RADICAL REACTION

ALKYLATION In 2005, Yoza and coworkers [256] reported one of the three isomeric structures of $La@C_{74}(C_6H_3Cl_2)$ adducts which was isolated from the extraction of soot with 1,2,4-trichlorobenzene (TCB) under reflux. Theoretical calculations suggested that the $La@C_{74}(D_{3h})$ would react with dichlorophenyl radicals which were produced during the extraction to form a stable closed-shell derivative. Similar adduct was obtained for the non-IPR endofullerene $La@C_{72}$ [91]. By heating $La@C_{82}(C_{2v})$ with 3-triphenylmethyl-5-oxazolidinone in toluene, a mixture of benzylated monoadducts $La@C_{82}(C_{2v})(CH_2C_6H_5)$ was obtained

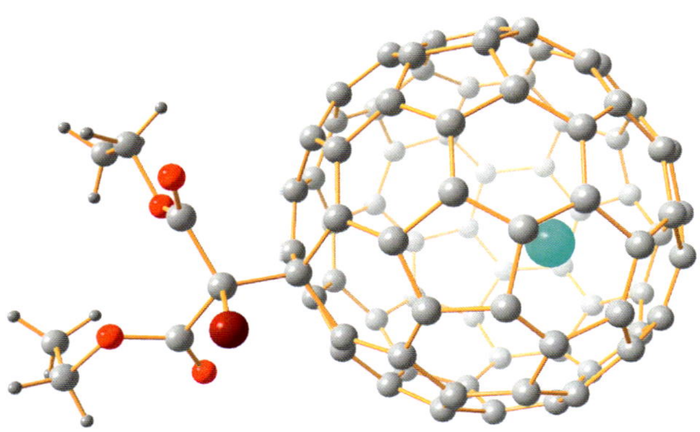

FIGURE 12.7 The structure of one isomer of $La@C_{82}(C_{2v})CBr(COOC_2H_5)_2$ [252]. *Reprinted with permission from [252]. Copyright 2005 American Chemical Society.*

[257]. Without the existence of 3-triphenylmethyl-5-oxazolidinone, the same monoadducts were also obtained by the photo-irradiation of $La@C_{82}(C_{2v})$ in toluene. However, the photo-irradiation of $La@C_{82}(C_{2v})$ in 1,2-dichlorobenzene in the presence of *R*,*R*,2,4-tetrachlorotoluene gave the monoadduct $La@C_{82}(C_{2v})(CHClC_6H_3Cl_2)$ also (Scheme 12.11) [257]. One isomer of the $La@C_{82}(C_{2v})(CHClC_6H_3Cl_2)$ was identified by single-crystal X-ray diffraction. Similarly a dibenzyl adduct $Sc_3N@C_{80}(CH_2C_6H_5)_2$ [258] was synthesized in 82% yield and high regioselectivity by irradiating a deoxygenated solution of $Sc_3N@C_{80}(I_h)$ and 1000 equiv. of benzyl bromide in toluene at 355 nm without cooling for 1 h. The structure of $Sc_3N@C_{80}(CH_2C_6H_5)_2$ was elucidated by NMR spectroscopy and X-ray diffraction.

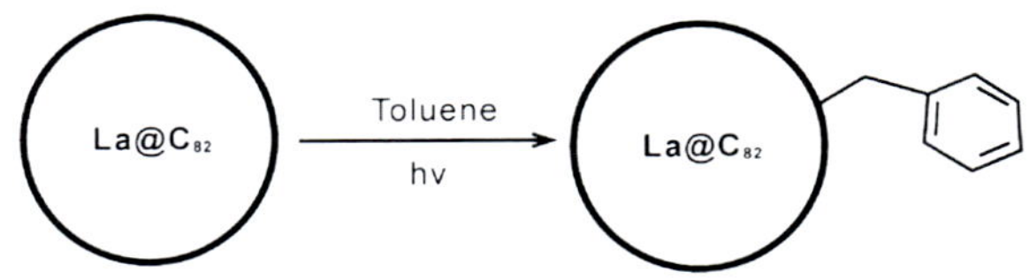

SCHEME 12.11 Alkylation reaction of $La@C_{82}(C_{2v})$ [257]. *Adapted with permission from [257]. Copyright (2008) American Chemical Society.*

PERFLUOROALKYLATION The perfluoroalkylation of endofullerenes was first carried out on $La@C_{82}(C_s)$ [259]. The degassed and cooled solution of $La@C_{82}(C_s)$ with an excess of perfluorooctyl iodide in toluene was irradiated with a UV lamp over 15 h. The product $La@C_{82}(C_s)(C_8F_{17})_2$ was analyzed by HPLC, mass spectrum, and EPR spectrum . By reacting a $Y@C_{82}$-enriched extract with $AgCF_3CO_2$ in a quartz reactor under dynamic vacuum at 400°C for 10 h, two stable isomers of $Y@C_{82}(CF_3)_5$ were obtained [260]. They were isolated and characterized by mass spectrometry and ^{19}F NMR spectrum. Using CF_3I gas as the fluoroalkylating agent at high temperature, a series of CF_3 derivatives of $Sc_3N@C_{80}$(I_h and D_{5h}) were synthesized. Two isomers ($Sc_3N@C_{80}(I_h)(CF_3)_2$ and $Sc_3N@C_{80}(D_{5h})(CF_3)_2$) were isolated by HPLC and characterized by ^{19}F NMR [261]. Recently, $Sc_3N@C_{80}(I_h)(CF_3)_{14}$ and $Sc_3N@C_{80}(I_h)(CF_3)_{16}$ were prepared by heating $Sc_3N@C_{80}(I_h)$ and $AgCF_3CO_2$ to 350°C in a sealed tube. Their structures, exemplified by $Sc_3N@C_{80}(I_h)(CF_3)_{14}$ shown in Fig. 12.8, were determined by single-crystal X-ray diffraction [262].

12.3.2.5. HYDROXYLATION

The first hydroxyl derivative of endofullerene was prepared in 1997 by Pei and coworkers [263]. A mixture of gadolinium endofullerenes was reacted with potassium metal in toluene solution under N_2 atmosphere for 2 h, and then the dry precipitate was hydrolyzed with water. A mixture of hydroxylated endofullerenes was obtained, in which the average number of additional hydroxyl group was predicted up to 20. In 1999,

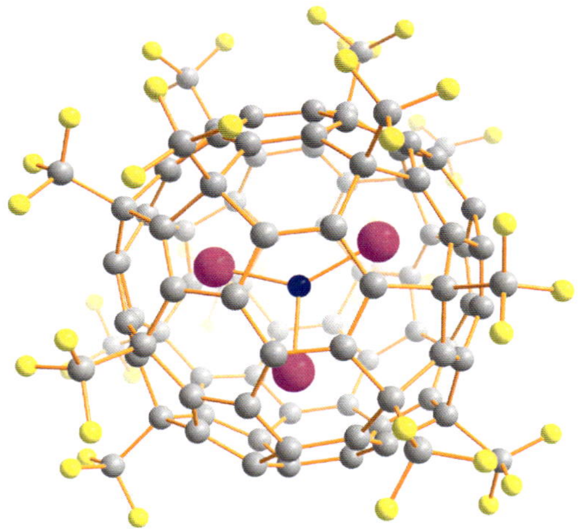

FIGURE 12.8 The structure of $Sc_3N@C_{80}(I_h)(CF_3)_{14}$ [262]. *Reprinted with permission from [262]. Copyright 2009 American Chemical Society.*

Wilson et al. reported the water-soluble radioactive EMFs $^{165}Ho@C_{82}(OH)_x$ ($x = 24-26$) which were obtained under alkaline phase-transfer conditions [264,265]. In the same year, water-soluble endohedral metallofullerenol $Pr@C_{82}O_m(OH)_x$ was prepared from pure $Pr@C_{82}$ and concentrated nitric acid, followed by hydrolysis [266]. By the similar strategies, a series of hydroxylated derivatives of endofullerenes have been synthesized, including $Gd@C_{82}$, $Gd_3N@C_{80}$, $Sc_pGd_{3-p}N@C_{80}$, and so on. Some of these derivatives have been investigated for potential applications in medicine and biotechnology [218–220]. However, the exact geometric structures of hydroxylated derivatives are still open for characterization, i.e., not fully characterized.

References

[1] H.W. Kroto, J.R. Heath, S.C. O'Brien, R.F. Curl, R.E. Smalley, Nature 318 (1985) 162.

[2] W. Kratschmer, L.D. Lamb, K. Fostiropoulos, D.R. Huffman, Nature 347 (1990) 354.

[3] S.J. Cyvin, E. Brendsdal, B.N. Cyvin, J. Brunvoll, Chem. Phys. Lett. 143 (1988) 377.

[4] H. Ajie, M.M. Alvarez, S.J. Anz, R.D. Beck, F. Diederich, K. Fostiropoulos, et al., J. Phys. Chem. 94 (1990) 8630.

[5] A.S. Koch, K.C. Khemani, F. Wudl, J. Org. Chem. 56 (1991) 4543.

[6] R.E. Haufler, J. Conceicao, L.P.F. Chibante, Y. Chai, N.E. Byrne, S. Flanagan, et al., J. Phys. Chem. 94 (1990) 8634.

[7] K. Tohji, A. Paul, L. Moro, R. Malhotra, D.C. Lorents, R.S. Ruoff, J. Phys. Chem. 99 (1995) 17785.

[8] B. Ahmad, M. Ahmad, M. Riaz, N. Ahmad, S. Ahmad, Mater. Chem. Phys. 94 (2005) 52.

[9] B. Ahmad, M. Riaz, M. Ahmad, S. Nawaz, S. Ahmad, Mater. Lett. 62 (2008) 3367.

[10] E.A. Rohlfing, D.M. Cox, A. Kaldor, J. Chem. Phys. 81 (1984) 3322.

[11] Y. Chai, T. Guo, C. Jin, R.E. Haufler, L.P.F. Chibante, J. Fure, et al., J. Phys. Chem. 95 (1991) 7564.

[12] L.P.F. Chibante, A. Thess, J.M. Alford, M.D. Diener, R.E. Smalley, J. Phys. Chem. 97 (1993) 8696.

[13] J.R. Heath, S.C. O'Brien, Q. Zhang, Y. Liu, R.F. Curl, F.K. Tittel, et al., J. Am. Chem. Soc. 107 (1985) 7779.

[14] S.Y. Xie, R.B. Huang, J. Ding, L.J. Yu, Y.H. Wang, L.S. Zheng, J. Phys. Chem. A 104 (2000) 7161.

[15] R.F. Curl, R.E. Smalley, Sci. Am. 265 (1991) 54.

[16] C.L. Fields, J.R. Pitts, M.J. Hale, C. Bingham, A. Lewandowski, D.E. King, J. Phys. Chem. 97 (1993) 8701.

[17] D. Laplaze, P. Bernier, G. Flamant, M. Lebrun, A. Brunelle, S. Della-Negra, Synth. Met. 77 (1996) 67.

[18] G. Flamant, J.F. Robert, S. Marty, J.M. Gineste, J. Giral, B. Rivoire, et al., Energy 29 (2004) 801.

[19] P. Gerhardt, S. Loeffler, K.H. Homann, Chem. Phys. Lett. 137 (1987) 306.

[20] J.B. Howard, J.T. McKinnon, Y. Makarovsky, A.L. Lafleur, M.E. Johnson, Nature 352 (1991) 139.

[21] P. Hebgen, A. Goel, J.B. Howard, L.C. Rainey, J.B. Vander Sande, Proc. Combust. Inst. 28 (2000) 1397.

[22] H. Takehara, M. Fujiwara, M. Arikawa, M.D. Diener, J.M. Alford, Carbon 43 (2005) 311.

[23] M. Bachmann, J. Griesheimer, K.H. Homann, Chem. Phys. Lett. 223 (1994) 506.

[24] J.M. Alford, C. Bernal, M. Cates, M.D. Diener, Carbon 46 (2008) 1623.

[25] H. Hepp, K. Siegmann, K. Sattler, Mater. Res. Soc. Symp. Proc. 359 (1995) 517.

[26] J. Ahrens, M. Bachmann, T. Baum, J. Griesheimer, R. Kovacs, P. Weilmuenster, et al., Int. J. Mass Spectrom. Ion Process. 138 (1994) 133.

[27] H. Richter, E. de Hoffmann, R. Doome, A. Fonseca, J.M. Gilles, J.B. Nagy, et al., Carbon 34 (1996) 797.

[28] H. Murayama, S. Tomonoh, J.M. Alford, M.E. Karpuk, Fuller Nanotub. Carbon Nanostr. 12 (2004) 1.

[29] T.M. Chang, A. Naim, S.N. Ahmed, G. Goodloe, P.B. Shevlin, J. Am. Chem. Soc. 114 (1992) 7603.

[30] R. Taylor, G.J. Langley, H.W. Kroto, D.R.M. Walton, Nature 366 (1993) 728.

[31] J. Osterodt, A. Zett, F. Voegtle, Tetrahedron 52 (1996) 4949.

[32] C. Crowley, H.W. Kroto, R. Taylor, D.R.M. Walton, M.S. Bratcher, P.C. Cheng, et al., Tetrahedron Lett. 36 (1995) 9215.

[33] C. Crowley, R. Taylor, H.W. Kroto, D.R.M. Walton, P.C. Cheng, L.T. Scott, Synth. Met. 77 (1996) 17.

[34] N.R. Conley, J.J. Lagowski, Carbon 40 (2002) 949.

[35] K.Y. Amsharov, M. Jansen, Carbon 45 (2007) 117.

[36] K.Y. Amsharov, K. Simeonov, M. Jansen, Carbon 45 (2007) 337.

[37] S.Y. Xie, R.B. Huang, L.J. Yu, J. Ding, L.S. Zheng, Appl. Phys. Lett. 75 (1999) 2764.

[38] S.Y. Xie, S.L. Deng, R.B. Huang, L.J. Yu, L.S. Zheng, Chem. Phys. Lett. 343 (2001) 458.

[39] S.Y. Xie, R.B. Huang, S.L. Deng, L.J. Yu, L.S. Zheng, J. Phys. Chem. B 105 (2001) 1734.

[40] S.W. McElvany, M.M. Ross, N.S. Goroff, F. Diederich, Science 259 (1993) 1594.

[41] Y. Tobe, N. Nakagawa, K. Naemura, T. Wakabayashi, T. Shida, Y. Achiba, J. Am. Chem. Soc. 120 (1998) 4544.

[42] Y. Tobe, N. Nakagawa, J.y. Kishi, M. Sonoda, K. Naemura, T. Wakabayashi, et al., Tetrahedron 57 (2001) 3629.

[43] Y. Rubin, T.C. Parker, S.I. Khan, C.L. Holliman, S.W. McElvany, J. Am. Chem. Soc. 118 (1996) 5308.

[44] Y. Tobe, H. Nakanishi, M. Sonoda, T. Wakabayashi, Y. Achiba, Chem. Commun. (1999) 1625.

[45] B. Gomez-Lor, O. de Frutos, A.M. Echavarren, Chem. Commun. (1999) 2431.
[46] M.M. Boorum, Y.V. Vasil'ev, T. Drewello, L.T. Scott, Science 294 (2001) 828.
[47] L.T. Scott, Pure Appl. Chem. 68 (1996) 291.
[48] L.T. Scott, P.C. Cheng, M.M. Hashemi, M.S. Bratcher, D.T. Meyer, H.B. Warren, J. Am. Chem. Soc. 119 (1997) 10963.
[49] S. Hagen, M.S. Bratcher, M.S. Erickson, G. Zimmermann, L.T. Scott, Angew. Chem., Int. Ed. Engl. 36 (1997) 406.
[50] L.T. Scott, H.E. Bronstein, D.V. Preda, R.B.M. Ansems, M.S. Bratcher, S. Hagen, Pure Appl. Chem. 71 (1999) 209.
[51] R.B.M. Ansems, L.T. Scott, J. Am. Chem. Soc. 122 (2000) 2719.
[52] L.T. Scott, Angew. Chem., Int. Ed. 43 (2004) 4994.
[53] L.T. Scott, M.M. Boorum, B.J. McMahon, S. Hagen, J. Mack, J. Blank, et al., Science 295 (2002) 1500.
[54] K. Amsharov, M. Jansen, Chem. Commun. (2009) 2691.
[55] K.Y. Amsharov, M. Jansen, J. Org. Chem. 73 (2008) 2931.
[56] G. Otero, G. Biddau, C. Sanchez-Sanchez, R. Caillard, M.F. Lopez, C. Rogero, et al., Nature 454 (2008) 865.
[57] R.E. Smalley, Acc. Chem. Res. 25 (1992) 98.
[58] H.W. Kroto, Nature 329 (1987) 529.
[59] P.W. Fowler, D.E. Manolopoulos, An Atlas of Fullerenes, Oxford University Press, Oxford, 1995.
[60] Y.Z. Tan, S.Y. Xie, R.B. Huang, L.S. Zheng, Nat. Chem. 1 (2009) 450.
[61] Z.Y. Gao, W.S. Jiang, D. Sun, S.Y. Xie, R.B. Huang, L.S. Zheng, Combust. Flame 157 (2010) 966.
[62] L.A. Paquette, Proc. Natl. Acad. Sci. U.S.A. 79 (1982) 4495.
[63] W. Qian, S.C. Chuang, R.B. Amador, T. Jarrosson, M. Sander, S. Pieniazek, et al., J. Am. Chem. Soc. 125 (2003) 2066.
[64] P.A. Troshin, A.G. Avent, A.D. Darwish, N. Martsinovich, A.a.K. Abdul-Sada, J.M. Street, et al., Science 309 (2005) 278.
[65] W. Qian, M.D. Bartberger, S.J. Pastor, K.N. Houk, C.L. Wilkins, Y. Rubin, J. Am. Chem. Soc. 122 (2000) 8333.
[66] J.C. Hummelen, B. Knight, J. Pavlovich, R. Gonzalez, F. Wudl, Science 269 (1995) 1554.
[67] B. Nuber, A. Hirsch, Chem. Commun. (1996) 1421.
[68] N. Tagmatarchis, K. Okada, T. Tomiyama, H. Shinohara, Synlett (2000) 1761.
[69] T. Zuo, L. Xu, C.M. Beavers, M.M. Olmstead, W. Fu, T.D. Crawford, et al., J. Am. Chem. Soc. 130 (2008) 12992.
[70] L.A. Paquette, R.J. Ternansky, D.W. Balogh, J. Am. Chem. Soc. 104 (1982) 4502.
[71] R.J. Ternansky, D.W. Balogh, L.A. Paquette, J. Am. Chem. Soc. 104 (1982) 4503.
[72] H. Prinzbach, A. Weller, P. Landenberger, F. Wahl, J. Worth, L.T. Scott, et al., Nature 407 (2000) 60.
[73] F. Wahl, A. Weiler, P. Landenberger, E. Sackers, T. Voss, A. Haas, et al., Chem. Eur. J. 12 (2006) 6255.
[74] C. Piskoti, J. Yarger, A. Zettl, Nature 393 (1998) 771.
[75] A. Koshio, M. Inakuma, T. Sugai, H. Shinohara, J. Am. Chem. Soc. 122 (2000) 398.
[76] A. Koshio, M. Inakuma, Z.W. Wang, T. Sugai, H. Shinohara, J. Phys. Chem. B 104 (2000) 7908.
[77] S.Y. Xie, F. Gao, X. Lu, R.B. Huang, C.R. Wang, X. Zhang, et al., Science 304 (2004) 699.
[78] Y.Z. Tan, J. Li, F. Zhu, X. Han, W.S. Jiang, R.B. Huang, et al., Nat. Chem. 2 (2010) 269.
[79] Y.Z. Tan, X. Han, X. Wu, Y.Y. Meng, F. Zhu, Z.Z. Qian, et al., J. Am. Chem. Soc. 130 (2008) 15240.
[80] Y.Z. Tan, Z.J. Liao, Z.Z. Qian, R.T. Chen, X. Wu, H. Liang, et al., Nat. Mater. 7 (2008) 790.
[81] C.R. Wang, Z.Q. Shi, L.J. Wan, X. Lu, L. Dunsch, C.Y. Shu, et al., J. Am. Chem. Soc. 128 (2006) 6605.
[82] X. Han, S.J. Zhou, Y.Z. Tan, X. Wu, F. Gao, Z.J. Liao, et al., Angew. Chem., Int. Ed. 47 (2008) 5340.
[83] C.R. Wang, T. Kai, T. Tomiyama, T. Yoshida, Y. Kobayashi, E. Nishibori, et al., Nature 408 (2000) 426.
[84] S. Stevenson, P.W. Fowler, T. Heine, J.C. Duchamp, G. Rice, T. Glass, et al., Nature 408 (2000) 427.
[85] M.M. Olmstead, H.M. Lee, J.C. Duchamp, S. Stevenson, D. Marciu, H.C. Dorn, et al., Angew. Chem., Int. Ed. 42 (2003) 900.
[86] Z.Q. Shi, X. Wu, C.R. Wang, X. Lu, H. Shinohara, Angew. Chem., Int. Ed. 45 (2006) 2107.
[87] S. Yang, A.A. Popov, L. Dunsch, Angew. Chem., Int. Ed. 46 (2007) 1256.
[88] H. Kato, A. Taninaka, T. Sugai, H. Shinohara, J. Am. Chem. Soc. 125 (2003) 7782.
[89] X. Lu, H. Nikawa, T. Nakahodo, T. Tsuchiya, M.O. Ishitsuka, Y. Maeda, et al., J. Am. Chem. Soc. 130 (2008) 9129.
[90] X. Lu, H. Nikawa, T. Tsuchiya, Y. Maeda, M.O. Ishitsuka, T. Akasaka, et al., Angew. Chem., Int. Ed. 47 (2008) 8642.
[91] T. Wakahara, H. Nikawa, T. Kikuchi, T. Nakahodo, G.M.A. Rahman, T. Tsuchiya, et al., J. Am. Chem. Soc. 128 (2006) 14228.
[92] M.D. Diener, J.M. Alford, Nature 393 (1998) 668.
[93] R. Ettl, I. Chao, F. Diederich, R.L. Whetten, Nature 353 (1991) 149.
[94] N.B. Shustova, I.V. Kuvychko, R.D. Bolskar, K. Seppelt, S.H. Strauss, A.A. Popov, et al., J. Am. Chem. Soc. 128 (2006) 15793.
[95] I.N. Ioffe, A.A. Goryunkov, N.B. Tamm, L.N. Sidorov, E. Kemnitz, S.I. Troyanov, Angew. Chem., Int. Ed. 48 (2009) 5904.
[96] S. Yang, A.A. Popov, L. Dunsch, J. Phys. Chem. B 111 (2007) 13659.
[97] F. Diederich, R.L. Whetten, C. Thilgen, R. Ettl, I. Chao, M.M. Alvarez, Science 254 (1991) 1768.
[98] K. Kikuchi, N. Nakahara, T. Wakabayashi, S. Suzuki, H. Shiromaru, Y. Miyake, et al., Nature 357 (1992) 142.
[99] N.B. Shustova, B.S. Newell, S.M. Miller, O.P. Anderson, R.D. Bolskar, K. Seppelt, et al., Angew. Chem., Int. Ed. 46 (2007) 4111.
[100] Y.Z. Tan, J. Li, T. Zhou, Y.Q. Feng, S.C. Lin, X. Lu, et al., J. Am. Chem. Soc. (2010). DOI:10.1021/ja102887t.
[101] A.A. Popov, M. Krause, S. Yang, J. Wong, L. Dunsch, J. Phys. Chem. B 111 (2007) 3363.
[102] C.M. Beavers, M.N. Chaur, M.M. Olmstead, L. Echegoyen, A.L. Balch, J. Am. Chem. Soc. 131 (2009) 11519.
[103] C.R. Wang, T. Sugai, T. Kai, T. Tomiyama, H. Shinohara, Chem. Commun. (2000) 557.
[104] S. Stevenson, G. Rice, T. Glass, K. Harlch, F. Cromer, M.R. Jordan, et al., Nature 401 (1999) 55.
[105] Y. Iiduka, T. Wakahara, T. Nakahodo, T. Tsuchiya, A. Sakuraba, Y. Maeda, et al., J. Am. Chem. Soc. 127 (2005) 12500.
[106] T. Zuo, M.M. Olmstead, C.M. Beavers, A.L. Balch, G. Wang, G.T. Yee, et al., Inorg. Chem. 47 (2008) 5234.
[107] H. Nikawa, T. Yamada, B. Cao, N. Mizorogi, Z. Slanina, T. Tsuchiya, et al., J. Am. Chem. Soc. 131 (2009) 10950.
[108] T. Akasaka, T. Wakahara, S. Nagase, K. Kobayashi, M. Waelchli, K. Yamamoto, et al., J. Am. Chem. Soc. 122 (2000) 9316.
[109] L. Feng, T. Tsuchiya, T. Wakahara, T. Nakahodo, Q. Piao, Y. Maeda, et al., J. Am. Chem. Soc. 128 (2006) 5990.
[110] T. Akasaka, T. Wakahara, S. Nagase, K. Kobayashi, M. Waelchli, K. Yamamoto, et al., J. Phys. Chem. B 105 (2001) 2971.
[111] M.M. Olmstead, A. de Bettencourt-Dias, S. Stevenson, H.C. Dorn, A.L. Balch, J. Am. Chem. Soc. 124 (2002) 4172.
[112] M.M. Olmstead, H.M. Lee, S. Stevenson, H.C. Dorn, A.L. Balch, Chem. Commun. (2002) 2688.

[113] B.Q. Mercado, C.M. Beavers, M.M. Olmstead, M.N. Chaur, K. Walker, B.C. Holloway, et al., J. Am. Chem. Soc. 130 (2008) 7854.
[114] N. Tagmatarchis, A.G. Avent, K. Prassides, T.J.S. Dennis, H. Shinohara, Chem. Commun. (1999) 1023.
[115] N.B. Tamm, L.N. Sidorov, E. Kemnitz, S.I. Troyanov, Chem. Eur. J. 15 (2009) 10486.
[116] C.R. Wang, T. Kai, T. Tomiyama, T. Yoshida, Y. Kobayashi, E. Nishibori, et al., Angew. Chem., Int. Ed. 40 (2001) 397.
[117] T.J.S. Dennis, T. Kai, T. Tomiyama, H. Shinohara, Chem. Commun. (1998) 619.
[118] I.N. Ioffe, C.B. Chen, S.F. Yang, L.N. Sidorov, E. Kemnitz, S.I. Troyanov, Angew. Chem., Int. Ed. 49 (2010). 10.1002/anie.201001082.
[119] C.M. Beavers, T. Zuo, J.C. Duchamp, K. Harich, H.C. Dorn, M.M. Olmstead, et al., J. Am. Chem. Soc. 128 (2006) 11352.
[120] G. Sun, M. Kertesz, Chem. Phys. 276 (2002) 107.
[121] S.I. Troyanov, N.B. Tamm, Chem. Commun. (2009) 6035.
[122] T. Zuo, C.M. Beavers, J.C. Duchamp, A. Campbell, H.C. Dorn, M.M. Olmstead, et al., J. Am. Chem. Soc. 129 (2007) 2035.
[123] H. Yang, C.M. Beavers, Z. Wang, A. Jiang, Z. Liu, H. Jin, et al., Angew. Chem., Int. Ed. 49 (2010) 886.
[124] I.E. Kareev, A.A. Popov, I.V. Kuvychko, N.B. Shustova, S.F. Lebedkin, V.P. Bubnov, et al., J. Am. Chem. Soc. 130 (2008) 13471.
[125] E. Kemnitz, S.I. Troyanov, Angew. Chem., Int. Ed. 48 (2009) 2584.
[126] N. Tagmatarchis, D. Arcon, M. Prato, H. Shinohara, Chem. Commun. (2002) 2992.
[127] H. Yang, C. Lu, Z. Liu, H. Jin, Y. Che, M.M. Olmstead, et al., J. Am. Chem. Soc. 130 (2008) 17296.
[128] Y. Che, H. Yang, Z. Wang, H. Jin, Z. Liu, C. Lu, et al., Inorg. Chem. 48 (2009) 6004.
[129] N.B. Tamm, L.N. Sidorov, E. Kemnitz, S.I. Troyanov, Angew. Chem., Int. Ed. 48 (2009) 9102.
[130] B.Q. Mercado, A. Jiang, H. Yang, Z. Wang, H. Jin, Z. Liu, et al., Angew. Chem., Int. Ed. 48 (2009) 9114.
[131] Q. Xie, E. Perez-Cordero, L. Echegoyen, J. Am. Chem. Soc. 114 (1992) 3978.
[132] J.W. Bausch, G.K.S. Prakash, G.A. Olah, D.S. Tse, D.C. Lorents, Y.K. Bae, et al., J. Am. Chem. Soc. 113 (1991) 3205.
[133] A.F. Hebard, M.J. Rosseinsky, R.C. Haddon, D.W. Murphy, S.H. Glarum, T.T.M. Palstra, et al., Nature 350 (1991) 600.
[134] A. Penicaud, J. Hsu, C.A. Reed, A. Koch, K.C. Khemani, P.M. Allemand, et al., J. Am. Chem. Soc. 113 (1991) 6698.
[135] J. Stinchcombe, A. Penicaud, P. Bhyrappa, P.D.W. Boyd, C.A. Reed, J. Am. Chem. Soc. 115 (1993) 5212.
[136] C. Bossard, S. Rigaut, D. Astruc, M.H. Delville, G. Felix, A. Fevrier-Bouvier, et al., J. Chem. Soc., Chem. Commun. (1993) 333.
[137] W.C. Wan, X. Liu, G.M. Sweeney, W.E. Broderick, J. Am. Chem. Soc. 117 (1995) 9580.
[138] V.M. Rotello, J.B. Howard, T. Yadav, M.M. Conn, E. Viani, L.M. Giovane, et al., Tetrahedron Lett. 34 (1993) 1561.
[139] M.F. Meidine, R. Roers, G.J. Langley, A.G. Avent, A.D. Darwish, S. Firth, et al., J. Chem. Soc., Chem. Commun. (1993) 1342.
[140] M. Tsuda, T. Ishida, T. Nogami, S. Kurono, M. Ohashi, J. Chem. Soc., Chem. Commun. (1993) 1296.
[141] J.A. Schlueter, J.M. Seaman, S. Taha, H. Cohen, K.R. Lykke, H.H. Wang, et al., J. Chem. Soc., Chem. Commun. (1993) 972.
[142] P. Belik, A. Guegel, J. Spickermann, K. Muellen, Angew. Chem. 105 (1993) 95.
[143] M. Maggini, G. Scorrano, M. Prato, J. Am. Chem. Soc. 115 (1993) 9798.
[144] M. Prato, M. Maggini, Acc. Chem. Res. 31 (1998) 519.
[145] S.H. Hoke II, J. Molstad, D. Dilettato, M.J. Jay, D. Carlson, B. Kahr, et al., J. Org. Chem. 57 (1992) 5069.
[146] Y. Nakamura, N. Takano, T. Nishimura, E. Yashima, M. Sato, T. Kudo, et al., Org. Lett. 3 (2001) 1193.
[147] S.R. Wilson, N. Kaprinidis, Y. Wu, D.I. Schuster, J. Am. Chem. Soc. 115 (1993) 8495.
[148] D.I. Schuster, J. Cao, N. Kaprinidis, Y. Wu, A.W. Jensen, Q. Lu, et al., J. Am. Chem. Soc. 118 (1996) 5639.
[149] X. Zhang, A. Romero, C.S. Foote, J. Am. Chem. Soc. 115 (1993) 11024.
[150] F. Diederich, L. Isaacs, D. Philp, Chem. Soc. Rev. 23 (1994) 243.
[151] J. Averdung, J. Mattay, D. Jacobi, W. Abraham, Tetrahedron 51 (1995) 2543.
[152] M. Yan, S.X. Cai, J.F.W. Keana, J. Org. Chem. 59 (1994) 5951.
[153] T. Akasaka, W. Ando, K. Kobayashi, S. Nagase, J. Am. Chem. Soc. 115 (1993) 1605.
[154] P.J. Fagan, P.J. Krusic, D.H. Evans, S.A. Lerke, E. Johnston, J. Am. Chem. Soc. 114 (1992) 9697.
[155] K. Komatsu, Y. Murata, N. Takimoto, S. Mori, N. Sugita, T.S.M. Wan, J. Org. Chem. 59 (1994) 6101.
[156] K. Komatsu, N. Takimoto, Y. Murata, T.S.M. Wan, T. Wong, Tetrahedron Lett. 37 (1996) 6153.
[157] C. Bingel, Chem. Ber. 126 (1993) 1957.
[158] A. Hirsch, I. Lamparth, H.R. Karfunkel, Angew. Chem. 106 (1994) 453.
[159] A. Hirsch, I. Lamparth, T. Groesser, H.R. Karfunkel, J. Am. Chem. Soc. 116 (1994) 9385.
[160] Y. Matsuo, E. Nakamura, Chem. Rev. 108 (2008) 3016.
[161] J.R. Morton, K.F. Preston, P.J. Krusic, S.A. Hill, E. Wasserman, J. Phys. Chem. 96 (1992) 3576.
[162] P.J. Krusic, E. Wasserman, B.A. Parkinson, B. Malone, E.R. Holler Jr., P.N. Keizer, et al., J. Am. Chem. Soc. 113 (1991) 6274.
[163] P.J. Krusic, E. Wasserman, P.N. Keizer, J.R. Morton, K.F. Preston, Science 254 (1991) 1183.
[164] C.N. McEwen, R.G. McKay, B.S. Larsen, J. Am. Chem. Soc. 114 (1992) 4412.
[165] I.V. Koptyug, A.G. Goloshevsky, I.S. Zavarine, N.J. Turro, P.J. Krusic, J. Phys. Chem. A 104 (2000) 5726.
[166] P.J. Fagan, P.J. Krusic, C.N. McEwen, J. Lazar, D.H. Parker, N. Herron, et al., Science 262 (1993) 404.
[167] S.I. Troyanov, A.A. Goryunkov, E.I. Dorozhkin, D.V. Ignat'eva, N.B. Tamm, S.M. Avdoshenko, et al., J. Fluorine Chem. 128 (2007) 545.
[168] I.E. Kareev, I.V. Kuvychko, S.F. Lebedkin, S.M. Miller, O.P. Anderson, S.H. Strauss, et al., Chem. Commun. (2006) 308.
[169] T. Mutig, S.M. Avdoshenko, E. Kemnitz, S.I. Troyanov, J. Fluorine Chem. 130 (2009) 241.
[170] A.A. Goryunkov, I.V. Kuvychko, I.N. Ioffe, D.L. Dick, L.N. Sidorov, S.H. Strauss, et al., J. Fluorine Chem. 124 (2003) 61.
[171] J.M. Wood, B. Kahr, S.H. Hoke II, L. Dejarme, R.G. Cooks, D. Ben-Amotz, J. Am. Chem. Soc. 113 (1991) 5907.
[172] F. Diederich, R. Ettl, Y. Rubin, R.L. Whetten, R. Beck, M. Alvarez, et al., Science 252 (1991) 548.
[173] K.M. Creegan, J.L. Robbins, W.K. Robbins, J.M. Millar, R.D. Sherwood, P.J. Tindall, et al., J. Am. Chem. Soc. 114 (1992) 1103.
[174] R. Taylor, J.P. Parsons, A.G. Avent, S.P. Rannard, T.J. Dennis, J.P. Hare, et al., Nature 351 (1991) 277.
[175] J.M. Wallis, J.K. Kochi, J. Am. Chem. Soc. 110 (1988) 8207.
[176] J.M. Hawkins, Acc. Chem. Res. 25 (1992) 150.
[177] M.S. Meier, B.R. Weedon, H.P. Spielmann, J. Am. Chem. Soc. 118 (1996) 11682.

[178] M.S. Meier, H.P. Spielmann, R.C. Haddon, R.G. Bergosh, M.E. Gallagher, M.A. Hamon, et al., Carbon 38 (2000) 1535.
[179] M.S. Meier, P.S. Corbin, V.K. Vance, M. Clayton, M. Mollman, M. Poplawska, Tetrahedron Lett. 35 (1994) 5789.
[180] A.D. Darwish, A.a.K. Abdul-Sada, G.J. Langley, H.W. Kroto, R. Taylor, D.R.M. Walton, J. Chem. Soc., Perkin Trans. 2 (1995) 2359.
[181] S. Ballenweg, R. Gleiter, W. Kraetschmer, Tetrahedron Lett. 34 (1993) 3737.
[182] C.C. Henderson, P.A. Cahill, Science 259 (1993) 1885.
[183] C. Ruechard, M. Gerst, J. Ebenhoch, H.D. Beckhaus, E.E.B. Campbell, R. Tellgmann, et al., Angew. Chem. 105 (1993) 609.
[184] M. Gerst, H.D. Beckhaus, C. Ruechardt, E.E.B. Campbell, R. Tellgmann, Tetrahedron Lett. 34 (1993) 7729.
[185] K. Shigematsu, K. Abe, M. Mitani, K. Tanaka, Chem. Express 8 (1993) 37.
[186] T. Osaki, T. Hamada, Y. Tai, React. Kinet. Catal. Lett. 78 (2003) 217.
[187] H. Selig, C. Lifshitz, T. Peres, J.E. Fischer, A.R. McGhie, W.J. Romanow, et al., J. Am. Chem. Soc. 113 (1991) 5475.
[188] K. Kniaz, J.E. Fischer, H. Selig, G.B.M. Vaughan, W.J. Romanow, D.M. Cox, et al., J. Am. Chem. Soc. 115 (1993) 6060.
[189] A.A. Gakh, A.A. Tuinman, J.L. Adcock, R.A. Sachleben, R.N. Compton, J. Am. Chem. Soc. 116 (1994) 819.
[190] O.V. Boltalina, A.Y. Borschevskii, L.N. Sidorov, J.M. Street, R. Taylor, Chem. Commun. (1996) 529.
[191] O.V. Boltalina, A.Y. Lukonin, A.A. Gorjunkov, V.K. Pavlovich, A.N. Rykov, V.M. Seniavin, et al., Proc. Electrochem. Soc. 97−14 (1997) 257.
[192] O.V. Boltalina, J.M. Street, R. Taylor, J. Chem. Soc., Perkin Trans. 2 (1998) 649.
[193] A.A. Goryunkov, V.Y. Markov, O.V. Boltalina, B. Zemva, A.K. Abdul-Sada, R. Taylor, J. Fluorine Chem. 112 (2001) 191.
[194] O.V. Boltalina, D.B. Ponomarev, A.Y. Borschevskii, L.N. Sidorov, J. Phys. Chem. A 101 (1997) 2574.
[195] O.V. Boltalina, V.Y. Markov, R. Taylor, M.P. Waugh, Chem. Commun. (1996) 2549.
[196] P.A. Troshin, O.V. Boltalina, N.V. Polyakova, Z.E. Klinkina, J. Fluorine Chem. 110 (2001) 157.
[197] H. Holloway John, E.G. Hope, R. Taylor, G.J. Langley, A.G. Avent, T.J. Dennis, et al., J. Chem. Soc., Chem. Commun. (1991) 966.
[198] O.V. Boltalina, A.a.K. Abdul-Sada, R. Taylor, J. Chem. Soc., Perkin Trans. 2 (1995) 981.
[199] H. Selig, K. Kniaz, G.B.M. Vaughan, J.E. Fischer, A.B. Smith III, Macromol. Symp. 82 (1994) 89.
[200] G.A. Olah, I. Bucsi, C. Lambert, R. Aniszfeld, N.J. Trivedi, D.K. Sensharma, et al., J. Am. Chem. Soc. 113 (1991) 9385.
[201] F.N. Tebbe, J.Y. Becker, D.B. Chase, L.E. Firment, E.R. Holler, B.S. Malone, et al., J. Am. Chem. Soc. 113 (1991) 9900.
[202] A.J. Adamson, J.H. Holloway, E.G. Hope, R. Taylor, Fuller. Sci. Technol. 5 (1997) 629.
[203] P.A. Troshin, O. Popkov, R.N. Lyubovskaya, Fuller. Nanotubes Carbon Nanostruct. 11 (2003) 165.
[204] P.R. Birkett, A.G. Avent, A.D. Darwish, H.W. Kroto, R. Taylor, D.R.M. Walton, J. Chem. Soc., Chem. Commun. (1993) 1230.
[205] I.V. Kuvychko, A.V. Streletskii, A.A. Popov, S.G. Kotsiris, T. Drewello, S.H. Strauss, et al., Chem. Eur. J. 11 (2005) 5426.
[206] P.A. Troshin, R.N. Lyubovskaya, I.N. Ioffe, N.B. Shustova, E. Kemnitz, S.I. Troyanov, Angew. Chem., Int. Ed. 44 (2005) 234.
[207] N.B. Shustova, A.A. Popov, L.N. Sidorov, A.P. Turnbull, E. Kemnitz, S.I. Troyanov, Chem. Commun., (2005) 1411.
[208] S.I. Troyanov, N.B. Shustova, A.A. Popov, L.N. Sidorov, E. Kemnitz, Angew. Chem., Int. Ed. 44 (2005) 432.
[209] F.N. Tebbe, R.L. Harlow, D.B. Chase, D.L. Thorn, G.C. Campbell Jr., J.C. Calabrese, et al., Science 256 (1992) 822.
[210] P.R. Birkett, P.B. Hitchcock, H.W. Kroto, R. Taylor, D.R.M. Walton, Nature 357 (1992) 479.
[211] A.L. Balch, V.J. Catalano, J.W. Lee, Inorg. Chem. 30 (1991) 3980.
[212] A.L. Balch, V.J. Catalano, J.W. Lee, M.M. Olmstead, J. Am. Chem. Soc. 114 (1992) 5455.
[213] P.J. Fagan, J.C. Calabrese, B. Malone, J. Am. Chem. Soc. 113 (1991) 9408.
[214] A.L. Balch, L. Hao, M.M. Olmstead, Angew. Chem., Int. Ed. Engl. 35 (1996) 188.
[215] M. Rasinkangas, T.T. Pakkanen, T.A. Pakkanen, M. Ahlgren, J. Rouvinen, J. Am. Chem. Soc. 115 (1993) 4901.
[216] H. Shinohara, Rep. Prog. Phys. 63 (2000) 843.
[217] S. Liu, S. Sun, J. Organomet. Chem. 599 (2000) 74.
[218] L. Dunsch, S. Yang, Small 3 (2007) 1298.
[219] L. Dunsch, S. Yang, Phys. Chem. Chem. Phys. 9 (2007) 3067.
[220] M.N. Chaur, F. Melin, A.L. Ortiz, L. Echegoyen, Angew. Chem., Int. Ed. 48 (2009) 7514.
[221] T. Akasaka, T. Kato, K. Kobayashi, S. Nagase, K. Yamamoto, H. Funasaka, et al., Nature 374 (1995) 600.
[222] E.B. Iezzi, J.C. Duchamp, K. Harich, T.E. Glass, H.M. Lee, M.M. Olmstead, et al., J. Am. Chem. Soc. 124 (2002) 524.
[223] H.M. Lee, M.M. Olmstead, E. Iezzi, J.C. Duchamp, H.C. Dorn, A.L. Balch, J. Am. Chem. Soc. 124 (2002) 3494.
[224] S. Stevenson, R.R. Stephen, T.M. Amos, V.R. Cadorette, J.E. Reid, J.P. Phillips, J. Am. Chem. Soc. 127 (2005) 12776.
[225] Y. Maeda, J. Miyashita, T. Hasegawa, T. Wakahara, T. Tsuchiya, T. Nakahodo, et al., J. Am. Chem. Soc. 127 (2005) 12190.
[226] S. Osuna, M. Swart, M. Sola, J. Am. Chem. Soc. 131 (2009) 129.
[227] T. Akasaka, S. Nagase, K. Kobayashi, T. Suzuki, T. Kato, K. Yamamoto, et al., J. Chem. Soc., Chem. Commun. (1995) 1343.
[228] T. Akasaka, T. Kato, S. Nagase, K. Kobayashi, K. Yamamoto, H. Funasaka, et al., Tetrahedron 52 (1996) 5015.
[229] T. Akasaka, S. Okubo, M. Kondo, Y. Maeda, T. Wakahara, T. Kato, et al., Chem. Phys. Lett. 319 (2000) 153.
[230] T. Wakahara, J.I. Kobayashi, M. Yamada, Y. Maeda, T. Tsuchiya, M. Okamura, et al., J. Am. Chem. Soc. 126 (2004) 4883.
[231] Y. Iiduka, O. Ikenaga, A. Sakuraba, T. Wakahara, T. Tsuchiya, Y. Maeda, et al., J. Am. Chem. Soc. 127 (2005) 9956.
[232] T. Wakahara, Y. Iiduka, O. Ikenaga, T. Nakahodo, A. Sakuraba, T. Tsuchiya, et al., J. Am. Chem. Soc. 128 (2006) 9919.
[233] M. Yamada, T. Wakahara, T. Tsuchiya, Y. Maeda, M. Kako, T. Akasaka, et al., Chem. Commun. (2008) 558.
[234] B. Cao, T. Wakahara, Y. Maeda, A. Han, T. Akasaka, T. Kato, et al., Chem. Eur. J. 10 (2004) 716.
[235] X. Lu, X. He, L. Feng, Z. Shi, Z. Gu, Tetrahedron 60 (2004) 3713.
[236] L. Feng, X. Lu, X. He, Z. Shi, Z. Gu, Inorg. Chem. Commun. 7 (2004) 1010.
[237] C.M. Cardona, A. Kitaygorodskiy, A. Ortiz, M.A. Herranz, L. Echegoyen, J. Org. Chem. 70 (2005) 5092.
[238] T. Cai, Z. Ge, E.B. Iezzi, T.E. Glass, K. Harich, H.W. Gibson, et al., Chem. Commun. (2005) 3594.
[239] C.M. Cardona, A. Kitaygorodskiy, L. Echegoyen, J. Am. Chem. Soc. 127 (2005) 10448.
[240] C.M. Cardona, B. Elliott, L. Echegoyen, J. Am. Chem. Soc. 128 (2006) 6480.
[241] L. Echegoyen, C.J. Chancellor, C.M. Cardona, B. Elliott, J. Rivera, M.M. Olmstead, et al., Chem. Commun. (2006) 2653.
[242] N. Martin, M. Altable, S. Filippone, A. Martin-Domenech, L. Echegoyen, C.M. Cardona, Angew. Chem., Int. Ed. 45 (2006) 110.

[243] M. Yamada, T. Wakahara, T. Nakahodo, T. Tsuchiya, Y. Maeda, T. Akasaka, et al., J. Am. Chem. Soc. 128 (2006) 1402.
[244] T. Cai, C. Slebodnick, L. Xu, K. Harich, T.E. Glass, C. Chancellor, et al., J. Am. Chem. Soc. 128 (2006) 6486.
[245] T. Cai, L. Xu, M.R. Anderson, Z. Ge, T. Zuo, X. Wang, et al., J. Am. Chem. Soc. 128 (2006) 8581.
[246] T. Cai, L. Xu, H.W. Gibson, H.C. Dorn, C.J. Chancellor, M.M. Olmstead, et al., J. Am. Chem. Soc. 129 (2007) 10795.
[247] Y. Maeda, Y. Matsunaga, T. Wakahara, S. Takahashi, T. Tsuchiya, M.O. Ishitsuka, et al., J. Am. Chem. Soc. 126 (2004) 6858.
[248] Y. Iiduka, T. Wakahara, K. Nakajima, T. Nakahodo, T. Tsuchiya, Y. Maeda, et al., Angew. Chem., Int. Ed. 46 (2007) 5562.
[249] T. Akasaka, T. Kono, Y. Matsunaga, T. Wakahara, T. Nakahodo, M.O. Ishitsuka, et al., J. Phys. Chem. A 112 (2008) 1294.
[250] T. Akasaka, T. Kono, Y. Takematsu, H. Nikawa, T. Nakahodo, T. Wakahara, et al., J. Am. Chem. Soc. 130 (2008) 12840.
[251] B. Cao, H. Nikawa, T. Nakahodo, T. Tsuchiya, Y. Maeda, T. Akasaka, et al., J. Am. Chem. Soc. 130 (2008) 983.
[252] L. Feng, T. Nakahodo, T. Wakahara, T. Tsuchiya, Y. Maeda, T. Akasaka, et al., J. Am. Chem. Soc. 127 (2005) 17136.
[253] O. Lukoyanova, C.M. Cardona, J. Rivera, L.Z. Lugo-Morales, C.J. Chancellor, M.M. Olmstead, et al., J. Am. Chem. Soc. 129 (2007) 10423.
[254] M.N. Chaur, F. Melin, A.J. Athans, B. Elliott, K. Walker, B.C. Holloway, et al., Chem. Commun. (2008) 2665.
[255] T. Cai, L. Xu, C. Shu, H.A. Champion, J.E. Reid, C. Anklin, et al., J. Am. Chem. Soc. 130 (2008) 2136.
[256] H. Nikawa, T. Kikuchi, T. Wakahara, T. Nakahodo, T. Tsuchiya, G.M.A. Rahman, et al., J. Am. Chem. Soc. 127 (2005) 9684.
[257] Y. Takano, A. Yomogida, H. Nikawa, M. Yamada, T. Wakahara, T. Tsuchiya, et al., J. Am. Chem. Soc. 130 (2008) 16224.
[258] C. Shu, C. Slebodnick, L. Xu, H. Champion, T. Fuhrer, T. Cai, et al., J. Am. Chem. Soc. 130 (2008) 17755.
[259] N. Tagmatarchis, A. Taninaka, H. Shinohara, Chem. Phys. Lett. 355 (2002) 226.
[260] I.E. Kareev, S.F. Lebedkin, V.P. Bubnov, E.B. Yagubskii, I.N. Ioffe, P.A. Khavrel, et al., Angew. Chem., Int. Ed. 44 (2005) 1846.
[261] N.B. Shustova, A.A. Popov, M.A. Mackey, C.E. Coumbe, J.P. Phillips, S. Stevenson, et al., J. Am. Chem. Soc. 129 (2007) 11676.
[262] N.B. Shustova, Y.S. Chen, M.A. Mackey, C.E. Coumbe, J.P. Phillips, S. Stevenson, et al., J. Am. Chem. Soc. 131 (2009) 17630.
[263] S. Zhang, D. Sun, X. Li, F. Pei, S. Liu, Fuller. Sci. Technol. 5 (1997) 1635.
[264] L.J. Wilson, D.W. Cagle, T.P. Thrash, S.J. Kennel, S. Mirzadeh, J.M. Alford, et al., Coord. Chem. Rev. 190-192 (1999) 199.
[265] D.W. Cagle, S.J. Kennel, S. Mirzadeh, J.M. Alford, L.J. Wilson, Proc. Natl. Acad. Sci. U.S.A. 96 (1999) 5182.
[266] D. Sun, H. Huang, S. Yang, Z.L.S. Liu, Chem. Mater. 11 (1999) 1003.

CHAPTER

13

Synthesis of Organometallic Compounds

Julia K.C. Abbott, Brenda A. Dougan, Zi-Ling Xue

The University of Tennessee, Knoxville, USA

Organometallic compounds refer to those containing at least one metal—carbon bond. They are in an interdisciplinary area between inorganic and organic chemistry. Studies of organometallic compounds have significantly advanced our understanding of chemical bonding, as these complexes show unique bonds and structures [1—16]. Organometallic compounds have played a critical role in catalysis and organic synthesis [1—16], often leading to more efficient use of reagents, higher yields of products, and less use of energy. Organometallic compounds have also been used as precursors in the preparation of nanomaterials and microelectronic materials such as thin films in integrated circuits [17,18]. Species containing M—C bonds have been found in biology as well. The vitamin B_{12} coenzyme contains a Co—C bond [19].

Synthesis of M—C-containing compounds plays a central role in the field of organometallic chemistry. Both the metals and ligands are diverse. The former include those of main group, transition metals, and lanthanide and actinide elements. The latter range from CO to multidentate organic molecules/groups. Given the range and diversity of these compounds, it is difficult to give a concise review of the preparation of organometallic compounds in one chapter. We intend here to discuss fundamental methods with typical examples.

Since organometallic compounds contain metals and ligands, the synthetic methods are in general grouped into two types: (1) reactions between metal species and preformed ligands or ligand precursors; and (2) reactions of ligands in organometallic compounds yielding new ligands. The former is used in the preparation of Grignard reagents (Eq. (13.1)), organolithium reagents, $M(CH_2Bu^t)_4$ (Eq. (13.2)) [20], and $MeCo(CO)_4$ (Eq. (13.3)) [21]. The latter is typified by the synthesis of Fischer carbene and carbyne complexes containing M=C and M≡C bonds, which will be discussed in Section 13.2.3.

$$R{-}X + Mg \rightarrow RMgX \qquad (13.1)$$

$$\begin{matrix} MCl_4 + 4LiCH_2Bu^t \rightarrow M(CH_2Bu^t)_4 \\ M = Ti,\ Zr,\ Hf \end{matrix} \qquad (13.2)$$

$$Na^+[Co(CO)_4]^- + MeI \rightarrow MeCo(CO)_4 \qquad (13.3)$$

Synthetic organometallic chemistry is discussed in many textbooks [1—11], monographs [22—24], book series [12—14,25—30], and reviews [15,16]. In particular, the 10 volume set *Synthetic Methods of Organometallic and Inorganic Chemistry* edited by Wolfgang A. Herrmann [25], the four volume set *Organometallic Syntheses* edited by R. Bruce King and John J. Eisch [26—29], and *Inorganic Syntheses* series, with Volume 35 published recently [30] are valuable sources that give detailed procedures for the synthesis of important organometallic and inorganic compounds. The monograph *Synthesis of Organometallic Compounds: A Practical Guide* edited by Sanshiro Komiya [22] provides the basic synthetic methods used in organometallic chemistry and detailed protocols. It gives an overview of the organometallic chemistry of each metal and the synthesis and handling of its important organometallic compounds. Another monograph *Synthetic Coordination and Organometallic Chemistry* edited by Alexander D. Garnovskii and Boris I. Kharisov illustrates the common ligands and synthetic methods for coordination and organometallic compounds [24]. It also provides a number of synthetic procedures for particular types

Modern Inorganic Synthetic Chemistry, DOI: 10.1016/B978-0-444-53599-3.10013-7

of complexes. It should be mentioned that there is a chapter in the book *An Introduction to Organometallic Chemistry* by A.W. Parkins and R.C. Poller that is devoted to the preparation of organometallic compounds [7]. The readers are encouraged to consult these valuable resources.

This chapter first discusses typical synthetic reactions to prepare organometallic complexes. These reactions include ligand substitution, oxidative addition and reductive elimination, insertion and elimination of ligands, and nucleophilic and electrophilic attacks on coordinated ligands. In the next section, preparation of several types of organometallic compounds is presented. These include metal carbonyls, complexes with M–C single and multiple bonds (carbene and carbyne), metal hydrides, and complexes. Some fundamental experimental techniques, especially those for handling air-sensitive compounds, are summarized. The readers hopefully will gain an understanding of the features of synthetic organometallic chemistry and will be able to find the detailed synthetic procedures in the references provided in this chapter.

13.1. SYNTHETIC REACTIONS

13.1.1. Ligand Substitution

Ligand substitutions are often the first (and last) step in a catalytic cycle which makes them a common reaction type in organometallic chemistry. This type of reaction usually occurs at the metal with no overall change in oxidation state or coordination number. Equation (13.4) shows the general reaction.

$$L_nML_1 + L_2 \rightarrow L_nML_2 + L_1 \quad (13.4)$$

A typical reaction is the substitution of a CO ligand by phosphine, PR_3. Here it is important to discuss the trans effect [31]. The trans effect, often observed in square-planar complexes, directs where the ligand substitution takes place. In other words, certain ligands are capable of facilitating the departure of another ligand, trans to the first. The more effective a ligand is at facilitating the departure, the higher its trans effect is. Ligands that form strong σ bonds (hydrides and alkyls) or are strong π acceptors (CN^-, CO, PR_3), bond strongly with the metal and cause preferential substitution of the trans metal–ligand bond. The preparation of *cis*-platin, $PtCl_2(NH_3)_2$, from $PtCl_4^{2-}$ and 2 equiv. NH_3 is a typical example (Eq. (13.5)). The second NH_3 replaces the Cl^- ligand cis to the first NH_3, because Cl^- ligand has a larger trans effect than NH_3 [6,31].

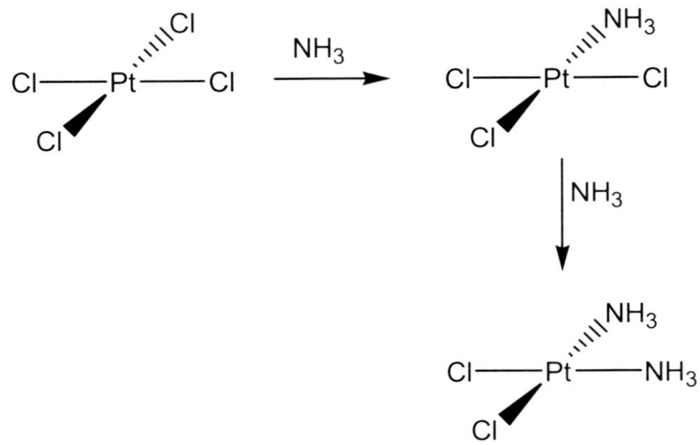

(13.5)

There are two main types of ligand substitution reactions: associative (*A*) and dissociative (*D*). Intermediate cases are also known (*I*: I_a if closer to the associative or I_d if closer to the dissociative mechanism).

Associative Substitution

In associative substitution, the incoming ligand first attacks restore, e.g., a 16e complex forming an 18e intermediate followed by a rapid expulsion of a ligand to form a 16e product (Eq. (13.6)).

$$L_nM \xrightarrow{A,\ k_1} L_nM\text{-}A \xrightarrow{-L,\ \text{fast}} L_{n-1}M\text{-}A \quad (13.6)$$

The overall reaction is often second order (Eq. (13.7)).

$$\text{Rate} = k_1[A][L_nM] \quad (13.7)$$

The reaction shown in Eq. (13.5) is an example of an associative mechanism. Associative substitution often takes place with d^8, square planar, 16e species. It is possible for an 18e complex to undergo associative substitution as well, but one of the ligands is usually rearranged in order to avoid a 20e configuration. A good example is nitrosyls that change from their linear, 3e to bent, 1e coordination as shown in the reaction in Eq. (13.8) [32]. The departure of a CO ligand returns the NO ligand to linear.

$$(OC)_4Mn(lin\text{-}NO) \xrightarrow{L,\ \text{slow}} (OC)_4LMn(bent\text{-}NO) \xrightarrow{-CO,\ \text{fast}} (OC)_3LMn(lin\text{-}NO) \quad (13.8)$$

Dissociative Substitution

The other common ligand replacement is the dissociative substitution. In this type, the first step is a slow loss of CO to generate a vacant site at the metal. The incoming ligand then binds to the metal. This is most commonly observed in 18e carbonyl complexes (Eq. (13.9)).

$$L_nM\text{-}CO \underset{+CO,\ k_{-1}}{\overset{-CO,\ k_1}{\rightleftharpoons}} L_nM\text{—}\square \xrightarrow{A,\ k_2} L_nM\text{-}A \quad (13.9)$$

The rate is dependant on the rate at which the CO ligand leaves. Thus it is independent of the concentration of the incoming ligand, or the type of ligand (Eq. (13.10)).

$$\text{Rate} = k_1[L_nM{-}CO] \quad (13.10)$$

Equation (13.11) is an example of a dissociative substitution, in which the intermediate rearranges and removes the second SR_2 ligand [33].

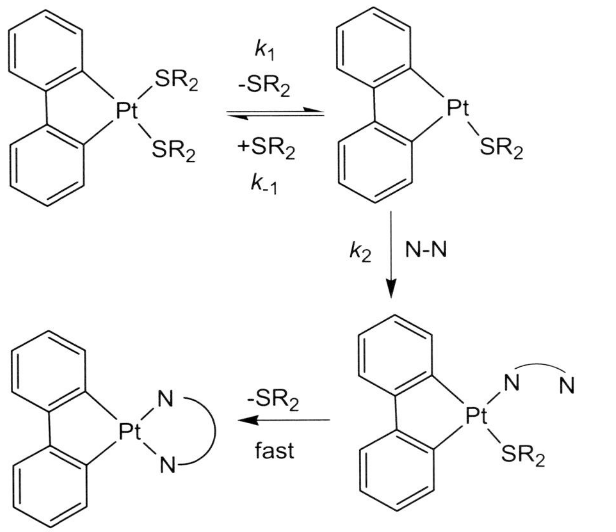

(13.11)

Interchange Type

There are cases in which it may be difficult to distinguish between associative or dissociative substitution. The difference is in how strongly the ligands bind to the metal in the intermediate. In I_a the ligands bind strongly to the metal, and in I_d the ligands bind weakly. Darensbourg gave a detailed review of the mechanism [34].

13.1.2. Oxidative Addition and Reductive Elimination

Oxidative addition (OA), and its reverse reaction, reductive elimination (RE), are important in many catalytic cycles and syntheses. Oxidative addition consists of breaking an A—B ligand and adding the A and B moieties to a metal. Both A and B usually gain an electron in the process to become anionic A^- and B^- ligands. The oxidation state, coordination number, and electron count of the metal in the complex all increase by two during the addition. Since reductive elimination is the reverse of oxidative addition, the pair of ligands, A^- and B^-, are removed from the metal to form A—B. The oxidation state, coordination number, and electron count of the metal are reduced by two in the reaction. A general reaction is given in Eq. (13.12).

$$L_nM + A\text{-}B \underset{RE}{\overset{OA}{\rightleftharpoons}} L_nM\genfrac{}{}{0pt}{}{A}{B} \quad (13.12)$$

Oxidative Addition

In oxidative addition, two ligands are added to the metal. The metal needs to have two open binding sites. Often the compound before the reaction is tetra-coordinated with 16 valence electrons. It is also necessary that the metal at the n oxidation can reach its $n+2$ oxidation state. Oxidative addition is favored by strong donor ligands because they tend to stabilize the oxidized state of the metal.

There are multiple mechanisms by which oxidative addition may occur. The Vaska's compound [35–37], a square-planar Ir(I) complex, offers a typical example of the mechanistic pathways. A few of these pathways are discussed below.

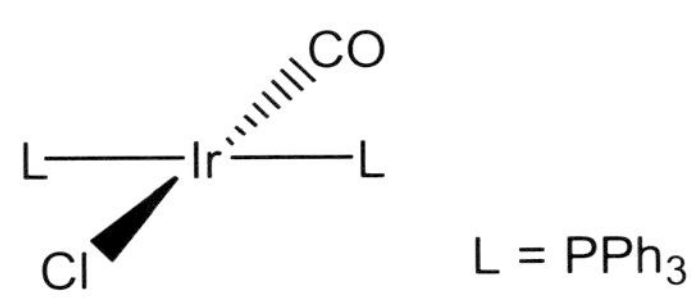

Vaska's compound

THREE-CENTER CONCERTED ADDITIONS

In a concerted addition, a nonpolar compound, such as H_2, binds as a σ complex, forming a three-membered ring transition state. Then the H—H bond is broken due in part to the strong back donation from the metal into the H—H σ* orbital, forming the final product. Sometimes the transition state is stable and the reaction stops there. Usually, however, the reaction proceeds to the end, yielding the addition product (Eq. (13.13)). In the example in Eq. (13.13), H_2 adds to the Vaska's complex, forming a trigonal bypyramid Ir(I) intermediate, which then undergoes the oxidation to give the Ir(III) dihydrides.

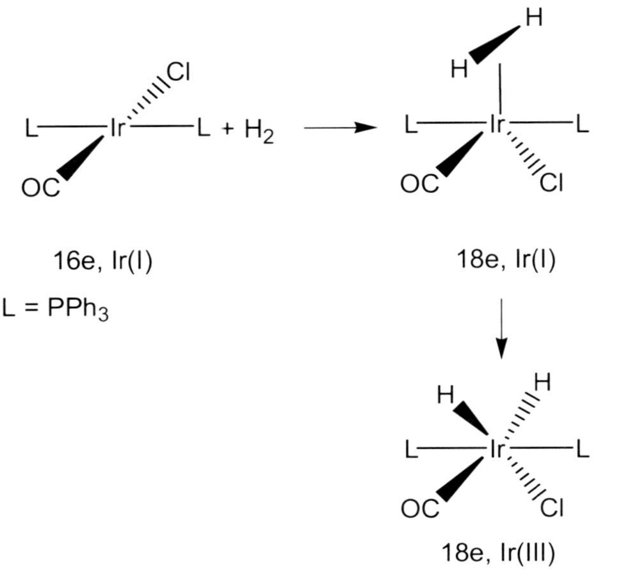

(13.13)

Several nonpolar reagents, including C—H or Si—H bonds and aryl halides, react through a concerted mechanism as well.

NUCLEOPHILIC OXIDATIVE ADDITION OF R—X

Nucleophilic oxidative addition is similar to S_N2 reactions in organic chemistry involving a polar compound such as alkyl halide. The incoming A—B compound is attacked at the least electronegative atom by L_nM (Eq. (13.14)). Similar to the concerted mechanism discussed above, this type is usually second-order reactions with negative entropies of activation, and their rates are accelerated by polar solvents.

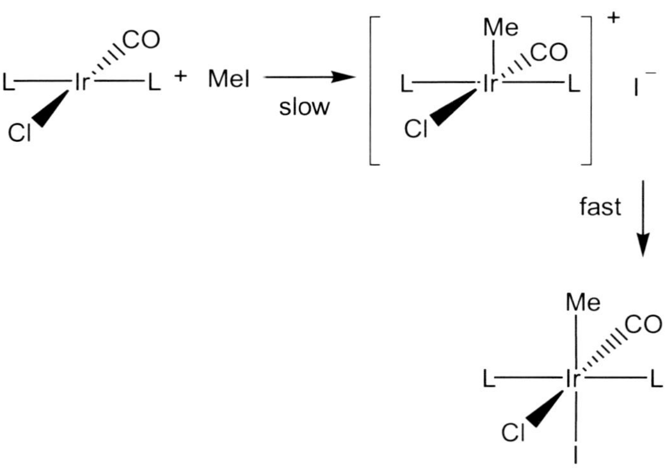

(13.14)

Another characteristic of nucleophilic oxidative addition is that, the more nucleophilic the metal is, the greater its reactivity is. Halide ions can increase the nucleophilicity of the metal and often enhance the rate of the addition. The reaction of the Vaska's complex with MeI in Eq. (13.14) is an example [36]. Such oxidative addition of MeI plays an important role in the catalytic carbonylation of methanol to acetic acid using $[RhI_2(CO)_2]^-$ and HI in the Monsanto Acetic Acid Process [2,37]. In the reaction, MeOH reacts with HI to form MeI and H_2O. MeI then undergoes carbonylation to form Rh-C(=O)Me, followed by hydration to give MeCOOH.

RADICAL MECHANISMS

Radical oxidative additions commonly occur in metals with an odd number of *d* electrons such as Co (II) and Rh(II). There are two types of radical pathways: chain and nonchain. The chain mechanism is similar to other radical chain mechanisms with initiation, induction, and termination steps. The reaction of the Vaska's complex below again offers a good example [38].

$$R\bullet + Ir^{I}Cl(CO)L_2 \longrightarrow RIr^{II}Cl(CO)L_2$$

$$RIr^{II}Cl(CO)L_2 + RX \longrightarrow RXIr^{III}Cl(CO)L_2 + R\bullet$$

$$2R\bullet \longrightarrow R{-}R$$

The second type, or nonchain mechanism, has a key feature in which one electron is transferred from M to the RX σ* to form M^+ and RX^-. This is followed by the transfer of X^- to M^+ to give R•, as shown in the example below [39,40].

$$PtL_2 + RX \xrightarrow{slow} \bullet PtL_2^{+} + \bullet RX^{-} \longrightarrow \bullet PtXL_2 + R\bullet$$

$$\bullet PtXL_2 + R\bullet \xrightarrow{fast} R{-}PtXL_2$$

IONIC MECHANISMS

Acids such as hydrogen halides and CF_3COOH are usually dissociated in solution. Their anions and protons add to the metal complex in separate steps. This is what occurs in an ionic oxidative addition. There are two types, and one is much more common than the other. In the more common type, the complex is basic enough to protonate, and then the anion binds to give the final product. This type is favored by basic ligands and low oxidation state metals. The other, less common type involves the initial halide attack, followed by protonation of the intermediate. This is favored by electron—acceptor ligands and by a net positive charge on the complex. An example of each type of ionic mechanism is given in Eqs (13.15) and (13.16) [41].

$$\mathrm{Pt(PPh_3)_4 + H^+ + Cl^- \xrightarrow{-PPh_3} [ClPt(PPh_3)_3]^- + H^+ \xrightarrow{-PPh_3} HPtCl(PPh_3)_2}$$

$$\mathrm{[Ir(COD)L_2]^+ + Cl^- + H^+ \longrightarrow [IrCl(COD)L_2] + H^+ \longrightarrow [IrHCl(COD)L_2]^+}$$

COD = 1,5-cyclooctadiene; L = $PMePh_2$

Reductive Elimination

As previously stated, reductive elimination (RE) is the reverse of oxidative addition. The oxidation state, coordination number, and electron count all decrease by two, as shown by the reaction in Eq. (13.17) [42].

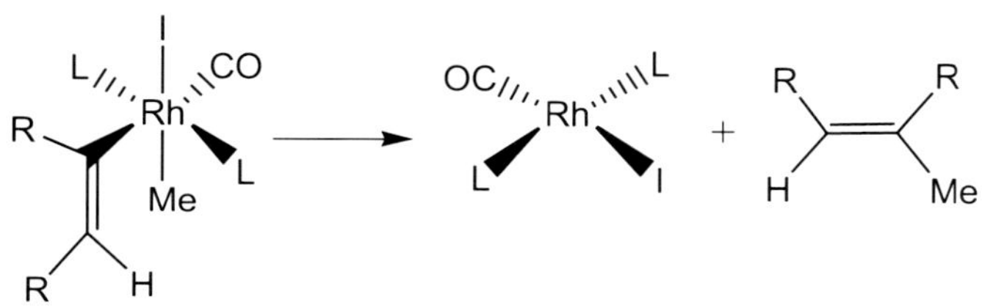

(13.17)

Several additional examples are given in Eqs (13.18)–(13.21). Certain groups are more easily eliminated than others. Usually complexes with bulky ligands are favored, since, with elimination, there is relief of steric hindrance and strain. Complexes with high oxidation states, and those with groups that will stabilize the reduced metal after ligand loss, are also favored. In addition, reactions that involve hydride ligands, such as those in Eqs (13.18) and (13.19), are fast because the transition state energy is lowered. The reaction is also efficient for intermediate oxidation states of late transition metals, including d^8 square-planar metals: Ni(II), Pd(II), and Au(III), as well as d^6 octahedral metals: Pt(IV), Pd(IV), Ir(III), and Rh(III).

$$\mathrm{L_nMRH \rightarrow L_nM + R{-}H} \quad (13.18)$$

$$\mathrm{L_nMH(COR) \rightarrow L_nM + RCHO} \quad (13.19)$$

$$\mathrm{L_nMR_2 \rightarrow L_nM + R{-}R} \quad (13.20)$$

$$\mathrm{L_nMR(SiR_3) \rightarrow L_nM + R{-}SiR_3} \quad (13.21)$$

Most reductive eliminations, including those above, seem to occur through a concerted three-centered process, the reverse of the concerted oxidative additions. In this process, the two leaving groups are usually cis to each other and there is retention of stereochemistry at the leaving group atom.

Overall, reductive elimination is not as well studied as oxidative addition mainly due to instability of complexes in reductive elimination reactions. It is, however, becoming better understood and widely used mainly due to its applications in organic synthesis. There is always a push for new methods to make C–C bonds. The last step in a catalytic cycle often involves combining two C-containing fragments, and reductive elimination of two C-containing ligands is a good method to accomplish this task.

13.1.3. Insertion and Elimination

Substitution and oxidative addition allow us to put ligands onto metals. Insertion and its reverse reaction, elimination, rearrange, transform, and combine these ligands intramolecularly. There are two main types, commonly called 1,1- or 1,2-insertions. Both types are given below.

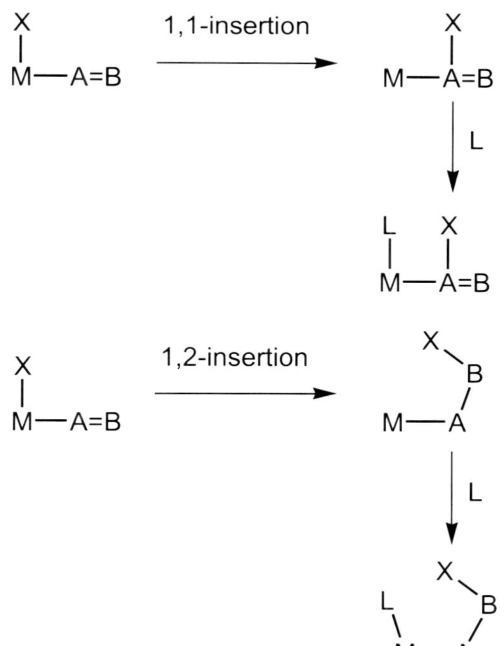

In both reactions, the ligands undergoing the insertion are cis (adjacent) to each other, and there is no change in the overall formal oxidation state. In the 1,1-insertion, the inserting ligand and the metal bind to the same α atom of the new ligand. In comparison, in 1,2-insertion, the inserting ligand and the metal bind to α and β atoms of the new ligands, respectively. The type of insertion depends on the nature of the inserting ligand. For example, CO usually undergoes 1,1-insertion, while ethylene typically favors 1,2-insertion. In general, η^1 and η^2-type ligands give 1,1- and 1,2-insertion, respectively.

1,1-Insertion

As previously stated, CO is a ligand that inserts into metal alkyl bonds, resulting in a 1,1-insertion. The following well-studied example is used to explain the reaction, mechanism, and kinetics (Eq. (13.22)) [43].

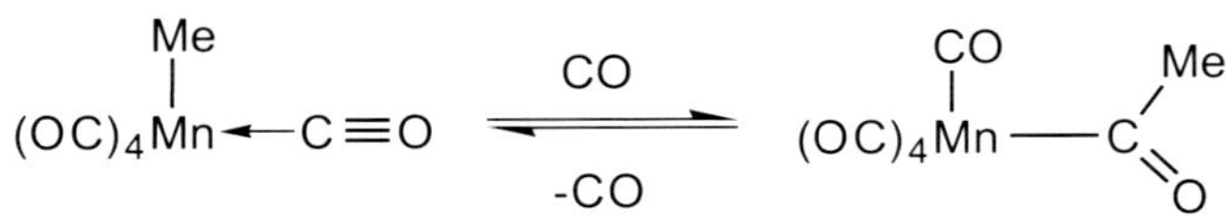

(13.22)

Mechanistic studies show that the inserted CO group is bound to the metal before the insertion occurs. This was confirmed by ^{13}C labeling studies in which ^{13}CO was added to the system, and the reaction gives a compound with the ^{13}CO ligand cis to an unlabeled acetyl group (Eq. (13.23)).

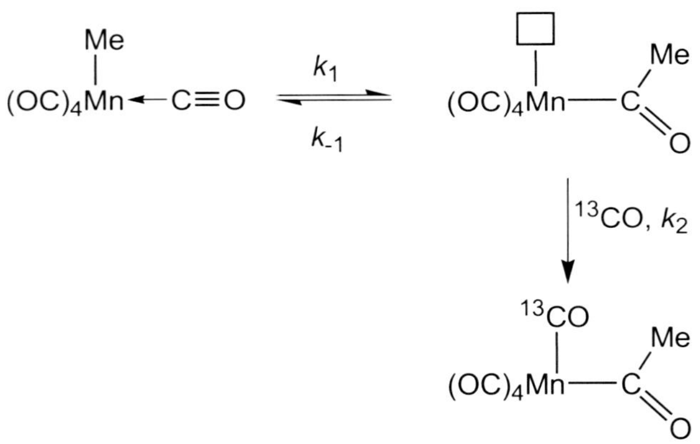

(13.23)

The rate of the reaction may be obtained by using the steady-state approach to give Eq. (13.24):

$$\text{Rate} = -\frac{d[S]}{dt} = \frac{k_1 k_2 [L][S]}{k_{-1} + k_2[L]} \qquad (13.24)$$

where S is the beginning compound, and L is the added ligand. The rate is dependant on the magnitudes of k_1, k_{-1}, k_2. Below are the three possible situations.

1. k_{-1} is very small relative to k_2[L], i.e., $k_{-1} \ll k_2$[L]. Since k_{-1} is small, the reaction stays in the intermediate phase, trapped by L, with the overall reaction based on k_1, in a first-order reaction (Eq. (13.25)).

$$\text{Rate} = -\frac{d[S]}{dt} = k_1[S] \qquad (13.25)$$

2. k_{-1} is very large relative to k_2[L]. Here, the intermediate goes back to the starting material, and the attack by L governs the overall rate, giving a second-order reaction (Eq. (13.26)).

$$\text{Rate} = -\frac{d[S]}{dt} = \frac{k_1 k_2 [L][S]}{k_{-1}} \qquad (13.26)$$

3. k_{-1} is comparable to k_2[L], causing the intermediate to be trapped by L at a rate similar to the reverse reaction. The equation thus becomes significantly more complicated (Eq. (13.27)).

$$\text{Rate} = -\frac{d[S]}{dt} = k_{obs}[S] \qquad (13.27)$$

$$\text{where } k_{obs} = \frac{k_1 k_2 [L]}{k_{-1} + k_2[L]}$$

CHANGE IN INSERTION RATES

The insertion rate of these reactions may be influenced by several factors. One contributor is steric bulk. If a ligand is bulky, the reaction releases steric hindrance when the bulky ligand leaves. The reaction in Eq. (13.22) is an example. The alkyl and the carbonyl ligands take two coordination sites, while in the product, the −C(=O)Me ligand takes one coordination site [44]. If a bulkier alkyl group replaces the Me ligand in this reaction (Eq. (13.22)), the reaction is expected to be faster.

Lewis acids also increase the rate of migration by stabilizing the metal in the transition state, and speeding up the trapping of the incoming ligand. Polar solvents also increase the rate, as shown by Wax and Bergman [45]. Oxidation of the metal and the use of an early transition metal have also been found to facilitate the insertion [45–47].

1,2-Insertion

As previously stated, 1,2-insertion with alkenes is an important reaction [48–50]. The alkene insertion into M−H bonds yields alkyls which are used in a variety of catalytic reactions. The intermediate is sometimes considered to have an agostic interaction (Eq. (13.28)) [1,51]. The intermediate needs to be coplanar for the reaction to be reversible [48]. The stereochemistry of the insertion is usually *syn*, so stereochemistry at both carbons is retained.

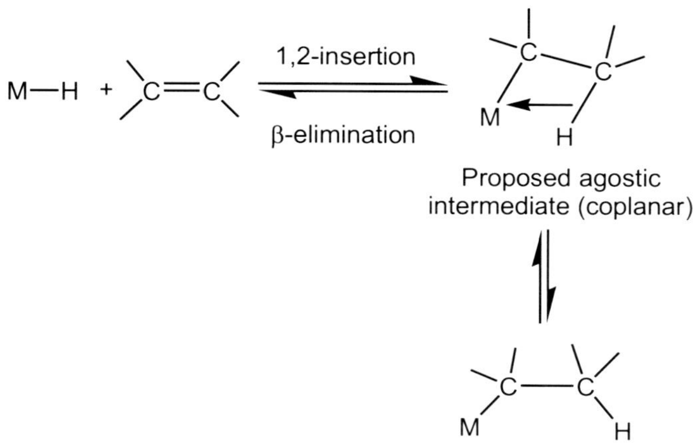

(13.28)

The reverse of this reaction is β-elimination, and the equilibrium between the two is decided by the thermodynamics of the system, and depends almost entirely on the alkene used in the reaction. Electron-withdrawing atoms such as F at the α- or β-alkyl positions lessen the tendency for β-elimination by strengthening the M−C bond, as shown in the example in Eq. (13.29) [49].

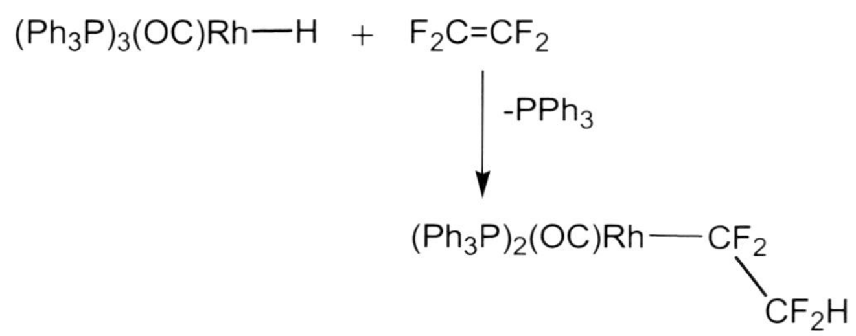

(13.29)

INSERTION INTO M−H VS M−R BONDS

In 1,1-insertion, CO usually inserts into M−R, but not M−H bonds due to unfavorable thermodynamics. The opposite is observed in 1,2-insertion. Brookhart et al. [52] have compared barriers for insertion of ethylene into M−H and M−Et bonds, and found that the barrier for the insertion into the M−H bond is significantly lower than that for the insertion into the M−Et bond.

OTHER INSERTIONS

In principle, an unsaturated ligand can undergo insertions. SO_2 ligand has been shown to undergo both 1,1- and 1,2-insertions. Other possible ligands for insertion include O_2 and CO_2. An interesting insertion of SO_2 is given below in a Pd complex where the new ligand bridges two metal atoms (Eq. (13.30)) [53].

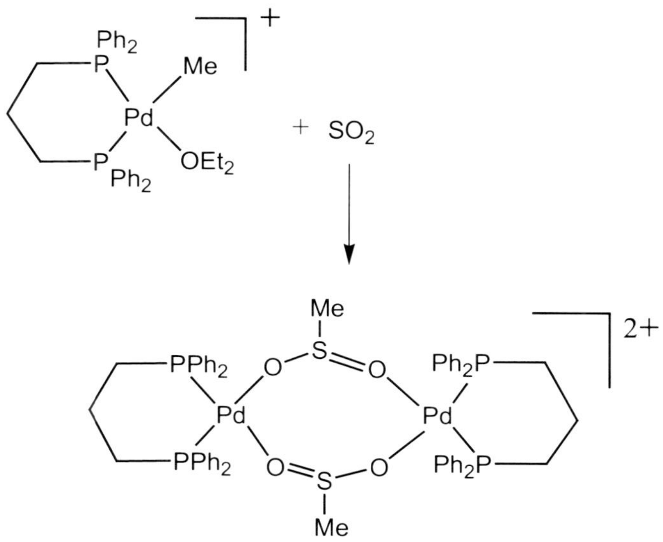

(13.30)

α, β, γ, δ Eliminations

The most common type of elimination is β-elimination. It is the main decomposition pathway for alkyls containing β-H atoms. In order for the reaction to occur, a vacant site on the metal is usually required. A coplanar M−C−C−H arrangement brings the β-H atom closer to the metal center (Eq. (13.31)). If the species is an 18e complex, it first loses a ligand to open up a site for elimination.

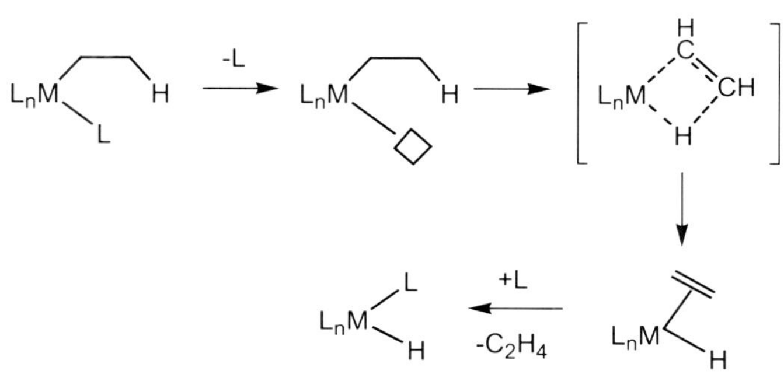

(13.31)

If there are no β-H atoms, a complex may go through α, γ, or δ elimination. These elimination reactions are not as common as β-H elimination. A simple α elimination is that of a methyl group, which has no β-H atoms and gives methylene hydride (Eq. (13.32)). Methylene hydride, though, is unstable, and often reacts further. γ and δ eliminations give cyclic products (Eq. (13.33)), but are also not common due to unfavorable sterics.

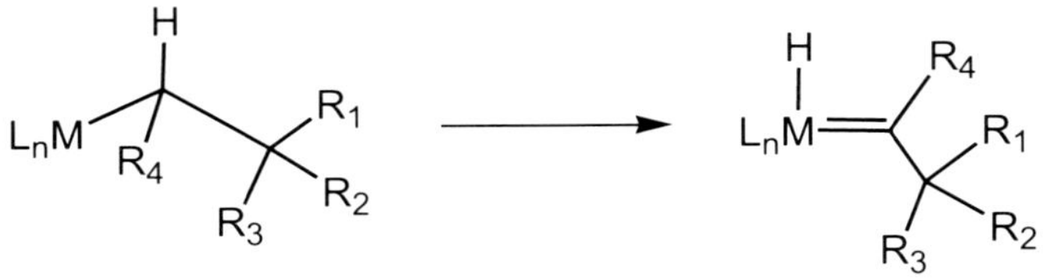

(13.32)

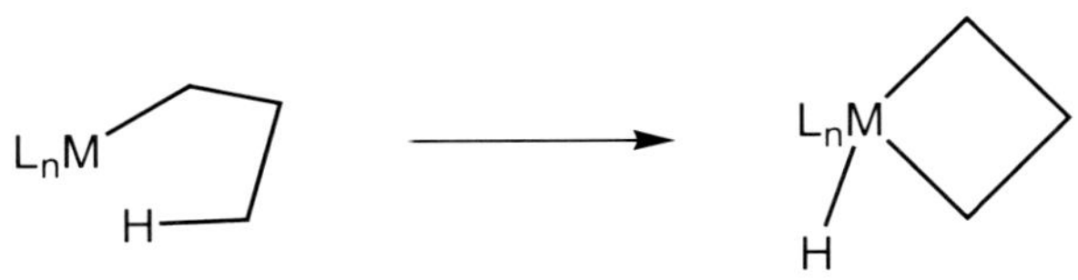

(13.33)

13.1.4. Nucleophilic and Electrophilic Attacks on Coordinated Ligands

In the previous sections, reactions involve ligands that react with each other or a reagent that reacts with the metal itself. In this section we discuss how a metal center activates a bound ligand, followed by the attack of the ligand by an external reagent.

The attacking reagent is usually an electrophile or a nucleophile. In the reaction, there is either an addition or abstraction, depending on the properties of the metal fragment. In the nucleophilic attack by Nu^-, the metal complex L_nM is usually a poor π base and a good σ acid, and the ligands L are electron withdrawing. The electron density on the metal complex L_nM or one of the ligand L is low, thus facilitating the nucleophilic attack. In the electrophilic attack by E^+, the complex L_nM is usually a good π base and a poor σ acid, and the ligands L are electron donating. There is high electron density on the metal or the ligands, thus making the complex prone to electrophilic attack. Within each category, there is addition or abstraction. A comparison of these reactions is given below.

$L_nM\text{-}L' + Nu^-$ → $[L_nM\text{-}L'\text{-}Nu]^-$ Nucleophilic Addition

$L_nM\text{-}L' + Nu^-$ → $L_nM^- + L'\text{-}Nu$ Nucleophilic Abstraction

$L_nM\text{-}L' + E^+$ → $[L_nM\text{-}L'\text{-}E]^+$ Electrophilic Addition

$L_nM\text{-}L' + E^+$ → $L_nM^+ + L'\text{-}E$ Electrophilic Abstraction

It should be pointed out that it is also possible for an attack on the metal. This reaction may result in a ligand substitution if the incoming ligand is a nucleophile, or oxidative addition if the incoming ligand is an electrophile. These types of attack are discussed below in more detail.

Nucleophilic Addition

One of the most common complexes to undergo nucleophilic addition is metal carbonyl. Normally, a carbonyl ligand (and other ligands discussed in this section) is not susceptible to nucleophilic attack because of its high electron density. When it is bound to an electron-deficient metal, however, the ligand becomes positively charged and open to attack by a nucleophile. A typical example of nucleophilic addition is the reaction of metal carbonyls with a nucleophilic lithium reagent, followed by an electrophilic addition to give a stable Fischer carbene (Eq. (13.34)) [54].

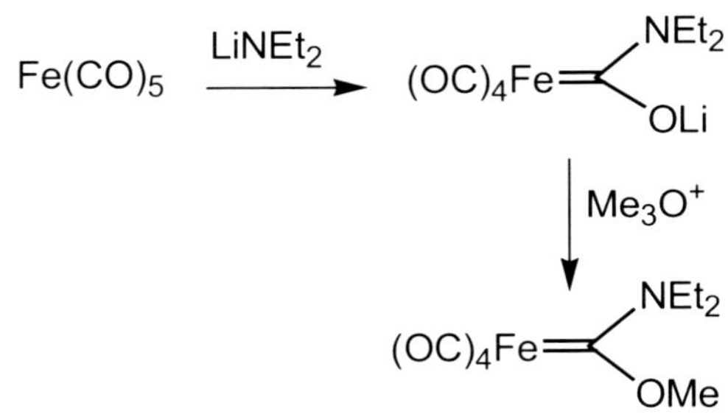

(13.34)

Sometimes a metal complex is so electron poor that even water attacks a coordinated CO ligand (Eq. (13.35)) [55]. The attack is followed by β-elimination to give a stable metal hydride. This example is also an important synthetic reaction because it removes CO from a metal complex, which is often difficult because CO usually binds to metals tightly.

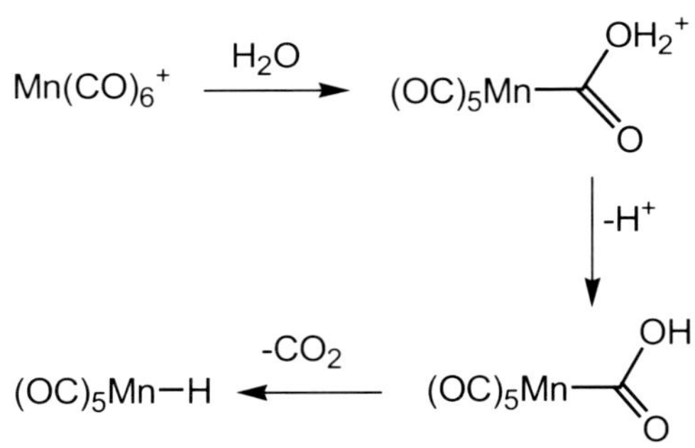

(13.35)

Isonitriles are similar to carbonyls in their chemical behavior. They undergo nucleophilic addition when bound to electron-deficient metal centers. Isonitriles tend to bind metals at higher oxidation states, and they are therefore especially susceptible to nucleophilic attack. In the example in Eq. (13.36), the resulting carbene is electron rich, and it thus resists any further nucleophilic attack [56].

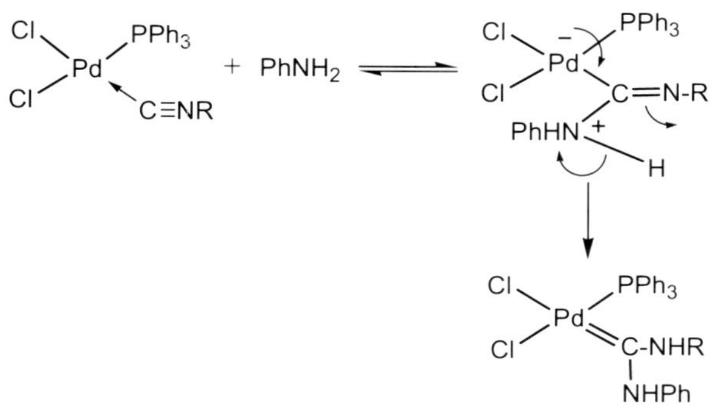

(13.36)

Nucleophilic addition also occurs on a number of π ligands including carbenes, carbynes, and alkenes/polyenes such as benzene and ethylene. It is not uncommon to have several polyenes or polyenyl ligands in one complex, and there is thus often selective attack of one site over another. The regiochemistry of these additions has been well studied and summarized by Davies et al. [57] into a set of rules that help predict the site of addition. The rules can be summarized as follows:

1. Neutral polyenes (such as butadine) react before anionic polyenyls (such as cyclopentadienl Cp^-);
2. Open ligands such as butadiene react before closed ligands such as benzene; and
3. For even, open polyenes, addition occurs at the terminal position. For odd, open polyenyls, addition occurs at the terminal position unless the metal fragment is strongly electron withdrawing.

When Rules 1 and 2 are in conflict, Rule 1 takes precedence. These rules and corresponding reactions are described in detail in organometallic and inorganic textbooks [1,4,6]. Thus only two examples are given here (Eqs (13.37) and (13.38)) [57].

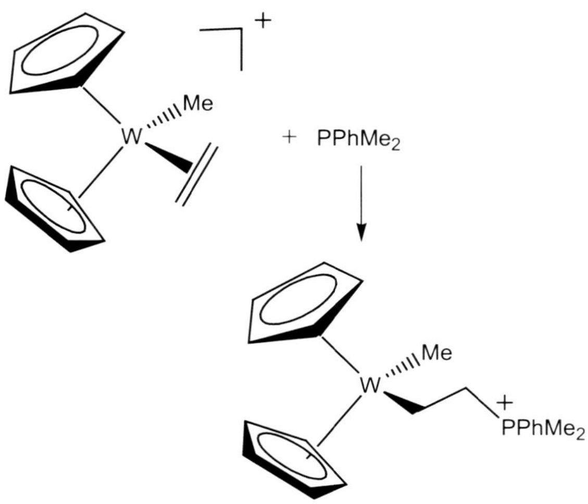

(13.37)

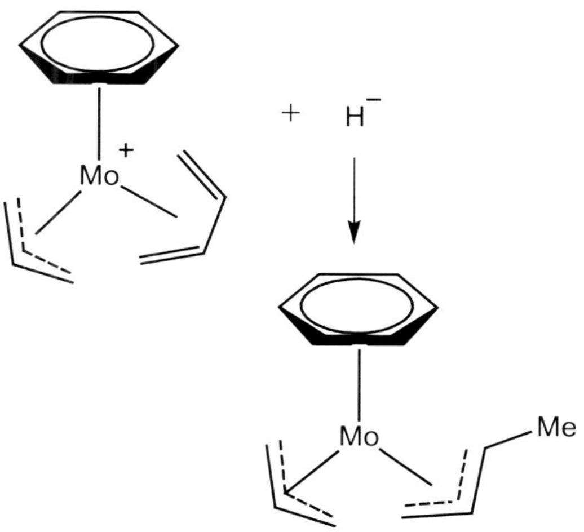

(13.38)

Alkene complexes that undergo nucleophilic addition to give metal alkyls often rearrange to give other products. This has been the basis for an important industrial process known as the Wacker [58,59] process to make aldehydes from alkenes using $PdCl_2$ as catalyst. In this reaction, $PdCl_2$ is reduced to Pd(0) which is then oxidized by $CuCl_2$ back to $PdCl_2$. $CuCl_2$ is reduced to CuCl in the process, and O_2 oxidizes CuCl back to $CuCl_2$. The overall reaction scheme is given below.

$$H_2C{=}CH_2 + 1/2\ O_2 \longrightarrow CH_3CHO$$

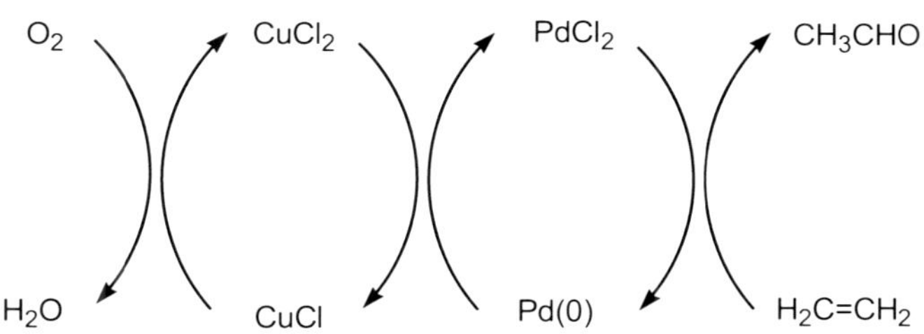

Nucleophilic Abstraction

In nucleophilic abstraction, the attacked ligand is partially or wholly removed along with the nucleophile. Such reactions are observed in hydrides, alkyls, and acyls. In a metal hydride, deprotonation of the metal hydride produces a nucleophilic metal anion (Eq. (13.39)). In the intermediate $MeReH_6L_2$, the driving force for methane loss is greater than that for H_2 loss. Thus the resulting compound is an unusual polyhydride ReH_5L_2 which reacts further with cyclopentadiene to give off H_2 [60].

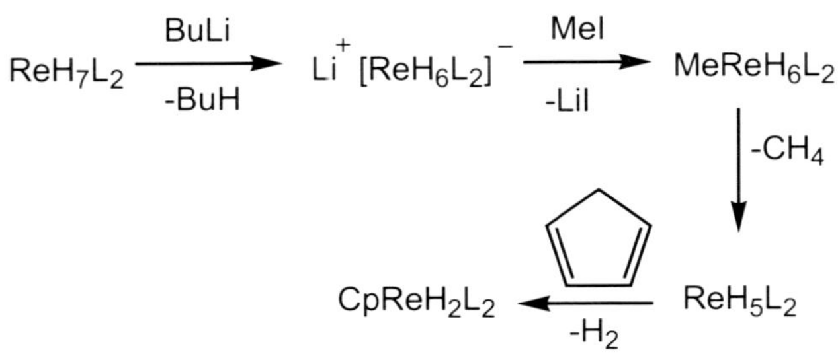

(13.39)

In metal alkyl complexes, an attack by another nucleophilic complex may lead to the transfer of the alkyl ligand and the inversion of the configuration at the stereogenic carbon center. For example, benzyl-*α*-*d* chloride (*S*)-(+) PhCH(D)Cl partially racemizes in the presence of $Pd(PPh_3)_4$ [61]. The racemization is attributed to a nucleophilic exchange equilibrium process during the oxidative addition of PhCH(D)Cl. Electron rich, zero-valent Pd attacks the chiral Pd(II) alkyl complex, abstracting the alkyl ligand as shown in Eq. (13.40). The Pd(0) by-product will then attack the inverted Pd(II) alkyl complex, resulting in another inversion of configuration. Acyl ligands are also readily abstracted by nucleophiles, especially when the metal is Pd.

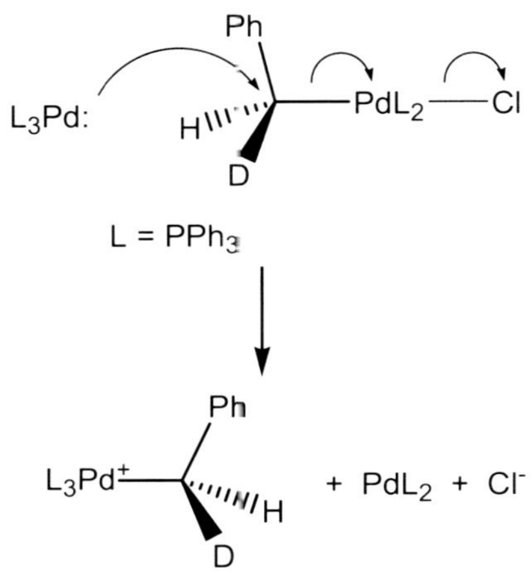

(13.40)

Electrophilic Addition

Attack by an electrophile such as H^+ or Me^+ can occur on the metal, ligand, or a metal−ligand bond that often leads to the addition of the electrophile to the metal complex [62−65].

In oxidative addition by S_N2 or the ionic mechanism, the first step is an electrophilic addition. An example of such an electrophilic addition to a metal center is given in Eq. (13.14) in the first step of the oxidative addition of MeI to the Vaska's complex. The addition of any zero-electron ligand, including $AlMe_3$, BF_3, and even CO_2 (when it binds by η^1) is also considered an electrophilic addition. Another common type of electrophilic addition to a metal center is protonation, as shown in Eq. (13.41) [63]. This is one principal method to prepare metal hydride complexes, as discussed later in Section 13.2.4.

$$\text{Ni(diphosphine)}_2 + H^+ \rightarrow \left[\text{HNi(diphosphine)}_2\right]^+ \quad (13.41)$$

Attack of an M−L bond is typified by the protonation of the carbyne ligand in $W(\equiv CBu^t)(PMe_3)_4Cl$ in Eq. (13.42) [64].

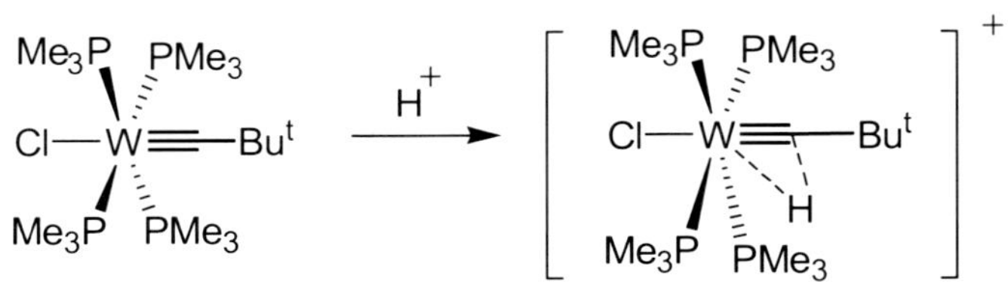

(13.42)

An example of the electrophilic attack of a ligand is protonation of nickelocene Cp_2Ni (Eq. (13.43)) [65]. This addition occurs at the exo position of the Cp ligand, and the product does not undergo scrambling of the deuterium label.

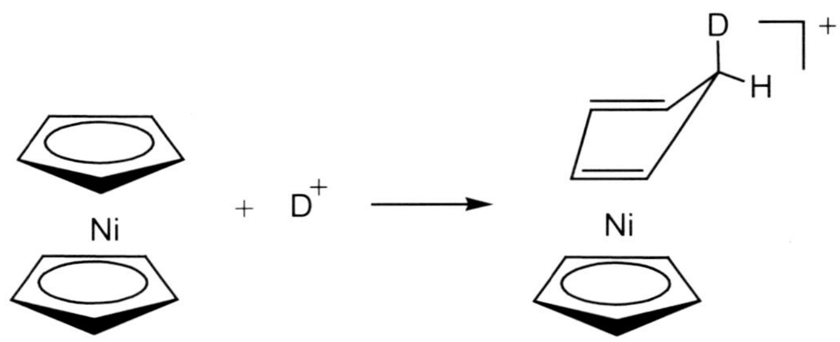

(13.43)

Electrophilic Abstraction

The most common and well-studied electrophilic abstraction is that of metal alkyls (Eqs (13.44)−(13.46) [66−68]. Electrophiles that undergo such abstraction include H^+, BAr_3, and Hg^{2+}. Protons are capable of cleaving metal alkyl bonds, especially when there is an electropositive metal, giving off an alkane in such protonation. An example is the reaction of HCl with Schrock-type carbyne complex $(Bu^tCH_2)_3W\equiv CBu^t$, leading to the release of CMe_4 and formation of $[Bu^tC\equiv WCl_3(DME)]$ (DME = dimethoxyethane, Eq. (13.44)) [66]. This carbyne complex and its reaction with HCl are used in the formation of the Schrock-type olefin metathesis catalysts, as discussed below (Eq. (13.81)). The protonation of the vinyl ligand in the Zr complex in Eq. (13.45) removes the vinyl ligand as an alkene with retention of the configuration [67]. Lewis acids such as $B(C_6F_5)_3$ and methyl aluminum oxide (MAO) have also acted as electrophiles to abstract alkyl ligands. One such reaction is shown in Eq. (13.46) [68]. The cationic products of the reactions are active catalysts for stereoselective olefin polymerization [69]. The electrophilic reactions with Lewis acids to cleave metal alkyl bonds have played a critical role in the development of olefin polymerization catalysts.

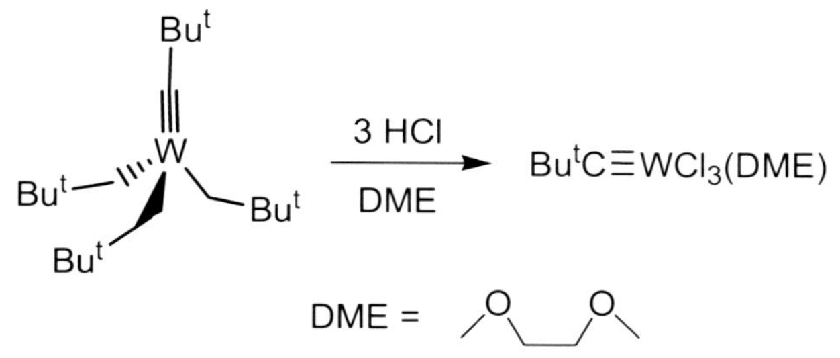

(13.44)

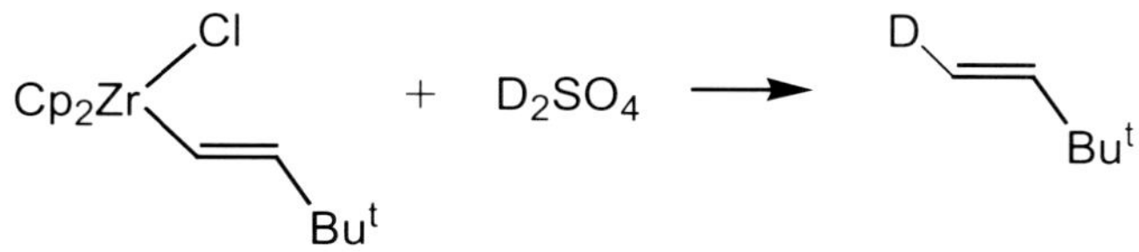

(13.45)

$$Cp_2ZrMe_2 + B(C_6F_5)_3 \rightarrow (Cp_2ZrMe)^+[B(C_6F_5)_3Me]^- \quad (13.46)$$

When metal ions such as Hg^{2+} cleave metal alkyl bonds in electrophilic abstraction reactions, two pathways are known [70]. In one pathway, the attack occurs at the metal or the M–C bond, and the configuration is retained. In the other pathway, attack occurs at the α carbon of the alkyl. The configuration at the carbon may be inverted in the reaction.

13.1.5. Reactions of Metal Vapors with Ligands or Ligand Precursors

This is a unique method to prepare coordination compounds. Vapors of metals react directly with ligands or their precursors such as alkenes, alkynes, arenes, and phosphines to give products. Metal vapors are more reactive than metals in the solid or liquid state. This is primarily because gaseous metal atoms have minimal steric effects and they are of higher energies. Metal vapors are usually obtained by heating the metals at high temperatures (1700–2200°C) and transporting in vacuum to the ligands or their precursors [71–73]. The ligands or their precursors may be in the gas phase or are kept at low temperatures of −263 to 0°C (10–273 K). One typical example of such reactions is the preparation of $Cr(\eta^6\text{-}C_6H_6)_2$ by the condensation of gaseous chromium metal and arene (Eq. (13.47)) [71].

$$Cr(g) + 2C_6H_6(g) \rightarrow Cr\left(\eta^6\text{-}C_6H_6\right)_2 \quad (13.47)$$

In some cases, thermally unstable compounds/species may be prepared in low-temperature matrices. For example, condensing Cu atoms and CO in an adamantine matrix at liquid nitrogen temperature of −196°C gives CuCO and $Cu(CO)_3$ (Eq. (13.48)), as identified by EPR [74]. Because low temperatures are usually used in the synthesis, the method has been called "cryosynthesis" [73].

$$\begin{aligned} Cu(g) &\xrightarrow[\text{adamantane matrix}]{-196^\circ C} Cu(s) \\ &\xrightarrow[-196^\circ C]{CO} CuCO + Cu(CO)_3 \end{aligned} \quad (13.48)$$

13.2. PREPARATION OF TYPICAL ORGANOMETALLIC COMPOUNDS

Organometallic compounds can be prepared by many different synthetic routes. Detailed descriptions of such preparation methods can be found in reference books [25–30]. The goal of this section is to briefly describe principal synthetic routes to obtain typical organometallic complexes.

The majority of organometallic complexes contain an M–C σ, π, and/or dative bond. Thus, a discussion of the preparation of M–C σ bond formation via the reaction of carbonyls, alkyls, and hydrides with metals will be given. The utilization of these complexes with M–C σ-bonds in the preparation of well-known Fisher and Schrock-type carbene and carbynes containing M–C multiple bonds will be explored. This section will include preparation of so-called π complexes: metal allyls, olefins, acetylenes, arenes, and polyenes containing dative bonds from C–C π bonds. The examples, illustrations, and discussion given below are representatives of typical synthetic procedures for organometallic compounds.

13.2.1. Metal Carbonyls

Many metal carbonyl complexes are commercially available and can serve as starting materials (Table 13.1). These complexes may be prepared by direct addition of CO, utilizing a reducing agent, or by starting from a reactive organic carbonyl compound. A brief overview of the preparation of metal carbonyls is presented.

Mononuclear metal carbonyls may be prepared by direct reaction of CO with the metals [76]. For instance,

TABLE 13.1 Commercially Available Metal Carbonyl Complexes [75]

Group 6	Group 7	Group 8	Group 9	Group 10
$Cr(CO)_6$ White crystals mp 150°C, dec.	$Mn_2(CO)_{10}$ Yellow solid 154°C	$Fe(CO)_5$ Yellow liquid mp −20°C; bp 103°C	$Co_2(CO)_8$ Dark orange crystals mp 51–52°C, dec.	$Ni(CO)_4$ Colorless liquid bp 42°C
$Mo(CO)_6$ White crystals mp 150°C, dec.	$Re_2(CO)_{10}$ White solid 170°C, dec.	$Fe_2(CO)_9$ Golden crystals 100°C, dec.	$Co_4(CO)_{12}$ Black solid mp 60°C, dec.	
$W(CO)_6$ White crystals mp 150°C, dec.		$Fe_3(CO)_{12}$ Black crystals 140°C, dec.	$Rh_4(CO)_{12}$ Dark red crystals >130°C dec.	
		$Ru_3(CO)_{12}$ Orange solid mp 154°C	$Rh_6(CO)_{16}$ Black crystals >300°C dec.	
		$Os_3(CO)_{12}$ Yellow crystals mp 224°C	$Ir_4(CO)_{12}$ Yellow crystals mp 195°C, dec.	

$Fe(CO)_5$ is synthesized under high pressure from iron metal and CO at 200°C. The reaction can proceed further by photolysis to produce dimer $Fe_2(CO)_9$ (Eq. (13.49)) [76].

$$Fe + 5CO \xrightarrow[200^\circ C]{200\ atm} Fe(CO)_5 \xrightarrow{hv} Fe(CO)_9 \quad (13.49)$$

While direct addition of CO to a metal may be the simplest method in preparing metal carbonyls, alternative methods are available. By introducing a metal halide or metal oxides to CO, formation of metal carbonyls such as $W(CO)_6$ [76], $Cr(CO)_6$ [76], $Ni(CO)_4$ [76], $Re_2(CO)_{10}$ [77], and $Ru_3(CO)_{12}$ [78] occurs due to reductive carbonylation (Eqs (13.50)–(13.54)). Reducing agents such as aluminum alkyls can be used when the initial metal complex is electropositive (e.g., WCl_6) at a high oxidation state. However, as shown in Eq. (13.53), CO can act as a reducing agent.

$$WCl_6 + 6CO + 2Et_3Al \xrightarrow[140\ atm]{50^\circ C} W(CO)_6 \quad (13.50)$$

$$CrCl_3 + Al + 6CO \xrightarrow[140^\circ C,\ 600\ atm]{AlCl_3} Cr(CO)_6 + AlCl_3 \quad (13.51)$$

$$NiSO_4 + 4CO + S_2O_4^{2-} \rightarrow Ni(CO)_4 \quad (13.52)$$

$$[NH_4][ReO_4] + CO + HAl(CH_2CHMe_2)_2 \xrightarrow{1\ atm} Re_2(CO)_{10} \quad (13.53)$$

$$RuCl_3 \xrightarrow[Na\ acetylacetonate,\ 165^\circ C]{30\ atm\ H_2,\ 120\ atm\ CO} Ru_3(CO)_{12} \quad (13.54)$$

Metal carbonyls serve as excellent building blocks for utilization in such areas as hydrogenation. $Fe(CO)_5$, prepared as shown in Eq. (13.55), undergoes a base reaction with NaOH to produce $[HFe(CO)_4]^-$ (Eq. (13.49)) [79]. The same method applies for $Fe_2(CO)_9$ and results in the formation of $[HFe_2(CO)_8]^-$. The anion $[HFe(CO)_4]^-$ reacts with an alkyl halide and CO to give carbonylate $RCOFeH(CO)_4$ (R = alkyl group) containing a functionalized aldehyde group [80].

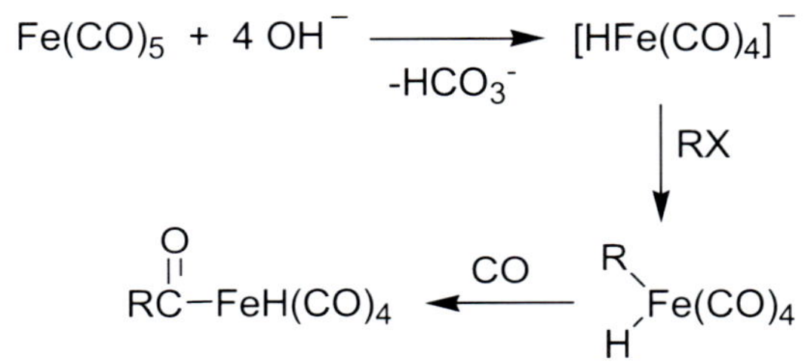

(13.55)

Equation (13.56) illustrates an alternative route for the synthesis of a functionalized metal carbonylate using $Co_2(CO)_9$ [81,82]. Thus, neutral and anionic metal carbonyls are utilized in a wide array of reactions. Reducing agents such as sodium amalgam Na/Hg or Na in liquid ammonia at −33°C are used in the formation of metal carbonyl anions $[Co(CO)_4^-$ and $Ni_5(CO)_{12}^{2-}$ (Eq. (13.57))] [83]. Sodium amalgam is an alloy of mercury with sodium metal, and it is safer to use as a reducing reagent than sodium metal.

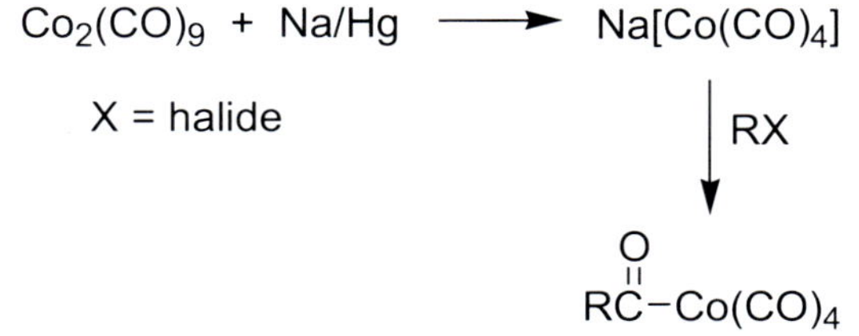

(13.56)

$$Ni(CO)_4 + Na \xrightarrow{NH_3\ (liquid)} Na_2[Ni_5(CO)_{12}] \quad (13.57)$$

13.2.2. Complexes with M—C σ Bonds

The formation of M—C σ bonds is one of the most fundamental aspects of organometallic chemistry. Preparation of complexes with M—C bonds has led to novel complexes and insightful chemistry. Alkylation usually refers to the reaction of a metal complex containing, e.g., a halide ligand (X) with a Grignard or alkyl lithium reagent to give an alkyl ligand (Eq. (13.58)).

$$M{-}X + R^- \rightarrow M{-}R + X^- \qquad (13.58)$$

The alkylating reagent often reacts with a metal halide or cationic metal complex to give a metal alkyl complex as shown in the preparation of $Zr(CH_2Ph)_4$ (Eq. (13.59)) [84] and WMe_6 (Eq. (13.60)) [85,86]. See also the preparation of $M(CH_2Bu^t)_4$ (M = Ti, Zr, Hf) in Eq. (13.2). It should be pointed out that WMe_6 adopts a trigonal prismatic structure (Eq. (13.60)) [86]. The square-planar Pt complex in Eq. (13.61) is alkylated with $(LiCH_2CH_2)_2$ to give a metallacyclopentane [87,88]. Maintaining proper reaction conditions (e.g., temperature) and molar ratios of the alkylating reagents are important in these reactions.

$$ZrCl_4 + 4PhCH_2MgCl \rightarrow Zr(CH_2Ph)_4 \qquad (13.59)$$

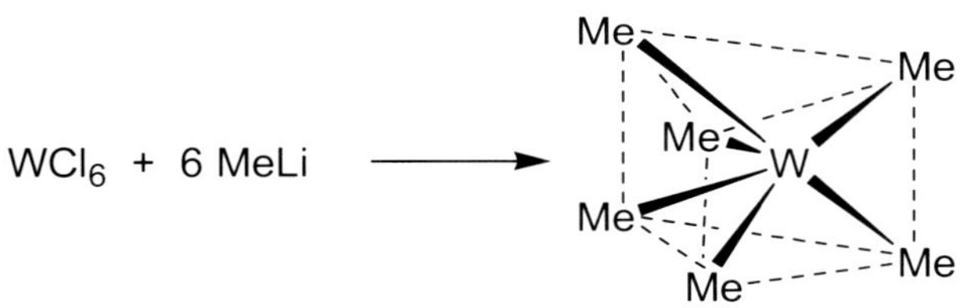

(13.60)

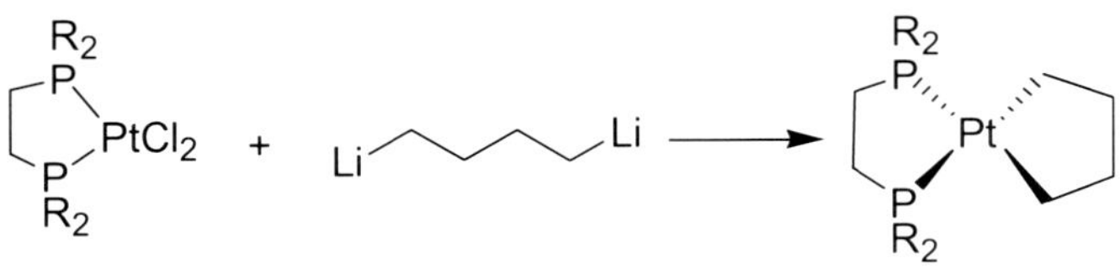

(13.61)

Alkyl lithium compounds have been used to substitute Grignard reagents in many reactions. These reactions are often slower than those of Grignard reagents. In addition, a few alkyl lithium reagents such as MeLi, Bu^nLi, and Bu^tLi are pyrophoric with limited life. Consequently, Grignard reagents are more popular and favorable to use.

In another method, a transition metal complex is treated with a reducing agent such as sodium metal to give an anionic metal species. It then reacts with alkyl halides to give metal alkyl complexes. For instance, $W(CO)_6$, synthesized by the reaction in Eq. (13.50), is treated with NaCp, followed by alkylation with EtI, to give $CpW(CO)_3Et$ [89]. Reduction of $Mn_2(CO)_{10}$ with sodium amalgam Na/Hg gives $[Mn(CO)_5]^-$ (Eq. (13. 63)) [90]. Reaction of the anion with MeI yields $MeMn(CO)_5$ (Eq. (13.63)). The alkylation here to give $MeMn(CO)_5$ is similar to that in the synthesis of $MeCo(CO)_4$ in Eq. (13.3). Similar reduction of $Cp(OC)_2Fe{-}Fe(CO)_2Cp$ with sodium amalgam yields $[Cp(OC)_2Fe]^-$ (also known as Fp^-) [91]. It is a widely used reagent to make many Fe complexes, as shown in the example in Eq. (13.64), and organic compounds [91].

$$W(CO)_6 + NaCp \longrightarrow Na[CpW(CO)_3] \xrightarrow{EtI} CpW(CO)_3Et \qquad (13.62)$$

$$Mn_2(CO)_{10} + Na/Hg \longrightarrow Na[Mn(CO)_5] \xrightarrow{MeI} MeMn(CO)_5 \qquad (13.63)$$

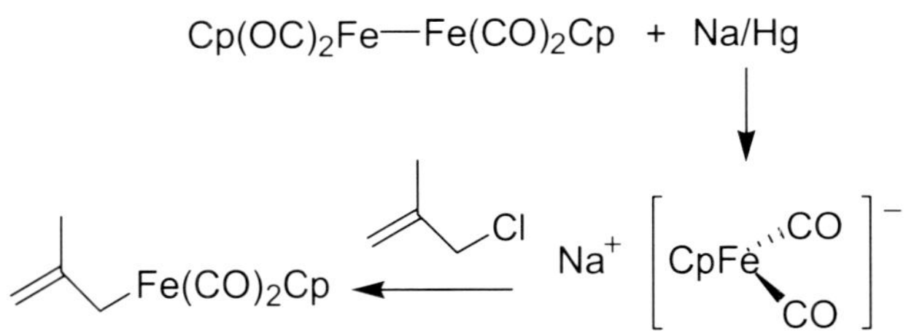

(13.64)

Though alkyllithiums and Grignard reagents are among the most common alkylating reagents, removal of magnesium or lithium halides (e.g., $MgCl_2$ or LiCl) may be difficult. Thus alternative alkylation methods have been developed. By employing alkyl complexes of metals such as Zn, Mg, or Li, transfer of an alkyl group from these metals to the new metal occurs in a process known as transmetallation (Eqs (13.65)–(13.68)) to give WMe_6 [92], $MeTiCl_3$ [93], Me_2NbCl_3 [94], and $(Bu^tCH_2)_3TaCl_2$ [95–98]. For WMe_6, see Eq. (13.60) for its preparation using MeLi [85,86].

$$WCl_6 + 2AlMe_3 \rightarrow WMe_6 \qquad (13.65)$$

$$TiCl_4 + AlMe_3 \rightarrow MeTiCl_3 \qquad (13.66)$$

$$NbCl_5 + ZnMe_2 \rightarrow Me_2NbCl_3 \qquad (13.67)$$

$$TaCl_5 + Zn(CH_2Bu^t)_2 \rightarrow (Bu^tCH_2)_3TaCl_2 \quad (13.68)$$

There are a few disadvantages in using alkylation to make M–C bonds. The main drawback is that the alkylating reagents may undergo nucleophilic attack on a ligand on the metal atom, as shown in the example in Eq. (13.69) [99], where the methyl group is inserted into the carbonyl group.

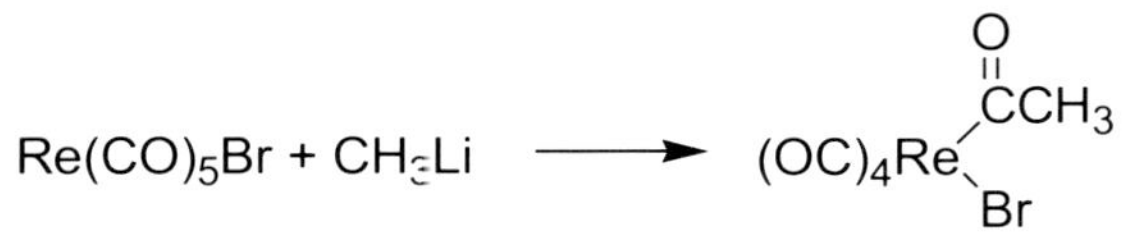

(13.69)

Another technique is hydrometallation by adding an alkene group (e.g., C_2H_4) to a transition metal complex [100,101]. Such reactions are commonly observed in metathesis and polymerization reactions, as shown in the examples in Eq. (13.70) [100] and Eq. (13.71) [101].

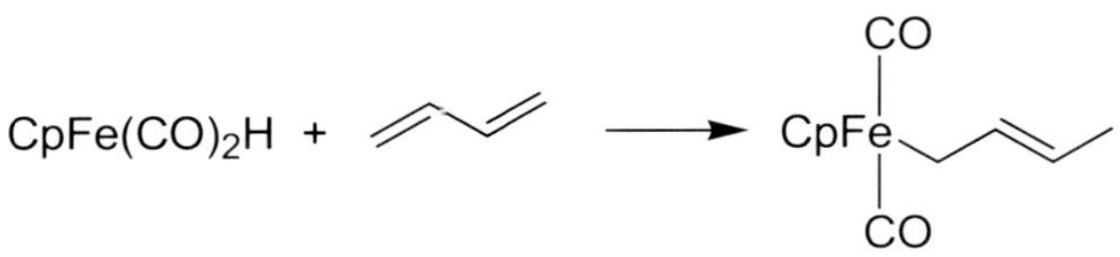

(13.70)

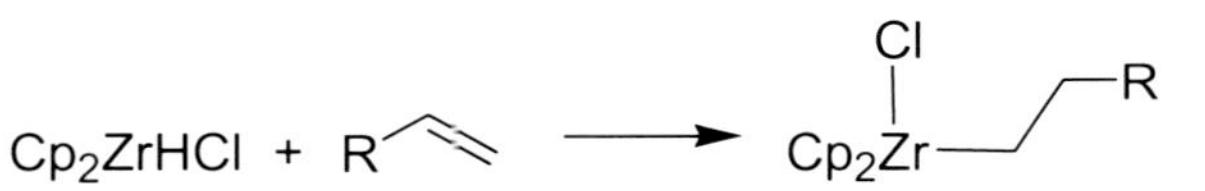

(13.71)

13.2.3. Complexes with M–C Multiple Bonds

Organometallic compounds containing an M=C double bond are called metal carbene complexes. When an M≡C triple bond is present, they are called metal carbyne complexes. Two types of carbene and carbyne complexes are known, and they are called Fischer and Schrock-type carbenes and carbynes. Fischer carbene and carbyne compounds, e.g., $(OC)_5W{=}C(OCH_3)Ph$ and $(OC)_4ClMo{\equiv}CEt$, illustrate low oxidation state metals, and the carbene/carbyne ligands are electrophilic. In contrast, Schrock carbene and carbyne complexes, often known as alkylidene and alkylidyne complexes such as $(Bu^tCH_2)_3Ta{=}CHBu^t$ and $(Me_3SiCH_2)_3W{\equiv}CSiMe_3$, consist of metal centers at high oxidation states, and the carbene/carbyne ligands are nucleophilic.

Carbenes

There are several methods to prepare Fischer and Schrock-type carbenes, more so than carbyne complexes. The work of Chauvin, Grubbs, and Schrock to use carbene complexes in olefin metathesis [102–104] led to their receiving of the 2005 Chemistry Nobel Prize, and has attracted intense interest in metal carbene chemistry. Within the last decade or so, research has been predominant in the preparation of N-heterocyclic carbenes, as demonstrated in the second-generation Grubbs olefin metathesis catalysts discussed below.

Typical preparation of Fischer and Schrock-type carbenes is given in Eq. (13.72) [105] and Eq. (13.73) [95–98], respectively. The Fischer carbenes $(OC)_5M{=}C(OMe)R$ prepared in Eq. (13.72) are often used to make Fischer carbynes, which are discussed below. Another example of the preparation of a Fischer carbene is given earlier in Eq. (13.34).

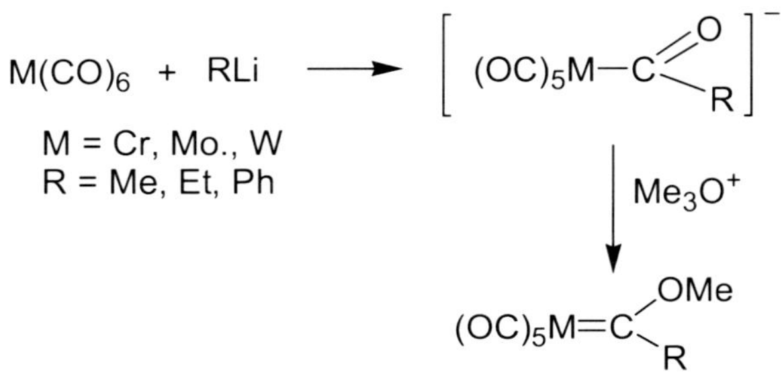

(13.72)

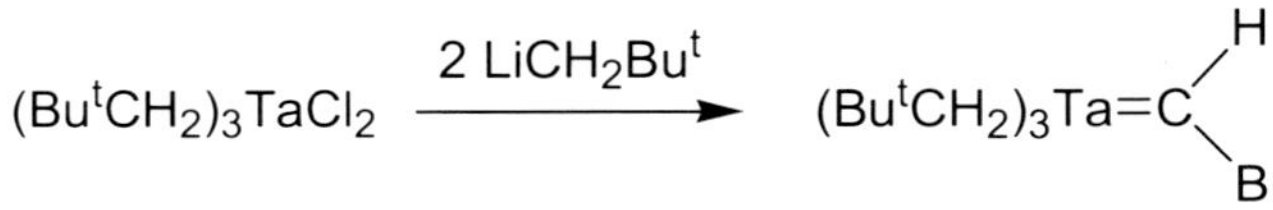

(13.73)

For Fischer carbenes, derivatives of the carbenes may be synthesized by reacting a carbene complex containing an –OR moiety on the carbene C atom with an alcohol, amine, or thiol to give a new carbene, as illustrated in Eqs (13.74) and (13.75) [106–108].

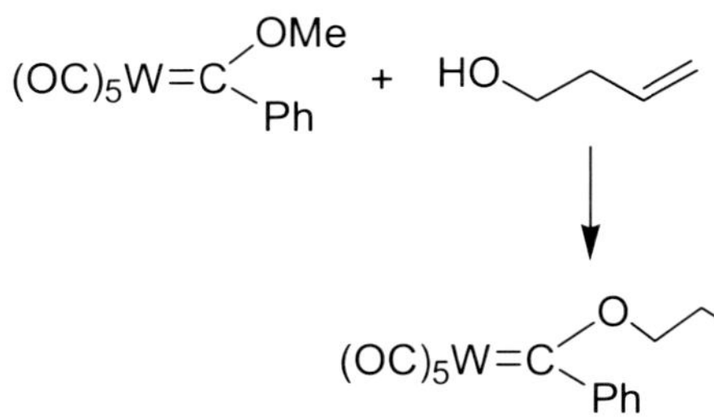

(13.74)

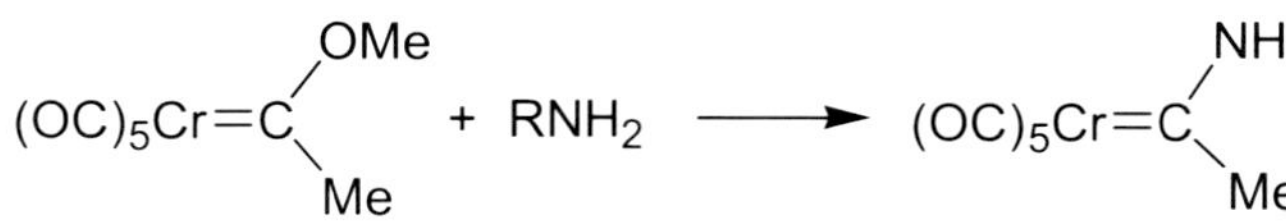

(13.75)

The first Schrock carbene $(Bu^tCH_2)_3Ta{=}CHBu^t$ may be prepared by directly adding neopentylmagnesium chloride in excess to $TaCl_5$ (Eq. (13.76)) [95–98]. Since both $TaCl_5$ and product $MgCl_2$ are insoluble, yield is low. An alternative route with higher yield uses Zn $(CH_2Bu^t)_2$ to alkylate $TaCl_5$ to $(Bu^tCH_2)_3TaCl_2$ (Eq. (13.68)), followed by reaction with neopentyllithium (Eq. (13.73)) [95–98]. The intermediate $(Bu^tCH_2)_3TaCl_2$ is highly soluble and can be isolated by sublimation. Subsequent mechanistic studies by Xue and coworkers reveal that the formation of the archetypical Schrock carbene $(Bu^tCH_2)_3Ta{=}CHBu^t$ involves unstable pentaneopentyltantalum $Ta(CH_2Bu^t)_5$ as an intermediate [97,98]. $NbCl_5$ is similarly treated with $Mg(CH_2Bu^t)_2$(dioxane) to give $(Bu^tCH_2)_3NbCl_2$, followed by addition of $LiCH_2$-Bu^t, to give $(Bu^tCH_2)_3Nb{=}CHBu^t$ (Eq. (13.77)) [96]. Another example is given in Eq. (13.78). Substitution of two chloride ligands in $(PhCH_2)_3TaCl_2$ by bulky Cp^- ligands leads to the α-H abstraction in part to relief steric constraints. α-H abstraction is the prevailing pathway in the formation of the Schrock carbene ligands.

(13.76)

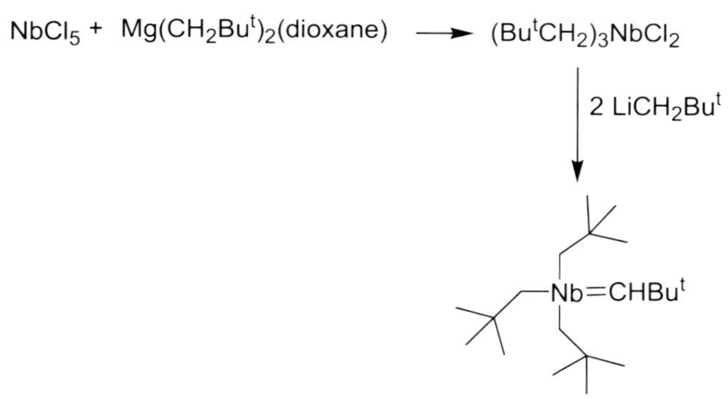

(13.77)

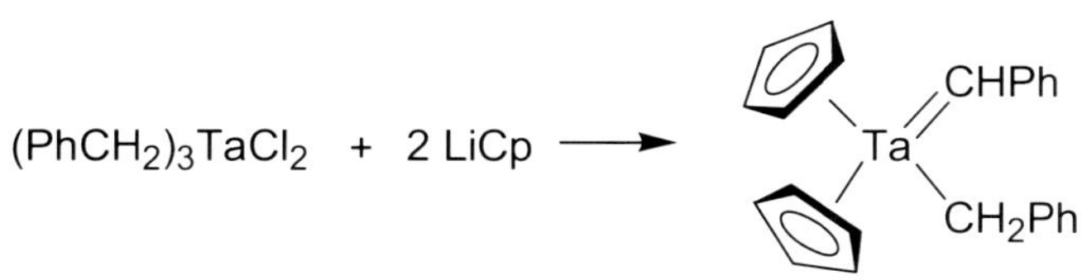

(13.78)

Silyl analogs of the alkyl alkylidenes such as $Ta({=}CHBu^t)(CH_2Bu^t)_2[Si(SiMe_3)_3]$ are prepared from $(Bu^tCH_2)_3TaCl_2$ and lithium silyl compound LiSi-$(SiMe_3)_3$ (Eq. (13.79)) [109]. Reaction of the tungsten species in Eq. (13.80) with 2 equiv. $LiCH_2SiMe_3$ yields an alkylidene complex [110]. α-H abstraction is the prevailing process in the formation of Schrock carbene complexes.

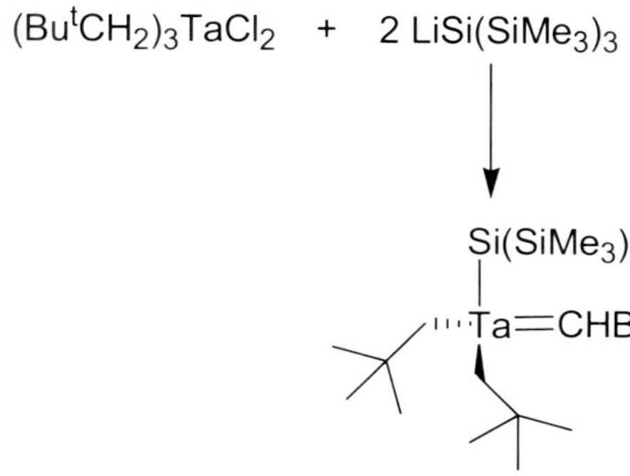

(13.79)

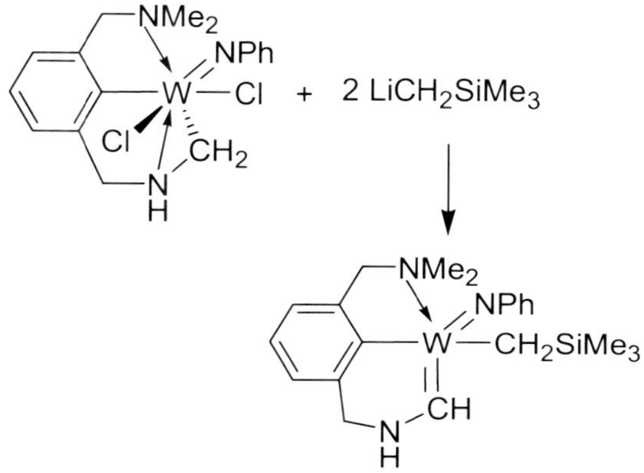

(13.80)

Since the synthesis of N-heterocyclic carbenes in the 1950s, these carbenes and their use as ligands in metal complexes have been an extensive area of research. In the late 1990s, active Ru catalysts containing N-heterocyclic carbene ligands, the so-called second-generation Grubbs catalysts, were developed for olefin metathesis and metathesis-based polymerization such as ring-opening metathesis polymerization (ROMP) and acyclic diene metathesis polymerization (ADMET) [103,111,112]. The work has been among the factors that have driven recent interests in N-heterocyclic carbene ligands.

The Grubbs catalysts are popular as they show remarkable functional group tolerance and are air stable. The preparation of the first-generation Grubbs catalysts is given in Eq. (13.81) [111,112]. These catalysts do *not* contain an N-heterocyclic carbene ligand. $RuCl_3$ is reacted with an excess amount of PPh_3 to give $RuCl_2(PPh_3)_3$, followed by addition of a diazoalkane ($N_2{=}CHR$; R = alkyl group), to give the Ru–carbene complex. PPh_3 may be replaced by PCy_3 (Cy = cyclohexyl) to increase the catalytic activities.

Replacing one of the phosphine ligands in the first-generation Grubbs catalysts by N-heterocyclic carbenes gives the second-generation Grubbs catalysts, as illustrated in Eq. (13.82) [113]. They often show higher activities in the olefin metathesis reactions [113]. It should be noted that the heterocyclic carbene ligand here donates its two electrons to the Ru atom through a dative bond. This is similar to the bonding of phosphine ligands. There is another Ru carbene bond Ru=CHPh that is directly involved in olefin metathesis. Compared to phosphines, N-heterocyclic carbene ligands are stronger σ donors and much less labile. Thus the N-heterocyclic carbenes in the complexes are not able to readily dissociate, and they stabilize the electron-deficient intermediates. In addition, they enhance the dissociation of the labile trans PCy_3 ligand, a step needed to open a site for olefin metathesis. These effects make the second-generation Grubbs catalysts among the most active in olefin metathesis.

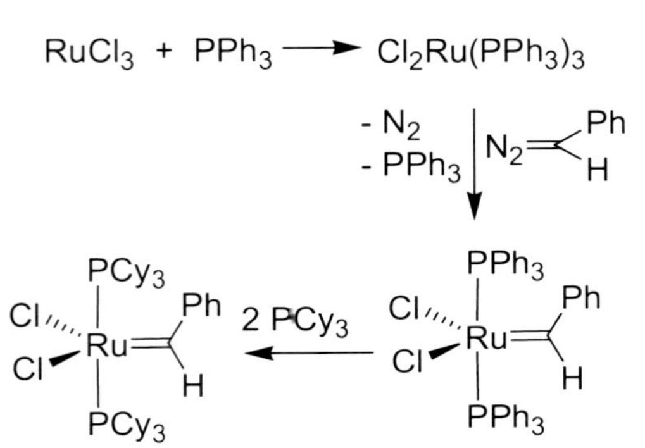

(13.81)

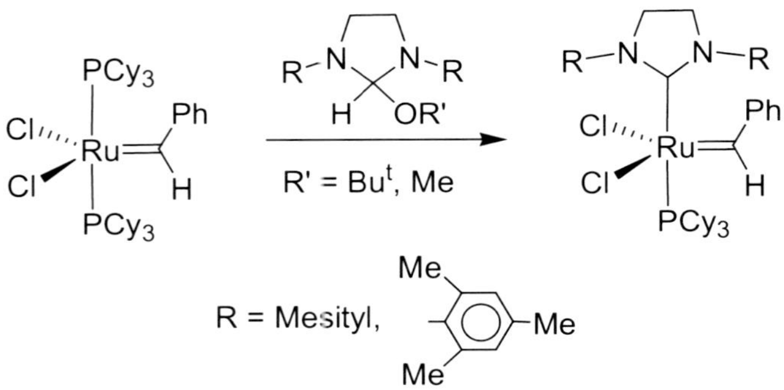

(13.82)

Schrock catalysts for olefin metathesis are usually W or Mo carbene complexes and contain an imido and two alkoxide ancillary ligands. A typical synthesis of Schrock catalysts is given in Eq. (13.83) [104,114,115]. It starts from Schrock-type carbyne $(Bu^tCH_2)_3W{\equiv}CBu^t$ [114], which is discussed below (Eq. (13.86)). The first step is acidification of the carbyne complex [71], which is discussed earlier in Eq. (13.44). The acidification is followed by removal of one Cl ligand by $Me_3SiNHAr$ to introduce an arylamide ligand ArNH− [115]. H migration from the amide to the carbyne ligand, catalyzed by NEt_3, yields each a carbene and an imide ligand. The last step is the substitution of the Cl ligands by alkoxides to give the tetra-coordinated catalysts [115]. Bulky imide ligand and the Bu^t group of the neopentylidene ligand prevent the carbene C atom against bimolecular reactions. Electron-withdrawing alkoxide ligands such as $-OCMe(CF_3)_2$ enhance the catalytic activities. Schrock catalysts are in general more active than Grubbs catalysts, but Grubbs catalysts are more tolerant of functional groups such as −OH and acids.

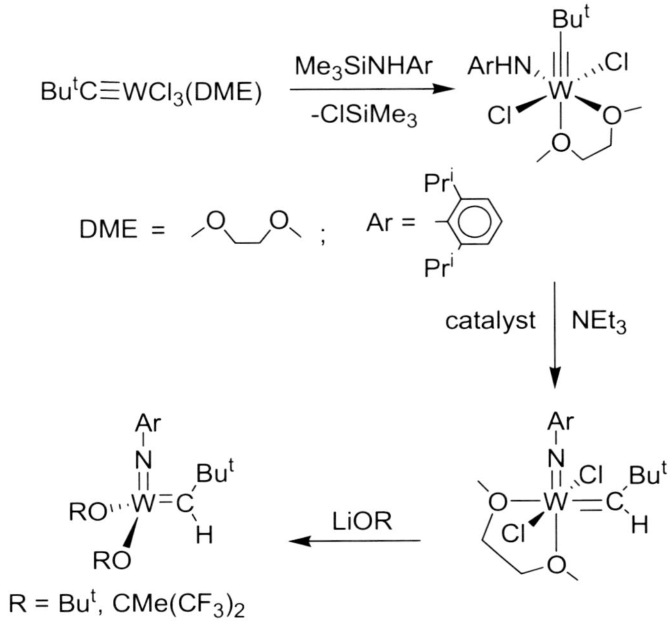

(13.83)

Carbynes

When treating Fischer-type carbene complex $(OC)_5M{=}C(OMe)R$ (M = Cr, Mo, W) with an electrophile, such as BCl_3, a Fischer carbyne is generated. This method is applied to several Fischer carbene complexes, as shown in Eq. (13.84) [116].

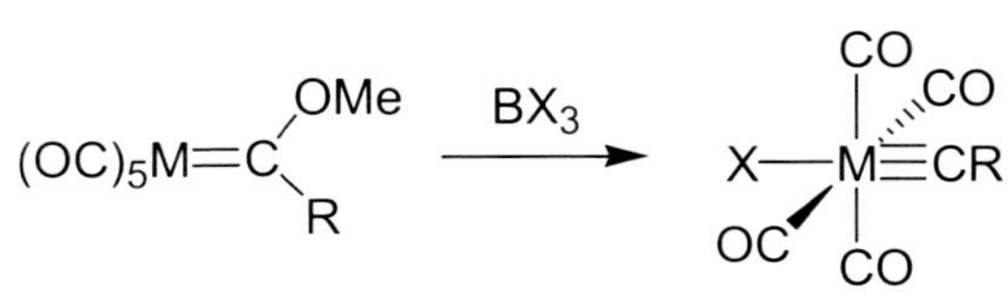

(13.84)

M = Cr, Mo, W; X = Cl, Br, I; R = Me, Et, Ph

Another interesting procedure to prepare Fischer carbynes is given in Eq. (13.85) [117]. $Mo(CO)_6$ is treated in sequence with *n*-butyllithium, oxalyl chloride, and pyridine (py) to form $Cl(py)_2(OC)_2Mo{\equiv}CBu^n$. Substitution of the pyridine and chloride ligands with NaCp gives $Cp(OC)_2Mo{\equiv}CBu^n$. Metals in Fischer-type carbynes are at low oxidation states such as +3 or +4 for Group 6 Cr, Mo, and W.

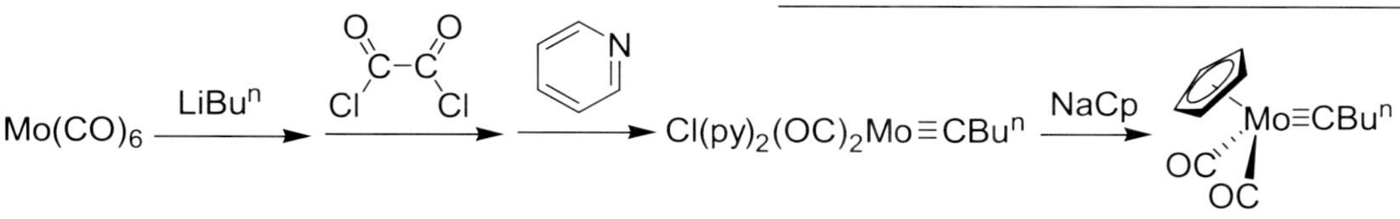

(13.85)

Schrock-type alkylidynes such as $(Bu^tCH_2)_3W{\equiv}CBu^t$ may be prepared by a one-step reaction [114,115]. For instance, when WCl_6 is added to an excess amount of Grignard reagent Bu^tCH_2MgCl (Eq. (13.86)) and Me_3SiCH_2MgCl, $(Bu^tCH_2)_3W{\equiv}CBu^t$ [114] and $(Me_3SiCH_2)_3W{\equiv}CSiMe_3$ [118] are formed, respectively. The yield in this one-step synthesis is low, in part because of the low solubility of WCl_6. Greater yields are obtained by preparing $Cl_3W(OMe)_3$ from WCl_6. It is then treated with either Me_3SiCH_2MgCl or Bu^tCH_2MgCl to give $(Me_3SiCH_2)_3W{\equiv}CSiMe_3$ (Eq. (13.87)) and $(Bu^tCH_2)_3W{\equiv}CBu^t$, respectively [114]. These two complexes are classic Schrock-type carbynes. $(Bu^tCH_2)_3W{\equiv}CBu^t$ is used to yield the Schrock olefin metathesis catalyst in Eq. (13.87).

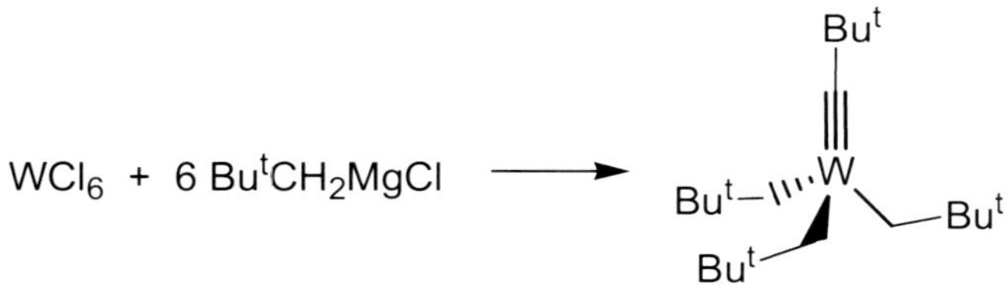

(13.86)

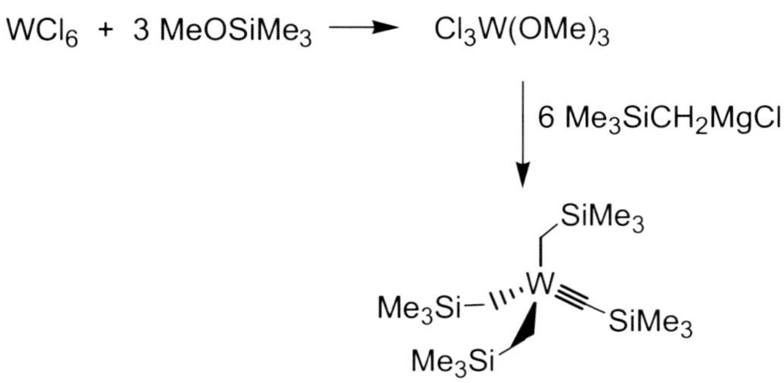

(13.87)

Schrock-type carbyne complexes can be produced by metathesis shown in Eq. (13.88). Reaction of $(Bu^tO)_3W{\equiv}W(OBu^t)_3$ with 2-butyne, via metathesis, gives Schrock-type carbyne $(Bu^tO)_3W{\equiv}CMe$. Subsequent metathesis with $Me_3SiC{\equiv}CSiMe_3$ gives a new carbyne $(Bu^tO)_3W{\equiv}CSiMe_3$ [119]. Substitution of the butoxide ligands by Grignard reagent Bu^tCH_2MgCl gives another W(VI) Schrock-type carbyne $(Bu^tCH_2)_3W{\equiv}CSiMe_3$ [120,121].

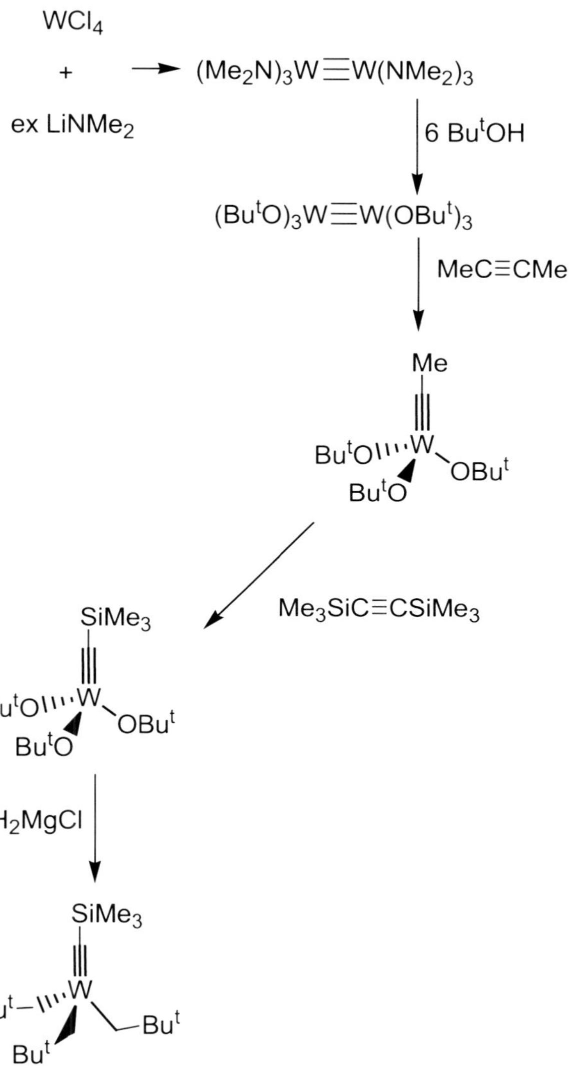

(13.88)

Carbyne ligands are often terminal. Bridging alkylidynes are, however, among the first known carbyne complexes. In 1971, Wilkinson and coworkers published two novel bridging metal carbynes, as shown in Eq. (13.89) [122–124]. By adding Grignard Me_3SiCH_2MgCl to a solution of MCl_5, the dimeric Nb and Ta complexes are formed. Crystals of both species confirm that they are dimers [122–124]. Subsequent mechanistic studies by Li et al. [98] revealed that pentaalkyl $Ta(CH_2SiMe_3)_5$ is a precursor that undergoes α-H abstraction to give carbene $(Me_3SiCH_2)_3Ta{=}CHSiMe_3$. The carbene subsequently undergoes second α-H abstraction, probably via a dimeric intermediate, to give the bridging carbyne.

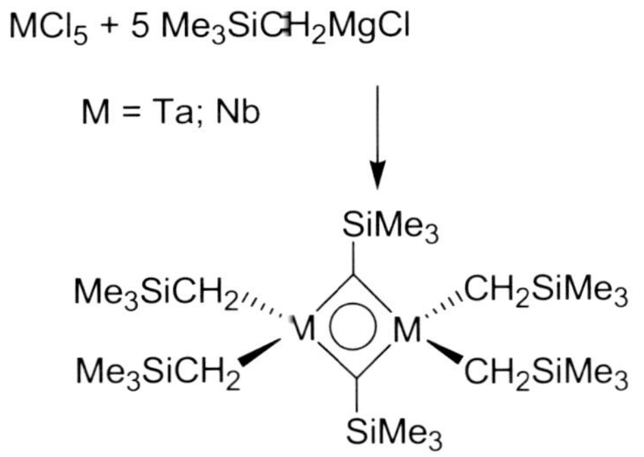

(13.89)

13.2.4. Metal Hydrides

Four popular methods have been used to prepare metal hydrides. One route is by direct addition of hydrogen gas. The second is the protonation of metal complexes by water or a weak acid. The third involves the substitution of ligands such as halides by hydrides in, e.g., $LiAlH_4$ or $NaBH_4$. The fourth method is β-H elimination from an alkyl ligand such as $M{-}CH_2CH_3$.

Direct addition of hydrogen gas to complexes leads to hydrides through oxidative addition [125,126]. The Vaska's complex $IrCl(CO)(PPh_3)_2$ reacts with H_2 to give $H_2IrCl(CO)(PPh_3)_2$ as discussed earlier in Eq. (13.13). Another example is the formation of $HCo(CO)_4$ from the oxidative addition of H_2 to dimeric $Co_2(CO)_8$ (Eq. (13.90)) [125]. These direct additions require low-valent and unsaturated metal complexes. Since a metal halide such as $W^{VI}Cl_6$ cannot be used in the formation of a metal hydride through oxidative addition, a W(0) complex may be used. When $W(CO)_3(PR_3)_2$ ($R = Pr^i$) is reacted with hydrogen gas, the reaction forms a dihydrogen complex shown in Eq. (13.91) [126].

$$Co_2(CO)_8 + H_2 \rightarrow 2HCo(CO)_4 \qquad (13.90)$$

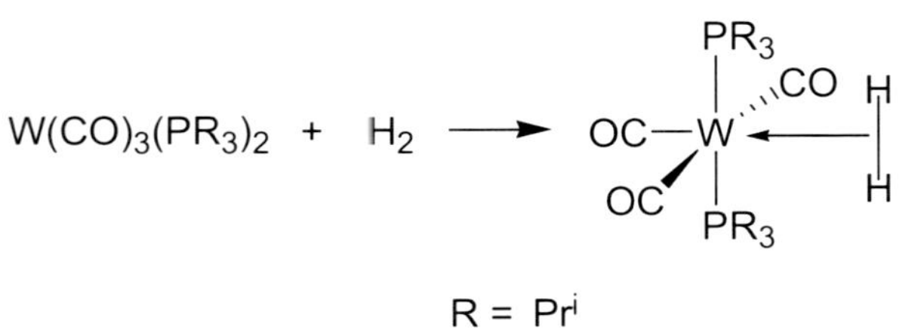

(13.91)

Water and acid can be served as a hydrogen donor/source in the synthesis of metal hydrides through protonation of, e.g., anionic species. Thus adding H^+ to $Na[Co(CO)_4]$ yields $HCo(CO)_4$ (Eq. (13.92)) [127]. As discussed earlier, $Na[Co(CO)_4]$ can be prepared from the reduction of $Co_2(CO)_8$ by sodium amalgam (Eq. (13.56)). The method in Eq. (13.92) is an alternative route to $HCo(CO)_4$ and does not require the use of gaseous H_2 (Eq. (13.90)) [125].

$$Na[Co(CO)_4] \xrightarrow{H^+} HCo(CO)_4 \qquad (13.92)$$

Neutral or basic metal complexes also react with either a weak or strong acid to give hydride complexes, as discussed earlier in "Electrophilic Addition" (Eq. (13.41)). Illustrated in Eq. (13.93), hyperchloric acid reacts with an Os(0) complex, yielding ionic species $[Os(CO)_3(PPh_3)_2H][ClO_4]$ [128]. Likewise when weak acid NH_4PF_6 is added to $CpCo(PMe_3)_2$, only one proton is added to the basic metal forming $[CpCo(PMe_3)_2H]$-$[PF_6]$ (Eq. (13.94)) [129].

$$Os(CO)_3(PPh_3)_2 + HClO_4 \rightarrow [Os(CO)_3(PPh_3)_2H]^+ [ClO_4]^- \qquad (13.93)$$

$$CpCo(PMe_3)_2 + NH_4PF_6 \rightarrow [CpCo(PMe_3)_2H]^+ [PF_6]^- \qquad (13.94)$$

Main-group hydrides (e.g., $LiAlH_4$ or $NaBH_4$) are used as a hydride source in the preparation of metal hydrides through substitution of halide ligands. The main-group hydrides are commercially available and their use is often relatively easy, making them favorable in metal hydride synthesis. WCl_6, a common starting material for W complexes, reacts with $LiEt_3BH$ in the presence of a phosphine or $NaBH_4$ and NaCp, yielding W(IV) hydride complexes $WH_4(PMePh_2)_4$ (Eq. (13.95)) and Cp_2WH_2 (Eq. (13.96)), respectively [130,131]. When $LiAlH_4$ or $NaBH_4$ is reacted with two similar Ir (III) complexes, they lead to two different Ir(III) hydrides as shown in Eqs (13.97) [132] and (13.98) [133]. The $LiAlH_4$ addition results in a typical six-coordinate Ir complex [132]. In contrast, the reaction involving $NaBH_4$ yields a bridging borane, most often observed in borane clusters [133].

$$WCl_6 + PMePh_2 + LiEt_3BH \rightarrow WH_4(PMePh_2)_4 \qquad (13.95)$$

$$WCl_6 + NaCp + NaBH_4 \rightarrow Cp_2WH_2 \qquad (13.96)$$

$$IrHCl_2(PPh_3)_3 + LiAlH_4 \rightarrow IrH_3(PPh_3)_3 \qquad (13.97)$$

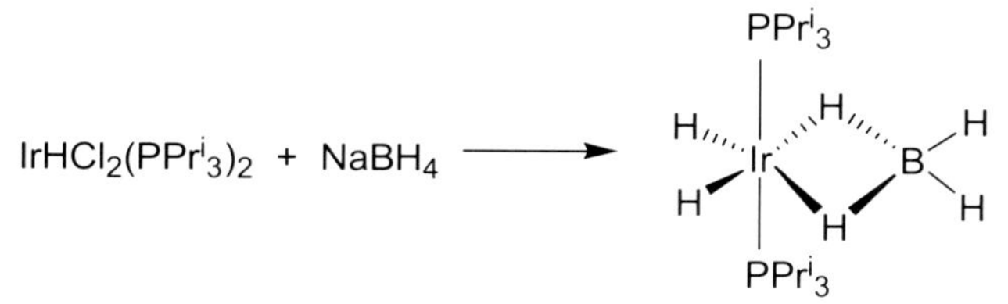

(13.98)

β-H elimination from an alkyl ligand is another method to prepare metal hydride complexes. The Pt(II) ethyl complex in Eq. (13.99) undergoes β-elimination to give metal hydride $PtHCl(PEt_3)_2$ [134]. In the example in Eq. (13.100), alkylation of $RuCl_3$ in the presence of phosphine PPh_3, followed by β-elimination, leads to the formation of a Ru dihydride.

$$PtEtCl(PEt_3)_2 \rightarrow PtHCl(PEt_3)_2 + C_2H_4 \quad (13.99)$$

$$RuCl_3 + PPh_3 + AlEt_3 \rightarrow RuH_2(PPh_3)_4 \quad (13.100)$$

13.2.5. π Complexes

Olefin and Acetylene Complexes

Olefin and acetylene complexes are π complexes involving the donation of the π electrons to the metal and the back donation of electron density on the metal to the π* anti-bonding orbitals in olefin and acetylene ligands. These complexes are often used in processes such as hydrogenation, polymerization, hydrocarbonylation, hydroformylation, and oxidation such as the Wacker process discussed in Section 13.1.4.

Many alkene complexes are prepared by displacement of a carbonyl or halide ligand by an alkene. For example, in Eq. (13.101), one of the four chloride ions in the square-planar $[PtCl_4]^{2-}$ is displaced by ethylene (C_2H_4) [135–137]. In Eq. (13.102), the two COD (cyclooctadiene) ligands in the $Pt(COD)_2$ moiety are displaced by three ethylene groups [138]. When $Fe(CO)_5$ is reacted with 1,3-butadiene, a simple olefin complex is formed (Eq. (13.103)) [139].

$$K_2PtCl_4 + C_2H_4 \xrightarrow[180\ bar]{HCl_{(aq)}} K^+\left[PtCl_3(C_2H_4)\right]^- + KCl \quad (13.101)$$

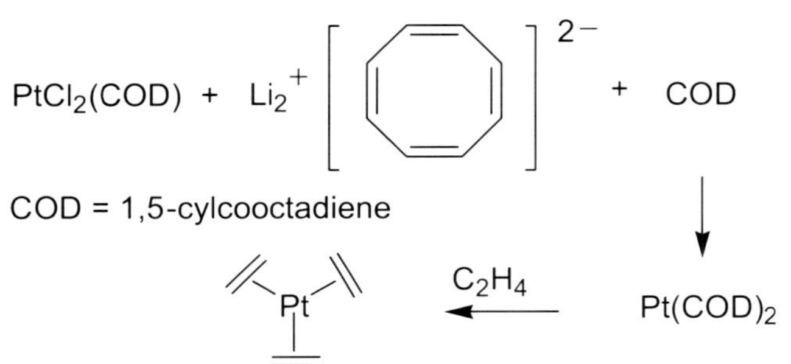

(13.102)

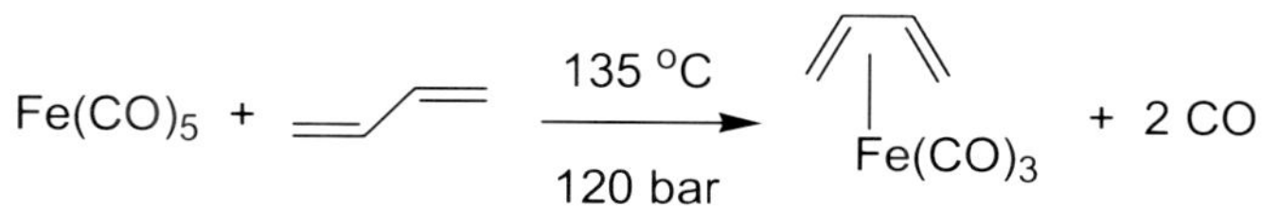

(13.103)

Recently metal fullerene derivatives such as Ir(CO)-Cl(η^2-C_{60})$(PPh_3)_2$ have been prepared by procedures similar to those used in the preparation of olefin complexes. For example, Vaska's complex, *trans*-Ir(CO)-Cl$(PPh_3)_2$, is used to form an ethylene complex in Eq. (13.104) [140]. When Vaska's complex is reacted with fullerene (C_{60}), Ir(CO)Cl(η^2-C_{60})$(PPh_3)_2$ is formed (Eq. (13.105)) [141].

$$Ir(CO)Cl(PPh_3)_2 + H_2C{=}CH_2 \rightarrow Ir(CO)Cl(\eta^2\text{-}C_2H_4)(PPh_3)_2 \quad (13.104)$$

$$Ir(CO)Cl(PPh_3)_2 + C_{60} \rightarrow Ir(CO)Cl(\eta^2\text{-}C_{60})(PPh_3)_2 \quad (13.105)$$

Due to steric effects, smaller and terminal alkene complexes are more stable. However, preparation of large bulky olefin complexes has been achieved by implementing a reducing reagent, liquid–vapor condensation, or hydride abstraction from an alkyl metal complex. When a reducing agent, such as aluminum alkyl, is treated with a metal halide and diene, a bulky metal olefin complex is formed (Eq. (13.106)) [142]. Butadiene undergoes trimerization to give the intermediate. The cyclic triene ligand is then replaced by COD to form $Ni(COD)_2$.

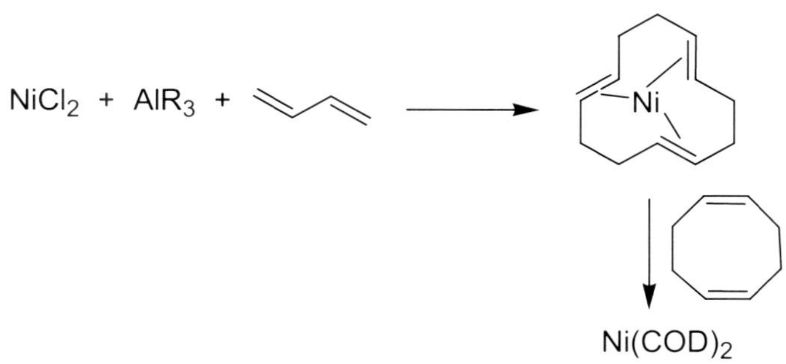

(13.106)

Transition metals readily react with alkynes. Yet, relatively few form simple, stable metal alkyne complexes mainly because many alkyne complexes are reactive with the addition of another alkyne or the formation of more elaborate complexes or organic products. Substitution of CO ligands by alkyne is a reaction used often to make alkyne complexes. An example is given below in Eq. (13.107) [143]. The reactivities of alkyne complexes are illustrated in the trimerization of alkyne to form benzene ligands, as shown in Eq. (13.112) discussed below in the preparation of arene complexes.

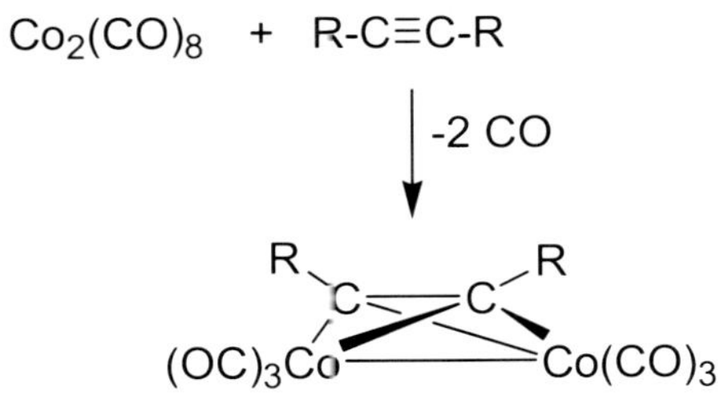

(13.107)

Cyclopentadienyl and Arene Complexes

Metallocenes are well known for their applications in hydrogenation, alkylation, and aminolation. Cp ligands are usually stable and their sterics may be adjusted by placing substituents on the cyclopentadienyl ligands. Cyclopentadienyl complexes are thus often used in asymmetric catalysis and regioselective synthesis. Metallocene derivatives of the Ziegler–Natta catalysts are a prime example of cyclopentadienyl (Cp) chemistry [69].

The general synthesis of the first-row Cp metal complexes is by reacting a metal halide with NaCp or a cyclopentadienyl derivative, shown in Eqs (13.108) and (13.109) [144]. In the case of the reaction with $CrCl_3$, NaCp serves not only as a ligand but also as a reducing agent. Alternative methods include direct use of cyclopentadiene in the presence of a base such as Et_2NH to deprotonate cyclopentadiene (Eq. (13.110)) [145].

$$MCl_2 + 2NaCp \rightarrow M(\eta^5\text{-}Cp)_2 \quad M = \text{Fe or Ni} \qquad (13.108)$$

$$CrCl_3 + NaCp \rightarrow Cr(\eta^5\text{-}Cp)_2 \qquad (13.109)$$

$$MCl_2 + C_5H_6 + Et_2NH \rightarrow M(\eta^5\text{-}Cp)_2 \qquad (13.110)$$

The η^6-arene metal complexes are often prepared by reacting a metal complex or metal halide with an aromatic hydrocarbon (e.g., C_6H_6) through displacement of ligands such as CO (Eq. (13.111)) [146] or cyclotrimerization of acetylenes (Eq. (13.112)) [147]. Microwave irradiation assists the displacement of CO in the reaction in Eq. (13.111) [146]. In the reaction involving $MnCl_2$ (Eq. (13.112)), PhMgBr reduces Mn(II) to Mn(0) [147].

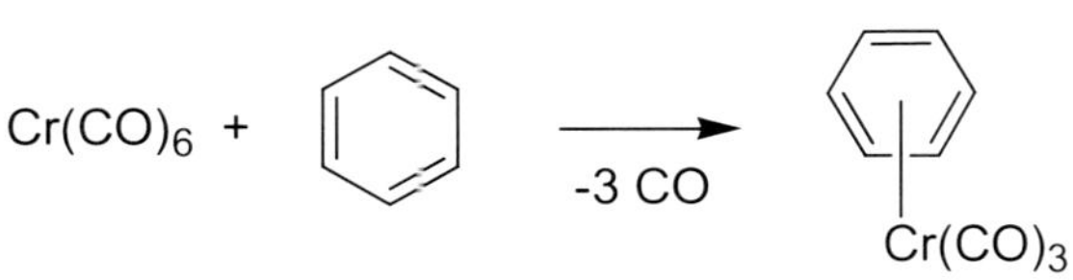

(13.111)

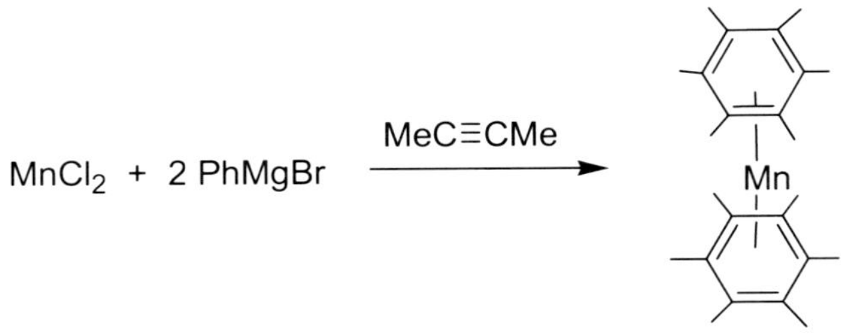

(13.112)

Alternatively, $Cr(\eta^6\text{-}C_6H_6)_2$ can be prepared by the condensation of gaseous chromium metal and arene, as discussed earlier in Eq. (13.47) [71]. Use of a metal gas has also been used in the preparation of other arene complexes [72,73]. It is, however, a less popular technique.

Allylic and Polyene Complexes

Several techniques have been developed to prepare π-allyl complexes. Popular methods include the removal of a halide ligand by allyl anion, substitution of the halide in $CH_2{=}CHCH_2Cl$ by an anionic metal complex, conversion of an alkene ligand to an allyl and a hydride ligand, and reactions involving diene or polyene ligands. Additional synthetic methods are available. Those discussed here are among the most widely used.

In the examples in Eqs (13.113) and (13.114), the metal center undergoes a nucleophilic halide substitution by an allyl Grignard reagent [148,149]. Anionic metal complexes such as $Mn(CO)_5^-$ acts as a nucleophile to replace chloride in $CH_2{=}CHCH_2Cl$ to give an η^1-allyl complex (Eq. (13.115)), which under heating, converts to an η^3-allyl complex with the elimination of CO. This is in essence an electrophilic attack of $CH_2{=}CHCH_2Cl$ on the metal [150].

$$ZrCl_4 + 4C_3H_5MgCl \xrightarrow[-78\ ^\circ C]{} Zr(C_3H_5)_4 \qquad (13.113)$$

$$NiBr_2 + 2C_3H_5MgBr \xrightarrow[-40\ ^\circ C]{} Ni(C_3H_5)_2 \qquad (13.114)$$

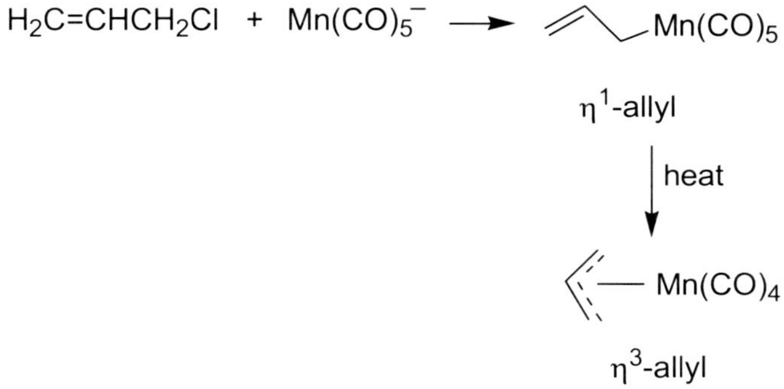

(13.115)

Alkene complexes such as $Mo(dppe)_2(\eta^2\text{-}CH_2{=}CHCH_3)$ (dppe = $Ph_2PCH_2CH_2PPh_2$) undergo C–H oxidative addition to give an allyl hydride complex (Eq. (13.116)) [151]. The conversion is reversible.

(13.116)

Alkene, diene, or polyene metal complexes are also precursors to allyl complexes through addition or removal of a proton or hydride. For example, deprotonation of (cycloheptatriene)$Fe(CO)_3$ gives a deep red nucleophilic anion $(C_7H_7)Fe(CO)_3^-$ containing an antiaromatic cycloheptatrienide anion [152]. Two limiting bonding alternatives have been considered for the anion: one delocalizes the negative charge in the ring as an allylic carbanion and second places the charge density on the metal center (Eq. (13.117)). The insertion of allene into the Pt-H bond in $[PtH(acetone)(PPh_2Me)_2]^+$ gives the allyl ligand in $[Pt(allyl)(PPh_2Me)_2]^+$ (Eq. (13.118)) [153].

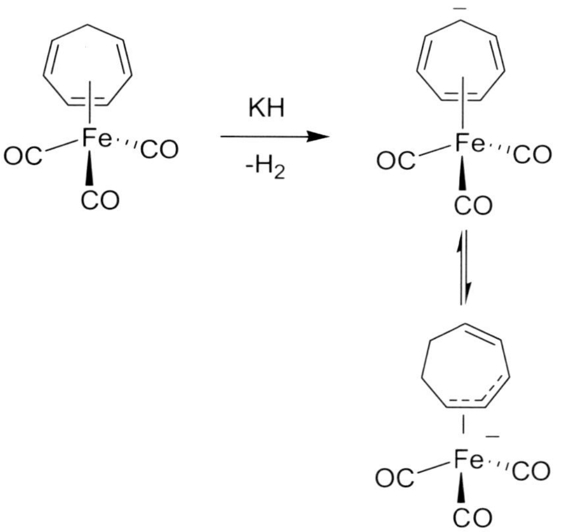

(13.117)

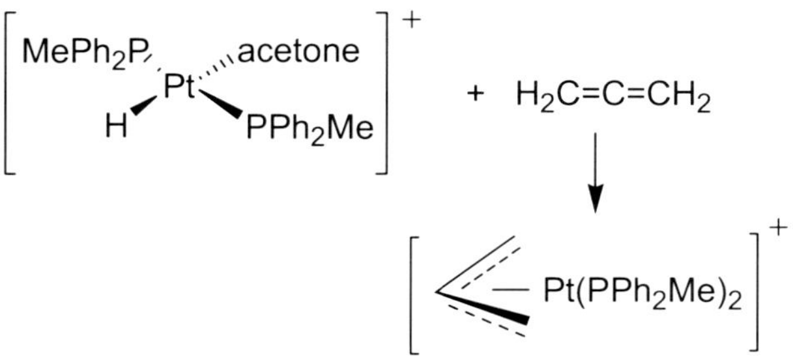

(13.118)

13.3. EXPERIMENTAL TECHNIQUES

One key feature of synthetic organometallic chemistry is the use of inert gas techniques. Many organometallic compounds or their precursors, especially those of earlier transition metals, are air-sensitive, and thus require their synthesis or handling be conducted under an inert gas, typically nitrogen or argon. A variety of air-sensitive techniques such as Schlenk flask, glove box, vacuum manifold/inert gas line, and stopcock have been developed for such use. There are advanced monographs describing the air-sensitive techniques in detail. In particular, *The Manipulation of Air-Sensitive Compounds* by Duward F. Shriver and Mark A. Drezdzon, and *Advanced Practical Inorganic and Metalorganic Chemistry* by R. John Errington are two excellent books on those techniques [154,155]. The readers are encouraged to consult these books for valuable air-sensitive techniques. A summary of key features of air-sensitive techniques is given here. It should be noted that the Schlenk techniques are named after German chemist Wilhelm Johann Schlenk (1879–1943) who developed the glassware that is widely used today in handling air-sensitive techniques. Seyferth recently wrote a review of Schlenk's career and his organoalkali-metal chemistry [156].

Schlenk flasks are flasks attached to a side arm (Fig. 13.1). The sidearm is a stopcock or valve, and a glass or Teflon stopcock is often used. The side arm allows the user to pull vacuum on the flask, followed by refilling it with an inert gas. There is a ground joint on the Schlenk flask, and it is either plugged with a stopper or used to connect the Schlenk flask to another apparatus. The flask is often under a slightly positive pressure maintained by a Schlenk line that is also called inert gas/vacuum double manifold.

The basic design of a Schlenk line (Fig. 13.2) allows the atmosphere inside the Schlenk flask to be readily switched between vacuum and the inert gas by turning a three-way stopcock. One part of the double manifold is under vacuum maintained by a pump. A cold trap is often placed between the pump and the manifold in order to condense volatile species such as solvents from the Schlenk flask and to protect the pump. Another part of the double manifold is under a slightly positive atmosphere of the inert gas. This part is attached to the source of the inert gas such as a gas cylinder. Often a gas purification apparatus is placed between the cylinder and the manifold to remove trace amounts of oxygen and moisture in the inert gas. Usually four or five taps are placed on one manifold so that several Schlenk flasks or other items can be connected to the manifold at the same time.

Solvents used in air-sensitive organometallic reactions need to be dried to remove trace amounts of

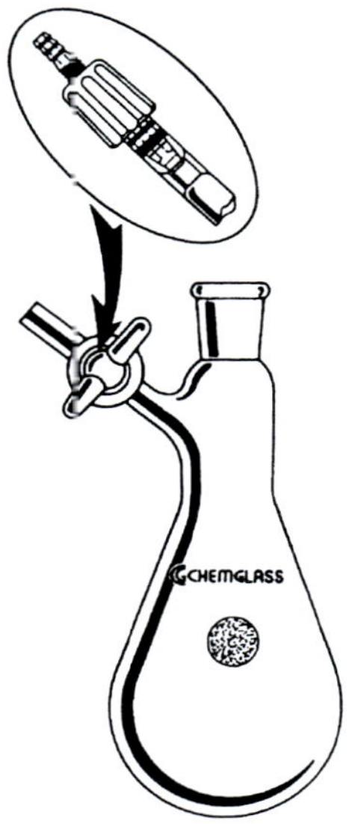

FIGURE 13.1 Scheme of a typical Schlenk flask. The stopcock is made of either a solid glass plug or a high vacuum Teflon® plug (as shown in the insert). *Courtesy of Chemglass Life Sciences (www.chemglass.com). Images reproduced with permission by Chemglass Life Sciences.*

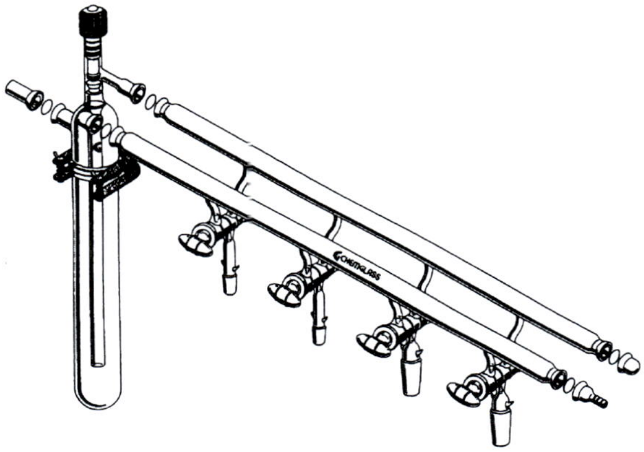

FIGURE 13.2 Scheme of a typical Schlenk line. *Courtesy of Chemglass Life Sciences. Images reproduced with permission by Chemglass Life Sciences.*

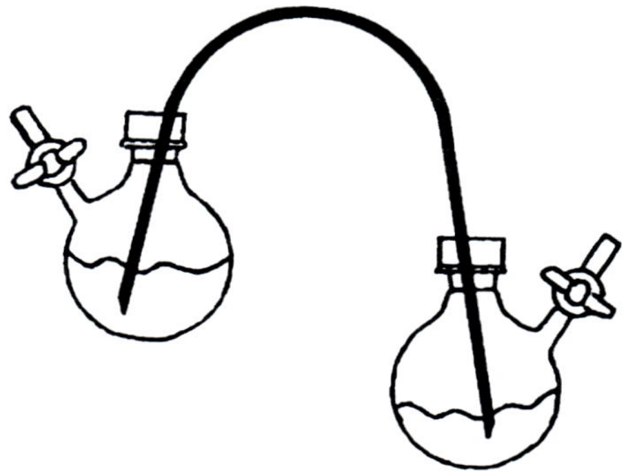

FIGURE 13.3 Transfer of solvent under an inert atmosphere. *Courtesy of Chemglass Life Sciences. Images reproduced with permission by Chemglass Life Sciences.*

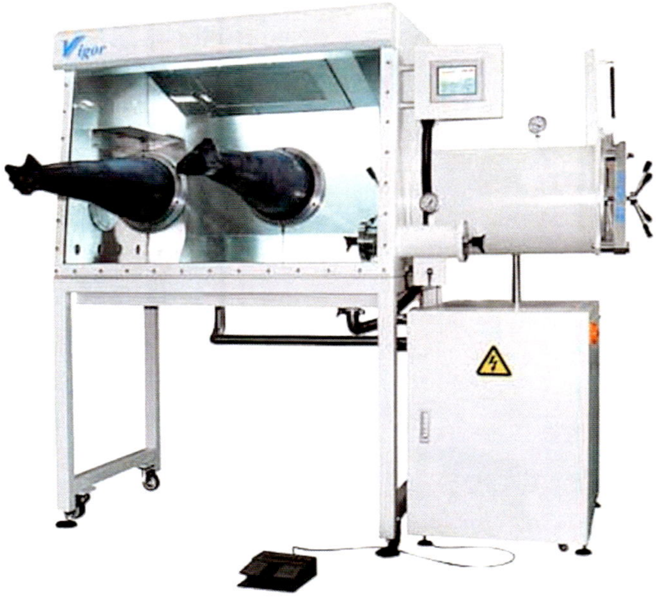

FIGURE 13.4 Photo of a glove box.

oxygen and moisture. New, column purification-based systems that Grubbs and colleagues developed [157] have gradually replaced distillations, often from sodium or potassium, and vacuum transfers. This is a nonhazardous alternative and requires no undue supervision or cooling, providing a rapid collection of large quantities of pure solvents on demand. The purified solvents are stored in Schlenk-type solvent bottles under an inert atmosphere. Such purification-based systems are now commercially available.

The transfer of a purified solvent from its storage bottle to a Schlenk flask is often conducted by the use of a syringe or more commonly a cannula. A cannula is a long Teflon or metal tube. Its two ends are pointed so that the ends can be inserted through rubber septa on the Schlenk flask and the solvent bottle. The pressure in the solvent bottle is slightly higher than that in the receiving Schlenk flask, thus leading to the flow of the solvent (Fig. 13.3).

Often one needs to transfer air-sensitive solids or prepare an NMR or IR sample of the solids. This is usually conducted in a glove box which is kept under an inert gas (nitrogen or argon). A glove box is essentially a gas-tight container from which air (and thus oxygen and moisture) is excluded (Fig. 13.4). The user operates through gloves. To bring a Schlenk flask into the box, it is first pumped in an antechamber attached to the glove box. Following refilling the antechamber with the inert gas, the Schlenk flask is brought into the box by opening a side door between the box and the antechamber. The inert gas inside the box is circulated, passing over an O_2-removing catalyst and drying agents in order to keep oxygen and moisture levels at minimum. It should be pointed out that more and more research groups now conduct reactions inside glove boxes. There is thus no need to use Schlenk glassware (flasks, double manifold, etc.) and such a practice reduces the chances of air-leak into a Schlenk flask. The O_2-removing catalyst and drying agents for the glove box require, however, much more frequent regenerations.

In essence the air-sensitive techniques are to keep the samples, solvents, and operations under an inert atmosphere excluding air. Once one learns these techniques, he/she will benefit greatly and will be able to prepare not only air-sensitive organometallic compounds, but also many other types of compounds such as organic compounds and polymer materials that may require the use of the techniques during their syntheses.

References

[1] R.H. Crabtree, The Organometallic Chemistry of the Transition Metals, fourth ed., Wiley, New York, 2005.
[2] J.F. Hartwig, Organotransition Metal Chemistry. From Bonding to Catalysis, University Science Books, Sausalito, CA, 2010.
[3] F.A. Cotton, G. Wilkinson, C.A. Murillo, M. Bochman, Advanced Inorganic Chemistry Wiley, New York, 1999, pp. 1165.
[4] J.P. Collman, L.S. Hegedus, J.R. Norton, R.G. Finke, Principles and Applications of Organotransition Metal Chemistry, University Science Books, Mill Valley, CA, 1987.
[5] C. Elschenbroich, Organometallics, third ed., Wiley-VCH, 2006.
[6] G.O. Spessard, G.L. Miessler, Organometallic Chemistry, Prentice-Hall, Upper Saddle River, N.J, 1997.
[7] A.W. Parkins, R.C. Poller, An Introduction to Organometallic Chemistry, Macmillan, London, 1986.
[8] R.C. Mehrotra, A. Singh, Organometallic Chemistry, Wiley Eastern, New Delhi, 1991.
[9] C.E. Housecroft, A.G. Sharpe, Inorganic Chemistry, third ed, Pearson Education Ltd, Essex, England, 2008.
[10] J.E. Huheey, E.A. Keiter, R.L. Keiter, Inorganic Chemistry. Principles of Structure and Reactivity, fourth ed., HarperCollins College Publishers, New York, 1993.
[11] A. Yamamoto, Organotransition Metal Chemistry. Fundamental Concepts and Applications, Wiley, New York, 1986.
[12] Comprehensive Organometallic Chemistry III, Elsevier, Amsterdam, 2007.
[13] Comprehensive Organometallic Chemistry II, Pergamon, New York, 1995.
[14] Comprehensive Organometallic Chemistry, Pergamon Press, New York, 1982.
[15] Advances in Organometallic Chemistry, Academic Press.
[16] Topics in Organometallic Chemistry, Springer, Berlin.
[17] A.C. Jones, M.L.E. Hitchman, Chemical Vapour Deposition. Precursors, Processes and Applications, Royal Society of Chemistry, London, 2009.
[18] Precursor chemistry of advanced materials, in: R.A. Fischer (Ed.), Topics in Organometallic Chemistry, CVD, ALD and nanoparticles, vol. 9, Springer, Berlin, 2005.
[19] R. Banerjee, S.W. Ragsdale, Annual Review of Biochemistry 72 (2003) 209.
[20] P.J. Davidson, M.F. Lappert, R. Pearce, Journal of Organometallic Chemistry 57 (1973) 269.
[21] R.E. Nitzschmann, Chemische Berichte 97 (1964) 2098.
[22] S.E. Komiya, Synthesis of Organometallic Compounds: a Practical Guide, Wiley, New York, 1997.
[23] Experimental organometallic chemistry, A practicum in synthesis and characterization, in: A.L. Wayda, M.Y. Darensbourg (Eds.), ACS Symposium Series, 1987. vol. 357.
[24] A.D. Garnovskii, B.I.E. Kharisov, Synthetic Coordination and Organometallic Chemistry, Marcel Dekker, New York, 2003.
[25] W.A. Herrmann (Ed.), Synthetic Methods of Organometallic and Inorganic Chemistry, Thieme Verlag, New York, 1996–2002.
[26] R.B. King, vol. 1, Organometallic Syntheses, Academic Press, New York, 1965.
[27] J.J. Eisch, vol. 2, Nontransition-Metal Compounds, Academic Press, New York, 1981.
[28] R.B. King, J.J. Eisch, vol. 3, Organometallic Syntheses, Elsevier, Amsterdam, 1986.
[29] R.B. King, J.J. Eisch, vol. 4, Organometallic Syntheses, Elsevier, Amsterdam, 1988.
[30] Inorganic Syntheses, Wiley, New York,
[31] J.E. Huheey, E.A. Keiter, R.L. Keiter, Inorganic Chemistry fourth ed., HarperCollins College Publishers, New York, 1993, p. 543.
[32] H. Wawersik, F. Basolo, Journal of the American Chemical Society 89 (1967) 4626.
[33] M.R. Plutino, L.M. Scolaro, R. Romeo, A. Grassi, Inorganic Chemistry 39 (2000) 2712.
[34] D.J. Darensbourg, Advances in Organometallic Chemistry 21 (1982) 113.
[35] L. Vaska, Accounts of Chemical Research 1 (1968) 335.
[36] W.H. Thompson, C.T. Sears, Inorganic Chemistry 16 (1977) 769.
[37] C.M. Thomas, G. Suss-Fink, Coordination Chemistry Reviews 243 (2003) 125.
[38] J.A. Labinger, J.A. Osborn, Inorganic Chemistry 19 (1980) 3230.
[39] M.F. Lappert, P.W. Lednor, Advances in Organometallic Chemistry 14 (1976) 345.
[40] T.L. Hall, M.F. Lappert, P.W. Lednor, Journal of the Chemical Society-Dalton Transactions (1980) 1448.
[41] R.H. Crabtree, J.M. Quirk, T. Fillebeenkhan, G.E. Morris, Journal of Organometallic Chemistry 181 (1979) 203.
[42] J. Schwartz, D.W. Hart, J.L. Holden, Journal of the American Chemical Society 94 (1972) 9269.
[43] F. Calderazzo, Angewandte Chemie International Edition in English 16 (1977) 299.
[44] L. Gonsalvi, H. Adams, G.J. Sunley, E. Ditzel, A. Haynes, Journal of the American Chemical Society 121 (1999) 11233.
[45] M.J. Wax, R.G. Bergman, Journal of the American Chemical Society 103 (1981) 7028.
[46] R.H. Magnuson, S. Zulu, W.M. Tsai, W.P. Giering, Journal of the American Chemical Society 102 (1980) 6887.
[47] G. Erker, F. Rosenfeldt, Angewandte Chemie International Edition in English 17 (1978) 605.
[48] R.H. Crabtree, Accounts of Chemical Research 12 (1979) 331.
[49] G. Yagupsky, C.K. Brown, G. Wilkinso., Journal of the Chemical Society A –Inorganic Physical Theoretical (1970) 1392.
[50] J. Yu, J.B. Spencer, Journal of the American Chemical Society 119 (1997) 5257.
[51] M. Brookhart, M.L.H. Green, G. Parkin, Proceedings of the National Academy of Sciences of the United States of America 104 (2007) 6908.

[52] M. Brookhart, A.F. Volpe, D.M. Lincoln, I.T. Horvath, J.M. Millar, Journal of the American Chemical Society 112 (1990) 5634.
[53] D.P. Gates, P.S. White, M. Brookhart, Chemical Communications (2000) 47.
[54] E.O. Fischer, Advances in Organometallic Chemistry 14 (1976) 1.
[55] D.J. Darensbourg, .A. Froelich, Journal of the American Chemical Society 99 (1977) 4726.
[56] B. Crociani, P. Uguagliati, U. Belluco, Journal of Organometallic Chemistry 117 (1976) 189.
[57] S.G. Davies, M.L.H Green, D.M.P. Mingos, Tetrahedron 34 (1978) 3047.
[58] J. Smidt, W. Hafner, R. Jira, J. Sedlmeier, R. Sieber, R. Ruttinger, et al., Angewandte Chemie. International Edition 71 (1959) 176.
[59] J.M. Takacs, X.T. Jiang, Current Organic Chemistry 7 (2003) 369.
[60] N. Aktogu, D. Baudry, D. Cox, M. Ephritikhine, H. Felkin, R. Holmessmith, J. Zakrzewski, Bulletin de la Societe Chimique de France (1985) 381.
[61] K.S.Y. Lau, P.K. Wong, J.K. Stille, Journal of the American Chemical Society 98 (1976) 5832.
[62] R.A. Henderson, Angewandte Chemie International Edition in English 35 (1996) 947.
[63] C.J. Curtis, A. Miedaner, W.W. Ellis, D.L. DuBois, Journal of the American Chemical Society 124 (2002) 1918.
[64] S.J. Holmes, D.N. Clark, H.W. Turner, R.R. Schrock, Journal of the American Chemical Society 104 (1982) 6322.
[65] G.K. Turner, W. Kaui, M. Scotti, H. Werner, Journal of Organometallic Chemistry 102 (1975) C9.
[66] R.R. Schrock, D.N Clark, J. Sancho, J.H. Wengrovius, S.M. Rocklage, S.F. Pedersen, Organometallics 1 (1982) 1645.
[67] J.A. Labinger, D.W. Hart, W.E. Seibert, J. Schwartz, Journal of the American Chemical Society 97 (1975) 3851.
[68] X. Yang, C.L. Stern, T.J. Marks, Journal of the American Chemical Society 113 (1991) 3623.
[69] Frontiers in metal-catalyzed polymerization (special issue), Chemical Reviews 100 (2000) 1167–1682.
[70] D. Dong, D.A. Slack, M.C. Baird, Inorganic Chemistry 18 (1979) 188.
[71] P.L. Timms, Journal of Chemical Education 49 (1972) 782.
[72] K.J. Klabunde, Chemistry of Free Atoms and Particles, Academic Press, New York, 1980.
[73] B.I. Kharisov, A.D. Garnovskii, L.M. Blanco, A.S. Burlov, A. Garcia-Luna, Journal of Coordination Chemistry 49 (1999) 113.
[74] J.H.B. Chenier, C.A. Hampson, J.A. Howard, B. Mile, Journal of Physical Chemistry 93 (1989) 114.
[75] C.E. Housecroft, A.C. Sharpe, Inorganic Chemistry, third ed. Pearson Education Ltd, Essex, England, 2008. 816.
[76] R.B. King, vol. 1, Organometallic Syntheses, Academic Press, New York, 1965, p. 82.
[77] S. Top, P. Morel, H Pankowski, G. Jaouen, Journal of the Chemical Society, Dalton Transactions (1996) 3611.
[78] B.F.G. Johnson, J. Lewis, Inorganic Syntheses, vol. 13, Wiley, New York, 1971, p. 92.
[79] R.B. King, vol. 1, Organometallic Syntheses, Academic Press, New York, 1965, p. 96.
[80] T. Mitsudo, Y. Watanabe, M. Yamashit, Y. Takegami, Chemistry Letters (1974) 1385.
[81] I. Wender, H.W. Sternberg, M. Orchin, Journal of the American Chemical Society 75 (1953) 3041.
[82] M. Roper, M. Schieren, B.T. Heaton, Journal of Organometallic Chemistry 299 (1986) 131.
[83] G. Longoni, P. Chini, L.D. Lower, L.F. Dahl, Journal of the American Chemical Society 97 (1975) 5034.
[84] U. Zucchini, E. Albizzat, U. Giannini, Journal of Organometallic Chemistry 26 (1971) 357.
[85] A.J. Shortlan, G. Wilkinso, Journal of the Chemical Society, Dalton Transactions (1973) 872.
[86] V. Pfennig, K. Seppelt, Science 271 (1996) 626.
[87] K. Clauss, C. Beermann, US Patent 2,951,085 (1958).
[88] H.J. Berthold, G. Groh, Zeitschrift fur Anorganische und Allgemeine Chemie 319 (1963) 230.
[89] T.S. Piper, G. Wilkinson, Journal of Inorganic & Nuclear Chemistry 3 (1956) 104.
[90] J.A.M. Andersen, J.R. Moss, Organometallics 13 (1994) 5013.
[91] J.R. Green, e-EROS Encyclopedia of Reagents for Organic Synthesis, Wiley, 2001.http://www.mrw.interscience.wiley.com/eros/articles/rs060/frame.html, 2001.
[92] A.L. Galyer, G. Wilkinson, Journal of the Chemical Society, Dalton Transactions (1976) 2235.
[93] M.T. Reetz, B. Raguse, T. Seitz, Tetrahedron 49 (1993) 8561.
[94] G.W.A. Fowles, D.A. Rice, J.D. Wilkins, Journal of the Chemical Society-Dalton Transactions (1973) 961.
[95] R.R. Schrock, Journal of the American Chemical Society 96 (1974) 6796.
[96] R.R. Schrock, J.D. Fellmann, Journal of the American Chemical Society 100 (1978) 3359.
[97] J.K.C. Abbott, L.T. Li, Z.L. Xue, Journal of the American Chemical Society 131 (2009) 8246.
[98] L.T. Li, M.L. Hung, Z.L. Xue, Journal of the American Chemical Society 117 (1995) 12746.
[99] K.P. Darst, C.M. Lukehart, Journal of Organometallic Chemistry 171 (1979) 65.
[100] T.A. Shackleton, M.C. Baird, Organometallics 8 (1989) 2225.
[101] D.W. Hart, J. Schwartz, Journal of the American Chemical Society 96 (1974) 8115.
[102] Y. Chauvin, Angewandte Chemie International Edition 45 (2006) 3740.
[103] R.H. Grubbs, Angewandte Chemie International Edition 45 (2006) 3760.
[104] R.R. Schrock, Angewandte Chemie International Edition 45 (2006) 3748.
[105] H. Fischer, In Transition Metal Carbene Complexes, Verlag Chemie, Deerfield Beach, FL., 1983, p. 1.
[106] C.P. Casey, A.J. Shusterman, Journal of Molecular Catalysis 8 (1980) 1.
[107] J.A. Connor, E.O. Fischer, Journal of the Chemical Society A – Inorganic Physical Theoretical (1969) 578.
[108] P.E. Baikie, E.O. Fischer, O.S. Mills, Chemical Communications (1967) 1199.
[109] L.T. Li, J.B. Diminnie, X.Z. Liu, J.L. Pollitte, Z.L. Xue, Organometallics 15 (1996) 3520.
[110] J.A.M. Brandts, E. Kruiswijk, J. Boersma, A.L. Spek, G. van Koten, Journal of Organometallic Chemistry 585 (1999) 93.
[111] P. Schwab, R.H. Grubbs, J.W. Ziller, Journal of the American Chemical Society 118 (1996) 100.
[112] P. Schwab, M.B. France, J.W. Ziller, R.H. Grubbs, Angewandte Chemie International Edition in English 34 (1995) 2039.
[113] M. Scholl, S. Ding, C.W. Lee, R.H. Grubbs, Organic Letters 1 (1999) 953.
[114] R.R. Schrock, J. Sancho, S.F. Pederson, Inorganic Syntheses, vol. 26, Wiley, New York, 1989, p. 45.
[115] C.J. Schaverien, J.C. Dewan, R.R. Schrock, Journal of the American Chemical Society 108 (1986) 2771.
[116] E.O. Fischer, G. Kreis, C.G. Kreiter, J. Muller, G. Huttner, H. Lorenz, Angewandte Chemie International Edition in English 12 (1973) 564.
[117] K.E. Torraca, I. Ghiviriga, L. McElwee-White, Organometallics 18 (1999) 2262.

[118] R.A. Andersen, M.H. Chisholm, J.F. Gibson, W.W. Reichert, I.P. Rothwell, G. Wilkinson, Inorganic Chemistry 20 (1981) 3934.
[119] M.L. Listemann, R.R. Schrock, Organometallics 4 (1985) 74.
[120] M.H. Chisholm, J.C. Huffman, J.A. Klang, Polyhedron 9 (1990) 1271.
[121] K.G. Caulton, M.H. Chisholm, W.E. Streib, Z.L. Xue, Journal of the American Chemical Society 113 (1991) 6082.
[122] F. Huq, W. Mowat, A.C. Skapski, G. Wilkinson, Journal of the Chemical Society D – Chemical Communications (1971) 1477.
[123] G. Yagupsky, W. Mowat, A. Shortland, G. Wilkinson, Journal of the Chemical Society D – Chemical Communications (1970) 1369.
[124] F. Huq, W. Mowat, A. Shortlan., A.C. Skapski, G. Wilkinso, Journal of the Chemical Society D – Chemical Communications (1971) 1079.
[125] F. Ungvary, Journal of Organometallic Chemistry 36 (1972) 363.
[126] G.J. Kubas, R.R. Ryan, B.I. Swanson, P.J. Vergamini, H.J. Wasserman, Journal of the American Chemical Society 106 (1984) 451.
[127] H.W. Sternberg, I. Wender, M. Orchin, Inorganic Syntheses, vol. 5, Wiley, New York, 1957, p. 192.
[128] K.R. Laing, W.R. Roper, Journal of the Chemical Society A – Inorganic Physical Theoretical (1969) 1889.
[129] H. Werner, W. Hofmann, Chemische Berichte–Recueil 110 (1977) 3481.
[130] G.G. Hlatky, R.H. Crabtree, Coordination Chemistry Reviews 65 (1985) 1.
[131] M.L.H. Green, P.J. Knowles, Journal of the Chemical Society, Perkin Transactions 1 (1973) 989.
[132] R.G. Hayter, Journal of the American Chemical Society 83 (1961) 1259.
[133] H. Werner, M. Schulz, M.A. Esteruelas, L.A. Oro, Journal of Organometallic Chemistry 445 (1993) 261.
[134] R. Romeo, G. Alibrandi, L.M. Scolaro, Inorganic Chemistry 32 (1993) 4688.
[135] W.C. Zeise, Poggendorff's Annalen der Physik und Chemie 21 (1831) 497.
[136] D. Seyferth, Organometallics 20 (2001) 2.
[137] R.A. Love, T.F. Koetzle, G.J.B. Williams, L.C. Andrews, R. Bau, Inorganic Chemistry 14 (1975) 2653.
[138] J.L. Spencer, Inorganic Syntheses, vol. 19, Wiley, New York, 1979, p. 213.
[139] O.S. Mills, G. Robinson, Proceedings of the Chemical Society of London (1960) 421.
[140] L. Vaska, R.E. Rhodes, Journal of the American Chemical Society 87 (1965) 4970.
[141] A.L. Balch, V.J. Catalano, J.W. Lee, Inorganic Chemistry 30 (1991) 3980.
[142] G. Wilke, Angewandte Chemie International Edition 2 (1963) 105.
[143] R.S. Dickson, P.J. Fraser, Advances in Organometallic Chemistry, vol. 12, Academic Press, 1974, p. 323.
[144] R.B. King, Organometallic Syntheses, vol. 1, Academic Press, New York, 1965, p. 64.
[145] J.M. Birmingham, D. Seyferth, G. Wilkinson, Journal of the American Chemical Society 76 (1954) 4179.
[146] Y.T. Lee, S.Y. Choi, S.I. Lee, Y.K. Chung, T.J. Kang, Tetrahedron Letters 47 (2006) 6569.
[147] M. Tsutsui, H. Zeiss, Journal of the American Chemical Society 83 (1961) 825.
[148] J.K. Becconsa, B.E. Job, S. Obrien, Journal of the Chemical Society A – Inorganic, Physical, Theoretical (1967) 423.
[149] J. Terao, Y. Naitoh, H. Kuniyasu, N. Kambe, Chemical Communications (2007) 825.
[150] W. McClellan, E.L. Muetterties, B.W. Howk, H.H. Hoehn, H.N. Cripps, Journal of the American Chemical Society 83 (1961) 1601.
[151] J.W. Byrne, H.U. Blaser, J.A. Osborn, Journal of the American Chemical Society 97 (1975) 3871.
[152] G.M. Williams, D.E. Rudisill, B.A. Barnum, K. Hardcastle, R.H. Heyn, C.J. Kozak, J.W. McMillan, Journal of the American Chemical Society 112 (1990) 205.
[153] H.C. Clark, H. Kurosawa, Inorganic Chemistry 11 (1972) 1275.
[154] D.F. Shriver, M.A. Drezdzon, The Manipulation of Air-Sensitive Compounds, second ed. Wiley, New York, 1986.
[155] R.J. Errington, Advanced Practical Inorganic and Metalorganic Chemistry, Blackie Academic & Professional, London, 1997.
[156] D. Seyferth, Organometallics 28 (2009) 2.
[157] A.B. Pangborn, M.A. Giardello, R.H. Grubbs, R.K. Rosen, F.J. Timmers, Organometallics 15 (1996) 1518.

CHAPTER

14

Synthesis and Assembly Chemistry of Inorganic Polymers

Xiaozhen Tang, Xiaobin Huang

Shanghai Jiao Tong University, China

One of the indicators of the progress in human civilization is the type of materials that are accessible to society. In spite of the wide-ranging applications and ubiquitous presence of organic polymer materials, it does not have the capability to fulfill all the demands [1]. Most of the organic polymers are not suitable for applications at extreme temperatures. They may become very brittle at low temperatures while oxidative at high temperatures. Organic polymers are normally easy to burn because of their flammability. Their main limitation is that the basic petroleum feed stocks are not going to last for ever [2]. However, organic polymers are so diverse in terms of their structure and property, that it's possible to find new types of polymeric systems with completely different properties by combining with inorganic compounds. It would greatly motivate scientists to synthesize and assemble a series of polymer systems that are based on inorganic polymers, oligomers, or elements [3–7]. Organic polymers could be prepared in a number of ways taking advantage of the rich functional group chemistry of organic molecules. Unfortunately almost all of the inorganic polymers require more difficult and different synthetic strategies [2]. Among inorganic polymers, the polymers based on silicon element are well known for their superior properties, rigorous monomers, and hard preparation. In this chapter, we will deal with the synthesis and assembly chemistry of the novel inorganic polymers known as polyphosphazenes in continuation of our work in recent times. They can form the hybrid systems of organic and inorganic polymers including blends and copolymers with stepped-up excellent properties. Their synthesis, assembly, functionalization, and possible utilization will be involved. The most recently development of the silicones, especially on the polysilaethers bearing Si−H bond will also be introduction.

14.1. POLYPHOSPHAZENES

Phosphazenes are the intriguing substrates having chemical structures based on −P=N− repetitive units from low-molecular weight cyclic derivatives to more than 10,000 molecular weight polymers (Fig. 14.1). More interesting thing is that the two substituent groups which are chosen among a great number of different candidates are attached to the phosphorus atoms, thereby producing a series of new materials that are able to cover an unbelievably large number of practical applications [8–9].

The first polyphosphazene compound, polydichlorophosphazene, was prepared in crosslinked form by Stokes at the end of the nineteenth century by the thermal ring-opening polymerization (ROP) of the cyclic trimer $[PNCl_2]_3$ [10]. The material referred to as "inorganic rubber" and remained a chemical curiosity due to its intractability and hydrolytic instability until the mid-1960s when it was shown by Allcock and Kugel

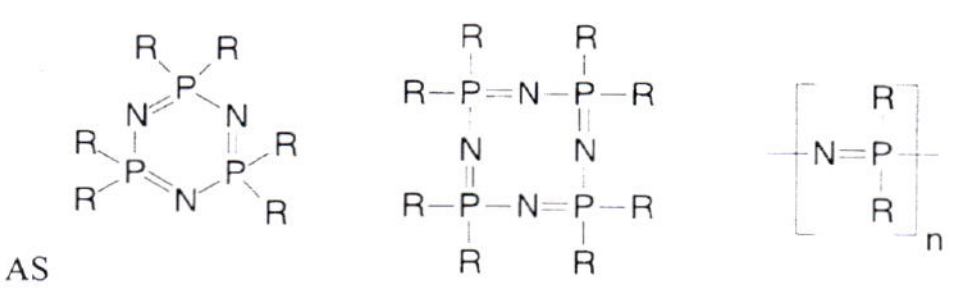

FIGURE 14.1 General structures of cyclo- and linear polyorganophosphazenes. *Reprinted from Progress in Polymer Science, Vol. 23, Roger De Jaeger, Mario Gleria, Poly(organophosphazene)s and related compounds: Synthesis, properties and applications, Page179, Copyright (1998), with permission from Elsevier.*

Modern Inorganic Synthetic Chemistry, DOI: 10.1016/B978-0-444-53599-3.10014-9

that uncrosslinked polydichlorophosphazene which was soluble in organic solvents was formed if the ROP of pure cyclic trimer was carried out carefully [11]. Subsequent reaction of this highly reactive polymeric species with nucleophiles brought out a great deal of hydrolytically stable polyorganophosphazenes by the pioneering work of Allcock and coworkers [12,13]. After that, the high-tech functional applications of the polyphosphazenes compounds were found which greatly stimulated a major research effort involving several industrial and academic laboratories over the past 40 years. The present work was focused on the synthesis of novel phosphazenes compounds and the assembly of cyclomatrix polyphosphazene micro and nanomaterials.

14.2. SYNTHESIS AND ASSEMBLY CHEMISTRY OF CYCLOPHOSPHAZENE

Cyclomatrix polymers containing crosslinked cyclophosphazene units are not so interesting as compared to other members of the phosphazene polymer family. It has not had much success, though $N_3P_3Cl_6$ itself would be an ideal monomer for this purpose as it contained six reactive groups. The basic structure of cyclomatrix polyphosphazene was shown in Fig. 14.2. It was a high degree crosslinked thermosetting material and difficult to process. It can only be used as adhesives or fillers to improve the thermal and flame-retardant properties of the matrix, which baffle the development of the cyclomatrix polyphosphazenes [12,13].

Until 2006, based on the special stereo structure of cyclophosphazene and the beamed substitution of chlorine bond, a series of controllable shaped micro- and nanomaterials were firstly obtained by one pot preparation [14–19]. The pioneering work by Tang and his coworkers revealed a new synthesis and assembly mechanism of cyclophosphazene which greatly pushed the cyclomatrix polyphosphazenes forward. Cycolphosphazene played the core role in the synthesis and assembly process of cyclomatrix polyphosphazenes

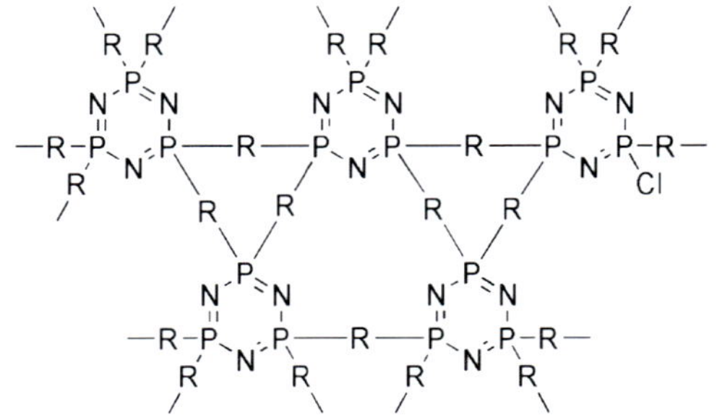

FIGURE 14.2 The structure of cyclomatrix polymers. *Reprinted with permission from [12]. Copyright 1994 American Chemical Society.*

and provided a new way to prepare the shape-controlled micro- and nanomaterials.

Polymeric nanotubes have been explored for application in bionanotechnology field, quantum devices, electronic, or electrochromic devices. Without question, it would be a challenge to develop facile approaches to polymer nanotubes with controlled dimensions under a mild condition [20–28].

The above-mentioned method to prepare cyclomatrix polyphosphazenes was called one pot method because the shaped materials could be formed during the polymerization reaction and without any postprocessing, which would overcome both the multistep processing and the difficulties to separate products from the templates. Up to now, the reported techniques of using one pot method to prepare the cyclomatrix polyphosphazene micro- and nanomaterials include four types: in situ template, self-assembly of primary particles, in situ skeleton convert technique, and surface polymerization on the nano- or microparticles.

14.2.1. In Situ Template Method

Formation mechanism of cyclomatrix nanotubes via in situ template approach is shown in Fig. 14.3.

The strategy was inspired by the rod-shaped structure and the high surface energy of triethylammonium chloride (TEACl) crystals, a byproduct of the reaction. The TEACl nanocrystals could be formed, then

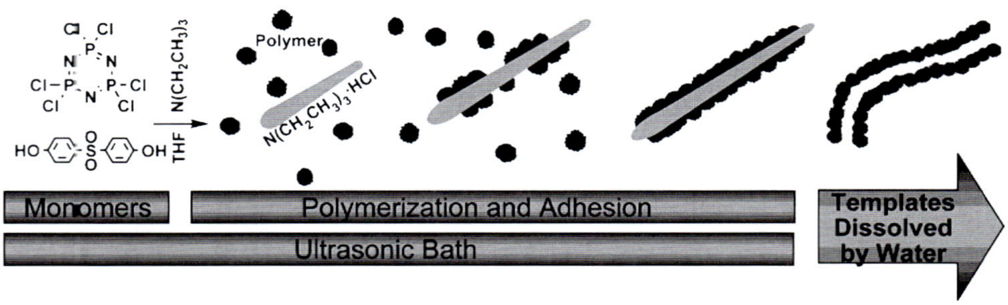

FIGURE 14.3 Formation mechanism of cyclomatrix nanotubes via in situ template approach.

separated from the reaction systems and suspended in it in situ and play a template role in the polymer nanotubes assembly. Crosslinked poly(cyclotriphosphazene-co-4,4′-sulfonyldiphenol) (PPZS) was prepared through the polymerization of cyclotriphosphazene (HCCP) and 4,4′-sulfonyldiphenol (BPS). The polymerization was performed in an ultrasonic bath in a solution of tetrahydrofuran (THF) with excess TEA as an acid acceptor. The polycondensation of HCCP with equimolar BPS generated prepolymer to form the primary polymer particles and hydrogen chloride (HCl). Triethylamine absorbed the HCl to afford TEACl, which accelerated the polymerization. During the polymerization, TEACl was precipitated out and the nano-sized rod-like crystals were thus obtained. As the polymerization proceeded under ultrasonic irradiation, the primary polymer particles with low surface energy settled down onto the TEACl nanocrystal surface gradually to get a core/shell structure material. With the TEACl nanocrystals growing along their axes, the crosslinking reaction among the polymers particles was carried out continually on the surfaces of the nanocrystals, and thus, core/shell structures assembled well. In the last step, the TEACl nanocrystals inside the core/shell materials were removed by washing with water to obtain the fine and uniform PPZS nanotubes. Figure 14.4 shows the reaction procedure of cyclophosphazene and 4,4′-sulfonyldiphenol.

The whole polymerization process is shown in Fig. 14.5. One can clearly see the formation of primary particles, the crystal templates and the surface deposit (a–c). The in-depth growth of the crystals was during the polymerization (d, e), the flexible polymer nanotubes were obtained after washing out the TEACl crystals with water (g, h). The degree of polymerization

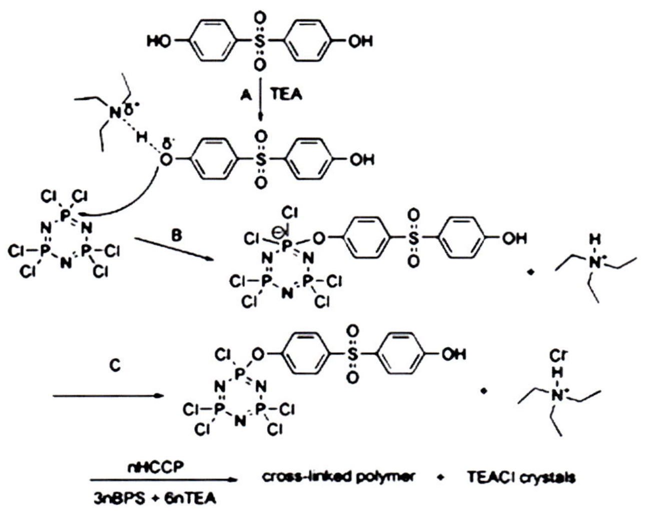

FIGURE 14.4 The reaction process of cyclophosphazene and BPS.

was evaluated by the EDX analysis (f, i). The obtained nanotubes were amorphous. Controlling the size of the nanocrystals by choosing the solvent system, the dimensions of the tubes could be easily controlled.

14.2.2. Self-assembly of Primary Particles

Using the in situ template method, it was very easy to obtain cyclomatrix polyphosphazene nanotubes but difficult to obtain the other morphology materials like fiber, porous particles, sphere, etc. By the recently reported method for self-assembly of primary particles, the micro- and nanofiber, sphere, and porous particles could also be obtained during the polymerization process of cyclophosphazene and BPS. Figure 14.6 shows the self-assembly process based on the special stereo structure of cyclophosphazene and the beamed substitute of chlorine bond and Fig. 14.7 were the images of obtained cyclomatrix polyphosphazene nanofibers. The inset scale bar in Fig. 14.7a is 50 nm, in Fig. 14.7b it is is 500 nm.

Different morphologies of the cyclomatrix polyphosphazene nanofibers and nanospheres [29,30] are shown in Fig. 14.8.

14.2.3. In Situ Skeleton Convert Technique

A more interesting, in situ skeleton convert technique, to prepare the special shaped micro- and nanoparticles was built in the research on the cyclomatrix polyphosphazenes. During test, thermal properties of the nanotubes gotten by the manner mentioned above, Tang and coworkers found that the char yields of the tubes were very high even at high temperature of 900 °C (Fig. 14.9). SEM and TEM images further showed that the residue power retained the tube morphology (Figs 14.10 and 14.11) and more exciting test results by EDX analysis gave out the element difference of the cyclomatrix polyphosphazene nanotubes between heat treatment at 800 °C for 2 h before and after (Fig. 14.12). Major component elements of the powder was carbon, which demonstrated that the skeleton has converted from the −N=P-bond to the −C−C-bond while keeping the same size of the tubes. The micro- or nanofiber and sphere had the similar properties to convert to corresponding carbon fiber and carbon sphere. It may be a novel way to prepare micro- or nanocarbon tubes, carbon fibers, carbon sphere, etc.

For application it was necessary to modify carbon tubes, fibers, and spheres, by introducing functional groups or elements on the particle surface, which are extremely difficult to achieve normally. The most challenging character of the in situ skeleton convert technique than that of the traditional ones is to greatly

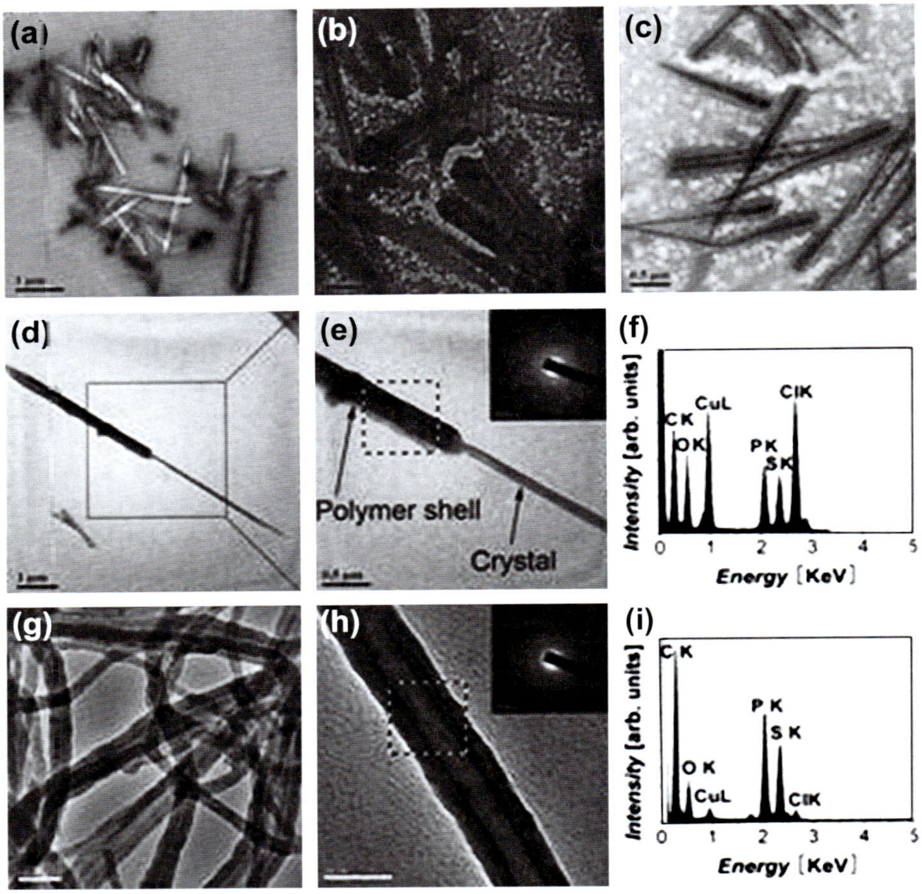

FIGURE 14.5 The detailed polymerization and assembly process.

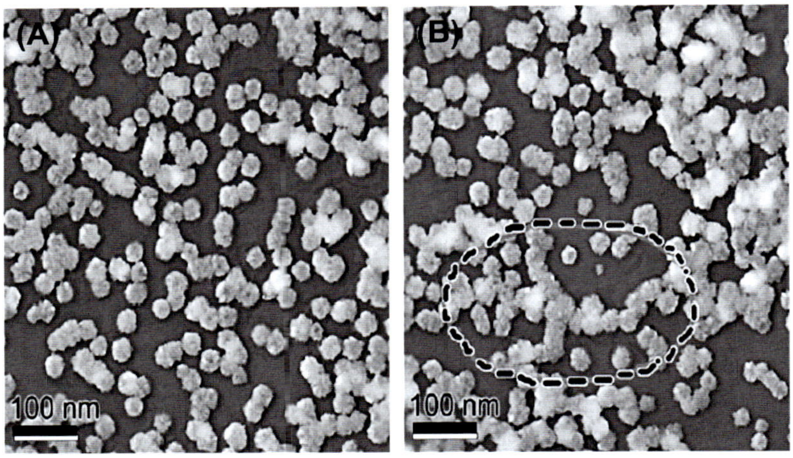

FIGURE 14.6 Self-assembly process of primary cyclomatrix particles.

increase its application value. We can keep a series of groups or elements in the particles during the preparation process by controlling the treatment conditions. In the EDX analysis of heat treated nanotubes (Fig. 14.12), it was very significant to find that there were traces of oxygen, phosphorus, and sulfur in the carbon nanotubes in spite of carbon being the main. These elements had either reacted amongst themselves or converted to the other reactive groups, which modified the carbon nanotubes. It would be beneficial to introduce the active surface cyclomatrix polyphosphazene micro- and nanomaterials or the converted carbon micro- and

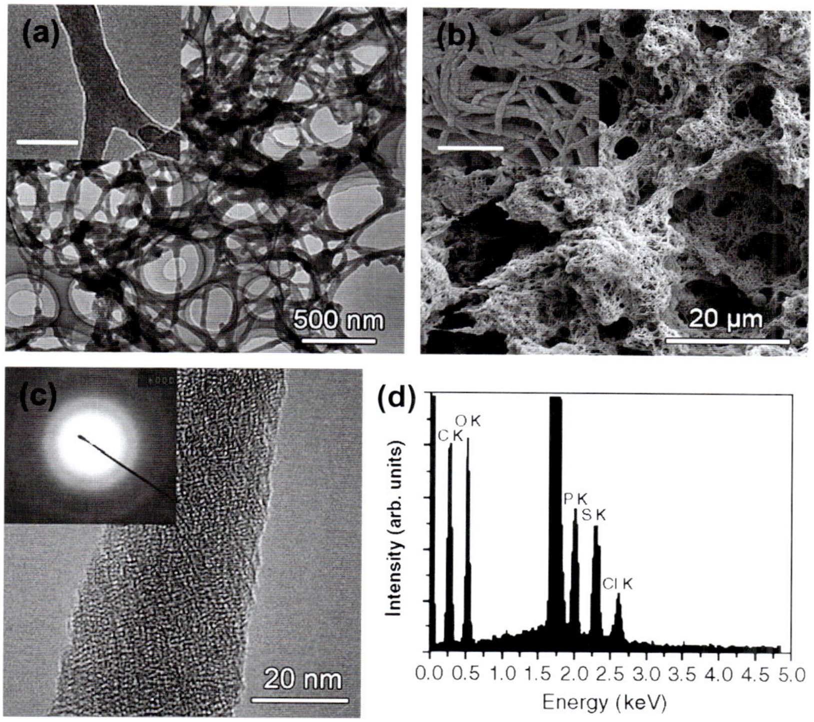

FIGURE 14.7 TEM and SEM images of PPZS nanofiber matrices.

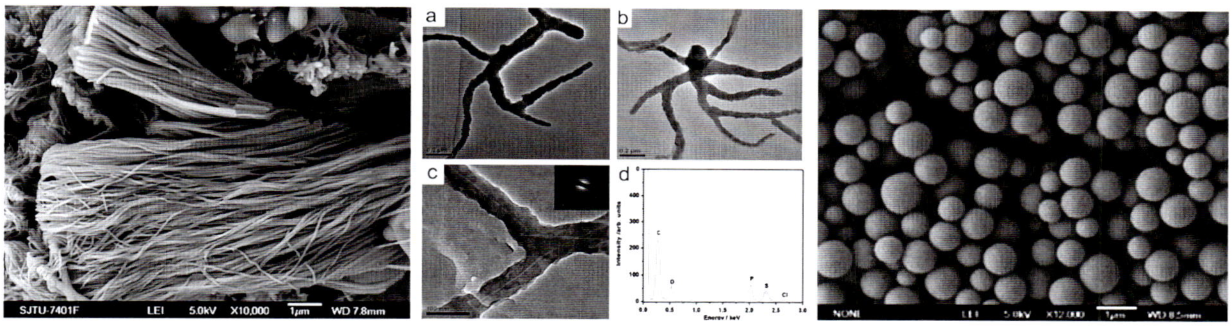

FIGURE 14.8 The different morphologies of the cyclomatrix polyphosphazene nanofibers and spheres.

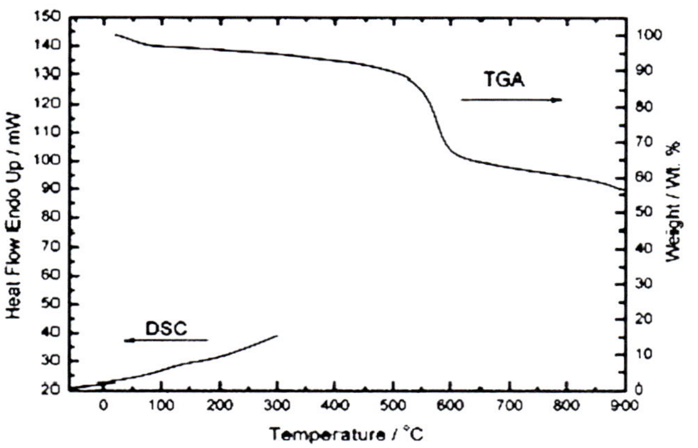

FIGURE 14.9 TGA and DSC curves of the cyclomatrix polyphosphazene nanotubes.

nanomaterials into the macromolecule network or chain to obtain the completely new polymer systems.

Just by controlling the ratio of BPS to cyclophosphazene, Tang and coworkers easily obtained the hydroxyl cyclomatrix polyphosphazene nanotubes as shown in Fig. 14.13. Using the hydroxyl cyclomatrix polyphosphazene nanotubes as crosslinkers, a novel modified polyurethane network made up of the molecules and particles were obtained (Fig. 14.14).

14.2.4. Surface Polymerization on the Nano- or Microparticles

The cyclomatrix polyphosphazene primary particles can quickly undergo polymerization on the higher

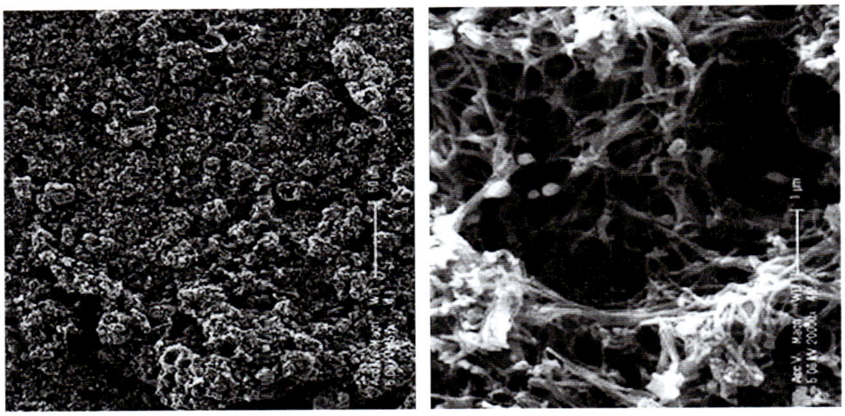

FIGURE 14.10 SEM images of the residue cyclomatrix polyphosphazene treated half an hour at 900°C.

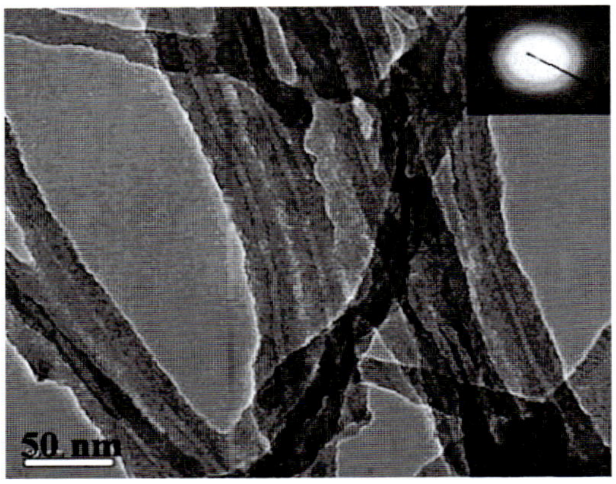

FIGURE 14.11 TEM image of the residue cyclomatrix polyphosphazene treated half an hour at 900 °C nanotubes.

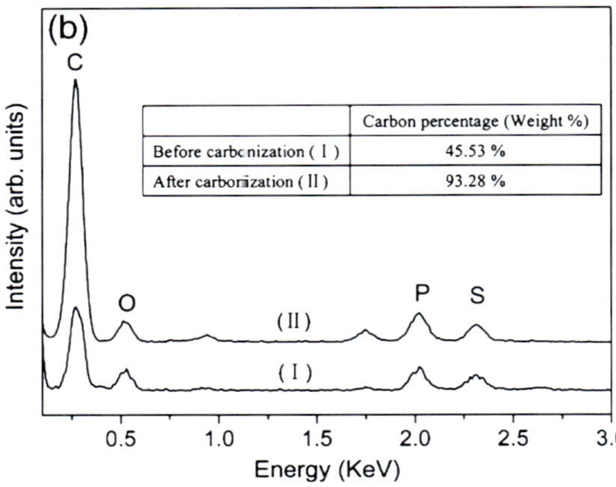

FIGURE 14.12 The EDX analysis of the cyclomatrix polyphosphazene. *Reprinted from Materials Letters, Vol. 62, Jianwei Fu, Yawen Huang, Yang Pan, Yan Zhu, Xiaobin Huang, Xiaozhen Tang, An attempt to prepare carbon nanotubes by carbonizing polyphosphazene nanotubes with high carbon content, Copyright (2008), with permission from Elsevier.*

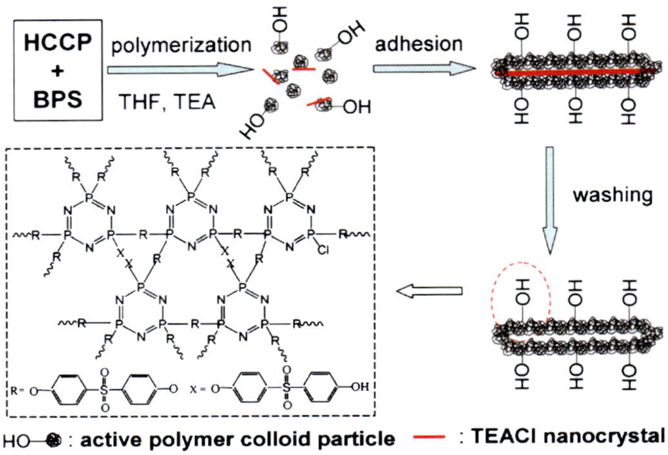

FIGURE 14.13 Formation process of hydroxyl cyclomatrix polyphosphazene nanotubes. *Reprinted from Applied Surface Science, Vol. 255, Jianwei Fu, Xiaobin Huang, Yan Zhu, Yawen Huang, Xiaozhen Tang, Facile fabrication of novel cyclomatrix-type polyphosphazene nanotubes with active hydroxyl groups via an in situ template approach. Copyright (2009), with permission from Elsevier.*

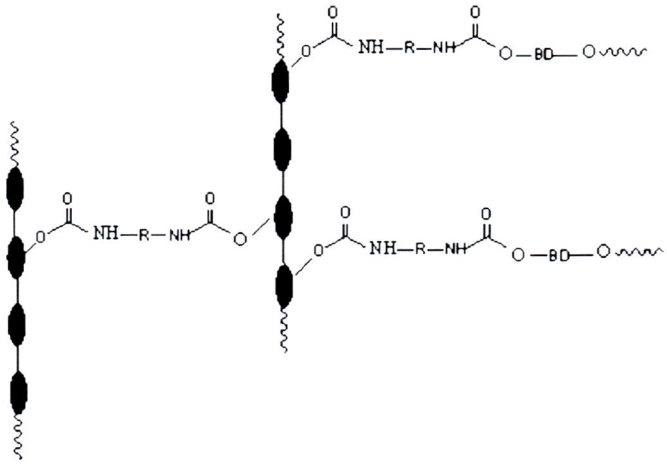

FIGURE 14.14 A modified polyurethane network (black ellipses represented the nanotubes).

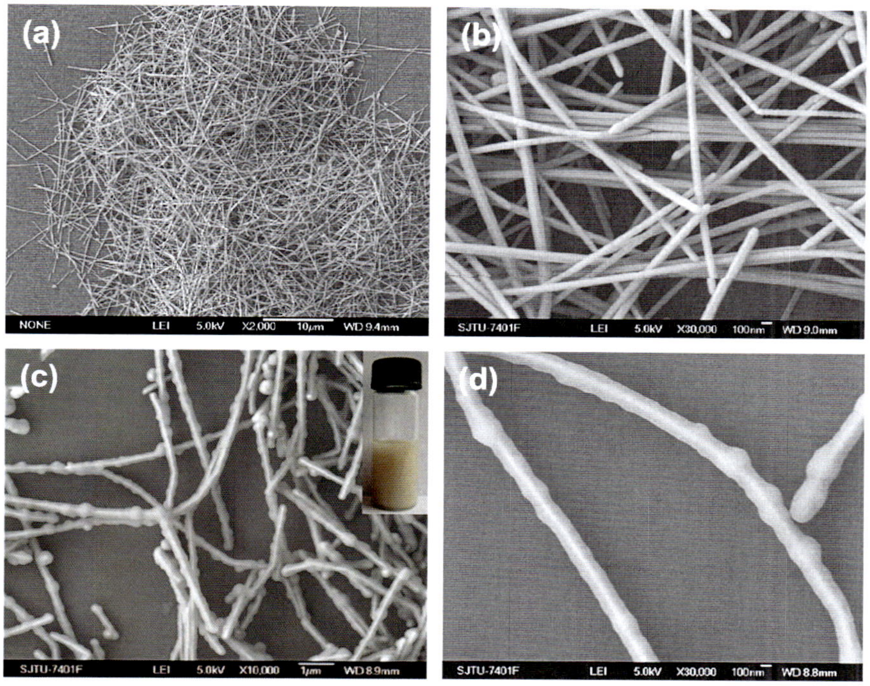

FIGURE 14.15 The cyclomatrix polyphosphazene modified silver nanowire.

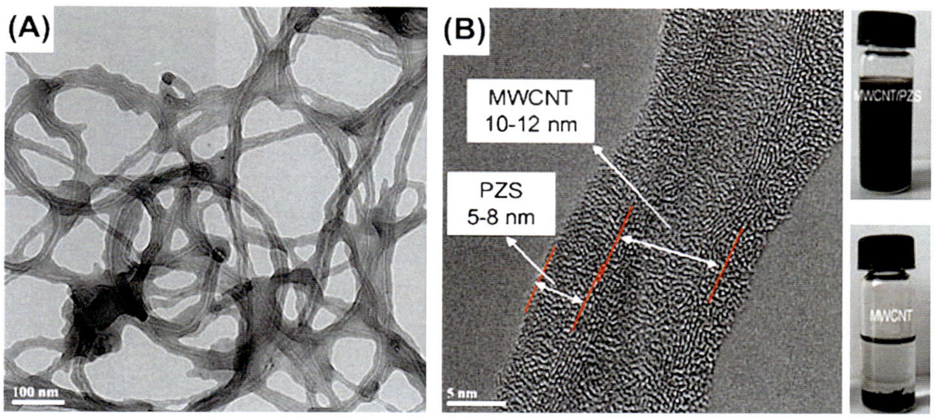

FIGURE 14.16 The cyclomatrix polyphosphazene modified carbon nanotubes. *Reproduced with permission from The Royal Society of Chemistry.*

surface energy particles due to their lower surface energy to form shaped nano- or micro-core/shell particles. It would be an easy and effective manner to produce high performance or functional particles. As some typical and useful samples, Figs 14.15–14.17 showed the cyclomatrix polyphosphazene-modified silver nanowire, carbon nanotubes, and magnetic nanotubes.

Recently, by reacting different monomers with cyclophosphazene, the similar novel cyclomatrix polyphosphazene nanomaterials could also be successfully obtained. Using 4,4′-diaminodiphenyl ether (ODA) to replace BPS, the cyclomatrix polyphosphazene spheres with active amino groups were prepared as shown in Fig. 14.18. The active amino sphere-modified carbon nanotubes could reduce gold ion in the solution directly (Fig. 14.19) [31].

Using 4,4′-(hexafluoroisopropylidene)diphenol (BPAF) to replace BPS, the superhydrophobic cyclomatrix polyphosphazene sphere was obtained (Fig. 14.20) [19].

14.3. APPLICATIONS OF CYCLOMATRIX POLYPHOSPHAZENES

14.3.1. Biomedical Phosphazenes

As promising biomaterial candidates, polyorganophosphazenes possess inherent advantages: satisfactory

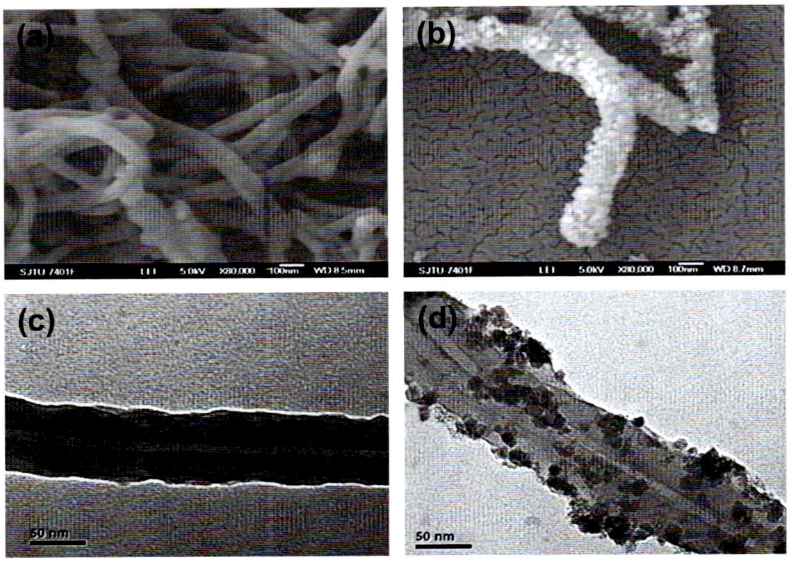

FIGURE 14.17 The magnetic cyclomatrix polyphosphazene nanotubes.

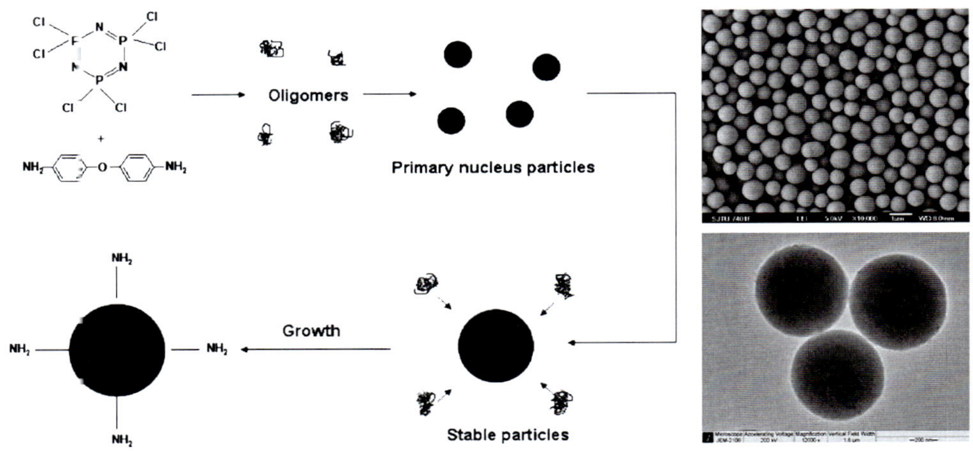

FIGURE 14.18 The amino active nanospheres made of ODA.

biocompatibility with human body, to have their relatively fast degradation rate and can be controlled by varying the side substituents [32], to know phosphate, ammonia as the degradation residues of the materials, and to choose the side groups which are either nontoxic when being present in small quantities or easily metabolized by human body [33,34]. Drug delivery, biological membranes, coatings, and polymeric medical devices and components such as prosthetics and implants are often their application fields.

Our researches showed that the obtained cyclomatrix polyphosphazene nanomaterials had no biotoxity and could absorb many medical molecules. For example, it can absorb doxorubicin and release it slowly. The cyclomatrix polyphosphazene nanomaterials had potential applications in controllable drug release. Based on the formation mechanism, we thought that the cyclomatrix polyphosphazene could be used to modify many ordinary nanomaterials or form multifunctional composite materials. Here, several new biodegradable polyphosphazenes with well-defined architecture such as micro-crosslinked, star-shaped, and penta-armed will be introduced.

14.3.1.1. *Micro-crosslinked Polyphosphazenes*

Linear polyphosphazenes known as fine biodegradable materials could undergo rapid hydrolysis degradation but exhibited poor mechanical properties. The inorganic backbone of polyphosphazenes was naturally too flexible to obtain necessary mechanical properties.

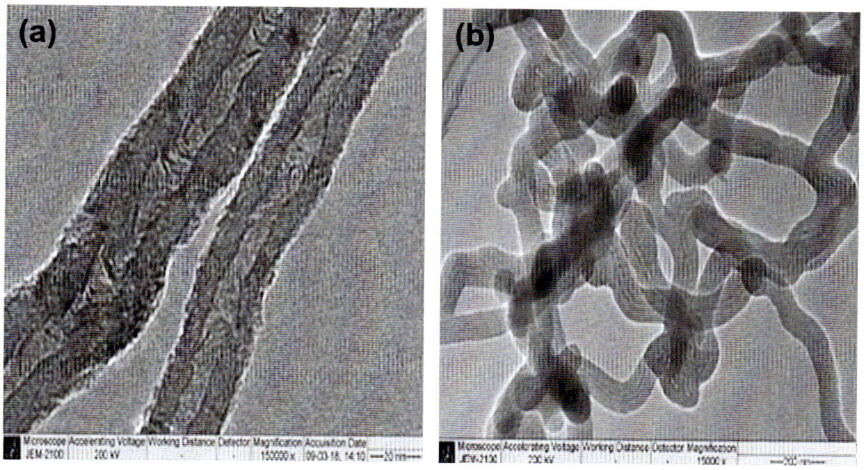

FIGURE 14.19 Active amino-modified carbon nanotubes.

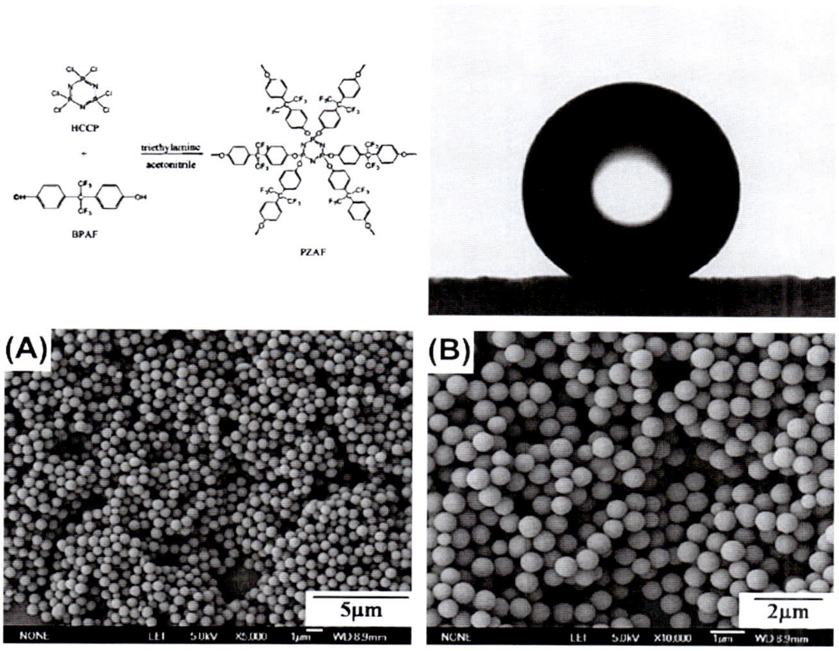

FIGURE 14.20 Superhydrophobic cyclomatrix polyphosphazene spheres. *Reproduced with permission from The Royal Society of Chemistry.*

To overcome their inherent limitation, the modified polyphosphazenes blended with biodegradable polyesters [35–38], polyanhydrides [38], and inorganic particles [39] were prepared. The materials did strengthen their mechanical properties but gave a rise to slower degradation rate [40]. Co-substituted polyorganophosphazenes with unsaturated side groups had formed interpenetrating polymer networks with acrylonitrile, styrene, acrylic acid (AA), and methyl methacrylate using sequential interpenetrating method [41] to balance the mechanical properties and the degradation rate of the materials. Here, two kinds of micro-crosslinked polyphosphazenes were further designed and prepared to give new and convenient way to obtain better results. Allylamine was an important monomer whose polymerized products had widespread applications in microencapsulation and gene transfer vectors as cationic polymers [42–45]. The synthesis route of poly(glycino ethyl ester)(allyl amino)phosphazenes is shown in Fig. 14.21 [46].

Through the crosslink reaction initiated by BPO, crosslinked polyphosphazenes with modified hydrolytic sensitivity and surface hydrophilicity could be obtained. These modified properties are based on the degree of crosslink dominated by the ratio of glycino ethyl ester and allylamine. Co-substitution of allyl

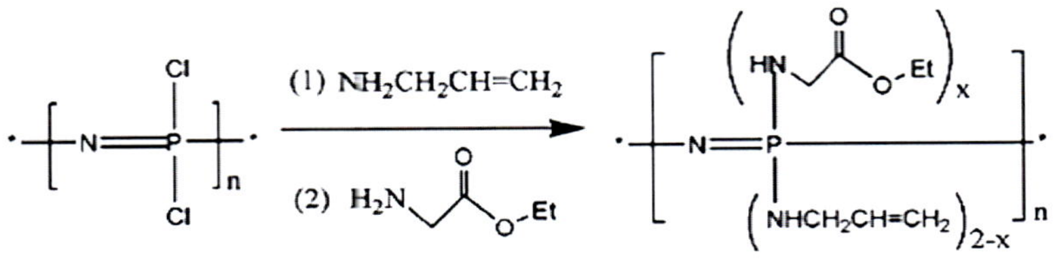

FIGURE 14.21 Synthesis route of the polymer.

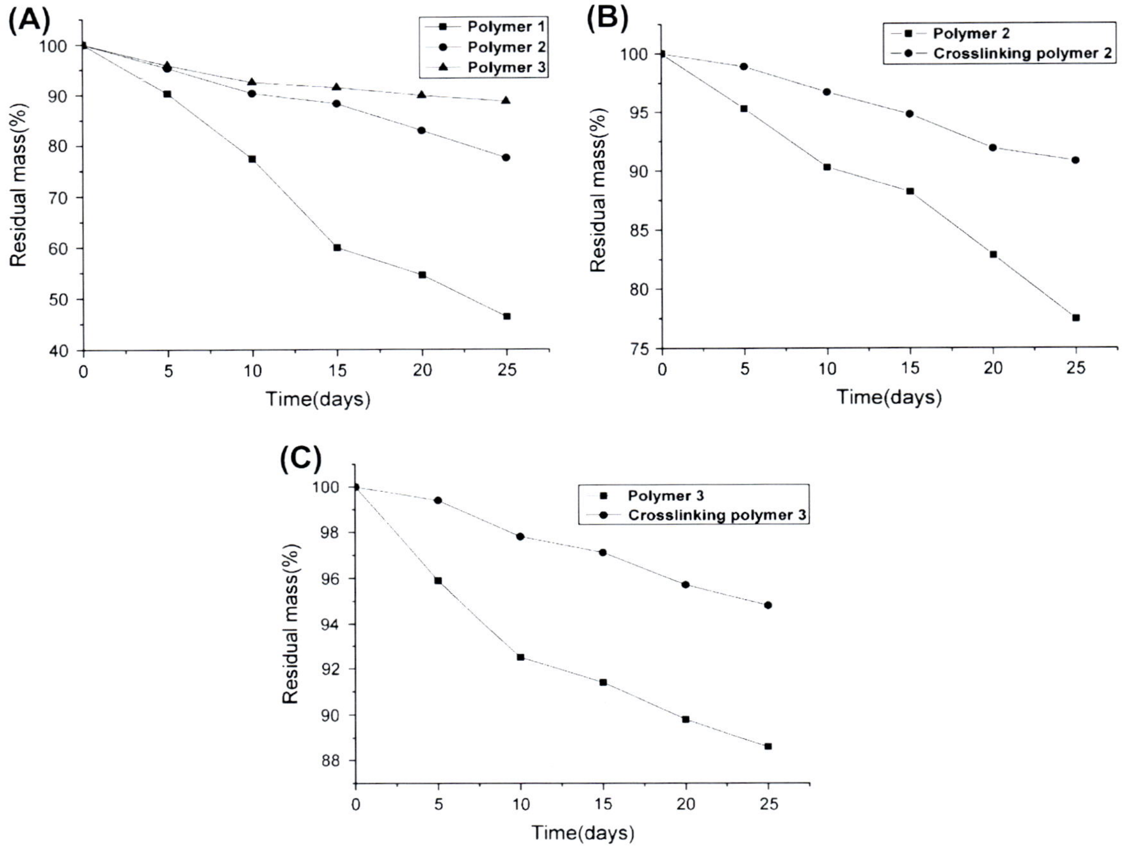

FIGURE 14.22 Weight loss of (A) the polymers 1–3, (B) the linear polymer and crosslinked polymer 2 and (C) the linear and crosslinked polymer 3 in phosphate buffer of pH 7.4. *Reprinted from Polymer Degradation and Stability, Vol. 92, Lin Yin, Xiaobin Huang, Xiaozhen Tang, Synthesis, characterization and hydrolytic degradation of linear and crosslinked poly[(glycino ethyl ester)(allyl amino)phosphazene], Page 795, Copyright (2007), with permission from Elsevier.*

amino groups and glycine ethyl ester units increased the surface hydrophobicity and increased the contact angles from 52° to 69° to result in the decrease of hydrolytic rate. Compared to the linear ones, crosslinked polymer films had hydrophobic surfaces and gave the contact angles of 91°and 96°. The rate of degradation decreased as the allylamine increased at the pendant groups (see in Fig. 14.22, the ratio of glycino ethyl ester to allyl amino in polymers 1,2, and 3 was 1:0, 2:1 and 1:2, respectively).

2-Hydroxyethyl methacrylate (HEMA) had satisfactory biocompatibility and bifunctionality. It was attached to the side chain along with glycine ethyl ester to form a precursor with unsaturated substituents. The co-substituted polyorganophosphazenes was blended with HEMA or AA followed by a free radical polymerization to prepare micro-crosslinked polyorganophosphazenes. The synthesis route of the micro-crosslinked polyorganophosphazenes is shown in Fig. 14.23 [40].

The mechanical strength and hydrolysis degradation profiles are depicted in Table 14.1 and Fig. 14.24, respectively. The more ethenyl side groups were introduced in the precursors, the better mechanical properties were thus obtained. HEMA crosslinked polyphosphazenes showed an approximately 13–17 times increase in terms of modulus of elasticity when compared to the linear counterpart. AA-crosslinked polyphosphazenes showed an approximately 11–14 times increase in terms of modulus of elasticity when compared to the linear counterpart. The polymers hydrolyzed at a rather slow rate when HEMA were used as crosslinkers. AA-crosslinked polyorganophosphazenes underwent moderate hydrolysis degradation.

TABLE 14.1 The Mechanical Properties of the Polyphosphazenes

Polymers	Modulus of elasticity (KPa)	Failure stress (KPa)	Failure strain $\delta L/L$
2	17.2 ± 2.6	23.5 ± 2.9	2.15 ± 0.32
6	212.8 ± 13.9	248.0 ± 30.6	3.08 ± 0.27
7	287.5 ± 33.7	294.5 ± 39.4	3.90 ± 0.25
8	191.3 ± 10.9	229.6 ± 17.3	2.97 ± 0.35
9	239.6 ± 27.6	242.5 ± 18.9	3.12 ± 0.28

14.3.1.2. Star-shaped Polyphosphazenes

The well-defined macromolecular architecture such as star-shaped polymers [47–50], star block polymers [51], comb-like polymers [52], and hyperbranched polymers [53,54] have attracted considerable interest of scientists due to their various functions and properties based on their special structures. Among them, star-shaped polymers gained attention because of their structure and unique physicochemical properties which are different from those of their linear counterparts [55,56]. Generally, star-shaped polymers can be synthesized by two different routes: one was the arm-first method and the other was the core-first approach using multifunctional initiators [57–59]. So far, the core-first approach has proven to be a very efficient strategy to form expected structure polymers by ring-opening polymerization. Cyclotriphosphazenes exhibited useful thermal properties and excellent biocompatibility [60,61]. Moreover, cyclotriphosphazenes could be used as multifunctional initiators [62]. Star-shaped poly(L-lactide) and poly(D,L-lactide) with hydroxyl-terminated cyclotriphosphazene core were synthesized by ring-opening polymerization using stannous octanoate as catalyst. The synthetic routes are shown in Fig. 14.25 [63].

The compound of $N_3P_3(OC_6H_4\text{-}p\text{-}CH_2OH)_6$ was obtained from the reaction of $N_3P_3(OC_6H_4\text{-}p\text{-}CHO)_6$ with sodium borohydride at room temperature [64]. Molecular weight of star-shaped poly(L-lactide) and poly(D,L-lactide) linearly increases with the molar ratio of monomer to initiator. The star-shaped poly(L-lactide) possessed low melting point, crystallinity, and onset decomposition temperature. Star-shaped poly(D,L-lactide) also possessed the best degradability while linear poly(L-lactide) possessed the poorest degradability. And the degradation of star-shaped polylactide was divided into two steps, which was different from the degradation process of linear polylactide. Highly branched architecture of star-shaped PCL resulted in interrupted crystallization form and subsequently lower melting

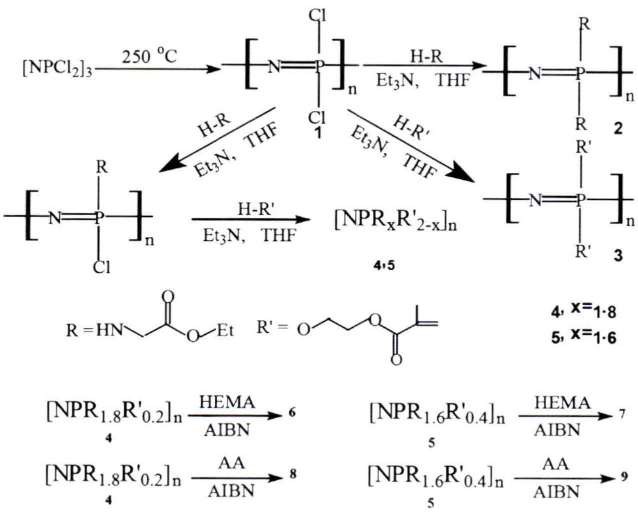

FIGURE 14.23 Synthesis route of micro-crosslinked polyphosphazenes. *Reprinted from Biomaterials, Vol. 25, Yanjun Cui, Xian Zhao, Xiaozhen Tang, Yinpei Luo, Novel micro-crosslinked poly(organophosphazenes) with improved mechanical properties and controllable degradation rate as potential biodegradable matrix, Page 451, Copyright (2004), with permission from Elsevier.*

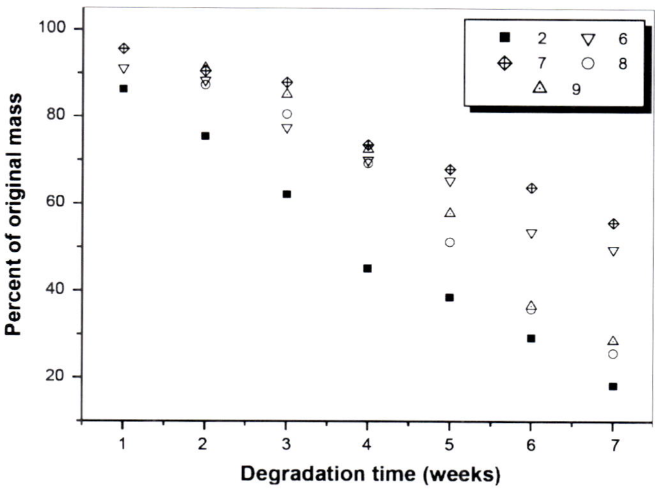

FIGURE 14.24 The hydrolysis degradation rate of the polymers 2 and 6–9. *Reprinted from Biomaterials, Vol. 25, Yanjun Cui, Xian Zhao, Xiaozhen Tang, Yinpei Luo, Novel micro-crosslinked poly(organophosphazenes) with improved mechanical properties and controllable degradation rate as potential biodegradable matrix, Page 451, Copyright (2004), with permission from Elsevier.*

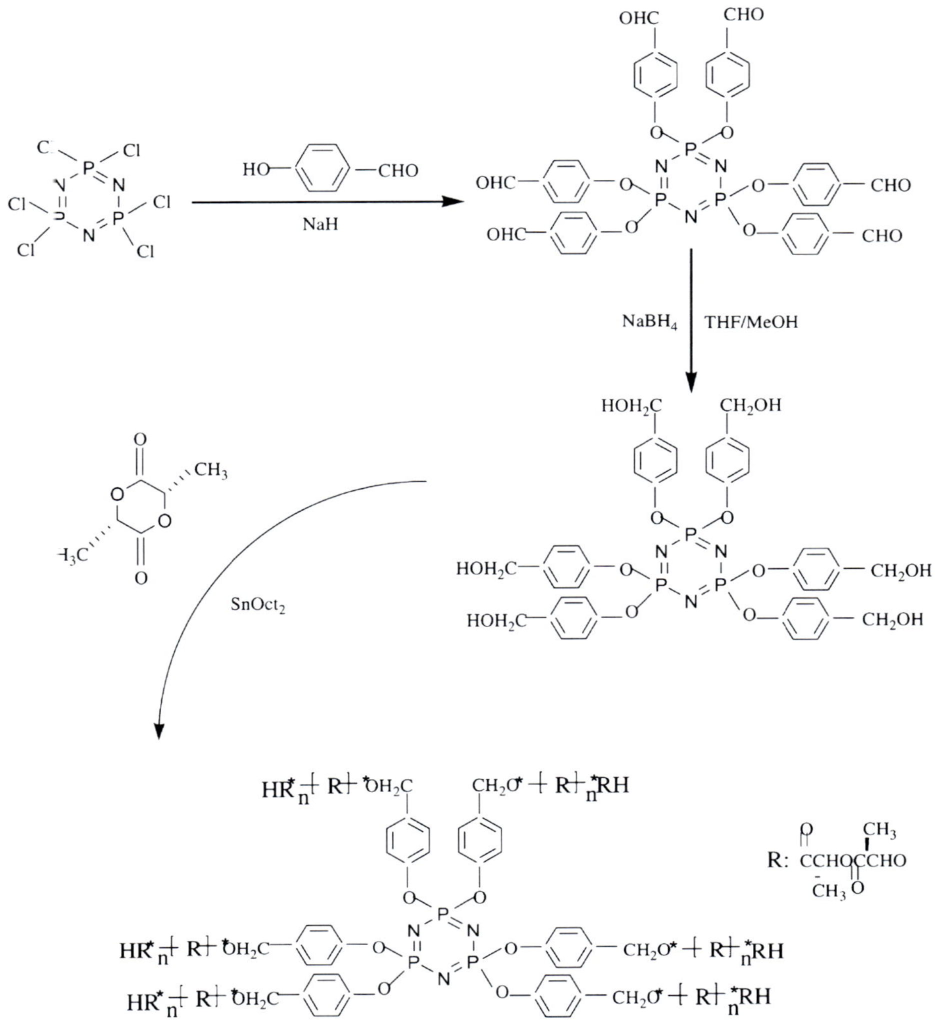

FIGURE 14.25 Synthesis of star-shaped poly(L-lactide). *Reprinted from Polymer Degradation and Stability, 87, Weizhong Yuan, Lu Zhu, Xiaobin Huang, Sixun Zheng, Xiaozhen Tang, Synthesis, characterization and degradation of hexa-armed star-shaped poly(L-lactide)s and poly(D,L-lactide)s initiated with hydroxyl-terminated cyclotriphosphazene, pp. 503–509, Copyright (2005), with permission from Elsevier.*

temperature. Thermogravimetric analysis (TGA) carried out on the star-shaped PCL suggested that the introduction of phosphazenes ring strengthened the thermal stability of the resulting polymers. The star-shaped PCL-*b*-D,L-PLAGA copolymer could combine, excellent permeability of PCL and higher biodegradability of the D,L-PLAGA block. The synthesis routes are shown in Fig. 14.26.

14.3.1.3. Penta-armed Polyphosphazenes

N-Dichlorophosphoryl-*P*-trichlorophosphazene (NDT) was the simplest linear short-chain polyphosphazenes. Its derivatives possessed good biocompatibility and biodegradability because of their structures. Using the hydroxyl groups end capped derivative of NDT as initiator, the star-shaped penta-armed asymmetric poly(ε-caprolactone)s or poly(L-lactide)s with lower crystallinity, faster degradation rate was obtained by ring-opening polymerization of ε-caprolactone or L-lactide. The synthesis routes are shown in Figs 14.27 and 14.28 [65].

The degradation behavior of linear and penta-armed PCL (Mn ≈ 20,000) was investigated by immersing them in pH 7.4 PBS at 37°C. The weight losses of linear and penta-armed PCL during the degradation period are shown in Fig. 14.29. The degradation rate of linear PCL was very slow (for example, the residual weight was 98.1% after degradation for 33 days), which was attributed to its high degree of crystallinity (54.7%). The degradation rate of penta-armed PCL was obviously faster than that of the linear one. This is due to the lower degree of crystallinity of penta-armed PCL (42.6%). The degradation of polyester took place by random hydrolysis of ester bonds. The penta-armed

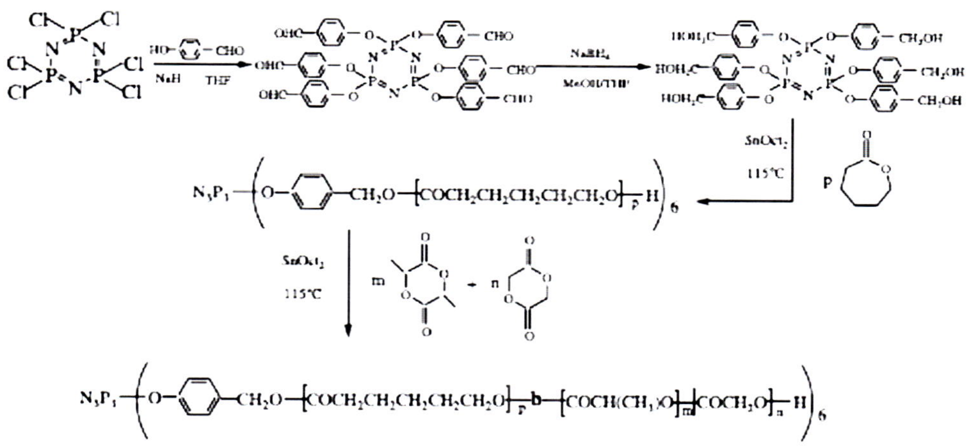

FIGURE 14.26 Synthetic route of star-shaped copolymer phosphazenes.

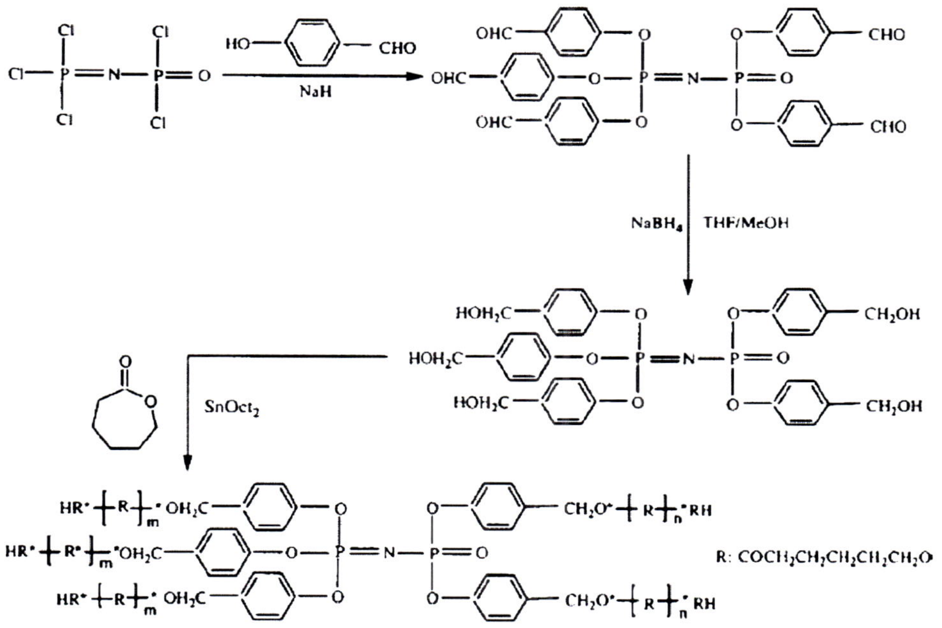

FIGURE 14.27 Synthetic route of penta-armed poly(ε-caprolactone)s.

PCL possessed a branched structure which led to the lowering of crystallinity and the rate of amorphous of polymer. Therefore, the water molecule pervaded the molecule of penta-armed PCL faster than that of the linear one, which promoted the hydrolytic degradation of polymer. The degradation of penta-armed PCL included two steps: the PCL chains were cleaved randomly, and then short PCL chains were detached from phosphazene cores.

14.3.2. Polyphosphazenes with Nonlinear Optical Properties

The development of nonlinear optical (NLO) chromophore-containing polymeric materials has drawn considerable attention due to their potential photonics applications over the last decade [66–69]. The applications included reversible optical data storage [70–74], fabrication of diffractive elements with specific polarization properties [75,76], optical switching and/or slab waveguides [77,78], and many other purposes. In recent years, the significant interest existed in the design and development of organic materials with large molecular hyperpolarizability values to improve their optical transparency and good thermal stability. It has been theoretically and experimentally shown by Fichou and others that heteroaromatic-based chromophore [79,80] and multidipolar chromophore [81,82] had larger macroscopic nonlinearities, higher thermal stability, and better orientation order and stability than one-dimensional,

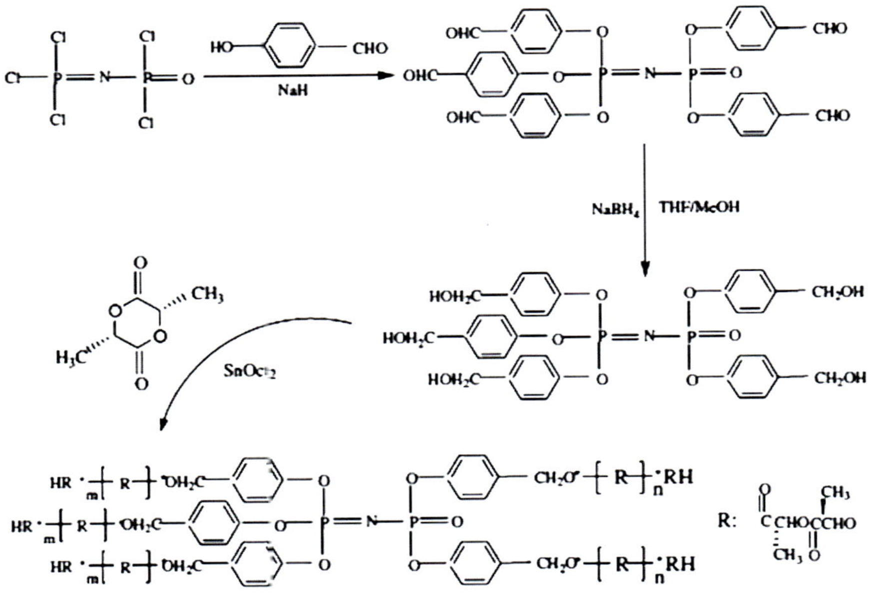

FIGURE 14.28 Synthetic route of penta-armed poly(L-lactide)s. *Reprinted from European Polymer Journal, 41, Weizhong Yuan, Lu Zhu, Xiaobin Huang, Sixun Zheng, Xiaozhen Tang, Synthesis and properties of pentaarmed poly(l-lactide)s on N-dichlorophosphoryl-P-trichlorophosphazene derivative core, pp. 1867–1873, Copyright (2005), with permission from Elsevier.*

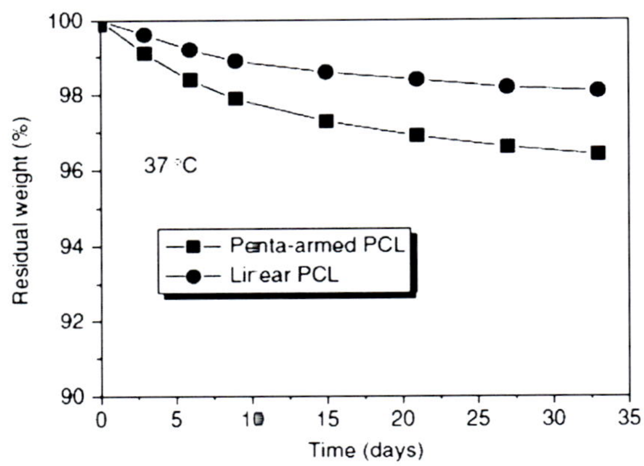

FIGURE 14.29 The weight losses of linear and penta-armed PCL in phosphate buffer at 37 °C.

donor–acceptor disubstituted *p*-conjugated molecules. How to provide general laws predicting NLO properties of multidipolar chromophore was still a big challenge. This problem should be solved further to push the development of NLO polymeric materials forward and bring this kind of promising material into practical applications. Polyphosphazenes presented a number of useful features for practical devices, such as excellent flexibility, high thermal and oxidative stability, optical transparency from 220 nm to the near-IR region, and controlled incorporation of covalent chromophore, which can be easily accomplished over a broad concentration range. A combination of the advantages of heteroaromatic based and multidipolar chromophore could produce a new type of chromophore that presented high electro-optical coefficients and good temporal stability. Via a post-azo coupling reaction an easy two-step method, some new polymers with imidazole-based and multiple charge-transfer Y-shaped chromophore were synthesized. The synthetic routes are shown in Fig. 14.30 [83].

Chromophore contents can reach 14 mol%. The polymers had good optical transparency. The birefringence was about 10^{-2}. The maximum absorption λ_{max} was 354–375 nm and the absence of absorption was 425–465 nm. They provided a wide transparent window and dozens of blue shift compared to the other materials and had excellent photorefractivity. The polymers' T_g were 150–180 °C, which ensured the stability when the polymers were in the stable melt state. It was the first time that polyphosphazene polymers were observed to present photorefractivity without adscititious electric field or advanced polarization in room temperature. These polyphosphazene polymers were the new kinds of materials which were called all-optical photorefractive materials. To these materials, the possible damage of the materials caused by adscititious electric field or advanced polarization needed not to be considered. Membrane of the polymers could be made thicker to get better performance when there was no restriction of adscititious electric field or advanced polarization.

14.3.3. Phosphazene Separation Membranes

The pervaporation (PV) was considered to be a promising alternative of conventional energy intensive technologies such as extractive or azeotropic distillation in liquid mixture separation. There has been growing research interest in the applications of pervaporation in

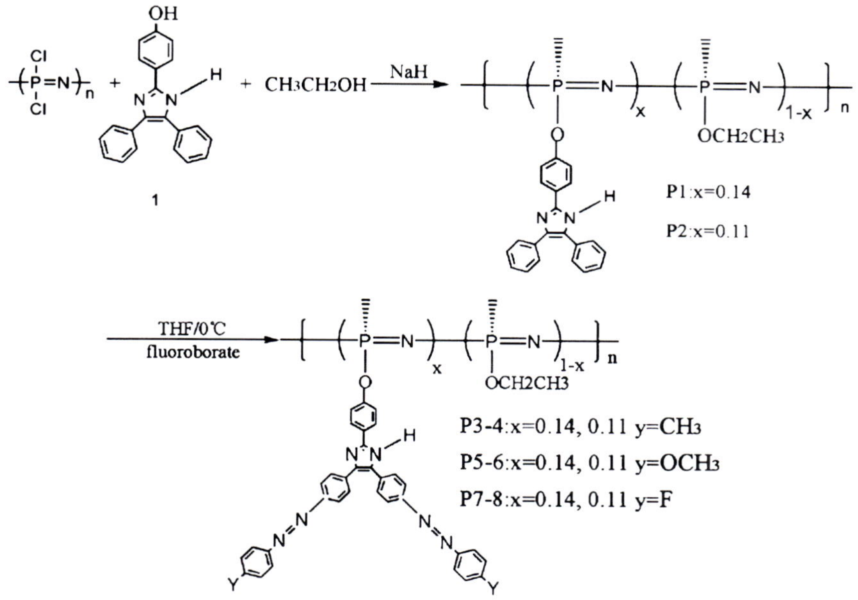

FIGURE 14.30 Synthetic routes of the NOL. *Reprinted from Synthetic Metals, 159, Yang Pan, Xiaozhen Tang, Synthesis and characterization of new photorefractive polymers with high glass-transition temperatures, pp. 1796–1799, Copyright (2009), with permission from Elsevier.*

biotechnology areas, for an example, recovery of ethanol from fermentation broths [84–86]. Success of PV in ethanol–water separation depended on the development of a membrane material that had high permeability, high permselectivity and good film-forming properties. The polymers that have been used in ethanol-permselective membranes included polydimethylsiloxane (PDMS) [87,88] and poly[1-(trimethylsilyl)-1-propyne] (PTMSP) [89]. Polyorganophosphazenes were the important organic–inorganic hybrid polymers and were potential membrane materials in the mentioned applications owing to their flexible main chain and good chemical and physical tailoring properties.

Three important points when polyorganophosphazenes membranes are used to remove ethanol from water are to be followed: (1) the flexibility of pendant groups should be tuned to proper degree to meet necessary mechanical properties and enough high diffusivity; (2) the hydrophobicity of polymer should be properly high to reduce interchain force which results in easy diffusion of ethanol and water in membrane and low swelling degree; and (3) the polymer should have enough affinity for ethanol. Single functional polyorganophosphazenes obviously can't achieve the requirement. Tang and coworkers synthesized a series of phosphazene heteropolymers with three substituting groups. The three groups are 2-allylphenoxy, phenoxy, and hydrophobic groups, respectively. 2-Allylphenoxy group provided both crosslinking sites, whereas phenoxy groups were used to increase the strength of membrane and help to keep hydrophobicity of materials. Hydrophobic groups including $-OC_2H_5$, $-OCH_2CF_3$, and $-OCH_2CF_2CF_2CF_2CF_2H$ groups were incorporated into polyorganophosphazene membranes (the polymers were sequentially called PAEPP, 2CF, and 5CF) to compare the pervaporation performance. The synthesis routes of PAEPP are shown in Fig. 14.31 and that of 2CF and 5CF were similar to PAEPP [90].

2CF membranes had the highest affinity to ethanol, PAEPP membranes had the highest affinity for water, 5CF membranes had the best hydrophobic properties and fluorine-containing materials had good sorption selectivity for ethanol. To achieve good separation performance of ethanol/water mixture, pendant groups of polyphosphazenes should have good affinity to ethanol and low affinity to water. It should also provide properly high diffusivity of ethanol.

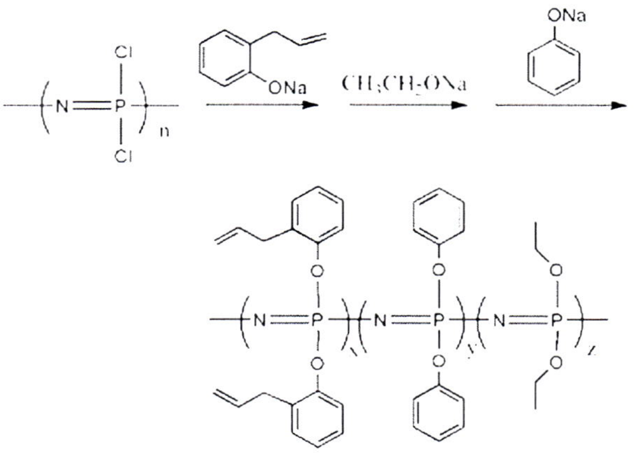

FIGURE 14.31 Synthetic route of PAEPP.

Polymer systems containing dispersed nanoparticles represented a class of materials with a combination of ideal properties generally not obtainable in conventional polymers. For the pervaporation technique, incorporated mixed matrix membranes have been considered as innovative and attractive membranes. The various nanocomposite membranes have been prepared to be applied in pervaporation technology [91–94]. However, the pervaporation performance of CNT–polymer hybrid membranes for ethanol/water and other volatile organic compound (VOC)/water mixture has not been studied till now possibly due to their poor affinity of CNT and general VOCs. Besides, the bad dispersibility of CNTs in solvents and polymers caused inconvenience in membrane preparation. The cost of CNTs was higher, which also limited their applications. Polyphosphazene nanotubes (PZSNTs) [14] were a novel type of nanotubes which can be easily prepared by the reaction between HCCP and BPS as shown before. PZSNTs were a highly crosslinked polymer products, resulting in high stability in solvent. Compared with CNTs, PZSNTs had following merits: preparation of PZSNTs was easy and convenient and sizes of PZSNTs can be easily tailored by changing the preparation conditions. Polyphosphazene microspheres (PZSMSs) having the same chemical structure with PZSNTs can also be easily prepared. It was convenient to recognize the relationship between morphology of fillers and pervaporation performance. PZSNTs were organic–inorganic hybrid polymers and can be well dispersed in most general solvents. In addition, there were nanopores on the wall of PZSNTs [95] which may reduce diffusion resistance. By incorporating the PZSNTs in PDMS, the PDMS/PZSNTs nanocomposite membranes were obtained. [96] Incorporation of PZSNTs in PDMS membranes increased the permeability and selectivity of membranes for water/ethanol mixtures. As the permeation of PZSNTs contents increased from 0% to 10%, both flux and selectivity increased. According to our successful research polyphosphazene nanotubes had potential applications as the filler of pervaporation membranes for ethanol/water. It would also improve membrane performance to separate other VOC–water mixtures. Also, many functional groups could be attached on PZSNTs which can further enhance their affinity to VOCs.

14.3.4. Phosphazenes Modified Polyurethane

Polyurethanes are important and versatile class of polymer materials, which are receiving steadily increasing attention and are excellent engineering materials due to their high abrasion resistance, wearability, chemical resistance, high impact strength, and outstanding damping ability. However, traditional polyurethanes have relatively poor thermal, flame resistant, and hydrophobic properties, which could not meet the special requirements in high-tech application fields. Polyphosphazenes have so far proven to be of limited in their wider applications because polyphosphazenes of high molecular weight had high production costs and were difficult to produce. But phosphazene oligomers were easily obtained by one new pot method in Tang's labs [97]. Therefore, phosphazene oligomers were used to improve thermal, cold, flame resistant, and hydrophobic properties of polyurethanes without any dramatic damage to their mechanical properties. Tang and coworkers synthesized a series of hydroxyl-terminated phosphazene oligomers and used them as different components, chain extenders [98], crosslinkers [99,100], and polyols [101], of polyurethanes to synthesize the novel polyphosphazene/polyurethane copolymerization materials.

14.3.4.1. Phosphazene Oligomer as Chain Extenders

A series of thermoplastic poly(oligophosphazene-urethane)s (POUs) by using the phosphazene oligomers as chain extenders were successfully prepared. The synthesis routes are shown in Fig. 14.32.

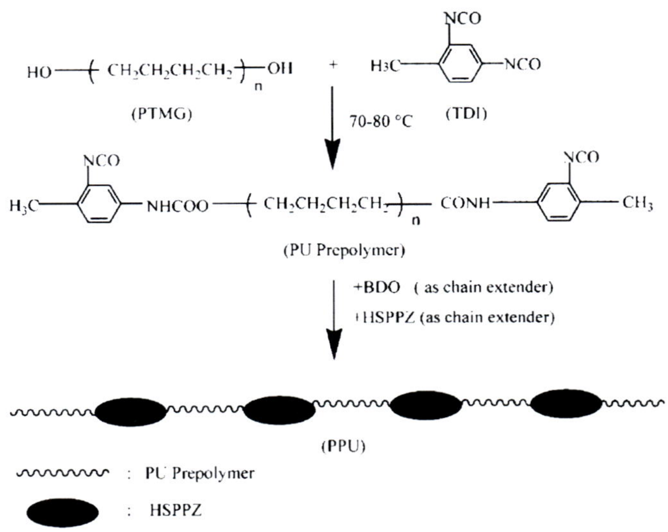

FIGURE 14.32 Synthetic routes of POUs.

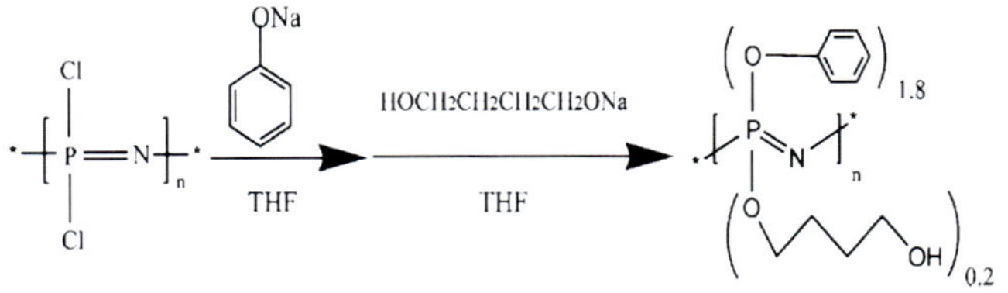

FIGURE 14.33 Synthetic route of OPHBP.

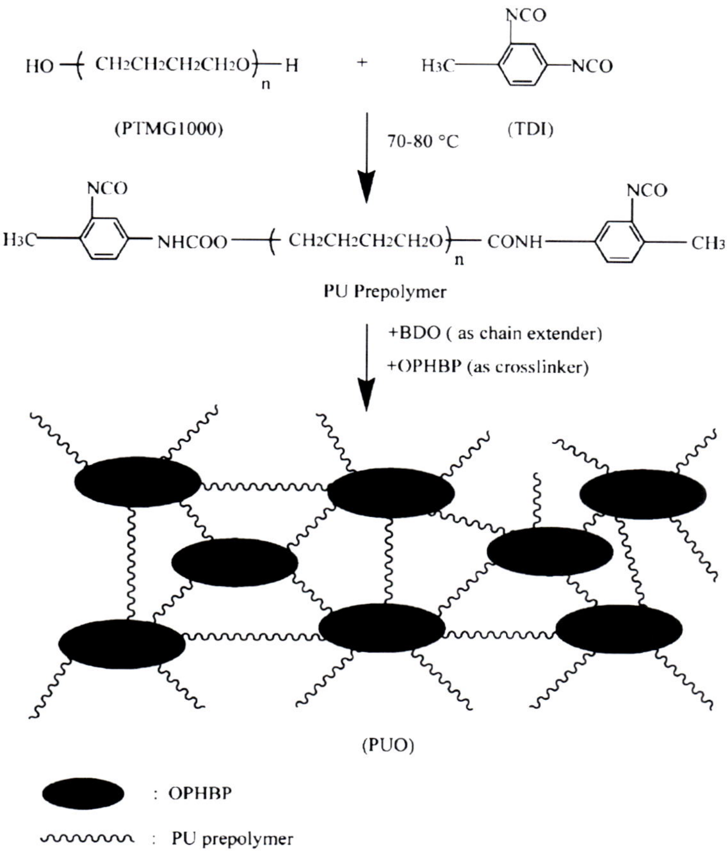

FIGURE 14.34 Synthetic route of PUOs.

Compared to the ordinary polyurethanes, POUs were completely amorphous and increased the onset degradation temperature by about 22°C. Their glass transition temperatures (T_g) decreased from −20.6 to −32.3°C and their water contact angles increased from 78.6° to 104.5°.

14.3.4.2. Phosphazenes Oligomer as Chain Crosslinkers

Chemical crosslinking was one of the most popular techniques to improve thermal stability of the polymer materials. The phenoxyphosphazene oligomers containing hydroxyl groups (OPHBP, see Fig. 14.33) were prepared and used as crosslinker to prepare thermosetting polyurethanes (PUOs). The thermosetting polyurethanes were prepared in a two-step method, in which polyurethane prepolymers were crosslinked by phenoxyphosphazene oligomers (see Fig. 14.34). PUOs were amorphous and their onset degradation temperature increased by about 78°C. Their glass transition temperatures decreased from −13.6°C to −32.4°C and their water contact angles increased from 74.8° to 94.6°.

14.3.4.3. Phosphazene Oligomers as Polyols

To further improve the mechanical strength of the phosphazene oligomers, the oligomers with active hydroxyl groups were reacted with isocyano group. In comparison with other methods, this approach could be performed at mild temperature, which may keep the functional groups of phosphazene oligomers from thermal degradation. The method would overcome requirement of unsaturated groups and multistep processes. Furthermore, high polar and rigid urethane oligomer can enhance mechanical properties of the elastomer (see Fig. 14.35) [102].

14.3.5. Phosphazene Solid Polymer Electrolytes

The development of polymer electrolytes with high ionic conductivity has become one of the main goals in polymer research due to the potential applications in solid-state batteries [103−105]. The main advantages of polymer electrolytes are favorable mechanical properties, ease in fabrication of thin film of desirable size, and an ability to form effective electrode−electrolyte contacts. Armand et al. [106] have claimed that the crystalline complexes formed from alkali metal salts with polyethylene oxide (PEO) are capable of demonstrating significant ionic conductivity and have highlighted their possible application as battery electrolytes. In order to increase the low ionic conductivity and lithium ion

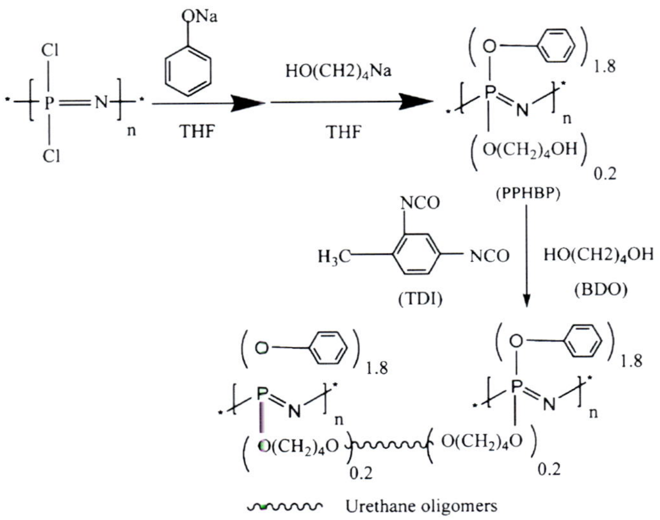

FIGURE 14.35 Synthetic route of phosphazene as polyols.

transference number of solid polymer electrolytes at room temperature, studies have been made via various approaches such as incorporation of ceramic fillers into the polymer matrix [107]. The Lewis acid–base interaction plays a vital role in the enhancement of ionic conductivity, electrochemical stability, and interfacial stability of solid composite polymer electrolytes. But the ceramic fillers are easy to aggregate because of the poor compatibility between polymer and fillers.

Recently, the successful preparations of fully cross-linked poly(cyclotriphosphazene-*co*-4,4′-sulfonyldiphenol) (PZS) microspheres and nanotubes viaprecipitation polymerization in our group offer a new kind of molecular-level inorganic–organic hybrid micro- and nanoscale materials for solid composite polymer electrolytes [108–111]. The structure of PZS microspheres and PZSNTs consist of alternating organic segments (benzene rings and C, O, H atoms) and inorganic segments (inorganic HCCP rings). Compared with traditional ceramic fillers which are easy to aggregate when highly loaded, PZS microspheres and PZSNTs have better compatibility with PEO chains due to the presence of same elements as PEO on the hybrid inorganic–organic structure and it could combine the advantages of organic and inorganic fillers. Furthermore, the active groups such as hydroxyl groups and chlorine atoms can be retained selectively on PZS microspheres and PZSNTs by controlling the ratio of the reactants for further modification. The rigid hybrid structure of PZS microspheres and PZSNTs could also act as anion receptors and form the complex with anions, causing a decrease in the share of ion pairs and immobilization of anions, resulting in an increase in the lithium ion transference number [112–116]. With all the excellent properties, PZS microspheres and PZSNTs are a new class of fillers in the development of solid composite polymer electrolytes.

14.3.5.1. Polyphosphazene Microspheres Incorporated in PEO-based Solid Composite Polymer Electrolytes

Solid composite polymer electrolytes based on poly (ethylene oxide) (PEO) by using $LiClO_4$ as doping salts and PZS microspheres as fillers could enhance the electrochemical and thermal properties of PEO-based polymer electrolytes significantly. Differential scanning calorimetry (DSC) results showed that there was a decrease in the glass transition temperature of the electrolytes and the crystallinity of the samples in the presence of the fillers, which is shown in Table 14.2. The decrease in T_g and X_c indicates that the addition of PZS microspheres could increase the flexibility of PEO chains and the ratio of amorphous state PEO because of the good compatibility between PZS microspheres and PEO chains.

Figure 14.36 shows the conductivity of PEO_{10}–$LiClO_4$, PEO_{10}–$LiClO_4$/10%SiO_2 and PEO_{10}–$LiClO_4$/10%PZSMSs as a function of temperature. In the temperature range studied the conductivity of PEO_{10}–$LiClO_4$/10%PZSMSs is higher than PEO_{10}–$LiClO_4$/10%SiO_2. The chemical elements of PZS microspheres surface

TABLE 14.2 Thermal Properties of PEO_{10}–$LiClO_4$ and PEO_{10}–$LiClO_4$/x%PZSMSs Composite Polymer Electrolytes Obtained from DSC Analysis

Sample	Glass point T_g (°C)	Melting point T_m (°C)	Melting enthalpy[a] δH_m (J g^{-1})	Crystallinity[b] X_c (%)
PEO_{10}–$LiClO_4$	−34.9	60.5	94.0	44.0
PEO_{10}–$LiClO_4$/5%PZSMSs	−38.4	58.7	79.7	37.3
PEO_{10}–$LiClO_4$/10%PZSMSs	−42.9	57.8	67.3	31.5
PEO_{10}–$LiClO_4$/15%PZSMSs	−45.4	56.5	63.7	29.8
PEO_{10}–$LiClO_4$/20%PZSMSs	−45.6	54.9	61.3	28.7

[a] *The data have been normalized to the weight of the PEO matrix.*
[b] $X_c = (\delta H_m^{sample}/\delta H_m^*) \times 100$, *where* $\delta H_m^* = 213.7$ *(J g*$^{-1}$*).*

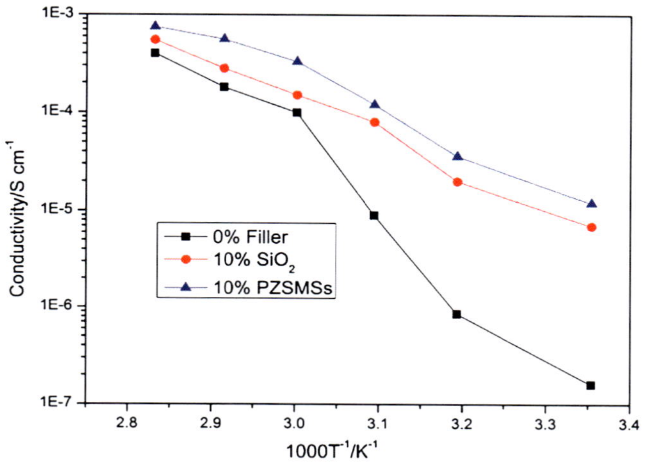

FIGURE. 14.36 Ionic conductivity of composite SPE as a function of temperature.

include S, O, and N atoms which can interact with Li^+ ions and enhance the dissolvability of $LiClO_4$ in PEO matrix more efficiently than SiO_2. As a result, more free Li^+ ions are released and then the conductivity and t_{Li}^+ are increased more by PZSMSs than SiO2. Maximum ionic conductivity values of 1.2×10^{-5} S cm^{-1} at ambient temperature and 7.5×10^{-4} S cm^{-1} at 80 °C were obtained and lithium ion transference number was 0.29.

The linear voltage sweep curves (Fig. 14.37) show that the maximum working voltage (V_{max}) of PEO_{10}−$LiClO_4$ only extends to about 4.5 V versus Li and the addition of PZSMSs widens the electrochemical stability window. The V_{max} of PEO_{10}−$LiClO_4$/10%PZSMSs exceeds to 4.8 V, indicating that PZSMSs obviously improve the electrochemical stability.

Compared with traditional ceramic fillers, the addition of PZS microspheres increased the ionic conductivity of the electrolytes and led to remarkable enhancement in lithium ion transference number and electrochemical stability.

PZS microspheres with active hydroxyl groups were also incorporated in PEO, using $LiClO_4$ as a dopant salt, to form a novel composite polymer electrolyte (CPE). Compared with plain PZS microspheres, Fig. 14.38 showed that the PZS microspheres with active hydroxyl groups can enhance the ionic conductivity and increase the lithium ion transference number of PEO-based electrolytes more effectively. Maximum ionic conductivity values of 3.36×10^{-5} S cm^{-1} at ambient temperature and 1.35×10^{-3} S cm^{-1} at 80 °C with 10 wt % content of active PZS microspheres were obtained and the lithium ion transference number was 0.34. The OH groups on the surface of PZSMSs-OH can act as Lewis acid sites to enhance the dissolvability of $LiClO_4$ and PZSMSs-OH could induce more free ions and the ionic conductivity is enhanced higher by PZSMSs-OH.

14.3.5.2. PEO-based Solid Composite Polymer Electrolytes with Inorganic−organic Hybrid Polyphosphazene Nanotubes as Fillers

Compared with PZS microspheres, PZSNTs have higher surface area and the pores on the surface make PZSNTs a porous structure. This characteristic endures PZSNT as a more promising material in the development of solid composite polymer electrolytes. After the addition of PZS nanotubes, solid polymer electrolytes had increased ionic conductivity, lithium ion transference number, and electrochemical stability window.

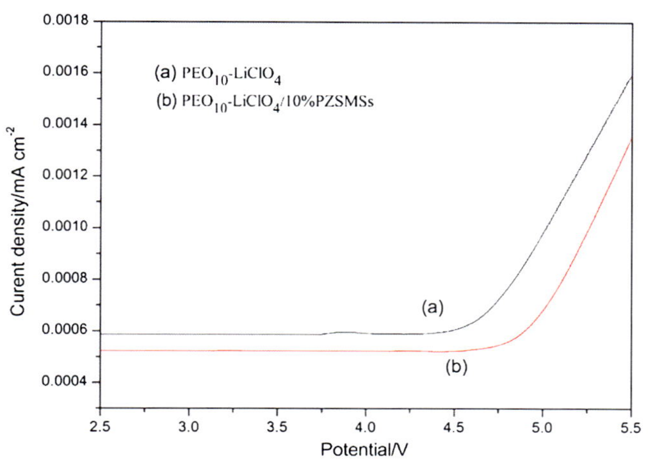

FIGURE 14.37 Linear voltage sweep curves of composite SPE. *With kind permission from Springer Science+Business Media: Journal of Applied Electrochemistry, Novel PEO-based solid composite polymer electrolytes with inorganic−organic hybrid polyphosphazene microspheres as fillers, DOI: 10.1007/s10800-010-0126-6, 2010, Jiawei Zhang, Figure 1.*

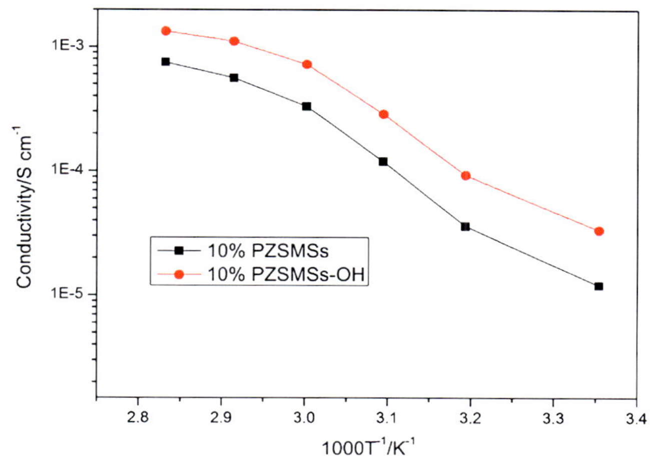

FIGURE 14.38 Ionic conductivity of PZSMSs and PZSMSs-OH composite SPE. *Reprinted from Materials Chemistry and Physics, 121, Jiawei Zhang, Xiaobin Huang, Jianwei Fu, Yawen Huang, Wei Liu, Xiaozhen Tang, Novel PEO-based composite solid polymer electrolytes incorporated with active inorganic–organic hybrid polyphosphazene microspheres, pp. 511−518, Copyright (2010), with permission from Elsevier.*

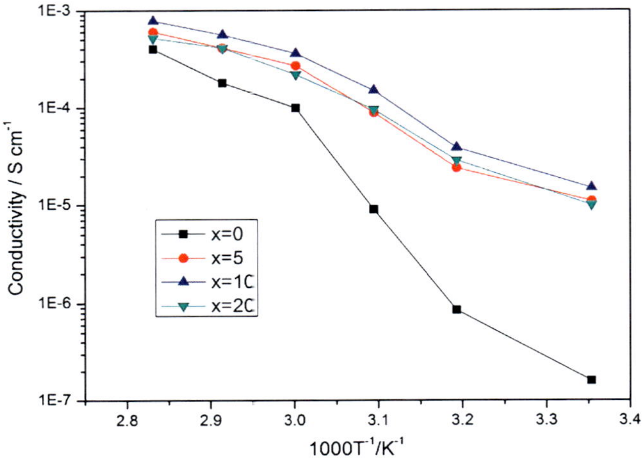

FIGURE 14.39 Ionic conductivity of PEO_{10}–$LiClO_4$/x%PZSNT SPE composites.

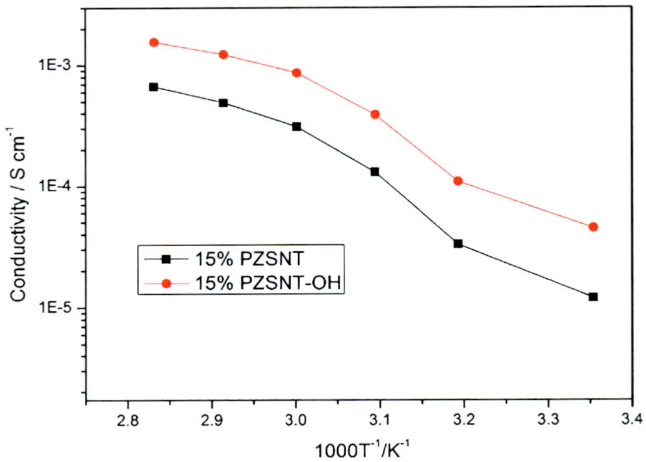

FIGURE 14.40 Ionic conductivity of PZSNTs and PZSNTs-OH composites SPE.

Figure 14.39 shows the conductivity of PEO_{10}–$LiClO_4$/x%PZSNTs at different temperatures. Maximum ionic conductivity values of 1.5×10^{-5} S cm^{-1} at ambient temperature and 7.8×10^{-4} S cm^{-1} at 80 °C were obtained with 10 wt% content of PZS nanotubes and the lithium ion transference number was 0.35. Compared with PZS microspheres, the ions can transport through the pores on the porous surface of PZS nanotubes and then the conductivity, the lithium transference number, and electrochemical stability window of the composite polymer electrolyte could be enhanced. Another important reason for the increased ionic conductivity and lithium transference number is the entrapment of ClO_4^- anion by the rigid hybrid structure of PZSNTs, similar to what is observed in boron-based systems and additives [117]. The rigid hybrid structure of PZSNTs could act as anion receptors and form the complex with anions, causing a decrease in the share of ion pairs and immobilization of anions, resulting in an increase in the lithium ion transference number [118,119]. The PZS nanotubes serve as excellent fillers to polymer electrolytes for its porous inorganic–organic hybrid structure.

PZS nanotubes with active hydroxyl groups were incorporated in PEO to form a novel composite polymer electrolyte (CPE) and the incorporation of active PZS nanotubes in PEO–$LiClO_4$ polymer electrolytes enhances ionic conductivity and facilitates salt dissociation. The experiment results in Fig. 14.40 show that the inorganic–organic hybrid PZSNTs with active hydroxyl groups can enhance the ionic conductivity and increase the lithium ion transference number of PEO-based electrolytes more effectively. Ionic conductivity values of 3.6×10^{-5} S cm^{-1} at ambient temperature and 1.34×10^{-3} S cm^{-1} at 80 °C were obtained when the content of active PZS nanotubes was 15 wt% with lithium ion transference number 0.37.

The reason is believed to be caused by Lewis acid–base interactions among active OH groups at the filler surface and the ether oxygen atoms of PEO chains and $LiClO_4$, which could weaken the interactions between Li^+ and ClO_4^-, enhance the dissolvability of $LiClO_4$ and release more free ions [120]. On the other hand, the addition of PZSNTs-OH fillers could provide a highly conducting layer at the electrolyte/filler interface [121–125]. This interface layer could be an amorphous polymer layer surrounding PZSNTs-OH and a space-charge layer [126,127]. The existence of the conductive layers between the inorganic–organic hybrid surface of PZSNTs-OH and PEO chains could act as the special conducting pathway of the charge carriers and enhance the lithium ion transference number with the addition of PZSNTs-OH fillers.

The addition of PZSNTs-OH also widens the electrochemical stability window and the V_{max} of PEO_{10}–$LiClO_4$/15%PZSNTs-OH exceeds to 5.0 V and obviously improves the electrochemical stability of the composite polymer electrolytes.

The good chemical properties of the solid-state composite polymer electrolytes suggested that the PZS microspheres and PZSNTs had a promising use as fillers in SPEs and the composite solid composite polymer electrolyte can be used as candidate material for lithium polymer batteries.

14.4. SILICONES

Polysiloxane and polysilane are two most important kinds of silicones and have been given widespread attention since their commercial introduction in the 1940s. There were over 20,000 publications (approximately half of which were patents) covering various

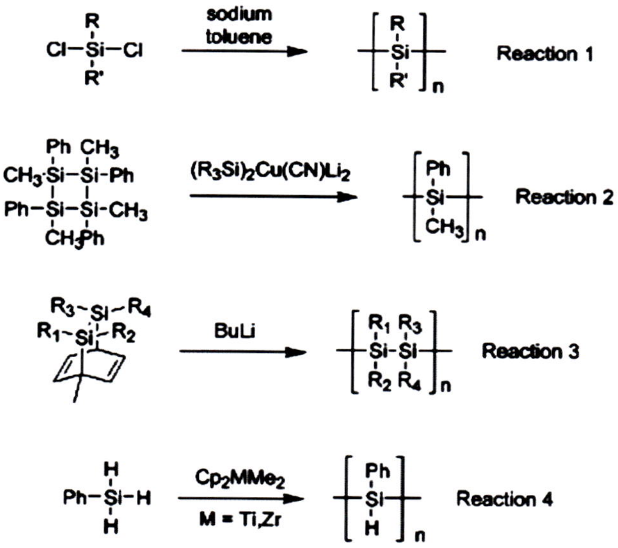

FIGURE 14.41 Synthesis of polysilane.

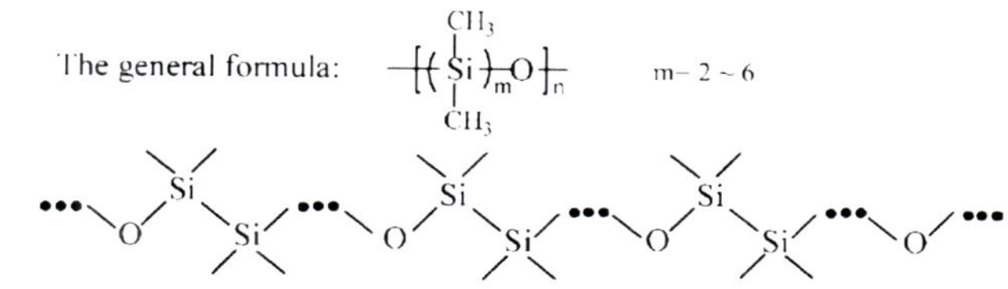

FIGURE 14.42 The general formula of polysilaethers. *Reprinted from Polymer, Vol. 46, Mengzhong Cui, Zhuyun Li, Xiaobin Huang, Jingyu Xi, Xiaozhen Tang, Sixun Zheng, Polysilaethers bearing Si—H and its functionalization via hydrosilylation with acrylic acid, Page 9162, Copyright (2005), with permission from Elsevier.*

aspects of the technology of silicone elastomers and their applications [128–139].

14.4.1. Polysiloxanes and Polysilane

Inorganic polymers, polysiloxanes contain alternate silicon and oxygen atoms as their backbone. The silicon atom has two side groups while the oxygen atom carries none. In this regard they are similar to polyphosphazenes [128,129]. The pioneering work on polysiloxanes dates back to 1863–1871, to the studies of Friedel, Crafts, and Ladenburg. But Kipping and his coworkers first demonstrated the polymeric siloxane structures in the early 1900s. The detailed account of the historical background on polysiloxane chemistry can be seen in the literature [128–132]. The most widely studied polymer of this family is poly(dimethylsiloxane) (PDMS) which contains methyl groups as the substituents on silicon. PDMS has the unusual properties like extremely low T_g (−123 °C), very high hydrophobicity, high thermal stability, low surface tension, low variation of viscosity with temperature, high gas permeability, and nontoxicity [131–133]. Despite their outstanding properties, however, their mechanical properties are relatively poor.

A very effective and important way to improve the mechanical strength of a weak, rubbery polymer, without chemical crosslinking, is to prepare the block (AB or ABA) or the segmented [(AB)*n*] copolymers where the soft, rubbery components are chemically linked to the glassy or the crystalline segments. Based on the unique properties of siloxanes, a wide variety of block or segmented copolymers containing PDMS as the soft segments and various thermoplastics containing PDMS as the hard segments have been synthesized and characterized [129].

Polysilanes are linear polymers with a backbone of silicon atoms that are usually substituted with aryl and/or alkyl groups. They possess excellent optoelectronic properties, such as strong ultraviolet (UV) absorption and relatively high electric conductivity, etc. [129]. There are four known procedures of the polysilane synthesis (Fig. 14.41), the Wurtz-type reductive-coupling of dichloroorganosilanes (Reaction 1), the ring-opening of cyclosilanes (Reaction 2), the anionic polymerization of "masked" disilenes (Reaction 3), and the catalytic dehydrogenation of primary silanes

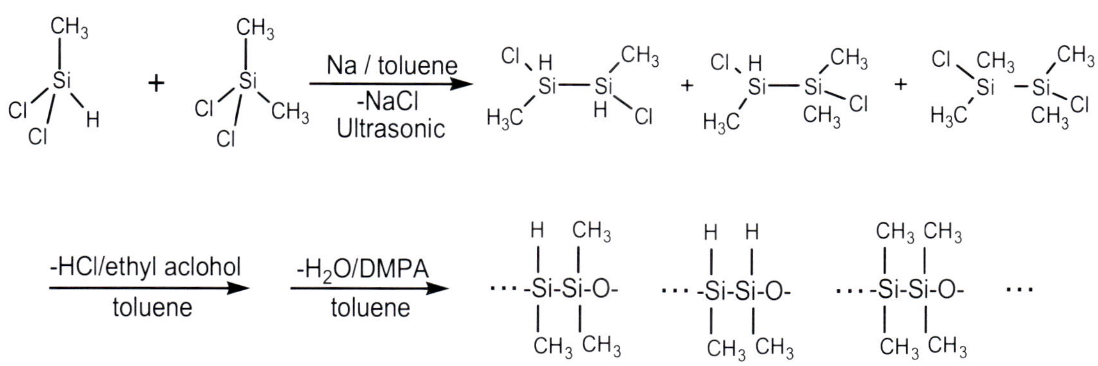

FIGURE 14.43 Synthesis route of PSEMH.

TABLE 14.3 The Raw Material Components and the Yields

Sample	Na/Me_2SiCl_2 (molar ratio)	$MeHSiCl_2$:Me_2SiCl_2 (molar ratio)	Yields (%)
A	1:1	1:0	72.6
B	1:1	1:1	71.7
C	1:1	0.4:0.6	72.3
D	1:1	0.2:0.8	70.8

A:PSEMH containing 50% Si–H bond; B:PSEMH containing 25% Si–H bond; C:PSEMH containing 20% Si–H bond; D:PSEMH containing 10% Si–H bond;

(Reaction 4) [129,140–143]. Detailed descriptions of these procedures can be seen in recent reviews [144].

14.4.2. Polysilaethers Bearing Si–H Bond

Polysilanes have relatively poor mechanical properties although they have excellent optoelectronic properties, which adversely affect their processability and limit their exploitation. By introducing Si–O–Si and Si–H into the macromolecular backbones of polysilanes, they could significantly improve the processability and functionalizability of the polymers while their optoelectronic properties will be basically reserved. Chojnowski and coworkers reported the synthesis of polysilaethers with permethyl-substituted groups (e.g., poly[oxy(multidimethylsilylene)s]) [145,146]. At the macromolecular backbones of permethyl polysilaethers, Si–Si bonds were regularly intercalated by oxygen atoms (Fig. 14.42) and the polymers were highly crystalline [147]. Permethyl-substituted polysilaethers can be synthesized via polycondensation of α,ω-functionalized linear oligosilanes or oligosiloxanes [148,149] and the ring-opening polymerization (ROP) of cyclic silaethers [150,151], respectively.

It is expected that the polysilaethers will exhibit some novel features owing to the electronic effect of hydrogen when the Si–H bonds are introduced into the macromolecular backbones. In addition, the presence of Si–H

TABLE 14.4 The Determination Results of the Synthesized Polymers

sample	Before polycondensation $\overline{Mn}/10^4$	PDI	After polycondensation[a] $\overline{Mn}/10^5$	PDI	IR[b] (cm^{-1}) ν_{Si-H}; $\nu_{Si-O-Si}$	δ/NMR (ppm) ^{29}Si[c]	^1H	H% (wt%) Cal.	Found
A	2.49	1.38	2.83	1.74	399; 2163,2113	−35.18[c]	0.1−0.16; 4.96−5.03	1.92	1.91
B	2.02	1.32	3.28	1.81	383; 2160,2095	−17.23,−35.18,−37.30	0.1−0.16; 5.0−5.05	0.85	0.84
C	2.16	1.41	3.11	1.52	381; 2159,2082	−17.23,−35.18,−37.11	0.01−0.13; 5.07	0.66	0.63
D	2.51	1.90	3.41	2.01	380; 2158,2075	−17.23,−35.36,−37.11	0.01~0.35; 5.08	0.31	0.30

[a] *They were further polycondensed at 110 °C.*
[b] *The split of ν_{Si-Si} and ν_{Si-H}.*
[c] *End groups were not detected perhaps due to the high molecular weight.*

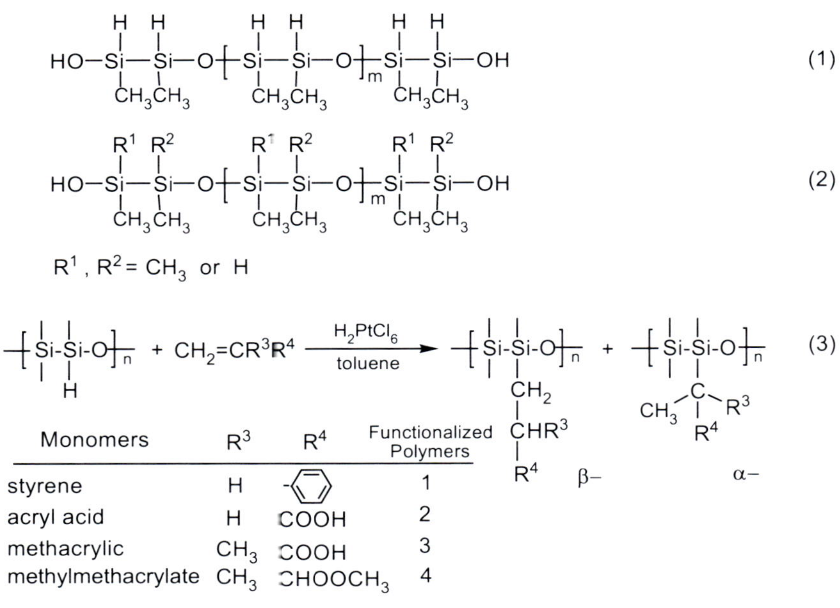

FIGURE 14.44 The different substituted polysilaethers.

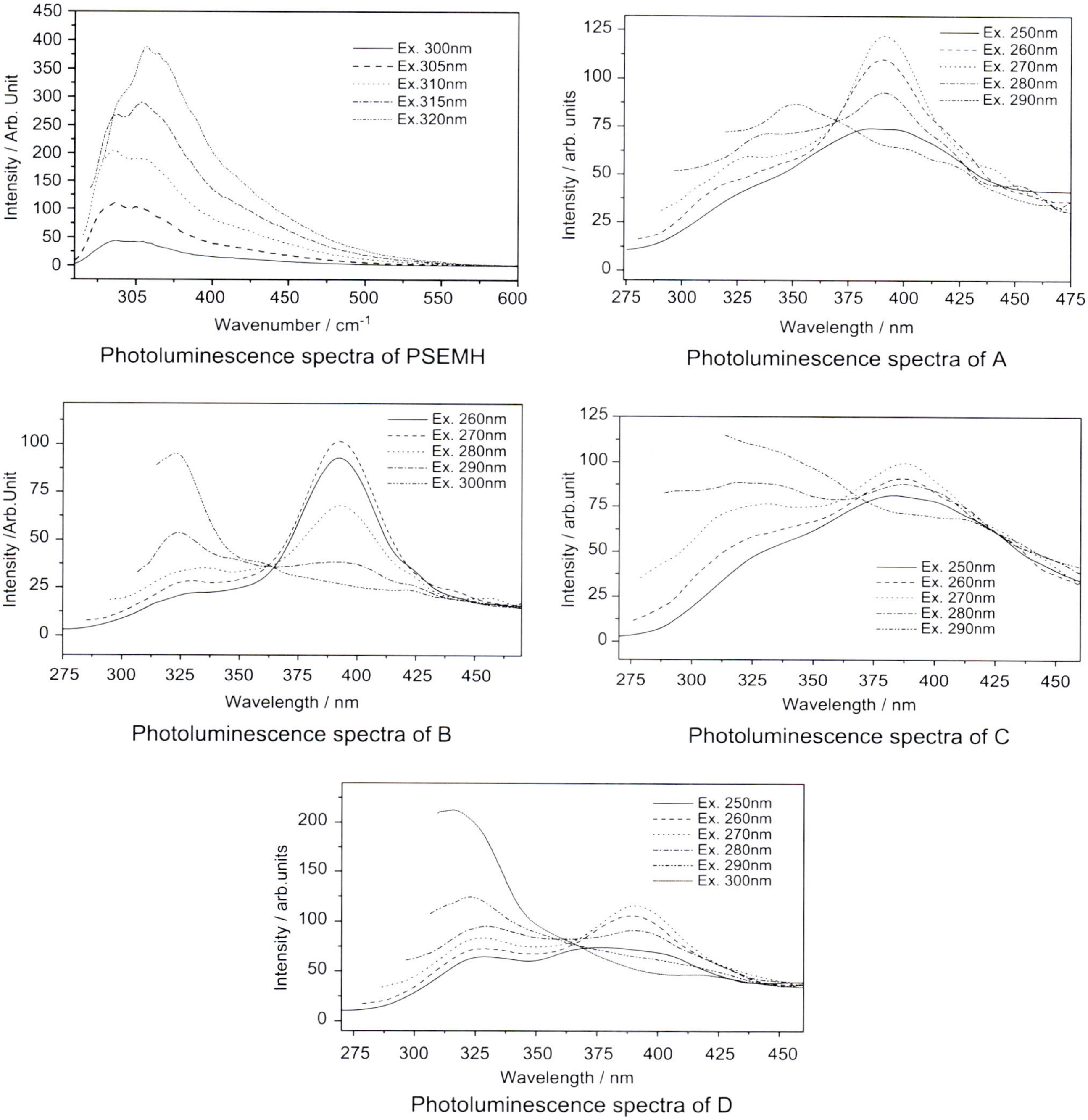

FIGURE 14.45 The photoluminescence spectra of the substituted polysilaethers.

bonds will provide tremendous potentials to prepare a variety of functionalized derivatives through hydrosilylation. Kumada et al. first reported the synthesis of the simplest silicon analogues of polyethers fully substituted with hydrogen in 1968. Unfortunately, the polymers were very unstable because each Si atom was connected with two reactive Si—H bonds at the backbone [152].

Recently, Cui reported the synthetic route of polysilaethers containing moiety Si—H bonds at the side chain (PSEMH) which allowed access to hitherto inaccessible oxygen interrupt polysilanes [153,154]. The synthetic route is shown in Fig. 14.43. By a Wurtz reductive-coupling reaction, an equimolar ratio of dichloromethylsilane to alkali metal yielded dichlorodisilane. The alcoholysis of Wurtz coupling resultants was performed

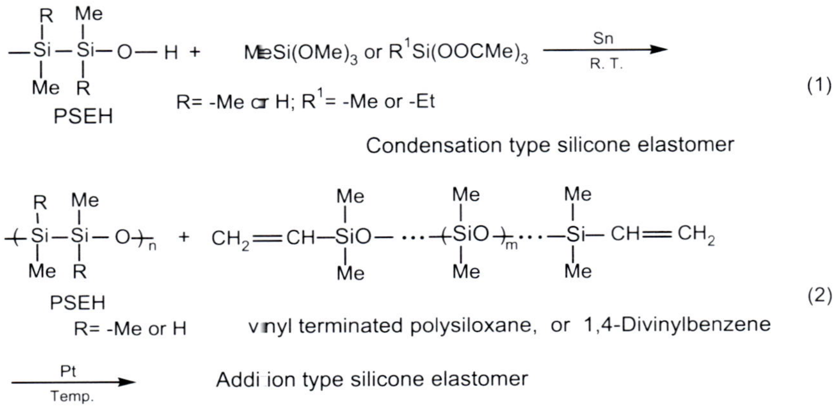

FIGURE 14.46 Synthesis route of functional silicone polymers.

in situ and the polycondensation and hydrolysis occurred simultaneously in the presence of a small amount of *N*,*N*-(dimethylamino)pyridine. The Si—H bonds at the macromolecular side chains provided tremendous potentials to prepare a variety of functionalized polysilaethers via hydrosilylation.

Using the above method, the yields can reach to 70%. The ratio of monomers used in synthesis of PSEMH and the results of characterization data are summarized in Table 14.3. The IR, ^{29}Si NMR, content of Si—H bond of the synthesized polymers is shown in Table 14.4.

The side groups of PSEMH had good reactivity. In base water solution, alcohol water solution (pH $\geq$ 8), Si—H bond was very easily degradable and can react with C=C, C≡C, C=O, OH, =NH and C≡N bond. It can also be grafted by organic polymer. Figure 14.44 showed the styrene, acrylic acid, methacrylic and methyl methacrylate-substituted polysilaethers and Si—H bond played an important role. The very interesting thing was that the photoluminescence spectra of the substituted polysilaethers changed dramatically (see Fig. 14.45).

Novel silicone elastomers can be obtained by reaction of linear polymer of polysilaethers bearing Si—H bond with silane-coupling agent, crosslinking vinyl or polysiloxane-containing vinyl groups (see Fig. 14.46). These novel elastomers kept unique properties of ordinary silicone elastomers and gave new functional properties like photoluminescence, etc. which would pioneer the new way to synthesize a series of functional silicone polymers.

References

[1] H.R. Allcock, F. Lampe, J. Mark, Contemporary Polymer Chemistry, third ed., Prentice-Hall, Englewood Cliffs, 2004.
[2] H.R. Allcock, Chemistry and Applications of Polyphosphazenes, John Wiley & Sons, Hoboken, 2003.
[3] J.E. Mark, R. West, H.R. Allcock, Inorganic Polymers, Prentice-Hall, Englewood Cliffs, 1992.
[4] R.D. Archer, Inorganic and Organometallic Polymers, Wiley-VCH, Weinheim, 2001.
[5] I. Manners, Angew. Chem. Int. Ed. Engl. 35 (1996) 1602—1621.
[6] P. Nguyen, P. Gomez-Elipe, I. Manners, Chem. Rev. 99 (1999) 1515—1548.
[7] I. Manners, Synthetic Metal Containing Polymers, Wiley-VCH, Weinham, 2004.
[8] T. Ishikawa, Superbases for Organic Synthesis: Guanidines, Amidines, Phosphazenes and Related Organocatalysts, John Wiley & Sons, Ltd, Chichester, 2009.
[9] R. De Jaeger, M. Gleria, Prog. Polym. Sci. 23 (1998) 179—276.
[10] C.W. Allen, Coord. Chem. Rev. 130 (1994) 137—173.
[11] H.R. Allcock, R.L. Kugel, J. Am. Chem. Soc. 87 (1965) 4216—4217.
[12] H.R. Allcock, Chem. Mater. 6 (1994) 1476—1491.
[13] H.R. Allcock, Adv. Mater. 6 (1994) 106—115.
[14] L. Zhu, Y. Xu, W. Yuan, J. Xi, X. Huang, X. Tang, et al., Adv. Mater. 18 (2006) 2997—3000.
[15] X.Y. Zhang, X.B. Huang, X.Z. Tang, New J. Chem. 33 (2009) 2426—2430.
[16] J.W. Fu, X.B. Huang, Y.W. Huang, Y. Pan, Y. Zhu, X.Z. Tang, J. Phys. Chem. C 112 (2008) 16840—16844.
[17] X.Y. Zhang, X.B. Huang, X.Z. Tang, J. Mater. Chem. 19 (2009) 3281—3285.
[18] J.W. Fu, X.B. Huang, Y.W. Huang, J.W. Zhang, X.Z. Tang, Chem. Commun. (2009) 1049—1051.
[19] W. Wei, X.B. Huang, X.L. Zhao, P. Zhang, X.Z. Tang, Chem. Commun. 46 (2010) 487—489.
[20] N. Kol, L. Adler-Abramovich, D. Barlam, R.Z. Shneck, E. Gazit, I. Rousso, Nano. Lett. 5 (2005) 1343—1346.
[21] X.Y. Gao, H. Matsui, Adv. Mater. 17 (2005) 2037—2050.
[22] D. Tománek, Curr. Appl. Phys. 2 (2002) 47—49.
[23] X.Y. Zhang, S.K. Manohar, J. Am. Chem. Soc. 127 (2005) 14156—14157.
[24] S.I. Cho, D.H. Choi, S.H. Kim, S.B. Lee, Chem. Mater. 17 (2005) 4564—4566.
[25] M.H. Jin, X.J. Feng, L. Feng, T.L. Sun, J. Zhai, T.J. Li, et al., Adv. Mater. 17 (2005) 1977—1981.
[26] Y. Cui, C. Tao, S.P. Zheng, Q. He, S.F. Ai, J.B. Li, Macromol. Rapid Commun. 26 (2005) 1552—1556.
[27] Z.C. Wang, C.J. Medforth, J.A. Shelnutt, J. Am. Chem. Soc. 26 (2004) 15954—15955.
[28] M.J. Mulvihill, B.L. Rupert, R.R. He, A. Hochbaum, J. Arnold, P.D. Yang, J. Am. Chem. Soc. 127 (2005) 16040—16041.
[29] J.W. Fu, X.B. Huang, Y.W. Huang, L. Zhu, Y. Zhu, X.Z. Tang, Macromol. Mater. Eng. 293 (2008) 173—177.

[30] Y. Zhu, X.B. Huang, J.W. Fu, G. Wang, X.Z. Tang, Mater. Sci. Eng. B 153 (2008) 62–65.
[31] P. Zhang, X.B. Huang, J.W. Fu, Y.W. Huang, Y. Zhu, X.Z. Tang, Macromol. Chem. Phys. 210 (2009) 792–798.
[32] H.R. Allcock, S.R. Pucher, A.G. Scopelianos, Macromolecules 27 (1994) 1–5.
[33] H.R. Allcock, S.R. Pucher, A.G. Scopelianos, Macromolecules 27 (1994) 1071–1075.
[34] H.R. Allcock, T.J. Fuller, K. Matsumura, Inorg. Chem. 21 (1982) 515–521.
[35] S.E. Ibim, M.A. Ambrosio, M.S. Kwon, H.R. Allcock, L.C.T. Laurencin, Biomaterials 18 (1997) 1565–1569.
[36] H.R. Allcock, K.B. Visscher, Chem. Mater. 4 (1992) 1182–1187.
[37] M.A. Ambrosio, H.R. Allcock, D.S. Katti, C.T. Laurencin, Biomaterials 23 (2002) 1667–1672.
[38] L.Y. Qiu, K.J. Zhu, Polym. Int. 49 (2000) 1283–1288.
[39] C.S. Reed, K.S. Tenhuisen, P.W. Brown, H.R. Allcock, Chem. Mater. 8 (1996) 440–447.
[40] Y.J. Cui, X. Zhao, X.Z. Tang, Y.P. Luo., Biomaterials 25 (2004) 451–457.
[41] H.R. Allcock, K.B. Visscher, Y.B. Kim, Macromolecules 29 (1996) 2721–2728.
[42] O. Boussif, T. Delair, C. Brua, L. Veron, A. Pavirani, H.V.J. Kolbe, Bioconjugate Chem. 10 (5) (1999) 877–883.
[43] E. Carretti, L. Dei, P. Baglioni, R.G. Weiss, J. Am. Chem. Soc. 125 (17) (2003) 5121–5129.
[44] E.S. Forzani, V.M. Solis, E.J. Calvo, Anal. Chem. 72 (21) (2000) 5300–5307.
[45] M.Z. Lu, H.L. Lan, F.F. Wang, S.J. Chang, Y.J. Wang, Biotechnol. Bioeng. 70 (5) (2000) 479–483.
[46] L. Yin, X.B. Huang, X.Z. Tang, Polym. Degrad. Stabil. 92 (2007) 795–801.
[47] K.Y. Baek, M. Kamigaito, M. Sawamoto, Macromolecules 34 (2001) 215–221.
[48] Y.L. Zhao, Q. Cai, J. Jing, X.T. Shuai, J.Z. Bei, F.C. Chen, Polymer 43 (2002) 5819–5825.
[49] T. Beiela, A. Duda, K. Rode, H. Pasch, Polymer 44 (2003) 1851–1860.
[50] Q. Cai, Y.L. Zhao, J.Z. Bei, F. Xi, S.G. Wang, Biomacromolecules 44 (2003) 828–834.
[51] C.M. Dong, K.Y. Qiu, Z.W. Gu, X.D. Feng, Macromolecules 34 (2001) 4691–4696.
[52] X.X. Kong, T. Kawei, J. Ake, T. Iyoda, Macromolecules 34 (2001) 1837–1844.
[53] D.A. Tomaila, P.R. Dvornic, Nature 372 (1994) 617–618.
[54] N. Higashi, T. Koga, M. Niwa, Adv. Mater. 12 (2000) 1373–1375.
[55] Y.J. Cui, X.Z. Tang, X.B. Huang, Y. Chen, Biomacromolecules 4 (2003) 1491–1494.
[56] Y.L. Zhao, X.T. Shuai, C.F. Chen, F. Xi, Chem. Mater. 15 (2003) 2836–2843.
[57] J. Roovers, L. Zhou, P.M. Toporowski, M. Zwan, H. Iatrou, N. Hadjichristidis, Macromolecules 26 (1993) 4324–4331.
[58] J. Ueda, M. Matsuyama, M. Kamigaito, M. Sawamoto, Macromolecules 31 (1998) 557–562.
[59] J. Xia, X. Zhang, K. Matyjaszewski, Macromolecules 32 (1999) 4482–4484.
[60] S.B. Lee, S.C. Song, J.I. Jin, Y.S. Sohn, Macromolecules 34 (2001) 7565–7569.
[61] R.J. Waltman, D.J. Pocker, H. Deng, N. Kobayashi, Y. Fujii, T. Akad, et al., Chem. Mater. 15 (2003) 2362–2375.
[62] J.Y. Chang, H.J. Ji, M.J. Han, S.B. Rhee, S.K. Cheong, M.J. Yoon, Macromolecules 27 (1994) 1376–1380.
[63] W.Z. Yuan, L. Zhu, X.B. Huang, S.X. Zheng, X.Z. Tang, Polym. Degrad. Stabil. 87 (2005) 503–509.
[64] H.R. Allcock, P.E. Austain, Macromolecules 14 (1981) 1616–1622.
[65] W.Z. Yuan, L. Zhu, X.B. Huang, S.X. Zheng, X.Z. Tang, Eur. Polym. J. 41 (2005) 1867–1873.
[66] M.S. Wong, C. Bosshard, F. Pan, P. Gunter, Adv. Mater. 8 (1996) 677–680.
[67] W. Moerner, A. Jepsen, C. Thompson, Annu. Rev. Mater. Sci. 27 (1997) 585–623.
[68] S.R. Marder, B. Kippelen, A.K.-Y. Jen, N. Peyghambarian, Nature 388 (1997) 845–851.
[69] L. Yu, J. Polym, Sci. Part A: Polym. Chem. 39 (2001) 2557–2564.
[70] M. Yang, B. Champagne, J. Phys. Chem. A 17 (2003) 3942–3951.
[71] V. Ostroverkhov, R.G. Petschek, K.D. Singer, R. Twieg, J. Chem. Phys. Lett. 340 (2001) 109–115.
[72] S.V. Elshocht, T. Verbiest, M. Kauranen, L. Ma, H. Cheng, K. Musick, et al., Chem. Phys. Lett. 309 (1999) 315–320.
[73] H. Pan, X.L. Gao, Y. Zhang, P.N. Prasad, Chem. Mater. 7 (1995) 816–821.
[74] A.K. Jen, Y.M. Cai, P.V. Bedworth, S.R. Marder, Adv. Mater. 9 (1997) 132–135.
[75] H.R. Allcock, C. Cameron, T. Skloss, S. Meyers, J. Haw, Macromolecules 29 (1996) 233–238.
[76] J. Zyss, Nonlinear Opt. 1 (1991) 3–18.
[77] P. Wang, P.W. Zhu, W. Wu, H. Kang, C. Ye, Phys. Chem. Chem. Phys. 1 (1999) 3519–3525.
[78] B. Cho, S. Lee, K. Kim, Y. Son, J. Doo, G. Lee, et al., Chem. Mater. 13 (2001) 1438–1440.
[79] E.M. Breitung, C.F. Shu, R.J. McMahon, J. Am. Chem. Soc. 122 (2000) 1154–1160.
[80] Z.Y. Hu, A. Fort, M. Barzoukas, A.K.-Y. Jen, S. Barlow, S.R. Marder, J. Phys. Chem. B 108 (2004) 8626–8630.
[81] R.M. Christopher, S. Ermer, M.L. Steven, I.H. McComb, D.S. Leung, R. Wortmann, et al., J. Am. Chem. Soc. 118 (1996) 12950–12955.
[82] J. Zyss, J. Chem., Phys. 98 (1993) 6583–6599.
[83] Y. Pan, X.Z. Tang, Synth. Met. 159 (2009) 1796–1799.
[84] D.J. O'Brien, L.H. Roth, A.J. McAloon, J. Membr. Sci. 166 (2000) 105–111.
[85] M.D. Luccio, C.P. Borges, T.L.M. Alves, Desalination 147 (2002) 161–166.
[86] L.M. Vane, J. Chem. Technol. Biotechnol. 80 (2005) 603–629.
[87] L. Li, Z.Y. Xiao, S.J. Tan, L. Pa, Z.B. Zhang, J. Membr. Sci. 243 (2004) 177–187.
[88] F.J. Xiangli, Y.W. Chen, W.Q. Jin, N.P. Xu, Ind. Eng. Chem. Res. 46 (2007) 2224–2230.
[89] T. Masuda, M. Takatsuka, B.Z. Tang, T. Higashimura, J. Membr. Sci. 49 (1990) 69–83.
[90] Y.W. Huang, J.W. Fu, Y. Pan, X.B. Huang, X.Z. Tang, Sep. Sci. Technol 66 (2009) 504–509.
[91] F.B. Peng, F.S. Pan, H.L. Sun, L.Y. Lu, Z.Y. Jiang, J. Membr. Sci. 300 (2007) 13–19.
[92] L.M. Vane, V.V. Namboodiri, T.C. Bowen, J. Membr. Sci. 308 (2008) 230–241.
[93] T.C. Bowen, R.G. Meier, L.M. Vane, J. Membr. Sci. 298 (2007) 117–125.
[94] I.F.J. Vankelecom, D. Depre, S.D. Beukelaer, J.B. Uytterhoeven, J. Phys. Chem. 99 (1995) 13193–13197.
[95] J.W. Fu, X.B. Huang, L. Zhu, X.Z. Tang, Scripta Mater. 58 (2008) 1047–1049.
[96] Y.W. Huang, P. Zhang, J.W. Fu, Y.B. Zhou, X.B. Huang, X.Z. Tang, J. Membr. Sci. 339 (2009) 85–92.
[97] A.C. Gabino, G.I.A. Francisco, G.E. Paloma, J.I. Fidalgo, J.L. Garcia, Chem. Eur. J. 9 (2003) 3833–3836.
[98] Y.B. Zhou, X.B. Huang, X.Z. Tang., Polym. Int. 58 (2009) 710–714.

[99] Y.B. Zhou, X.B. Huang, X.Q. Kang, J.W. Fu, Y. Zhu, X.Z. Tang., Macromol. Mater. Eng. 294 (2009) 605–610.
[100] Y.B. Zhou, X.B. Huang, Y.W. Huang, X.Z. Tang, J. Appl., Polym. Sci. 113 (2009) 473–478.
[101] Y.B. Zhou, X.B. Huang, X.Z. Tang, Polym. Plast. Technol. Eng. 48 (2009) 950–952.
[102] P. Król, B. Pilch-Pitera, Polymer 44 (2003) 5075–5101.
[103] J. Xi, X. Qiu, X. Ma, M. Cui, J. Yang, X. Tang, et al., Solid State Ionics 176 (2005) 1249–1260.
[104] K. Edelmann, B. Sandner, Solid State Ionics 170 (2004) 225–232.
[105] J. Paulsdrof, M. Burjanadze, K. Hagelschur, H.D. Wiemhöfer, Solid State Ionics 169 (2004) 25–33.
[106] M.B. Armand, J.M. Chabagno, M. Duclot, et al., Fast Ion Transport in Solid, Elsevier, New York, 1979.
[107] F. Croce, G.B. Appetecchi, L. Persi, B. Scrosati, Nature 394 (1998) 456–458.
[108] H.R. Allcock, Phosphorus Sulfur 179 (2004) 661–671.
[109] A. Singh, N.R. Krogman, S. Sethuraman, L.S. Nair, J.L. Sturgeon, P.W. Brown, et al., Biomacromolecules 7 (2006) 914–918.
[110] L. Zhu, Y. Zhu, Y. Pan, Y. Huang, X.B. Huang, X.Z. Tang, Macromol. React. Eng. 1 (2007) 45–52.
[111] Y. Zhu, X.B. Huang, W.Z. Li, J.W. Fu, X.Z. Tang, Mater. Lett. 62 (2008) 1389–1392.
[112] L.F. Li, H.S. Lee, H. Li, X.Q. Yang, K.W. Nam, W.S. Yoon, et al., J. Power Sources 184 (2008) 517–521.
[113] B. Xie, H.S. Lee, H. Li, X.Q. Yang, J. McBreen, L.Q. Chen, Electrochem. Commun. 10 (2008) 1195–1197.
[114] E. Zygadło-Monikowska, Z. Florjańczyk, A. Tomaszewska, M. Pawlicka, N. Langwald, R. Kovarsky, et al., Electrochim. Acta 53 (2007) 1481–1489.
[115] N.S. Choi, S.W. Ryu, J.K. Park, Electrochim. Acta 53 (2008) 6575–6579.
[116] P.Y. Pennarun, S. Papaefthimiou, P. Yianoulis, P. Jannasch, Sol. Energy Mater. Sol. Cells 91 (2007) 330–341.
[117] J. McBreen, H.S. Lee, X.Q. Yang, X. Sun, J. Power Sources 89 (2000) 163–167.
[118] T. Mizumo, K. Sakamoto, N. Matsumi, H. Ohno, Electrochim. Acta 50 (2005) 3928–3933.
[119] T. Itoh, Y. Mitsuda, T. Ebina, T. Uno, M. Kubo, J. Power Sources 189 (2009) 531–535.
[120] T.A. Luther, F. Stewart, J.L. Budzien, R.A. LaViolette, W.F. Bauer, M.K. Harrup, et al., J. Phys. Chem. 107 (2003) 3168–3176.
[121] M. Salomon, M. Xu, E.M. Eyring, S. Petrucci, J. Phys. Chem. 98 (1994) 8234–8244.
[122] C-W. Nan, D.M. Smith, Mater. Sci. Eng. B 10 (1991) 99–106.
[123] C-W. Nan, Prog. Mater. Sci. 37 (1993) 1–116.
[124] W. Wieczorek, D. Raducha, A. Zalewska, J. Phys. Chem. B 102 (1998) 8725–8731.
[125] W. Wieczorek, A. Zalewska, D. Raducha, Z. Florjanczyk, J.R. Stevens, Macromolecules 29 (1996) 143–155.
[126] J. Maier, Prog. Solid State Chem. 23 (1995) 171–263.
[127] N. Sata, K. Eberman, K. Eberl, J. Maier, Nature 408 (2000) 946–949.
[128] I. Yilgor, J.E. McGrath, Adv. Polym. Sci. 86 (1988) 1–86.
[129] J.H. Simon, G.J. Richard, in: F. Ganachaud, et al. (Eds.), Silicon Based Polymers, Springer Science+Business Media B.V., 2008, pp. 249–277.
[130] A. Muzafrov, E. Rebrov, in: P.R. Dvornic, M.J. Owen (Eds.), Silicon-Containing Dendritic Polymers, Advances in Silicon Science 2, Springer Science + Business Media B.V, 2009, pp. 21–30.
[131] V. Chandrasekhar, Inorganic and Organometallic Polymers, Springer, New York, 2005.
[132] W. Noll, Chemistry and Technology of Silicones, Academic Press, New York, 1968.
[133] M.G. Voronkov, V.P. Mileshkevich, Y.A. Yuzhelevskii, The Siloxane Bond, Consultants Bureau, New York, 1978.
[134] E.G. Rochow, Silicon and Silicones, Springer-Verlag, Berlin, Heidelberg, 1987.
[135] P.F. Bruins, Silicone Technology, Wiley, New York, 1970.
[136] C. Eaborn, Organosilicon Compounds, Butterworths Scientific Publications, London, 1960.
[137] I. Goodman, Developments in Block Copolymers, Applied Science Publishers, London, 1982.
[138] A. Petrov, B. Mironov, V. Ponomorenko, E. Chernyshev, Synthesis of Organosilicon Monomers, Consultants Bureau, New York, 1964.
[139] W. Lynch, Handbook of Silicone Rubber Fabrication, Van Nostrand and Reinhold, New York, 1978.
[140] J.E. Mark, H.R. Allcock, R. West, Inorganic Polymers, Prentice-Hall, New Jersey, 1992.
[141] M. Cypryk, J. Chrusciel, E. Fossum, K. Matyjaszewski, Makromol. Chem. Macromol. Symp. 73 (1993) 167–176.
[142] K. Sakamoto, K. Obata, H. Hirata, M. Nakajima, H. Sakurai, J. Am. Chem. Soc. 111 (1989) 7641–7643.
[143] C. Aitken, J.F. Harrod, E. Samuel, J. Organomet. Chem. 279 (1985) C11–C13.
[144] R.G. Jones, W. Ando, J. Chojnowski, Silicon-Based Polymers: The Science and Technology of their Synthesis and Applications, Kluwer Academic Publishers, Dordrecht, 2000.
[145] J. Chojnowski, M. Cypryk, J. Kurjata, Prog. Polym. Sci. 28 (2003) 691–728.
[146] T. Ganicz, W.A. Stańczyk, Prog. Polym. Sci. 28 (2003) 303–329.
[147] J. Chojnowski, J. Kurjata, S. Rubinsztajn, M. Ścibiorek, Synthesis and property of some polysilaethers–polyoxymultisilylenes, in: A.R. Bassindale, P.P. Gaspar (Eds.), Frontiers of Organosilicon Chemistry, The Royal Society of Chemistry, Cambridge, 1991, pp. 70–80.
[148] J. Chojnowski, J. Kurjata, Macromolecules 27 (1994) 2302–2309.
[149] J. Kurjata, M. Ścibiorek, W. Fortuniak, J. Chojnowski, Organometallics 18 (1999) 1259–1266.
[150] H. Stüger, M. Eibl, E. Hengge, I. Kovacs, J. Organomet. Chem. 431 (1992) 1–15.
[151] J. Kurjata, J. Chojnowski, Makromol. Chem. 194 (1993) 3271–3286.
[152] M. Kumada, K. Tamao, Adv. Organomet. Chem. 6 (1968) 19–117.
[153] M.Z. Cui, Zh.Y. Li, X.B. Huang, J.Y. Xi, X.Z. Tang, S.X. Zheng, Polymer 46 (2005) 9162–9169.
[154] M.Z. Cui, Zh.Y. Li, J.Y. Xi, S.X. Zheng, X.Z. Tang, J. Polym. Sci. A Polym. Chem. 43 (2005) 2476–2482.

CHAPTER

15

Synthetic Chemistry of Nonstoichiometric Compounds

Jilin Zhang, Guangyan Hong

Changchun Institute of Applied Chemistry, Chinese Academy of Sciences, China

In solid-state chemistry, nonstoichiometric compounds constitute a special type of solid-state compounds. These compounds possess definite structure and thermodynamics characteristic, which differ from the stoichiometric counterpart and a mixture. These nonstoichiometric compounds exhibit different properties such as conductivity, magnetism, catalytic nature, color, and other unique solid-state properties which have important technological applications. Moreover, some unusual information on solid-state chemistry, stability, and dynamics can be revealed through exploring their structure and characteristic. Therefore, study on the nonstoichiometric compounds has become an active research field in solid-state chemistry and defect chemistry.

In chemistry, the law of definite proportions states that a chemical compound always contains exactly the same proportion of elements by mass. An equivalent statement is the law of constant composition, which states that all samples of a given chemical compound have the same elemental composition. According to this law, the proportion of elements by mass for all compounds should not have any change irrespective of their origin, existence form, preparation, and determination methods. All chemical compounds obeying the law of definite proportions are called as stoichiometric compounds. These are also referred to as Daltonides, which came from John Dalton who advocated theories of the law of definite proportions in the nineteenth century. Although the law of definite proportions is very useful in the foundation of modern chemistry, it is not true universally. For crystalline compounds the precise stoichiometric composition is the exception rather than the rule because real crystals always contain defects.

As opposed to the stoichiometric compounds or Daltonides, the nonstoichiometric compounds are also known as Berthollides from the name of chemist's Claude Louis Berthollet. Nonstoichiometric compounds are the chemical compounds deviated from stoichiometry, namely their elemental composition cannot be represented by a ratio of well-defined natural numbers, and therefore violate the law of definite proportions. The composition of a nonstoichiometric compound usually varies in a continuous manner over a narrow range. Wustite, ferrous oxide, is an example of a nonstoichiometric compound with chemical formula $Fe_{1-x}O$, where x may have values varying from 0.05 to 0.15 and its physical and chemical properties vary depending on the value of x [1].

Nonstoichiometry is very pervasive existence amongst the solid compounds of transition metals containing lanthanide and actinide metals (e.g., $Fe_{1-x}O$, $Ni_{1-x}O$, CeO_{2-x}, UO_{2+x}) [2]. In fact, nonstoichiometry is becoming a rule rather than exception and represents a challenge and new opportunity in chemistry. It can be foreseen that a discussion on the synthesis and characterization of the nonstoichiometric compounds is an important task to explore their specific properties and potential applications.

In this chapter, definition and classification of the nonstoichiometric compounds as well as point defects in the nonstoichiometric compounds are introduced. In the next part, some important synthesis methods of the nonstoichiometric compounds are summarized. In the last part, characterization methods on the nonstoichiometric compounds are presented.

Modern Inorganic Synthetic Chemistry, DOI: 10.1016/B978-0-444-53599-3.10015-0

15.1. NONSTOICHIOMETRIC COMPOUNDS

15.1.1. Definition of Nonstoichiometric Compounds

In solid-state chemistry, the nonstoichiometric compound represents a uniform physical phase in which unit cell parameters vary with its composition in a continuous manner, and free energy is function of composition and temperature of the system. That is to say, "it is to be defined in strictly operational terms; the criterion is that a crystalline compound, in equilibrium with its environment, behaves as a thermodynamically bivariant system. The chemical potentials of the components are a continuous function of the composition of the solid, over an observable range of composition." [1] The structural and thermodynamical criteria on the nonstoichiometric compounds are in agreement with their nature. Structure is related to microcosmic characteristic, while thermodynamics deals with macroscopical system.

Structurally, though there exists a broad-range change in composition for a nonstoichiometric compound, the X-ray diffraction pattern always possesses characteristic of the parent crystallography cell, and the unit cell parameters continuously vary with its composition [3]. This is different from a crystalline stoichiometric compound or a mixture composed of crystalline stoichiometric compounds for which all have own characteristic X-ray diffraction patterns and unit cell parameters.

Schottky and Wagner pointed out that at any temperature above 0 K, every solid-state compound exists in a single phase in the range of composition change. Thus, the stoichiometrically ideal crystal has no special status. A nonstoichiometric solid-state substance may be stated according to the following two aspects: (1) the stoichiometric compound defined by pure chemistry is a uniform physical phase in which nonstoichiometric composition can be determined by means of chemical analysis, X-ray diffraction analysis, equilibrium vapor pressure, etc., for instance, FeO_{1+x} and FeS_{1+x}; and (2) it is difficult to examine the concentration of lattice defects by measurement of chemical analysis or X-ray diffraction analysis due to the small composition deviation. But they can be studied by measuring their optical, electrical, and magnetic properties [4].

Thermodynamically, taking G (free energy)-x (composition) relation of two component systems at a fixed temperature for example, we explain the concept of what a nonstoichiometric compound is. In a system, nonstoichiometric phase AB is produced by coexistence of two linear phases A_2B and AB_2, as shown in Fig. 15.1. Dots a and b stand for free energies of the pure phases A

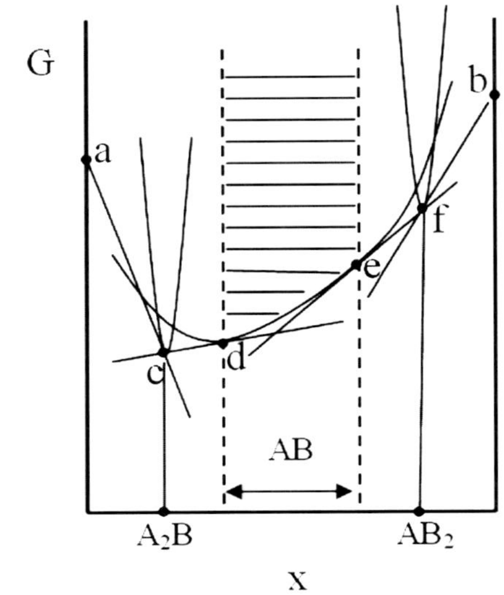

FIGURE 15.1 Relationship curve of the nonstoichiometric compound AB and the linear phases A_2B and AB_2.

and B, respectively, and dots c and f do free energies of accurate stoichiometric compounds A_2B and AB_2, respectively. If the linear phases A_2B or AB_2 deviate from their stoichiometric compositions to form an A-rich or B-rich single phase, the corresponding free energies will become very high, as two sharp wedge-shaped curves shown in Fig. 15.1. If coexistence of the pure A phase and the linear phase A_2B is permitted, the corresponding free energy will be simple addition of the G(A) and $G(A_2B)$, which is denoted using a tangent ac. Similarly, the corresponding free energy for addition of the G(B) and $G(AB_2)$ is denoted using a tangent bf. Asymmetry line de indicates that free energy of nonstoichiometric phase AB gently varies with its composition in a continuous manner due to easy formation of point defects. But the type of its crystal structure in the homogenetic region of existence of nonstoichiometric compound remains unaltered with change in its composition. This is key characteristic and general phenomenon of a nonstoichiometric phase which is produced by the point defects of thermal equilibrium [5]. In contrast, some radiation-induced defects (e.g., a pair of defects (an anion vacancy and an anion interstitial) in the colored NaCl formed after ionizing radiation) are not those of thermal equilibrium and disappear upon annealing. Strictly speaking, compound such as colored NaCl often is considered as defect compound, not a nonstoichiometric compound [6].

Most nonstoichiometric compounds can be expressed by such formulas as WO_{3-x}, $Co_{1-x}O$, or $Zn_{1+x}O$ $(0 < x < 1)$, because they have compositions that are close to those of stoichiometric compounds. But in certain cases apparent nonstoichiometry results from the existence of homologous series of stoichiometric compounds, such as the series of molybdenum oxides having the formulas Mo_nO_{3n-1} ($n = 8, 9, 10, 11, 12$, and 14); a true nonstoichiometric compound MoO_{3-x} would show a continuous variation of x within some range of values, and no discrete species would be detectable [7]. Similarly, the niobium oxides $NbO_{2.500}$, $NbO_{2.4906}$, $NbO_{2.4800}$, $NbO_{2.4681}$, $NbO_{2.4545}$, and $NbO_{2.467}$ do not represent compositions within some range of stoichiometry, but are a succession of discrete, crystallographically defined compounds $Nb_{28}O_{70}$ (Nb_2O_5), $Nb_{53}O_{132}$, $Nb_{25}O_{62}$, $Nb_{47}O_{116}$, $Nb_{22}O_{54}$, and $Nb_{12}O_{29}$ [1]. Therefore, the term "nonstoichiometric compound" can be occasionally used loosely.

15.1.2. Classification of Nonstoichiometric Compounds

Most of the nonstoichiometric compounds are transition metal oxides, but also include hydrides, carbides, nitrides, sulfides, tellurides, fluorides, and so on [5,8,9]. The nonstoichiometric compounds existing only in the condensed state are often classified in terms of element composition.

The simplest way to classify nonstoichiometric compounds is to consider which element is in excess and how this excess is brought about. For example, a perovskite-type nonstoichiometric oxide such as oxygen-deficient ABO_{3-x}, A-deficient $A_{1-x}BO_3$, or B-deficient $AB_{1-x}O_3$ often appears. In the first case, an oxygen vacancy would be formed, and in the second and third cases deviation from stoichiometric composition (A:B = 1:1) would result in the formation of some lattice imperfections. Based on the criterion, nonstoichiometric compounds can be divided into four types [10]: (1) anion vacancy type (e.g., TiO_{2-x}, ZrO_{2-x}); (2) cation vacancy type (e.g., $Fe_{1-x}O$, $Cu_{2-x}O$); (3) anion interstitialcy type (e.g., UO_{2+x}); (4) cation interstitialcy type (e.g., $Zn_{1+x}O$, $Cd_{1+x}O$).

On the basis of composition range, the nonstoichiometric compounds can be classified into two groups: (1) stoichiometric deviations in nonstoichiometric compounds are very small, but they are detectable with special analytical experiments or methods based on direct determination of carrier concentrations. Examples are PbS_{1+x}, Cu_2O_{1+x}, ZnO_{1-x}, alkali halides containing F-centers, etc. It is reasonable to assume that the stoichiometric deviation is small enough for the lattice defects to be isolated and randomly distributed. (2) The crystals show gross departure from ideal stoichiometry (e.g., the nonstoichiometric rare earth oxides, etc). Their lattice defects are not regarded as randomly distributed due to existence of the interaction between point defects. The nonstoichiometry is due to assimilation of vacancies or interstitial atoms as structure elements of the crystal [11]. The former is referred to as narrow-range nonstoichiometric compounds which deviate only slightly from the stoichiometric compounds, while the latter is so-called wide-range nonstoichiometric compounds which grossly departure from ideal stoichiometry [5].

15.1.3. Point Defects in the Nonstoichiometric Compounds

The existence of nonstoichiometric compounds indicates the presence of defects in the lattice structures of crystalline substances, such as the absence of ions from sites that would normally be occupied. But it should be noted that there is no absolute defect-free crystals at $T > 0$ K, so that the presence of defects is not by itself an indication of nonstoichiometry. For example, a sodium chloride crystal that lacks a sodium ion and a chloride ion is defective, but still stoichiometric because the numbers of sodium ions and chloride ions are the same; if the sodium ion site, however, is filled by a neutral sodium atom, which then gives up its valence electron to fill the chloride ion site, the crystal defect is remedied, but the crystal is now nonstoichiometric because it contains more sodium ions than chloride ions [7].

The defects are mainly crystallographic point defects such as interstitial atoms and vacancies resulting from excess or deficiency of a component element, respectively. In fact, point defects are an important cause of formation of the nonstoichiometric compounds. Moreover, ionic and electronic point defects determine fundamental properties of functional materials composed of the nonstoichiometric compounds.

The point defects can be formed by creating vacancy and interstitial up to a certain concentration lead to decrease in free energy. The concepts of vacancy and interstitial are as follows: if an atom is not present on the site that it should occupy in a perfect crystal, then a vacancy is located at that site, while if an atom is present on any site that would be unoccupied in a perfect crystal then that atom is an interstitial.

The nonstoichiometric compounds include two types of important point defects: Frenkel defect and Schottky defect. A nearby pair of a vacancy and an interstitial are often called a Frenkel defect or Frenkel pair. The phenomenon is named after the previous Soviet physicist Yakov Frenkel, who discovered it in 1926. When there is a large difference in size between the cations and anions, the crystals with low coordination numbers and large interstitial sites will tend to favor Frenkel defects by displacing the smaller ion, for example Al_2O_3. For large cations (U^{4+}, Ce^{4+}, and Th^{4+}), the defects formed in UO_2, CeO_2, and ThO_2 are anion Frenkel defects. A vacancy (or pair of vacancies in an ionic solid) is sometimes called a Schottky defect named after

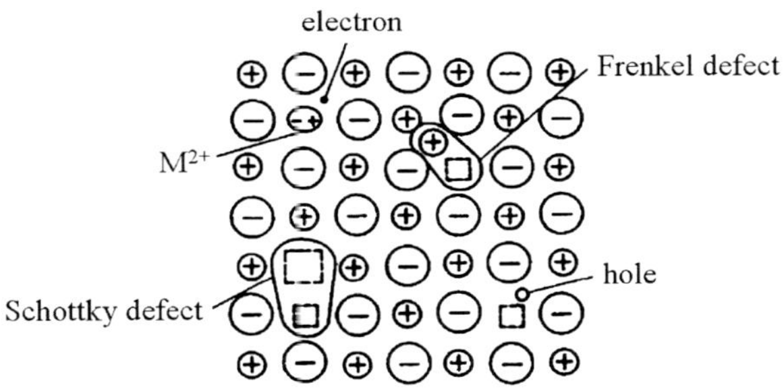

FIGURE 15.2 Schematic of point defects in ionic crystal lattice.

Walter H. Schottky. It can be formed when oppositely charged ions leave their lattice sites, creating vacancies. Schottky defects in crystals with high coordination numbers such as MgO are found to easily form. Vacancies may be also present on different sites for the same type of ion. For example, $NiFe_2O_4$ is an inverse spinel $[Fe(NiFe)O_4]$; Fe^{3+} ions sit in both tetrahedral and octahedral sites. Thus, we can have Fe^{3+} vacancies on the tetrahedral or octahedral sites, but these two types of vacancies are not the same [12]. Normally, Schottky defects will bring about a decrease in the density of the crystal. The point defects in ionic crystal lattice can be illustrated schematically with a two-dimensional diagram (see Fig. 15.2).

For historical reasons, many point defects especially in ionic crystals are called "centers"; for example, the vacancy in many ionic solids is called an F-center. The origin of the name, F-center, originates from the German Farbzentrum. F-center is a type of crystallographic defect in which an anionic vacancy in a crystal is filled by one or more electrons, depending on the charge of the missing ion in the crystal. That is to say, these F-centers are due to an electron or a hole trapped at an anion vacancy site. Electrons in such a vacancy tend to absorb light in the visible spectrum. Such a material that is usually transparent becomes colored. The color is the complement of what is absorbed by the color center. Colors are produced in most of the alkali halides following irradiation with ionizing radiation and are characteristic of the crystal, not the radiation source used. Several different types of color centers have been identified and some of the common ones are illustrated in Fig. 15.3 for alkali halides [12].

The point defects still employ symbols of the Kröger–Vink defect notation, which has been adopted by the IUPAC [13]. In this notation, structural elements are denoted as S_P^C. Here S represents the symbol of point defect, P its position in the lattice, and C the effective charge of the defect relative to the perfect crystal lattice. For this part of notation, dots ($^\bullet$), dashes ($'$), and ($^\times$) represent effective positive charges, effective negative

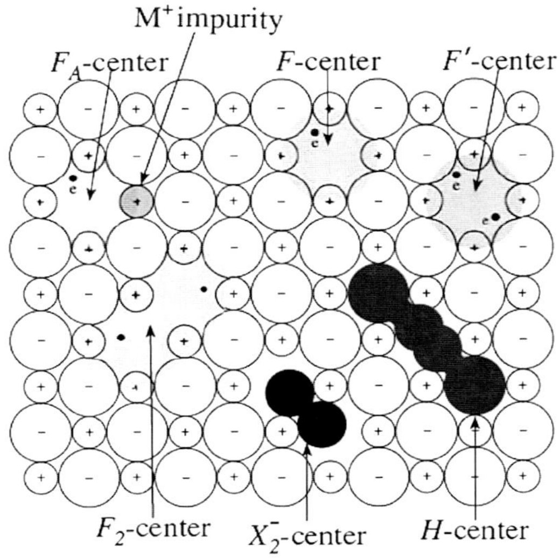

FIGURE 15.3 Schematics of color centers in alkali halides [12].

charges, and an effective zero charge, respectively. Examples of the symbols of intrinsic point defects are:

■ Vacancy	V
■ Metal ion (M^+) vacancy	V_M
■ Metal ion (M^+) interstitial	$M_i^{\cdot}$
■ Empty interstitial site	$V_i^{\times}$
■ Oxide ion vacancy	$V_O^{\cdot\cdot}$
■ Magnesium ion (Mg^{2+}) vacancy	V_{Mg}''
■ Magnesium ion (Mg^{2+}) on a regular lattice site	$Mg_{Mg}^{\times}$

It is the above-mentioned intrinsic defect of nonstoichiometric compounds that endow their specific electrical, optical, magnetic, chemical, and mechanical properties and make them functional materials.

15.2. PREPARATION OF NONSTOICHIOMETRIC COMPOUNDS

15.2.1. Formation Mechanism of Nonstoichiometric Compounds

An ideal crystal with a periodic, three-dimensional and completely ordered ensemble of particles is only a physical model of perfect solid. Real crystals always have various types of defects. The presence of defects means that an exact stoichiometric composition of crystalline compounds is more an exception than a rule [9]. It is the defect in a crystal that leads to the formation of a nonstoichiometric compound. Cr_2O_3 is an example for narrow nonstoichiometric oxide. The alternative responses of intrinsic defects in the Cr_2O_3 to chemical

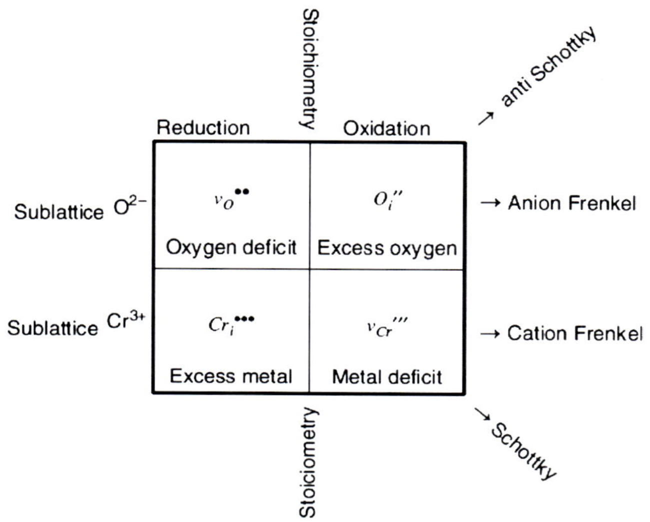

FIGURE 15.4 Possible structural defect compensations for chemical reduction (to the left of the center) and oxidation (to the right) of a pure, stoichiometric Cr_2O_3. *Reprinted from Journal of Solid State Chemistry, Vol. 179, Pavel Karen, Nonstoichiometry in oxides and its control, Page 3167, Copyright (2006), with permission from Elsevier [6].*

oxidation and reduction are outlined in Fig. 15.4 [6]. When the precisely stoichiometric Cr_2O_3 is reduced, a deficit of oxygen or an excess of metal can be created. Upon oxidation, either an excess of oxygen or a deficit of metal may be the response of structure.

At a given temperature there is an equilibrium concentration of point defects in a crystal. This means that a crystal containing point defects must have a lower free energy (ΔG) than the corresponding crystal without any defect. The ΔG minimum corresponds to equilibrium concentration of point defects in the crystal. Therefore, formation of a nonstoichiometric compound can be understood from minimization of the free energy and diffusive equilibrium of the point defects.

Based on thermodynamics, the "point defects" are created in a solid lattice due to increase in the configurational entropy, which gives rise to minimization of free energy of the system [4–6,12]. The configurational entropy ($kT \ln \Omega$) is a driving force behind the vacancy creation. Where $\Omega (=(N_0 + n)!/(N_0!n!))$ is the number of permutation ways calculated as n vacancies can be arranged in the $N_0 + n$ sites. Gibbs free energy originates from both entropy and enthalpy contributions and can be calculated by using expression $\Delta G = \Delta H - \Delta S$. For formation of n vacancies with a low-vacancy concentration, change of the Gibbs energy is $\Delta G = n(\Delta H - T\Delta S_v) - kT \ln \Omega$. ΔH deals with the enthalpy of the vacancy formation by breaking of some bonds, while the ΔS_v is the vibration entropy associated with the formation of the vacancy. ΔG in the system depends on two terms: $n(\Delta H - T\Delta S_v)$ and $\mathrm{k}T \ln \Omega$. Both factors show different behaviors with increase of the defect concentration in the system, as shown in Fig. 15.5. It can be seen from the curves that the introducing vacancies can decrease the free energy of the crystal until an equilibrium concentration corresponding to a small equilibrium amount of vacancies ($n_{eq.}$) is reached. This is because the configurational entropy term is nonlinear, therefore, adding too many vacancies gives rise to increase in ΔG. The $n_{eq.}$ value can be evaluated by differentiation according to n:

$$\frac{d}{dn}\left(n\Delta H - nT\Delta S_v - kT \ \mathrm{In}\frac{(N_0 + n)!}{N_0!n!}\right) = 0.$$

For large value of x the factorial term can be simplified by application of the Stirling's formula, $\ln(x!) \approx x \ln x - x$:

$$\Delta H - T\Delta S_v - kT \ \mathrm{In}\left(\frac{N_0 + n_{eq}}{n_{eq}}\right) = 0.$$

Note that the term involving $n_{eq.}$ is the inverse of the fractional concentration of the vacancies formed. A rearrangement shows that the fractional concentration of vacancies is a sole function of temperature, with two parameters (ΔH and ΔS_v), involved in a mass action-type term:

$$\frac{n_{eq}}{N_0 + n_{eq}} = \exp\left(\frac{\Delta S_v}{k} - \frac{\Delta H}{kT}\right).$$

It can be seen from the above equation that at higher temperatures the equilibrium number of vacancies $n_{eq.}$ increases and leads to formation of the stable nonstoichiometric phase. This implies that the nonstoichiometric phase at relatively higher temperature is more stable than one at low temperature due to possessing a lower free energy in a higher defect concentration (see Fig. 15.6). If such equilibrium in the crystal without

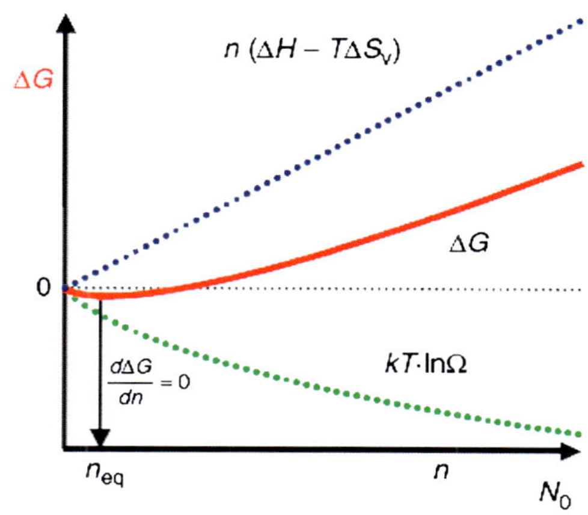

FIGURE 15.5 Gibbs energy of formation of n vacancies in an ideal crystal of N_0 sites. *Reprinted from Journal of Solid State Chemistry, Vol. 179, Pavel Karen, Nonstoichiometry in oxides and its control, Pages 3167, Copyright (2006), with permission from Elsevier [6].*

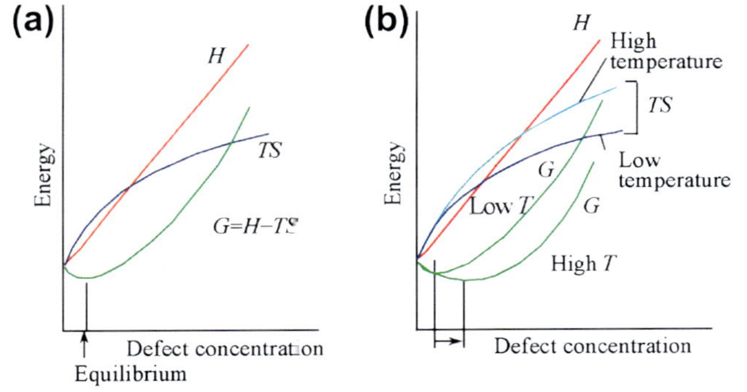

FIGURE 15.6 Change of G minimum and the equilibrium concentration of defects in crystal lattices with temperature increase [14].

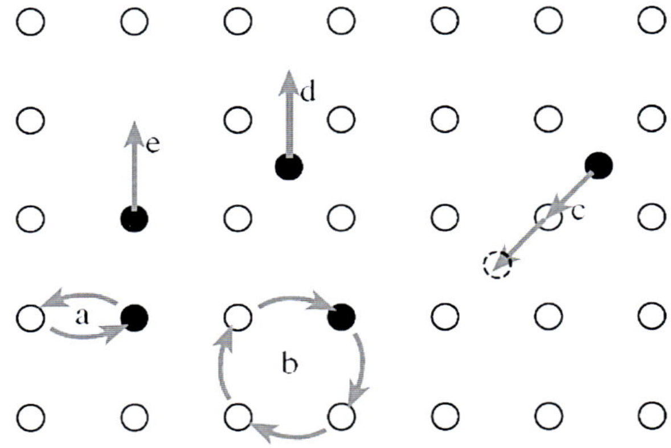

FIGURE 15.7 Possible mechanisms occurring during diffusion. (a) Exchange, (b) ring, (c) vacancy, (d) interstitial, and (e) knock-on [12].

defect clusters is achieved at a relatively high temperature, the equilibrium concentrations of noninteracting point defects can be treated by the mass action law. Since these defects are in thermodynamic equilibrium, annealing or other thermal treatments cannot remove them. Thus, the point defects in the nonstoichiometric compounds are mainly responsible for formation of the nonstoichiometric compounds.

In addition, the compounds containing changeable valence state ions are in favor of formation of the nonstoichiometric compounds, because the vacancy and nonstoichiometry are closely related to a peculiar electron configuration and inscrutable relationship between the valence instability and the electron transfer (e.g., rare earth oxides). Meanwhile, it should be pointed out that nonstoichiometric compounds are basically highly ionic compounds, as formation of the point defects in a covalent compound crystal usually needs very high energy [5].

Based on kinetics, diffusion which occurs by atomic defects moving through the crystal is a thermally activated process. There are four basic mechanisms that can, in principle, occur. These are illustrated in Fig. 15.7 [12].

- Direct exchange (difficult and not energetically probable)
- Ring mechanism (cooperative motion: possible but not demonstrated)
- Vacancy mechanism (most common, dominant in metals)
- Movement of interstitials (occurs when you can put interstitials in); an example is Zn in ZnO.

The diffusion coefficient (D) has a strong dependence on temperature (T). Large numbers of experimental results show that the higher the system temperature is, the more remarkable the diffusion phenomenon is; the lesser the activation energy is, the larger the diffusion coefficient is.

$$D = D_0 \exp\left(\frac{-Q}{RT}\right)$$

Here R is gas constant and Q is activation energy (an activation energy barrier to diffusion).

The following examples are applied to illustrate special features of diffusion in ceramics.

- $Zn_{1+x}O$ is a metal-rich oxide with two different charges in which excess zinc is present as interstitials. As the partial pressure of Zn (p_{Zn}) increases, the density of interstitials increases and the diffusion of Zn through the structure increases.
- The group of binary oxides with rocksalt structure (e.g., FeO, NiO, CoO, and MnO) is metal-deficient. The diffusion depends on pO_2 [12]. The most important of these is $Fe_{1-x}O$ because FeO is unstable and can contain as many as 15% vacancies due to the equilibrium concentration of Fe^{3+} cations. $Fe_{1-x}O$ is special because point defects are always present in FeO and the cation is always present in two charge states.
- CuO_{1-x} is an example of a group of oxygen-deficient oxides (MO_{1-x}), where the nonstoichiometry is accommodated by oxygen vacancies. With increase of the pO_2, the diffusion coefficient will decrease. Comparison of CuO_{1-x} and $Zn_{1+x}O$ showed different crystal structures and point defect chemistry in spite of same cationic charges, 2 + or 1 + for both metals.

In a word, as long as there exists concentration gradient due to inhomogeneous distribution of the atoms or ions in a solid, there must exist chemical potential gradient which will result in directional flow and mutual mixing of the atoms and ions till diffusion finally eliminates the chemical potential gradient and reaches uniform distribution of the component concentration in the system [4].

15.2.2. Synthesis Condition of Nonstoichiometric Compounds

In the MX crystal, $M_{1+x}X$ (or $M_{1-x}X$) represents metal-rich (or metal-deficient) nonstoichiometric compound; while MX_{1+x} (or $MX)_{1-x}$ represents X-rich (or X-deficient) nonstoichiometric compound. The value $1 \pm x$ has a homogeneity range in which the compound still maintains stable and almost unchanged structure. For a nonstoichiometric compound, compositional range depends mainly on interaction energy between defects, temperature, and the intrinsic out-of-order distribution fraction (N^0/N_i, N^0 is number of the intrinsic defects; N_i is number of the interstitial M atoms). There are three conditions resulting in easy formation of a nonstoichiometric compound: (1) the energy used for produced point defects is not large, namely, change of the free energy brought about by the x alteration is small; (2) it is easy for atomic valence alteration of the M due to smaller energy difference between various oxidation states of the M; and (3) difference in the atomic radii between various oxidation states is not too large. Many transition metal compounds such as oxides, hydrides, carbides, nitrides, sulfides, and tellurides can meet the above conditions to form corresponding nonstoichiometric compounds. But nonstoichiometry in metal oxides has been of great interest because it sensitively affects physical and chemical properties of materials. If one gas component (e.g., O_2) takes part in reaction and atoms can transfer between crystal and gas phase to form vacancies and interstitials in the crystal, we will obtain the corresponding nonstoichiometric compounds. So an important approach to prepare nonstoichiometric oxides is through adding or removing O_2 molecule in the metal oxide crystal [15].

Optimization of synthesis conditions of the nonstoichiometric compound with precise composition can be chosen using a stable region phase diagram. The experimental technique which determines the stable region of the nonstoichiometric compounds is described as follows. (1) A compound which is close to the desired composition is kept in a synthesis reactor in which the temperature and oxygen partial pressure can be well controlled for a long time. Equilibrium at an elevated temperature can be reached from two opposite directions: oxidation and reduction. The equilibrium is related to the oxygen exchange between the sample and the surrounding gas atmosphere. The oxygen-exchange reaction is composed of two processes. "The first is the surface reaction where O_2 splits into two oxygen anions and two holes or vice versa. This is the actual chemical oxidation (or reduction when in the opposite direction), because the holes represent oxidized specie of the transition metal. The subsequent process is a diffusion-driven homogenization during which the surface oxygen excess (or its lack) homogeneously extends across the bulk of the sample." [6] (2) The sample is taken out after the system temperature is quenched to 0°C or so. "Appropriate cooling regime is critical to all techniques of nonstoichiometry control. During cooling, high-temperature gradients may be generated between the surface and the center of the sample, resulting in oxidation of the surface at the expense of the center. For best results, very rapid cooling should be adopted when the sample has access to oxygen; certainly in all open systems. Very slow cooling should be adopted for the low $p(O_2)$ getter techniques where no O_2 is left to oxidize the sample." [6] (3) We determine its composition using an exact chemical analysis method and analyze its substance phase by X-ray analysis method. Through the above operation steps repeated again and again, we can find the relationship of temperature, oxygen partial pressure, and composition (T, $p(O_2)$, x), and hence get to the bottom of the stable region. (4) We can obtain optimization parameters of the synthesis condition from the above experimental results.

The phase relation in metal–oxygen system is a basic factor in designing the advanced materials with high performance. For example, Fig. 15.8A illustrates the superposition of stability region of σ-phase on the phase diagram of Pr-O system [16]. The difference in stability

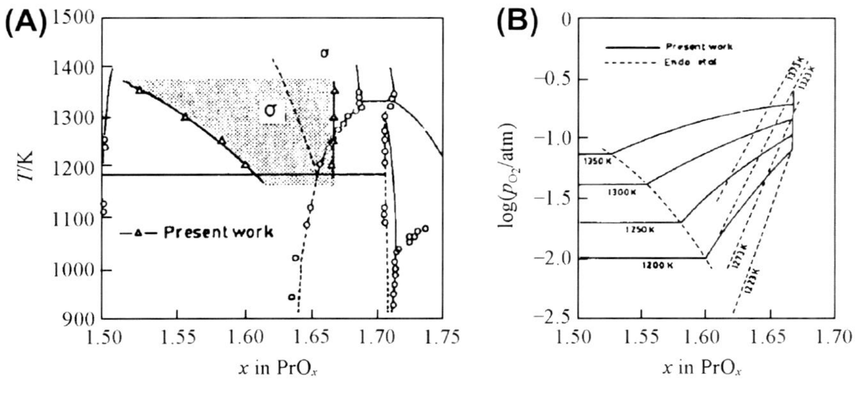

FIGURE 15.8 Stability region of the σ-phase (shadow area) superposed on the phase diagram of Pr-O system (A) and Logarithm of the equilibrium oxygen partial pressure ($\log p(O_2)$) of PrO_x as a function of composition (x) (B) [16].

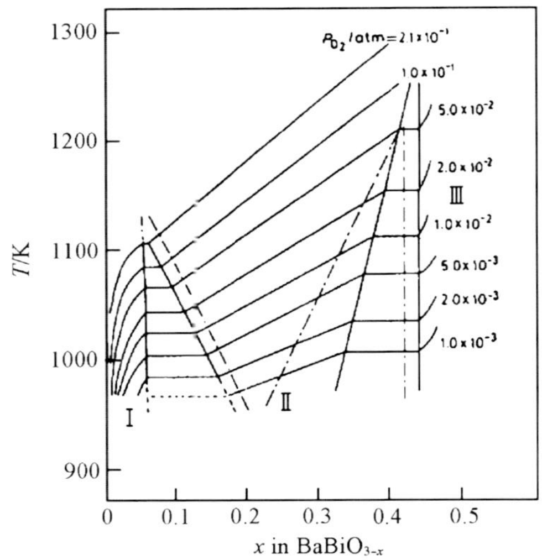

FIGURE 15.9 Relation between temperature (T) and composition (x) in $BaBiO_{3-x}$ at the oxygen partial pressures indicated. The dashed and chained lines indicate two-phase regions and phase boundaries, respectively [16].

region of σ-phase is clearly shown and the existence of line phase, $PrO_{1.667}$, is conclusively indicated. The stable region of the σ-phase (shadow area) is clearly shown. In Fig. 15.8B, log $p(O_2)$ is plotted against x and the results obtained by Endo et al. are also plotted in dotted lines for comparison [16]. It reveals the relationship of the composition and oxygen partial pressure. Figure 15.9 exhibits the relation between temperature (T) and composition (x) in $BaBiO_{3-x}$ at the oxygen partial pressures in the temperature range of 975−1300 K [16]. At 1050 K, two-phase regions lie in the range of $x = 0.05-0.11$, whereas two-phase regions of II and III are found at higher temperatures and larger nonstoichiometry (i.e., $x = 0.37-0.44$ at 1100 K).

By the way, the atmosphere with a high accuracy oxygen partial pressure can be obtained using a buffering gas mixing method and a simple gas mixing method [15,17]. There are several control parameters that may be used to maintain the high-temperature equilibrium and establish the desirable oxygen content in the sample. A more detailed discussion of these methods may be found in Ref. [6].

The oxygen partial pressure can be measured using an electrochemistry method [15,17]. The cell structure in Fig. 15.10 is described as follows:

$$Pt|O_2(p^*(O_2))|(ZrO_2)_{0.85}\bullet(CaO_2)_{0.15}|CO_2\text{-}H_2(p^*(O_2))|Pt$$
$$\text{In}(p(O_2)/p^*(O_2)) = -4EF/RT$$

Here E, F, and $p^*(O_2)$ stand for electromotive force, Faraday constant, and the known oxygen partial pressure, respectively, in the counter electrode (e.g., $p^*(O_2) = 1$ atm.). The sensor is used to measure the $p(O_2)$ in automobile exhaust gases as well as gases present in the carburizing process used to harden steel components.

15.2.3. Synthesis of Nonstoichiometric Compound

Preparing nonstoichiometric compounds with predetermined composition, structure, and properties is of great interest due to following reasons. First of all, design and tailoring of materials with specific electronic, magnetic, and optical properties depend on effective and precise control of the composition, defect, and structure. Second, presence of certain nonstoichiometric defects is necessary and desirable, such as in ionic and/or electronic conductors in batteries, fuel cells, gas separating membranes, and chemical sensors. Third, some functional properties such as magnetic coupling or valence mixing are coupled to specific valence states of the metal atoms in these oxides, but a small undesirable nonstoichiometry may frustrate the desired properties.

As mentioned above, the stable nonstoichiometric phase of a transition metal oxide is usually formed at high temperature and corresponding defect equilibrium is established in reasonable time. At the same time, the phase composition of the nonstoichiometric compounds depends strictly on temperature and oxygen partial pressure as well as annealing or quenching process. Therefore, for the nonstoichiometric compounds a dominant synthesis technique is high-temperature synthesis, including solid-state reactions, thermal decomposition, and gas−solid synthesis at elevated temperatures. But then, precursors of the nonstoichiometric compounds

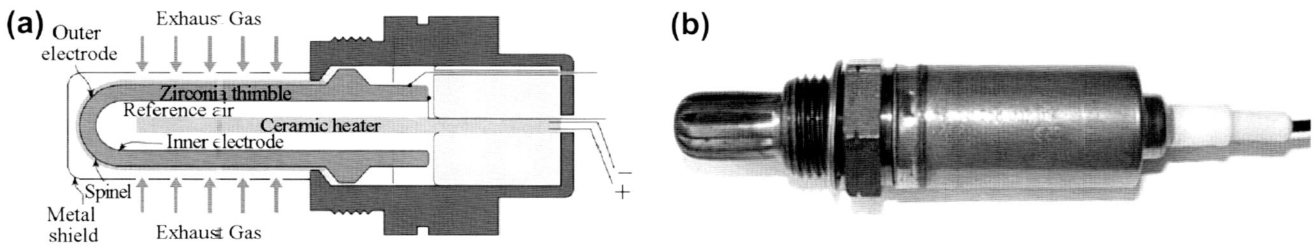

FIGURE 15.10 Schematic of a ZrO_2-based oxygen sensor and an actual sensor unit [12].

may be prepared by various chemistry-based synthetic methods such as coprecipitation method, sol–gel techniques, and hydrothermal and solvothermal processing.

15.2.3.1. High-Temperature Solid-State Reaction Method

(1) $YBa_2Cu_3O_{7-x}$

The high-temperature superconductor $YBa_2Cu_3O_{7-x}$ is the first material to become superconducting above 77 K (the boiling point of nitrogen). All materials developed before 1986 became superconducting only at temperatures near the boiling points of liquid helium or liquid hydrogen ($T_b = 20.28$ K)—the highest being Nb_3Ge at 23 K. Several high-temperature superconducting devices used for magnetic resonance imaging, magnetic levitation, and Josephson junctions have realized commercial application.

Structural feature of the $YBa_2Cu_3O_{7-x}$ is illustrated in Fig. 15.11 [18]. $YBa_2Cu_3O_{7-x}$ can be viewed as a variant of the perovskite family of materials. For the cuprates, Y + Ba occupy "A sites", whereas Cu occupies the "B sites". The structure of the $YBa_2Cu_3O_{7-x}$ depends on oxygen content. When $x = 1$, O(1) sites in the Cu(1) layer are vacant and the structure is tetragonal $YBa_2Cu_3O_6$, which is insulating and does not superconduct. Increasing the oxygen content slightly causes more of the O(1) sites to become occupied. For $x < 0.65$, Cu–O chains along the b-axis of the crystal are formed. Elongation of b-axis changes the structure to orthorhombic, with lattice parameters of $a = 3.82$, $b = 3.89$, and $c = 11.68$ Å. Optimum superconducting properties occur when $x \sim 0.07$ and all of the O(1) sites are occupied with few vacancies.

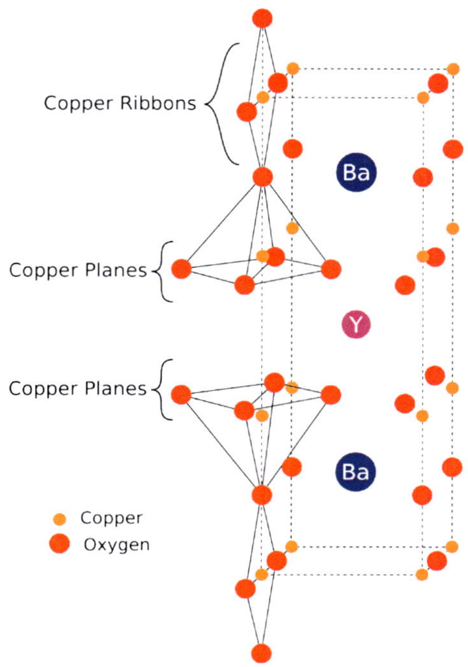

FIGURE 15.11 The structure unit of $YBa_2Cu_3O_{7-x}$ [18]

The oxygen in the $YBa_2Cu_3O_{7-x}$ plays a vital role in the structure and superconductivity of $YBa_2Cu_3O_{7-x}$ materials and is dependent on the value of x (i.e., its oxygen content). In addition, the annealing procedure in the presence of oxygen is a key to obtain better superconducting properties, too. For these reasons, T-$p(O_2)$-x phase equilibrium diagram, as shown in Fig. 15.12 [16], is necessary for us to optimize synthetic parameters of $YBa_2Cu_3O_{7-x}$ superconductor.

According to T-$p(O_2)$-x phase equilibrium diagram, polycrystalline samples of $YBa_2Cu_3O_{7-x}$ superconductor may be prepared either by powder mixing method or by coprecipitation technique. For the former, high purity oxides, carbonates, and/or nitrates powders are mixed in an appropriate ratio followed by subsequent heating at desirable high-temperature for a suitable reaction time in air [19]. For the latter, three nitrates solutions of each constituent metal are mixed in an appropriate ratio. The ammonium oxalate solution was then used to form precipitation from the mixed solution which are filtrated, dried, and calcined at desirable temperature in air. In annealing process, the resultant samples were reground, pressed into pellets, and then sintered in air.

The samples with different oxygen contents were prepared by the following method: the sintered pellet samples were equilibrated with an oxygen partial pressure for a time, depending on

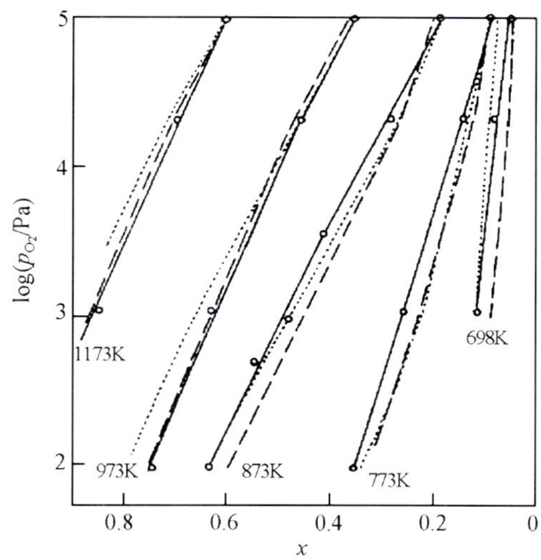

FIGURE 15.12 Variation of the oxygen nonstoichiometry x in $YBa_2Cu_3O_{7-x}$ with the oxygen partial pressure $p(O_2)$ at different temperatures [16].

temperature and oxygen partial pressure, and then cooled rapidly to room temperature in an ice bath. The oxygen partial pressure was controlled with O_2/air or air/Ar mixed gas.

The atomic compositions of Ba and Y in $YBa_2Cu_3O_{7-x}$ can be chemically analyzed by precipitating Ba as $BaSO_4$ and Y as Y_2O_3, respectively. The atomic composition of Cu was determined by titration using $Na_2S_2O_3$ solution. The equilibrium value of oxygen nonstoichiometry can be determined through thermogravimetric measurement in conjunction with chemical analysis. The homogeneity of the samples can be confirmed by X-ray diffraction analysis [19].

The samples of $YBa_2Cu_3O_{7-x}$ superconductor film can be prepared by combination of sol–gel method and high-temperature annealed technique. There are two routes to be used: according to route I, $Y(NO_3)_3 \cdot 5H_2O$, $Cu(NO_3)_2 \cdot H_2O$, and $Ba(NO_3)_2$ agents are used as starting materials. These chemical agents with an appropriate ratio are dissolved in ethylene glycol to form a uniform mixed solution, which is then refluxed at an appropriate temperature (e.g., 130–180°C) and the gel is gradually created as the solvent is vaporized out of the reaction system. The resultant gel can be coated on the surface of a matrix and is sintered at a high temperature (e.g., 950°C) in oxygen atmosphere with a desirable oxygen partial pressure, which results in $YBa_2Cu_3O_{7-x}$ film sample with orthogonal structure. In route II, organometallic compounds $Y(OC_3H_7)_3$, $Cu(O_2CCH_3)_2 \cdot H_2O$, and $Ba(OH)_2$ with an appropriate ratio are dissolved under heating with vigorous stirring in ethylene glycol. After formation of a uniform mixed solution, the solvent is vaporized to obtain a gel. The resultant gel is then coated on surface of a matrix (e.g., [110] plane of sapphire, [100] plane of $SrTiO_3$ single crystal, and [001] plane of ZrO_2 single crystal). In order to obtain 10–100 μm of uniform $YBa_2Cu_3O_{7-x}$ film with good superconducting properties, two annealing ways have been employed. In the first way, the film is heated to 400°C in the presence of oxygen using a program control means at the rate of 2°C per minute, then to 950°C at the rate of 5°C per minute with subsequent cooling to room temperature at the rate of 3°C per minute. This step is repeated 2–3 times and the film is sintered at 950°C in oxygen atmosphere for 12 h, and finally is cooled to room temperature at the rate of 3°C per minute. In the second way, the film is first calcined at 950°C for 10 min in air. Then, the film is coated again and is subsequently calcined again and the step is repeated several times. Finally, the film is annealed at 550–950°C in oxygen atmosphere for 5–12 h [20].

(2) TiO_{2-x}

Titanium dioxide is often regarded as an oxygen-deficit nonstoichiometric compound (TiO_{2-x}) whose nonstoichiometry (x) is closely related to the concentration of both ionic and electronic defects. Recent studies have shown that TiO_2 not only exhibit n-type semiconductor property, but also have p-type semiconductor property (i.e., formula of the p-type TiO_2 may be expressed as TiO_{2+x}). Therefore, the general formula, which represents both n- and p-type TiO_2, is $TiO_{2\pm x}$ [21]. Titanium dioxide has been considered as the most promising candidate for photoelectrochemical water-splitting devices and photocatalytic water purification. Figure 15.13 gives structure unit of rutile TiO_2 [22].

There are mainly five kinds of O atoms: bridging (O_b), sub-bridging (O_{sb}), sub-bridging-2 (O_{sb2}), surface (O_s), and subsurface (O_{ss}), as shown in Fig. 15.13a. There are two types of Ti atoms on the rutile TiO_2 (110) surface. One is the sixfold-coordinated Ti between the two bridging O atoms, and the other is a fivefold-coordinated Ti in the subsurface.

For oxides the defect-related properties depend on the specimens obtained by following well-known processing conditions at elevated temperatures, including the temperature, oxygen activity, and the rate of cooling. The gas/solid equilibrium for the oxide–oxygen system (e.g., TiO_2–O_2 system) may be established at elevated temperatures.

At lower temperatures, oxidation of metal oxides is mainly limited to physical adsorption. At moderate temperatures, the physically adsorbed species (molecules and/or atoms) become ionized, leading to the formation of several chemisorbed species, such as O^- and O^{2-}. Catalytic and photocatalytic properties of oxides are closely related to the concentration of these species. At elevated temperatures, oxidation of metal oxides results in oxygen incorporation into the lattice. This leads to a change in the ability of the oxide crystal to donate or accept electrons during chemical reactions [21]. The defect diagram for undoped TiO_2 at 1173 K and 1123 K is shown in Fig. 15.14. The defect diagram shows the effect of oxygen activity on the concentration of both ionic and electronic defects for TiO_2 at 1173 K and 1123 K [21,23].

An example on the synthesis of TiO_{2-x} using a high-temperature solid-state reaction method is described as follows [24]. Samples of nonstoichiometric cubic titanium monoxide TiO_y ($y = 2 - x$) with different oxygen contents y ($0.74 < y < 1.26$) were synthesized by solid-phase sintering from a mixture of powders of metallic titanium and titanium dioxide TiO_2 in a vacuum of 10^{-3} Pa and at a temperature of 1770 K. The samples

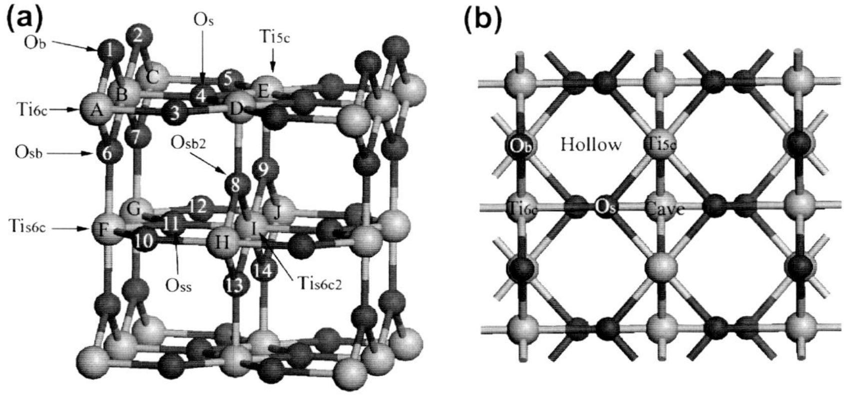

FIGURE 15.13 Structure of rutile TiO_2 (110) surface: (a) side view and (b) top view. The large light and small dark spheres represent Ti and O atoms, respectively. The numbers and the block letters are used for identification purposes. *Reprinted with permission from [22]. Copyright 2009 American Chemical Society.*

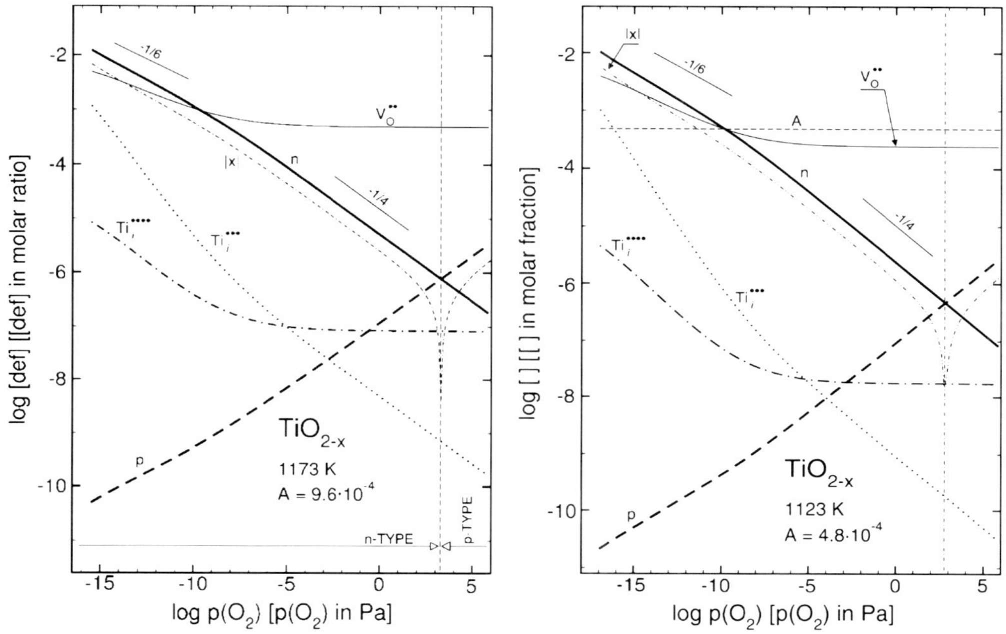

FIGURE 15.14 Defect disorder diagram for undoped TiO_2, showing the effect of oxygen activity on the concentration of ionic and electronic defects at 1173 K and 1123 K, where A corresponds to the equilibrium concentration of titanium vacancies. *Reprinted with permission from [21,23]. Copyright 2008 American Chemical Society. Sheppard, L. R.; Nowotny, M. K.; Bak, T.; Nowotny, J.: Effect of cooling on electrical conductivity of TiO2. Physica Status Solidi (b). 2008. Volume 245, Page 1816. Copyright Wiley-VCH Verlag GmbH & Co. KGaA. Reproduced with permission.*

were transferred to disordered state by calcination for 3 h at a temperature of 1330 K, after which the ampoules with the samples were quenched in water with a rate of 200 K/s. To obtain the ordered state, quenched samples of disordered titanium monoxide were calcined in evacuated quartz ampoules for 3 h at a temperature of 1330 K, followed by slow cooling from 1330 to 300 K with a rate of 10 K/h. Calcination produced ordered samples, only with a monoclinic structure. Structural and chemical characterizations of the samples showed that as the oxygen content in titanium monoxide increases ($TiO_{0.74}$–$TiO_{1.26}$), the number of vacancies in the oxygen sublattice decreases from 25 to 3 at% and the number of titanium vacancies grows from 8 to 23 at%.

15.2.3.2. High-Temperature Thermal Decomposition Method

(1) UO_{2+x}

Uranium oxides are known as nonstoichiometric compounds. Their compositions alter with external conditions such as temperature and oxygen partial pressure. Uranium dioxide can adsorb additional oxygen in the lattice but adsorption decreases below 300°C. Phase diagrams of UO_2–U_3O_8 (left) and UO_2–U_4O_9 (right) systems are shown in Fig. 15.15 [25]. At low temperatures, the stable phases are UO_2, the pseudocubic U_4O_9 phase, two (or more) tetragonal phases, and the orthorhombic U_3O_8 phase. The densities of U_4O_9 and the two tetragonal

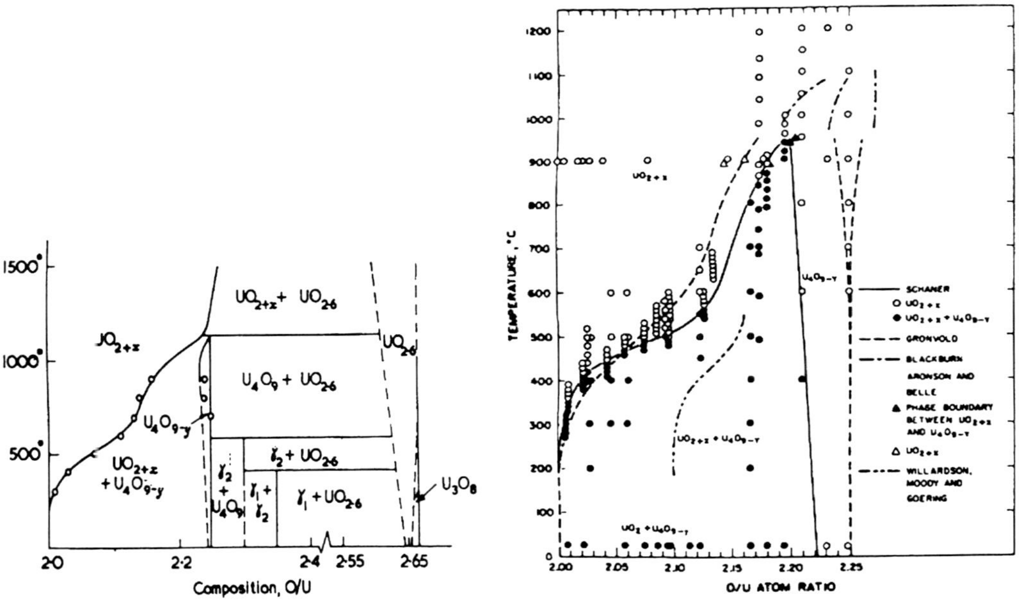

FIGURE 15.15 Phase diagrams of UO_2–U_3O_8 (left) and UO_2–U_4O_9 (right) systems [25].

phases are higher than that of UO_2 due to the additional oxygen being in interstitial positions. Above 300°C, there is a genuine solubility of oxygen in the UO_2 structure, the unit cell contracting linearly as the oxygen concentration increases. It is also evidenced that nonstoichiometric UO_{2+x} phase contains interstitial oxygen.

The nonstoichiometric UO_{2+x} can be prepared by the direct oxidation of stoichiometric uranium dioxide obtained by thermal decomposition of uranyl iodide [25].

(2) FexO

Iron(II) oxide is a black powder with the chemical formula FeO. Its mineral form is known as Wustite. This is a typical example of a nonstoichiometric compound with a known stability range from $x = 0.83$ to 0.96 above 560°C and contains at least 5% vacancies on the cation sublattice. So the formula should be written as FexO [or better $Fe(II)_{1-2\delta}Fe(III)_{\delta}O$]. Figure 15.16 gives iron oxygen phase diagram [5,26].

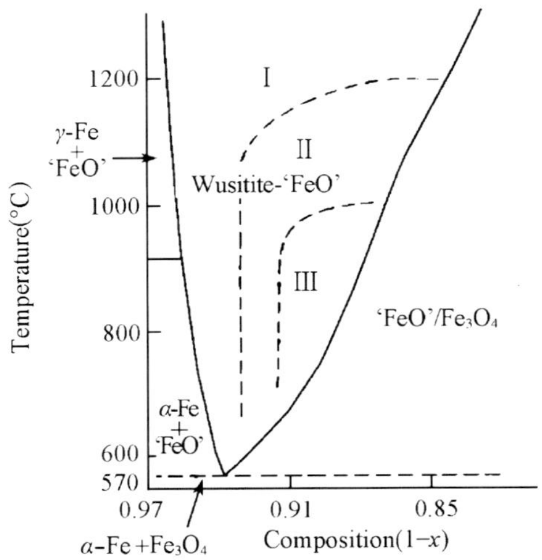

FIGURE 15.16 The iron oxygen phase diagram. *Reprinted from Journal of Physics and Chemistry of Solids, Vol. 30, B.E.F. Fender, F.D. Riley, Thermodynamic properties of $Fe_{1-x}O$. Transitions in the single phase region, Page 793, Copyright (1969), with permission from Elsevier [5,26].*

FexO has a defect rocksalt structure with an ordered distribution of iron vacancies (see Fig. 15.17) [17,27]. Fe_xO is paramagnetic at room temperature and antiferromagnetic or weakly ferrimagnetic below the Néel temperature T_N of about 183 K or 198 K due to a transition from the cubic to a rhombohedral or a monoclinic structure. The transition is strongly related to the defect structure of Wustite.

For the bulk, Wustite Fe_xO was typically prepared by heating iron and magnetite in sealed vessels based on the iron oxygen phase diagram. This is because it is stable only above 560–570°C, while below this temperature it decomposes via a two-step mechanism into α-Fe and magnetite, Fe_3O_4. Recently, Redl et al. [28] reported the synthesis of nanocrystalline Wustite Fe_xO ($0.84 < x < 0.95$) with controllable size and shape using a high-temperature thermal decomposition method. This method is based on the decomposition of iron acetylacetonate, iron acetate, and iron pentacarbonyl precursors in the presence of organic

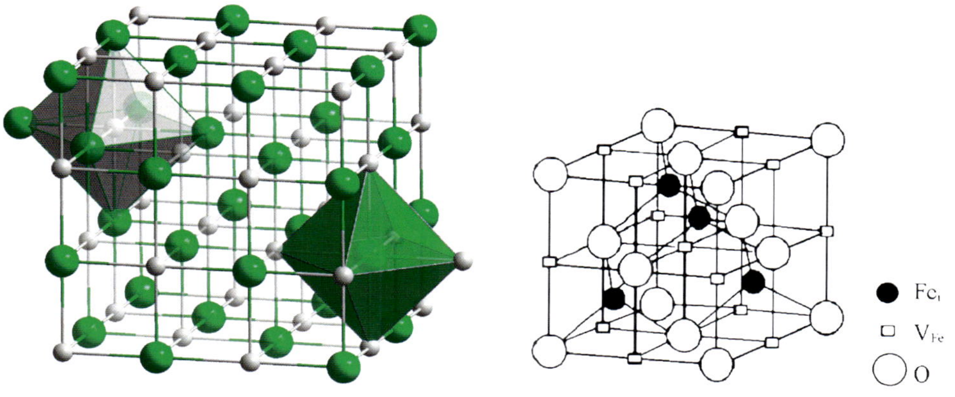

FIGURE 15.17 Crystal structure of Wustite (Fe_xO) [17,27].

solvents with high boiling temperatures. Tight control over temperature allows the syntheses of cubic or faceted Fe_xO nanocrystals with narrow size distributions by thermolysis of iron(II) acetate or a selective oxidation route of iron pentacarbonyl with pyridine N-oxide. The more detailed reaction conditions and results of products may be found in Ref. [28].

15.2.3.3. High-Temperature and High-Pressure Synthesis Method

(1) $Fe_{3-x}O_4$

Fe_3O_4 belongs to a family of inverse spinel ferrites (see Fig. 15.18). It consists of a close-packed face-centered cube of large O^{2-} anions. The smaller cations (Fe^{2+} and Fe^{3+}) are located in the interstitial sites of the anion lattice. All of the Fe^{2+} ions occupy half of the octahedral sites and the Fe^{3+} are distributed evenly across the remaining octahedral sites and tetrahedral sites. That is to say, there exist two kinds of cation sites in the crystal: (i) tetrahedrally coordinated to oxygen (A site), which is occupied only by Fe^{3+} ions; and (ii) octahedrally coordinated to oxygen (B site), which is occupied by equal numbers of Fe^{2+} and Fe^{3+} ions.

Fe_3O_4 is a half-metallic oxide suitable as a source of spin-polarized electrons for spintronic devices. This is because it has a very high degree of spin polarization (half metallicity) at the Fermi level that possesses a high Curie temperature ($\sim$859 K). There is a phase transition at 120 K, the so-called Verwey transition where there is a discontinuity in the structure, conductivity, and magnetic properties. In other words, at room temperature the electron hopping between the Fe^{2+} and Fe^{3+} ions occupying B site results in a high conductivity of manganite. Below $\sim$120 K, the conductivity of magnetite drastically reduces by over two orders of magnitude. The Verwey transition is due to an ordering of the extra electron at the Fe^{2+} B site.

Nonstoichiometric magnetite can be represented by the formula $Fe_{3-x}O_4$, where x can, in principle, range from zero (stoichiometric magnetite) to 0.333 (stoichiometric maghemite). The variations in magnetic properties are related to different parameters, such as cationic distribution and vacancies, nonstoichiometry, spin-canting, or surface contribution. If one considers that hopping occurs only between pairs of ferrous and ferric ions, it can also be represented by the formula [30]:

$$Fe^{3+}\left[{Fe^{2+}}_{1-3x}{Fe^{3+}}_{1-3x}{Fe^{3+}}_{5x-x}\right]O_4$$

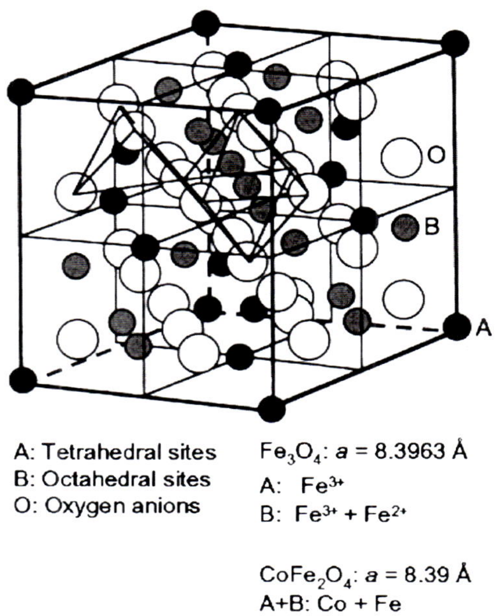

FIGURE 15.18 Schematic model of the spinel unit cell structure. *Reprinted with permission from [29]. Copyright 2004 American Chemical Society.*

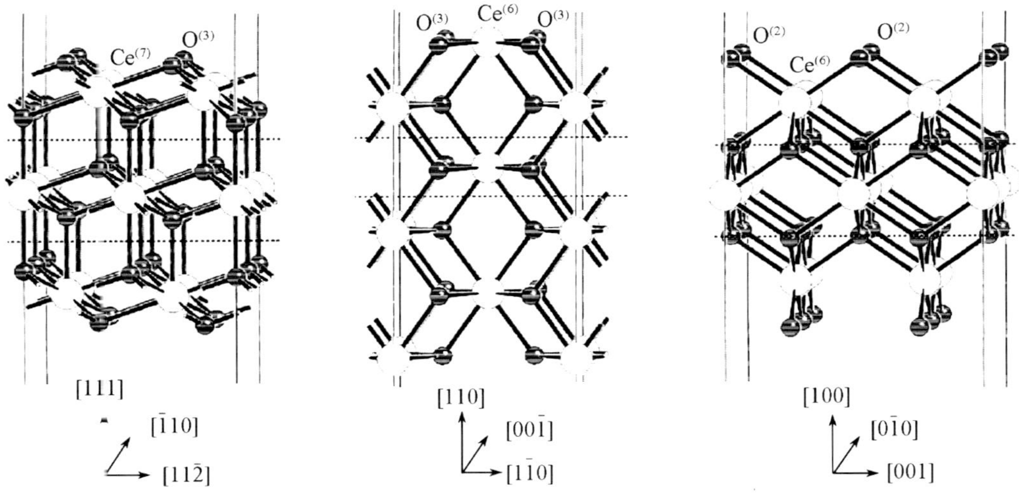

FIGURE 15.19 Bulk truncated CeO_2 (111), (110), and (100) surfaces. The dotted lines indicate the repeating unit (1L) in the direction perpendicular to the surface Small and large circles represent the oxygen and cerium atoms, respectively. *Reprinted from Surface Science Reports, Vol. 62, M. Verónica Ganduglia-Pirovano, Alexander Hofmann, Joachim Sauer, Oxygen vacancies in transition metal and rare earth oxides: Current state of understanding and remaining challenges, Page 219, Copyright (2007), with permission from Elsevier [31].*

Recently, preparation of nonstoichiometric magnetite nanoparticles ($Fe_{2.95}O_4$) was reported by Daou et al. [3]. The nanosized nonstoichiometric magnetite was synthesized by coprecipitation at 70°C from ferrous Fe^{2+} and ferric Fe^{3+} ions by an ($N(CH_3)_4OH$) solution, followed by hydrothermal treatment at 250°C. After analyzing and calculating Mossbauer spectrum data in detail, authors thought that there would be oxidation of Fe^{2+} on the B site in Fe^{3+}accompanied by vacancy formation.

(2) CeO_{2-x}

Pure stoichiometric CeO_2 has a structure of calcium fluoride (fluorite) type with space group Fm3m over the temperature range from room temperature to the melting point. It consists of a cubic array of fourfold coordinated oxygen ions with the metal ions occupying half of the eightfold coordinated cationic interstice. The relative stability of the low-index ceria surfaces decreases in the (111) > (110) > (100) order. Figure 15.19 gives bulk truncated CeO_2 (111), (110), and (100) surfaces [31]. In recent years, research on synthesis of the ceria (i.e., CeO_{2-x}) nanoparticles has received significant attention because of its extensive applications such as excellent automobile exhaust catalysts, additives in ceramics, phosphors, high-energy efficiency fuel cells, and polishing materials [32].

When exposed to an O_2-deficient atmosphere at high temperatures, ceria can lead to the formation of several cerium oxide phases of the type CeO_{2-x} with a range of possible compositions ($0 \leq x \leq 0.5$) due to removal of oxygen from ceria, depending on partial pressure of oxygen. There are five different point-defect species in the CeO_{2-x}: the reduced cerium ion Ce'_{Ce}; two types of oxygen vacancies at different charge states, $V_O^{\cdot\cdot}$ and $V_O^{\cdot}$; and two types of defect pairs formed by the association of Ce'_{Ce} with the appropriate oxygen vacancies, $(CeV_O)^{\cdot}$ and $(CeV_O)^{\times}$. Figure 15.20 shows schematically a planar projection of the CeO_2 lattice containing the above-mentioned defect species [33].

Figure 15.21 shows the phase diagram of CeO_{2-x} constructed by Mogensen et al. [34]. They concluded that CeO_2 can be reduced to nonstoichiometric compositions, CeO_{2-x}, where $1.7 \leq 2-x \leq 2$.

For $T > 921$ K (648°C) and $2 > 2-x > 1.818$, only the α-phase exists. For $T < 722$ K (449°C), CeO_{2-x} forms a discrete set of compositions. The relation between pO_2 and x (in CeO_{2-x}) obtained by thermogravimetry in the temperature range 900−1400°C is within the α-phase region.

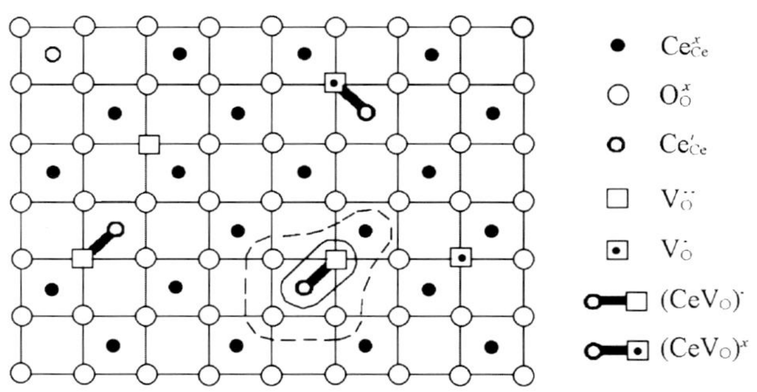

FIGURE 15.20 Schematic of a planar projection of the CeO_2 lattice containing various species of point defects complexes. *Reprinted figure with permission from [33] as follows: S. Ling, Phys. Rev. B, volume 49, page 864, 1994. Copyright (2010) by the American Physical Society.*

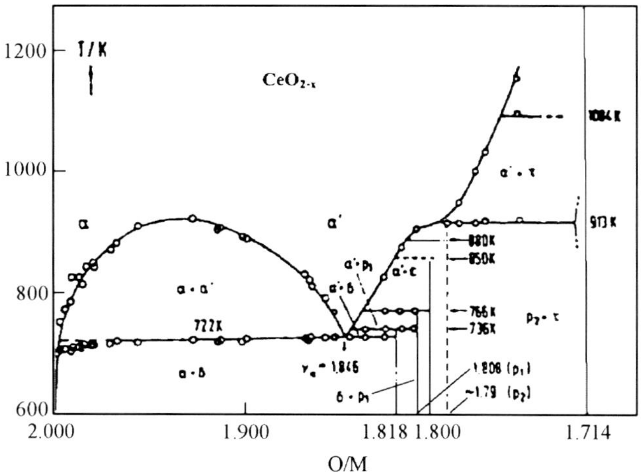

FIGURE 15.21 Phase diagram of CeO_{2-x}. *Reprinted from Solid State Ionics, Vol. 129, Mogens Mogensen, Nigel M. Sammes, Geoff A. Tompsett, Physical, chemical and electrochemical properties of pure and doped ceria, Page 63, Copyright (1994), with permission from Elsevier [34].*

Very recently, solvothermal/hydrothermal technique is used for synthesis of the CeO_{2-x} nanomaterials. During the solvothermal treatment, the pressure in the autoclave is able to be readily adjusted by using different solvents, which have different vapor pressures at a given temperature. There are two examples to be described as follows:

(1) An alcohothermal method with the addition of bases (KOH or NaOH) is used to fabricate uniform ceria nanocrystals with sizes ranging from 2.6 to 6.9 nm. The alcohothermal mechanism is investigated to follow a hydrolytic route [32]: (i) the formation of Ce^{3+} and Ce^{4+} ions when Ce(III) and Ce(IV) nitrates are dissolved; (ii) the formation of polymers of cerium hydroxide coordinated with ethanol; (iii) the transition of hydroxides to hydrated and alcoholized CeO_{2-x}; and (iv) CeO_{2-x} nanocrystals obtained through alcohothermal treatment.

(2) The CeO_{2-x} nanotubes were synthesized by two successive stages: precipitation and aging [35]. In a typical synthesis, at the precipitation stage, 0.9 g of cerium nitrate ($Ce(NO_3)_3 \cdot 6H_2O$) was added to 10 mL of deionized water and heated at 100°C. Once a large amount of vapor formed, 7 mL of 5% ammonia hydroxide solution was added. Very fine yellowish precipitates were formed immediately. After boiling the precipitates for 3 min, the solution was quickly transferred and cooled at 0°C. After 45 days aging, the CeO_{2-x} nanotubes were obtained. The formation of the tubular structure strongly depends on precipitation temperature and aging time.

15.2.3.4. *γ-Irradiation Synthesis Method*

$Cu_{2-x}Se$ is a *p*-type extrinsic semiconductor with an energy gap of 1.1–1.29 eV, which is near the optimum value for solar cell applications. It is stable over a wide range of departures from stoichiometry ($0 < x < 0.15$).

Recently, γ-irradiation technique has become an important method, which can be applied at room temperature and ambient pressure to prepare nanoscale materials. It can provide high energy and an evenly distributed irradiation field. The synthesis of nanocrystalline $Cu_{2-x}Se$ in aqueous solution using sodium selenosulfate as selenium source under γ-irradiation in the field of a 2.59×10^{15} Bq ^{60}Co g-ray source with absorbed dose rate of 1×10^2 Gy min^{-1} at room temperature and ambient pressure was reported by Qiao et al. [36]. The effect pH of value on the solutions and some coordination of copper ions (Cl^- and NH_3) on the $Cu_{2-x}Se$ nanocrystals were studied in detail, while the mechanism of the formation of nanocrystalline selenide in the γ-irradiation process is not very clear up to now.

15.3. CHARACTERIZATION OF NONSTOICHIOMETRIC COMPOUNDS

Characterization of nonstoichiometric compounds mainly deals with determination of their compositions, substance phase, and defects. This is because it is not enough for us to merely consider composition of a nonstoichiometric compound, we must also combine its composition with its structure and properties which depends on the defects in the crystal to think over. For example, composition can be determined by means of chemical analysis; substance phase and structure can be examined through X-ray, electron, and neutron diffraction analyses; and relationship of the composition, structure, and properties can be detected using various physicochemistry methods. So and so only, we can really confirm that the compound is a nonstoichiometric compound rather than stoichiometric compound.

15.3.1. Composition Analysis

Compositions of some nonstoichiometric compounds can be determined through chemical methods. Iodometric and cerimetric redox titrations are suitable for the determination of significant levels of nonstoichiometry in metal oxides. There are two types of element nonstoichiometry to be suitable for iodometric analysis: (1) those in highly oxidizing valence states that reduce to stable nonoxidizing ions in aqueous solutions, the type includes any nonstoichiometry above the following oxidation states such as trivalent Cr, divalent Mn, divalent Co, divalent Ni, trivalent Ce, trivalent Tb, and trivalent Pr; (2) those compounds which are stabilized by formation of soluble or insoluble iodide complex after reduction. An example of the type is determination of any nonstoichiometry above monovalent Cu. Alternatively, cerimetric analysis is more versatile for similar nonstoichiometry determination. The key technique that makes the above titrations simple and applicable to analyses of nonstoichiometry in various solid oxides is closed-ampoule digestion [6].

For the wide-range nonstoichiometric oxides, their composition with a sufficiently high precision can be obtained by the above chemical titration methods. However, for the narrow-range nonstoichiometric compounds, it is very difficult and even impossible to determine their compositions by the chemical titration methods. This is because error for the common quantitative analysis methods is $\pm10^{-3}$, while the nonstoichiometry is generally less than 10^{-3}. But then, for some special nonstoichiometric oxides with different oxidizing valence states such as ZnO_{1-x} ($Zn^{2+}O + Zn^0$) and FeO_{1-x} ($Fe^{2+}O - Fe_2^{3+}O_3$), the metal-excess or metal-deficiency can be determined using the chemical titration methods such as Iodometry and Cerimetry. In addition, ZnO_{1-x} can also be determined by an electroanalysis method [4].

As the ratio of intensities of two six-line spectra composing the complex Mössbauer spectrum of magnetite depend sensitively on x, Volenik et al. [37] reported the applicability of Mössbauer spectroscopy and X-ray diffraction methods for the determination of the deviation of magnetite from stoichiometry. The results showed that among the data obtainable by both methods above, the ratio of intensities of two partial spectra composing the Mössbauer spectrum of magnetite enables to evaluate the deviation of magnetite from stoichiometry quantitatively. Therefore, the measurement of the ratio of intensities of two partial Mössbauer spectra is a suitable method for determining the deviation x of magnetite from stoichiometry.

15.3.2. Structure Determination [4,38]

Undoubtedly, X-ray diffraction techniques are one of the most important methods to confirm phase structure. Polycrystal powder sample can be used for the determination of the phase structure and unit cell parameters. Single crystal sample can directly measure its crystal structure and study its crystal integrality. However, it is difficult to determine the structures of the homologous series phases (e.g., rare earth oxides) by X-ray diffraction due to the weak scattering of X-rays by oxygen atoms.

Neutron diffraction turned out to be the best method to determine the structures of these ordered oxygen-deficient phases. But there exist the following two problems: (1) it is almost impossible to prepare a sample consisting of a single phase because the oxygen content of the sample will easily vary with temperature and oxygen partial pressure; and (2) the refinement of the neutron diffraction data from a powder sample needs proposed models. Those models might come from the knowledge about a structural principle for the homologous series phases.

To overcome the above problems, high-resolution electron microscopy (HRTEM) has become a powerful tool for unraveling the structural features in nonstoichiometric compounds (e.g., rare earth oxides). HRTEM is able to give an insight in the real structure of the rare earth oxides and therefore overcomes the problem of multidomains, even with different compositions. HRTEM opens up the possibility to simultaneously obtain diffraction data in the back focal plane and images showing the projection of the atoms arrangement of a structure at atomic scale in the image plane of the objective lens. The beam size, which can be changed from 100 μm to 1 nm, restricts the area on a sample where the diffraction data and the structure image come from. It is possible to record an electron diffraction pattern and a structure image at time intervals of 1/16 of a second. If the sample is located in an environmental cell of the electron microscope it is possible to record even the valence variation using EELS and a CCD camera. This means that the transition of different homologous series phases either at reduction or oxidation processes can be observed. Therefore HRTEM is a key method to study the higher oxides of the binary rare earths.

15.3.3. Defect Detection

In principle, all physical properties of a crystal are derived from its crystal composition and structure. Moreover, some physical properties such as density, electrical conductivity, optical, and optoelectronic properties strongly depend on kinds and concentration of the defects in the crystal. Accordingly, we can determine existence and concentration of the defects using the physical properties. For instance, because Schottky defects can bring about density decrease of the crystal, we can combine density determination method with X-ray diffraction technique (i.e., the lattice parameter with a high precision can be determined using X-ray diffraction and the precise lattice-parameter method) to confirm concentration of the defects in the crystal. Also, in situ thermogravimetry is one of the most universal methods of nonstoichiometry analysis. The kinds and concentration of the defects in the crystal can also be obtained with high precision thermobalances [4]. Similarly, the charge states of bulk defects can be inferred indirectly from physical and electrical measurements. For example, measurements of TiO_2 electrical conductivity versus oxygen partial pressure have been used to develop defect models that account for charged vacancies and interstitials. The ionic conductivity is proportional to the oxygen partial pressure raised to an exponent that depends upon oxygen vacancy charge state [39].

The electron–positron annihilation method is a promising and universally recognized experimental approach to investigation of solids with point, linear, extended, and bulk defects (e.g., positrons injected into a semiconductor become preferentially trapped in features such as vacancies, vacancy clusters, and negative charge centers, where they subsequently annihilate from the trapped state in the defect) is a promising and universally recognized experimental approach to investigation of solids with point, linear, extended, and bulk defects. As an extra merit, this method of exploring local electronic structure of solids is also sensitive to even very small concentrations of structural vacancies in a solid, from 10^{-6} to 10^{-3} vacancies per atom. The results obtained by the method identify the defects contained in a solid and provide information on the presence or absence of structural vacancies in it, establish the nature of the vacancies—monovacancies, divacancies, or clusters of vacancies, as well as permit estimation of their electronic state throughout the volume of the solid under study [24].

It is very difficult to see individual point defects in a material. It is not due to their small size (i.e., transmission electron microscopy (TEM) has the resolution), but is attributed to the interference caused by presence of surrounding atoms [12]. At present, X-ray absorption near-edge structure (XANES) spectroscopy is one of the few (and probably most straightforward) techniques for the studies on individual cation vacancy and layer-specific hole concentrations: it definitely probes local concentration of holes as the X-ray absorption spectrum is determined by electronic transitions from a selected atomic core level to the unoccupied electronic states near the Fermi level [40].

In addition, electron paramagnetic resonance (EPR) spectra provide detailed structural information about semiconductor's defects, including their symmetry, atomic, and lattice configurations. EPR is equally useful for defect clusters as it is for isolated point defects [41]. Similarly, laser Raman spectroscopy (LRS) is also a powerful means to investigate the structural information on microenvironment of center ions because LRS is sensible to the slight change of crystalline structure. Raman spectroscopy measurement is an efficient way for studying the local atomic arrangement change [41].

References

[1] J.S. Anderson, Proceedings of the Indian Academy of Sciences – Chemical Sciences 93 (1984) 861.

[2] K. Naito, T. Tsuji, T. Matsui, Journal of Radioanalytical and Nuclear Chemistry 143 (1990) 221.

[3] T.J. Daou, G. Pourroy, S. Begin-Colin, J.M. Greneche, C. Ulhaq-Bouillet, P. Legare, et al., Chemistry of Materials 18 (2006) 4399.

[4] M.Z. Su, Introduction to Solid State Chemistry, Peking University Press, Beijing, 1987, pp. 140 & 340.

[5] Y.Q. Zhou, J.Z. Li, Inorganic Chemistry Series, vol. 14, Academic Press, Beijing, 1997, p. 367.

[6] P. Karen, Journal of Solid State Chemistry 179 (2006) 3167.

[7] http://www.britannica.com/EBchecked/topic/417848/nonstoichiometric-compound.

[8] V.N. Lipatnikov, L.V. Zueva, A.I. Gusev, A. Kottar, Physics of the Solid State 40 (1998) 1211.

[9] A.I. Gusev, Uspekhi Fizicheskikh Nauk 170 (2000) 3.

[10] C.F. Yu, General Defect Chemistry, Wuhan University of Technology Press, Wuhan, 2006, pp.44–50.

[11] Binary Rare Earth Oxides, in: G. Adachi, N. Imanaka, Z.C. Kang (Eds.), Springer, Amsterdam, 2004, p. 1.

[12] C.B. Carter, M.G. Norton, Ceramic Materials-Science and Engineering, Springer Press, New York, 2007, pp. 181–200.

[13] F. Manea, D. Perniud, J. Schoonman, Defect Chemistry of Sensor Materials: the Defect Chemistry (NATO Science for Peace and Security Series C: Environmental Security), Springer, Amsterdam, 2008, pp. 105–123.

[14] http://ocw.nctu.edu.twuploadsschemistrysschemistry_lecturenotesssch-5-1.pdf.

[15] Japanese Chemistry Society, Inorganic Solid State Reaction, in: W.T. Dong, S.J. Dong (Translaters), Science Press, Beijing, 1985, p. 140.

[16] J. Nowotny, W. Weppner, Non-Stoichiometric Compounds – Surfaces, Grain Boundaries and Structural Defects, Kluwer Academic Publishers, Netherlands, 1988, pp. 11 & 471.

[17] Japanese Chemistry Society, Inorganic Solid State Reaction, in: W.T. Dong, S.J. Dong (Translaters), Science Press, Beijing, 1985, p. 20.

[18] http://en.wikipedia.org/wiki/Yttrium_barium_copper_oxide.

[19] B.N. Onwuagba, Turkish Journal of Physics 26 (2002) 403.

[20] R.R. Xu, W.Q. Pang, Inorganic Synthesis and Preparation Chemistry, Higher Education Press, Beijing, 2001, p. 44.

[21] J. Nowotny, T. Bak, L.R. Sheppard, M.K. Nowotny, Journal of the American Chemical Society 130 (2008) 9984.

[22] R. Long, N.J. English, Journal of Physical Chemistry C 113 (2009) 9423.

[23] L.R. Sheppard, M.K. Nowotny, T. Bak, J. Nowotny, Physica Status Solidi B – Basic Solid State Physics 245 (2008) 1816.

[24] A.A. Valeeva, A.A. Rempel, W. Sprengel, H.E. Schaefer, Physics of the Solid State 51 (2009) 924.

[25] R.F. Gould (Ed.), Nonstoichiometric Compounds, The 142nd Meeting of the American Chemical Society, Washington, D.C, 1962, p. 58.

[26] http://web.chemistry.gatech.edu/~wilkinson/Class_notes/CHEM_3111_6170/Defects_nonstoichiometry_ionic_conductivity_solid_state.pdf.

[27] http://en.wikipedia.org/wiki/W%C3%BCstite.

[28] F.X. Redl, C.T. Black, G.C. Papaefthymiou, R.L. Sandstrom, M. Yin, H. Zeng, et al., Journal of the American Chemical Society 126 (2004) 14583.

[29] J. Li, H. Zeng, S.H. Sun, J.P. Liu, Z.L. Wang, Journal of Physical Chemistry B 108 (2004) 14005.

[30] G.M. Da Costa, E. De Grave, P.M.A. De Bakker, R.E. Vandenberghe, Clays and Clay Minerals 43 (1995) 656.

[31] M.V. Ganduglia-Pirovano, A. Hofmann, J. Sauer, Surface Science Reports 62 (2007) 219.
[32] Q. Yuan, H.H. Duan, L.L. Li, L.D. Sun, Y.W. Zhang, C.H. Yan, Journal of Colloid and Interface Science 335 (2009) 151.
[33] S. Ling, Physical Review B 49 (1994) 864.
[34] M. Mogensen, N.M. Sammes, G.A. Tompsett, Solid State Ionics 129 (2000) 63.
[35] W.Q. Han, L.J. Wu, Y.M. Zhu, Journal of the American Chemical Society 127 (2005) 12814.
[36] Z.P. Qiao, Y. Xie, J.G. Xu, X.M. Liu, Y.J. Zhu, Y.T. Qian, Canadian Journal of Chemistry – Revue Canadienne De Chimie 78 (2000) 1143.
[37] K. Volenik, M. Seberini, J. Neid, Czechoslovak Journal of Physics 25 (1975) 1063.
[38] G. Adachi, et al. (Eds.), Binary Rare Earth Oxides, Kluwer Academic Publishers, Amsterdam, 2004, p. 57.
[39] E.G. Seebauer, M.C. Kratzer, Charged Semiconductor Defects: Structure, Thermodynamics and Diffusion, Springer, London, 2009, pp. 39–61.
[40] M. Karppinen, H. Yamauchi, Chemical Design of Copper-Oxide Superconductors: Homologous Series and Oxygen Engineering, Springer Verlag, Berlin, 2005, pp. 255–294.
[41] J.F. Qu, W. Wang, Y. Chen, G. Li, X.G. Li, Physical Review B 73 (2006). 092518-1092518-4.

CHAPTER

16

Synthetic Chemistry of the Inorganic Ordered Porous Materials

Qisheng Huo
Jilin University, China

16.1. POROUS MATERIALS

Inorganic porous material has developed quickly in the past half century despite the fact that the inorganic porous solid materials can be crystalline or amorphous. The synthesis and property of porous material involve a lot of fundamental concepts. They have wide applications as adsorbent, catalyst, ion exchanger, energy material, environment material, life sciences, nanotechnology, etc. Open structure and high surface area (interior surface and outer surface) porous materials strengthen their catalytic and adsorption abilities. Many novel materials with desired structure and property have been developed. The inorganic porous materials include natural zeolites, synthetic zeolites (from low siliceous zeolites to high siliceous zeolites), pure silica molecular sieves, microporous phosphates, oxides ($AlPO_4$, GeO_2, etc.), and even organic–inorganic hybrid materials such as MOFs.

Porous materials are classified into microporous, mesoporous, and macroporous materials by their pore size. According to the International Union of Pure and Applied Chemistry (IUPAC) notation [1], microporous materials have pore diameters of less than 2 nm and macroporous materials have pore diameters of greater than 50 nm, whereas the mesoporous materials lie in the middle. Some illustrations of porous material with examples are shown in Fig. 16.1 [2].

The crystal structure of zeolite or molecular sieve results in the unique pore size and shape. Ordered mesoporous material such as MCM-41 have regular pore (shape and size), but its wall is amorphous at atomic level. The pore size and pore shape can be controlled by selecting or modifying different structures.

The pore property is the most important structural parameter for porous material. Nitrogen adsorption is one of the most important techniques in studying the surface area and porosity of porous materials [3]. Figure 16.2 illustrates this technique by showing the nitrogen adsorption–desorption isotherms (−196 °C) on three different porous materials: (a) CaA (calcium-exchanged zeolite type A) zeolite having only micropores, (b) MCM-48 having small mesopores, and (c) SBA-15 having large mesopores [4].

This chapter will focus only on the ordered porous inorganic and inorganic–organic hybrid materials and discuss the new trends in their synthetic chemistry. The structure, synthesis, and property of microporous materials (zeolite, microporous aluminophosphate, etc.) can be obtained from the official website of International Zeolite Association (IZA) [5]. Many books

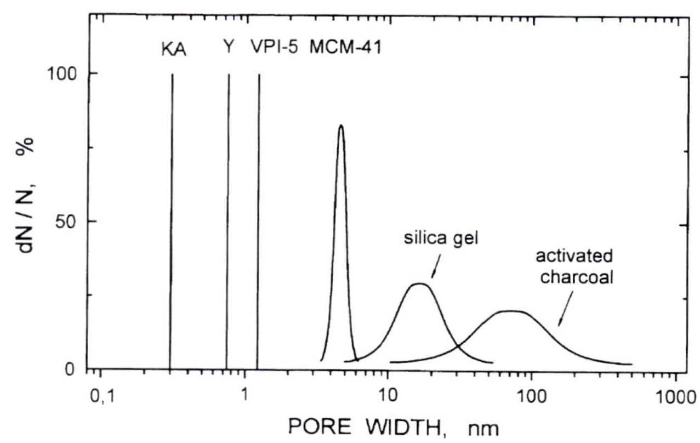

FIGURE 16.1 Comparison of pore size for typical porous materials [2].

Modern Inorganic Synthetic Chemistry, DOI: 10.1016/B978-0-444-53599-3.10016-2

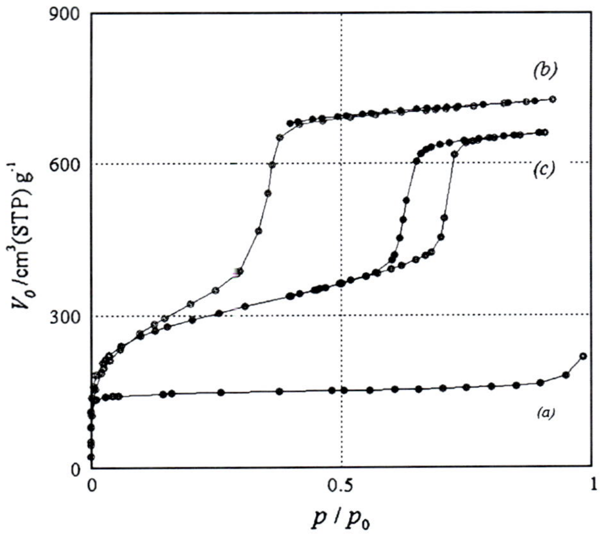

FIGURE 16.2 Nitrogen adsorption–desorption isotherms of (a) CaA zeolite; (b) MCM-48; and (c) SBA-15. *Images reproduced with permission [4].*

[2,6–13] and review papers [14–28] covering synthesis, properties, and applications of zeolite, microporous molecular sieves, and mesoporous materials are also available.

16.2. ZEOLITE AND ITS STRUCTURE

Zeolites are microporous, aluminosilicate minerals, and accommodate a wide variety of cations, such as Na^+, K^+, Ca^{2+}, Mg^{2+}, etc. The term "Zeolite" was originally coined in 1756 by Swedish mineralogist Axel Fredrik Cronstedt. The synthesis of zeolite can be traced up to mid-nineteenth century [14,15], but modern zeolite synthesis begin after World War II. Barrer studied the transformation of a natural zeolite, and obtained a new structure [29]. Milton of USA used active starting materials (fresh aluminosilicate gel) and synthesized zeolite A and zeolite X [30]. The industrial applications of zeolites (e.g., adsorbent, catalyst in petroleum industry, additive in detergent) are encouraging and promoting the research on synthesis, structure, properties, and applications of inorganic porous materials.

In 1932 J.W. McBain proposed the term "molecular sieve" to describe a class of materials that exhibited selective sorption properties for gases and liquids. These materials have the property of acting as sieves on a molecular scale. Common materials include natural aluminosilicate minerals, clays, porous glasses, microporous charcoals, zeolites, activated carbons, or synthetic compounds.

Among molecular sieves, zeolites and zeolite-like materials have occupied an extremely important position in modern material science and industry, because of their regular structures and various predictable properties. The large advantages of zeolites include: (1) adjustability of framework composition (incorporation of heteroelements in the structure); (2) very high surface area and capacity of adsorption; (3) various surfaces (from hydrophilic to hydrophobic); (4) controllablity of the intensity and density of the active sites (e.g., acidic sites); (5) uniform pore sizes with molecular dimensions which fit most gas and liquid molecules (0.5–1.2 nm); (6) strong electric fields inside pores; (7) various systems of pore, channel and cage, allow adsorbing selectively the reactants and results of reaction, and avoiding the side reaction; (8) exchange capacity of the cation; (9) molecular sieve capacity (based on the size, shape, polarity, and other properties of molecules); (10) thermal and hydrothermal stability; (11) highly chemical stability; and (12) regeneration with simple technologies (e.g., heating, reducing pressure, ion exchange).

In general, natural zeolites were named after their discoverer or location, while synthetic zeolites were named after their discoverer's affiliation. For example, ZSM-*n* is a series of zeolites that are found in Mobil Corporation. ZSM represents Zeolite Socony Mobil. There are nearly 1000 Zeolite-type materials (natural and synthetic) which may or may not be totally same, for instance over 20 materials such as ZSM-5, mutinaite, silicalite, and TS-1 have the same crystal structure. Nearly 200 types of zeolite framework structures have been confirmed. According to the IUPAC nomenclature, International Zeolite Association (IZA) assigned a "three letters" code to each framework type. For example, FAU represents Faujasite structure; MFI represents ZSM-5 structure. Till the beginning of 2010, 191 Framework Type Codes have been assigned [31]. One structure type can be used to describe the materials with different chemical compositions. For example, the materials with FAU structure can be natural Faujasite, zeolite X (synthetic Faujasite with low ratio of SiO_2 to Al_2O_3), zeolite Y (synthetic Faujasite with high ratio of SiO_2 to Al_2O_3), SAPO-37 (silicoaluminophosphate molecular sieves), or Zincophosphate X. IUPAC and IZA also have issued the detailed regulation to describe the zeolite composition. For example, zeolite X should be expressed as $|Na_{58}|[Al_{58}Si_{134}O_{384}]$-FAU or Na-[Al–Si–O]-FAU.

16.2.1. Basic Structural Unit of the Zeolite

The aluminosilicate structure of zeolite is negatively charged and attracts the positive cations that reside within. Zeolites consist of crosslinked TO_4 tetrahedra where TO_4 is $[SiO_4]^{4-}$ or $[AlO_4]^{5-}$. Each T atom occupies four connected vertices of a three-dimensional network and the oxygen occupies two connected positions between the four connected vertices. The O–T–O bond angle is close to the ideal tetrahedra bond angle of 109.5°. The T—O—T bond angle is much more flexible than the O—T—O bond angle and is usually around 140—165° [32]. The TO_4 tetrahedra are often referred to as the primary building units of zeolite structures. Primary building units are linked together to form secondary building units (SBUs) which are named based on their geometry. For example, a 6-ring (6R) SBU is a ring structure made from six oxygen atoms and six silicon atoms. The SBU is shown schematically in Fig. 16.3 [31]. Each corner in the secondary building units represents the center of tetrahedra.

16.2.2. Framework Structure of Zeolite and Molecular Sieve

SBU can be linked to form cages or channels within the structure. Connecting rings of different sizes leads to many different structures. For example, β-cage (sodalite cage) is a common unit which can be constructed with 4R and 6R SBUs. At least four classical zeolites (i.e., Sodalite (SOD), Zeolite A (LTA), Faujasite (FAU), and EMC-2 (EMT) are built from the sodalite cage. In SOD, the sodalite cages are directly joined on the 4R rings to form a cubic arrangement of cages. In FAU and EMT, the sodalite cages are joined by oxygen bridges on the 6R rings, resulting in a diamond arrangement of cages. In FAU, a large cage (called supercage, diameter of 1.18 nm, 0.74 nm window of 12-ring) is formed. LTA is a cubic arrangement of sodalite cages joined by oxygen bridges on the 4R rings. This leads into a larger cavity (α-cage) of minimum free diameter 11.4 Å with window of an 8-ring of 4.2 Å. Figure 16.4 describe these structures [33].

Another description of extended zeolite structures is sheet building unit. Some zeolite structures (e.g., ZSM-5) can be generated through attaching two-dimensional sheet units. The detailed zeolite structure description [33] can be found in the website of IZA [5].

For TO_4 in the typical zeolite structures, the cation T to oxygen radius ratio should fall within the range of 0.225—0.414 (Pauling's rule). The distribution of SiO_4 and AlO_4 is not random in zeolites and zeolite-type crystals. Lowenstein's rule governs the linking together of SiO_4 and AlO_4. Two tetrahedrals are linked by one oxygen bridge. There can be only one AlO_4, while the other must be SiO_4 or another TO_4 such as PO_4 with T of electrovalency 4 or more.

The porosity of zeolite-type frameworks can be expressed in terms of the framework density (FD), defined as the number of T atoms per 1000 $Å^3$. For the zeolites and zeolite-type materials, FD values range from 12.5 to 20.2. There is a clear gap in the FD values of zeolite and dense frameworks (FD = 20—22) [34].

Extra-framework cation can have a great influence on the pore (window or opening) of zeolite. For example, zeolite A has a cubic unit cell. Each unit cell contains eight α-cages or supercages and eight β or sodalite cages. The openings to the α-cages are 8-membered rings that are approximately 5 Å across. The presence of charge-balancing cations (Na/K for 3A, Na for 4A, and Na/Ca for 5A) reduces the effective pore size of the opening to 3 or 4 Å, depending on the type.

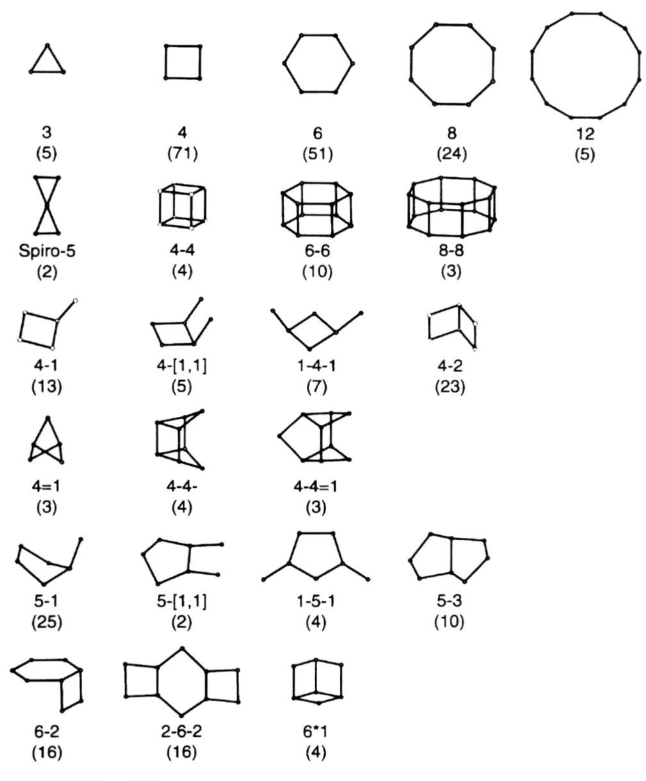

FIGURE 16.3 Secondary building units and their symbols (number in parenthesis is the frequency of occurrence) [31].

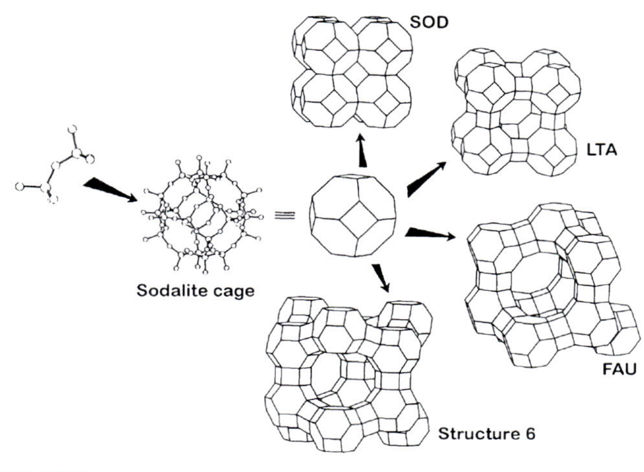

FIGURE 16.4 zeolite structures related to sodalite cage. *From Ref. [33]. Reprinted with permission from AAAS.*

16.2.3. Intergrowth in Zeolite

Commonly, the phase-pure zeolite is obtained only when the synthesis parameters are finely tuned. Zeolites can form different families due to common structure building units for the frameworks. Therefore they have a tendency to form intergrowths between different structures within the same family. For example, ZSM-2, ZSM-3, ZSM-20, VPI-6, CSZ-1 or ECR-30 are actually intergrowths of two end-members, cubic (FAU) and hexagonal (EMT) structures [35]. The Periodic Building Unit (PerBU) in the Faujasite family of zeolite frameworks (FAU and EMT) is composed of sodalite cages which are linked through double 6 rings (D6R) into a hexagonal layer. Neighboring PerBUs can be connected along [001] through D6R in two different ways: (1) the top layer is shifted over $1/3(-a+b)$ before connecting it to the bottom layer. The resulting connectivity exhibits inversion symmetry: (**i**) between successive layers; and (2) the top layer is rotated over 60° about [001] (followed by the shift vector $1/3(-a+b)$) before connecting it to the bottom layer. The connectivity now shows mirror symmetry (**m**) between successive layers. Once the distribution of the symmetry elements **i** and **m** between the layers stacked along [001] is known, the three-dimensional structure is defined. An example of an intermediate structure in the Faujasite family of zeolite frameworks is shown in Fig. 16.5 [5]. Pure EMT and FAU are obtained when neighboring PerBUs along the (hexagonal) [001] axis are exclusively related by **m** and **i**, respectively.

Gmelinite (GME) is a well-known zeolite having a structure in which the main feature is a large 12-ring channel. However, the adsorptive properties of gmelinite zeolites are similar to zeolites having smaller pores. The reason for this is that gmelinites have a propensity to intergrow with chabazite or related zeolites, which creates stacking faults that block and restrict access to the 12-ring channel of the gmelinite structure.

16.2.4. The Composition of Zeolite and Molecular Sieve

Most elements in the periodic table can be introduced into zeolite frameworks. From a view of composition, zeolite and zeolite-type materials include: (1) low silica zeolite (Si/Al ≤ 2), e.g., zeolite A and zeolite X, (2) intermediate silica zeolite (2 < Si/Al ≤ 5), such as zeolite Y, mordenite (MOR) and zeolite Ω (MAZ), (3) high-silica zeolite (Si/Al > 5), like ZSM-5 (MFI), ZSM-11 (MEL), zeolite β (BEA), ZSM-12 (MTW), and ZSM-35 (FER), (4) all-silica molecular sieves (Si/Al ~ ∞), e.g., Silicalite-1 (MFI), (5) silica clathrate, like ZSM-39 (MTN), (6) heteratom-substituted zeolites or silica molecular sieves, e.g., TS-1 (MFI), (7) aluminophosphate, such as AlPO-5, AlPO-11, and VPI-5, (8) metalloaluminophosphate and silicoaluminophosphate, e.g., MeAPO-34, SAPO-37, (9) metallo-phosphates, such as GaPO-*n*, ZnPO, MoPO, CoPO, (10) germium oxide or germate, (11) sulfides, (12) Mn oxides, titanates, and alumate, and (13) MOF.

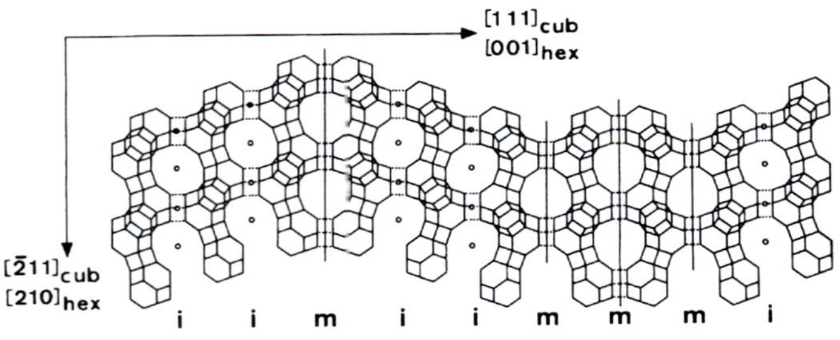

FIGURE 16.5 Connectivity sequence of PerBUs with **m** and **i** as symmetry elements [5].

16.3. THE SYNTHESIS OF ZEOLITE

16.3.1. Hydrothermal Zeolite Synthesis

Aluminosilicate zeolites are usually synthesized under hydrothermal conditions from reactive gels in alkaline media at temperatures between about 80 and 200°C [14,15].

The pioneering work in the synthesis of hydrothermal zeolites was done by Barrer and Milton over 60 years ago. Richard Barrer began his studies in the early 1940s, initially by investigating the conversion of known mineral phases under the action of strong salt solutions at fairly high temperatures (ca. 170–270°C) [36]. Robert Milton began his synthesis program in the laboratories of Linde Corporation in 1949. Use of more reactive starting materials (freshly precipitated aluminosilicate gels) enabled reactions to be carried out under milder conditions. Zeolite A [30] had been prepared, together with a crystalline impurity designated X. In 1950, pure zeolite X (FAU) was isolated [14,15].

Many classical zeolites such as the zeolites A, X, and Y are made in inorganic reactants synthesis system. However, in 1961 the range of reactants was expanded to include quaternary ammonium cations as templates. The introduction of organic constituents was to have a major impact upon zeolite synthesis and the key step followed quite rapidly. Subsequently here has been a large rise in the number of known synthetic zeolites leading to the discovery of new families of zeolite-like or zeolite-related materials. A major expansion of zeolite composition occurred in 1982 when a family of molecular sieves based on aluminophosphates was reported. Such materials display great compositional diversity and generally have frameworks unlikely for zeolites.

A typical hydrothermal zeolite synthesis [14,15] can be described as follows. (1) Reactants containing silica and alumina are mixed together with a cation source, usually in a high pH medium. (2) The reaction mixture is heated, often in a sealed autoclave for temperature above 100 °C. (3) For some time after rising to synthesis temperature, the reactants remain amorphous. (4) After the above "induction period", crystalline zeolite product can be detected. (5) Gradually, amorphous material is replaced by an approximately equal mass of zeolite crystals which are recovered by filtration, washing, and drying.

The verified synthesis (procedure) of typical zeolites and molecular sieves can be obtained from IZA website [5]. The purity and crystallinity of zeolite can be measured by X-ray diffraction. The XRD patterns for typical zeolites can be found in IZA website [5].

16.3.2. Zeolite Crystallization and Formation Mechanism

Despite many experimental studies that have been carried out on the synthesis of different zeolites, understanding is still lacking [14,15,37]. In fact, zeolites are thermodynamically metastable. The overall free energy change for a zeolite synthesis reaction is usually quite small, so that the outcome is most frequently kinetically controlled. Zeolite crystallization is a very complex phenomenon that cannot be adequately described by the classical variables of reactant composition, temperature, and pressure. Crystallization also involves polymerization–depolymerization, solution-precipitation, nucleation–crystallization, and other complex phenomena encountered in aqueous colloidal dispersions [38]. This process is a multiphase reaction crystallization, commonly involving at least one liquid phase and both amorphous and crystalline solid phases. Generally the liquid phases present in synthesis mixtures are not true solutions but sols.

Crystallization from solution generally occurs via the sequential steps, i.e., nucleation of the phase and growth of the nuclei to larger sizes. Nucleation and crystal growth rates typically are governed by a driving force related to the supersaturation. Most zeolite syntheses are preceded by the formation of an amorphous gel phase which dissolves to replace reagents consumed from the solution by crystal growth. In hydrothermal zeolite systems it is more difficult to identify a "supersaturation," because of the (i) myriad species present in the aluminosilicate solution, (ii) role of templates (structure-directing agents) in some cases, and (iii) the relative concentrations of these in a batch system change as the crystallization proceeds [39].

Studies on Synthesis Mechanism [14,15]

Understanding the synthesis mechanism of the microporous and mesoporous materials is one of the challenging chemical problems today. It is difficult to study the detailed mechanisms of the zeolitic formation due to the extreme complexity of hydrothermal crystallizations. Many analytical techniques have been used to study the formation of zeolite and other porous materials. For example, UV Raman spectroscopy can provide molecular information from aqueous solutions to solid phases of zeolite synthesis mixtures due to the low Raman scattering cross-section of water [40].

The most probable mechanistic pathways in zeolite formation are described in chronological sequence: induction period, nucleation, and crystal growth (see Fig. 16.6) [14,15,41].

The induction period is the time between the notional start of the reaction and the point at which crystalline product is first observed. When the reactants are initially mixed together, the primary gel or colloid is frequently formed. The primary amorphous phase represents the initial and immediate product from the reactants and is a nonequilibrium. After some time, the above mixture undergoes changes due to the equilibration reactions and is converted into the secondary amorphous phase. Then, the secondary amorphous phase is converted into the crystalline zeolite.

The nucleation mechanisms in liquid–solid systems include (1) primary nucleation: homogeneous or heterogeneous; and (2) secondary nucleation which is catalyzed by the presence of parent crystals of the same phase. The parent crystals might be added as seed crystals at the beginning of a synthesis, or grown in the original unseeded system.

Zeolite crystals appear to grow rather slowly (hours to days), compared to simple ionic crystals. The growth

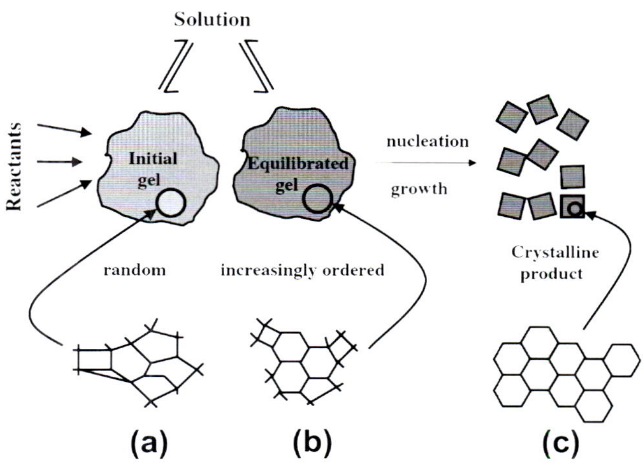

FIGURE 16.6 The evolution of order, from the primary amorphous phase (a) through the secondary amorphous phase (b) to the crystalline product (c). *Reprinted from Microporous and Mesoporous Materials, Vol. 82, Colin S. Cundy, Paul A. Cox, The hydrothermal synthesis of zeolites: Precursors, intermediates and reaction mechanism, Copyright (2005), with permission from Elsevier [15].*

by addition of monomers, low-molecular-weight species, or nanoparticles cannot be ruled out. Crystal growth can be described by an induction period followed by a sudden rapid growth which can be monitored. The resulting S-shaped growth curve of bulk crystallinity against time is the most commonly reported measurement of zeolite crystallization kinetics (see Fig. 16.7) [42].

The two extreme views on the zeolite synthesis mechanism [14,15] (see Fig. 16.8): (a) the amorphous gel is dissolved to yield small soluble species from which the zeolite crystals grow by a solution-mediated mechanism; and (b) the zeolite lattice is formed via an in situ rearrangement (ordering) of the gel, apparently in the absence of solution participation.

Recently, Cundy and Cox proposed a generalized mechanism (see Fig. 16.9) for zeolite synthesis to replace the two extremes [15] A fragment or domain of amorphous material (a) equilibrates with solution species (anions and cations) to develop elements of local order (b). In due course, the equilibration process leads to an area of sufficient order for a periodic structure to become established, i.e., nucleation to occur (c). The same equilibration reactions (T—O—T bond-making and bond-breaking) then allow the nascent crystal to grow and the amorphous areas to dissolve (d). The self-assembly process is mediated by the associated solvated cations, which act as coordination centers (templates) for the construction of the framework (central insert). These transformations most usually take place via a bulk solution phase, but may occur within a solvated layer at the surface of a "dry" solid (apparent solid-phase transformation) [15].

Templating

In zeolite synthesis, the formation of a specific framework type and a polyhedral building unit depends on one or two cation species. The cation specificity is strong

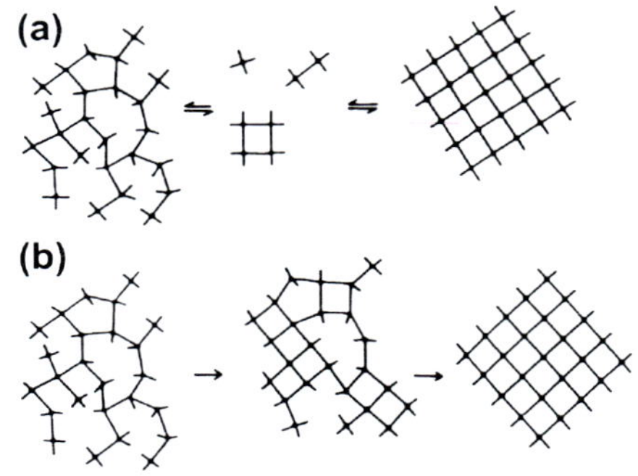

FIGURE 16.8 Two extreme views of the zeolite synthesis mechanism. *Reprinted with permission from [14,15]. Copyright 2003 American Chemical Society. Reprinted from Microporous and Mesoporous Materials, Vol 82, Colin S. Cundy, Paul A. Cox, The hydrothermal synthesis of zeolites: Precursors, intermediates and reaction mechanism, Page 1, Copyright (2005), with permission from Elsevier [14,15].*

for the α-cage, sodalite cage, gmelinite cage, and D4R unit. In some cases the cation (hydrated or anhydrous) is observed to fit nicely into the building unit.

Templates can be considered as cationic species added to synthesis media to aid or guide in the polymerization or organization of the anionic building blocks that form the framework [38]. It has been found that the kind of zeolite which crystallizes from alkaline gels can be strongly influenced by the type of cation present

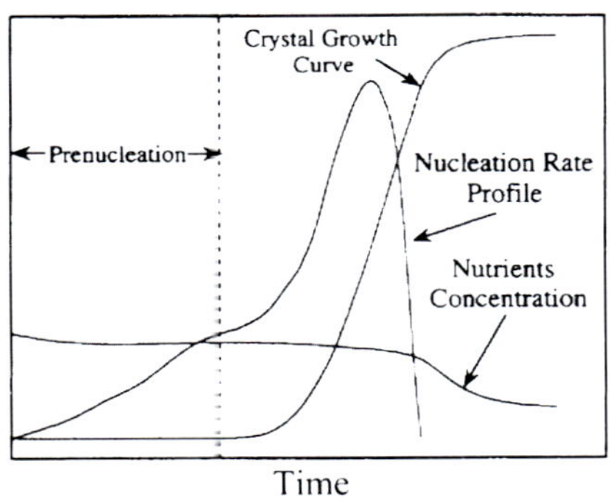

FIGURE 16.7 S-shaped growth curve of bulk crystallinity of zeolite. *Reprinted from Microporous and Mesoporous Materials, Vol 21, Vladimiros Nikolakis, Dionisios G. Vlacho, Michael Tsapatsis, Modeling of zeolite crystallization: the role of gel microstructure, Page 337, Copyright (1998), with permission from Elsevier [42].*

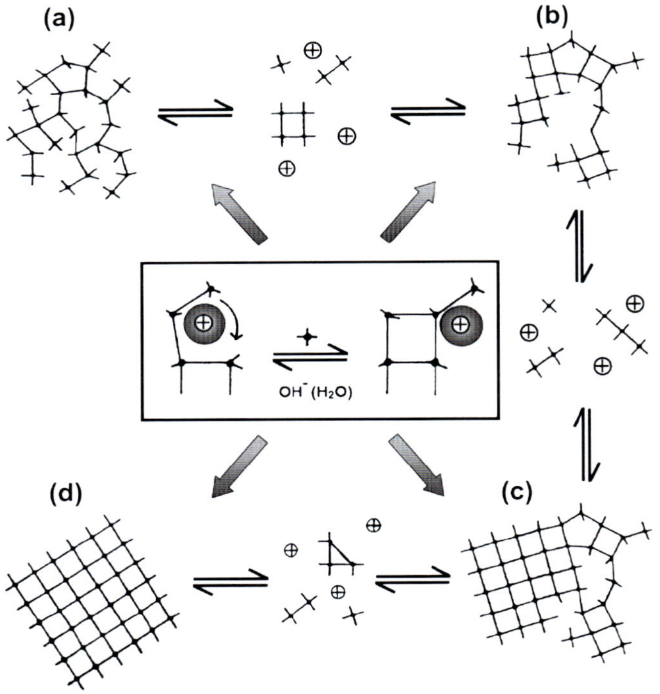

FIGURE 16.9 A generalized mechanism for zeolite synthesis. *Reprinted from Microporous and Mesoporous Materials, Vol. 82, Colin S. Cundy, Paul A. Cox, The hydrothermal synthesis of zeolites: Precursors, intermediates and reaction mechanism, Copyright (2005), with permission from Elsevier [15].*

[43]. For example, Na favored sodalite and cancrinite hydrates, Na−P, gmelinites, Faujasites, and zeolite A. Water associated with cations is at least in part displaced by silicate, aluminate, or aluminosilicate species in solution. The cations may thus serve as centers or templates. The cations are tending to function as templates having structure-directing role.

By addition of alkylammonium cations to the synthesis gel, not only known zeolite structures (i.e., LTA, MAZ, etc.) can be synthesized with an increased Si/Al ratio, but also many new structures (i.e., KFI, MFI, etc.) can be obtained. Starting with highly siliceous gels results in the formation of high-silica zeolites, including catalytically important zeolites ZSM-5, ZSM-11, etc.

The charge distribution and the size and geometric shape of the template are invoked to explain structure direction. There were three basic questions on the structure-directing effect of organic species [44]: (1) how can one template give rise to so many different structures? (2) how can so many templates, differing in size and shape, all direct the same structure? and (3) why do certain structures not form in the absence of a specific template molecule? The solution to this dilemma is the interplay of templating and "gel chemistry" where "gel chemistry" represents all the other reaction parameters governing the gel, such as oxide composition, temperature, time, reagent type, and pH. In a real sense the template is a necessary but not a sole condition for structure formation.

Davis and Lobo [27] classified templates into three different types: (1) "true" templates (direct the framework to adopt the geometric and electronic configurations unique to the templating molecule), (2) structure-directing agents (direct a specific product to form), and (3) space-filling species (many different organic molecules give one structure).

The fluoride anion has also shown a templating effect with its mineralizing effect in molecular sieve synthesis [45]. The structure-directing effect was found when F^- incorporated into D4R (double-four-ring).

In summary, a strong geometric match is often observed between the framework and the template [27,38]. Alkali cations are most effective in templating low Si/Al zeolites from basic media. Quaternary ammonium cations are best at templating medium to high Si/Al zeolites and AlPO-based molecular sieves. Amines have been used to template AlPO-based molecular sieves and high Si zeolites, and it is believed that the effective form of the amine is certainly the protonated form in AlPO-based synthesis and probably the protonated form even at the higher pH range typical of the high Si zeolites. The structure-directing role of the template is dominated by stereo-specific space-filling and stoichiometry between the template and the framework.

16.3.3. Optimization of Zeolite Synthesis

A great deal of effort has been made in establishing reliable and reproducible synthetic routes to known zeolites. Clearly, there are many different aspects to this goal and a variety of approaches to its attainment. There are many reports in the literature which discuss the possible improvements in zeolite synthesis systems. The zeolite synthesis is influenced by the reaction parameters (temperature, reaction composition, mass transfer effects, etc.). The crystallization behavior of a synthesis reaction mixture is often dependent upon the time−temperature history of the sample in advance of the principal heating step. Aging of mixture gel can be used as a means to control product phase purity and crystal size, as demonstrated in the synthesis of zeolite X.

Generally, a given zeolite crystallizes not only from one single composition but also there exists a composition domain inside which only more or less important parameters are observed for the composition of the zeolite, its structure remaining the same. Figure 16.10 shows the formation domains of zeolites in the ternary $Na_2O-SiO_2-Al_2O_3$ system [2].

In the zeolite synthesis, Ostwald's law of successive transformations is obeyed. The transformation proceeds from amorphous to metastable to more stable phase. For example, zeolite A converts to the more stable sodalite after longer crystallization time under the synthesis condition. Thus, time can be optimized in the synthesis of zeolites.

The alkalinity (pH) in a synthesis batch is one of the most important parameters for the control of the crystallization of zeolites. It determines their composition and is to a great extent responsible for the type of the crystallizing product [46,47]. The pH value of the solution is determined by the total alkali content and complicated buffering equilibrium of the silicate and aluminate species. The solubility of amorphous gel phase in reaction mixture also depends on the alkalinity. It assures the supersaturation for nucleation and growth processes. Generally, the different zeolite types crystallize within rather narrow ranges of pH.

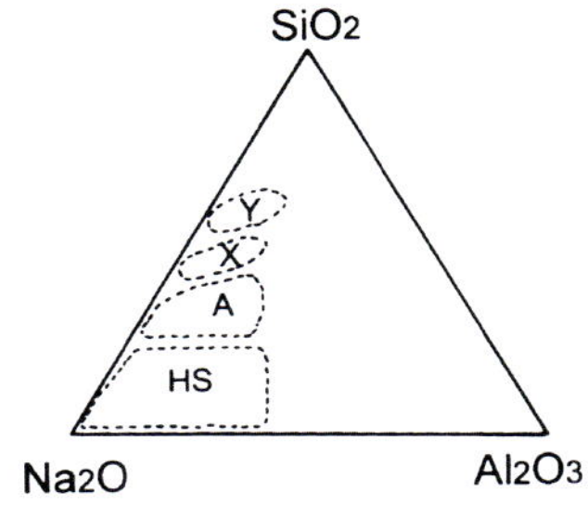

FIGURE 16.10 Formation domains of zeolites Y, X, A and HS in the ternary $Na_2O-SiO_2-Al_2O_3$ diagram [2].

16.4. ZEOTYPE: ZEOLITE-LIKE MATERIALS

Since the applications of aluminosilicate zeolites in the areas of ion exchange, separations, and catalysis [6], the research of porous materials has become one of the hot topics for chemists and material scientists for the last 50 years [48]. The explosive growth in the number of microporous and open-framework materials is largely due to many variable synthetic and structural parameters. Among these, the use of structure-directing agents with different charge, size, and shape is particularly effective in assisting the formation of oxide frameworks. Furthermore, the systematic and controlled variation of framework atoms has led to a variety of open-framework solids [25].

16.4.1. The Pure Silica Molecular Sieves and Clathrasil Compounds

Pure silica molecular sieves are highly thermally stable, structurally diverse, and chemically simple (silica polymorphs). Over 20 zeolite topologies can be made in pure silica form, i.e., the ITQ-*n* and SSZ-*n* series of materials. Pure silica molecular sieves are normally synthesized in the presence of organic structure-directing agents.

Silicalite (MFI) [49], Al-free ZSM-5, is one excellent example of pure silica molecular sieves. It was prepared by hydrothermal synthesis using alkylammonium (for example TPA) cations. The concept of silicalite was very significant as it brought about a number of important realizations, namely: (1) high-silica zeolites are essentially impure silica polymorphs; (2) high-silica zeolites are intrinsically hydrophobic and organophilic; and (3) for a given structure, there is usually a smooth transition in properties as the aluminum content is varied from zero to a limiting value.

Clathrasils are porous framework silicates with cage-like voids similar to zeolites [50]. Their window size is small (8 rings or less). Typical materials include pure silica sodalite (SOD), ZSM-39 (MTN), DD3H, and DDR-type clathrasils.

16.4.2. Aluminophosphate and Other Phosphate Molecular Sieves

The most remarkable breakthrough of zeotype material synthesis was the discovery of the $AlPO_4$ family in 1982 [51,52]. Flanigen and coworkers reported the synthesis of the first zeolite analogue, the microporous aluminophosphate solids ($AlPO_4$-*n*, where *n* refers to a structural type) which demonstrated that microporous materials need not contain any silica at all. Furthermore, there was subsequent finding that a large number of other elements could be substituted into $AlPO_4$ frameworks (giving the SAPO and MeAPO series of materials). The typical $AlPO_4$-*n* and SAPO-*n* materials include AlPO-5 (AFI), AlPO-11 (AEL), VPI-5 (VFI), MeAPO-5 (AFI), MeAlPO-11 (AEL), SAPO-34 (CHA), SAPO-37 (FAU), etc. Any three-dimensional and appreciably covalent framework could probably be persuaded to behave similarly. This pioneering zeolite-like family stimulated the discovery of novel inorganic porous materials [25], in addition to the family of phosphates with a large list of different metal ions [13,25].

16.4.3. Metal-Organic Frameworks (MOFs)

Along with the progress of purely inorganic porous solids, another innovation for the synthesis of porous materials emerged at the beginning of the 1990s with the introduction of organic molecules as constituents of the structure. Metal-organic frameworks (MOFs) [53–55] which are constructed from metal ions or metal ion clusters and bridging organic linkers, have recently emerged as an important family of porous materials [56]. In the beginning of the twenty-first century, thousands of publications describing MOFs with diverse topologies, pore sizes, shapes, and natures have been reported. Only zeolitic-structured MOF will be briefly discussed in this chapter. More details about MOF can be found in other chapters of this book.

Generally speaking, MOFs are formed by using transition metal ions as nodes and multidentate organic ligands containing O- or N-donors as linkers. One of the successful synthesis strategies is to obtain four-connected and porous MOFs with zeolite topology by designing inorganic and organic four-connected nodes [57–59]. The structures of MOFs with expanded zeolite topology such as MTN, ABW, BCT, SOD, and RHO would have larger pores and higher porosity than zeolites. Figure 16.11 shows the structure of rho-ZMOF [57].

Recently, the invention of zeolitic imidazolate frameworks (ZIFs) has provided materials based on simple zeolite structures [60]. ZIFs can be formulated as $T(Im)_2$ (Im = imidazolate and its derivatives, T = tetrahedrally bonded metal ion) and are similar to the TO_2 frameworks of zeolites; in particular the T–Im–T angle of 145° is close to the Si–O–Si angle typically found in zeolites.

MOFs with channel sizes up to mesoporous range can be prepared [61]. For example, $[Cd_3(bpdc)_3(DMF)]\cdot 5DMF\cdot 18H_2O$ (JUC-48) has ne-dimensional hexagonal nanotube-like channels of 24.5 Å × 27.9 Å and etb topology, constructed from rod-shaped cadmium carboxylate SBUs [62]. MOF materials actually bridge the gap between microporous materials and mesoporous

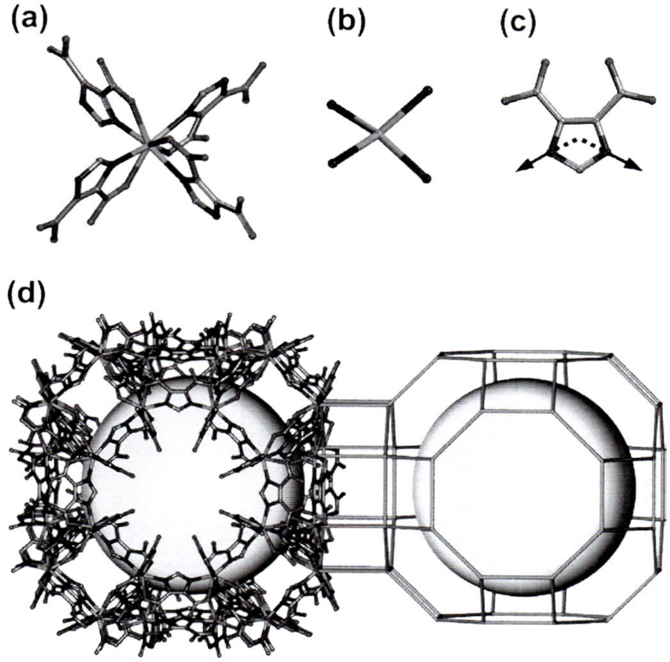

FIGURE 16.11 Structure of *rho*-ZMOF. *Images reproduced with permission [57].*

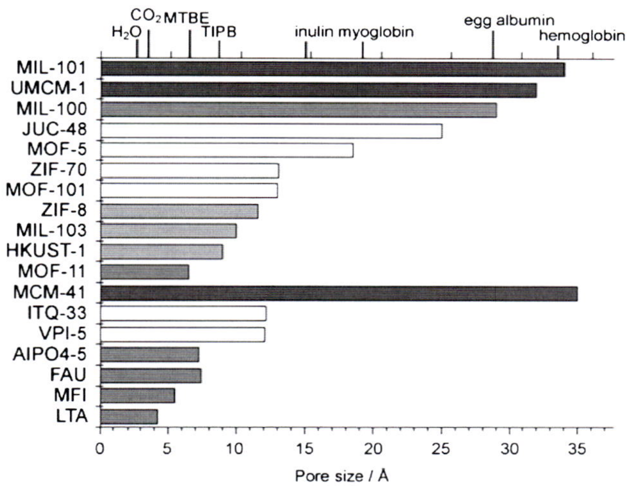

FIGURE 16.12 Cavity size of porous MOFs compared with standard zeolite, microporous molecular sieves and mesoporous materials. MTBE = methyl *tert*-butyl ether; TIPB = triisopropylbenzene [64].

materials as shown in the arbitrary selection in Fig. 16.12. In principle, the very large cavities of isoreticular MIL-101 (46 Å) can address triglycerides and small protein substrates (<10,000 Daltons) [63].

16.4.4. Open-framework Chalcogenides

Open-framework chalcogenides are inorganic solids analogous to zeolites, in which the O^{2-} anions have been replaced by chalcogens (S^{2-}, Se^{2-}, and Te^{2-}) [65]. By combining synthetic and structural concepts in chalcogenide clusters and porous materials, many microporous or open-framework crystalline solids containing spatially organized chalcogenide clusters have been created [66]. Because of their composition, these materials can also exhibit semiconducting, metallic, or semimetallic properties common to solids built up from the linkage of many of the main group metals or semimetals through sulfur, selenide, and telluride bridges.

In open-framework chalcogenides, chalcogenide clusters behave like large artificial tetrahedral atoms. To construct open-framework topology, tetrahedrally shaped clusters capable of forming linkages through their corners are preferable. Tetrahedral clusters are usually joined together with a single S^{2-} (or Se^{2-}) bridge. The three-dimensional framework of a chalcogenide includes zeolite structure types such as ABW and SOD.

16.4.5. Organic–inorganic Hybrid Zeolite and Microporous AlPO

The organic–inorganic hybridization of zeolites can be classified into two families: functionalization with terminal organic groups and containing organic framework [67].

Jones and coworkers introduced organosilyl groups in beta-zeolite frameworks by various terminal organic groups [68,69]. They employed organosilanes such as phenethyltrimethoxysilane and 3-mercaptopropyltrimethoxysilane as a part of silicon source. Through this hybridization, a new function can be added to a zeolite material.

Yamamoto and Tatsumi developed a series of ZOL [67] (zeolite with organic groups as lattice) materials from a synthetic system containing an organosilane such as bis(triethoxysilyl)methane (BTESM). In this synthesis, methylene groups bridge two silicon atoms ($Si-CH_2-Si$) and replace a part of oxygen atom ($Si-O-Si$) in the zeolite framework. ZOLs of various structures with the LTA, MFI, and *BEA topologies can be synthesized under synthetic conditions similar to those for conventional zeolites. The presence of framework methylene is demonstrated through the ^{13}C MAS NMR analyses. Although ZOL has physical properties similar to those of its inorganic counterpart, it shows higher hydrophobicity because of the presence of the framework organic group. The organic moieties in ZOL materials are thermally stable enough to be retained after the calcination for removal of templates.

Nitrogen-substituted zeolites are typically prepared by high-temperature (500–700 °C) treatment of the starting material with ammonia or amine [70]. However, it is extraordinarily difficult to synthesize zeolites with nitrogen substitutions without significant damage to the zeolite framework structure.

Yan et al. [71] synthesized a hybrid VFI-type $AlPO_4$ material, in which phenylphosphonic acid was employed as a part of phosphorus source. Maeda et al. [72] incorporated organosilyl groups in aluminophosphate frameworks. The organosilyl-substituted aluminophosphate materials have improved hydrophobicity and acidity.

16.5. NEW STRATEGIES AND NEW TRENDS OF ZEOLITE SYNTHESIS

Some topics such as zeolite film and membrane, synthesis by microwave heating, and structure simulation and design will be discussed in other chapters.

16.5.1. Ultralarge-pore Zeolites and Zeolite-like Materials

To manipulate larger molecules or support large moieties, larger micropore systems than those available in conventional zeolites are required. It appears that there is not a thermodynamic limitation to the synthesis of extra-large-pore zeolites [73]. However, for synthesizing zeolites with the largest pores to date, the presence of cations such as Ga, Be, and Ge has been required. Until recently, the pore openings of zeolites have been limited in size to 12 tetrahedral atoms. The largest pore and/or channel opening that has been observed in inorganic frameworks are 18-membered rings (18MR) in silicates [63,74], 24MR in metal phosphates [75–77], 26MR in metal phosphites [78], 30MR in ITQ-37 [73], and 30MR in germanates [79].

The syntheses of microporous aluminophosphate materials and their analogues have led to the discovery of new structures with pore openings larger than 12 tetrahedra, such as AlPO-8 (ATE) [52], VPI-5 (VFI) [80], cloverite (–CLO) [81], zinc phosphate with gigantic pores of 24 tetrahedra [75], gallium phosphate containing 24-ring channels [76], Nickel(II) phosphate VSB-5 with 24-ring tunnels [77], and helical metal phosphite NTHU-5 with 26R channel structure. Most of these materials contain mixed metal ion coordinations (octahedral and tetrahedral), terminal hydroxyl groups, or other nontetrahedral framework groups. Since these large-pore phosphate materials do not have the thermal stability of silicates, their applications have been limited.

New silicates, UTD-1F (DON) [82], CIT-5 (CFI) [83], and OSB-1 (OSO) [84] contain 14-ring openings. While the beryllium silicate, OSB-1, is thermally unstable, UTD-1 and CIT-5 represent the first thermally stable silicate zeolites having pore openings larger than 12-ring openings. The synthesis of zeolites with larger ring structures has been believed to be hindered by the low Si–O–Si bond angles available.

The thermally stable gallosilicate ECR-34 (ETR) [74] is prepared from a mixed alkali metal reaction gel containing tetraethylammonium (TEA) cations. It is not easy to obtain a pure phase. This material has 18-ring pores (diameter 10.1 Å) on the basis of a combination of simulated framework annealing and model building. Gas adsorption, ion exchange of TBA^+ cations, and FTIR data are consistent with the one-dimensional, 18-ring model. ECR-34 exhibited low micropore volume.

The silicogermanate zeolite ITQ-33 [63] exhibits straight large-pore channels with circular openings of 18 rings along the *c*-axis interconnected by a bidirectional system of 10-ring channels, yielding a structure with very large micropore volume. The conditions for synthesis are easily accessible, but are not typical, and were identified using high-throughput techniques.

The mesoporous germanium oxide, SU-M [79], with gyroidal channels (30 rings) separated by crystalline walls that lie about the *G* minimal surface as in the mesoporous MCM-48 has the largest primitive cell and lowest framework density of any inorganic material. Other examples of germinate-based solids with 30-ring windows include ITQ-37 [85] with extra-large pore and JLG-12 [86] containing Ge_7 and Ge_9 clusters prepared under solvothermal conditions.

Al-ITQ-37 zeolite [73] with the largest pore has been reported to date. Silicogermanate ITQ-37 is the first chiral mesoporous material with an interrupted zeolite framework. ITQ-37 was synthesized using the bulky diammonium ion. ITQ-37 has a very open framework (framework density: 10.3 T atoms per 1000 $Å^3$) and 30R pores (2.2×0.7 nm diameter) which result in a large accessible volume.

In the hybrid frameworks, however, the cavities were determined from pore structure analysis based on gas sorption studies. By using bulky organic units to co-construct the framework, even higher porosity or larger pores can be produced. For example, the pore size found in the framework MIL-101 [87] was up to 4.6 nm. A porous hybrid organic zinc phosphate with three-dimensional framework, NTHU-8 [88], contains nanometer-sized channels with intriguing bimodal porosity.

16.5.2. Large Crystal

The synthesis of large zeolite single crystals is of extensive interest for a large number of requirements, including single-crystal structure analysis, studies of crystal growth mechanisms, studies of adsorption and diffusion, electrical, magnetic or optical properties, and utilization of zeolite single crystals as matrices to create arrayed nanoclusters [89]. The synthesis of large zeolite crystals was achieved by: (1) addition of nucleation suppressor, i.e., triethanolamine for zeolite X; (2) using two kinds of silica sources [89], i.e., sodium silicate solution and dry silica powder being jointly used as silicon sources for mordenite (MOR) and ZSM-5 (MFI); (3) in the presence of F^-, there are many examples such as AlPO-5, 34 and ZSM-5. Large modenite (MOR) zeolite crystals can be easily obtained from the low-water content system by using $Al(OH)_3$ as an Al source or in the presence of NH_4F [90]; (4) from clear solution, for zeolites A and X; and (5) from nonaqueous system for pure silica sodalite and others [91].

16.5.3. Nanocrystal

Zeolite nanocrystals have much larger accessible external surface area, which could be exposed to more active sites than conventional zeolite crystals and facilitate the mass transfer and reaction process [92]. Colloidal suspensions of zeolite nanocrystals have also been used to construct hierarchical porous materials for many applications such as catalysts, membranes, and low-*k* thin films [93,94]. Classical low silica zeolites (i.e., zeolite L, A, X, Y, ZSM-2) can be made in nanocrystals easily [6,13]. For the colloidal suspensions of high-silica zeolite nanocrystals, hydrothermal procedures using clear aqueous solutions are usually used in the presence of organic templates.

Single zeolite A nanoscale single crystals (10–30 nm) was found in amorphous gel particles within 3 days at room temperature [95]. The gel particles were consumed during further crystal growth at room temperature, forming a colloidal suspension of zeolite A nanocrystals of 40–80 nm. Zeolite Y with an average crystal size of 80 nm was synthesized by optimizing synthesis composition and addition of NaCl [96].

Li et al. [97] have used a two-stage synthesis method to enhance the yield of zeolite nanocrystals. The method involves a low-temperature nucleation process (first stage) and a subsequent high-temperature crystallization step (second stage). The addition of methylene blue (MB) to the synthesis precursor solution is a quick and easy method to reduce the size of silicalite nanocrystals [98].

16.5.4. Phase Transition

Although not normally considered in terms of synthesis, zeolites undergo many solid-state transformations. Most of these relate to decomposition reactions, or changes in symmetry rather than overall structure (e.g., for ZSM-5, ZSM-39, and $AlPO_4$-5). However, in some cases a more complete phase transition from one zeolitic structure into another can take place, such as Na-EAB (EAB) into sodalite (SOD) at above 600°C in dry nitrogen. This transformation involves cleavage of selected T–O–T bridges followed by bond rotation to give TO_4 inversion. The examples for aluminophosphate family include $AlPO_4$-21 into $AlPO_4$-25, VPI-5 into $AlPO_4$-8, and JDF-20 into $AlPO_4$-5.

In some cases, zeolite can be formed by calcination of a layer-structured precursor. For example, the two-dimensional aluminosilicate PREFER transformed into the zeolite ferrierite on calcination at 550 °C [99]. The precursor appears to consist of ferrierite-type sheets in the *bc* plane, separated by molecules of the bulky template (4-amino-2,2,6,6-tetramethylpiperidine). On burning out the occluded organic molecule, the ferrierite layers progressively link together in the *a*-direction as new T–O–T linkages are formed by condensation reactions. Other materials can also be obtained from this two-dimensional precursor (see Fig. 16.13). Similar cases include MCM-22 family [100] and NU-6 [101].

16.5.5. Nonaqueous Synthesis: Solvothermal, Ionothermal, and Dry Gel Systems

Nonaqueous syntheses were carried out in nonaqueous solvents [91,102–104], ionic liquid, "dry gel" or "vapor phase".

The earliest synthesis in nonaqueous media is the preparation of an all-silica form of sodalite by Bibby and Dale in 1985 [102]. The reaction medium was ethylene glycol. No water was added, and no attempt was made to ensure an absolutely anhydrous system, and the reagents themselves (NaOH, silica, glycol) generated some water.

However, a variation of the method (adding the template as an amine–HF complex) later proved very successful for growing huge single crystals [91]. Application of nonaqueous solvent and mixed-solvent techniques for the synthesis of phosphates have produced a richer variety of results [13], including the discovery of some novel phases such as the 20-ring aluminophosphate JDF-20 [105].

The nonaqueous system was extended to ionothermal synthetic system [106] in which an ionic liquid or eutectic mixture is the reaction solvent and also the structure-directing agent (SDA) during the synthetic process. One most important feature of this strategy is the removal of competition between the template–framework and solvent–framework interactions. Applicable material system includes zeolite, phosphate, and MOF.

In 1990, Xu et al. [107] reported the first example of a zeolite (ZSM-5) synthesis in which some of the reaction components were supplied from the vapor phase. A prepared "dried sodium aluminosilicate gel" was suspended above liquid in an autoclave and subjected to

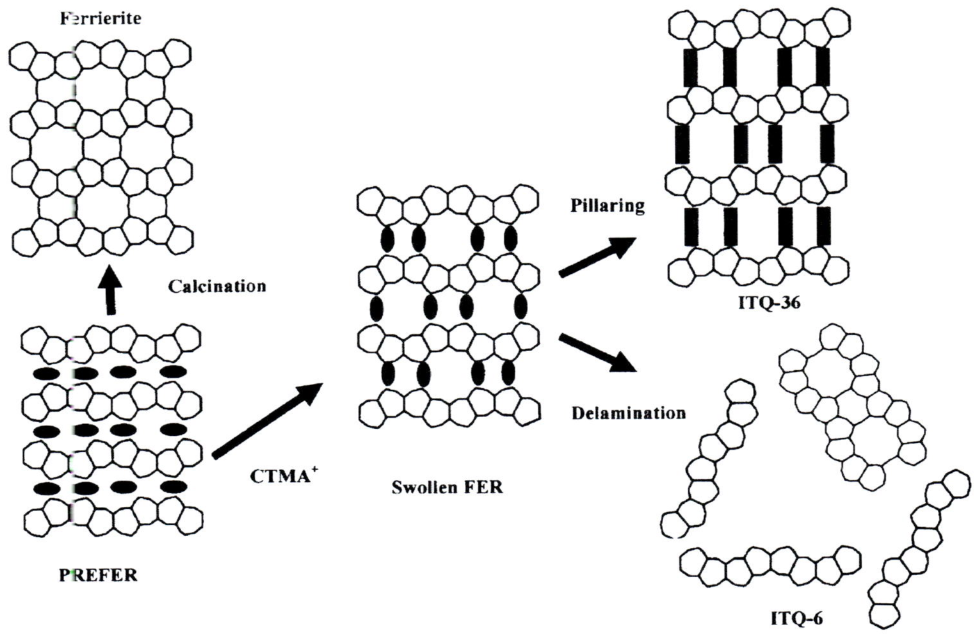

FIGURE 16.13 The family of materials derived by delamination and reassembly from the laminar precursor PREFER ($CTMA^+$ = cetyltrimethylammonium). *Reprinted with permission from [99]. Copyright 2000 American Chemical Society.*

the mixed vapor of ethylenediamine, triethylamine, and water at elevated temperature and pressure. Subsequent work by several groups has provided many examples of this type of process, which is described variously as the dry gel conversion or vapor phase transport technique [108]. The apparent lack of a liquid phase does not mean that there is no water (or other polar solvent) present.

16.5.6. Mesoporous Zeolites

Hierarchical zeolites (zeolites with intracrystal mesopores) are of considerable current interest [109–111], because the creation of mesoporosity in ordinary zeolite crystals strongly improves the catalytic performance in reactions that tend to suffer from diffusion limitations. The formation of mesopores in zeolites can be realized by the following techniques: (1) dealumination, such as steaming and chemical treatments (acid, alkaline, EDTA, etc.) can create disordered mesopores (5–50 nm) in zeolite crystals, (2) desilication [112], the recrystallization of mesoporous materials into zeolites in the presence of template like tetrapropylammonium [113], (3) the use of amphiphilic organosilanes to prepare zeolite [114], and (4) nanocasting synthesis [115,116]. The hard templates for preparation of mesoporous zeolites include nanostructured carbons (particles, nanotubes, mesoporous carbon CMKs, etc.) mesoscale cationic polymers, mesoscale organosilanes, and nanosized inorganic materials (e.g., nanosized $CaCO_3$).

16.5.7. F^- as Mineralizer

In typical zeolite synthesis, hydroxide ion acts as mineralizer in high pH systems. However, fluoride ion can be used as mineralizing agents to mobilize reaction components [117], in particular silica-rich zeolite synthesis systems in slightly acidic media (pH 5).

The most common and preferred fluoride sources are NH_4F, NH_4HF_2, or HF. Fluoride may also combine with the source of framework elements such as in $(NH_4)_2SiF_6$. When the crystallization is carried out in the presence of an organic cation, fluoride is generally occluded in the pores of the solid as a compensating negative charge, in addition to the negative framework charge, of the organic cations. Fluoride is completely removed on calcination.

The fluoride route has led also to novel structures, e.g., ITQ-*n* series [118]. The 20-T-ring gallophosphate cloverite [81] has never been made in the absence of fluoride. The unambiguous location of fluoride by diffraction or MAS NMR methods has shown the structure-directing agent role of the F^- ion in the formation of the double 4-ring (D4R) unit [117].

The crystals are usually of good quality and the size generally exceeds the values obtained in alkaline-type synthesis. The fluoride route also allows the synthesis of large crystals of purely siliceous molecular sieves [91] and consequently facilitates the crystal structure determination from single-crystal X-ray diffraction. By working at a pH below 10 and 11, most nonbridging $\equiv SiO^-$ defects are avoided.

16.5.8. Synthesis Challenges from new Pentasil Zeolite Minerals

Three new pentasil zeolite minerals are gottardiite [119] (NES, the natural counterpart of NU-87), mutinaite [120] (MFI, the natural counterpart of ZSM-5), tschernichite [121] (the natural counterpart of zeolite beta), and terranovaite [120] (no synthetic analogue). Although gottardiite is indeed one of the highest Si/Al ratio found in natural zeolites, the Si:Al ratios for gottardiite and mutinaite are much lower than their synthetic counterparts. These minerals are rich in calcium. This suggests both that their compositional range is larger than had been thought and also that organic templates are probably unnecessary for their synthesis. The existence of these materials offers a fascinating challenge to the synthetic zeolite chemist.

16.6. BASIC OF ORDERED MESOPOROUS MATERIALS

The dimensions and accessibility of pores of zeolites and microporous solids are restrained to the subnanometer scale ($<\sim 1.5$ nm). This limits their applications to small molecules, rather than larger organic or biological molecules. Mesoporous materials overcome this limitation. The mesopores allow many reactions on ordered porous materials to be possible, such as modification by utilizing larger organic or biological molecules. Mesoporous materials also provide new opportunities for both fundamental research (e.g., gas adsorption modeling, biomolecular catalysis) and practical application (e.g., adsorption, separation and purification of gas and liquid, catalyst, biological material, semiconductor, optics component, semiconductor, sensors, drug delivery carrier, material for environmental protection, energy storage host, and so on).

The synthesis of the ordered mesoporous material was started as early on as the 1970s [122]. A research group in Japan also started synthesis work in 1990 [123]. Only the report of the M41S family (MCM-41, MCM-48, etc.) in 1992 [124,125] started to attract the attention of many scientists all over the world. Figure 16.14 shows M41S structures [126].

The synthesis of the M41S family of mesoporous silica breaks through the traditional principles of zeolite synthesis: use of single molecules or cations as templates. The organized assemblies of molecules (micelles) are used as templates for the synthesis of mesoporous materials.

The ordered porous materials, M41S and related families, have regular pore sizes of $\sim$20–200 Å. The materials are ordered, but not conventionally crystalline, since the pore walls are usually amorphous. Silicate, nonsilicate, aluminosilicate, and heterosubstituted varieties have been synthesized, usually at moderate temperatures (25–150 °C). The commonest template for mesoporous silica is amphilic molecule (surfactant). Surfactant molecules will assemble into micelles to maintain low energy in the solutions. Therefore, the micelles look like a huge soluble (hydrophilic) molecule which may be the real template for the synthesis of mesoporous silica.

The attractive parts of the ordered mesoporous material are that they possess some exclusive outstanding properties which other porous materials do not have. These properties include (1) well-defined pore system: high surface area, pore sizes and shape, narrow pore size distribution and adjustable in the range of $\sim$1.3 to $\sim$100 nm, existence of micropores in the amorphous wall (for those thicker wall materials); (2) various wall (framework) compositions obtained from direct synthesis, or posttreatment or modification; (3) high thermal and hydrothermal stability if properly prepared or treated; (4) various controllable regular morphologies in different scale from nanometers to micrometers; and (5) application potentials such as large-molecule catalysis, biological process, selective adsorption, functional materials.

Since the discovery of M41S materials in 1992 [124,125], there are many achievements in the synthetic chemistry. The main results include the synthesis of mesoporous silica in extreme acidic system [127–129], the use of block copolymers as soft templates and discovery of large-pore mesoporous materials such

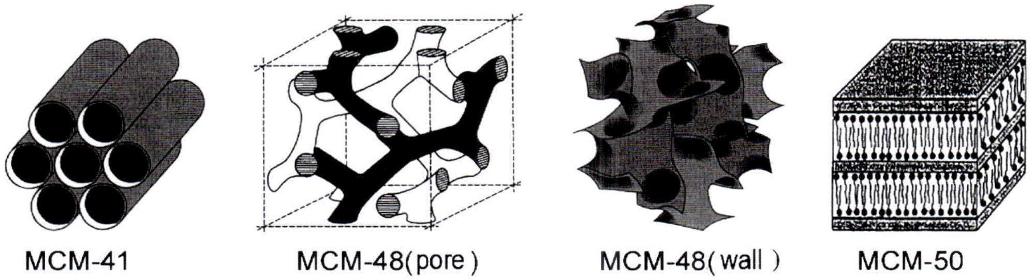

FIGURE 16.14 Structures of M41S family [126].

as SBA-15 [130,131], the self-assembly of silsesquioxanes and the synthesis of mesoporous organosilicas [22,132], nanocasting synthesis of mesoporous carbons and other nonsilica materials using hard templates [133,134], polymer synthesis system using block copolymers as soft templates [23,135], and the fabrication of mesoporous materials with crystalline walls [136–138].

16.7. UNDERSTANDING THE SYNTHESIS OF MESOPOROUS MATERIALS

16.7.1. Synthetic System

Procedure for synthesis of mesoporous material is simple and synthesis parameters can be controlled easily. The simple procedure does not mean that the reactions or interaction among reactants in the synthetic system are simple. The typical synthesis mixture for mesoporous materials contains four major components: inorganic precursors, organic template molecules, solvent, and acid or base catalyst. The formation of a material with a desired structure and morphology depends on a delicate interplay between several basic processes, whose relative rates determine the structure and properties of the final structure. These are the self-assembly of the organic template molecules to form organized structures that serve as templates, the sol–gel chemistry that generates the inorganic network, and the specific interaction at the interface between the organic assemblies and the inorganic oligomers. In the real synthetic system, these interactions affect each other, either enhance or restrict. Figure 16.15 shows the interactions between the three main components [139].

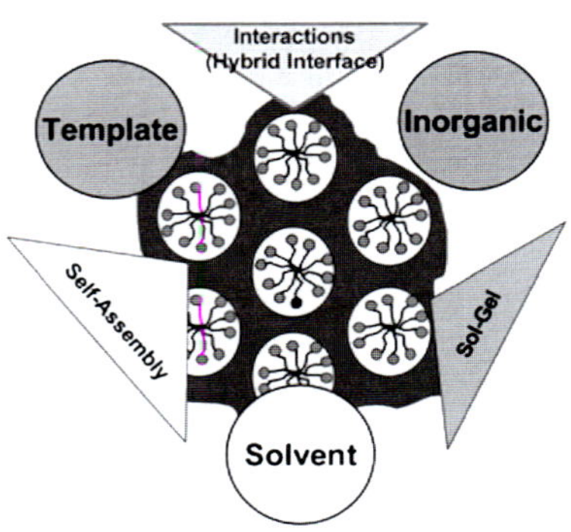

FIGURE 16.15 The relationships between various components in synthetic system. *Reprinted from Current Opinion in Colloid & Interface Science, Vol 8, Galo J. de A. A. Soler-Illia, Eduardo L. Crepaldi, David Grosso, Clément Sanchez, Block copolymer-templated mesoporous oxides, Page 109, Copyright (2003), with permission from Elsevier [139].*

16.7.2. Formation Mechanism of Mesostructure: Liquid Crystal Template and Cooperative Self-assembly

Since the report on MCM-41, the formation mechanism of MCM-41 and other M41S members have been an important study topic. Even now, we do not have a complete understanding of formation mechanism.

Initially, two possible formation pathways were proposed by the Mobil group (see Fig. 16.16) [125]. In the first, silicate species interact with a pre-existing liquid crystal phase to form the composite mesophase. The alternative model is based on the cooperative interaction of inorganic species with surfactant micelles in solution to generate the ordered composite structure. All other mechanisms are derived from the initial Mobil mechanisms.

Since most syntheses are carried out at surfactant concentrations well below those necessary for the formation of liquid crystalline phases. The second mechanistic pathway, cooperative formation mechanism, in original Mobil mechanism believes that the liquid crystal mesophase is the template for the formation of MCM-41, but the liquid crystal mesophase is formed after addition of inorganic precursor. The formation of organic–inorganic mesostructure is a cooperative self-assembly of the ammonium surfactant and the silicate species. However, there are different ways to describe the details for the cooperative self-assembly of the surfactant and the silicate species. The two mechanisms proposed by Chen et al. [140] and Stucky [127] are more noticeable.

Stucky and coworkers [128,141] developed the more general "cooperative assembly" mechanism to describe the detailed cooperative self-assembly process based on systematic syntheses under different conditions and NMR studies. This mechanism emphasizes the role of specific interactions between surfactant headgroups and inorganic species.

Recently, new mechanisms proposed for the formation of particles or fibers were based on the nucleation of a silicatropic liquid crystals by a cooperative assembly followed by a classical growing mechanism or by a colloidal aggregation [142]. Based on dynamic light scattering (DLS) result, Mesa et al. [143] have shown that liquid crystal nuclei in certain acidic medium systems are neither observed during syntheses with a cationic surfactant nor with a triblock copolymer surfactant. The building blocks are composite nanoparticles formed during the first stage of the reaction by the hydrolysis of TEOS and by migration of the resulting siliceous species into the hydrophilic corona of the triblock micelles [144]. These nanoparticles can be isolated and are stable when a silane is added to terminate the reaction [145].

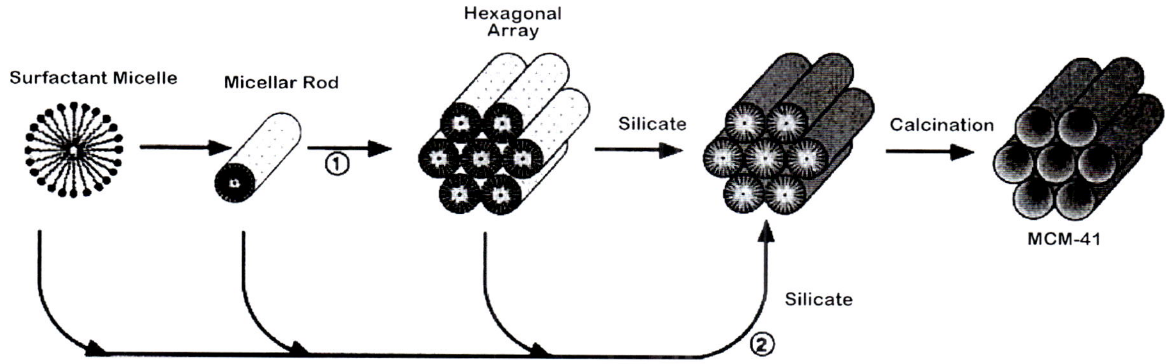

FIGURE 16.16 The initial LTC mechanism for the formation of MCM-41: (1) liquid crystal template (2) co-assembly template. *Reprinted with permission from [125]. Copyright 1992 American Chemical Society.*

16.7.3. Interaction Between Organic Template and Inorganic Species

The interaction between the organic template and the inorganic species (e.g., charge matching [146]) is the key for the mesoporous material synthesis. It guides the synthesis process [147]. Any kind of interactions between inorganic and organic is feasible [128]. Figure 16.17 shows the various types of silica–surfactant interfaces (synthesis pathways) according to the type of interactions between surfactants and the inorganic precursors.

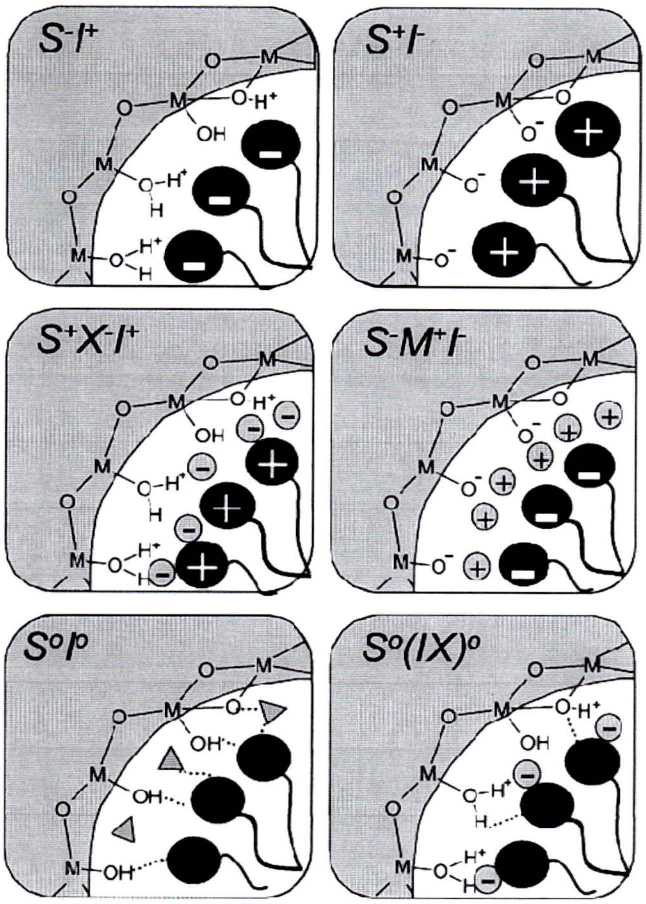

FIGURE 16.17 Schematic representation of the different types of silica–surfactant interfaces. Solvent molecules are not shown, except for the S^0I^0 case (triangles); dashed lines correspond to H-bonding interactions. *Reprinted with permission from [22]. Copyright 2002 American Chemical Society.*

16.7.4. The Surfactant Packing Parameter

The packing parameters of ionic surfactants are widely used in predicting and explaining the final mesostructures. The calculation of g value is simple but of great significance and guidance: $g = V/(a_0 l)$. Here, V is the total volume of surfactant hydrophobic chains plus any cosolvent (organic molecules) between the chains, a_0 is the effective hydrophilic headgroup area at the aqueous–micelle surface, and l is the kinetic surfactant tail length. The expected mesophase sequence as a function of g value is cubic (*Pm3n*, etc.) and three-dimensional hexagonal ($P6_3/mmc$) with $g < 1/3$, two-dimensional hexagonal (*p6 mm*) with $1/3 < g < 1/2$, cubic (*Ia3d*) with $1/2 < g < 2/3$, and lamellar with $g \approx 1$.

Many experimental results can be explained by the change of g values, such as the factor of surfactant molecular structure, the effects of inorganic and organic additives, and phase transformation. Similarly, the hydrophilic/hydrophobic volume ratios (VH/VL) are suggested especially for nonionic surfactant templating systems to account for the formation of different mesophases.

16.8. TYPICAL MESOSTRUCTURES AND MESOPOROUS MATERIALS

Since the discovery of ordered mesoporous silica M41S (MCM-41, 48, 50) [124,125], a large variety of ordered mesoporous materials with different mesostructures (two-dimensional hexagonal p6 mm, three-dimensional hexagonal *P6₃/mmc*, cubic *Pm3m*, *Pm3n*, *Fd3m*, *Fm3m*, *Im3m*, bicontinuous cubic *Ia3d*, etc.) and compositions (silica, metal oxides, metal sulfides, metals, and even polymers and carbons) have been synthesized. For example, FSM-16 [123], SBA family [129,131], FDU family [148,149], KIT family [150], AMS family [151], HOM family [152], MSU [153,154], and HMS [155] have been synthesized under a wide synthetic range from highly basic to strongly acidic conditions using cationic, anionic, neutral, or nonionic surfactants.

16.8.1. Two-dimensional Hexagonal Structure: MCM-41, SBA-15, FSM-16 and SBA-3

The two-dimensional hexagonal materials have honeycomb arrays of nonintersecting primary channels. Typical materials include well-known MCM-41 and SBA-15, SBA-3 made from acidic media, and FSM-16 made from kanemite.

MCM-41

The earliest MCM-41 synthesis is that of adding surfactant $C_{16}H_{33}(CH_3)_3OH/Cl$ solution with a solution of sodium silicate solution [124,125]. The typical pore size is about 4.0 nm. For the high-quality MCM-41 sample, XRD can give more than four *hk0* diffraction peaks. Figure 16.18 shows the XRD pattern of a high-quality MCM-41 made from an extreme low surfactant concentration system [156]. Figure 16.19 shows high-resolution TEM images of the calcined MCM-41 [157].

The typical N_2 adsorption isotherm for MCM-41 belongs to Type IV isotherm. A high-quality MCM-41 sample has a narrow pore size distribution, high surface area (>1000 m^2/g), and large-pore volume (>0.7 cm^3/g). Shown in Fig. 16.20 are argon and nitrogen adsorption isotherms acquired at 77 K for MCM-41 silicas with pore diameters from 2.4 to 6.5 nm. In cases of both adsorbates, the same pattern of adsorption–desorption behavior is observed. Multilayer adsorption that takes place at lower pressures is followed by the capillary condensation at pressures gradually and systematically increasing with the pore diameter. Adsorption–desorption isotherms for smaller pores are reversible, but adsorption–desorption hysteresis is observed for larger pore diameters [158].

SBA-3, the acidic version of MCM-41, was synthesized in extreme acidic media (1–7 M of acid, 2 M HCl is preferred). It is a typical example of $S^0X^-I^0$ synthetic pathway [128,129]. This hexagonal silica mesophase can be formed at room temperature within a few minutes. Longer reaction time and high synthesis temperature will improve the stability of SBA-3. The mesoporous silicas from acidic synthetic systems have

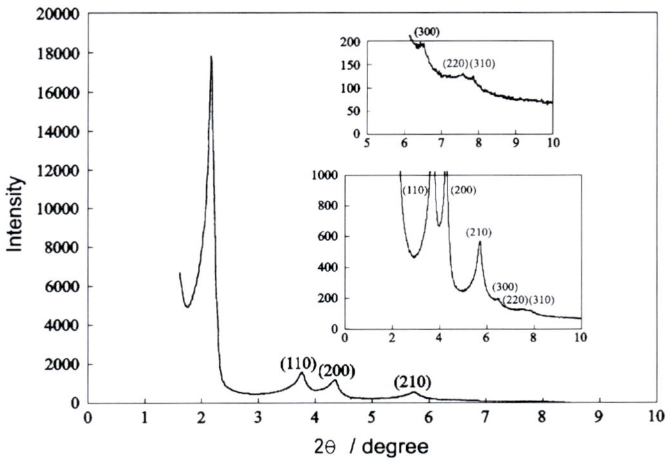

FIGURE 16.18 XRD pattern for a high quality MCM-41. *Reprinted from Microporous and Mesoporous Materials, Vol 32, Qiang Cai, Wen-Yong Lin, Feng-Shou Xiao, Wen-Qin Pang, Xi-Hua Chen, Ben-San Zou, The preparation of highly ordered MCM-41 with extremely low surfactant concentration, Page 1, Copyright (1999), with permission from Elsevier [156].*

regular "crystal" morphology, even curved shapes (see Fig. 16.21) [128].

SBA-15

In 1998, a new family of highly ordered mesoporous silica materials including SBA-15 and SBA-16 has been synthesized in an acidic medium by the use of commercially available non-ionic polyethylene oxide–polypropylene oxide–polyethylene oxide triblock copolymer (PEO–PPO–PEO) with large polyethylene oxide (PEO) and polypropyleneoxide (PPO) blocks [130,131]. The formation of this family of mesoporous silica is believed a $(S^0H^+)(X^-I^0)$ process.

SBA-15 may exhibit a large variety of morphologies depending on the synthetic conditions. The pore size can be from ~4 to ~30 nm by varying synthesis composition and condition (e.g., addition of swelling agent, TMB). The silica wall (~2 to 6 nm) is much thicker than that of MCM-41. Since the pore size (>4 nm) of SBA-15 is large, the nitrogen adsorption–desorption isotherm for SBA-15 sample shows clearly an H1 hysteresis loop. SBA-15 shows relatively high hydrothermal stability.

The typical SBA-15 contains significant amount of micropores in its framework (the wall for mesopore). The micropore volume is about 0.1 cm^3/g. This microporosity is the result of silica templating by PEO fingers forming a corona around each micelle. In other words, these micropores result from the insertion of hydrophilic PEO of surfactant into silica wall. The templating polymer within both mesopores and micropores can be removed stepwise [159]. The different amounts of micropores, changes in the pore wall thickness, and the

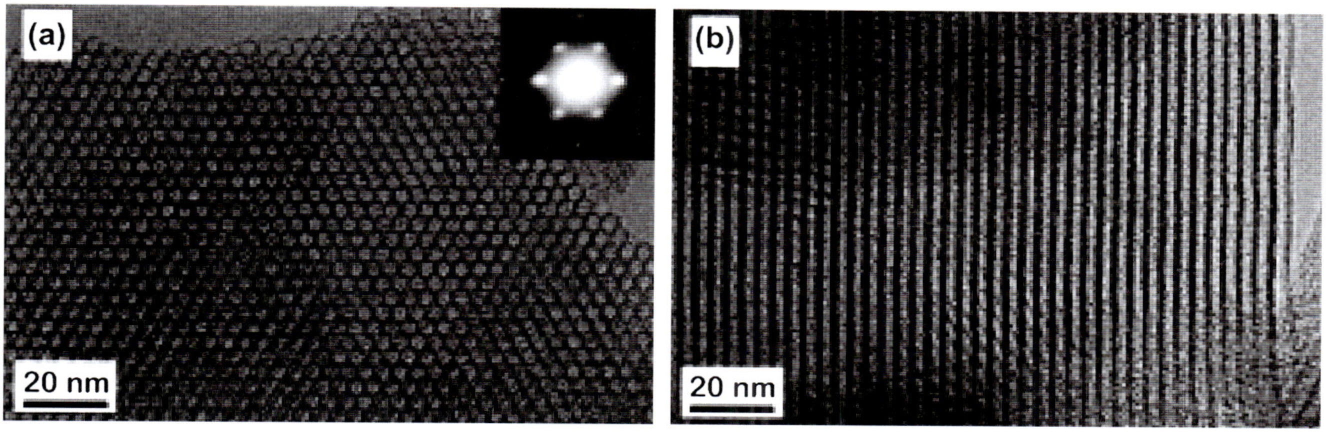

FIGURE 16.19 TEM images of calcined mesoporous MCM-41 (a) along the channel direction (*c* axis) and (b) perpendicular to it. *Liu, Z.; Sakamoto, Y.; Ohsuna, T.; Hiraga, K.; Terasaki, O.; Ko, C. H.; Shin, H. J.; Ryoo, R.: TEM Studies of Platinum Nanowires Fabricated in Mesoporous Silica MCM-41. Angewandte Chemie. 2000. Volume 39, Page 3107. Copyright Wiley-VCH Verlag GmbH & Co. KGaA. Reproduced with permission.[157].*

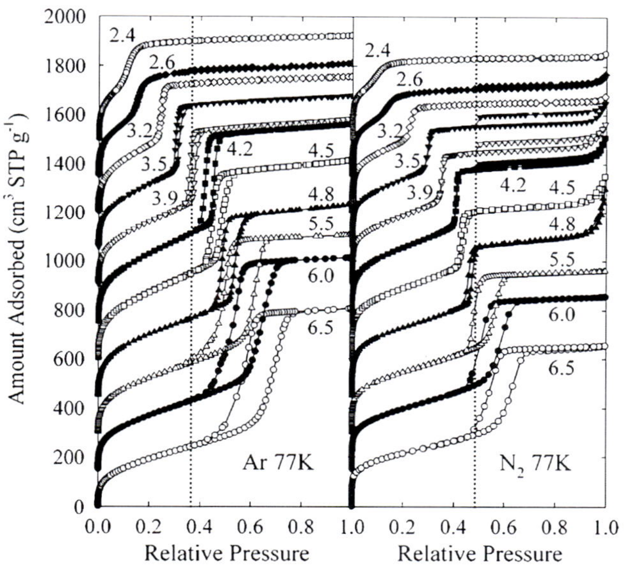

FIGURE 16.20 Argon and nitrogen adsorption–desorption isotherms measured at 77 K for a series of MCM-41 silicas with pores of diameter 2.4–6.5 nm. *Reprinted with permission from [158]. Copyright 2003 American Chemical Society.*

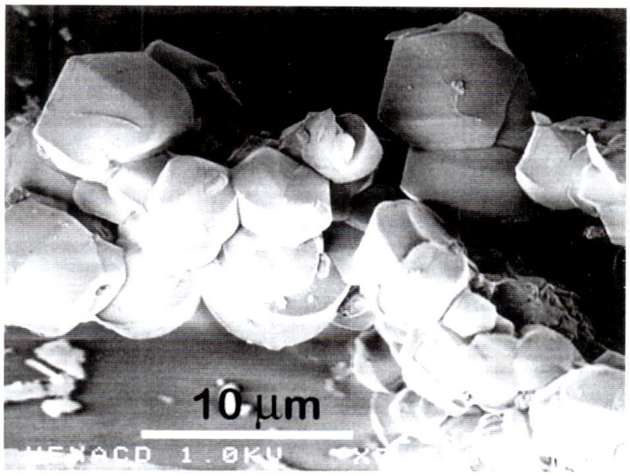

FIGURE 16.21 SEM of SBA-3 (scale bar 10 µm). *Reprinted with permission from Macmillan Publishers Ltd:* Nature *[128], Copyright (1994).*

morphology of SBA-15 could be obtained by changing template, post-synthesis treatment [139], or varying synthesis conditions [160–164], like temperature, the addition of additives such as co-surfactants, swelling agents, electrolytes, salts, etc.

16.8.2. Cubic Channel Mesostructures: MCM-48, FDU-5, and *Im3m* Materials

This family of mesostructure has an enantiomeric pair of three-dimensional channel systems (Q230). Till date, there are a few members in the structure family: *Ia3d* (FDU-5 [165], FDU-15 [166], KIT-6 [167]), *Im3m* and *Pn3m*. In view of material application, three-dimensional channel is more favorable for mass transfer than one-dimensional channel.

MCM-48

The initial proposed structure for MCM-48 by Monnier et al. [146], based on lyotropic liquid crystal model was in good agreement with XRD and TEM experimental results [168]. Figure 16.22 shows the high-resolution XRD (synchrotron X-ray resource) pattern of MCM-48. All diffraction peaks can be indexed with *Ia3d*, and all peaks (in low-angle region) are visible.

The basic concept for the syntheses of MCM-48 is to control the effective surfactant packing parameter, *g* is

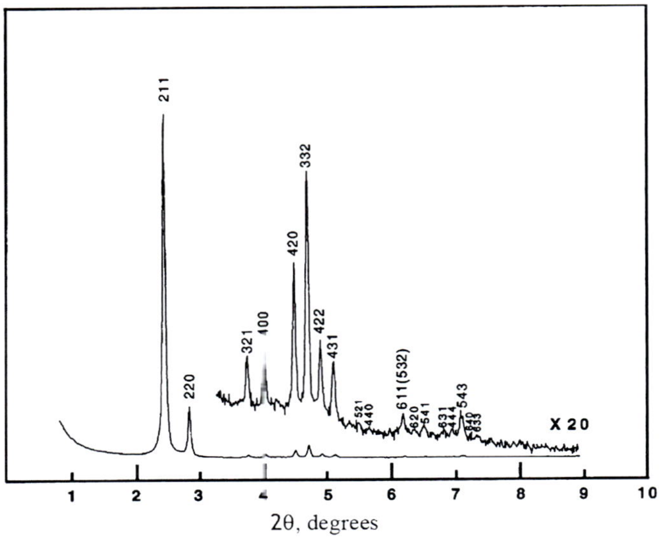

FIGURE 16.22 XRD pattern of MCM-48 (synchrotron X-ray resource, wavelength 0.17 nm). *From Ref. [146]. Reprinted with permission from AAAS.*

between 1/2 and 2/3, in another word, increase the palisade layer volume of micelle [13].

16.8.3. Cubic Caged Structures

The surfactants with large headgroup prefer the caged mesostructures. Typical structure includes *Pm3n* (SBA-1 [127,128], SBA-6 [169]), *Im3m* (SBA-16 [131]), and *Fm3m* (FDU-12 [170]). Typical three-dimensional hexagonal close packing (*hcp*, $P6_3/mmc$) and cubic close packing (*ccp*, *Fm3m*) intergrowth mesostructured materials include SBA-2, SBA-12, and FDU-1 [148,171].

SBA-2

SBA-2 was first produced using divalent quaternary ammonium surfactants, Cn-s-1 in both basic and acidic media [172]. SBA-2 was initially assigned a three-dimensional hexagonal cage mesostructure with a space group $P6_3/mmc$ on the basis of the powder XRD patterns and TEM images. $P6_3/mmc$ symmetry had not been previously reported for conventional liquid crystal phases before SBA-2 was reported [173]. Later, more extensive TEM studies [174] suggest that the SBA-2 is an intergrowth of *hcp* structure and *ccp* structure. The N_2 adsorption–desorption isotherm is type IV with an H2 hysteresis for even small pore SBA-2 samples (<2.5 nm). The window size of SBA-2 is strongly dependent on the conditions of its synthesis media [175].

SBA-16

SBA-16 [131] was synthesized using F127 ($EO_{106}PO_{70}EO_{106}$) or F108 ($EO_{132}PO_{50}EO_{132}$) as template. The mesostructure of SBA-16 can be described as a body-centered arrangement of cages with diameter 9.5 nm connected through large windows with diameter 2.3 nm along the [111] directions. The pore cage diameter (from ~4.5 to 9 nm) and entrance size in SBA-16 can be enlarged in a wide range not only by increasing the synthesis temperature and time, but also by using blends of Pluronic F127 with Pluronic P123 copolymer [176].

Small pore (~2.6 nm) thick-walled (7.7 nm) SBA-16 (ST-SBA-16) [177] can be synthesized by using oligomeric surfactant with ultra-long hydrophilic chains, Brij700. The mesoscopic ordering of ST-SBA-16 can be greatly improved by the addition of suitable amount of TMB. The pore size and surface area can be effectively tailored by changing the hydrothermal treatment time and the calcination temperature.

16.8.4. Deformed Mesophases, Low Ordered Mesostructures and Other Possible Mesophases

The initial as synthesized two-dimensional hexagonal films on substrate were composed of cylindrical micelles highly organized structure with the *c*-axis preferentially aligned parallel to the surface plane. During drying and calcination, unidirectional contraction of the initial mesostructure leads to two-dimensional centered rectangular (*c2mm*) mesoporous structures.

The bulk materials with deformed mesostructure have also been synthesized with a judicious choice of template or synthetic condition. For example, the deformed hexagonal mesostructured silica, SBA-8 [178], with a two-dimensional centered rectangular lattice (*cmm*) was synthesized by using a bolaform surfactant containing a rigid unit in the hydrophobic chain at room temperature. KSW-2 [179] with a mesostructure of rectangular arrangements of square or lozenge one-dimensional channels was synthesized by mild acid treatment of a layered alkyltrimethylammonium–kanemite complex. Three-dimensional low symmetry mesoporous silicas, FDU-11, and FDU-13 (tetragonal and orthorhombic structure) [180], were obtained from tetra-headgroup rigid bolaform quaternary ammonium surfactant synthesis systems.

16.8.5. Siliceous Mesostructured Cellular Foams (MCFs)

MCFs [181] can be obtained by adding a sufficiently large amount of an organic additive (e.g., TMB) in the SBA-15 synthetic system. MCFs are microemulsion templated. MCFs have well-defined ultralarge spherical mesopores (22–42 nm in diameter) interconnected by windows (10 nm in diameter). The pore and window sizes can be controlled by changing the TMB/block copolymer ratio and adding ammonium fluoride. In addition, the windows can be enlarged by postsynthesis treatment in hot water.

16.8.6. New Mesostructure

New structure is still possible after nearly 20 year extensive synthetic study. Han et al. [182] have developed the first tri-continuous mesoporous material, IBN-9, with a silica pore wall following a hexagonal minimal surface using a unique surfactant template, *N,N*-dimethyl-L-phenylalanine. This surfactant has a unique tunable headgroup as well as a long hydrocarbon tail that has variable levels of hydrophobic qualities. IBN-9 consists of three identical continuous interpenetrating channels, which are separated by a silica wall that follows a hexagonal minimal surface. This completely new porous structure previously has been predicted only mathematically.

16.9. SYNTHESIS STRATEGIES FOR MESOPOROUS SILICA

16.9.1. Template

Surfactant template control of pore system and mesophase topology has been a fruitful approach to the design of mesoporous materials.

Cationic Surfactant

For the systematic investigation of the formation of mesoporous silica materials, a series of cationic surfactants were selected with and without organic additives, which favor a range of *g* values when used as templates to synthesize silica mesophases in different reaction conditions [13,23,129].

Anionic Surfactants and Costructure-directing Agent (CSDA)

The anionic surfactant can be used as template directly to synthesize mesoporous materials such as SnO_2, Al_2O_3, and Ga_2O_3 through an S^-I^+ interaction pathway. A new route ($S^-N^+I^-$) is the self-assembly of anionic surfactants and inorganic precursors by using aminopropylsiloxane or quaternized aminopropylsiloxane as the co-structure-directing agent (CSDA) [151,183]. The negatively charged headgroups of the anionic surfactants interact electrostatically with the positively charged ammonium sites of the CSDAs. The alkoxysilane groups of the CSDA co-condense with tetraalkoxysilane and are subsequently assembled to form the silica framework. Figure 16.23 is a scheme of this synthetic route.

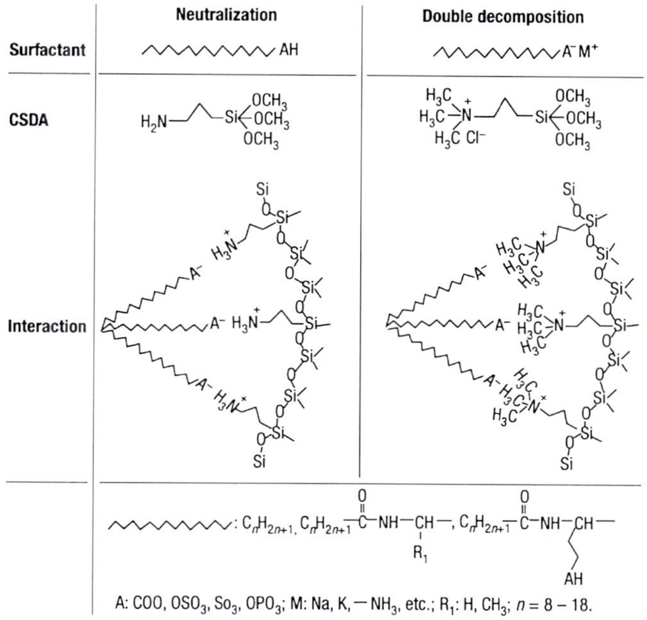

FIGURE 16.23 A scheme of CSDA. *Reprinted by permission from Macmillan Publishers Ltd:* Nature Materials *[183], Copyright (2003).*

In the synthetic system with anionic surfactant *N*-myristoyl-L-glutamic acid as template and *N*-trimethoxylsilylpropyl-*N,N,N*-trimethylammonium chloride as CSDA, the packing of the micelle was controlled by simply adjusting the neutralization degree of the C14GluA surfactant. Different mesophases ranging from tetragonal *P42/mnm* (cage type, AMS-9), cubic *Fd3m* (cage type, AMS-8), to two-dimensional hexagonal *p6mm* (cylindrical, AMS-3), and a bicontinuous double diamond cubic Pn-3m mesophase (AMS-10) were obtained by decreasing the amount of NaOH that was added into the reaction system [184]. AMS-10 may exhibit a lower curvature close to bicontinuous cubic *Ia3d* from the sequence of the mesophases. Changing the degree of ionization of the surfactant results in changes of the surfactant packing parameter *g*, which leads to different mesostructures. Furthermore, variation of the charge density of positively charged amino groups of the CSDA also gives rise to different values of *g* [185].

Nonionic (neutral) Surfactant

Most of nonionic surfactants as templates may be charged in extreme acidic conditions. The common surfactants used in synthesis of mesoporous silica materials include [142]: (1) long-chain amine surfactants such as $C_{12}NH_2$, (2) long-chain polyethylene oxide surfactants, (3) the molecule containing more than one amine group such as $NH_2(CH_2)_nNH_2$ ($n = 10-22$) [186] and $C_nH_{2n-1}NH(CH_2)_2NH_2$ ($n = 10$, 12, 14), and (4) block copolymers such as PEO–PPO–PEO and PI-*b*-PEO.

Mixed Surfactants

It is of practical interest to find a way to change the effective head- and tail-group sizes systematically and continuously in order to use them more efficiently. The use of mixed surfactants is the easiest way. Such control of the hydrophilic–hydrophobic balance is often impractical because surfactant molecules with various chain sizes that cover the range to be studied are not easily available.

Nonsurfactant Templating

A novel synthetic route for ordered mesoporous materials without the use of amphiphilic surfactants has been developed (see Fig. 16.24) [187]. Hoogsteen-bonded tetrads and pentamers are formed by a large variety of organic molecules through H-donor and acceptor groups capable of inducing self-organization to form columnar and hexagonal mesophases. On the basis of the supramolecular templating of stacked arrays of the tetramer-forming pterin groups of folic acid under a variety of synthetic conditions. Hexagonally ordered mesoporous structures with pores on the order of 25–30 Å in diameter and surface areas above 1000 m^2/g were obtained. More importantly circular dichroism studies revealed that the folate template possesses a chiral signature within the pores in the as-synthesized solid and that chirality is transferred from the folate template to pore surface via aminopropyltriethoxysilane costructure-directing agent used in supramolecular assembly.

16.9.2. Organic Additives [13]

Lyophilic organics can be solubilized within the micelles. Small molecules are preferentially located near the micelle–water interface, while large molecules are absorbed in the core. This results in the change of micellar shapes and *g* values and, in turn, phase transformation, enlargement of mesopores, or variation of morphologies of the final products. For example, when a hydrophobic, nonpolar organic additive such as TMB is added, it seeks the most hydrophobic region that is at the tail end of surfactant array and swells the micelle size. Both v and l are affected and the net result can be either a phase change or an increase in the effective pore or cage size. Thus, when TMB is added as a swelling agent, relatively large-pore size changes are observed. This approach has been used in large-pore MCM-41 frequently. A suitable polar additive is able to enter the hydrophilic–hydrophobic palisade region (first few carbon atoms) of the micelle, with a relative increase in the volume of the hydrophobic core to form surfactant molecular aggregates with lower curvature intersurfaces, e.g., from sphere to rod. Thus when *t*-amyl alcohol, a polar additive, is added into the synthesis mixture at basic synthetic media, the SBA-2 product is replaced by MCM-41 when C16-3-1 is used as template. Another example is the effect of additive such as EtOH on the formation of MCM-48.

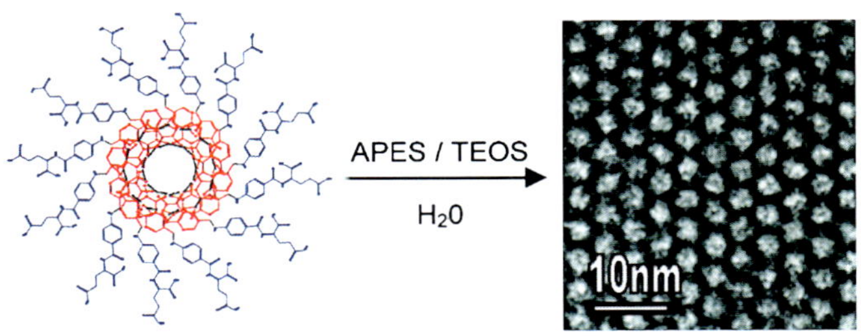

FIGURE 16.24 Mesoporous silica with nonsurfactant templating. *Reprinted with permission from Ref. [187]. Copyright 2009 American Chemical Society.*

16.9.3. Pore Size Control

There are many strategies available for changing pore size [13,188]. The basic principle is the same: change the size and volume of micelles. Most used methods include: (1) choice of surfactants with different chain lengths or molecular structures, or use of mixed surfactants; (2) use of low-polar or nonpolar organic swelling agents; (3) change of synthetic conditions, like temperature, pH, concentration of surfactant, and reaction time; (4) posthydrothermal treatment; (5) modification of pore surface with chemicals such as silane; and (6) use of nanometer-sized liquid or solid (e.g., emulsion or colloid) as template.

16.9.4. Postsynthesis Hydrothermal Treatment

Postsynthesis hydrothermal treatment [127,129] is simple and powerful technique to optimize or change mesostructure, or improve property of product. Here, the so-called treatment is a heating process of as-made material in water with or without additives. Phase transformations are frequently observed as consequence of mild treatments of as-made samples. Postsynthesis hydrothermal treatment can improve the properties and structural ordering (XRD quality, silica polymerization, stability, etc.) of most as-made mesostructured silica materials including the products from both of acidic and basic media. In some cases, the unit cell and pore size of materials are enlarged. For example, the as-made MCM-41 with mixed surfactants (C22-3-1 and $C18TMA^+$) as template and in pH ~12 at room temperature for a short reaction time (0.5–2 h) was treated water (pH ~ 7) at 100 °C for two weeks. The unit cell of hexagonal silica mesostructure increased from 5.5 to 7.9 nm, and the structural ordering improved (up to eight XRD peaks were observed). The postsynthesis ammonia hydrothermal treatment can simultaneously restructure pore size, nanochannel regularity, and morphology of SBA-3 [189]. ^{29}Si MAS NMR result indicates that the silica structure further condenses during ammoniacal hydrothermal treatment and makes it more thermally stable.

Sayari et al. [190] found the unprecedented expansion of the pore size (from ~3.15 nm to 25 nm) and pore volume (from 0.85 to 3.6 cm^3/g) of MCM-41 after hydrothermal treatment in the presence of certain amine such as *N,N*-dimethyldecylamine (DMDA). Figure 16.25 shows schematic representation of MCM-41 pore-expansion, selective extraction, and calcinations [190,191]. The treated material contains the surfactant template and the swelling agent. The swelling agent is then selectively extracted with ethanol, producing a mesoporous material with a surfactant layer on its surface and large void channels for the adsorbate molecules to diffuse freely. These materials are good adsorbents for different compounds [192].

16.9.5. Stabilization of Silica Mesophases

During last decades, several methods have been developed for improving the stabilities of mesoporous silica materials. A few effective methods [13]: (1) by increasing the polymerization level of silica wall by varying synthesis reaction equilibrium, adding inorganic salts, increasing synthesis or calcination temperature, and time; (2) chemical modification to build a protective layer; (3) increase in the wall thickness; and (4) change wall in composition and structures.

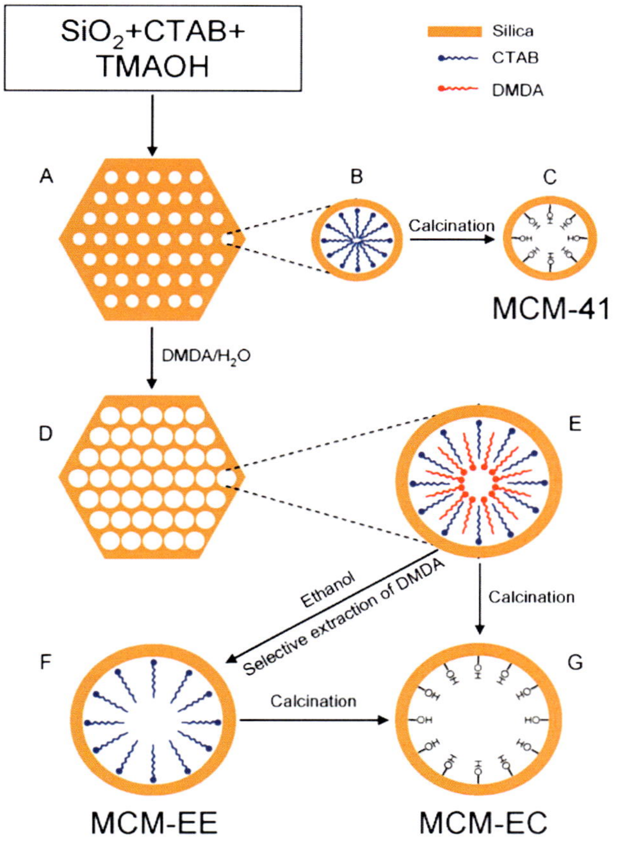

FIGURE 16.25 Schematic representation of MCM-41 pore-expansion, selective extraction, and calcination. *Reprinted with permission from [191]. Copyright 2007 American Chemical Society.*

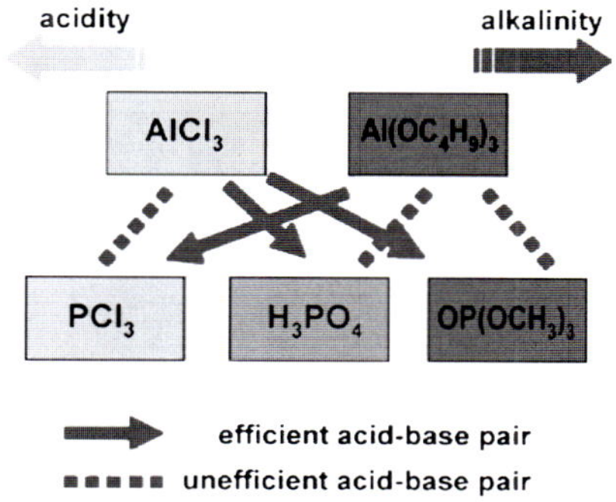

FIGURE 16.26 "Acid–base pair" routes for synthesizing mesostructured AlPOs. *Reprinted from Microporous and Mesoporous Materials, Vol. 67, Limin Wang, Bozhi Tian, Jie Fan, Xiaoying Liu, Haifeng Yang, Chengzhong Yu, Bo Tu, Dongyuan Zhao, Block copolymer templating syntheses of ordered large-pore stable mesoporous aluminophosphates and Fe-aluminophosphate based on an "acid–base pair" route, Page 123, Copyright (2004), with permission from Elsevier [194].*

16.9.6. Synthesis through Acid–base Pair

Wan et al. [193] described the self-adjusted inorganic–inorganic interplay between two or more inorganic precursors is guided by acid–base chemistry considerations. A wide variety of highly ordered, large pore, homogeneous, stable, and multicomponent mesostructured minerals, including metal phosphates and metal borates, as well as various metal oxides and mixed metal oxides, have been obtained based on this route. It takes a different perspective to address the influence of inorganic–inorganic interplay on the synthesis of mesoporous materials.

Block copolymer (e.g., P123 and F127) templating syntheses lead to the formation of stable mesoporous AlPOs with uniform large pores based on the "acid–base pair" routes [194]. As shown in Fig. 16.26, an appropriate combination of inorganic sources such as $AlCl_3/H_3PO_4$, $AlCl_3/OP(OCH_3)_3$, and $Al(OC_4H_9)_3/PCl_3$ is quite important for the formation of Al–O–P bonds.

16.9.7. Evaporation-induced Self-assembly (EISA) Process

Beginning with a homogeneous solution of soluble silica and surfactant prepared in ethanol–water solvent, preferential evaporation of ethanol concentrates the nonvolatile surfactant and silica species. The progressively increasing surfactant concentration drives self-assembly of silica–surfactant micelles and their further organization into liquid crystal mesophases.

The EISA process [195] enables the rapid production of mesophases or mesoporous materials in the form of films (see Chapter 22 for details), fibers, or powders. Through variation of the initial reaction mixture composition it is possible to form different final mesostructures. The EISA process is perfect for the synthesis of film and monolith and can be used for transition metal-containing silica materials and nonsilica mesoporous materials [196].

16.9.8. Chemical Modification: Grafting and Cocondensation

One advantage of mesoporous materials is their amorphous wall which can tolerate various elements or organic groups. Another advantage of mesoporous materials is their huge inner surface which can be easily modified. For modifying the mesoporous materials through covalent (or other kind of) linkage between functional groups and silica framework, the major pathways include grafting approach (the organic or inorganic groups are bound on the silica surface by a postsynthetic treatment), cocondensation of tetraalkoxysilanes and agents containing functional elements or groups (direct synthesis), substitution of inorganic wall with nonsilicon atom, ion exchange or loading-active species (e.g., Cu, Ni), and techniques related with periodic mesoporous organosilicates.

Grafting (postsynthetic functionalization) makes use of the silanol groups present on the silica surface of the ordered mesoporous silica, because they are the anchor sites for metal species or silane-coupling agents [197]. Various functional groups including amino, thiol, and alkyl groups can be introduced by direct reaction of organosilanes to silica surface. The distribution and concentration of functional groups are influenced by reactivity of the organosilane and their accessibility to surface silanols, which are limited by diffusion and steric factors.

Another major approach for functionalization of mesoporous silica is cocondensation method where organosilane is condensed or polymerized together with conventional silica sources such as TMOS and TEOS. The one-pot pathway of the cocondensation method provides several advantages such as homogeneous distribution of the functional groups, unique pore size, and short preparation time. Limitations of this approach are that the functional components must be compatible with the synthesis conditions, must not disrupt the mesophases' self-assembly process, and must distribute the functional species to the desired locations within the final product.

Grafting can be controlled easily. Cocondensation is simple and often give higher loading. When using the grafting approach, the diffusion limitations of the rather large grafting species within the pores may lead to an inhomogeneous distribution, meaning that they are often located in the vicinity of the pore openings. Using cocondensation methods, silsesquioxane precursors with terminal organic groups must be co-assembled to some extent with TEOS. A more homogeneous distribution of the organic functionalities can be realized, because the organic units are direct parts of the mesoporous framework. The inhomogeneity is mainly due to the different reaction rates (hydrolysis and condensation) of both components which can lead more likely to homo- than to cocondensations and the tendency of the (large) organic groups to protrude into the pores.

16.9.9. Periodic Mesoporous Organosilicas (PMOs)

Periodic mesoporous organosilicas (PMOs) [4,198], one of the most advanced organic–inorganic hybrid materials, have attracted much research attention because of their combined advantages of ordered mesoporous structure—fusion of organic and inorganic fragments within the pore wall. The PMO materials were invented independently by three groups: Inagaki group [199], Ozin group [200], and Stein group [196] in 1999. In PMOs, organic and inorganic components are integrated by means of chemical, bottom-up, and self-assembly approach. The organic functionalities are not only completely homogeneously distributed but also possess additionally ordered pore arrangements accompanied by sharp pore size distributions. To date, many kinds of organic groups including methylene, phenylene, biphenylene, thiophene, ferrocene, large heterocyclic bridging groups, etc. have been successfully incorporated in the framework of PMOs [201].

Among one of the unique finding, Inagaki et al. [136] successfully prepared crystal-like pore wall structure in PMO material from benzene-bridged organosilane, 1,4-bis(triethoxysilyl)-benzene, where hydrophilic silicate layers and hydrophobic benzene layers array alternately. The ordered benzene–silica hybrid material has a hexagonal array of mesopores and crystal-like pore walls that exhibit structural periodicity with a spacing of 7.6 Å along channel direction. Periodic wall structure results from alternating hydrophilic and hydrophobic layers, composed of silica and benzene, respectively.

16.9.10. Nanocasting

Nanocasting or repeat templating is the process in which a template with relevant structures on length scale of nanometers is filled with another material, and initial template is afterward removed. Generally, two kinds of templates, hard and soft templates, can be used for nanocasting processes. The nanocasting pathways [133,202] use hard templates (see Fig. 16.27) to create ordered replicas, which allow us to create mesoporous materials with new compositions, controllable structure, and specific functionality. Nanocasting is a powerful method for creating materials that are more difficult to synthesize by conventional processes. The template is actually used as a true mold to produce mesoporous materials with controllable pore size, pore shape, and distribution.

Nanocasting with mesoporous inorganic solids or mesoporous carbons: Ordered mesoporous silica seems to be an ideal hard template, which can be used as a mold for other mesostructures with various compositions, such as ordered mesoporous carbon and metal oxides. Silica is relatively easily dissolved in HF or NaOH. Alternatively, mesoporous carbons with solid skeleton structure are also of suitable choices as hard template due to their excellent structural stability on thermal or hydrothermal and chemical treatments. A pronounced advantage of carbon is the fact that it is much easier to remove than silica by simple combustion.

Highly ordered mesoporous silica can be regenerated from a mesoporous carbon CMK-3 that is a negative replica of mesoporous silica SBA-15, indicating reversible replication between carbon and inorganic materials [203]. This method is likely to be a valuable complement to the existing methods for the preparation of new mesoporous materials. The HUM-1 (*I4₁/a*) [204], a silica replica of mesoporous carbon, is a cubic mesoporous silica that is distinctly different from the original MCM-48 silica and represents a previously unreported new mesoporous silica material. The HUM-1 does not possess the two noninterconnecting channel systems found in the starting MCM-48 framework. It may not be possible to make this material using the current conventional surfactant assembly methods.

To obtain well-ordered porous products by nanocasting, two requirements should be satisfied for the precursor of the expected composition. (i) Precursor should have a high solubility in a suitable solvent or should exist in liquid state to maximize the loading amount and produce a sufficiently rigid skeleton in calcined solid to avoid collapse. Basically, a very high concentration of the precursor solution, if at all possible, the neat precursor, is highly recommended. (ii) Resulting material should have a melting point higher than the temperature at which the carbon templates are combusted if the mesoporous carbon is used as template. Silica has an excellent thermal stability, and TEOS as silica precursor allows a high filling degree of the pore system of the porous carbon. In addition, the moderate hydrolysis rate helps to avoid a too rapid and vigorous reaction. Alternatively, an aqueous Na_2SiO_3 solution can be used as silica precursor [205].

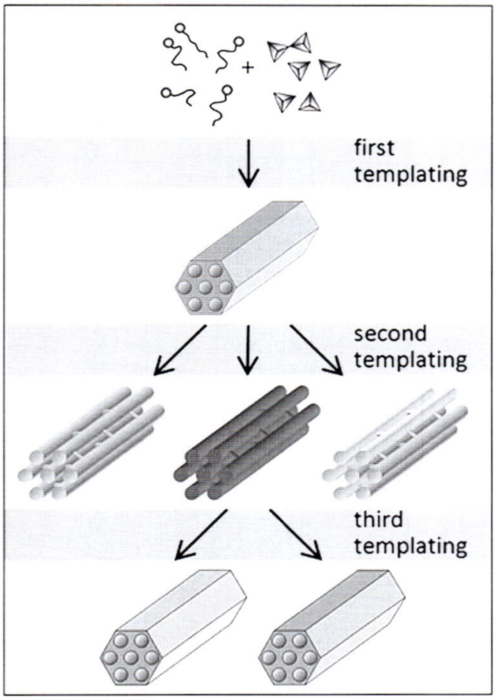

FIGURE 16.27 Schematic drawing of repeated templating for the synthesis of ordered mesoporous materials. *Reprinted with permission from Ref. [202]. Copyright 2008 American Chemical Society.*

16.9.11. Hierarchical Porous Materials

Recent interest is being devoted to the development of hierarchically ordered porous structures, in which different levels of porosity are integrated into a single body. The hierarchical materials have a number of potential advantages, e.g., improved overall mass transport characteristics [206], better hydrothermal stability [207], multifunctionality, etc. There are many studies reported on the bimodal micro–mesoporous [208], double mesoporous, macro–mesoporous, and trimodal porous materials. The most common synthetic way involves the secondary templates such as colloid crystals or emulsion or vesicle droplets for the formation of macropores. In this section, a brief introduction of micro–mesoporous bimodal materials is given due to the limited scope of this chapter.

Different porous materials that combine micro- and mesopores have been developed in the last decade such as SBA-15, plugged hexagonal-templated silica, mesoporous materials build up from zeolite nanoparticles or precursors, zeolite crystals or particles with mesopores, etc.

The plugged hexagonal-templated silica, PHTS [209,210] (see Fig. 16.28), has been obtained by increasing the silica over surfactant ratio in the synthesis of SBA-15. PHTS possesses extra microporous amorphous nanoparticles (plugs) in uniform mesoporous channels. There are two types of micropores: originating from the walls and the nanocapsules, respectively. Micropore volumes have a high value (up to 0.3 cm^3/g) and the total pore volume exceeds 1 cm^3/g. The pillaring effect of the nanoparticles shows that PHTS is much more stable than the conventional micellar-templated mesostructures. The phenomenon of amorphous silica plugs has been observed in other materials such as cubic *Fd3m* mesoporous silica monolith [211].

Mesoporous materials with zeolitic properties [212]: One simple synthesis approach is the deposition of a clear solution containing zeolitic precursor particles in large-pore mesoporous silica materials like siliceous mesostructured cellular foams [213] or the impregnation of zeolitic precursor solutions within mesoporous materials like SBA-15.

The direct synthetic method is to use zeolitic precursor solutions as silica source in the normal mesoporous material synthetic system with CTAB, P123, etc. as template. There are several successful examples for the self-assembly of preformed zeolite primary and secondary structural building units (nanoclusters) with surfactant micelle. A series of mesoporous materials build up from Faujasite, zeolite beta, ZSM-5, titanium silicalite-1, etc. precursor particles have been reported

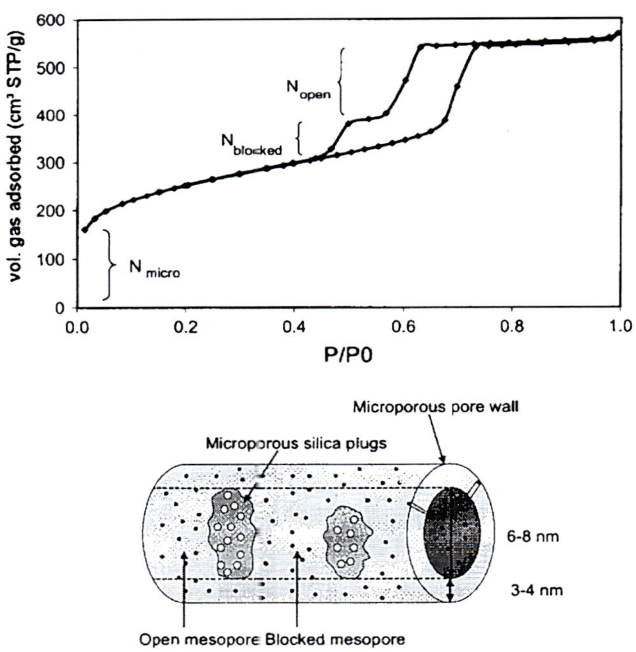

FIGURE 16.28 Nitrogen sorption isotherms at −196 °C for PHTS. Schematic representation of the different pore structures. *Reprinted with permission from [209]. Copyright 2002 American Chemical Society.*

by Pinnavaia and coworkers [214,215], Xiao [138], and other groups. This method is effective in introducing Al or other metals such as Ti into ordered mesoporous silica materials. These materials are steaming stable and have strongly acidic sites.

16.10. MORPHOLOGY CONTROL IN MESOPOROUS MATERIALS

The control of morphology in mesoporous materials is thought to be governed by kinetic effects as the self-assembly of surfactant molecules and inorganic species. Synthetic conditions, i.e., inorganic source or additive such as organoalkoxysilanes, temperature, stirring rate, ionic strength, acidity, and reactant ratio, affect the mesostructures and macrostructures of materials. Mesoporous inorganic materials have been synthesized with various mesophase structures and macroscopic morphologies (e.g., powders, various morphologies in microscale [216], mesostructures within nanoconfinement [217], "Single Crystal" [149], hollow spheres [218], nanoparticles [156], beads in milometer scale [219], ribbons [220], fibers [221,222], thin films [223,224], and monoliths [225]).

16.10.1. Thin Film

Mesoporous thin films have potential applications as low dielectric constant films, low refractive index films, hydrogen sensors, biomolecular sensors, nanostructured magnetic materials, photomodulated mass-transport layers, nanostructured solar cells, and nanostructured thermoelectrics.

An ordered mesoporous thin film can be obtained by a combination of sol–gel and supramolecular chemistry using a micelle-templated self-assembly process. The evaporation-induced self-assembly (EISA) [195,224,226] is a very effective method to make thin film, which is reviewed in Chapter 22. The EISA technique is more simple and easily controllable than other techniques such as film growth on interface from reaction solution [223].

Pai and Watkins [227] have used selective deposition of silica in a block copolymer film dialyzed by supercritical carbon dioxide. Unlike EISA, this strategy allows pre-organization of the template by decoupling the metal oxide condensation and the template self-assembly. Mesoporous silica films with well-ordered nanochannels oriented parallel to the substrate have been also produced by this method [228].

16.10.2. Spheres and Nanoparticles

Mesoporous silica nano- and microspheres have been intensively investigated for controllable drug delivery [230–233], biosignal probing [234], gene transport and expression [235], biomarking [180], other important bioapplications [236,237], as building blocks for structured materials [238], macromolecular separations, catalytic supports, and template agents for photonic crystals. Several synthetic strategies to control the sizes of mesoporous nanoparticles from hundreds of nanometers to tens of micrometers have been reported [239–241].

Among the synthetic strategies leading to the small mesoporous silica particles, a modified Stöber route is used with adding the structure-directing agent to the water–alcohol–ammonia-tetraalkoxysilane system [242–244].

The hollow spherical mesoporous silicas are fabricated either via hard templating such as inorganic or polymer beads [245,246] or soft templating methods such as emulsion [247].

Lu et al. [248] reported a rapid, aerosol-based process for synthesizing solid, well-ordered spherical particles with stable pore mesostructures of hexagonal and cubic topology, as well as layered (vesicular) structures. This method relies on evaporation-induced interfacial self-assembly confined to spherical aerosol droplets.

Mesoporous silica MCM-41 nanoparticles with a diameter of ~20 nm and without deformation of the hexagonal mesostructure have been successfully synthesized [239]. In this synthesis, a nonionic surfactant (F127) was used as a suppressant in grain growth to achieve a balance between the ordered mesostructure and the nanoparticle.

The nanoscaled mesoporous silica spheres with controlled particle size in highly diluted aqueous solution were synthesized, in which diluted surfactant concentration was used to overcome the particle aggregation [249–251]. However, it was very difficult to isolate the product, and the high dilution technique often resulted in low yields.

A series of monodispersed mesoporous silica spheres with tunable particle size were synthesized by using self-assembly of cationic gemini surfactant C18-3-1 and silica source under basic condition [252]. The particle sizes of these mesoporous silica spheres with radially oriented pore structure were controlled between 120 and 490 nm, and the pore sizes changed from 2.2 to 3.4 nm by varying the initial synthesis conditions. With increasing the concentration of ethanol, the sphere diameter varied from 70 to 460 nm.

Bein and coworkers [253–255] reported a high-yield synthetic procedure to synthesize the suspensions of colloidal mesoporous silica with 50–100 nm particle size by addition of triethanolamine (TEA). TEA was thought to act as a complexing agent for silicate species and additionally as an encapsulator for mesoporous particles, limiting the growth and aggregation of particles.

In the synthesis system with cetyltrimethylammonium chloride (CTAC) as template, the nanoparticle size can be controlled from ~25 nm to ~200 nm by adding suitable additive agents (e.g., inorganic bases, alcohols) to control the hydrolysis and condensation of silica precursor TEOS in the solution [256]. A certain acid–base buffer capacity of the reaction mixture in the range of pH 6–10 is essential for the formation of mesoporous silica nanoparticles.

16.10.3. Fiber and Rod

Huo et al. [222] described the synthesis of mesoporous silica fibers by the general procedure for one-step, two-phase synthesis at room temperature. The silica fibers, which are 1–5 μm diameter and up to 5 cm long, have excellent long-range order. Control over the morphology and dimensions of the fibers is demonstrated through control of the self-assembly and growth mechanisms, which leads to orientationally ordered hexagonal mesopores within the fibers and waveguide properties. The mesopores in silica fibers run in a circular direction around the axis [257]. The kinetics of fiber formation and the product distribution vary, being strongly dependent on the silica source and the amount of additional oil [221].

Mesoporous silica nanofibers [258] can be synthesized by a one-phase route under strongly acidic conditions with cationic surfactants. The diameters of the fibers range from 50 to 250 nm, and the length is up to millimeters (see Fig. 16.29). The nanofibers can have either a longitudinal pore architecture, in which the pore channels are aligned parallel to the fiber axis, or a circular pore architecture, in which the pore channels are wound circularly around the fiber axis. The pore channels in both types of nanofibers are hexagonally packed. The cross-sections of the nanofibers with longitudinal pore architectures are hexagonal and those of the nanofibers with circular pore architectures are circular. Generally, low temperatures produce nanofibers with longitudinal pore architectures and high temperatures produce nanofibers with circular pore architectures. The transition temperatures for CTAC and CTAB surfactants are around 75 °C.

Monodispersed SBA-15 materials with ~100% rod-like morphologies (~1 to 2 μm long) have been obtained by using P123 as template under static conditions without using salts [163]. This method produced rod-like particles with temperature-dependent pore sizes ranging from 5.8 to 12.5 nm but with similar external dimensions. In order to prepare short monodispersed SBA-15 rods, the absence of stirring is essential, and the temperature of the first stage should not exceed 60°C.

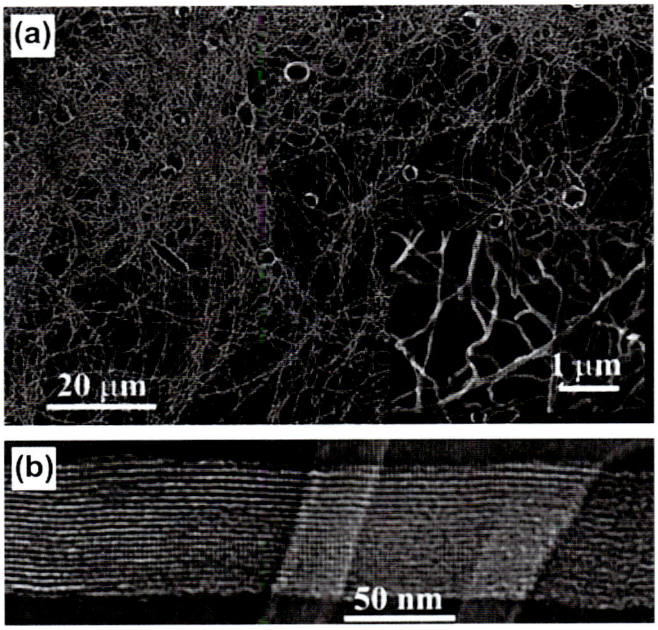

FIGURE 16.29 (a) SEM image of as-made silica nanofibers. (b) TEM images of as-made nanofibers. *Reprinted with permission from [258]. Copyright 2004 American Chemical Society.*

16.10.4. Monolith

The direct liquid crystal templating pathway is effective to make mesostructured silica monolith. For example, Feng et al. [259] synthesized monolithic mesoporous silica in ternary and microemulsion-type quaternary systems while preserving long-range ordering at the mesoscopic level.

Large-sized, crack-free silica monoliths with highly ordered mesostructure are prepared by a fast and easy way via liquid-paraffin-medium protected solvent evaporation [260]. By employing the inert liquid paraffin, cracks of the materials can be successfully avoided. The block copolymer–silica composite monoliths are transparent and crack-free with a large size.

16.11. NEW COMPOSITIONS: NONSILICA-BASED MESOPOROUS MATERIALS [13,22]

In addition to the preparation of various mesoporous silica structures, incorporation of heteroatoms such as Cu, Zn, Al, B, Ga, Fe, Cr, Ti, V, and Sn into mesoporous silica framework has been widely investigated. Furthermore, it is possible to synthesize mesoporous structures of materials other than silica by using the similar supramolecular assembly of surfactants or nanocasting method [133,139,261–264]. One of the main difficulties arouse from the organization of two different units on the surface of surfactant assemblies. Hydrogen-bonded and ligand-assisted synthetic pathways were proposed for the formation of various nonsilica-based mesoporous materials and used surfactants were easily removed by extraction. This research area covers the synthesis, structural characterization, morphological control (e.g., film, fiber, and monolith), and alignment control of mesochannels [23,193,264–267]. Mesoporous materials now include a variety of inorganic-based materials, for example, aluminophosphate, transition metal oxides, carbon compounds, inorganic–organic hybrid materials, polymers, and even metals.

16.11.1. Mesostructured and Mesoporous AlPO-based Materials

The surfactant-templated synthesis is effective for the preparation of mesoporous AlPOs by keeping the AlPO frameworks amorphous. Crystallized AlPO frameworks have high charge densities, which result in the formation of lamellar phases containing a larger amount of ionic surfactants.

Block copolymer is good template for mesoporous AlPO with two-dimensional or three-dimensional hexagonal phase [268]. The periodicities of mesopores are lost for almost all the mesoporous AlPOs after the removal of surfactants though the uniformities of the mesopores are somehow retained. Morphological controls such as ordered mesoporous AlPO films based on the acid–base pair route and mesoporous AlPO spheres by the nanocasting technique (hard templating) have been achieved recently [269].

16.11.2. Metal Oxides and Other Inorganic Materials

Nonsiliceous mesostructured systems, such as transition metal oxides (TMOs), are often of greater interest than silicate mesostructures due to varied framework properties which are potentially useful for (photo)catalysis, sensing, optics, energy conversion, etc. However, there are two additional challenges for TMO synthesis [264]. First, metal oxide precursors (generally chlorides or alkoxides) are much more reactive than silica-based analogues; uncontrolled condensation yields macroscopic phase segregation. Strategies for controlling the hydrolysis–condensation rates of TMO precursors include utilizing specific pH ranges, stabilizing ligands, nonaqueous media, preformed nanoclusters, controlled hydrolysis, or some combination thereof. Second, redox reactions, phase transformations, and crystallization can collapse the framework at elevated temperatures. To address this problem, block-copolymer SDAs that yield thicker inorganic walls and careful heat treatment are often employed.

An important development in the synthesis of mesoporous TMOs was the work by Antonelli and Ying, who developed a modified sol–gel route using acetylacetonate chelators [270]. The resulting mesostructured TiO_2 and Nb_2O_5 were thermally stable but with amorphous pore walls.

Evaporation-induced self-assembly (EISA) has proven to be an extremely useful process for both controlling macroscopic form (thin films, membranes, and monoliths) and enabling the synthesis of mesostructured TMOs [162,195,271]. Porous TMOs with semicrystalline frameworks were formed by utilizing EISA with inorganic precursors in highly acidic alcoholic solutions [162].

Using a surfactant as a template to synthesize metal oxides often leads to the loss of the ordered structure after removal of the template. Using mesoporous silica as a hard template to create an ordered metal oxide is an alternative effective method.

The porous chromium oxide single crystals were obtained by using aminopropyltriethoxysilane functionalized SBA-15 as the template and $H_2Cr_2O_7$ as the chromium precursor, which was chemically adsorbed. Subsequent thermal treatment resulted in the formation of chromium oxide and the template was removed by an aqueous hydrofluoric acid (HF) solution [272]. In this synthesis, calcination temperatures exceeding 350°C were needed in order to create the crystalline phase.

Zhao has demonstrated that three-dimensional mesoporous silica can be used as hard template and hydrated metal nitrates as source to fabricate various mesostructured crystalline metal oxides (Co_3O_4, Mn_2O_3, CeO_2, and In_2O_3) [166,273]. The nanocasting pathway can be extended to other nonsilica compositions which are not accessible by solution-based methods. In addition, metal sulfide nanowires, such as CdS, ZnS, and In_2S_3, with ordered mesostructure were casted using mesoporous SBA-15 as a hard template [274].

However, when silica is used as the hard template, the leaching process with strong alkaline or HF is not compatible with many different oxides. Carbon templates provide an alternative to silica.

Monodisperse and high-surface-area mesoporous inorganic spheres of various compositions including metal oxides, mixed oxides, and metal phosphates [269] have been prepared by templating mesoporous carbon spheres replicated from spherical mesoporous silica. Due to the rigid and thermally stable framework of carbon template, the crystalline phases of the obtained metal oxide spheres can be readily tailored by controlling crystalline temperatures.

16.11.3. The Mesoporous Metals and Related Metal-based Nanomaterials [275]

Mesoporous metals hold promises for a wide range of potential applications, in electronic devices, magnetic recording media, and metal catalysts. Mesoporous metals with highly ordered networks and narrow pore size distributions have been produced by using nanocasting method with mesoporous silica as a hard template or the direct-template a synthesis approach from lyotropic liquid crystals (LLCs) made of nonionic surfactants at high concentrations. Moreover, mesoporous alloys of various compositions can also be designed by controlling the composition.

Various types of nanostructured metals, including three-dimensional networks, nanoparticles, nanowires, and nanonecklaces, have been created by using mesoporous silicas or organic–inorganic hybrid mesoporous materials. For example, Shin et al. [276] reported nanoporous Pt with three-dimensional networks by the impregnation of a Pt precursor followed by H_2 reduction within the mesopores of the MCM-48 ($Ia3d$) template.

Many mesoporous metals have been prepared by the chemical or electrochemical reduction of metal salts dissolved in aqueous LLC domains or hard template. As a soft template, LLCs are more versatile and therefore more advantageous than hard templates. Moreover, the hard-template method has been widely applied for the creation of metal nanowire films by utilizing electrochemical processes (Fig. 16.30) [277,278]. The electrodeposition method, which uses an external power source, is an efficient technique for depositing metals in mesoporous silica films. Wang et al. [277] fabricated Pd nanowire thin films through the reduction of Pd ions within the mesochannels of two-dimensional hexagonally ordered mesoporous silica films. After the removal of the mesoporous silica template, Pd nanowire thin films were successfully prepared. Several three-dimensional nanowire networks can be prepared by using mesoporous silica films with three-dimensional mesoporous structures. The metal nanowires gradually grow from the bottom surface of the conductive substrate by electrodeposition. The electrodeposition process can easily prevent metal deposition in the outer part of the silica template by controlling the electrodeposition time.

16.12. POROUS CARBON MATERIALS

Porous carbon materials are commonly used as adsorbent, catalytic supports, etc. [135,279]. While many porous carbon materials are known to exhibit periodic structures resulting from the uniform stacking of graphene sheets and periodic arrangement of atoms within these sheets. Carbon materials with periodic

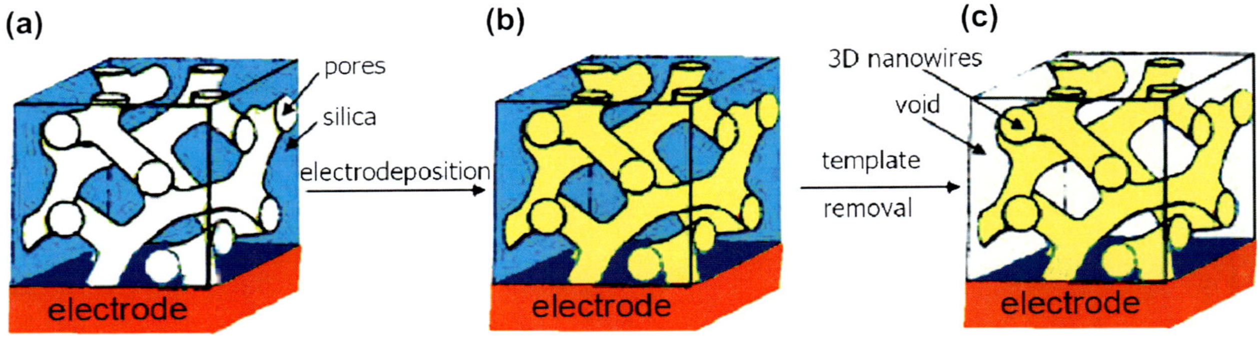

FIGURE 16.30 Schematic representation of the formation of three-dimensional continuous macroscopic metal or semiconductor nanowire networks by a template-electrodeposition technique. *Reprinted with permission Journals: Wang, D.; Luo, H.; Kou, R.; Gil, M. P.; Xiao, S.; Golub, V. O.; Yang, Z.; Brinker, C. J.; Lu, Y.: A General Route to Macroscopic Hierarchical 3D Nanowire Networks. Angewandte Chemie International Edition. 2004. Volume 43, Page 6169. Copyright Wiley-VCH Verlag GmbH & Co. KGaA. Reproduced with permission [278].*

microporous or mesoporous structures have been reported only recently. The effective synthesis methods include nanocasting (see Fig. 16.31 [279]) and the carbonization of mesoporous polymer.

16.12.1. Microporous Carbons

Zeolites were already employed as template in the synthesis of microporous carbon with ordered structures [280]. Zeolite Y template can cast a microporous carbon with a three-dimensional nano-array structure which is identical to those of supercages in zeolite Y crystal [281]. Carbon has a surprisingly high BET surface area (3600 m^2/g) with almost no mesoporosity. Zeolite EMC-2 (EMT) allows the formation of faithful carbon replica exhibiting up to three well-resolved XRD peaks [282].

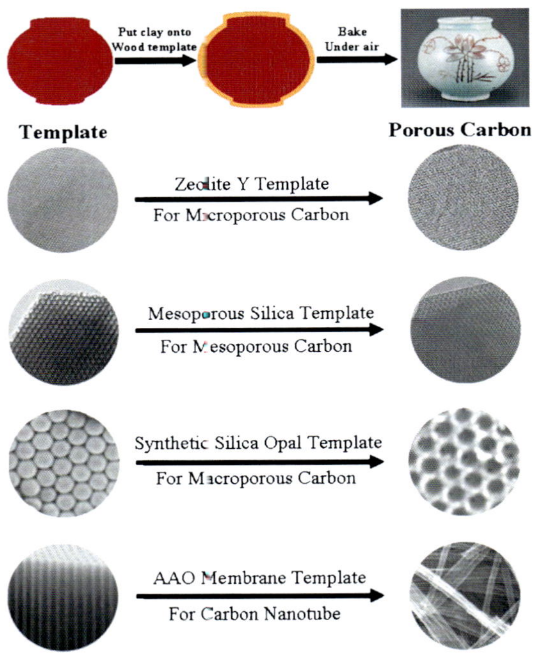

FIGURE 16.31 a) Schematic representation showing the concept of template synthesis. (b) Microporous, (c) mesoporous, and (d) macroporous carbon materials, and (e) carbon nanotubes were synthesized using zeolite, mesoporous silica, a synthetic silica opal, and an AAO membrane as templates, respectively. *Lee, J.; Kim, J.; Hyeon, T.: Recent Progress in the Synthesis of Porous Carbon Materials.. Advanced Materials. 2006. Volume 18, Page 2073. Copyright Wiley-VCH Verlag GmbH & Co. KGaA. Reproduced with permission [279].*

16.12.2. Mesoporous Carbon Materials [283]

It is very difficult to obtain mesoporous carbon materials with an ordered structure via a sol–gel process involving a surfactant templating strategy. The discovery of ordered mesoporous silica materials opened new opportunities in the synthesis of periodic carbon structures using the templating approach. By employing mesoporous silica structures as hard templates, ordered mesoporous carbon replicas have been synthesized from a nanocasting strategy. The synthesis is quite tedious and involves two main steps: (i) preparation and calcination of the silica mesophase; and (ii) filling the silica pore system by a carbon precursor, followed by the carbonization and selective removal of the silica framework. Various carbon precursors such as sucrose [284], furfuryl alcohol [134,285], acetylene gas, phenol–formaldehyde, phenol resin, mesophase pitch [286], polydivinylbenzene, acrylonitrile, and pyrrole can be used for the synthesis of ordered mesoporous carbons. These organic substances after carbonization formed rigid carbon frameworks in the mesopores of silica template. Following formation of the carbon replica the silica substrate can be removed by dissolution in sodium hydroxide or HF solution and the structural order of the carbon frameworks can be retained. These ordered mesoporous carbons exhibit high specific surface areas (typically 1300–2000 $m^2 g^{-1}$), uniform pore diameters (2–6 nm), large adsorption capacities

($1-2\ cm^3\ g^{-1}$), and high thermal, chemical, and mechanical stabilities. For the low temperatures (700–1000 °C) synthesis, the resulting carbon materials are usually amorphous because of no graphitization.

The carbon materials exhibit wide varieties of pore shapes, connectivity, and pore wall thickness, depending on silica templates that are synthesized with various structures and pore diameters. The first mesoporous carbon examples are CMK-1 [284] with symmetry cubic $I4_132$ or $I4_1/a$ from sucrose and CMK-4 [287] from acetylene with symmetry $Ia3d$ by using MCM-48 as template. Figure 16.32 shows the XRD patterns of CMK-3 (keep original symmetry) and CMK-1 (create new symmetry).

After the discovery of CMK-1, many studies were carried out to synthesize mesoporous carbons with ordered structures [289,290]. Till date, many mesoporous silicas, such as SBA-3, HMS, MSU-H, MCM-41, MCM-48, SBA-15, KIT-6, FDU-5, FDU-12, SBA-16, and mesocellular silica foam, have been used as templates for periodic mesoporous carbons.

It is noteworthy that different types of carbons with SBA-15 as template but different framework configurations, rod- and tube-types, were obtained. The rod-type *p6mm* mesoporous carbon CMK-3 was synthesized by a complete filling of mesoporous cylindrical channels of the SBA-15 silica. The CMK-5 carbon resulted from an incomplete filling of the SBA-15 channels, which after silica dissolution gave interconnected nanopipes. NCC-1 carbon, which is essentially similar to CMK-5, has a bimodal pore size distribution (two pore systems: the pores left by the silica template and the pores in the inner part of the nanotubes) [291]. The crucial factors for the synthesis of such carbons are an aging temperature of 140°C for the template SBA-15, a relatively low concentration of furfuryl alcohol (25 vol.%), and a carbonization temperature higher than 750°C [292].

Figure 16.33 shows the XRD pattern of CMK-5, TEM image, and schematic structure [134]. The structural model is provided to indicate that the carbon nanopores are rigidly interconnected into a highly ordered hexagonal array by carbon spacers. The outside diameter of the carbon structures is controllable by the choice of a template SBA-15 aluminosilicate with suitable diameter; the inside diameter is controllable by the amount of the carbon source. The (10) diffraction peak is lower than (11) in intensity, owing to the diffraction interference between the walls and the spacers interconnecting adjacent cylinders.

Nanoporous carbon with narrow pore size distribution can be prepared directly via a one-step nanocasting technique by carbonization of cyclodextrin–silica organic–inorganic hybrid composite [293]. The method consists of preparing a cyclodextrin-templated silica mesophase via soft chemistry, followed by direct carbonization of the occluded cyclodextrins. The nanocasting procedure provides granules (in millimeter scale) or monoliths (in centimeter scale).

The ordered mesoporous carbon can also be synthesized with a catalytic CVD method [294]. The ordered carbons possess bimodal pores: the pores arise from the replica of frameworks of the template and the pores correspond to carbon nanotubes formed in the channels of the template.

Meng et al. [295] fabricated highly ordered mesoporous phenolic resin polymers by using the low-molecular-weight and soluble polymers of phenolic and formaldehyde as organic precursors and amphiphilic triblock copolymers as templates via an EISA approach. Subsequent carbonization at 600–1400 °C under

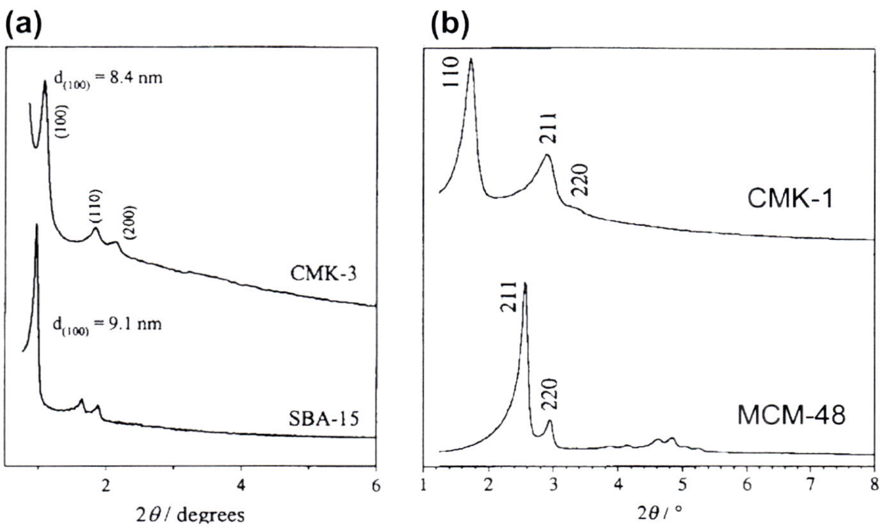

FIGURE 16.32 Powder XRD patterns of: (a) CMK-3 carbon and SBA-15 silica used as template for the CMK-3 synthesis. *Reprinted with permission from [288]. Copyright 2000 American Chemical Society.* (b) template MCM-48 and CMK-1. *Reprinted with permission from [284]. Copyright 1999 American Chemical Society.*

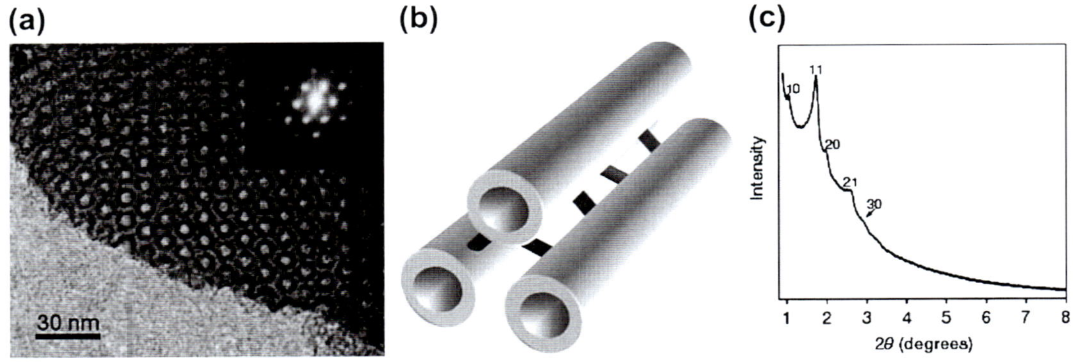

FIGURE 16.33 CMK-5 (a) TEM image viewed along the direction of the ordered nanoporous carbon and the corresponding Fourier diffractogram. (b) Schematic model for the carbon structure. (c) XRD pattern indicating the hexagonal order between carbon cylinders. *Reprinted by permission from Macmillan Publishers Ltd:* Nature *[134], Copyright (2001).*

nitrogen directly transforms the polymers to carbon counterparts. The synthesis approach is quite similar to that of the mesoporous silica. The key issue for the successful organization of organic–organic mesostructures is the preformed polymer guest, which has plenty of hydroxyl groups and strongly interacts with the amphiphilic triblock copolymer host via hydrogen bonds. The amphiphilic surfactant can be easily removed by solvent extraction or calcination in N_2.

16.13. ORDERED MACROPOROUS MATERIALS

Ordered macroporous materials [296] with pore sizes in the submicrometer range have applications in low dielectric constant materials and lightweight structural materials. Several single, binary, and tertiary macroporous oxides have been prepared using sol–gel chemistry: SiO_2, TiO_2, ZrO_2, Al_2O_3, Fe_2O_3, Sb_4O_6, WO_3, Eu_2O_3, Nd_2O_3, Sm_2O_3, $YZrO_2$, V–P–O; $PbTiO_3$, etc.

Colloidal crystals consisting of three-dimensional ordered arrays of monodispersed spheres can be used as templates for the preparation of ordered macroporous materials that exhibit precisely controlled pore sizes and highly ordered three-dimensional porous structures. The macroscale templating approach typically consists of three steps [297]. First, the interstitial voids of the monodisperse sphere arrays are filled with precursors of various classes of materials, such as ceramics, semiconductors, metals, and monomers. In the second step, the precursors condense to form a solid framework around the spheres. Finally, the spheres are removed by either calcination or dissolution with a solvent leaving a periodic array of holes in the inorganic matrix. For example, a macroporous silica [298] or TiO_2 [299] can be synthesized through a slow sedimentation of colloidal particles onto a template, the monomeric alkoxide precursors permeate the array of bulk polystyrene spheres and condense in air at room temperature. Close packed, open-pore structures with ~100 to ~2000 nm voids are obtained after calcination of the organic component at high temperatures.

Imhof and Pine [300] reported a microemulsion method for producing highly monodisperse macroporous materials with pore sizes ranging from 50 nm to several micrometers. Macroporous titania, silica, and zirconia were formed by using the nonaqueous monodisperse emulsion droplets of isooctane in formamide as templates around which material was deposited through a sol–gel process. Gelation and removal of the droplets led to a macroporous amorphous titania gel.

16.14. CHALLENGES FOR POROUS MATERIAL SCIENTIST

There is no doubt of the great importance of porous materials in the general subject of materials science, even though some of their essential features are not wholly understood [37]. Porous materials, especially zeolites, microporous molecular sieves, and ordered mesoporous solids, have many present and potential future applications for catalysis, adsorption and separation, life science, energy, environmental, functional materials for high technologies, and so on.

Considerable scientific efforts has been made on the preparation, characterization, and application of porous materials. Strategy aimed at the controllable synthesis has been focused on the control of micro-, meso-, and macroscale, including synthetic methods, architecture concepts, and fundamental principles that govern the rational design and synthesis. However, many challenges remain for the future: (1) the practical applications of known porous materials, to compete with current technologies. (2) The design of materials with fully controllable pore structures, framework composition, active sites, morphologies, and properties. The influence of changes in porous characteristics on the final properties in adsorption, catalysis, diffusion, etc. will facilitate in designing new, adjusted, or improved porous materials that can be implemented in industrial processes of various kinds. (3) The clear understanding of formation mechanisms and the interactions between templates and wall precursors. The knowledge of the growth mechanism and behavior of the zeolite nuclei in different environments will be of utmost importance. Formation mechanism of the mesophase needs to be fully understood. (4) The development of new and versatile synthetic methods. (5) The industrial preparation of porous materials should be ecologically friendly, low toxic, low-cost, and simple procedure.

References

[1] J. Rouquerol, D. Avnir, C.W. Fairbridge, D.H. Everett, J.M. Haynes, N. Pernicone, et al., Pure and Applied Chemistry 66 (1994) 1739.
[2] J. Weitkamp, L. Puppe, Catalysis and Zeolites: Fundamentals and Applications, Springer, 1999.
[3] M. Kruk, M. Jaroniec, Chemistry of Materials 13 (2001) 3169.
[4] P. Van der Voort, C. Vercaemst, D. Schaubroeck, F. Verpoort, Physical Chemistry Chemical Physics 10 (2008) 347.
[5] http://www.iza-online.org.
[6] D.W. Breck, Zeolite Molecular Sieves: Structure, Chemistry, and Use, Wiley, New York, 1973.
[7] R.M. Barrer, Hydrothermal Chemistry of Zeolites, Academic Press, London, 1982.
[8] J. Cejka, H. van Bekkum, A. Corma, et al., Introduction to Zeolite Molecular Sieves third ed. (Studies in Surface Science and Catalysis, Vol. 168), Elsevier, 2007.
[9] S.M. Auerbach, K.A. Carrado, P.K. Dutta, Handbook of Zeolite Science and Technology, CRC, 2003.
[10] C.J. Brinker, G.W. Scherer, Sol-Gel Science, Academic Press, New York, 1990.
[11] R. Szostak, Molecular Sieves: Principles of Synthesis and Identification, Blackie Academic & Professional, London, 1998.
[12] H.G. Karge, J. Weitkamp, Molecular Sieves: Science and Technology, vol. 1, Synthesis, Springer-Verlag, Berlin, 1998.
[13] R. Xu, W. Pang, J. Yu, Q. Huo, J. Chen, Chemistry of Zeolites and Related Porous Materials: Synthesis and Structure, Wiley-Interscience, 2007.
[14] C.S. Cundy, P.A. Cox, Chemical Reviews 103 (2003) 663.
[15] C.S. Cundy, P.A. Cox, Microporous and Mesoporous Materials 82 (2005) 1.
[16] X. He, D. Antonelli, Angewandte Chemie International Edition 41 (2002) 214.
[17] F. Schuth, Angewandte Chemie International Edition 42 (2003) 3604.
[18] A. Stein, Advanced Materials 15 (2003) 763.
[19] F. Schuth, W. Schmidt, Advanced Materials 14 (2002) 629.
[20] J.Y. Ying, C.P. Mehnert, M.S. Wong, Angewandte Chemie International Edition 38 (1999) 56.
[21] A. Corma, Chemical Reviews 97 (1997) 2373.
[22] G.J.D. Soler-illia, C. Sanchez, B. Lebeau, J. Patarin, Chemical Reviews 102 (2002) 4093.
[23] Y. Wan, D.Y. Zhao, Chemical Reviews 107 (2007) 2821.
[24] J.H. Yu, R.R. Xu, Accounts of Chemical Research 36 (2003) 481.
[25] A.K. Cheetham, G. Ferey, T. Loiseau, Angewandte ChemieInternational Edition 38 (1999) 3268.
[26] T.J. Barton, L.M. Bull, W.G. Klemperer, D.A. Loy, B. McEnaney, M. Misono, et al., Chemistry of Materials 11 (1999) 2633.
[27] M.E. Davis, R.F. Lobo, Chemistry of Materials 4 (1992) 756.
[28] M.E. Davis, Nature 417 (2002) 813.
[29] R.M. Barrer, Journal of the Chemical Society (1948) 127.
[30] D.W. Breck, W.G. Eversole, R.M. Milton, T.B. Reed, T.L. Thomas, Journal of the American Chemical Society 78 (1956) 5963.
[31] C. Baerlocher, L.B. McCusker, D.H. Olson, Atlas of Zeolite Framework Types, sixth ed., Elsevier, 2007.
[32] R.F. Lobo, Handbook of Zeolite Science and Technology, Marcel Dekker Inc., 2003.
[33] J.M. Newsam, Science 231 (1986) 1093.
[34] G.O. Bnmner, W.M. Meier, Nature 337 (1989) 146.
[35] C.N.R. Rao, J.M. Thomas, Accounts of Chemical Research 18 (1985) 113.
[36] G.T. Kerr, Science 140 (1963) 1412.
[37] J. Coronas, Chemical Engineering Journal 156 (2010) 236.
[38] S.T. Wilson, in: H. Robson (Ed.), Verified Synthesis of Zeolitic Materials,second revised edition, Elsevier, Amsterdam, 2001, p. 27.
[39] R.W. Thompson, in: H. Robson (Ed.), Verified Synthesis of Zeolitic Materials, second revised edition, Elsevier, Amsterdam, 2001, p. 21.
[40] F.T. Fan, Z.C. Feng, C. Li, Accounts of Chemical Research 43 (2010) 378.
[41] M.A. Nicolle, F. Di Renzo, F. Fajula, P. Espiau, T.d. Courieres, in: J.B.H.R. von Ballmoos, M.M.J. Treacy (Eds.), Proc. 9th Int. Zeolite Conf., Butterworth-Heinemann, Boston, 1993, p. 313.
[42] V. Nikolakis, D.G. Vlacho, M. Tsapatsis, Microporous and Mesoporous Materials 21 (1998) 337.
[43] R.M. Barrer, Zeolites 1 (1981) 130.
[44] B.M. Lok, T.R. Cannan, C.A. Messina, Zeolites 3 (1983) 282.
[45] H.K.J.L. Guth, P. Caullet, J. Hazm, A. Merrouche, J. Patarin, in: R. von Bailmoos, J.B. Higgins (Eds.), Proc. 9th Int. Zeolite Conf., Butterworth-Heinemann, 1993, p. 215.
[46] H. Lechert, P. Staelin, C. Kuntz, Zeolites 16 (1996) 149.
[47] H. Van Bekkum, E.M. Flanigen, J.C. Jansen, Introduction to Zeolite Science and Practice, Elsevier, Amsterdam, 1989.
[48] A. Corma, Chemical Reviews 95 (1995) 559.
[49] E.M. Flanigen, J.M. Bennett, R.W. Grose, J.P. Cohen, R.L. Patton, R.M. Kirchner, et al., Nature 271 (1978) 512.
[50] H. Gies, Journal of Inclusion Phenomena and Macrocyclic Chemistry 2 (1984) 275.
[51] E.M. Flanigen, B.M. Lok, R.L. Patton, S.T. Wilson, Pure and Applied Chemistry 58 (1986) 1351.
[52] S.T. Wilson, B.M. Lok, C.A. Messina, T.R. Cannan, E.M. Flanigen, Journal of the American Chemical Society 104 (1982) 1146.
[53] H. Li, M. Eddaoudi, M. O'Keeffe, O.M. Yaghi, Nature 402 (1999) 276.
[54] A. Phan, C.J. Doonan, F.J. Uribe-Romo, C.B. Knobler, M. O'Keeffe, O.M. Yaghi, Accounts of Chemical Research 43 (2010) 58.
[55] N.W. Ockwig, O. Delgado-Friedrichs, M. O'Keeffe, O.M. Yaghi, Accounts of Chemical Research 38 (2005) 176.
[56] S. Qiu, G. Zhu, Coordination Chemistry Reviews 253 (2009) 2891.
[57] Y. Liu, V.C. Kravtsov, R. Larsen, M. Eddaoudi, Chemical Communications (2006) 1488.
[58] Y. Liu, V. Kravtsov, M. Eddaoudi, Angewandte Chemie International Edition 47 (2008) 8446.
[59] X.C. Huang, Y.Y. Lin, J.P. Zhang, X.M. Chen, Angewandte Chemie International Edition 45 (2006) 1557.

[60] R. Banerjee, A. Phan, B. Wang, C. Knobler, H. Furukawa, M. O'Keeffe, et al., Science 319 (2008) 939.
[61] K. Koh, A. Wong-Foy, A. Matzger, Angewandte Chemie International Edition 47 (2008) 677.
[62] Q.R. Fang, G.S. Zhu, Z. Jin, Y.Y. Ji, J.W. Ye, M. Xue, et al., Angewandte Chemie International Edition 46 (2007) 6638.
[63] A. Corma, M.J. Diaz-Cabanas, J.L. Jorda, C. Martinez, M. Moliner, Nature 443 (2006) 842.
[64] D. Farrusseng, S. Aguado, C. Pinel, Angewandte Chemie International Edition 48 (2009) 7502.
[65] C.L. Bowes, W.U. Huynh, S.J. Kirkby, A. Malek, G.A. Ozin, S. Petrov, et al., Chemistry of Materials 8 (1996) 2147.
[66] P. Feng, X. Bu, N. Zheng, Accounts of Chemical Research 38 (2005) 293.
[67] K. Yamamoto, T. Tatsumi, Chemistry of Materials 20 (2008) 972.
[68] C.W. Jones, K. Tsuji, M.E. Davis, Nature 393 (1998) 52.
[69] C.W. Jones, M. Tsapatsis, T. Okubo, M.E. Davis, Microporous and Mesoporous Materials 42 (2001) 21.
[70] K.D. Hammond, M. Gharibeh, G.A. Tompsett, F. Dogan, A.V. Brown, C.P. Grey, et al., Chemistry of Materials 22 (2010) 130.
[71] W. Yan, E.W. Hagaman, S. Dai, Chemistry of Materials 16 (2004) 5182.
[72] K. Maeda, Y. Mito, T. Yanagase, S. Haraguchi, T. Yamazaki, T. Suzuki, Chemical Communications (2007) 283.
[73] J.X. Jiang, J.H. Yu, A. Corma, Angewandte Chemie International Edition 49 (2010) 3120.
[74] K.G. Strohmaier, D.E.W. Vaughan, Journal of the American Chemical Society 125 (2003) 16035.
[75] G.Y. Yang, S.C. Sevov, Journal of the American Chemical Society 121 (1999) 8389.
[76] C.H. Lin, S.L. Wang, K.H. Lii, Journal of the American Chemical Society 123 (2001) 4649.
[77] N. Guillou, Q. Gao, P.M. Forster, J.S. Chang, M. Nogues, S.E. Park, et al., Angewandte Chemie International Edition 40 (2001) 2831.
[78] Y.L. Lai, K.H. Lii, S.L. Wang, Journal of the American Chemical Society 129 (2007) 5350.
[79] X. Zou, T. Conradsson, M. Klingstedt, M.S. Dadachov, M. O'Keeffe, Nature 437 (2005) 716.
[80] M.E. Davis, C. Saldarriaga, C. Montes, J. Garces, C. Crowdert, Nature 331 (1988) 698.
[81] M. Estermann, L.B. McCusker, C. Baerlocher, A. Merrouche, H. Kessler, Nature 352 (1991) 320.
[82] C.C. Freyhardt, M. Tsapatsis, R.F. Lobo, K.J. Balkus, M.E. Davis, Nature 381 (1996) 295.
[83] P.A. Barrett, M.J. Díaz-Cabañas, M.A. Camblor, R.H. Jones, Journal of the Chemical Society, Faraday Transactions 94 (1998) 2475.
[84] T. Cheetham, H. Fjellåg, T.E. Gier, K.O. Kongshaug, K.P. Lillerud, G.T. Stucky, Studies in Surface Science and Catalysis 135 (2001) 788.
[85] J. Sun, C. Bonneau, A. Cantin, A. Corma, M.J. Diaz-Cabanas, M. Moliner, et al., Nature 458 (2009) 1154.
[86] X. Ren, Y. Li, Q. Pan, J. Yu, R. Xu, Y. Xu, Journal of the American Chemical Society 131 (2009) 14128.
[87] A. Sonnauer, F. Hoffmann, M. Froba, L. Kienle, V. Duppel, M. Thommes, et al., Angewandte Chemie International Edition 48 (2009) 3791.
[88] S.H. Huang, C.H. Lin, W.C. Wu, S.L. Wang, Angewandte Chemie International Edition 48 (2009) 6124.
[89] S. Qiu, J. Yu, G. Zhu, O. Terasaki, Y. Nozue, W. Pang, et al., Microporous and Mesoporous Materials 21 (1998) 245.
[90] B. Lu, Y. Oumi, T. Sano, Journal of Crystal Growth 291 (2006) 521.
[91] A. Kuperman, S. Nadimi, S. Oliver, G.A. Ozin, J.M. Garces, M.M. Olken, Nature 365 (1993) 239.
[92] S.C. Larsen, The Journal of Physical Chemistry C 111 (2007) 18464.
[93] L. Huang, Z. Wang, J. Sun, L. Miao, Q. Li, Y. Yan, et al., Journal of the American Chemical Society 122 (2000) 3530.
[94] Z. Wang, H. Wang, A. Mitra, L. Huang, Y. Yan, Advanced Materials 13 (2001) 746.
[95] S. Mintova, N.H. Olson, V. Valtchev, T. Bein, Science 283 (1999) 958.
[96] G. Zhu, S. Qiu, J. Yu, Y. Sakamoto, F. Xiao, R. Xu, et al., Chemistry of Materials 10 (1998) 1483.
[97] Z.J. Li, S. Li, H.M. Luo, Y.S. Yan, Advanced Functional Materials 14 (2004) 1019.
[98] C.M. Lew, Z. Li, S.I. Zones, M. Sun, Y. Yan, Microporous and Mesoporous Materials 105 (2007) 10.
[99] A. Corma, U. Diaz, M.E. Domine, V. Fornes, Journal of the American Chemical Society 122 (2000) 2804.
[100] Y.J. He, G.S. Nivarthy, F. Eder, K. Seshan, J.A. Lercher, Microporous and Mesoporous Materials 25 (1998) 207.
[101] S. Zanardi, A. Alberti, G. Cruciani, A. Corma, V. Fornes, M. Brunelli, Angewandte Chemie International Edition 43 (2004) 4933.
[102] D.M. Bibby, M.P. Dale, Nature 317 (1985) 157.
[103] R.E. Morris, S.J. Weigel, Chemical Society Reviews 26 (1997) 309.
[104] Q.S. Huo, R.R. Xu, S.H. Feng, Journal of the Chemical Society, Chemical Communications, (1988) 1486.
[105] Q. Huo, R. Xu, S. Li, Z. Ma, J.M. Thomas, R.H. Jones, et al., Journal of the Chemical Society, Chemical Communications (1992) 875.
[106] E.R. Parnham, R.E. Morris, Accounts of Chemical Research 40 (2007) 1005.
[107] W. Xu, J. Dong, J. Li, W. Li, F. Wu, Journal of the Chemical Society, Chemical Communications (1990) 755.
[108] M. Matsukata, M. Ogura, T. Osaki, P.R.H.R. Rao, M. Nomura, E. Kikuchi, Topics in Catalysis 9 (1999) 77.
[109] C.H. Christensen, K. Johannsen, E. Tonqvist, I. Schmidt, H. Topsoe, C.H. Christensen, Catalysis Today 128 (2007) 117.
[110] J. Perez-Ramirez, C.H. Christensen, K. Egeblad, C.H. Christensen, J.C. Groen, Chemical Society Reviews 37 (2008) 2530.
[111] Y. Tao, H. Kanoh, L. Abrams, K. Kaneko, Chemical Reviews 106 (2006) 896.
[112] S. Gopalakrishnan, A. Zampieri, W. Schwieger, Journal of Catalysis 260 (2008) 193.
[113] H. Vinh-Thang, Q. Huang, A. Ungureanu, M. Eic, D. Trong-On, S. Kaliaguine, Langmuir 22 (2006) 4777.
[114] H. Xin, A. Koekkoek, Q. Yang, R.v. Santen, C. Li, E.J.M. Hensen, Chemical Communications (2009) 7590.
[115] W. Fan, M.A. Snyder, S. Kumar, P.S. Lee, W.C. Yoo, A.V. McCormick, et al., Nat Mater 7 (2008) 984.
[116] X. Meng, F. Nawaz, F.S. Xiao, Nano Today 4 (2009) 292.
[117] P. Caullet, J.L. Paillaud, A. Simon-Masseron, M. Soulard, J. Patarin, Comptes Rendus Chimie 8 (2005) 245.
[118] A. Corma, M.E. Davis, Chem. Phys. Chem. 5 (2004) 304.
[119] S. Ori, G. Vezzalini, E. Galli, Microporous and Mesoporous Materials 126 (2009) 171.
[120] T. Kol'tsova, Inorganic Materials 42 (2006) 658.
[121] A. Alberti, G. Cruciani, E. Galli, R. Millini, S. Zanardi, The Journal of Physical Chemistry C 111 (2007) 4503.
[122] F. Di Renzo, H. Cambon, R. Dutartre, Microporous Materials 10 (1997) 283.
[123] S. Inagaki, Y. Fukushima, K. Kuroda, Journal of the Chemical Society, Chemical Communications (1993) 680.

[124] C.T. Kresge, M.E. Leonowicz, W.J. Roth, J.C. Vartuli, J.S. Beck, Nature 359 (1992) 710.
[125] J.S. Beck, J.C. Vartuli, W.J. Roth, M.E. Leonowicz, C.T. Kresge, K.D. Schmitt, et al., Journal of the American Chemical Society 114 (1992) 10834.
[126] P. Selvam, S.K. Bhatia, C.G. Sonwane, Industrial & Engineering Chemistry Research 40 (2001) 3237.
[127] Q. Huo, D.I. Margolese, U. Ciesla, D.G. Demuth, P. Feng, T.E. Gier, et al., Chemistry of Materials 6 (1994) 1176.
[128] Q. Huo, D.I. Margolese, U. Ciesla, P. Feng, T.E. Gier, P. Sieger, et al., Nature 368 (1994) 317.
[129] Q. Huo, D.I. Margolese, G.D. Stucky, Chemistry of Materials 8 (1996) 1147.
[130] D. Zhao, J. Feng, Q. Huo, N. Melosh, G.H. Fredrickson, B.F. Chmelka, et al., Science 279 (1998) 548.
[131] D. Zhao, Q. Huo, J. Feng, B.F. Chmelka, G.D. Stucky, Journal of the American Chemical Society 120 (1998) 6024.
[132] F. Hoffmann, M. Cornelius, J. Morell, M. Fröba, Angewandte Chemie International Edition 45 (2006) 3216.
[133] A.H. Lu, F. Schuth, Advanced Materials 18 (2006) 1793.
[134] S.H. Joo, S.J. Choi, I. Oh, J. Kwak, Z. Liu, O. Terasaki, et al., Nature 412 (2001) 169.
[135] Y. Wan, Y. Shi, D. Zhao, Chemistry of Materials 20 (2008) 932.
[136] S. Inagaki, S. Guan, T. Ohsuna, O. Terasaki, Nature 416 (2002) 304.
[137] X. Lai, X. Li, W. Geng, J. Tu, J. Li, S. Qiu, Angewandte Chemie International Edition 46 (2007) 738.
[138] Z. Zhang, Y. Han, L. Zhu, R. Wang, Y. Yu, S. Qiu, et al., Angewandte Chemie International Edition 40 (2001) 1258.
[139] G.J.d.A.A. Soler-Illia, E.L. Crepaldi, D. Grosso, C. Sanchez, Current Opinion in Colloid & Interface Science 8 (2003) 109.
[140] C.Y. Chen, S.L. Burkett, H.X. Li, M.E. Davis, Microporous Materials 2 (1993) 27.
[141] A. Firouzi, D. Kumar, L.M. Bull, T. Besier, P. Sieger, Q. Huo, et al., Science 267 (1995) 1138.
[142] Y. Wan, Y. Shi, D. Zhao, Chemical Communications (2007) 897.
[143] M. Mesa, L. Sierra, J.L. Guth, Microporous and Mesoporous Materials 112 (2008) 338.
[144] L. Sierra, S. Valange, J.L. Guth, Microporous and Mesoporous Materials 124 (2009) 100.
[145] Q. Huo, J. Liu, L.Q. Wang, Y. Jiang, T.N. Lambert, E. Fang, Journal of the American Chemical Society 128 (2006) 6447.
[146] A. Monnier, F. Schuth, Q. Huo, D. Kumar, D. Margolese, R.S. Maxwell, et al., Science 261 (1993) 1299.
[147] P. Behrens, Angewandte Chemie International Edition in English 35 (1996) 515.
[148] C. Yu, Y. Yu, L. Miao, D. Zhao, Microporous and Mesoporous Materials, 44-45 (2001) 65.
[149] C. Yu, B. Tian, J. Fan, G.D. Stucky, D. Zhao, Journal of the American Chemical Society 124 (2002) 4556.
[150] R. Ryoo, J.M. Kim, C.H. Ko, C.H. Shin, The Journal of Physical Chemistry 100 (1996) 17718.
[151] T. Yokoi, T. Tatsumi, Journal of the Japan Petroleum Institute 50 (2007) 299.
[152] S. El-Safty, Journal of Porous Materials 15 (2008) 369.
[153] S.A. Bagshaw, E. Prouzet, T.J. Pinnavaia, Science 269 (1995) 1242.
[154] S.S. Kim, T.R. Pauly, T.J. Pinnavaia, Chemical Communications (2000) 835.
[155] P.T. Tanev, T.J. Pinnavaia, Science 267 (1995) 865.
[156] Q. Cai, W.Y. Lin, F.S. Xiao, W.Q. Pang, X.H. Chen, B.S. Zou, Microporous and Mesoporous Materials 32 (1999) 1.
[157] Z. Liu, Y. Sakamoto, T. Ohsuna, K. Hiraga, O. Terasaki, C.H. Ko, et al., Angewandte Chemie 39 (2000) 3107.
[158] M. Kruk, M. Jaroniec, Chemistry of Materials 15 (2003) 2942.
[159] A.H. Lu, W. Schmidt, B. Spliethoff, F. Schuth, Chemistry – A European Journal 10 (2004) 6085.
[160] W.H. Zhang, L. Zhang, J. Xiu, Z. Shen, Y. Li, P. Ying, et al., Microporous and Mesoporous Materials 89 (2006) 179.
[161] M. Kruk, M. Jaroniec, C.H. Ko, R. Ryoo, Chemistry of Materials 12 (2000) 1961.
[162] P. Yang, D. Zhao, D.I. Margolese, B.F. Chmelka, G.D. Stucky, Nature 396 (1998) 152.
[163] A. Sayari, B.H. Han, Y. Yang, Journal of the American Chemical Society 126 (2004) 14348.
[164] A. Galarneau, H. Cambon, F.D. Renzo, R. Ryoo, M. Choi, F. Fajula, New Journal of Chemistry 27 (2003) 73.
[165] X. Liu, B. Tian, C. Yu, F. Gao, S. Xie, B. Tu, et al., Angewandte Chemie International Edition 41 (2002) 3876.
[166] B. Tian, X. Liu, L.A. Solovyov, Z. Liu, H. Yang, Z. Zhang, et al., Journal of the American Chemical Society 126 (2004) 865.
[167] T.W. Kim, F. Kleitz, B. Paul, R. Ryoo, Journal of the American Chemical Society 127 (2005) 7601.
[168] A. Carlsson, M. Kaneda, Y. Sakamoto, O. Terasaki, R. Ryoo, S.H. Joo, Journal of Electron Microscopy (Tokyo) 48 (1999) 795.
[169] Y. Sakamoto, M. Kaneda, O. Terasaki, D.Y. Zhao, J.M. Kim, G. Stucky, et al., Nature 408 (2000) 449.
[170] J. Fan, C. Yu, F. Gao, J. Lei, B. Tian, L. Wang, et al., Angewandte Chemie International Edition 42 (2003) 3146.
[171] J.R. Matos, M. Kruk, L.P. Mercuri, M. Jaroniec, L. Zhao, T. Kamiyama, et al., Journal of the American Chemical Society 125 (2003) 821.
[172] Q. Huo, R. Leon, P.M. Petroff, G.D. Stucky, Science 268 (1995) 1324.
[173] M. Clerc, Journal de Physique II 6 (1996) 961.
[174] W. Zhou, H.M.A. Hunter, P.A. Wright, Q. Ge, J.M. Thomas, The Journal of Physical Chemistry B 102 (1998) 6933.
[175] A.E. Garcia-Bennett, S. Williamson, P.A. Wright, I.J. Shannon, Journal of Materials Chemistry 12 (2002) 3533.
[176] T.W. Kim, R. Ryoo, M. Kruk, K.P. Gierszal, M. Jaroniec, S. Kamiya, et al., The Journal of Physical Chemistry B 108 (2004) 11480.
[177] L. Wang, J. Fan, B. Tian, H. Yang, C. Yu, B. Tu, et al., Microporous and Mesoporous Materials 67 (2004) 135.
[178] D. Zhao, Q. Huo, J. Feng, J. Kim, Y. Han, G.D. Stucky, Chemistry of Materials 11 (1999) 2668.
[179] T. Kimura, T. Kamata, M. Fuziwara, Y. Takano, M. Kaneda, Y. Sakamoto, et al., Angewandte Chemie International Edition 39 (2000) 3855.
[180] Y.S. Lin, C.P. Tsai, H.Y. Huang, C.T. Kuo, Y. Hung, D.M. Huang, et al., Chemistry of Materials 17 (2005) 4570.
[181] P. Schmidt-Winkel, W.W. Lukens, P. Yang, D.I. Margolese, J.S. Lettow, J.Y. Ying, et al., Chemistry of Materials 12 (2000) 686.
[182] Y. Han, D. Zhang, L.L. Chng, J. Sun, L. Zhao, X. Zou, et al., Nat Chem 1 (2009) 123.
[183] S. Che, A.E. Garcia-Bennett, T. Yokoi, K. Sakamoto, H. Kunieda, O. Terasaki, et al., Nature Materials 2 (2003) 801.
[184] C. Gao, Y. Sakamoto, K. Sakamoto, O. Terasaki, S. Che, Angewandte Chemie International Edition 45 (2006) 4295.
[185] C. Gao, H. Qiu, W. Zeng, Y. Sakamoto, O. Terasaki, K. Sakamoto, et al., Chemistry of Materials 18 (2006) 3904.
[186] T. Sun, J.Y. Ying, Angewandte Chemie International Edition 37 (1998) 664.
[187] R. Atluri, N. Hedin, A.E. Garcia-Bennett, Journal of the American Chemical Society 131 (2009) 3189.
[188] A. Sayari, Y. Yang, M. Kruk, M. Jaroniec, The Journal of Physical Chemistry B 103 (1999) 3651.
[189] H.P. Lin, C.Y. Mou, S.B. Liu, Advanced Materials 12 (2000) 103.

[190] A. Sayari, S. Hamoudi, Y. Yang, Chemistry of Materials 17 (2005) 212.
[191] R. Serna-Guerrero, A. Sayari, Environmental Science & Technology 41 (2007) 4761.
[192] Y. Belmabkhout, R. Serna-Guerrero, A. Sayari, Industrial & Engineering Chemistry Research 49 (2010) 359.
[193] Y. Wan, H. Yang, D. Zhao, Accounts of Chemical Research 39 (2006) 423.
[194] L. Wang, B. Tian, J. Fan, X. Liu, H. Yang, C. Yu, et al., Microporous Mesoporous Mater. 67 (2004) 123.
[195] C.J. Brinker, Y. Lu, A. Sellinger, H. Fan, Advanced Materials 11 (1999) 579.
[196] B.J. Melde, B.T. Holland, C.F. Blanford, A. Stein, Chemistry of Materials 11 (1999) 3302.
[197] S. Zheng, L. Gao, Q.H. Zhang, J.K. Guo, Journal of Materials Chemistry 10 (2000) 723.
[198] Q. Yang, J. Liu, L. Zhang, C. Li, Journal of Materials Chemistry 19 (2009) 1945.
[199] S. Inagaki, S. Guan, Y. Fukushima, T. Ohsuna, O. Terasaki, Journal of the American Chemical Society 121 (1999) 9611.
[200] T. Asefa, M.J. MacLachlan, N. Coombs, G.A. Ozin, Nature 402 (1999) 867.
[201] W. Whitnall, L. Cademartiri, G.A. Ozin, Journal of the American Chemical Society 129 (2007) 15644.
[202] M. Tiemann, Chemistry of Materials 20 (2008) 961.
[203] A.H. Lu, W. Schmidt, A. Taguchi, B. Spliethoff, B. Tesche, F. Schuth, Angewandte Chemie International Edition 41 (2002) 3489.
[204] J.Y. Kim, S.B. Yoon, J.S. Yu, Chemistry of Materials 15 (2003) 1932.
[205] M. Kang, S.H. Yi, H.I. Lee, J.E. Yie, J.M. Kim, Chemical Communications (2002) 1944.
[206] C.H. Kim, D.K. Lee, T.J. Pinnavaia, Langmuir 20 (2004) 5157.
[207] A. Karlsson, M. Stocker, R. Schmidt, Microporous and Mesoporous Materials 27 (1999) 181.
[208] V. Meynen, P. Cool, E.F. Vansant, Microporous and Mesoporous Materials 104 (2007) 26.
[209] P. Van Der Voort, P.I. Ravikovitch, K.P. De Jong, M. Benjelloun, E. Van Bavel, A.H. Janssen, et al., The Journal of Physical Chemistry B 106 (2002) 5873.
[210] M. Kruk, M. Jaroniec, S.H. Joo, R. Ryoo, The Journal of Physical Chemistry B 107 (2003) 2205.
[211] S.A. El-Safty, F. Mizukami, T. Hanaoka, Journal of Materials Chemistry 15 (2005) 2590.
[212] R. Garcia, I. Diaz, C. Marquez-Alvarez, J. Perez-Pariente, Chemistry of Materials 18 (2006) 2283.
[213] T.O. Do, A. Nossov, M.A. Springuel-Huet, C. Schneider, J.L. Bretherton, C.A. Fyfe, et al., Journal of the American Chemical Society 126 (2004) 14324.
[214] Y. Liu, W. Zhang, T.J. Pinnavaia, Journal of the American Chemical Society 122 (2000) 8791.
[215] Y. Liu, T.J. Pinnavaia, Chemistry of Materials 14 (2002) 3.
[216] D. Zhao, P. Yang, Q. Huo, B.F. Chmelka, G.D. Stucky, Current Opinion in Solid State and Materials Science 3 (1998) 111.
[217] Y. Wu, G. Cheng, K. Katsov, S.W. Sides, J. Wang, J. Tang, et al., Nature Materials 3 (2004) 816.
[218] S. Schacht, Q. Huo, I.G. Voigt-Martin, G.D. Stucky, F. Schüth, Science 273 (1996) 768.
[219] Q. Huo, J. Feng, F. Schuth, G.D. Stucky, Chemistry of Materials 9 (1997) 14.
[220] J. Wang, C.K. Tsung, R.C. Hayward, Y. Wu, G.D. Stucky, Angewandte Chemie International Edition 44 (2005) 332.
[221] F. Kleitz, F. Marlow, G.D. Stucky, F. Schuth, Chemistry of Materials 13 (2001) 3587.
[222] Q. Huo, D. Zhao, J. Feng, K. Weston, S.K. Buratto, G.D. Stucky, et al., Advanced Materials 9 (1997) 974.
[223] H. Yang, A. Kuperman, N. Coombs, S. Mamiche-Afara, G.A. Ozin, Nature 379 (1996) 703.
[224] Y. Lu, R. Ganguli, C.A. Drewien, M.T. Anderson, C.J. Brinker, W. Gong, et al., Nature 389 (1997) 364.
[225] N.A. Melosh, P. Lipic, F.S. Bates, F. Wudl, G.D. Stucky, G.H. Fredrickson, et al., Macromolecules 32 (1999) 4332.
[226] P. Innocenzi, L. Malfatti, T. Kidchob, P. Falcaro, Chemistry of Materials 21 (2009) 2555.
[227] R.A. Pai, J.J. Watkins, Advanced Materials 18 (2006) 241.
[228] H.T. Chen, T.A. Crosby, M.H. Park, S. Nagarajan, V.M. Rotello, J.J. Watkins, Journal of Materials Chemistry 19 (2009) 70.
[229] T.C. Wei, H.W. Hillhouse, Langmuir 23 (2007) 5689.
[230] I.I. Slowing, B.G. Trewyn, V.S.Y. Lin, Journal of the American Chemical Society 129 (2007) 8845.
[231] N.K. Mal, M. Fujiwara, Y. Tanaka, Nature 421 (2003) 350.
[232] Y.Z. You, K.K. Kalebaila, S.L. Brock, D. Oupicky, Chemistry of Materials 20 (2008) 3354.
[233] M. Vallet-Regi, F. Balas, D. Arcos, Angewandte Chemie International Edition 46 (2007) 7548.
[234] I.I. Slowing, B.G. Trewyn, S. Giri, V.S.Y. Lin, Advanced Functional Materials 17 (2007) 1225.
[235] I.I. Slowing, J.L. Vivero-Escoto, C.W. Wu, V.S.Y. Lin, Advanced Drug Delivery Reviews 60 (2008) 1278.
[236] Y. Wang, F. Caruso, Chemistry of Materials 17 (2005) 953.
[237] B.G. Trewyn, S. Giri, I.I. Slowing, V.S.Y. Lin, Chemical Communications (2007) 3236.
[238] Y. Wang, A.D. Price, F. Caruso, Journal of Materials Chemistry 19 (2009) 6451.
[239] K. Suzuki, K. Ikari, H. Imai, Journal of the American Chemical Society 126 (2004) 462.
[240] G. Derrien, C. Charnay, J. Zajac, D.J. Jones, J. Roziere, Chemical Communications (2008) 3118.
[241] T. Nakamura, M. Mizutani, H. Nozaki, N. Suzuki, K. Yano, The Journal of Physical Chemistry C 111 (2006) 1093.
[242] C.X. Lin, S.Z. Qiao, C.Z. Yu, S. Ismadji, G.Q. Lu, Microporous and Mesoporous Materials 117 (2009) 213.
[243] Y. Zhu, J. Shi, W. Shen, H. Chen, X. Dong, M. Ruan, Nanotechnology 16 (2005) 2633.
[244] M. Liong, J. Lu, M. Kovochich, T. Xia, S.G. Ruehm, A.E. Nel, et al., ACS Nano 2 (2008) 889.
[245] M.M. Titirici, A. Thomas, M. Antonietti, Advanced Functional Materials 17 (2007) 1010.
[246] Y.D. Xia, R. Mokaya, Advanced Materials 16 (2004) 886.
[247] H. Djojoputro, X.F. Zhou, S.Z. Qiao, L.Z. Wang, C.Z. Yu, G.Q. Lu, Journal of the American Chemical Society 128 (2006) 6320.
[248] Y. Lu, H. Fan, A. Stump, T.L. Ward, T. Rieker, C.J. Brinker, Nature 398 (1999) 223.
[249] R.I. Nooney, D. Thirunavukkarasu, Y. Chen, R. Josephs, A.E. Ostafin, Chemistry of Materials 14 (2002) 4721.
[250] C.E. Fowler, D. Khushalani, B. Lebeau, S. Mann, Advanced Materials 13 (2001) 649.
[251] Q. Cai, Z.S. Luo, W.Q. Pang, Y.W. Fan, X.H. Chen, F.Z. Cui, Chemistry of Materials 13 (2001) 258.
[252] Qianru Chen, Lu Han, C. Gao, S. Che, Microporous and Mesoporous Materials 128 (2010) 203.
[253] K. Moller, J. Kobler, T. Bein, Advanced Functional Materials 17 (2007) 605.
[254] Y. Hoshikawa, H. Yabe, A. Nomura, T. Yamaki, A. Shimojima, T. Okubo, Chemistry of Materials 22 (2010) 12.
[255] J. Kobler, T. Bein, ACS Nano 2 (2008) 2324.
[256] Z.A. Qiao, L. Zhang, M. Guo, Y. Liu, Q. Huo, Chemistry of Materials 21 (2009) 3823.

[257] F. Marlow, B. Spliethoff, B. Tesche, D. Zhao, Advanced Materials 12 (2000) 961.
[258] J. Wang, C.K. Tsung, W. Hong, Y. Wu, J. Tang, G.D. Stucky, Chemistry of Materials 16 (2004) 5169.
[259] P. Feng, X. Bu, G.D. Stucky, D.J. Pine, Journal of the American Chemical Society 122 (2000) 994.
[260] H. Yang, Q. Shi, B. Tian, S. Xie, F. Zhang, B. YanTu, et al., Chemistry of Materials 15 (2003) 536.
[261] F. Schuth, Chemistry of Materials 13 (2001) 3184.
[262] M.G. Kanatzidis, Advanced Materials 19 (2007) 1165.
[263] A. Sayari, P. Liu, Microporous Mater. 12 (1997) 149.
[264] S.W. Boettcher, J. Fan, C.K. Tsung, Q. Shi, G.D. Stucky, Accounts of Chemical Research 40 (2007) 784.
[265] Y. Yamauchi, M. Sawada, M. Komatsu, A. Sugiyama, T. Osaka, N. Hirota, et al., Chemistry – An Asian Journal 2 (2007) 1505.
[266] J.D. Bass, D. Grosso, C. Boissiere, E. Belamie, T. Coradin, C. Sanchez, Chemistry of Materials 19 (2007) 4349.
[267] C.M. Yang, H.A. Lin, B. Zibrowius, B. Spliethoff, F. Schuth, S.C. Liou, et al., Chemistry of Materials 19 (2007) 3205.
[268] B. Tian, X. Liu, B. Tu, C. Yu, J. Fan, L. Wang, et al., Nature Materials 2 (2003) 159.
[269] A. Dong, N. Ren, Y. Tang, Y. Wang, Y. Zhang, W. Hua, et al., Journal of the American Chemical Society 125 (2003) 4976.
[270] D.M. Antonelli, J.Y. Ying, Angewandte Chemie International Edition in English 35 (1996) 426.
[271] D. Grosso, F. Cagnol, G.J.D.A. Soler-Illia, E.L. Crepaldi, H. Amenitsch, A. Brunet-Bruneau, et al., Advanced Functional Materials 14 (2004) 309.
[272] K. Zhu, B. Yue, W. Zhou, H. He, Chemical Communications (2003) 98.
[273] B. Tian, X. Liu, H. Yang, S. Xie, C. Yu, B. Tu, et al., Advanced Materials 15 (2003) 1370.
[274] X. Liu, B. Tian, C. Yu, B. Tu, Z. Liu, O. Terasaki, et al., Chemistry Letters 32 (2003) 824.
[275] Y. Yamauchi, K. Kuroda, Chemistry – An Asian Journal 3 (2008) 664.
[276] H.J. Shin, R. Ryoo, Z. Liu, O. Terasaki, Journal of the American Chemical Society 123 (2001) 1246.
[277] D. Wang, W.L. Zhou, B.F. McCaughy, J.E. Hampsey, X. Ji, Y.B. Jiang, et al., Advanced Materials 15 (2003) 130.
[278] D. Wang, H. Luo, R. Kou, M.P. Gil, S. Xiao, V.O. Golub, et al., Angewandte Chemie International Edition 43 (2004) 6169.
[279] J. Lee, J. Kim, T. Hyeon, Advanced Materials 18 (2006) 2073.
[280] T. Kyotani, Z. Ma, A. Tomita, Carbon 41 (2003) 1451.
[281] Z. Ma, T. Kyotani, Z. Liu, O. Terasaki, A. Tomita, Chemistry of Materials 13 (2001) 4413.
[282] F.O.M. Gaslain, J. Parmentier, V.P. Valtchev, J. Patarin, Chemical Communications (2006) 991.
[283] C. Liang, Z. Li, S. Dai, Angewandte Chemie International Edition 47 (2008) 3696.
[284] R. Ryoo, S.H. Joo, S. Jun, The Journal of Physical Chemistry B 103 (1999) 7743.
[285] M. Kruk, M. Jaroniec, R. Ryoo, S.H. Joo, The Journal of Physical Chemistry B 104 (2000) 7960.
[286] Z. Li, M. Jaroniec, The Journal of Physical Chemistry B 108 (2004) 824.
[287] M. Kaneda, T. Tsubakiyama, A. Carlsson, Y. Sakamoto, T. Ohsuna, O. Terasaki, et al., The Journal of Physical Chemistry B 106 (2002) 1256.
[288] S. Jun, S.H. Joo, R. Ryoo, M. Kruk, M. Jaroniec, Z. Liu, et al., Journal of the American Chemical Society 122 (2000) 10712.
[289] F.B. Su, Z.C. Zhou, W.P. Guo, J.J. Liu, X.N. Tian, X.S. Zhao, Chemistry and Physics of Carbon, vol. 30, CRC Press-Taylor & Francis Group, Boca Raton, 2008, p. 63.
[290] C.D. Liang, Z.J. Li, S. Dai, Angewandte Chemie International Edition 47 (2008) 3696.
[291] A.H. Lu, W. Schmidt, B. Spliethoff, F. Schuth, Advanced Materials 15 (2003) 1602.
[292] A.H. Lu, W.C. Li, W. Schmidt, F. Schuth, Microporous and Mesoporous Materials 80 (2005) 117.
[293] B.H. Han, W.Z. Zhou, A. Sayari, Journal of the American Chemical Society 125 (2003) 3444.
[294] W.H. Zhang, C. Liang, H. Sun, Z. Shen, Y. Guan, P. Ying, et al., Advanced Materials 14 (2002) 1776.
[295] Y. Meng, D. Gu, F. Zhang, Y. Shi, H. Yang, Z. Li, et al., Angewandte Chemie International Edition 44 (2005) 7053.
[296] A. Stein, R.C. Schroden, Current Opinion in Solid State & Materials Science 5 (2001) 553.
[297] M.A. Carreon, V.V. Guliants, European Journal of Inorganic Chemistry 2005 (2005) 27.
[298] O.D. Velev, T.A. Jede, R.F. Lobo, A.M. Lenhoff, Chemistry of Materials 10 (1998) 3597.
[299] J.E. Wijnhoven, W.L. Vos, Science 281 (1998) 802.
[300] A. Imhof, D.J. Pine, Nature 389 (1997) 948.

CHAPTER

17

Assembly Chemistry of Anion-intercalated Layered Materials

Xue Duan, Jun Lu, David G. Evans
Beijing University of Chemical Technology, China

17.1. STRUCTURE OF ANION-INTERCALATED LAYERED MATERIALS

17.1.1. Introduction to Layered Double Hydroxides (LDHs)

Layered double hydroxides (LDHs) are a group of anion-intercalated inorganic functional materials which are also known as hydrotalcite-like compounds or anionic clays. In LDHs, positively charged metal hydroxide host layers stack alternately with interlayer anions to form a sandwich structure. The metal ions in the host layer are octahedrally coordinated divalent M^{II} and trivalent M^{III} cations, whereas the interlayer A^{n-} anions interact with the host layers by ionic bonding, hydrogen bonding, and van der Waals forces. Much of the interest in LDHs arises from their wide range of possible compositions which give rise to the possibility of tailoring the properties of the material in several different ways [1] to realize the microscopic controllability of layer chemical composition, the microscopic controllability of the nature and amount of interlayer anions, and the mesoscopic controllability of crystallite size and size distribution.

Many examples of minerals with the LDH structure are found in nature [2–4] and these are classified as the sjögrenite–hydrotalcite group [5]. The formula of one such mineral, hydrotalcite $[Mg_6Al_2(OH)_{16}]CO_3 7\cdot 4H_2O$ was first determined almost 100 years ago and its crystal structure was determined in the 1960s [6–8]. From then onward, great efforts have been focused on detailed studies of the structure of LDHs including the possible range of stoichiometry and composition, the extent of ordering of metal cations within the layers, the stacking arrangement of the layers, and the distribution of interlayer anions and water molecules. The literature up to 2005 describing the wide range of experimental and theoretical studies of the structure of LDHs has been reviewed [9]. In this chapter, we focus on some more recent advances in our understanding of the structural chemistry of LDHs.

17.1.2. Structure of LDHs

17.1.2.1. Brucite-like Layers

The layer structure of LDHs is based on brucite $[Mg(OH)_2]$ which consists of Mg^{2+} ions surrounded octahedrally by hydroxide ions. These octahedral units form infinite layers by sharing edges, with the O–H bond perpendicular to the plane of the layers [10]. In the lattice of LDHs, trivalent M^{III} cations substitute some of the Mg^{2+} cations in the brucite layers resulting in positively charged host layers, and the interlayer A^{n-} anions are located in hydrated interlayer galleries, whereas the host layers stack on top of one another to form a three-dimensional (3D) structure, as shown in Fig. 17.1. In brucite itself, the hydroxyl anions are close-packed in the two-dimensional (2D) planes with triangular symmetry and the metal cations occupy the octahedral holes between alternate pairs of OH planes and thus occupy a triangular lattice which is the same as that occupied by the OH^- ions. The local geometry of the metal (D_{3d}) is strongly distorted away from the idealized O_h arrangement due to compression of the octahedra along the stacking axis [11]. All brucite-like metal hydroxides show this type of distortion [12,13] which results in hexagonal symmetry (space group $P\bar{3}m1$). The weak forces between the layers of brucite have been attributed to contributions from dispersion forces and hydrogen bonding [10,14–16].

Modern Inorganic Synthetic Chemistry, DOI: 10.1016/B978-0-444-53599-3.10017-4

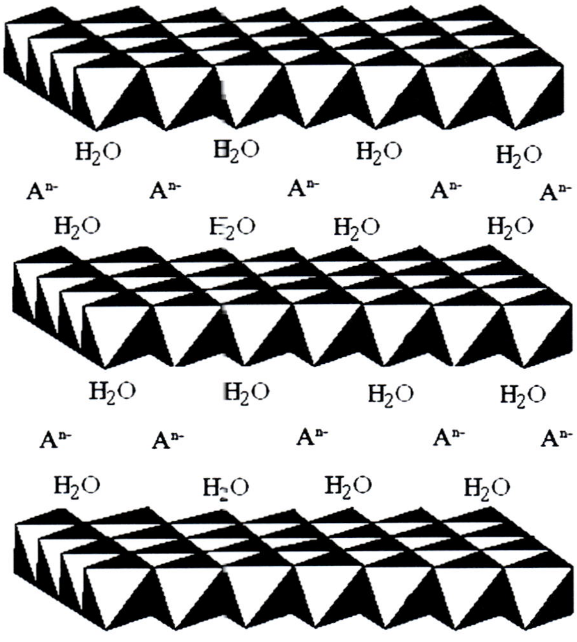

FIGURE 17.1 Structure of layered double hydroxides.

17.1.2.2. LDH Stacking Polytypes

The brucite-like layers in LDHs may be stacked in various ways, which gives rise to a variety of possible polytype structures. If the opposing OH groups lie vertically above one another an interlayer with a trigonal prismatic arrangement results, whilst if they are offset, an octahedral arrangement is formed [17–19]. There are three possible two-layer polytypes with hexagonal stacking of the layers, denoted as $2H_1$, $2H_2$, and $2H_3$, where the $2H_1$ polytype has all prismatic interlayers and the $2H_2$ polytype has all octahedral interlayers, whilst both types of interlayers are present in the $2H_3$ polytype. There are nine possible three-layer polytypes, of which two have rhombohedral symmetry—$3R_1$ and $3R_2$—whilst the remaining seven ($3H_1$–$3H_7$) have hexagonal symmetry. The $3R_1$ polytype has all prismatic interlayers; $3R_2$, $3H_1$, and $3H_2$ have all octahedral interlayers and other polytypes involve both types. The various possible six-layer polytypes with rhombohedral (6**R**) and hexagonal symmetry (6**H**) have also been described [17–19]. Hydrotalcite (MgAl–CO_3^{2-}-LDH) has the $3R_1$ polytype structure, which most other carbonate-containing LDHs also adopt [8,19], since the prismatic arrangement of hydroxyl groups facilitates hydrogen bonding with the oxygen atoms of the CO_3^{2-} anion and the interlayer water molecules. Vucelic et al. [20] have shown that MgAl-LDHs containing intercalated benzoate or terephthalate anions also have the $3R_1$ structure. Newman et al. [21] have reported that in the $3R_2$ polytype of an LDH containing interlayer hydroxide anions, the hydroxyl groups of adjacent layers and the interlayer hydroxide ions form a cubic close-packed array.

It is difficult to distinguish between the various polytypes, since the intensities of reflections are sensitive to the interlayer anion. Simulated XRD patterns and single crystal XRD indicate that both manasseite and sjögrenite (one polytype of $[Mg_6Fe_2(OH)_{16}]CO_3\cdot4\cdot5H_2O$) have the $2H_1$ structure [8]. Hydrotalcite-like minerals containing interlayer sulfate ions are sometimes found with prismatic ($3R_1$) interlayers although octahedral ($3R_2$, $3H_2$) interlayers are more prevalent [18]. For most LDHs it is not possible to unambiguously identify the polytype present, because the powder XRD patterns contain only a limited amount of structural information. The reflections in the XRD pattern of an LDH can be divided into three types:

(1) A series of strong basal 00*l* reflections at low Bragg angles allow the basal spacing normal to the (00*l*) plane (c_o), which equals the thickness of one brucite-like layer plus one interlayer, to be determined. The unit cell parameter $c = 3c_o$ for a 3**R** polytype, while $c = 2c_o$ for a 2*H* polytype [19,22–24].

(2) The position of the 110 reflection allows the value of the lattice parameter a_o ($a_o = 2d_{110}$) to be determined, which corresponds to the closest distance between two metal cations in the layer.

(3) The positions of the 01*l* and/or 10*l* reflections can, at least in principle, be used to determine the stacking pattern of the layers. Rhombohedral and hexagonal structures can be distinguished by systematic absences [23].

The XRD reflections of hydrotalcite $[Mg_6Al_2(OH)_{16}](CO_3)\cdot4H_2O$ (JCPDS 22–700) can be indexed in a 3**R** polytype with rhombohedral symmetry (space group R 3^- m, $a_o = 0.306$ nm and $c = 2.34$ nm) [6–8]. The mineral manasseite (JCPDS 14–525) has the same formula as hydrotalcite and the reflections can be indexed in a 2**H** polytype with hexagonal symmetry (space group $P6_3/mmc$, $a_o = 0.306$ nm and $c = 1.56$ nm) [25]. Hydrotalcite and manasseite are often found intergrown [26,27].

For some synthetic LDHs, the stacking sequence varies with temperature. A ZnAl–Cl-LDH undergoes the polytype transformation $3R_1 \rightarrow 2H_1$ at 150°C, whilst an $Ni^{II}Co^{III}$–CO_3-LDH undergoes the transformation $3R_1 \rightarrow 1H$ at 200°C [28]. $[Ca_2Al(OH)_6]X\cdot2H_2O$ LDHs exist as rhombohedral polytypes [6**R** (X = Cl), 6**R** or 3**R** (X = Br), and 3**R** (X = I)] at high temperature and undergo a phase transition to a monoclinic polytype on cooling [29,30]. The polytype of the type $[LiAl_2(OH)_6]Cl\cdot nH_2O$ varies with the extent of hydration n, with the $n = 2$ polytype having a rhombohedral form [31] and that with $n = 1$ having a hexagonal polytype [32]. In contrast, $[LiAl_2(OH)_6]OH\cdot nH_2O$ has a random stacking of layers [32]. Different types of disorder may be introduced during the crystallization of LDHs leading to poor crystallinity [33–38]. Loss of regular stacking results in turbostratic disorder, which is reflected in increased broadening of the *h0l* reflections, whilst the *hk0* (100 and 110) reflections adopt a characteristic "shark's fin" shape [38,39–42]. Radha et al. [43] have investigated the conservation of order, disorder, and "crystallinity" during anion exchange in ZnAl-LDHs. The anion-exchange reactions yield a solid with a greater degree of order if the incoming ion is CO_3^{2-} or Cl^-. In contrast, incoming NO_3^- ions yield an interstratified phase, whereas incoming SO_4^{2-} ions generate turbostratic disorder.

17.1.3. Metal Ions in LDH Layers

17.1.3.1. Cation Substitution in LDH Layers

The identity and proportions of the di- and trivalent cations as well as the identity of the interlayer anions can be varied to form a family of LDH materials with the general formula $[M^{II}_{1-x}M^{III}_x(OH)_2]^{x+}[A^{n-}]_{x/n}\cdot yH_2O$, where M^{II} = Mg, Zn, Mn, Fe, Co, Ni, Cu, Ca, Cd, etc. and M^{III} = Al, Co, Fe, Mn, Cr, Ga, In, etc. Many ternary and quaternary LDHs involving mixtures of different M^{II} and/or M^{III} cations can also be prepared [44] and this family can also include layered materials containing monovalent ions with the formula $[LiAl_2(OH)_6]^+[A^{n-}]_{1/n}\cdot yH_2O$ [32].

It has been speculated that the ionic radii of M^{II} and M^{III} and the difference between them [4,45–48], as well as the solubility products of $M^{II}(OH)_2$ and $M^{II}CO_3$ [49], play roles in defining the boundaries of the range of metal cations that can form LDHs. It was once believed that Cu^{2+}, which is subject to a Jahn–Teller distortion, does not form LDHs unless diluted by other cations [2,50,51], but $[Cu_{0.69}Cr_{0.31}(OH)_2]Cl_{0.31}\cdot 0.61H_2O$ [52,53] has since been well characterized, although it does show a corrugation of the sheets associated with the Jahn–Teller distortion around individual Cu^{II} centers. Recently rare earth ions such as Eu^{3+} and Tb^{3+}, which tend to adopt a coordination number higher than six, have been incorporated into LDH layers [54–56].

Many researchers have suggested [2,4,47,48] that pure LDH phases can only be formed for stoichiometries in the range $0.20 < x < 0.33$, corresponding to M^{II}/M^{III} ratios in the range 2–4. For $x > 0.33$, the presence of M^{III}–O–M^{III} linkages is unavoidable and this is energetically unfavorable due to the strong repulsion between the adjacent trivalent cations [57–59]. However, the Cr^{III}–O–Cr^{III} unit does exist in the pure Zn_7Cr_4–CO_3-LDH phase [60]. LDHs of formula $[M^{II}Al_4(OH)_{12}](NO_3)_2\cdot nH_2O$ (M^{II} = Co, Ni, Cu, Zn) have been prepared [61] by reaction of an activated gibbsite with $M^{II}(NO_3)_2$ solutions. Only half of the cation vacancies in gibbsite are filled with M^{II} cations, so that although the M^{II}/Al ratio is 1:4, there are still cation vacancies, whereas all the cation sites are occupied in the brucite-like layers of conventional LDHs. Monovalent Li^I cations can fill all of the vacancies in gibbsite giving LDH-like materials containing cation-ordered $[LiAl_2(OH)_6]^+$ layers [32]. Furthermore, the Al^{III} cations can be substituted by Fe^{III} to give $[LiFe_xAl_{2-x}(OH)_6](CO_3)_{0.5}\cdot nH_2O$ phases in which the ordering of M^I and M^{III} cations in the layers is maintained [62].

17.1.3.2. Long-range Cation Order–disorder

Since LDHs are used as catalysts [63] or as precursors to catalysts [64,65], ferrimagnetic spinels [66–68], or other functional materials, they should ideally have a homogeneous distribution of cations without segregation of "lakes" of separate cations. Diffraction techniques are not well suited for studies of the extent of cation disorder in many of the most common LDHs, due to the indistinguishability of cations such as Mg and Al by X-ray diffraction, the difficulties in synthesizing fully deuterated LDHs for high-resolution neutron diffraction studies, and the turbostratic disorder and stacking faults existing within many LDHs [42,57]. Pauling's rules [69] suggest that M^{III} cations should not occupy adjacent sites (unless accompanied by vacant cation sites [70]): this implies that the minimum possible M^{II}/M^{III} ratio is 2 and that when M^{II}/M^{III} is exactly equal to 2 there is also a *long-range cation order* with a superlattice involving an ordered array with each M^{III} cation having six M^{II} nearest neighbors and each M^{II} cation having three M^{II} and three M^{III} nearest neighbors [59]. The superlattice has a cell parameter $a\sqrt{3}$ where a is the cell parameter of the disordered brucite-like layers [42].

17.1.3.2.1. MG_2AL-LDHS WITH LONG-RANGE CATION ORDER

Solid-state nuclear magnetic resonance (NMR) studies of LDHs are a potential source of structural information [71]. Sideris [72] reported that by using extremely fast magic angle spinning frequencies of up to 60 kHz the homonuclear coupling is effectively eliminated in ^{1}H NMR spectra and allows information about the Mg and Al distribution in MgAl–NO_3-LDHs to be obtained. They used MgAl–NO_3-LDH samples containing 19%, 25%, and 33% Al (denoted MgAl-19, MgAl-25, and MgAl-33) as representatives of LDHs with low, medium, and high Al content, allowing the degree of cation ordering to be explored as a function of Al content. Since each layer hydroxyl group is coordinated to three metal cations, if the metal cations are arranged randomly in the layers there are four possible hydroxyl sites (Mg_3-OH, Mg_2Al-OH, $MgAl_2$-OH, and Al_3-OH) as shown in Fig. 17.2A. The calculated ratios of the

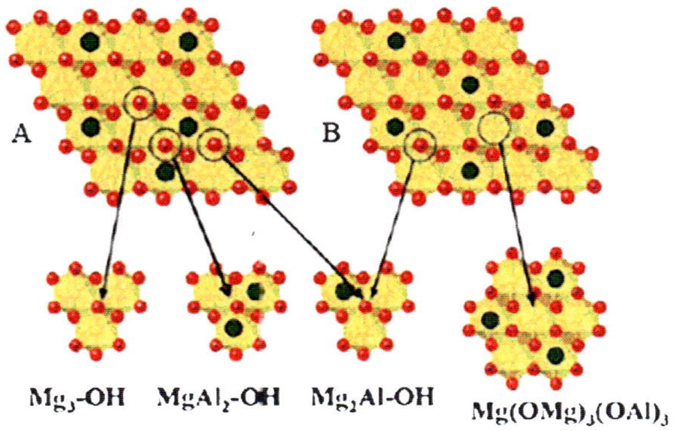

FIGURE 17.2 The metal hydroxide sheets of an LDH with an Mg/Al ratio of 2:1 [MgAl-33] are shown with (A) random and (B) ordered cation distributions. Three major classes of hydroxyl groups are present in (A) (Mg_3-OH, Mg_2Al-OH, and $MgAl_2$-OH), whereas only one hydroxyl environment (Mg_2Al-OH) and one Mg local environment [$Mg(OMg)_3(OAl)_3$] are present in (B). Small red circles and large green circles are hydroxyl and aluminum atoms, respectively. *From [72]. Reprinted with permission from AAAS.*

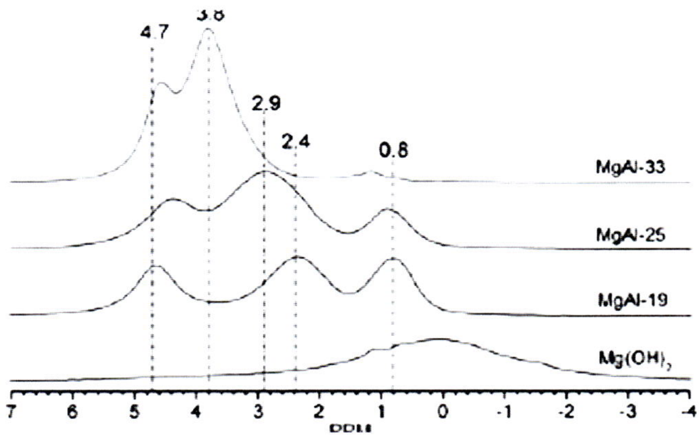

FIGURE 17.3 The effect of Al content on the single-pulse ^{1}H MAS NMR spectra collected at a spinning speed of 60 kHz. Hydrotalcite-like MgAl-LDHs containing 19%, 25%, and 33% Al are shown. *From [72]. Reprinted with permission from AAAS.*

populations of the sites for the different MgAl−NO_3-LDHs shown in Table 17.1 indicate that at least three hydroxyl peaks (the population of Al_3-OH sites is probably too low to give an observable signal) are expected in the ^{1}H MAS NMR spectrum. However, the spectrum of MgAl-19 (Fig. 17.3) shows only two hydroxyl ^{1}H resonances at 0.8 and 2.4 ppm (the third signal at 4.7 ppm can be attributed to interlayer water [73]). The resonance at 0.8 ppm can be assigned to Mg_3-OH by analogy with the spectrum of brucite [74]. The resonance at 2.4 ppm can therefore be assigned to Mg_2Al-OH. No third peak at lower field which is assigned to $MgAl_2$-OH units can be observed in the spectra of MgAl-19 (or in the spectra of MgAl-25 and MgAl-33) (Fig. 17.3). This strongly suggests that the cations are ordered with no adjacent Al atoms (Fig. 17.2B) since 2D magnetization exchange experiments [75] confirmed that the exchange processes involving different OH and H_2O sites were too slow to affect the chemical shift positions and intensities of the ^{1}H resonances, and ^{1}H−^{27}Al transfer of population in double resonance (TRAPDOR) NMR spectra [76] confirmed that no additional resonances (such as that due to $MgAl_2$-OH) at higher frequencies were obscured by the water resonance. The intensity ratios of the peaks assigned to Mg_3-OH and Mg_2Al-OH sites show a good correlation with those calculated for an ordered arrangement of cations for the three different LDHs (Table 17.1), further supporting the conclusion that the cations are ordered.

17.1.3.2.2. $M^{II}_2M^{III}$-LDHS WITH LONG-RANGE CATION ORDER

Zn_nCr−Cl-LDHs can only be prepared for $n=2$, indicative of the presence of ordered layers [52]. Mg_2Ga-LDH has a supercell of $a\sqrt{3}$ as indicated by a strong (100) reflection (0.463 nm) in its XRD pattern [77]. Single crystals of $[Ca_2Al(OH)_6]X$ LDHs ($X=CO_3^{2-}$ [78,79], NO_3^- [80,81], and Cl^- [82,83]) show complete ordering of cations, which is believed to result from the large size and the anisotropy of the seven-coordinate Ca^{2+} ion [84]. Ni_2Al−CO_3-LDH [85] also shows a $(\sqrt{3}\times\sqrt{3})R30^\circ$ superlattice. Both X-ray and neutron powder diffraction [32,86] show that $[LiAl_2(OH)_6]X$ LDHs always have an ordered array of cations with a hexagonal supercell ($a=a_o\sqrt{3}=0.532$ nm), which originates from Li^+ cations filling the ordered pattern of

TABLE 17.1 A Comparison of Relative Concentrations of Hydroxyl Groups (%) Calculated for a Random and an Ordered Distribution of Cations[a]

	Random cation distribution				Ordered model/experimental data	
	Mg_3OH	Mg_2AlOH	$MgAl_2OH$	Al_3OH	Mg_3OH	Mg_2AlOH
MgAl-19	54	37	8	0.6	43/38(3)	57/62(3)
MgAl-25	42	42	14	2	25/20(3)	75/80(3)
MgAl-33	30	44	22	4	1/3(1)	99/97(1)

[a] *Assuming Al−Al avoidance, with values determined experimentally from the deconvolution and integration of single-pulse ^{1}H MAS NMR spectra obtained at a spinning speed of 60 kHz. Resonances in the 0.8−1.5, and 2.4−3.6 ppm ranges contribute to the Mg_3-OH and Mg_2Al-OH local environments, respectively.*

vacancies in the dioctahedral gibbsite lattice [87,88]. The mineral shigaite, $[Mn_6Al_3(OH)_{18}][(SO_4)_2\{Na(OH_2)_6\}(H_2O)_6]$, shows an ordered array of cations in a $(\sqrt{3} \times \sqrt{3})R30°$ array with a **3R** polytype [89]. The mineral nikischerite (M = Ni) [90] is isostructural with shigaite, and motukoreaite, $[Mg_{5.6}Al_{3.4}(OH)_{18}][(SO_4)_{1.3}(CO_3)Na_{0.6}(H_2O)_{12}]$, also has essentially the same structure [91]. The mineral zaccagnite, $[Zn_4Al_2(OH)_{12}]CO_3 \cdot 3H_2O$ adopts a $\mathbf{2H_1}$ polytype stacking with a supercell ($a = a_o\sqrt{3} = 0.462$ nm) [92]. The minerals $[M^{II}_4Al_2(OH)_{12}]CO_3 \cdot 3H_2O$ (charmarite for M = Mn, quintinite for M = Mg, and caresite for M = Fe) show a superlattice of the type $(2\sqrt{3} \times 2\sqrt{3})R30°$ [93]. A single crystal X-ray study [94] on a mineral with the composition $[Mg_4Al_2(OH)_{12}]CO_3 \cdot 3H_2O$ revealed that it stacked in a **2H** polytype with an ordered cation distribution giving rise to a $(\sqrt{3} \times \sqrt{3})R30°$ superlattice ($a = a_o\sqrt{3} = 0.5283$ nm and $c = 1.5150$ nm). Cation order has also been found in synthetic materials with the formula $[Fe^{II}_4Fe^{III}_{(2-6y)}Al^{III}_{6y}(OH)_{12}]SO_4 \cdot 8H_2O$ for $y = 0$ (so-called "green rust" materials) and $y = 0.08$ [95].

17.1.3.3. Short-range Cation Order

EXAFS is a useful means for studying *short-range cation ordering* and has provided strong evidence for the local cation ordering in LDHs. It has been shown that Fe^{III} cations in $Mg_nFe^{III}-CO_3$-LDHs ($n = 2$, 3, and 4) never occupy neighboring sites [59]. The Fourier transform of the EXAFS oscillation at the cation K edges in Zn_2Cr-Cl- and Cu_2Cr-Cl-LDHs can be simulated with a six-shell model in which the first peak arises from the O atoms coordinated to the metal whilst the other five peaks result from metallic neighbors at a, $a\sqrt{3}$, $2a$, $a\sqrt{7}$, and $3a$ [96]. Local cation ordering was observed in a Cu_2Cr-LDH with a corrugated sheet due to the Jahn–Teller distortion of the Cu^{2+} ion [96], and in a Zn_2Cr-LDH with a flat sheet [52,97]. EXAFS studies at the Zn, Co, and Fe K edges have shown the ordered nature of the cations in Zn_2Al-Cl-LDHs [98,99] and in $[Co_2(Fe_yAl_{1-y})(OH)_6]Cl \cdot nH_2O$ $(0 \le y \le 1)$ [100]. The presence of local cation ordering in Mg_2Ga-CO_3- and Mg_5Ga-CO_3-LDHs has also been shown by EXAFS [77].

It has been proposed that the nature of the vibrations in the 800–250 cm^{-1} region of the IR spectra of LDHs gives an indication of the extent of lattice ordering. When there are five sharp vibrations in this region, as observed for Mg_2Al-CO_3-, Ni_2Al-CO_3-, and Mg_2Fe-CO_3-LDHs, this has been taken to indicate an ordered array of $M^{II}_2Al(OH)$ units [101]. In the Raman spectrum of $Mg_{2.3}Al-CO_3/NO_3$-LDH [102] and stichtite ($[Mg_6Cr_2(OH)_{16}]CO_3 \cdot 4H_2O$) [103] distinct hydroxyl stretching bands can be attributed to the individual Mg_3-(OH), Mg_2Al-(OH)/Mg_2Cr-(OH) and $MgAl_2$-(OH)/Cr_3-(OH) units. Separate hydroxyl stretching bands attributable to Mg_3OH, Zn_3OH, and Al_3OH moieties were also observed in the Raman spectra of the ternary $[Mg_xZn_{6-x}Al_2(OH)_{16}](CO_3) \cdot 4H_2O$ ($x = 0$–6) LDHs [104]. The Raman spectra of $[Cu_xZn_{6-x}Al_2(OH)_{16}]CO_3 \cdot 4H_2O$ indicate that when $x < 2$, the cations are ordered, whilst when $x > 2$ there is some evidence for segregation of different cations [105].

In so-called green rust materials, the extent of cation ordering has been correlated with anion ordering [106]. For an LDH with layers of $[Fe^{II}_{1-x-y}Mg_yFe^{III}_x(OH)_2]^{x+}$, the Mössbauer spectrum is consistent with an ordered arrangement of cations in which each Fe^{III} cation is surrounded by six M^{II} cations. The Mössbauer spectra of $[Fe^{II}_4Fe^{III}_{2-y}Al_y(OH)_{12}](SO_4) \cdot nH_2O$ [95,107] are also consistent with an ordered array of Fe^{III}/Al^{III} cations around Fe^{II} sites. In the case of Zn_2Cr- and Zn_3Cr-LDHs, it has been suggested that the UV–visible diffuse reflectance spectra are consistent with the presence of $Cr^{III}O_6$ sites surrounded by six Zn^{2+} ions without any $Cr^{III}-O-Cr^{III}$ linkages [108,109].

17.1.3.4. Non-octahedral Coordination of Layer Cations

The structure of LDHs generally involves octahedrally coordinated M^{II} and M^{III} cations. The presence of seven-coordinate cations has been observed and confirmed by single crystal XRD of LDHs containing the Ca^{II} ion [8,29,30,78,80,81,83,110]. The seventh coordinated oxygen may arise from interlayer anions or interlayer water molecules, and in $[Ca_4Al_2(OH)_{12}]CO_3 \cdot 5H_2O$, it is one of the carbonate oxygen atoms. FT-IR and ^{13}C CP/MAS NMR spectra indicate that there may also be a coordinate bond between sulfonate oxygen atoms and Al^{3+} ions in sulfonate-intercalated LDHs [111]. On heating, some interlayer anions can become grafted to the layers in LDHs, with the anions substituting one of the hydroxyl groups, thus maintaining the octahedral coordination of the cation. Such grafting has been reported for carboxylate, sulfonate, and phosphonate anions [112], oxometalate [113], and polyoxometalate anions [114–116], as well as for carbonate anions [117].

17.1.3.5. LDHs and Related Structures without M^{III} Cations

Recently, a variety of LDHs containing M^{II} cations together with M^{IV} or even M^V or M^{VI} cations have been reported. Examples include $ZnSi-CO_3$-LDH [118] and other $M^{II}M^{IV}$-LDHs [119–123], CoV^V- [124], $ZnMo^{VI}$- [125], and MgAlZr-LDHs [126]. In contrast, the α-hydroxides of nickel and cobalt [40,127] and $Zn_5(OH)_8(NO_3)_2 \cdot 2H_2O$ [128], $Ni_{1-x}Zn_{2x}(OH)_2(CH_3CO_2)_{2x} \cdot nH_2O$ [129,130] and $[Co_7(OH)_{12}](O_3SC_2H_4SO_3) \cdot 2H_2O$ [131] also have structures that are related to LDHs, although they only contain M^{II} cations. The mineral lawsonbauerite $[(Mn,Mg)_9Zn_4(OH)_{22}](SO_4)_2 \cdot 8H_2O$ also contains only

M^{II} cations and has vacancies in two-ninths of the cation sites in a brucite-like lattice with O atoms of the vacant sites and SO_4^{2-} ions coordinated to tetrahedral Zn^{II} ions [132].

17.1.4. Interlayer Anions of LDHs

17.1.4.1. Interlayer Hydration and Hydrogen Bonding

The maximum amount of water which can be accommodated in the interlayer galleries of an LDH with $[M^{II}_{1-x}M^{III}_x(OH)_2]^{x+}$ layers is given by $(1 - Nx/n)$ where N is the number of sites occupied by an anion of charge n [133]. The general formula of an LDH containing interlayer carbonate anions is therefore:

$$\left[M^{II}_{1-x}M^{III}_x(OH)_2\right]\left(CO_3\right)_{x/2}(H_2O)_{1-3x/2-\Delta},$$

where Δ represents the number of vacant interlayer sites [8]. In the case of hydrotalcite, there is one vacant layer site per $[Mg_6Al_2(OH)_{16}]CO_3 \cdot 4H_2O$ formula unit. In synthetic LDHs, the experimentally determined amount of water can exceed $(1 - Nx/n)$ because there is *extrinsic* water adsorbed on external surfaces [134]. Hou et al. [135] have shown that the swelling and water sorption behavior of LDHs depend markedly on the interlayer anion. Molecular dynamics (MD) calculations [136] have shown that interlayer water molecules have on average less than four nearest neighbor water molecules. This differs from bulk water, which involves distorted tetrahedral arrangements similar to that in ice (I_h). Thermodynamic measurements have indicated [137] that interlayer water in LDHs has a higher entropy than that of ice, but lower than that of bulk water. The interlayer water in $MgAl-CO_3$-LDHs is more strongly bonded than physisorbed water by ca. 12 kJ mol^{-1} [138]. Increasing the M^{II}/M^{III} ratio in the layers leads to a decreasing number of anions and an increase in the diffusion coefficient [139]. The extent of disorder decreases with increasing charge on the interlayer anions [140]. The presence of an extensive hydrogen-bonding network involving the interlayer water molecules and the layer hydroxyl groups has also been confirmed by IR and Raman spectroscopy and inelastic neutron scattering [141].

17.1.4.2. Arrangement of Interlayer Anions and Water

17.1.4.2.1. INTERLAYER ALIGNMENT FROM X-RAY DIFFRACTION ANALYSIS

If a metal complex anion is incorporated into the interlayer galleries of LDHs, the second basal reflection of the XRD pattern is generally [142] stronger than the first [143–147] and sometimes the first basal reflection may be too weak to be observed, as is the case of $[Mo_7O_{24}]^{6-}$-intercalated LDHs [133]. When a tetrasulfonated perylene dye is intercalated into an MgAl-LDH, two maxima in the electron density in the interlayer gallery are observed which presumably correspond to the positions of the sulfur atoms, and the observed two minima can be taken to locate the position of the C–N single bonds [148]. In stearate-intercalated MgAl-LDHs, the electron density distribution confirmed the presence of a bilayer of guest molecules [149–151]. The packing sequences and interlayer structures in $[Fe^{II}_4Fe^{III}_2(OH)_{12}]X \cdot nH_2O$ are different for ($X = Cl_2$, $n = 4$) [152] and ($X = SO_4$, $n \approx 8$) [153]. Mohanambe et al. [154] found that in the carboxymethyl-β-cyclodextrin (β-CMCD)-intercalated MgAl-LDH, the intercalation resulted in an increase in the interlayer spacing from 8.9 Å to 24.5 Å. In view of the dimensions of β-CMCD, two possible arrangements of intercalated β-CMCD can be proposed, as shown in Fig. 17.4. A combination of X-ray diffraction, ^{13}C NMR spectroscopy, infrared and Raman spectroscopy, augmented with molecular dynamics calculations, showed that the actual arrangement was that in Fig. 17.4b, which allows guest species to access the wider opening of the β-CMCD species, since this is located in the center of the interlayer galleries.

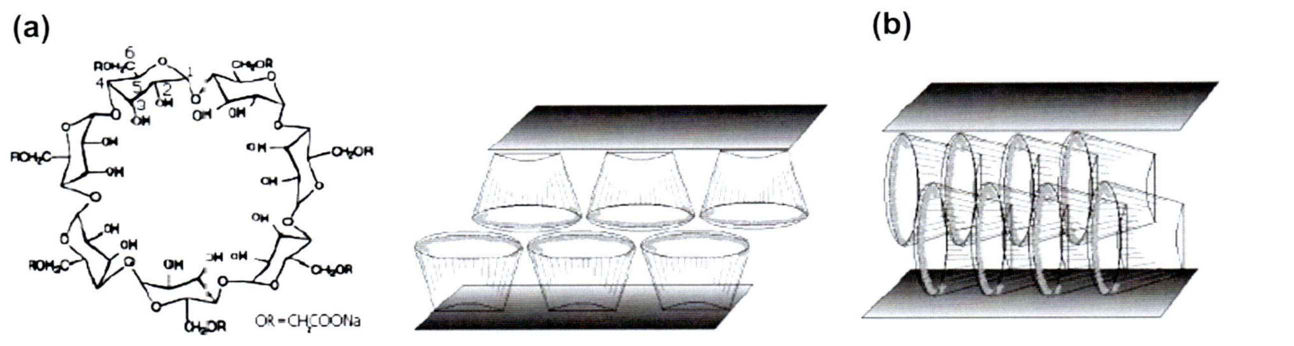

FIGURE 17.4 (a) Structure of the β-CMCD molecule and (b) possible orientations of β-CMCD molecules within the MgAl-LDH galleries. *Reprinted with permission from [154]. Copyright 2005 American Chemical Society.*

17.1.4.2.2. INTERLAYER MOLECULAR MODELING AND MOLECULAR DYNAMICS SIMULATIONS

Mohanambe et al. [155] studied MgAl-LDHs intercalated with three drug molecules, ibuprofen, diclofenac, and indomethacin which all adopt a double layer arrangement in the interlayer galleries. Spectroscopy and molecular dynamics simulations showed that there is a change in the geometry of intercalated diclofenac and indomethacin, which is believed to originate from the electrostatic interaction between the Cl group on the drug molecule and the positively charged layers of the LDH. Yan et al. [156] studied the orientation of the 4-vinylbenzenesulfonate anion (VBS)-intercalated NiAl-LDH by MD simulations. Three orientation angles for the VBS and water molecules were defined as shown in Fig. 17.5. The VBS molecules have a tendency to tilt toward a vertical arrangement with respect to the layers as the amount of interlayer water increases. For the water molecules, the H······H vector tends toward becoming parallel to the layers and the plane of the water molecule becomes more tilted with increasing amounts of interlayer water molecules. An ultrathin film (UTF) of an LDH containing intercalated *tris*(1,10-phenanthroline-4,7-diphenylsulfonate)ruthenium(II) anions has also been studied [157]. The most significant change on intercalation is that the ruthenium complex anion undergoes a dramatic degradation in symmetry.

17.1.4.2.3. SPECTROSCOPIC MEASUREMENTS

X-ray absorption near-edge spectroscopy (XANES) and EXAFS data at the V K-edge indicate that ion-exchange of a ZnAl–Cl-LDH with ammonium metavanadate may give either discrete $[V_{10}O_{28}]^{6-}$ anions or polymeric anions based on VO_4 tetrahedra in the interlayer galleries, depending on the reaction pH [158,159]. The variation in the arrangement of interlayer benzoate anions with extent of hydration of the LDH has been studied by XANES [160,161].

Vibrational spectroscopy has been widely employed in the study of LDHs [162,163]. The ν_3 band in the C–O stretching region of their IR spectra is typically split into two bands separated by 30–60 cm^{-1} [164], which suggests a lowering of the symmetry to C_{2v} or C_s, associated with hydrogen bonding of the CO_3^{2-} to hydroxyl groups and/or interlayer water molecules [162]. The positions of the two peaks arising from the ν_3 band depend on the nature of the divalent cation in the hydroxyl layer [164–166]. Varying the trivalent cation has a similar effect [167]. For $[Mg_{1-x}Al_x(OH)_2](NO_3)_x \cdot nH_2O$ LDHs, it has been reported that when $x < 0.26$, the NO_3^- anions lie in the center of the interlayer gallery with their planes parallel to the layers [168]; when $x > 0.26$, alternate NO_3^- anions are shifted up and down the *c*-axis with their planes remaining parallel to the layers so that they are strongly hydrogen bonded to one layer and do not interact significantly with the other, in a "stick-lying" structure [169,170]. For intercalated octahedral $[Fe(CN)_6]^{4-}$ anions with their C_3-axis perpendicular to the sheets [142,171–173], the expected one IR-active band is split into two bands, which can be attributed to a reduction in symmetry to D_{3d} [171,172,174].

Solid-state ^{13}C MAS NMR spectroscopy can detect the presence of co-intercalated carbonate (and/or bicarbonate) anions [175,176]. The dynamics and orientation of terephthalate anions in $[Mg_{1-x}Al_x(OH)_2][\{p\text{-}C_6D_4(COO)_2\}_y(CO_3)_{1-y}]_{x/2} \cdot nH_2O$ have been studied by ^2H solid-state NMR [177]. A ^{35}Cl NMR study of ClO_4^--intercalated MgAl- and LiAl-LDHs has shown that the anion is rigidly held at low relative humidities and temperatures, but undergoes rapid isotropic reorientation on raising the humidity and/or temperature [178]; the isostructural interlayer SeO_4^{2-} anion, however, did not undergo isotropic reorientation at any relative humidity as shown by ^{77}Se NMR spectroscopy [179]. A first-order phase transition near 6°C of $[Ca_4Al_2(OH)_{12}]Cl_2 \cdot 4H_2O$ can be detected by ^{35}Cl NMR, involving a transformation of the Cl^- anions located in triaxial sites to uniaxial sites [180]. Dynamically disordered water molecules in $[LiAl_2(OH)_6]Cl \cdot xH_2O$ have been detected by ^1H NMR [181].

The absorption spectrum of 9-anthracenecarboxylate-intercalated LDHs shows a band around 490 nm, not present in the free anion, which has been attributed to an aggregate of the guest anions [182]. Reaction of $[LiAl_2(OH)_6]Cl$ with 4-nitrohippuric acid (NHA) leads

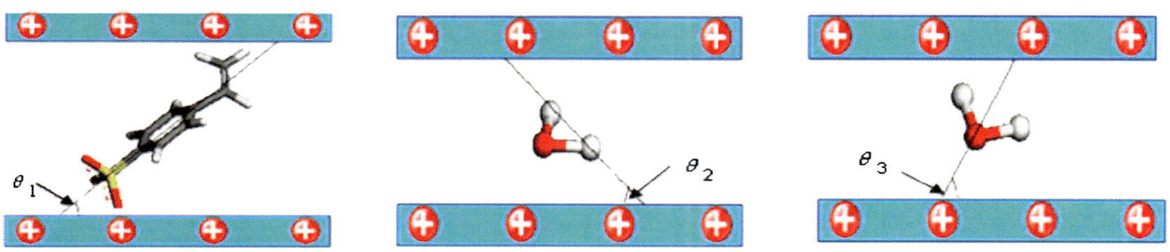

FIGURE 17.5 Definition of the three orientation angles, θ_1, θ_2, and θ_3 for the VBS/LDH system, standing for the C_3-axis of the sulfonate group, the H······H vector and the dipole vector of the H_2O molecule relative to the host layers, respectively. *Reprinted with permission from [156]. Copyright 2008 American Chemical Society.*

to an unusual intercalation of the neutral organic molecule without loss of Cl^-. The resulting material exhibits frequency-doubling characteristics, indicating an ordered array of intercalated guest molecules [183]. By using the polarized fluorescence method, Shi et al. [184] have studied the molecular orientation of Zn_2Al-LDH intercalated with α-naphthalene acetate and β-naphthalene acetate.

17.1.4.2.4. GUEST–GUEST INTERACTIONS

The photo-induced [2 + 2] dimerization of crystalline *trans*-cinnamic acid requires both close contacts of approximately 0.4 nm and a specific orientation of the unsaturated bonds, and can be used to explore the arrangement of substituted cinnamate ions in the interlayer galleries of LDHs [47,185]. Head-to-head (HH) cyclodimers were formed selectively on irradiation of LDHs containing cinnamate ions, indicating that the arrangement of the intercalated anions inhibits the formation of the head-to-tail (HT) dimer. Furthermore, the ratio of *anti*-HH to *syn*-HH dimers increases as the Mg/Al ratio increases. It was suggested that larger MgAl ratios result in longer guest–guest distances which favor the formation of the *anti*-HH isomer [186]. Intercalation of (Z,Z)-2,4-hexadienedioic acid into a $LiAl_2$-LDH host affords a vertical arrangement of interlayer guest anions which results in the 1,3-diene moieties being arranged in a face-to-face manner which facilitates their polymerization on photoirradiation [187].

In MgAl-LDHs intercalated with $p\text{-}CH_3C_6H_4CO(CH_2)_nCO_2^-$ ($n = 4$–10), the chains become increasingly interdigitated as their length increases [188]. Aloisi et al. [182] prepared an LDH containing co-intercalated coumarin-3-carboxylic acid (3-CCA) and 9-anthracene carboxylic acid (9-ACA) to probe the energy transfer between the two confined guest anions. Yan et al. [189] investigated the orientation of 3,4,9,10-perylene tetracarboxylate (PTCB) intercalated in an MgAl-LDH, and found the existence of PTCB aggregates of H-type and J-type dimers.

17.1.4.3. *Interlayer Anion Ordering*

Ordering of the interlayer anions may be observed in LDHs. This may result from either an ordered arrangement of cations or a specific pattern of hydrogen bonding in the interlayer galleries. The XRD pattern of Mg_2Al–benzoate-LDH can be indexed to a superlattice with $a = a_0\sqrt{3}$ characteristic of ordered benzoate anions [20]. A similar conclusion was drawn for an oxalate-intercalated $Mg_{2.2}Al$-LDH [190]. The ordering of interlayer silicate anions in a Zn_3Al-LDH with a superlattice (with $a = a_0\sqrt{3} = 0.5305$ nm) has been explained [191] in terms of the symmetry match between the hexagonal condensed silicate anions and the layers. Sulfate anions tend to be ordered when the interlayer gallery is expanded to ca. 1.1 nm by inclusion of additional water and/or hydrated cations [18,192]. The lattice periodicity may be $a_0\sqrt{3}$ (for 2:1 LDHs) as in $[Mg_{3.96}(Al_{1.98}Fe^{III}_{0.06})(OH)_{12}][(SO_4)_{1.30}Na_{0.56}(H_2O)_{7.3}]$ [18,192] or $3a_o$ for shigaite [90]. The synthesis of $[Zn_{0.67}Cr_{0.33}(OH)_2][(SO_4)_{0.22}Na_{0.11}(H_2O)_{1.25}]$ LDH adopts a two-layer polytype with an $a_0\sqrt{3}$ superlattice [193]. High-intensity XRD, $[Ni_{1-x}Al_x(OH)_2](CO_3)_{x/2}\cdot yH_2O$, shows to have ordering of cations for the case of $x = 0.33$ [85], whereas for $x = 0.30$, the carbonate anions form an ordered $(\sqrt{13}\times\sqrt{13})R13.90°$ superlattice [85,194]. Single crystals of $[Mg_4Al_2(OH)_{12}]CO_3\cdot 3H_2O$ exhibited an ordered distribution of interlayer anions [94]. The interlayers are composed of alternate layers of carbonate anions and a net of water molecules which gives a hexagonal structure ($P\bar{6}2m$, $a = a_0\sqrt{3} = 0.5283$ nm, $c = 1.5150$ nm).

Atomic force microscopy (AFM) images of the crystal surface of $[Mg_6Al_2(OH)_{16}](CO_3)_{0.5}Cl\cdot 2H_2O$ in contact with an aqueous solution of Na_2SO_4 show 2D hexagonal periodicity, which results from the hydroxyl anions in the basal layer plane [195]. Scanning tunneling microscopy (STM) image of the same sample in air showed a 2D lattice ($a = 0.75 \pm 0.04$ nm, $b = 1.10 \pm 0.03$ nm, and $\alpha = 70 \pm 3°$). STM images of the crystal in contact with aqueous solutions of $[Fe(CN)_6]^{n-}$ ($n = 3$, 4) showed 2D ordered arrays of the anions with different lattice parameters, which was also reported for $[Fe(CN)_6]^{n-}$ anions adsorbed on $[Mg_6Al_2(OH)_{16}](CO_3)_{0.75}(Cl)_{0.5}\cdot 2H_2O$ [196]. Studies have suggested that ordering of organic anions is more common than for inorganic anions [197].

17.1.4.4. *Staging Structures*

An increasing number of LDH materials with more than one type of interlayer, giving rise to so-called *staged structures* have been reported [198]. A mineral $[Mg_4Al_2(OH)_{12}]CO_3\cdot 3H_2O$ has alternate interlayers of carbonate anions and water molecules [94]. The mineral coalingite $[Mg_{10}Fe_2(OH)_{24}]CO_3\cdot 3H_2O$ has interlayers containing carbonate and water with alternate vacant interlayers [199]. On dehydration, the mineral motukoreaite $[Mg_{5.6}Al_{3.4}(OH)_{18}](SO_4)_{1.3}(CO_3)Na_{0.6}\cdot 12H_2O$ first reaches an intermediate staged phase with basal spacing of 1.98 nm and alternate hydrated and dehydrated interlayers [200]. A mineral with the approximate formula $[Mg_4Al_2(OH)_{12}](SO_4)_{0.5}(CO_3)_{0.5}\cdot nH_2O$ has alternate interlayers of carbonate (0.756 nm) and sulfate (1.098 nm) anions with the $6\mathbf{R}_4$ polytype [192].

At intermediate temperatures, an interstratified intermediate of terephthalate-intercalated MgAl-LDHs containing alternate layers of vertical and horizontal anions can be isolated with a 6**R** repeat unit [20]. Staging has also been reported during reactions of terephthalate-intercalated MgAl-LDHs with inorganic anions giving partial exchange of terephthalate anions [201]. Staging in LDHs containing alternate interlayers of inorganic and organic anions such as (4-phenylazophenyl)acetate [151] or succinate or tartrate [96,202] anions has also been reported. The staged intermediates can be observed in exchange reactions of LDHs by in situ energy-dispersive XRD experiments [203,204]. Second-stage intermediates were revealed during the anionic exchange of Cl^- in Cu_2Cr-Cl-LDHs by tartrate and succinate anions, resulting from the hydrophobic binding effect of the organic moieties [96]. A carboxyethyl-substituted azacrown ether derivative (CSAE) was intercalated into an MgAl-LDH with a vertical orientation [205]. At 100°C, a staged phase with basal spacing of 2.33 nm appeared, attributed to anion exchange of CO_3^{2-} and CSAE. Another 2.08 nm staging phase appeared at 150°C, due to the regular stacking of the 1.18 and 0.77 nm phases. To date, all staging examples observed in LDHs have involved second-stage materials, which can be explained by the Rüdorff model [206], rather than the Daumas and Hérold model involving buckling of the layers usually invoked for graphite [198]. The layers have been shown to be capable of buckling, in species such as $[SmW_{10}O_{36}]^{9-}$ intercalated LDHs, however [44,207], suggesting that buckling of the layers cannot be ruled out.

LDHs are not necessarily easy to prepare as a pure single phase and there still remain unresolved issues concerning their structural chemistry. These include (i) the range and the relative amounts of possible cations incorporated in a pure LDH phase; (ii) the stacking fashion of the brucite layers; (iii) the extent of ordering of cations, both long-range and short-range; and (iv) the detailed arrangements in the interlayer galleries. Highly crystalline LDHs or even single crystals are required to obtain the most accurate structural information [78,114,208–211]. Technical advances including high-intensity XRD sources [28,85], area detectors [212], and image plate technology [149] as well as improved methods of modeling static and dynamic disorder [213] may also facilitate detailed studies of the structure of LDHs. The rapid improvements in both hardware and software for the theoretical study of materials should also enhance our theoretical understanding of the structure of LDHs.

17.2. PREPARATIVE CHEMISTRY OF ANION-INTERCALATED LAYERED MATERIALS

17.2.1. Methods of Preparation of LDHs

17.2.1.1. Coprecipitation Methods

The coprecipitation method has been extensively employed for the one-pot direct synthesis of LDHs containing different layer cations and interlayer anions [214], and can easily be scaled up to produce large quantities of LDHs in a commercial process. Many LDHs, such as Zn_nAl-CO_3-LDH [48], may be prepared with a range of M^{II}/M^{III} ratios using this method, while some LDHs can be obtained only with a narrow range of ratios, irrespective of the M^{II}/M^{III} ratio employed in the precursor solution. This synthetic route is often chosen as the method for preparing organic anion-containing LDHs which are difficult to synthesize by other ways. A supersaturated state is necessary to ensure the simultaneous precipitation of two or more cations; this is generally implemented by controlling the pH value of the solution. The coprecipitation should be carried out at a pH value no less than that at which the most soluble hydroxide is precipitated [2]. A procedure using thermal treatment following coprecipitation is often necessary to increase yields and/or the crystallinity of amorphous or poorly crystallized materials [48]. Two main coprecipitation methods have been commonly employed: precipitation at constant pH value, and at variable pH value [48]. LDHs have a high affinity for carbonate anions and coprecipitation reactions are therefore generally carried out under N_2 in order to avoid competition from CO_3^{2-} resulting from atmospheric CO_2.

17.2.1.1.1. PRECIPITATION AT CONSTANT pH

In *the constant pH method*, the pH value of the reaction system remains the same throughout the whole preparation process, leading to the coprecipitation of the two metallic ions at low supersaturation [48]. Pure intercalated LDHs with high crystallinity can be obtained. The target-intercalated anion should be present in excess and have a high affinity for the LDH layers. Metal nitrate and chloride salts are commonly used as precursors because of the low affinity of LDHs for these anions.

Many anions have been intercalated into LDHs by precipitation at constant pH, including CO_3^{2-} [134,215–218], NO_3^- [219], naphthalene-2,6-disulfonate [220], tetraphenyl porphyrins [221], $[Ni(EDTA)]^{2-}$ [222], an anionic azobenzene derivative (AzAA) [151], alizarin red S [111], amino acids [223–225], nucleoside monophosphates, deoxyribonucleic acid (DNA) [226], and vitamins (A, C, E) [227]. A perylene chromophore has been intercalated into an LDH host in order to increase its stability when used as a pigment [148]. Catalytically active species, including polyoxometalate anions such as $PW_{12}O_{40}^{3-}$ or $SiW_{12}O_{40}^{4-}$, have been introduced into the interlayer galleries of LDHs giving active catalysts or catalyst precursors [228]. Zr-containing LDHs prepared by this method possess excellent catalytic performance for the selective hydroxylation of phenol to catechol in the liquid phase [229,230].

17.2.1.1.2. PRECIPITATION WITH VARIABLE pH

The variable pH method involves addition of a mixed M^{II}/M^{III} salt solution to an alkaline solution containing the target interlayer anion under intensive agitation, which leads to a continuous change in solution pH value, followed by an aging step. This method generally gives poorly crystalline products, due to the large number of crystallized nuclei obtained at high supersaturation. The formation of metal hydroxide impurity phases is common, and thus an LDH product with an M^{II}/M^{III} ratio different from that in the precursor mixture often results.

Using this method many LDHs including MgAl-, NiAl-, CoMnAl-, CoMnMgAl-, and NiCrAl$-CO_3$-LDHs containing interlayer CO_3^{2-} anions [231], MgAl$-OH^-$, $-Cl^-$, $-SO_4^{2-}$ [232], or $-W_7O_{24}^{6-}$ [233] intercalated LDHs and (Cu, Zn)Al$-CO_3/SO_4$-LDHs [234] have been prepared. Alkoxide-intercalated derivatives of LDHs have been synthesized using coprecipitation in methanol or ethanol using the variable pH method [235]. LDHs containing both Fe^{II} and Fe^{III} cations including $MFe^{II}Fe^{III}$ (M = Mg, Co, and Ni) [67,68], $MgFe^{III}$ [236], and $[Fe(CN)_6]^{3-}$ intercalated $MgFe^{III}$-LDHs have also been prepared by this method [237]. Organic anions, including caprate [238], indole-2-carboxylate (ICTA) [239], 4′-chloro-4-stilbenecarboxylate [240], terephthalate [241,242], amino acids [243], dicarboxylates [175], anti-inflammatory drugs [244,245], and L-aspartic acid [246,247] have also been incorporated into the interlayer galleries of LDHs by coprecipitation at variable pH. LDHs with a broad range of M^{II}/M^{III} ratios can generally be prepared by this method. Preferential intercalation of specific isomers of sulfonated anthraquinone ions has been observed using this method [248].

17.2.1.1.3. SEPARATE NUCLEATION AND AGING STEPS (SNAS) METHOD

It is difficult to control the particle size and size distribution of LDHs using traditional coprecipitation methods at either constant or variable pH [214]. A method involving *separate nucleation and aging steps (SNAS)* has been devised [249], in which a very rapid mixing and nucleation process in a modified colloid mill is followed by a separate aging process. The design of the colloid mill is illustrated in Fig. 17.6. The LDH products formed using the SNAS method have a slightly higher crystallinity than those produced using the coprecipitation at a constant pH value [2,134,250–253]. Moreover, the SNAS method affords smaller crystallites with a higher aspect ratio, having a very narrow distribution of crystallite size, as can be seen from Fig. 17.7 [249]. In the colloid mill, the mixing and nucleation are complete in a very short time giving nuclei of all approximately the same size; these are then subjected to a separate aging process. This results in well-formed crystallites with a similarly narrow range of diameters. The SNAS method has proved to be very amenable to scale-up [254] and is now employed to produce LDHs commercially in China.

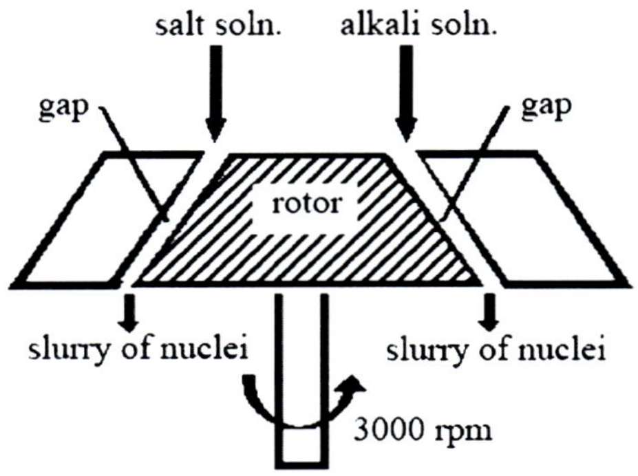

FIGURE 17.6 Schematic illustration of a colloid mill. [253] With kind permission from Springer Science+Business Media:< Preparation of Layered Double Hydroxides, Struct Bond (2006) 119: 89–119, Jing He, MinWei, Bo Li, Yu Kang, David G Evans and Xue Duan>.

The *non-equilibrium aging method* was developed [255] as a complement to the SNAS method. After a period of aging, another portion of salt and alkali solution is added to ensure that the metal ions are always supersaturated. A high temperature and slow rate of addition of the reactants favor the adsorption of the added metal ions on the preformed crystal particles, rather than the formation of new nuclei [256]. MgAl$-CO_3$-LDHs with increased crystallinity can be prepared by this method [255].

17.2.1.1.4. UREA HYDROLYSIS METHOD

Urea is a very weak Brønsted base ($pK_b = 13.8$) and can be employed as a reagent for "homogeneous" precipitation from solution [257]. The hydrolysis of urea results in a solution with pH ≈ 9, depending on the temperature, which is suitable for precipitating many metal hydroxides and a variety of LDHs. The *urea hydrolysis method* is suitable for the preparation of MgAl-LDHs with high layer charge densities [258]. If the $M^{II}(OH)_2$ is more insoluble than $Mg(OH)_2$, the LDH obtained has a composition near to that in the original precursor solutions (e.g., ZnAl- and NiAl-LDHs). This method results in LDH crystallites with a relatively large size (microns) and a well-defined hexagonal shape due to a low supersaturation during precipitation [259]. A prolonged aging time and decreased total metal concentration allows MgAl$-CO_3$-LDHs with high crystallinity to be obtained [260]. The particle size can be controlled by altering the reaction temperature which affects the hydrolysis rate of urea and larger particles are formed at lower temperatures due to the lower nucleation rate [261].

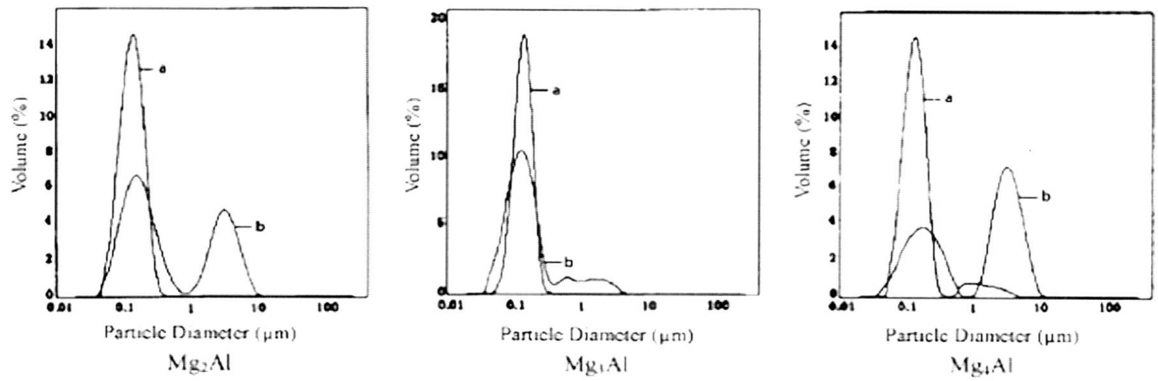

FIGURE 17.7 Distribution of particle size for MgAl–CO_3-LDHs with different Mg/Al ratios prepared using (a) the SNAS method and (b) conventional coprecipitation at constant pH value. *Reprinted with permission from [249]. Copyright 2002 American Chemical Society.*

The as-synthesized LDHs usually contain carbonate ions which generally cannot be deintercalated. However, treatment with gaseous or aqueous HCl [210] or HCl/NaCl mixtures [262] can lead to replacement of carbonate with chloride anions, which are labile and can be replaced by bulky organic anions. However, the urea hydrolysis method is not suitable for preparing LDHs containing Cu^{II} or Cr^{III} [210]. Hexamethylenetetramine hydrolyzes at high temperature with the release of ammonia, which makes the solution alkaline [263] and has been used as an alternative to urea for the homogeneous preparation of LDHs. A simplified synthesis of MgAl–NO_3-LDHs with various MgAl ratios using $NH_3 \cdot H_2O$ as a base has been developed [264]. No incorporation of CO_3^{2-} was observed using this approach even in ambient atmosphere.

17.2.1.2. *Anion-exchange Method*

The anion-exchange method is based on the exchangeable properties of the interlayer anions [265]. This method is especially useful when the coprecipitation method is inapplicable, for example, the divalent or trivalent metal cations or the anions involved are unstable in alkaline solution, or the direct reaction between metal ions and guest anions is more favorable, or there is no suitable soluble salt of the guest anions. In thermodynamic terms, anion exchange in LDHs depends mainly on the electrostatic interactions between the host sheets and the exchanging anions and, to a lesser extent, on the free energy associated with the changes of hydration [265,266].

Anions with higher negative charges tend to exchange and replace those with lower charge in the interlayer galleries. The ion-exchange ability of common anions is as follows:

$$CO_3^{2-} > HPO_4^{2-} > SO_4^{2-}, OH^- > F^- > Cl^- > Br^- > NO_3^- > I^-.$$

Therefore, NO_3^- or Cl^--intercalated LDHs are usually used as the precursors for ion exchange [267].

The anion-exchange method is especially useful for the preparation of LDHs containing anions other than carbonate. A large number of organic and inorganic anions have been incorporated into LDHs using the ion-exchange process, such as carboxylates [265], anionic surfactants [268], phosphonates [203], β-cyclodextrin derivatives [269], pharmaceutically active species [270,271], nucleoside and DNA molecules [272], amino acids [273], glyphosate anions [274], dye anions [263,275], polyoxometalates [252,276], metal complex ions [144,147,277–285], polyoxirane derivatives [286, 287], and polystyrene sulfonate [288].

In general, the ion-exchange process can follow one of two pathways shown schematically as follows:

$$(1)\ \mathrm{LDH}\cdot A^{m-} + X^{n-} \rightarrow \mathrm{LDH}\cdot(X^{n-})_{m/n} + A^{m-}$$
$$\text{or } (2)\ \mathrm{LDH}\cdot A^{m-} + X^{n-} + mH^+ \rightarrow \mathrm{LDH}\cdot(X^{n-})_{m/n} + H_mA$$

In the first case, the LDH precursor generally contains intercalated univalent anions ($A^- = Cl^-$, NO_3^-, ClO_4^-) which have a weak electrostatic interaction with the host layers. By using different MgAl-LDH precursors, OsO_4^{2-} [289], $W_7O_{24}^{6-}$ [233], naproxen [270], and Eu $(EDTA)^-$ [282] intercalated MgAl-LDHs have been prepared by this method.

In the second case, the LDH precursor contains anions susceptible to acid attack, such as carbonate or carboxylates (e.g., terephthalate). By using this method under acidic conditions, inorganic anions such as Cl^-, Br^-, NO_3^-, SO_4^{2-} [290], heteropolyoxometalate anions containing vanadium [291], and organic anions such as adipate, terephthalate, succinate, dodecyl sulfonate, p-hydroxybenzoate, and benzoate [292,293] intercalated LDHs can be prepared. Chisem et al. [294] have determined the surface acidity of Cl^- and NO_3^- intercalated LiAl-LDHs using the Hammett indicator adsorption method, which showed that the acidity of the intercalation products is related to the nature of the interlayer anions, namely: carbonate < nitrate < chloride < decavanadate.

It has been verified that the favorability of the anion-exchange process is related to the following five main factors. (1) The exchange ability of incoming anions increases with increasing charge and decreasing ionic radius. The co-intercalation of a second anion was found to have no effect on the order of anion-exchange preference [295]. (2) Appropriate choice of solvent will favor the swelling and anion exchange of an LDH precursor [296]. (3) The pH value should generally be greater than 4, in order to preserve the host hydroxyl layer against damage. A low pH value favors liberation of anions of weak conjugate acids and incorporation of a less basic anion [297]. (4) In some cases, the chemical composition of the LDH host layer affects the anion-exchange process. (5) Generally, higher temperatures favor anion exchange [298].

The interlayer arrangement of organic anions depends strongly on the area available to each anion [9]. In some cases, shape-selective intercalation of isomers [4,299,300] has been observed in the anion-exchange process. *Second staging* structures as intermediate phases during the intercalation of organic guests by anion exchange [151,202,204] have been observed, as noted above. Phase segregation during the intercalation of hexacyanoferrate(III) anions in the LDH-chloride system [301] has also been observed in an anion-exchange process. After intercalation of a variety of bifunctional organic aromatic anions in LDHs using ion-exchange, subsequent thermal treatment resulted in a grafting reaction between the hydroxyl groups of the host layers and either the anionic functionality or a hydroxyl group on the aromatic anions [112]. An alternative method of anion exchange in LDHs based on the formation of a salt between an anionic and a cationic surfactant was reported by Crepaldi et al. [302,303]. The exchange is achieved by the formation of a water-insoluble surfactant salt. The exchange of dodecyl sulfate by different anions (chloride, carbonate, terephthalate, cholate and a sulfonated Cu-phthalocyanine) was achieved within 30 min with efficiency higher than 98.5%.

17.2.1.3. Calcination/reconstruction Methods

Calcination of LDHs at intermediate temperatures results in a mixed metal oxide (MMO) [304]. Exposure of the as-prepared MMO to an aqueous solution of an anion results in regeneration of the layered structure, with intercalation of the anion concerned [25,305–307]; this is sometimes described in the literature as the *structural memory effect* of LDHs (Fig. 17.8). It is an important method of synthesizing both inorganic and organic anion-intercalated LDHs [182,308–328]. The appropriate calcination temperature should be selected based on the composition of the LDH precursor. Generally, it is possible to reconstruct the LDH structure when the calcination temperature (T_c) is below 500°C. The products of calcination of MgAl-LDHs at 600°C and above contain the dense spinel phase, which cannot be reconverted to the LDH structure. An inert N_2 atmosphere is required during the rehydration process when an anion other than carbonate is to be incorporated. An anion such as tartrate, which has a strong tendency to complex metal cations, cannot intercalate into an LDH by the reconstruction method [41,323]. It is difficult to obtain pure crystalline intercalated products because the lamellar structure of LDHs can often only be partially restored during the reconstruction stage [324].

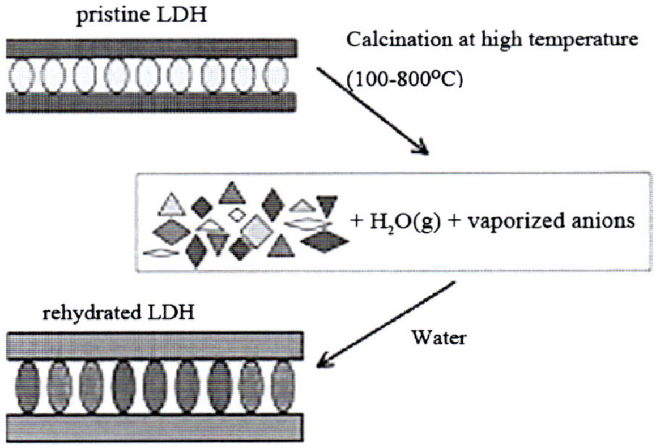

FIGURE 17.8 A schematic diagram of the calcination/reconstruction method [328]. *Reprinted from Progress in Organic Coatings 51 (2004), F. Wong, R.G. Buchheit, Utilizing the structural memory effect of layered double hydroxides for sensing water uptake in organic coatings, pages 91–102, Copyright (2004), with permission from Elsevier.*

It has been found that the extent of intercalation observed using this method depends on the reaction medium, the composition of the host layer, and the geometric and electronic structures of the anions. Organic chromophores [182], surfactants [311], ε-caprolactam [312], a sulfonated spiropyran dye [315], the herbicide glyphosate [316], naphthalene disulphonates [319], metal complex anions [318,320,325], amino acids and peptides [321], hexose [322], and even non-ionized guest pentoses [314] have been incorporated into LDHs using this method. $W_7O_{24}^{6-}$ and $[Cr(C_2O_4)_3]^{3-}$ intercalated MgAl-LDHs have been prepared by del Arco et al. [278]. Li et al. [326,327] prepared benzoate, phthalate, isophthalate, and terephthalate anion-intercalated MgAl- and ZnAl-LDHs. A study of $[LiAl_2(OH)_6](CO_3)_{0.5}\cdot nH_2O$ showed that the rate of recovery of the LDH by hydration of a sample calcined at 220°C was faster than that of a material calcined at 820°C [328].

17.2.1.4. *Hydrothermal Methods*

Hydrothermal synthesis has been shown to be effective for the intercalation of organic guest species with low affinity for LDHs. When insoluble metal hydroxides—for example $Mg(OH)_2$ and $Al(OH)_3$—are used as the inorganic precursors, there is no chance of competition for intercalation from anions in the precursor, ensuring that the desired anion occupies the interlayer galleries [329]. Single crystals of $[Ca_4Al_2(OH)_{12}]CO_3\cdot 5H_2O$ LDH were synthesized by a hydrothermal process using Ca$(OH)_2$, $Al(OH)_3$, and $CaCO_3$ as precursors [79]. This method has also been used to prepare a mesoporous and fibrous LDH containing tetrahedrally coordinated Al^{III} ion in the presence of dodecyl sulfate [330].

17.2.1.5. *Pre-pillaring Method*

The *pre-pillaring method* (also known as the *pre-intercalation method* or the *secondary assembly method*) is an anion-exchange method, and is a useful way to prepare LDHs intercalated with bulky guest anions or those with low charge density [331–333]. In such cases, pre-intercalation by smaller guests is an effective way to enlarge the interlayer space, and then the desired anions can be subsequently intercalated into the interlayer galleries by anion exchange [258].

C_{60} molecules [334] and polyethylene oxide [332] have been introduced into LDH interlayer galleries by using dodecyl sulfate pre-pillared LDH precursor. A three-step process has been used for the immobilization of penicillin G acylase (PGA) in the interlayer galleries of LDHs [331]. Inorganic polyoxometalate-pillared LDH derivatives have also been prepared using organic anion-intercalated LDHs as precursors [241,302,303,333]. A dioxomolybdenum(VI) complex $MoO_2Cl_2(THF)_2$ [99] has also been intercalated into the interlayer galleries of a ZnAl-LDH by using a 2,2′-bipyridine-5,5′-dicarboxylate pillared LDH precursor. Tagaya et al. [335] achieved co-intercalation of a photosensitizer, indole spirobenzopyransulfonate (SP-SO_3^-) and *p*-toluenesulfonate LDHs by this method. The pyran ring opening/closing interconversion of SP-SO_3^- (stable in non-polar conditions) and its photochemical product, merocyanine (MC) anions, can be realized in this co-intercalated LDH under UV irradiation.

17.2.1.6. *Intercalation by Dissolution and Re-coprecipitation*

The dissolution and re-coprecipitation method involves dissolution of the LDH in a solution of an organic acid to form a transparent solution, which is then added to an alkali solution resulting in the synthesis of the organic carboxylate-intercalated LDH. Using this method, competitive intercalation by carbonate formed from atmospheric CO_2 is not observed and pure and crystalline products can be prepared without protection by N_2 or another inert atmosphere. This method is not only suitable for the preparation of bulky anion-intercalated LDHs, but is also an effective means of pre-intercalation of the host layers of LDHs. Glutamate [331], citrate, oxalate, tartrate, and malate intercalated [336] LDHs have been prepared by this procedure.

17.2.1.7. *Other Methods*

Besides the methods for preparation discussed above, various other methods have been developed for the synthesis of LDHs. *The salt–oxide (or salt–hydroxide) method* can be employed in the preparation of ZnCr–Cl- and CuCr–Cl- [337,338], MgAl- [327,339,340], NiAl- and NiFe- [341] LDHs. Using the *microwave* or *ultrasound aging methods* [342], MgAl- and MgGa-LDHs can be prepared rapidly and conveniently [343–345]. $Ni^{II}M^{III}$ ($M^{III} = Al^{III}$, Cr^{III}, Mn^{III}, and Fe^{III}) [346] and $Fe^{II}Fe^{III}$–CO_3-LDHs [347] have been prepared by *electrochemical synthesis*. *In situ oxidation* can be employed to prepare $Mg_{0.3}Co^{II}_{0.6}Co^{III}_{0.2}$–$NO_3$-LDH [348], $MgFe^{III}$- [349,350], and $MgMn^{III}$- LDHs [208]. The *template synthesis* method involves the growth of LDH crystallites in a solution of an organic anion in the absence of CO_2, and is suitable for the intercalation–assembly of polymer anions (e.g., polyethylene sulfonate, polypropylene sulfonate, and polystyrene sulfonate) [250,266]. Water-in-oil emulsions have been used as a soft template to synthesize MgAl–CO_3-LDHs with a floccule or fiber-like morphology [351]. The *sol–gel method* using nickel acetylacetonate and aluminum isopropoxide as precursors has been employed in the preparation of $Ni_{2.5}$Al-LDHs with high surface areas [218,352]. The *"chimie douce"* method employs a highly crystalline-layered γ-oxyhydroxide precursor and gives very crystalline LDHs with controllable M^{II}/M^{III} ratios. LDHs of the type $[Ni_{1-x}Co_x(OH)_2]X_{x/n}\cdot mH_2O$ ($X^{n-} = CO_3^{2-}, SO_4^{2-}, NO_3^{-}, OH^{-}$) [114,353] have been prepared using this method.

17.2.2. Control of the Chemical Composition of LDHs

17.2.2.1. *The Identities and Ratios of M^{II}/M^{III} in the Brucite Layers*

A variety of naturally occurring minerals have the LDH structure, such as the carbonate-containing LDHs hydrotalcite and manasseite (Mg_3Al-), sjogrenite and pyroaurite (Mg_3Fe-), stichtite (Mg_3Cr-), takovite (Ni_3Al-), reevesite (Ni_3Fe-), and sulfate-containing LDHs like woodwardite (Cu_2Al-). An often quoted empirical generalization states that only six-coordinate M^{II} and M^{III} cations with their radius close to that of the Mg^{II} ion (72 pm [354]) can enter the octahedral voids formed by the hydroxyl groups and generate the brucite-like layers of LDHs [2–4,39,46]. Ca^{II} ion is an exception (ionic radius, 100 pm), which can form a stable LDH layer with Al^{III} [110,355]. Of the trivalent metal ions having an ionic radius in the range 54–80 pm, only Al^{III}, Cr^{III}, Fe^{III}, Co^{III}, and Ga^{III} can form stable binary LDH layers with divalent metal ions. Table 17.2 shows some combinations of metal ions which are known to form LDH layers [2,3]. Although they are unable to form a stable binary LDH with one other metal ion, almost all of the divalent and trivalent ions in the fourth and the fifth periods, and some in the sixth period of the periodic table may be introduced into the hydroxide layers to a small extent in combination with a third (and sometimes fourth) ion to form ternary or quaternary LDHs. For example, the molar ratio of Cu^{II} to the other divalent metal ion should generally be no more than 1, in order to generate a pure LDH structure [356–359], although there are exceptions as noted earlier [52,53].

In general [2,3], the M^{II}/M^{III} ratio of a pure LDH phase ranges from 2 to 4, i.e., $0.25 < x < 0.33$. When $x < 0.1$ or $x > 0.33$, $M^{II}(OH)_2$ or $M^{III}(OH)_3$ impurity phases will be precipitated, respectively [58,59,77,360]. Wider or narrower ranges of x for which pure phase LDHs can be synthesized have been suggested for some compositions of the layers, e.g., MgAl–OH-LDHs (0.25–0.44) [361], NiAl–CO_3-LDHs (0.17–0.33), MgAl–ClO_4-LDHs (0.20–0.33) [70], MgAl–CO_3-LDHs (0.10–0.34) [2], and ZnCr–CO_3-LDHs (0.25–0.35) [362]. As discussed above, an increasing number of LDHs containing tetravalent metal cations such as Sn^{IV} and Zr^{IV} have also been reported [229,363]. Fe^{II}, Co^{II}, and Mn^{II} cations are easily oxidized to the corresponding trivalent metal cations, and ternary LDHs containing metal cations with different valences can be obtained by controlling the synthesis conditions: examples include $MgCo^{II}Co^{III}$-LDH [348], $MgMn^{II}Mn^{III}$-LDH [364], and $CoCu^{II}Fe^{III}$-LDH [51].

17.2.2.2. Incorporation of Noble Metal Ions into LDHs

Introducing a small amount of noble metal cations into the layers of a ternary or quaternary LDH should not seriously affect its stability. Basile et al. [365] synthesized a series of MMgAl-LDHs (M = Rh^{III}, Ir^{III}, Ru^{III}, Pd^{II}, and Pt^{II}, 0.04–5.00 at.%) using coprecipitation at constant pH 10.0. Apart from Pt^{II}, the other four noble metal ions quantitatively entered the LDH layers according to their proportion in the synthesis mixture. Qi et al. [366] prepared $NiAlM^{III}$-LDHs ($Ni/(Al + M^{III}) = 4$, M^{III} = Au 1.2 at.%, Ir or Rh 2.4 at.%) by using the same method. Melo et al. [367] prepared ternary $M^{II}MgAl$-LDHs ($M^{II} = Pd^{II}$ or Ru^{II} 0.4 at.%) at pH 13. By calcination and/or reduction, noble metal catalysts can be prepared using these ternary LDH precursors. In these catalysts, the noble metal particles are uniformly dispersed, and the thermal stability and specific surface area of the catalysts are significantly higher than that of similar catalysts prepared by the impregnation method [365]. The resulting loaded noble metal/LDH catalysts were active in reactions such as CO_2 reforming [368], methane oxidation [369], Heck reactions [370], Guerbet condensation [371], hydrogenation [372], and methanol steam reforming for hydrogen production [366].

TABLE 17.2 Combinations of M^{II} and M^{III} Cations which give Stable LDH Structures (The Shaded Boxes)

	Mg^{II}	Fe^{II}	Co^{II}	Ni^{II}	Cu^{II}	Zn^{II}	Ca^{II}	Li^{I}
Al^{III}	■	■	■	■	■	■	■	■
Cr^{III}			■	■	■	■	■	
Fe^{III}	■	■	■		■	■	■	
Co^{III}		■	■	■				
Ni^{III}				■				
Ga^{III}	■	■	■	■	■	■	■	
In^{III}				■			■	
Ti^{IV}			■					

17.2.2.3. Metal Complexes in the Interlayer Galleries

Another strategy to introduce noble metals into LDHs is to intercalate their anionic complexes into the interlayer galleries. Highly dispersed catalysts such as M/$MgAlO_x$ (M = Ru, Rh, Ir) with high catalytic activity and CO selectivity in methane reforming were prepared by calcination/reduction of a variety of M(EDTA) chelate anion-intercalated MgAl–M(EDTA)-LDHs [373]. Carpentier et al. [374] obtained $[PdCl_2(OH)_2]^{2-}$-intercalated MgAl-LDHs using the coprecipitation method at pH 10. MgAlPd-LDHs calcined at 290°C exhibited high catalytic activity and CO selectivity in toluene oxidation. With the same method (pH 8.6), Sabbar et al. [375] obtained materials with co-intercalated metal oxalate complexes such as MgAl–$[Cu(C_2O_4)_2/Pt(C_2O_4)_2]$-LDHs. Francová et al. [372] obtained the co-intercalation of (hydroxycitrato)palladium(II) anion and $CO_3{}^{2-}$ in MgAl–$Pd[(C_6O_7H_3)OH]$–CO_3-LDHs by the anion-exchange method. With the same method, Zhang et al. [376] prepared a rhodium phosphine complex anion-intercalated ZnAl–$[RhCl(CO)(P(m\text{-}C_6H_4SO_3{}^-)_3)_2]$-LDH. Other transition metals can also be intercalated into the interlayer galleries of LDHs, giving materials such as the bis(oxalato)copper(II)-intercalated MgCuAl-LDH, (citrato)nickel(II)-intercalated [377], and (nitriloacetato)zinc(II)-intercalated MgAl-LDH [378].

17.2.3. Control of Mesomorphology of LDHs

17.2.3.1. Control of Particle Size

Controlled rapid nucleation and slow growth is the key to prepare LDHs with small particle size. LDHs prepared by the variable pH method have a small particle size, typically on the nanoscale, due to the fast nucleation resulting from coprecipitation at high supersaturation. LDHs with slightly larger crystallite size can be prepared by the constant pH method due to the slow generation of new nuclei resulting from a low supersaturation state at all times. The SNAS method gives the smallest particle sizes due to the high-speed (almost instantaneous) nucleation step. If LDHs with small particle size are required, crystallization should be carried out at low temperature, and the crystallization time should be limited.

The urea hydrolysis method is an ideal way to prepare large-size LDH crystallites. The pH of the solution gradually increases and eventually reaches a constant value of about 9, which is very suitable for the growth of many different LDHs. The rate of decomposition of urea can be controlled by adjusting the reaction temperature, ensuring that the reaction system is in a state of low supersaturation and slow nucleation, and that most of the reactants are consumed in grain growth, rather than in the formation of new nuclei. This method is suitable for preparation of high charge density (low M^{II}/Al^{III} ratio) LDHs [258]. When the solubility of the hydroxide $M^{II}(OH)_2$ is significantly less than that of $Mg(OH)_2$, LDHs such as ZnAl-LDHs and NiAl-LDHs can be obtained with high layer charge densities by using precursor solutions with low M^{II}/Al ratios. The urea hydrolysis method is not suitable for preparation of LDHs precipitating at relatively high or low pH value (e.g., CoAl-LDHs, MnAl-LDHs, or CoCr-LDHs) [379].

The LDHs produced by hydrothermal synthesis with slightly soluble metal oxides or hydroxides as precursors have a large particle size (a few microns), resulting from heterogeneous nucleation and growth on the surface of the metal oxide or hydroxide. The surface of the metal oxide or hydroxide gradually hydrolyzes to form a local supersaturation. Kovanda et al. [380] subjected Mg_2Al-CO_3-LDHs to hydrothermal treatment at 180°C for 8 h, and the crystallite size of the product increased to 800–1000 nm. In the non-equilibrium crystallization method, the reactants are replenished during the thermal crystallization process which results in a non-equilibrium crystallization and large-size LDH crystallites [255].

17.2.3.2. *Control of Crystallinity*

The degree of crystallinity of LDHs can be controlled by adjusting the supersaturation during the nucleation and crystal growth processes. The main factors affecting the supersaturation degree are temperature, pH value, and reactant concentration. In the variable pH method, a high supersaturation state leads to the generation of new nuclei at a faster rate than crystal growth, which gives products with low crystallinity containing impurities. The constant pH method ensures an appropriate and stable degree of supersaturation, and the M^{II}/M^{III} ratio varies over a narrow range and can be modified within a certain range by regulating the pH value. The optimal pH values for precipitation vary widely. For example, ZnAl-LDHs can be prepared in the pH range 7–9, whereas ZnCr-LDHs and NiCr-LDHs require pH values of 4.5–5 and ~11.5, respectively. Hibino et al. [217] prepared MgAl–CO_3-LDHs using the constant pH method (pH 10, 70°C) without any treatment at later stage, and obtained a very poorly crystalline flaky powder with a grain diameter of 25–50 nm. When precipitated at the same pH and temperature of 40°C, followed by aging at 70°C for 40 h, hexagonal LDH platelets with a size of 40–120 nm were obtained [134]. In general, when several metal ions coexist in solution, the pH value for coprecipitation is different from the respective pH values for the precipitation of each metal ion alone. The pH value to prepare MgAl–OH-LDHs ranges from 7.7 to 8.5 [381], different from that for $Mg(OH)_2$ (9.5) and $Al(OH)_3$ (4.0–4.5) [2]. Table 17.3 shows the pH values required for the coprecipitation of some cations to form LDHs. It is very important for the synthesis of highly crystalline LDHs to select the appropriate ratio of metal ions and anion to be intercalated and their concentrations. Suitable ratios for the value of x are typically in the range 0.17–0.34, and the ratio of the concentration of the intercalated anion to that of M^{III} (A^{n-}/M^{III}) is $(1/n - 1)$ [359]. The concentration of metal ions can range from 1 mmol/L to 1 mol/L, and lower concentrations of metal ions result in larger LDH particle sizes.

TABLE 17.3 The pH Values Required for Coprecipitation of Some LDHs

Metal cations	Interlayer anions	pH	Molar ratio $R = M^{II}/M^{III}$
Zn^{II}, Al^{III}	Cl^-	7.0	$1.0 \leq R \leq 5.0$
Zn^{II}, Al^{III}	Cl^-	10.0	$1.0 \leq R \leq 3.0$
Zn^{II}, Al^{III}	CO_3^{2-}	9.0	$1.7 \leq R \leq 2.3$
Mg^{II}, Al^{III}	CO_3^{2-}	8.0	$1.0 \leq R \leq 3.0$
Ni^{II}, Al^{III}	ClO_4^-	10.0	$1.0 \leq R \leq 3.0$
Zn^{II}, Cr^{III}	Cl^-	4.5	$R \approx 2.0$
Zn^{II}, Cr^{III}	Cl^-	10.0	$2.0 \leq R \leq 3.0$
Ni^{II}, Cr^{III}	Cl^-	11.5	$1.0 \leq R \leq 3.0$
Ni^{II}, Cr^{III}	CO_3^{2-}	13.0	$1.0 \leq R \leq 2.0$
Cu^{II}, Cr^{III}	Cl^-	5.5	$1.6 \leq R \leq 2.3$
Mg^{II}, Fe^{III}	CO_3^{2-}	–	$2.7 \leq R \leq 5.6$
Co^{II}, Fe^{III}	Cl^-	9.0	$1.8 \leq R \leq 4.0$
Co^{II}, Fe^{III}	CO_3^{2-}	9.0	$1.0 \leq R \leq 3.0$

17.2.4. Control of the Macromorphology of LDHs

17.2.4.1. Three-dimensional Framework Structure

With polystyrene (PS) microspheres as the template, Géraud et al. [382–384] prepared open-framework structures of LDHs by coprecipitation. In the SEM image shown in Fig. 17.9, the regular arrangement of macroscopic pores obtained after removal of the template can be clearly observed. The resulting 3D macroporous LDHs exhibit higher activity for the photodegradation of 2,6-dimethylphenol than LDHs prepared by conventional routes.

17.2.4.2. Hollow and Solid Spherical Particles

Recently, Li et al. [385] reported the preparation of hollow spherical particles of LDHs with PS microspheres as the sacrificial template (Fig. 17.10). In the preparation process, the delaminated MgAl-LDH monolayers were assembled with poly(sodium 4-styrene sulfonate) (PPS) using layer-by-layer (LBL) techniques on the PS microspheres to achieve a PS@(LDH/PPS)$_{20}$ composite with a core–shell structure which was calcined at 480°C to remove the organic species and then rehydrated in humid air to reconstruct the LDH structure.

Gunawan et al. [386] prepared dodecyl sulfonate (DS)-intercalated MgAl–DS-LDH microspheres under rapid agitation in an ethylene glycol/methanol mixed solvent. The product had a unique coral-like solid spherical morphology (Fig. 17.11), which is related to the self-assembly of the LDH nanosheets under the action of the solvent mixture.

Wang et al. [387] reported the preparation of LDH solid spheres with diameters of tens of microns by spray-drying methods (Fig. 17.12). The method is simple and can easily be scaled up for industrial production of LDHs.

17.2.5. Fabrication of LDH Films

17.2.5.1. Colloidal Deposition Methods

Various ways of preparation and deposition of LDH colloidal nanoparticles with a narrow size distribution have been devised [388–391]. Itaya et al. [388] ultrasonically agitated LDH particles prepared by a coprecipitation method followed by high-speed centrifugation. The as-prepared LDH nanoparticles homogenously dispersed in the supernatant were deposited on SnO_2 electrodes to form an LDH film with a thickness of ca. 100 nm. Lee et al. [389,390] ultrasonically dispersed LDH nanoparticles treated under hydrothermal conditions in a butanol/isopropanol mixture, and then fabricated compact monolayer LDH films on an Si substrate (Fig. 17.13). 1,10-Sebacate anions may be intercalated into the LDH film by a solvothermal anion-exchange reaction, to give a hybrid coating. With a coprecipitation method, Roto et al. [391] and Qiu and Villemure [392] prepared NiAl-LDHs and NiFe-LDHs, which were deposited on the surface of an Au electrode to form a type of clay-modified electrode.

17.2.5.2. Solvent Evaporation Method

The *solvent evaporation method* is a common colloidal deposition method for the preparation of LDH films [235,393,394]. Gardner et al. [235] reported that when alkoxide-intercalated LDH gels with a particle size of 8–10 nm were subjected to solvent evaporation at room temperature, continuous transparent LDH films were formed. Iyi et al. [395] coated exfoliated acetate-intercalated MgAl-LDH colloids on a substrate to form

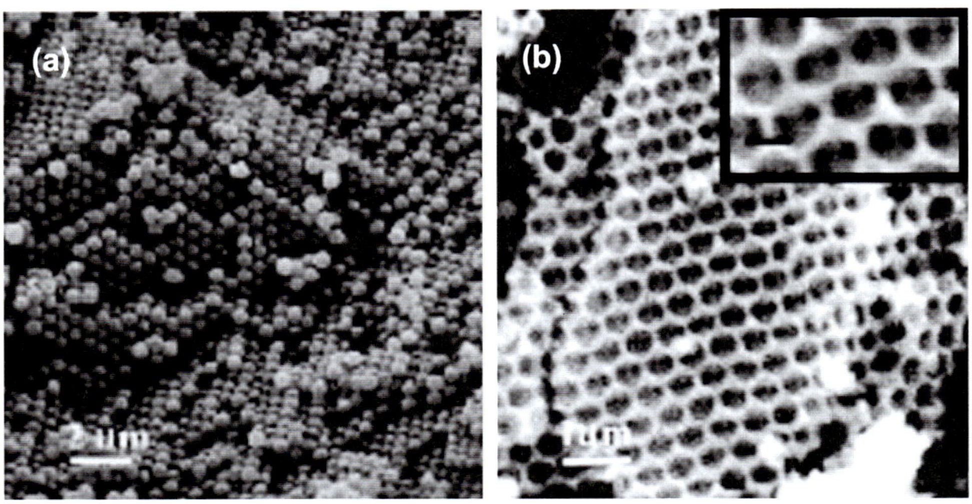

FIGURE 17.9 SEM micrographs of LDH-PS composite materials (a), and the resulting macroporous LDH obtained after removal of the PS template (b). *Reprinted with permission from [383]. Copyright 2006 American Chemical Society.*

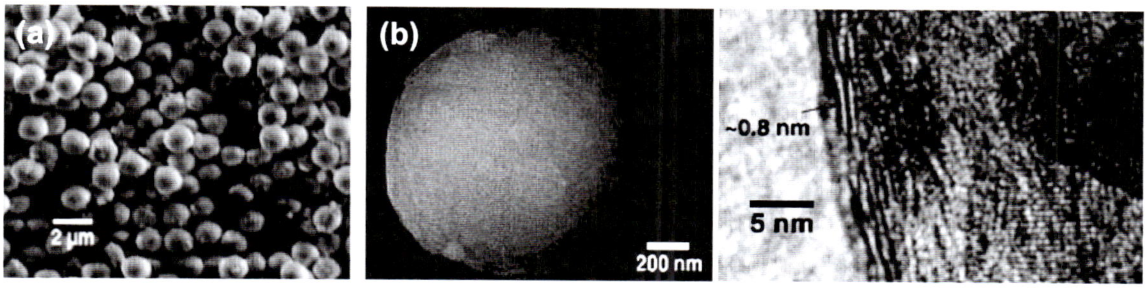

FIGURE 17.10 SEM micrographs of PS@(LDH/PPS)$_{20}$ (left, center) and a high-resolution TEM micrograph of the LDH hollow shell (right). *From* [385]. *Reproduced with permission from The Royal Society of Chemistry.*

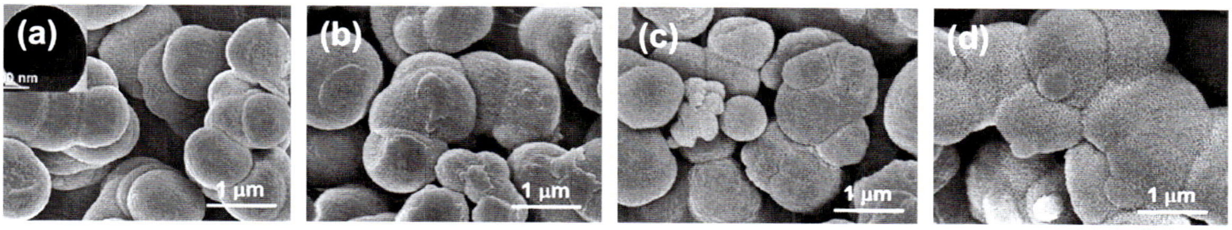

FIGURE 17.11 SEM micrographs of MgAl−DS-LDHs prepared under different crystallization conditions: (a) 100 °C, 18 h; (b) 120 °C, 18 h; (c) 150 °C, 4 h; and (d) 150 °C, 8 h. *[386] Reproduced by permission of The Royal Society of Chemistry.*

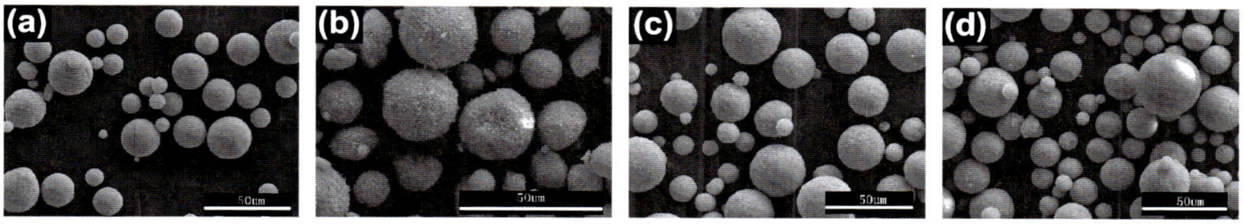

FIGURE 17.12 SEM micrographs of LDH solid spheres with different compositions produced by spray-drying: (a) NiAl-LDHs, (b) ZnAl-LDHs, (c) CuZnAl-LDHs ((Cu + Zn):Al = 2) and (d) CuZnAl-LDHs ((Cu + Zn):Al = 3). *Reprinted with permission from [387]. Copyright 2008 American Chemical Society.*

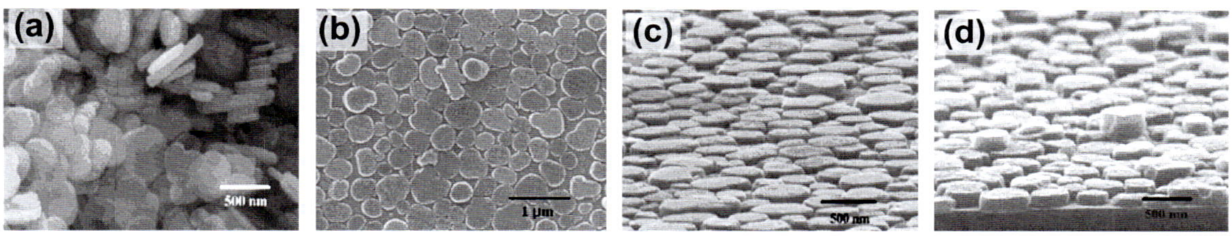

FIGURE 17.13 (a) MgAl-LDH powder; (b, c) MgAl-LDH monolayer film; (d) MgAl-LDH film after intercalation by 1,10-sebacic acid. *Reprinted with permission from [389]. Copyright 2004 American Chemical Society.*

an MgAl−AcO-LDH film after vacuum drying. Self-supporting MgAl−AcO-LDH films (areas of up to 10−20 cm^2, and thickness of 10−25 μm), with a certain extent of flexibility, could be obtained by exfoliation from the hydrophilic substrates such as polyethylene and polypropylene. Wang et al. [396] prepared a suspension of colloidal LDH nanoparticles using the SNAS method, which after deposition on glass afforded a transparent and continuous LDH thin film with high degree of orientation. Li et al. [397] subsequently showed that large area and continuous composite metal oxide films and macroporous spinel films with a (111) orientation could be obtained by calcination of these (00*l*)-oriented NiAl−NO_3-LDH films at different temperatures.

17.2.5.3. *Layer-by-layer Assembly Method*

Electrostatic layer-by-layer (LBL) assembly is one of the most widely applied technologies to fabricate ultrathin film (UTF). UTFs can be fabricated on various shapes and sizes of substrates, and the film composition and thickness can be controlled [398–402]. The positively charged host layers of LDHs can be exfoliated to obtain dispersed nanosheets [403], which are ideal building blocks for constructing composite UTFs. LDH nanosheets have been deposited alternately with polystyrene sulfonate (PSS), or polyacrylic acid (PAA), on a negatively charged substrate, resulting in LDH/polyelectrolyte hybrid UTFs. Liu et al. [404] obtained (CoAl-LDH/PSS) composite UTFs showing a magneto-optical effect by LBL assembly. $(Mg_{2/3}Al_{1/3}(OH)_2/Ti_{0.91}O_2)_n$ and $(Mg_{2/3}Al_{1/3}(OH)_2/Ca_2Nb_3O_{10})_n$ inorganic composite UTFs were also prepared by LBL self-assembly [405]. Zhang et al. [406] assembled the positively charged LDH nanosheets obtained by exfoliating a lactic acid-intercalated LDH, with the negatively charged layered MnO_2 nanosheets to form $(LDH/MnO_2)_n$ composite UTFs. Wang et al. [407] deposited exfoliated CoAl-LDH nanosheets on indium tin oxide (ITO) glass to obtain an electrode material. Yan et al. [408] used the electrostatic LBL method to assemble sulfonated derivatives of poly(*p*-phenylene) (APPP) and poly(*p*-phenylenevinylene) (APPV) with exfoliated MgAl-LDH nanosheets in order to fabricate APPP/LDH UTFs showing blue luminescence and APPV/LDH UTFs showing orange luminescence (Fig. 17.14). The presence of inorganic LDH monolayers improves the thermal and optical stability of the interlayer anionic polymers. Yan et al. also demonstrated an example of

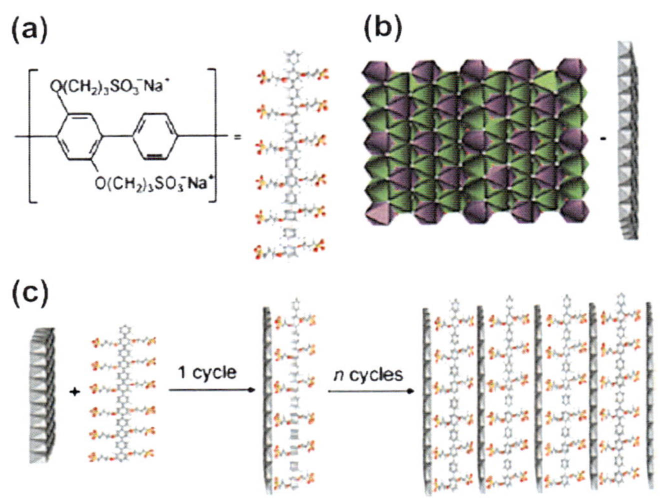

FIGURE 17.14 (a) The chemical formula of APPP. (b) A representation of one sheet of an Mg–Al layered double hydroxide (Mg–Al-LDH; purple: $Al(OH)_6$ octahedra; green: $Mg(OH)_6$ octahedra). (c) Assembly process for $(APPP/LDH)_n$ UTFs.

ordered periodically stacked UTFs based on the anionic complex tris(1,10-phenanthroline-4,7-diphenylsulfonate)ruthenium(II), $[Ru(dpds)_3]^{4-}$, and LDH monolayers by using the LBL method. The stepwise and regular deposition procedure resulted in almost equal amounts of $[Ru(dpds)_3]^{4-}$ being incorporated in each cycle [157].

17.2.5.4. *In Situ Growth Method (surface synthesis)*

In the *in situ growth method*, the substrate not only serves as a matrix for attaching the LDH particles, but is also directly involved in the reaction as a precursor for the LDH. The resulting LDH film is linked to the substrate by strong chemical bonding. In 1983, Mok et al. [409] synthesized NiAl-LDHs by in situ synthesis on the surface of α-Al_2O_3 with urea or ammonia as a precipitator. Schaper et al. [410] obtained NiAl-LDHs in the pores of an Al_2O_3 support. Clause et al. [411,412] prepared CoAl-, NiAl-, or ZnAl-LDHs on γ-Al_2O_3 by regulating the pH value through bubbling ammonia into the solution. Mao et al. [413] and Zhang et al. [414] prepared NiAl-LDHs in situ on the surface and in the pores of spherical γ-Al_2O_3.

The in situ growth of LDH films can be achieved on an anodized aluminum surface. Anodizing leads to a coating of porous anodic alumina (PAO) with regular vertical channels. The $Al(OH)_4^-$ ions adsorbed on the alumina surface constitute the active centers, which induce the heterogeneous nucleation and growth of the LDH on the surface. The in situ growth of LDH on the alumina surface ensures a strong chemical interaction between the alumina and LDH films. Using such in situ growth techniques, Chen et al. [415] showed that an anodized aluminum matrix acted as both substrate and aluminum source in the preparation of a highly oriented NiAl–CO_3-LDH film with the (00*l*) planes of the LDH perpendicular to the substrate (Fig. 17.15). LDH films with controllable nano- and micrometer composite structures can be achieved. Zhang et al. [416] prepared a ZnAl–NO_3-LDH precursor film on the same PAO/Al substrate. Laurate (La) anions can be intercalated into the ZnAl-LDH interlayer galleries to afford ZnAl–La-LDH thin films with superhydrophobic properties (Fig. 17.16). The resulting LDH film gives a high corrosion resistance. Standard tests showed that the film is strongly bound to the Al matrix, so that the film has the required long-term stability for practical application as a corrosion resistant coating.

Leggat et al. [275,417] prepared an $[LiAl_2(OH)_6]NO_3$ thin film on an aluminum alloy surface. Gao et al. [418] fabricated ZnAl-LDH films on a glass substrate coated with an aluminum layer. Lei et al. [419] obtained LDH thin films by hydrothermal in situ crystallization on

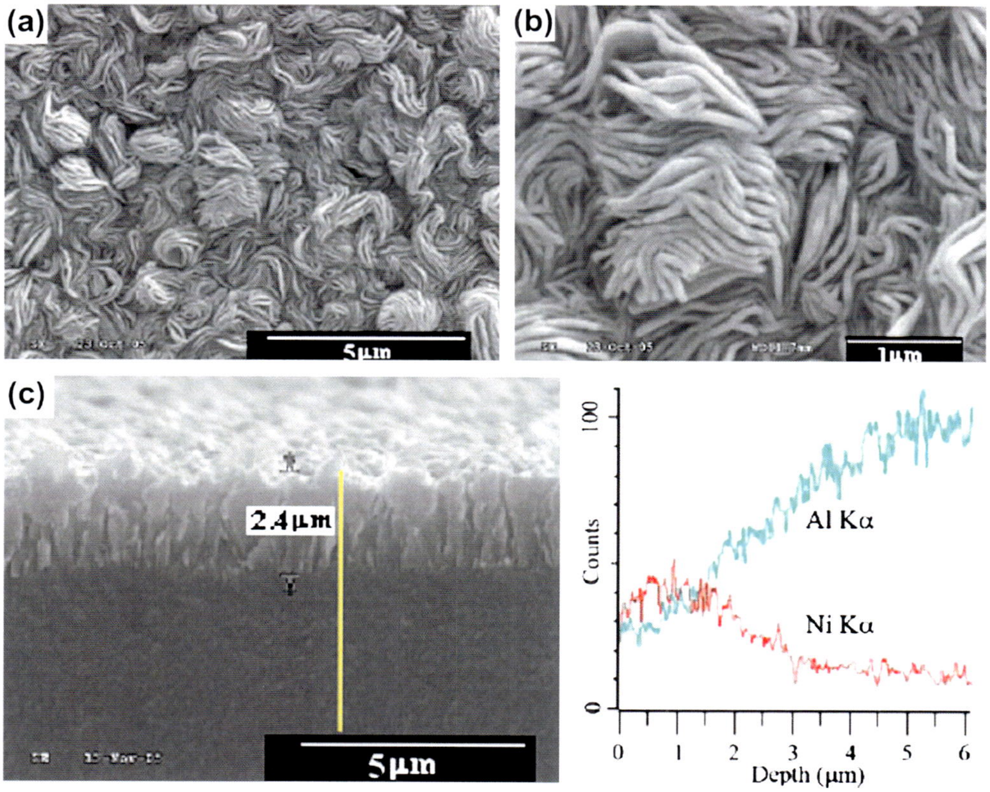

FIGURE 17.15 SEM micrographs of the in situ growth of an NiAl-LDH film on PAO/Al: (a) top view, (b) enlarged view of (a), (c) cross-section view and the corresponding depth-scanning energy dispersive spectra (EDS).

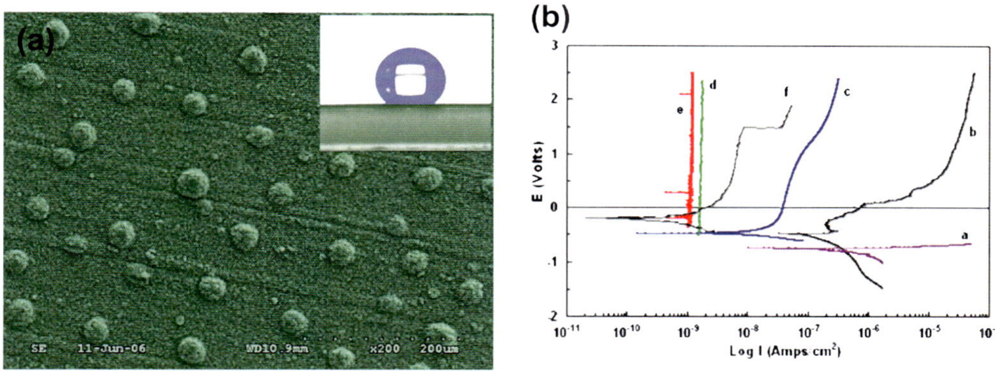

FIGURE 17.16 (a) SEM image of the ZnAl–La-LDH/Al film and the contact angle photograph of a water droplet on the surface; (b) the polarization curve (vs SCE) of the ZnAl–La-LDH/Al film in 3.5% NaCl aqueous solution at room temperature for 30 min: (a) bare Al substrate, (b) PAO/Al substrate, (c) ZnAl–NO_3-LDH film, d) PAO/Al–La film, e) ZnAl–La-LDH film, f) ZnAl–La-LDH film after immersion for 21 days.

the surface of sulfonated polystyrene. Lei et al. [420] also reported the fabrication of uniformly distributed LDH thin films with an upright platelet morphology on the surface of sulfonated polystyrene. The as-prepared LDH thin film has strong adhesion to the substrate and a high degree of orientation (Fig. 17.17). The mechanism of the growth of MgAl–CO_3-LDH film on the negatively charged surface of the polystyrene substrate has been analyzed in detail [420]. The growth of the film can be divided into three stages: the adsorption and enrichment of metal ions, supersaturation and nucleation of the LDH, and the gradual growth of LDH nuclei. The surface sulfonate groups play an important role in both the heterogeneous nucleation and the film growth of the LDH on the surface of the substrate. The sulfonate group becomes the interlayer anion of the LDH layers

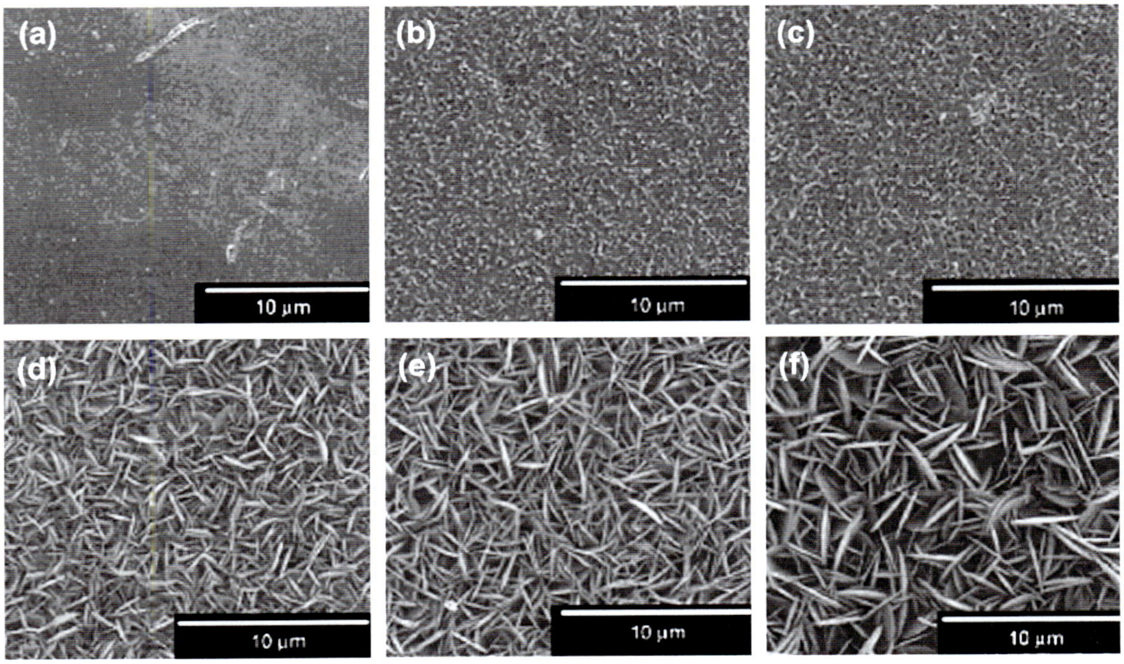

FIGURE 17.17 SEM micrographs showing the in situ growth of an MgAl-LDH thin film on a polystyrene substrate: (A) 1 h; (B) 24 h; (C) 48 h; (D) 72 h; (E) 120 h; and (F) 168 h. [420]. *Reprinted from Chemical Engineering Science 62 (2007). Zhi Lü, Fazhi Zhang, Xiaodong Lei, Lan Yang, David G. Evans, Xue Duan. Microstructure-controlled synthesis of oriented layered double hydroxide thin films: Effect of varying the preparation conditions and a kinetic and mechanistic study of film formation, pages 6069–6075, Copyright (2007), with permission from Elsevier.*

and is the bridge between the polystyrene matrix and the LDH and ensures a strong interaction. However, this preparation method for preparing LDH films gives relatively low yields, because of the homogeneous precipitation of a large amount of LDH powder.

Both LDHs and their calcined products show good catalytic performances in a variety of reactions such as epoxide polymerization, aldol condensations, alkoxylation reactions, and transesterification reactions. Lü et al. [421] prepared an MgAl-LDH film on an anodic aluminum oxide (AAO)/Al substrate by this method. After a calcination/rehydration process, the film showed excellent catalytic activity in the self-condensation reaction of acetone (Fig. 17.18). The method provides a new approach to integrate LDHs as an active component into a structured catalytic reactor.

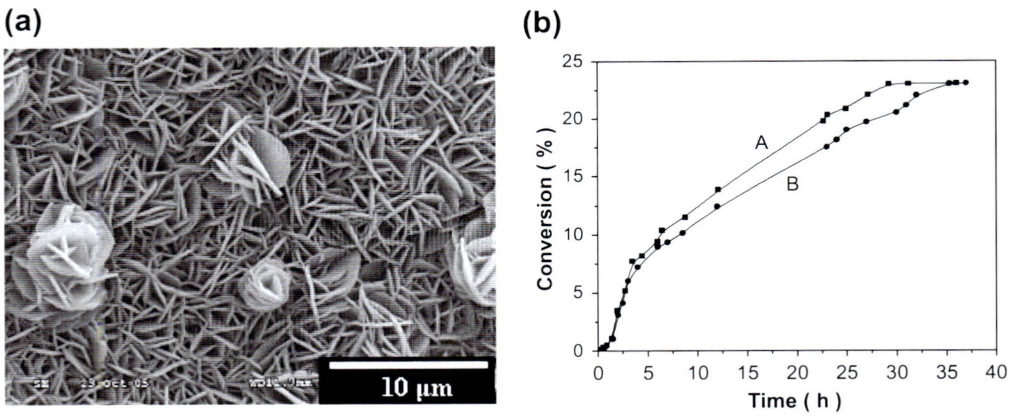

FIGURE 17.18 (a) SEM image of the product obtained after rehydration of the MgAl–CO_3-LDH film calcined at 500 °C for 8 h, and (b) the conversion rate in the self-condensation of acetone at 0 °C with freshly activated (A), and five-times cycled (B) rehydrated LDH/AAOs (acetone/LDH films = 2 mol/0.5 m^2) as a catalyst [421]. *Reprinted from Chemical Engineering Science 63 (2008), Zhi Lü, Fazhi Zhang, Xiaodong Lei, Lan Yang, Sailong Xu, Xue Duan, In situ growth of layered double hydroxide films onanodic aluminum oxide/aluminum and its catalytic feature in aldol condensation of acetone, pages 4055–4062, Copyright (2008), with permission from Elsevier.*

17.2.5.5. *Langmuir–Blodgett (L–B) Method*

Langmuir–Blodgett technology [422–427] is based on the adsorption and arrangement of amphiphilic molecules at a gas–liquid interface, and molecular combinations with a specific spatial order can be realized at the substrate surface [422,423,428]. He et al. obtained amphiphilic Ru^{II} complex/LDH composite L–B films at a gas–liquid interface and on the surface of mica [426]. He et al. [427] also deposited stearate/MgAl–CO_3-LDHs films on a mica substrate.

17.2.5.6. *Spin-coating Method*

The spin-coating method is a simple, convenient, and rapid method for the preparation of uniform films with large area and varying thickness on any flat substrate. Zhang et al. [429] deposited exfoliated colloidal LDH nanoparticles on a magnesium alloy surface by this method and the thickness of the LDH film increased linearly with increasing concentration of the LDH colloid.

17.3. ASSEMBLY CHEMISTRY OF ANION-INTERCALATED LAYERED MATERIALS

17.3.1. Intercalation–assembly Concepts

17.3.1.1. *Selection of Intercalated Guests*

In the LDH structure, the host layers are positively charged, and must be balanced by the interlayer anions to maintain charge neutrality. LDHs have the ability to incorporate simple organic (e.g., carboxylate, sulfonate anions), simple inorganic, hetero-/iso-polyoxometalates (e.g., $Mo_7O_{24}{}^{6-}$, $V_{10}O_{28}{}^{6-}$, $PW_{11}CuO_{39}{}^{6-}$, $SiW_9V_3O_{40}{}^{7-}$, etc.), metal complex anions (e.g., $Zn(BPS)_3{}^{4-}$, $Ru(BPS)_3{}^{3-}$ (BPS = 4,7-diphenyl-1,10-phenanthrolinedisulfonate), etc.), polymer anions (e.g., polyacrylic acid, polyvinyl sulfonate, polyaniline sulfonate, etc.), and even biological guests, such as drug molecules, amino acids, nucleotides, DNA, di-peptides, enzymes and proteins with appropriate charge density and ionic volume. The ion-exchange capacity (IEC) of anions has a positive correlation with their charge density; when the charge density is the same, the smaller the diameter of the anion, the larger the IEC. For instance, some bulky organic dye anions with sulfonate groups can still be intercalated into LDHs by the ion-exchange method due to their high charge densities [430].

17.3.1.2. *Driving Force for Intercalation–assembly*

The main driving force for intercalation–assembly is the host–guest interaction between the hydroxyl layers and interlayer anions. In early studies, the intercalation–assembly process was mainly regarded as being driven by the host–guest electrostatic interactions. Since the LDH host layer is positively charged, in theory all anions (including organic, inorganic, and biological anions) can interact with the LDH through electrostatic interaction. For organic anions, the protonation which takes place under acidic conditions will reduce the electrostatic interaction between the anion and LDH layers, making the organic anions de-intercalate from the interlayer galleries.

There is an extensive network of hydrogen-bonding interactions between the guest anions and water molecules in the interlayer galleries and/or between the guest anions and the metal hydroxyl groups of the layers [431]. For example, H_2O molecules and O atoms of $CO_3{}^{2-}$ anions are distributed near the axis of the hydroxyl group of the adjacent layers [2] and $CO_3{}^{2-}$ and –OH are connected by hydrogen bonding [8,432]. The thickness of the interlayer hydrogen-bonding zone is reported to be ca. 0.27 ± 0.1 nm [41], and this value has often been used to derive a model of an intercalated structure [247,433]. Studies have shown that covalent bonds can form between the LDH hydroxyl layers and the guest anions, which thus become grafted to the layers. For example, ZnAl-LDHs can be structurally recovered in aqueous solution from calcined LDHs, and then become hydrophobic after surface modification through esterification [317,434].

17.3.2. LDHs as Molecular Vessels and Molecular Reactors

17.3.2.1. *Molecular Vessels*

LDHs are biocompatible and have been used as a matrix for stabilization and controlled release of pharmaceuticals such as antipyretic, analgesic, and anti-inflammatory drugs [435]. In general, most biological molecules and drugs are chiral molecules, and only one chiral enantiomer has the desired performance and efficacy, while the other is ineffective and may even have undesirable side-effects. Therefore, it is important to preserve the chiral characteristics of these substances. When chiral guest anions are intercalated into the interlayer galleries of LDHs, their racemization can be inhibited significantly, and their thermo- and photostability can be enhanced greatly due to the host–guest and guest–guest interactions. Furthermore, the host layers of LDHs can block UV irradiation, which further enhances the photostability and preserves the optical activity of guest anions. Therefore, the intercalated LDH systems can be employed as a "molecular vessel" to enhance the stability of guest molecules with respect to racemization and decomposition by protecting them from the environment [223,273]. L-Tyrosine (4-hydroxyphenylalanine, represented as L-Tyr) is a non-essential amino acid, the lack of which in humans

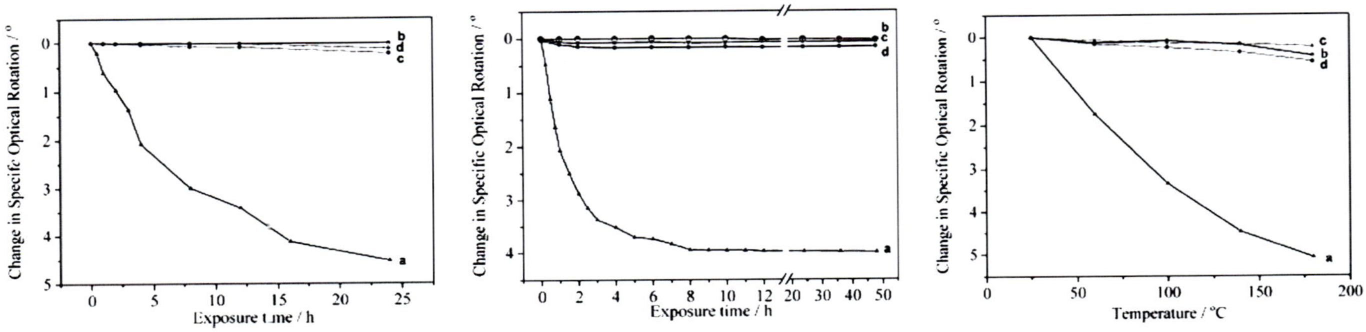

FIGURE 17.19 Left plot: The relationship between optical activity and exposure time to sunlight. (a) *L*-tyrosine; (b) NiAl−*L*-tyrosine-LDH; (c) MgAl−*L*-tyrosine-LDH; (d) ZnAl−*L*-tyrosine-LDH. Middle plot: the relationship between optical activity and time of exposure to heat. (a) *L*-tyrosine; (b) NiAl−*L*-tyrosine-LDH; (c) MgAl−*L*-tyrosine-LDH; (d) ZnAl−*L*-tyrosine-LDH. Right plot: The relationship between optical activity and temperature. (a) *L*-tyrosine; (b) NiAl−*L*-tyrosine-LDH; (c) MgAl−*L*-tyrosine-LDH; (d) ZnAl−*L*-tyrosine-LDH [436]. *Reproduced by permission of The Royal Society of Chemistry.*

can lead to the symptoms of dysthymia and mental depression. We have studied the influence on the rate of racemization when L-Tyr is intercalated in NiAl-, MgAl-, and ZnAl-LDHs by the coprecipitation method [436]. Compared with free L-Tyr molecules, the rate of racemization of the intercalated L-Tyr was significantly reduced on exposure to sunlight, ultraviolet irradiation, or high temperatures, as shown in Fig. 17.19. LDHs can therefore effectively preserve the properties and performance of the intercalated guest species, and can be regarded as "molecular vessels".

Generally, to have maximum performance, drug molecules should be continuously transported to the target site with a controlled rate, and should be administered in a form which maintains an appropriate therapeutic drug dose over a period of time [437]. Intercalation of drug molecules (e.g., diclofenac sodium, ibuprofen, naproxen, etc.) in LDHs affords an effective controlled release system, in addition to stabilizing the drug during storage [271,438,439].

Various other studies have shown how the thermal stability of a species may be increased by intercalation in LDHs. Wei et al. intercalated naproxen drug molecules (Fig. 17.20A) into the interlayers of MgAl-LDHs by the ion-exchange method [270]. The intercalated naproxen began to decompose at 250°C and underwent a structural change at 300°C. In contrast, the free naproxen decomposed at 170°C.

Meng et al. [440] have intercalated the herbicide glyphosate anion (gly, Fig. 17.20B) in Mg-Al LDHs by a coprecipitation method, in order to obtain a glyphosate-release agent. When the intercalated MgAl−gly-LDHs was exposed to the atmosphere, the intercalated glyphosate was gradually deintercalated from the interlayer galleries because the ubiquitous carbonate ions have stronger affinity for LDHs than glyphosate, and result in a slow release of the glyphosate. Zhang et al. [441] intercalated the antihypertensive drug, captopril (Fig. 17.20C) in MgAl-LDHs. The thermal stability of the intercalated product was enhanced significantly compared with the free drug, due to the host−guest interactions in the LDH intercalate.

The molecular vessel characteristics of LDHs are also highlighted when a second host species is intercalated in the interlayer galleries, when the recognition and interaction between the host and guest molecules can take place within a limited space, and the stability and selectivity of the host molecule are enhanced. Cyclodextrin is a cyclic oligosaccharide that contains a donut-shaped non-polar ring. Cyclodextrin and its derivatives can accommodate various organic or inorganic guest molecules in their cavity [442−444], and form inclusion compounds. Wei et al. [269] studied the molecular structure and thermal decomposition behavior of the sulfonated cyclodextrin (CD-SO_3^-) intercalated in LDHs. The CD-SO_3^- anions aligned to form a single-layer, with the direction of their cavity axis perpendicular to the host layer of LDHs and the sulfonate groups of adjacent CD moieties alternately interacting with the upper and lower host layer of the LDH. The thermostability of the intercalated CD-SO_3^- was significantly enhanced, with its combustion temperature being increased by about 250°C compared to the free CD-SO_3^- salt. Cyclodextrin and its derivatives intercalated in LDHs can be used as a new stationary phase for liquid chromatography [445], which has a high degree

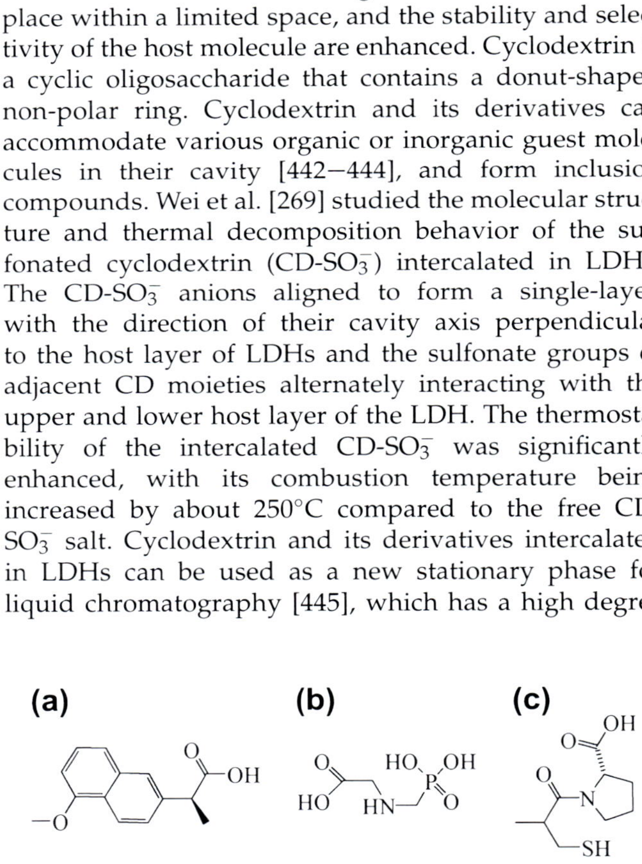

FIGURE 17.20 Molecular structrues of (A) naproxen, (B) glyphosate, (C) captopril

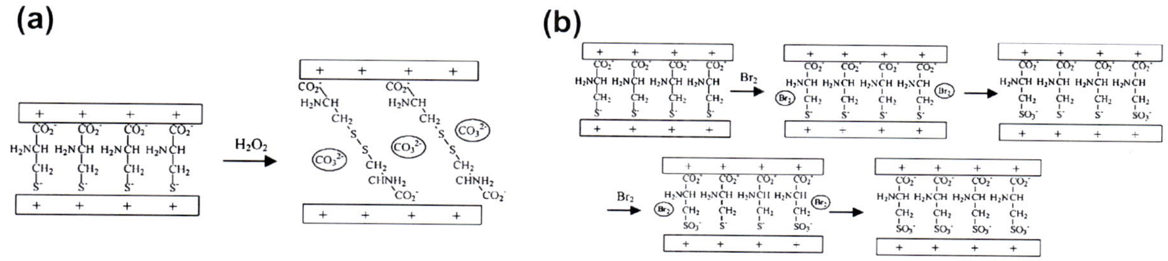

FIGURE 17.21 Oxidation of L-cysteine intercalated in an LDH host by H_2O_2 (A) and Br_2 (B). *Reprinted with permission from [446]. Reproduced with permission from The Royal Society of Chemistry*

of chiral selectivity and low leaching. This is due to the interlayer confinement effect of the LDHs, the rigidity and stability of the layers and the molecular recognition properties of the intercalated CD-SO_3^- guest, which can enhance its working temperature range and enable it to be applied to a wider range of chiral analysis problems.

DNA was successfully intercalated into Mg_2Al-NO_3-LDH through an anion-exchange reaction to form a DNA−LDH nanohybrid as a DNA vessel [272]. Circular dichroism (CD) spectroscopic studies indicate that the B-type DNA molecules are electrostatically bound in the interlayer of LDHs, and the intercalated DNA molecules are more or less twisted due to the charge mismatch between anionic DNA and cationic LDH. Since the as-prepared LDH particles (80, 150, 300 nm) are smaller in size than the DNA molecules, some parts of the intercalated DNA chains dangle outside the host LDH layer, and are decomposed quickly by DNase I. The remaining DNA strands protected by LDH layers can be recovered by treating with an acidic solution to dissolve the LDH host.

17.3.2.2. Molecular Reactors

L-Cysteine and L-cysteine intercalated MgAl-LDHs have been prepared by coprecipitation. It was found that L-cysteine is oxidized to L-cystine by some oxidants (e.g., H_2O_2), just as in solution, whereas the intercalated L-cysteine was oxidized by Br_2 to the single product of cysteic acid, irrespective of the amount of oxidant, which is contrast to the case in solution, where the oxidation product of L-cysteine is related to the consumption of Br_2. This discrepancy is closely related to interlayer diffusion process of Br_2 molecules, which is the rate-limiting step of the oxidation reaction. The proposed reaction mechanism is shown in Fig. 17.21 [446].

Wei et al. [447] intercalated methyl methacrylate (MMA) into the interlayer of MgAl-LDHs by a coprecipitation method and heated the intercalated product to obtain the polyanion-intercalated LDH by in situ thermal polymerization in the interlayer galleries (see Fig. 17.22).

Under N_2 protection, the intercalated *m*-aminobenzenesulfonate (*m*-$NH_2C_6H_4SO_3^-$) anion can be thermally polymerized in situ at 300°C, which results in polyaniline (PANI)/LDH nanocomposites [448]. After polymerization at 300°C, the UV−visible absorption spectra of the composites showed a strong absorption band at 296 nm corresponding to the $\pi_B-\pi_B*$ transition absorption of the benzene-type polyaniline. After polymerization at 320°C, the $\pi_B-\pi_Q*$ absorption of the quinone-type of PANI appeared at 550 nm which showed that the intercalated polyaniline existed in its highest oxidation state. Semi-empirical PM3 calculations suggested that two types of dimer structure, namely (α, β) and (*N*, *N*) (Fig. 17.23), are closest to the experimental results. Shichi [449] intercalated the vinyl benzoate, *p*-benzene diacrylate into the interlayer of LDHs by the anion-exchange method, and achieved interlayer polymerization of the organic anions.

A wide variety of methods are available for the synthesis of LDHs. For example, hydrothermal synthesis gives pure phases with high crystallinity allowing

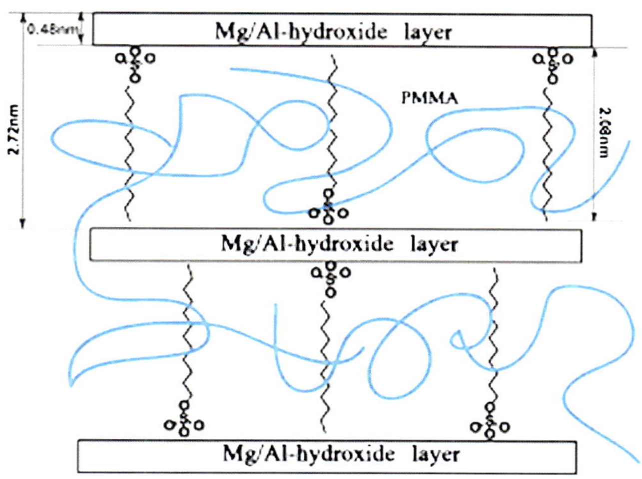

FIGURE 17.22 The in situ polymerization product of PMMA/MgAl-LDHs. [447]. *Reprinted from Solid State Communications 130 (2004), Wei Chen, Li Feng, Baojun Qu, In situ synthesis of poly(methyl methacrylate)/MgAl layered double hydroxide nanocomposite with high transparency and enhanced thermal properties, pages 259−263, Copyright (2004), with permission from Elsevier.*

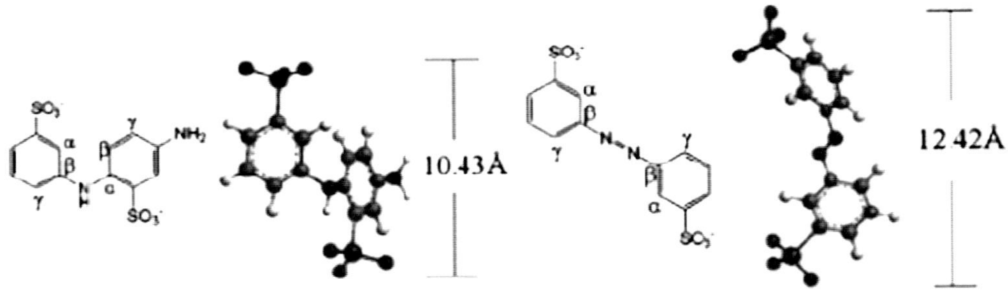

FIGURE 17.23 Models of the polymerization of *m*-aminobenzenesulfonate intercalated in LDHs. Left (α, β-), right (*N*, *N*-) linkage.

accurate structural information to be obtained, whereas coprecipitation is generally more appropriate for preparing LDHs for use as catalysts or catalyst precursors. Colloidal deposition, solvent evaporation, and layer-by-layer assembly methods can be employed for the fabrication of LDH films. The structure of LDHs allows them to be used as "molecular vessels" and "molecular reactors".

LDHs are one of the most technologically promising inorganic layered materials, due to the microscopic controllability of their layer chemical composition, the microscopic controllability of the nature and amount of interlayer anions, and the mesoscopic controllability of their crystallite size and size distribution. Even though much work of academic and commercial interest on LDH materials has been carried out, still more problems remain to be solved in order to fully exploit their potential applications.

References

[1] X. Duan, F.Z. Zhang, M. Pu, H. Yan, J. Lu, L. Jin, et al., Wuji Chaofenzi Cailiao de Chaceng Zuzhuang Huaxue, Science Press, Beijing, 2009.
[2] F. Cavani, F. Trifirò, A. Vaccari, Catal. Today 11 (1991) 173–301.
[3] V. Rives, in: Layered Double Hydroxides – Present and Future, Nova Science Publishers, New York, 2001.
[4] A.I. Khan, D. O'Hare, J. Mater, Chem. 12 (2002) 3191–3198.
[5] R.V. Gaines, H.C.W. Skinner, E.E. Foord, B. Mason, A. Rosenzweig, Dana's New Mineralogy, eighth ed., Wiley, New York, 1997.
[6] R. Allmann, Acta Crystallogr. B 24 (1968) 972–977.
[7] H.F.W. Taylor, Miner. Mag. 37 (1969) 338–342.
[8] H.F.W. Taylor, Miner. Mag. 39 (1973) 377–389.
[9] D.G. Evans, R.C.T. Slade, Struct. Bond. 119 (2005) 1–87.
[10] M. Catti, G. Ferraris, S. Hull, A. Pavese, Phys. Chem. Miner. 22 (1995) 200–206.
[11] C. Greaves, M.A. Thomas, Acta Crystallogr. B 42 (1986) 51–55.
[12] R.C. Peterson, R.J. Hill, G.V. Gibbs, Can. Mineral. 17 (1979) 703–711.
[13] G.W. Brindley, C.C. Kao, Phys. Chem. Miner. 10 (1984) 187–191.
[14] J.B. Parise, K. Leinenweber, D.J. Weidner, K. Tan, R.B. von Dreele, Am. Mineral. 79 (1994) 193–196.
[15] T. Nagai, T. Hattori, T. Yamanaka, Am. Mineral. 85 (2000) 760–764.
[16] K. Shinoda, N. Aikawa, Phys. Chem. Miner. 25 (1998) 197–202.
[17] A.S. Bookin, V.A. Drits, Clays Clay Miner. 41 (1993) 551–557.
[18] A.S. Bookin, V.I. Cherkashin, V.A. Drits, Clays Clay Miner. 41 (1993) 558–564.
[19] V.A. Drits, A.S. Bookin, Chapter 2 in [3].
[20] M. Vucelic, G.D. Moggridge, W. Jones, J. Phys. Chem. 99 (1995) 8328–8337.
[21] S.P. Newman, W. Jones, P. O'Connor, D.N. Stamires, J. Mater. Chem. 12 (2002) 153–155.
[22] C. Vaysse, L. Guerlou-Demourgues, A. Demourgues, F. Lazartigues, D. Fertier, C. Delmas, J. Mater. Chem. 12 (2002) 1035–1043.
[23] L. Guerlou-Demourgues, C. Denage, C. Delmas, J. Power Sources 52 (1994) 269–274.
[24] G.S. Thomas, M. Rajamathi, P.V. Kamath, P. Vishnu, Clays Clay Miner. 52 (2004) 693–699.
[25] T.S. Stanimirova, G. Kirov, E. Dinolova, Mater. Sci. Lett. 20 (2001) 453–455.
[26] C. de la Calle, C.H. Pons, J. Roux, V. Rives, Clays Clay Miner. 51 (2003) 121–132.
[27] G. Parthasarathy, M.L. Kantam, B.M. Choudary, C.V. Reddy, Micropor. Mesopor. Mater. 56 (2002) 147–152.
[28] D.R. Hines, S.A. Solin, U. Costantino, M. Nocchetti, Phys. Rev. B 61 (2000) 11348–11358.
[29] G. Renaudin, J.P. Rapin, E. Elkaïm, M. François, Cement Concrete Res. 34 (2004) 1845–1852.
[30] J.P. Rapin, G. Renaudin, E. Elkaïm, M. François, Cement Concrete Res. 32 (2002) 513–519.
[31] A.M. Fogg, A.J. Freij, G.M. Parkinson, Chem. Mater 14 (2002) 232–234.
[32] A.V. Besserguenev, A.M. Fogg, R.J. Francis, S.J. Price, D. O'Hare, V.P. Isupov, et al., Chem. Mater. 9 (1997) 241–247.
[33] A. Ennadi, A. Legrouri, A. de Roy, J.P. Besse, J. Mater. Chem. 10 (2000) 2337–2341.
[34] S. Guggenheim, D.C. Bain, F. Bergaya, M.F. Brigatti, V.A. Drits, D.D. Eberl, et al., Clays Clay Miner. 50 (2002) 406–409.
[35] A.V. Radha, P.V. Kamath, G.N. Subbanna, Mater. Res. Bull. 38 (2003) 731–740.
[36] C. Delmas, C. Tessier, J. Mater. Chem. 7 (1997) 1439–1443.
[37] M. Rajamathi, P.V. Kamath, R. Seshadri, J. Mater. Chem. 10 (2000) 503–506.
[38] T.N. Ramesh, R.S. Jayashree, P.V. Kamath, Clays Clay Miner. 51 (2003) 570–576.
[39] A.S. Prakash, P.V. Kamath, M.S. Hegde, Mater. Res. Bull. 35 (2000) 2189–2197.
[40] Z.P. Xu, H.C. Zeng, Int. J. Inorg. Mater. 2 (2000) 187–196.
[41] V. Prévot, C. Forano, J.P. Besse, Inorg. Chem. 37 (1998) 4293–4301.
[42] W. Hofmeister, H. von Platen, Cryst. Rev. 3 (1992) 3–29.
[43] A.V. Radha, P.V. Kamath, C. Shivakumara, J. Phys. Chem. B 111 (2007) 3411–3418.

[44] T.R. Park, J. Korean Phys. Soc. 44 (2004) 398–402.
[45] G. Alberti, U. Costantino, Solid state supramolecular chemistry: Two- and three-dimensional inorganic networks, in: G. Alberti, T. Bein (Eds.), Comprehensive Supramolecular Chemistry, vol. 7, Pergamon and Elsevier Science, Oxford, 1996. Chapter 1.
[46] K. Takagi, N. Saito, T. Shichi, Y. Sawaki, Chem. Lett. 28 (1999) 275–276.
[47] S.P. Newman, W. Jones, in: W. Jones, C.N.R. Rao (Eds.), Supramolecular Organization and Materials Design, Cambridge University Press, Cambridge, 2001.
[48] A. de Roy, C. Forano, J.P. Besse, Chapter 1 in [3].
[49] J.T. Kloprogge, D. Wharton, L. Hickey, R.L. Frost, Am. Mineral. 87 (2002) 623–629.
[50] A. Alejandre, F. Medina, P. Salagre, X. Correig, J.E. Sueiras, Chem. Mater. 11 (1999) 939–948.
[51] A.H. Iglesias, O.P. Ferreira, D.X. Gouveia, A.G. Souza Filho, J.A.C. de Paiva, J.M. Filho, et al., J. Solid State Chem. 178 (2005) 142–152.
[52] H. Roussel, V. Briois, E. Elkaïm, A. de Roy, J.P. Besse, J. Phys. Chem. B 104 (2000) 5915–5923.
[53] L. Bigey, C. Depège, A. de Roy, J.P. Besse, J. Physique IV 7 (1997) 949–950.
[54] T. Stumpf, H. Curtius, C. Walther, K. Dardenne, K. Ufer, T. Fanghaenel, Environ. Sci. Tech. 41 (2007) 3186–3191.
[55] H. Curtius, K. Ufer, Clays Clay Miner. 55 (2007) 354–360.
[56] P. Gunawan, R. Xu, J. Phys. Chem. C 113 (2009) 17206–17214.
[57] A.V. Radha, P.V. Kamath, C. Shivakumara, Acta Crystallogr. B 63 (2007) 243–250.
[58] G. Bourrié, F. Trolard, P. Refait, F. Feder, Clays Clay Miner. 52 (2004) 382–394.
[59] M. Vucelic, W. Jones, G.D. Moggridge, Clays Clay Miner. 45 (1997) 803–813.
[60] N. Gutmann, B. Müller, J. Solid State Chem. 122 (1996) 214–220.
[61] A.M. Fogg, G.R. Williams, R. Chester, D. O'Hare, J. Mater. Chem. 14 (2004) 2369–2371.
[62] K. Piotr, W. Agnieszka, R.-L. Alicja, P.-J. Agnieszka, D. Roman, Clays Clay Miner. 53 (2005) 18–27.
[63] D. Kim, C. Huang, H. Lee, I. Han, S. Kang, S. Kwon, et al., Appl. Catal. A-Gen 249 (2003) 229–240.
[64] F. Basile, A. Vaccari, Chapter 10 in [3].
[65] A. Monzón, E. Romeo, A.J. Marchi, Chapter 11 in [3].
[66] U. Chellam, Z.P. Xu, H.C. Zeng, Chem. Mater. 12 (2000) 650–658.
[67] J.J. Liu, F. Li, D.G. Evans, X. Duan, Chem. Commun. (2003) 542–543.
[68] F. Li, J.J. Liu, D.G. Evans, X. Duan, Chem. Mater 16 (2004) 1597–1602.
[69] L. Pauling, J. Am. Chem. Soc. 51 (1929) 1010–1026.
[70] G.W. Brindley, S.M. Kikkawa, Am. Mineral. 64 (1979) 836–843.
[71] A. Vyalikh, D. Massiot, U. Scheler, Solid State Nucl. Magn. Res. 36 (2009) 19–23.
[72] P.J. Sideris, U.G. Nielsen, Z. Gan, C.P. Grey, Science 321 (2008) 113–117.
[73] J.P. Yesinowski, H. Eckert, G.R. Rossman, J. Am. Chem. Soc. 110 (1988) 1367–1375.
[74] U. Sternberg, E. Brunner, J. Magn. Res. Ser. A 108 (1994) 142–150.
[75] J. Jeener, B.H. Meier, P. Bachmann, R.R. Ernst, J. Chem. Phys. 71 (1979) 4546–4553.
[76] C.P. Grey, A.J. Vega, J. Am. Chem. Soc. 117 (1995) 8232–8242.
[77] M. Bellotto, B. Rebours, O. Clause, J. Lynch, D. Bazin, E. Elkaïm, J. Phys. Chem. 100 (1996) 8527–8534.
[78] M. François, G. Renaudin, O. Evrard, Acta Crystallogr. C 54 (1998) 1214–1217.
[79] G. Renaudin, M. François, O. Evrard, Cement Concrete Res. 29 (1999) 63–69.
[80] G. Renaudin, J.P. Rapin, B. Humbert, M. François, Cement Concrete Res. 30 (2000) 307–314.
[81] G. Renaudin, M. François, Acta Crystallogr. C 55 (1999) 835–838.
[82] G. Renaudin, F. Kubel, J.P. Rivera, M. François, Cement Concrete Res. 29 (1999) 1937–1942.
[83] A. Terzis, S. Filippakis, H.J. Kuzel, H. Burzlaff, Z. Kristallogr. 181 (1987) 29–34.
[84] I. Rousselot, C. Taviot-Guého, F. Leroux, P. Léone, P. Palvadeau, J.P. Besse, J. Solid State Chem. 167 (2002) 137–144.
[85] S.A. Solin, D.R. Hines, G.T. Seidler, M.M.J. Treacey, J. Phys. Chem. Solids 57 (1996) 1043–1048.
[86] C.J. Serna, J.L. Rendon, J.E. Iglesias, Clays Clay Miner. 30 (1982) 180–184.
[87] I. Sissoko, E.T. Iyagba, R. Sahai, P. Biloen, J. Solid State Chem. 60 (1985) 283–288.
[88] X. Hou, R.J. Kirkpatrick, Inorg. Chem. 40 (2001) 6397–6404.
[89] M.A. Cooper, F.C. Hawthorne, Can. Mineral. 34 (1996) 623–630.
[90] D.M.C. Huminicki, F.C. Hawthorne, Can. Mineral. 41 (2003) 79–82.
[91] J. Rius, F. Plana, Neues Jahrb. Miner. Monatsh. (1986) 263–272.
[92] S. Merlino, P. Orlandi, Am. Miner. 86 (2001) 1293–1301.
[93] G.Y. Chao, R.A. Gault, Can. Miner. 35 (1997) 1541–1549.
[94] A.V. Arakcheeeva, D.Y. Pushcharovskii, R.K. Rastsvetaeva, D. Atencio, G.U. Lubman, Kristallografiya 41 (1996) 1024–1034. Crystallogr. Rep. 41(1996) 972–982.
[95] C. Ruby, M. Abdelmoula, R. Aissa, G. Medjahdi, M. Brunelli, M. François, J. Solid State Chem. 181 (2008) 2285–2291.
[96] C. Taviot-Guého, F. Leroux, C. Payen, J.P. Besse, Appl. Clay Sci. 28 (2005) 111–120.
[97] H. Roussel, V. Briois, E. Elkaïm, A. de Roy, J.P. Besse, J.P. Jolivet, Chem. Mater. 13 (2001) 329–337.
[98] F. Leroux, M. Adachi-Pagano, M. Intissar, S. Chauvière, C. Forano, J.P. Besse, J. Mater. Chem. 11 (2001) 105–112.
[99] S. Gago, M. Pillinger, A.A. Valente, T.M. Santos, J. Rocha, I.S. Gonçalves, Inorg. Chem. 43 (2004) 5422–5431.
[100] M. Intissar, R. Segni, C. Payen, J.P. Besse, F. Leroux, J. Solid State Chem. 167 (2002) 508–516.
[101] M.J. Hernandez-Moreno, M.A. Ulibarri, J.L. Rendon, C.J. Serna, Phys. Chem. Miner. 12 (1985) 34–38.
[102] J.T. Kloprogge, L. Hickey, R.L. Frost, Appl. Clay Sci. 18 (2001) 37–49.
[103] R.L. Frost, K.L. Erickson, Spectrochim. Acta A 60 (2004) 3001–3005.
[104] T.E. Johnson, W. Martens, R.L. Frost, Z. Ding, J.T. Kloprogge, J. Raman Spec. 33 (2002) 604–609.
[105] R.L. Frost, Z. Ding, W. Martens, T.E. Johnson, J.T. Kloprogge, Spectrochim. Acta A 59 (2003) 321–328.
[106] J.M.R. Génin, C. Ruby, Solid State Sci. 6 (2004) 705–718.
[107] R. Aissa, C. Ruby, A. Gehin, M. Abdelmoula, J.M.R. Génin, Hyperfine Interact. 156/7 (2004) 445–451.
[108] J.W. Boclair, P.S. Braterman, J. Jiang, S. Lou, F. Yarberry, Chem. Mater. 11 (1999) 303–307.
[109] J.W. Boclair, P.S. Braterman, Chem. Mater. 10 (1998) 2050–2052.
[110] R. Allmann, Neues Jahrb. Miner. Monatsh. (1977) 136–144.
[111] E. Kanezaki, J. Incl. Phenom. Macrocycl. Chem. 46 (2003) 89–95.
[112] V. Prévot, C. Forano, J.P. Besse, Appl. Clay Sci. 18 (2001) 3–15.
[113] V. Rives, M.A. Ulibarri, Coord. Chem. Rev. 181 (1999) 61–120.
[114] K.S. Han, L. Guerlou-Demourgues, C. Delmas, Solid State Ionics 84 (1996) 227–238.

[115] M. Ménétrier, K.S. Han, L. Guerlo-Demourgues, C. Delmas, Inorg. Chem. 36 (1997) 2441–2445.
[116] F. Malherbe, L. Bigey C. Forano, A. de Roy, J.P. Besse, J. Chem. Soc., Dalton Trans. (1999) 3831–3839.
[117] C. Vaysse, L. Guerlou-Demourgues, C. Delmas, Inorg. Chem. 41 (2002) 6905–6913.
[118] S. Osama, T. Hideyuki, J. Porous Mater. 16 (2009) 81–89.
[119] E.M. Seftel, E. Popovici, M. Mertens, G. van Tendeloo, P. Cool, E.F. Vansant, Micropor. Mesopor. Mater. 111 (2008) 12–17.
[120] O. Saber, J. Phys. Conf. Ser. 61 (2007) 825–830.
[121] O. Saber, B. Hatano, H. Tagaya, J. Incl. Phenom. Macrocycl. Chem. 51 (2005) 17–25.
[122] O. Saber, J. Mater. Sci. 42 (2007) 9905–9912.
[123] O. Saber, J. Colloid Interf. Sci. 297 (2006) 182–189.
[124] O. Saber, H. Tagaya, Rev. Adv. Mater. Sci. 10 (2005) 59–63.
[125] O. Saber, H. Tagaya, J. Porous Mater. 14 (2007) 481–484.
[126] H. Curtius, K. Ufer, K. Dardenne, Radiochim. Acta 97 (2009) 423–428.
[127] P.V. Kamath, G.H.A. Therese, J. Gopalakrishnan, J. Solid State Chem. 128 (1997) 38–41.
[128] W. Stählin, H.R. Oswald, Acta Crystallogr. B 26 (1970) 860–863.
[129] J.H. Choy, Y.M. Kwon, K.S. Han, S.W. Song, S.H. Chang, Mater. Lett. 34 (1998) 356–363.
[130] J.H. Choy, Y.M. Kwon, S.W. Song, S.H. Chang, Bull. Korean Chem. Soc. 18 (1997) 450–453.
[131] P.M. Forster, M.M. Tafoya, A.K. Cheetham, J. Phys. Chem. Solids 65 (2004) 11–16.
[132] A.H. Treiman, D.R. Peacor, Am. Miner. 67 (1982) 1029–1034.
[133] T. Hibino, A. Tsunashima, Chem. Mater. 9 (1997) 2082–2089.
[134] S.K. Yun, T.J. Pinnavaia, Chem. Mater. 7 (1995) 348–354.
[135] X.Q. Hou, D.L. Bish, S.L. Wang, C.T. Johnston, R.J. Kirkpatrick, Am. Miner. 88 (2003) 167–179.
[136] J. Wang, A.G. Kalinichev, R.J. Kirkpatrick, Geochim. Cosmochim. Acta 68 (2004) 3351–3365.
[137] R.K. Allada, A. Navrotsky, J. Boerio-Goates, Am. Miner. 90 (2005) 329–335.
[138] L. Pesic, S. Salipurovic, V. Markovic, D. Vucelic, W. Kagunya, W. Jones, J. Mater. Chem. 2 (1992) 1069–1073.
[139] S.P. Newman, S.J. Williams, P.V. Coveney, W. Jones, J. Phys. Chem. B 102 (1998) 6710–6719.
[140] V. Mehrotra, E.P. Giannelis, J. Appl. Phys. 72 (1992) 1039–1048.
[141] W. Kagunya, R. Baddour-Hadjean, F. Kooli, W. Jones, Chem. Phys. 236 (1998) 225–234.
[142] J.W. Boclair, P.S. Braterman, B.D. Brister, F. Yarberry, Chem. Mater. 11 (1999) 2199–2204.
[143] J. Inacio, C. Taviot-Guého, S. Morlat-Thérias, M.E. de Roy, J.P. Besse, J. Mater. Chem. 11 (2001) 640–643.
[144] P. Beaudot, M.E. de Roy, J.P. Besse, Chem. Mater. 16 (2004) 935–945.
[145] E. Kanezaki, K. Maeda, Nippon Kagaku Kaishi (2002) 393–397.
[146] P. Beaudot, M.E. de Roy, J.P. Besse, J. Solid State Chem. 177 (2004) 2691–2698.
[147] P. Beaudot, M.E. de Roy, J.P. Besse, J. Solid State Chem. 161 (2001) 332–340.
[148] J. Bauer, P. Behrens, M. Speckbacher, H. Langhals, Adv. Funct. Mater. 13 (2003) 241–248.
[149] T. Itoh, N. Ohta, T. Shichi, T. Yui, K. Tagaki, Langmuir 19 (2003) 9120–9126.
[150] T. Kanoh, T. Shichi, K. Takagi, Chem. Lett. 28 (1999) 117–118.
[151] N. Iyi, K. Kurashima, T. Fujita, Chem. Mater. 14 (2002) 583–589.
[152] P.H. Refait, M. Abdelmoula, J.M.R. Génin, Corros. Sci. 40 (1998) 1547–1560.
[153] L. Simon, M. François, P. Refait, G. Renaudin, M. Lelaurin, J.M.R. Génin, Solid State Sci. 5 (2003) 327–334.
[154] L. Mohanambe, S. Vasudevan, Langmuir 21 (2005) 10735–10742.
[155] L. Mohanambe, S. Vasudevan, J. Phys. Chem. B 109 (2005) 15651–15658.
[156] D. Yan, J. Lu, M. Wei, H. Li, J. Ma, F. Li, et al., J. Phys. Chem. A 112 (2008) 7671–7681.
[157] D. Yan, J. Lu, M. Wei, J. Ma, D.G. Evans, X. Duan, Chem. Commun. (2009) 6358–6360.
[158] V. Rives, Mater. Chem. Phys. 75 (2002) 19–25.
[159] C. Barriga, W. Jones, P. Malet, V. Rives, M.A. Ulibarri, Inorg. Chem. 37 (1998) 1812–1820.
[160] G.D. Moggridge, P. Parent, G. Tourillon, Clays Clay Miner 42 (1994) 462–472.
[161] G.D. Moggridge, P. Parent, G. Tourillon, Physica B: Condens. Matter 209 (1995) 269–270.
[162] J.T. Kloprogge, R.L. Frost, Chapter 5 in [3].
[163] J.T. Kloprogge (Ed.), The Application of Vibrational Spectroscopy to Clay Minerals and Layered Double Hydroxides, Clay Minerals Society, Aurora, 2005.
[164] J.T. Kloprogge, R.L. Frost, J. Solid State Chem. 146 (1999) 506–515.
[165] R.L. Frost, W. Martens, Z. Ding, J.T. Kloprogge, T.E. Johnson, Spectrochim. Acta A 59 (2003) 291–302.
[166] J.T. Kloprogge, L. Hickey, R.L. Frost, J. Raman Spec. 35 (2004) 967–974.
[167] J. Pérez-Ramírez, G. Mul, F. Kapreijn, J.A. Moulijn, J. Mater. Chem. 11 (2001) 2529–2536.
[168] R. Xu, H.C. Zeng, Chem. Mater. 15 (2003) 2040–2048.
[169] Z.P. Xu, H.C. Zeng, J. Phys. Chem. B 105 (2001) 1743–1749.
[170] Z.P. Xu, H.C. Zeng, Chem. Mater 13 (2001) 4555–4563.
[171] J.W. Boclair, P.S. Braterman, B.D. Brister, Z. Wang, F. Yarberry, J. Solid State Chem. 161 (2001) 249–258.
[172] J.T. Kloprogge, M. Weier, I. Crespo, M.A. Ulibarri, C. Barriga, V. Rives, et al., J. Solid State Chem. 177 (2004) 1382–1387.
[173] H.C.B. Hansen, C.B. Koch, Clays Clay Miner 42 (1994) 170–179.
[174] P.S. Braterman, C.Q. Tan, J.X. Zhao, Mater. Res. Bull. 29 (1994) 1217–1221.
[175] L. Raki, D.G. Rancourt, C. Detellier, Chem. Mater. 7 (1995) 221–224.
[176] M. del Arco, S. Gutiérrez, C. Martín, V. Rives, J. Rocha, J. Solid State Chem. 177 (2004) 3954–3962.
[177] R.S. Maxwell, R.K. Kukkadapu, J.E. Amonette, H. Cho, J. Phys. Chem. B 103 (1999) 5197–5203.
[178] X.Q. Hou, R.J. Kirkpatrick, Chem. Mater. 14 (2002) 1195–1200.
[179] X.Q. Hou, R.J. Kirkpatrick, Chem. Mater. 12 (2000) 1890–1897.
[180] R.J. Kirkpatrick, P. Yu, X.Q. Hou, Y. Kim, Am. Miner. 84 (1999) 1186–1190.
[181] S.G. Kozlova, S.P. Gabuda, V.P. Isupov, L.E. Chupakhina, J. Struct. Chem. 44 (2003) 198–205.
[182] G.G. Aloisi, U. Costantino, F. Elisei, L. Latterini, C. Natalia, M. Nocchetti, J. Mater. Chem. 12 (2002) 3316–3323.
[183] S. Cooper, P.K. Dutta, J. Phys. Chem. 94 (1990) 114–118.
[184] W. Shi, M. Wei, J. Lu, F. Li, J. He, D.G. Evans, et al., J. Phys. Chem. C 112 (2008) 19886–19895.
[185] H.C. Greenwell, W. Jones, S.P. Newman, P.V. Coveney, J. Mol. Struct. 647 (2003) 75–83.
[186] T. Shichi, K. Takagi, Y. Sawaki, Chem. Commun. (1996) 2027–2028.
[187] S.W. Rhee, D.Y. Jung, Bull. Korean Chem. Soc. 23 (2002) 35–40.
[188] K. Takagi, E. Harata, T. Shichi, T. Kanoh, Y. Sawaki, J. Photochem. Photobiol. A Chem. 105 (1997) 47–54.
[189] D. Yan, J. Lu, M. Wei, J. Ma, D.G. Evans, X. Duan, Phys. Chem. Chem. Phys. 11 (2009) 9200–9209.

[190] J.C.A.A. Roelofs, J.A. van Bokhoven, A.J. van Dillen, J.W. Geus, K.P. de Jong, Chem. Eur. J. 8 (2002) 5571–5579.
[191] C. Depège, F.Z. El Metoui, C. Forano, A. de Roy, J. Dupuis, J.P. Besse, Chem. Mater. 8 (1996) 952–960.
[192] V.A. Drits, T.N. Sokolova, G.V. Sokolova, V.I. Cherkashin, Clays Clay Miner. 35 (1987) 401–417.
[193] M. Khaldi, A. de Roy, M. Chaouch, J.P. Besse, J. Solid State Chem. 130 (1997) 66–73.
[194] D.R. Hines, G.T. Seidler, M.M.J. Treacy, S.J. Solin, Solid State Commun. 101 (1997) 835–839.
[195] K. Yao, M. Taniguchi, M. Nakata, M. Takahashi, A. Yamagishi, Langmuir 14 (1998) 2410–2414.
[196] K. Yao, M. Taniguchi, M. Nakata, A. Yamagishi, J. Electroanal. Chem. 458 (1998) 249–252.
[197] K.R. Franklin, E. Lee, C.C. Nunn, J. Mater. Chem. 5 (1995) 565–569.
[198] A.M. Fogg, J.S. Dunn, D. O'Hare, Chem. Mater. 10 (1998) 356–360.
[199] J. Pastor-Rodriguez, H.F.W. Taylor, Miner. Mag. 38 (1971) 286–294.
[200] G.W. Brindley, Miner. Mag. 43 (1979) 337–340.
[201] M. Kaneyoshi, W. Jones, Chem. Phys. Lett. 296 (1998) 183–187.
[202] J. Pisson, C. Taviot-Guého, Y. Israëli, F. Leroux, P. Munsch, J.P. Itié, et al., J. Phys. Chem. B 107 (2003) 9243–9248.
[203] G.R. Williams, A.J. Norquist, D. O'Hare, Chem. Mater. 16 (2004) 975–981.
[204] G.R. Williams, A.J. Norquist, D. O'Hare, Chem. Commun. (2003) 1816–1817.
[205] S. Ma, C. Fan, L. Du, G. Huang, X. Yang, W. Tang, et al., Chem. Mater. 21 (2009) 3602–3610.
[206] G. Timp, M.S. Dresselhaus, J. Phys. C Solid State Phys. 17 (1984) 2641–2651.
[207] T.R. Park, T.Y. Park, H. Kim, P. Min, J. Phys. Condens. Matter 14 (2002) 11687–11700.
[208] H.C.B. Hansen, R.M. Taylor, Clay Miner. 26 (1991) 507–525.
[209] U. Costantino, N. Coletti, M. Nocchetti, G.G. Aloisi, F. Elisei, Langmuir 15 (1999) 4454–4460.
[210] U. Costantino, F. Marmottini, M. Nocchetti, R. Vivani, Eur. J. Inorg. Chem. (1998) 1439–1446.
[211] U. Costantino, M. Casciola, L. Massinelli, M. Nocchetti, R. Vivani, Solid State Ionics 97 (1997) 203–212.
[212] N.S.P. Bhuvanesh, J.H. Reibenspies, J. Appl. Crystallogr. 36 (2003) 1480–1481.
[213] F. Izumi, Solid State Ionics 172 (2004) 1–6.
[214] E.L. Crepaldi, P.C. Pavan, J.B. Valim, J. Braz. Chem. Soc. 11 (2000) 64–70.
[215] M.A. Aramendia, V. Borau, C. Jimenez, J.M. Marinas, J.R. Ruiz, F.J. Urbano, J. Solid State Chem. 168 (2002) 156–161.
[216] A. Corma, V. Fornes, R.M. Martin-Aranda, F. Rey, J. Catal. 134 (1992) 58–65.
[217] T. Hibino, Y. Yamashita, K. Kosuge, A. Tsunashima, Clays Clay Miner. 43 (1995) 427–432.
[218] F. Prinetto, G. Ghiotti, P. Graffin, D. Tichit, Micropor. Mesopor. Mater. 39 (2000) 229–247.
[219] M. Meyn, K. Beneke, G. Lagaly, Inorg. Chem. 29 (1990) 5201–5207.
[220] E. Kanezaki, Mater. Res. Bull. 34 (1999) 1435–1440.
[221] S. Bonnet, C. Forano, A. de Roy, J.P. Besse, Chem. Mater. 8 (1996) 1962–1968.
[222] A.I. Tsyganok, K. Suzuki, S. Hamakawa, K. Takehira, T. Hayakawa, Chem. Lett. 30 (2001) 24–25.
[223] S. Aisawa, S. Takahashi, W. Ogasawara, Y. Umetsu, E. Narita, J. Solid State Chem. 162 (2001) 52–62.
[224] A. Fudala, I. Palinko, B. Hrivnak, I. Kiricsi, J. Therm. Anal. Calorim. 56 (1999) 317–322.
[225] S. Aisawa, S. Takahashi, W. Ogasawara, Y. Umetsu, E. Narita, Clay Sci. 11 (2000) 317–328.
[226] S.Y. Kwak, Y.J. Jeong, J.S. Park, J.H. Choy, Solid State Ionics 151 (2002) 229–234.
[227] S.H. Hwang, Y.S. Han, J.H. Choy, Bull. Korean Chem. Soc. 22 (2001) 1019–1022.
[228] W. Kagunya, Z. Hassan, W. Jones, Inorg. Chem. 35 (1996) 5970–5974.
[229] S. Velu, R. Veda, A. Ramani, B.M. Chanda, S. Sivasanker, Chem. Commun. (1997) 2107–2108.
[230] S. Velu, D.P. Sabde, N. Shah, S. Sivasanker, Chem. Mater. 10 (1998) 345–350.
[231] French Patent 2091785 to BASF AG, 1971.
[232] V.R.L. Constantino, T.J. Pinnavaia, Inorg. Chem. 34 (1995) 883–892.
[233] M. del Arco, D. Carriazo, S. Gutiérrez, C. Martín, V. Rives, Inorg. Chem. 43 (2004) 375–384.
[234] J.A.M. Correa, in: M. Pecchio, F.R.D. Andrade, L.Z. D'Agostino, H. Kahn, L.M. Sant'Agostino, M.M.M.L. Tassinari (Eds.), Applied Mineralogy Proceedings, 1, ICAM-BR, Sao Paulo, Brazil, 2004, p. 399.
[235] E. Gardner, K.M. Huntoon, T.J. Pinnavaia, Adv. Mater. 13 (2001) 1263–1266.
[236] W. Meng, F. Li, D.G. Evans, X. Duan, Mater. Res. Bull. 39 (2004) 1185–1193.
[237] W. Meng, F. Li, D.G. Evans, X. Duan, Mater. Chem. Phys. 86 (2004) 1–4.
[238] S. Carlino, M.J. Hudson, J. Mater. Chem. 5 (1995) 1433–1442.
[239] M.Z.B. Hussein, C.W. Long, Mater. Chem. Phys. 85 (2004) 427–431.
[240] R. Sasai, N. Shin'ya, T. Shichi, K. Takagi, K. Gekko, Langmuir 15 (1999) 413–418.
[241] M.A. Drezdzon, Inorg. Chem. 27 (1988) 4628–4632.
[242] Z.P. Xu, H.C. Zeng, J. Phys. Chem. 104 (2000) 10206–10214.
[243] N.T. Whilton, P.J. Vickers, S. Mann, J. Mater. Chem. 7 (1997) 1623–1629.
[244] B.X. Li, J. He, D.G. Evans, X. Duan, Appl. Clay Sci. 27 (2004) 199–207.
[245] B.X. Li, J. He, D.G. Evans, X. Duan, Int. J. Pharm. 287 (2004) 89–95.
[246] Q. Yuan, M. Wei, Z.Q. Wang, G. Wang, X. Duan, Clays Clay Miner 52 (2004) 40–46.
[247] Q. Yuan, M. Wei, D.G. Evans, X. Duan, J. Phys. Chem. B 108 (2004) 12381–12387.
[248] W.K. Kuk, Y.D. Huh, J. Mater. Chem. 7 (1997) 1933–1936.
[249] Y. Zhao, F. Li, R. Zhang, D.G. Evans, X. Duan, Chem. Mater. 14 (2002) 4286–4291.
[250] J.J. Bravo-Suárez, E.A. Páez-Mozo, S.T. Oyama, Chem. Mater. 16 (2004) 1214–1225.
[251] Y.J. Feng, D.Q. Li, C.X. Li, Z.H. Wang, D.G. Evans, X. Duan, Clays Clay Miner. 51 (2003) 566–569.
[252] X.R. Jiang, F. Li, D.G. Evans, X. Duan, Huaxue Xuebao 62 (2004) 16–21.
[253] J. He, M. Wei, B. Li, Y. Kang, D.G. Evans, X. Duan, Struct. Bond. 119 (2006) 89–119.
[254] X. Duan, Q. Jiao, Chinese Patent 00132145.5 to Beijing University of Chemical Technology, 2000.
[255] Y. Zhao, Q.Z. Jiao, F. Li, D.G. Evans, X. Duan, Chin. J. Inorg. Chem. 17 (2001) 830–834.
[256] M.R. Shedam, A.V. Rao, Mater. Chem. Phys. 52 (1998) 263–266.
[257] A.I. Vogel, Quantitative Inorganic Chemical Analysis, fifth ed., Longman, Harlow, Essex, UK, 1989, p. 424.
[258] J.J. Bravo-Suárez, E.A. Páez-Mozo, S.T. Oyama, Quim. Nova 27 (2004) 601–614.

[259] M. Adachi-Pagano, C. Forano, J.P. Besse, J. Mater. Chem. 13 (2003) 1988–1993.
[260] J.M. Oh, S.H. Hwang, J.H. Choy, Solid State Ionics 151 (2002) 285–291.
[261] M. Ogawa, H. Kaiho, Langmuir 18 (2002) 4240–4242.
[262] U. Costantino, N. Coletti, M. Nocchetti, G.G. Aloisi, F. Elisei, L. Latterini, Langmuir 16 (2000) 10351–10358.
[263] K. Yao, Y. Imai, L.Y. Shi, E. Abe, Y. Adachi, K. Nishikubo, et al., Chem. Lett. 33 (2004) 1112–1113.
[264] J. Olanrewaju, B.L. Newalkar, C. Mancino, S. Komarneni, Mater. Lett. 45 (2000) 307–310.
[265] N. Morel-Desrosiers J. Pisson, Y. Israëli, C. Taviot-Guého, J.P. Besse, J.P. Morel, J. Mater. Chem. 13 (2003) 2582–2585.
[266] Y. Israëli, C. Taviot-Guého, J.P. Besse, J.P. Morel, N. Morel-Desrosiers, J. Chem. Soc., Dalton Trans. (2000) 791–796.
[267] S.P. Newman, W. Jones, J. Solid State Chem. 148 (1999) 26–40.
[268] Z.P. Xu, P.S. Braterman, J. Mater, Chem. 13 (2003) 268–273.
[269] J. Wang, M. Wei, R. Rao, D.G. Evans, X. Duan, J. Solid State Chem. 177 (2004) 366–371.
[270] M. Wei, S. Shi, J. Wang, Y. Li, X. Duan, J. Solid State Chem. 177 (2004) 2534–2541.
[271] A.I. Khan, L. Lei, A.J. Norquist, D. O'Hare, Chem. Commun. (2001) 2342–2343.
[272] J.H. Choy, S.Y. Kwak, J.S. Park, Y.J. Jeong, J. Portier, J. Am. Chem. Soc. 121 (1999) 1399–1400.
[273] A. Fudala, I. Palinko, I. Kiricsi, Inorg. Chem. 38 (1999) 4653–4658.
[274] F. Li, L.H. Zhang, D.G. Evans, C. Forano, X. Duan, Thermochim. Acta 424 (2004) 15–23.
[275] R.B. Leggat, S.A. Taylor, S.R. Taylor, Colloids Surf. A Physicochem. Eng. Aspects 210 (2002) 69–81.
[276] E.A. Gardner, S.K. Yun, T. Kwon, T.J. Pinnavaia, Appl. Clay Sci. 13 (1998) 479–494.
[277] J.C. Villegas, O.H. Giraldo, K. Laubernds, S.L. Suib, Inorg. Chem. 42 (2003) 5621–5631.
[278] M. del Arco, S. Gutiérrez, C. Martín, V. Rives, Inorg. Chem. 42 (2003) 4232–4240.
[279] V. Prévot, C. Forano, J.P. Besse, J. Solid State Chem. 153 (2000) 301–309.
[280] S. Bhattacharjee, T.J. Dines, J.A. Anderson, J. Catal. 225 (2004) 398–407.
[281] A.I. Tsyganok, T. Tsunoda, S. Hamakawa, K. Suzuki, K. Takehira, T. Hayakawa, J. Catal. 213 (2003) 191–203.
[282] C. Li, G. Wang, D.G. Evans, X. Duan, J. Solid State Chem. 177 (2004) 4569–4575.
[283] V.P. Isupov, L.E. Chupakhina, R.P. Mitrofanova, K.A. Tarasov, A.Y. Rogachev, V.V. Boldyrev, Solid State Ionics 265 (1997) 101–103.
[284] F. Malherbe, J.P. Besse, J. Solid State Chem. 155 (2000) 332–341.
[285] I. Carpani, M. Berrettoni, B. Ballarin, M. Giorgetti, E. Scavetta, D. Tonelli, Solid State Ionics 168 (2004) 167–175.
[286] Q.Z. Yang, D.J. Sun, C.G. Zhang, X.J. Wang, W.A. Zhao, Langmuir 19 (2003) 5570–5574.
[287] F. Leroux, P. Aranda J.P. Besse, E. Ruiz-Hitzky, Eur. J. Inorg. Chem. (2003) 1242–1251.
[288] E.M. Moujahid, J.P. Besse, F. Leroux, J. Mater. Chem. 12 (2002) 3324–3330.
[289] B.M. Choudary, N.S. Chowdari, K. Jyothi, M.L. Kantam, J. Am. Chem. Soc. 124 (2002) 5341–5349.
[290] D.L. Bish, Bull. Miner. 103 (1980) 170–175.
[291] J. Guo, T. Sun, J.P. Shen, J.G. Liu, D.Z. Jiang, E.Z. Min, Wuji Huaxue Xuebao 11 (1995) 134–139.
[292] Y.S. Sun, Q.Z. Jiao, Y. Zhao, D.G. Evans, X. Duan, Wuji Huaxue Xuebao 17 (2001) 414–418.
[293] Y.S. Sun, Q.Z. Jiao, Y. Zhao, D.G. Evans, X. Duan, Yingyong Huaxue 18 (2001) 781–784.
[294] I.C. Chisem, W. Jones, I. Martín, C. Martín, V. Rives, J. Mater. Chem. 8 (1998) 1917–1925.
[295] R.P. Bontchev, S. Liu, J.L. Krumhansl, J. Voigt, T.M. Nenoff, Chem. Mater. 15 (2003) 3669–3675.
[296] S.P. Newman, W. Jones, New J. Chem. (1998) 105–115.
[297] K.K. Ravi, S.W. Marc, E.A. James, Chem. Mater. 9 (1997) 417–419.
[298] Y. Xing, D.Q. Li, L.L. Ren, D.G. Evans, X. Duan, Huaxue Xuebao 61 (2003) 267–272.
[299] A.M. Fogg, J.S. Dunn, S.G. Shyu, D.R. Cary, D. O'Hare, Chem. Mater. 10 (1998) 351–355.
[300] S.W. Rhee, J.H. Lee, D.Y. Jung, J. Colloid Interf. Sci. 245 (2002) 349–355.
[301] M. Jobbágy, A.E. Regazzoni, J. Phys. Chem. B 109 (2005) 389–393.
[302] E.L. Crepaldi, P.C. Pavan, J.B. Valim, Chem. Commun. (1999) 155–156.
[303] E.L. Crepaldi, P.C. Pavan, J.B. Valim, J. Mater. Chem. 10 (2000) 1337–1343.
[304] R.E. Johnsen, P. Norby, J. Phys. Chem. C 113 (2009) 19061–19066.
[305] K.L. Erickson, T.E. Bostrom, R.L. Frost, Mater. Lett. 59 (2005) 226–229.
[306] A.J. Marchi, C.R. Apesteguía, Appl. Clay Sci. 13 (1998) 35–48.
[307] J. Rocha, M. del Arco, V. Rives, M.A. Ulibarri, J. Mater. Chem. 9 (1999) 2499–2503.
[308] A. Vaccari, Appl. Clay Sci. 14 (1999) 161–198.
[309] F. Kooli, V. Rives, M.A. Ulibarri, Inorg. Chem. 34 (1995) 5114–5121.
[310] F. Kooli, C. Depege, A. Ennaqadi, A. de Roy, J.P. Besse, Clays Clay Miner. 45 (1997) 92–98.
[311] Y.W. You, H.T. Zhao, G.F. Vance, J. Mater. Chem. 12 (2002) 907–912.
[312] S.I. Katahira, K. Yasue, J. Mater. Res. 14 (1999) 1178–1180.
[313] H. Tagaya, A. Ogata, T. Kuwahara, S. Ogata, M. Karasu, J.I. Kadokawa, et al., Micropor. Mater. 7 (1996) 151–158.
[314] S. Aisawa, H. Hirahara, K. Ishiyama, W. Ogasawara, Y. Umetsu, E. Natira, J. Solid State Chem. 174 (2003) 342–348.
[315] T. Kuwahara, H. Tagaya, K. Chiba, Micropor. Mater. 4 (1995) 247–250.
[316] M.J.S. Martín, M.V. Villa, M. Sanchez-Camazano, Clays Clay Miner. 47 (1999) 777–783.
[317] H. Morioka, T.M. Karasu, J. Kadokawa, K. Chiba, J. Solid State Chem. 117 (1995) 337–342.
[318] A. Corma, V. Fornés, F. Rey, A. Cervilla, E. Llopis, A. Ribera, J. Catal. 152 (1995) 237–342.
[319] E. Kanezaki, J. Mater. Sci. 30 (1995) 4926–4929.
[320] C.A.S. Barbosa, A.M.D.C. Ferreira, V.R.L. Constantino, A.C.V. Coelho, J. Incl. Phenom. Macrocyclic Chem. 42 (2002) 15–23.
[321] H. Nakayama, N. Wada, M. Tsuhako, Int. J. Pharm. 269 (2004) 469–478.
[322] S. Aisawa, H. Hirahara, S. Takahashi, Y. Umetsu, E. Narita, Chem. Lett. 33 (2004) 306–307.
[323] A. Vyalikh, F.R. Costa, U. Wagenknecht, G. Heinrich, D. Massiot, U. Scheler, J. Phys. Chem. C 113 (2009) 21308–21313.
[324] C.O. Oriakhi, I.V. Farr, M.M. Lerner, Clays Clay Miner. 45 (1997) 194–202.
[325] K. Chibwe, W. Jones, J. Chem. Soc., Chem. Commun. (1989) 926–927.
[326] L. Li, D. Mo, Q.S. Luo, Wuji Huaxue Xuebao 20 (2004) 256–260.
[327] L. Li, Q.S. Luo, X. Duan, J. Mater. Sci. Lett. 21 (2002) 439–441.

[328] F. Wong, R.G. Buchheit, Prog. Org. Coat. 51 (2004) 91–102.
[329] M. Ogawa, S. Asai, Chem. Mater. 12 (2000) 3253–3255.
[330] M. de J. Martínez-Ortiz, E. Lima, V. Lara, J.M. Vivar, Langmuir 24 (2008) 8904–8911.
[331] L.L. Ren, J. He, S.C. Zhang, D.G. Evans, X. Duan, R.Y. Ma, J. Mol. Catal. B Enzym. 18 (2002) 3–11.
[332] G.A. Bubniak, W.H. Schreiner, N. Mattoso, F. Wypych, Langmuir 18 (2002) 5967–5970.
[333] E.D. Dimotakis, T.J. Pinnavaia, Inorg. Chem. 29 (1990) 2393–2394.
[334] W.Y. Tseng, J.T. Lin, C.Y. Mou, S. Cheng, S.B. Liu, P.P. Chu, et al., J. Am. Chem. Soc. 118 (1996) 4411–4418.
[335] H. Tagaya, S. Sato, T. Kuwahara, J. Mater. Chem. 4 (1994) 1907–1912.
[336] J. Zhang, F. Zhang, L.L. Ren, D.G. Evans, X. Duan, Mater. Chem. Phys. 85 (2004) 207–214.
[337] H.P. Boehn, J. Steinle, C. Vieweger, Angew. Chem. Int. Ed. 16 (1977) 265–266.
[338] K. El Malki, A. de Roy, J.P. Besse, Eur. J. Solid State Inorg. Chem. 26 (1989) 339–351.
[339] J. Zhang, D.Q. Li, L.L. Ren, D.G. Evans, X. Duan, Chin. J. Inorg. Chem. 20 (2004) 1208–1212.
[340] V.P. Isupov, L.E. Chupakhina, R.P. Mitrofanova, J. Mater. Synth. Proc. 8 (2000) 251–253.
[341] A.V. Radha, P.V. Kamath, Bull. Mater. Sci. 26 (2003) 661–666.
[342] A. de la Hoz, Á. Díaz-Ortiz, A. Moreno, Chem. Soc. Rev. 34 (2005) 164–178.
[343] D. Tichit, A. Rolland, F. Prinetto, G. Fetter, M.J. Martínez-Ortiz, M.A. Valenzuela, et al., J. Mater. Chem. 12 (2002) 3832–3838.
[344] G. Fetter, A. Botello, V.H. Lara, P. Bosch, J. Porous Mater (2001).
[345] Y. Seida, Y. Nakano, Y. Nakamura, Clays Clay Miner 50 (2002) 525–532.
[346] L. Indira, M. Dixit, P.V. Kamath, J. Power Sources 52 (1994) 93–97.
[347] L. Legrand, M. Abdelmoula, A. Géhin, A. Chaussé, J.M.R. Génin, Electrochim. Acta 46 (2001) 1815–1822.
[348] H.C. Zeng, Z.P. Xu, M. Qian, Chem. Mater 10 (1998) 2277–2283.
[349] H.C.B. Hansen, C.B. Koch, Appl. Clay Sci. 10 (1995) 5–19.
[350] H.C.B. Hansen, C.B. Koch, R.M. Taylor, J. Solid State Chem. 113 (1994) 46–53.
[351] J. He, B. Li, D.G. Evans, X. Duan, Colloid Surf. A Physicochem. Eng. Aspects 251 (2004) 191–196.
[352] M. Jitianu, M. Bălăuoiu, M. Zaharescu, A. Jitianu, A. Ivanov, J. Sol-Gel Sci. Tech. 19 (2000) 453–457.
[353] C. Delmas, Y. Borthomieu, J. Solid State Chem. 104 (1993) 345–352.
[354] R.D. Shannon, Acta Crystallogr. A 32 (1976) 751–767.
[355] F. Millange, R.I. Walton, L. Lei, D. O'Hare, Chem. Mater 12 (2000) 1990–1994.
[356] G.J. Ross, H. Kodama, Amer. Mineral 52 (1967) 1036–1047.
[357] P. Courty, C. Marcilly, A scientific approach to the preparation of bulk mixed oxide catalysts, in: G. Poncelet, P. Grange, P.A. Jacobs (Eds.), Preparation of Catalysts III—Scientific Bases for the Preparation of Heterogeneous Catalysts, Elsevier, Amsterdam, 1983, pp. 485–519.
[358] Y.J. Feng, D.Q. Li, C.X. Li, Z.G. Wang, D.G. Evans, X. Duan, Huaxue Xuebao 61 (2003) 78–83.
[359] J. He, M. Wei, X. Duan, in: X. Duan, F.Z. Zhang (Eds.), Chaceng Zuzhuang yu Gongneng Cailiao, Chemical Industry Press, Beijing, 2007, p. 17.
[360] D.L. Bish, G.W. Brindley, Am. Mineral 62 (1977) 458–464.
[361] I. Pausch, H.H. Lohse, K. Schurmann, R. Allmann, Clays Clay Miner. 34 (1986) 507–510.
[362] G. Del Piero, M. Di Conca, F. Trifirò, A. Vaccari, in: P. Barret, L.C. Dufour (Eds.), Reactivity of Solids, Elsevier, Amsterdam, 1985, p. 1029.
[363] S. Velu, K. Suzuki, T. Osaki, F. Ohashi, S. Tomura, Mater. Res. Bull. 34 (1999) 1707–1717.
[364] J.M. Fernández, C. Barriga, M.A. Ulibarri, F.M. Labajos, V. Rives, J. Mater. Chem. 4 (1994) 1117–1121.
[365] F. Basile, G. Fornasari, M. Gazzano, A. Vaccari, Appl. Clay Sci. 16 (2000) 185–200.
[366] C. Qi, J.C. Amphlett, B.A. Peppley, Catal. Lett. 104 (2005) 57–62.
[367] F. Melo, N. Morlanés, Catal. Today 133–135 (2008) 374–382.
[368] F. Basile, G. Fornasari, E. Poluzzi, A. Vaccari, Appl. Clay Sci. 13 (1998) 329–345.
[369] F. Basile, G. Fornasari, M. Gazzano, A. Vaccari, Appl. Clay Sci. 18 (2001) 51–57.
[370] T.H. Bennur, A. Ramani, R. Bal, B.M. Chanda, S. Sivasanker, Catal. Commun. 3 (2002) 493–496.
[371] C. Carlini, M. Marchionna, M. Noviello, A.M.R. Galletti, G. Sbrana, F. Basile, et al., J. Mol. Catal. A 232 (2005) 13–20.
[372] D. Francová, N. Tanchoux, C. Gérardin, P. Trens, F. Prinetto, G. Ghiotti, et al., Micropor. Mesopor. Mater. 99 (2007) 118–125.
[373] A.I. Tsyganok, M. Inaba, T. Tsunoda, S. Hamakawa, K. Suzuki, T. Hayakawa, Catal. Commun. 4 (2003) 493–498.
[374] J. Carpentier, J.F. Lamonier, S. Siffert, E.A. Zhilinskaya, A. Aboukaïs, Appl. Catal. A-Gen 234 (2002) 91–101.
[375] E.M. Sabbar, M.E. de Roy, F. Leroux, Micropor. Mesopor. Mater 103 (2007) 134–141.
[376] X. Zhang, M. Wei, M. Pu, X.J. Li, H. Chen, D.G. Evans, et al., J. Solid State Chem. 178 (2005) 2701–2707.
[377] L.Y. Wang, G.Q. Wu, D.G. Evans, Mater. Chem. Phys. 104 (2007) 133–140.
[378] Y.C. Zhang, L.Y. Wang, G.Q. Wu, A. Chang, D.G. Evans, J. Phys. Chem. Solids 69 (2008) 1095–1097.
[379] P.P. Yang, M.P. Su, X.W. Yang, G.Z. Liu, J.F. Yu, T.H. Wu, et al., Wuji Huaxue Xuebao 19 (2003) 485–489.
[380] F. Kovanda, D. Koloušek, Z. Cílová, V. Hulínský, Appl. Clay Sci. 28 (2005) 101–109.
[381] G.J. Ross, H. Kodama, Am. Mineral 52 (1967) 1037–1040.
[382] E. Géraud, V. Prévot, F. Leroux, J. Phys. Chem. Solids 67 (2006) 903–908.
[383] E. Géraud, V. Prévot, J. Ghanbaja, F. Leroux, Chem. Mater. 18 (2006) 238–240.
[384] E. Géraud, S. Rafqah, M. Sarakha, C. Forano, V. Prévot, F. Leroux, Chem. Mater. 20 (2008) 1116–1125.
[385] L. Li, R. Ma, N. Iyi, Y. Ebina, K. Takada, T. Sasaki, Chem. Commun. (2006) 3125–3127.
[386] R. Gunawan, R. Xu, J. Mater. Chem. 18 (2008) 2112–2120.
[387] Y.C. Wang, F.Z. Zhang, S.L. Xu, X.Y. Wang, D.G. Evans, X. Duan, Ind. Eng. Chem. Res. 47 (2008) 5746–5750.
[388] K. Itaya, H.C. Chang, I. Uchida, Inorg. Chem. 26 (1987) 624–626.
[389] J.H. Lee, S.W. Rhee, D.Y. Jung, Chem. Mater. 16 (2004) 3774–3779.
[390] J.H. Lee, S.W. Rhee, D.Y. Jung, Chem. Commun. (2003) 2740–2741.
[391] R. Roto, A. Yamagishi, G. Villemure, J. Electroanal. Chem. 572 (2004) 101–108.
[392] J.B. Qiu, G. Villemure, J. Electroanal. Chem. 395 (1995) 159–166.
[393] J.H. Lee, S.W. Rhee, D.Y. Jung, Chem. Mater. 18 (2006) 4740–4746.
[394] J.A. Gursky, S.D. Blough, C. Luna, C. Gomez, A.N. Luevano, E.A. Gardner, J. Am. Chem. Soc. 128 (2006) 8376–8377.
[395] N. Iyi, Y. Ebina, T. Sasaki, Langmuir 24 (2008) 5591–5598.

[396] L.Y. Wang, C. Li, M. Liu, D.G. Evans, X. Duan, Chem. Commun. (2007) 123–125.
[397] C. Li, L.Y. Wang, M. Wei, D.G. Evans, X. Duan, J. Mater. Chem. 18 (2008) 2666–2672.
[398] F. Hua, Y.M. Lvov The New Frontiers of Organic and Composite Nanotechnology, Elsevier, Amsterdam, 2008, pp. 1–44.
[399] G. Decher, Science 277 (1997) 1232–1237.
[400] M. Lahav, G. Tobias, A.N. Shipway, I. Willner, J. Am. Chem. Soc. 121 (1999) 258–259.
[401] L.Z. Wang, N. Sakai, Y. Ebina, K. Takada, T. Sasaki, Chem. Mater 17 (2005) 1352–1357.
[402] J.H. Kim, S. Fujita, S. Shiratori, Thin Solid Films 499 (2006) 83–89.
[403] K. Okamoto, T. Sasaki, T. Fujita, N. Iyi, J. Mater. Chem. 16 (2006) 1608–1616.
[404] Z.P. Liu, R.Z. Ma, M. Osada, N. Iyi, Y. Ebina, K. Takada, et al., J. Am. Chem. Soc. 128 (2006) 4872–4880.
[405] L. Li, R. Ma, Y. Ebina, K. Fukuda, K. Takada, T. Sasaki, J. Am. Chem. Soc. 129 (2007) 8000–8007.
[406] X. Zhang, Y. Wang, X. Chen, W. Yang, Mater. Lett. 62 (2008) 1613–1616.
[407] Y. Wang, W.S. Yang, C. Chen, D.G. Evans, J. Power Sources 184 (2008) 682–690.
[408] D.P. Yan, J. Lu, M. Wei, J.B. Han, J. Ma, F. Li, et al., Angew. Chem. Int. Ed. 48 (2009) 3073–3076.
[409] K.B. Mok, J.R.H. Ross, R.M. Sambrook, Thermally and mechanically stable catalysts for steam reforming or methanation, a new concept in catalyst design, in: G. Poncelet, P. Grange, P.A. Jacobs (Eds.), Preparation of Catalysts III—Scientific Bases for the Preparation of Heterogeneous Catalysts, Elsevier, Amsterdam 1983, pp. 291–299.
[410] H. Schaper, R.B.M. Doesburg, J.M.C. Quartel, L.L. van Reijen, Synthesis of methanation catalysts by deposition- precipitation, in: G. Poncelet, P. Grange, P.A. Jacobs (Eds.), Preparation of Catalysts III—Scientific Bases for the Preparation of Heterogeneous Catalysts, Elsevier, Amsterdam, 1983, pp. 301–309.
[411] J.B. d'Espinose de la Caillerie, M. Kermarec, O. Clause, J. Am. Chem. Soc. 117 (1995) 11471–11481.
[412] E. Merlen, P. Gueroult, J.B. d'Espinose de la Caillerie, B. Rebours, C. Bobin, O. Clause, Appl. Clay Sci. 10 (1995) 45–56.
[413] Y.B. Mao, D.Q. Li, F.Z. Zhang, D.G. Evans, X. Duan, Wuji Huaxue Xuebao 20 (2004) 596–601.
[414] R. Zhang, D.Q. Li, F.Z. Zhang, D.G. Evans, X. Duan, Fudan Xuebao 42 (2003) 333–338.
[415] H.Y. Chen, F.Z. Zhang, S.S. Fu, X. Duan, Adv. Mater. 18 (2006) 3089–3093.
[416] F.Z. Zhang, L.L. Zhao, H.Y. Chen, S.L. Xu, D.G. Evans, X. Duan, Angew. Chem. Int. Ed. 47 (2008) 2466–2469.
[417] R.B. Leggat, S.A. Taylor, S.R. Taylor, Colloids Surf. A: Physicochem. Eng. Aspects 210 (2002) 83–94.
[418] Y.F. Gao, M. Nagai, Y. Masuda, F. Sato, W.S. Seo, K. Koumoto, Langmuir 22 (2006) 3521–3527.
[419] X.D. Lei, L. Yang, F.Z Zhang, D.G. Evans, X. Duan, Chem. Lett. 34 (2005) 1610–1611.
[420] Z. Lü, F.Z. Zhang, X.D. Lei, L. Yang, D.G. Evans, X. Duan, Chem. Eng. Sci. 62 (2007) 6069–6075.
[421] Z. Lü, F.Z. Zhang, X.D. Lei, L. Yang, S.L. Xu, X. Duan, Chem. Eng. Sci. 63 (2008) 4055–4062.
[422] K.B. Blodgett, J. Am. Chem. Soc. 56 (1934). 495-495.
[423] K.B. Blodgett, I. Langmuir, Phys. Rev. 51 (1937) 964–982.
[424] R.M. Leblanc, Curr. Opin. Chem. Biol. 10 (2006) 529–536.
[425] A.P. Girard-Egrot, S. Godoy, L.J. Blum, Adv. Colloid Interf. Sci. 116 (2005) 205–225.
[426] J.X. He, K. Kobayashi, M. Takahashi, G. Villemure, A. Yamagishi, Thin Solid Films 397 (2001) 255–265.
[427] J.X. He, S. Yamashita, W. Jones, A. Yamagishi, Langmuir 18 (2002) 1580–1586.
[428] M.C. Petty, Langmuir–Blodgett Film Structure: an Introduction to Langmuir–Blodgett Films, Cambridge University Press, London, 1996.
[429] F.Z. Zhang, M. Sun, S.L. Xu, L.L. Zhao, B.W. Zhang, Chem. Eng. J. 141 (2008) 362–367.
[430] J. He, M. Wei, X. Duan, in: X. Duan, F.Z. Zhang (Eds.), Chaceng Zuzhuang yu Gongneng Cailiao, Chemical Industry Press, Beijing, 2007, p. 48.
[431] A. de Roy, C. Forano, K. El Malki, J.P. Besse, Anionic clays: trends in pillaring chemistry, in: M.L. Occelli, H.E. Robson (Eds.), Synthesis of Microporous Materials, vol. II., Expanded Clays and Other Microporous Solids, New York, Van Nostrand Reinhold, 1992, pp. 108–169.
[432] R. Allmann, Chimia 24 (1970) 99–107.
[433] M. Wei, J. Wang, J. He, D.G. Evans, X. Duan, Micropor. Mesopor. Mater. 78 (2005) 53–61.
[434] H. Tagaya, S. Ogata, H. Morioka, J.I. Kadokawa, M. Karasu, K. Chiba, J. Mater. Chem. 6 (1996) 1235.
[435] V. Ambrogi, G. Fardella, G. Grandolini, L. Perioli, Int. J. Pharm 220 (2001) 23–32.
[436] M. Wei, Q. Yuan, D.G. Evans, Z. Wang, X. Duan, J. Mater. Chem. 15 (2005) 1197–1203.
[437] M. Partyka, B.H. Au, C.H. Evans, J. Photochem. Photobiol. A Chem. 140 (2001) 67–74.
[438] H. Zhang, K. Zou, H. Sun, X. Duan, J. Solid State Chem. 178 (2005) 3485–3493.
[439] K. Zou, H. Zhang, X. Duan, Chem. Eng. Sci. 62 (2007) 2022–2031.
[440] J.H. Meng, H. Zhang, D.G. Evans, X. Duan, Gaodeng Xuexiao Huaxue Xuebao 24 (2003) 1315–1319.
[441] H. Zhang, K. Zou, S. Guo, X. Duan, J. Solid State Chem. 179 (2006) 1792–1801.
[442] M.L. Bender, M. Komiyama, Cyclodextrin Chemistry, Springer, Berlin, 1978.
[443] Y. Inoue, T. Hakushi, Y. Liu, L.H. Tong, B.J. Shen, D.S. Jin, J. Am. Chem. Soc. 115 (1993) 475–481.
[444] X. Wang, M.L. Brusseau, Environ. Sci. Technol 29 (1995) 2632–2635.
[445] J.M. Oh, S.Y. Kwak, J.H. Choy, J. Phys. Chem. Solids 67 (2006) 1028–1031.
[446] M. Wei, Z. Shi, D.G. Evans, X. Duan, J. Mater. Chem. 16 (2006) 2102–2109.
[447] C. Wei, F. Li, B. Qu, J. Solid State Commun. 130 (2004) 259–263.
[448] M. Wei, X. Tian, J. He, M. Pu, G. Rao, H. Yang, et al., Eur. J. Inorg. Chem. (2006) 3442–3450.
[449] T. Shichi, S. Yamashita, K. Takagi, Supramol. Sci. 5 (1998) 303–308.

CHAPTER

18

Host–guest Functional Materials

Jiesheng Chen

Shanghai Jiao Tong University, China

A composite material is usually superior in property as compared to the individual components from which the composite is formed, and therefore it is preferable to obtain composites through various strategies. Among the composite materials, those formed from a porous host and guest species have been attracting increasing interest because of their unique functions and their great potential in practical applications. Depending on the nature of the porous host, the technique to introduce a guest into the host is varied. Within the pores of the host, the guest species may interact with each other or with the pore walls of the host in a way distinct from that outside the pores. Furthermore, it is possible to obtain new entities with unusual physical–chemical properties and these entities cannot be prepared outside the host pores.

Porous materials include conventional zeolites which are crystalline aluminosilicates with a pore diameter less than 1.5 nm. Many zeolite-related materials including crystalline porous aluminophosphates, and mesoporous materials have also been invented in the past after the discovery of zeolites. These porous materials as well as the conventional zeolites have been successfully used in a variety of fields such as ion exchange, gas separation, adsorption, and catalysis. However, in the past decades many efforts have been directed toward the use of zeolites and related materials as host to prepare host–guest composites. Especially the mesoporous molecular sieves announced in 1990s provided new hosts for host–guest assembly. The compositions and structures of zeolitic materials are versatile and in principle, it is possible to encapsulate a particular guest species using a special zeolite host. In return, through variation of the guest species and the assembly approaches, various guest matters, which may exhibit versatile chemical–physical properties, can be prepared. As the diameters of the channels and cages in zeolites fall into the category of nanometer scale, the guest particles confined in a zeolite should also be nanometer sized and this lays foundation for the chemical preparation of guest species with quantum size effects [1,2]. In general, the host–guest composite materials based on porous hosts can be classified as five major categories. The first category includes composites of zeolites with encapsulated metal clusters or metal ion clusters and occasionally in the clusters there are nonmetal ligands such as carbonyl groups. The second category concerns the host–guest composite materials of zeolites and dye molecules. The third category involves polymers or carbons such as fullerenes and carbon nanotubes in zeolitic materials. Whereas the fourth category consists of inorganic semiconductor particles in the channels and cages of zeolites and the fifth includes metal complexes encapsulating zeolites. These host–guest materials not only exhibit interesting chemical–physical properties but also show great application potential in various areas. For example, when metal complexes and clusters are bonded to zeolite supports, they may combine the technological advantages of solid catalysts with the selectivity of soluble molecular catalysts [3].

18.1. METAL CLUSTERS IN ZEOLITES

18.1.1. Preparation Approaches of Metal Clusters

Two general approaches can be used to prepare metal clusters in zeolites: (i) involves direct evaporation and deposition of metals into the pores of zeolites; and (ii) is to load the metal-containing precursors to the zeolite pores followed by formation of metal clusters in the zeolite pores through decomposition or reduction [4].

Modern Inorganic Synthetic Chemistry, DOI: 10.1016/B978-0-444-53599-3.10018-6

According to the preparation techniques used, these two approaches can be further divided into following sub-approaches.

18.1.1.1. Metal Salt Impregnation and Reduction

A solution of the metal salt is prepared and thoroughly mixed with the corresponding zeolite. In this way, the metal salt will penetrate into the zeolite pores. After impregnation the solid zeolite is dried, calcined, and reduced in gaseous hydrogen to form metal clusters in zeolite pores. The metal salts used are usually chlorides, nitrates, or carboxylates. The impregnation method is suitable for the following cases: (a) zeolites which lack or have low ion exchange capacity; (b) the loaded metals (such as Mo) that are present as anions that are not able to be ion exchanged into the zeolite pores; and (c) generation of protons needs to be avoided because H_2-reduction after ion exchange usually produces protons in the zeolite channels. In some cases, the combination of the impregnation and the ion exchange methods may reduce the reduction temperature rendering the formation of metal clusters easier.

18.1.1.2. Adsorption and Decomposition of Metal Compounds

Organometallic and metal-carbonyl compounds can be introduced into the zeolite pores through vapor deposition or through solution adsorption. After entering into the pores, these precursor compounds can be decomposed through thermal treatment or other physical–chemical methods to form metal clusters in the zeolite pores. This technique requires that the precursor molecules should be smaller than the pore diameter so that they are able to diffuse inside the pores. In addition, the vapor pressure of the precursors should not be too low because otherwise the evaporation cannot proceed smoothly. During decomposition, formation of intermediates which can diffuse outside the zeolite channels should be avoided and the decomposition of organic ligands should not lead to easily depositing species to block the zeolite channels.

18.1.1.3. Preparation through Metal Vapor

For metals with a relatively high vapor pressure, it is possible to load the metal clusters into the zeolite pores through direct evaporation. Rabo et al. first reported the preparation of neutral or ionic sodium clusters through contact of carefully dehydrated zeolites with sodium vapor in vacuum. In fact, besides metal itself, some compounds such as NaN_3 can also be used as the metal sources. The apparatus for the preparation of zeolite-encapsulated metal clusters through the vapor preparation method [5] usually consists of the vacuum pumping system, the valves, the joints, and the tube reactor.

18.1.1.4. Ion Exchange and Reduction

This technique is commonly used for preparation of metal clusters in zeolite pores. First, the metal ions are exchanged into the zeolite pores and then these metal ions are reduced through various chemical–physical approaches to form metal clusters. The reduction of the ion-exchanged zeolites is usually realized/accomplished through thermal treatment in H_2 atmosphere and to avoid diffusion of the reduced metal atoms to condense outside the zeolite pores, the reduction temperature should be as low as possible depending on the reducibility of the metal ions. In some cases, in vacuum or in an inert atmosphere, thermal treatment will lead to spontaneous reduction of the exchanged metal ions to form metal atoms. The presence of extra-framework molecules or ions may facilitate this spontaneous reduction.

18.1.2. Alkali Metal Clusters

18.1.2.1. Preparation of Alkali Metal Clusters

Previously, the formation of ion clusters in zeolites was realized through physical approaches, for instance, γ-ray irradiation or high-frequency ultraviolet irradiation of cations in zeolites. In 1965, Kasai et al. reported that after γ-ray irradiation of dehydrated NaY zeolite in vacuum, the obtained pink sample showed electron spin resonance signals. The signal splitting indicates the presence of Na_4^{3+} metal clusters in the zeolite. Later on, Rabo and coworkers discovered that the exposure of NaY to sodium vapor at 580°C resulted in the same species. But in this case the sample was bright red in color, suggesting that the metal cluster concentration was relatively high.

In 1984, Edwards et al. reported the formation of Na_4^{3+} in zeolites Y and A and the formation of K_4^{3+} in K^+-exchanged zeolite Y. Later on, they found that if the zeolite was NaY, the formed metal ion clusters are Na_4^{3+} no matter what (Na or K) the reaction vapor was, whereas if the zeolite used was K-Y, the obtained metal clusters will be K_4^{3+}. That is, the formed metal cluster species is not related with the vapor but simply depends on the type of the cations in the zeolite used. After the alkali metals enter the channels or cages of zeolites to form metal ion clusters, the electrons on the original metal atoms will be released to be shared by more than one metal atom. It has been confirmed that these free electrons actually occupy the holes formed by the metal atoms (ions). Therefore, these electrons are also called solid solvated electrons (in analogy with the solvated electrons formed by alkali metals in solvents such as liquid ammonia [6]) and the formed compounds are called solid electrides.

18.1.2.2. *Location of Alkali Metal Ion Clusters*

Either in sodalite, zeolite A or in zeolite Y, the Na_4^{3+} formed gives rise to identical ESR spectra, indicating that the Na_4^{3+} ion cluster is located in the same microenvironment as in the sodalite cage. The sodalite cage of zeolites A and X can also accommodate K_3^{2+} ion cluster. Relevant studies indicate that other ion clusters can also form in the sodalite cage. However, alkali metals with a larger radius such as Rb and Cs would exist in different structural forms when they form ion clusters. For instance, Cs_4^{3+} clusters in the sodalite cages of zeolite A may be connected with one another to form a chain structure. It is also proposed that in the sodalite cages the clusters may not be isolated but interact with the ions or ion clusters in neighboring cages to form ion cluster groups. When the content of Na metal in a zeolite is very high, the composite becomes black. Under this circumstance, the ESR fine structure disappear and only one broad signal with $g = 2$ shows up. This phenomenon arises from the electron exchange because the metal ion clusters in adjacent sodalite cages of the zeolite get very close to one another. Potassium has been loaded in zeolites LiX and LiA [7] and the as-prepared host–guest composites have been elucidated via ^{133}Cs NMR and ESR spectroscopy. It is found that at low Cs loadings, the isolated Cs cations are present whereas at higher loadings Cs clusters are formed in the cages of the zeolites.

18.1.2.3. *Metal Ion Cluster Crystals and Cation Continuum*

For samples with a sodium concentration which is not too low, the ESR spectrum of Na/NaY shows no fine structure of Na_4^{3+} but a single line without fingerprints. Initially, Edwards et al. thought that this single line arises from the metal nanoparticles in the zeolite. But the subsequent study indicated that this single line might originate from the interaction between adjacent Na_4^{3+} ion clusters. Through *ab initio* molecular dynamic modeling calculations, Ursenbach et al. confirmed this postulation. The calculation suggested that the extent of interactions between Na_4^{3+} clusters in adjacent sodalite cages of NaY is comparable to the reciprocal of observed ^{23}Na ultrafine coupling constant of Na_4^{3+}. In this case, if each ion cluster has at least one adjacent ion cluster, the disappearance of the ultrafine structure of ESR spectrum becomes obvious. On the basis of detailed investigation in the structures of metal ion-containing zeolites, Armstrong and coworkers pointed out that there exist Na_4^{3+} interaction networks in zeolites.

The distance between metal ion clusters and the orientation of the clusters in zeolite cages depends on the zeolite host and as a result the three-dimensional arrays can be described in different ways. They can be called cluster crystals or superlattices. The Na/NaY cluster crystal is formed by accommodation of one Na_4^{3+} cluster in each sodalite cage. Because the Na_4^{3+} cluster can be present in zeolites with different structures [8–11], it is possible to prepare cluster crystals with various geometric arrangements. In these cluster crystals, the centered cubic array formed by the encapsulation of sodalite framework is the simplest and it exists in the host–guest compound designated "black sodalite" [12,13].

18.1.2.4. *Properties of Alkali Metal Clusters*

Na_4^{3+} cluster can reversibly absorb O_2 and form O_2^- ion in NaY [14]. Other molecules such as SO_2, N_2, Ar, Kr, CO, CO_2, NH_3, benzene, *n*-hexane, and haloalkanes also interact with these metal clusters [15,16] and as a result, the metal clusters may catalyze the chemical reactions involving these molecules. In principle, the metal clusters in the zeolite channels and cages may act as basic catalysts. Martens et al. investigated [17] the catalytic performance of sodium clusters in zeolites A, X, and Y for the isomerization reaction of *cis*-2-butene and for the hydrogenation reaction of *cis*-2-butene, acetylene, and benzene. Later on, Hannus et al. [18] also reported the catalytic performance of sodium clusters in a series of Y zeolites for 1-butene isomerization and they proposed that in this catalytic reaction process the butane isomerization intermediate is a carbanion because of the basicity of the metal clusters. Simon et al. reported that [19] in NaX the Na_6^{5+} clusters could facilitate the isomerization of cyclopentene to propene. The metal ion clusters formed by pyrolysis of NaN_3 in the cages of zeolite can catalyze branch-alkylation of toluene and the aldehyde condensation of ketones.

Because cation-containing zeolites such as NaX can adsorb a large amount of alkali metals to form metal clusters, a large number of extra electrons in unit volume can be produced in the compounds. When the concentration of these extra electrons reaches a certain value, it is possible for the metal cluster-containing host–guest material to undergo transition from insulator to conductor. In particular, when there are crystallographically one-dimensional channels, the one-dimensional metal clusters formed may become quantum wires [20–-23]. In 1997, Anderson et al. reported that [21] when more and more extra electrons are introduced into the zeolite hosts, some zeolites indeed show increase in electric conductance. The K/K-A host–guest compound exhibits interesting ferromagnetic behavior [24] and this magnetism may be related with the formation of superlattice [25].

18.1.3. Metal Clusters of Silver

In 1962, Ralek et al. [26] reported that after dehydration/rehydration treatment, the silver-exchanged zeolite

A changed from white (hydrated) to yellow to orange and finally to brick red in color. In 1977, Kim and Seff [27] observed similar phenomenon for Ag-A single crystal when carefully treated at 400°C and on the basis of XRD analysis they revealed that after dehydration there exist Ag_6^0 metal clusters in the Ag-A zeolite. Later, Jacobs et al. [28] revealed that Ag_3^{2+} clusters may be present in the zeolite. In the absence of coordination water molecules, isolated Ag^0 atoms do not actually exist because isolated silver atoms are paramagnetic and theoretically they may be detected through ESR spectroscopy. Nevertheless, the compounds show no ESR signals. Therefore, it is concluded that the Ag^0 atoms formed through reduction interact with one another to form metal clusters [29]. In zeolites, silver clusters can also be obtained through reduction of the silver ions in the zeolites using alkali metals. Through H_2-reduction, silver atom clusters or ion clusters with various sizes can be prepared using Ag-A or AgNa-A zeolites. In some cases, these clusters show ESR signals indicating that there exist unpaired electrons. By means of physical methods such as ray irradiation, paramagnetic silver metal clusters may also be acquired at low temperatures [30,31].

Because in zeolites the silver clusters change their color when in contact with water, it has been suggested that these compounds may be used as water vapor sensing materials. Ozin et al. [32] observed the vapor-pressure chromic, cathode-ray chromic, water chromic, photochromic, and thermochromic properties for silver sodalite and they proposed that these compounds hold promise for use as sensors. In addition, the silver clusters in zeolites are also sensitive to other molecules and therefore they have great potential as chemical-sensing materials. Silver nanoparticles have also been confined within the channels of selectively grafted mesoporous silica SBA-15 [33]. The as-formed Ag-SBA-15 exhibits an excellent electrocatalytic activity toward the reduction of hydrogen peroxide.

18.1.4. Noble Metal Clusters

Differing from alkali metal ion clusters, the zero-valence noble metal clusters prepared through encapsulation in zeolites not only vary considerably in size range but also in the number of atoms contained in the cluster. For example, small-angle X-ray scattering (SAXS) and transmission electron microscopic (TEM) analyses indicate that the Pt-Y zeolite prepared through hydrogen reduction at 300°C contains Pt metal clusters with a diameter of 0.6–1.3 nm [34,35]. Furthermore, TEM structure analysis reveals that almost all the Pt clusters are smaller than the size (1.2 nm) of the supercage of zeolite Y and the particles are homogeneously distributed in the zeolite. It is reasonable to believe that these Pt clusters are located in the supercages of the Y zeolite [36]. The number of atoms in each cluster is probably 40 and the cluster is accommodated in the faujasite supercage in the form of a truncated tetrahedron. Of course, under different preparation conditions, other Pt clusters may also form in the sodalite cages of the Y zeolite or on the external surface of the crystal. It is claimed that a high dispersion of platinum to predominantly monoatomic particles in the cavities of NaX zeolite has been realized through impregnation and thermal decomposition of the Pt(II)-acetylacetonate [37].

Pd^{2+} ion and its reduced atom have a mobility in zeolite higher than the corresponding Pt^{2+} and Pt. As a result, it is difficult for all the prepared Pd metal clusters to be located in the supercages of the host–guest Pd-Y zeolite. In general, smaller Pd metal clusters with a diameter of 0.6 nm can be formed in the supercages of zeolite Y under mild conditions. Pd-exchanged NaY zeolites have been prepared [38] and used for catalytic Heck coupling reactions. Other noble metal (Ru, Rh, Os, Ir) clusters in zeolite Y prepared through ion exchange and hydrogen reduction method are usually smaller than the Pt cluster prepared under similar conditions. Some noble metal (Ru, Ir) clusters can also be obtained through loading of metal-carbonyl compounds in zeolite pores followed by pyrolysis.

The carbonyl metal cluster anions of Ru such as $[Ru_6C(CO)_{16}]^{2-}$ and $[H_2Ru_{10}(CO)_{25}]^{2-}$ can be loaded in mesoporous molecular sieves through direct absorption from organic solvents. Because the hydroxyls in the mesopores are able to interact with the carbonyl group of the cluster anions through H-bonding, the two species are attracted to each other and the cluster anions are bound in the mesopores tightly. High-resolution electron microscopy clearly shows that the metal cluster anions are arrayed in the MCM-41 channels [39]. Shephard et al. [40] prepared a bimetal complex of Ag and Ru and introduced this complex into the channels of MCM-41 through adsorption. The subsequent thermal treatment of the composite material converts the bimetal complex to metal nanoparticles. It is revealed that the metal nanoparticles are attached onto the channel walls of the mesoporous host with the silver atoms being closer to the SiO_2 wall. It is found that the bimetal cluster particles in the mesoporous channels are catalytically active for hydrogenation reactions. A series of bimetallic nanoparticles have been anchored within mesoporous silica hosts and the host–guest composites exhibit high catalytic activities and high selectivities [41] in a number of hydrogenations at low temperatures.

The reaction of trimethoxysilane with the SBA-15 molecular sieve leads to formation of a silane-modified mesoporous host and through impregnation metallic rhodium particles may be introduced [42] into the channels of the mesoporous host (Fig. 18.1). The metal particles are firmly bound to the mesoporous host and the as-

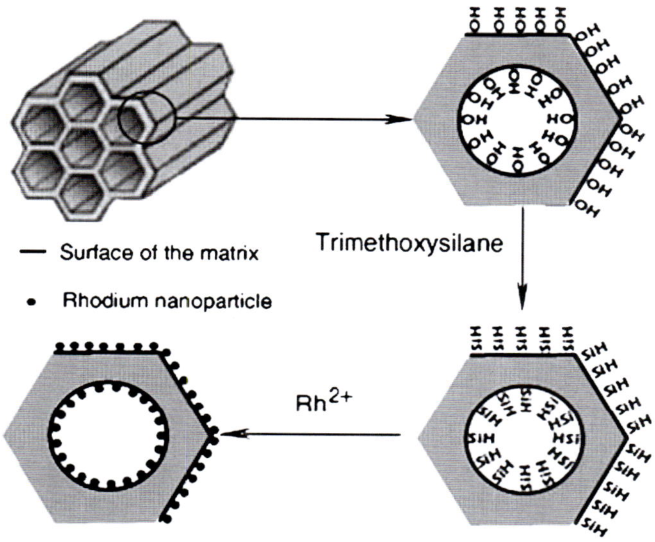

FIGURE 18.1 Schematic representation for the Rh nanoparticles attached on the external and internal surfaces of SBA-15 mesoporous host [42].

formed composite material shows high catalytic activity for the Heck coupling reaction. Gao and coworkers [43] grafted a branched polymer onto the walls of the mesoporous SBA-15 and they used this polymer-grafted SBA-15 to trap Pd particles. They discovered that the nanocomposite is a good catalyst for hydrogenation and the catalytic selectivity of the material is much superior to the conventional Pd/Al_2O_3 catalyst. Furthermore, the composite catalytic system shows excellent stability and recyclability. After storing for several months or many cycles of catalytic testing, the catalytic performance of the material remains almost unchanged. The polymer molecules grafted on the mesoporous channels play a role not only in the stabilizing of the Pd nanoparticles but also in facilitating their catalytic performance.

18.1.5. Other Metal Clusters

The mercury-exchanged zeolite X can adsorb a large quantity of mercury vapor to form Hg_2^{2+} and/or $Hg_{(x+2)}^{2+}$ species in the cages of the zeolites. When Cd-A reacts with cadmium vapor, Cd_n^{2+} ($n = 2-4$) and Cd_5^{4+} clusters [44–46] form in the zeolite, as revealed by structure and spectroscopic analysis. Cadmium and zinc react with H-Y to form Cd_2^{2+} and Zn_2^{2+} cationic clusters [47,48]. The reports on other zeolite–metal cluster host–guest compounds are very rare. On the basis of single crystal X-ray diffraction analysis, Heo et al. pointed out that [49,50] it is possible to grow In_5^{8+} and In_3^{2+} ion clusters in zeolite A. Univalent Zn^+ ions [51] in the SAPO-CHA pores can be prepared when zinc vapor reacts with the small-pore H-SAPO-CHA (chabazite structure) molecular sieve. Furthermore, trisulfur anionic radicals, which are chromogenic, have been formed through reaction of vaporized sulfur with univalent zinc in the Zn^+-containing SAPO-CHA (Fig. 18.2). It is found [52] that the sulfur radicals are accessible to guest molecules and respond uniquely to water molecules and as a result the radical-containing composite material may be used to detect trace amounts of H_2O efficiently on the basis of color variation. It is revealed that when water molecules are allowed to interact with the trisulfur radicals, the electrons on the radicals are transferred to the water molecules and correspondingly the color of the radicals is changed from deep blue to light yellow. The adsorption–desorption of water in the SAPO–CHA material with sulfur radicals is reversible so that a convenient system based on the composite material for water detection may be fabricated.

After grinding of the mixture of hydrated cobalt nitrate $[Co(NO_3)_2 \cdot 6H_2O]$ and NaX, the cobalt nitrate can easily enter the pores of the X zeolite. In cobalt nitrate-loaded NaX zeolite upon heating at 60°C, the cobalt nitrate decomposes to form Co_3O_4. If this host– guest material is further heated in H_2 atmosphere at 550°C, the cobalt oxide located in the zeolite would be reduced to cobalt atoms which undergo coarsening to form cobalt metal clusters [53]. However, X-ray diffraction and electron microscopy analysis indicate that the particle size of the cobalt clusters reaches 15–25 nm, much larger than the supercage size of the X zeolite.

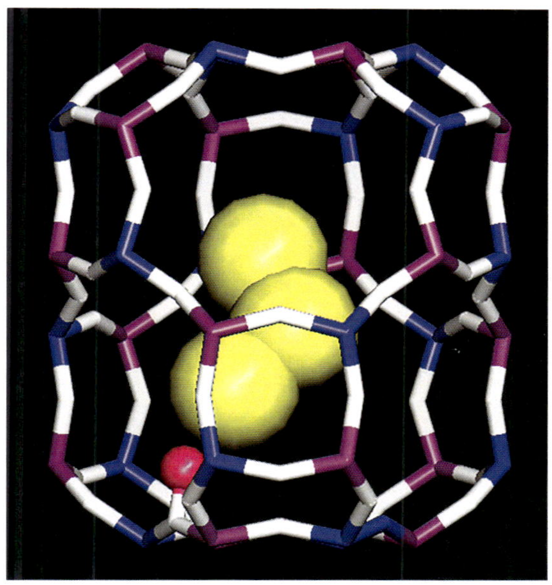

FIGURE 18.2 Schematic representation of encapsulation of a S_3 molecule in the cage of Zn-containing SAPO-CHA. *Reproduced with permission from The Royal Society of Chemistry [52].*

Further inspection reveals that the cobalt metal clusters are not dispersed on the external surface of the zeolite crystals but homogeneously distributed in the bulk of the crystals. The only explanation on this phenomenon is that there exist a large amount of defects in the zeolite crystals and the cobalt metal clusters are accommodated in these defects.

18.1.6. Clusters of Metal Oxides

Lanthanide-exchanged zeolite Y can be used as high-performance petroleum refinery catalysts after calcination. The thermal stability of this catalyst is very high and this high thermal stability may originate from the charged O-bridged lanthanide metal clusters formed in the sodalite cages of the zeolite [54,55]. On the basis of XRD analysis, Park and Seff [56] pointed out that in the sodalite and supercages of completely dehydrated, partially dehydrated, and anhydrous LaX zeolites there exist various metal–oxygen cluster ions. A typical example is the $La_4O_4^{4+}$ cluster located in the sodalite cage. Similar Pb_4O_4 cluster can also be formed in zeolite A. In fact, this Pb_4O_4 cluster can further interact with Pb^{2+} or Pb^{4+} ions to form larger clusters such as $Pb_8O_4^{P+}$. It has been evidenced that in sodalite there exists $Pb_2(OH)(H_2O)_3^{3+}$ [57]. When zeolite Y with adsorbed $Fe(CO)_5$ is thermally treated under O_2 atmosphere, Fe_6O_n clusters are formed and in these clusters, the Fe occupies the vertex of an octahedron [58]. The interaction of dehydrated Zn-A with Zn vapor results in body-centered cubic $Zn_9O_n^{P+}$ clusters. In fact, in these clusters there may exist Zn–Zn bonds because of O deficiency [59].

Li and Shi [60] successfully prepared a host–guest composite formed by incorporation of nanometer-sized Mg–Al layered double hydroxide (LDH) in a mesoporous material. This composite exhibits high catalytic activity as a basic catalyst. The size of the LDH nanometer particles is not larger than 9 nm and the particles contain rich active sites which render the composite material superior in catalytic performance.

18.2. ENCAPSULATION OF DYES IN ZEOLITES

In solutions, dye molecules aggregate even at very low concentrations. After aggregation, the energy of excited dye molecules can be released easily through thermal relaxation and therefore, their optically active properties cannot be realized. If the dye molecules are dispersed in the zeolite pores, the aggregation of molecules can be avoided and as a result the dye molecules may exhibit excellent optical activities such as lasing [61].

There are four approaches to load dyes in porous molecular sieves: (i) direct ionic exchange for cationic dyes [62], (ii) vapor phase deposition [63], (iii) crystallization encapsulation [64], and (iv) precursor in situ synthetic method [65]. Hoppe et al. [66] successfully loaded methyl blue into zeolite NaY through ionic exchange and crystallization encapsulation methods. X-ray structure analysis indicates that in the host–guest material prepared through ionic exchange, the dye molecule is located at the center of the supercage of the zeolite; whereas in the compound prepared via the crystallization encapsulation technique the dye molecule is close to the opening window of the supercage.

If detemplated $AlPO_4$-5 crystals are directly mixed with solutions of dye molecules, the dye molecules will enter the channels of the molecular sieve with the solvent molecules. After separation and drying of the obtained solid material, the solvent molecules will vaporize and the dye molecules will be attached in the channels of the microporous crystals. Rurack et al. investigated the loading of 2,2′-dibyridine-3,3′-diphenol dye molecules into $AlPO_4$-5 molecular sieve and the spectral properties of the formed host–guest composite [67]. It is found that the host–guest material exhibits intense fluorescence and the crystal displays optical anisotropy. The optical anisotropy indicates that the guest dye molecules loaded in the host channels are arranged linearly. Nevertheless, there is a tilting angle between the axis of the guest dye molecule and the running direction of the crystal channel. Proflavin is a commonly used luminescent dye molecule. It is somewhat basic and as a result this molecule is able to interact with proton to form a salt. Protonated proflavin can be exchanged into the pores of zeolites Y and L [68]. The fluorescent intensity of the proflavin dye molecules loaded in a zeolite increases with loading amount but decreases when the concentration reaches a certain value because when the dye molecules get too close to one another, self-quenching occurs. In zeolite L the proflavin molecules are present in both monoprotonated and di-protonated forms and the luminescent intensity is high, whereas in zeolite Y, the proflavin molecules are mainly monoprotonated.

By adding a small amount of pyridine 2 dye molecules into the synthetic system for $AlPO_4$-5, Vietze et al. [69] obtained a composite of $AlPO_4$-5 crystals with the encapsulated dye molecules. The diameter of the dye molecule is about 0.6 nm and the molecule can just be accommodated in the one-dimensional channel of the $AlPO_4$-5 crystal. It is believed that the chain direction of the dye molecules is parallel with the axis of the zeolite molecular sieve, i.e., the dye molecules are arranged along the microporous channels of the molecular sieve. The zeolite-dye host–guest composite obtained through this method exhibits apparent optical anisotropy and the crystal displays thermoelectrical properties, suggesting that the dye molecules in the $AlPO_4$-5 channels are oriented not only along the channel direction but also in a head-to-tail manner. When pumped with an Nd:YAG laser, the dye-containing $AlPO_4$-5 exhibits intense fluorescence. Spectral analysis indicates that the fluorescent emission peak is narrow and in some cases several lines appear, whereas some crystals give only one single line. The lasing behavior of the dye/$AlPO_4$-5 composite suggests that this host–guest composite can be used as solid laser [70].

This approach of preparation of microlaser through loading of dyes in regular channels is also valid for mesoporous materials. Yang et al. [71] and Wirnsberger and Stucky [72] used a similar method and successfully loaded dye molecules in mesoporous silica to form microlaser material. With block copolymer as template, Vogel et al. [73] successfully loaded rhodamine 6G dye molecules into mesoporous titania. Due to dispersion effect of the block copolymer, the loaded 6G molecules in the mesoporous material overcame aggregation and as a result, they exhibited superior laser emission properties. These mesoporous host–guest materials can form films and through lithography they can be patterned. Therefore, these materials may find applications in fabrication of microlasers and other optically active devices. Besides dye molecules, other luminescent matters can also be loaded into zeolite channels to form laser materials. For instance the loading of Nd^{3+} complexes in nanosized zeolite cages leads to a host–guest composite material that emits intensely in the near-infrared region [74]. The loading of rhodamine dye molecules in microporous and mesoporous molecular sieves results in sensing material. For example, after grafting onto the channel walls of the mesoporous MCM-41, rhodamine B sulfonate (Rh B-sulfo) shows fluorescent spectrum very sensitive to SO_2 molecule [75].

Calzaferri et al. assembled various dye molecules in zeolite L taking advantage of the wide channels of this zeolite and they found that these multi-component dyes exhibit efficient energy transfer properties as guests in the channels. They also called these host–guest systems zeolite antennae [76,77]. In another work, Calzaferri and coworkers [78] prepared oriented monolayers of zeolite L crystals which have their channels perpendicular to the surface of the substrate (Fig. 18.3). Subsequently, they inserted dyes into the open channel ends and tested the energy transfer of the host–guest composite. The external acceptor or donor stopcock dye at the channel ends traps electronic excitation energy from donor molecules or injects it to an acceptor inside the channels. The well-organized host–guest material offers possibilities for developing photonic devices.

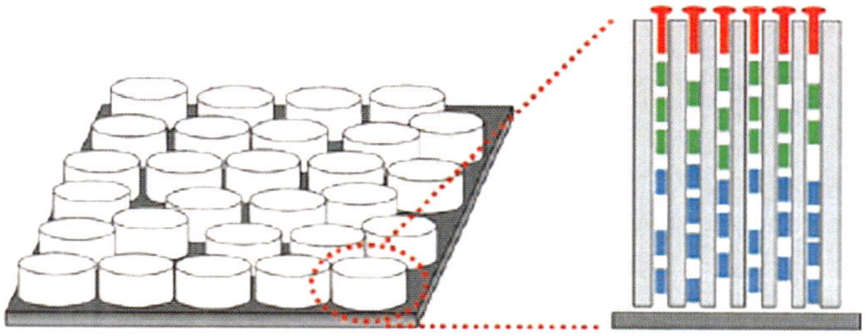

FIGURE 18.3 Organization of host–guest functional dye–zeolite crystals in oriented monolayers [78].

In another approach, pyronine molecules were loaded in the central part of zeolite L microcrystal whereas oxinine molecules were accommodated at the ends of the zeolite channels [79]. In this case, because the absorption and emission wavelengths of pyronine are shorter than those of oxinine, after excitation of pyronine upon light absorption, the energy will be transferred among different pyronine molecules along the zeolite channel and finally the energy will be transferred to the oxinine molecules at the ends of the zeolite crystal where the energy will be released through spectral emission. Calzaferri et al. also suggested that if the double-dye-molecule-loaded zeolite crystal is attached onto the surface of a semiconductor, it is possible that the energy will be transferred from the dye to the semiconductor, achieving the goal of exciting semiconductor with short wavelength light.

Bussemer et al. reported the loading of neutral dye molecules in the channels of zeolite L and performed quantum chemical study on the host–guest interactions [80]. The guest dye molecules show emission properties in zeolite L comparable to that of solution. Matrix rank analysis of the emission spectra reveals that there are three contributing pieces, whereas quantum chemical calculations provide different conformations depending on the Si/Al distribution of the framework and the extraframework cations and protons. Hashimoto et al. [81] applied fluorescence microscopy and confocal laser scanning microscopy techniques for the observation of space-resolved pictures of the distribution and chemical reaction of fluorescent dyes inside zeolite crystals.

In some cases, when the size of dye molecules is similar or larger than that of the pores of a microporous crystal, it is difficult to realize the host–guest assembly through direct loading. Nevertheless, an alternative approach is to load the dye molecules into a larger pore zeolite followed by alteration of the host zeolite structure through chemical–physical techniques and in this way larger dye molecules can be loaded into smaller-pore zeolites. For instance, under mild conditions methyl blue and perylene can be loaded into VPI-5 which is converted to $AlPO_4$-8 after thermal treatment at 120°C. This thermal treatment of the host–guest material will lead to an $AlPO_4$-8 with the guest dye molecules tightly confined in the framework of the newly formed host [82].

Microcrystalline $AlPO_4$-5 possesses one-dimensional circular channels that are suitable for loading chain organic molecules. When optically active organic molecules are orderly arranged in the $AlPO_4$-5 channels, they may exhibit special properties and functions. *para*-nitroaniline has relatively large dipole moment and it shows nonlinear optical properties. However, the crystal formed by *p*-NA molecules possesses symmetry center and therefore, the crystal exhibits no nonlinear optical property. To take advantage of the special feature of *p*-NA molecule, it is necessary to array these molecules in the three-dimensional space in such a way that no symmetry center arises. Zeolite Y has a symmetry center and therefore, loading of *p*-NA in zeolite Y cannot lead to nonlinear optical materials. However, $AlPO_4$-5 crystal structure has no symmetry center and if *p*-NA is loaded in $AlPO_4$-5 it is possible to form a host–guest material that shows nonlinear optical properties. This has been proved true. Through a vapor transport deposition technique, *p*-NA can be loaded in the one-dimensional channels of $AlPO_4$-5 [83,84]. The guest molecules can be arranged in the channels in two different manners. One is that the *p*-NA molecules are positioned head-to-tail to form a chain structure in the channel, whereas in another manner, there are weak interactions among the *p*-NA molecules and the molecules are similar to those in a molten state of *para*-nitroaniline. However, in the second case, the *p*-NA molecules in the $AlPO_4$-5 channels are still arrayed rather orderly. If the loading amount of *p*-NA in $AlPO_4$-5 is small, there are no H-bonds between the *p*-NA molecules and the obtained host–guest composite shows no second-order nonlinear optical properties (second harmonic generation, abbreviated as SHG). When the loading amount reaches a certain value, the host–guest composite exhibits apparent second harmonic generation and infrared spectroscopy indicates that in this case there are strong H-bonding interactions between the *p*-NA molecules in the channels. *p*-NA/$AlPO_4$-5 host–guest composite can exhibit not only nonlinear optical properties, but also thermoelectricity [85].

The loading of dye molecules in zeolites can also result in spectral hole-burning material. Hole burning is defined as the phenomenon of absorption intensity decrease at the irradiation laser frequency when a material is irradiated with a laser light. Spectral hole burning can be observed at low temperatures when thionine and methyl blue are loaded into zeolite X or Y through ion exchange [86]. The hole-burning frequency of the thionine/Y host–guest composite is $605.05\ cm^{-1}$ and the highest hole-burning temperature is 13 K. A higher hole-burning temperature (about 80 K) has been observed for the composite formed by loading of phthalocynine dye in zeolites [87].

18.3. POLYMERS AND CARBON MATTERS IN ZEOLITES

18.3.1. Polymers in Zeolites

The organic monomers adsorbed into zeolite pores can easily undergo polymerization to form polymers under proper conditions [88,89]. Attention has been attracted by polymeric materials with conducting properties in microporous and mesoporous molecular sieves. These polymers may exist as single chains due to the confinement of zeolite pores and therefore they are very important for investigation in physico-chemical properties of polymers and for miniaturization of electronic devices. When acetylene molecules are adsorbed into zeolites, they may polymerize under appropriate conditions to form polymeric fragments with conjugated double bonds [90]. In zeolites, the polymerization of monomers sometimes requires the presence of oxidizing agents. For instance, Cu^{2+} and Fe^{3+} exchanged into the pores of zeolites Y and mordenite can act as an oxidant to facilitate the polymerization of the successively adsorbed pyrrole or thiophene, forming polypyrrole or polythiophene [91–93]. Further oxidation of polypyrrole will render the polymer chains electrically conductive. Another common oxidant for polymerization is the water-soluble persulfate. Usually oxidation polymerization occurs when a zeolite with adsorbed monomers is mixed with persulfate. Through this approach polyaniline [94] can be prepared in zeolites mordenite and Y. The electrically conductive properties of polyaniline are closely related with the extent to which the molecule is oxidized and protonated and as a result, the structure and composition of the zeolite host may affect the properties considerably.

Using zeolites as a host template, acrylonitrile monomers can polymerize to form polyacrylonitrile in the zeolite channels [95]. First, the zeolite is dehydrated through evacuation followed by contact of the dehydrated zeolite with the vapor of liquid acrylonitrile monomers and then the monomers will be adsorbed into the zeolite channels through diffusion. Polyacrylonitrile encapsulated in the zeolite framework can be obtained by appropriately heating the mixture of the acrylonitrile monomer-containing zeolite, persulfate, and sulfite solutions. After formation of the host–guest material, the zeolite framework can be removed through dissolution in HF solution and the remaining polymer is characteristic of the normal polyacrylonitrile on the basis of analysis results. It has been found that the polymerization of acrylonitrile differs in zeolites with different structures. Moller et al. [96] investigated the formation of polymethylmethacrylate (PMMA) through polymerization of methylmethacrylate (MMA) in channels of microporous crystals including NaY, mordenite, zeolite β, ZSM-5, and mesoporous molecular sieves such as MCM-41 and MCM-48. MMA in zeolites can also polymerize as acrylonitrile and with the increase in the host channel size the polymerization degree is enhanced. Electron microscopic observation indicates that the polymerization reaction proceeds mainly inside the zeolite channels, because almost no polymers are observable on the external surface of the zeolite particles. This is further confirmed by the fact that the polymer/zeolite composite lacks the characteristic glass transition temperature for a bulk polymer.

When the suspension of dehydrated H-Y zeolite in dichloromethane is mixed with vinyl ether, the latter will enter the channels of the zeolite rapidly and undergo polymerization to various degrees [97]. Protons are present in H-Y and these protons may destroy the ether bonds of the vinyl ether, resulting in the formation of polyethylene cations on the polymer chains. The presence of the polyethylene cations causes the host–guest composites to show various colors. The Na/H ratio in H-Y and the substituent groups on the vinyl ether influence the behavior of the polymerization reaction and the length of the polyethylene cations to a great extent.

18.3.2. Preparation of Carbon Matters in Zeolites

Porous carbon can be prepared by using zeolites as the host materials. First, organic species such as acrylonitrile [95], polyacrylonitrile [98], polyfurfuryl alcohol [99], or phenolic resin [100] are loaded into zeolite channels and then the loaded organic species are carbonized through pyrolysis; the carbonized composites are mixed with an acid solution to dissolve the inorganic substrate and carbon molecular sieves are formed consequently. The zeolites used include zeolites Y, mordenite, β, and L. By varying organic species and preparatory conditions various porous carbon materials can be obtained. Except for a few examples, the pore structures of the porous carbons prepared through this approach are not uniform and therefore, these porous carbons cannot be used as molecular sieves as the zeolites do. Through similar techniques, porous carbons can also be achieved using non-one-dimensional pore mesoporous molecular sieves, but in this case the obtained porous carbons have a pore size falling into the range 5–10 nm. Furthermore, unlike the porous carbons obtained from microporous zeolites, the porous carbon materials obtained from mesoporous molecular sieves possess uniform pore structures and sizes [101–103]. Actually this type of porous carbons can be classified as molecular sieves because of their pore uniformity and they may exhibit chemical–physical properties different from those of other porous materials. For instance, platinum nanoparticles can be homogeneously dispersed onto the internal surface of these uniform-pore carbon molecular sieves [104].

If one-dimensional channel mesoporous molecular sieves are used as host for preparation of carbon matters, the obtained carbon is usually of linear structure because the carbon matter grows only in the one-dimensional channels and their interconnection to form a three-dimensional network is not possible. Wu and Bein [105] reported evaporation of acrylonitrile monomers into mesoporous molecular sieves at room temperature followed by polymerization of the monomers in the presence of a catalyst. The host–guest composite after polymerization underwent pyrolysis at various temperatures, forming one-dimensional graphitic matter distributed in the mesoporous channels. These graphitic species can be regarded as one-dimensional conducting wiles and the experimental results indicate that the one-dimensional carbon/MCM-41 possesses electrical conductivity. The zeolite/polymer host–guest precursor for preparation of porous carbon can also be obtained through direct contact of monomers in a carrier gas with zeolite molecular sieves followed by polymerization. For example, propylene can enter into zeolite Y under the carriage of N_2 and polymerize to form polypropylene. After pyrolysis, the polypropylene undergoes carbonization and the host zeolite framework of the carbonization product can be removed by dissolution in acids, leaving carbon material with characteristic pores [106].

It is also possible to prepare photoluminescent carbon nanoparticles in zeolite host materials. Using a magnesium-substituted aluminophosphate with a chabazite structure (MAPO-44) as the host, highly photoluminescent carbon nanoparticles have been obtained [107] through thermal decomposition of the occluded template or loaded organic molecules. The resulting composite containing carbon nanoparticles (C-MAPO-44) can be excited by a broad range of light in the ultraviolet region and the emission wavelength is tunable through varying the thermal treatment condition. It is demonstrated that the emission wavelength is dependent on the carbon content in the composite material and the higher the content, the longer the emission wavelength. Correspondingly, upon excitation at a single UV wavelength, the emission color is tunable from violet to orange red for samples with various carbon contents. Spectroscopic characterization reveals that heterogeneity or defects are prevalent in the carbon nanoparticles of the composite and it is these defects that form surface states responsible for the photoluminescence of the material. In contrast with the previously reported carbon phosphors which are coated with organic molecules, the C-MAPO-44 phosphor material is thermally stable because the material has been subjected to elevated temperature during the preparation process.

18.3.3. Fullerenes Encapsulated in Zeolites

The van der Waals diameter of C_{60} is about 1 nm, larger than the pore opening of most microporous crystals. Therefore, it is rather difficult to load C_{60} into zeolite molecular sieves, although the adsorption of fullerenes on the external surface of hydrophobic zeolites [108] is easily realized. However, VPI-5 is a one-dimensional channel microporous crystal with a relatively larger (about 1.25 nm) pore diameter. Its pore size is large enough to accommodate C_{60} molecules. Hamilton et al. [109] successfully loaded C_{60} molecules into VPI-5 channels through interactions of C_{60} benzene solution with VPI-5 molecular sieve under 50 atm at 50°C. Because of the confinement of the molecular sieve framework, the C_{60} in VPI-5 exhibits properties very different from those of the parent C_{60}. First of all, the C_{60} can prevent VPI-5 from conversion to $AlPO_4$-8 under thermal treatment. C_{60} can also be loaded into the channels of mesoporous materials through solution impregnation. Drljaca et al. [110] mixed the toluene solution of C_{60} with mesoporous SiO_2 and obtained a host–guest composite containing solvated C_{60} molecules in the channels. The color of this solid material is similar to that of the C_{60} toluene solution (pale purple). However, when the toluene solvent in the mesoporous channels is evaporated through thermal treatment, the sample becomes yellow. The appearance of this yellow color and the corresponding UV–vis diffuse reflectance spectrum indicate that the C_{60} molecules in the channels undergo aggregation. If the yellow-colored host–-guest material is contacted with toluene again, the original pale purple color reappears, suggesting that the aggregation and dissolution of the C_{60} molecules located in the mesoporous channels is a reversible process. If hydrophobic cyclodextrin molecules with strong encapsulating ability are added into the mesoporous SiO_2 solid containing toluene/C_{60}, these cyclodextrin molecules easily capture the C_{60} molecules in the mesoporous channels to form C_{60} aggregates in the space encapsulated by the cyclodextrin. In this case, the host–guest solid material becomes brown yellow. C_{60} can also be directly loaded into the channels of mesoporous molecular sieves through thermal diffusion.

Chen et al. [111] mixed C_{60} and MCM-41 material with various features under vacuum and they observed the dehydroxylation behavior of the latter. They discovered that the C_{60} apparently facilitates the dehydroxylation process. MCM-41 possesses a large amount of hydroxyls on the inner walls after detemplating and these hydroxyls exist as either H-bonded or isolated forms [112]. It has been revealed that when the loading amount of C_{60} is small, the isolated hydroxyls are lost more readily than the H-bonded ones, and the dehydroxylation behavior in the presence of C_{60} differs for MCM-41 materials with different hydroxyl amounts or features. Infrared spectroscopy analysis indicates that during dehydroxylation, the C_{60} captures the hydroxyl groups of the MCM-41 silanols to form C–H and C–OH bonds followed by condensation at higher temperatures to be removed as H_2O molecules.

18.3.4. Carbon Nanotubes Grown in Zeolites

Since Iijima reported [113] carbon nanotube preparation and its microstructure in 1991, carbon nanotubes have attracted enormous attention. Because of their unique structures, these materials exhibit a variety of interesting physico-chemical properties and they may play an important role in the manufacture of nanodevices. Among the carbon nanotubes that can be capped by semisphere of C_{60} [114], those with a zigzag structure have an integer pair of (9,0), whereas those with an armchair structure have an integer pair of (5,5). Carbon nanotubes can be prepared through various techniques. The technique reported by Iijima is similar to the one for the preparation of C_{60} and other fullerenes, i.e., graphite arc-discharge method. Pyrolysis of organic molecules such as benzene in the presence of H_2 has also been used to prepare multiwall carbon nanotubes. Using porous alumina as an external template, carbon nanotubes can be grown in the channels. However, the pore size of porous alumina is large, falling into macropore (diameter > 50 nm) category and therefore, the diameter of the prepared carbon nanotubes is also large and they are usually multiwalled.

Although there are many techniques for preparation of carbon nanotubes, the preparation of single-wall carbon nanotubes with a uniform diameter is usually not easy. Carbon nanotubes with a diameter smaller than 1 nm are even difficult to prepare. Because the properties of single-wall carbon nanotubes are different from those of multiwall ones and these properties can be easily elucidated, it is more important to prepare single-wall carbon nanotubes. Theoretically, microporous crystals with one-dimensional channels are able to be used as external templates for preparation of carbon nanotubes. Of course, the pore size of these microporous crystals should not be too small because otherwise even the thinnest carbon nanotube cannot be accommodated in the micropores. Tang and coworkers [115,116] tried the growth of carbon nanotubes in $AlPO_4$-5 and they succeeded. First, they prepared perfect $AlPO_4$-5 single crystals hydrothermally using tripropylamine (TPA) as the template in the presence of F^- ions. Second, they placed the $AlPO_4$-5 single crystals in a vacuum system for pyrolysis at 500−800°C. The pyrolyzed $AlPO_4$-5 single crystals appeared deep black and they exhibited anisotropy for polarized light absorption. When the polarization direction of the polarized light coincides with the crystal channel running direction, the absorption reaches the maximum, whereas when the polarization direction is perpendicular to the channel axis the absorption is the weakest [117].

Although in principle the formation of carbon nanotubes in $AlPO_4$-5 is possible, it is rather difficult to confirm the existence of carbon nanotubes in the pyrolyzed product in the $AlPO_4$-5 channels. The carbon matter obtained through dissolution of the pyrolyzed $AlPO_4$-5 crystals in hydrochloric acid is very easily damaged and form graphite fragments under high-resolution electron microscope. However, after close inspection, it is found that tiny single-wall carbon nanotubes are indeed present [118]. These observed carbon nanotubes may be more stable than others because they have fewer defects. The diameter of the carbon nanotube is about 0.42 nm, larger than the lattice spacing (0.34 nm) for the graphite fragments in the same TEM image. The evidence for the presence of carbon nanotubes in the pyrolyzed $AlPO_4$-5 crystal is from Raman spectrum. The Raman signals corresponding to the template molecules appear in the spectrum of the as-synthesized TPA-containing $AlPO_4$-5 crystals. When the sample is treated at 400°C, the Raman spectral peaks disappear completely and in this case the template molecules decompose to form an amorphous matter. If the crystals are calcined in air or in O_2 atmosphere, the carbon matter is removed completely and in this case there are no Raman signals. Nevertheless, when the crystals are pyrolyzed in vacuum at 500−550°C, the Raman spectrum shows a series of new peaks which are different from either TPA or graphite (diamond) signals. This phenomenon indicates that a new allotrope of carbon is formed. On the basis of the signal positions, it is inferred that this new carbon matter possesses the features of a carbon nanotube [119]. Similar to the UV−vis absorption, the Raman spectrum of the carbon nanotube containing $AlPO_4$-5 also exhibits anisotropy.

The powder X-ray diffraction patterns for the as-synthesized $AlPO_4$-5 crystals, the pyrolyzed carbon nanotube-containing $AlPO_4$-5 crystals, and the $AlPO_4$-5 crystals after complete template removal show no distinct change. However, the relative intensities of the diffraction peaks of the three materials differ and with the removal of the template the diffraction intensities increase [120]. In addition, the diffraction positions are moved slightly toward higher angles. From the X-ray diffraction patterns it is concluded that the template removal leads to slight lattice contraction of the $AlPO_4$-5 crystals, but the formation of carbon nanotubes does not damage the basic structure of $AlPO_4$-5.

Owing to their structural features, carbon nanotubes exhibit unique electrical conduction properties. The carbon nanotubes prepared through common methods have a larger diameter and usually they show metallic or semiconducting properties. In $AlPO_4$-5 crystals, upon pyrolysis the carbon nanotubes can grow along the micro-crystal channels and they array orderly. As a result, it is convenient to measure their conduction properties. The experimental results show that the template-containing $AlPO_4$-5 and the completely detemplated $AlPO_4$-5 crystals are typically insulators, whereas the carbon nanotube/$AlPO_4$-5 crystal exhibits unique conduction properties. At room temperature, the measured conductivity is in 0.1 $\Omega\,cm^{-1}$ order of magnitude. The magnitude of this conductivity is smaller than that of metallic carbon nanotubes but is similar to that of semiconducting carbon nanotubes. As the temperature is decreased, the conductivity drops as well, indicating that the carbon nanotubes have semiconducting behavior. However, when the temperature is lowered below 20 K, the carbon nanotube/$AlPO_4$-5 crystals exhibit Meissner effect and this effect implies that the carbon nanotubes in the $AlPO_4$-5 crystals are one-dimensional superconductor at temperatures below 20 K [121].

18.4. SEMICONDUCTOR NANOPARTICLES IN ZEOLITES

Since late 1980s, increasing interest has been focused on chemical fabrication of zero-dimensional semiconductor clusters (or quantum dots) because these species exhibit unusual opto-electronic properties [122]. As the pore diameter of a zeolite is usually smaller than 1.5 nm, the semiconductor particles grown in a zeolite microporous crystal host will exhibit apparent quantum size effect.

Metal organic chemical vapor deposition (MOCVD) technique can be used to prepare semiconductor nanoparticles in the pores of zeolites effectively. A typical example is the formation of compound semiconductor particles in zeolite Y through grafting of organometallic species followed by reaction with gases. This approach can be used to obtain II−VI, IV−VI, and III−V type semiconductor compounds encapsulated in zeolite Y [123−125]. Due to the presence of charge-balancing protons, it is very easy for H-Y to react with the organometallic species in the channels. Quantitatively, the number of protons in zeolite channels corresponds to the number of organometallic molecules that can be grafted into the zeolite channels to form methane molecules. After formation of the grafting host–guest material, the grafted organometallic groups may react with the H_2S introduced into the system to form hydrogen sulfide compounds. Afterward, the host–guest material is thermally treated at a certain temperature and the hydrogen sulfide species in the microporous channels will condensate to form sulfide clusters (Sn_4S_6). Of course, other types of metal sulfide nanoclusters such as Cd_6Se_4 and Zn_6S_4 can also be obtained through variation of the metal in the organometallic molecule or through reaction with hydrogen sulfide. In some cases, the newly formed protons can be reacted with further introduced organometallic compounds and as a result, the amount of the loaded metal sulfide can be increased further. Through this approach, not only II–VI but also III–V type semiconductor, such as GaP, clusters can be prepared. These semiconductor clusters show distinct quantum size effects and apparent blue shift is observed for the band edge of their electronic transition absorptions.

In 1990, Canham reported [126] the room temperature luminescence of porous silicon. This important discovery has attracted much attention for the investigation of silicon with quantum size effect. Through chemical vapor deposition, it is also possible to load silicon nanoparticles into zeolite channels and pores. First, disilane is reacted with the protons in a zeolite and is grafted to the zeolite walls. Second, mild thermal treatment leads to the decomposition of the grafted disilane molecules and the decomposed product reacts with the grafted disilane to form clusters. Finally, clusters containing as many as 60 Si atoms are formed in the supercages of zeolite Y [127,128]. These clusters located in the zeolite cages emit orange–red light at room temperature. With the decrease of temperature, the emission intensity of the host–guest material is further enhanced. Gao et al. used single crystal silicon as the silicon source and prepared giant crystals of silicalite-I and they deposited silicon nanoparticles in the zeolite channels through reaction of silane with the detemplated silicalite-I single crystals [129]. Further investigation indicates that the nano-silicon particles located in the silicalite-I single crystals emit strong red light and the luminescence spectral wavelength is 570 nm at room temperature, whereas it shows blue shift (551 nm) at 10 K. Besides silicon nanoclusters, through the vapor deposition technique it is also possible to prepare germanium and silicon–germanium nanoclusters in zeolite Y. Similar to compound semiconductor nanoclusters, these element semiconductor nanoclusters also exhibit apparent quantum size effects and their band gap is distinctly blue-shifted in comparison with the corresponding bulk matters.

Through chemical vapor deposition, not only semiconductor nanoclusters in microporous zeolites can be prepared, but also larger semiconductor clusters in mesoporous molecular sieves can be obtained. Detemplated mesoporous silica possesses rich silanols on the channel walls and these silanols easily react with organometallic molecules so as to graft the latter onto the mesoporous channel walls. It has been demonstrated that in detemplated MCM-41 200 wt% disilane can be loaded [127]. After thermal treatment, these grafted disilane species decompose to form silicon nanoclusters. Because the silicon content is high, the silicon nanoclusters in the mesoporous channels are actually able to connect one another to form nanowires. The case for nano-semiconductors in microporous crystals, the nano-semiconductor clusters located in mesoporous molecular sieves also exhibit quantum size effects. Their band gaps and emission energies are correlated with the loading amount of semiconductor and the particle size. Through chemical vapor deposition, Ge nanowires can also be formed in mesoporous channels [130].

It has been shown that the reaction of Cd^{2+}-ion-exchanged zeolites with H_2S gas results in the formation of cadmium sulfide nanoparticles. In the sodalite cage of faujasite nanoclusters such as Cd_4S_4 can be formed as confirmed by extended X-ray fine structure analysis and X-ray powder diffraction structural analysis [131]. Owing to the quantum size effects of nanoparticles, the UV–vis diffuse reflectance spectrum of the cadmium sulfide nanoparticles loaded in zeolite microporous channels is distinctly different from that of bulk CdS. The band gap of bulk CdS semiconductor corresponds to the absorption edge of electronic transition. However, in microporous zeolite crystals, if the concentration of the Cd_4S_4 nanoparticles formed is small, the related electronic absorption spectrum markedly exhibits blue shifts, indicating that as the nanoparticles are formed the semiconductor band gap is widened. When the amount of Cd^{2+} ions in the system is increased, the concentration of the formed Cd_4S_4 nanoparticles is increased as well. In this case, because of the interactions between semiconductor nanoparticles located in adjacent sodalite cages, the absorption edge shifts gradually toward longer wavelength region. Nevertheless, for molecular sieves with different structures, the degree to which the nanoparticles interact with one another varies.

In the treatment of zeolite with H_2S many mesoporous defects will be produced and these mesoporous defects may accommodate larger cadmium sulfide clusters. Therefore, in the microporous crystal with a particular structure, there may exist many different cadmium sulfide particles with various sizes. However, the content of these cadmium sulfide clusters located defects is usually limited and as a result, they have little effects on the electronic spectral properties of the composite. Detailed composition and structural analysis indicates that the nanoclusters contain not only Cd and S, but also O. In fact, these clusters can be written as $Cd_4(S,O)_4$. CdS and ZnS clusters have also been incorporated within the pore structure of mordenite and clinoptilolite. It is demonstrated that the metal sulfide clusters interconnect to form well-defined aggregates with a behavior intermediate between that of discrete clusters and the bulk semiconductor. Furthermore, the dynamics and the optical absorption are strongly affected by the interface and the host [132].

Shi and coworkers [133] reported the preparation of mesoporous molecular sieve membranes using a block copolymer as the template. After formation of the molecular sieve membrane, Cd^{2+} cations were introduced into the molecular sieve channels through ion exchange, followed by passing gaseous H_2S to form CdS nanoparticles. The amount of CdS formed in the channels is so large that the optical property of the composite membrane is varied to a great extent. The third-order nonlinear optical coefficient of the as-prepared membrane is increased by two orders of magnitude in comparison with that previously reported for other CdS–polymer membranes. The enhancement of the nonlinear optical performance is attributed to the increase of CdS amount, the homogeneity of CdS particle size distribution and the localized field effect arising from the matrix around the CdS particles. Besides cadmium sulfide, other semiconductor particles can also be loaded in zeolite microporous crystals. Moller et al. prepared cadmium sulfide nanoparticles in zeolite Y through a similar approach [134]. Nevertheless, structural analysis indicates that the formed cluster particles are actually rather complex and apart from cadmium sulfide clusters, there exist other nanoclusters such as Cd_4O_4 or Cd_2O_2Se in the channels of zeolite Y. These nanoclusters are not isolated and they strongly interact with the framework oxygen of the zeolite.

The low temperature phase of bulk silver sulfide is a semiconductor with a monoclinic structure and its room temperature band gap is 1 eV. Because of its unique luminescent property, silver sulfide nanoparticles have attracted much interest recently. The technique to grow silver sulfide nanoparticles usually involves formation of Ag^+-loaded zeolite through ion exchange of Ag^+ ions into a particular zeolite (such as zeolite A). Silver sulfide nanoparticles can be obtained through reaction of the loaded zeolite with H_2S gas. The silver sulfide particle size in zeolite A may be controlled through varying the amount of Ag^+ ions initially exchanged into the zeolite channels. Nevertheless, if the silver sulfide nanoclusters are located only in the α-cages of zeolite A, their size should not exceed the diameter of the cage (about 1.5 nm). At higher concentrations, the silver nanoclusters in adjacent cages may interact with one another, leading to variation of the UV–vis absorption and emission spectra [135,136].

PbI_2 exhibits semiconducting properties. This compound may be loaded into zeolite channels through vapor transportation. When dehydrated Na-A zeolite is sealed with PbI_2 in an evacuated quartz tube followed by thermal treatment at 420°C for 24 h, the PbI_2 will enter into Na-A through vapor deposition process [137]. The loading amount of the guest in the host can be varied through controlling the ratio of zeolite over PbI_2 in the quartz tube. UV–vis diffuse reflectance spectroscopy reveals that as the loading amount of the guest is increased, PbI_2 forms $(PbI_2)_4$ clusters initially in the zeolite supercages followed by growth to $(PbI_2)_5$ clusters. HgI_2 is also a typical semiconductor material and it is also easy to vaporize after thermal treatment. Therefore, HgI_2 may be loaded into various zeolites through vapor transportation approach. Due to the confinement of zeolite framework, HgI_2 in zeolites may exhibit distinct quantum size effects. The electronic transition absorption spectrum of HgI_2 loaded in $AlPO_5$-5 single crystals shows apparent blue shift and in the meantime, the composite exhibits anisotropic property [138]. This anisotropy suggests that the HgI_2 in $AlPO_4$-5 one-dimensional channels are probably present as chains.

Chen et al. [139] described the preparation of ZnO nanoarrays from the one-dimensional channels (diameter 0.9 nm) of VSB-1 nickel phosphate crystals. These ZnO nanoarrays exhibit an unusual ultraviolet exciton absorption which is located at about 266 nm, a wavelength much smaller than that of the exciton absorption for bulk zinc oxide. In addition, the absorption peak and the corresponding photoluminescent emission peak of the ZnO nanoarrays are both quite sharp. The blue shift of the absorption is attributable to the quantum confinement effect, whereas the sharpness of the peaks indicates that the ZnO nanoparticles are uniform in size distribution. Because the diameter of the VSB-1 channels is 0.9 nm, the diameter of the ZnO nanoparticles formed inside the channels of the VSB-1 host should not exceed 0.9 nm. Nevertheless, the Bohr radius of the ZnO exciton is 1.25 nm, distinctly larger than the diameter of the VSB-1 channel diameter (or the diameter of the ZnO nanoparticle), accounting for the observation of the intense exciton absorption peak. Zhang et al. [140] grafted amino groups onto the channel walls of MCM-41 and using this modified mesoporous material as a precursor they prepared a Zn^{2+}-containing MCM-41 through adsorption. Zinc oxide nanoclusters were formed in the MCM-41 channels upon calcination of the Zn^{2+}-containing MCM-41 material. Spectroscopic and electron microscopic characterization demonstrates that the zinc oxide particles are located in the mesoporous channels and the absorption peak of the nanoparticles shows remarkable blue shift as compared to the bulk ZnO. Photoluminescence has been observed for the ZnO particles located in the channels. The emission wavelength is in the visible region and this visible emission may originate from defects in the ZnO particles. Through similar approaches it is possible to prepare cobalt oxide nanoparticles in MCM-41 channels, but the corresponding nickel and copper oxide nanoparticles fail to form using the same preparation technique. Abe et al. [141] successfully loaded Fe_2O_3 nanoparticles in MCM-41 and the band gap of the loaded particles is increased from that (2.1 eV) of the corresponding bulk material to 4.1 eV.

Element semiconductors include Se, Te, Ge, Si, etc. Selenium has a low melting point (230°C) and at lower temperatures it gives high vapor pressure. Therefore, it is possible to load selenium in a zeolite through vapor deposition at a temperature as low as 150°C when selenium is sealed in a vacuum system with the dehydrated zeolite. Parise et al. investigated the incorporation of Se into zeolites A, X, Y, $AlPO_4$-5 and mordenite [142]. Element, selenium, exists as different forms in different channel structures. In linear channel molecular sieves ($AlPO_4$-5 and mordenite), it is present as chains running along the channel direction [143–145], whereas in the smaller-cage zeolite A, it is located in the cages as Se_8 ring molecules. However, in larger cage zeolites such as X and Y, there exist both selenium ring molecules and spiral chains. Selenium/zeolite Y nanocomposites can be readily prepared through sorption of the liquid or gaseous chalcogen at moderate temperatures. The host–guest interaction renders it possible to tune the structural, electronic, and thermodynamic properties of the composite materials, whereas the molecular character of Se can be shifted from chains to rings through interactions with the extraframework cations [146]. The distance between the adjacent selenium atoms in each selenium chain located in a zeolite is shorter than that of the Se–Se bond length of bulk selenium crystal, because the interactions between selenium chains are weakened when they are accommodated in zeolite channels. Poborchii et al. through polarized Raman spectroscopic study on Se-incorporated $AlPO_4$-5 single crystals [147], discovered that in the main channels there are a small amount of Se chains besides the spiral Se single chains and Se_8 ring molecules. In mordenite, the state of selenium in the channels varies to a certain extent depending on the types of cations in the zeolite and the preparation method of the host–guest material [148]. When the Na is replaced by K in mordenite, the ratio of chained Se over Se_6 ring in the loaded guest species is decreased and in the meantime, if the guest is loaded with the vapor deposition method, the obtained guest Se is more ordered than the guest molecules obtained through liquid Se injection. The case for Te is similar to that for Se, i.e., the guest Te may exist in the host as either chains or as ring forms depending on the host channel structure. Polarized Raman spectroscopy and electronic transition spectroscopy indicate that [149] the spiral S and Te chains are able to be loaded in mordenite through vapor deposition technique and in the meantime, S_6, S_8, and Te_6 elemental ring molecules may also exist.

III–V type compound semiconductors have been gaining much more attention recently. It has been reported that III–V type semiconductor nanoparticles or nanowires can be grown in mesoporous channels through chemical vapor deposition. The preparation of III–V semiconductor nanoclusters follows the same principle as for silicon and germanium clusters. First, the organometallic compound of Al, Ga, or In (such as trimethylindium) is grafted onto the mesoporous channel walls through vapor deposition reaction and second phosphine is introduced to react with the alkyl metal to form III–V semiconductor compound particles. Because the reaction of phosphine proceeds at a rather high temperature (about 300°C), there is no need to pyrolyze the reaction product [150]. Spectroscopic and high-resolution electron microscopic analysis indicates that the III–V semiconductor clusters prepared through this technique are distributed not only in the mesoporous channels but also partly deposited on the external surface of the mesoporous molecular sieve. Through reaction of adsorbed $(CH_3)_3Ga$ with PH_3, it is possible to grow GaP nanoclusters in zeolite pores and channels. Srdanov et al. investigated the assembly of GaAs in mesoporous MCM-41 and the optical properties of the assembled host–guest composite material [151]. They used *tert*-butyl arsine and trimethylgallium as the arsenic and gallium sources, respectively, and deposited gallium arsenide directly in the MCM-41 channels at 700°C through organometallic chemical vapor deposition. The electronic transition absorption spectrum of the host–guest composite compound formed through deposition shows apparent blue shift, indicative of quantum size effect. The luminescent property of the composite depends on pore size of the host MCM-41 material. Further analysis indicates that the particle size distribution of the deposited GaAs nanoparticles is wide and the particles are present not only in the MCM-41 channels but also on the external surface of the mesoporous molecular sieve. The GaAs particles on the external surface of mesoporous material are larger in size.

GaN dispersed in mesoporous channels can also be prepared using MCM-41 as a host material. Winkler et al. [152] reported the preparation of the composite material MCM-41/GaN nanoparticles through passing ammonia into the organogallium-containing MCM-41 precursor. This composite shows characteristic luminescent property of GaN. The introduction of GaN into the MCM-41 channels decreases the X-ray diffraction intensity of the latter and correspondingly the surface area and pore volume of the MCM-41 host are reduced to a considerable extent. However, the high-resolution TEM reveals that the mesoporous structure of the MCM-41 host remains unchanged after loading of the GaN nanoparticles, indicating that the nanoparticles are indeed located in the mesoporous channels.

18.5. METAL COMPLEXES IN MOLECULAR SIEVES

Metal complexes encapsulated in zeolite molecular sieves exhibit physical–chemical properties different from those in solutions and in solid state [153]. The molecular sieve frameworks protect the metal complexes and the latter show enhanced thermal stability and anti-oxidation property. The methods for preparation of metal complexes in zeolites are versatile, but they may be classified as the following: (a) direct synthesis of porous framework compounds encapsulating metal complexes using the metal complex as a template; (b) through the so-called ship-in-bottle approach, small ligands complex metal ion helps to form larger metal complex in the zeolite channels; (c) volatile metal complexes can be loaded into zeolite channels through vapor transportation; and (d) grafting of complexes onto the walls of molecular sieves and this approach is suitable for preparation of complex/mesoporous molecular sieve host–guest composites.

18.5.1. Incorporation of Metal–pyridine Ligand Complexes

18.5.1.1. Pyridine Complexes

Copper–pyridine complex was used as a homogeneous catalyst for oxidation-coupling reactions and the reactants included acetylene [154], phenols [155], and so on. Copper(II) tetra–pyridine complex may catalyze the oxidation of 2,6-dimethylphenol to oxides of phenols in the presence of O_2. If the complex is loaded or dispersed on a solid support or into a porous crystal, it may be possible to obtain composite heterogeneous catalysts with superior catalytic performance. Ukisu et al. [156] successfully incorporated the copper complex of substituted pyridine into the pores of zeolite Y and they investigated the state and catalytic performance of the loaded complex. They first dealuminated zeolite Y to form Si–OH enriched microporous crystal and then they used 2-(2-trichlorosilanyl)ethylpyridine to react with the Si–OH groups. The former is thus grafted onto the walls of the Y zeolite, forming Py–NaY composite. When this Py–NaY composite is contacted with Cu^{2+}-containing solution, Cu–Py–NaY composite is formed. X-ray absorption fine structure analysis indicates that the oxidation state of the oxidized copper is +2 whereas that of the reduced copper is +1. Dai and Lunsford [157] ever reported that the Cu^{2+} of $[Cu(Py)_4]^{2+}$ located in the cages of zeolite Y is reduced to Cu^+ after evacuation to dehydrate, but the Cu^+ is able to be oxidized to Cu^{2+} again in the presence of O_2.

Böhlmann et al. [158] loaded copper pyridine complex into mesoporous MCM-41 and investigated the electron paramagnetic property of the composite. The loading method they used was ion exchange, i.e., the cation complex of Cu^{2+} and pyridine was first prepared followed by ion exchange of the complex into Al-containing MCM-41 channels. Pöppl and Kevan [159] incorporated copper–pyridine complex into Al-free MCM-41 channels and they found that in this case the interactions between the complex and the MCM-41 walls are weak. Yamada [160] investigated the substitution reaction of copper(II) bis(acetylacetonate) loaded in zeolite NaX by pyridine. ESR signal variation was clearly observed after the substitution reaction and this substitution reaction could not proceed in solution or on silica gel. The static electric field generated by the zeolite cations tends to reduce the stability of the Cu $(acac)^{2+}$ complex.

The complexes of pyridine and ruthenium possess unique luminescent and catalytic properties and therefore their assemblage as guest molecules has been extensively investigated. Dewilde et al. [161] first reported the incorporation of tris(2,2′-bipyridine)ruthenium(II) complex into zeolite Y through the ship-in-bottle approach. They first introduced $[Ru(III)(NH_3)_6]^{3+}$ into zeolite Y through ion exchange to form $[Ru(III)(NH_3)_6]$-Y and then heated the mixture of bipyridine and $[Ru(III)(NH_3)_6]$-Y. In this case, the Ru(III) was reduced to Ru(II) and $[Ru(II)(bpy)_3]$-Y, which has spectral properties completely different from that in solution, was formed. After incorporation, the complex exhibits versatile opto-physical behaviors which depend on hydration degree, and loading amount. Later on, Quayle and Lunsford [162] reported the oxidation of the $[Ru(II)(bpy)_3]$-Y to $[Ru(III)(bpy)_3]$-Y using Cl_2. The Ru(III) complexes loaded in zeolites are able to decompose water into O_2.

Lainé et al. [163] described the encapsulation of $[Ru(bpy)_3]^{2+}$ in the supercages of zeolite Y. They found that generally the first ship-in-bottle reaction may introduce $[Ru(bpy)_3]^{2+}$ into 50% of the supercages of zeolite Y, whereas multiple loadings may incorporate the complex into 65% of the supercages. Attempt to load $[Ru(bpy)_3]^{2+}$ into more than 65% of the supercages would lead to the formation of other complexes such as $[Ru(bpyn(NH_3)_{6-2n})]^{2+}$. Ledney and Dutta [164] investigated, in detail the catalytic performance of [Ru(III)(bpy)$_3$]$^{3+}$ loaded in zeolite for the oxidation of water and it was found that when the pH of the system was lower than 4, no O_2 was formed. They believed that the Ru(III) complex first binds H_2O to form OH radicals which evolve to H_2O_2 and the latter then interacts with the un-reacted Ru(III) to form O_2. The same research team [165] prepared [Ru(bpy)$_2$(H_2O)$_2$]-Y and oxidized it to [(bpy)$_2$(Oz)Ru(IV)=O]$^{2+}$ in air (where Oz stands for zeolite framework O). They also prepared Ru(bpy)$_3$]$^{2+}$–zeolite composite material with different loading amounts and investigated the interactions between the guest species located in different supercages [166].

Maruszewki et al. [167] characterized the multipyridine complex of ruthenium loaded in zeolite Y cages using absorption and resonance Raman spectroscopies. The complexes involved included [Ru(bpy)$_2$]$^{2+}$, [Ru(bpy)(bpz)]$^{2+}$, [Ru(byp)$_2$(dmb)]$^{2+}$, and so on. Photoelectron transfer reaction may occur [168] between [Ru(bpy)$_3$]$^{2+}$ loaded in zeolite cages and methyl violet. After excitation by light illumination, the excited state *[Ru(II)(bpy)$_3$]$^{2+}$ generated through charge transfer transition transfers its electron to the methyl violet in an adjacent cage to form MV$^{+\bullet}$ ion radical, whereas the complex itself is converted to [Ru(III)(bpy)$_3$]$^{3+}$ ion. To maintain the charge balance with the zeolite framework, for every electron transferred, an Na^+ ion must be moved simultaneously. Therefore, the mobility of the Na^+ ion affects the electron transfer rate and the recovery of the electron. The lifetime of the MV$^{+}\cdot$ cationic radical is several hours. Raman spectroscopy indicates that the radical has strong interactions with the zeolite framework. Sykora and Kincald [169] incorporated the complex of ruthenium with bipyridine, 2,2′-bipiperazine (abbreviated as bpz), Ru(bpy)$_2$bpz, ruthenium 5-methyl-2,2′-bipyridine (abbreviated as mmb) complex, and *N,N′*-trimethyl-2,2′bipyridine (abbreviated as DQ$_{55}$$^{2+}$) into the adjacent supercages of zeolite Y to form a photochemical system in which the [Ru(II)(bpy)$_2$(bpz)]$^{2+}$ acts as a photosensor. Upon illumination with light, an electron is promoted from the ground state to the excited state to form *[Ru(bpy)$_2$(bpz)]$^{2+}$and then this excited state loses an electron which is transferred to the DQ$_{55}$$^{2+}$, the latter being reduced to DQ$_{55}$$^{+}$, whereas the complex being oxidized to [Ru(bpy)$_2$(bpz)]$^{3+}$. The oxidized complex may obtain another electron from the [Ru(mmb)$_3$]$^{2+}$ in an adjacent cage to form its reduced state. In this way, the [Ru(mmb)$_3$]$^{2+}$ does not interact with DQ$_{55}$$^{2+}$ directly to undergo redox reaction and the electron transfer process is accomplished through the photosensor-separated electron donor–acceptor system. After reduction, the electron acceptor DQ$_{55}$$^{2+}$ interacts with the propylviologen sulphonate (PVS) in the solution and reduces the latter. The PVS-reduction efficiency of this double-complex system is higher than that of the monocomplex-zeolite Y system formed from [Ru(mmb)$_3$]$^{2+}$ or [Ru(bpy)$_2$(bpz)]$^{2+}$ alone.

Innocenzi et al. incorporated [Ru(bpy)$_3$]$^{2+}$ into a silica gel membrane prepared through the sol–gel technique and investigated the luminescent property [170] of the composite. They hydrolyzed tetraethyl orthosilicate mixed with [Ru(bpy)$_3$]$Cl_2\cdot 6H_2O$ and directly obtained the silica film encapsulating the complex molecules. It was found that the luminescence arising from recovery of ground state from excited state which corresponds to metal–ligand charge transfer transition in the complex on the sol–gel film is red-shifted, whereas after thermal treatment the luminescence is blue-shifted. However, when the heating temperature is over 200°C, the luminescence is red-shifted again. With the drying temperature increase, the fluorescence lifetime is increased, whereas the luminescence efficiency is decreased.

Besides forming unique complexes with ruthenium, bipyridine can also form functional coordination compounds with other metals. Kim et al. investigated [171] the incorporation of bipyridine complex of manganese (II) into mesoporous MCM-41 and they found that the composite is an effective catalyst for the oxidation of styrene. Through electron spin resonance spectroscopy (ESR) in combination with UV–vis and infrared spectroscopic analysis, Luan et al. [172] revealed the state and related property of the manganese–bipyridine complex loaded in MCM-41 channels. They discovered that when the loading amount is low, the guest complex exists as monomolecules in the channels, whereas if the loading concentration exceeds a particular value, the guest complex molecules undergo aggregation and the ESR signal fine structure disappears. The manganese ion in the guest complex molecule may undergo reversible redox reaction and the oxidation state may vary between +2 and +4. This lays the foundation for the host–guest composite material to be used as a redox catalyst. Knops-Gerrits et al. [173] investigated the luminescence property of the zeolite X, Y and EMT-encapsulated complex formed from Mn(II) and bipyridine or *o*-phenanthroline (abbreviated as phen). The structure of zeolite X is identical with that of zeolite Y but the numbers of negative charges contained in the framework and the cations (Na^+) are different for these two zeolites. The topology of EMT is different from that of zeolite X or Y and the number of its framework negative charges and cations is also less, but its pore opening size is comparable with that of zeolite X or Y. [Mn(bpy)$_3$]$^{2+}$ in all zeolites shows charge transfer transition absorption (495 nm), whereas the charge transfer transition absorption for [Mn(phen)$_3$]$^{2+}$ is broadened because of stereic hindrance which causes distortion. The luminescence efficiency of [Mn(bpy)$_3$]$^{2+}$ incorporated in zeolite X is enhanced, whereas the luminescence wavelength is red-shifted; when this molecule is loaded in the more spacious EMT, its luminescence wavelength is red-shifted and the corresponding luminescence efficiency is reduced. The luminescence of $[Mn(phen)_3]^{2+}$ in zeolites is weak and the signal is widened.

Quayle et al. [174] exchanged Fe(II) into NaY under the protection of N_2 and then mixed the exchanged product with bipyridine and heated the mixture to form $[Fe(II)(bpy)_3]$-Y composite compound. When the content of Fe(II) is smaller than one Fe(II) per supercage, the coordination is the most effective. Whereas if the Fe (II) content is too high, the coordination reaction will drive the excess Fe(II) to be located in between two complex molecules and to bind the ligands through extra π-bondings. When the Fe(II) loading amount is low, chlorine gas may oxidize $[Fe(bpy)_3]^{2+}$ to $[Fe(bpy)_3]^{3+}$and the oxidation conversion is 90%. At high loading amount, this oxidation is much incomplete, probably because the chlorine gas cannot enter into the zeolite pores easily or because there is not enough room for the formation of chloride ion. Umemura et al. [175] used various techniques to investigate the assembly and states of $[FeL_3]^{2+}$ in zeolite Y (L = ethylenediamine, 2-(aminoethyl)pyridinie, 2,2′-bipyridine, 1,10-*o*-phenanthroline, 4,4′-dimethyl-2,2′-bipyridine, 5,6-dimethyl-1,10-*o*-phananthroline). It was found that the former four ligands can form guest complexes with the iron ions in the zeolite channels because their volumes are small, whereas the latter two are not able to form coordination compounds with the Fe(II) ion in the zeolite Y cages because of their larger sizes. The structure of the guest complexes formed by ethylenediamine and 2-(aminoethyl)pyridine undergoes no distortion, whereas in the guest complexes formed by 2,2′-bipyridine and 1,10-*o*-phenanthroline the Fe(II) is in its low-spin state, suggesting that the squeezing of the zeolite framework enhances the coordination ability of the ligand toward the Fe(II) center.

It has been demonstrated [176] that electrochemiluminescent cells can be manufactured on the basis of incorporation of ruthenium tris-bipyridyl in zeolite Y supercages. The cell contains an active layer consisting of the Y-encapsulated Ru complex and the operation of the cell requires the addition of polyethylene glycol as the solid electrolyte. The cell also exhibits electrical conductivity behavior typical of a semiconductor. Ion exchange of sodium by cesium and vapor deposition of calcium metal inside the zeolite pores distinctly enhances the electrochemiluminescent efficiency of the cell.

18.5.2. Incorporation of Metal−Schiff Base Complexes

N,N′-di(salicylal)ethylenediimine is a typical Schiff base formed through condensation of salicylaldehyde and ethylenediamine and this compound is abbreviated as SALEN. SALEN is a widely used chelating ligand; not only its two N atoms are capable of coordinating but also its two OH groups are able to participate in coordination.

Balkus et al. [177] successfully prepared Rh(SALEN) complex in zeolites X and Y. They incorporated Rh(III) into the zeolite through ion exchange and heated the mixture of the zeolite and SALEN at 140°C for 13 h. The reaction product was cooled and washed copiously with chloroform and brown yellow Rh(SALEN)-Y or Rh (SALEN)-X was obtained. In 1991, Bedioui et al. [178] reported incorporation of Co(III)−SALEN complex into zeolite Y and investigated the electrochemical property of the composite. Co(SALEN)-Y has two electrochemical signals, corresponding to Co(III)/Co(II) and Co(II)/Co(I) redox pairs, respectively. These potential values are the same as those of the signals of the monomer $Co(SALEN)^{3+}$ in solution. In addition, there is another pair of redox signal for $Co(SALEN)^{3+}$ between the typical signals for Co(III)/Co(II) and Co (II)/Co(I). Gaillon et al. attributed this pair of signals to the Co(III)/Co(II) redox reaction of $Co(SALEN)^{3+}$ which has strong interactions with the zeolite framework. In other words, $Co(SALEN)^{3+}$ exists in zeolite Y in two different coordination states but the detailed coordination modes are not clear. It was revealed that the assembly product Co(SALEN)-Y may be used as a good oxidation−reduction catalyst. Gaillon et al. [179] also prepared $[Mn(III)SALEN]]^{+}$ and [Fe(III) $SALEN]^{+}$ complexes in zeolite Y. Electrochemical analysis results indicated that depending on preparation methods, two types of complexes with different coordination modes are present in the zeolite channels. In addition, the incorporated complex is accessible by small molecules, especially molecular oxygen, in solution and the latter may get activated.

Bessel and Rolison [180] reported the electrochemical behavior of $[Co(SALEN)]^{2+}$ and $Fe(bpy)_3]^{2+}$ in zeolite Y. They prepared an electrode using the complex-zeolite composite and carbon powder and tested the electrochemical behaviors of the electrode and the composite dispersed in a solution. It was found that the electrochemical behaviors of these two materials differ to a great extent. After several cycles, the former loses all the electrochemical signals, whereas the latter continuously shows the signals. They believed that the electrochemical signals arise from the complex attached onto the zeolite external surface (defects or external supercages), whereas the complex inside the zeolite channel does not participate in electron transfer of the electrochemical process. In fact, there has been dispute on whether the electrochemical signals arise from electron transfer in zeolite channels or those on the zeolite external surface. Both views can find experimental supports [181,182].

Pd(SALEN) exhibits good catalytic activity when used as homogeneous and heterogeneous hydrogenation catalysts, with limited selectivity. To solve this problem, Kowalak et al. [183] incorporated Pd(SALEN) into zeolites X and Y and used the composite for selective hydrogenation catalyst. They discovered that the composite shows high selectivity for the hydrogenation of alkenes the main product obtained through hydrogenation of hexene is hexane and *trans*-2-hexene, whereas the by-product cyclohexane is not observed in the product. This suggests that channels of zeolite X or Y limit the formation of larger molecules so that the reaction selectivity is enhanced.

A series of transition metal (M = Cu, Co, Fe, and Mn) tetrahydro-Schiff base complexes (denoted as M-[H_4] Schiff with Schiff = salen, salpn, and salicyhexen) have been successfully prepared and encapsulated in zeolite Y [184] through in situ reaction approach. The prepared M-[H_4]Schiff/Y catalysts show higher activities in the oxidation of cyclohexane than the corresponding M-Schiff analogues encapsulated in zeolite Y. But the catalytic performance of the M-[H_4]Schiff/Y catalysts also depends on the transition metals because of the different redox nature of transition metals and the different oxidative dehydrogenation activity of the formed complexes toward oxygen. Among the as-prepared composite materials, the Cu-[H_4]Schiff/Y catalysts are highly active and stable for the oxidation of cycloalkanes. It has been revealed that a radical mechanism predominates the catalytic process.

18.5.3. Encapsulation of Porphyrin and Phthalocyanine Complexes

Nakamura et al. [185] successfully synthesized tetramethylporphin (TMP) complexes of iron and manganese in NaY zeolite. They investigated the catalytic properties of the composite for oxidation of cyclohexane in the presence of H_2O_2. The results indicate that the catalytic activity of [Fe(TMP)]-Y and [Mn(TMP)]-Y is enhanced in comparison with the corresponding Fe(II) and Mn(II) exchanged Y zeolite and the catalytic product mainly consists of cyclohexanol and cyclohexanone. Liu et al. [186] silylated the mesoporous MCM-41 channel walls using 3-aminopropyltriethoxysilane and then introduced the ruthenium complex of porphyrin carbonyl. The porphyrin they used was tetra(4-chlorobenzene)porphyrin. The four N atoms of the porphyrin and one carbonyl occupy five coordination sites of the ruthenium and the remaining coordination site was occupied by a solvent molecule and the solvent molecule can be easily replaced by the NH_2 group of the silylation agent, as a result the complex is attached onto the MCM-41 channel walls tightly. The assembly composite compound possesses good catalytic property for oxidation reactions. The research results also indicate that the catalytic performance is optimal if the Ru weight content is in the range of 0.1–0.8%. Unlike in solution, the incorporated complex does not form dimers because it is attached on the channel wall. This is one of the reasons that the catalytic activity of the composite material is high.

Wang et al. synthesized tetrachlorotetramethylporphine (TCTMP) and tetrabromotetramethylporphine (TBTMP) [187] the size of which is comparable with the inner diameter of NaX zeolite. They encapsulated the porphyrin into Co^{2+} ion-exchanged NaY supercages through a solid–liquid phase step-by-step synthetic technique and found that the decomposition temperature of the encapsulated porphyrin molecules is increased by about 70°C. For the oxidation reaction of styrene by H_2O_2 [188], the porphyrin-loaded zeolite exhibits a catalytic conversion 12 times that of the pristine metal porphyrin. Through step-by-step sealed synthesis, tetraphenyltetrabenzoporphine may also be incorporated into NaY supercages. Wang et al. investigated the catalytic activities of tetraphenyltetrabenzoporphine zinc(II) (TPTBP-Zn(II)), NaY, and the porphyrin-loaded NaY (TPTBP-Zn(II)-NaY) for the oxidation reaction of styrene by H_2O_2 [189] and they found that tetraphenyltetrabenzoporphine zinc(II) has no catalytic activity, whereas the zeolite shows some activity but the composite of zeolite and the porphyrin exhibits much enhanced activity. The fact that the catalytic activity of the metal porphyrin loaded in a zeolite is enhanced may be because the reactants in the zeolite cages are electrostatically interacted by both the metal porphyrin and the zeolite so that the styrene activation degree is increased. It was also found experimentally that as a catalyst, the amount of the porphyrin molecules should not be high because there should be enough free space for the reactants to get in contact with the catalytic sites to lower the activation energy of the reaction and to increase the conversion.

The size of phthalocyanine is larger than that of porphine and there is still dispute on whether phthalocyanine is able to enter the pores of microporous crystals. Nevertheless, it is possible for faujasite, which has an inner cage diameter of 1.3 nm, to accommodate phthalocyanine molecule and there have been occasional reports in the literature on the incorporation of phthalocyanine metal complexes into faujasite cages [190]. Paez-Mozo et al. [191] investigated the incorporation of cobalt phthalocyanine complex into zeolite Y and the physicochemical properties of the composite in detail. However, they found that besides the cobalt phthalocyanine complex, un-removable impurity compounds, which are tightly bound to the acidic sites of the zeolite, are also present in the zeolite cages. Balkus et al. [192] also prepared and characterized the complexes of Co

(II) and Cu(II) with hexadecafluorophthalocyanine (MF16Pc) incorporated in zeolites. They used two incorporation methods: (i) ion-exchange reaction and (ii) direct synthesis of zeolites in the presence of the complexes. Middle-infrared, UV−vis spectroscopies and powder X-ray diffraction in combination with elemental analysis indicate that the MF16Pc complex is located inside the zeolite cages. Electrochemical analysis also shows Co(II)/Co(I) and Cu(II)/Cu(I) redox signals, which are not observable in solution.

Metal phthalocyanine compounds are large in size so that the incorporation of these molecules into microporous molecular sieves is limited to a certain degree. The appearance of mesoporous molecular sieves laid the foundation for incorporation of metal phthalocyanine complexes into molecular sieves. The incorporation of metal phthalocyanine compounds into mesoporous molecular sieves may be realized through two pathways: (i) by addition of the metal phthalocyanine complex into the synthetic system of the mesoporous material, the formation of which directly occludes the metal phthalocyanine with the surfactant template in the mesoporous channels; and (ii) by introducing the metal phthalocyanine into the mesoporous channels through impregnation of the metal phthalocyanine solution with detemplated mesoporous material. Depending on the incorporation technique and condition, the existing state and dispersion degree of metal phthalocyanine in mesoporous channels vary to a certain extent [193−195]. Zinc phthalocyanine (ZnPc) is dispersed in the template medium of mesoporous channels as single molecules after incorporation into MCM-41 through the direct synthesis approach, because its absorption spectrum is very similar to that of the complex dissolved in DMF solution. Through impregnation, the incorporation of ZnPc into mesoporous channels exists not only as single molecules but also as molecular dimers.

18.5.4. Encapsulation of Other Metal Complexes

Through solution and solid-state ion exchange, it is possible to exchange Cu^{2+} into mesoporous molecular sieve MCM-41 [196] to form Cu−MCM-41 composite. Pöppl et al. used electron spin resonance and electron spin echo modulation techniques to investigate the coordination state of Cu^{2+} in Cu−MCM-41 and its interactions with adsorbates D_2O and NH_3. The research results indicate that locations of the copper ions in samples obtained through solution exchange differ to a great extent from those in samples obtained through solid exchange.

Ethylenediamine is a common aliphatic chain chelate and it forms very stable chelating compounds with a variety of transition metals. Howe and Lunsford [197] incorporated Co(II)−ethylenediamine complex into zeolites X and Y to form composite compounds which can adsorb oxygen. In both zeolite cages the oxygen adduct $[Co(II)])(en)_2O_2]^{2-}$ may be formed and this complex adduct is stable up to 70°C in the presence of oxygen. The ESR parameters of the adduct are similar to those for the adduct in solution.

Nonaromatic N-containing heterocyclic compounds have attracted much interest recently as ligands. The aromaticity of these macrocyclic compounds is weaker than aromatic heterocyclic compounds so that their configurations can be varied greatly upon coordination and their coordination modes are also versatile. Triazacyclononane is a typical nonaromatic N-containing heterocyclic compound. Its three N atoms may participate in coordination to metal ions. De Vos et al. [198] incorporated $[Mn(tmtacn)]^{2+}$, the Mn(II) complex of 1,4,7-trimethyl-1,4,7-triazacyclononane (tmtacn) into zeolite Y supercages and characterized the existing state of the complex using ESR spectroscopy. They found that the incorporated complex is suitable for use as a catalyst for epoxidation reaction with H_2O_2 as the oxidant.

Dioxotetramine macrocyclic compounds are another type of extensively investigated N-containing nonaromatic macrocyclic ligands. The complexes formed by these ligands with metal ions exhibit various unique properties. Taking advantage of the large channel size of MCM-41, the Cu(II) complexes (14OCu and 14T2Cu) formed by the dioxotetramine macrocyclic ligand 1,4,8,11-tetrazacyclotetradecane-12,14-dione (abbreviated 14O) and the substituted dioxotetramine macrocyclic ligand 4,8-bis(2-thienymethyl)-1,4,8,11-tetraazacyclotetradecane-12,14-dione (abbreviated 14T2) may be loaded into the channels of silica [199]. Diffuse reflectance UV−vis and ESR spectroscopies indicate that after loading, the absorption peak of 14OCu remains unchanged whereas that of 14T2Cu blue shifts by 19 nm, suggesting that the interaction of 14T2Cu with MCM-41 is stronger than that of 14OCu. After incorporation, the ESR spectra for both 14OCu and 14T2Cu show anisotropy. Salavati-Niasari [200] reported the synthesis of a series of nickel(II) complexes with a 14-membered hexaaza macrocyclic ligand in the cages of zeolite Y. The in situ formation of the complexes was realized through direct reactions of formaldehyde and ethylenediamine with amine within the cavities of the zeolite and the as-prepared host−guest materials were characterized by chemical analysis and spectroscopic techniques. It is demonstrated that the encapsulated complexes do not experience extensive distortions in the supercage and that chemical ligation to the zeolitic surface is minimal.

Recently, Salavati-Niasari et al. [201] described the synthesis of copper(II) complexes with tetraoxo dithia tetraaza macrocyclic ligands $[Cu([18]aneN_4S_2)]^{2+}$, $[Cu([20]aneN_4S_2)]^{2+}$, $[Cu(Bzo_2[18]aneN_4S_2)]^{2+}$, and

$[Cu(Bzo_2[20]aneN_4S_2)]^{2+}$) in the supercages of zeolite Y. The obtained host–guest nanocomposite materials have been thoroughly characterized via a variety of spectroscopic techniques.

Cobaltocene cation (Cp_2Co^+) is rather rigid and it is stable even under hydrothermal conditions. Using cobaltocene cation as template zeolites nonasil and ZSM-51 (NON) can be synthesized. The size of cobaltocene matches with that of the cage of NON structure [202,203], therefore this template is tightly encapsulated by the zeolite framework of NON structure. Cobaltocene can also act as a template for the synthesis of $AlPO_4$-16 and $AlPO_4$-5 [204]. Through using methylated cobaltocene cation ($Cp^*_2Co^+$) as template, UTD-1, a totally new structure zeolite has been synthesized. This is also the first high-silica zeolite with 14-membered rings [205]. The methylcobaltocene cations in UTD-1 can be removed through washing. Honma and Zhou [206] reported the synthesis of mesoporous M41S silica using a ferrocene-quaternary ammonium derivative (ferrocenyl–$(CH_2)_{11}$–$N^+(CH_3)_3$) as template.

Enzymes in biological system are constructed from protein and many enzymes contain transition metals. These polypeptide chains bound or coordinated metal ions play unique roles in catalysis. Therefore, it has been continuously attempted to synthesize metal amino acid complexes to mimic natural metal enzymes. Weckhuysen et al. [207] incorporated copper(II) histidine complex into zeolite Y and they found that the incorporated complex exhibits excellent catalytic performance for oxidation. Differing from the commonly used ion exchange followed by coordination method, the technique they adopted is to synthesize the $Cu(His)_2^{2+}$ complex first and then directly ion exchange it into NaY zeolite. Through ESR analysis, they found that the amino N, carboxylate O, and imidazole ring N of one histidine participate in coordination to the Cu(II) ion, whereas only the amino N and the carboxylate O of another histidine coordinate to the metal. The sixth coordination site of the Cu(II) ion may accept extra ligand. In catalytic oxidation reactions, this sixth coordination site may activate the oxidant. It has been revealed that using *tert*-butyl peroxide as the oxidant and composite as catalyst, the conversions of 1-pentanol, benzyl alcohol, and cyclohexene are 12%, 56%, and 28%, respectively. The main products are pentanoic acid, phenylacetylaldehyde, and 1,2-cyclohexanediol and the selectivity is rather high. Therefore, the $[Cu(His)_2]^{2+}$-Y composite can be regarded as an effective enzyme-mimicking compound.

Inspired by enzymes [208], a mononuclear copper(II) complex with one MIm2Pr ligand [MIm_2Pr = 3,3-bis (1-methylimidazol-2-yl)propionate] in the supercages of zeolite Y has been prepared by Weckhuysen et al. (Fig. 18.4). The preparation procedure leaded to two compounds (labeled **1** and **2**), which differ in molecular structure and chemical composition. In **1**, the copper is five-coordinated, with one MIm_2Pr ligand in a facial-type NNO coordination toward copper, the other two coordination sites being occupied by oxygen atoms from either the zeolite framework and/or a water molecule. The total charge of **1** is 1+. But in **2**, the copper is surrounded by two MIm_2Pr ligands, both in a facial-type coordination mode as found in the homogeneous $Cu(MIm_2Pr)_2$ complex. **2** is neutral in charge and can be easily washed out of the zeolite, whereas the mononuclear species **1** remains inside the zeolite material upon washing. The catalytic activity for 3,5-di-*tert*-butylcatechol and benzyl alcohol oxidation of **1** is comparable with that of the zeolite-immobilized Cu(histidine) complexes but differed from that of the homogeneous $Cu(MIm_2Pr)_2$ compound. Therefore, the encapsulation in zeolite offers a route to stabilize a fivefold-coordinated copper complex with novel catalytic properties.

The structures and compositions of microporous crystals are becoming more versatile and this has laid foundations for incorporation of complexes with various structures and functionalities in microporous crystals. The previously reported complex incorporation was mainly limited to the use of zeolites as the hosts. The aluminophosphate microporous crystal family discovered in the 1980s possesses versatile structures as well, but the incorporation of complexes in aluminophosphate hosts has been rarely reported. It is not surprising

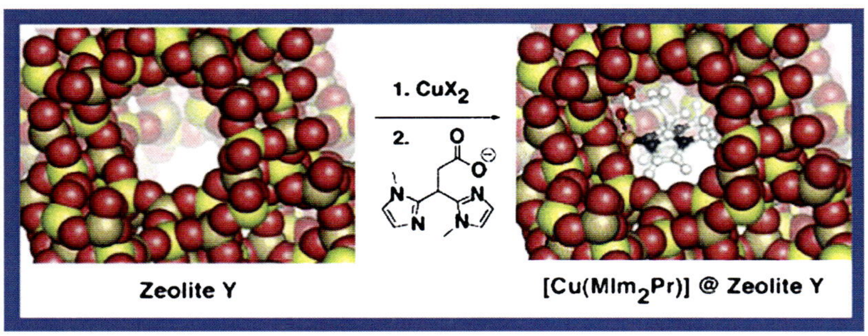

FIGURE 18.4 Schematic representation for the encapsulation of $Cu(MIm_2Pr)_2$ in zeolite Y. *Reprinted with permission from [208]. Copyright 2006 American Chemical Society.*

that complex incorporation in microporous aluminophosphates will extend the porous host–complex guest assembly chemistry. Mesoporous molecular sieves M41S (including MCM-41 and MCM-48) have the advantages of large pore size (>1.5 nm) and being capable of accommodating large molecules. Complex molecules with a large size may enter or be loaded into the channels or cages of M41S mesoporous molecular sieves to form composite materials with special functionalities [172] such as high catalytic performance. The channel size of mesoporous molecular sieves is large and after introduction of complex molecules there is still enough room for guest molecules to pass through and as a result, diffusion in these materials may not be affected when they are used as catalysts. Therefore, it is envisioned that mesoporous molecular sieves can be widely used to accommodate complex molecules to form high-performance catalysts. Evans et al. [209] reported the grafting of aminosilane onto the walls of mesoporous silica and it was found that after grafting the amino groups of the aminosilane exhibit strong coordination ability and they may coordinate to many metal ions such as Mn^{2+}, Cu^{2+}, Co^{2+}, and Zn^{2+} to form complexes. Evans et al. investigated the physical–chemical properties of the complex/mesoporous silica host–guest composite materials and their use as catalysts for the oxidation of aromatic amines. It was discovered that the manganese-containing host–guest material showed the highest catalytic activity and the activity of the copper-containing compound is the second highest followed by those of the cobalt- and zinc-containing compounds. There is an apparent induction period for the reaction involving the latter two compounds as the catalysts.

It is also of significance to incorporate complex molecules into microporous crystals to form photochemically or photophysically active centers. Because of the separation by the host framework, the complexes located in the channels or cages of microporous crystals are isolated. If the isolated centers with oxidation or reduction features are loaded in the connected and adjacent cages of a microporous crystal, redox pairs may be formed. Electron transfer may occur on these redox pairs under the excitation of light and therefore photochemical reactions may proceed effectively. This is important for the utilization of solar energy. In addition, this type of assembly system may also be used to simulate the electron transfer process of oxidation–reduction in biological systems.

Apart from forming catalysts for photochemical reactions, some rare earth ion complexes may also form efficient luminescent materials after incorporation into microporous crystals. Alvaro et al. [210] loaded europium complex into zeolite Y, mordenite, and ZSM-5. Because of the confinement of zeolite framework, the chance for the luminescent centers to decay nonradiationally is reduced and as a result the lifetime is increased in comparison with that in solution. In the meantime, upon formation of complex, the luminescent intensity of Eu^{3+} ion is distinctly increased. Therefore, it is possible to prepare valuable composite luminescent materials using microporous crystals as hosts and complexes as guests.

References

[1] G.D. Stucky, J.E. MacDougall, Science (247) (1990) 669.
[2] G.A. Ozin, Adv. Mater. (4) (1992) 612.
[3] J. Guzman, B.C. Gates, Dalton Trans. (2003) 3303–3333.
[4] P. Gallezot, in: Molecular Sieves, vol. 3, Springer-Verlag, Berlin Heidelberg, 2002, p. 257.
[5] P.P. Anderson, in: Molecular Sieves, vol. 3, Springer-Verlag, Berlin Heidelberg, 2002, p. 307.
[6] P.P. Edwards, P.A. Anderson, J. Thomas, Acc. Chem. Res. (29) (1996) 23.
[7] C.J. Reinhold, P.A. Anderson, P.P. Edwards, V.V. Terskikh, C.I. Ratcliffe, J.A.J. Ripmeester, Phys. Chem. C (112) (2008) 17796–17803.
[8] P.J. Kasai, J. Chem. Phys. (43) (1965) 3322.
[9] W.G. Hodgson, J.S. Brinen, E.F. Williams, J. Chem. Phys. (47) (1967) 3719.
[10] P.P. Edwards, M.R. Harrison, J. Klinowski, et al., J. Chem. Soc. Chem. Commun. (1984) 982.
[11] M.R. Harrison, P.P. Edwards, J. Klinowski, et al., J. Solid State Chem. (54) (1984) 330.
[12] R.M. Barrer, J.M. Cole, J. Phys. Chem. Solids (29) (1968) 1755.
[13] N.P. Blake, V.I. Srdanov, G.D. Stucky, et al., J. Chem. Phys. (104) (1996) 8721.
[14] J.A. Rabo, C.L. Angell, P.H. Kasai, et al., Discuss. Faraday Soc. (41) (1966) 328.
[15] T.Y. Ben, C. Naccache, M. Che, et al., Chem. Phys. Lett. (24) (1974) 41.
[16] U. Westphal, G. Geismar, Z. Anorg. Allg. Chem. (508) (1984) 165.
[17] L.R.M. Martens, P.J. Grobert, P.A. Jacobs, Nature (315) (1985) 568.
[18] I. Hannus, I. Kirisci, A. Beres, et al., Stud. Surf. Sci. Catal. (98) (1995) 81.
[19] M.W. Simon, J.C. Edwards, S.L. Suib, J. Phys. Chem. (99) (1995) 4698.
[20] P.A. Anderson, R.G. Bell, C.R.A. Catlow, et al., Chem. Mater. (8) (1996) 2114.
[21] P.A. Anderson, A.R. Armstrong, A. Porch, et al., J. Phys. Chem. B (48) (1997) 9892.
[22] M.J. Kelly, J. Phys. Condens. Matter (7) (1995) 5507.
[23] P.A. Anderson, A.R. Armstrong, P.P. Edwards, Angew. Chem. Int. Ed. (33) (1994) 641.
[24] Y. Ikemoto, T. Nakano, Y. Nozue, et al., Mater. Sci. Eng. B (48) (1997) 116.
[25] Y. Maniwa, H. Kira, F. Shimuzu, et al., J. Phys. Soc. Jpn. (9) (1999) 2902.
[26] M. Ralek, P. Jiru, O. Grubner, et al., Collect. Czech. Chem. Commun. (27) (1962) 142.
[27] Y. Kim, K. Seff, J. Am. Chem. Soc. (99) (1977) 7055.
[28] P.A. Jacobs, J.B. Uytterhoeven, H.K. Beyer, J. Chem. Soc. Faraday Trans. I (75) (1979) 56.
[29] T. Sun, K. Seff, Chem. Rev. (94) (1994) 857.
[30] B. Xu, L. Kevan, J. Phys. Chem. (95) (1991) 1147.
[31] N. Narayana, L. Kevan, J. Chem. Phys. (76) (1982) 3999.

[32] G.A. Ozin, J. Godber, A. Stein, US Patent 4 942 119 (1990).
[33] D.H. Lin, Y.X. Jiang, Y. Wang, S.G. Sun, J. Nanomater. (2008). Article ID 473791, doi:10.1155/2008/473791.
[34] P. Gallezot, I. Mutin, G. Dalmai-Imelik, et al., J. Microscop. Spectrosc. Electron. (1) (1976) 1.
[35] P. Gallezot, D.A. Alarcon, J.A. Dalmon, et al., J. Catal. (39) (1975) 334.
[36] G. Bergeret, P. Gallezot, in: Proceedings of the 8th International Congress Catalysis, vol. 5, Dechema, Frankfurt, 1984, p. 659.
[37] S. Mentus, Z. Mojovic, V. Radmilovic, J. Serb. Chem. Soc. (74) (2009) 1113–1123.
[38] L. Djakovitch, K.J. Koehler, Am. Chem. Soc. (123) (2001) 5990–5999.
[39] W.Z. Zhou, J.M. Thomas, D.S. Shephard, et al., Science (280) (1998) 705.
[40] D.S. Shephard, T. Maschmeyer, B.F.G. Johnson, et al., Angew. Chem. Int. Ed. (36) (1997) 2242.
[41] J.M. Thomas, B.F.G. Johnson, R. Raja, G. Sankar, P.A. Midgley, Acc. Chem. Res. (36) (2003) 20–30.
[42] L. Li, J. Shi, Adv. Synth. Catal. (350) (2008) 667–672.
[43] Y. Jiang, Q.J. Gao, Am. Chem. Soc. (128) (2006) 716–717.
[44] L.B. McCusker, K. Seff, J. Am. Chem. Soc. (101) (1979) 5235.
[45] S.B. Jang, S. Kim U, Y. Kim, et al., J. Phys. Chem. (98) (1994) 3796.
[46] A. Goldbach, P.D. Barker, P.A. Anderson, et al., Chem. Phys. Lett. (292) (1998) 137.
[47] T. Sprang, A. Seidel, M. Wark, et al., J. Mater. Chem. (7) (1997) 1429.
[48] A. Seidel, F. Rittner, E. Boddenberg, J. Phys. Chem. (102) (1998) 7176.
[49] Y. Tian, G.D. Li, J.S. Chen, J. Am. Chem. Soc. (125) (2003) 6622.
[50] N.H. Heo, S.H. Kim, H.C. Choi, et al., J. Phys. Chem. B (101) (1997) 5531.
[51] N.H. Heo, S.H. Kim, H.C. Choi, et al., J. Phys. Chem. B (102) (1998) 17.
[52] Q. Gao, Y. Xiu, G.D. Li, J.S. Chen, J. Mater. Chem. (20) (2010) 3307.
[53] I. Hussain, I. Gameson, P.A. Ander erson, et al., J. Chem. Soc., Dalton Trans. (1996) 775.
[54] J.A. Rabo, P.H. Kasai, Prog. Solid State Chem. (9) (1975) 1.
[55] J.V. Smith, J.M. Bennett, E.M. Flanigen, Nature (215) (1967) 241.
[56] H.S. Park, K. Seff, J. Phys. Chem. B (104) (2000) 2224.
[57] K. Sahl, Z. Kristallogr. (152) (1980) 13.
[58] A. Carlsson, T. Oku, J.O. Bovin, et al., Chem. Eur. J. (5) (1999) 244.
[59] J.E. Readman, I. Gameson, J.A. Hriljac, et al., Chem. Commun. (2000) 595.
[60] L. Li, J. Shi, Chem. Commun. (2008) 996–998.
[61] G. Schulz-Ekloff, D. Wohrle, B. van Duffel, et al., Microp. Mesop. Mater. (51) (2002) 91.
[62] G. Calzaferri, N. Gfeller, J. Phys. Chem. (96) (1992) 3428.
[63] S.D. Cox, T.E. Gier, G.D. Stucky, et al., J. Am. Chem. Soc. (110) (1988) 2986.
[64] S. Wohlrab, R. Hoppe, G. Schulz-Ekloff, et al., Zeolites (12) (1992) 862.
[65] N. Herron, G.D. Stucky, C.A. Tolman, J. Chem. Soc. Chem. Commun. (1986) 1521.
[66] R. Hoppe, G. Schulz-Ekloff, D. Wohrle, et al., Adv. Mater. (7) (1995) 61.
[67] K. Rurack, K. Hoffmann, W. Al-Soufi, et al., J. Phys. Chem. B. (106) (2002) 9744.
[68] V. Ganesan, R. Ramaraj, J. Lumin. (92) (2001) 167.
[69] U. Vietze, O. Krauβ, F. Laeri, et al., Phys. Rev. Lett. (81) (1998) 4628.
[70] G. Ihlein, F. Schuth, L. Benmohammadi, et al., Adv. Mater. (10) (1998) 1117.
[71] P. Yang, G. Wirnsberger, H.C. Huang, et al., Science (287) (2000) 465.
[72] G. Wirnsberger, G.D. Stucky, Chem. Mater. (12) (2000) 2525.
[73] R. Vogel, P. Meredith, I. Kartini, et al., Chem. Phys. Chem. (4) (2003) 595.
[74] Y. Wada, T. Okubo, M. Ryo, et al., J. Am. Chem. Soc. (122) (2000) 8583.
[75] M. Ganschow, M. Wark, D. Wohrle, et al., Angew. Chem. Int. Ed. (39) (2000) 160.
[76] M.M. Yatskou, M. Meyer, S. Huber, et al., Chem. Phys. Chem. (4) (2003) 567.
[77] M. Pauchard, S. Huber, R. Meallet-Renault, et al., Angew. Chem. Int. Ed. (40) (2001) 2839.
[78] A.Z. Ruiz, H.R. Li, G. Calzaferri, Angew. Chem. Int. Ed. (45) (2006) 5282.
[79] G. Calzaferri, in: M. Anpo (Ed.), Photofunctional Zeolites, NOVA Science Publishers, New York, 2000, p. 205.
[80] B. Bussemer, D. Munsel, H. Wulnscher, G.J. Mohr, U.-W. Grummt, J. Phys. Chem. B (111) (2007) 8–15.
[81] S. Hashimoto, H.R. Moon, K.B. Yoon, Microp. Mesop. Mater. (101) (2007) 10–18.
[82] Y.M. Jin, H. Chon, Chem. Commun. (1996) 135.
[83] S.D. Cox, T.E. Gier, G.D. Stucky, Chem. Mater. (2) (1990) 609.
[84] W. Hill, F. Marlow, J. Kornatowski, Appl. Spectrosc. (48) (1994) 224.
[85] G.J. Klap, S.M. van Klooster, M. Wubbenhorst, et al., J. Phys. Chem. B (102) (1998) 9518.
[86] M. Ehrle, H.W. Kindervater, F.W. Deeg, et al., J. Phys. Chem. (98) (1994) 11756.
[87] M. Ehrle, F.W. Deeg, C. Brauchle, et al., J. Phys. Chem. (98) (1994) 47.
[88] C. Pereira, G.T. Kokotailo, R.J. Gorte, J. Phys. Chem. (95) (1991) 705.
[89] S.D. Cox, G.D. Stucky, J. Phys. Chem. (95) (1991) 710.
[90] P. Tsai, R.P. Cooney, J. Heaviside, et al., Chem. Phys. Lett. (59) (1978) 510.
[91] T. Bein, P. Enzel, Angew. Chem. (101) (1989) 1737.
[92] P. Enzel, T. Bein, J. Chem. Soc. Chem. Commun. (1989) 1326.
[93] D.E. De Vos, P.A. Jacobs, in: H. van Bekkum, E.M. Flanigen, P.A. Jacobs, et al. (Eds.), Introduction to Zeolite Science and Practice, Elsevier, Amsterdam, 2001, p. 957.
[94] C. Wu, T. Bein, Science (264) (1994) 1757.
[95] P. Enzel, T. Bein, Chem. Mater. (4) (1992) 819.
[96] K. Moller, T. Bein, R.X. Fischer, Chem. Mater. (10) (1998) 1841.
[97] A. Graeser, S. Spange, Chem. Mater. (10) (1998) 1814.
[98] T. Kyotani, T. Nagai, S. Inoue, et al., Chem. Mater. (9) (1997) 609.
[99] Z. Ma, T. Kyotani, Z. Liu, et al., Chem. Mater. (13) (2001) 4413.
[100] S.A. Johnson, E.S. Brigham, P.J. Olliver, et al., Chem. Mater. (9) (1997) 2448.
[101] R. Ryoo, S. Joo, J. Jun, J. Phys. Chem. B (103) (1999) 7743.
[102] J.W. Lee, S.H. Yoon, T.H. Hyeon, et al., Chem. Commun. (1999) 2177.
[103] S.B. Yoon, J.Y. Kim, J.S. Yu, Chem. Commun. (2001) 559.
[104] S.H. Joo, S.J. Choi, I. Oh, et al., Nature (412) (2001) 169.
[105] C.G. Wu, T. Bein, Science (266) (1994) 1013.
[106] J. Rodriguez-Mirasol, T. Cordero, L.R. Radovic, et al., Chem. Mater. (10) (1998) 550.
[107] Y. Xiu, Q. Gao, G.D. Li, K.X. Wang, J.S. Chen, Inorg. Chem. (49) (2010) 5859.
[108] E.H. Ellison, J. Phys. Chem. B (110) (2006) 11406–11414.
[109] B. Hamilton, J.S. Rimmer, M. Anderson, et al., Adv. Mater. (5) (1993) 583.

[110] A. Drljaca, C. Kepert, L. Spiccia, et al., Chem. Commun. (1997) 195.
[111] J.S. Chen, Q.H. Li, H. Ding, et al., Langmuir (13) (1997) 2050.
[112] J.S. Chen, Q.H. Li, R. Xu, et al., Angew. Chem. Int. Ed. (34) (1995) 2694.
[113] S. Iijima, Nature (354) (1991) 56.
[114] P.J.F. Harris, Carbon Nanotubes and Related Structures, University Press, Cambridge, 1999, p. 4.
[115] Z.K. Tang, H.D. Sun, J. Wang, et al., Appl. Phys. Lett. (73) (1998) 2287.
[116] G.D. Li, Z.K. Tang, N. Wang, J.S. Chen, Carbon (40) (2002) 917.
[117] Z.M. Li, Z.K. Tang, H.J. Liu, et al., Phys. Rev. Lett. (87) (2001) 127401.
[118] N. Wang, Z.K. Tang, G.D. Li, J.S. Chen, Nature (408) (2000) 50.
[119] H.D. Sun, Z.K. Tang, J. Chen, G. Li, Appl. Phys. A (69) (1999) 381.
[120] G.D. Li, Z.K. Tang, N. Wang, et al., Stud. Surf. Sci. Catal. (135) (2001) 296.
[121] Z.K. Tang, L.Y. Zhang, N. Wang, et al., Science (292) (2001) 2462.
[122] C.L. Bowes, A. Malek, G.A. Ozin, Chem. Vapour Deposit. (2) (1996) 97.
[123] M.R. Steele, P.M. Macdonald, G.A. Ozin, J. Am. Chem. Soc. (115) (1993) 7285.
[124] G.A. Ozin, M.R. Steele, A.J. Holmes, Chem. Mater. (6) (1994) 999.
[125] J.E. MacDougall, H. Eckert, G.D. Stucky, et al., J. Am. Chem. Soc. (111) (1989) 8006.
[126] L.T. Canham, Appl. Phys. Lett. (57) (1990) 1046.
[127] E. Chomski, Ö. Dag, A. Kuperman, G.A. Ozin, Adv. Mater. (7) (1995) 72.
[128] Ö. Dag, A. Kuperman, G.A. Ozin, Adv. Mater. (6) (1994) 147.
[129] F.F. Gao, G.S. Zhu, X.T. Li, et al., J. Phys. Chem. B (105) (2001) 12704.
[130] R. Leon, D. Margolese, P.M. Petroff, G.D. Stucky, Phys. Rev. B (52) (1995) R2285.
[131] N. Herron, Y. Wang, M.M. Eddy, et al., J. Am. Chem. Soc. (111) (1989) 530.
[132] F. Iacomi, Surf. Sci. (532) (2003) 816–821.
[133] F. Qin, J.L. Shi, C.Y. Wei, et al., J. Mater. Chem. (18) (2008) 634–636.
[134] K. Moller, M.M. Eddy, G.D. Stucky, et al., J. Am. Chem. Soc. (111) (1989) 2564.
[135] R. Seifert, A. Kunzmann, G. Calzaferri, Angew. Chem. Int. Ed. (37) (1998) 1521.
[136] D. Bruhwiler, R. Seifert, G. Calzaferri, J. Phys. Chem. B (103) (1999) 6397.
[137] Z.K. Tang, Y. Nozue, O. Terasaki, T. Goto, Mol. Cryst. Liq. Cryst. (218) (1992) 61.
[138] Z.K. Tang, M.M. Loy, J.S. Chen, R.R. Xu, Appl. Phys. Lett. (70) (1997) 34.
[139] Z. Chen, Q.M. Gao, M.L. Ruan, et al., Appl. Phys. Lett. (87) (2005) 093113.
[140] W.H. Zhang, J.L. Shi, L.Z. Wang, et al., Chem. Mater. (12) (2000) 1408–1413.
[141] T. Abe, Y. Tachibana, T. Uematsu, et al., J. Chem. Soc. Chem. Commun. (1995) 1617–1618.
[142] J.B. Parise, J.E. MacDougall, N. Herron, et al., Inorg. Chem. (27) (1988) 221.
[143] K. Tamura, S. Hosokawa, H. Endo, et al., J. Phys. Soc. Jpn. (55) (1986) 528.
[144] Z.K. Tang, M.M.T. Loy, T. Goto, et al., Solid State Commun. (101) (1997) 333.
[145] I.L. Li, J.P. Zhai, P. Launois, S.C. Ruan, Z.K. Tang, J. Am. Chem. Soc. (127) (2005) 16111.
[146] A. Goldbach, M.-L. Saboungi, Acc. Chem. Res. (38) (2005) 705–712.
[147] V.V. Poborchii, A.V. Kolobov, J. Caro, et al., Chem. Phys. Lett. (280) (1997) 17.
[148] V.V. Poborchii, A.V. Kolobov, H. Oyanagi, et al., Chem. Phys. Lett. (280) (1997) 10.
[149] V.V. Poborchii, Chem. Phys. Lett. (251) (1996) 230.
[150] J.R. Agger, M.W. Anderson, M.E. Pemble, et al., J. Phys. Chem. B (102) (1998) 3345.
[151] V.I. Srdanov, I. Alxneit, G.D. Stucky, et al., J. Phys. Chem. (102) (1998) 3341.
[152] H. Winkler, A. Birkner, V. Hagen, et al., Adv. Mater. (11) (1999) 1444–1448.
[153] X.Z. You, Q.J. Meng, W.S. Han, Progress in Coordination Chemistry, Higher Education Press, Beijing, 2000, p. 222.
[154] A.S. Hay, J. Org. Chem. (25) (1960) 1275.
[155] A.S. Hay, H.S. Blanchand, G.F. Endres, et al., J. Am. Chem. Soc. (81) (1959) 6335.
[156] Y. Ukisu, A. Kazusaka, M. Nomura, J. Mol. Catal. (70) (1991) 165.
[157] P.S.E. Dai, J.H. Lunsford, Inorg. Chem. (19) (1980) 262.
[158] W. Böhlmann, K. Schandert, A. Pöppl, et al., Zeolites (19) (1997) 297.
[159] A. Pöppl, L. Kevan, Langmuir (11) (1995) 4486.
[160] Y. Yamada, Bull. Chem. Soc. Jpn. (45) (1972) 64.
[161] W. Dewilde, G. Peeters, J.H. Lunsford, J. Phys. Chem. (84) (1980) 2306.
[162] W.H. Quayle, J.H. Lunsford, Inorg. Chem. (21) (1982) 97.
[163] P. Lainé, M. Lans, G. Calzaferri, Inorg. Chem. (35) (1996) 3514.
[164] J.M. Ledney, P.K. Dutta, J. Am. Chem. Soc. (117) (1995) 7687.
[165] P.K. Dutta, S.K. Das, J. Am. Chem. Soc. (119) (1997) 4311.
[166] M. Sykora, J.R. Kincail, P.K. Dutta, N.B. Castagnola, J. Phys. Chem. B (103) (1999) 309.
[167] K. Maruszewki, D. Strommen, K. Handrich, et al., Inorg. Chem. (30) (1991) 4579.
[168] P.K. Dutta, J.A. Incavo, J. Phys. Chem. (91) (1987) 4443.
[169] M. Sykora, J.R. Kincald, Nature (387) (1997) 162.
[170] P. Innocenzi, H. Kozuka, T.F. Yoko, J. Phys. Chem. B (101) (1997) 2285.
[171] S.S. Kim, W. Zhang, T.J. Pinnavaia, Catal. Lett. (43) (1997) 149.
[172] Z.H. Luan, J. Xu, L. Kevan, Chem. Mater. (10) (1998) 3699.
[173] Knops-Gerrits, F.C. De Schryver, M. van de Auweraer, et al., Chem. Eur. J. (2) (1996) 592.
[174] W.H. Quayle, G. Peeters, G.L. De Roy, et al., Inorg. Chem. (21) (1982) 2226.
[175] Y. Umemura, Y. Minai, T. Tominaga, J. Phys. Chem. B (103) (1999) 647.
[176] M. Alvaro, J.F. Cabeza, D. Fabuel, A. Corma, H. Garcia, Chem. Eur. J. (13) (2007) 3733–3738.
[177] K.J. Balkus Jr., A.A. Welch, B.E. Gnade, Zeolites (10) (1990) 722.
[178] F. Bedioui, E. De Boysson, J. Devynck, et al., J. Chem. Soc. Faraday Trans. (87) (1991) 3831.
[179] Z. Gaillon, N. Sajot, F. Bedioui, et al., J. Electroanal. Chem. (345) (1993) 157.
[180] C.A. Bessel, D.R. Rolison, J. Phys. Chem. B (101) (1997) 1148.
[181] D.R. Rolison, C.A. Bessel, M.D. Baker, et al., J. Phys. Chem. (100) (1996) 8610.
[182] F. Bedioui, J. Devynck, K.J. Balkus Jr., J. Phys. Chem. (100) (1996) 8607.
[183] S. Kowalak, R.C. Weiss, K.J. Balkus Jr., J. Chem. Soc. Chem. Commun. (1991) 57.
[184] C. Jin, W.B. Fan, Y.J. Jia, B.B. Fan, J.H. Ma, R.F. Li, J. Mol. Catal. A. Chem. (249) (2006) 23–30.

[185] M. Nakamura, T. Tatsumi, H.-O. Tominaga, Bull. Chem. Soc. Jpn. (63) (1990) 3334.

[186] C.J. Liu, S.G. Li, W.Q. Pang, et al., Chem. Commun. (1997) 65.

[187] X.Q. Wang, Y.X. Liang, Y.W. Liu, et al., Chem. J. Chin. Univ. (14) (1993) 19.

[188] X.Q. Wang, S. Gao, Y.W. Liu, et al., Chem. J. Chin. Univ. (15) (1994) 789.

[189] X.Q. Wang, S. Gao, C.S. Cao, et al., Chin. J. Catal. (17) (1996) 343.

[190] F. Bedioui, Coord. Chem. Rev. (144) (1995) 39.

[191] E. Paez-Mozo, N. Gabriunas, F. Lucaccioni, et al., J. Phys. Chem. (97) (1993) 12819.

[192] K.J. Balkus Jr., A.G. Gabrielov, S.L. Bell, et al., Inorg. Chem. (33) (1994) 67.

[193] M. Ganschow, D. Wohrle, G. Schulz-Ekloff, J. Porphyrins Phthalocyanines (3) (1999) 299.

[194] M. Wark, A. Ortlam, M. Ganschow, et al., Ber. Bunsenges Phys. Chem. (102) (1998) 1548.

[195] R. Hoppe, A. Ortlam, J. Rathousky, et al., Microp. Mater. (8) (1997) 267.

[196] A. Pöppl, M. Newhouse, L.E. Kevan, J. Phys. Chem. (99) (1995) 10019.

[197] R.F. Howe, J.H. Lunsford, J. Phys. Chem. (79) (1975) 1836.

[198] D.E. De Vos, J.L. Meinershagen, T. Bein, Angew. Chem. Int. Ed. (35) (1996) 2211.

[199] X.C. Cao, G.D. Li, J.S. Chen, et al., Chem. J. Chin. Univ. (20) (1999) 25.

[200] M. Salavati-Niasari, Inorg. Chem. Commun. (7) (2004) 963–966.

[201] M. Salavati-Niasari, S.N. Mirsattari, K. Saberyan, Bull. Korean. Chem. Soc. (30) (2009) 348–354.

[202] G. van de Goor, B. Lindlar, P. Behrens, J. Felsche, J. Chem. Soc. Chem. Commun. (1995) 2559.

[203] P. Behrens, G. van de Goor, C.C. Freyhardt, Angew. Chem. Int. Ed. (34) (1995) 2680.

[204] K.J. Balkus Jr., A.G. Gabrielov, S. Shepelev, Microp. Mater. (3) (1995) 489.

[205] C.C. Freyhardt, M. Tsapatsis, R.F. Lobo, et al., Nature (381) (1996) 295.

[206] I. Honma, H.S. Zhou, Adv. Mater. (10) (1998) 1532.

[207] B.M. Weckhuysen, A.A. Verberckmoes, I.P. Vannijvel, et al., Angew. Chem. Int. Ed. (34) (1995) 2652.

[208] K. Kervinen, P.C.A. Bruijnincx, A.M. Beale, J.G. Mesu, G. van Koten, R.J.M.K. Gebbink, et al., J. Am Chem. Soc. (128) (2006) 3208–3217.

[209] J. Evans, A.B. Zaki, M. El-Sheikh, et al., J. Phys. Chem. B (104) (2000) 10271.

[210] M. Alvaro, V. Forues, S. Garcia, et al., J. Phys. Chem. B (102) (1998) 8744.

CHAPTER

19

Chemical Preparation of Advanced Ceramic Materials

Jingkun Guo, Jiang Li, Huamin Kou

Shanghai Institute of Ceramics, Chinese Academy of Sciences, China

The history of ceramics is intertwined with human history. From the first use of flint and obsidian during the Stone Age, the formation of vessels from clay, the use of refractories in the iron and steel industry, to the fabrication of optical fibers for high-speed communication, ceramics have revolutionized the society and technology in many ways. Novel ceramic materials are always surprising us. So how to prepare advanced ceramics has become the key point. In this chapter we will explore the topic of chemical preparation of advanced ceramic materials. We will mention some of the most recent developments in the field of advanced ceramics. The chapter has been divided into four sections. Section I: Fabrication of nanoceramics; Section II: Making of ceramic matrix composites; Section III: Integration of structures and functions of ceramics; and Section IV: Fabrication of some transparent ceramics such as laser ceramics, ceramic scintillators, upconversion ceramics, and optical ceramics for windows.

19.1. NANOCERAMICS

Conventionally nanostructured ceramics are defined as inorganic materials composed of structural units with a size scale of less than 100 nm in any dimension [1]. Based on dimensions, nanoceramics are classified as zero-dimensional nanocrystals, one-dimensional nanowires and nanotubes, two-dimensional nanofilms and nanowalls, and three-dimensional bulk materials with at least one nanocrystalline phase [2]. Nanoceramics, in particular, bulk nanoceramic materials, exhibit superior and unique properties in comparison to conventional ceramics with coarser structured units. A major challenge in the research on bulk nanoceramics and nanoceramic composites is concerned with the aspect of processing. The basic steps in nanoceramic fabrication mainly involve obtaining unagglomerated nanosized powders with uniform size distribution and sintering to near theoretical density without grain growth.

19.1.1. Preparation of Nanoscaled Powders

The synthesis of nanocrystalline powders is an essential and first step in the processing of bulk nanoceramics. Efforts have been made to ensure appropriate control of particle size, surface contamination, and degree of agglomeration. Various chemical methods have been adapted to synthesize nanocrystalline powders. The main advantages of chemical synthesis method lie in their ability to produce a large variety of compositions and ensure homogeneous (atomic level) mixing of the constituent particles.

19.1.1.1. Chemical Vapor Deposition (CVD)

Chemical vapor deposition (CVD), in which a precursor is converted to nanoparticles, is one of the conventional and widely used techniques to synthesize ceramic nanopowders. This method is mainly used for fabrication of oxide, nitride, and carbide nanopowders. For example, TiO_2 nanopowder [3] and TiO_2/Al_2O_3 nanocomposite [4] have been synthesized by Shi et al. CVD has also been utilized for synthesis of a technologically important material—carbon nanotube [5]. Laser-induced chemical vapor deposition(LICVD) is a recently developed method to prepare amorphous or crystalline nanopowders with controllable particle size and homogeneous size distribution. A similar process, inert gas condensation technique, has been used to synthesize

Modern Inorganic Synthetic Chemistry, DOI: 10.1016/B978-0-444-53599-3.10019-8

nanosized TiO_2 powders, involving formation of unltrafine Ti particles and subsequent in situ oxidation [6].

19.1.1.2. Chemical Precipitation

The nanosized ceramic powders are routinely synthesized by chemical precipitation and reaction routes. In such process, a solution containing precipitating agent, commonly hydroxide, ammonium acid carbonate, or oxalic acid, is added to the solution containing cation of the desired oxide. This is followed by firing of the precipitates (hydroxides, carbonates, or oxalates). Chemical precipitation routes include direct precipitation, coprecipitation and homogeneous precipitation. In the direct precipitation process, only one cation is contained in the solution. For example, alumina precursor $NH_4Al(OH)_2CO_3$ has been produced by direct precipitation using $NH_4Al(SO_4)_2\cdot 12H_2O$ as starting material, NH_4HCO_3 as precipitating agent, and nanosized α-Al_2O_3 as seed [7]. α-Al_2O_3 powders with average particle size of 150 nm can be obtained by washing, drying, and firing the precursor. It is found that the temperature of nucleation can be reduced and the transformation rate can be increased effectively by adding a small quantity of α-Al_2O_3 seeds [8].

In addition, the effects of fluoride additions on the phase transition of α-Al_2O_3 formation have also been investigated. The addition of 2% LiF and AlF_3 decreases the transformation temperature by 300°C and well-dispersed α-Al_2O_3 powders with average particle size of ~2 μm were obtained [9]. The LiF and AlF_3 additives prove to be effective in enhancing the phase formation of α-Al_2O_3 because an intermediate compound, AlOF, can be formed in the case of the phase transformation and AlOF can accelerate the mass transportation from transient phase to stable α-Al_2O_3 phase [10].

For the coprecipitation method, multications are present in the mixed solution. Nanosized YAG powders with average particle size of 60 nm have been synthesized by a coprecipitation method from a mixed solution of $NH_4Al(SO_4)_2$ and $Y(NO_3)_3$, with NH_4HCO_3 as the precipitant [11]. Composition of this precursor will be the result of competition between OH^- and the carbonate species generated by the following chemical reactions during combination with metal cations:

$$NH_4HCO_3 + H_2O \Leftrightarrow NH_4OH + H_2CO_3 \quad (19.1)$$

$$NH_4OH \Leftrightarrow +NH_4^+ + OH^- \quad (19.2)$$

$$H_2CO_3 \Leftrightarrow H^+ + HCO_3^- \quad (19.3)$$

$$HCO_3^- \Leftrightarrow H^+ + CO_3^{2-} \quad (19.4)$$

As mentioned earlier, Al^{3+} may precipitate as AlOOH or $NH_4Al(OH)_2CO_3$. On the other hand, Y^{3+} may most likely precipitate as normal carbonate of $[Y_2(CO_3)_3\cdot nH_2O$ $(n = 2 \pm 3)]$ [12] or basic carbonate of $[Y(OH)CO_3]$ [13] from the carbonate anions containing NH_4HCO_3 solution. Also, homogeneous distribution nano-mulite powders (~100 nm) have been prepared by hydrolysis and coprecipitation method using TEOS and $AlCl_3\cdot 6H_2O$ as starting materials [14].

Compared to direct precipitation and coprecipitation method, homogeneous precipitation has the advantage of excellent homogeneity of nucleation and precipitation. In the homogeneous precipitation process, urea is usually used as precipitant. For example, Eu, Li-codoped ZnO:(Eu, Li) nanopowders have been synthesized by the homogeneous coprecipitation method using $Zn(NO_3)_2\cdot 2H_2O$, $CO(NH_2)_2$, $Eu(NO_3)_3\cdot 6H_2O$, and $LiNO_3\cdot 2H_2O$ as raw materials. The ZnO:(Eu, Li) nanopowder calcined at 700°C for 2 h is a pure hexagonal wurtzite structure and the particle size distribution is uniform in the range of 20–40 nm [15]. Using the same urea homogeneous precipitation method, monophase ytterbium aluminum garnet ($Yb_3Al_5O_{12}$) powders with the particle size of 20–30 nm have also been obtained at the low calcination temperature of 900°C [16].

Spherical anatase microparticles (15–40 nm) with good crystallinity have been synthesized by homogeneous precipitation under mild conditions (83–100°C), employing ammonium fluorotitanate as the titanium source and urea as the precipitant [17]. Urea decomposes at a temperature higher than 80°C during the urea precipitation process, as shown in Eq. (19.5).

$$2(NH_2)_2CO + 5H_2O \rightarrow 4NH_4^+ + OH^- + HCO_3^- + CO_3^{2-} \quad (19.5)$$

The NH_4^+ and O–H$^-$ ions generated then react readily with ammonium fluorotitanate to yield TiO_2 according to Eq. (19.6).

$$(NH_4)_2TiF_6 + 4NH_4^+ + 4OH^- \rightarrow TiO_2 + 6NH_4F + 2H_2O \quad (19.6)$$

Figure 19.1 shows the particle morphologies of the powders obtained after 30 min at three typical temperatures of 83, 90, and 100°C. The resultant particles are microspheres in each case, irrespective of the reaction temperature. Furthermore, these microparticles show good dispersion and no apparent aggregation is observed.

19.1.1.3. Sol–gel Method

Among various synthetic routes, sol–gel processing is also one of the most widely used techniques for nanoparticle formation. In this process, organic or inorganic precursors are gelled to form an interconnected three-dimensional network after hydrolysis and

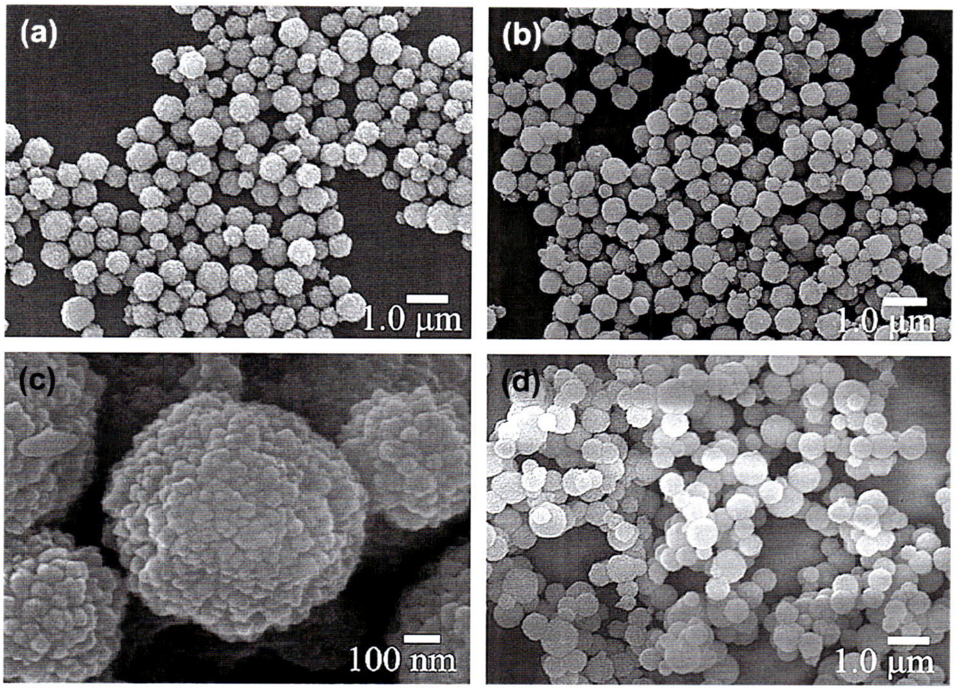

FIGURE 19.1 The particle morphologies of the anatase powders obtained after 30 min at different temperatures (a) 80°C, (b, c) 90°C, (d) 100°C [17].

polycondensation in organic solvents. This is followed by dehydration and calcination of the gel to form nanopowders. The sol–gel approach is of particular interest which allows compositional and microstructural tailoring through controlling the precursor chemistry and processing conditions. Sol–gel processing provides an excellent chemical homogeneity and the possibility of deriving unique metastable structures at low reaction temperatures. TiO_2 nanocrystals have been derived via sol–gel hydrolysis precipitation of titanium isopropoxide $Ti(OC_3H_7)_4$, followed by calcination or hydrothermal treatment [18].

Simultaneous control over particle size, morphology, and surface is one of the pertinent advantages of using sol–gel process for nanopowder preparation. However, high costs of alkoxide precursors possess a serious limitation to this process.

Gel combustion is a method, which combines the advantages of sol–gel and combustion method. The synthesis method includes two steps: the first is gelation and then combustion reaction. The gelation process from solution can guarantee a precise control of cationic stoichiometry and homogeneous mixture of metal ions at the atomic level. The following gel combustion reaction can avoid the rigorous condition in the dry process. Ultrafine α-Al_2O_3 powders with average particle size of ~200 nm have been synthesized by the gel combustion method using a citrate polymeric precursor derived from aluminum nitrate and citric acid mixed solution [19–21]. Also, nanosized Nd:YAG powders have been prepared at low temperature by a gel combustion method with citric acid as fuel and nitrate as oxidizer [22]. By altering the citrate-to-nitrate ratio, a variety of Nd:YAG powders with different agglomerate structures could be obtained [23].

Gel combustion for synthesizing fine powders is of significant interest, primarily due to the overall simplicity and high efficiency of the technique. However, the powders produced are not as uniform as those from the sol–gel process: they often contain hard agglomerates.

19.1.1.4. *Microemulsions*

Since Boutonnet et al. put forward using microemulsions to prepare nanomaterials in 1982 [24], the microemulsion method has been paid more and more attention. The development of the microemulsion method offers a good way to obtain the nanocrystals with controlled surface and size [25]. Usually, the preparation of nanocrystal in the microemulsion process involves two steps. One is to prepare nanoparticles in a water-in-oil (W/O) microemulsion, which is a transparent, isotropic liquid medium with nanosized water droplets dispersed in a continuous oil phase and stabilized by surfactant molecules at the water/oil interface. These surfactant-covered water pools offer a unique microenvironment for the formation of nanoparticles. They act as microreactors for processing reactions and inhibit the excess aggregation of particles. As a result, the particles obtained in such a microenvironment are generally very fine and monodispersed [26]. The other step in the process is to crystallize the nanoparticles by increasing the annealing temperature. The properties of nanocrystal, such as crystalline size, shape, strain, and stacking faults, are governed by the chemical nature of the "microreactor" and the annealing process of the nanoparticles. Using the microemulsion method, cubic phase CeO_2 nanocrystals with size of 6–8 nm have been prepared by Zhang et al. [27]. For another example, nanosized 3 mol% Y_2O_3 stabilized tetragonal ZrO_2 powders with average particle size of 14 nm have also been prepared by microemulsion processing [28]. The obtained particles are weakly agglomerated into polyhedral or spherical-shaped powders.

The microemulsion method has many advantages in synthesizing nanopowders, however, the method still has limitation in application. A mass of surfactant is needed in the water-in-oil (W/O) microemulsion system, so the cost is too high. The second disadvantage is the low yield due to the small amount of water solvent in the microemulsion. In addition, the existence of surfactant also has some influence on the purity and properties of the obtained nanopowders.

19.1.1.5. *Hydrothermal/solvothermal Method*

In a sealed vessel, solvents can be brought to temperatures well above their boiling points by the increase in autogenous pressures resulting from heating. Performing a chemical reaction under such supercritical conditions is referred to as solvothermal processing or, in the case of water as solvent, hydrothermal processing. Hydrothermal/solvothermal processing allows many inorganic materials to be prepared at temperatures substantially below those required by traditional solid-state reactions. Unlike the cases of coprecipitation and sol–gel methods, which also allow for substantially reduced reaction temperatures, the products of hydrothermal/solvothermal reactions are usually crystalline and do not require postannealing treatments [29,30].

Highly crystallized, well-dispersed perovskite-type $BaTiO_3$ crystallites with very fine (<100 nm) particles have been produced by the hydrothermal reaction of a newly prepared $Ti(OH)_4$ gel with the $Ba(OH)_2$ solution [31], and the crystallographic properties of hydrothermal $BaTiO_3$ crystallites have also been investigated [32].

The synthesis of nanocrystalline TiO_2, which is an important photocatalyst for the decomposition of toxic chemicals, is one of the more thoroughly investigated hydrothermal/solvothermal reactions. The preparation of anatase by hydrothermally processing hydrous titania prepared by the controlled hydrolysis of $Ti(OEt)_4$ in ethanol has been reported by Oguri et al. in 1988 [33]. More recently, a method for preparing nanoparticulate, phase-pure rutile from aqueous $TiCl_4$ by a hydrothermal process has been developed by Zhang and Gao [34]. By adding citric acid to stabilize the TiO_2 nanoparticles and hydrothermally heating the precursors in the presence of KCl or NaCl mineralizers, 2–10 nm crystallites of monodispersed, phase-pure anatase has been produced by Yin et al. [35]. Furthermore, TiO_2 nanocrystals can be prepared by the low-temperature reaction of low-valent organometallic precursor [36]. Pure anatase TiO_2 have bee prepared by the hydrolysis of $TiCl_4$ with ethanol at 273 K followed by calcination at 360 K for 3 days [37]. Nanocrystals of the transition metal dichalcogenides (ME2; M = Fe, Co, Ni, Mo; E = S or Se) with diameters in the range of 4–200 nm have also been prepared by a hydrothermal route [38].

Compared with hydrothermal reaction method, solvothermal decomposition of organometallic precursors is more common for preparation of metal oxide nanocrystals. The use of cupferron complexes as precursors to prepare γ-Fe_2O_3, Cu_2O, and Mn_3O_4 nanocrystals has been described by Rockenberger et al. [39]. Metallic ReO_3 nanocrystals with diameters in the 8.5–32.5 nm are obtained by the solvothermal decomposition of R_2O_7-dioxane under solvothermal condition [40]. Figure 19.2a shows a TEM image of ReO_3 nanocrystals of 17 nm average diameter with the size distribution histogram as an upper inset. The lower inset shows an high resolution transmission electron microscopy (HRTEM) image of 8.5 nm nanocrystal. The nanocrystals exhibit a surface plasmon band around 520 nm which undergoes blue-shifts with decrease in size (Fig. 19.2b). Nanocrystals of metal chalcogenides are also prepared by the reaction of metal salts with an appropriate sulfiding or seleniding agent under solvothermal conditions. Toluene-soluble CdSe nanocrystals with a dia-meter of 3 nm have been prepared solvothermally by reacting cadmium stearate with elemental Se in toluene in the presence of tetralin [41].

Unlike the cases of the precipitation and sol–gel methods, hydrothermal/solvothermal method can obtain agglomeration free (or less agglomeration), high crystallinity and high sintering activity nanosized powders with postannealing treatment [42]. However, this method still has its intrinsic limitation, such as long reaction period, rigorous temperature, and pressure.

19.1.2. Sintering of Nanoceramics

There are two challenges in nanoceramic sintering: achieving full density and maintaining nanosized grains. While a number of techniques are capable of achieving full density, maintaining nanosized grains

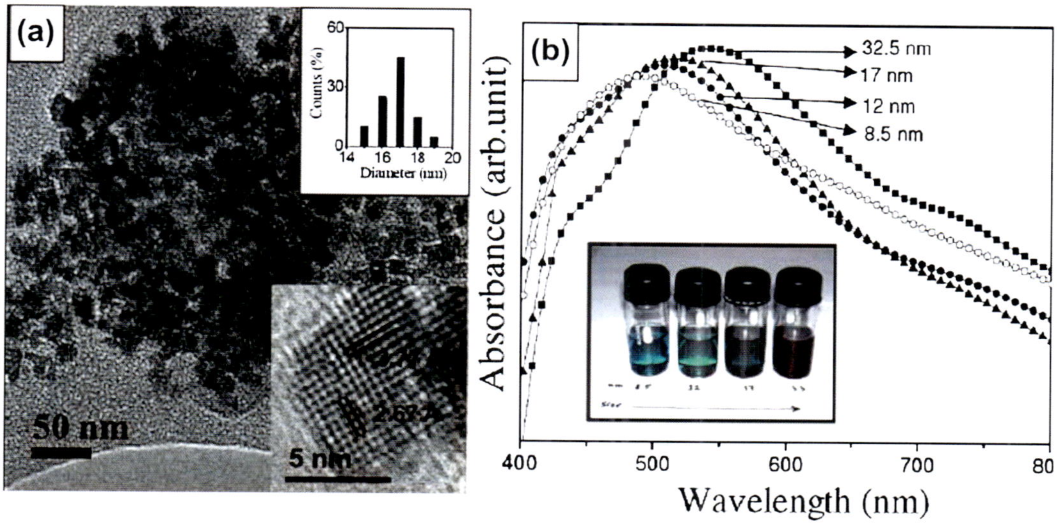

FIGURE 19.2 (a) TEM image of ReO_3 nanocrystals of average diameter 17 nm. Upper inset shows the size distribution histogram. Lower inset shows the HRTEM image of a single 8.5 nm nanocrystal. (b) Optical absorption spectra of ReO_3 nanocrystals with average diameters of 8.5, 12, 17, and 32.5 nm. Inset in (b) shows a picture of four different sizes of ReO_3 nanocrystals dissolved in CCl_4. *Reprinted with permission from [40]. Copyright 2006 American Chemical Society.*

remains to be a tough problem to solve. When both full densification and nanosized grains are desired, innovative sintering approach becomes a must. The success in consolidation of nanopowders to yield nanoceramics lies with the precise control of the coarse-densification competition during the sintering process, i.e., enhancing densification while suppressing coarsening. In the following section, some nanoceramic sintering methods will be introduced.

19.1.2.1. Pressureless Sintering

In the case of conventional pressureless sintering, due to the absence of any external driving force, higher temperature and longer time are required for obtaining near theoretical densification. The enhanced duration of elevated temperature results in significant grain growth, especially in the last stage of sintering [43]. For nanoparticle sintering, there are three aspects that can be utilized to improve pressureless sintering [44]. The first is devising the sintering schedules to control the relative contribution of densification and grain growth in nanoceramics. The second is attainment of high nanoparticle packing (green density). And the third is to introduce second phase, reducing the grain boundary mobility while encouraging grain boundary diffusion. The application of a two-step sintering process, which involves lowering the peak sintering temperature and holding the sample at the lower temperature for the final stage of densification, can be employed to prepare nanoceramics via pressureless sintering. The success of this process is first demonstrated by Y_2O_3 nanoparticle sintering [45,46]. Two different sintering schedules have been adopted to prepare Y_2O_3 while using nanosized Y_2O_3 powders with particle size of 10–15 nm as raw material. For pure Y_2O_3 initially heated to 1310°C, full density was achieved after holding at 1150°C for 20 h, during which there is no grain growth. As another example, with the first step reaching 1250°C, full densification without grain growth is achieved after holding at 1150°C for 20 h. For 1 wt% MgO-doped Y_2O_3, fully dense ceramic with average grain size of ~60 nm has been obtained by sintering at 1000°C for 20 h, after reaching 76% relative density in an initial firing to 1080°C, as shown in Fig. 19.3. According to the explanation provided by Chen and Wang [45], the kinetics of grain boundary mobility is reduced at the lower holding temperature, while maintaining sufficient grain boundary diffusion kinetics for elimination of residual pores. This approach is further corroborated by $BiTiO_3$ and Ni–Cu–Zn systems [47].

In sintering, besides the starting nanoparticle size and size distribution, high packing density is a critical factor to consider. Due to the rapid grain growth tendency of nanoceramics, one natural approach in maintaining the nanosized grains is to reduce the necessary sintering shrinkage by starting with high green density. The low-density samples formed under convention compaction show rapid densification but at the expense of rapid grain growth. For the high green density samples compacted under superhigh pressure, there is no excessive grain growth. More importantly, the final sintered densities are always higher for the higher green density samples. Sixty percent dense Y_2O_3–ZrO_2 sample

FIGURE 19.3 Microstructure of fully dense Y_2O_3 doped with 1 wt% MgO. *Reprinted by permission from Macmillan Publishers Ltd: Nature [45], Copyright (2000)*

compacted under 3 GPa pressure can be sintered at 1050°C while maintaining the grain size at about 80 nm [48]. γ-Al_2O_3 is a metastable phase and vermicular structure will be formed during γ- to α-phase transformation. So it is very difficult to get dense alumina ceramics by traditional pressureless sintering of γ-Al_2O_3 nanopowders. However, when the γ-Al_2O_3 nanoparticles are compacted at >3 GPa pressure, there is no formation of vermicular structure and nearly full density is obtained with grain size at 150 nm [49]. As another example involves phase transformation during sintering, 99% dense TiO_2 nanoceramic with grain size at 60 nm has been obtained from the titania hydrosol heated at 600°C [50].

In order to obtain fully dense nanoceramics by pressureless sintering, dopants or lower melting temperature species that form liquid phase at sintering temperature are introduced to reduce the grain boundary mobility and encourage grain boundary diffusion. Y_2O_3 and Al_2O_3 are used in the Y_2O_3–ZrO_2 and Al_2O_3–ZrO_2 systems. The Al_2O_3 particles have 10 nm size and Y_2O_3 particles have <20 nm size. The ZrO_2 matrix particles also have 10 nm size. After 1100°C sintering, the ZrO_2 composites have 30 nm grain size with full density. Such small grain size is believed to result from a homogeneous distribution of Y_2O_3 and Al_2O_3 particles hindering ZrO_2 grain boundary mobility and thus grain growth by pinning the grain boundaries [51]. Empirically, the suppression of grain growth during sintering by pinning the grain boundaries with second phase is successful when the starting particle size does not exceed 30 nm. The desired condition for controlled grain growth is that secondary phase is well separated and homogeneously dispersed in the matrix phase [52].

By any means, pressureless sintering of nanoceramics requires careful control of sintering schedule, nanoparticle size, nanoparticle packing, and sintering additives in order to enhance sintering densification while suppressing grain growth. Even with the best efforts, pressureless sintering of nanoceramics remains to be a challenging task. So pressure sintering and other new sintering techniques are more prevalent in sintering of nanoceramics.

19.1.2.2. Pressure Sintering

Pressure sintering can drastically accelerate nanoceramic densification and achieve nanosized grains by providing external sintering driving force. As a result, lower sintering temperature and shorter sintering time can be used while achieving nanoceramic full densification. Pressure sintering mainly includes hot pressing, hot isostatic pressing, and sinter forging. Hot pressing and sinter forging involve the application of uniaxial pressure during elevated temperature exposure, in the presence or absence of constraining die, respectively. MgO of 10 nm particle size is hot pressed to 99.5% at 790°C, 150 MPa and <74 nm grain size is obtained [53]. When additives are used as sintering aids, liquid-phase amount needs to be carefully adjusted for combined liquid phase sintering and hot pressing. For example, dense SiC ceramics with a mean grain size of <100 nm can be obtained through combined hot pressing and liquid-phase sintering [54]. In this process, SiC nanoparticles of 30 nm are used as raw materials with Al_2O_3 and Y_2O_3 as sintering aids. The optimal liquid phase for SiC matrix is 10 wt%. It is clear that the most desirable approach is still hot pressing the nanopowders with additives. As mentioned before, phase transformation during sintering will hinder nanoceramics full densification. However, one special type of hot pressing, whose pressure is in the order of a few gigapascals, can desirably utilize phase transformation to assist consolidation. Pressure-induced phase transformation of anatase TiO_2 has been sintered at 1.5 GPa for 1 h, and the sintered compact is 98% dense rutile phase with a mean grain size of 36 nm [55]. The anatase to rutile phase transformation that begins at 650°C can be greatly lowered as the pressure is raised. The increased pressure during the phase transformation assisted sintering will lead to the grain refinement of the rutile phase [56].

Sinter forging is a combined superplastic deformation and sintering process. Sinter forging of ZrO_2 nanoparticles at 950°C and 300 MPa achieves 45 nm grain size and full density [57]. Uchic et al. [58] were able to densify nanocrystalline TiO_2 powders to near theoretical density at 650°C via sinter forging and obtained ceramic retains grain size of 50–60 nm. Sinter forging can achieve fast densification. However, there is little theoretical understanding about the microstructure evolution. Compared with other sintering techniques, sintering forging is not a mainstream sintering method for nanoceramics.

In the case of hot isostatic pressing, the pressure is uniform from all directions and it is applied to densify the ceramic body via fluid medium, such as inert gas. The application of nanostructured oxides is best demonstrated by 10 vol% ZrO_2–Al_2O_3 nanocomposite with the intent of using ZrO_2 nanoparticles to reduce Al_2O_3 grain growth. The nanoparticle mixture is cold isostatically compacted at 495 MPa and presintered at 1200°C for 2 h before it is isostatically pressed at 1200°C and 247 MPa. The sintered sample is 99% dense with <100 nm grain size [59]. However, hot isostatic pressing is not widely applied to oxide nanoceramics as the relatively low pressure cannot provide sufficient sintering driving force for grain growth control. For nonoxides, hot isostatic pressing has been utilized to carbide nanoceramics, as well as their composites. Dense and nanostructured SiC ceramic with average grain size of about 100 nm has been obtained at 1850°C and 200 MPa pressure after sintering for 1 h [60]. Si_3N_4/SiC nanocomposite is prepared by using pretreated Si_3N_4 and SiC nanoparticles; dense and homogeneous nanostructure with an average grain size of 50 nm is obtained at 1750°C and 150 MPa for 1 h. Hot isostatic pressing has the advantages of uniform densification in all directions and homogeneous microstructures. However, it has its own disadvantages, such as the complexity and high cost of the equipment, low pressure, and lack of container materials suitable for nanoceramic sintering [61].

19.1.3.3. Spark Plasma Sintering

Spark plasma sintering is also named as plasma-activated sintering, pulse electric current sintering, and pulse discharge pressure sintering [62]. In spark plasma sintering, sintering is realized by subjecting the green compact to arc discharge generated by a pulsed electric current. An electric discharge process takes place on a microscopic level and accelerates material diffusion. The externally applied pressure is generally 30–200 MPa and the heating rate is 100–1000°C min^{-1}[63]. Spark plasma sintering is the most effective densification technique for nanoceramics. One of the most remarkable features of this technique is that the small grain size can be maintained while achieving full densification. For oxides such as MgO, nanoparticles are spark plasma sintered between 700 and 825°C under applied pressure of 100–150 MPa. The obtained MgO transparent ceramic is fully dense and with 52 nm average grain size [64]. For nitrides such as Si_3N_4, nanoparticles of 10 nm size are spark plasma sintered at 1600°C for 5 min at a heating rate of 300°C min^{-1}. The sample is fully densified with homogeneous nano Si_3N_4 grains of 70 nm size [65]. Besides the single-phase nanoceramics mentioned above, SPS is also used for nanoceramic or nanocomposite fabrication. Hydroxyapatite–3 mol% Y_2O_3–ZrO_2 composite has been sintered by spark plasma sintering at 1150°C and 50 MPa with a heating rate of 200°C min^{-1}. The grain size of Y–TZP is only about 50 nm and hydroxyapatite grain size is about 100 nm [66].

A wide variety of nanoceramics have been successfully densified by the SPS process. There are several advantages that accrue from this process, such as fast heating rate, fast cooling rate, and very short holding time. These combined advantages offer very limited grain growth. However, the sintering mechanism of spark plasma sintering is not very clear, so the fundamental study is still needed.

Finally, it should be stated that, in spite of the progress made in the processing of bulk nanoceramics, this remains an emerging field and further research is required to optimize processing parameters, develop processing techniques for large scale production, gain further understanding of their properties, and render nanoceramics and ceramic nanocomposites viable for further novel applications.

19.2. CERAMIC MATRIX COMPOSITES (CMCS)

Ceramic matrix composite is a kind of material which uses ceramic as matrix and fiber, whisker, pellet, or particle as reinforcement. CMC is also named as multiphase composite. CMCs have the advantages of high temperature resistance, high hardness, corrosion resistance, and high wearing resistance. In addition, the intrinsic brittleness can be improved and the unexpected destruction can be avoided by introducing the reinforcer into the ceramic matrix. CMCs have been widely used in high technology, aerospace, national defense, and national economic fields. Generally speaking, CMCs include fiber (or whisker) reinforced ceramic matrix composites, particle dispersion-strengthened ceramic matrix composites, in situ growth ceramic matrix composites, and ceramic nanocomposites.

The characteristic of atomic arrangement in ceramic structure leads to the lack of plastic deformation like ability in metal. When broken, ceramic has no other energy absorption mechanisms, except to form new broken surfaces [67]. Usually, the toughening mechanisms involve dispersion toughening, microcrack toughening, bridging crack toughening, crack deflection toughening, fiber extraction toughening, compressive prestress toughening, and stress-induced phase transformation toughening [68]. By the proper design of composite, the brittleness of ceramic materials can be greatly improved. In this section, some typical ceramic matrix composites with important use and applications will be introduced.

19.2.1. Fiber-reinforced Ceramic Matrix Composites

19.2.1.1. C_f/SiO_2 Composites

Fused silica (SiO_2) has very low density, thermal expansion coefficient, electric conductivity, relatively high mechanical property, heat resistance, thermal-shock resistance, corrosion resistance, and dielectric constant, so it is extensively used in optical and opto-electronics devices, microwave dielectric materials, and refractory materials. However, compared with other materials, fused silica is a kind of typical brittle material with low strength and toughness, which influence its application and reliability. So the introduction of reinforced fiber to form composite is the main approach to improve its mechanic properties. Carbon fiber-reinforced fused silica (C_f/SiO_2) composite materials have been developed by Guo and Yen [69]. Compared with silica glass, the strength and toughness of C_f/SiO_2 composite have been greatly improved. The strength has been increased to 11 times and the work of fracture has been increased to two orders of magnitude. The high temperature, lightweight C_f/SiO_2 composite have been used.

19.2.1.2. C_f/C Composites

The carbon/carbon (C_f/C) composites contain carbon fiber and carbon matrix. Among the thermostructural CMCs, C_f/C composite is the most lightweight material with excellent properties. The advantageous properties of C_f/C composites such as their high temperature strength, creep and corrosion resistances, high toughness, resistance to shocks, fatigue and damage, and reliability make them ideal candidates for the military and aerospace applications. The Young's modulus of high-performance carbon fiber is over 700 MPa and the strength is at the range of 3–4 GPa. Carbon fibers in C_f/C composites can be continuous or discontinuous. By adjusting the orientation and weave of carbon fibers, the strength and modulus of C_f/C composites at different directions can be controlled.

Based on CVI, chemical liquid vapor infiltration (CLVI), which utilizes liquid hydrocarbon as precursor, is introduced as a new method to fabricate C_f/C composites [70,71]. By a rapid CLVI processing, C_f/C composites have been fabricated with the deposition temperatures in the range of 850–1400°C and under the system pressure of ~0.1 MPa [72]. By introducing microwave heating into the densification process of carbon fiber-reinforced CMCs, a new method named microwave pyrolysis chemical vapor infiltration (MPCVI) has been developed. C_f/C composites have been prepared by MPCVI with different deposition temperatures [73].

C_f/C composites are full carbon materials and their main disadvantage is the degradation in an oxidizing atmosphere beyond 400°C, resulting in the need for an external oxidation protection. It is known from oxidation kinetics that increasing the final heat treatment temperature (for example by a graphitization step) results in an improved oxidation resistance of the C-fibers. Therefore, the oxidation resistance is also improved by reinforcements with high modulus (HM) or ultrahigh modulus (UHM) carbon fibers in comparison to high tenacity (HT) fibers. Long-term oxidation protection requires multilayer protection coatings, where the C_f/C composites are protected for example with SiC layers and additional self-healing glass forming layers, based on oxides like mullite, alumina, or silicon[74]. The silicon carbide layer can be performed via pack cementation, but superior oxidation resistance can be achieved with pure β-SiC layers, deposited via the CVD process. SiC and modified SiC coatings have been produced by a pack cementation technique on surface of C_f/C composites [75]. The results show that the anti-oxidation resistance of the modified SiC coating is greatly improved at 1773 K. In another example, precoated carbon layers on the surface of C_f/C composites have also been prepared first by slurry and high-temperature treatment for preparing the anti-oxidation coating with SiC concentration gradient [76].

High-melting $MoSi_2$ or WSi_2 with superior oxidation resistance has been used as one of the best materials to protect C_f/C composites [77–79]. However, single WSi_2 is too hard and brittle and $MoSi_2$ has high-temperature plasticity. These disadvantages influence their high-temperature mechanical properties and confine/limit their uses in high-temperature materials. Compared with single WSi_2 or $MoSi_2$, $WSi_2/MoSi_2$ composite has better property. An $SiC-WSi_2/MoSi_2$ multilayer oxidation protective coating for C_f/C composites has been prepared by pack cementation [80]. A stable and dense glassy SiO_2 film can be formed on the coating surface with 10.0 at% (W and Mo) during the oxidation test. The coated specimen with 10.0 at% (W and Mo) has no weight loss after oxidation in air for 315 h at 1500°C and no cracks after thermal cycling between 1500°C and room temperature for 18 times, which implies that the multicoating has excellent oxidation protective ability and thermal shock resistance.

19.2.1.3. C_f/SiC Composites

The applications of carbon fiber-reinforced silicon carbide (C_f/SiC) lie in fields, where conventional materials, due to their insufficient mechanical properties at high temperatures or limited damage tolerance behavior can no longer be considered and includedin all principal areas of lightweight construction. The carbon fibers generally decreases the brittleness of SiC considerably so that the damage tolerance of C_f/SiC components lies in the same order of magnitude as for grey cast iron. Due to the excellent mechanical properties, C_f/SiC composites have been a new and potential generation of high-temperature structural material used in the aerospace field. Nowadays, the traditional preparation methods for C_f/SiC composites contain hot pressing (HP) [81], chemical vapor infiltration (CVI) [82], reactive melt infiltration (RMI) [83], and precursor infiltration and pyrolysis (PIP)[84].

Using carbon fibers as reinforcer, C_f/SiC composites have been prepared by hot pressing at 1800–1880°C and 20 MPa [85]. By chemical vapor infiltration, two-dimensional C_f/SiC composites used for hinge bearing have been prepared [86]. Excellent wear resistance and load-carrying ability has been demonstrated by low wear and especially small deformation. The hot-pressing method, the preparation temperature for C_f/SiC composites, is relatively high and it is only suitable for component with simple configuration.

Through chemical vapor infiltration (CVI) combined with liquid melt infiltration (LMI), the C_f/SiC brake materials have been prepared [87]. The carbon fiber preform is fabricated with the three-dimensional needling method. The C_f/SiC composites exhibited excellent toughness and the main toughening mechanism of the brake materials is fiber pull-out, interfacial debonding, bridging of fibers, and deflection of crack (Fig. 19.4). Chemical vapor infiltration is a traditional method for C_f/SiC composition fabrication.

Compared with hot pressing and chemical vapor infiltration, precursor immersion pyrolysis (PIP) is a comparably novel method. Using this method, large-scaled, complex-shaped, and large batch C_f/SiC compositions can be prepared at low temperature. C_f/SiC compositions have been prepared by PIP utilizing polycarbosilane as the precursor for matrix formation and two-dimensional layered 0°/90° carbon fabrics as the reinforcement [84].

Regarding the advantages of CVI and PIP, carbon fiber needling preform-reinforced C_f/SiC compositions are prepared by "CVI+PIP" combined process [88]. The results show that the densification efficiency of

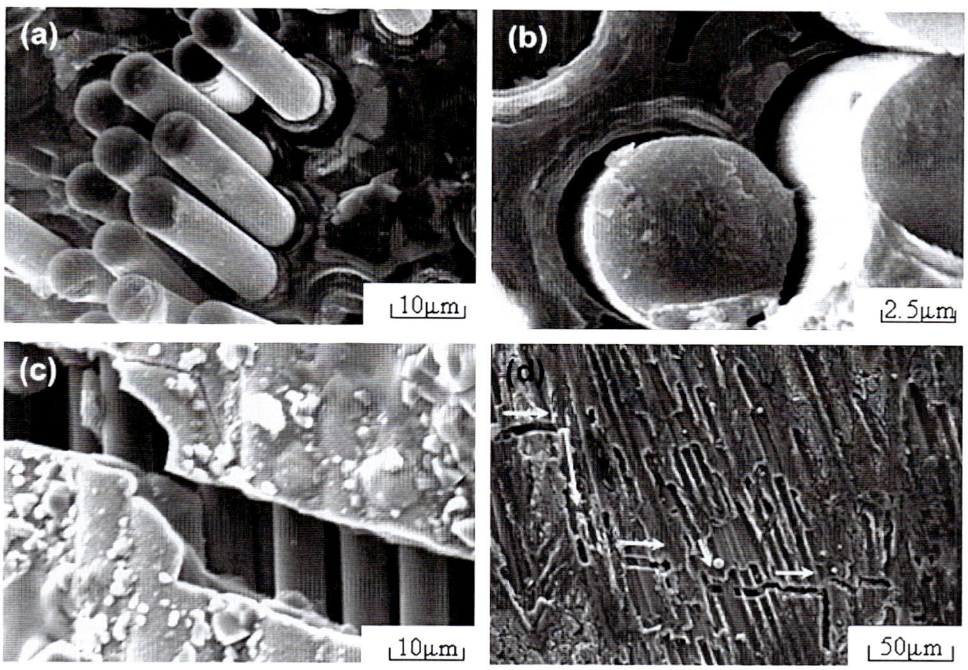

FIGURE 19.4 Typical SEM micrograph of the samples after tensile test (a) fiber pull-out, (b) interfacial debonding, (c) bridging of fibers, and (d) deflection of crack. *Reprinted from Composites Science and Technology, Vol. 67, Shangwu Fan, Litong Zhang, Yongdong Xu, Laifei Cheng, Jianjun Lou, Junzhan Zhang, Lin Yu, Microstructure and properties of three-dimensional needle-punched carbon/silicon carbide brake materials, Copyright (2007), with permission from Elsevier [87].*

carbon fiber needling preform is higher than that of three-dimensional woven preform. The composites reinforced with carbon fiber needling preform exhibit good mechanical properties with flexural strength of 237 MPa and shear strength of 26 MPa, respectively. In addition, C_f/SiC compositions fabricated by the reaction melt infiltration (RMI) method show some excellent advantages such as moderate density (about 2.0 g/cm^3), low open porosity (<5%), and good oxidation resistance in comparison with C_f/C composites [83,89].

Due to the anisotropic coefficient of thermal expansion (CTE), C_f/SiC composites, the oxidation protection is more difficult than it is for nonreinforced carbon or graphite bulk materials. The mismatch of CTE between the SiC coating and the carbon fiber reinforcement creates cracks in the SiC coating during the cooling-down period after deposition [90]. At the same time, residual pores will inevitably remain during the fabrication process of C_f/SiC composites [91]. The existence of cracks and pores provides the diffusion channels for oxygen and leads to the oxidation of carbon fiber. Thus, the properties of C_f/SiC composites will be greatly depressed. SiC coating has become the commonly used method to protect carbon fiber from being oxidized, because SiC can be oxidized into SiO_2 with very low oxygen diffusion coefficient and the crack can be enveloped. The oxidation behavior of C_f/SiC composites with three SiC protection layers has been investigated by Cheng et al. [92]. It is found that, at the temperature range from room temperature to 1500°C, the oxidation degree of interface carbon and carbon is the most serious at 700°C. Considering the bad oxidation resistance of SiC coating at the intermediate and low temperatures, B_4C [93] system has been introduced into the SiC coating to protect C_f/SiC composites. B_4C and SiBC can be oxidized into B_2O_3 with low melting point (450°C) and good fluidity. The cracks in the SiC coating can be closed up and oxygen can be prevented from diffusing into the fibers. However, the B coating is usually used as middle healing coating on C_f/SiC composites due to volatility of B_2O_3. Modification of SiC coating has been carried out with B and BC_x prepared by CVD. SiC/B/SiC and SiC/BC_x/SiC multilayer self-healing coatings are prepared on a two-dimensional C_f/SiC composite [94]. The main reactions involved during preparation of B and BC_x middle layers by CVD are as follows:

$$3BCl_3 + 3H_2 \xrightarrow{1100^\circ C} 3B + 6HCl$$

$$6BCl_3 + 2xC_3H_6 + (9-6x)H_2 \xrightarrow{1000^\circ C} 6BC_x + 18HCl$$

The results show that the B_2O_3 glass produced by oxidation of B or BC_x can seal the cracks that existed in the coating well, the kinetics of oxidation is controlled by diffusion of oxygen through the mocrocracks and B_2O_3 glass layer. SiC/BC_x/SiC$-C_f$–SiC composites have low weight loss during the oxidation and higher strength retention after oxidized for 10 h.

Long-carbon-fiber-reinforced SiC composites are thoroughly investigated and the materials have been widely utilized in high-temperature structural elements for aerospace equipments like rocket nozzles, aeronautic jet engines, and leading edges. In the recent years, the research on short carbon fiber-reinforced silicon carbide (C_{sf}/SiC) has progressed [95,96]. C_{sf}/SiC composites have been fabricated by hot pressing at 1800°C and 25 MPa using Si as the main sintering aid [97]. It is found that the maximum oxidation weight loss of the composite occurs at 900°C [98]. Below 900°C, the oxidation weight loss of the composites is caused by the carbon fibers oxidation. Between 900 and 1450°C, short carbon fibers on the composite surface are oxidized and the SiC matrix forms a protecting film. The inner short carbon fibers still have the strengthening effect on the oxidation layer and matrix protecting them from oxidation, and the composite have good self-resistance oxidation property. Above 1600°C, the composite surface is destroyed badly and loses the resistance oxidation capacity.

19.2.1.4. SiC$_f$/CMCs

Silicon carbide fiber-reinforced ceramic matrix composites (SiC_f/CMCs) are developed initially for military and aerospace applications. Now they are being introduced into new fields and their range of applications will grow when their cost is lowered drastically. The advantageous properties of SiC_f/CMCs such as their excellent high-temperature strength, creep and corrosion resistances, low density, high toughness, resistance to shocks, fatigue and damage, and reliability make them ideal candidates for the replacement of metals and ceramics in many engineering applications involving loads, high temperatures, and aggressive environments [99,100]. Compared with carbon fibers, SiC fibers have better high temperature stability. Even at the high temperature of 1200°C, there is still no obvious decrease in tensile strength and Young's modulus of SiC fibers. So SiC_f/CMCs are of great potential application and used as high-temperature structural materials. The preforms of SiC fibers have been developed from discontinuous structure to one-dimensional, two-dimensional, three-dimensional, and multi-dimensional overall structures [101].

As to the SiC_f/CMCs system, the main matrices include Al_2O_3, ZrO_2, $ZrSiO_4$, Si_3N_4, and SiC. The interest in the development of SiC_f/Si_3N_4 composites stems from the fact that: (1) SiC and Si_3N_4 are thermodynamically compatible and stable at temperature to 1700°C, (b) Si_3N_4 processing temperature can be tailored to avoid fiber degradation, and (c) Si_3N_4 matrix microstructure can be controlled to improve composite properties such as cracking strength and thermal conductivity [102]. The fabrication of SiC_f/Si_3N_4 composites begins by the use of monofilaments, fiber tows, and textile processed to form multifiber bundles or tows of ceramics fibers into two-dimensional and three-dimensional fiber weave preforms that meet the product size and shape requirements. For assuring crack deflection between the fibers and final matrix, a thin fiber coating or interphase material with a mechanically weak microstructure, such as boron nitride (BN) or pyrolytic carbon is then applied to the fiber surfaces by chemical vapor infiltration (CVI).Silicon nitride-based matrices are then formed within the coated preforms by a variety of processes such as CVI, reaction-bonded silicon nitride (RBSN), polymer infiltration and pyrolysis (PIP), hot pressing (HP), and hot-isostatic pressing (HIP). The pressureless-sintering method commonly used for the fabrication of monolithic silicon nitride ceramics cannot be successfully adapted for the fabrication of SiC fiber-reinforced silicon nitride matrix composites because of significant degradation strength of currently available fibers at the high temperatures required for sintering, and also because of the problems associated with retardation of densification of the matrix material by the presence of already dense fibers. Fabrication of SiC_f/Si_3N_4 composites by CVI was not successful because of the difficulty in maintaining stoichiometry of silicon nitride, and infiltrating into thick sections of the preforms. Only SiC monofilament and SiC fiber tow-reinforced Si_3N_4 matrix composites fabricated by reaction-bonding, or hot pressing, or hot-isostatic pressing have been successfully prepared [103].

The SiC_f/SiC composites consist of an SiC-based matrix reinforced by SiC fibers. Usually, the SiC_f/SiC composites can be prepared by chemical vapor infiltration (CVI). In very simple terms, the SiC-based matrix is deposited from gaseous reactants onto a heated substrate of fibrous preforms (SiC). CVI is a slow process, and the obtained composite materials possess some residual porosity and density gradients. Despite these drawbacks, the CVI process presents a few advantages [104]: (i) the strength of reinforcing fibers is not affected during composite manufacture, (ii) the nature of the deposited material can be changed easily, simply by introducing the appropriate gaseous precursors into the infiltration chamber, and (iii) a large number of components and large complex shapes can be produced. The SiC_f /SiC composites prepared by CVI exhibit good mechanical properties at room and high temperatures that depend on the fiber/matrix interface. Pyrocarbon (PyC) has proven to be an efficient interphase to control fiber/matrix interactions and composite mechanical behavior. But pyrocarbon is sensitive to oxidation at temperatures above 450°C. A few versions of high temperature resistant SiC_f/SiC composites are produced. In order to protect the PyC interphase against oxidation, multilayered interphases and matrices have been developed. Multilayered matrices contain phases which produce sealants at high temperatures preventing oxygen from reaching the interphase. This composite is referred to as CVI SiC_f/Si−B−C. Oxidation-resistant interphases such as BN or multilayered materials can also be coated on the fibers. An "oxygen getter" can be added to the matrix to scavenge oxygen that might ingress into the matrix (enhanced CVI SiC_f/SiC).

19.2.2. Whisker-reinforced Ceramic Matrix Composites

Using carbothermic reduction reactions of low-cost silica and carbon precursors, industrial-scale quantities of SiC whiskers have been produced. At the present time, commercial sale of whiskers is limited to a few companies in the world with SiC whiskers, the most predominant [105]. In addition, whiskers of TiC, TiN, Al_2O_3, mullite, Si_3N_4, and B_4C have also been produced and studied. However, SiC remains the most interesting and commercially successful whisker.

19.2.2.1. SiC_w/Al_2O_3 Composites

The ideal SiC whisker for reinforcement of ceramic composites would have several characteristics that would also depend on the matrix phase. In every case, the ideal whisker would have no internal structural imperfections, be relatively smooth, and possess high strength. For alumina matrix where the difference in thermal expansion is higher between the matrix and the whisker, the whisker diameter would be in the range of 1.5−2.0 μm [106]. For Silicon carbide whisker-reinforced alumina (SiC_w/Al_2O_3) composite, it was determined that the presence of excess surface carbon on the whiskers resulted in a weak interfacial bond and produced materials with high toughness and strength. In the fabrication of SiC_w/Al_2O_3 composites, the SiC whiskers are normally received from commercial sources in an agglomerated form. Deagglomeration and dispersion of the SiC whiskers have been found to be extremely important in the fabrication of composites with good mechanical properties. The methods used to disperse the SiC whiskers and mix them with the matrix powders include ultrasonic homogenization, high shear mixing, ball milling, and turbomilling[107]. Densification methods fall into two general categories: (1) pressureless-sintering and (2) pressure-assisted. Pressureless sintering is the most economical and is capable of producing complex shaped parts. However, whisker loadings are limited where high sintered densities cannot be achieved. The pressure-assisted techniques include hot-pressing and hot isostatic pressing

(HIP).These methods produce high densities at high whisker loadings.

19.2.2.2. SiC_w/Si_3N_4 Composites

Si_3N_4 ceramic is a structural material with excellent mechanical properties, thermal shock resistance, oxidation resistance, and low coefficient of thermal expansion. It is one of the most promising material used as the component of ceramic engine. However, the component will be catastrophically destroyed during the application course due to the intrinsic brittleness of Si_3N_4 ceramic that prevents its wide utility. In order to improve the mechanical property of Si_3N_4 ceramic, SiC whisker is used as the reinforcement. SiC_w/Si_3N_4 composites have been prepared by hot pressing at 1780°C for 1 h and under the pressure of 25 MPa [108]. Compared to Si_3N_4 ceramic, the bending strength and fracture toughness of the obtained SiC_w/Si_3N_4 have been improved to a certain extent. As to the property of thermal shock resistance, the SiC_w/Si_3N_4 composite is also better than Si_3N_4 ceramic. In the system of SiC_w/Si_3N_4 composite, debonding along the Si_3N_4 matrix–SiC whisker interface and microcrack propagation are both attributed to the increase of fracture toughness.

19.2.3. Particle Dispersion-strengthened Ceramic Matrix Composites

19.2.3.1. Ni/Al_2O_3 (Co/Al_2O_3) Composites

Introducing ductile metal phase into the brittle alumina ceramic matrix has been proved to be a promising way to improve the toughness [109,110]. In the metal/alumina systems, Ni/Al_2O_3 has been widely investigated. For example, Ni/Al_2O_3 composites have been prepared by hydrogen reduction and hot pressing of powder mixtures of NiO and Al_2O_3 obtained by two processes, namely, the ball-milled mixing process and the coating process [111]. In the ball-milled mixing process, NiO nanopowders with average particle size of 8 nm are prepared by chemical precipitation method using analytically pure $Ni(NO_3)_2 \cdot 6H_2O$ and NH_4HCO_3 as starting materials [112]. The reaction can be represented in the form of following equation:

$$3Ni^{2+} + 6HCO_3^- + H_2O = NiCO_3 \cdot 2Ni(OH)_2 \cdot 2H_2O\downarrow + 5CO_2\uparrow$$

However, after ball milling, the obtained $NiO-Al_2O_3$ composite is severely agglomerated. In the coating method, Ni-coated Al_2O_3 powders are prepared by the heterogeneous precipitation method using Al_2O_3 with diameter of 350 nm, $Ni(NO_3)_2 \cdot 6H_2O$ and NH_4HCO_3 as raw materials [113]. Figure 19.5 shows TEM micrographs of $Ni-Al_2O_3$ composite powders prepared by the above-mentioned two processes. It can be seen that spherical Ni particles of size 10–106 nm are obtained by ball-milled mixing and then hydrogen reduction. By heterogeneous precipitation, calcination, and reduction method, spherical Ni particles of size 10–55 nm can be obtained [111].

Dense $Ni-Al_2O_3$ composites are prepared by hot pressing at 1300–1500°C for 1 h in argon atmosphere with an applied pressure of 20 MPa [111]. The $Ni-Al_2O_3$ composite prepared by direct mixing method, many small Ni particles of size 100–600 nm are located with Al_2O_3 grain, and the relatively large Ni particles are dispersed at grain boundaries. For the heterogeneous precipitation method, most Ni particles are located at grain boundaries, and few small ones are entrapped within Al_2O_3 grains. In addition, $Ni-Al_2O_3$ composites have also been prepared by a chemical method and hot-pressing technique using $Al(NO)_3 \cdot 9H_2O$ and NiO as starting materials [114]. It is revealed from microstructure observation that Ni grains are uniformly dispersed in the Al_2O_3 matrix.

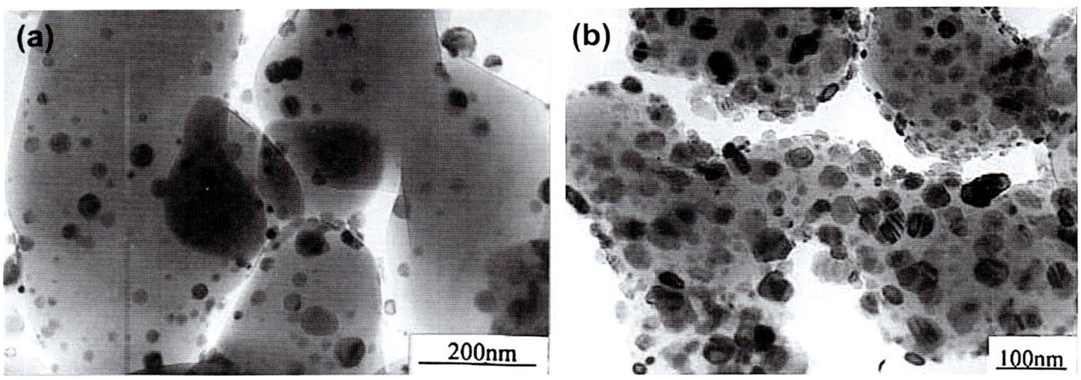

FIGURE 19.5 TEM micrographs of $Ni-Al_2O_3$ composite powders prepared by (a) direct mixing, and (b) heterogeneous precipitation. *Reprinted from Materials Science and Engineering: A, Vol 352, Guo-jun Li, Xiao-xian Huang, Jing-kun Guo, Fabrication and mechanical properties of Al_2O_3–Ni composite from two different powder mixtures, Copyright (2003), with permission from Elsevier [111].*

Besides $Ni-Al_2O_3$ composites, $Co-Al_2O_3$ ceramic matrix composites have also been prepared and investigated. For example, $Co-Al_2O_3$ nanocomposite powders have been prepared by coprecipitation method using $Al(NO_3)_3 \cdot 9H_2O$ and $Co(NO_3)_2 \cdot 6H_2O$ as raw materials, NH_4HCO_3 as precipitation agent and PEG as dispersant [115]. The resulting precipitates are washed, filtered, dried, calcined, and then in situ selectively reduced to form $Co-Al_2O_3$ nanocomposite powders with average particle size of about 30 nm. $Co-Al_2O_3$ ceramic matrix composites can be prepared by hot pressing at 1400°C in Ar atmosphere for 30 min under the pressure of 30 MPa. It is observed that the obtained 15 vol% $Co-Al_2O_3$ composite contains submicron-sized cobalt particles disperse at the alumina matrix grain boundaries. Most of the triple junctions are occupied by Co particles.

19.2.3.2. SiC_p/Si_3N_4 Composites

As mentioned before, Si_3N_4 ceramics are structural materials with excellent mechanical property, which can be prepared by pressureless sintering, hot pressing, and hot isostatic-pressing [116–118]. However, the brittleness of Si_3N_4 ceramics prevents its use in wide applications. Besides fiber and whisker-reinforced Si_3N_4 ceramic matrix composites, particle reinforcement is also used to improve the mechanical property. SiC_p/Si_3N_4 composites are representational materials in the particle dispersion-strengthened ceramic matrix composites and have been used in cutting ceramics, bearing balls, ceramic dies, and components of plunger pump. In contrast to fibers and whiskers, particles have less reinforcement effect. However, particle-reinforced ceramic matrix composites have wide application potential due to their prominent advantages of homogeneous mixing of raw material, better densification behavior, and easy to obtain products with complex shapes. SiC_p/Si_3N_4 composites have been prepared from organic precursor [119]. The bending strength of the obtained 5 wt% SiC_p/Si_3N_4 composite is as high as 970 MPa, which is 25% higher than that of Si_3N_4 ceramic prepared at the same conditions.

19.2.4. In Situ Growth Ceramic Matrix Composites

19.2.4.1. In Situ Growth Whisker-reinforced Ceramic Composites

Whisker-reinforced ceramic matrix composites have been widely investigated. However, introducing whiskers into ceramics to fabricate whisker-reinforced ceramic matrix composites still has following disadvantages [120]: (a) difficult to disperse whiskers homogeneously, (b) expensive and bad for health, (c) easily polluted in air, and (d) physicochemical change between whisker and matrix during sintering and application at high temperature. In situ growth whisker is likely to overcome the mentioned disadvantages. Whiskers can be homogeneously dispersed in the matrix because they grow uniformly in the powders. Whiskers and composites are prepared at a certain protective atmosphere, which prevents them from being polluted. Because whiskers grow in the ceramic matrix at high temperature, both whiskers and matrix are in the state of thermodynamic and phase equilibrium. In addition, the possible bonding between whiskers and matrix can improve their interface bonding property. At the same time, the process of in situ growth whisker-reinforced ceramic composites can decrease the fabrication cost.

In the process of in situ growth whisker-reinforced ceramic composites, whisker growth initiator is preintroduced into the matrix and homogeneously dispersed. After the green body is prepared, it is treated and densified at a certain condition and finally, in situ growth whisker-reinforced ceramic composite can be obtained. The main preparation methods of in situ growth whisker include carbothermic or aluminothermic method of reducing and synthetic reaction method.

For the carbothermic method, SiC_w/Si_3N_4 composite will be taken as an example. At first, graphite powder or carbon black, SiO_2 and Si_3N_4 powders are mixed by ball milling. Then the slurry will be dried, sieved and reduced at N_2 protective atmosphere. Additives are added into the formed SiC_w/Si_3N_4 powder mixtures and calcined in air to remove the residual carbon. The SiC_w/Si_3N_4 composite can be obtained by forming and then hot pressing at N_2 or Ar atmosphere. The main reactions in the SiO_2 carbothermic reduction process can be described as follows [121]:

$$\begin{aligned}
&SiO_2(s) + C(s) \rightarrow SiO(g) + CO(g) \\
&SiO_2(s) + CO(g) \rightarrow SiO(g) + CO_2(g) \\
&SiO(g) + C(s) \rightarrow SiC(s) + CO(g) \\
&SiO(g) + 3CO(g) \rightarrow SiC(s) + 2CO_2(g) \\
&CO_2(g) + C(s) \rightarrow 2CO(g)
\end{aligned}$$

In addition, in situ growth SiC_w/Al_2O_3 composite can be fabricated by aluminothermic method of reduction [122]:

$$4Al + 3SiO_2 + 3C \rightarrow 2Al_2O_3 + 3SiC_w$$

For the synthetic reaction method, Si_3N_4 or SiC whiskers can be synthesized in situ by silicon nitrification [123] or carbonization reaction [124]. The main reactions are as follows:

$$3Si(l, s) + 2N_2(g) \rightarrow \beta\text{-}Si_3N_4(s)$$

$$Si(l) + C(s) \rightarrow SiC_w(s)$$

In situ growth whisker-reinforced ceramic composites have some unique advantages. However, only some exploratory work has been done; systemic and deep investigation is still needed.

19.2.4.2. *In Situ Growth Particle-reinforced Ceramic Composites*

The research on TiN_p/Al_2O_3 composite has attracted much attention, due to its excellent properties and wide applications. The introduction of TiN particles into Al_2O_3 matrix can increase not only its strength, toughness, and resistance to abrasion, but also its electrical conductivity. Usually, $TiN-Al_2O_3$ composite is prepared by ball milling TiN and Al_2O_3 powders, forming, and then sintering. However, this method will cause the agglomeration of the additive or local composition segregation. Ultimately, the mechanical property and electrical conductivity of the $TiN-Al_2O_3$ composite will be reduced. A novel in situ preparation method has been provided by Li et al. Nanosized $TiN-Al_2O_3$ composite powders are prepared by in situ nitridation using $TiO_2-Al_2O_3$ nanopowders synthesized by a coprecipitation method [125]. Experimental results show that the nitridation reaction begins at 700°C and completes at 900°C for 5 h. Nanosized TiN with particle size of about 50–70 nm are homogeneous in the Al_2O_3 matrix. In another example, in situ formation of zirconia–alumina–spinel–mullite ceramic multicomposites has been prepared by a simple one-step reaction sintering of commercial-grade zircon, alumina, and magnesium carbonate according to the equation given below [126]:

$$2ZrSiO_4 + (3 + x)Al_2O_3 + xMgO \rightarrow 2ZrO_2 + 3iAl_2O_3 \cdot 2SiO_2 + xMgAl_2O_4$$

19.2.5. Ceramic Nanocomposites

As pioneered by Niihara [127], the significant improvement in mechanical properties, in particular, strength, hardness, and fracture toughness, can be achieved in ceramics by nanocomposite design. Ceramic nanocomposites can also be classified on the basis of their microstructural design, depending on whether the matrix or the reinforcement or both are nanocrystalline, and distribution of the nanocrystalline reinforcement. It is to be noted that the development of ceramic materials by nanocomposite design not only results in the improvement of mechanical properties like hardness, strength, and reliability at room temperature but also in high temperature mechanical properties like hot hardness, strength retention, creep, and fatigue fracture resistance [128].

19.3. INTEGRATION OF STRUCTURES AND FUNCTIONS

With the development of our society, the application requirements of materials have been increasingly improved. The reinforcement mechanisms in structural ceramics are put forward to improve the mechanical properties of functional ceramics, because functional ceramics need to be used at rigorous environment/conditions. In addition, by incorporating functional composites into the structural ceramics, the optical, electric, and magnetic applications have been gradually developed. Utilizing the perfect ceramic fabrication technology and accurately controlling fabrication process, it is observed that some structural ceramics themselves are excellent functional materials. Among the earliest materials that have realized the integration of structures and functions, the most widely used structural ceramics is alumina ceramic. In the early 1960s, translucent alumina ceramics have been developed by General Electric Co. The obtained alumina ceramics of high transmission, high strength, and high corrosion resistance are widely used as high voltage sodium lamp envelope. In order to realize the integration of structures and functions, reasonable design and controllable fabrication of materials are necessary. Transparent ceramics, such as laser ceramics and ceramic scintillators, are typical examples, which will be discussed in detail in the next section "Advanced ceramic materials". In this section, $CNTs/SiO_2$ and $CNTs/BaTiO_3$ composites will be mainly recommended as the examples of structural–functional integration.

19.3.1. $CNTs/SiO_2$ Composites

Due to the unique structure and outstanding properties of carbon nanotube, CNT-reinforced ceramic matrix composites are expected to realize the integration of structures and functions. However, in the previous works, the research on carbon nanotube–ceramic composites mainly focuses on fabrication process and mechanical properties [129–132]. The investigations on functional characteristics were seldom involved.

By introducing CNTs in SiO_2 matrix, the mechanical properties of $CNTs/SiO_2$ composite can be improved and some functional properties can be developed. As to $CNTs/SiO_2$ composites, usually, there are two different fabrication methods. One is direct mixing method [133] and the other is in situ composite method [134]. Direct mixing is a relative simple and common used method. However, in the in situ composite method, surfactants are generally used to disperse CNTs in the SiO_2[135,136]. Using the in situ composite method, $CNTs/SiO_2$ composite with high bending strength and fracture toughness has also been prepared by hot pressing. The obtained $CNTs/SiO_2$ composites have higher thermal conductivity [137], high electrical and dielectric properties [138–140], and high complex permittivities [138]. It is also suggested that the fused silica matrix with appropriate characteristic impedance and MWNTs is suitable to design low reflecting and high absorbing efficiency sandwich gradient electromagnetic absorbing materials [141].

At the same time, strong and broad-band optical limiting has been observed in single-walled and multi-walled nanotube suspension [142,143] or solution [144], as well as inorganic films [145]. But most of these investigations are carried out in suspensions or solutions, which have many practical drawbacks. So it is urgent and also interesting to prepare CNTs–ceramic composite with improved strength and stability for optical limiter. Transparent $MWNTs/SiO_2$ composites have been prepared by rapid sol–gel process and hot-pressing technology [146]. It is considered that $MWNTs/SiO_2$ composite is a good candidate for optical limiting application.

19.3.2. CNTs/$BaTiO_3$ Composites

Among the ceramic matrices, barium titanate ($BaTiO_3$) is a very important functional material, which has been applied in various fields such as ferroelectrics, piezoelectrics, and thermoelectrics. Incorporating MWNTs into $BaTiO_3$ has been adopted to obtain novel composite materials [147]. A suitable scheme has been developed to prepare well-dispersed carbon nanotube–$BaTiO_3$ composite powders and then further fabricate novel $MWNTs/BaTiO_3$ nanocomposite ceramics. At first, an effective one-step scheme is introduced for immobilization of rutile TiO_2 on the side wall of MWNTs [148]. $MWNTs/BaTiO_3$ nanocomposite ceramics with varied contents of MWNTs have been obtained by hot pressing at 1200°C for 1 h. All the samples are sintered densely and the sizes of $BaTiO_3$ grains are statistically reduced when adding MWNTs [147]. A maximal toughness value of 1.65 MPa $m^{1/2}$ has been obtained in 1 wt% $MWNTs/BaTiO_3$ nanocomposite, which is about 240% that of the carbon-free $BaTiO_3$ ceramic, whose toughness is only 0.68 MPa $m^{1/2}$.

For the $MWNTs/BaTiO_3$ nanocomposites prepared by SPS method, it is found that the incorporation of carbon nanotubes can effectively accelerate the densification process of MWNTs–$BaTiO_3$ composites. Excellent electrical and thermal conductivity of MWNTs are proposed to account for this unexpected sintering behavior [149]. The resistivity of $MWNTs/BaTiO_3$ nanocomposites is found higher than that of MWNTs-free $BaTiO_3$ ceramics. And the temperature dependence of resistivity shows abnormal metallic conductance behavior in the 1 wt% $MWNTs/BaTiO_3$ nanocomposite. It is proposed to be due to the additional subgrain boundary between MWNTs and $BaTiO_3$ matrix, which leads to transformation from an impurity-scattering mechanism to a dominant lattice-scattering mechanism in the $MWNTs/BaTiO_3$ nanocomposites as the content of carbon nanotubes increases [150]. Based on observed unique electrical properties, bilayer ceramics stacked by one layer of MWNTs-free $BaTiO_3$ and another layer of $MWNTs/BaTiO_3$ nanocomposite are fabricated and show excellent rectification property. These electrical properties of $MWNTs/BaTiO_3$ nanocomposites shed light on the future application of carbon nanotubes in functional bulk ceramics.

19.4. TRANSPARENT CERAMICS

Since translucent alumina ceramics (Lucalox) were first prepared by Coble in 1962 [151], some other transparent (or translucent) ceramic systems, such as MgO [152], Y_2O_3[153], $MgAl_2O_4$[154], PLZT [155], AlON [156], SiAlON [157], AlN [158], and YAG [159], were gradually developed. In one side, these ceramics have good transparency, and in the other side, they are of high strength, hardness, corrosion resistance, and high-temperature resistance like structural ceramics. So transparent ceramics are suitable for windows and lens materials. For example, transparent $MgAl_2O_4$ and AlON ceramics, which have higher strength and hardness than glasses, have been utilized in windows of military vehicle, airplane and missile, windows of bar code scanner, and scratch-resistant lenses. What's more important is that the functional characteristics of transparent ceramics have been paid more and more attention. Transparent ceramics have been used as solid-state laser host [160], upconversion materials [161], and ceramic scintillators for advanced medical X-ray detectors [162,163]. The following section involves the fabrication of optical ceramics for windows, laser ceramics, upconversion nanoparticles and ceramics, and ceramic scintillators.

19.4.1. Optical Ceramic for Windows

19.4.1.1. Alumina Transparent Ceramics

Transparent alumina ceramics, discovered by Coble in the early 1960s [164], have been developed for infrared domes and high-intensity discharge lamps because of their high heat resistance and high chemical durability. It is widely known that the elimination of scattering centers such as residual pores, optical anisotropy due to birefringence, secondary-phase inclusions, and rough surfaces is essential for improving transparency. Residual pores, in particular, reduce transmission drastically. A porosity of ≪1% is needed for translucent polycrystalline alumina (PCA) ceramics [165]. As alumina has hexagonal structure, optical birefringence at grain boundaries also affects transmission. The birefringence is caused by the differences in the refractive indices for different polarizations due to the crystal structure. The refractive index of Al_2O_3 is 1.768 for ordinary rays (perpendicular to *c*-axis) and 1.760 for extraordinary rays (parallel to *c*-axis) [166]. The scattering of birefringence is closely related to the PCA microstructure. It has been reported that the in-line transmission of PCA increases with a decrease in grain size [167].

To obtain alumina ceramics with high transparency, two approaches have been utilized in order to reduce the source of light scattering and absorption, such as grain boundaries and pores. One approach is sintering at very high temperatures (>1700°C) for a long time which mainly reduces the area of the grain boundaries [168]. High-temperature sintering, however, causes severe grain growth, and hence, poor mechanical properties. Furthermore, the in-line transmission of sintered coarse-grained alumina is typically lower than 10%. For example, translucent alumina ceramics have been prepared by gelcasting in air and pressureless sintering in hydrogen atmosphere. The grain size of sintered alumina is in the range of 10–20 mm. The in-line transmission increases gently from 9% to 10% in the wavelength range of 190–1100 nm for the sample with a thickness of 0.5 mm [169]. The low in-line transmission probably results from residual pores which are not sufficiently eliminated by conventional pressureless sintering. However, using a magnetic-field-assisted slip-casting method, transparent polycrystalline alumina ceramics with orientated optical axes have been demonstrated [170]. The photo and transmission of the obtained alumina ceramicsare shown in Fig. 19.6. The samples are placed 8 mm above the paper. It can be seen that the sample cast in the magnetic field exhibits better transparency than that without using a magnetic field. The transmission of the sample cast in the magnetic field is much higher than that of the sample prepared without a magnetic field. Furthermore, the transmission maintains a high level as the wavelength shifts toward the UV range. For example, a transmission of 44% is observed at 300 nm. The XRD results revealed that grains in the alumina ceramics have been orientated successfully by the magnetic-field-assisted slip casting method.

For transparent alumina ceramic fabrication, the other approach is hot isostatic pressing (HIP) at low temperatures (1200–1300°C), which sufficiently eliminates residual pores with suppressed grain growth [171]. By using HIP, the porosity can easily be reduced to lower than 0.05% and the grain size can be suppressed to less than 1 μm. The fully dense and fine-grained alumina has good mechanical properties and the in-line transmission exceeds typically 50%. This is the reason why many studies have recently been conducted on transparent alumina prepared by using HIP [172]. For example, Krell et al. [171] have manufactured flat tiles and hollow bodies by gelcasting and HIP. The results illustrated that transparent alumina ceramics with complicated shapes could be produced by the combination of gelcasting and HIP methods. Highly translucent PCA with various grain sizes has been fabricated by HIP. The effects of residual pores and optical birefringence on transmission through translucent polycrystalline alumina have been studied with the Mie scattering theory [173]. It has been revealed by the model that porosity has a significant effect on both the total forward scattering and the in-line transmission, whereas grain size affects only in-line transmission. The calculated transmittance is in good agreement with the experimental values. The in-line transmission gradually increased with a decrease in grain size, and the effect of birefringence on the total forward transmission is negligible in the visible spectrum. The total forward

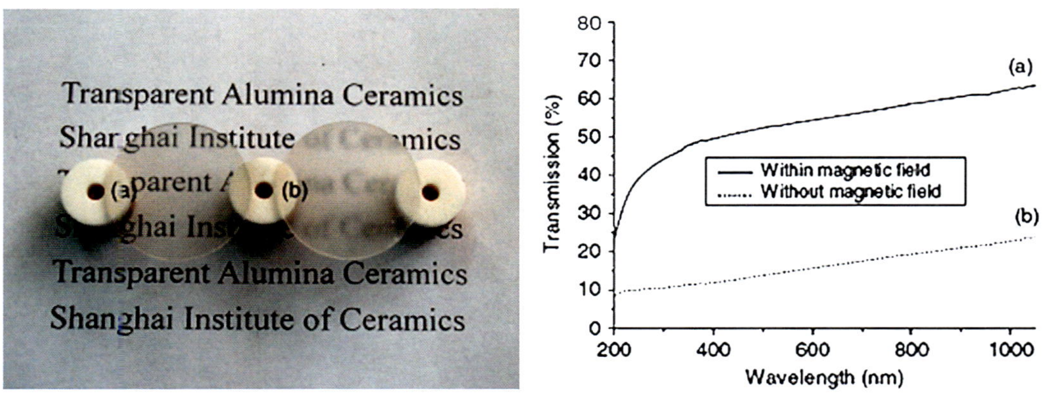

FIGURE 19.6 Photo and transmission of polycrystalline alumina ceramics slip cast (a) in and (b) outside a magnetic field [170]. *Reprinted with permission from Wiley.*

transmission in the visible spectrum is mainly governed by residual pores.

Another available candidate for obtaining transparent alumina is spark plasma sintering (SPS). Transparent alumina ceramic with a fine grain size of 270 nm (Fig. 19.7) has been obtained by controlling the heating rate during SPS processing [174]. The alumina sintered at 1150°C with a heating rate of 8°C/min has a residual porosity of 0.03% and an in-line transmission of 47% for a wavelength of 640 nm. It is also shown that a low heating rate has an effect on the densification and transparency of alumina for sintering at 1150°C.

In another case, highly IR transparent alumina with a maximum transmittance of 85% has also been successfully sintered at 1300° C for 5 min using the SPS method [175]. It is suggested that annealing as well as sintering for a longer time or at a slower heating rate can reduce the residual porosity, which contributes to the improved transmission. The localized residual stress at grain boundaries may be another factor influencing the optical transmission among other factors such as grain size, porosity, and oxygen vacancy concentration for the SPS-sintered alumina.

19.4.1.2. $MgAl_2O_4$ Transparent Ceramics

Polycrystalline magnesium aluminate spinel ($MgAl_2O_4$) is one of the most promising optically transparent ceramics that exhibits a unique combination of optical and mechanical properties. Its advantage lies in the optically isotropic cubic structure, so that light scattering at grain boundaries is less critical than with optically anisotropic systems. To fabricate $MgAl_2O_4$ transparent ceramics, it is very important to obtain high-quality $MgAl_2O_4$ powders. Ultrafine $MgAl_2O_4$ spinel powders have been prepared by various methods to enhance their sinterability. These methods include hydroxide coprecipitation [176,177], ammonium bicarbonate precipitation [178], freeze drying [179], and hydrolysis alkoxides [180]. However, $MgAl_2O_4$ spinel is one of the hard-to-sinter ceramics, and hence a transparent spinel ceramic is difficult to fabricate directly from a high-purity powder using conventional techniques such as pressureless sintering processing. In most studies on transparent $MgAl_2O_4$ spinel ceramics, fine powders doped with sintering aids, such as CaO, LiF, and B_2O_3, are utilized to enhance the sinterability. Furthermore, the powders have usually been consolidated into dense bulks at high temperatures using hot pressing (HP), hot isostatic pressing (HIP), or spark plasma sintering (SPS) techniques. For example, a translucent polycrystalline $MgAl_2O_4$ ceramic has been prepared from finely divided coprecipitated spinel, in which a small amount of CaO, added as a sintering aid, is uniformly distributed [154]. The CaO promotes densification through the formation of a liquid phase at the sintering temperatures. Depending on the sintering treatment, the relative density of the sintered spinel is 99.7−100% of theoretical. The in-line optical transmission was >10% from 0.3 to 6.5 μm. Total transmission in the visible region was between 67% and 78%. In another case, B_2O_3 is used as sintering aid to prepare transparent $MgAl_2O_4$ spinel ceramic by hot isostatic pressing of the sintered materials composed of fine grains [181]. It is found that a small addition of B_2O_3 is very effective to restrain anomalous grain growth, which is important for transparency. The samples added with 0.15 wt% B_2O_3 show high optical transmittance of 80%. $MgAl_2O_4$ spinel doped with LiF can be hot pressed to densities above 98% [182], but the residual porosity must be removed by a subsequent process step such as hot-isostatic pressing. Considering that various steps are needed for formation and densification, there exist many possible processing variations, including rates of heating and cooling, rates of pressure application and removal, hold times, and the maximum temperatures and pressures. It has been indicated that the distribution of LiF within the powder compact before sintering is essential in utilizing its effectiveness [183]. The removal of LiF before complete densification is further necessary as it otherwise scatters light. This makes hot pressing with LiF challenging, because the detailed chemical reactions and subsequent vaporization involved in this process are sensitive to temperature and applied pressure. Therefore, it is of interest to control the distribution and resulting behavior of LiF and any associated reaction products throughout the entire sintering process of spinel [184]. Using an HP technique at 1550°C for 2 h, Reimanis et al. [185] have successfully attained an

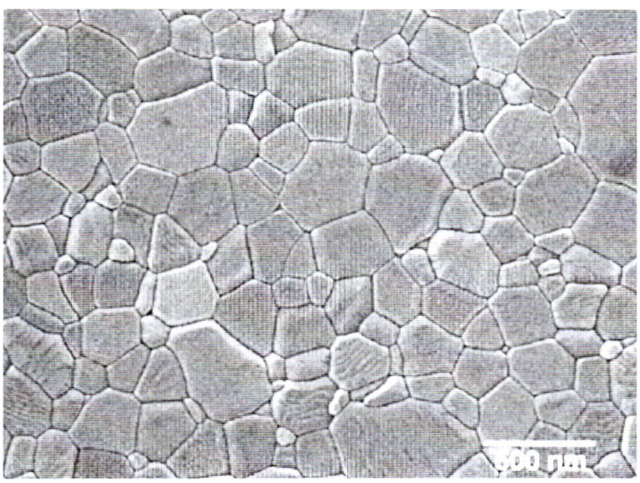

FIGURE 19.7 Microstructure of the transparent alumina SPS-sintered at a heating rate of 8°C/min. *Reprinted from Scripta Materialia, Vol 57, Byung-Nam Kim, Keijiro Hiraga, Koji Morita, Hidehiro Yoshida, Spark plasma sintering of transparent alumina, Copyright (2007), with permission from Elsevier [174].*

excellent visible transmittance of 85% in a 1 wt% LiF-doped spinel. Utilizing superhigh pressure of 2–5 GPa and nanopowders, transparent $MgAl_2O_4$ spinel nanoceramics can be sintered at relatively low temperature of 500–700°C using a hydrostatic press[186].

Besides HP and HIP, spark plasma sintering (SPS) technology has also been utilized to fabricate transparent $MgAl_2O_4$ spinel nanoceramics. The SPS technique has a significant advantage over the HIP and HP techniques, because it can complete the powder consolidation within a short sintering time owing to the available high heating rates of typically ≥50°C min^{-1}. Frage et al. [187] have reported a light transmission of <50% and ~70% by SPS processing at the low heating rate of 10°C min^{-1} in highly pure and LiF-doped materials, respectively. However, the grain size of the transparent spinel exceeded 10 μm, probably because of the high sintering temperature of 1600°C; grain coarsening became more pronounced owing to the existence of sintering aids. In addition to transparency, another important point regarding optical applications is the mechanical properties, high strength, and toughness are required for window materials. For brittle ceramic materials, as fine-grained materials are expected to possess better mechanical properties than coarse-grained ones, a fine-grained spinel is preferable for engineering applications. According to Morita et al.'s [188] work, transparent $MgAl_2O_4$ spinel ceramic can be fabricated without any sintering aids by spark plasma sintering (SPS) processing for only a 20 min soak at 1300°C. The spinel exhibits an in-line transmission of 47% for a visible-wavelength of 550 nm and a fracture strength of 500 MPa.

19.4.1.3. AlON Transparent Ceramics

The acronym "AlON" is used for an aluminum oxynitride, of composition $Al_{(64+x)/3}O_{32-x}N_x$ $(0 \leq x \leq 8)$—a pseudo-binary solid solution of AlN in Al_2O_3, which may be regarded as a nitrogen stabilized cubic γ-Al_2O_3 phase (spinel). It has interesting optical, mechanical, and chemical properties, making it potentially useful as a high-performance ceramic material [189]. The material, AlON, has been processed into fully dense transparent materials suitable for use in infrared and visible window applications. It is regarded that four different spinel AlON phases exist at molar ratios of approximately 10 mol% AlN (δ-AlON), 17 mol% AlN (ø′-AlON), 21 mol% AlN (γ′-AlON), and 36 mol% AlN (γ-AlON) [190]. Gamma AlON exists in the widest stability range, which is a function of temperature with eutectoid decomposition into Al_2O_3 and AlN below about 1640°C [191]. Dense AlON ceramic has excellent optical properties with useful transmission from about 0.2 μm in the UV through the visible range to about 6.0 μm in the near-IR [192]. While AlON is denser than AlN, its cubic structure allows it to be made transparent, even at considerable thickness, while maintaining decent mechanical properties [193]. Usually, there are two methods to fabricate transparent AlON ceramics. A single-step preparation method has been used to make transparent AlON ceramics, by mixing Al_2O_3 and AlN powders in the molar ratio of 5:9 and subsequent reactive sintering at 2025°C in nitrogen atmosphere [156]. In order to inhibit grain growth of AlON, Zeta alumina ($LiAl_5O_8$) is reaction sintered with α-Al_2O_3 and AlN to produce transparent LiAlON ceramic [194]. Microwave reactive sintering methodhas also been used to fabricate highly transparent AlON ceramic using high purity α-Al_2O_3 and AlN powders as starting materials and Y_2O_3 as sintering aid to improve the densification and transparency of the sintered body [195]. Another conventional method to fabricate transparent AlON ceramics involves using presynthesized AlON powder to form a green body, followed by sintering in a nitrogen atmosphere at high temperatures. Carbothermal nitridation of alumina is a well-established preparation procedure. The reaction process can be described as follows:
$(\frac{64+x}{3})Al_2O_3 + 3xC + xN_2(g) \rightarrow 2Al_{(64+x)/3}O_{(32-x)}N_x + 3CO(g)$ Using the prepared AlON powders, transparent AlON ceramics can be obtained by sintering.

19.4.2. Laser Ceramics

19.4.2.1. YAG Transparent Ceramics

Solid-state lasers are widely used in metal processing, medical applications, such as eye surgery, red–green–blue (RGB) light sources in laser printers and projectors, environmental instrumentation measurements and optical transmission systems, and have potential for future nuclear-fusion applications [196]. Conventionally, single crystal or glass is used as the gain medium in solid-state lasers, the first example being the ruby laser, devised by Maiman in 1960 [197]. Since the success in generating continuous-wave (c.w.) laser oscillation using an Nd:YAG single crystal at room temperature in 1960 [198], designs of solid-state lasers using single crystal have continually progressed. However, there remains the challenge of finding an approach that can overcome the technological and economical issues of conventional single-crystal laser gain media. Attempts to synthesize solid-state laser materials from polycrystalline ceramics such as Nd-doped Y_2O_3–ThO_2 ceramics have been reported by Greskovich [199,200], and attempts from YAG ceramics by de With and Van Dijk [159], Mudler and de With [201] and Sekita et al. [202]. In the experiments of both de With and Sekita et al., laser output can be realized due to relatively large scattering loss in the obtained YAG transparent ceramics. In 1995, Ikesue et al. first demonstrated the possibility of fabricating transparent Nd:YAG ceramics of sufficient quality for

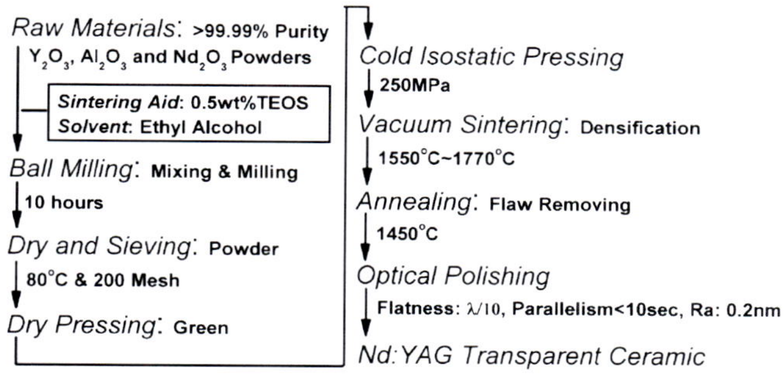

FIGURE 19.8 Solid-state reaction fabrication process of Nd:YAG transparent ceramics. *Reprinted from Optical Materials, Vol 31, Jiang Li, Yusong Wu, Yubai Pan, Wenbin Liu, Liping Huang, Jingkun Guo, Fabrication, microstructure and properties of highly transparent Nd:YAG laser ceramics, Copyright (2008), with permission from Elsevier [211].*

solid-state lasers with reasonable efficiency [160,203]. Since then, Nd:YAG laser ceramics have attracted much attention because the optical quality has been improved greatly and highly efficient laser oscillations can be obtained whose efficiencies are comparable or superior to those of Nd:YAG single crystals [204–207].

As compared to laser single crystal, one of the prominent advantages of laser ceramics is the ease in achieving composite structure. According to the manufacturing process of ceramics, multilayer ceramic laser materials can be fabricated together, whereas this is impossible with the single-crystal growth method. This advantage provides more freedom in the design of laser systems. Usually, there are two typical methods to fabricate high-quality YAG transparent ceramics. One is solid-state reaction method and the other is wet-chemical precipitation and vacuum sintering technology. Transparent polycrystalline Nd:YAG ceramics with different concentrations have been fabricated by solid-state reaction method by sintering a mixture of commercial Al_2O_3,Y_2O_3, and Nd_2O_3 powders [208–210], and the fabrication process is shown in Fig. 19.8[211].

In the Al_2O_3–Y_2O_3 system, three phases are known, monoclinic phase (YAM, $Y_4Al_2O_9$), perovskite phase (YAP, $YAlO_3$), and cubic phase (YAG, $Y_3Al_5O_{12}$), and the reaction temperature of each phase has been reported by Kinsman and McKittrick [212]. Figure 19.9 shows the photo and optical transmittance of the mirror-polished 2.0 at%Nd:YAG transparent ceramic (Φ59 mm × 3 mm) sintered at 1770°C for 50 h. It can be seen that the sample is very homogeneous and transparent. The optical transmittance of the sample is 82% at the lasing wavelength of 1064 nm.

Using the solid-state reaction method, highly transparent Yb:YAG[213], Tm:YAG [214], and Er:YAG [215] transparent ceramics have also been fabricated. Using end-pump method, high efficient laser outputs have been obtained. For example, the obtained 6 at% Tm:YAG ceramic slab was end-pumped by a laser diode at 792 nm. The maximum output power of 4.5 W has been obtained with a slope efficiency of 20.5% at 2015 nm. In addition, Cr, Nd:YAG [216,217] and Cr, Yb:YAG [218] transparent ceramics can be prepared by this method as self-Q-switched laser materials.

Besides solid-state reaction method, wet-chemical method is also a primary way to fabricate YAG laser ceramics [219]. Nd:YAG nanopowders have been prepared by coprecipitation method using aluminum, yttrium, and neodymium chlorides as starting materials and ammonium hydrogen carbonate as precipitator and highly transparent Nd:YAG ceramics can be obtained by vacuum sintering [220].

Method for preparation of YAG laser ceramics using nanopowders obtained by calcining of precursors deposited from nitrate solutions with ammonium bicarbonate as a precipitant has also been developed [221]. High-pressure colloidal slip-casting (HPCSC) method is used for nanopowders compactions, and green density up to 60%. Application of HPCSC compaction method reduces the requirement of starting nanopowders. Transparent YAG ceramics have been obtained

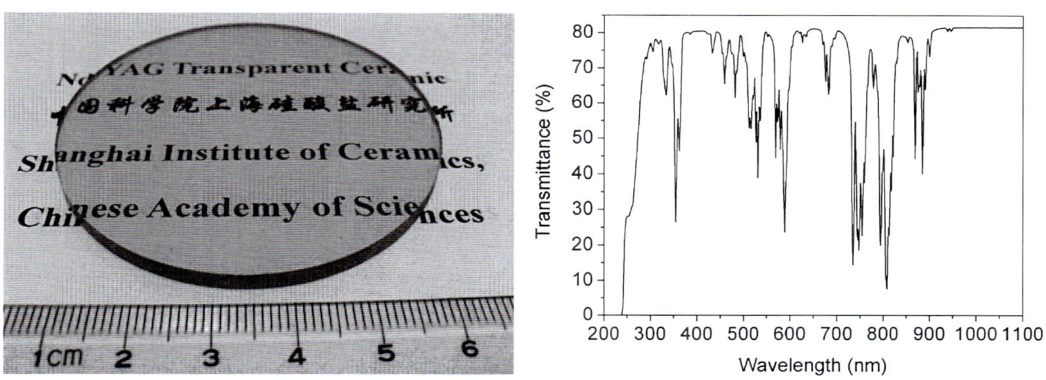

FIGURE 19.9 Photo and optical transmittance of the mirror-polished 2.0 at% Nd:YAG transparent ceramic (Φ59 mm × 3 mm) sintered at 1770°C for 50 h.

by sintering of compacts at 1700–1800°C in vacuum. Laser action has been achieved in plate ceramic samples with longitudinal pumping by semiconductor laser.

The above-mentioned two fabrication processes are mainly used to prepare polycrystalline YAG laser ceramics. As compared to laser single crystal, one of the prominent advantages of laser ceramics is ease of achieving composite structure. According to the manufacturing process of ceramics, laser ceramic composites can be easily fabricated, whereas this is impossible with the single-crystal growth method [222,223]. This advantage provides more freedom in the design of laser systems. Figure 19.10 shows an all-ceramic composite with a layer-by-layer (YAG/Nd:YAG/YAG) structure and its interferometry image. The sample is completely transparent and, as can be seen in the photograph, the interferometry shows straight fringes, meaning that the crystal is optically homogeneous. Although the laser oscillation greatly depends on the resonator system, the slope efficiency of this composite laser reaches 58% when the reflectance of the output mirror was 88.5%. This layer-by-layer composite is designed and fabricated for the application of a high-power laser because the thermal conductivity of the pure YAG bonded on both surfaces of the Nd:YAG is about 15 W/mK and these two layers (pure YAG) act as a heat-sink system.

19.4.2.2. Sesquioxide Transparent Ceramics

The sesquioxides scandia (Sc_2O_3), yttria (Y_2O_3), and lutetia (Lu_2O_3) can easily be doped with rare-earth ions and exhibit higher heat conductivity than yttrium aluminum garnet (YAG) [224]. These properties make the sesquioxides attractive for high-power solid-state lasers. However, growth of large crystals from the melt is a difficult task due to the high melting point >2400°C [225].

With the development of nanopowder technology and vacuum sintering method, the optical quality of sesquioxide ceramics has been improved greatly. The ceramic lasers can be obtained with a high efficiency that is comparable with that of single-crystal lasers. To prepare sesquioxide transparent ceramics, nanopowders of Y_2O_3, Sc_2O_3 and Lu_2O_3 are synthesized by wet-chemical method. For example, Y_2O_3 nanopowders can be prepared via hydroxyl precipitation and yttria transparent ceramics can be obtained by vacuum sintering [226]. In another case, hydroxyl-type Sc_2O_3 precursors have been synthesized via precipitation at 80° C with hexamethylenetetramine as the precipitant [227]. Hard-aggregated precursors (γ-$ScOOH \cdot 0.6H_2O$) are formed with scandium nitrate, which is converted to Sc_2O_3 at temperatures >400°C, yielding nanocrystalline oxides of low surface area. The use of sulfate leads to a loosely agglomerated basic sulfate powder having an approximate composition of $Sc(OH)_{2.6}(SO_4)_{0.2}\,H_2O$. The powder transforms to Sc_2O_3 via dehydroxylization and desulfurization at temperatures up to 1000°C. Well-dispersed Sc_2O_3 nanopowders (~64.3 nm) of high purity have been obtained by calcining the basic sulfate at 1000°C for 4 h.

Based on sesquioxide nanopowders, a series of laser ceramics, such as $Nd:Y_2O_3$[228], $Yb:Y_2O_3$[229,230], $Yb:Sc_2O_3$[231], and $Nd:Lu_2O_3$[232], have been produced. As the ceramics fabrication technique overcomes the shortcomings of single-crystal growth method, large-size high-quality $Nd:Y_2O_3$ ceramics have been fabricated and highly efficient CW laser oscillation has also been demonstrated under LD end-pumping. Due to its remarkable advantages, such as refractory nature, stability, ruggedness, optical clarity over a broad spectral region, broad emission band, and high thermal conductivity, $Nd:Y_2O_3$ transparent ceramic is a very promising laser media that can be used in many fields, such as high-power industrial lasers and laser fusion system, $Yb:Y_2O_3$ ceramics, a new solid-state laser material, have been fabricated and highly efficient CW laser oscillation has been achieved [233]. The broadband luminescence of Yb^{3+} ions in this novel laser material also makes it possible to generate the femtosecond pulses from diode pumped solid-state laser (DPSSL). Demonstrated promising results and nonexhausted potential of advanced ceramic technology indicate that this novel laser material can be applied in many fields, such as femtosecond lasers, tunable lasers, and high-power solid-state lasers.

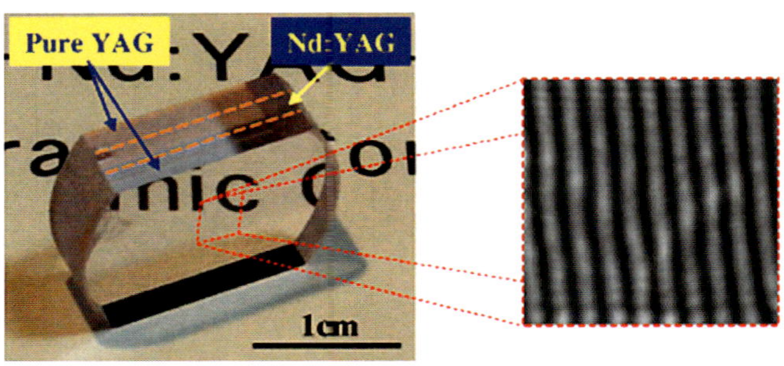

FIGURE 19.10 Appearance of a layer-by-layer YAG/Nd:YAG/YAG laser ceramic composite and its wave front image [222]. *Reprinted with permission from Wiley.*

19.4.3. Upconversion Nanoparticles and Ceramics

Upconversion luminescent materials can convert infrared radiation into visible light, which have a variety of potential applications, such as upconversion lasers, three-dimensional volumetric displays, fluorescent labels, and fiber optic amplifiers[234–236]. Most of the upconversion luminescent materials that had been developed are rare-earth-doped halides, oxyhalides, and oxysulfides existing in the forms of crystals, glasses, or ceramics [237]. Recently, oxides and oxynitrides are attracting much attention because of their good chemical, photochemical stability, and mechanical properties. Rare-earth-doped oxides such as Y_2O_3[238], Lu_2O_3[239] and yttrium aluminum garnet (YAG) [240] show very useful visible emissions, and have great potential application in upconversion devices. Er^{3+}:Y_2O_3 and Yb^{3+},Er^{3+}:Y_2O_3 nanocrystalline powders have been prepared via reverse-strike coprecipitation method using nitrates and ammonia as raw materials [241]. The obtained powders are of cubic-phase structure of Y_2O_3 and the particle size is in the range of 60–80 nm. Strong red ($^4F_{9/2}$–$^4I_{15/2}$) and ($^2H_{11/2}$/$^4S_{3/2}$–$^4I_{15/2}$) upconversion luminescence can be observed in all samples when excited with a 980 nm continuous wave (CW) diode laser. Yb^{3+}, Ho^{3+}:Lu_2O_3 nanocrystalline powders with the particle size of 30–50 nm have also been synthesized by the same method using NH_4HCO_3 and $NH_3 \cdot H_2O$ as precipitators [242]. Under the excitation of a 980-nm continuous wave diode laser, green and red emissions centered around 548 and 667 nm, respectively, can be observed and the green emission dominates the upconversion spectrum. Yb^{3+},Er^{3+}:YAG pure phase nanopowders can be prepared by a coprecipitation method in which ammonium hydrogen carbonate is used as precipitator [243]. Upconversion emissions are investigated on the nanopowders pumped by a 980-nm continuous wave diode laser, and red emission centered at 660 nm and green emission centered at 563 nm can be observed. In addition, Yb^{3+},Tm^{3+}:YAG nanopowders with crystallite size of 40 nm have been prepared using the same method. Strong blue emission can be seen with by the naked eye under the excitation of 940 nm [244].

Oxynitrides, such as γ-AlON with spinel structure, could also be a promising host for upconversion application because of their excellent mechanical and optical properties suitable for infrared/visible window and γ-AlON has low phonon energy[245]. The low photon energy leads to low nonradiative transition rates due to multiphoton relaxation and high radiative transition rates, which increase the quantum efficiency from excited states of active ion. Er-doped γ-AlON phosphors with different Er^{3+} concentrations for upconversion luminescence have been prepared by a combination of carbothermal reduction and nitridation and solid-state synthesis methods [246]. Under the excitation of 980 nm diode laser, green emission centered at 548 nm and red emission centered at 666 nm can be observed, which are ascribed to the transitions of $^4S_{3/2}/^2H_{11/2} \rightarrow {}^4I_{15/2}$ and $^4F_{9/2} \rightarrow {}^4I_{15/2}$ of Er^{3+} ions, respectively. Using the same method, γ-AlON:Yb^{3+},Tm^{3+} ceramic phosphors are prepared and upconversion emission properties have also been investigated under single-wavelength diode laser excitation of 980 nm [247].

Although highly efficient upconversion emissions have been achieved in rare-earth-doped oxide and oxynitride powders, it is obvious that bulk transparent materials have more broad practical applications than those of the powders. Er^{3+}:Y_2O_3 transparent ceramics have been fabricated successfully via nanocrystalline powder technology and hydrogen atmosphere sintering method[248]. Under the excitation of 980 nm diode laser, very strong green emission centered at 562 nm and red emission centered at 660 nm are observed, which can be assigned to the transitions of $^4S_{3/2}$/$^2H_{11/2}$–$^4I_{15/2}$ and $^4F_{9/2}$–$^4I_{15/2}$ energy levels of Er^{3+} ions, respectively. That the excited-state absorption mechanism played an important role in the upconversion processes and cooperative energy transfer (CET) processes between two nearby Er^{3+} ions cannot be ignored. By changing the doping concentration of Er^{3+} ions, the color of upconversion emissions can be tuned from green to red, gradually. In another case, Yb^{3+},Er^{3+}:YAG transparent ceramics are fabricated by solid-state reaction method and vacuum sintering by using high-purity Y_2O_3, Al_2O_3, Yb_2O_3, and Er_2O_3 powders as starting materials [249]. When the obtained ceramics are excited by a CW 980 nm laser diode, strong green emission centered at 523 and 559 nm and red emission centered at 669 nm can be observed.

19.4.4. Ceramic Scintillators

Inorganic scintillators play an important role in radiation detection in many sectors of research concerning almost all medical diagnostic imaging modalities that use X-rays or gamma rays, dosimetry, nuclear medicine, high-energy physics, and also in many industrial measuring systems [250]. In different applications, the scintillator is essentially a luminescent material that absorbs high-energy photons and then emits visible light [163]. The common scintillator requirements are fast response time (10–100 ns), high light yield, high density, and high atomic number [251]. Scintillation optical ceramic materials based on slow 4*f*–4*f* transitions of rare-earth ions appeared about 20 years ago. Now ceramic scintillators have become competitive to single crystal ones in medical imaging, namely, in computer tomography [252], where some of them are widely used today. Luminescence and scintillation characteristics of the fast scintillation ceramics based on the Ce^{3+} -doped $Y_3Al_5O_{12}$ (Ce:YAG) have been reported by Zych and coworkers [253,254], followed by a few other reports [255]. For example, Ce:YAG transparent ceramics have been fabricated by a solid-state reaction method and vacuum sintering using commercial α-Al_2O_3, Y_2O_3, and CeO_2 as starting material [256].

Transparent Ce:YAG ceramics show a substantial advantage with respect to single crystal. In particular, there is an absence of antisite defects (Y at Al site) and related electron traps [257], allowing, in principle, a much faster and more efficient energy transfer to the Ce^{3+} emission centers. Consequently, there still is a considerable space for optimization of such ceramic scintillators.

Lutetium aluminum garnet (LuAG) exists in cubic form and has a garnet structure. Recently, because of its high density (6.73 g/cm^3, 94% of bismuth germanate (BGO)) and other physical properties such as shock resistivity, chemical radiation stability, LuAG has been known to be a promising host structure for scintillating materials. This host lattice containing rare-earth ions as luminescent activators especially cerium (Ce^{3+}), yields fast decay and is an efficient high-response scintillator [258]. On the other hand, Tm^{3+}-, Ho^{3+}-, or Yb^{3+}-doped lutetium aluminates are very promising for microchip laser or neutrino physics. However, LuAG single crystals are almost entirely grown from the melt (Czochralski method) by slow cooling in an expensive iridium crucible. They are very expensive, and it is difficult to produce large-sized LuAG single crystals. As an alternative, polycrystalline LuAG can be used if high density and transparency are acquired using ceramic processing techniques. Ce:LuAG transparent ceramics have been prepared by either solid-state reaction [259,260] or coprecipitation routes [261]. For example, using the Lu_2O_3 powders synthesized by a precipitation method and commercial Al_2O_3 and CeO_2 powders as starting materials, Ce:LuAG transparent ceramic has been prepared by solid-state reactive sintering at 1770°C for 10 h under vacuum and then annealed at 1450°C for 20 h in air to remove oxygen vacancy.

In another case, Ce:LuAG nanopowders with a uniform particle size of about 30 nm have been synthesized by an alternative wet-chemical route, namely, the urea homogeneous precipitation method [262]. The fabrication of transparent Ce:LuAG ceramics has been attempted by vacuum sintering of the obtained powders without any additives. The transmittance in the visible light region of the transparent Ce:LuAG ceramics fabricated at 1800°C for 10 h reaches 70% [263]. Optical absorption of Ce:LuAG transparent ceramic shows an additional light scattering loss due to nanovoids and/or refraction index inhomogeneities. Even though radioluminescence intensity of the optical ceramics prepared by this method exceeds that of single crystal, its scintillation light yield is lower due to the charge carrier retrapping in the process of energy transfer toward the Ce^{3+} emission centers [264]. The obtained Ce:LuAG transparent ceramic does not show the presence of the LuAl antisite defects which decreases the scintillation figure of merit in single crystals. Furthermore, using the ceramic technology, one can achieve both the homogeneous doping and higher Ce concentration. Similar to Ce:YAG, the transparent ceramics may therefore become competitive to the Ce:LuAG single crystals, if trapping phenomena at other defects can be further reduced.

References

[1] A. Mukhopadhyay, B. Basu, International Materials Reviews 52 (2007) 257.

[2] C.N.R. Rao, S.R.C. Vivekchand, K. Biswasa, A. Govindaraja, Dalton Transactions (2007) 3728.

[3] L.Y. Shi, C.Z. Li, H.C. Gu, Acta Metallurgica Sinica 36 (2000) 295.

[4] L.Y. Shi, C.Z. Li, D.Y. Fang, H.C. Gu, Y.H. Zhu, A.P. Chen, Chinese Journal of Inorganic Chemistry 16 (2000) 683.

[5] S. Iijima, Nature 354 (1991) 56.

[6] R.W. Siegel, S. Ramasamy, H. Hahn, Z.Q. Li, T. Lu, R. Gronsky, Journal of Materials Research 3 (1988) 1367.

[7] J. Li, Y.B. Pan, F.G. Qiu, L.P. Huang, J.K. Guo, Materials Science and Engineering A – Structural Materials Properties Microstructure and Processing 435 (2006) 611.

[8] J. Li, Y.B. Pan, J.W. Ning, L.P. Huang, J.K. Guo, Journal of Inorganic Materials 19 (2004) 1293.

[9] J. Li, Y.S. Wu, Y.B. Pan, W.B. Liu, J.K. Guo, Ceramics International 33 (2007) 919.

[10] K. Daimon, E. Kato, Journal of Crystal Growth 75 (1986) 348.

[11] J. Li, Y.B. Pan, J.J. Zhang, Journal of the Chinese Ceramic Society 31 (2003) 490.

[12] N. Saito, S. Matsuda, T. Ikegami, Journal of the American Ceramic Society 81 (1998) 2023.

[13] D.J. Sordelet, M. Akinc, Journal of Colloid and Interface Science 122 (1988) 47.

[14] X.H. Jin, L. Gao, J.K. Guo, Journal of Inorganic Materials 16 (2001) 555.

[15] W.Y. Hu, D.Y. Yang, X. Liu, Journal of the Chinese Ceramic Society 37 (2009) 1419.

[16] Y.S. Wu, J. Li, Y.B. Pan, Q. Liu, J.K. Guo, Ceramics International 35 (2009) 25.

[17] S.H. Liu, X.D. Sun, J.G. Li, X.D. Li, Z.M. Xiu, D. Huo, European Journal of Inorganic Chemistry (2009) 1214.

[18] C.C. Wang, J.Y. Ying, Chemistry of Materials 11 (1999) 3113.

[19] J. Li, Y.B. Pan, C.S. Xiang, Q.M. Ge, J.K. Guo, Ceramics International 32 (2006) 587.

[20] J. Li, Y.S. Wu, Y.B. Pan, J.K. Guo, Ceramics International 33 (2007) 735.

[21] J. Li, Y.S. Wu, Y.B. Pan, W.B. Liu, Y. Zhu, J.K. Guo, Ceramics International 34 (2008) 1539.

[22] J. Li, Y.B. Pan, F.G. Qiu, Y.S. Wu, W.B. Liu, J.K. Guo, Ceramics International 33 (2007) 1047.

[23] J. Li, Y.B. Pan, F.G. Qiu, Y.S. Wu, J.K. Guo, Ceramics International 34 (2008) 141.

[24] M. Boutonnet, J. Kizling, P. Stenius, Colloids and Surfaces 5 (1982) 209.

[25] M.J. Schwuger, K. Stickdorn, R. Schomacker, Chemical Reviews 95 (1995) 849.

[26] D.H. Chen, S.H. Wu, Chemistry of Materials 12 (2000) 1354.

[27] J. Zhang, X. Ju, Z.Y. Wu, T. Liu, T.D. Hu, Y.N. Xie, et al., Chemistry of Materials 13 (2001) 4192.

[28] L. Gao, H.C. Qiao, H.B. Qiu, D.S. Yan, Journal of the European Ceramic Society 16 (1996) 437.

[29] C.T. Xia, E.W. Shi, W.Z. Zhong, J.K. Guo, Journal of the European Ceramic Society 15 (1995) 1171.
[30] B.L. Cushing, V.L. Kolesnichenko, C.J. O'Connor, Chemical Reviews 104 (2004) 3893.
[31] C.T. Xia, E.W. Shi, W.Z. Zhong, J.K. Guo, Journal of Crystal Growth 166 (1996) 961.
[32] E.W. Shi, C.T. Xia, W.Z. Zhong, B.G. Wang, C.D. Feng, Journal of the American Ceramic Society 80 (1997) 1567.
[33] Y. Oguri, R.E. Riman, H.K. Bowen, Journal of Materials Science 23 (1988) 2897.
[34] Q.H. Zhang, L. Gao, Langmuir 19 (2003) 967.
[35] H.B. Yin, Y. Wada, T. Kitamura, T. Sumida, Y. Hasegawa, S. Yanagida, Journal of Materials Chemistry 12 (2002) 378.
[36] J. Tang, F. Redl, Y.M. Zhu, T. Siegrist, L.E. Brus, M.L. Steigerwald, Nano Letters 5 (2005) 543.
[37] G.S. Li, L.P. Li, J. Boerio-Goates, B.F. Woodfield, Journal of the American Chemical Society 127 (2005) 8659.
[38] X.H. Chen, R. Fan, Chemistry of Materials 13 (2001) 802.
[39] J. Rockenberger, E.C. Scher, A.P. Alivisatos, Journal of the American Chemical Society 121 (1999) 11595.
[40] K. Biswas, C.N.R. Rao, Journal of Physical Chemistry B 110 (2006) 842.
[41] U.K. Gautam, M. Rajamathi, F. Meldrum, P.E.D. Morgan, R. Seshadri, Chemical Communications (2001) 629.
[42] E.W. Shi, C.T. Xia, B.G. Wang, Journal of Inorganic Materials 11 (1996) 193.
[43] M.J. Mayo, D.C. Hague, D.J. Chen, Materials Science and Engineering A – Structural Materials Properties Microstructure and Processing 166 (1993) 145.
[44] K. Lu, International Materials Reviews 53 (2008) 21.
[45] I.W. Chen, X.H. Wang, Nature 404 (2000) 168.
[46] X.H. Wang, P.L. Chen, I.W. Chen, Journal of the American Ceramic Society 89 (2006) 431.
[47] X.H. Wang, X.Y. Deng, H.L. Bai, H. Zhou, W.G. Qu, L.T. Li, et al., Journal of the American Ceramic Society 89 (2006) 438.
[48] L. Gao, W. Li, H.Z. Wang, J.X. Zhou, Z.J. Chao, Q.Z. Zai, Journal of the European Ceramic Society 21 (2001) 135.
[49] E.J. Gonzalez, B. Hockey, G.J. Piermarini, Materials and Manufacturing Processes 11 (1996) 951.
[50] K.N.P. Kumar, K. Keizer, A.J. Burggraaf, T. Okubo, H. Nagamoto, S. Morooka, Nature 358 (1992) 48.
[51] U. Betz, A. Sturm, J.F. Loffler, W. Wagner, A. Wiedenmann, H. Hahn, Materials Science and Engineering A – Structural Materials Properties Microstructure and Processing 281 (2000) 68.
[52] M.J. Mayo, International Materials Reviews 41 (1996) 85.
[53] D. Ehre, E.Y. Gutmanas, R. Chaim, Journal of the European Ceramic Society 25 (2005) 3579.
[54] D. Sciti, J. Vicens, N. Herlin, J. Grabis, A. Bellosi, Journal of Ceramic Processing Research 5 (2004) 40.
[55] S.C. Liao, K.D. Pae, W.E. Mayo, Nanostructured Materials 8 (1997) 645.
[56] B.H. Kear, W.E. Mayo, Advanced Materials Processing Ii 437–438 (2003) 399.
[57] G. Skandan, Nanostructured Materials 5 (1995) 111.
[58] M. Uchic, H.J. Hofler, W.J. Flick, R. Tao, P. Kurath, R.S. Averback, Scripta Metallurgica et Materialia 26 (1992) 791.
[59] S. Bhaduri, S.B. Bhaduri, E. Zhou, Journal of Materials Research 13 (1998) 156.
[60] S.M. Dong, D.L. Jiang, S.H. Tan, J.K. Guo, Journal of Materials Science Letters 16 (1997) 1080.
[61] S. Seal, S.C. Kuiry, R. Georgieva, A. Agarwal, MRS Bulletin 29 (2004) 16.
[62] S.W. Wang, L.D. Chen, T. Hirai, J.K. Guo, Journal of Inorganic Materials 16 (2001) 1055.
[63] S.J. Wu, L.C. DeJonghe, M.N. Rahaman, Journal of the American Ceramic Society 79 (1996) 2207.
[64] R. Chaim, Z.J. Shen, M. Nygren, Journal of Materials Research 19 (2004) 2527.
[65] X. Xu, T. Nishimura, N. Hirosaki, R.J. Xie, Y.C. Zhu, Y. Yamamoto, et al., Journal of the American Ceramic Society 88 (2005) 934.
[66] W. Li, L. Gao, Biomaterials 24 (2003) 937.
[67] J.K. Guo, Journal of the Chinese Ceramic Society (1987) 385.
[68] J.X. Deng, Z.Q. Li, X. Ai, Materials Reviews 5 (1994) 72.
[69] J.K. Guo, T.S. Yen, Microstructure and Properties of Ceramic Materials, Science Press, Beijing, 1984, p. 281.
[70] I. Golecki, Materials Science & Engineering R-Reports 20 (1997) 37.
[71] Y.H. Zhang, J.P. Wang, G.J. Qiao, Journal of the Chinese Ceramic Society 35 (2007) 389.
[72] W.C. Sun, H.J. Li, S.P. Chen, S.Y. Zhang, K.Z. Li, Journal of Inorganic Materials 18 (2003) 121.
[73] J.Z. Zou, X.R. Zeng, X.B. Xiong, Journal of the Chinese Ceramic Society 35 (2007) 1062.
[74] J.F. Huang, X.R. Zeng, H.J. Li, X.B. Xiong, Y.W. Fu, Carbon 42 (2004) 1517.
[75] G.S. Jiao, H.J. Li, K.Z. Li, Journal of the Chinese Ceramic Society (2007) 721.
[76] Y.L. Zhang, H.J. Li, Q.G. Fu, K.Z. Li, H.B. Ouyang, Journal of Inorganic Materials 24 (2009) 125.
[77] S.J. Park, M.K. Seo, Carbon 39 (2001) 1229.
[78] X.R. Zeng, H.J. Li, Z. Yang, Journal of the Chinese Ceramic Society 27 (1999) 8.
[79] L.F. Cheng, Y.D. Xu, L.T. Zhang, X.W. Yin, Carbon 37 (1999) 977.
[80] D.S. Hou, K.Z. Li, H.J. Li, Q.G. Fu, Y.L. Zhang, Journal of Inorganic Materials 23 (2008) 213.
[81] X.M. Zhang, X.B. He, C.R. Zhang, Journal of the Chinese Ceramic Society 27 (1999) 611.
[82] S.Q. Guo, Y. Kagawa, Journal of the American Ceramic Society 84 (2001) 2079.
[83] F. Xu, L.T. Zhang, X.K. Meng, Y.F. Chen, Y.D. Xu, S.W. Fan, et al., Journal of Inorganic Materials 23 (2008) 233.
[84] Y.Z. Zhu, Z.R. Huang, S.M. Dong, M. Yuan, D.L. Jiang, Journal of Inorganic Materials 22 (2007) 685.
[85] Y.S. Ding, S.M. Dong, L. Gao, P. He, X.Y. Zhang, D.L. Jiang, Journal of Inorganic Materials 23 (2008) 1151.
[86] Y.N. Zhang, L.T. Zhang, L.F. Cheng, Y.D. Xu, Journal of the American Ceramic Society 90 (2007) 1139.
[87] S.W. Fan, L.T. Zhang, Y.D. Xu, L.F. Cheng, H.J. Lou, J.Z. Zhang, et al., Composites Science and Technology 67 (2007) 2390.
[88] L.S. Yan, H. Cui, K.Z. Li, H.J. Li, T. Wang, M.L. Song, Journal of Inorganic Materials 23 (2008) 223.
[89] J.Z. Zhang, Y.D. Xu, L.T. Zhang, L.F. Cheng, International Journal of Applied Ceramic Technology 4 (2007) 463.
[90] F. Lamouroux, G. Camus, J. Thebault, Journal of the American Ceramic Society 77 (1994) 2049.
[91] X.W. Yin, L.F. Cheng, L.T. Zhang, Y.D. Xu, C. You, Materials Science and Engineering A – Structural Materials Properties Microstructure and Processing 290 (2000) 89.
[92] L.F. Cheng, Y.D. Xu, L.T. Zhang, X.W. Yin, Journal of the American Ceramic Society 85 (2002) 989.
[93] S. Goujard, L. Vandenbulcke, H. Tawil, Thin Solid Films 245 (1994) 86.
[94] W.H. Zhang, L.F. Cheng, L.T. Zhang, W.B. Yang, Y.S. Liu, Y.D. Xu, Journal of Inorganic Materials 23 (2008) 774.
[95] K. Walter, International Journal of Ceramic Technology 1 (2004) 188.
[96] X.L. He, Y. Zhou, D.C. Jia, Y.K. Guo, Ceramics International 32 (2006) 929.

[97] H.L. Tang, X.R. Zeng, X.B. Xiong, Journal of the Chinese Ceramic Society 35 (2007) 1057.
[98] H.L. Tang, X.R. Zeng, X.B. Xiong, L. Li, J.Z. Zou, Journal of Inorganic Materials 24 (2009) 305.
[99] C.A. Lewinsohn, L.A. Giannuzzi, C.E. Bakis, R.E. Tressler, Journal of the American Ceramic Society 82 (1999) 407.
[100] H. Tenailleau, X. Bourrat, R. Naslain, R.E. Tressler, L.A. Giannuzzi, Journal of the American Ceramic Society 81 (1998) 2037.
[101] R. Raj, Journal of the American Ceramic Society 76 (1993) 2147.
[102] R.T. Bhatt, in: N.P. Bansal (Ed.), Handbook of Ceramic Composites, Kluwer Academic Publishers, London, 2005, p. 150.
[103] N.D. Corbin, G.A. Rossetti Jr., S.D. Hartline, Ceramic Engineering and Science Proceedings 10 (1989) 1083.
[104] J. Lamon, in: N.P. Bansal (Ed.), Handbook of Ceramic Composites, Kluwer Academic Publishers, London, 2005, p. 56.
[105] T. Tiegs, in: N.P. Bansal (Ed.), Handbook of Ceramic Composites, Kluwer Academic Publishers, London, 2005, p. 312.
[106] T.N. Tiegs, Handbook on Discontinuously Reinforced Ceramic Matrix Composites, Am. Ceram. Soc., Westerville, 1995, p. 225.
[107] T.N. Tiegs, K.J. Bowman, Handbook on Discontinuously Reinforced Ceramic Matrix Composites, Am. Ceram. Soc., 1995, p. 91.
[108] Y.F. Zhang, X.X. Huang, J.K. Guo, Journal of Inorganic Materials 7 (1992) 329.
[109] W.H. Tuan, R.J. Brook, Journal of the European Ceramic Society (1990) 31.
[110] T. Sekino, K. Niihara, Journal of Materials Science 32 (1997) 3943.
[111] G.J. Li, X.X. Huang, J.K. Guo, Materials Science and Engineering A – Structural Materials Properties Microstructure and Processing 352 (2003) 23.
[112] G.J. Li, X.X. Huang, Y. Shi, J.K. Guo, Materials Letters 51 (2001) 325.
[113] G.J. Li, X.X. Huang, J.K. Guo, Materials Research Bulletin 36 (2001) 1307.
[114] G.J. Li, X.X. Huang, J.K. Guo, Materials Research Bulletin 38 (2003) 1591.
[115] X.M. Shi, Y.B. Pan, J.K. Guo, Ceramics International 33 (2007) 1509.
[116] J.D. Walton, American Ceramic Society Bulletin 53 (1974) 255.
[117] S. Immelmann, E. Welle, W. Reimers, Materials Science and Engineering A – Structural Materials Properties Microstructure and Processing 238 (1997) 287.
[118] M.J. Hoffmann, A. Geyer, R. Oberacker, Journal of the European Ceramic Society 19 (1999) 2359.
[119] P.Z. Gu, W.R. Zhang, Y. Li, Journal of the Chinese Ceramic Society 23 (1995) 266.
[120] Y.Q. Wu, Y.F. Zhang, J.K. Guo, Materials Reviews 14 (2000) 20.
[121] S.T. Song, G.Z. Liu, Journal of the Chinese Ceramic Society 21 (1993) 1.
[122] D. Bandyopadhyay, L.C. Pathak, I. Mukherjee, S.K. Das, R.G. Ganguly, P. Ramachandrarao, Materials Research Bulletin 32 (1997) 75.
[123] Y.D. Xu, L.T. Zhang, Z. Zhang, Acta Materiea Compositae Sinica 12 (1995) 43.
[124] W.W. Chen, Z.S. Zou, T.M. Wang, Journal of the Chinese Ceramic Society 26 (1998) 127.
[125] J.G. Li, L. Gao, J.K. Guo, Journal of Inorganic Materials 17 (2002) 437.
[126] M. Awaad, M.E. Zawrah, N.M. Khalil, Ceramics International 34 (2008) 429.
[127] K. Niihara, Journal of the Ceramic Society of Jpn 99 (1991) 974.
[128] K. Niihara, Y. Suzuk, Materials Science and Engineering A261 (1999) 6.
[129] J. Sun, L. Gao, W. Li, Chemistry of Materials 14 (2002) 5169.
[130] L.N. An, W.X. Xu, S. Rajagopalan, C.M. Wang, H. Wang, Y. Fan, et al., Advanced Materials 16 (2004) 2036.
[131] C. Balazsi, Z. Shen, Z. Konya, Z. Kasztovszky, F. Weber, Z. Vertesy, et al., Composites Science and Technology 65 (2005) 727.
[132] J. Wang, H.M. Kou, X.J. Liu, Y.B. Pan, J.K. Guo, Ceramics International 33 (2007) 719.
[133] J.W. Ning, J.J. Zhang, Y.B. Pan, J.K. Guo, Materials Science and Engineering A – Structural Materials Properties Microstructure and Processing 357 (2003) 392.
[134] Y. Zhu, Y.B. Pan, H. Xu, J.K. Guo, Journal of Non-Crystalline Solids 355 (2009) 785.
[135] J.W. Ning, J.J. Zhang, Y.B. Pan, J.K. Guo, Ceramics International 30 (2004) 63.
[136] Y. Zhu, Y.B. Pan, H. Xu, C.S. Xiang, H.M. Kou, J.K. Guo, Chemistry Letters 36 (2007) 1098.
[137] J.W. Ning, J.J. Zhang, Y.B. Pan, J.K. Guo, Journal of Materials Science Letters 22 (2003) 1019.
[138] C.S. Xiang, X.M. Shi, Y.B. Pan, J.K. Guo, High-Performance Ceramics Iii, Pts 1 and 2, 280–283 Trans Tech Publications Ltd, Switzerland, (2005), 123.
[139] C.S. Xiang, Y.B. Pan, X.J. Liu, Journal of Nanoscience and Nanotechnology 6 (2006) 1.
[140] C.S. Xiang, Y.B. Pan, J.K. Guo, Ceramics International 33 (2007) 1293.
[141] C.S. Xiang, J. Yang, Y. Zhu, Y.B. Pan, J.K. Guo, Journal of Inorganic Materials 22 (2007) 101.
[142] L. Vivien, E. Anglaret, D. Riehl, F. Bacou, C. Journet, C. Goze, M. Andrieux, et al., Chemical Physics Letters 307 (1999) 317.
[143] P. Chen, X. Wu, X. Sun, J. Lin, W. Ji, K.L. Tan, Physical Review Letters 82 (1999) 2548.
[144] J.E. Riggs, D.B. Walker, D.L. Carroll, Y.P. Sun, Journal of Physical Chemistry B 104 (2000) 7071.
[145] X. Sun, R.Q. Yu, G.Q. Xu, T.S.A. Hor, W. Ji, Applied Physics Letters 73 (1998) 3632.
[146] H. Xu, Y.B. Pan, Y. Zhu, H.M. Kou, J.K. Guo, Journal of Alloys and Compounds 481 (2009) L4.
[147] Q. Huang, L. Gao, J. Sun, Journal of the American Ceramic Society 88 (2005) 3515.
[148] Q. Huang, L. Gao, Journal of Materials Chemistry 13 (2003) 1517.
[149] Q. Huang, L. Gao, Y.Q. Liu, J. Sun, Journal of Materials Chemistry 15 (2005) 1995.
[150] Q. Huang, L. Gao, Applied Physics Letters 86 (2005) 123104.
[151] R.L. Coble, Journal of the American Ceramic Society 45 (1962) 123.
[152] G.D. Miles, R.A.J. Sambell, J. Rutherfo, G.W. Stephens, Transactions of the British Ceramic Society 66 (1967) 319.
[153] R.A. Lefever, J. Matsko, Materials Research Bulletin 2 (1967) 865.
[154] R.J. Bratton, Journal of the American Ceramic Society 57 (1974) 283.
[155] G.H. Haertling, C.E. Land, Journal of the American Ceramics Society 54 (1971) 303.
[156] J.W. McCauley, N.D. Corbin, Journal of the American Ceramics Society 62 (1979) 476.
[157] M. Mitomo, Y. Moriyoshi, T. Sakai, T. Ohsaka, M. Kobayashi, Journal of Materials Science Letters 1 (1982) 25.
[158] N. Kuramoto, H. Taniguchi, Journal of Materials Science Letters 3 (1984) 471.
[159] G. Dewith, H.J.A. Van Dijk, Materials Research Bulletin 19 (1984) 1669.
[160] A. Ikesue, T. Kinoshita, K. Kamata, K. Yoshida, Journal of the American Ceramic Society 78 (1995) 1033.
[161] M. Liu, S.W. Wang, J. Zhang, Journal of Rare Earths 24 (2006) 732.

[162] C. Greskovich, S. Dominic, H. David, American Ceramic Society Bulletin 71 (1992) 1120.
[163] C. Greskovich, S. Duclos, Annual Review of Materials Science 27 (1997) 69.
[164] R.L. Coble, US Patent 3,026,210, 1962.
[165] J.G.J. Peelen, R. Metselaa, Journal of Applied Physics 45 (1974) 216.
[166] M.E. Thomas, S.K. Andersson, R.M. Sova, R.I. Joseph, Infrared Physics & Technology 39 (1998) 235.
[167] Y.T.O, J.B. Koo, K.J. Hong, J.S. Park, D.C. Shin, Materials Science and Engineering A – Structural Materials Properties Microstructure and Processing 374 (2004) 191.
[168] G.C. Wei, A. Hecker, D.A. Goodman, Journal of the American Ceramic Society 84 (2001) 2853.
[169] X.J. Mao, S.Z. Shimai, M.J. Dong, S.W. Wang, Journal of the American Ceramic Society 91 (2008) 1700.
[170] X.J. Mao, S.W. Wang, S. Shimai, J.K. Guo, Journal of the American Ceramic Society 91 (2008) 3431.
[171] A. Krell, P. Blank, H.W. Ma, T. Hutzler, M.P.B. van Bruggen, R. Apetz, Journal of the American Ceramic Society 86 (2003) 12.
[172] R. Apetz, M.P.B. van Bruggen, Journal of the American Ceramic Society 86 (2003) 480.
[173] I. Yamashita, H. Nagayama, K. Tsukuma, Journal of the American Ceramic Society 91 (2008) 2611.
[174] B.N. Kim, K. Hiraga, K. Morita, H. Yoshida, Scripta Materialia 57 (2007) 607.
[175] D. Jiang, D.M. Hulbert, U. Anselmi-Tamburini, T. Ng, D. Land, A.K. Mukherjee, Journal of the American Ceramic Society 91 (2008) 151.
[176] R.J. Bratton, American Ceramic Society Bulletin 48 (1969) 759.
[177] C.R. Bickmore, K.F. Waldner, D.R. Treadwell, R.M. Laine, Journal of the American Ceramic Society 79 (1996) 1419.
[178] J.G. Li, T. Ikegami, J.H. Lee, T. Mori, Y. Yajima, Journal of the European Ceramic Society 21 (2001) 139.
[179] C.T. Wang, L.S. Lin, S.J. Yang, Journal of the American Ceramic Society 75 (1992) 2240.
[180] T. Shiono, K. Shiono, K. Miyamoto, G. Pezzotti, Journal of the American Ceramic Society 83 (2000) 235.
[181] K. Tsukuma, Journal of the Ceramic Society of Japan 114 (2006) 802.
[182] M.C.L. Patterson, D.W. Roy, G. Gilde, in: J.W. McCauley, A. Crowson, W.A. Gooch Jr., A.M. Rajendran, S.J. Bless, K.V. Logan, et al. (Eds.), Ceramic Transactions, vol. 134, The American Ceramic Society, Westerville, 2002, p. 595.
[183] K. Rozenburg, I.E. Reimanis, H.J. Kleebe, R.L. Cook, Journal of the American Ceramic Society 90 (2007) 2038.
[184] K. Rozenburg, I.E. Reimanis, H.J. Kleebe, R.L. Cook, Journal of the American Ceramic Society 91 (2008) 444.
[185] I.E. Reimanis, H.J. Kleebe, R.L. Cook, A. DiGiovanni, Proceedings of the International Society for Optical Engineering (SPIE) Defense and Security Symposium (2004) 30.
[186] T.C. Lu, X.H. Chang, J.Q. Qi, X.J. Luo, Q.M. Wei, S. Zhu, et al., Applied Physics Letters 88 (2006) 213120.
[187] N. Frage, S. Cohen, S. Meir, S. Kalabukhov, M.P. Dariel, Journal of Materials Science 42 (2007) 3273.
[188] K. Morita, B.N. Kim, K. Hiraga, H. Yoshida, Scripta Materialia 58 (2008) 1114.
[189] G. Long, L.M. Foster, Journal of the American Ceramic Society 44 (1961) 255.
[190] N.D. Corbin, Journal of the European Ceramic Society 5 (1989) 143.
[191] H.X. Willems, M.M.R.M. Hendrix, G. de With, Journal of the European Ceramic Society 10 (1992) 339.
[192] T.M. Hartnett, S.D. Bernstein, E.A. Maguire, SPIE 3060 (1997) 284.
[193] E.A. Maguire, J.K. Rawson, R.W. Tustison, SPIE 2286 (1994) 26.
[194] D. Clay, D. Poslusny, M. Flinders, S.D. Jacobs, R.A. Cutler, Journal of the European Ceramic Society 26 (2006) 1351.
[195] J.P. Cheng, D. Agrawal, Y.J. Zhang, R. Roy, Journal of Materials Science Letters 20 (2001) 77.
[196] A. Ikesue, Y.L. Aung, Nature Photonics 2 (2008) 721.
[197] T.H. Maiman, Nature 187 (1960) 493.
[198] T.H. Maiman, Physical Review Letters 4 (1960) 564.
[199] J.E. Geusic, H.M. Marcos, L.G. Vanuitert, Applied Physics Letters 4 (1964) 182.
[200] C. Greskovich, K.N. Woods, American Ceramic Society Bulletin 52 (1973) 473.
[201] C.A.M. Mulder, G. Dewith, Solid State Ionics 16 (1985) 81.
[202] M. Sekita, H. Haneda, T. Yanagitani, S. Shirasaki, Journal of Applied Physics 67 (1990) 453.
[203] A. Ikesue, I. Furusato, K. Kamata, Journal of the American Ceramic Society 78 (1995) 225.
[204] J. Lu, M. Prahu, J. Xu, Applied Physics Letters 77 (2000) 3707.
[205] J. Lu, M. Prahu, J. Xu, Journal of Applied Physics Letters 40 (2001) L552.
[206] J. Lu, T. Murai, K. Takaichi, Applied Physics Letters 78 (2001) 3586.
[207] H. Yagi, T. Yanagitani, K. Takaichi, K. Ueda, A.A. Kaminskii, Optical Materials 29 (2007) 1258.
[208] J. Li, Y.S. Wu, Y.B. Pan, L.Q. An, J. Zhang, S.W. Wang, et al., Journal of Inorganic Materials 22 (2007) 798.
[209] J. Li, Y.S. Wu, Y.B. Pan, Journal of the Chinese Ceramic Society 35 (2007) 1600.
[210] S.H. Lee, S. Kochawattana, G.L. Messing, J.Q. Dumm, G. Quarles, V. Castillo, Journal of the American Ceramic Society 89 (2006) 1945.
[211] J. Li, Y.S. Wu, Y.B. Pan, W.B. Liu, L.P. Huang, J.K. Guo, Optical Materials 31 (2008) 6.
[212] K.M. Kinsman, J. McKittrick, Journal of the American Ceramic Society 77 (1994) 2866.
[213] Y.S. Wu, J. Li, Y.B. Pan, J.K. Guo, B.X. Jiang, Y. Xu, et al., Journal of the American Ceramic Society 90 (2007) 3334.
[214] W.X. Zhang, Y.B. Pan, J. Zhou, W.B. Liu, J. Li, B.X. Jiang, et al., Journal of the American Ceramic Society 92 (2009) 2434.
[215] J. Zhou, W.X. Zhang, J. Li, B.X. Jiang, W.B. Liu, Y.B. Pan, Ceramics International 36 (2010) 193.
[216] J. Li, Y.S. Wu, Y.B. Pan, J.K. Guo, Journal of Non-Crystalline Solids 352 (2006) 2404.
[217] J. Li, Y.S. Wu, Y.B. Pan, H.M. Kou, Y. Shi, J.K. Guo, Ceramics International 34 (2008) 1675.
[218] Y.S. Wu, J. Li, F.G. Qiu, Y.B. Pan, Q. Liu, J.K. Guo, Ceramics International 32 (2006) 785.
[219] T. Tachiwaki, M. Yoshinaka, K. Hirota, T. Ikegami, O. Yamaguchi, Solid State Communications 119 (2001) 603.
[220] J.R. Lu, K. Ueda, H. Yagi, T. Yanagitani, Y. Akiyama, A.A. Kaminskii, Journal of Alloys and Compounds 341 (2002) 220.
[221] Y.L. Kopylov, V.B. Kravchenko, S.N. Bagayev, V.V. Shemet, A.A. Komarov, O.V. Karban, et al., Optical Materials 31 (2009) 707.
[222] A. Ikesue, Y.L. Aung, Journal of the American Ceramic Society 89 (2006) 1936.
[223] J. Li, Y.S. Wu, Y.B. Pan, W.B. Liu, L.P. Huang, J.K. Guo, International Journal of Applied Ceramic Technology 5 (2008) 360.
[224] P.H. Klein, W.J. Croft, Journal of Applied Physics 38 (1967) 1603.
[225] K. Petermann, G. Huber, L. Fornasiero, S. Kuch, E. Mix, V. Peters, et al., Journal of Luminescence 87–89 (2000) 973.
[226] T. Ikegami, J.G. Li, T. Mori, Y. Moriyoshi, Journal of the American Ceramic Society 85 (2002) 1725.

[227] J.G. Li, T. Ikegami, T. Mori, Y. Yajima, Journal of the American Ceramic Society 87 (2004) 1008.
[228] J. Lu, J. Lu, T. Murai, Japanese Journal of Applied Physics 40 (2001) L1277.
[229] M. Tokurakawa, K. Takaichi, A. Shirakawa, K.I. Ueda, H. Yagi, T. Yanagitani, et al., Applied Physics Letters 90 (2007) 071101.
[230] J. Zhang, L.Q. An, M. Liu, S. Shimai, S.W. Wang, Journal of the European Ceramic Society 29 (2009) 305.
[231] J. Lu, J.F. Bisson, K. Takaichi, T. Uematsu, A. Shirakawa, M. Musha, et al., Applied Physics Letters 83 (2003) 1101.
[232] J. Lu, K. Takaichi, T. Uematsu, A. Shirakawa, M. Musha, K. Ueda, et al., Applied Physics Letters 81 (2002) 4324.
[233] J.R. Lu, K. Takaichi, T. Uematsu, A. Shirakawa, M. Musha, K. Ueda, et al., Japanese Journal of Applied Physics Part 2-Letters 41 (2002) L1373.
[234] E. Downing, L. Hesselink, J. Ralston, R. Macfarlane, Science 273 (1996) 1185.
[235] A.J. Silversmith, W. Lenth, R.M. Macfarlane, Applied Physics Letters 51 (1987) 1977.
[236] G.S. Yi, B.Q. Sun, F.Z. Yang, D.P. Chen, Y.X. Zhou, J. Cheng, Chemistry of Materials 14 (2002) 2910.
[237] J. Silver, M.I. Martinez-Rubio, T.G. Ireland, G.R. Fern, R. Withnall, Journal of Physical Chemistry B 105 (2001) 948.
[238] J.A. Capobianco, J.C. Boyer, F. Vetrone, A. Speghini, M. Bettinelli, Chemistry of Materials 14 (2002) 2915.
[239] J.A. Capobianco, F. Vetrone, J.C. Boyer, A. Speghini, M. Bettinelli, Optical Materials 19 (2002) 259.
[240] L.A. Diaz-Torres, E. De la Rosa, P. Salas, H. Desirena, Optical Materials 27 (2005) 1305.
[241] J. Zhang, S.W. Wang, T.J. Rong, L.D. Chen, Journal of the American Ceramic Society 87 (2004) 1072.
[242] L.Q. An, J. Zhang, M. Liu, S.W. Wang, Journal of the American Ceramic Society 88 (2005) 1010.
[243] M. Liu, S.W. Wang, J. Zhang, L.Q. An, L.D. Chen, Optical Materials 29 (2007) 1352.
[244] M. Liu, S.W. Wang, J. Zhang, L.Q. An, L.D. Chen, Optical Materials 30 (2007) 370.
[245] J. Zheng, B. Forslund, Journal of the European Ceramic Society 15 (1995) 1087.
[246] F. Zhang, S.W. Wang, X.J. Liu, L.Q. An, X.Y. Yuan, Journal of Applied Physics 105 (2009) 093542.
[247] F. Zhang, L.Q. An, X.J. Liu, G.H. Zhou, X.Y. Yuan, S.W. Wang, Journal of the American Ceramic Society 92 (2009) 1888.
[248] J. Zhang, S.W. Wang, L.Q. An, M. Liu, L.D. Chen, Journal of Luminescence 122 (2007) 8.
[249] L. Min, S.W. Wang, Z. Ban, L.Q. An, Journal of Rare Earths 24 (2006) 732.
[250] C.W.E. van Eijk, Nuclear Instruments & Methods in Physics Research Section A – Accelerators Spectrometers Detectors and Associated Equipment 460 (2001) 1.
[251] C.W.E. vanEijk, Nuclear Instruments & Methods in Physics Research Section A – Accelerators Spectrometers Detectors and Associated Equipment 392 (1997) 285.
[252] C.W.E. van Eijk, Physics in Medicine and Biology 47 (2002) R85.
[253] E. Zych, C. Brecher, A.J. Wojtowicz, H. Lingertat, Journal of Luminescence 75 (1997) 193.
[254] E. Zych, C. Brecher, H. Lingertat, Journal of Luminescence 78 (1998) 121.
[255] T. Yanagida et al., IEEE Trans. Nucl. Sci. 52, 1836 (2005); M. Nikl, Meas. Sci. Technol. 17, R37 (2006).
[256] Y. Shi, Y.B. Pan, X.Q. Feng, J. Li, J.K. Guo, Journal of Inorganic Materials 25 (2010) 125.
[257] E. Mihokova, M. Nikl, J.A. Mares, A. Beitlerova, A. Vedda, K. Nejezchleb, et al., Journal of Luminescence 126 (2007) 77.
[258] M. Nikl, E. Mihokova, J.A. Mares, A. Vedda, M. Martini, K. Nejezchleb, et al., Physica Status Solidi A – Applied Research 181 (2000) R10.
[259] H.L. Li, X.J. Liu, L.P. Huang, Journal of the American Ceramic Society 88 (2005) 3226.
[260] H.L. Li, X.J. Liu, L.P. Huang, Journal of Inorganic Materials 21 (2006) 1161.
[261] H.L. Li, X.J. Liu, R.J. Xie, Y. Zeng, L.P. Huang, Journal of the American Ceramic Society 89 (2006) 2356.
[262] H.L. Li, X.J. Liu, L.P. Huang, Optical Materials 29 (2007) 1138.
[263] H.L. Li, X.J. Liu, R.J. Xie, G.H. Zhou, N. Hirosaki, X.P. Pu, et al., Japanese Journal of Applied Physics 47 (2008) 1657.
[264] M. Nikl, J.A. Mares, N. Solovieva, H.L. Li, X.J. Liu, L.P. Huang, et al., Journal of Applied Physics 101 (2007) 033515.

CHAPTER

20

Amorphous Materials

Zhuang-Qi Hu

Institute of Metal Research, Chinese Academy of Sciences, China

Amorphous materials are amongst one of the important types of nonequilibrium materials. Traditionally, the solid physics refers to crystal physics, usually in equilibrium state in which the atomic arrangement is of long-range periodicity. In recent years, the metastable materials at far from the equilibrium state have become one of the most active research fields. The reasons being: (i) many new preparation techniques are developed for producing various kinds of metastable materials; and (ii) due to rapid progress of high technology a lot of new materials having various exceptionally peculiar properties and functions are required. The deep theoretical understanding of the amorphization and nonequilibrium state guides and promotes the research and development of amorphous materials.

This chapter briefly summarizes the structure, formation criteria, and preparation technology of amorphous materials.

20.1. AMORPHOUS STRUCTURE

Crystalline and amorphous materials are all real solid, possessing basic characteristics of solid materials. The main discrepancy is the difference in atomic-scale structure. In crystal, the atomic equilibrium locations are arranged in a long-range periodicity and on the contrary in the amorphous there is a lack of long-range periodicity and they have a disordered arrangement as shown in Fig. 20.1. In this case the amorphous material can also be termed as metallic glass material.

20.1.1. Morphology of Noncrystal

Figure 20.2 demonstrates the path from gaseous state to solid state. During cooling, the gaseous atoms transform to liquid atoms at boiling temperature T_b, and the liquid volume continuously drops. The slope of

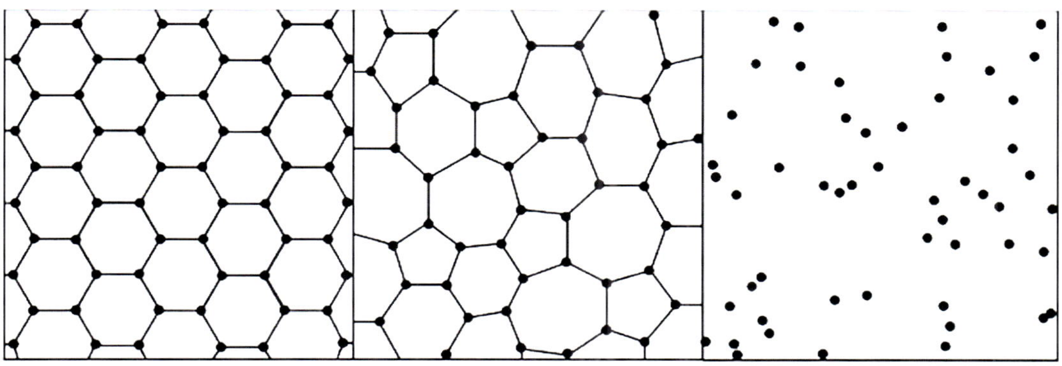

FIGURE 20.1 Schematic representation of atomic arrangements in crystal, amorphous and gas. (a) crystalline, (b) amorphous, and (c) gas. *R. Zallen, The Physics of Amorphous Solids, Wiley-Interscience Publication, Copyright 1983 Wiley-VCH Verlag GmbH & Co. KGaA. Reproduced with permission.*

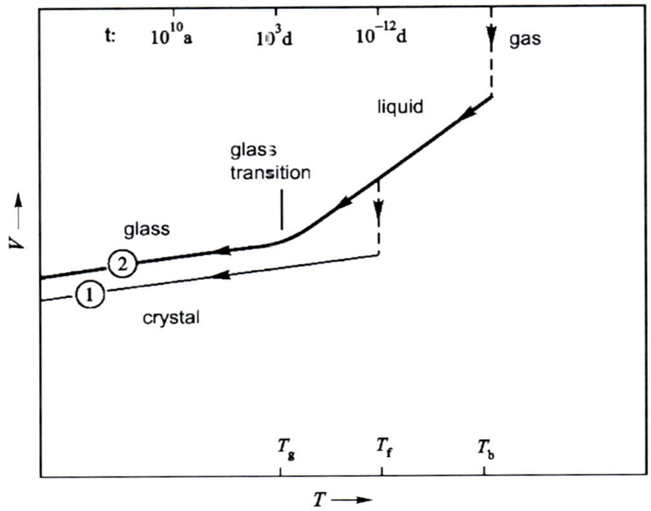

FIGURE 20.2 Cooling paths from gas to solid. 1 Path—route to crystal and 2 Path—route to glass by rapid quenching.

$V(T)$ curve is the thermal expansion coefficient of the liquid. When the temperature is lowered to the melting point T_f, the liquid transforms to solid state (except liquid helium). One of the characteristics of solid is the smaller slope of $V(T)$ curve. The transition of liquid to solid can be expressed by the abrupt volume contraction and discontinuity of $V(T)$ curve. If the cooling rate is rapid enough, the liquid state can be maintained right up to the glass transition temperature T_g. It is the second solidification phenomenon, that is, the transition from liquid state to amorphous state. Here volume change discontinuity is absent. Large number of experiments prove that there is a close relationship between the glass transition temperature and cooling rate, that is, the kinetic nature of the glass transition temperature. Generally, an ordered change in the cooling rate can alter the glass transition temperature for several degrees. If the cooling process is longer, the glass transition moves to a lower temperature because of the relation between atomic relaxation time τ and the temperature.

The atoms are frozen in an amorphous state, that must favor to the condition that $\tau(T)$ has to exceed the practical cooling time.

For a long time, it is considered that only a few materials can be obtained in amorphous state. Sometimes some oxide glasses and organic polymer compounds are also named as glassy solid. Most of the condensed matters have glass-forming ability. It is a generalized character. All the materials probably can be transformed into amorphous state, only if the cooling rate is high enough and the cooling temperature is low. Table 20.1 lists the bonding modes and glass transition temperatures of some amorphous materials, including metals, alloys, oxides, and organic compounds [1].

Relative to the lowest energy of thermodynamic equilibrium state of crystalline materials, the noncrystalline is metastable. It is correct. However, it is noticed that dynamically it is hard to revert back to the crystalline state, for example, once the glass forms, it can remain for a long time. Under standard temperature and pressure, graphite is a stable equilibrium phase, but diamond can remain in metastable state forever.

20.1.2. Long-range Disorder of Noncrystal

The radial distribution function is used to characterize the structure of amorphous alloy, in which there still exists short-range order having the nearest and subnearest coordinations, therefore in the radial distribution function curve there are clear first and second peaks. Also the amorphous alloy has no long-range order distribution function curve. Due to the random distribution in the space, in the system of the average number of atoms $\bar{n}$ per unit volume, radial distribution function curve can be obtained by the number of atoms in a shell volume of $4\pi r^2 dr$. Figure 20.3 shows the typical radial distribution function curves of crystal, noncrystal, and gas. When radius r is large, the part of radial distribution function curve approaching the dotted curve gives information related to average density. Sometimes it can be also expressed by the reduced radial distribution function $g(r)$. When r becomes higher, $g(r)$ approaches zero. The radial distribution function can be obtained by diffraction experiment after Fourier transformation. Therefore, the diffraction experiment is an important means to study the amorphous structure.

TABLE 20.1 Bonding and Glass Transition Temperature of Some Amorphous Materials [1]

Glass	Bonding	T_g (K)
SiO_2	Covalent	1430
GeO_2	Covalent	820
Si, Ge	Covalent	—
$Pd_{0.4}Ni_{0.4}P_{0.2}$	Metallic	580
BeF_2	Ionic	570
As_2S_3	Covalent	470
Polystyrene	Polymerized	370
Se	Polymerized	310
$Au_{0.8}Si_{0.2}$	Metallic	290
H_2O	Hydrogen	140
C_2H_5OH	Hydrogen	90
Isopentane	Van der Waals	65
Fe, Co, Ni	Metallic	—

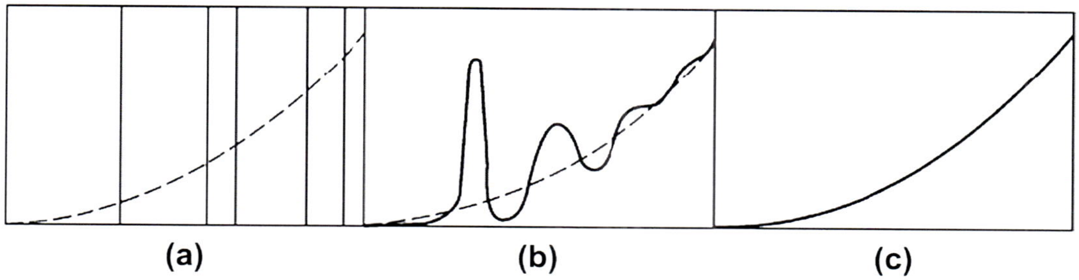

FIGURE 20.3 Radial distribution function curves: (a) crystalline solid, (b) amorphous solid, and (c) gas.

From the diffraction data of X-ray, electron or electron scattering one-dimensional description of the atomic-scale amorphous structure can be calculated. Recently developed extended X-ray absorption fine structure (EXAFS) can further overcome the shortage of scattering technique which only describes an average surrounding of an atom in the solid state, suitable for a pure metal, but for an alloy composed of different atoms, it is not suitable due to the presence of bonding between different atoms. Figure 20.4 depicts a typical X-ray diffraction intensity curve of amorphous $Fe_{80}P_{13}C_7$ alloy. In the figure there is a diffraction curve of polycrystalline alloy with same composition for comparison. Figure 20.5 shows the electron scattering intensity curves of amorphous Si and crystalline Si [2].

Because of the long-range disorder, the amorphous alloys have a large peculiarity in optical and electrical properties. The reflectivity in the electron excitation range of crystalline, amorphous, and liquid Ge is compared in Fig. 20.6 [4]. The low-frequency behavior of liquid Ge is different from that of crystalline and amorphous Ge. When the electron energy of liquid Ge approaches zero, the reflectivity is nearly 100% but it is about 36% for amorphous and crystalline Ge, this explains why the amorphous material is of long-range disorder and short-range order. Figure 20.7 shows the imaginary part of dielectric constant of crystalline Si and amorphous Si. It can be seen that the band of crystalline Si has a characteristic of full structure, whereas that of amorphous Si is smooth [5]. The above-mentioned result once again proves that the amorphous material has long-range disorder which has to increase the electrical resistivity as shown in Fig. 20.8 [6].

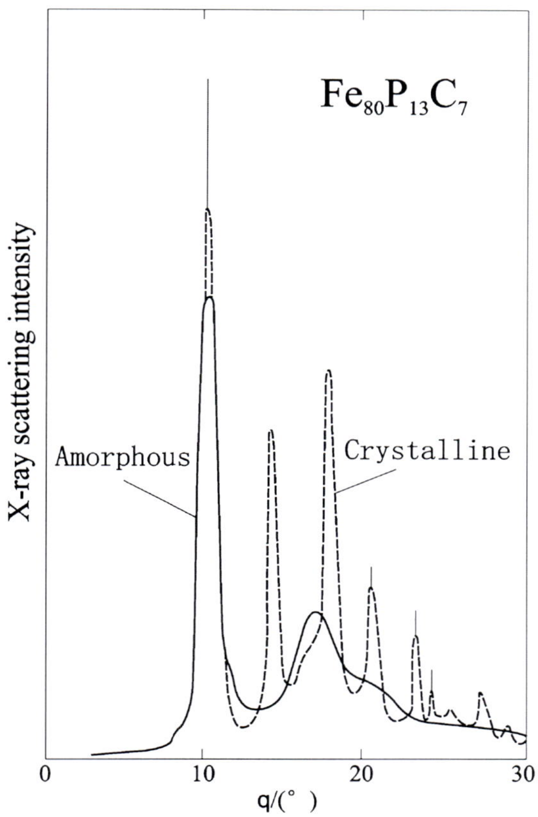

FIGURE 20.4 X-ray scattering results of metallic glass (thick line) compared to crystalline alloy of the same sample after crystallization (fine line) [3]. *With kind permission from Springer Science + Business, Media: Zeitschrift für Physik B Condensed Matter, An X-ray diffraction study of the structure of amorphous Cu_{57}-Zr_{43} alloy, volume 21, 1975, page 235, Y. Waseda, figure 1, and any original (first) copyright notice displayed with material.*

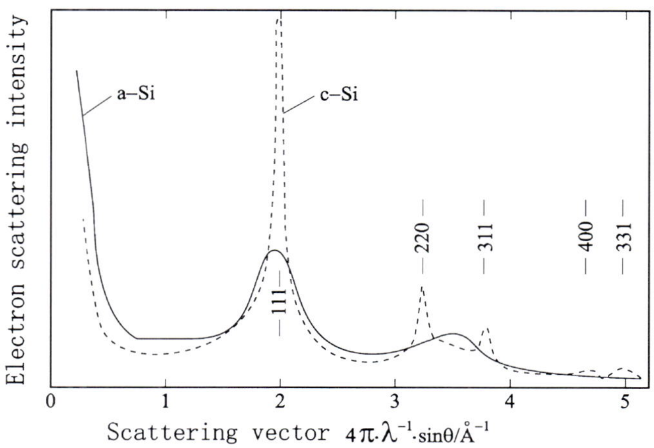

FIGURE 20.5 Electron diffraction pattern of amorphous Si (thick line) compared to partially crystallized film (fine line) [2].

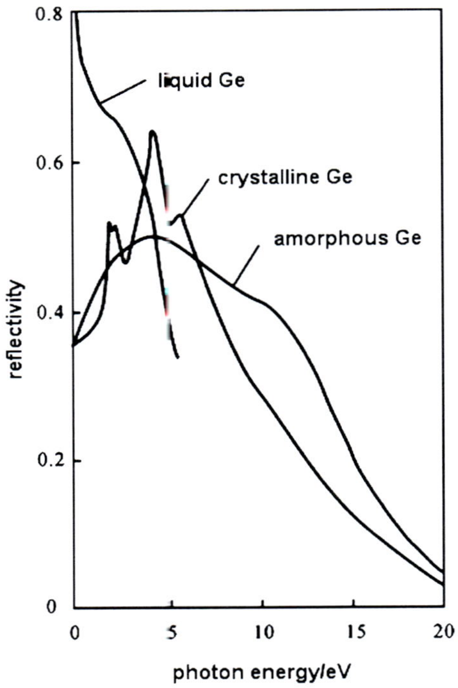

FIGURE 20.6 Reflectivity in electron excitation range of crystalline, amorphous, and liquid Ge [4].

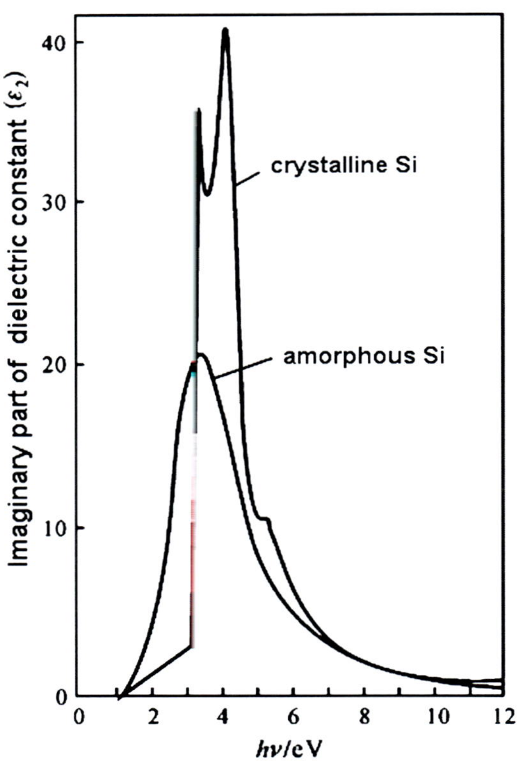

FIGURE 20.7 A comparison of imaginary part of dielectric constant between crystalline and amorphous Si [5].

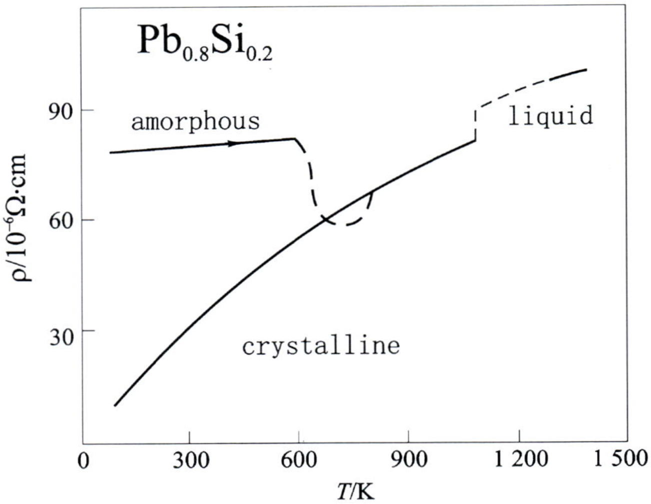

FIGURE 20.8 Resistivity of $Pb_{0.8}Si_{0.2}$ metallic glass [6].

20.1.3. Molecular Dynamics Computer Simulation

Computer simulation of electronic and atomic levels has been developed to an advanced stage. The advanced theories combined with supercomputer push the profound understanding of the materials' behaviors with ever thorough details and believable accuracy about electronic and atomic levels, that finally creates a new interdisciplinary research field. By the help of this epochal progress, it is possible not only to simulate experiments, but also to design new alloys and predict their properties.

The applications of molecular dynamics calculation on materials science, particularly on rapid solidification as well as on the phase transformation during rapid heating or rapid cooling, already made great achievements. For example, Ni_3Al is an intermetallic compound vastly studied nowadays, which cannot be amorphized by conventional rapid cooling technologies (usually less than 10^7 K/s cooling rate). Now the computational materials science can help us to know at what cooling rate Ni_3Al can be amorphized. Figure 20.9 represents the total two-body distribution function of Ni_3Al at 4×10^{13} K/s cooling rate. It can be seen that the three former peaks become higher as the temperature goes down. The lowering of peak valley means that the atomic arrangement tends to short-range ordering. At 300 K final simulated state, the second peak is split, which is the indication of amorphization. Again from Fig. 20.10, the number of 1551 bond pair indicating amorphous structure is of majority at $R_{c1}(4 \times 10^{13}$ K/s) and $R_{c2}(1 \times 10^{13}$ K/s) cooling rates. On the contrary, the 1551 bond pair is absent at $R_{c3}(2.5 \times 10^{12}$ K/s) and $R_{c4}(4 \times 10^{11}$ K/s) cooling rates, which means that under these conditions it cannot be amorphized [7].

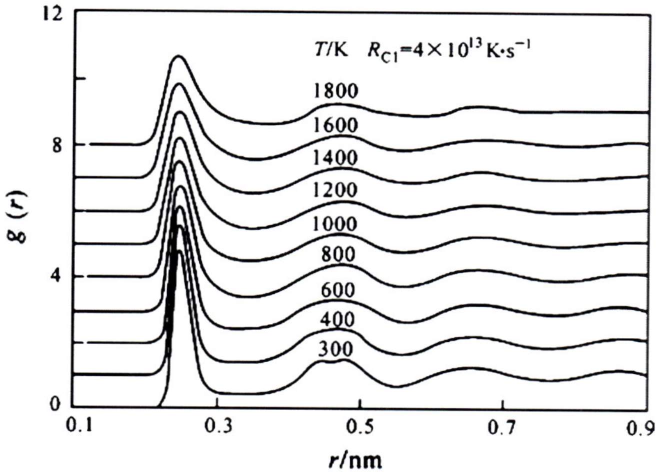

FIGURE 20.9 Total two-body distribution function $g(r)$ of Ni_3Al during rapid solidification at different temperatures [7].

It is very difficult to perform high-pressure experiment, but by using pseudo-potential theory under the approximation of free atoms, prediction of high-pressure phase stability can be made. Jin [8] studied the phase transformation of Mg at high pressure. Figure 20.11(a) indicates that Mg has hexagonal close-packed structure at normal pressure after rapid solidification at 1×10^{12} K/s and 6×10^{12} K/s cooling rates. But it is amorphous structure, if the cooling rate is 1.2×10^{13} K/s. At high pressure (45 GPa) at cooling rates of 8×10^{12} K/s and 5×10^{13} K/s, Mg has a body-centered cubic (bcc) structure, but Mg becomes amorphized at cooling rate of 1×10^{14} K/s (Fig. 20.11(b)). Table 20.2 compiles the numbers of different bond pairs [8]. Liquid Mg and amorphous Mg have numerous 1551 and 1541 bond pairs at normal and high pressures. Increase of pressure can enhance the 1551 bond pairs, which renders the second peak of $g(r)$ to split.

Wang et al. [9] calculated the two-body distribution function of Au and Ni at liquid, supercooled, and solid states. The calculated data are very close to the experimental ones. The splitting of second peak goes gradually and completes at a rather low temperature, which is usually the intrinsic characteristic of amorphous formation. According to Abraham theory, the amorphous transformation temperature T_g can be obtained (Fig. 20.12). By using molecular dynamics, the temperatures for amorphization of Au and Ni are 787 K and 1010 K, respectively, and the cooling rates should exceed 3.8×10^{13} K/s and 4.0×10^{13} K/s, respectively. During rapid solidification, the main point is the increase of icosahedrons number. The dynamic factor governs the structure evolution. The difference between

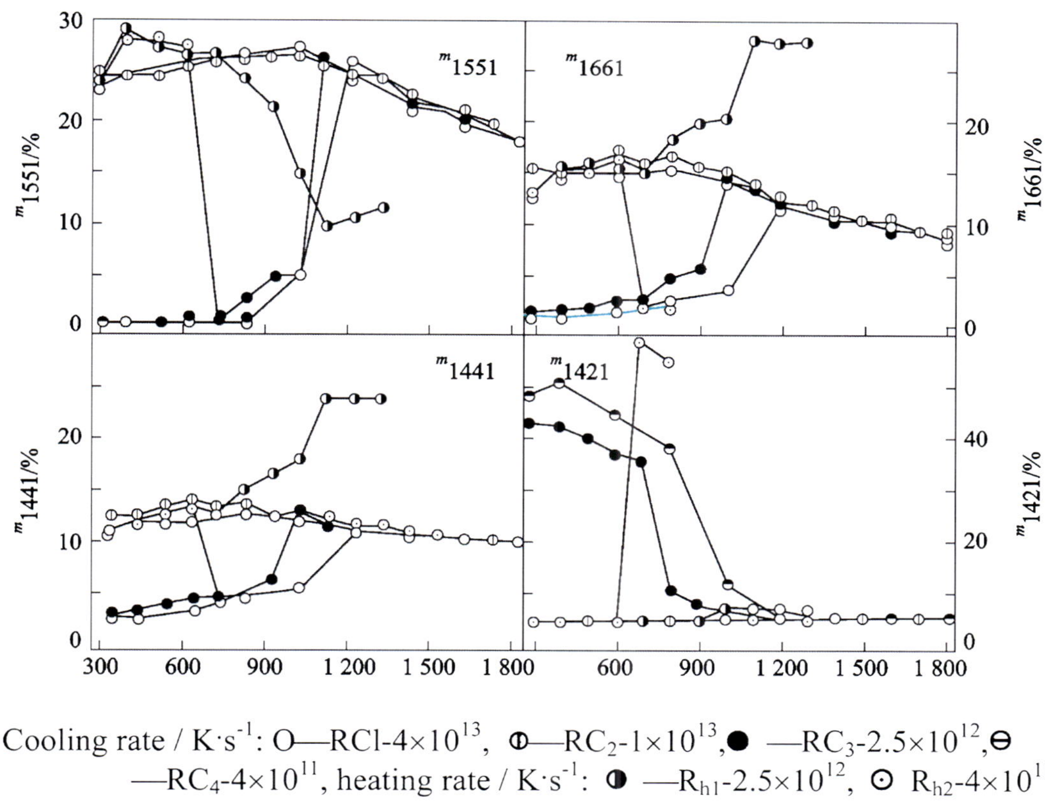

FIGURE 20.10 Influence of temperature on relative number m of different bond pairs of Ni_3Al [7].

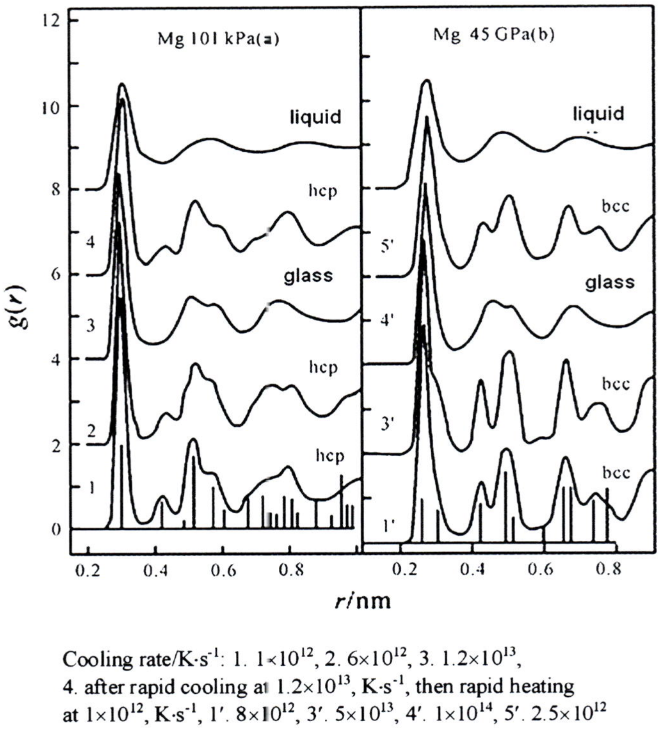

FIGURE 20.11 Total two-body distribution $g(r)$ Mg at normal pressure and 45 GPa high pressure [8].

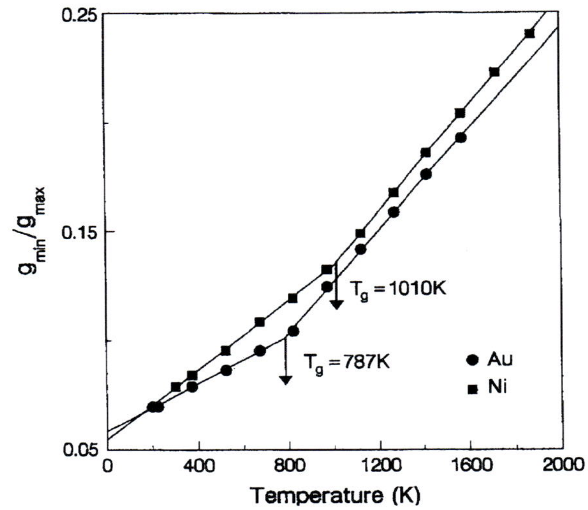

FIGURE 20.12 Relationship between Abraham coefficient g_{min}/g_{max} and temperature. *Reprinted from Physica B: Condensed Matter, Vol 239, Luhong Wang, Haozhe Liu, Kuiying Chen, Zhuangqi Hu, The local orientational orders and structures of liquid and amorphous metals Au and Ni during rapid solidification. Copyright (1997), with permission from Elsevier [9].*

TABLE 20.2 Relative Number of Typical Bond Pairs of Liquid Mg, Amorphous Mg, and Crystalline Mg at Normal and High Pressures [8]

P/MPa	State	1551	1541	1421	1422	1431	1661	1441
0.101	Liquid	0.115	0.135	0.039	0.078	0.211	0.04	0.044
	Amorphous	0.178	0.224	0.095	0.116	0.216	0.044	0.026
	Crystalline (hcp)	0	0.052	0.459	0.330	0.04	0.002	0.008
45×10^3	Liquid	0.213	0.143	0.018	0.036	0.154	0.094	0.080
	Amorphous	0.332	0.205	0.032	0.056	0.156	0.109	0.071
	Crystalline (bcc)	0.039	0.072	0.011	0.005	0.04	0.458	0.350

the supercooled liquid and amorphous structure is that the amorphous structure has more icosahedra, while the supercooled liquid has a broad distribution of local ordering, exhibiting a higher disordering.

20.1.4. Atomic Diffusion in Amorphous Alloy

Diffusion is the transport process of solute atoms or matrix atoms driven by thermal activation. There are many important properties of amorphous alloys closely related with the diffusion, for example, self-diffusion influences, viscosity, creep, time-dependent paramagnetism, electrical resistivity, internal friction, crystallization rate, etc., chemical diffusion (transport of extraneous atoms), control over oxygen dynamics, intermetallic compound formation, degradation of diffusion barrier of amorphous film, etc. Besides the vacancy mechanism and interstitial mechanism describing the atomic diffusion in amorphous alloys, the cooperative movement of adjacent atoms cluster is a common mechanism. The hydrogen movement in amorphous alloy is definitely governed by the interstitial mechanism. For those larger atoms, it is controlled by vacancy mechanism. Due to lack of definite vacancy in amorphous alloy, it can be imagined as a vacancy-like, and the diffusion for large atoms can be operated by a cooperative movement of nearby atoms, that can be described as the redistribution of many small vacancies to reform into several large vacancies easy for diffusion of atom.

The atom diffusion in amorphous alloys is influenced by matrix structure, chemical composition, atom species, etc. The influence of relaxation on diffusion depends on the thermal history of the preparation methods. Generally, the effect of relaxation on structure change is very small that makes the determination very hard. In those preparation technologies in which there happens self-relaxation already, there is no pronounced influence of relaxation on diffusion.

The influence of plastic deformation on diffusion of atom in amorphous alloy can be explained by the change of free volume. The free volume can be increased by plastic deformation that enhances the atom diffusion.

At a certain temperature, the diffusion coefficient becomes smaller in a following series: deformed state > quenched state > relaxed state. Radiation can lower the diffusion coefficient in amorphous alloy. Cahn et al. [10] measured the diffusion coefficient of Au in amorphous alloy $Ni_{64}Zr_{36}$ before and after fast neutron radiation, they found that the diffusion coefficient was lowered by radiation despite the increase in average atomic volume by radiation because the radiation reinforces the chemical short-range disordering that in turn results in volume contraction and lowering of diffusion coefficient of Au. Besides these factors, other factors which influence diffusion in amorphous alloys are solute atom concentration [11] and environmental pressure [12].

The methods of measuring the atomic diffusion in amorphous alloys are mostly related to the ionic beam, such as the measurement of Ag diffusion in amorphous $Pd_{81}Si_{19}$ alloy. Ag was implanted in a thin film of the amorphous alloy. After diffusion annealing, the surface film was thinned by Ar bombardment layer by layer. The profile of Ag concentration versus depth can be obtained by measuring the intensity of isotope of silver. The secondary ion mass spectroscope (SIMS) also needs ion peeling, but this time stable isotope was used. By applying this method, the diffusion of Al in $Fe_{78}Si_{19}B_{13}$ amorphous alloy was measured as shown in Fig. 20.13 [13]. The nondestructive Rutherford back scattering (RBS) method can also be used. Another method is nuclear reaction method, which is complex, accurate, and nondestructive, with the help of nuclear reaction between diffusive atoms and detected ions.

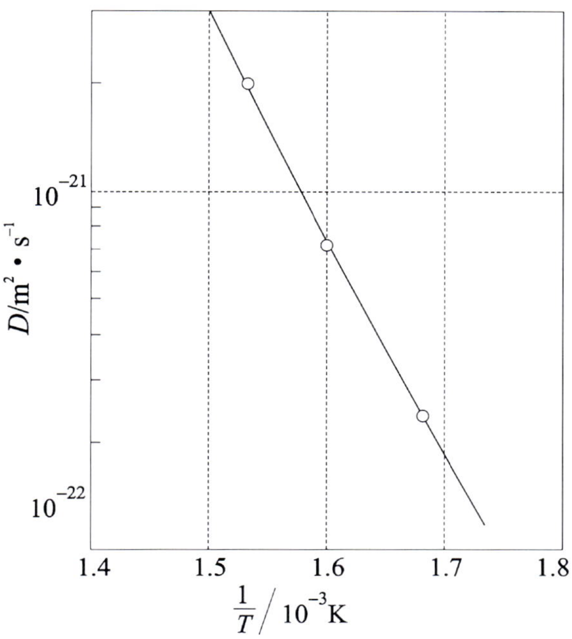

FIGURE 20.13 Diffusion coefficient of Al in $Fe_{78}Si_9B_{13}$ amorphous alloy [13].

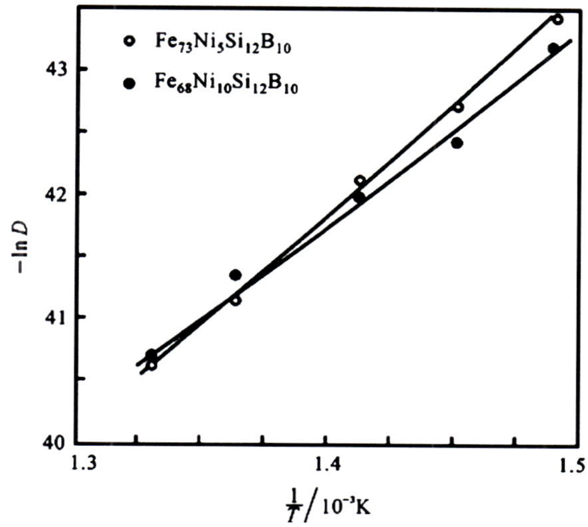

FIGURE 20.14 Diffusion coefficient of Si in (Fe, $Ni)_{78}Si_{12}B_{10}$ amorphous alloy. *Reprinted from Materials Letters, Vol 17, H.G. Jiang, B.Z. Ding, H.Y. Tong, J.T. Wang, Z.Q. Hu, Diffusivity of Si in two amorphous alloys of (Fe, Ni)$_{78}$Si$_{12}$B$_{10}$., Copyright (1993), with permission from Elsevier [14].*

The diffusion in amorphous alloys can be also studied by indirect methods, i.e., to measure the physical or chemical factors governed by diffusion. There is no need to measure the in-depth solute composition directly. The indirect methods to measure the diffusion coefficient in amorphous alloy must have the following prerequisites: first, it must have a physical or chemical process relating to the diffusion; second, it can be expressed by a definite mathematical formula; and finally, the corresponding physical or chemical values can be accurately measured, for example, crystallization kinetics, internal friction controlled by solute migration (to measure hydrogen diffusion), magnetic resonance spectrum, X-ray diffraction of multilayer film, viscous flow, or surface segregation dynamics. Figure 20.14 shows the Si diffusion coefficient in (Fe, $Ni)_{78}Si_{12}B_{10}$ amorphous alloy determined by surface segregation dynamics method [14,15].

20.2. FORMATION RULE OF AMORPHOUS ALLOY

20.2.1. Principles for Formation of Amorphous Alloy

The glass-forming abilities of various metals and alloys are quite different, for example, S and Se can become amorphous only at a moderate cooling rate and other typical metals need a cooling rate higher than 10^{10} K/s to suppress the nucleation. Adequate alloying can lower the cooling rate less than 10^6 K/s to be amorphized. For three alloys $Pd_{77.5}Cu_6Si_{16.5}$, $Pd_{60}Cu_{20}P_{20}$, and

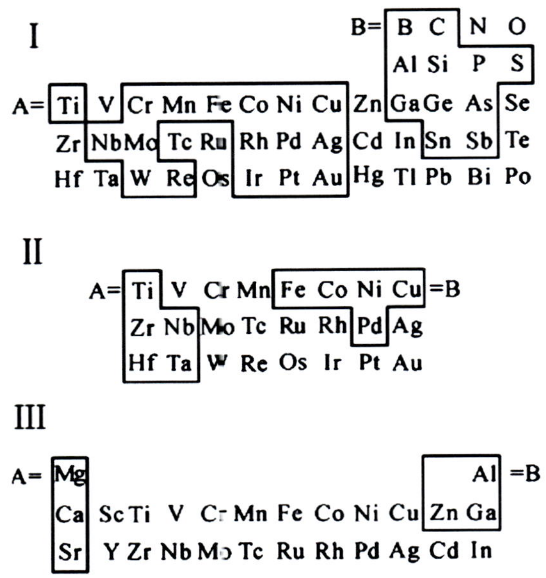

FIGURE 20.15 Basic constituents of three categories of A_xB_y amorphous alloys [16].

$Pd_{56}Ni_{24}P_{20}$, the cooling rate can be as low as 10^2 K/s to form millimeter-size bulk metallic glass.

Recently the known formation rules of binary alloys are as follows as shown in Fig. 20.15 [16].

(1) Base metals (inert or late transition metals) are alloyed with about 20 at% semimetal (B, C, Si, P, etc.) to form amorphous material like $Fe_{80}B_{20}$, $Au_{75}Si_{25}$, $Pd_{80}Si_{20}$, etc. (Fig. 20.15I).

(2) Late transition metals such as Fe, Co, Ni, Pd (right side of periodic table) together with Cu as base metal, are alloyed with early transition metals (Ti, Zr, Nb, Ta, etc.) on the left side of periodic table, to form amorphous materials like $Ni_{50}Nb_{50}$, $Cu_{60}Zr_{40}$, etc. (Fig. 20.15II).

(3) The binary alloys consisting of alkali-earth metals (Mg, Ca, Sr) of II group in periodic table and solute atoms (Al, Zn, Ga) are easily amorphized, e.g., $Mg_{70}Zn_{30}$, $Ca_{35}Al_{65}$ (Fig. 20.15III).

(4) Alloy compositions near their eutectics are prone to be amorphized (Fig 20.16), the shaded areas in the rectangular blocks under the six binary phase diagrams, are the composition ranges near eutectic compositions, in which the alloys can easily be amorphized (glass formation).

20.2.2. Semi-empirical Criteria of Metallic Glass Formation

There are several simple semi-empirical criteria to predict the metallic glass formation: First criterion is T_g/T_m, where T_g is the amorphization temperature and T_m is the alloy melting temperature. The larger the ratio, the easier the glass formation. At the "nose" of temperature–time transformation curve the higher the alloy viscosity, the higher the T_g/T_m value. The second criterion is the value of $(T_x - T_g)$, where T_x is crystallization temperature. If $(T_x - T_g)$ value is higher, the alloy is amorphized easily. Based on the parameter $(T_x - T_g)$, the other two criteria were derived, that is: $K_{gl} = \frac{T_x - T_g}{T_m - T_z}$ and $S = \frac{(T_p - T_x)(T_x - T_g)}{T_g}$, where T_p is the crystallization peak temperature. The greater the K_{gl} or S, the easier the formation of metallic glass. The third one is the thermodynamic state functions ΔH and ΔS. Increasing ΔS or lowering ΔH can lower the homogeneous nucleation rate and crystallization growth rate, this is beneficial for the metallic glass formation. Similarly, enhancing the solid/liquid interfacial energy can also promote the amorphization.

Marcus and Turnbull [17] proposed a normalized parameter $\frac{\Delta T}{T_1^0}$, where ΔT is the deviation of T_1 away from the ideal liquidus temperature T_1^0, T_1^0 can be calculated from the following formula:

$$T_1^0 = \frac{\Delta H_f^A T_m^A}{\Delta H_f^A - R\ln(1-x)T_m^A}, \tag{20.1}$$

where ΔH_f^A = melting enthalpy of solvent metal; T_m^A = melting point of solvent metal; and x = mole fraction of solvent metal. If the $\frac{\Delta T}{T_1^0}$ is a big positive value, it denotes the glass formation ability (GFA) is good. This criterion is suitable to predict GFA of the alloy system of late transition metal with semimetal, but it is not suitable on the binary alloy system of late transition metal with early transition metal, because in this case $\frac{\Delta T}{T_1^0}$ is negative value.

Donald and Davies [18] suggested Eq. (20.2) to describe GFA of eutectic alloys of metal–metal systems and metal–semimetal systems:

$$\Delta T^* = (T_1^{mix} - T)_1 T_1^{mix}, \tag{20.2}$$

$$T_1^{mix} = \sum^{n} x_i T_m^i, \tag{20.3}$$

where x_i is mole fraction of I component in the alloy and T_m is the melting point.

Another empirical criterion is that the difference in atomic radii of two components of binary alloy must exceed 15% and the difference of the two group numbers in the periodic table (Δn) must be greater than five. However, the above-mentioned amorphization criterion is not obeyed in some systems, for example, $\Delta n = 2$ for Be-(Ti, Zr) and $\Delta n = 0$ for Ca–Mg, which are easy to be amorphized.

The key aim is that the amorphization model should successfully predict composition and cooling rate which can suppress nucleation. For this reason, a profound knowledge on the crystallization kinetics is necessary.

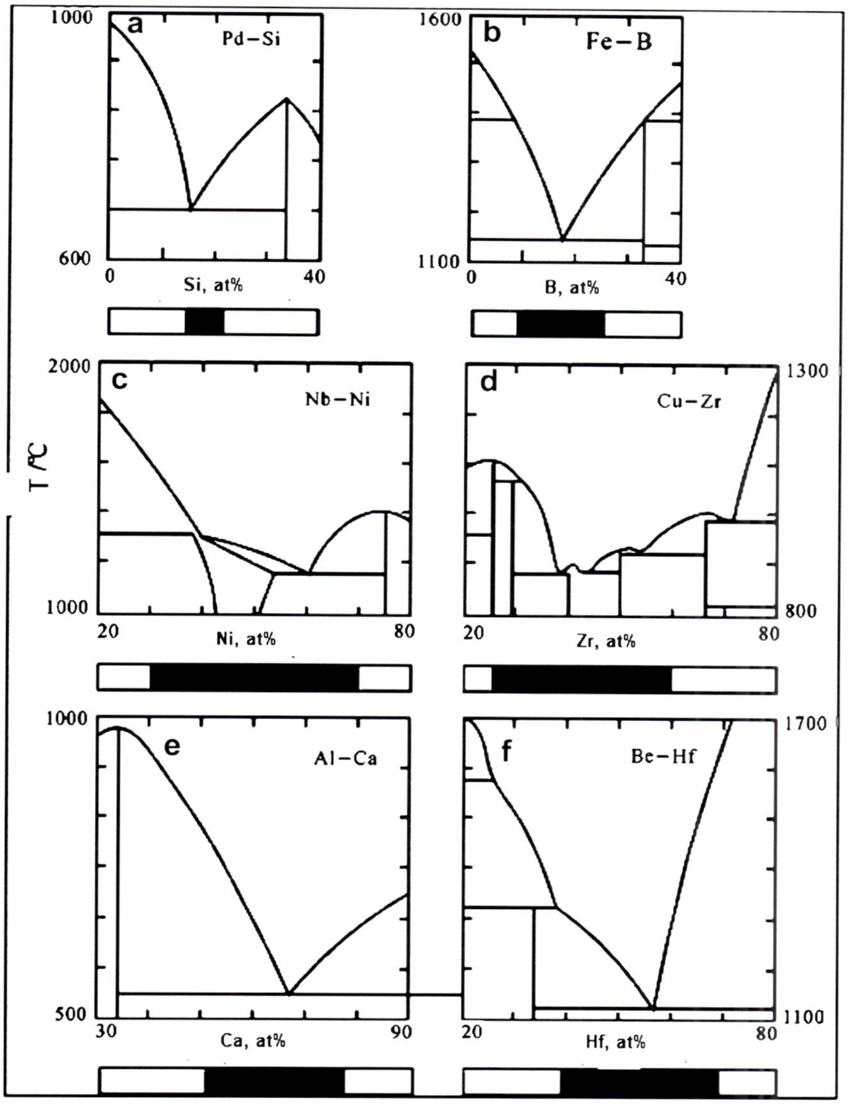

FIGURE 20.16 Composition ranges easy to be amorphized [16] (expressed by the shaded areas).

The crystallization fraction x and nucleation frequency I (nuclei formed in unit time t and unit volume) are related to the growth rate R:

$$x = \frac{1}{3}\pi I R^3 t^4. \tag{20.4}$$

For homogeneous nucleation:

$$I = N\nu \exp\left(\frac{-16\pi\,\gamma^3 T_f^2}{3L^2 kT \Delta T^2}\right). \tag{20.5}$$

For metal having typical low melting entropy:

$$R \approx a_0 v\left[1 - \exp\left(-\frac{L\Delta T}{kTT_f}\right)\right], \tag{20.6}$$

where ν—atom jump frequency, approximately D/a_0^2; N—number of atoms in unit volume; γ—solid/liquid interface energy; T_f—equilibrium solidification temperature; L—melting latent heat of unit volume; k—Boltzmann constant; T—temperature; δT—supercooling ($T_f - T$); and a_0—atomic spacing.

By using above three equations (Fig. 20.17), one can predict the cooling rate required for the amorphization.

Liu studied the effects of crystal structure, atomic size and electro-negativity on glass formation ability after systematic experiments of ion mixing, and proposed a new criterion of maximum possible amorphization range (MPAR) to predict the glass forming possibility. MPAR is measured as 100% minus two maximum solid solubilities on both sides of equilibrium binary phase diagram. The larger the MPAR, the broader the composition range, in which the alloy can be amorphized [19].

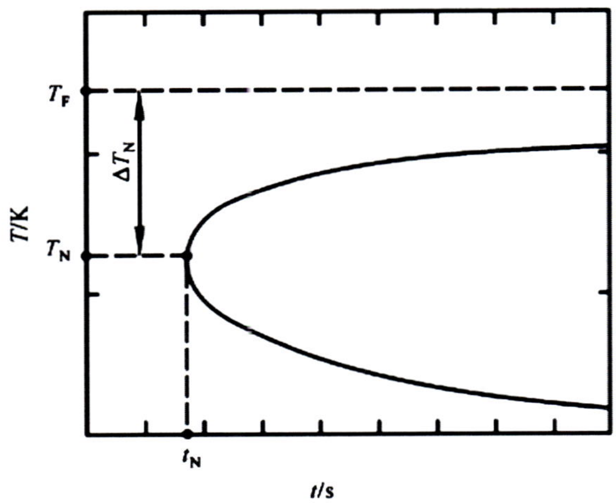

FIGURE 20.17 Map of temperature versus time to predict the beginning of crystallization (T_F—crystallization temperature, T_N—nucleation temperature).

Liu used the combination of two characteristic parameters, ΔH_f and MPAR, to predict the glass forming ability. Figure 20.18 is the map of δH_f—MPAR, in which experimental data of 54 binary alloys were summed up. It is concluded that:

(a) MPAR < 20% of the binary alloys is hard to be amorphized (HGF);

(b) MPAR > 20% of the binary alloys can be amorphized (PGF); and

(c) MPAR > 65% and $\delta H_f < 0$ of the binary alloys can be easily amorphized (RGF).

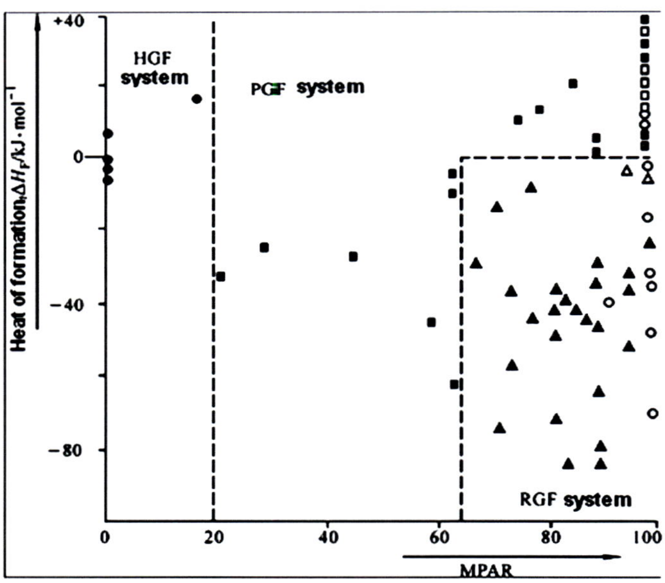

FIGURE 20.18 Map of δH_f—MPAR diagram [21], ▲—readily glass forming, ■—possibly glass forming, ●—hardly glass forming (the empty symbols denote those predicted alloy systems have not been proved experimentally) *B.X. Liu and O. Jin, Physica Status Solidi A – Applied Research, 161 (1997) 3.Copyright Wiley-VCH Verlag GmbH & Co. KGaA. Reproduced with permission.*

According to three glass-formation abilities, it can be classified into three categories: hardly glass-forming system (HGF), possibly glass-forming system (PGF), and readily glass-forming system (RGF) [20].

Liu et al. first pointed out the important function of interface in multilayer film and estimated the interface free energy. The calculation results indicate that for systems of $\Delta H_f > 0$, the interface energy is increased with the increase of atom fraction at the interface, which intersects the convex free energy curve of amorphous state. It means that the interacted composition parts on both sides of phase diagram are now possible to be amorphized. Continuing to increase the interface fraction, it is possible to make the initial energy state of multilayer completely over the amorphous curve, therefore the composition near the peak of the free energy curve can also transform into amorphous state, that is realized by experiments. For those systems having $\Delta H_f < 0$, the effect of interface energy on alloying is rather small, therefore thermodynamic models fitting for $\Delta H_f < 0$ and $\Delta H_f > 0$ binary alloy systems, that can predict the amorphization tendency by multilayer ion mixing [21].

Fan et al. [22,23] proposed a dialectical concept. In the preparation of amorphous alloy by mechanical alloying, the mechanical ball milling of amorphous ribbon can have two effects: (i) destruction of the short-range order and diminishing of the nuclei; and (ii) to create a new volume and accelerating the crystal growth. The competition between these two mechanisms governs the thermal stability of amorphous $Fe_{80}B_{20}$ alloy. It is proved that low-energy ball milling can improve the thermal stability. The crystallization temperature T_p, crystallization latent heat δH, and crystallization activation energy of $Fe_{80}B_{20}$ amorphous alloy are increased with the prolongation of milling time, as compiled in Table 20.3. By using mechanical alloying technique, $Ti_{100-x}Al_x$ alloy can be synthesized from Al powder (200 mesh, 99.9%)

TABLE 20.3 Influence of Low-energy Ball Milling on Crystallization Temperature, Latent Heat, and Activation Energy of $Fe_{80}B_{20}$ Amorphous Alloy [23]

Milling time (h)	T_p (K)	ΔH (cal/g)	E_x (eV)
0	751.8	128.3	2.78
5	752.4	128.0	2.81
10	754.3	135.7	2.95
20	755.4	140.9	3.01
40	755.6	143.7	3.06

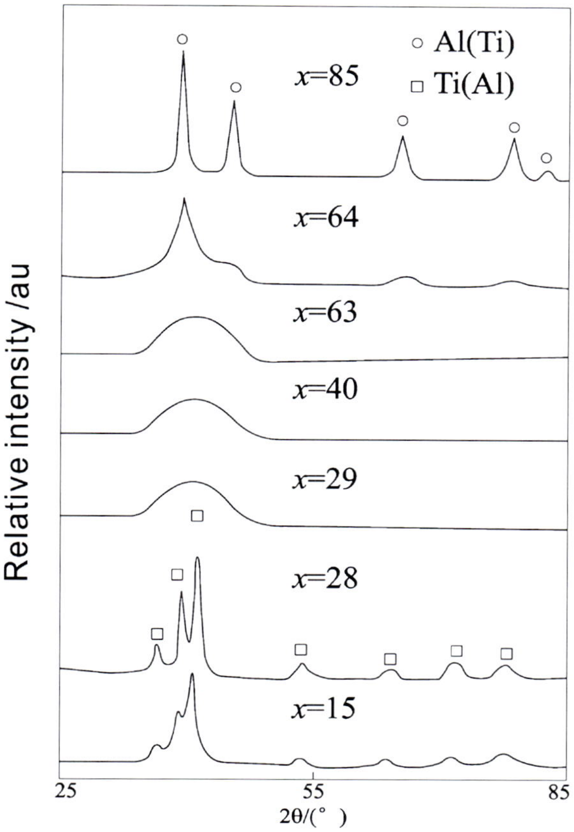

FIGURE 20.19 X-ray diffraction pattern of $Ti_{100-x}Al_x$ alloys produced by mechanical alloying. *Reprinted from Scripta Metallurgica et Materialia, Vol 32, G. J. Fan, M. X. Quan, Z. Q. Hu, Mechanically driven phase transformation from crystal to glass in Ti–Al binary system, Copyright (1995), with permission from Elsevier [24].*

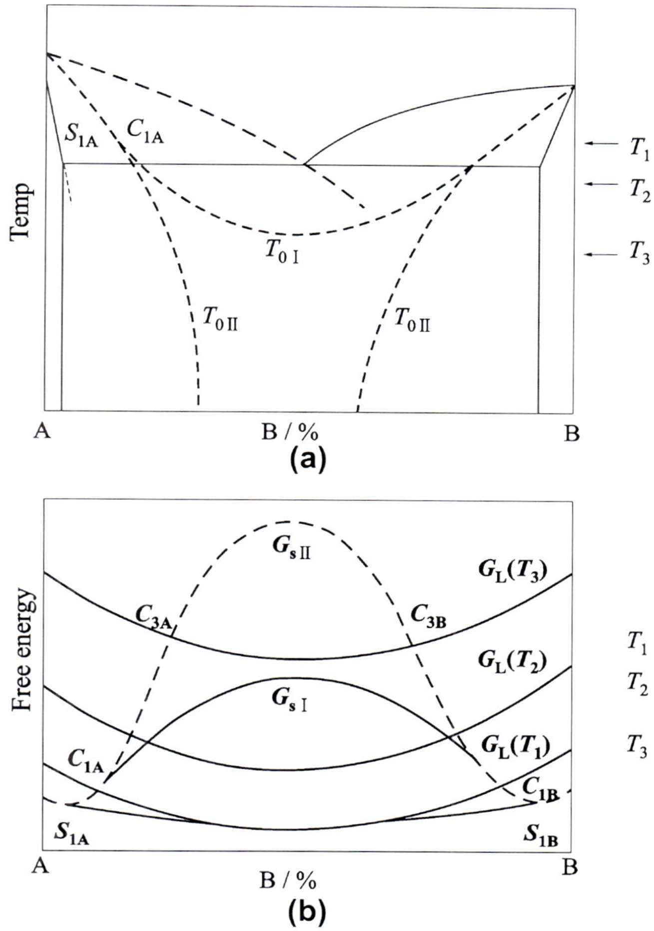

FIGURE 20.20 Effect of rapid solidification on T_o curve [26]. (a) Extension of T_o curve in the phase diagram and (b) different T_o from different G_s.

and Ti powder (360 mesh, 99.9%). It is found that at $x \geq 64\%$, Ti(Al)supersaturated solid solution is formed; at $x \leq 28\%$, Al (Ti) supersaturated solid solution is formed, only at x, in the range between 29% and 63%, the alloy can be amorphized as shown in Fig. 20.19 [24].

20.2.3. Criterion of Thermodynamic T_o Curve

An explicit analysis on thermodynamic T_o curve has been given by Baker and Cahn [25], which has important significance on the formation of amorphous alloy. In the binary phase diagram, T_o curve is the locus of equal value of liquid mole free energy (G_L) and solid mole free energy (G_S).

T_o curve always locates between the liquidus and solidus and is the maximum limit of liquid composition and temperature in diffusionless solidification. Therefore, T_o curve can be obtained by alloy thermodynamic calculation.

In rapid solidification, T_o curve can be extended like the extension of liquidus and solidus, as shown in Fig. 20.20 [26,27]. The T_o curve in Fig. 20.20(a) is the locus of the intersection of G_s and G_L in Fig. 20.20(b). Different G_s curves can get different T_o curves due to the movement of the intersection points like T_{oI} curve and T_{oII} curve. For T_{oI} curve, a continuous metastable solid solution can be produced after rapid solidification. On the contrary, it is impossible to get this continuous metastable solid solution in the case of T_{oII} curve.

Now we can discuss with some real examples. Figure 20.21(a) shows a continuous T_o curve of Ag–Cu system; therefore it can form α-continuous solid solution after rapid solidification [27]. There is a cross-intersection of two T_o curves in Fig. 20.21(b) like Al–Al_6Fe system, which shows that it cannot form a continuous solid solution. In Fig. 20.21(c) the two T_o curves neither intersect nor connect continuously, like $Pd_{73}Si_{12}Cu_{15}$–Pd_9Si_2 system. It indicates that the alloy cannot undergo partitionless solidification or diffusionless solidification. From Fig. 20.22 [28], it can be seen that it cannot undergo diffusionless solidification or partitionless solidification because the

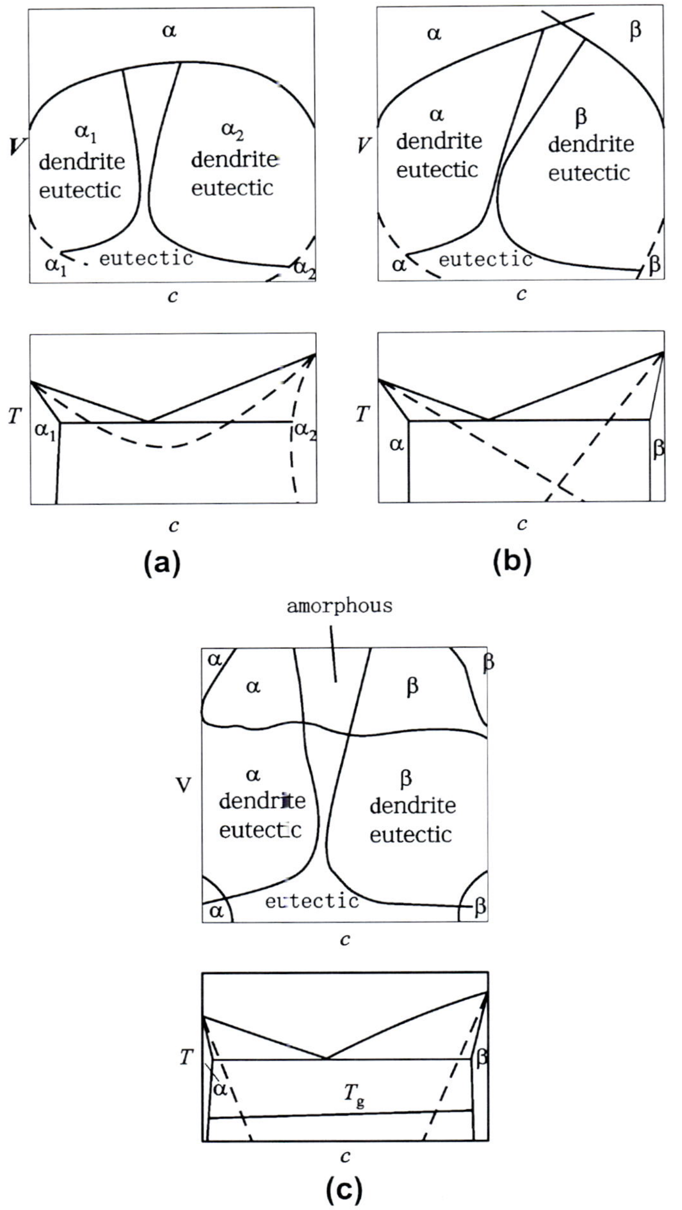

FIGURE 20.21 Three kinds of T_o curves [28].

alloy composition cannot meet the T_o curve (it cannot crystallize whether the curve of transformation from liquid to solid extends to a very low temperature). In this occasion, the alloy can be easily amorphized. Figure 20.21(c) shows that the alloy can turn into amorphous state after the solidification rate reaches to a certain value.

20.2.4. Formation of Amorphous Alloy Under High Pressure

Since the discovery of metallic glass, researchers and technicians are trying to produce it in a bulk in order to utilize its promising properties in real applications. One type of method is to increase the crystal energy that helps the crystalline alloy to transform spontaneously into amorphous at a temperature lower than the corresponding crystallization temperature. The solid reaction amorphization at high pressure is a typical example. High pressure induces formation of new high-energy metastable crystalline phase in which the energy is higher than the noncrystalline phase. Below the room temperature, it spontaneously transforms or dissociates into noncrystalline material. Another kind of method (vapor deposition, melt quenching, etc.) is to freeze the disorder state (such as liquid). The starting point of previous research is to let the atoms no time to move for long-range diffusion and retain original melt disorder state. However, to freeze the atoms needs a very high cooling rate that limits the melt quenching or vapor deposition to produce bulk metallic glass. High-pressure technology is an effective way to overcome the above-mentioned drawback. Due to the introduction of high pressure, the atomic spacing is decreased, atoms are hard to diffuse, not to have long-range diffusion, then it is possible to retain the whole block in a disorder state, and finally to form a bulk amorphous alloy.

Li et al. [29] studied the formation of Cu−Ti amorphous alloy. $Cu_{60}Ti_{40}$ alloy is cooled from 1573 K at a cooling rate of 300 K/s under 5.5 GPa, that is, basically transformed into amorphous state proved by X-ray diffraction. Figure 20.23 is the electron diffraction ring of amorphous $Cu_{60}Ti_{40}$ alloy. If the cooling rate drops to 50 K/s, only a little indication of the amorphous presence remains. It means the higher the cooling rate, the easier the high temperature disordered state to be retained When the $Cu_{60}Ti_{40}$ alloy is cooled from 1373 K instead of 1573 K at a cooling rate of 300 K/s under 5.5 GPa, a part of the sample can be amorphized. Where the amorphous halo is still clear, it becomes fully crystallized as it is cooled from 1473 K at a cooling rate of 300 K/s under 3 GPa. It indicates that the melt quenching under high pressure exists at a critical pressure. Below this critical pressure, it is impossible to get the amorphous alloy. For $Cu_{60}Ti_{40}$ alloy quenched at 300 K/s, the critical pressure is around 3−4 GPa. Cd−Sb system can also be amorphized by high-pressure melt quenching, as shown in Fig. 20.24 [30,31]. Heat the alloy to 670°C under 9 GPa, keep it for 5 min and quench to room temperature at a cooling rate of 10^2 K/s by liquid nitrogen. Figure 20.24(a) is the X-ray diffraction pattern of original crystalline $Cd_{43}Sb_{57}$ alloy, and Fig. 20.24(b−f) shows the presence of the high-pressure metastable phase after holding for 12 h, 24 h, 36 h, 48 h, and 60 h, respectively. Metastable γ-phase is formed after high-pressure melt quenching, which is a simple hexagonal structure with $a = 0.3182$ nm and $c = 0.2939$ nm. With

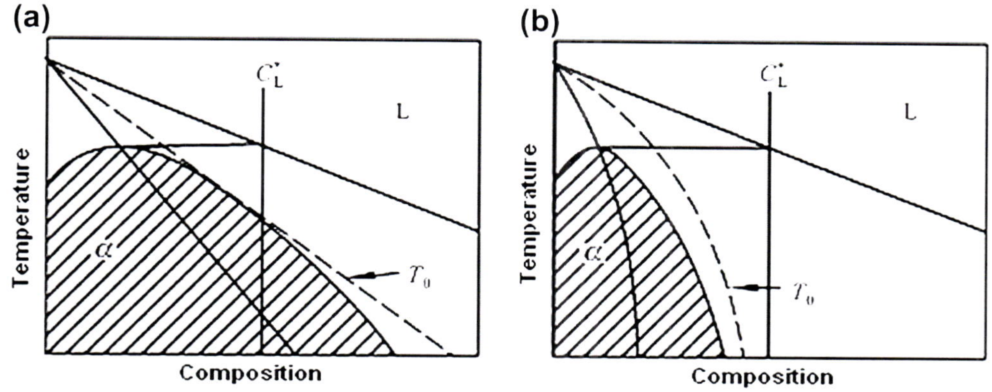

FIGURE 20.22 Requisite condition for partitionless solidification [28]. (a) If the composition cuts T_o curve, it cannot be amorphized. (b) If the composition cannot cut T_o curve, it can be amorphized.

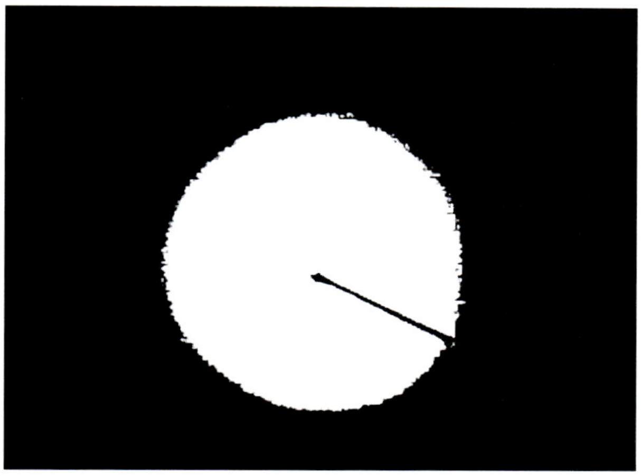

FIGURE 20.23 Electron diffraction pattern of amorphous $Cu_{60}Ti_{40}$ alloy [29].

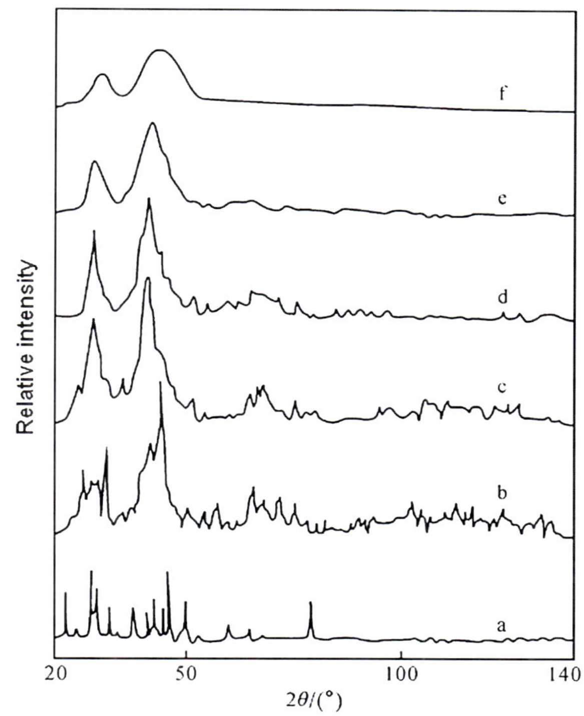

FIGURE 20.24 X-ray diffraction patterns (CuK_{α}) of high pressure metastable phase in $Cu_{43}Sb_{57}$ alloy [30,31]. a—crystal; b to f—high-pressure metastable phase at room temperature held for 12 h, 24 h, 36 h, 48 h, and 60 h, respectively.

the prolongation of holding time, metastable γ-phase disappears and fully transforms into amorphous phase (Fig. 20.24(f)).

The pressure has influence on melt viscosity and density [32]. η_0 is the melt viscosity at normal pressure, and the viscosity at pressure p is:

$$\eta(p) = \eta_0 \exp\left[(E + pVN_A)/kT\right], \tag{20.7}$$

where, E—activation energy of viscous flow; V—volume; N_A—Avogadro constant; k—Boltzmann constant; and T—temperature.

The melt viscosity is increased with the rise of pressure, and solid viscosity is also increased with the rise of pressure. However, the solid volume change is less than melt volume change with pressure. Although the solid viscosity at normal pressure is greater than the melt viscosity, the melt viscosity increases with increase in pressure much more than that of solid, therefore, there is an intersection of the two curves expressing the pressure effect on melt or solid viscosity. The intersection point is actually the critical pressure P_c. At the same time, the densities of melt and solid have similar trend. At normal pressure, the melt density is smaller than the solid density. The melt volume is expansible. When the pressure elevates, both the melt density and solid density increase, but the rate of melt density rise is faster than that of solid density rise.

The melting temperature decreases with the increase in pressure. When the slope of melting curve in $T-P$ map is negative, the solidification process involves volume expansion. Pressure suppresses the transformation from molten state to crystalline state and favors the disordered molten state to be retained at room temperature. The condition for high-pressure melt quenching to form amorphous is the system that has a negative slope of the melting curve. It must be higher than the critical pressure, that is, the melting temperature is decreased with the pressure rise. It means that the glass-forming ability depends upon the critical pressure. The system having large compressibility possesses good glass-forming ability but the system having low compressibility is hard to be amorphized due to the high critical pressure.

At high pressure, the glass-forming temperature T_g is closely related to the pressure [33].

$$T_g = T_g^*(E + W + pVN_A)/(E^* + W^*), \tag{20.8}$$

where T_g^* is the glass-forming temperature at normal pressure, W^* is the nucleation barrier at normal pressure, and W is the nucleation barrier at pressure p.

From Eq. (20.8), it can be roughly verified that T_g is increased with the pressure rise due to $W^* \ll E^*$

(1) For systems $\Delta V_f = V_L - V_s > 0$, $dT_m/dp > 0$, the melting temperature T_m increases with increase in pressure. Although T_g is enhanced with the pressure rise, the important GFA criterion T_g/T_m changes a little with the pressure variation.

(2) For systems $\Delta V_f = V_L - V_s < 0$, $dT_m/dp < 0$, the melting temperature T_m decreases with increase in pressure, while T_g is increased with the pressure rise. Both reasons make T_g/T_m rise, that reinforces the glass-forming ability.

It can be concluded that the systems possessing $\Delta V_f = V_L - V_s < 0$ can satisfy the glass formation by melt quenching under high pressure. The larger the $|\Delta V_f|$, the less the pressure required and the easier the glass formation.

20.3. PREPARATION TECHNOLOGY OF AMORPHOUS MATERIALS

The preparation technology of amorphous materials can be summarized as following:

1. melt quenching
   (1) melt quenching
   (2) atomization
   (3) laserglazing
2. large supercooling of pure melt
   (1) droplet emulsion method
   (2) flux method
   (3) drop tube method
3. physical and chemical vapor deposition
   (1) evaporation
   (2) sputtering
   (3) laser chemical vapor deposition
   (4) plasma-excited chemical vapor deposition
4. radiation
   (1) ion bombardment
   (2) electron bombardment
   (3) neutron bombardment
   (4) ion implantation
   (5) ion mixing
5. chemical methods
   (1) hydrogenation
   (2) electrode position
   (3) chemical plating
6. mechanical methods
   (1) high-energy milling
   (2) mechanical alloying
7. reaction
   (1) solid-state reaction
   (2) dissociation of solid solution
8. high pressure

20.3.1. Melt Quenching

The schematic diagram of melt quenching is shown in Fig. 20.25, in which (a), (b), (c), and (d) represent two-piston method, single-roller method, melt spinning

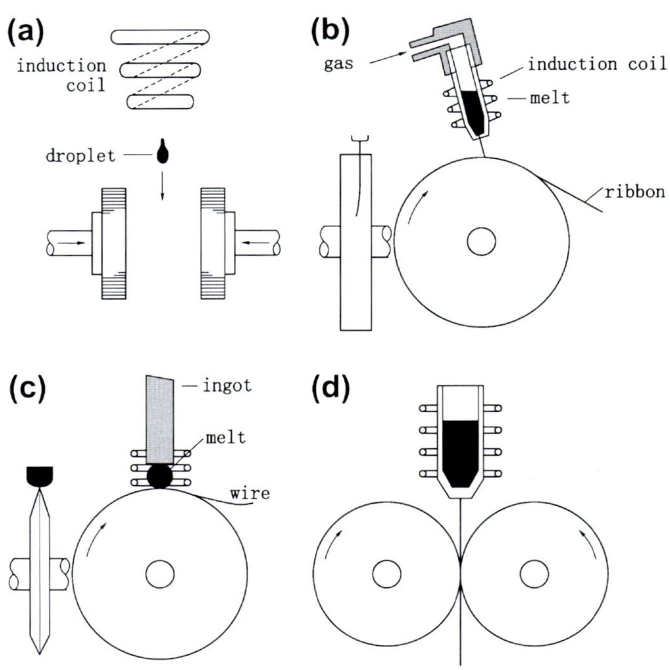

FIGURE 20.25 Schematic diagram of melt quenching. (a) two-piston method, (b) single-roller method, (c) melt spinning method, and (d) double-roller method.

method, and double-roller method, respectively. All the above-mentioned methods belong to a kind of rapid solidification that the hot melt directly smashes onto the cool metallic surface. By using this melt quenching, the melt stream can directly touch the outer metallic surface as well as to the inner metallic surface of the roller. This subdivides the single-roller method into two methods: one is free jet method in which the melt is freely sprayed on the roller surface rotated at high speed, and the other is planar flow method in which the puddle of melt flow directly touches the rotating roller when the melt jet nozzle is placed very close to the roller surface. The latter method can damp the perturbation of melt flow, improve the geometrical accuracy of ribbon size which in turn guarantees whole section of ribbon subjected to have same cooling rate in order to obtain homogeneous microstructure.

Based on heat convection mechanism, the interface heat transfer coefficient is an important parameter, reaching $10^6\ N\,(m^2\,K)^{-1}$, that can roughly estimate the moving velocity of solid/liquid interface to be $1\ m\,s^{-1}$. The cooling rate mainly depends on the ribbon thickness, as shown in Eq. (20.9). Generally, the cooling rate can be as high as 10^5–$10^6\ K\,s^{-1}$.

$$R = \frac{h(T - T)_0}{l\rho C}, \tag{20.9}$$

where R is cooling rate, h is interface heat transfer coefficient, T is melt temperature, T_0 is roller temperature, l is ribbon thickness, ρ is metal density, and C is metal heat capacity.

20.3.2. Atomization

Figure 20.26 is the schematic diagram of atomization. In the subsonic range, the melt stream converts into atomized droplets in order to overcome the low tangential drag. For those high property and easy oxidized material, a method of argon gas atomization is usually selected.

But the gas content is still high at about 10 or $20\ \mu g\,g^{-1}$. The cooling rate of this atomization is also not high, usually at 10^2–$10^3\ K\,s^{-1}$. The atomized powder quality is not good enough due to high gas porosity and low density. Moreover, the powder has satellite structure (small particles sticking on large particles) that makes the microstructure nonhomogeneous which makes the powder screening difficult and prone to gas contamination. Afterward, a new atomization technology, forced convection centrifugal atomization by helium gas, is developed which increases the cooling rate up to $10^5\ K\,s^{-1}$. The cooling rate can be elevated to an order of magnitude by helium gas instead of argon gas. Recently a new ultrasonic atomization is developed, which involves applying a pulsed helium or argon gas flow having a velocity of 2–2.5 Mach (680–850 $m\,s^{-1}$) and a frequency of 20,000–100,000 Hz directly on the melt stream to form very fine atomized powder particles. The principle is to use a cone nozzle of Hartmann shock wave tube, and the transmission of ultrasonic wave in liquid passes through by standing wave mechanism that creates compression and depression periodically. During depression, vacuum cavity is formed which is closed during compression and exerts a large shock wave at a level of a few megapascals which convert the melt stream into enormous droplets. The higher the frequency, the smaller the droplet size. The cooling rate can reach $10^5\ K\,s^{-1}$. Table 20.4 summarizes the cooling rate and powder quality of various atomization processes. The cooling rate can be calculated according to the following equation:

$$R = \frac{3h(T - T_0)}{r\rho C}, \tag{20.10}$$

where r is the droplet radius.

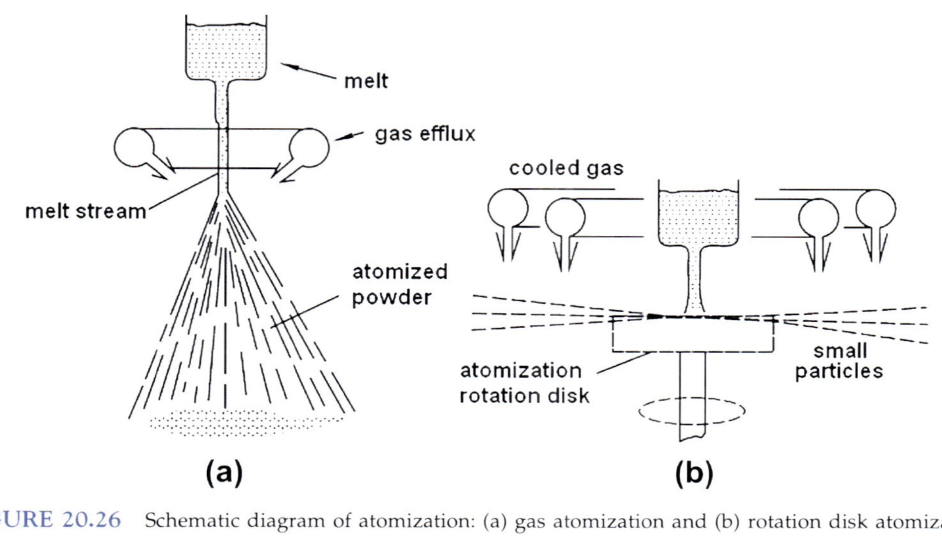

FIGURE 20.26 Schematic diagram of atomization: (a) gas atomization and (b) rotation disk atomization.

TABLE 20.4 Cooling Rate and Powder Quality of Various Atomization Processes

Process	Powder size (μm)	Average size (μm)	Cooling rate ($K\,s^{-1}$)	Gas entrapment	Powder quality
Subsonic atomization	1 to >500	50–70	10^0–10^2	Yes	Spherical having satellite
Ultrasonic atomization	1–250	20	10^4–10^5	No	Spherical with less satellite
Rotation electrode atomization	100–600	200	10	No	Spherical without satellite
Centrifugal atomization	1 to >500	70–80	10^5	No	Spherical with less satellite
Soluble gas atomization	1 to >500	40–70	10^2	No	Irregular having satellite
Electrohydrodynamic atomization	10^{-3}–40	10^{-1}–10^{-2}	10^7–10^8	No	Spherical without satellite
Electrosparking ablation atomization	10^{-3}–75	10^{-1}–10^{-2}	10^7–10^8	No	Spherical without satellite

20.3.3. Laserglazing

The schematic diagram of laserglazing is shown in Fig. 20.27. This technology uses laser beam of very high energy density (about 10^7 W cm^{-2}) and for a very short time period (10^{-3}–10^{-12} s) to interact with metal. Such high energy is enough to heat the metal surface to thousands of degrees to melt the surface metal, even to evaporate it. Afterward, the melt is rapidly quenched by the substrate in cold state with a cooling rate of 10^5–10^9 K s^{-1}, e.g., by using pulsed laser, the pulse energy is 100 J and pulse width is 2–8 ms, the peak power density can reach 400–1700 kW cm^{-2}. For a 2-kW continuous laser, the power density is about 70 kW cm^{-2}. The future direction is to shorten the pulse width from millisecond to picosecond. Besides, one can apply laser rotation lens for scanning, so that the scan width can increase to 20 mm or so.

There are two ways to increase cooling rate of laserglazing, one is to increase the absorbed heat flux density, and the other one is to shorten the interaction time. For example, by using 10^{-12} s pulsed laser, it can make silicon to be amorphous. Roughly speaking, if the absorbed heat flux density is increased 10 times, or the interaction time is decreased 100 times, it is equivalent to shortening the melt depth 10 times, enhancing the solidification rate 10 times, increasing the thermal gradient at liquid/solid interface 10 times, or increasing the cooling rate 100 times.

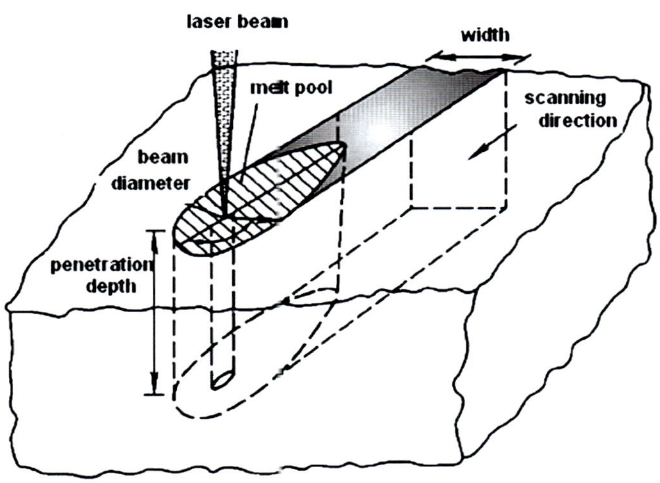

FIGURE 20.27 Schematic diagram of laserglazing.

When the laser absorption length is smaller than the heat diffusion length, the heat source can be imagined as a surface heat source, the cooling rate can be approximately calculated as:

$$R = \frac{(1 - R^*)\, I_0}{\rho C (2\alpha_s t_p)^{\frac{1}{2}}}, \tag{20.11}$$

where R^* is reflectivity of laser beam to metal surface; I_o is the output of laser power; α_s is the heat diffusivity; t_p is the pulse width.

When the laser absorption length is larger than the heat diffusion length, the influence of heat diffusion can be ignored. The cooling rate can be roughly estimated by the following equation:

$$R = \frac{2\alpha_s\,(1 - R^*)\, I_0 a^3 t_p}{\rho\, C}, \tag{20.12}$$

where a is the absorption coefficient.

Figure 20.28 demonstrates the relationship among cooling rate, surface melt depth, and absorbed power density in laserglazing [34]. Here, the rules are the same, i.e., the thinner the surface melt depth, the larger the cooling rate. The higher the absorbed power density, the larger the cooling rate.

It has to be emphasized that various laser beams interacting on a same material with various conditions can get different absorption coefficients (Fig. 20.29).

20.3.4. Emulsion Droplet Method

One way to increase the cooling rate is to increase the supercooling. The homogeneous nucleation needs larger

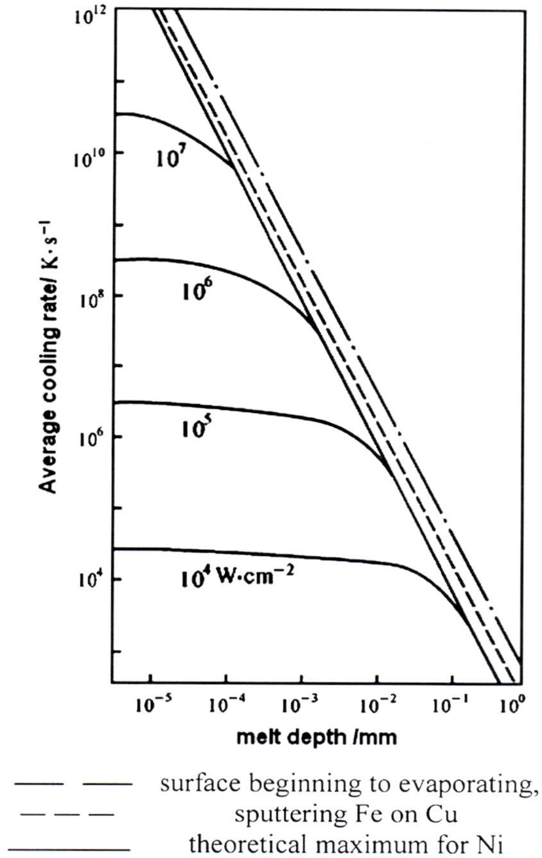

FIGURE 20.28 During laserglazing, the relationship among cooling rate, melt depth, and absorbed power density [34].

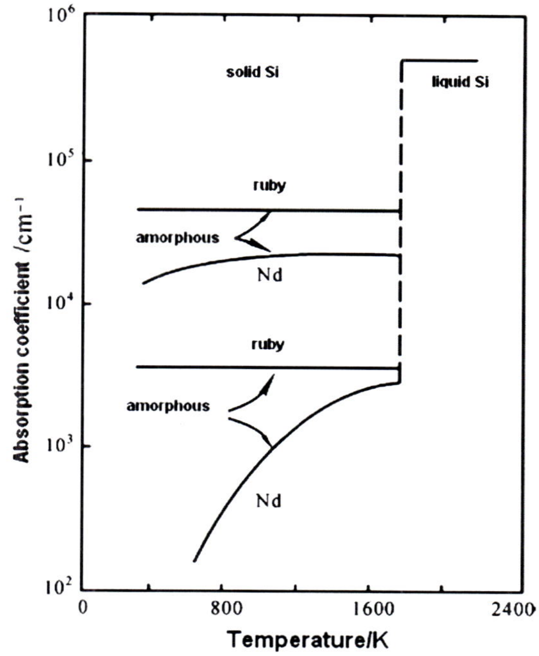

FIGURE 20.29 Absorption coefficient and temperature of amorphous, solid, and liquid Si laserglazed by ruby laser ($\lambda = 0.69$ μm) and Nd glass laser ($\lambda = 1.06$ μm) [35].

undercooling than heterogeneous nucleation. The maximum undercooling is usually about 20% of the absolute temperature of the melting temperature of metal. Using the emulsion droplet method it can dramatically increase up to 30–40%. The scientists hope that later it will be increased to two-third of the absolute temperature of the melting temperature of metal. Figure 20.30 is the schematic diagram of emulsion droplet method [36]. The droplet is dispersively distributed in a special solution. For high purity metal, only a limited number of droplets contain nucleants that can create very large undercooling. One of the important factors to define the undercooling is the particle size. For example, if the average diameter of Sn droplet is 275 μm, the undercooling is 48°C; and if the average diameter drops to 4 μm, the undercooling rises to 187°C.

Flemings [36] put Sn–Pb alloy in polyphenylether emulsifier, which has very high boiling temperature, about 520°C. In order to prevent the sticking of droplets, several drops of oxidizer were added (usually phthalic acid or organic peroxide). The additional amount is 0.05 g phthalic acid per 1 g metal. Heat the alloy to a temperature of 25°C higher than the melting

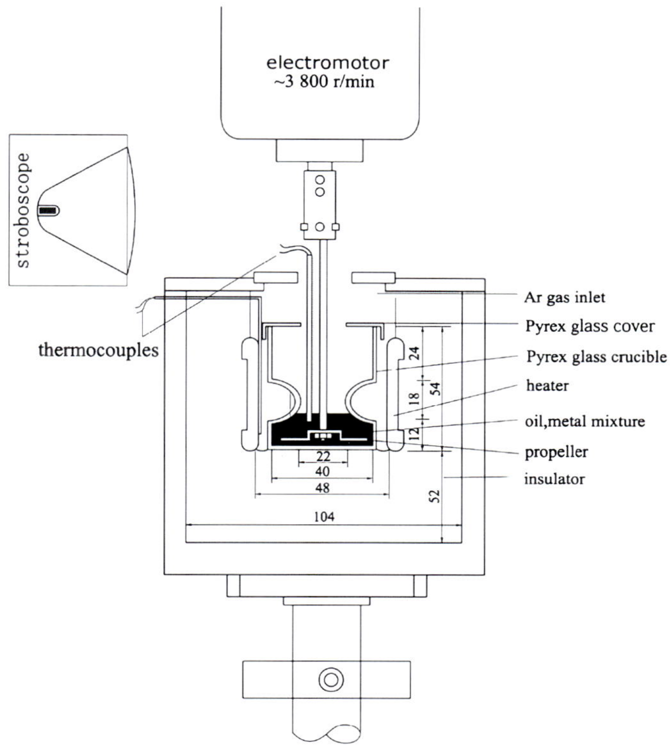

FIGURE 20.30 Schematic diagram of experimental setup for emulsion droplets to study undercooling of low-melting point metal in oil [36].

temperature. Mix for 40 min under argon atmosphere. The rotating speed is 4000 r min^{-1} that disintegrates molten alloy in carrier into 2–30 μm droplets; usually a ratio of 1 g metal and 5 mL emulsifier carrier can obtain 10^8 droplets with the diameter of 5–30 μm.

20.3.5. Mechanic Method

By using high-energy ball miller to study Se amorphous formation, it can be transformed into amorphous state after milling for 5 h. The purity of crystalline Se reaches to five nine that is milled under Ar protection at a rate of 0.8 mL s^{-1}. The mass ratio of ball and metal is 10. If the milling is carried out in a mixture of dry ice, methanol, and liquid nitrogen at temperature (-100 ± 5)°C, only 2 h is required to transform Se into amorphous as shown in Fig. 20.31 [37].

Shen et al. [38] mechanically ball milled Ge and S powders, which can also be transformed into $Ge_{1-x}S_x$ ($x = 0.61$, 0.67, and 0.72) semiconductor amorphous alloy. For Ni–Nb alloy at 1000°C, 3.5% Ni can be solid solutioned into Nb, while 4.2% Nb can be solid solutioned into Ni. After mechanical alloying, both can be extended to 10%, for Ni_xNb_{1-x} alloy ($0.24 < x < 0.79$) which can be transformed into amorphous by mechanical alloying [39,40]. Another system of Ni_xZr_{1-x} ($0.24 < x < 0.85$) can also be amorphized by mechanical alloying [41].

The following two factors can be considered as the driving force of amorphization: (i) the rapid elevation of free energy by moving the composition to nonstoichiometry; and (ii) increase of defect concentration. Another criterion fitted for thin-film diffusion couple method can be depicted. If we want two pure metals to be amorphized, there must be a very large negative mixing heat and a large difference between mutual diffusions which is also fitted for mechanical alloying. Some alloys first form intermetallic before amorphization as an intermediate product, such as $Nb_{75}Ge_{25}$ alloy and $Ni_{75}Sn_{25}$ alloy, respectively. They first form Nb_3Ge or Nb_3Sn with A15 structure, then transform into amorphous state. In the process of formation of $Cu_{71}Ni_{11}P_{18}$ ternary amorphous alloy by mechanically alloying, Cu, Ni, and P powders are refined and diffused mutually in the first milling stage, Cu_3P intermetallic is formed in the intermediate stage without the formation of Ni_3P as its activation energy is higher than Cu_3P. In the third stage, it forms $Cu_{71}Ni_{11}P_{18}$ amorphous alloy from Cu_3P and Ni.

Binary systems like Fe–Nb, Cu–Ti, and Ti–Fe having larger negative mixing enthalpy are able to be amorphized by ball milling method. The two requisites to produce amorphous alloys by mechanical alloying of the two metals mixed powders are: (i) there should be large negative mixing enthalpy between the two metals and (ii) one metal can quickly diffuse through another metal. The former provides the driving force for the amorphization reaction, and the latter guarantees the amorphization formation rate. Lou et al. [42] carried out an interesting experiment, in which they mechanically milled $Cu_{60}Ti_{40}$ and $Fe_{50}Nb_{50}$ amorphous alloy powders and quickly crystallized them into nanometer-sized solid solution. But if they mechanically milled the four metal powders with an atomic ratio of 38.4Cu–25.6Ti–18Fe–18Nb, it can be amorphous in the early milling stage because in the former experiment, the two amorphous alloys have already released the energy that makes the negative mixing enthalpy very small that is not beneficial to amorphization. However, in the latter experiment, the negative mixing enthalpies of the mixture of above four metals are

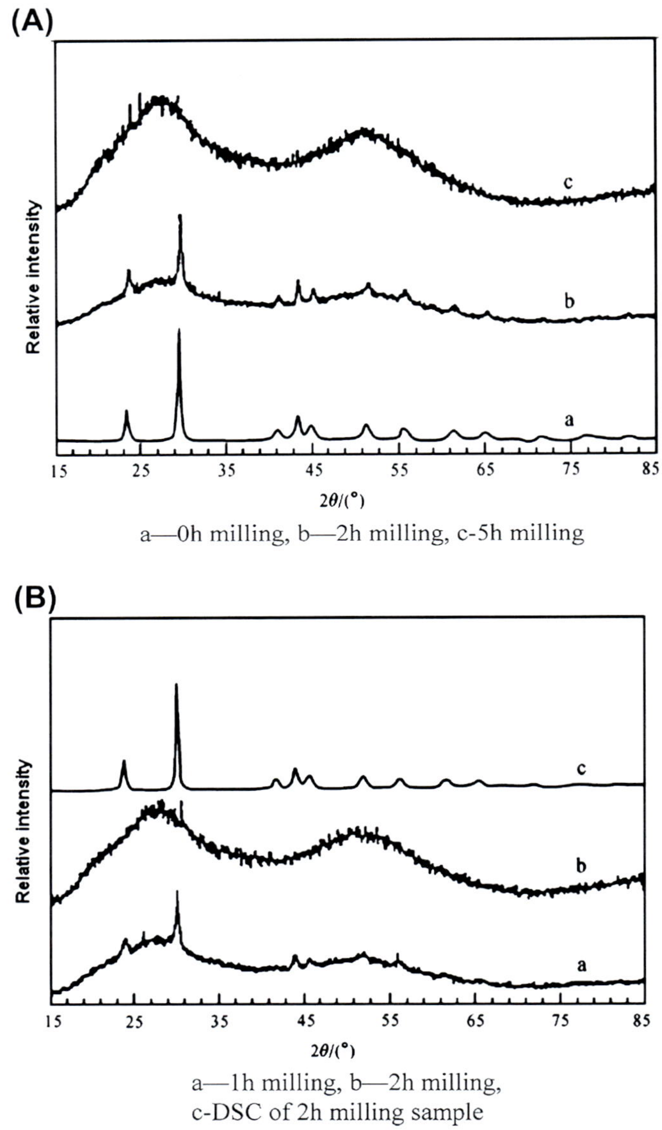

FIGURE 20.31 XRD of Se [37] milled (A) at room temperature and (B) at −100°C.

large, so as to have larger driving force which is beneficial for amorphization.

Similar to mechanical alloying method, the multiphase powder rolling method can be also applied for the amorphization, as shown in Fig. 20.32 [43]. For example, the Ag-70 at% Cu alloy begins amorphization only after seven passes reversely. After 30 times of reverse rolling, the amorphization becomes much evident in XRD examination. Shen et al. carried out reverse rolling Ni/Ti-enveloped powders and annealing with a constant heating rate, that can form amorphous alloy [44].

20.3.6. Solid-state Reaction Method

Besides melt rapid quenching to produce amorphous alloy, it is possible to form amorphous alloy by solid-state mutual diffusion reaction below melting point. For example, if Au and La multilayer film is annealed below 125°C, the product is amorphous alloy. The fundamental condition for forming amorphous alloy by solid-state reaction is that the free energy of noncrystalline state should be lower than the corresponding metastable crystalline state. It is the thermodynamical driving force of amorphization.

Comparison between solid-state reaction amorphization and melt rapid solidification deduces the following peculiarities:

(1) The composition range for amorphous formation becomes wider.

(2) It is not restricted by the melting points of system components and mutual solubilities.

The solid-state reaction method becomes frequently used method for producing amorphous alloys, because it cannot be limited by the cooling rate which gives the possibility to produce bulk amorphous alloys. The mechanism of amorphous formation is controlled by atom diffusion. The kinetic prerequisite of amorphous alloy formation is the existence of very large difference in mutual diffusion coefficients of the A—B system components. In other words, A metal component possesses a very high diffusion coefficient in B metal, which in turn has a low diffusion coefficient in A.

The main research interest on mutual diffusion of solid-state polycrystalline thin film is at first focused on the metallization of semiconductor integrated circuit. In order to enhance the stability and service life of semiconductor integrated circuit, it is necessary to understand the chemical reaction and mutual diffusion between the metallized films. It is different from bulk material, inside the solid-state polycrystalline thin film, wherein exist high-density grain boundaries and dislocations. It becomes very special that the mutual diffusion and chemical reaction can be evolved even at low temperature. In recent years, the study on metastable materials has become an active area of research. Using solid-state reaction and mutual diffusion of polycrystalline thin film to produce amorphous and other metastable materials has been studied enormously. Schwarz and Johnson [45] first discovered that after vacuum annealing of Au—La polycrystalline thin film, it can form metastable amorphous alloy via solid-state reaction which is not only used to produce amorphous alloys, but also can be used for modeling theoretically

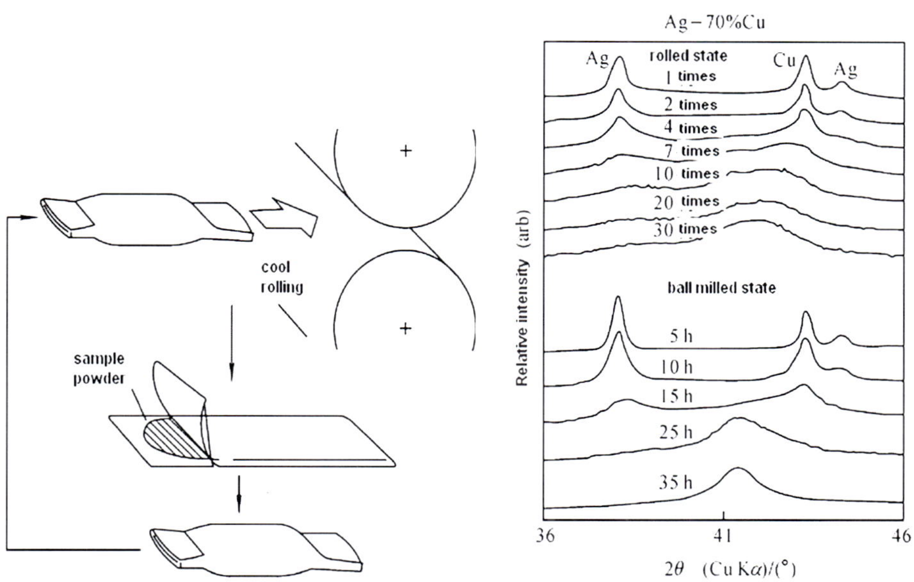

FIGURE 20.32 Reverse powder rolling method to amorphize Ag-70% Cu alloy and XRD patterns after repeated rolling times and various durations of ball milling [43] (a) reverse powder rolling method and (b) XRD patterns.

the thermodynamic conditions and kinetic process of the solid-state reaction (such as mechanical ball milling).

20.3.7. Radiation Method

Replacement of electron beam with laser beam can also be used to produce amorphous alloys. The electron beam reacts with nuclei and electrons of the substrate. The collision with nuclei is elastic. Same as laser radiation, the transfer of energy is rapid to the atomic lattice as heat energy. If the heating process can be approximately considered as quasi-adiabatic, the thermal conductivity effect can be omitted. The temperature distribution curve follows the electron energy loss curve of the substrate interior.

The energy maximum appears at a certain depth. The depth increases and the distribution width is broadened with the rise in electron energy. In laser radiation on a homogeneous medium, the maximum energy concentrated on the surface; in this case the surface structure is very sensitive to absorption and reflectivity.

Figure 20.33 shows the three main differences between electron beam and laser radiation during heating [46,47]. For Al, the main difference between ruby laser ($\lambda = 0.69$ μm) and electron beam (10–50 keV) is the cooling rate (see Fig.20.33(c)), which is due to the different energy absorption depth (see Fig.20.33(b)). For example, the 20 keV electron beam having 50 ns typical pulse width and about 1 μm energy absorption depth to radiate on Al, the quenching rate is mainly governed by the energy absorption depth. Such depth determines the thermal gradient in solid and heat dissipation time. It is similar to the solid cooling rate just after the solidification. To a specified material, the short pulse width does not always create a rapid cooling rate, only if the energy absorption depth is small. The energy absorption depth also governs the minimum melting depth caused by the pulse electron beam. The heat flow calculation gives the 2.6 μm melting depth, about 500 ns melting time, about 8 m s^{-1} resolidification interface rate, if the pulse is 50 ns and 1.5 J cm^{-2}.

The main differences between electron beam and laser radiation can be summarized as follows:

(1) The molten layer by electron beam radiation is thicker.
(2) The energy deposition range by electron beam radiation is larger than that by laser radiation.
(3) The liquid temperature after electron beam radiation is lower than that after laser radiation, therefore, the thermal gradient is also lower.
(4) The cooling rate by electron beam radiation is lower.
(5) The electron beam radiation has a lower regrowth rate, roughly an order of magnitude lower than the laser radiation.

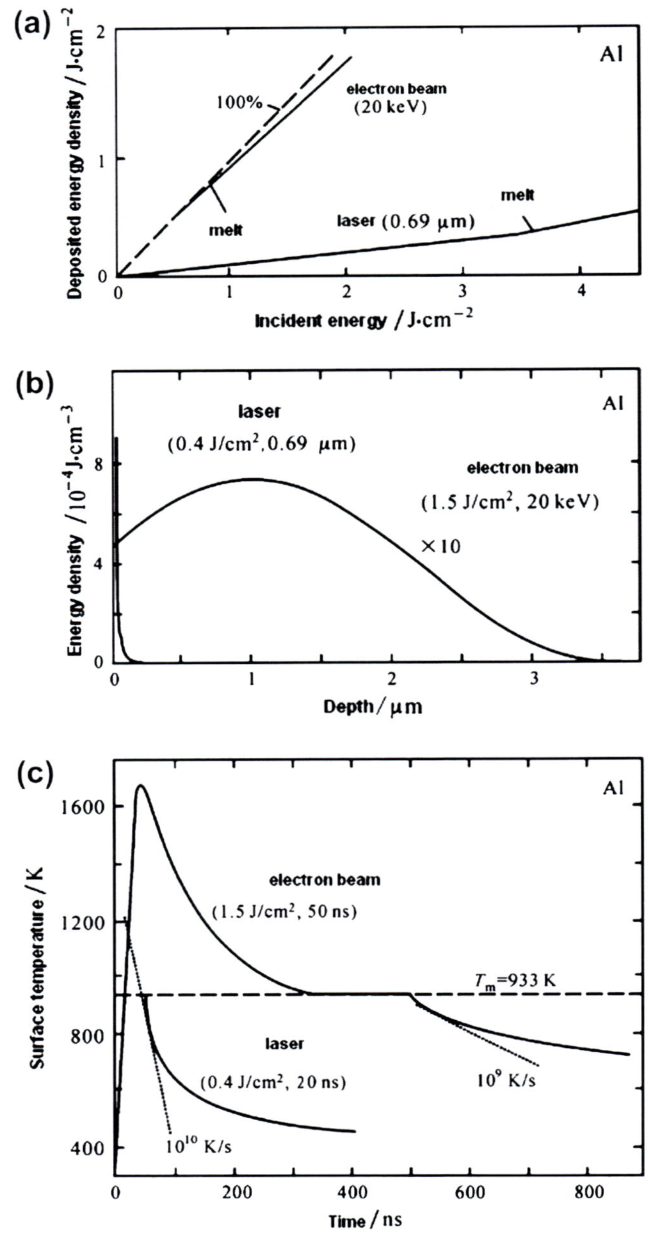

FIGURE 20.33 A comparison of pulse heating Al between electron beam (20 keV) and ruby laser [46–48]. (a) Absorbed energy; (b) distribution curve between absorbed energy density and depth; and (c) calculated surface temperature.

The ion beam radiation has been used for pulse heating of metal which has an energy absorption distribution curve qualitatively similar to that of electron beam radiation. Typically, during the initial few microseconds, the curve is approximately in a Gaussian distribution. Besides, both the electron beam and ion beam radiation deposit energy in the electron excitation and lattice excitation, but the ion beam energy entering lattice excitation has a larger percentage. On the contrary the laser

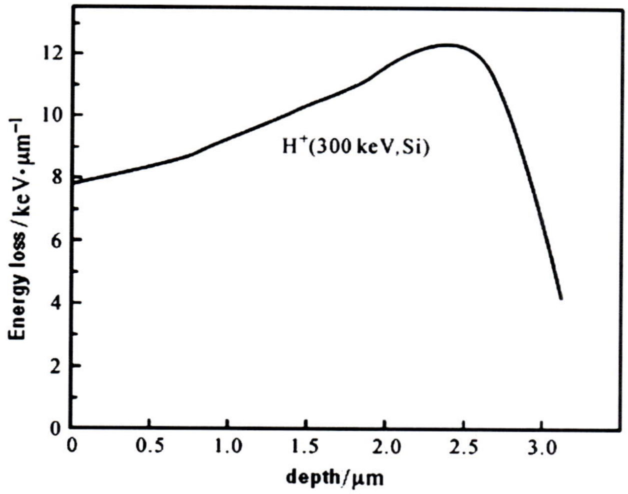

FIGURE 20.34 Energy distribution curve of 300 keV proton beam bombardment on Si [35].

energy first deposits in electrons initially, that has a maximum intensity at the surface, but decreases with the depth exponentially.

During ion beam radiation, ions transfer energy to nuclei and electrons. The transfer predominantly depends on the ion velocity, bombarding particles as well as atomic number and atomic weight. The ion beam radiation can deposit energy at the substrate surface in a short time that can make the surface melt [48].

By using 300 keV proton beam to bombard Si, the deposited energy distribution curve is shown in Fig. 20.34. In the range from sample surface to the depth of 2.5 μm, the deposited energy distributed homogeneously [35].

20.4. BULK AMORPHOUS ALLOY-BASED COMPOSITE MATERIALS

The bulk amorphous alloys have a large potential to be used as a structural material due to their peculiar properties of bulk amorphous alloys, such as high strength, high hardness, large elastic limit, good corrosion resistance, and superplasticity at high temperature. However, the room temperature ductility of the bulk amorphous alloy is limited to the formation and propagation of single band or only a few shear bands, and ruptures suddenly without any macroscopic plastic deformation. In order to improve the ductility of bulk amorphous alloy, shear band must be taken into account, as single shear band may govern the plastic deformation behavior of bulk amorphous alloy, to control the initiation and propagation of shear band and increase in number of shear bands can enhance the amorphous alloy ductility. Therefore, it is considered that the formation of amorphous alloy-based composite material by introducing nanocrystalline, microcrystalline ductile particles, or adding secondary phases (ceramic particles, fiber, or dendrite grains) can alter the geometrical constraint during deformation.

The interaction between the secondary particles and shear bands can obstruct the shear band propagation and induce the formation of multiple shear bands that can enhance the room temperature plastic deformation. The emergence of amorphous alloy matrix composite material gives a broad prospect of enlarged applications of amorphous alloy.

20.4.1. Preparation of Amorphous Alloy Matrix Composite Material

The amorphous alloy matrix composite material is prepared by introducing secondary phase into amorphous matrix to improve the comprehensive properties. The reinforcing particles can be in situ precipitated or added from outside, and can be nanosized or micrometer-sized.

1. Methods to add secondary particles or fibers
   (1) Liquid phase process (infiltration)
   Maintain the temperature above liquidus, mix thoroughly the particles into melt and quickly quench, for example, wire, steel wire, or carbon fiber-reinforced Zr-based amorphous matrix composite material can be produced by this process.
   (2) Powder metallurgy process
   Produce amorphous powders by atomization or mechanical alloying, mix thoroughly the amorphous powders, secondary-phase particles, and hot-extrude in the undercooled temperature range, but the as-prepared materials have some microporosity, low density, and low strength as compared with the material prepared by melt quenching. Now it is prepared by the combination of powder metallurgy process and melt quenching process, that is induction heat and melt the as-pressed slab, then cast into Cu mold. Using this new combined process amorphous Zr-based [49,50], Cu-based [51,52], and Mg-based [53,54] matrix composite materials reinforced with metallic particles (W, Nb, Ta, Mo, etc.), and ceramic particles (SiC, ZrC, TiC, TiB, WC, intermetallics, etc.) can be produced.
2. Methods to introduce in situ secondary particles
   (i) Partial crystallization or in situ precipitation of nanocrystals:

By isothermal annealing of the bulk amorphous alloy in the undercooled liquid region in order to produce partial crystalline phase, a nanocrystalline-reinforced amorphous matrix composite material.

(ii) In situ precipitation of micron-size secondary phase or dendrites:

By adding refractory metal Nb, Ta, etc. in Zr-based, Cu-based, or Ti-based amorphous alloys to precipitate β-Ti type particles or dendrites with bcc structure, it finally forms amorphous alloy-based composite material. For the Pd-based and La-based amorphous systems, the alloy compositions at deviated eutectics are selected and the composites are formed by in situ precipitation of dendrite phases.

20.4.2. Zr-based Amorphous Matrix Composite

1. Two-phase reaction dynamics

To select $Zr_{55}Al_{10}Ni_5Cu_{30}$ alloy having good glass formation ability and to study the interface between amorphous matrix and reinforcing phase are very important because the characteristics of this region dominate the load transfer and crack obstruction during deformation of the metal matrix composite material. However, the interface between crystalline metal matrix and ceramic phase has been studied since 1960, but the study on the interface between amorphous matrix and ceramic phase is not much studied, because the interest in bulk metallic glass started a few years ago, and only one point that can be generally accepted is to increase the wettability, to control the interface reaction, and to reduce the oxide formation, the interface strength, is to be optimized. The interface can effect mechanical adherence or chemical bonding between matrix and reinforcement. Only when the interface strength is higher than the surface tension of liquid metal, the wetting between solid and liquid is effective. The wettability can be obtained by measuring the contact angle.

Figure 20.35 is an SEM micrograph of $Zr_{55}Al_{10}Ni_5Cu_{30}$ amorphous alloy reinforced with 60 vol.% W fibers, which are uniformly distributed in the glassy matrix. The bonding quality between matrix and W fiber is influenced by many factors. If the criteria are the interface reaction and matrix amorphization, there are two important factors: one is infiltration temperature and the other is holding time during infiltration.

2. Dynamics of two-phase wetting

Figure 20.36 is the wetting dynamic curve between SiC and $Zr_{55}Al_{10}Ni_5Cu_{30}$ alloy. At 1173 K (m.p.

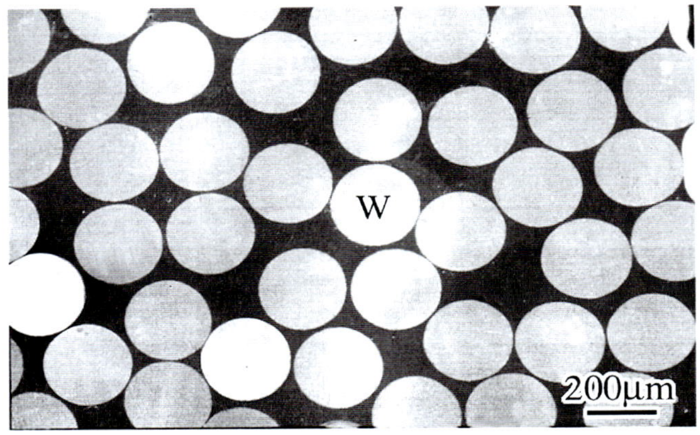

FIGURE 20.35 SEM micrograph of bulk amorphous matrix with 60 vol.% W fibers [49].

1108 K), it is only 65 K higher than the melting temperature, so the superheating is low, viscosity of alloy melt is high, flowability is bad, and the contact angle is as high as 74° at the equilibrium condition. When the temperature increases to 20 K, the wetting dynamic curve does not change much but the contact angle drops a bit. When the temperature is 1223 K, the dynamic curve changes to flat form and the superheating is 115 K and the time to reach equilibrium contact angle is rather longer. When the temperature is 1373 K, the liquid alloy possesses a very high superheating (265 K) that requires only a short time to reach equilibrium.

The relation between temperature and contact angle of a $Zr_{55}Al_{10}Ni_5Cu_{30}$ liquid alloy droplet on SiC substrate is demonstrated in Fig. 20.37. It can be summarized that the curve trend is similar, and for better wettability increasing temperature is quicker and better than prolonging time, for example, the

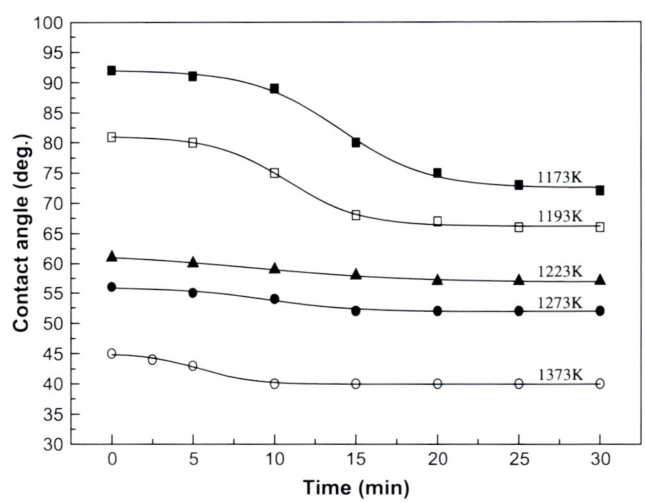

FIGURE 20.36 Relation between time and contact angle of $Zr_{55}Al_{10}Ni_5Cu_{30}$ alloy droplet on SiC substrate [49].

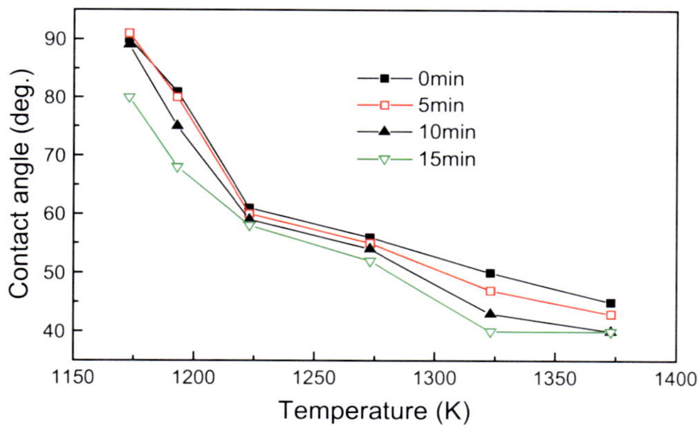

FIGURE 20.37 Relation between temperature and contact angle of $Zr_{55}Al_{10}Ni_5Cu_{30}$ alloy droplet on SiC substrate [49].

contact angle at 1323 K (15 min) and at 1373 K (10 min) is similar. In general, the higher the temperature, the shorter the time needed to reach equilibrium.

References

[1] R. Zallen, The Physics of Amorphous Solids, Wiley-Interscience Publication (1983).
[2] S.C. Moss, J.F. Graczyk, Physical Review Letters 23 (1969) 1167.
[3] Y. Waseda, T. Masumoto, Zeitschrift Fur Physik B – Condensed Matter 22 (1975) 121.
[4] J. Tauc, Amorphous and Liquid Semiconductors, Plenum (1974).
[5] D.T. Pierce, W.E. Spicer, Physical Review B 5 (1972) 3017.
[6] P. Duwez, Phase Stability in Metals and Alloys, New York, 1967.
[7] Z.Q. Hu, L.H. Wang, Y. Liu, Chinese Journal of Materials Research 12 (1988) 1.
[8] Z.H. Jin, A Theoretical Study of Liquid Metals and Alloys, Institute of Metal Research, Chinese Academy of Sciences, Shenyang, 1996.
[9] L.H. Wang, H.Z. Liu, K.Y. Chen, Z.Q. Hu, Physica B 239 (1997) 267.
[10] R.W. Cahn, B. Toloui, D. Akhtar, M. Thomas, Radiation damage in a nickel-zirconium (Ni64Zr36) glass, in: T. Masumoto, K. Suzuki (Eds.), Proceedings of 4th International Conference on Rapidly Quenched Metals, vol. 1, Japan Institute of the Metals, Sendai, 1982, p. 749.
[11] K. Yamada, Y. Iijima, K. Fukamichi, Journal of Materials Research 8 (1993) 2231.
[12] Y. Limoge, Acta Metallurgica et Materialia 38 (1990) 1733.
[13] H.G. Jiang, B.Z. Ding, J.T. Wang, Acta Metallurgica Sinica (English Edition), Series B 5 (1992) 255.
[14] H.G. Jiang, B.Z. Ding, H.Y. Tong, J.T. Wang, Z.Q. Hu, Materials Letters 17 (1993) 69.
[15] H.G. Jiang, B.Z. Ding, J.T. Wang, Z.Q. Hu, Chinese Science Bulletin 38 (1993) 1276.
[16] H. Jones, Rapid Solidification of Metals and Alloys, Institute of Metallurgists, Monograph Series (1982).
[17] M. Marcus, D. Turnbull, Materials Science and Engineering 23 (1976) 211.
[18] I.W. Donald, H.A. Davies, Journal of Non-Crystalline Solids 30 (1978) 77.
[19] B.X. Liu, E. Ma, J. Li, L.J. Huang, Nuclear Instruments & Methods in Physics Research Section B: Beam Interactions with Materials and Atoms 19-2 (1987) 682.
[20] B.X. Liu, Progress in Physics 13 (1993) 38.
[21] B.X. Liu, O. Jin, Physica Status Solidi A: Applied Research 161 (1997) 3.
[22] G.J. Fan, M.X. Quan, Z.Q. Hu, Journal of Applied Physics 80 (1996) 1.
[23] G.J. Fan, M.X. Quan, Z.Q. Hu, Applied Physics Letters 68 (1996) 319.
[24] G.J. Fan, M.X. Quan, Z.Q. Hu, Scripta Metallurgica et Materialia 32 (1995) 247.
[25] J.C. Baker, J.W. Cahn, in: "Solidification", ASM, Metals Park, Ohio, 1971.
[26] B.C. Giessen, R.H. Willem, Materials Science and Technology vol. III (1990).
[27] R. Mehrabian, International Metals Reviews 27 (1982) 186.
[28] W.J. Boettinger, Growth kinetic limitations during rapid solidification, in: B.H. Kear, B.C. Giessen, M. Cohen (Eds.), Rapidly Solidified Amorphous and Crystalline Alloys, vol. 8, Proc. Materials Research Society, 1982, p. 15.
[29] D.J. Li, J.T. Wang, B.Z. Ding, et al., Chinese Journal of High Pressure Physics 8 (1994) 74.
[30] Z.Q. Hu, W.Q. Guo, D.J. Li, et al., Nonequilibrium solidification of metastable materials and formation of metastable phases, 94 Autumn C-MRS, Part 2, Low-Dimension Materials, vol. 2, Chemical Industry Publishing House, Beijing, 1995, p. 1.
[31] D.J. Li, J.T. Wang, B.Z. Ding, Scripta Metallurgica et Materialia 26 (1992) 621.
[32] D.J. Li, Phase Transformation Mechanism of Metastable Phase under High Pressure, Institute of Metal Research, Chinese Academy of Sciences, Shenyang, 1993.
[33] V.M. Glazov, V.B. Koltsov, Zhurnal Fizicheskoi Khimii 55 (1981) 2759 (in Russian).
[34] E.M. Breinan, B.H. Kear, Rapid solidification laser processing of materials for control of microstructure and properties, in: R. Mehrabian, B.H. Kear, M. Cohen (Eds.), Rapid Solidification Processing, Principles and Technologies, Proceedings of International Conference on Rapid Solidification Processing, 1978, p. 87.
[35] J.M. Poate, C. Foi, D.C. Jacobson, Surface Modification and Alloying by Laser, Ion and Electron Beams, Plenum Press, 1983.
[36] M.C. Flemings, Experimental Program on Nucleation and Structure in Undercooled Melts, Materials Processing Center, Annual Report 1981, School of Engineering, MIT, 1981.
[37] G.J. Fan, F.Q. Guo, Z.Q. Hu, M.X. Quan, K. Lu, Physical Review B 55 (1997) 11010.
[38] T.D. Shen, K.Y. Wang, M.X. Quan, Z.Q. Hu, Applied Physics Letters 63 (1993) 1637.
[39] C.C. Koch, Materials Transactions JIM 36 (1995) 85.
[40] P.Y. Lee, C.C. Koch, Journal of Non-Crystalline Solids 94 (1987) 88.
[41] P.Y. Lee, C.C. Koch, Journal of Materials Science 23 (1988) 2837.
[42] T.P. Lou, F.S. Li, J. Hong, et al., Chinese Journal of Materials Science 12 (1998) 320.
[43] P.H. Shingu, K.N. Ishihara, Materials Transactions JIM 36 (1995) 96.
[44] T.D. Shen, M.X. Quan, J.T. Wang, Z.Q. Hu, Mater. Sci. Prog. 7 (1993) 189.
[45] R.B. Schwarz, W.L. Johnson, Physical Review Letters 51 (1983) 415.
[46] J.A. Knapp, D.M. Follstaedt, Pulsed electron beam melting of iron, in: B.R. Appleton, C.K. Celler (Eds.), Laser and Electron Beam Interactions with Solids, vol. 4, Proceedings of Materials Research Society, 1982, p. 407.
[47] P.S. Peercy, D.M. Follstaedt, S.T. Picraux, W.R. Wampler, Defects and aluminium antimonide precipitate nucleation in laser irradiated aluminium, in: B.R. Appleton, C.K. Celler (Eds.), Laser and Electron Beam Interactions with Solids, vol. 4, Proceedings of Materials Research Society, 1982, p. 401.
[48] R.T. Hodgson, J.E.E. Baglin, R. Pal, J.M. Neri, D.A. Hammer, Applied Physics Letters 37 (1980) 187.

[49] K.Q. Qiu, The Glass-Forming Ability, Composites and Properties of Bulk Metallic Glasses, Institute of Metal Research, Chinese Academy of Sciences, 2002.

[50] W.B. Dong, Zirconium Based Bulk Metallic Glasses and Their Composites, Institute of Metal Research, Chinese Academy of Sciences, Shenyang, 2006.

[51] J. Das, M.B. Tang, K.B. Kim, R. Theissmann, F. Baier, W.H. Wang, et al., Physical Review Letters 94 (2005).

[52] H.M. Fu, Cu-based Bulk Metallic Glasses and Their Composites, Institute of Metal Research, Chinese Academy of Sciences, Shenyang, 2006.

[53] D.G. Pan, Preparation and Properties of Mg-Based Bulk Metallic Glass and Composite, Institute of Metal Research, Chinese Academy of Sciences, Shenyang, 2007.

[54] W.R. Liu, Mg-Er Based Bulk Amorphous Alloys and Their Composites, Institute of Metal Research, Chinese Academy of Sciences, Shenyang, 2007.

CHAPTER

21

Synthetic Chemistry of Nanomaterials

Shi Zhang Qiao, Jian Liu, Gao Qing (Max) Lu

The University of Queensland, Australia

21.1. BASICS OF THE SYNTHETIC CHEMISTRY OF NANOMATERIALS

We are currently witnessing the advancement and development of a new interdisciplinary scientific field—Nanoscience. The synthesis of nanoscale materials, the foundation of nanoscience and nanotechnology, has become one of the most popular research topics in recent years. Nanoscale materials, or so-called nanomaterials, can be defined as materials with at least one dimension that in the nanometer scale (1 nm = 10^{-9} m), and with its properties influenced by this dimensional confinement [1–4]. Nanomaterials possess unique chemical and physical properties such as size, surface, and interface effect, compared to their corresponding bulk or isolated atoms and molecules. The technological importance of these nanomaterials is well demonstrated in innumerous applications, including their use in catalysis, biotechnology, photonic, and energy-storage [5–30]. Relative to their bulk counterparts, the thermodynamic, dynamic, mechanical, optical, electronic, magnetic, and chemical properties of nanomaterials can be significantly altered [31–56]. These properties are dependent not only on size but also on morphology and spatial organization [4–6]. For example, the color or absorption spectrum changes dramatically with size when the size is small, compared to the de Broglie wavelength or the Bohr exciton radius of the electron. When a metal particle such as gold is smaller than 10 nm, it essentially exists in a state that is neither liquid nor solid. The shape of the particle is in a constant state of flux. When a common liquid such as water is confined to a space that is only a few nanometers in dimension, its properties are significantly different from those of the liquid water and solid ice that we are familiar with [6]. Therefore, in addition to the tremendous potential in various applications of nanomaterials, the fascinating properties of nanoscale materials will remain a strong motivation for scientific discovery and exploration.

The synthesis of nanomaterials is one of the most active fields in nanoscience and nanotechnology [12–56]. It is clear that in order for nanomaterials to realize their full potential, evolutional progress and revolutionary breakthroughs are needed in the synthesis of nanomaterials, and in the understanding of the fundamental properties of such materials. Nanochemistry is a science to utilize synthetic chemistry to make nanoscale building blocks of different size and shape, composition and surface structure, charge, and functionality [3,4]. Nanochemistry aims to extend the traditional length scales of synthetic chemistry and exploit the collective properties of organized assemblies. There are numerous methods for synthesizing nanomaterials with various characteristics. Typically, two drastically different approaches are involved: the top-down approach and the bottom-up approach. The bottom-up approach involving self-assembly of molecular species, with controllable chemical reactions, is more efficient and flexible. Nanomaterials with well-controlled shape, surface properties, and developed porous structures can be conveniently prepared through such chemical synthesis techniques—bottom-up method [3,4,40]. Synthesis of nanoparticles with narrow size distributions has allowed the construction of a superlattice or artificial atomic structures using different assembly techniques.

Structurally organized porous nanomaterials and zero-, one-, two-, and three-dimensional (0D, 1D, 2D and 3D) nanomaterials have attracted much attention for emerging applications (e.g., catalysis, storage and controlled release systems, smart fillers, and biotechnologies) and become a focal area in nanochemistry [24–30]. To constitute a complete toolbox for the "top-down" and "bottom-up" approaches in nanotechnology, the ability of making functional materials into various

Modern Inorganic Synthetic Chemistry, DOI: 10.1016/B978-0-444-53599-3.10021-6

artificial nanokits and nanoparts is essential [4,31–56]. This chapter will mainly focus on the synthetic chemistry of nanomaterials. The nanomaterials to be covered in this chapter include nanocrystals, nanowires, hollow nanostructures, and core–shell nanomaterials. This chapter is aimed at early career researchers who require nanofabrication for their research, or who are interested in learning how to fabricate nanostructures in practice. It will cover many of the techniques employed in the nanofabrication of hard materials (semiconductors, metals, and ceramics) and soft materials (polymers and polymer–biological), with reference to applications in biotechnology, catalysis, sensors, clean energy, nanomedcine, lithium-batteries, nanophotonics, and nanoelectronics. A key aim of the chapter is to assist researchers in choosing the appropriate nanofabrication techniques for their particular project, as well as information on how to access these capabilities. This chapter is organized as follows. In the first section we systematically survey the synthetic approaches for nanomaterials, and the available strategies are broadly categorized into two themes, top-down, and bottom-up approaches, and each includes several well-established methods. By choosing some typical references as examples, we will present a comprehensive overview of synthetic strategies for nanomaterials. We then explicitly discuss a number of strategies that are being developed for synthesis of the nanocrystals, nanowires, hollow nanostructures, and core–shell nanostructures. The novel properties and applications of these nanostructures will be surveyed. Finally, we will present our thoughts on the future development of the synthetic chemistry of nanomaterials. It is the intention of this contribution to provide a brief account of these research activities and some of our recent works in this research area.

21.2. SYNTHETIC METHOD FOR NANOMATERIALS

The synthesis of nanomaterials is an essential component of nanoscience and nanotechnology. Advances in this field largely depend on the ability to synthesize nanomaterials with different components, sizes, and morphologies [7,8]. It is well recognized that properties of metal nanoparticles highly depend on the size, shape, composition, structure, and crystallinity [32–37]. Therefore, various approaches have been developed to control these parameters and, hence, meet the requirements for various applications. Although nanomaterials can be generated by physical (top-down) approaches [20,22] (such as evaporation and laser ablation), the chemical (bottom-up) approach is the more preferred and efficient method for fabricating a wide variety of nanoparticles. The most important nanomaterials synthesis method includes nanolithography techniques [21,23], templated-directed synthesis [25–31], the vapor-phase method [38,55], vapor–liquid–solid (VLS) methods [38,40], solution–liquid–solid (SLS) methods [38,40], the sol–gel process [56–58], micelle and emulsion [51–55], vapor deposition [34–36], hydrothermal and solvothermal methods [47,48], self-assembly [59], layer-by-layer assembly [60–63], spin-coating, spraying, and some other special synthesis methods [49,64–68] as shown in Fig. 21.1. It is generally accepted that photolithography, which includes chemical vapor deposition (CVD) or metal oxide CVD (MOCVD), is the most popular method used in the semiconductor industries, which has already been reviewed in Chapter 7. The bottom-up synthesis methods of thermal decomposition, hydrothermal, and solvothermal synthesis have been reviewed in Chapters 2 and 4. The template method, widely used in the synthesis of nanoporous materials, has been introduced in Chapter 16. The special synthesis methods, including microwave irradiation, photochemical synthesis, and bioinspired synthesis, are mentioned and illustrated in Chapters 6 and 23, respectively. The wide array of synthesis techniques pursued to prepare nanomaterials is well beyond the scope of this chapter. We will therefore focus on a few types of routes that are being frequently used to prepare nanomaterials.

21.2.1. Top-down Methods

Top-down strategies involve two categories: (1) using macroscopic tools to transfer a computer-generated

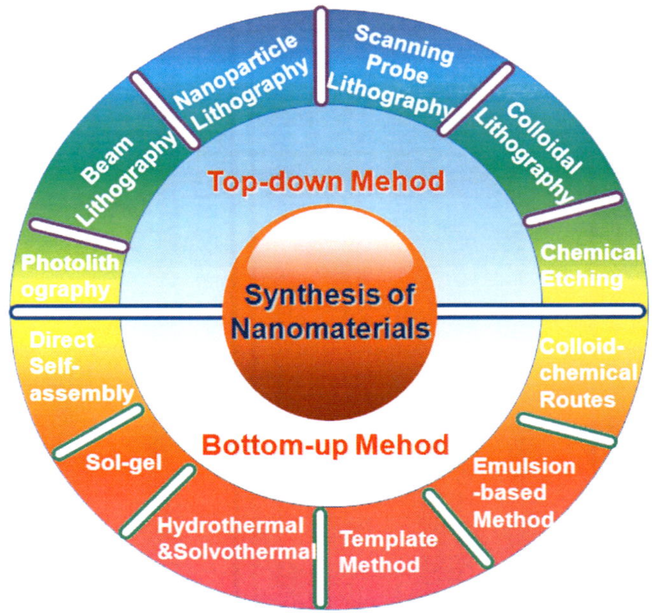

FIGURE 21.1 (a) Schematic illustration of the synthesis method of nanomaterials.

pattern onto a larger piece of bulk material first, and then "sculpting" a nanostructure by physically removing material (e.g., through wet/dry etching); and (2) using macroscopic tools to directly add/rearrange materials on a substrate. In the first category, the most common techniques are based on photolithography, which is cost-effective and relatively fast, but its resolution is ultimately limited by optical diffraction effects to typically 0.2–0.5 μm. Electron and ion-based lithographic methods, in contrast, permit the creation of ordered nanostructured arrays with high resolution (i.e., 50 nm features and/or spacing) and allow very good control over particle shape and spacing; their throughput, however, is limited and line-by-line pattern generation (a serial technique) is considered slow when compared with a parallel technique (such as photolithography), in which the entire surface is simultaneously patterned at once. The second category of top-down approaches includes scanning probe lithographic (SPL) techniques (e.g., dip-pen nanolithography (DPN) and scanning tunneling microscopy (STM)), microcontact printing (μCP), and nanoimprint lithography (NIL). Currently, patterns generated using DPN can be as small as 15 nm, whereas STM offers the unparalleled capability to position individual atoms to pattern structures with ultrahigh, subnanometer precision. However, DPN and STM are also serial techniques and are therefore not suitable for high-volume manufacturing technologies, although this is a drawback that may eventually be overcome by the introduction of massively parallel microfabricated probe tip arrays.

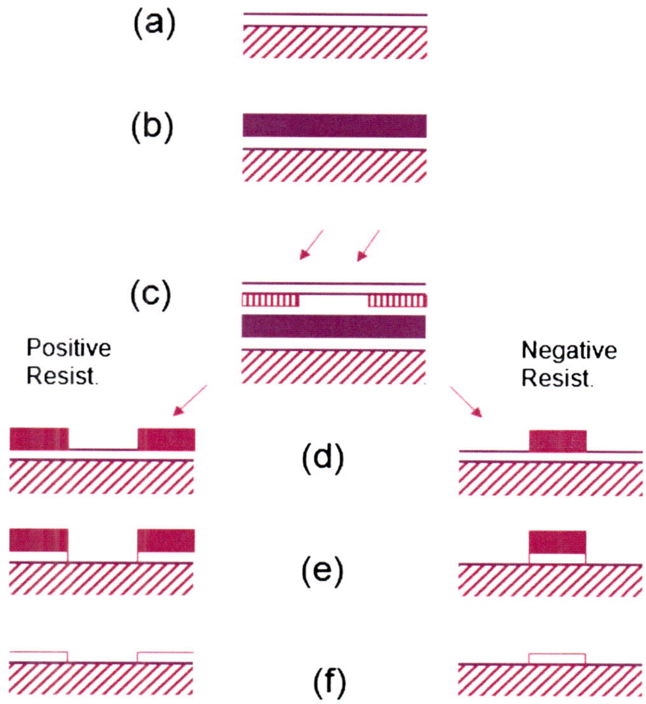

FIGURE 21.2 Schematic illustration of lithographic techniques to prepare nanomaterials in various feature sizes.

21.2.1.1. *Photolithography*

Photolithography is the basic technique used to define the shape of nanomaterials [6,9,10]. The technique is essentially the same as that used in the microelectronics industry. Figure 21.2 lists lithographic techniques reported to produce various feature sizes such as a thin film of some materials (e.g., SiO_2) on the substrate of other materials (e.g., a silicon wafer). It is desirable that the thin film can be selectively removed so that it only remains in particular areas on the silicon wafer (Fig. 21.2f). In this technique, first a mask is prepared. This is typically a chromium pattern on a glass plate. The wafer is then coated with a polymer which is sensitive to ultraviolet light (Fig. 21.2b), it is so-called a photoresist. Ultraviolet light is then shone through the mask onto the photoresist (Fig. 21.2c). The photoresist is then developed which transfers the pattern on the mask to the photoresist layer (Fig. 21.2d). Micropatterns can be readily produced using well-known techniques. Most reported methods of micropatterns include photolithography and micromachining. These techniques can produce micro features as small as possible. Recent advances in X-ray, electron- and ion-beam lithography, nanoparticle lithography, soft lithography, colloidal lithography, and microcontact printing (μCP) and nanocontact printing have broken the wavelength barrier to prepare patterns as small as 300 nm. To distinguish between micro and nanolithography, techniques capable of producing feature sizes of 100 nm or smaller are defined as nanolithography.

21.2.1.2. *X-ray, Electron- and Ion-beam Lithography*

In conventional photolithography, a projection configuration is used, where a collimated light passes through a mask and illuminates the resist surfaces [9]. Depending on the optics, various magnifications could be projected onto the resists. There are, however, no ideal materials available for X-ray projection. Therefore, shadow mask techniques were developed, which consist of a thin membrane of SiC, or Si, and heavy metals such as Ta, W, or Au to absorb X-ray photons. These masks are produced using microfabrication/processing and electron-beam lithography.

Figure 21.3 illustrates the fundamental approaches in electron- and ion-beam lithography. In electron-beam lithography, a focused electron-beam is achieved by converting commercial SEM or TEM instruments, which operate at 50–100 kV [27–29]. The electron-beam has very high energy, and is focused on a tiny spot (a few nanometers). The spot can be moved by the electron

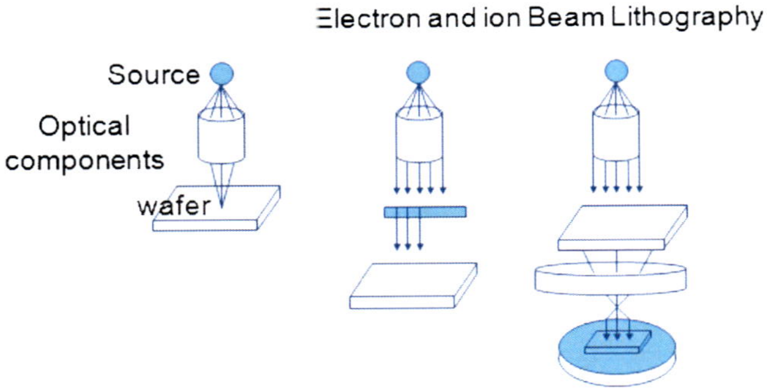

FIGURE 21.3 Three basic approaches adopted by electron- and ion-beam lithography and subsequent methods.

optics following a programmed trajectory. The resolution is not determined by the spot size, but rather by electron scattering, which impacts both the exposed and nearby areas (proximity effect). The correction and improvement of proximity effect has been intensely studied.

21.2.1.3. *Nanoparticle Lithography*

To produce periodic arrays of nanostructures, a relatively inexpensive and simple method of nanoparticle (or nanosphere) lithography has attracted much attention [21,23,69]. Monodispersed latex particles are first synthesized and purified. Particles can assemble into ordered and closely packed 2D and 3D structures under carefully controlled conditions. These assemblies are used as templates, where the void space is filled with the materials of interest. Particles can then be removed by either calcination or solvent dissolution. Arrays of nanostructures are produced, which include 2D and 3D arrays of metals, metal oxides and silica, porous membranes of polyurethane, as well as 3D opal or reversed opal structures of photonic band-gap materials such as SiO_2 and TiO_2.

21.2.1.4. *Scanning Probe Lithography*

Scanning probe microscopy such as scanning tunneling microscope (STM) and atomic focal microscopy (AFM) are well known for their capability to visualize surfaces of materials with the highest spatial resolution. In terms of structural characterizations, AFM can attain a lateral resolution of 0.1 Å and vertical resolution of 0.05 Å [10]. STM resolution is intrinsically five times higher than that of AFM. These high resolutions are reached for crystalline systems. For noncrystalline surfaces or soft-and-sticky surfaces, it is more difficult to achieve high resolution. AFM-based lithography is a very active area of research because of the flexibility and simplicity of the technique. AFM tips may be used to carry catalysts to selectively induce surface reactions, or as a pen to attach molecules on surfaces in DPN and its derivative techniques. AFM tips may also be used as an electrode to direct local oxidation on surfaces.

21.2.1.5. *Colloidal and Soft Lithography*

Colloidal lithography provides another inexpensive method for creating nanoscale topographies. This technique allows for the production of surfaces with controlled heights and diameters. Colloidal lithography involves the use of nanocolloids as an etching mask [21,23,70]. These nanocolloids are dispersed as a monolayer and are electrostatically self-assembled over a surface. Directed reactive ion-beam bombardment or film evaporation can then be used to etch the area surrounding the nanocolloids, as well as the nanocolloids themselves.

Self-assembled monolayers (SAMs) for patterning and shaping materials over different length scales and dimensions were first developed by Whitesides. SAMs embrace concepts of self-assembly, templating, and crystal engineering, together with soft lithographic techniques of microcontact printing and micromolding. It is an alternative low-cost, nonphotolithographic. SAMs provides access to topologically complex 3D objects for building micro- or nanomachines. Chemists are interested in experiment with reliable, convenient, and inexpensive methods for patterning planar or cured surfaces with organic, inorganic, polymer, liquid crystal, ceramic, or biological structures. μCP uses a polydimethylsiloxane (PDMS) stamp to "print" a chemical "ink" on a target substrate. This technique has been widely used to pattern SAMs on gold because alkanethiol molecules can be transferred from a PDMS stamp to a gold surface to form patterns of SAMs (Fig. 21.4). Using this technique, nanoscale patterns of SAMs are printed and the patterned molecules are used along with chemical etching to generate metallic nanostructures. SAMs can act as an etching resist to protect the underlying gold surface from chemical etching. Thus, when a gold film patterned with SAMs is placed into solution of cyanide-free gold etch (Transene), the exposed areas of gold are removed to reveal a pattern resembling the relief features of the PDMS stamp (Fig. 21.4, top). Archival-grade gold CD-Rs were used to produce more complex metallic nanostructures. Using a PDMS stamp from an Aluminum master, some researchers printed SAMs on the gold-CD surface. Molecules were transferred to the gold-CD only in the regions where the lines of the stamp were in contact with the lines on the gold-CD. When the lines on the stamp were nearly perpendicular to the lines of the CD, a checkerboard pattern of SAMs was produced. After students placed this substrate into the gold etching, the gold regions not protected by the SAM were etched away to reveal a checkerboard gold pattern (Fig. 21.4, bottom).

21.2.1.6. *Chemical Etching*

Chemical etching is a means of producing nanoscale features on the surface of a material by soaking it in an

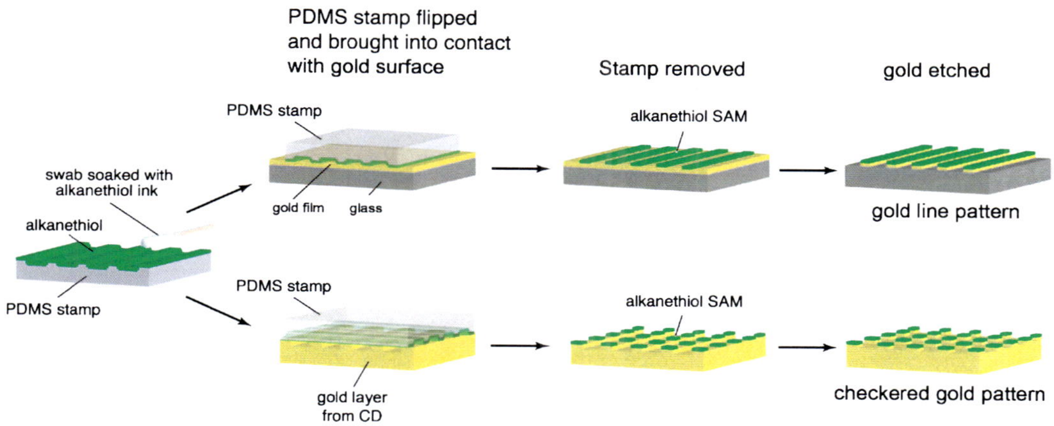

FIGURE 21.4 Procedure for micro-contact printing and etching of a gold film (top) and a gold-CD (bottom).

etchant. Typical etchants are hydrofluoric acid (HF) and sodium hydroxide (NaOH). As the material is etched away, the surface is roughened creating pits and protrusions at the nanometer scale. This process is essentially a surface treatment and cannot create structures with any prescribed geometry or organization. It can, however, provide a very quick, easy, and inexpensive means of creating a nanostructured surface by changing the scale of the roughness on the material surface.

Silicon wafers can be etched using HF to create nanometer-scale roughness on the surface of the wafer. Etching time can control the roughness with longer etching times leading to rougher surfaces (i.e., smaller surface pits and protrusions). Very recently, Yin and coworkers [71–74] described the formation of hollow silica nanoparticles by reacting silica particles with $NaBH_4$ or NaOH. Polyvinyl pyrrolidone (PVP) or polyacrylic acid (PAA) is selected as the surface-protecting agent because its carbonyl groups can form strong hydrogen bonds with the hydroxyl groups on the silica surface. Because the silica surface is protected by PVP, OH^- ions cannot dissolve the surface layer rapidly. Instead, the small ions diffuse into the interiors of the silica spheres, leading to a relatively higher etching rate at the center, eventually producing hollow spheres upon continued etching (Fig. 21.5) [71,72]. Further etching removes materials from the particle cores, making the appearance of hollow structures more pronounced (Fig. 21.5d). Simultaneously, the shells come to appear rougher and less homogeneous in transmission contrast, which can be attributed to partial and local etching of the remaining shell material. Park and Kim [75] described that the spontaneous hollowing of silica nanoparticles can occur in a very mild basic solution of typical bases used in the synthesis of silica nanoparticles, demonstrating that the hollowing effect is a general phenomenon that can commonly occur in silica nanoparticles. They also show that the shape evolution of silica particles is size-dependent, reinforcing the importance

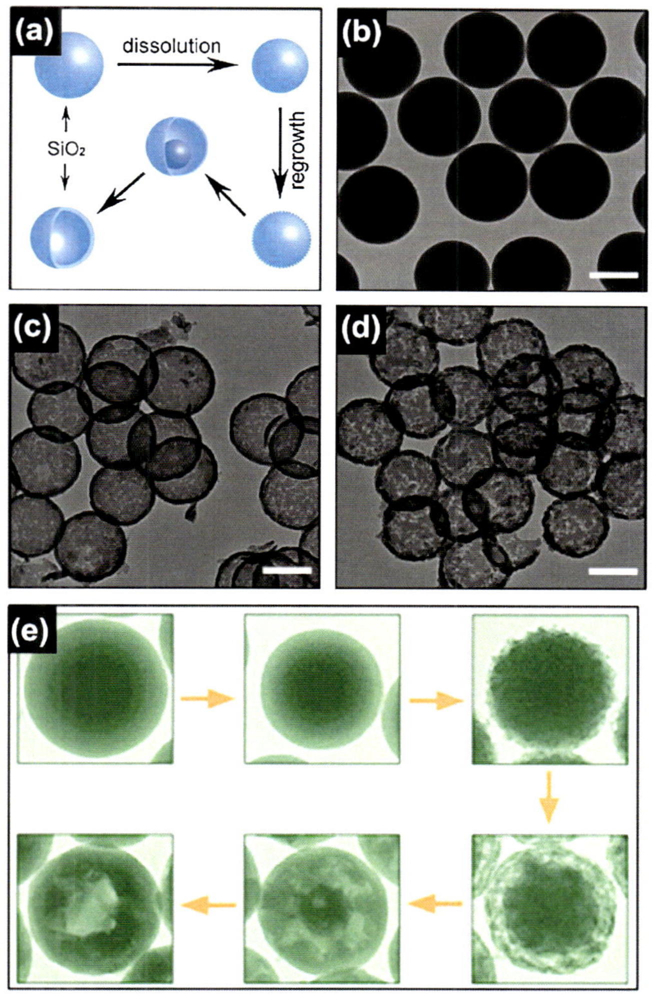

FIGURE 21.5 (a) Schematic illustration of the spontaneous formation of hollow SiO_2 spheres. (b) TEM images of as-prepared SiO_2 spheres. (c and d) Samples after reacting with 0.06 g/mL $NaBH_4$ for 5 h at 51°C (c), and 5 h at 56°C (d). Scale bars are 200 nm. (e) Schematic illustration of the spontaneous formation of hollow SiO_2 spheres at room temperature reacting [72].

of evaluating chemical reactivity and structural transformation as well as physical properties for different size ranges.

Top-down methods can produce large quantities of nanomaterials. However, the size control of nanomaterials is very difficult. The bottom-up approach takes advantage of physicochemical interactions for the hierarchical synthesis of ordered nanoscale structures through the self-assembly of basic building blocks. The particle morphology and size can be easily controlled.

21.2.2. Bottom-up Methods

Bottom-up nanofabrication is driven by the self-assembling processes of component molecules, where various examples based on supramolecular chemistry have been intensively investigated. Currently, the most common types of bottom-up fabrication procedures are those based on the use of a templating substrate, such as chemically or topologically patterned surfaces, inorganic mesoporous structures, and organic supramolecular complexes (mainly block copolymer (BCP) systems). Topographically/chemically patterned surfaces have shown to be particularly effective in directing the nucleation and growth of 1D/2D colloidal crystals or highly uniform colloidal aggregates with well-controlled sizes, shapes, and structures. BCPs provide another versatile route for templating the self-assembly of ordered nanoscale structures (metallic nanoparticles and other inorganic materials), particularly on surfaces. The periodicity of the microphase-separated domains in BCP systems is typically in the range of 10–200 nm. However, the main disadvantage to the use of block copolymer templates is the requirement, in some synthesis cases, of highly specialized polymeric components that may not be readily or commercially available.

21.2.2.1. Direct Self-assembly

One of the major technological challenges in nanoscience and nanotechnology is the self-assembly of tiny nano-building units (nanokits and nanoparts) into larger organized conformations and geometrical architectures for device applications. Self-assembly phenomena have been used both to form templates for fabricating nanostructures and to assemble the nanostructure itself [59]. Self-assembled processes are molecular processes. Self-assembly also can be used to describe the spontaneous formation process for highly ordered assemblies from disordered components by noncovalent interactions. During the self-assembly, the system minimizes its free energy thereby evolving toward equilibrium. It balances attractive and repulsive colloidal and intermolecular forces that include van der Waals interactions, electrostatic interactions, hydrophobic/hydrophilic interactions, surface tension, capillary forces, steric forces, and hydrogen bonds such as the hybridization of DNA. For the preparation of thin films, the Langmuir–Blodgett (LB) technique [59,76] and the SAMs [77] method are usually used. Direct self-assembly of building blocks, which may involve molecular and organic structure-directing additives, in addition to the constituent building units, is considered to be distinct from spontaneous self-assembly. Using the self-assembly method, various nanomaterials including nanoclusters, nanorods, nanotubes, nanowires, nanoporous materials, and block copolymer nanostructures have been synthesized. Particle morphology will thus influence the collective properties of the assembly structure through various kinds of translational and orientational ordering, and the interaction and coupling of the individual nanoparticle building blocks. Self-assembled superlattice structures have modified properties compared to isolated particles due to interparticle interactions. Very exciting results have been obtained with self-assembly using crystalline surfactants and polymer liquid crystalline structures, with proteins and other biomolecules. Assembling, patterning, and integration of nanoparticles in functional and ordered networks (on suitable surfaces) are of great importance for the fabrication of efficient electronic, photonic, or sensor devices. However, the self-assembly approach still needs to demonstrate the large-scale control required for functional devices, and needs to find ways to make connections and interfaces between the devices and the systems.

21.2.2.2. Layer-by-layer Assembly

Based on the charge theory, the layer-by-layer (LbL) adsorption technique provides an easy and inexpensive process for multilayer formation [60–62]. Therefore, the LbL assembly method can be regarded as a versatile bottom-up nanofabrication technique. Since its introduction in 1991 [78], the LbL assembly technique has rapidly expanded to become a premier method for the preparation of nanoscale films with tailored properties. Typically, the LbL process begins with the adsorption of a charged species onto a substrate with opposite charge, thereby reversing the substrate surface charge. Further layers are then deposited by the alternate adsorption of oppositely charged species onto the substrate, until the desired thickness is achieved. The versatility of the LbL approach has allowed a broad range of materials (e.g., polymers, nanoparticles, lipids, proteins, dye molecules) to be assembled on various substrates, on the basis of not only electrostatic interactions but also hydrogen bonding, hydrophobic interactions, covalent bonding, and complementary base pairing. The LbL approach offers a number of key advantages when compared to other methods for surface modification. An important feature is the precision with which the layer thicknesses can be controlled. This control can be achieved by varying: (a) the specific materials being used, (b) the number of layers assembled, and/or (c) the specific adsorption conditions used. For instance, by using variables such as the temperature of adsorption, the ionic strength of the adsorption and rinse solutions, and the solvent polarity, the thickness of each adsorbed layer can be tuned within nanometer resolution. The flexibility in the LbL films' structure allows researchers to fabricate higher-dimensional structures from functional components. One of the most outstanding strategy modifications for LbL assembly preparation involves invisible particles and subsequent hollow capsule formation, which is reported by Caruso et al. [63] in their pioneering work (Fig. 21.6). In this strategy, the LbL films were assembled sequentially, similar to the conventional assemblies, on a colloidal core. Dissolution of the central particle core upon exposure of the particles to appropriate solvents results in hollow capsules. A similar template synthesis was applied to nanotube synthesis. For example, Hou et al. [79] reported fabrication of nanotubes using LbL assembly between 1,10-decanediyl-bis(phosphonic acid) and zirconium ion on porous alumina template. Highly versatile LbL assembly methods can be combined with the template synthesis approach to yield complex assemblies such as free-standing and nanoporous films, capsules, nanotubes, nanoporous particles, and macroporous and biomimetic structures. Their structures and properties may be readily controlled through the choice of the LbL components and the template.

21.2.2.3. *Template Method*

"Templating" essentially involves the replication of one structure into another under structural inversion. In the case of direct templating, the templated material is a 1:1 copy of the template structure, and no changes in order or length scale of the template structure occur (Fig. 21.7). Templating therefore is a versatile technique for the formation of nanostructured materials, especially for nanoporous materials, as size and shape of the resulting pore structures can be easily tuned by choosing the appropriate template structures [80]. As described in Chapter 16, templating is one of the most frequently used methods of synthesizing materials with structural units ranging from nanometers to micrometers, as in the case of zeolites and photonic crystals, respectively.

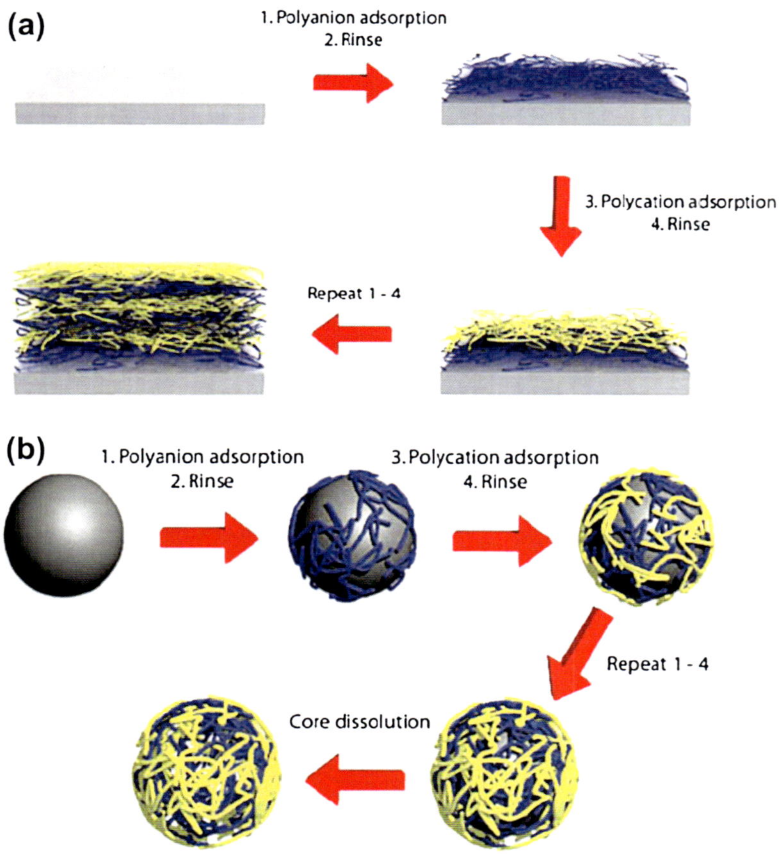

FIGURE 21.6 Layer-by-layer (LbL) assembly of polyelectrolytes on (a) planar and (b) colloidal substrates. A substrate with inherent charge is first exposed to an oppositely charged polyelectrolyte, followed by thorough rinsing. Reversal of the surface charge then facilitates further adsorption steps. The process is continued until the desired layer number (or thickness) is achieved. In (b), the core can be dissolved to give hollow polyelectrolyte capsules. *From Ref. [62]. Reproduced with permission from The Royal Society of Chemistry.*

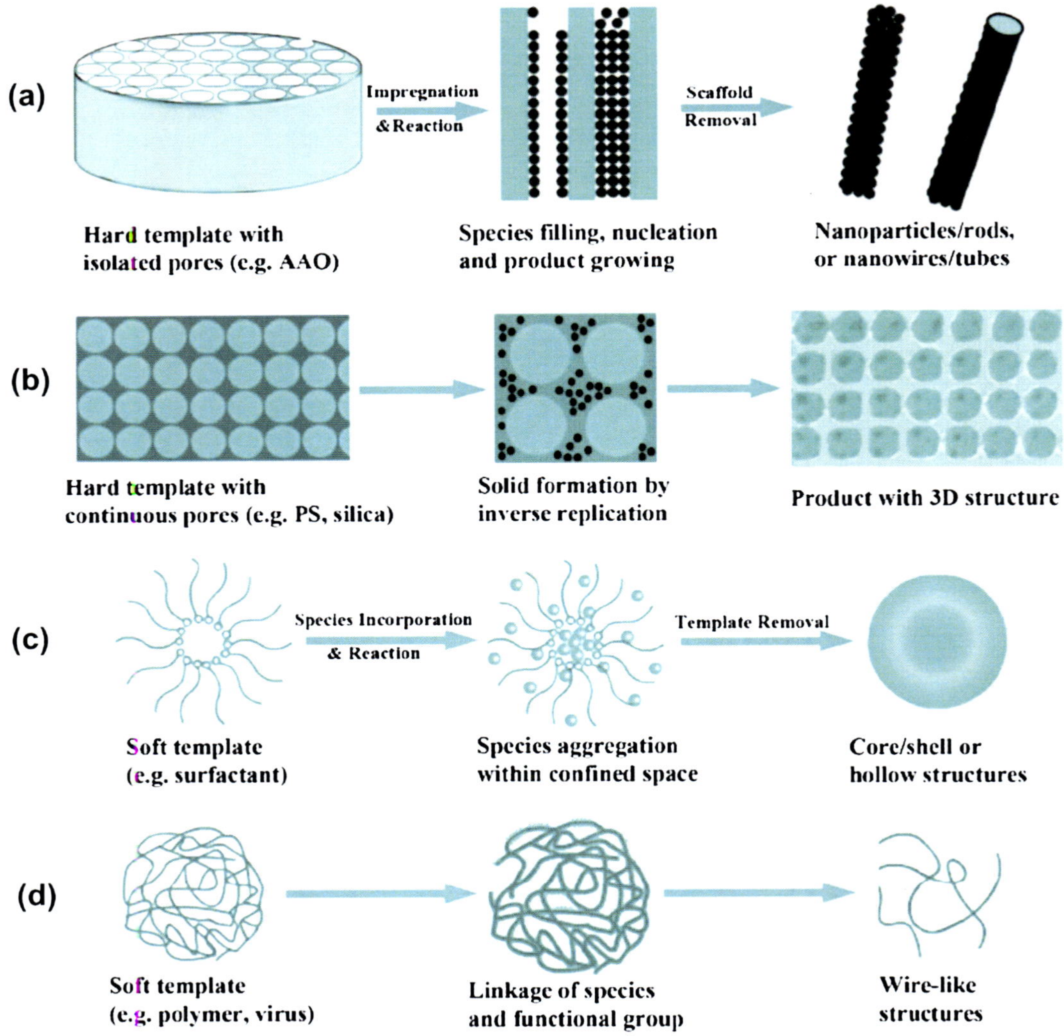

FIGURE 21.7 Schematic depiction showing some typical examples of the hard and soft template synthesis of electrode materials with diverse morphologies. *Reprinted with permission from Ref. [80]. Copyright 2008 American Chemical Society.*

The templates can be in a variety of forms including molecules, supramolecular aggregates, nanoparticles, colloids, nanoporous matrix, and even biological materials such as viruses. Generally, the employed materials in template synthesis can be simply classified into hard templates and soft templates. Usually, organic molecules or supramolecular aggregates are used as soft templates to impart structural features to the porous solid formed. Conversely, the resulting porous solids can be used as hard templates to structure the guests, often being inverse replicas of the hosts. The key to successful templating synthesis is to select a template that ensures the formation of a desired nanostructure and can be easily removed without damaging it. For using nanoporous solid as hard templates, generally, the connectivity of the pores or channels strongly influences the structure of the resulting solid product. The hard template can be anodic aluminum oxide (AAO) membranes, colloid beads, ordered mesoporous materials, zeolites, ice, etc. For soft templates, supramolecules function as structure-directing agents that assist in the assembly of reacting species. It is generally accepted that soft templates are facile for controllable synthesis of nanomaterials because of either the unique anisotropic structures or the functionality of their subunit groups. Typical examples are surfactants, long-chain polymers, micelles, and supramolecular aggregates. Except them, the biological species such as bacteria, textiles/paper, hair, cells, insect wings, living cells, mushroom gills, plant leaves, diatoms, spider silk, wool, and wood have also been used as templates for synthesis of nanomaterials. In the template method, a new concept called repeated templating provides a versatile method to synthesize various nanomaterials. The general principle of repeated templating is very straightforward. The matrices are themselves created by templating procedures; thus the entire process can be envisaged as "repeated templating". For example, a porous solid material is used as a rigid matrix (template), that is, its pores are filled with one or more precursor species, which react in situ to form the desired material. The matrix is finally removed to yield the product as its negative replica; the replica can be further used as template to synthesize nanomaterials.

Overall, hard template-based synthetic approaches suffer from several intrinsic disadvantages, which range from the inherent difficulty of achieving high product yields from the multistep synthetic process to the lack of structural robustness of the shells upon template removal. These difficulties can be partly overcome by the use of soft templates. Compared to solid templates, the soft templates can be more efficient and easily removed. Additionally, soft templates allow facile and efficient encapsulation of functional species like biomolecules.

21.2.2.4. Microemulsion- or Nanoemulsion-based Synthesis Methods

Microemulsion- or nanoemulsion-based synthesis has proven to be a versatile route to synthesize a variety of nanomaterials. A micro-emulsion is a thermodynamically stable dispersion of two immiscible liquids in the presence of an emulsifier or surfactant. They are characterized by ultralow interfacial tension, large interfacial area, and capacity to solubilize both water and oil components. Micro-emulsions are of use in oil recovery, pharmaceutics, cosmetics, detergency, and lubrication, among other processes. They are categorized as water-in oil (W/O) microemulsions when the water is dispersed homogeneously in an organic media with the help of the surfactant and oil-in-water (O/W) micro-emulsions, where oil is dispersed in water. The water-in-sc-CO_2 (sc = supercritical) micro-emulsion is another class of microemulsion developed recently. In W/O micro-emulsions, the aqueous phase is dispersed as microdroplets (typically 1–50 nm in diameter) surrounded by a monolayer of surfactant molecules in the continuous hydrocarbon phase. The size of the reverse micelle is determined by the molar ratio of water to surfactant. By mixing two identical W/O microemulsions containing the desired reactants, the microdroplets will continuously collide, coalesce, and break again, and finally a precipitate forms in the micelles. The precipitate can be extracted by filtering or centrifuging the mixture by the addition of solvent, such as acetone or ethanol, to the microemulsions. In this sense, a micro-emulsion can be used as a nanoreactor for the formation of nanoparticles. Microemulsion-based synthesis is a powerful method where expensive or specialized instruments are not needed, contrasting to the case for several physical methods such as plasma synthesis, ball milling, and chemical vapor deposition. The product obtained is micro homogeneous as the desired stoichiometry is maintained inside the water pools. Metallic nanoparticles, semiconductor quantum dots, polymeric nanoparticles, metal oxides nanoparticles, ceramics, core–shell nanostructures, etc. are a few examples of nanomaterials synthesized using reverse micelles. The reverse micelles collide among themselves to exchange the reactants and then again break apart. This coalescence process is critical since it is only through this mechanism that the reactants, solubilized in individual reverse micelles (nanoreactors), come in close contact and undergo homogeneous mixing. While decoalescence ensures the presence of the protective coating of the amphiphile for the controlled nucleation and growth, it also prevents aggregation. On mixing the microemulsions, the reverse micelles containing the reactants collide with each other to form water channel, resulting in a transient dimer. Once such a dimer is formed, intermicellar exchange of the reactants takes place and thus nucleation starts at the micellar edges with the well-known growth process from the boundary to core. The intermicellar exchange rate can be characterized by the type of microemulsion chosen. Along with intermicellar exchange time, the time required for the chemical reaction, *r* (occurring inside the reverse micelles), is also critical. The ratio *r/ex* determines the kinetics of the chemical reaction inside the micelle. An encounter rate factor, depending on the film flexibility, affects the exchange rate constant k_{ex}.

First, for the synthesis of spherical metal nanoparticles in the 1980s, the emulsion-based synthesis method has grown considerably, many highly complex and multifunctional nanostructures have been synthesized by this method today [81]. Understanding of the subject has benefited tremendously due to the availability of several new techniques to follow the dynamics of the processes, underlying the synthesis carried out in microemulsions. The future of microemulsion-based synthesis appears bright because this method can be applied in several applications in high growth industries such as cosmetics, food, and pharmaceuticals. However, there are some challenges that will dominate the directions of this research area in the next decade. The major challenge is the utilization and recycling of the used solvents involved in these microemulsion systems. There is no doubt that the microemulsion-based synthesis will always be more appropriate for the preparation of nanomaterials with stringent size and shape restrictions.

21.2.2.5. Colloid Chemical Routes

The colloid chemical method is popular for the synthesis of a wide variety of metal and metal oxide nanoparticles. In the seeding growth, fine metal particles are first produced by reducing metal ions with a suitable reducing agent. These fine metal particles are so-called seed particles, since in a subsequent step they are added to growth solutions containing the same or different metal ions along with other additives such as dopants and ligands. In this approach, a small quantity of relatively strong reducing agent is introduced into the reaction medium, in which a mild reducing agent is already present. The purpose of the strong reducing agent is to initiate homogeneous nucleation in the reaction medium. The growth of the seed created in situ from the reduction of metal ions by the strong reducing agent is then carried forward by the mild reducing agent present in the reaction medium. Nanorods/nanowires, rectangle-, cube- or tetrapod-like gold nanoparticles, and silver nanoplates and disks have been synthesized in aqueous solution at room temperature by this method [82]. Colloid–chemical preparative processes do not always deliver monodispersed particles. The polydispersity of the final product (in terms of either particle sizes or shapes) may be acceptable only in certain cases. Therefore, size and shape separations are necessary.

21.2.2.6. Sol–gel Method

During the twentieth century, the sol–gel synthesis processes were developed both in academic research and in industry, producing glasses by straightforward polymerization of molecular precursors in solution. Basically, sol–gel synthesis is carried out in a liquid medium. Typical examples are silica gels which have been synthesized in 1846, since then many ceramics materials have been synthesized through the sol–gel method [56–58]. This process involves the evolution of inorganic networks through the formation of a colloidal suspension which is called sol and gelation of the sol to form a network in a continuous liquid phase which is denoted as gel. Three reactions generally describe the sol–gel process: (1) hydrolysis reaction, (2) alcohol condensation process, and (3) water condensation process. The sol–gel approach also provides an alternative and usual way for synthesis of nanomaterials. Combined with chemical nanotechnologies, remarkable progress has been achieved and sol–gel techniques have taken their place as a fundamental approach to the development of new nanomaterials. The sol–gel synthesis method has been used for the production of metal, metal oxide, and ceramic nanoparticles with high purity and good homogeneity. If an organic surfactant is added to the sol as the structuring agent, it is even possible to obtain an ordered porous structure in two dimension or three dimension.

The sol–gel process is considered as a low-temperature synthesis method that gives pure, homogeneous nanoparticles with good size distribution in the design of complex nanoarchitectures. These nanoarchitectures can be further functionalized using organic molecules or polymers as the gel developed or even after as the gel has dried. Furthermore, many kinds of nanoparticles including oxides, sulfides, metals, and semiconductors with nanoporous structures can be synthesized through a precise heat treatment (Fig. 21.8). The versatility of the process is largely due to the rich and varied chemistry of organometallic precursors, combined with the low processing temperature.

21.2.3. Special Synthetic Method for Nanomaterials

21.2.3.1. Biomimetic and Bioinspired Method

Nature is indeed a school for nanomaterials synthesis and the high degree of sophistication and miniaturization found in natural materials provides inspiration for scientists to synthesize nanomaterials. A biomimetic and bioinspired approach to materials is one of the most promising scientific and technological challenges in the coming years. Bioinspired selective multifunctional materials with associated properties such as separation, adsorption, catalysis, sensing, biosensing, imaging, and multitherapy will appear in the near future. Chapter 23 provides detailed synthesis examples to introduce this intriguing method [18–20].

21.2.3.2. Photochemical and Radiation–chemical Method

Synthesis of nanomaterials is associated with the generation of highly active strong reducers like electrons, radicals, and excited species [64–66]. Photosynthesis is characterized by energies below ~60 eV, whereas radiolysis uses energies of 103–104 ev. Optical irradiation can provide a relatively easy, controlled reduction of metal salts. Photochemical reduction involves UV irradiation of the metal-precursor solution in the presence of an

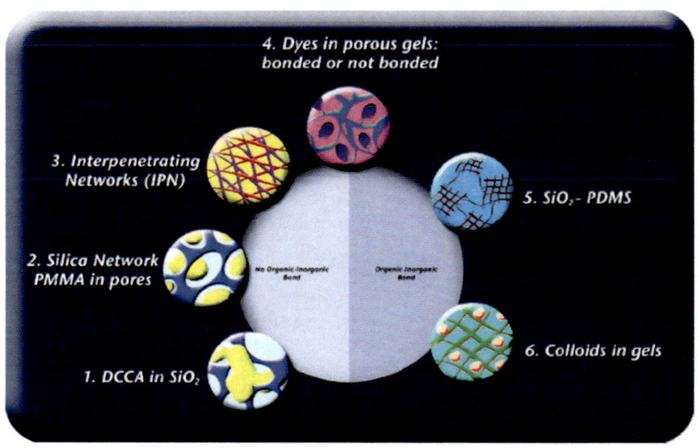

FIGURE 21.8 Various sol–gel-derived nanoarchitectures for sol–gel organic–inorganic hybrids. *Reprinted with permission from Ref. [58]. Copyright 2009 American Chemical Society.*

electron-donating reagent. Photochemical and radiation–chemical reduction methods make it possible to produce nanoparticles under solid-state conditions and at low temperatures. Photoreduction of silver nitrate in the presence of polycarboxylic acids allows the methods of controlling the shape and size of particles to be developed. Spherical and rod-like silver particles are obtained. Synthesis of silver nanoparticles in nanoemulsions by radiation reduction is also conducted and described. A more rapid variant of the photochemical method is a mixed chemical–photochemical reduction of metal precursors, where an active reducing agent is also present in reaction solution.

21.2.3.3. *Ultrasonic-assisted Synthetic Method*

Sonochemistry is one of the earliest techniques used to prepare nanomaterials [67]. Sonochemistry is the research area in which molecules undergo a chemical reaction due to the application of powerful ultrasound radiation (20 kHz–10 MHz). The physical phenomenon responsible for the sonochemical process is acoustic cavitation. A number of theories have been developed in order to explain how 20-kHz sonic radiation can break chemical bonds. Nanomaterials can be easily prepared using the ultrasonic method. Furthermore, the sonochemical method is superior to all other techniques in the following topics related to nanomaterials: (1) preparation of amorphous products; (2) insertion of nanomaterials into mesoporous materials; (3) deposition of nanoparticles on ceramic and polymeric surfaces; and (4) morphology control of nanomaterials.

21.2.3.4. *Microwave Synthetic Method*

Microwave (MW)-assisted rapid heating has received considerable attention as a new promising method for the one-pot synthesis of nanostructured materials in solutions [24,64]. Conceptually, advantageous applications of this method have been demonstrated in the preparation of single-crystalline Ag, Au, Pt, and Au–Pd nanostructures or zeolites, mesoporous materials with various morphologies, such as spherical nanoparticles, polygonal plates, sheets, rods, wires, tubes, and dendrites, can be prepared within a few minutes under MW heating conditions. In general, nanostructures with smaller sizes, narrower size distributions, and a higher degree of crystallization have been obtained under MW heating compared to those in conventional oil-bath heating. The MW synthetic method could also be another avenue for green synthesis of nanomaterials. The advantages of green nanosynthesis include shorter reaction time, reduced energy consumption, and better product yields.

21.2.3.5. *Ionic liquid-assisted Synthesis Method*

Water and traditional organic solvents are generally involved in synthesis of traditional inorganic nanomaterials. These synthetic approaches have their advantages and disadvantages, regarding both fundamental and application purposes, and new synthetic routes with other solvents need to be developed [49,68].

Room-temperature ionic liquids (ILs) are organic salts, composed entirely of ions, with low melting points of below 100°C, sometimes as low as 96°C. The first finding of an IL with a melting point of 12°C was reported in 1914 [68]. After that, a succession of the interesting analogous compounds has been discovered. In recent years, the advantages of ILs in synthesis of inorganic nanomaterials have been gradually realized and have received more and more attentions due to their unique physical and chemical properties. ILs have many advantages including environmental friendly, negligible vapor pressures, good thermal stability, wide electrochemical potential windows, and tunable solubility for organic and inorganic substances. Using ILs as additives, various approaches have been reported to prepare inorganic nanomaterials with unprecedented and unique structures and properties. IL-assisted synthesis method can produce not only conventional inorganic nanomaterials but also new inorganic nanomaterials with properties that cannot or only with great difficulty be made via conventional processes.

Since the inception of the IL-based system for the synthesis of inorganic nanomaterials in 2000 [83], it has witnessed the fast growth in this research area. ILs have been applied in the synthesis of many nanomaterials, including metal nanostructures, mesoporous silicas, organosilicas, zeolites, metal oxides, metal chalcogenides, metal salts, and open-framework structures. In many examples, ILs provided new synthesis pathways for nanomaterials with special morphologies and controlled crystalline phases. Using an ILs, 1-ethyl-3-methylimidazolium bis(trifluoromethylsulfonyl)imide ([EMim][NTf_2]), instead of water as the solvent, Dai's group first reported the synthesis of SiO_2 aerogel in 2000 [83]. Dupont et al. [84] prepared uniform Ir nanoparticles in ILs media. Hollow TiO_2 microspheres were prepared by Nakashima and Kimizuka [85] at the interface between a toluene droplet and an IL. Similarly, using an IL as solvent, Zhou and Antonietti obtained spherical TiO_2 nanoaggregates [86]. By using an IL precursor, Taubert brought forth an "all-in-one" concept for the synthesis of CuCl nanoplatelets [87]. Later, ionthermal synthesis concept was proposed by Cooper et al. [88] for the preparation of aluminum phosphate open-frameworks.

The synthesis of inorganic nanomaterials in ILs is a rather new field, which has roughly emerged over the last 10 years [89]. The unique adaptability and flexibility of ILs provide us with a new, flexible, and powerful tool for the fabrication of interesting and sophisticated nanostructures via chemical approaches. However, the application of the ILs in synthesis of inorganic nanomaterials is still in its infant stages. Further research on developing the approach and examining the breadth of the IL application is highly desirable. The use of ILs may lead to new nanomaterials with interesting morphologies, structures, and properties that are not accessible by using conventional synthesis method.

21.2.3.6. *Electrochemical Synthesis*

Inducing chemical reactions in an electrolyte solution via the use of an applied voltage are called the electrochemical nanomaterials synthesis method. This technique has the advantage of being compatible with a wide variety of materials. Metallic nanoparticles have been prepared by electrochemical pathways with and without the use of nanoporous hard templates. Hard templates consist of nanoporous materials such as track-etched polymers, anodized-aluminum oxides, and others. Electrochemical deposition is carried out by coating one face of the membrane with a metal film, which acts as a cathode for electroplating. Appropriate metal ions are then electrochemically reduced and deposited within the pores of the template membranes. The size and morphology of the nanoparticles can be controlled by varying the electro-deposition parameters, such as potential, number of coulombs passed, temperature, deposition time, appropriate surfactants, or soluble polymers, during the deposition process. Electrochemical synthesis is a powerful method for fabricating multicomponent nanowires with different metals. Electrospinning and electrodeposition are also simple, versatile techniques for generating nanofibers from a rich variety of materials including polymers, composites, and ceramics. Electrospinning has exhibited a strong ability to generate polymeric nanofibers in the past decade. Combined with calcination or carbonation, ceramic or other inorganic nanofibers could also be synthesized using the electrospinning technique.

21.3. SYNTHESIS OF NANOMATERIALS

Instead of compiling all the literature that would by far exceed the scope of this chapter, we try to present typical and representative examples, and the most recent reports for the discussion of nanomaterials' synthetic pathway. The materials covered here include magnetic nanoparticles, nanocrystals with large percentages of reactive facets, semiconductor nanowires, porous and hollow nanomaterials, and core–shell nanocomposites. Many of these materials represent the most exciting, and cutting edge research in the recent years.

21.3.1. Gold and Silver Nanocrystals

The synthesis of small, monodisperse nanoparticles is a major challenge in nanotechnology research. Due to diminished surface energy, it is easy for small particles to aggregate. A protective coating or capping is necessary during synthesis to keep them in a highly dispersed state. The synthesis of Au and Ag nanocrystals with well-defined shapes appears to be one of the most popular subjects of research over the past decade judging from the total number of publications per year. Au nanocrystals represent an ideal platform for chemical and biological sensing as well as applications in nanomedicine. Ag nanocrystals have been widely used as substrates for surface enhanced Raman spectroscopy (SERS), optical labeling, and near-field optical probing. Both Au and Ag nanoparticles have been finding use as a catalyst in various oxidation and oxidative coupling reactions.

There have been many reported methods for controlling the shape of Au and Ag nanocrystals including the well-known polyol reduction method. As pioneered by the Murphy group, the seeded growth has proven to be extremely powerful in the synthesis of Au nanocrystals [90]. In their method, first, by reduction with a strong reducing agent, small Au nanoparticles were prepared. These Au seeds are then used to initiate further nanocrystal growth when an additional precursor is added and then slowly reduced in the presence of a surfactant or polymeric stabilizer. In the presence of citrate-capped twinned Au seeds and cetyltrimethylammonium bromide (CTAB), $HAuCl_4$ is reduced with ascorbic acid to produce anisotropic Au nanorods and nanowires. As a variant of these seeded growth methods, Habas et al. [91] extended their heteroepitaxial method to prepare Au nanorods. In this example, single-crystal Pt nanocubes were used to seed Au nanocrystal growth.

Recently, Xia's group reported a very rapid route to Ag nanocubes, in which oxidative etching was not necessarily involved [92]. In this case, sulfide or hydrosulfide is added to ethylene glycol at parts per million level prior to the introduction of PVP and $AgNO_3$. Upon $AgNO_3$ addition, Ag_2S clusters or nanocrystallites are formed, which can serve as primary sites for nucleation by catalyzing the reduction of Ag^+ [93]. In a related study, Tao et al. [94] also produced single-crystal Ag nanocubes in a matter of minutes, in this case by adding trace $CuCl_2$ to a 1,5-pentanediol-based polyol synthesis.

21.3.2. Magnetic Nanoparticles

Magnetic nanoparticles are of great interest for researchers from a wide range of disciplines, including magnetic fluids, catalysis, biotechnology/biomedicine, magnetic resonance imaging, data storage, and environmental remediation [33–37]. While a number of suitable methods have been developed for the synthesis of magnetic nanoparticles of various different compositions, successful application of such magnetic nanoparticles is highly dependent on the stability of the particles under a range of different conditions. Magnetic nanoparticles have been synthesized with a number of different compositions and phases that include iron oxides, such as magnetite (Fe_3O_4) and maghemite (γ-Fe_2O_3); pure metals, such as Fe and Co; or alloys such as $CoPt_3$ and FePt; as well as spinel-type ferromagnets, such as $MgFe_2O_4$; and $CoFe_2O_4$ alloys. In the last decades, much research has been devoted to the synthesis of magnetic nanoparticles with particle size being smaller than 20 nm. Among the practiced magnetic nanoparticle synthesis methods, coprecipitation, thermal decomposition, emulsion, and hydrothermal technologies are mentioned here. To date, magnetic nanoparticles prepared from coprecipitation and thermal decomposition are the most widely studied. Some recent examples are introduced as follows.

21.3.2.1. Coprecipitation

Coprecipitation is a very effective method to obtain iron oxides with a structure of either Fe_3O_4 or γ-Fe_2O_3. It was prepared in aqueous solutions by coprecipitation of Fe^{2+}/Fe^{3+} salt under alkaline condition at ambient temperature or at elevated temperature with the protection of inert gas. It has been found that a number of experimental factors have significant impacts for the synthesized magnetic NPs, such as the type of salts used (e.g., chlorides, sulfates, nitrates), the Fe^{2+}/Fe^{3+} molar ratio, the reaction temperature, the pH value, and ionic strength of the media. The particle size, shape, and composition of the prepared magnetic NPs can be controlled by changing the experimental conditions. Recently, significant advances in preparing monodisperse magnetite nanoparticles with different sizes have been made by the use of organic additives as stabilization and/or reducing agents [95].

21.3.2.2. Thermal Decomposition

Inspired by the synthesis of high-quality semiconductor nanocrystals and oxides in nonaqueous media by thermal decomposition [96–104], monodisperse magnetic nanoparticles can be synthesized through the thermal decomposition of organometallic compounds in boiling organic solvents that contain stabilizing surfactants [96]. The ratios of the starting reagents, including the organometallic compounds, surfactant, and solvent, together with the reaction temperature and reaction time, are the main parameters for controlling nanoparticle synthesis. For example, nearly monodisperse iron oxide crystals with sizes adjustable over a wide size range (3–50 nm) have been prepared [96–98]. The reaction system is composed of iron fatty acid salts, fatty acids, hydrocarbon solvents, and activation reagents. Monodisperse magnetic nanocrystals with smaller size can essentially be synthesized through the thermal decomposition [96,99,101].

Jana et al. [102] reported a general decomposition approach for the synthesis of size- and shape-controlled magnetic oxide nanocrystals based on the pyrolysis of metal fatty acid salts in nonaqueous solution. The reaction system was generally composed of the metal fatty acid salts, the corresponding fatty acids (decanoic acid, lauric acid, myristic acid, palmitic acid, oleic acid, stearic acid), a hydrocarbon solvent (octadecene (ODE), *n*-eicosane, tetracosane, or a mixture of ODE and tetracosane), and activation reagents. Nearly monodisperse Fe_3O_4 nanocrystals, with sizes adjustable over a wide size range (3–50 nm) can be synthesized, with controlled shapes, including dots and cubes. Park et al. [101] have also used a similar thermal decomposition approach for the preparation of monodisperse iron oxide nanoparticles (Fig. 21.9). They used nontoxic and inexpensive iron(III) chloride and sodium oleate to generate an iron oleate complex in situ, which was then decomposed at temperatures between 240 and 320°C in different solvents, such as 1-hexadecene, octyl ether, 1-octadecene, 1-eicosene, or trioctylamine. Particle sizes were in the range of 5–22 nm, depending on the decomposition temperature and aging period. In this synthesis, aging was found to be a necessary step for the formation of iron oxide nanoparticles. The nanoparticles obtained were dispersible in various organic solvents including hexane and toluene. The above-mentioned nanoparticles are dispersible in organic solvents. Recently, synthesis of water-soluble magnetic nanoparticles has attracted much attention because of their desirable applications in biotechnology. Using $FeCl_3 \cdot 6H_2O$ as an iron source and 2-pyrrolidone as coordinating solvent, Li et al. [103] prepared water soluble Fe_3O_4 nanocrystals under reflux (245°C). The mean particle size can be controlled at 4, 12, and 60 nm, respectively, when the reflux time was 1, 10, and 24 h. With increasing reflux time, the shapes of the particles

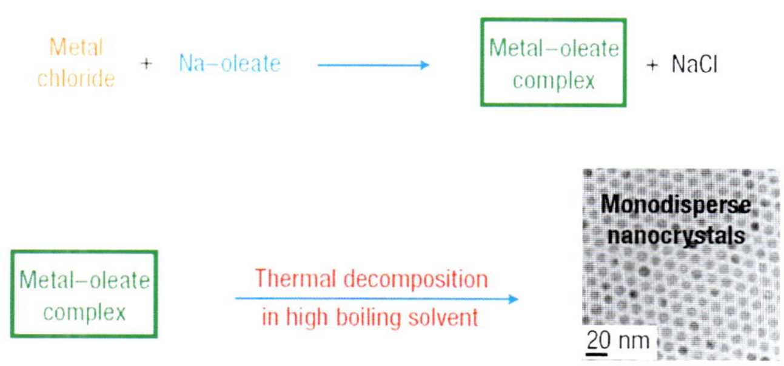

FIGURE 21.9 The overall scheme for the ultralarge-scale synthesis of monodisperse nanocrystals. Metal–oleate precursors were prepared from the reaction of metal chlorides and sodium oleate. The thermal decomposition of the metal–oleate precursors in high boiling solvent produced monodisperse nanocrystals. *Reprinted with permission from Macmillan Publishers Ltd: Nature Materials [101], copyright (2004).*

changed from spherical at early stage to cubic morphologies for longer times More recently, the same group developed a one-pot synthesis of water-soluble magnetite nanoparticles prepared under similar reaction conditions by the addition of α,ω-dicarboxyl-terminated poly(ethylene glycol) as a surface-capping agent [104]. These nanoparticles can potentially be used as magnetic resonance imaging contrast agents for cancer diagnosis.

21.3.2.3. *Microemulsion- or Nanoemulsion-based Synthesis Method (emulsion)*

As we mentioned above, the microemulsion method is very powerful for synthesis of nanomaterials. The magnetic particles can be prepared by dispersing magnetic elements in an organic phase containing an organo-soluble initiator and/or monomers, mixing the dispersion with an aqueous solution made of water and emulsifier, homogenizing the mixture to give droplets of organic phase with submicrometer size, and finally polymerizing the homogenized mixture after the addition of monomers, if necessary [105]. The distribution of the droplets (and hence of the polymer particles) is a function of the proportion of emulsifier present in the aqueous solution, and the ratio of the organic to the aqueous phase. The size distribution of the polymer particles obtained by this process is generally wide. By applying a purification process to the initial polydisperse crude emulsion, it is possible to obtain a set of highly monodisperse samples with high magnetic content (up to 70%) [106]. The purification method is analogous in principle to a fractionated crystallization process.

21.3.3. Semiconductor Nanowires and Ultrathin Nanowires

As testified by the number of citations, this field has accrued and the recent years have seen an explosive growth of research in 1D nanostructures [107–116]. As customary, its blossom was enabled by chemical breakthroughs that allowed the reproducible and affordable synthesis of such structures (Fig. 21.10). Through controlled growth and organization, semiconductor nanowires can be synthesized. They are a broad class of materials, and have led to a number of novel nanoscale photonic and electronic devices. The creation of semiconductor nanowires through a bottom-up approach is heavily dependent on the controlled synthesis of 1D, single-crystalline, high-optical-quality materials. The "vapor–liquid–solid" (VLS) growth, which promotes seeding and oriented growth by introducing a catalytic liquid alloy phase that can rapidly adsorb a vapor to supersaturation levels, has achieved the most success in producing various semiconductor nanowires in relatively large quantities.

21.3.3.1. *Vapor–liquid–solid (VLS) Method for Semiconductor Nanowire Growth*

The VLS mechanism was first proposed in 1964 as an explanation for silicon whisker growth from the gas phase in the presence of a liquid gold droplet placed upon a silicon substrate [117]. The vapor–liquid–solid method (VLS) is a mechanism for the growth of 1D nanostructures, such as nanowires. Growth of a crystal through direct adsorption of a gas phase onto a solid

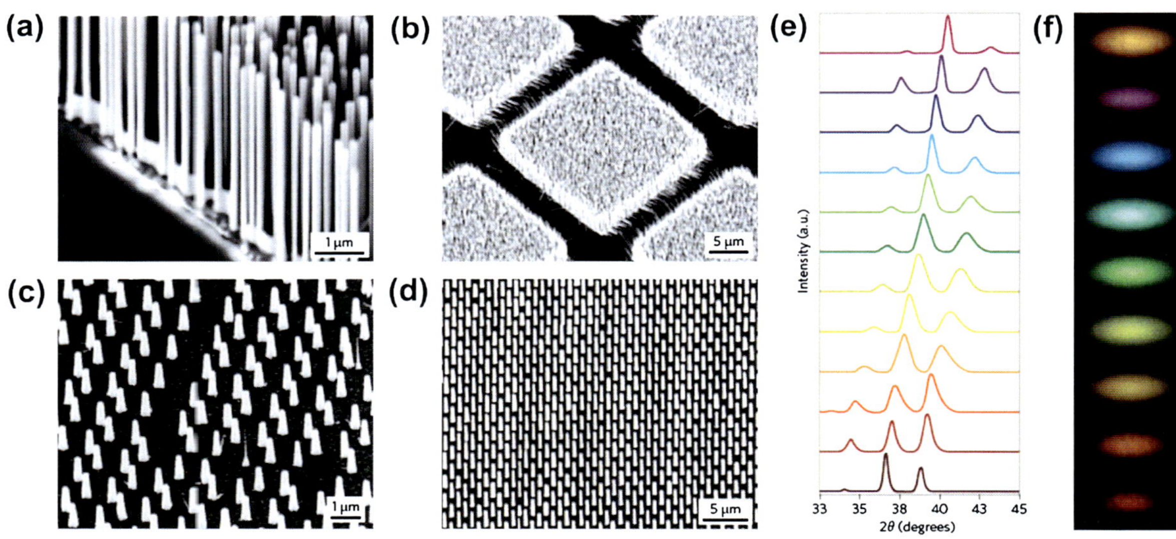

FIGURE 21.10 Semiconductor nanowire growth. (a–d) Scanning electron microscope images of ZnO (a), GaN (b), InP (c), and InP/InAs/InP core–multishell (d) nanowire arrays. (e) X-ray diffraction for InGaN nanowires with different concentrations, increasing downwards. (f) CCD camera image of the visible photoluminescence emission of InxGa1 − xN nanowire ($x = 0$–0.6). *Reprinted by permission from Macmillan Publishers Ltd: Nature Photonics [132], copyright (2009).*

surface is generally very slow. The VLS mechanism circumvents this by introducing a catalytic liquid alloy phase, which can rapidly adsorb a vapor to supersaturation levels, and from which crystal growth can subsequently occur from nucleated seeds at the liquid–solid interface [118–123]. The physical characteristics of nanowires grown in this manner depend, in a controllable way, upon the size and physical properties of the liquid alloy (Fig. 21.3). The VLS growth mechanism has been re-examined and developed over the past two decades by Lieber and coworkers [124–129], Yang and coworkers [130–139], and many other research groups. Understanding of the VLS mechanism makes the researchers to precisely control nanowire growth over length, diameter, growth direction, morphology, and composition. Nanowire diameter, typically from several nanometers to hundreds of nanometers, is determined by the size of the metal alloy droplets. Nanowire arrays with uniform size can be readily obtained by using monodispersed metal nanoparticles. The length of these nanowires can be easily controlled from micrometers to millimeters. Precise orientation control during nanowire growth can be achieved by applying conventional epitaxial crystal growth techniques to this VLS process. Recently, Tian et al. focused on using semiconductor nanowires as building blocks for photovoltaic devices on basis of investigations at the single nanowire level [43]. The use of single nanowires as photovoltaic elements presents several key advantages, which may be leveraged to produce high-efficiency, robust, integrated nanoscale photovoltaics (PV) power sources. First, the principle of bottom-up design allows the rational control of key nanomaterial parameters, which will determine PV performance, including chemical/dopant composition, diode junction structure, size, and morphology. Second, single or interconnected nanowire PV elements could be seamlessly integrated with conventional electronics and/or future nanoscale electronics to provide energy for low-power applications. Third, studies of PV properties at the single nanowire level will permit determination of the intrinsic limits, areas of improvement, and potential benefits of nanoenabled PV.

21.3.3.2. Solution–liquid–solid (SLS) Growth Process

Based on an analogy to the VLS process, Trentler et al. [140] developed a solution–liquid–solid (SLS) method for the synthesis of highly crystalline semiconductor nanowires at relatively low temperatures. In a typical procedure, a metal with a low melting point was used as the catalyst, and the desired material was generated through the decomposition of organometallic precursors [140–143]. The product was essentially single-crystalline whiskers or filaments with lateral dimensions of 10–150 nm and lengths up to several micrometers. In principle, one could reduce the operation temperature to a value below the boiling points of commonly used aromatic solvents. Recently, through a novel solution–liquid–solid approach, our group [143] reported synthesis of manganese-doped cadmium selenide (Mn-CdSe) colloidal nanowires which exhibit ferromagnetism and good conductivity without any change in the optical properties.

21.3.3.3. CVD Growth Process for Ultrathin Nanowires

Ultrathin nanowires are considered to be less than 10 nm in diameters in most of the literature, although increasing interest is emerging in nanowires below 2–3 nm. Such ultrathin nanowires are of great interest, as they have been demonstrated to show quantum conductance and ballistic conduction. Negative magnetoresistance, localization, low thermal conductivity, ferromagnetism, and quantum size effects are other properties that are of great importance for a variety of applications. The templating, ligand control, and oriented attachment methods have been used to synthesize ultrathin nanowires. The templating strategy involves the use of a hard or soft structure, called the template, as the directing agent of the growth of the nanowires. Recently, a few papers have appeared in the literature regarding Au ultrathin nanowires. In the first report, Xia's group described the reduction of gold chloride salts through oleylamine and silver nanocrystals [144]. The mechanism they propose is the formation of an $[(\text{oleylamine})\text{AuCl}]_n$ inorganic polymer, which templates the formation of ultrathin nanowires with uniform thickness. Later, Wang's group [123] reported the electrical characterization of samples obtained in very similar conditions. Au single-crystal nanowires of 9 nm diameter were found to have a good conductivity and a high failure current density. This work demonstrates that ultrathin nanowires can circumvent the electromigration problems of nanowires, which leads to failure. Ligand-control strategy for the synthesis of nanowires was an offspring of the work that was being done on shape control of nanocrystals. The first controllable synthesis of CdSe nanorods demonstrated that using mixtures of ligands on anisotropic structures (like CdSe wurtzite) allowed for the growth of anisotropic structures [145]. Metal oxide nanowires were also produced in ultrathin diameters by Hyeon's group. Sm_2O_3 nanowires had a rectangular cross section of 1–2 unit cells, corresponding to 1.1–2.2 nm [146]. Oriented attachment is maybe the most fascinating growth strategy for ultrathin nanowires, where single-crystalline nanowires are produced by oriented aggregation of faceted nanocrystals. Ultrathin CdSe nanowires were obtained by oriented attachment by reacting cadmium acetate ($CdAc_2$) and selenourea in long-chain amines [147].

21.3.4. Single Nanocrystals with a Large Percentage of Reactive or High-energy Facets

The surface stability and reactivity of inorganic single crystals have long been thought to be dominated by their surface chemistry, whose effect on the equilibrium morphology is critical for the synthesis of single crystals with high reactivity [148].

For anatase TiO_2, both theoretical and experimental studies found that the minority {001} facets in the equilibrium state were especially reactive. Anatase single crystals were synthesized using chemical transport reactions, but the process had a long reaction time and the crystals were of low purity and had no {001} facets. High-purity anatase single crystals with a high percentage of reactive {001} facets have promising applications in solar cells, photonic and optoelectronic devices, sensors and photocatalysis. Recently, anatase single crystals with 47% of the highly reactive {001} facets, using hydrofluoric acid (HF) as a capping agent through the hydrothermal synthesis method, were reported by our group (Fig. 21.11a, b) [148]. High-quality anatase TiO_2 single-crystal nanosheets, mainly dominated by {001} facets, have also been prepared using a water-2-propanol solvothermal synthetic route [149]. The anatase TiO_2 single-crystal nanosheets with 64% {001} facets show superior photoreactivity (more than 5 times), compared to P25 as a benchmarking material (Fig. 21.11c–f). Following this, a facile nonaqueous synthetic strategy was developed by Han et al. [150] to prepare nanostructured anatase TiO_2 with a large percentage of exposed high-energy facets (i.e., {001}, {010}). Wu et al. [151] used a similar strategy to synthesize anatase TiO_2 nanosheets with 89% exposed {001} facets and excellent photocatalytic efficiency. Very recently, Xia's group [152] employed an electrospinning process to manipulate the hydrolysis of a sol–gel precursor and obtain well-defined anatase TiO_2 nanocrystals with exposed, chemically active {001} facets. A micro-sheet anatase TiO_2 single crystal photocatalyst with a remarkable 80% level of reactive {001} facets was synthesized via a microwave-assisted hydrothermal route involving titanium tetrafluoride and a tetrafluoroborate-based ionic liquid [153]. Liu and coworkers [154–156] developed a facile and new route for one-pot synthesis of visible light responsive nitrogen-doped anatase TiO_2 sheets with dominant {001} facets. Tetragonal faceted-nanorods of single-crystalline anatase TiO_2 with a large percentage of higher-energy {100} facets have been synthesized by hydrothermal transformation of alkali titanate

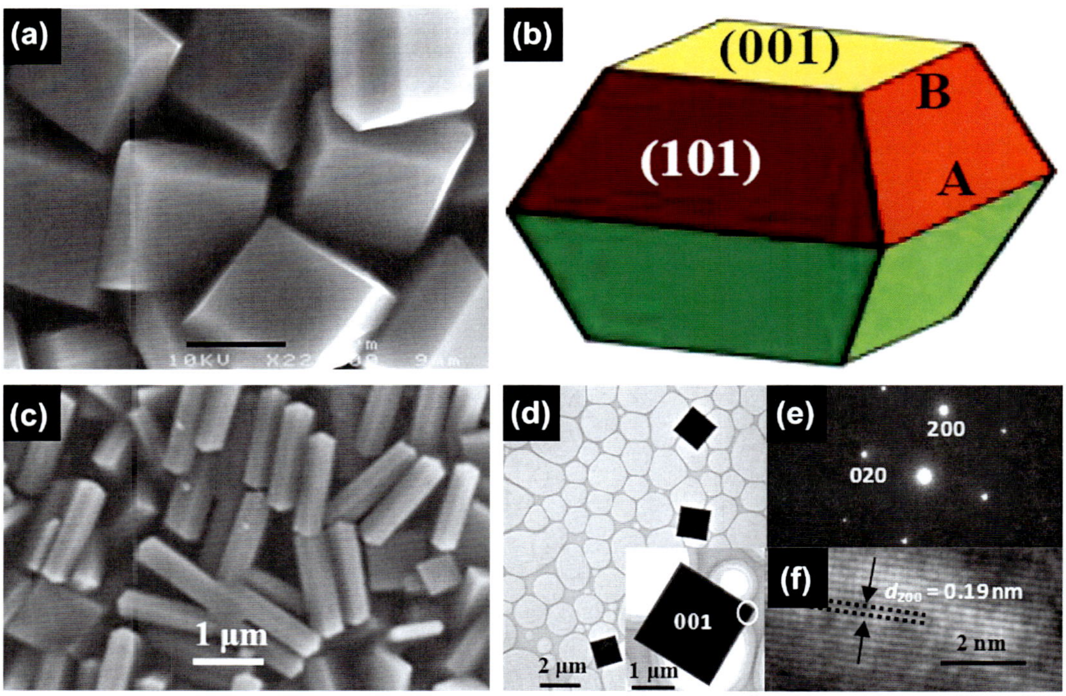

FIGURE 21.11 (a) SEM images and (b) schematic illustration of crystal orientation of anatase TiO_2 single microcrystals. (c) SEM image; (d) TEM image of anatase TiO_2 single-crystal nanosheets; and (e and f) SAED pattern and an HRTEM image of anatase TiO_2 single crystal nanosheets. *Reprinted with permission from Ref. [149]. Copyright 2009 American Chemical Society. Reprinted with permission from Macmillan Publishers Ltd: Nature [148], copyright (2008).*

nanotubes in basic solution [157–160]. At the same time, anatase TiO_2 single crystals with exposed {001} and {110} facets have been successfully synthesized using a modified hydrothermal technique in the presence of hydrogen peroxide and hydrofluoric acid solution [161].

These studies on TiO_2 single crystals motivate researchers to further explore the hydrothermal synthetic method for the preparation of other metallic oxides with a high percentage of reactive facets [162–168], which have promising applications as gas sensors, photocatalysts, solar cells, and photonic and optoelectronic devices. Very recently, octahedral SnO_2 particles with exposed high-energy {221} facets were synthesized by a simple hydrothermal route with the assistance of HCl and PVP [162]. The octahedral SnO_2 particles show excellent gas-sensing performance due to the high chemical activity of the exposed {221} facets. Elongated tetrahexahedral Au nanocrystals have been synthesized in high yields via seed-mediated growth [163]. These nanocrystals are single crystals enclosed by 24 high-index {037} facets. Electrochemical measurements show that they are more chemically active than octahedral Au nanocrystals that are enclosed by low-index {111} facets. Wang's group [164] reported an electrochemical method for the synthesis of tetrahexahedral (THH) Pt nanocrystals at high purity. The THH shape is bounded by 24 facets of high-index planes {730} and vicinal planes such as {210} and {310}. The synthesized THH Pt nanocrystals show enhanced catalytic activity in electro-oxidation of small organic fuels of formic acid and ethanol. Recently, Shen and coworkers [165,166] reported that Co_3O_4 nanorods with predominantly {110} planes exhibited surprisingly high catalytic activity for CO oxidation at temperatures as low as $-77°C$. Well-defined m-$BiVO_4$ nanoplates with exposed {001} facets by an easy and straightforward hydrothermal route were successfully synthesized, without the use of any template or organic surfactant [169]. The m-$BiVO_4$ nanoplates with exposed {001} facets exhibit greatly enhanced activity in the visible-light photocatalytic degradation of organic contaminants and photocatalytic oxidation of water for O_2 generation.

21.3.5. Hollow Porous Nanostructures

Hollow porous nanostructures are of great interest in many current and emerging areas of technology, and display new crystalline ordering and new phenomena. Consequently, they have been the subject of many recent investigations. Hollow porous nanospheres combining the merits of hollow and porous structures present improved flow performance and high surface areas [47,48,170]. They have an interior void as a storage space or a reaction chamber that can encapsulate various substances, while the shell structure contains paths affording control release systems. These materials have greatly impacted on various fields with many applications in diversified areas, including the encapsulation and controlled release of substances (e.g., drugs, genes, dyes, inks, cosmetics, pesticides, food stuffs), protection of biologically active species, removal of pollutants, catalysis, and sensing. Typically, such hollow porous nanospheres can be fabricated from hard templating, sacrificial templating, soft templating, and template-free methods (spray drying, aerosol-assisted self-assembly approach, ultrasonic method, galvanic replacement, and through the Kirkendall effect, etc.). The most popular method to synthesize the siliceous hollow nanospheres (SHNs) is template-mediated approach. Typically, a thin coating of the desired material is formed on the template to create a core–shell composite; subsequent removal of the template core (by calcination or selective etching) generates hollow structures whose inner diameter is largely determined by the size of the template (Fig. 21.12A). The principle of the synthetic strategy is straightforward, involving three main steps: (1) preparing the original template, (2) depositing the target shell material onto the surface of the template, and (3) removing the original template. In this case, the template simply serves as a scaffold, around which a desired material is directed to grow into a nanostructure with its morphology complementary to that of the template. Colloid crystals are widely used as hard template materials because of easy preparation with different particle size and easy functionalization. The second step of shell formation is usually the most challenging part of the process due to the unavoidable and usually undesirable homogeneous nucleation and growth of the shell materials. There are various methods to coat materials on these hard templates: layer-by-layer assembly [63], chemical deposition [171], adsorption [172], casting [173], and atomic layer deposition [174]. Various soft templates have been also employed for the synthesis of silica hollow nanostructures, including microemulsions formed in a two-phase solution, micelles, vesicles, and gas bubbles in liquid assembled by surfactants or supramolecules. Since the production cost of many well-defined sacrificial templates is usually high, large-scale synthesis for industrial applications using these methods has been impractical. Because of this, much effort has been devoted to establishing self-templating or template-free methods to synthesize hollow nanostructures, where no additional templates are needed.

21.3.5.1. Porous Hollow Nanostructured Silica

Hard template method can be used for preparation of porous hollow silica spheres [171–180]. Monodisperse spherical hollow nanoparticles of mesoporous silica featuring mesopores with a radial orientation in the silica shell were synthesized via a dual-templating

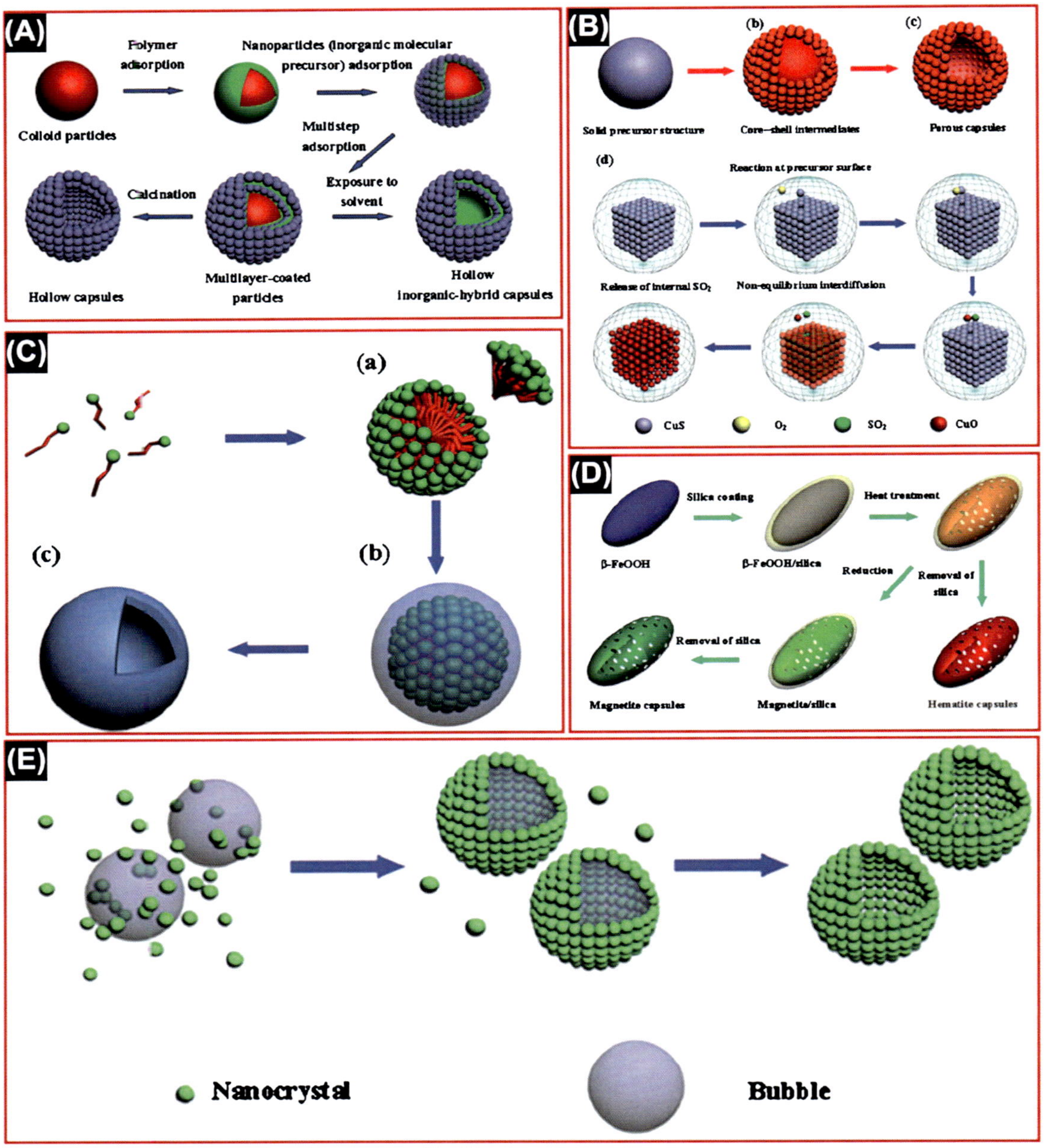

FIGURE 21.12 (A) Schematic illustration of colloid-templated LBL electrostatic assembly for the preparation of hollow inorganic capsules. (B) Schematic procedure used for fabrication of porous CuO capsules (C) Schematic illustration of emulsion/phase separation procedures for the preparation of hollow inorganic capsules. (a) Formation of a spherical micelle; (b) formation of the desired material in water phase that the exterior surface of a micelle serves as physical template; and (c) removal of surfactant molecules with an appropriate solvent (or by calcinations) to obtain an individual capsule. (D) Schematic illustration of procedure for the synthesis of uniform and water-dispersible porous iron oxides (magnetite and hematite) nanocapsules. (E) Schematic illustration of the gas-bubble templating process for preparing porous capsules. *From Ref. [170]. Reproduced with permission from The Royal Society of Chemistry.*

method. Specifically designed polystyrene latexes with anionic or cationic surface charges acted as the core templates, while cetyltrimethylammonium bromide (CTAB) served as a cotemplate to structure the mesopore formation during tetraethyl orthosilicate (TEOS) hydrolysis/condensation [175]. Similarly, hollow silica spheres with mesoporous shells were successfully synthesized by a modified Stöber method using CTAB and PS spheres as dual templates [175–177]. Uniform hollow mesoporous silica spheres and ellipsoids were synthesized by using hematite as hard template [178]. The outer diameter of the spherical mesoporous silica particles can be well adjusted in a submicrometric range (typically around 100–200 nm) by selecting suitable hematite particles as core templates. Silica hollow spheres were also synthesized using nano-sized calcium carbonate as inorganic templates and CTAB as organic templates [179]. Darbandi et al. [180] synthesized porous hollow silica nanoparticles with hole diameters of about 6.5 nm and wall thicknesses of 15 nm by using luminescent CdSe/ZnS cores as templates. Ammonia was used as the catalyst for the polymerization of silica and to dissolve the CdSe/ZnS cores at once.

The emulsion template method is another interesting approach for the preparation of mesoporous silica hollow spheres (MSHS). Schacht et al. [181] first demonstrated that MSHS could be prepared by interfacial reactions conducted in O/W emulsions. Later, it was reported that mesoporous silica hollow spheres could be synthesized via O/W, W/O, or the water/oil/water (W/O/W) interfacial reaction [49,53–82]. However, in most cases, the shell of the silica hollow spheres has disordered pore arrangement. For example, Sun et al. [182] described a facile emulsion method synthesis of hollow silica spheres with multilamellar structure and worm-like pores (~4 nm) on the shell using a triblock copolymer P123-based emulsion in combination with the commercially sodium silicate. In their later work, gigantic hollow silica spheres (20 μm) with a hierarchical intra- and interparticle porosity were obtained from an EO_{76}–PO_{29}–EO_{76}/butanol/ethanol/H_2O quaternary emulsion system [183]. Zhang et al. [184] demonstrated that the morphosynthesis of gigantic MSHS comprising a thin shell with a hole, and gigantic bowl-like hollow structures composed of small spheres, can be readily realized in the water–ethanolammonia–TEOS–CTAB system. They proposed that gigantic hollow shells were produced when the unhydrolyzed and partially hydrolyzed TEOS existed predominantly as emulsion droplets that acted as templates for the formation of hollow shells. Lin et al. [185] designed a sol–gel approach, using a water-in-oil microemulsion as the template, for the synthesis of silica hollow nanospheres (~40 nm) containing a luminescent agent and superparamagnetic Fe_3O_4 nanoparticles. MSHS were also prepared through a W/O emulsion formed from Span 80 and kerosene [186]. Fornasieri et al. [187] and Yu et al.'s [188] group reported the synthesis of MSHS from W/O, respectively. Porous silica nanocapsules have been fabricated by the combination of stabilizing condensation and dynamic self-assembly [189]. Fujiwara et al. reported the preparation of MSHS by an interfacial reaction using a W/O/W emulsion system, in which the reaction mechanism was proposed [190]. MSHS with large mesopore wall structures have been synthesized via CO_2-in-water emulsion templating in the presence of PEO–PPO–PEO block copolymers under supercritical fluid conditions [191,192]. The pore size and morphology of the MSHS can be controlled by varying the CO_2 operation pressure. Mann's group reported that hollow microspheres with ordered mesoporous walls could be synthesized under ambient conditions by a simple procedure involving dilution and neutralization of an aqueous TEOS–CTAB emulsion reaction mixture. By controlling the rate of TEOS hydrolysis specifically at the droplet/water interface, intact microspheres with uniform wall thickness and thermal stability can be routinely synthesized.

The synthesis of highly ordered mesoporous silicas with hollow interiors still remains a challenging topic in materials science. The combination of surfactant-assisted assembly and the emulsion chemistry has provided the possibility for simultaneous control of the mesostructure and morphology of the mesoporous materials [193–210]. In this method, the emulsified droplets act as templates for the hollow spheres, while cylindrical micelles work as templates for the ordered pore channels on the shells. This kind of material may combine some advantages of ordinary molecular sieves (e.g., MCM-41) and hollow spheres. Li et al. [193], for example, synthesized hollow spheres of aluminosilicate with ordered mesoporous shells by combining emulsion droplet templating and surfactant templating. The O/W emulsion is formed by adding TEOS into aqueous $Al_2(SO_4)_3$ solution. The surfactant (CTAB) molecules are responsible for stabilization of the droplets and templating the mesoporous shell formation [194–196]. Lin et al. [197] reported that mesoporous MCM-41 silica of hierarchical order was assembled from a solution of C_{14}TMAB, butanol, and sodium silicate by careful control of the acidification rate. The products consist of a high yield (~98%) of uniform-sized hollow spheres of diameter 5.0 nm. The inside of the micron-sized spheres has an intricate pillar structure possessing various topological genus ranks. The walls of these microspheres consist of a hexagonal arrangement of nanochannels in latitudinal directions. Liu et al. [199] report the fabrication of silica hollow spheres with highly ordered hexagonal arrangement of mesopore in the shell via a simple O/W [2,2,4-trimethylpentane (TMP)/water] emulsion template method in the presence of $(EO)_{20}(PO)_{70}(EO)_{20}$ under buffer solution (NaAc–HAc, pH = 4.4) using tetramethoxysilane (TMOS) as the silica source. The mesostructure of the shell transforms from a highly ordered 2D hexagonal structure to mesocellular foam with initial synthesis temperature increasing from 15 to 40°C. Zhou et al. [200] reported that highly ordered ethane bridged periodic mesoprous organosilicas (PMOs) hollow microspheres with ultralarge pore sizes up to 14.7 nm can be synthesized using a triblock copolymer F127 as a template and 1,3,5-trimethylbenzene (TMB) as a swelling agent at a temperature of 15°C. Randomly ordered MSHS of 20–50 μm in diameter with distinguished characteristics of interconnected porosity (7–14 nm) in their thin outer walls are prepared using dedecane as a swelling agent. Very recently, ordered MSHS were synthesized by a sol–gel/emulsion (oil-in-water/ethanol) method. The diameters of the hollow spheres are controllable in the range from 210 to 720 nm by changing the ratios of ethanol-to-water. The shell thicknesses are tunable by varying the concentration of CTAB used in the emulsion system [201]. Han et al. [202] synthesized monodisperse and uniform hollow silica spheres with ordered mesopores on the nano-sized shells using water/*n*-heptane/CTAB nanoemulsions induced by compressed CO_2 as a template. Wang et al. [203] prepared hierarchical morphology of budded mesoporous silica hollow spheres, which contained a wormhole-like mesoporous shell with protruding lamellar mesostructured buds by a single step emulsion-templating method. The anionic surfactant Sar–Na was used as both a surfactant and an oil phase after acidification.

It is challenging to obtain uniform emulsion droplets with a small size < 100 nm because of droplet coalescence and Ostwald ripening. Employing self-assembled supramolecular micelles/vesicles as soft templates can overcome some of these challenges and as a result has been attracting more attention. Polymer micelles are used for the synthesis of silica hollow spheres with diameters less than 100 nm, since they are desired in many potential applications, especially in drug and gene delivery systems. Generally, spheric micelles can be formed from an amphilic block copolymer in a selective solvent. The unique core–corona structure of the polymeric micelles allows one to encapsulate hydrophobic compounds within its hydrophobic core prior to silica shell formation. However, the template micelles are unstable during the synthesis and the formation of second-order or higher-order aggregates cannot be avoided. Special triblock copolymer micelles (such as ABC triblock copolymer micelles with a core–shell-corona structure) are needed for the synthesis of silica hollow spheres with an average outer diameter of 30 nm [211]. Cationic diblock copolymer 2-(dimethylamino)ethyl methacrylate micelles have been used as templates for the formation of core–shell hollow hybrid copolymer silica at ambient temperature and neutral pH. [212] Huo et al. [213] reported a new class of robust, ultrafine silica core–shell nanoparticles formed from silica cross-linked, individual block copolymer micelles. A highly acidic environment was employed to protonate the PEO chains of the polymeric micelles such that the PEO chains became the adsorption sites for the silica precursors. To prevent intermicellar aggregation and particle growth, diethoxydimethylsilane [$(Me_2Si(OEt)_2$] was needed to terminate the silicate condensation. Recently, Liu et al. [214] reported that the ethane–silica hollow nanospheres were facilely synthesized through condensation of 1,2-bis(trimethoxysilyl)ethane (BTME) around an inorganic-electrolyte-stabilized F127 micelle under a mild buffer condition (NaH_2PO_4–Na_2HPO_4, pH $\approx$ 7.0). In their latest work, they found that the approach can be also applicable to synthesize pure silica hollow nanospheres and various functionalized silica hollow nanospheres [215–218]. Silica and organosilicas containing various organic groups such as ethylene-, phenylene-, and 1,4-diethylphenylene hollow nanospheres with particle sizes less than 25 nm have been successfully fabricated using a similar method under an acidic medium [218]. Similarly, a stable aqueous dispersion of hybrid silica nanocapsules encapsulating fluorescent conjugated polymers has been synthesized via a facile and highly benign approach of templated condensation of silica precursors at the core–corona interface of PEO-based block copolymer micelles [219]. Furthermore, Feng et al. [220] reported a facile synthesis of nano-sized hollow mesoporous silica spheres with tunable size through this micelle templating route. (−)-*N*-Dodecyl-*N*-methylephedrinium bromide (DMEB) was employed as a dual template to direct the formation of hollow core and mesostructure of silica capsules (Fig. 21.12B). Du and coworkers [221,222] reported the synthesis of MSHS through the micelle template form from PVP and 1,12-diaminodedecane. Very recently, hollow silica nanospheres with mesoporous shells were fabricated with a one-pot strategy using a thermosensitive polymer, poly(*N*-isopropylacrylamide) (PNIPAm), as a reversible micelle template without the need of further calcination or chemical etching [223].

Silica hollow nanostructures can be synthesized by vesicles formed from surfactant. There are two types of approaches: morphosynthesis, which is based on the double-layer vesicle as reactor; and transcriptive synthesis, which is based on the sol–gel technique. Pinnavaia and coworkers [224,225] prepared ultrastable silica vesicles with the shells constructed of lamellar sheets. The method is based on supramolecular assembly through hydrogen bonding between neutral gemini surfactants of the type $C_nH_{2n+1}NH(CH_2)_2NH_2$ and silica precursors derived from TEOS. Highly ordered multilayer silica hollow nanospheres were prepared using thermodynamically stable vesicles formed by double chain (DODAB) and single-chain surfactant [dodecyltrimethylammonium bromide (DTAB)] mixtures as a template [226]. Using a vesicle template formed from CTAB-sodium dodecyl sulfate (SDS)-P123 in dilute silicate solution (pH = 5.0), Lin and coresearchers synthesized hollow silica spheres with disordered mesostructured shells with a pore size of approximately 5.5–7.5 nm [227]. Through a vesicle and a liquid crystal "dual templating" approach, Djojoputro et al. [228] reported the synthesis of periodic mesoporous organosilica (PMO) hollow spheres with a particle size around 200–800 nm and a highly ordered mesoporous structure in the shell. In their further work, various functional organosilica hollow spheres were synthesized [229]. Nanosized multilayered silica vesicles have been synthesized through a dual template way by using CTAB and $C_3F_7O(CFCF_3CF_2O)_2CFCF_3CONH(CH_2)_3N^+(C_2H_5)_2CH_3I^-$ (FC-4) as the co-templates and TEOS as the siliceous precursor. Multilamellar vesicles (MLVs), ethane–silica and silica hollow nanoparticles, were synthesized using a cationic surfactant with a partially fluorinated tail [230–232]. Zhang et al. [233] reported the synthesis of MLVs through a single-templating approach by using Pluronic P85 ($EO_{26}PO_{39}EO_{26}$) as a structure directing agent and 1,2-bis(triethoxysilyl) ethane (BTEE) as an organosilica source at mild conditions ($pH = 5.5$). Recently, through a combination of micelle and vesicle template, raspberry-like hierarchical silica hollow spheres were synthesized using a block copolymer $EO_{39}BO_{47}EO_{39}$ (commercial product known as B50-6600, Dow Company) as a template and TEOS as a silica source [234,235].

21.3.5.2. *Other Inorganic Hollow Nanostructures*

Xue and Liu have demonstrated a general top-down thermal oxidation strategy to synthesize uniform porous transition-metal oxide hollow capsules [236]. The thermal oxidation strategy is based on a nonequilibrium interdiffusion (the Kirkendall effect) process, and is schematically displayed in Fig. 21.12B. First, through a solvothermal method, they succeeded in synthesizing nearly monodisperse CuS and Cu_2S with different morphologies (Fig. 21.12C-a). Subsequently, these solid precursors were thermally oxidized in air at 700°C. Core–shell-structured intermediates formed simultaneously (Fig. 21.12C-b). Since the diffusion rate of inner sulfides is larger than that of atmospheric oxygen during oxidation reactions, voids are thus generated, which eventually result in a hollow cavity. Owing to the volume loss and release of internally born sulfur dioxide (SO_2) in the interdiffusion process, porous CuO hollow capsules, preserving the precursors' profile, were finally obtained (Fig. 21.12C-c). The detailed thermal oxidation process is shown in Fig. 21.12C. Xu and Wang [237] have recently prepared novel multishelled Cu_2O hollow spheres with single-crystalline shell structure via a simple CTAB vesicle templating route in aqueous solution. Depending on the concentration of CTAB in the range of 0.1–0.15 M, the Cu_2O hollow spheres were found to be dominantly single-, double-, triple-, or quadruple-shelled in the product. Piao et al. [238] reported the fabrication of water-dispersible hollow iron oxide (hematite and magnetite) nanocapsules from the nanostructural transformation of readily synthesized b-FeOOH nanoparticles. These obtained nanocapsules could be used not only as a novel controlled-release drug delivery vehicle, but also as a magnetic-resonance-imaging contrast agent. The overall wrap–bake–peel process (Fig. 21.12D) was composed of wrapping of b-FeOOH nanoparticles with silica coating, heat treatment, and etching of silica layer to produce hollow iron oxide nanocapsules.

The bubble-template mechanism has been considered as a reasonable interpretation for the formation of hollow inorganic materials [239–246]. The formation process based on this mechanism is illustrated in Fig. 21.12E. In the reaction process, monomers grow into nanocrystals after initial nucleation. These nanocrystals have a tendency to aggregate due to their high surface energy, and, at the same time, lots of bubbles produced during the reaction provide aggregation centers. Driven by the minimization of interfacial energy, small nanocrystals can aggregate around the gas/liquid interface between bubble and solvent and finally assemble hollow capsules form. Recently, this gas bubble templating was successfully applied to fabricating hollow capsules of metal (chainlike Co [240]), oxides (hierarchical ZnO, robust TiO_2, ferromagnetic Fe_3O_4, shaped VOOH), sulfides (nanometer-sized ZnS, CuS, and PbS), and selenides (micrometer-sized CdSe and ZnSe). The attachment of nanoparticles to a gas bubble is a complex and random process, which is affected by many factors, such as reaction system conditions, nanoparticle size, and properties. The past years have witnessed a certain degree of development in this area, however, it is still hard to control the size, monodispersity, and surface structure of final hollow capsules. ZnSe hollow spheres have been synthesized under hydrothermal conditions with hydrazine as the reducing agent, in which the resultant N_2 gas bubbles are proposed to play the role of soft templates [241,242]. Gu et al. [243] prepared ZnS hollow nanospheres via aggregation of small nanoparticles around the evolved H_2S bubbles produced from decomposition of thioacetamide. Xie and coworkers synthesized hollow spheres of TiO_2, Co_2P, and VOOH by templating against gas bubbles of O_2 [244], PH_3 and CH_4 [245], and N_2 [246], respectively. Recently, chains of Co hollow submicrospheres were synthesized via aggregation of PVP-stabilized primary nanoparticles, presumably assisted by the gas bubbles liberated from the reaction [240].

Gas bubble templating has also been suggested as a means for preparing hollow spheres of sulfides (ZnS, CuS) and oxides (SnO_2, TiO_2, Fe_3O_4). It should be pointed out that the gas bubble templating mechanism remains highly speculative, especially in systems involving highly soluble gases (CO_2, NH_3) resulting from decomposition of organic molecules such as urea and thiourea.

21.3.6. Silica-coated Core–shell Nanostructures

21.3.6.1. Silica-coated Metal and Quantum Dots Core–shell Nanostructures

Silica-coated metal core–shell nanoparticles (M@ SiO_2) have become increasingly important in the last decade for many promising catalytic and biomedical applications. Adopting the well-established Stöber method, the seminal silica coating of citrate-reduced gold nanoparticles conducted by Liz-Marzn, Mulvaney, and coworkers first involves the weak surface attachment with bi-functional (3-aminopropyl)trimethoxysilane in aqueous solution [247–250]. Recently, Han and coworkers have successfully synthesized highly monodisperse silica-coated Au [251], Ag [252], and Pt [253] nanoparticles of ~50 nm citrate-poststabilized gold nanoparticles with various controlled silica thickness through the Stöber method. Meanwhile (reverse) microemulsion has also been widely used for encapsulating nanoparticles and the produced silica shell has structural robustness, fabrication flexibilities, and handling convenience. Silica coating of different noble metal nanoparticles (Au, Ag, Pt, Rh) and magnetic metal nanoparticles (Fe, Ni, and so forth) has been successfully achieved.

To increase the colloidal and chemical stability, the composite magnetic luminescent spheres were additionally coated with an outer layer of silica. To elucidate how the silica coating affects the photoluminescence, spectra of the composite particles were measured before and after the deposition of the outer silica shell, observing a blue shift of the emission spectra (10 nm). Yi et al. [254] have developed a facile strategy to encapsulate hydrophobic QDs (CdSe) and magnetic nanoparticles (Fe_2O_3) or both QDs and MPs within a silica shell using reverse microemulsion. The as-synthesized hydrophobic trioctylphosphine oxide (TOPO)-capped CdSe QDs can be coated with silica in a direct one-pot reverse microemulsion method, employing Igepal as the nonionic surfactant. The hydrophobic TOPO-capped QDs are encapsulated within a reverse micelle and then in a microemulsion. Besides the use of small molecules, synthetic polymers can also play important roles in protecting metal nanoparticles by means of a steric stabilization mechanism for successful silica coating. One prominent example is PVP, an amphiphilic polymer available commercially in different molecular weights, that has been proven to be efficient for the sufficient coverage/protection of various metal nanoparticles by Graf et al. [255], followed by thorough removal of the excessive PVP molecules in solution through delicate centrifugation or longtime dialysis prior to silica coating. The integration of mesoporous silica with metal particles to form core–shell composite microspheres is undoubtedly of great interest for practical applications. Cationic surfactant such as cetyltrimethylammonium bromide (CTAB) were employed to not only serve as the stabilizing secondary surfactant for the transfer of the nanocrystals to the aqueous phase but also the organic template for the formation of the highly uniform mesoporous silica spheres [91]. By using similar methods, Habas et al. [91] detailed the preparation of thermally stable mesoporous silica-coated platinum catalysts of ~14 nm in size for high-temperature CO oxidation reactions. The core–shell catalysts can stand up to 750°C treatment, which is advantageous as this can drastically prolong the catalysts' lifetime for multiple cycles of catalysis.

21.3.6.2. Mesoporous Silica-coated Magnetic Iron Oxide Core–shell Nanostructures

In order to avoid contact between the magnetic core and the tissue while maintaining the colloidal suspension stability within the biological environment, synthesis of a magnetic core with mesoporous submicrospheres as the shell is the best choice. Mesoporous materials with unique magnetic properties have great potential in biological applications such as bioseparation, drug delivery carriers, cell sorting, diagnostic analysis, and so on [256–274].

Mesoporous silica-coated magnetic materials were first reported by Wu et al. [256]. These magnetic nanocomposites were achieved by deposition of mesoporous silica film on the magnetic cores using CTAC micelles as molecular templates. Ying's group [257] developed a reverse microemulsion method for the synthesis of mesoporous SiO_2-coated Fe_2O_3, using homogeneous SiO_2-coated Fe_2O_3 (SiO_2/Fe_2O_3) nanoparticles as a template.

Combining self-assembling, phase transfer, and microemulsion techniques, using CTAB as structural agent and co-agent for transforming magnetic particles from oil to water, several groups successfully encapsulated the magnetic into the mesoporous silica nanospheres [258–264].

By using phase transfer method, Hyeon's group [258] first reported magnetic fluorescent delivery vehicle using uniform mesoporous silica spheres embedded with monodisperse magnetic and semiconductor nanocrystals. In this approach, hydrophobic ligand-capped magnetic nanocrystals were first transferred from the organic phase to the aqueous phase by cetyltrimethylammonium bromide (CTAB). The subsequent sol–gel reaction of tetraethyl orthosilicate (TEOS) in an aqueous solution containing CTAB (and oleic acid)-stabilized magnetite nanocrystals resulted in magnetic mesoporous silica core shell structures. CTAB served as not only the stabilizing secondary surfactant for the transfer of the nanocrystals to the aqueous phase but also as the organic template for the formation of the mesoporous silica spheres. Later, the same group [259] reported synthesis of discrete and monodisperse mesoporous silica nanoparticles (NPs) consisting of single Fe_3O_4 nanocrystal core and mesoporous silica shell. The size of the mesoporous silica NPs can be easily controlled below 100 nm. The multifunctional bioapplications of the core–shell NPs for simultaneous magnetic resonance (MR) and fluorescence imaging, and for drug delivery were also demonstrated. Liong et al. [260] encapsulated superparamagnetic iron oxide nanocrystals into mesostructured silica spheres, then labeled with fluorescent dye molecules and coated with hydrophilic groups to prevent aggregation. Fluorescence imaging and cell viability assays show that the nanoparticles deliver water-insoluble drug molecules into cells and have increased specificity toward cancer cells.

We have synthesized the magnetic silica nanospheres (MSNs) with a highly ordered mesostructure [261]. Different from previous studies, the produced mesoporous magnetic silica nanospheres possess not only a uniform nanosize (90–140 nm) but also a highly ordered mesostructure. More importantly, the pore size can be tuned from 2.4 to 3.4 nm and the saturation magnetization values can be tuned from 0.43 emu g^{-1} to 7.62 emu g^{-1}. Binary adsorption and desorption of proteins cytochrome *c* (cyt *c*) and bovine serum albumin (BSA) demonstrate that MSNs are an effective and highly selective adsorbent for proteins with different molecular sizes. It is worthy to mention that magnetic nanocrystals have been successfully encapsulated into meosporous silicas with different morphologies such as hollow spheres [262], helical nanorods [263] recently by our group.

Recently, Lin et al. [264] used ammonium hydroxide as a basic catalyst and one-pot synthesis conditions to produce size tunable, sub-100-nm diameter, well-ordered multifunctional mesoporous magnetic silica nanocompomposties with sufficient magnetic loading. The PEG-modified multifunctional mesoporous magnetic silica nanocomposites exhibit excellent colloidal stability in both water and PBS solutions.

By hydrolysis of TEOS in oil-in-water reverse microemulsion, Lin et al. [265] reported the synthesis of uniform magnetic/luminescent functional mesoporous silica nanoparticles with well-ordered porous structure and good aqueous dispersity. The resulted magnetic silica nanocomposites were formed by fusing one or two $Fe_3O_4@SiO_2$ and one mesoporous silica nanoparticle, and they were able to internalize into NIH3T3 cells without any other uptake enhancing techniques. Ruiz-Hernandez et al.'s group [266] also presented an aerosol-assisted route to synthesize mesoporous silica microspheres encapsulating magnetic nanoparticles.

Fabricating uniform magnetic nanospheres with a large magnetic core/mesoporous silica shell structure has drawn intense scientific and technological interests because they possessed a wide range of applications in magnetic resonance imaging, catalysis, environmental, and biomedical fields [267–274]. Uniform magnetic silica nanospheres (MSNs) with a diameter of 270 nm were prepared by a sol–gel processing of silica on hematite particles followed by H_2 reduction [267,268]. The material had a high magnetization value, but its ferromagnetic property may limit its practical application in some areas. On the basis of this work, Kim et al. [269] further introduced nickel acetate into the mesopores of the MSNs by the wet impregnation method. The mesopores played an important role in loading large amounts of Ni^{2+} salts. After reduction in a flowing mixture of H_2 and N_2, magnetite core with Ni nanoparticles on the surface were formed. The exposed NiO nanoparticles provide for a selective adsorption of His-tagged protein from the mixed-protein solution, as well as *Escherichia coli* (*E. coli*) lysate, and the magnetic core allows the particles to be separated from the solution by applying an external magnetic field.

More recently, sandwich structured magnetic silica nanocomposites with large nanosphere magnetite core (100–200 nm) and ordered mesoporous silica shells were subject to extensive research due to their high magnetic response [270–274]. In particular, synthesis of superparamagnetic microspheres with Fe_3O_4 core and perpendicularly aligned mesoporous SiO_2 shell and their application for water treatment were reported by Zhao's group [270]. Later, the same group [271] reported a hematite ($-Fe_2O_3$) core–shell mesoporous nanocomposite hydrothermal synthesis of hematite with a mean size of 180–360 nm and octahedron-like morphology. The nanocomposites possess a uniform mesostructured layer with a thickness of 28 nm, and ordered perpendicular mesochannels. Using the similar method, Yang et al. [272] further functionalized the outer silica shell by the deposition of YVO4:Eu^{3+} phosphors, realizing a sandwich structured material with mesoporous, magnetic and luminescent properties. In their further work [273], monodisperse core–shell-structured $Fe_3O_4@nSiO_2@mSiO_2@NaYF_4:Yb^{3+}$, Er^{3+}/Tm^{3+} nanocomposites with mesoporous, up-conversion luminescent, and magnetic properties were successfully synthesized. This multifunctional composite was applied to the loading and controlled release of ibuprofen.

By using special surfactant 3-(trimethoxysily)propyl-octadecyldimethyl-ammonium chloride (TPODAC) with reactive group $-Si(OCH_3)_3$ on its hydrophilic groups, Stucky's group [274] independently reported synthesis of magnetic core (200 nm) mesoporous silica shell nanocomposites. The cooperative assembly of surfactant with silica precursors solidifies micelle arrays in the mesostructured silica framework, leading to the confinement of large amounts of surfactant micelles in a small volume. The results showed that magnetic permanently confined micelle arrays (Mag-PCMAs) are reusable sorbents for fast, convenient, and highly efficient removal of hydrophobic organic compounds (HOCs) from contaminated media.

21.4. CONCLUDING REMARKS

The ability to engineer materials at a nanometer length scale has sparked interest across many scientific disciplines and has enabled direct investigation into the fundamental size-dependent properties of matter. The successes in synthesis of nanomaterials have provided opportunities to tune their mechanical, optical, electrical, chemical, and other properties. In the past decade, considerable progress has been made in synthesis and applications of nanomaterials. This chapter has summarized the synthetic chemistry in the field of nanomaterials. The synthetic strategies for nanomaterials can be broadly categorized into two groups: (1) the top-down approach and (2) the bottom-up technique, the latter is by far the more common and effective for morphology controlled nanomaterials. As evidenced by the described method and discussed literature, a basic understanding of the mechanism and recent progress in these processes have enabled the fabrication and preparation of various nanomaterials with novel functionalities, compositions, and architectures. In addition, many of the bottom-up methods for synthesis of nanomaterials are based on solution synthesis, in which the concentration of precursors is usually very low, typically in the millimolar range. Scaling-up these syntheses to produce commercial-scale quantities for applications is expected to introduce significant challenges for the control of size, morphology, and structure. This aspect of nanomaterials synthesis, though still in its infancy, offers exciting opportunities for newcomers to the field.

A pertinent question after the discovery of nanomaterials for many years is what next? Beyond the obvious extensions already outlined in the previous chapters, one of the new frontiers will likely involve higher levels of complexity in the form of hierarchical architectures. This is a "global" way of thinking about nanomaterials, where, for example, nanomaterials with a particular form such as clusters, colloids, wires, rods, spirals, rings, and sheets can be self-assembled to hierarchical structures through specific chemical bonding and interfacial interactions to create purposeful constructs, and then to investigate the structure–function relationship.

Many of the approaches for constructing nanomaterials have not yet been refined and much remains to be accomplished with respect to fabricating these particles. It is envisioned that future research will, to a large extent, continue to be focused on optimizing existing approaches, as well as developing new procedures, in order to create multifunctional nanomaterials, to fabricate ultra-small nanoparticles (<2 nm), and to assemble nanomaterials into hierarchical architectures. Further into the future, we expect the top-down and bottom-up approaches to open many unique approaches for the synthesis of nanomaterials. With the methods and tools to control synthesis of nanomaterials in hand, the focus should be on engineering applications and the fabrication of devices.

References

[1] C.N.R. Rao, A. Müller, A.K. Cheetham, Nanomaterials Chemistry: Recent Developments and New Directions, Wiley-VCH, Weinheim, Germany, 2007.
[2] C.N.R. Rao, A. Müller, A.K. Cheetham, The Chemistry of Nanomaterials: Synthesis, Properties and Applications, Wiley-VCH, Weinheim, Germany, 2004.
[3] G.A. Ozin, A.C. Arsenault, Nanochemistry: A Chemical Approach to Nanomateials, RSC Publishing, Cambridge, 2005.
[4] L. Cademartiri, G.A. Ozin, Concepts of Nanochemistry, Wiley-VCH, Weinheim, 2009.
[5] P.D. Yang, The Chemistry of Nanostructured Materials, World Scientific Publishing, Singapore, 2003.
[6] G. Cao, Nanostructures & Nanomaterials, Synthesis, Properties & Applications, Imperial College Press, London, 2004.
[7] G.B. Sergeev, Nanochemistry, Elsevier, Amsterdam, 2006.
[8] E. Ruiz-Hitzky, K. Ariga, Y. Lvov, Bio-inorganic Hybrid Nanomaterials. Strategies, Syntheses, Characterization and Applications, Wiley-VCH, Weinheim, 2008.
[9] K.J. Klabunde, Nanoscale Materials in Chemistry, John Wiley & Sons, Inc., New York, USA, 2001.
[10] J.Z. Zhang, Z.L. Wang, J. Liu, S.W. Chen, G.Y. Liu, Self-assembled Nanostructures, Kluwer Academic Publishers, New York, USA, 2004.
[11] A. Sayari, M. Jaroniec, Nanoporous Materials, World Scientific Publishing, Singapore, 2008.
[12] C.C. Koch, Nanostructured Materials Processing, Properties, and Applications, William Andrew Inc, 2007.
[13] G.A. Ozin, Advanced Materials 4 (1992) 612.
[14] G.A. Somorjai, H. Frei, J.Y. Park, Journal of the American Chemical Society 131 (2009) 16589.
[15] J.Y. Ying, Chemical Engineering Science 61 (2006) 1540.
[16] Y. Yin, A.P. Alivisatos, Nature 437 (2005) 664.
[17] M. Grzelczak, J. Perez-Juste, P. Mulvaney, L.M. Liz-Marzan, Chemical Society Reviews 37 (2008) 1783.
[18] Q.H. Yang, J. Liu, in: Y. Gao (Ed.), Research Signpost, Kerala, 2010.
[19] S. Mann, Nature Materials 8 (2009) 781.
[20] S. Mann, Angewandte Chemie International Edition 47 (2008) 5306.
[21] Y.N. Xia, G.M. Whitesides, Angewandte Chemie International Edition 37 (1998) 551.
[22] W.H. Suh, K.S. Suslick, G.D. Stucky, Y.H. Suh, Progress in Neurobiology 87 (2009) 133.
[23] Y. Xia, Advanced Materials 16 (2004) 1245.
[24] I. Bilecka, M. Niederberger, Nanoscale, 2010.
[25] Q.H. Yang, J. Liu, L. Zhang, C. Li, Journal of Materials Chemistry 19 (2009) 1945.
[26] Y. Wan, H.F. Yang, D.Y. Zhao, Accounts of Chemical Research 39 (2006) 423.

[27] Y. Wan, D.Y. Zhao, Chemical Reviews 107 (2007) 2821.
[28] F. Schuth, W. Schmidt, Advanced Materials 14 (2002) 629.
[29] F. Schuth, Chemistry of Materials 13 (2001) 3184.
[30] J.Y. Ying, C.P. Mehnert, M.S. Wong, Angewandte Chemie International Edition 38 (1999) 56.
[31] M.E. Davis, Nature 417 (2002) 813.
[32] S.G. Kwon, T. Hyeon, Accounts of Chemical Research 41 (2008) 1696.
[33] J. Park, J. Joo, S.G. Kwon, Y. Jang, T. Hyeon, Angewandte Chemie International Edition 46 (2007) 4630.
[34] T. Hyeon, Chemical Communications (2003) 927.
[35] Y.W. Jun, J.S. Choi, J. Cheon, Chemical Communications (2007) 1203.
[36] Y.W. Jun, J.S. Choi, J. Cheon, Angewandte Chemie International Edition 45 (2006) 3414.
[37] A.H. Lu, E.L. Salabas, F. Schuth, Angewandte Chemie International Edition 46 (2007) 1222.
[38] Y.N. Xia, P.D. Yang, Y.G. Sun, Y.Y. Wu, B. Mayers, B. Gates, et al., Advanced Materials 15 (2003) 353.
[39] X. Wang, Y.D. Li, Inorganic Chemistry 45 (2006) 7522.
[40] C.M. Lieber, Z.L. Wang, MRS Bulletin 32 (2007) 99.
[41] L. Cademartiri, G.A. Ozin, Advanced Materials 21 (2009) 1013.
[42] A.I. Hochbaum, P.D. Yang, Chemical Reviews 110 (2010) 527.
[43] B. Tian, T.J. Kempa, C.M. Lieber, Chemical Society Reviews 38 (2009) 16.
[44] P.D. Yang, Dalton Transactions (2008) 4387.
[45] X. Duan, Y. Huang, Y. Cui, C.M. Lieber, in: M.A. Reed, T. Lee (Eds.), American Scientific Publishers, 2003, p. 199.
[46] A.L. Briseno, S.C.B. Mannsfeld, S.A. Jenekhe, Z. Bao, Y. Xia, Materials Today 11 (2008) 38.
[47] Y. Zhao, L. Jiang, Advanced Materials 21 (2009) 3621.
[48] X.W. Lou, L.A. Archer, Z.C. Yang, Advanced Materials 20 (2008) 3987.
[49] Z. Ma, J.H. Yu, S. Dai, Advanced Materials 22 (2010) 261.
[50] C.M. Cobley, Y.N. Xia, Elements 5 (2009) 309.
[51] Y. Xia, Y.J. Xiong, B. Lim, S.E. Skrabalak, Angewandte Chemie International Edition 48 (2009) 60.
[52] B. Wiley, Y.G. Sun, B. Mayers, Y.N. Xia, Chemistry – A European Journal 11 (2005) 454.
[53] S.E. Skrabalak, J.Y. Chen, Y.G. Sun, X.M. Lu, L. Au, C.M. Cobley, et al., Accounts of Chemical Research 41 (2008) 1587.
[54] B. Wiley, Y.G. Sun, Y. Xia, Accounts of Chemical Research 40 (2007) 1067.
[55] Y.J. Xiong, B.T. Mayers, Y.N. Xia, Chemical Communications (2005) 5013.
[56] C.J. Brinker, G.W. Scherer, Sol-gel Science, The Physics and Chemistry of Sol-gel Processing, Academic Press, Harcourt Brace & Company, Publishers, 1988.
[57] R. Corriu, N.T. Anh, Molecular Chemistry of Sol-Gel Derived Nanomaterials, Wiley-VCH, Weinheim, 2009.
[58] J.D. Mackenzie, E.P. Bescher, Accounts of Chemical Research 40 (2007) 810.
[59] K. Ariga, H.S. Nalwa, American Scientific Publishers, Los Angeles, 2009.
[60] Y.J. Wang, A.D. Price, F. Caruso, Journal of Materials Chemistry 19 (2009) 6451.
[61] Y. Wang, A.S. Angelatos, F. Caruso, Chemistry of Materials 20 (2008) 848.
[62] J.F. Quinn, A.P.R. Johnston, G.K. Such, A.N. Zelikin, F. Caruso, Chemical Society Reviews 36 (2007) 707.
[63] F. Caruso, R.A. Caruso, H. Mohwald, Science 282 (1998) 1111.
[64] Y.N. Xia, J.A. Rogers, K.E. Paul, G.M. Whitesides, Chemical Reviews 99 (1999) 1823.
[65] M. Haga, K. Kobayashi, K. Terada, Coordination Chemistry Reviews 251 (2007) 2688.
[66] M.A.M. Gijs, F. Lacharme, U. Lehmann, Chemical Reviews 110 (2010) 1518.
[67] A. Gedanken, Ultrasonics Sonochemistry 11 (2004) 47.
[68] Z.G. Li, Z. Jia, Y.X. Luan, T.C. Mu, Current Opinion in Solid State & Materials Science 12 (2008) 1.
[69] D. Qin, Y.N. Xia, G.M. Whitesides, Nature Protocols 5 (2010) 491.
[70] V. Meenakshi, Y. Babayan, T.W. Odom, Journal of Chemical Education 84 (2007) 1795.
[71] T.R. Zhang, Q. Zhang, J.P. Ge, J. Goebl, M.W. Sun, Y.S. Yan, et al., Journal of Physical Chemistry C 113 (2009) 3168.
[72] T.R. Zhang, J.P. Ge, Y.X. Hu, Q. Zhang, S. Aloni, Y.D. Yin, Angewandte Chemie International Edition 47 (2008) 5806.
[73] Q. Zhang, T.R. Zhang, J.P. Ge, Y.D. Yin, Nano Letters 8 (2008) 2867.
[74] J.P. Ge, Q. Zhang, T.R. Zhang, Y.D. Yin, Angewandte Chemie International Edition 47 (2008) 8924.
[75] S.J. Park, Y.J. Kim, Langmuir 24 (2008) 12134.
[76] K. Ariga, T. Kunitake, Accounts of Chemical Research 31 (1998) 371.
[77] J.C. Love, L.A. Estroff, J.K. Kriebel, R.G. Nuzzo, G.M. Whitesides, Chemical Reviews 105 (2005) 1103.
[78] G. Decher, J.D. Hong, Berichte Der Bunsen-Gesellschaft-Physical Chemistry Chemical Physics 95 (1991) 1430.
[79] S.F. Hou, C.C. Harrell, L. Trofin, P. Kohli, C.R. Martin, Journal of the American Chemical Society 126 (2004) 5674.
[80] F. Cheng, Z. Tao, J. Liang, J. Chen, Chemistry of Materials 20 (2008) 667.
[81] Z. Yin, D. Ma, X.H. Bao, Chemical Communications 46 (2010) 1344.
[82] T.K. Sau, A.L. Rogach, F. Jäckel, T.A. Klar, J. Feldmann, Advanced Materials 22 (2010) 1805.
[83] S. Dai, Y.H. Ju, H.J. Gao, J.S. Lin, S.J. Pennycook, C.E. Barnes, Chemical Communications (2000) 243.
[84] J. Dupont, G.S. Fonseca, A.P. Umpierre, P.F.P. Fichtner, S.R. Teixeira, Journal of the American Chemical Society 124 (2002) 4228.
[85] T. Nakashima, N. Kimizuka, Journal of the American Chemical Society 125 (2003) 6386.
[86] Y. Zhou, M. Antonietti, Journal of the American Chemical Society 125 (2003) 14960.
[87] A. Taubert, Angewandte Chemie International Edition 43 (2004) 5380.
[88] E.R. Cooper, C.D. Andrews, P.S. Wheatley, P.B. Webb, P. Wormald, R.E. Morris, Nature 430 (2004) 1012.
[89] P. Wasserscheid, T. Welton, in: P. Wasserscheid, T. Welton (Eds.), Wiley-VCH, Weinheim, 2008.
[90] N.R. Jana, L. Gearheart, C.J. Murphy, Journal of Physical Chemistry B 105 (2001) 4065.
[91] S.E. Habas, H. Lee, V. Radmilovic, G.A. Somorjai, P. Yang, Nature Materials 6 (2007) 692.
[92] S.E. Skrabalak, L. Au, X.D. Li, Y. Xia, Nature Protocols 2 (2007) 2182.
[93] A.I. Kryukov, A.L. Stroyuk, N.N. Zin'chuk, A.V. Korzhak, S.Y. Kuchmii, Journal of Molecular Catalysis A – Chemical 221 (2004) 209.
[94] A. Tao, P. Sinsermsuksakul, P.D. Yang, Angewandte Chemie International Edition 45 (2006) 4597.
[95] B.L. Cushing, V.L. Kolesnichenko, C.J. O'Connor, Chemical Reviews 104 (2004) 3893.
[96] S.H. Sun, C.B. Murray, D. Weller, L. Folks, A. Moser, Science 287 (2000) 1989.
[97] S.H. Sun, H. Zeng, Journal of the American Chemical Society 124 (2002) 8204.
[98] H. Zeng, J. Li, J.P. Liu, Z.L. Wang, S.H. Sun, Nature 420 (2002) 395.

[99] S.H. Sun, H. Zeng, D.B. Robinson, S. Raoux, P.M. Rice, S.X. Wang, et al., Journal of the American Chemical Society 126 (2004) 273.
[100] C. Wang, H. Daimon, T. Onodera, T. Koda, S.H. Sun, Angewandte Chemie International Edition 47 (2008) 3588.
[101] J. Park, K.J. An, Y.S. Hwang, J.G. Park, H.J. Noh, J.Y. Kim, et al., Nature Materials 3 (2004) 891.
[102] N.R. Jana, Y.F. Chen, X.G. Peng, Chemistry of Materials 16 (2004) 3931.
[103] Z. Li, Q. Sun, M.Y. Gao, Angewandte Chemie International Edition 44 (2005) 123
[104] F.Q. Hu, L. Wei, Z. Zhou, Y.L. Ran, Z. Li, M.Y. Gao, Advanced Materials 18 (2006) 2553.
[105] A. Bee, R. Massart, S. Neveu, Journal of Magnetism and Magnetic Materials 149 (1995) 6.
[106] T. Ishikawa, S. Kataoka, K. Kandori, Journal of Materials Science 28 (1993) 2693.
[107] D. Kim, N. Lee, M. Park, B.H. Kim, K. An, T. Hyeon, Journal of the American Chemical Society 131 (2009) 454.
[108] M. Bruchez, M. Moronne, P. Gin, S. Weiss, A.P. Alivisatos, Science 281 (1998) 2013.
[109] V.F. Puntes, K.M. Krishnan, A.P. Alivisatos, Science 291 (2001) 2115.
[110] H.M. Zheng, R.K. Smith, Y.W. Jun, C. Kisielowski, U. Dahmen, A.P. Alivisatos, Science 324 (2009) 1309.
[111] X.F. Duan, Y. Huang, R. Agarwal, C.M. Lieber, Nature 421 (2003) 241.
[112] L.J. Lauhon, M.S. Gudiksen, C.L. Wang, C.M. Lieber, Nature 420 (2002) 57.
[113] M.S. Gudiksen, L.J. Lauhon, J. Wang, D.C. Smith, C.M. Lieber, Nature 415 (2002) 617
[114] Y. Huang, X.F. Duan, Y. Cui, L.J. Lauhon, K.H. Kim, C.M. Lieber, Science 294 (2001) 1313.
[115] J.F. Wang, M.S. Gudiksen, X.F. Duan, Y. Cui, C.M. Lieber, Science 293 (2001) 1455.
[116] Y. Cui, Q.Q. Wei, H.K. Park, C.M. Lieber, Science 293 (2001) 1289.
[117] R.S. Wagner, W.C. Ellis, Applied Physics Letters 4 (1964) 89.
[118] Y.C. Lu, J. Zhong, in: T.E. Steiner (Ed.), Artech House, Inc., Norwood, MA., 2004, p. 191.
[119] R.S. Wagner, A.P. Levitt, Whisker Technology, Wiley-Interscience, New York, 1975.
[120] M.H. Huang, Y.Y. Wu, H. Feick, N. Tran, E. Weber, P.D. Yang, Advanced Materials 13 (2001) 113.
[121] J.T. Wang, Springer Verlag, Berlin, 2002.
[122] B. Bhushan, Springer Handbook of Nanotechnology, Spinger-Verlag, Berlin.
[123] C. Wang, Y.J. Hu, C.M. Lieber, S.H. Sun, Journal of the American Chemical Society 130 (2008) 8902.
[124] B.Z. Tian, P. Xie, T.J. Kempa, D.C. Bell, C.M. Lieber, Nature Nanotechnology 4 (2009) 824.
[125] B.Z. Tian, X.L. Zheng, T.J. Kempa, Y. Fang, N.F. Yu, G.H. Yu, et al., Nature 449 (2007) 885.
[126] J. Xiang, A. Vidan, M. Tinkham, R.M. Westervelt, C.M. Lieber, Nature Nanotechnology 1 (2006) 208.
[127] J. Xiang, W. Lu, Y.J. Hu, Y. Wu, H. Yan, C.M. Lieber, Nature 441 (2006) 489.
[128] C. Yang, Z.H. Zhong, C.M. Lieber, Science 310 (2005) 1304.
[129] F. Patolsky, B.P. Timko, G.F. Zheng, C.M. Lieber, MRS Bulletin 32 (2007) 142.
[130] A.L. Briseno, T.W. Holcombe, A.I. Boukai, E.C. Garnett, S.W. Shelton, J.J.M. Frechet, et al., Nano Letters 10 (2010) 334.
[131] R.X. Yan, P. Pausauskie, J.X. Huang, P.D. Yang, Proceedings of the National Academy of Sciences of the United States of America 106 (2009) 21045.
[132] R.X. Yan, D. Gargas, P.D. Yang, Nature Photonics 3 (2009) 569.
[133] E.C. Garnett, Y.C. Tseng, D.R. Khanal, J.Q. Wu, J. Bokor, P.D. Yang, Nature Nanotechnology 4 (2009) 311.
[134] B.D. Yuhas, P.D. Yang, Journal of the American Chemical Society 131 (2009) 3756.
[135] Y.J. Hwang, A. Boukai, P.D. Yang, Nano Letters 9 (2009) 410.
[136] E.C. Garnett, P.D. Yang, Journal of the American Chemical Society 130 (2008) 9224.
[137] A.R. Tao, J.X. Huang, P.D. Yang, Accounts of Chemical Research 41 (2008) 1662.
[138] A.I. Hochbaum, R.K. Chen, R.D. Delgado, W.J. Liang, E.C. Garnett, M. Najarian, et al., Nature 451 (2008) 163.
[139] T. Kuykendall, P. Ulrich, S. Aloni, P. Yang, Nature Materials 6 (2007) 951.
[140] T.J. Trentler, K.M. Hickman, S.C. Goel, A.M. Viano, P.C. Gibbons, W.E. Buhro, Science 270 (1995) 1791.
[141] Z. Li, O. Kurtulus, N. Fu, Z. Wang, A. Kornowski, U. Pietsch, et al., Advanced Functional Materials 19 (2009) 3650.
[142] J.W. Sun, W.E. Buhro, L.W. Wang, J. Schrier, Nano Letters 8 (2008) 2913.
[143] Z. Li, L.N. Cheng, Q. Sun, Z.H. Zhu, M.J. Riley, M. Aljada, et al., Angewandte Chemie International Edition 49 (2010) 9033.
[144] X.M. Lu, M.S. Yavuz, H.Y. Tuan, B.A. Korgel, Y.N. Xia, Journal of the American Chemical Society 130 (2008) 8900.
[145] X.G. Peng, L. Manna, W.D. Yang, J. Wickham, E. Scher, A. Kadavanich, et al., Nature 404 (2000) 59.
[146] T. Yu, J. Joo, Y.I. Park, T. Hyeon, Journal of the American Chemical Society 128 (2006) 1786.
[147] N. Pradhan, H.F. Xu, X.G. Peng, Nano Letters 6 (2006) 720.
[148] H.G. Yang, C.H. Sun, S.Z. Qiao, J. Zou, G. Liu, S.C. Smith, et al., Nature 453 (2008) 638.
[149] H.G. Yang, G. Liu, S.Z. Qiao, C.H. Sun, Y.G. Jin, S.C. Smith, et al., Journal of the American Chemical Society 131 (2009) 4078.
[150] X.G. Han, Q. Kuang, M.S. Jin, Z.X. Xie, L.S. Zheng, Journal of the American Chemical Society 131 (2009) 3152.
[151] B.H. Wu, C.Y. Guo, N.F. Zheng, Z.X. Xie, G.D. Stucky, Journal of the American Chemical Society 130 (2008) 17563.
[152] Y.Q. Dai, C.M. Cobley, J. Zeng, Y.M. Sun, Y.N. Xia, Nano Letters 9 (2009) 2455.
[153] D.Q. Zhang, G.S. Li, X.F. Yang, J.C. Yu, Chemical Communications (2009) 4381.
[154] G. Liu, H.G. Yang, X.W. Wang, L.N. Cheng, J. Pan, G.Q. Lu, et al., Journal of the American Chemical Society 131 (2009) 12868.
[155] G. Liu, H.G. Yang, X.W. Wang, L.N. Cheng, H.F. Lu, L.Z. Wang, et al., Journal of Physical Chemistry C 113 (2009) 21784.
[156] G. Liu, C.H. Sun, H.G. Yang, S.C. Smith, L.Z. Wang, G.Q. Lu, et al., Chemical Communications 46 (2010) 755.
[157] F. Amano, T. Yasumoto, O.O. Prieto-Mahaney, S. Uchida, T. Shibayama, B. Ohtani, Chemical Communications (2009) 2311.
[158] F. Amano, O.O. Prieto-Mahaney, Y. Terada, T. Yasumoto, T. Shibayama, B. Ohtani, Chemistry of Materials 21 (2009) 2601.
[159] Y. Alivov, Z.Y. Fan, Journal of Physical Chemistry C 113 (2009) 12954.
[160] J.M. Li, D.S. Xu, Chemical Communications 46 (2010) 2301.
[161] M. Liu, L.Y. Piao, L. Zhao, S.T. Ju, Z.J. Yan, T. He, et al., Chemical Communications 46 (2010) 1664.
[162] X.G. Han, M.S. Jin, S.F. Xie, Q. Kuang, Z.Y. Jiang, Y.Q. Jiang, et al., Angewandte Chemie International Edition 48 (2009) 9180.
[163] Y.Y. Ma, Q. Kuang, Z.Y. Jiang, Z.X. Xie, R.B. Huang, L.S. Zheng, Angewandte Chemie-International Edition 47 (2008) 8901.
[164] N. Tian, Z.Y. Zhou, S.G. Sun, Y. Ding, Z.L. Wang, Science 316 (2007) 732.

[165] X.W. Xie, Y. Li, Z.Q. Liu, M. Haruta, W.J. Shen, Nature 458 (2009) 746.
[166] X.W. Xie, W.J. Shen, Nanoscale 1 (2009) 50.
[167] D.B. Fan, P.J. Thomas, P. O'Brien, Journal of the American Chemical Society 130 (2008) 10892.
[168] T. Ming, W. Feng, Q. Tang, F. Wang, L.D. Sun, J.F. Wang, et al., Journal of the American Chemical Society 131 (2009) 16350.
[169] G.C. Xi, J.H. Ye, Chemical Communications 46 (2010) 1893.
[170] J. Liu, F. Liu, K. Gao, J.S. Wu, D.F. Xue, Journal of Materials Chemistry 19 (2009) 6073.
[171] M. Chen, L.M. Wu, S.X. Zhou, B. You, Advanced Materials 18 (2006) 801.
[172] I. Tissot, J.P. Reymond, F. Lefebvre, E. Bourgeat-Lami, Chemistry of Materials 14 (2002) 1325.
[173] M. Sanles-Sobrido, V. Salgueirino-Maceira, M.A. Correa-Duarte, L.M. Liz-Marzan, Small 4 (2008) 583.
[174] Y. Wan, S.H. Yu, Journal of Physical Chemistry C 112 (2008) 3641.
[175] G.S. Zhu, S.L. Qiu, O. Terasaki, Y. Wei, Journal of the American Chemical Society 123 (2001) 7723.
[176] S.B. Yoon, J.Y. Kim, J.H. Kim, S.G. Park, C.W. Lee, J.S. Yu, Current Applied Physics 6 (2006) 1059.
[177] Y. Le, J.F. Chen, J.X. Wang, L. Shao, W.C. Wang, Materials Letters 58 (2004) 2105.
[178] W.R. Zhao, M.D. Lang, Y.S. Li, L. Li, J.L. Shi, Journal of Materials Chemistry 19 (2009) 2778.
[179] J.F. Chen, H.M. Ding, J.X. Wang, L. Shao, Biomaterials 25 (2004) 723.
[180] M. Darbandi, R. Thomann, T. Nann, Chemistry of Materials 19 (2007) 1700.
[181] S. Schacht, Q. Huo, I.G. VoigtMartin, G.D. Stucky, F. Schuth, Science 273 (1996) 768.
[182] Q.Y. Sun, P.J. Kooyman, J.G. Grossmann, P.H.H. Bomans, P.M. Frederik, P. Magusin, et al., Advanced Materials 15 (2003) 1097.
[183] Q. Sun, P. Magusin, B. Mezari, P. Panine, R.A. van Santen, N. Sommerdijk, Journal of Materials Chemistry 15 (2005) 256.
[184] H.J. Zhang, J. Wu, L.P. Zhou, D.Y. Zhang, L.M. Qi, Langmuir 23 (2007) 1107.
[185] Y.S. Lin, S.H. Wu, C.T. Tseng, Y. Hung, C. Chang, C.Y. Mou, Chemical Communications (2009) 3542.
[186] W.J. Li, X.X. Sha, W.J. Dong, Z.C. Wang, Chemical Communications (2002) 2434.
[187] G. Fornasieri, W. Badaire, R. Backov, O. Mondain-Monval, U. Zakri, P. Poulin, Advanced Materials 16 (2004) 1094.
[188] C.Z. Yu, B.H. Tian, J. Fan, G.D. Stucky, D.Y. Zhao, Chemistry Letters (2002) 62.
[189] H.M. Chen, J.H. He, H.M. Tang, C.X. Yan, Chemistry of Materials 20 (2008) 5894.
[190] M. Fujiwara, K. Shiokawa, Y. Tanaka, Y. Nakahara, Chemistry of Materials 16 (2004) 5420.
[191] J.W. Wang, Y.D. Xia, W.X. Wang, R. Mokaya, M. Poliakoff, Chemical Communications (2005) 210.
[192] J.W. Wang, Y.D. Xia, W.X. Wang, M. Poliakoff, R. Mokaya, Journal of Materials Chemistry 16 (2006) 1751.
[193] Y.S. Li, J.L. Shi, Z.L. Hua, H.R. Chen, M.L. Ruan, D.S. Yan, Nano Letters 3 (2003) 609.
[194] C.E. Fowler, D. Khushalani, S. Mann, Chemical Communications (2001) 2028.
[195] C.E. Fowler, D. Khushalani, S. Mann, Journal of Materials Chemistry 11 (2001) 1968.
[196] C.E. Fowler, D. Khushalani, B. Lebeau, S. Mann, Advanced Materials 13 (2001) 649.
[197] H.P. Lin, Y.R. Cheng, C.Y. Mou, Chemistry of Materials 10 (1998) 3772.
[198] H.P. Lin, C.Y. Mou, S.B. Liu, C.Y. Tang, Chemical Communications (2001) 1970.
[199] J. Liu, C.M. Li, Q.H. Yang, J. Yang, C. Li, Langmuir 23 (2007) 7255.
[200] X.F. Zhou, S.Z. Qiao, N. Hao, X.L. Wang, C.Z. Yu, L.Z. Wang, et al., Chemistry of Materials 19 (2007) 1870.
[201] R. Schiller, C.K. Weiss, J. Geserick, N. Husing, K. Landfester, Chemistry of Materials 21 (2009) 5088.
[202] S.H. Han, W.G. Hou, J. Xu, Z.M. Li, Colloid and Polymer Science 282 (2004) 1286.
[203] J.G. Wang, Q. Xiao, H.J. Zhou, P.C. Sun, Z.Y. Yuan, B.H. Li, et al., Advanced Materials 18 (2006) 3284.
[204] B. Peng, M. Chen, S.X. Zhou, L.M. Wu, X.H. Ma, Journal of Colloid and Interface Science 321 (2008) 67.
[205] L. Li, J. Ding, J.M. Xue, Chemistry of Materials 21 (2009) 3629.
[206] Y.J. Zhao, J.L. Zhang, W. Li, C.X. Zhang, B.X. Han, Chemical Communications (2009) 2365.
[207] S.P. Naik, A.S.T. Chiang, R.W. Thompson, F.C. Huang, Chemistry of Materials 15 (2003) 787.
[208] W. Pan, J.W. Ye, G.L. Ning, Y. Lin, J. Wang, Materials Research Bulletin 44 (2009) 280.
[209] Z.G. Teng, Y.D. Han, J. Li, F. Yan, W.S. Yang, Microporous and Mesoporous Materials 127 (2010) 67.
[210] M.P. Kapoor, A. Vinu, W. Fujii, T. Kimura, Q.H. Yang, Y. Kasama, et al., Microporous and Mesoporous Materials 128 (2010) 187.
[211] A. Khanal, Y. Inoue, M. Yada, K. Nakashima, Journal of the American Chemical Society 129 (2007) 1534.
[212] J.J. Yuan, O.O. Mykhaylyk, A.J. Ryan, S.P. Armes, Journal of the American Chemical Society 129 (2007) 1717.
[213] Q.S. Huo, J. Liu, L.Q. Wang, Y.B. Jiang, T.N. Lambert, E. Fang, Journal of the American Chemical Society 128 (2006) 6447.
[214] J. Liu, Q.H. Yang, L. Zhang, H.Q. Yang, J.S. Gao, C. Li, Chemistry of Materials 20 (2008) 4268.
[215] J. Liu, F. Fan, Z. Feng, L. Zhang, S. Bai, Q. Yang, et al., Journal of Physical Chemistry C 112 (2008) 16445.
[216] J.S. Gao, J. Liu, S.Y. Bai, P.Y. Wang, H. Zhong, Q.H. Yang, et al., Journal of Materials Chemistry 19 (2009) 8580.
[217] J.T. Tang, J. Liu, P.Y. Wang, H. Zhong, Q.H. Yang, Microporous and Mesoporous Materials 127 (2010) 119.
[218] J. Liu, S.Y. Bai, H. Zhong, C. Li, Q.H. Yang, Journal of Physical Chemistry C 114 (2010) 953.
[219] H. Tan, N.S. Liu, B.P. He, S.Y. Wong, Z.K. Chen, X. Li, et al., Chemical Communications (2009) 6240.
[220] Z.G. Feng, Y.S. Li, D.C. Niu, L. Li, W.R. Zhao, H.R. Chen, et al., Chemical Communications (2008) 2629.
[221] L. Du, H.Y. Song, S.J. Liao, Applied Surface Science 255 (2009) 9365.
[222] L. Du, S.J. Liao, H.A. Khatib, J.F. Stoddart, J.I. Zink, Journal of the American Chemical Society 131 (2009) 15136.
[223] B.Y. Du, Z. Cao, Z.B. Li, A.X. Mei, X.H. Zhang, J.J. Nie, et al., Langmuir 25 (2009) 12367.
[224] S.S. Kim, W.Z. Zhang, T.J. Pinnavaia, Science 282 (1998) 1302.
[225] P.T. Tanev, Y. Liang, T.J. Pinnavaia, Journal of the American Chemical Society 119 (1997) 8616.
[226] T. Ogura, H. Shibata, K. Sakai, H. Sakai, M. Abe, Chemistry Letters 38 (2009) 120.
[227] Y.Q. Yeh, B.C. Chen, H.P. Lin, C.Y. Tang, Langmuir 22 (2006) 6.
[228] H. Djojoputro, X.F. Zhou, S.Z. Qiao, L.Z. Wang, C.Z. Yu, G.Q. Lu, Journal of the American Chemical Society 128 (2006) 6320.
[229] S.Z. Qiao, C.X. Lin, Y.G. Jin, Z. Li, Z.M. Yan, Z.P. Hao, et al., Journal of Physical Chemistry C 113 (2009) 8673.
[230] J. Liu, S.B. Hartono, Y.G. Jin, Z. Li, G.Q.M. Lu, S.Z. Qiao, Journal of Materials Chemistry 20 (2010) 4595.

[231] B. Tan, H.J. Lehmler S.M. Vyas, B.L. Knutson, S.E. Rankin, Advanced Materials 17 (2005) 2368.
[232] X. Gu, C.L. Li, X.H. Liu, J.W. Ren, Y.Q. Wang, Y.L. Guo, et al., Journal of Physical Chemistry C 113 (2009) 6472.
[233] Y. Zhang, M.H. Yu, L. Zhu, X.F. Zhou, Q.F. Zhao, H.X. Li, et al., Chemistry of Materials 20 (2008) 6238.
[234] M.H. Yu, H.N. Wang, X.F. Zhou, P. Yuan, C.Z. Yu, Journal of the American Chemical Society 129 (2007) 14576.
[235] M.H. Yu, J. Zhang, P. Yuan, H.N. Wang, N. Liu, Y.H. Wang, et al., Chemistry Letters 38 (2009) 442.
[236] J. Liu, D.F. Xue, Advanced Materials 20 (2008) 2622.
[237] H.L. Xu, W.Z. Wang, Angewandte Chemie International Edition 46 (2007) 1489.
[238] Y. Piao, J. Kim, H. Bin Na, D. Kim, J.S. Baek, M.K. Ko, et al., Nature Materials 7 (2008) 242.
[239] J.G. Wang, F. Li, H.J Zhou, P.C. Sun, D.T. Ding, T.H. Chen, Chemistry of Materials 21 (2009) 612.
[240] L. Guo, F. Liang, X.G. Wen, S.H. Yang, L. He, W.Z. Zheng, et al., Advanced Functional Materials 17 (2007) 425.
[241] Q. Peng, Y.J. Dong, Y.D. Li, Angewandte Chemie International Edition 42 (2003) 3027.
[242] C.L. Jiang, W.Q. Zhang, G.F. Zou, W.C. Yu, Y.T. Qian, Nanotechnology 16 (2005) 551.
[243] F. Gu, C.Z. Li, S.F. Wang, M.K. Lu, Langmuir 22 (2006) 1329.
[244] X.X. Li, Y.J. Xiong, Z.Q. Li, Y. Xie, Inorganic Chemistry 45 (2006) 3493.
[245] H.W. Hou, Q. Peng, S.Y. Zhang, Q.X. Guo, Y. Xie, European Journal of Inorganic Chemistry (2005) 2625.
[246] C.Z. Wu, Y. Xie, L.Y Lei, S.Q. Hu, C.Z. OuYang, Advanced Materials 18 (2006) 1727.
[247] S.H. Liu, M.Y. Han, Chemistry – An Asian Journal 5 (2010) 36.
[248] L.M. LizMarzan, M. Giersig, P. Mulvaney, Langmuir 12 (1996) 4329.
[249] Y. Lu, Y.D. Yin, Z.Y. Li, Y.A. Xia, Nano Letters 2 (2002) 785.
[250] F. Caruso, M. Spasova V. Saigueirino-Maceira, L.M. Liz-Marzan, Advanced Materials 13 (2001) 1090.
[251] S.H. Liu, M.Y. Han, Advanced Functional Materials 15 (2005) 961.
[252] S.H. Liu, Z.H. Zhang, M.Y. Han, Analytical Chemistry 77 (2005) 2595.
[253] S.H. Liu, Y. Wong, Y.B. Wang, D.S. Wang, M.Y. Han, Advanced Functional Materials 17 (2007) 3147.
[254] D.K. Yi, S.T. Selvan, S.S. Lee, G.C. Papaefthymiou, D. Kundaliya, J.Y. Ying, Journal of the American Chemical Society 127 (2005) 4990.
[255] C. Graf, D.L.J. Vossen, A. Imhof, A. van Blaaderen, Langmuir 19 (2003) 6693.
[256] P.G. Wu, J.H. Zhu, Z.H. Xu, Advanced Functional Materials 14 (2004) 345.
[257] D.K. Yi, S.S. Lee, G.C. Papaefthymiou, J.Y. Ying, Chemistry of Materials 18 (2006) 614.
[258] J. Kim, J.E. Lee, J. Lee, J.H. Yu, B.C. Kim, K. An, et al., Journal of the American Chemical Society 128 (2006) 688.
[259] J. Kim, H.S. Kim, N. Lee, T. Kim, H. Kim, T. Yu, et al., Angewandte Chemie International Edition 47 (2008) 8438.
[260] M. Liong, J. Lu, M. Kovochich, T. Xia, S.G. Ruehm, A.E. Nel, et al., ACS Nano 2 (2008) 889.
[261] L. Zhang, S.Z. Qiao, Y.G. Jin, H.G. Yang, S. Budihartono, F. Stahr, et al., Advanced Functional Materials 18 (2008) 3203.
[262] L. Zhang, S.Z. Qiao, Y.G. Jin, Z.G. Chen, H.C. Gu, G.Q. Lu, Advanced Materials 20 (2008) 805.
[263] L. Zhang, S.Z. Qiao, L.N. Cheng, Z.F. Yan, G.Q. Lu, Nanotechnology 19 (2008) 5608.
[264] Y.S. Lin, C.L. Haynes, Chemistry of Materials 21 (2009) 3979.
[265] Y.S. Lin, S.H. Wu, Y. Hung, Y.H. Chou, C. Chang, M.L. Lin, et al., Chemistry of Materials 18 (2006) 5170.
[266] E. Ruiz-Hernandez, A. Lopez-Noriega, D. Arcos, I. Izquierdo-Barba, O. Terasaki, M. Vallet-Regi, Chemistry of Materials 19 (2007) 3455.
[267] W.R. Zhao, J.L. Gu, L.X. Zhang, H.R. Chen, J.L. Shi, Journal of the American Chemical Society 127 (2005) 8916.
[268] W.R. Zhao, J.L. Shi, H.R. Chen, L.X. Zhang, Journal of Materials Research 21 (2006) 3080.
[269] J. Kim, Y. Piao, N. Lee, Y. Il Park, I.H. Lee, J.H. Lee, et al., Advanced Materials 22 (2010) 57.
[270] Y. Deng, D. Qi, C. Deng, X. Zhang, D. Zhao, Journal of the American Chemical Society 130 (2008) 28.
[271] X.H. Guo, Y.H. Deng, D. Gu, R.C. Che, D.Y. Zhao, Journal of Materials Chemistry 19 (2009) 6706.
[272] P.P. Yang, Z.W. Quan, C.X. Li, X.J. Kang, H.Z. Lian, J. Lin, Biomaterials 29 (2008) 4341.
[273] S.L. Gai, P.P. Yang, C.X. Li, W.X. Wang, Y.L. Dai, N. Niu, et al., Advanced Functional Materials 20 (2010) 1166.
[274] P. Wang, Q.H. Shi, Y.F. Shi, K.K. Clark, G.D. Stucky, A.A. Keller, Journal of the American Chemical Society 131 (2009) 182.

CHAPTER

22

Preparation Chemistry of Inorganic Membranes

Jerry Y.S. Lin, Shriya K. Seshadri
Arizona State University, USA

22.1. INORGANIC MEMBRANES AND THEIR MAJOR CHARACTERISTICS

Membranes are thin films allowing for selective transport of mass species, such as gases, liquids, or ions. According to the definition of IUPAC, in terms of the material pore size [1], membranes include macroporous (pore diameter $d_p > 50$ nm), mesoporous (2 nm $< d_p <$ 50 nm), and microporous ($0 < d_p < 2$ nm) membranes, corresponding to microfiltration (MF), ultrafiltration (UF), and nanofiltration (NF) membranes, respectively. Membranes with a pore size smaller than 0.3 nm through which helium cannot permeate are referred to as dense membranes. With respect to the materials used, membranes include polymeric, inorganic, and biological membranes. Both polymeric and inorganic membranes have found industrial applications for separation and chemical reaction, with polymeric membranes playing a major role in liquid separation (desalination) and gas separation due to their low membrane costs and easy fabrication into membrane modules with high packing density.

The major advantages of inorganic membranes as compared to polymeric membranes are their better thermal, chemical, and mechanical stability, and higher permselectivity [2]. Conversely, the disadvantages are higher membrane costs and an increased difficulty toward making membrane modules with high packing densities. Inorganic membranes will find applications that require high permselectivity and good chemical and thermal stability beyond what can be offered by polymeric membranes. For example, microporous zeolite membranes provide extremely high water to organic selectivity and can be used for solvent dehydration [3]. High temperature dense mixed-conducting ceramic membranes can be used in air separation to produce warm oxygen for power generation, and steel and glass production [4]. Many chemical reactions are operated at high temperatures, hence inorganic membranes are ideally suited in membrane reactors for chemical reactions to improve reaction conversion or product selectivity [5,6].

Porous inorganic membranes are typically prepared in an asymmetrical structure, consisting of a thick, large pore support and a thin, smaller pore separation layer. In most cases, macroporous and mesoporous membrane separation layers are prepared by submicron or nanometer-sized particles in which the size of the particles determines the pore size. Microporous inorganic membranes include crystalline (mainly zeolites) and amorphous (mainly SiO_2 and carbon) materials. The membranes are prepared as thin films on porous inorganic supports that provide mechanical strength. The thickness of the microporous film varies from a few ten nanometers to a few microns. In most cases, the microporous membrane film is coated on the surface of a mesoporous inorganic membrane layer with desired surface smoothness and pore size. Disk and single-tube are the most common geometries of the microporous inorganic membranes although microporous membranes are also prepared on hollow fiber and flat-sheet supports.

Dense membranes include ceramic oxides and metallic membranes. Dense ceramic oxide membranes are represented by a large number of fast ion conducting or mixed-conducting metal oxides of various crystalline structures such as fluorite and perovskite. Oxygen or hydrogen ions can transport through these dense membrane via hopping through lattice vacancies or interstitial sites. Hydrogen can form metal hydrides with many metals, and atomic hydrogen can permeate through crystalline metals interstitially under a hydrogen gradient driving force. For dense membranes, the surface reaction steps are very important, as permeation through these membranes involve multiple steps including

Modern Inorganic Synthetic Chemistry, DOI: 10.1016/B978-0-444-53599-3.10022-8

energy intensive molecule dissociation/charge transfer. Therefore, the membrane material should have good catalytic properties, or permeation should be conducted at elevated temperatures. Most fast ion-conducting ceramic membranes are about 1–2 mm in thickness as it is difficult to prepare these membranes as thin, gas-tight films on porous supports. Also, these membranes of 1–2 mm thick offer acceptable gas permeation flux at high temperatures. Metal membranes or dense silica membranes are now prepared as thin films on a porous support.

Synthesis and fabrication of inorganic membranes involve both physical and chemical methods. Chemistry plays an important role in the synthesis of inorganic particles (powders) that are used to form these membranes. Chemistry also plays an important role in the direct preparation of thin membrane layers on porous supports, though most inorganic membrane thin films are fabricated by the physical methods such as slip-casting, spin coating, and sputter deposition. For dense inorganic membranes of a given material, the composition, homogeneity, and desired phase structure are important parameters to control. For microporous inorganic membranes, crystalline structure, crystal orientation, and intercrystalline gaps (for crystalline microporous membranes) and micropore structure (for amorphous microporous membranes) are strongly influenced by the chemistry of the membrane synthesis processes. Although most mesoporous and macroporous membranes are prepared by particle compaction, synthesis of certain mesoporous and macroporous membranes (such as ordered mesoporous membranes) and modification of these membranes are also controlled by the chemistry of the synthesis processes.

As summarized above, inorganic membranes include a large group of materials with different chemical compositions and structures, and the chemistry involved in preparing these inorganic membranes is quite diverse. A comprehensive review of the chemistry for inorganic membrane synthesis is difficult to be completely summarized in a book chapter. Therefore, this chapter aims to provide a concise review of the chemistry involved in the synthesis of selected groups of microporous and mesoporous inorganic membranes. The review will start with synthesis of microporous silica membranes by the sol–gel method and pore narrowing of microporous zeolite membranes by chemical vapor deposition. The chemistry plays a key role in obtaining microporous silica membranes. Microporous crystalline metal-organic framework membranes will be reviewed and discussed to illustrate the latest development in membrane science. Finally, the review will discuss chemistry involved in the synthesis of ordered mesoporous silica membranes.

22.2. SYNTHESIS OF MICROPOROUS INORGANIC MEMBRANES

22.2.1. Synthesis of Amorphous Silica Membranes

Microporous silica membranes can be prepared by several methods including chemical vapor deposition and liquid sol–gel methods. The best-known method of silica membrane preparation starts from a sol with silica polymers. These polymers are made by the hydrolysis and condensation of an alkoxide precursor, such as tetraethyloxosilicate (TEOS) under controlled conditions [7]. Microporous silica membranes are coated on mesoporous supports, which in most cases, are γ-alumina membranes. The permeability of the microporous membrane is low due to the small pore size. Therefore, the coated microporous silica layer should be as thin as possible (down to a few ten nanometers) in order to obtain high permeance. This creates two challenges, one is the control of the thickness of the deposited film and the other is the avoidance of mesoporous and macroporous defects and pinholes. Good quality microporous silica membranes can be obtained only when the microstructure of the gel can be controlled and good integrity of the coated film can be achieved.

The most common precursor for sol–gel synthesis of microporous silica membranes are tetraethylothosilicate (TEOS), $Si(OC_2H_5)_4$ and tetramethylothosilicate (TMOS), $Si(OCH_3)_4$. Table 22.1 lists the physical properties of these two precursors. TEOS and TMOS do not dissolve in pure water, but can dissolve in ethanol or an ethanol/water mixture. Typical sol composition (in molar ratio) is 3.8 for ethanol/TEOS, 6.5 for deionized water/TEOS, and 0.09 for nitric acid/TEOS [8–10]. Experimentally solutions of TEOS/ethanol and nitric acid/water are prepared separately and then mixed in a flask equipped with a reflux condenser. The mixture is rapidly heated to and kept at 90°C for 3 h while being vigorously stirred. Finally, the resulting sol is cooled naturally to room temperature.

The microstructure of the final membrane is determined, to a large extent, by the structure of the silica clusters formed during the hydrolysis and condensation processes, which include the following three representative reactions:

$$\begin{array}{c}\text{Hydrolysis } (=>)\\ \equiv Si - OR + H_2O <=> \equiv Si - OH + ROH\\ \text{Esterification } (<=)\end{array} \qquad (A)$$

$$\begin{array}{c}\text{Alcoxolation } (=>)\\ \equiv Si - OR + HO - Si\equiv\ <=> \equiv Si - O - Si\equiv + ROH\\ \text{Alcoholysis } (<=)\end{array} \qquad (B)$$

TABLE 22.1 Physical Properties of Silicon Alkoxide Precursors

Precursor	MW	bp (°C)	*d* (20 °C)	Dipole moment (*D*)	Solubility
TMOS	152	121	1.02	1.71	Alcohols
TEOS	208	169	0.93	1.63	Alcohols

$$\begin{array}{l}\text{Water condensation } (=>) \\ \equiv Si-OH + HO-Si\equiv \;<=>\; \equiv Si-O-Si\equiv + H_2O \\ \text{Hydrolysis } (<=)\end{array} \qquad (C)$$

In general, conditions that favor the condensation rate with respect to the hydrolysis rate lead to a polymeric sol with silica clusters of higher fractal dimension. The above reactions show that decreasing water concentration will lower hydrolysis rate and increase the condensation reaction, resulting in a silica sol of polymeric clusters. Therefore, compared to synthesis of particulate sols such as boehmite sol, which is used to prepare mesoporous membranes, water in lower quantity is used to prepare silica sols for microporous silica membranes.

Another factor that has a more pronounced effect on the structure of silica clusters in silica sol is the use of catalysts [10–15]. The hydrolysis reaction replaces alkyl groups with hydroxyl groups [16]. Subsequent condensation between silanol groups produces siloxane bonds. In most cases, a catalyst, e.g., a mineral acid (HNO_3) or a base (NH_3) is employed. In the case of acid-catalyzed reaction, the hydrolysis proceeds through an electrophilic attack of the H^+-ion alkoxy group as shown below:

$$\equiv Si\text{-}OR + H^+ \xrightarrow[\text{fast}]{\text{electrophilic attack}} \equiv Si\text{-}\underset{R}{O}\text{-}H^+ \qquad (D)$$

$$\equiv Si-\underset{R}{O}-H^+ + H_2O \xrightarrow[\text{slow}]{} \equiv Si-OH + ROH + H^+$$

As hydrolysis proceeds, the number of OR groups decreases, causing the hydrolysis rate to decrease. The probability of forming fully hydrolyzed silicon (Si $(OH)_4$) is small and the condensation reaction will start before hydrolysis has been completed. The presence of the catalyst does not have as significant an effect on condensation reactions (B) and (C) as it does on the hydrolysis. Thus, the acid catalyst inhibits the hydrolysis reaction while maintaining a relatively constant condensation rate, resulting in polymeric sol.

Silica membranes prepared by compacting the silica mass fractal clusters in the acid-catalyzed sol have a microporous structure with pores defined by the space between the silica skeletons. Figure 22.1 shows the pore size distribution of such a microporous silica membrane measured by nitrogen adsorption porosimetry. The

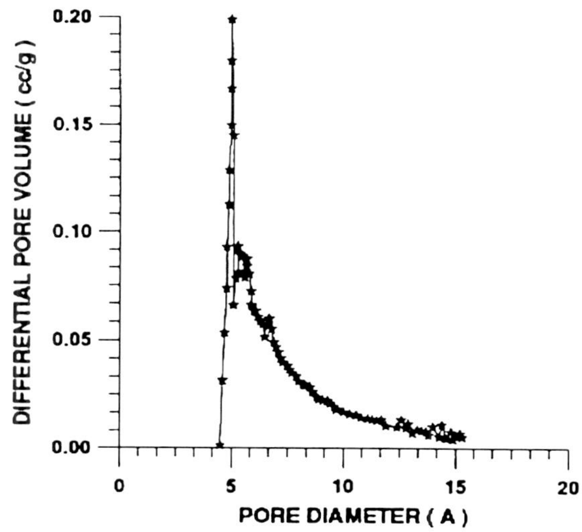

FIGURE 22.1 Pore size distribution of microporous silica membranes prepared in authors' laboratory.

membrane pores have a high surface area (S_{BET} = 920 m^2/g) and pore volume of about 0.46 cm^3/g. This particular membrane has pores ranging from about 0.4 nm to 1.5 nm in diameter. The mean pore diameter is about 0.6 nm.

The polymeric sol can be also prepared by hydrolysis and condensation in two steps [16–19]. The two-step sol–gel approach appears to provide greater pore size tailorability, allowing for superior gas separation performance over a single-step sol–gel method. da Costa et al. [20] compared microstructure and gas separation properties of microporous silica membranes prepared from silica sol of the same composition (molar composition of 1.0 TEOS, 3.8 ETOH, 6.0 H_2O, 0.1 1 M HNO_3), but synthesized by one-step and the two-step methods. In the two-step method, hydrolysis and condensation of TEOS were conducted first with smaller amounts of water and acid (molar composition of 1.0 TEOS, 3.8 ETOH, 1.0 H_2O, 7.0×10^{-4} 1 M HNO_3) at 60°C for 3 h. Additional water and acid were added to the system at room temperature. The two-step method gives a silica sol with weakly branched silica clusters, resulting in a silica membrane with a smaller pore size than that prepared by the one-step method. Thus, compared to the silica membrane prepared by the one-step method, the two-step sol–gel-derived silica membrane shows better gas separation properties, with a lower gas permeance and higher activation energy for diffusion of large gas molecules

22.2.2. Pore Narrowing of Zeolite Membranes

Zeolite is microporous crystalline alumina silicate with a uniform pore size, which is the main difference from the amorphous microporous silica membranes discussed above. Most zeolite membranes are synthesized by two approaches: in situ synthesis and seeded secondary growth. The essence of the in situ approach is to bring the surface of a porous support in contact with a zeolite synthesis solution (sol or gel) and keep the system under controlled conditions so that the zeolite can nucleate and grow to form a continuous film on the support surface. Several methods have been reported to grow zeolite films on porous supports by the in situ approach. In all cases, the experimental conditions, including the position of the support and the way the synthesis solution was brought in contact with the support surface, should be carefully controlled in order to create local supersaturation near the surface of the support. For example, for in situ synthesis of P-type zeolite membranes on alumina support in an autoclave [21], no or poor-quality P-zeolite films could be formed on the polished surface of the porous alumina disk when it was placed in the vertical or horizontal position. Continuous, high-quality P-type zeolite film can grow on the alumina support only when it is placed in a slanting position. Details on the synthesis chemistry and properties of zeolite membranes are available in several reviews [2,22,23].

Among the various zeolites, MFI-type zeolite, especially pure silica silicalite, has received the most attention in zeolite membrane research. This is because: (1) MFI zeolites have intermediate pores (about 0.55 nm) and find many applications in separation and reactions, (2) MFI-type zeolites (especially silicalite) have extremely high thermal and chemical stability, and (3) it is relatively easy to synthesize MFI-type zeolites. However, to use MFI-type zeolite membranes for separation of small gases, such as He, H_2, CO_2, CO, is not very effective since the pores of the MFI zeolite are still too large to offer sufficiently high diffusion-controlled permselectivity [24,25]. Since the sol–gel-derived microporous silica membranes are not hydrothermally stable, narrowing MFI zeolite pores to increase the membrane permselectivity is a promising and attractive method to prepare stable ultramicroporous inorganic membranes.

CVD of silica was applied to seal intercrystalline nonzeolitic pores of MFI-type zeolite membranes by Nomura et al. [27]. CVD was performed in the counter diffusion mode, with the membrane exposed to a silica precursor, such as TEOS (see Table 22.1) on one side and ozone (O_3) on the other side. TEOS and ozone interdiffused into the intercrystalline gaps and reacted, depositing silica and sealing the gaps or other nonzeolitic pores. Since the molecule of TEOS (0.95 nm) is larger than the size of the MFI-type zeolitic pores (0.55 nm), TEOS could only enter nonzeolitic pores resulting in deposition of silica to seal the defects. The CVD reaction was carried out between 200 and 400°C [25,27]. Similar CVD reactions were also used to seal intercrystalline gaps or nonzeolitic pores of DDR-type zeolite membranes [24]. In both cases, CVD modification resulted in an increase in permselectivity of the membranes to a certain extent, as a result of sealing of the intercrystalline nonzeolitic pores of the zeolite membranes. For example, the fresh DDR-type zeolite membrane (zeolitic pores of about 0.4 nm) shows hydrogen/helium permselectivity determined by the weight, not the size, of the permeating molecules (lighter, larger hydrogen is more permeable than heavier, smaller helium). The CVD-modified DDR-type zeolite membrane exhibits a permselectivity that is determined by the size of permeating molecules, i.e., helium is more permeable than hydrogen [24].

To further improve permselectivity of a zeolite membrane, especially for separation of nonadsorbing gases or at high temperatures, the size of the zeolitic pores of the membranes should be further reduced. Masuda et al. [26] reported an interesting CVD concept for narrowing zeolitic pores of MFI-type zeolite membranes. The process takes advantage of the phenomenon of coke deposition. A silane compound, which is smaller than the pores, can penetrate into the zeolitic pores and is chemisorbed on the active sites of the zeolite, such as the acid sites and metal cations. The chemisorbed species are decomposed by heating the zeolite membrane, leaving coke that contains Si atoms on the active sites in the zeolitic pores. The coke is burned in an air stream to yield an SiO_2 unit on each active site within the zeolitic pores. The size of the pores is reduced by the mono SiO_2 unit. This is similar to atomic layer deposition of SiO_2 on the surface of the mesopores of alumina [28]. Masuda et al. [26] demonstrated the concept by narrowing MFI-type zeolite containing Al and NH_4^+ in the framework with silica precursor of methyldiethoxysilane (MDES) and methyldimethoxysilane (MDMS), both of which are small enough to enter the zeolitic pores of MFI-type zeolite membranes. FTIR and TGA/chemisorption studies have identified a reaction mechanism that is shown in Fig. 22.2. The CVD modification resulted in about 10-fold reduction in hydrogen permeance but 30-fold increase in H_2/N_2 selectivity for the MFI-type zeolite membranes. The presence of proper active sizes in the zeolite framework and adequate CVD conditions are key to obtaining good results.

Hong et al. [29] used the same CVD approach to narrow the pore size of B-doped MFI-type zeolite and SAPO-34 membranes with MDES as the silica precursor. The CVD did not cause a change in the gas permeation properties of SAPO-34 membranes because MDES cannot enter the SAPO-34 zeolitic pores (about 0.4 nm). For B-MFI-type zeolite membranes, CVD modification increases the H_2/CO_2 permselectivity from about 1 to

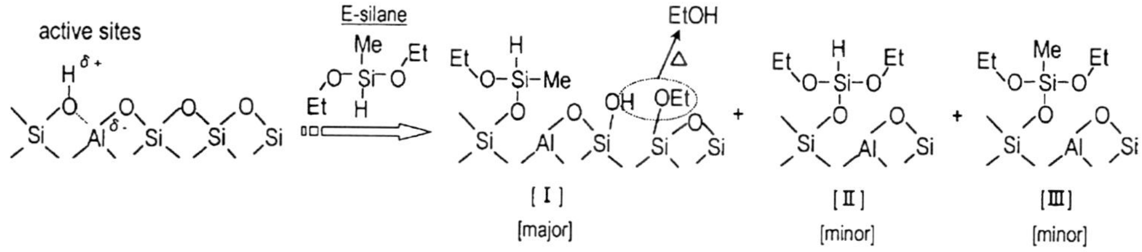

FIGURE 22.2 Chemical reaction mechanism for CVD of methyldiethoxysilane (E-silane) in the zeolitic pores of MFI-type zeolite membrane. *Reprinted from Microporous and Mesoporous Materials, Vol. 48, Takao Masuda, Naohiro Fukumoto, Masahiro Kitamura, Shin R. Mukai, Kenji Hashimoto, Tsunehiro Tanaka, Takuzo Funabiki, Modification of pore size of MFI-type zeolite by catalytic cracking of silane and application to preparation of H_2-separating zeolite membrane, Page 239, Copyright (2001), with permission from Elsevier [26].*

about 35. Similar to the work of Masuda et al. [26], CVD modification causes about 100-fold decrease in H_2 permeance for the membrane. Lin and Burggraf [30] analyzed the effect of pore narrowing of a membrane by chemical modification on gas permeation properties of the membrane and pointed out that such an approach usually resulted in an increase in selectivity at the expense of a substantial reduction in gas permeance due to the decrease in porosity and narrowing of the membrane pores over the entire membrane thickness. This analysis is consistent with the CVD results reported by these two research groups.

A major improvement in the CVD modification of zeolite membranes to enhance permselectivity was reported recently by Gu et al. [32]. CVD was conducted with MDES vapor carried by an H_2/CO_2 mixture on MFI-type zeolite prepared from a pure silica source. Though alumina was not used in the synthesis of the MFI zeolite membranes on alumina supports, the zeolite layer, in fact, contained aluminum from the support [33]. CVD modification improved H_2/CO_2 selectivity for the MFI zeolite membrane from roughly 3 to about 18, a sixfold improvement, at the expense of only a 40% reduction in H_2 permeance. The best result, however, was reported recently by Tang et al. [31]. They reported that CVD modification of the membrane improved H_2/CO_2 of MFI-type zeolite membrane (with Si/Al ratio of 111) by about 100-fold to 140-fold, while the H_2 permeance decreased by less than twofold. The IR spectra of the MFI-type zeolite were recorded before, during, and after CVD modification, as shown in Fig. 22.3. The silanol band in the unmodified MFI-type zeolite disappears and Si$-$OCH$_2$H$_5$, Si$-$CH$_3$, and Si$-$H bands appear after chemisorption of MDES vapor at 423 K. The Si$-$CH$_3$ and Si$-$H bands disappear while the Si$-$OCH$_2$H$_5$ band remains. The silanol band reemerges after heating to 573 K. Upon further increasing the temperature to 723 K, all bands disappear except for the silanol band. Gas chromatography$-$mass spectrometry (GC$-$MS) analysis of the effluent from the CVD reactor shows formation of H_2, CO_2, CO, CH_3, C_2H_4, H_2O, and C_2H_4OH. These results confirm the mechanism proposed by Masuda et al. [26] as shown in Fig. 22.2.

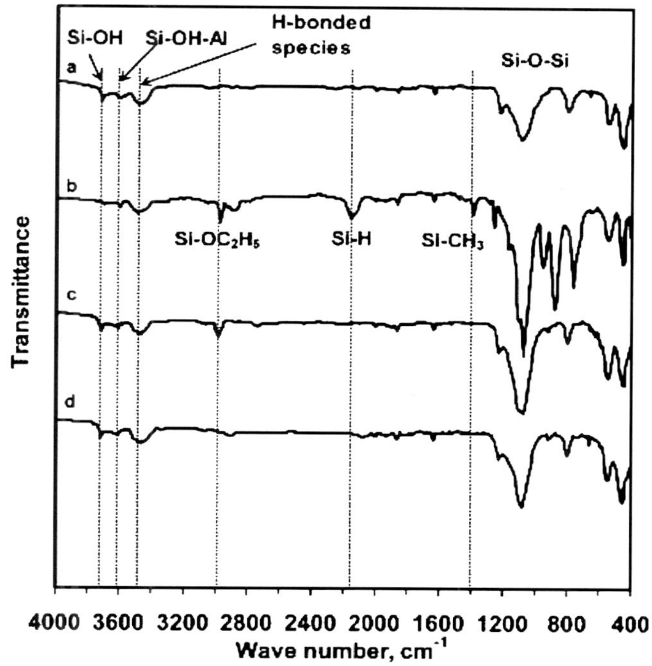

FIGURE 22.3 IR spectra of MFI-type zeolite samples (a) before CVD modification, (b) for MDES adsorbed at 424 K, (c) for MDES adsorbed at 573 K, and (d) modified by MDES at 723 K. *Reprinted with permission from [31]. Copyright 2009 American Chemical Society.*

In the work of Tang et al. [31], the presence of CO and H_2O in the effluent of the CVD reaction indicates a reverse water gas shift reaction at high temperature for the H_2/CO_2 carrier gas:

$$CO_2 + H_2 \Rightarrow CO + H_2O$$

This is one of the major differences between this work and that of Masuda et al. [26]. This might result in a faster CVD reaction, causing deposition of silica in a narrow region in the zeolite membrane layer [34,35]. Another possibility is that the zeolite membrane contains a small amount of acid sites that are scattered throughout the cross-section of the zeolite membrane. Silica is deposited in the zeolite pores in a discrete manner along the membrane diffusion direction. As a result, the zeolitic pores are reduced only in the narrow region. This improves membrane selectivity significantly with a minimum reduction in gas permeance. More studies are needed in order to understand the mechanism and to improve reproducibility of this CVD approach for membrane modification.

22.2.3. Metal-organic Framework Membranes

Metal-organic frameworks (MOFs) represent a new family of microporous materials that are formed by a network of transition metal ions linked by organic ligands [36]. These materials can have a nearly endless variety of ordered structures, functional groups, and very high porosity [37]. As a result, MOFs offer potential applications in gas separation and storage [38], catalysis [39] and membranes. To date, much attention has been focused on the synthesis of novel structures generated by the selection of different organic ligands and inorganic molecular building blocks to construct tailored frameworks [40] and their sorption properties for hydrogen, methane, and carbon dioxide [41]. With much progress made on the development of microporous [42] and mesoporous inorganic membranes for gas and liquid separation [43], MOFs have recently attracted attention in the materials and membrane communities as thin films (on dense substrates) or membranes (on porous substrates) for various applications [44].

A MOF membrane consists of a continuous film of a MOF material formed on a porous substrate which provides mechanical strength and allows gas or liquid to permeate with minimum resistance. Several research groups have reported synthesis of membranes of the following types of MOFs on porous supports: MOF-5 (or IRMOF-1), Cu-BTC ($Cu_3(BTC)_2$ or HKUST-1), MMOF, ZIF (ZIF-8 and ZIF-7), and $Mn(HCO_2)_2$; these are summarized in Table 22.2. MOF-5 has a cubic structure with each corner of $(Zn_4O)^{6+}$ linked by 1,4-benzenedicarboxylate (BDC) [38]. MOF-5 has a cavity size of about 1.2 nm and aperture of 0.8 nm. Cu-BTC has a face-center cubic structure with metal corners consisting of Cu^{2+} ions coordinated to 1,3,5-benzenetricarboxilate [45], and has a channel opening of about 0.9 nm. MMOF is built upon a copper paddle-wheel and a V-shaped dicarboxylate ligand, with a layered structure and one-dimensional pore channels of about 0.4 nm in pore size [46]. ZIF-8 and ZIF-7 resemble the zeolite SOD structure, with Zn replacing Si (or Al) and 2-methylimidazolate (for ZIF-8) or 2-benzimidazolate (for ZIF-7) replacing oxygen in the zeolite structure [47]. The pore openings of ZIF-8 and ZIF-7 are about 0.34 nm and 0.30 nm, respectively.

The majority of studies on MOF membranes have been published in the past year, as summarized in Table 22.2. Early efforts on MOF membranes were conducted by Arnold et al. [48] and Gascon et al. [49]. These two research groups reported on the synthesis of oriented microporous manganese formate $Mn(HCO_2)_2$, and Cu-BTC on porous graphite and alumina supports, respectively. Lai and coworkers [50,51] reported in situ synthesis by conventional and microwave heating of MOF-5 membranes about 25 μm thick on alumina supports. It was found that the gas permeation of CO_2, N_2, CH_4, and H_2 through MOF membranes is determined by the Knudsen mechanism. These data are consistent with theoretical prediction of the conditions that favor Knudsen diffusion. They also agree with the diffusivity data measured on MOF-5 crystals [52]. Guo et al. [53] reported on the synthesis of Cu-BTC membranes in a copper-net support. However, an H_2/CO_2 permselectivity of about 7 was reported for the Cu-BTC membrane at 25°C. These separation data [53] are inconsistent with the known transport mechanism through the membranes with pores in this size range [42,43].

TABLE 22.2 Summary of Recent Studies on Mof Membranes on a Porous Support

MOF	Pore diam. (nm)	Support	Synthesis method	Typical results and Reference
Manganese formate [$Mn(HCO_2)_2$]	0.45	In situ	Porous α-Al_2O_3, graphite	200–300 μm thick, poor membrane quality, no gas permeation data [48]
Cu-BTC (HKUST-1) [$Cu_3(BTC)_2$]	0.9	In situ, secondary growth	Porous α-Al_2O_3, copper-net	Noncontinuous 2–5 μm thick films, or 60 μm thick with unusual gas permeation data (H_2 permeance of $\sim 1 \times 10^{-6}$ mol/m^2 Pa s, with α_{H_2/CO_2} ~7 at 25 °C) [49,53]
MOF-5 [$Zn_4O(BDC)_3$]	0.8	In situ, microwave	Porous α-Al_2O_3	60 μm thick, gas permeation ($1–5 \times 10^{-6}$ mol/m^2 Pa s) governed by Knudsen diffusion [50,51]
MMOF [Cu(hfipbb)(H_2hfiobb)$_{0.5}$]	0.32–0.35	In situ, secondary growth	Porous α-Al_2O_3	20–40 μm thick, low permeance ($1–100 \times 10^{-10}$ mol/m^2 Pa s); α_{H_2/N_2} ~25 at 200 °C [54]
ZIF-8 [Zn-(mim)$_2$], ZIF-7 [Zn-(bim)$_2$]	0.34, 0.30	Secondary growth	Porous titania, α-Al_2O_3	30–50 μm thick (thinner for ZIF-7), H_2 permeance $6–7 \times 10^{-8}$ mol/m^2 Pa s, α_{H_2/CO_2} or α_{CO_2/CH_4} of 4–7 [55–57]

MOF-5 and Cu-BTC have pores larger than 12-ring FAU-type zeolites. These MOF membranes are unlikely to show molecular sieving properties for small gas molecules under normal conditions.

Very recently, Ranjan and Tsapatsis [54], Caro and workers [55,56], and Venna and Carreon [57] reported the synthesis of microporous MOF membranes with a pore diameter smaller than 0.4 nm. MMOF membranes of about 20–40 μm in thickness were grown on alumina supports in situ, as well as by the seeded secondary growth method [54]. The membranes showed H_2/N_2 selectivity of about 25 at 200°C. However, the hydrogen permeance was low (about 10^{-9} mol/m^2 Pa s), as expected from the small pore size of MMOF material and the thickness of the membranes. Membranes of ZIF-7 and ZIF-8 with pore diameter of about 0.30 and 0.34 nm were grown by the seeded growth method on porous alumina supports [55–57]. Both membranes showed an H_2/CO_2 selectivity in the range of 4–7, about the value of the Knudsen selectivity (4.7). Hydrogen permeance for the ZIF membranes is in the range of about 10^{-8} mol/m^2 Pa s, which is slightly larger than that for the MMOF membranes. The permselectivity of these microporous MMOF and ZIF membranes is comparable to the defect-free microporous MFI-type zeolite membrane, but the latter has higher hydrogen permeance than the former [25,58].

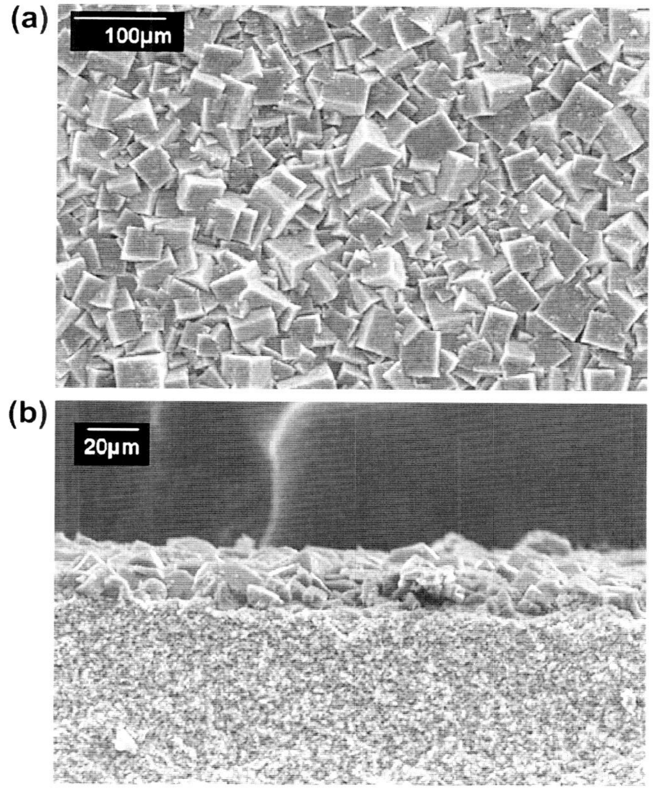

FIGURE 22.4 SEM images of the surface and cross-section of MOF-5 membranes on alumina support. *Reprinted from Microporous and Mesoporous Materials, Vol. 123, Yeonshick Yoo, Zhiping Lai, Hae-Kwon Jeong, Fabrication of MOF-5 membranes using microwave-induced rapid seeding and solvothermal secondary growth, Page 7, Copyright (2009), with permission from Elsevier [51].*

Similar to synthesis of zeolite membranes, MOF membranes were prepared by both in situ and seeded secondary growth methods. In the in situ synthesis, a porous alumina support is brought in contact with the MOF synthesis solution. Liu et al. [50] reported the following synthesis procedure. First, terephthalic acid (1,4-benzenedicarboxylic acid or H_2BDC, 0.38 mmol) was dissolved in degassed *N,N*-dimethylformamide (DMF) (10 mL) and the mixture was loaded into a 45 mL Teflon lined autoclave. A dry alumina support disk was placed into the solution with smooth surface upward and kept in the solution for half an hour. Next, dehydrated $Zn(NO_3)_2$ was added to the solution, and the mixture containing the support was sealed in the autoclave, heated to 105°C and kept at this temperature for 1–2 days. MOF-5 was nucleated on the support surface, and the crystals grew into a continuous MOF film. Figure 22.4 shows surface and cross-sectional morphology of the MOF-5 membrane grown by the in situ method.

The secondary growth method includes the synthesis of MOF crystals and suspension, formation of the seeded layer on the support, and secondary growth to obtain continuous MOF membrane. A typical procedure to prepare MOF-5 crystals starts with the preparation of synthesis solution by adding $Zn(NO_3)_2 \cdot 6H_2O$ (5.60 mmol) and terephthalic acid (1,4-benzenedicarboxylic acid or H_2BDC, 2.12 mmol) to degassed dimethylformamide (DMF) solvent (40 mL). The synthesis solution is put in a 100 mL container, which is sealed and heated to 130°C, and held at that temperature for 4 h under autogenous pressure. Cubic-like crystals of colorless powder are formed, and carefully collected, washed and dried under controlled conditions. Figure 22.5 shows an SEM image of the MOF-5 crystals prepared by this method. Cu-BTC crystals of similar size can also be prepared by the established method as reported in the literature [45,46]. A typical procedure for synthesizing Cu-BTC crystals starts with the preparation of a 0.3 M $Cu(NO_3)_2$/water solution and 0.16 M trimesic acid (1,3,5-benzenetricarboxylic acid/ethanol solution) by dissolving the corresponding salt and acid in water and ethanol, respectively. The two solutions are mixed in 1:1 ratio and placed in an autoclave and heated under hydrothermal conditions at 120°C for 12 h.

All of the good-quality large pore MOF membranes reported are fairly thick (20–300 μm), with thicknesses much larger than most zeolite membranes (typically in the range of 1–5 μm). This can be attributed to the size

FIGURE 22.5 SEM image of MOF-5 cubic crystals of about 40 μm in size prepared in the authors' laboratory.

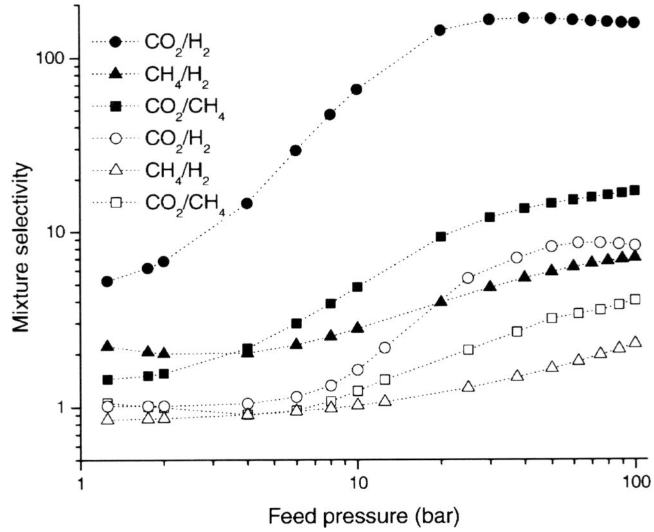

FIGURE 22.6 Equal molar binary gas mixture permselectivity by Cu-BTC (closed symbols) and MOF-5 (open symbols) membranes obtained by molecular simulation. *Reprinted with permission from [52]. Copyright 2009 American Chemical Society.*

of the MOF crystallites, which are mostly in the range of 10–50 μm, and generally much larger than zeolite crystallites. Thus, an MOF film on a porous support should be thicker than the crystallite size in order to achieve the desired crystallinity and continuity of the film. Reducing the thickness of the MOF membranes, while maintaining high quality, is key to improving the permeance and reducing the material cost. Perez et al. [59] reported the synthesis of 100 nm-sized MOF-5 nanocrystals for use in mixed-matrix membranes. The synthesis of MOF-5 nanocrystals includes the following differences and/or additional steps needed in comparison to the synthesis of micron-sized crystals described above: (1) $Zn(NO_3)_2 \cdot 6H_2/DMF$ solution is first prepared and 4A zeolite is added to the solution to remove excess water, before BDC acid is added to obtain the final synthesis solution; and (2) the synthesis solution is heated to 70°C (instead of 130°C), and after addition of triethylamine (TEA), the solution is maintained at that temperature under strong agitation for 10 min (instead of 4 h). However, such approach has not been used to prepare thin MOF membranes.

Much work has been done on molecular simulation of diffusion and adsorption of gases in MOF materials, in particular, Cu-BTC [60–62]. In 2009, Sholl and coworkers published several papers on the simulation of gas permeation and separation properties of MOF membranes [63–65]. They reported simulation results of room temperature single-gas permeation and binary gas separation of CO_2, H_2, CH_4, and N_2 for MOF-5 membranes by atomic simulation, and found that MOF-5 membranes offer essentially no selectivity for CO_2/H_2, CO_4/N_2, and CH_4/H_2 mixtures at low pressures (1 bar). The permselectivity for CO_2/H_2 and CO_4/N_2 increased to about 9 at higher pressures (about 50–60 bar). Lower pressure data are similar to the experimental results [50,51]. Room-temperature permeation and separation of CO_2/H_2, CH_4/H_2, and CO_2/CH_4 through a Cu-BTC membrane were also calculated by atomic simulation up to 100 bar [65]. For CO_2/H_2, the simulation result showed that the Cu-BTC membrane is more permselective toward CO_2 for similar gas mixtures than MOF-5 membranes, with CO_2/H_2 selectivity increasing from 5 to over 100 as the pressure increased from 1 to 100 bar. Figure 22.6 shows the simulation results. For both membranes, the simulation results show that the permselectivity is dominated by adsorption selectivity. Clearly, the results for Cu-BTC membrane are opposite from the experimental data for Cu-BTC membranes reported by Guo et al. [53]. The results indicate that the smaller pore MMOF membrane offers a much higher CO_2/CH_4 permselectivity (10^4–10^5) than the Cu-BTC membrane [63]. This reflects the molecular sieving effect of MMOF. However, the simulated permeance for these gases through the MMOF membrane was not reported by this group.

22.3. SYNTHESIS OF ORDERED MESOPOROUS MEMBRANES

22.3.1. General Characteristics of Ordered Mesoporous Materials

As discussed above, microporous membranes include amorphous silica, crystalline zeolite and MOF membranes. Similarly, mesoporous membranes include disordered mesoporous materials such as ultrafiltration γ-alumina or zirconia membranes whose pores are random and formed by packing small alumina or zirconia crystalline particles. Ordered mesoporous materials contain pores ranging from 2 to 50 nm. Unlike disordered mesoporous materials, these materials have pores arranged in an ordered fashion, with a pore wall made of amorphous material. Mesoporous materials were developed in an effort to increase the pore size of zeolites. While single molecules are used as structure-directing agents for zeolites, larger molecular assemblies of surfactants are used as directing agents for mesopores. The first ordered mesoporous material synthesized with the use of surfactant molecules was by the Mobil research group in 1992 [66]. Since this discovery, a number of groups have successfully synthesized various ordered mesoporous structures and morphologies.

Ordered mesoporous materials are characterized by having ordered pore structures with a narrow pore size distribution, high surface areas and a number of different pore connectivities that can be achieved by modulating the synthesis conditions and surfactants used. These properties make them particularly favorable for a number of membrane applications such as membrane reactors and catalytic reactors.

The pore ordering in mesoporous materials is derived from surfactant templates. During synthesis, metal alkoxides undergo either hydrolysis or condensation around preexisting liquid-crystal templates or cooperative self-assembly to form a material with long-range order. The surfactant can then be removed to obtain the final ordered mesoporous structure. The steps in the formation process of an ordered mesoporous system are illustrated in Fig. 22.7 [67]. While the chemistry of metal oxide hydrolysis and condensation is similar to that in sol–gel synthesis, the addition of surfactants introduces more complexities. The reaction rates for polymeric metal oxide formation need to be controlled so as to allow enough time for long-range order formation. The synthesis solution is usually made up of four components: (1) the inorganic precursor; (2) the structure-directing surfactant template; (3) solvent; and (4) catalyst.

There are three main processes that take place during synthesis, inorganic hydrolysis condensation, assembly of surfactant templates, and the interaction between the organic and inorganic species [67]. The delicate interplay between these processes results in the formation of a well-ordered system. A number of mechanisms have been proposed for mesopores formation [68], the discussion of which is beyond the scope of this chapter. However, the final morphology of the given product is essentially governed by the interaction between the charged inorganic species and the surfactant. Depending on the synthesis conditions used, a number of morphologies can be attained.

Huo et al. [69] have developed a number of materials using the concept of charge matching between the template and inorganic species, which is reviewed in Chapter 16. Ordered mesoporous silicate synthesis has been extensively studied in the past decade and these

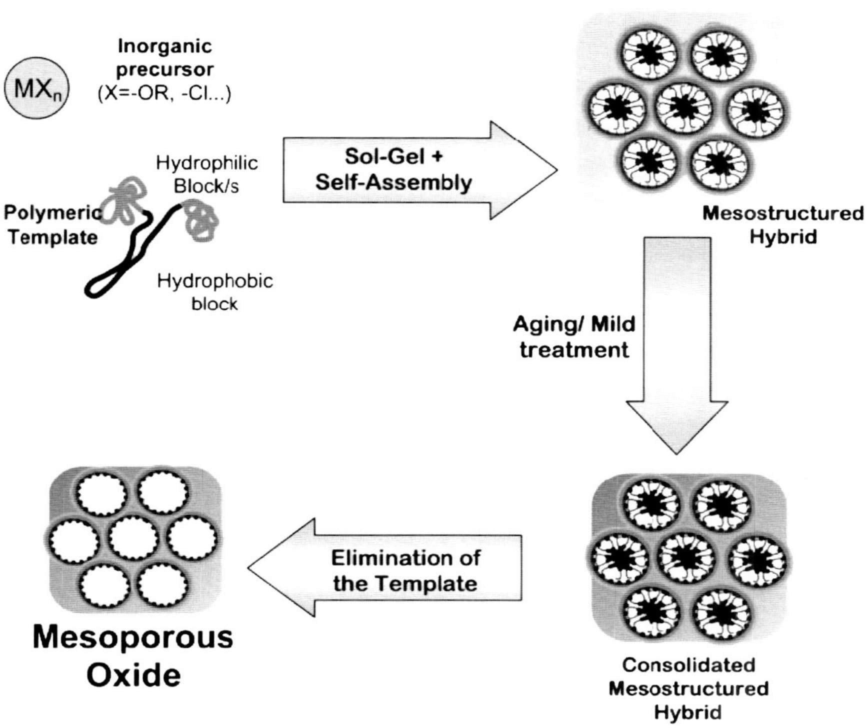

FIGURE 22.7 Schematic representation of ordered mesopores formation steps. *Reprinted from Current Opinion in Colloid & Interface Science, Vol. 8, Galo J. de A. A. Soler-Illia, Eduardo L. Crepaldi, David Grosso, Clément Sanchez, Block copolymer-templated mesoporous oxides, Page 109, Copyright (2003) with permission from Elsevier [67].*

materials can now be prepared by using a wide range of precursors under a vast number of reaction conditions (pH, temperature, etc) [70]. The precursors that are most commonly used are tetraethyl orthosilicate (TEOS) as silica source, water (and ethanol in some cases) is used as the solvent, and NaOH or an acid halide is usually used as the catalyst. This extensive array of synthesis conditions has led to a wide range of possible pore symmetries that can be formed. Table 22.3 summarizes the various pore systems that can be attained under various synthesis conditions. In a typical synthesis procedure for ordered mesoporous silica, the surfactant is mixed with water and catalyst, the silica source is then added to this mixture and stirred from anywhere between 30 min to 2 h. The system is heated to ~100°C in the oven for 3–6 days depending on the procedure chosen. The product is then washed and filtered to obtain the organic–inorganic composite. The surfactant can then be removed by liquid extraction or calcination to obtain the final product [71].

22.3.2. Synthesis of Ordered Mesoporous Membranes

Given the numerous morphologies and ordering that can be achieved, these materials have garnered considerable interest for fabrication into supported thin films. Various methods have been utilized to fabricate these membranes, the most predominant of them being hydrothermal synthesis, dip coating, and spin coating. Hydrothermal synthesis is widely used for production of MCM-48 type membranes. Here, ordered mesopore formation and membrane synthesis take place in a coupled process. The support for membrane synthesis is placed in the synthesis sol, and once the ordered gel is formed, it precipitates and deposits onto the support forming a membrane. The initial hydrolysis of silica and formation of the mesophase begin at room temperature. The system is then heated, and further ordering of the system, condensation–polymerization of the silicate species and the precipitation of the ordered silica–organic composite that form a film on porous support all take place at this elevated temperature. The final step involves removal of the surfactant to reveal an ordered mesoporous membrane.

In the first reported synthesis by Nishiyama et al. [72], the precursors used were tetraethyl orthosilicate (TEOS), quaternary ammonium surfactant ($C_{16}H_{33}(CH_3)_3NBr$; CTAB), NaOH, and deionized water in the following mole ratio 0.59 CTAB:1.0 TEOS:0.5 NaOH:61 H_2O. Macroporous alumina disks were used as the support for this particular membrane. In a typical procedure, the support was soaked in TEOS. Afterward, the remaining reactants were mixed together and the solution was

TABLE 22.3 Various Inorganic Precursor and Surfactant Interactions and Corresponding Mesoporous Structures Formed

Inorganic Precursor (I)	Surfactant (S)	Counter ion (X/F/M)	Interaction	Possible structures formed
Anionic	Cationic	—	I^-S^+ electrostatic force	Hexagonal MCM-41 and FSM-16, cubic MCM-48, hexagonal–cubic SBA-2, lamellar and hexagonal W2O3, lamellar, hexagonal and cubic antimony (V) Oxide, lamellar tin sulfide, and aluminophosphate
Anionic	Anionic	M^+	$I^-M^+S^-$ electrostatic force	AMS-*n*, lamellar zinc oxide, alumina
Cationic	Cationic	X^-	$I^-X^-S^+$ electrostatic force	Cubic SBA-1 (Pm3n), hexagonal–cubic SBA-2, hexagonal SBA-3, lamellar and hexagonal zirconia, hexagonal titanium dioxide, zinc phosphate
Cationic	Anionic	—	I^+S^- electrostatic force	Lamellar magnesium, aluminum, gallium, manganese oxides, hexagonal alumina, hexagonal gallium oxide, hexagonal tritium dioxide, hexagonal tin oxide
Cationic	Nonionic	X^-	$I^+X^-(S^0H^+)$	Hexagonal SBA-15
Cationic	Nonionic	F^-	$I^+F^-N^0$	Hexagonal silica
Neutral	Cationic	F^-	$I^0F^-S^+$	Hexagonal silica
Neutral	Nonionic	—	I^0S^0 hydrogen bond	Near hexagonal HMS
Neutral	Nonionic	N^0 (amine)	I^0N^0 hydrogen bond	Near hexagonal HMS MSU-X; hexagonal titanium aluminum, zirconia, tin oxides
Neutral	Nonionic	M^+	$I^0(S^0\ M^{n+})$	Hexagonal and cubic metal-containing silica
No charge	No Charge	—	IS covalent complex	Silica, Nb, Ta oxides, hexagonal Ta

added to the TEOS. After mixing for 90 min, the mixture was then transferred to an autoclave with the support held horizontally at the bottom. The system was heated to 90°C for 96 h. Table 22.4 summarizes the various studies conducted to fabricate membranes via the hydrothermal synthesis procedure.

Various methods can be used to remove the surfactant from the as-synthesized membrane. The most common one is involved in calcining the material at 500°C for 4 h. Upon calcination, the uncondensed silanol groups in the silica matrix condense to form siloxane bonds, hence resulting in the contraction of mesopores [73]. This shrinkage can lead to defects in the membranes. To avoid this, liquid extraction of surfactants has been carried out. The surfactant can be extracted using an EtOH/HCl solution containing 250 mL EtOH and 3.5 g of 37% HCl at 100°C for 24 h [74]. This modification has been observed to improve the quality of the membrane as was evidenced by improved gas separation properties. Due to the long synthesis time and large pore supports used, EDS and XRD analyses have shown that there is often considerable silica penetration into the supports. Various measures have been taken to minimize such infiltration. Kumar et al. [75] used asymmetric porous supports, covering one side with tape, to avoid the infiltration to some degree of success.

Other popular techniques for membrane synthesis are dip and spin coating. These two are similar, in that, both techniques employ solvent evaporation to induce structure formation by increasing reactant concentration within the system. These systems usually contain surfactants at concentrations well below critical micellar concentration (cmc) in the dip coating solution. This technique, called evaporation-induced self-assembly (EISA), was first reported by Luet al. in 1997 [76]. In a typical synthesis, TEOS ($Si(OC_2H_5)_4$), ethanol, water, and HCl (mole ratios $1:3:8:5 \times 10^{-5}$) were refluxed at 60°C for 90 min. Second, water and HCl were added, increasing the concentration of HCl to 7.34 mM. After stirring at 25°C for 15 min, the sols were aged at 50°C for 15 min and diluted with ethanol. Finally, CTAB was added in quantities corresponding to concentrations in the range 0.03–0.11 M (1.5–5.0 wt%). The final reactant mole ratios were: 1 TEOS:22 C_2H_5OH: $5H_2O$:0:004 HCl:0:054–0:18 CTAB. These synthesis systems are often acid catalyzed since the isoelectric point of silica is around a pH of 2. At the isoelectric point, condensation is minimized. Maintaining the system at this pH helps avoid condensation and structure formation before dip coating.

It has been observed that the optimum ordering is obtained when the number of Q1 species in the prehydrolyzed silica is the greatest. In sols where Q3 species are apparent ordering within the system is lost in the final dip-coated film [82]. Once the solution is dip coated, the more volatile components (alcohol, water, and acid) evaporate rapidly increasing local surfactant and silica concentrations. This leads to rapid self-assembly of the surfactant and silica, forming the liquid-crystal mesophase at the air/liquid interface. Similarly, at the substrate/liquid interface, local concentrations of surfactant are high due to formation of cylindrical micelles and hemimicelles which lead to the formation of ordered mesopores at the interfaces. Figure 22.8 reveals the approximate pathway that is traversed during synthesis. The ordering of the final structure is strongly dependent on the relative rate of three competing reactions: organic–inorganic interface co-assembly (k_{inter}), organic assembly (k_{org}), and inorganic condensation (k_{inorg}). In general, the rates of these reactions need to be controlled such that the $k_{inter} > k_{org} > k_{inorg}$ in order to synthesize well-ordered mesophases.

A number of different phase structures and pore orientations can be achieved by carefully controlling

TABLE 22.4 Hydrothermally Synthesized Membrane: Synthesis Solutions used and Membrane Properties

Support used	Precursors used	Pore size (nm)	Thickness (μm)	Reference
Stainless steel	1 TEOS:0.5 NaOH:60 H_2O:0.6 CTAB	—	500	Nishiyam et al. [72]
Asymmetric α-alumina	1 TEOS:0.5 NaOH:61 H_2O:0.59 CTAB	—	—	Nishiyam et al. 2001 [77]
α-Alumina	1 TEOS:0.5 NaOH:61 H_2O:0.59 CTAB	2.0	7	McCool et al. [78]
α-Alumina	1 TEOS:0.5 NaOH:61 H_2O:0.59 CTAB	2.5	15	Kumar et al. [74]
α-Alumina (tube)	1 TEOS:0.5 NaOH:61 H_2O:0.59 CTAB	2.7	6	Liu et al. [79]
Asymmetric alumina (porous)	1 TEOS:0.35–0.55 NaOH:55–65H_2O: 0.40–0.65 CTAB	2	0.5	Sakamoto et al. [80]
Symmetric zirconia	1 TEOS:0.46 NaOH:56 H_2O:0.48 CTAB	—	2–3	Ji et al. [81]
Asymmetric α-alumina	1 TEOS:0.5 NaOH:61 H_2O:0.59 CTAB	2.5	7	Kumar et al [75]

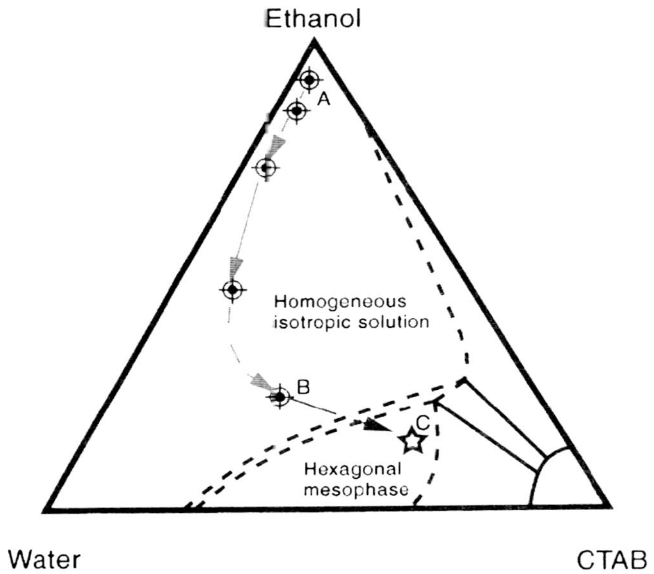

FIGURE 22.8 Steady-state film thinning profile established during dip coating of a complex fluid comprising soluble silica, surfactant, alcohol, and water. Approximate trajectory taken in ethanol/water/CTAB phase space during dip coating. Point A corresponds to the initial composition of solution, Point B is near the drying line, and Point C corresponds to the dried film. *Brinker, C. J.; Lu, Y. F.; Sellinger, A.; Fan, H. Y.: Evaporation-induced self-assembly: Nanostructures made easy. Advanced Materials. 1999. Volume 11, Page 579, Copyright Wiley-VCH Verlag GmbH & Co. KGaA. Reproduced with permission [83].*

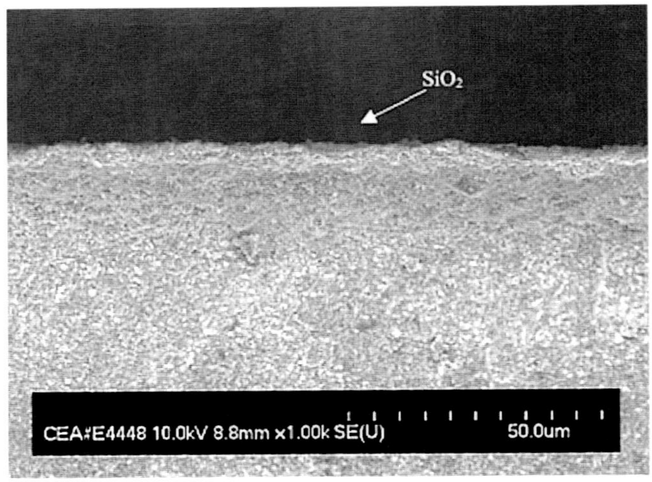

FIGURE 22.9 Cross-sectional view of a dip-coated mesoporous silica membrane. *Reprinted from Journal of Membrane Science, Vol. 218, B. A. McCool, N. Hill, J. DiCarlo, W. J. DeSisto, Synthesis and characterization of mesoporous silica membranes via dip coating and hydrothermal deposition techniques, Page 55, Copyright (2003), with permission from Elsevier [78].*

the precursor concentration and the humidity of the reaction environment. It is possible to control the pore structure of the final material by fine-tuning the composition of the initial reaction mixture. Toward this, synthesis of mesoporous thin films has been widely studied. McCool et al. [78] first reported on the synthesis of mesoporous MCM-48 membranes by dip coating. Here, a silica sol was prepared by mixing TEOS, ethanol, water, and HCl in the molar ratio of 1 TEOS:3 EtOH:1 H_2O:5 × 10^{-5} HCl. The sol was refluxed at 60°C for 1 h and then cooled down to the room temperature. The dip-coating solution was then prepared by mixing the prehydrolyzed sol with water, HCl, and CTAB. The final molar ratios of the dipping sol were: 1TEOS:20EtOH:5H_2O:0.004 HCl:0.14 CTAB. The dipping solution was aged for 1 week prior to use. Supports were then dipped in the solution in a Class 100 laminar flow cabinet. The dipping rate was approximately 3 cm/s. After dip coating, the membranes were dried in the laminar flow cabinet for 1 h. The membranes were calcined in air at 500°C for 4 h with heating and cooling rates of 1°C/min. The dip-coated membrane was seen to be ~2 μm thick. An SEM micrograph of the dip-coated membrane is seen in Fig. 22.9.

Table 22.5 lists the different studies that were conducted on membrane synthesis using dip-coating techniques, including the composition of the final sol precursor prepared. While most studies have focused on using ionic surfactants, nonionic pluronic surfactants have also been successfully used to synthesize mesoporous membranes [84,85]. These membranes showed larger pores (4.7 nm) than those prepared with the ionic surfactants (2.5 nm). For membranes synthesized using nonionic pluronic surfactants, a typical synthesis solution consisted of the following molar ratio: 1 TEOS:5.2 EtOH:12 H_2O:0.015 HCl:0.021 Pluronic P123. The thickness of the synthesized membrane is reported to be about 1 μm and has been reported to be of good quality.

Impregnation of silica sol into the support has also been observed in the case of dip-coated membranes. Boffa et al. [86] have sought to control silica impregnation into the support by increasing the viscosity of the sol with help of a urea-based additive called BYK-420. By rendering the sol thixotropic, the group was able to decrease silica penetration into the support. The addition of BYK-420 also resulted in an increased pore diameter. Beyond a maximum concentration, complete collapse of the pores was observed.

The sol used in the spin coating of ordered mesoporous membranes is similar to those used in dip coating. In spin coating, the sol is dropped on a substrate which is then spun at a fixed revolutions per minute (rpm). The sol spreads radially over the support due to centrifugal forces. Excess silica sol is spun off during this procedure. The volatile components then evaporate from the system, increasing local silica and surfactant concentration to form an ordered mesoporous structure. Relatively few membranes have been prepared by the spin-coating technique, as extremely smooth surfaces are required to

TABLE 22.5 Mesoporous Silica Membranes Synthesized by Dip and Spin Coating: Synthesis Solutions used and Membrane Property

Technique	Support used	Precursors used	Pore size (nm)	Thickness (nm)	Reference
Dip coating, spin coating	γ-Alumina (support by α-alumina)	1 TEOS:6.5 1-propanol:2.6 2-Butanol:10 H_2O:0.01 HCl:0.08 CTAB	2.8–3.4	20–30	Chowdhury et al. [87]
Dip coating	α-Alumina	1 TEOS:20 EtOH:5H_2O:0.004 HCl:0.14 CTAB	2.0	2000	McCool et al. [78]
Dip coating, spin coating	α-Alumina	1 TEOS:6.5 1-propanol:2.6 2-Butanol:10 H_2O:0.01 HCl:0.08 CTAB	2.3–2.6	65–70	Schmuhl et al. [88]
Spin Coating	Porous alumina	1 TEOS:6.5 1-propanol:2.6 2-Butanol:10 H_2O:0.01 HCl:0.08 CTAB	0.8	430	Nakagawa et al. [89]
Dip coating	α-Alumina and γ-alumina (supported by α-alumina)	1 TEOS:6.5 1-propanol:2.6 2-Butanol: 10 H_2O: 0.01 HCl:0.08 CTAB	–	≈600	Chowdhury et al. [90]
Dip coating	Asymmetric α-alumina		4.7	≈2000–3000	Higgins et al. [84]
Dip Coating	Anodic alumina	1 TEOS:8.7 EtOH:6 H_2O:0.004 HCl:0.25 Brij-56	5–20	—	Yoo et al. [91]
Dip coating	α-alumina	1 TEOS:1-propanol 15:15.5H_2O:0.14 HNO_3:0.13 CTAB	1.8–4.2	1000	Boffa et al. [86]
Spin coating	Assymetric alumina	1 TEOS:7.6 EtOH:5H_2O:0.005 HCl:0.1 CTAB	2.0	300	Sakamoto et al. [80]
Dip coating	Nafion 117	1 TEOS:20 EtOH:5H_2O:0.0087 HCl:0.0097 Pluronic P123	–	10,000–20,000	Lin et al. [92]
Dip coating	Anodic alumina	1 TEOS:5.2 EtOH:12 H_2O:0.015 HCl:0.021 Pluronic P123	4.7	1000	Higgins et al. [85]

assure proper coverage. The presence of macropores in the supports makes this requirement difficult to achieve. The composition of the sols used for spin coating is listed in Table 22.5.

22.3.3. Alternate Techniques for Preparation and Pore Orientation of Mesoporous Membranes

A number of alternative techniques have been developed for the synthesis of mesoporous membranes. These techniques include using shear force, magnetic fields, and aerosol-assisted deposition. Some of these techniques seek to fill the pores in the support, hence reducing the pore size and bringing it down into the mesopores range. There has also been considerable interest shown toward developing oriented pores, especially those which are oriented perpendicular to the support surface. Such orientation would allow for greater accessibility for molecule transport. Confinement techniques have developed into extremely promising techniques for the formation of oriented pores. Here, the structure-directing effects of the interface were sought to control mesostructure formation.

The first successful study was conducted by Yamaguchi et al. [93], in which a mixture of ethanol, TEOS, and 1 mL HCl aqueous solution was refluxed at 60°C for 90 min. To this, hydrolyzed sol, ethanol, HCl solution, and CTAB were added, and the solution was stirred for 30 min. Porous alumina membrane supports (pore diameter = 200 nm) were set in an ordinary membrane filtration apparatus, and the precursor solution was dropped onto the alumina membrane. Moderate aspiration was applied so that the precursor solution penetrated into the columnar alumina pores. The alumina membrane, including the precursor solution, was then dried in air at room temperature.

The resulting mesopores formed within the pores of the anopore support can be seen in Fig. 22.10. The mesopores were seen to have channels that run in a perpendicular direction to the support surface. The channels, however, did not run the entire thickness of the support. The authors ascribed the formation of the hexagonal structure in the direct vicinity of the pore wall to the adsorption of cationic CTAB to the alumina wall. Lu et al. [94] reported the synthesis of mesoporous SBA-15 having a pore diameter of 6 nm within alumina anopore supports by dipping it in a sol solution for 20 h.

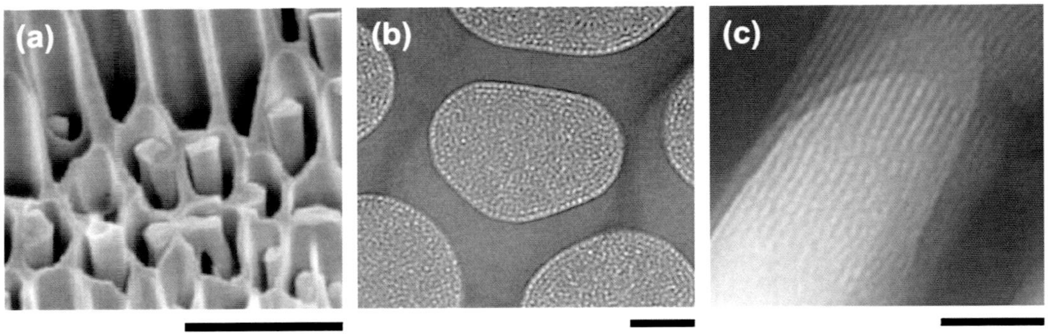

FIGURE 22.10 (a) Alumina anopore membrane with columnar mesoporous structures inside pores (scale bar 1 μm), (b) TEM image of the top of the fibers (scale bar 50 nm), and (c) side view of fibers (scale bar 50 nm). *Reprinted by permission from Macmillan Publishers Ltd: Nature Materials [93], Copyright (2004).*

Both these methods can be thought of as a combination of the EISA method and support confinement technique. Wu et al. [95] have published a comprehensive study on the effect of confinement on the mesophase formed in anopore supports. In situ grazing incidence small-angle X-ray scattering (GISAXS) techniques, in combination with ex situ transmission electron microscopy conducted by Platschek et al. [96], have revealed that in the case of ionic CTAB surfactant, contrary to previous belief, the formation of structure starts very late in the evaporation process. It has been proposed that the critical micelle concentration is reached first, followed by structure formation in the vicinity of the support wall, acting as a heterogenous nucleation center. Nonionic surfactants (i.e.: Brij-56 and Pluronic P123) where observed to form circular hexagonal structures first with the mesopores aligned around the circumference of the AAM channels. The circular structure subsequently transforms into a columnar hexagonal (P123 surfactant), or a mixture of columnar hexagonal and curved lamellar phases with the lamellae oriented parallel to the walls of the AAM channels (Brij-56 surfactant). It has been reported that these transformations occur after complete solvent evaporation and, therefore, differ from simple evaporation-induced phase formation.

In a different approach to oriented membrane formation, Seshadri et al. [97] have proposed a counter diffusion self-assembly (CDSA) method. A number of membrane supports have been studied. They have reported success with polycarbonate track-etch membranes supports which were held at the interface of an oil phase (tetrabutylorthosilicate, TBOS) and water phase (water/HCl/CTAB). The TBOS is allowed to diffuse into the water phase and undergo hydrolysis and condensation around micelles to form mesostructured fibers within support pores. They have identified precursor transport, surface chemistry, and solvent evaporation as controlling factors for good-quality membrane formation.

22.3.4. Functionalized Mesoporous Membranes

Functionalization of zeolites and sol–gel silica has been a topic of great interest for some time. Given that ordered mesoporous silica materials have high surface areas, availability of terminal silanol groups, and large pore sizes (which lead to better pore accessibility), they are expected to have better performance than zeolites and sol-gel silica in applications such as adsorbents, catalysis, sensing, and membrane reactors. Functionalization of the silica matrix can be achieved either in the synthesis stage by cocondensation or in a postsynthesis grafting step. The advantage of functionalizing during synthesis is that it can be accomplished with a one-step synthesis procedure. Additionally, these methods have been shown to provide more stable composite materials with a homogenous distribution of functional groups [98]. The main disadvantage is that the synthesis system will need to be modified to accommodate the additional reactants added to the synthesis mixture to maintain the delicate charge balance within the system. The loading of the material also needs to be kept at less than 25% to avoid structural collapse of the mesopores.

Postsynthesis functionalization, on the other hand, has the advantage of not requiring remodeling of the ordered material synthesis process. The functionalization of mesopores can be relatively easy to achieve due to the presence of uncondensed siloxane bonds within the system. These siloxane bonds act as anchoring sites for functional molecules. It has been suggested that the population density and the quality of the functionalized monolayers on mesoporous materials are greatly affected by the population of silanol groups and the number of adsorbed water molecules on the mesoporous silica surface. The calcining step used in preparing mesoporous silica causes condensation of the silanol groups present, thus resulting in poor surface coverage [99]. This can usually be overcome by introducing a small amount of water into the system to recover some of the lost silanol groups. Other drawbacks in these systems are the narrowing of the pores due to the addition of the grafted materials and difficulty in achieving uniform dispersion of the grafted molecule in the mesostructure. Furthermore, the pore openings are often blocked, which leads to low loading within the pores of the membrane.

Early studies conducted on functionalized materials and thin films have been seen to be quite successful. In membrane systems, however, fewer studies have been conducted. Most of these studies have focused on employing the postsynthesis grafting technique. Grafted materials include, but are not limited to long chain hydrocarbons such as 3-aminopropyltrimethoxysilane, polyethyleneimine, aminoproply triethoxysilane, and octadecyldimethylchlorosilane. These grafted membranes were used to study various gas separation properties. Park et al. [100] postfunctionalized MCM-48 membranes with trimethylchlorosilane to enhance hydrothermal stability and hydrophobicity. The separation properties of this membrane for ethanol, methylethyl ketone (MEK), and ethyl acetate (EA) in an aqueous were studied. Functionalization using trimethyl silane and triethylsilane resulted in enhanced separation properties, EtOH/water separation properties in the membrane [101]. Araki et al. [102] have shown that high dispersion of platinum into the silica matrix can be brought about with the use of sol–gel methods. Iron oxide nanocrystallites [103] and iron oxide clusters [104] have also been successfully incorporated into SBA-15 materials. These modifications can greatly enhance the potential of membranes in separation with catalytic reactor.

22.4. CONCLUSIONS

The most important properties of inorganic membranes include permeance and selectivity. These properties of microporous and mesoporous membranes are determined by the pore size, porosity, and membrane thickness. Chemistry plays a very important role in controlling these properties. Microporous amorphous silica membranes can be prepared by the acid catalyzed sol–gel method using an alkoxide precursor. The intercrystalline defects and zeolitic pores of crystalline zeolite membranes can be eliminated or narrowed by chemical vapor deposition through proper control of the surface chemistry of the zeolite framework. The last few years have seen progress in the synthesis of metal–organic framework membranes. The membranes are very thick because synthesis chemistry gave large MOF crystals. Synthesis chemistry controls the pore size and structure of ordered mesoporous materials as well as orientation of the pore of the ordered mesoporous membranes.

References

[1] W.J. Koros, Y.H. Ma, T. Shimidzu, Pure and Applied Chemistry 68 (1996) 1479.
[2] Y.S. Lin, Separation and Purification Technology 25 (2001) 39.
[3] T. Sano, S. Ejiri, K. Yamada, Y. Kawakami, H. Yanagishita, Journal of Membrane Science 123 (1997) 225.
[4] J.G. Sanchez Marcano, T.T. Tsotsis, Catalytic Membranes and Membrane Reactors, Wiley-VCH, Weinheim, 2002.
[5] R. Bredesena, K. Jordal, A. Bolland, Chemical Engineering and Processing 43 (2004) 1129.
[6] K. Li, Ceramic Membranes for Separation and Reaction, Wiley, New York, 2007.
[7] B.N. Nair, K. Keizer, T. Okubo, S.I. Nakao, Advanced Materials 10 (1998) 249.
[8] R.J.R. Uhlhorn, K. Keizer, A.J. Burggraaf, Journal of Membrane Science 66 (1992) 271.
[9] R.J.R. Uhlhorn, V.T. Zaspalis, K. Keizer, A.J. Burggraaf, Journal of Materials Science 27 (1992) 538.
[10] G.P. Fotou, Y.S. Lin, S.E. Pratsinis, Journal of Materials Science 30 (1995) 2803.
[11] C.J. Brinker, A.J. Hurd, P.R. Schunk, G.C. Frye, C.S. Ashley, Journal of Non-Crystalline Solids 147 (1992) 424.
[12] C.J. Brinker, T.L. Ward, R. Sehgal, N.K. Raman, S.L. Hietala, D.M. Smith, et al., Journal of Membrane Science 77 (1993) 165.
[13] R.S.A. Delange, J.H.A. Hekkink, K. Keizer, A.J. Burggraaf, Microporous Materials 4 (1995) 169.
[14] R.S.A. Delange, J.H.A. Hekkink, K. Keizer, A.J. Burggraaf, Journal of Membrane Science 99 (1995) 57.
[15] R.S.A. Delange, K. Keizer, A.J. Burggraaf, Journal of Membrane Science 104 (1995) 81.
[16] C.J. Brinker, G.W. Scherer, Sol-gel Science, Academic Press, London, 1990.
[17] C.J. Brinker, R. Sehgal, S.L. Hietala, R. Deshpande, D.M. Smith, D. Loy, et al., Journal of Membrane Science 94 (1994) 85.
[18] C.J. Brinker, N.K. Raman, M.N. Logan, R. Sehgal, R.A. Assink, D.W. Hua, et al., Journal of Sol-Gel Science and Technology 4 (1995) 117.
[19] R. Schafer, M. Noack, P. Kolsch, S. Thomas, A. Seidel-Morgenstern, J. Caro, Separation and Purification Technology 25 (2001) 3.
[20] J.C.D. da Costa, G.Q. Lu, V. Rudolph, Y.S. Lin, Journal of Membrane Science 198 (2002) 9.
[21] J. Dong, Y.S. Lin, Industrial & Engineering Chemistry Research 37 (1998) 2404.
[22] J. Caro, M. Noack, P. Kolsch, R. Schafer, Microporous and Mesoporous Materials 38 (2000) 3.
[23] J. Caro, M. Noack, Microporous and Mesoporous Materials 115 (2008) 215.
[24] M. Kanezashi, J. O'Brien-Abraham, Y.S. Lin, K. Suzuki, AIChE Journal 54 (2008) 1478.
[25] M. Kanezashi, Y.S. Lin, Journal of Physical Chemistry C 113 (2009) 3767.
[26] T. Masuda, N. Fukumoto, M. Kitamura, S.R. Mukai, K. Hashimoto, T. Tanaka, et al., Microporous and Mesoporous Materials 48 (2001) 239.
[27] M. Nomura, T. Yamaguchi, S. Nakao, Industrial Engineering & Chemistry Research 36 (1997) 4217.
[28] C.A. Cooper, Y.S. Lin, Journal of Membrane Science 195 (2002) 35.
[29] M. Hong, J.L. Falconer, R.D. Noble, Industrial & Engineering Chemistry Research 44 (2005) 4035.
[30] Y.S. Lin, A.J. Burggraaf, Journal of Membrane Science 79 (1993) 65.
[31] Z. Tang, J.H. Dong, T.M. Nenoff, Langmuir 25 (2009) 4848.

[32] X.H. Gu, Z. Tang, J.H. Dong, Microporous and Mesoporous Materials 111 (2008) 441.
[33] M. Kanezashi, J. O'Brien, Y.S. Lin, Microporous and Mesoporous Materials 103 (2007) 302.
[34] G. Xomeritakis, Y.S. Lin, Chemical Engineering Science 49 (1994) 3909.
[35] G. Xomeritakis, Y.S. Lin, AIChE Journal 44 (1998) 174.
[36] O.M. Yaghi, M. O'Keeffe, N.W. Ockwig, H.K. Chae, M. Eddaoudi, J. Kim, Nature 423 (2003) 705.
[37] H.K. Chae, D.Y. Siberio-Perez, J. Kim, Y. Go, M. Eddaoudi, A.J. Matzger, et al., Nature 427 (2004) 523.
[38] J.L.C. Rowsell, O.M. Yaghi, Microporous and Mesoporous Materials 73 (2004) 3.
[39] R.Q. Snurr, J.T. Hupp, S.T. Nguyen, AIChE Journal 50 (2004) 1090.
[40] J.J. Perry, J.A. Perman, M.J. Zaworotko, Chemical Society Reviews 38 (2009) 1400.
[41] M. Eddaoudi, J. Kim, N. Rosi, D. Vodak, J. Wachter, M. O'Keeffe, et al., Science 295 (2002) 469.
[42] Y.S. Lin, I. Kumakiri, B.N. Nair, H. Alsyouri, Separation and Purification Methods 31 (2002) 229.
[43] V.V. Guliants, M.A. Carreon, Y.S. Lin, Journal of Membrane Science 235 (2004) 53
[44] D. Zacher, O. Shekhah, C. Woll, R.A. Fischer, Chemical Society Reviews 38 (2009) 1418.
[45] S.S.Y. Chui, S.M.F. Lo, J.P.H. Chaemant, A.G. Orpen, I.D. Williams, Science 283 (1999) 1148.
[46] L. Pan, M.B. Sander, X.Y. Huang, J. Li, M. Smith, E. Bittner, et al., Journal of the American Chemical Society 126 (2004) 1308.
[47] H. Hayashi, A.P. Cote, H. Furukawa, M. O'Keeffe, O.M. Yaghi, Nature Materials 6 (2007) 501.
[48] M. Arnold, P. Kortunov, D.J. Jones, Y. Nedellec, J. Karger, J. Caro, European Journal of Inorganic Chemistry (2007) 60.
[49] J. Gascon, S. Aguado, F. Kapteijn, Microporous and Mesoporous Materials 113 (2008) 132.
[50] Y.Y. Liu, Z.F. Ng, E.A. Khan, H.K. Jeong, C.B. Ching, Z.P. Lai, Microporous and Mesoporous Materials 118 (2009) 296.
[51] Y. Yoo, Z.P. Lai, H.K. Jeong, Microporous and Mesoporous Materials 123 (2009) 100.
[52] Z.X. Zhao, Z. Li, Y.S. Lin, Industrial & Engineering Chemistry Research 48 (2009) 10015.
[53] H.L. Guo, G.S. Zhu, I.J. Hewitt, S.L. Qiu, Journal of the American Chemical Society 131 (2009) 1646.
[54] R. Ranjan, M. Tsapatsis, Chemistry of Materials 21 (2009) 4920.
[55] H. Bux, F.Y. Liang, Y.S. Li, J. Cravillon, M. Wiebcke, J. Caro, Journal of the, American Chemical Society 131 (2009) 16000.
[56] Y.S. Li, F.Y. Liang, H. Bux, A. Feldhoff, W.S. Yang, J. Caro, Angewandte Chemie International Edition 49 (2010) 548.
[57] S.R. Venna, M.A. Carreon, Journal of the American Chemical Society 132 (2010) 76.
[58] H. Verweij, Y.S. Lin, J.H. Dong, MRS Bulletin 31 (2006) 756.
[59] E.V. Perez, K.J. Balkus, J.P. Ferraris, I.H. Musselman, Journal of Membrane Science 328 (2009) 165.
[60] A.I. Skoulidas, Journal of the American Chemical Society 126 (2004) 1356.
[61] A.I. Skoulidas, D.S. Sholl, Journal of Physical Chemistry B 109 (2005) 15760.
[62] Q.Y. Yang, C.Y. Xue, C.L. Zhong, J.F. Chen, AIChE Journal 53 (2007) 2832.
[63] T. Watanabe, S. Keskin, S. Nair, D.S. Sholl, Physical Chemistry Chemical Physics 11 (2009) 11389.
[64] S. Keskin, D.S. Sholl, Industrial & Engineering Chemistry Research 48 (2009) 914.
[65] S. Keskin, J.C. Liu, J.K. Johnson, D.S. Sholl, Microporous and Mesoporous Materials 125 (2009) 101.
[66] C.T. Kresge, M.E. Leonowicz, W.J. Roth, J.C. Vartuli, J.S. Beck, Nature 359 (1992) 710.
[67] G. Soler-Illia, E.L. Crepaldi, D. Grosso, C. Sanchez, Current Opinion in Colloid & Interface Science 8 (2003) 109.
[68] K.J. Edler, S.J. Roser, International Reviews in Physical Chemistry 20 (2001) 387.
[69] Q.S. Huo, D.I. Margolese, U. Ciesla, P.Y. Feng, T.E. Gier, P. Sieger, et al., Nature 368 (1994) 317.
[70] G.J.D. Soler-illia, C. Sanchez, B. Lebeau, J. Patarin, Chemical Reviews 102 (2002) 4093.
[71] Q.S. Huo, D.I. Margolese, G.D. Stucky, Chemistry of Materials 8 (1996) 1147.
[72] N. Nishiyama, A. Koide, Y. Egashira, K. Ueyama, Chemical Communications (1998) 2147.
[73] P.Y. Feng, X.H. Bu, D.J. Pine, Langmuir 16 (2000) 5304.
[74] P. Kumar, J. Ida, S. Kim, V.V. Guliants, J.Y.S. Lin, Journal of Membrane Science 279 (2006) 539.
[75] P. Kumar, J.C. Ida, V.V. Guliants, Microporous and Mesoporous Materials 110 (2008) 595.
[76] Y.F. Lu, R. Ganguli, C.A. Drewien, M.T. Anderson, C.J. Brinker, W.L. Gong, et al., Nature 389 (1997) 364.
[77] N. Nishiyama, D.H. Park, A. Koide, Y. Egashira, K. Ueyama, Journal of Membrane Science 182 (2001) 235.
[78] B.A. McCool, N. Hill, J. DiCarlo, W.J. DeSisto, Journal of Membrane Science 218 (2003) 55.
[79] C.Y. Liu, L.Q. Wang, W.Z. Ren, Z.H. Rong, X.Q. Wang, J.Q. Wang, Microporous and Mesoporous Materials 106 (2007) 35.
[80] Y. Sakamoto, K. Nagata, K. Yogo, K. Yamada, Microporous and Mesoporous Materials 101 (2007) 303.
[81] H. Ji, Y.Q. Fan, W.Q. Jin, C.L. Chen, N.P. Xu, Journal of Non-Crystalline Solids 354 (2008) 2010.
[82] M. Klotz, A. Ayral, C. Guizard, L. Cot, Journal of Materials Chemistry 10 (2000) 663.
[83] C.J. Brinker, Y.F. Lu, A. Sellinger, H.Y. Fan, Advanced Materials 11 (1999) 579.
[84] S. Higgins, R. Kennard, N. Hill, J. DiCarlo, W.J. DeSisto, Journal of Membrane Science 279 (2006) 669.
[85] S. Higgins, W. DeSisto, D. Ruthven, Microporous and Mesoporous Materials 117 (2009) 268.
[86] V. Boffa, J.E. ten Elshof, D.H.A. Blank, Microporous and Mesoporous Materials 100 (2007) 173.
[87] S.R. Chowdhury, R. Schmuhl, K. Keizer, J.E. ten Elshof, D.H.A. Blank, Journal of Membrane Science 225 (2003) 177.
[88] R. Schmuhl, S.R. Chowdhury, J.E. ten Elshof, A. van den Berg, D.H.A. Blank, Journal of Sol-Gel Science and Technology 31 (2004) 249.
[89] K. Nakagawa, H. Matsuyama, T. Maki, M. Teramoto, N. Kubota, Separation and Purification Technology 44 (2005) 145.
[90] S.R. Chowdhury, A.M. Peters, D.H.A. Blank, J.E. ten Elshof, Journal of Membrane Science 279 (2006) 276.
[91] S.J. Yoo, D.M. Ford, D.F. Shantz, Langmuir 22 (2006) 1839.
[92] Y.H. Lin, H.D. Li, C.P. Liu, W. Xing, X.L. Ji, Journal of Power Sources 185 (2008) 904.
[93] A. Yamaguchi, F. Uejo, T. Yoda, T. Uchida, Y. Tanamura, T. Yamashita, et al., Nature Materials 3 (2004) 337.
[94] Q.Y. Lu, F. Gao, S. Komarneni, T.E. Mallouk, Journal of the American Chemical Society 126 (2004) 8650.
[95] Y.Y. Wu, G.S. Cheng, K. Katsov, S.W. Sides, J.F. Wang, J. Tang, et al., Nature Materials 3 (2004) 816.
[96] B. Platschek, R. Kohn, M. Doblinger, T. Bein, Langmuir 24 (2008) 5018.

[97] S.K. Seshadri, H.M. Alsyouri, Y.S. Lin, Microporous and Mesoporous Materials 129 (2010) 228.
[98] S.L. Burkett, S.D. Sims, S. Mann, Chemical Communications (1996) 1367.
[99] X. Feng, G.E. Fryxell, L.Q. Wang, A.Y. Kim, J. Liu, K.M. Kemner, Science 276 (1997) 923.
[100] D.H. Park, N. Nishiyama, Y. Egashira, K. Ueyama, Microporous and Mesoporous Materials 66 (2003) 69.
[101] D.H. Park, N. Nishiyama, Y. Egashira, K. Ueyama, Industrial & Engineering Chemistry Research 40 (2001) 6105.
[102] S. Araki, T. Satoh, H. Doi, H. Yano, Y. Miyake, Desalination and Water Treatment 7 (2009) 12.
[103] J.S. Li, X.T. Wei, Y.S. Lin, D. Su, Journal of Membrane Science 312 (2008) 186.
[104] K. Bachari, A. Touileb, Solid State Sciences 11 (2009) 1549.

CHAPTER

23

The Frontier of Inorganic Synthesis and Preparative Chemistry (I)—Biomimetic Synthesis

Kesong Liu[†], *Lei Jiang*[†, ‡]

[†] College of Chemistry and Environment, Beijing University of Aeronautics and Astronautics, China,
[‡] Institute of Chemistry, Chinese Academy of Sciences, China

After four and a half billion years of evolution, creatures in nature possess almost perfect structures and properties, exhibit the harmonization, and unification between structure and function. The term biomimetics (also known as bionics) emerged in the 1960s, which originates from Greek "Bios" (life, nature) and "Mimesis" (imitation, copy) [1]. It can be defined as the investigation of the structures and functions of biological materials that allows possible future design and synthesis of new and improved materials based on the principles of natural materials [2]. Biomimetic principles provide new method and approach for the construction of novel structural and functional materials. Learning from nature will give us important inspiration to develop new methods to construct artificial advanced materials [3]. This "learning" may be through inspiration in design, function, or a combination of both. Usually, this inspiration derives from a novel attribute of a biological system that suggests new and important insights into structure and function for materials science applications [1].

Inspired by nature, recently, much attention has been paid to biomimetic or bio-inspired design and synthesis of structural and functional materials [4–7]. Being that completeness was not an option, in order to give a fresh perspective on the field, in this chapter, we focus on the biomimetic synthesis of inorganic materials, particularly on biomineralization and its mimetic inorganic materials, biotemplated inorganic materials, biomimetic synthesis of inorganic chiral materials, and bio-inspired multiscale inorganic materials, illustrating by appropriate recent key examples.

Section 23.1 describes the natural biomineralization mechanism, the structures and formation mechanism of diatom, and nacre biominerals and their corresponding biomimetic mineralization materials. Section 23.2 provides an overview of the biomimetic synthesis of inorganic materials by using DNA, bacteria, insect wings, shell membranes, and other natural materials as the biotemplates. Some typical biomimetic synthesis of inorganic chiral materials, such as zeolites, molecular sieves, SiO_2, and metals, is covered in Section 23.3. Some representative bio-inspired multiscale inorganic materials, such as special wettability materials (superhydrophobicity, superoleophobicity), hollow-structured materials, nacre-inspired ultra-strong and super-tough layered materials, gecko-inspired high adhesive inorganic materials, and other materials are presented in Section 23.4. Due to the space constraints, we cannot review all the significant and interesting work in this active field. We hope that the chapter presented here will offer some inspirations to the graduate students and researchers in the fields of chemistry, materials science, biology, and nanoscience and will be helpful in the future development of biomimetics.

23.1. BIOMINERALIZATION AND ITS MIMETIC INORGANIC MATERIALS

23.1.1. Introduction

After billions of years of evolution, biological organisms produce mineralized tissues such as bone, teeth,

Modern Inorganic Synthetic Chemistry, DOI: 10.1016/B978-0-444-53599-3.10023-X

diatoms, corals, and shells. Biominerals are well-known composites of inorganic and organic materials in the form of fascinating shapes and high-ordered structures. The biological formation of minerals by living organisms is commonly called biomineralization [1,7–12]. The most common mineral components are calcium carbonate, calcium phosphate (hydroxyapatite), and amorphous silica, although over 20 minerals (with principal elements being Ca, Mg, Si, Fe, Mn, P, S, C, and the light elements H and O) have been identified [4]. The main types and functions of biominerals are shown in Table 23.1. The major mineralized tissues such as bone and teeth, and shells are composed of calcium phosphate and carbonate minerals, respectively, in combination with a complex organic macromolecular matrix of proteins, polysaccharides, and lipids [13].

Natural biominerals generally possess hierarchical structures organized at the nano, micro, meso, and macro levels, which exhibit optimized functions and high environmental adaptability as a result of long-time evolution. Biomineralization offers valuable ideas and inventive principles for materials scientists, chemists, and engineers to learn how to create superstructures resembling naturally existing biominerals with their unusual shapes and complexity. Learning from nature, a biomimetic/bio-inspired approach copying/applying strategies from biomineralization processes to inorganic or organic–inorganic hybrid materials with controlled mineralization analogous to those produced by nature has recently received much attention owing to their potential applications in industry [14–16]. Compared with conventional methods of materials production, biomimetic/bio-inspired mineralization is facile, environmentally benign, and economic, which is a multidisciplinary subject related to materials science, chemistry, biology, nanotechnology, physics, etc. During the past decade, a large variety of inorganic materials with unusual structural specialty and complexity have been synthesized [17]. The construction of inorganic materials or organic–inorganic hybrid materials with hierarchical structures through mimicking the biomineralization process has become an important branch in the broad area of biomimetics [18].

This section will focus on the biomineralization and its mimetic inorganic materials, especially on the natural biomineralization process, the diatom and nacre biominerals, and their corresponding biomimetic mineralization of inorganic materials.

TABLE 23.1 Types and Functions of the Main Inorganic Solids Found in Biological Systems [13]

Mineral	Formula	Organism/function
CALCIUM CARBONATE		
Calcite	$CaCO_3{}^a$	Algae/exoskeletons, Tribolites/eye lens
Aragonite	$CaCO_3$	Fish/gravity device, Molluscs/exoskeleton
Vaterite	$CaCO_3$	Ascidians/spicules
Amorphous	$CaCO_3 \cdot nH_2O$	Plants/Ca store
CALCIUM PHOSPHATE		
Hydroxyapatite	$Ca_{10}(PO_4)_6(OH)_2$	Vertebrates/endoskeletons, teeth, Ca store
Octa-calcium phosphate	$Ca_8H_2(PO_4)_6$	Vertebrates/precursor phase in bone?
Amorphous	?	Mussels/Ca store, Vertebrates/precursor phase in bone
CALCIUM OXALATE		
Whewellite	$CaC_2O_4 \cdot H_2O$	Plants/Ca store
Weddellite	$CaC_2O_4 \cdot 2H_2O$	Plants/Ca store
GROUP IIA METAL SULFATES		
Gypsum	$CaSO_4$	Jellyfish larvae/gravity device
Barite	$BaSO_4$	Algae/gravity device
Celestite	$SrSO_4$	Acantharia/cellular support
SILICON DIOXIDE		
Silica	$SiO_2 \cdot nH_2O$	Algae/exoskeletons
IRON OXIDE		
Magnetite	Fe_3O_4	Bacteria/magnetotaxis, Chitons/teeth
Goethite	α-FeOOH	Limpets/teeth
Lepidocrocite	γ-FeOOH	Chitons (mollusca)/teeth
Ferrihydrite	$5Fe_2O_3 \cdot 9H_2O$	Animals and plants/Fe storage proteins

[a]*A range of magnesium-substituted calcites are also formed.*
Reproduced with permission from The Royal Society of Chemistry.

23.1.2. Biomineralization

Biomineralization is a sophisticated process and a widespread phenomenon in nature leading to the formation of a variety of inorganic minerals by living organisms [19]. These biologically produced minerals with interesting properties and controlled hierarchical structures are inorganic–organic hybrid composites formed by self-assembled bottom-up processes under mild conditions (aqueous solutions and temperatures below 100°C) [20]. In nature, matrix macromolecules exert a strong influence on crystal nucleation, growth, and self-organization into hierarchical architectures and then control the biomineralization processes [21–23]. As proposed by Mann [13,24], the process of biomineralization can be divided into four key constructional stages, i.e., supramolecular preorganization, interfacial molecular recognition, vectorial regulation, and cellular processing. These four stages act cooperatively and are integrated within feedback systems. A number of studies have demonstrated that the ability to fine-tune both the rates of inorganic precipitation and the self-assembly of organic molecules at the same time within a reaction medium is the key to the synthetic fabrication of organized organic–inorganic materials. Here, we will briefly introduce the four constructional processes [13,24–26].

23.1.2.1. *Supramolecular Preorganization*

The first stage in biomineralization involves the construction of an organized reaction environment prior to mineral deposition. Two general approaches have evolved: (i) the self-assembly of enclosed protein cages and lipid vesicles in which the molecular construction of the compartment is mainly based on the balancing of hydrophobic–hydrophilic interactions that exist for amphiphilic molecules in aqueous environments; and (ii) the facilitated construction of extended protein–polysaccharide networks as organized polymeric environments for extracellular mineralization reactions. The former is prevalent in intracelluar biomineralization, whereas the latter is predominant in the extracellular (intercellular) spaces generated by multicellular organisms. Both systems produce architectures representative of first-order molecular tectonics.

23.1.2.2. *Interfacial Molecular Recognition*

The second constructional stage in biomineralization involves the controlled nucleation of inorganic clusters from aqueous solutions. The control of nucleation of inorganic clusters into the framework built in the first stage by supramolecular preorganization is one of the major points in biomineralization processes. In general view, these preorganized organic architectures consist of somehow functionalized surfaces that serve as molecular blueprints (templates) for site-directed inorganic nucleation. In this stage, the first-order molecular construction of organic supramolecular systems (such as vesicles, ferritin micelles, and polymeric networks) provides a framework for the second-order assembly of the inorganic phase. The primary function of organic surface is to reduce the activation energy of nucleation and/or influence the collision frequencies of the nucleation rate [27]. It is considered that the assembly of mineral nuclei is generally governed by electrostatic, structural, and stereochemical complementarities at the inorganic–organic interfaces [22]. Weissbuch et al. describe the auxiliary molecules which promote or inhibit crystal nucleation depending on their composition [28]. This aspect of molecular tectonics, in which interfacial molecular recognition assists the construction of nuclei (often with specific crystallographic structures and orientation), is not only a central feature of controlled biomineralization, but has important generic implications in synthetic materials chemistry [24].

23.1.2.3. *Vectorial Regulation*

The third stage of biomineralization is associated with the assembly of mineral phase through vectorial regulation of crystal growth and termination. These processes result in the characteristic ornamentation of many biominerals, such as diatom shells, coccolith scales, and acantharian skeletons. In each case, the patterning of mineralization process arises from the shaping of associated membrane vesicles subjected to a cellular stress field [13]. The vectorial regulation of biomineral shapes often reflects a compromise between the constraints of crystal physics and the genetic manipulation of morphology.

23.1.2.4. *Cellular Processing*

Biomineralization does not stop with the formation of small particles, but proceeds with the construction of higher-order architectures with elaborate ultra- and microstructural properties. The final stage of biomineralization is associated with a variety of constructional processes involving larger scale cellular activity and the production of higher-order biomineralized architectures with special properties. In the absence of further cellular intervention, mineral nuclei continue to grow within the confines of their supramolecular hosts. The resulting particles are constrained in size but have normal crystallographic structures and morphologies. In many organisms, however, a third order of tectonic complexity can be established by cellular processing, in which biominerals are endowed with unusual textures and shapes. At this level of construction, the disparity between biological and synthetic materials is most pronounced. Construction of these higher-order architectures within the cellular space can be based on either static or dynamic principles. An example of such an organized ultrastructure is the relatively rigid and spatially fixed linear assembly of a chain of membrane-bound magnetite crystals in magnetotactic bacteria [29]. Another example is the macroscopic organized architectures found in the nacreous layer of shells with sheet-like organic assemblies which are secreted periodically beyond the mineralization front [30]. The details of the recognition and organizational processes involved in the construction of these higher-order biomineralized architectures are currently unknown; it is probable that the phenomenon is a consequence of what one might describe as cellular logistics. The cellular processing of biomineral shapes often reflects a compromise between the constraints of crystal physics and the genetic manipulation of morphology.

23.1.3. Diatoms and Their Mimetic Mineralization Materials

23.1.3.1. Background

In nature, the biomineralization process yields intricately patterned silica shells. Amorphous silica is the second most abundant biomineral on Earth, and is mainly produced by diatoms [7,11,12]. Diatoms are single-celled eukaryotic algae that are found throughout the world's oceans and freshwater environments and adapted to living in soils and ice [31]. The food chain in the marine ecosystem is mainly based on the photosynthetic capacity of diatoms. Diatoms are estimated to contribute to 20% of global photosynthetic CO_2 fixation and 40% of primary biological production in the oceans, showing enormous ecological importance [32]. In addition to current interest in diatoms as mediators of global carbon flux, data indicated that in the past climate change has affected diatom evolution as well as changes in the global phytoplankton community structure that have affected the climate [33–35].

Diatom cell is encapsulated by a unique silica cell wall called the frustules, which are consisted of two symmetrical sides with a split between them. Diatom frustules are biosilica structures with genetically controlled three-dimensional ordered shapes on a nanometer to micrometer scale (Fig. 23.1). Research indicated that silica cell wall can serve as a protective armor against phytoplankton predators. Micromechanical force measurements in combination with finite element calculation revealed the exceptional mechanical stability of diatom cell walls, sufficient to withstand attack by the crushing tools of most diatom predators [36]. Diatom

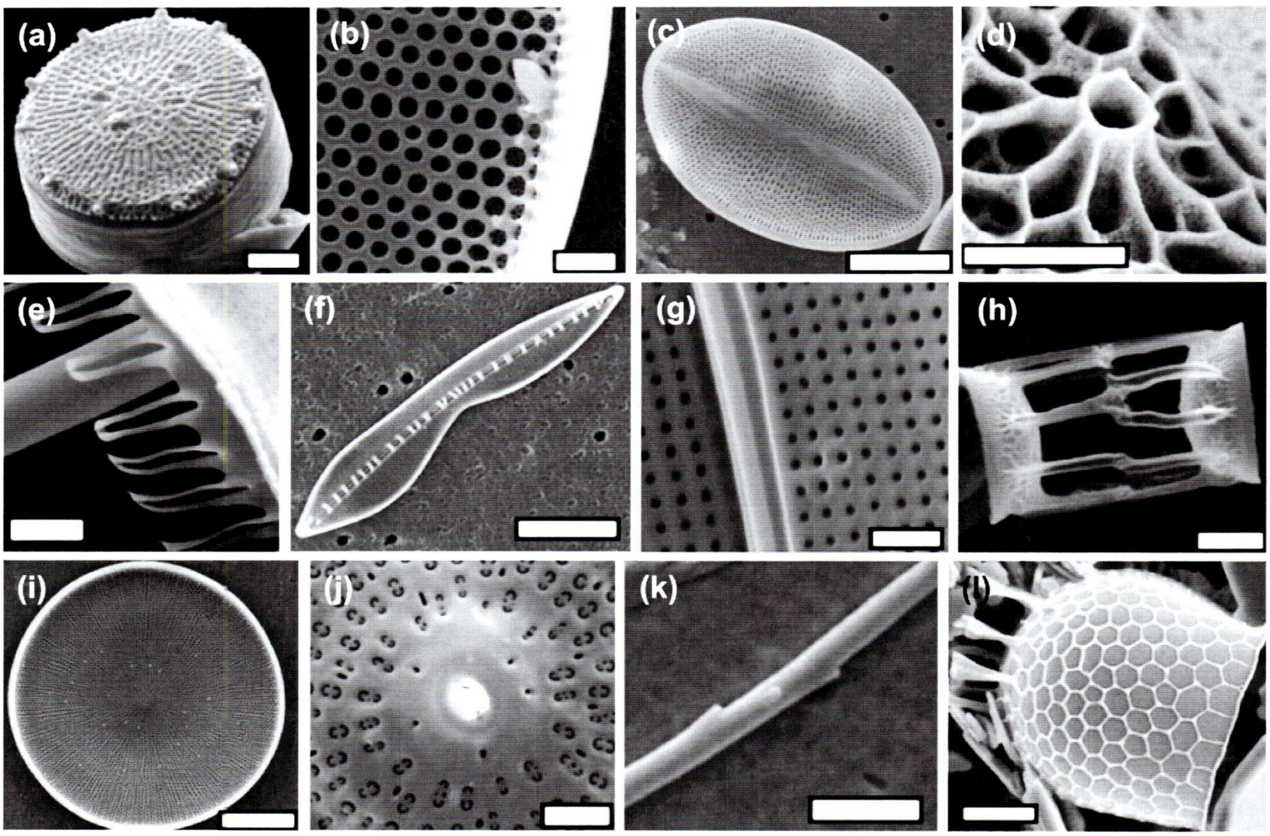

FIGURE 23.1 Diversity of diatom silica structures. (a) *Thalassiosira pseudonana*, bar = 1 μm, (b) *Coscinodiscus wailesii*, bar = 5 μm, (c) *Cocconeis* sp., bar = 10 μm, (d) rimoportula from *Thalassiosira weissflogii*, bar = 500 nm, (e) corona structure of *Ditylum brightwellii*, bar = 2 μm, (f) *Bacillaria paxillifer*, bar = 10 μm, (g) close up of pores in *Gyrosigma balticum*, bar = 2 μm, (h) *Skeletonema costatum*, bar = 2 μm, (i) valve of *C. wailesii*, bar = 50 μm, (j) close up of pores in *D. brightwellii*, bar = 2 μm, (k) seta of *Chaetoceros gracilis*, bar = 1 μm, and (l) *Stephanopyxis turris*, bar = 10 μm. *Reprinted with permission from [31]. Copyright 2008 American Chemical Society.*

species are classified according to their cell wall shapes and structures, and the estimated number of species is at least in the tens of thousands and perhaps more [12].

23.1.3.2. *Mechanisms*

Apart from their ecological significance, diatoms are well known for the intricate structures and spectacular patterns of their silica-based cell walls, which hold considerable promise for biological or biomimetic fabrication of nanostructured materials and devices. These elaborate patterns are species specific. They are precisely reproduced in each generation documenting a genetic control of this biomineralization process [37]. In recent years, much progress has been made in the study of the biomineralization processes in diatoms [38]. The understanding of complex mechanisms involved in diatom biomineralization processes is not only a fascinating biological problem, but also a great interest for nanomaterials science and technology [39].

Investigations into the cellular mechanism of silica biomineralization in diatoms were pioneered by electron microscopists during the early 1960s [11]. The hierarchically structured cell walls of diatoms contain amorphous silica and special biomolecules. Generally, diatom silica formation processes occur in the membrane-bound silica deposition vesicle (SDV) [31]. The SDV membrane, called the silicalemma, has distinct properties compared with other cellular membranes, which plays a significant role in diatom silica structure formation. Volcani and coworkers first engaged in biochemical studies on silicon metabolism and silica biomineralization in diatoms and many of their studies have been conducted with *Cylindrotheca fusiformis* [40]. Recently, diatom species *Thalassiosira pseudonana* (Fig. 23.1a) has been established as the model organism for silica biomineralization research. It is the first diatom species, whose complete genome sequence has been determined with the identification of components of its silica forming machinery. Over the past decade, three different classes of such biomolecules have been identified from diatom *T. pseudonana*: (i) the silaffins, highly posttranslationally modified peptides/proteins [41–43], (ii) long-chain polyamines (LCPAs) [44], and (iii) the highly acidic silacidins [45]. Silaffins (proteins with high silica affinity) and LCPAs are tightly associated with diatom silica, which have been shown to promote silica formation from monosilicic acid in vitro. The silacidins in diatom cell walls serve as the polyanion required in vivo for silica formation directed by polyamines (and/or silaffins). Most recently, a network-like chitin-based scaffold was found for the first time in the cell walls of diatom species *T. pseudonana* that resembles the size and shape of the biosilica. These scaffolds are composed of interconnected fibers with an average diameter of 25 nm that contain other unknown biomolecules apart from chitin [46]. In addition, a phase separation model was proposed by Sumper to explain the structure formation of diatom cell walls of the genus *Coscinodiscus*. In this model, the amphiphilic polyamine phase separates and forms an emulsion of micro- and nanodroplets within the flat SDV. The repeated phase separation events are assumed to result in the formation of self-similar silica patterns in smaller and smaller scales [47].

23.1.3.3. *Biomimetic Materials*

The ordered porous structures of diatom frustules irrefutably demonstrate precision and brilliance of natural design at the micro- and nanoscale. These unique structures have attracted much attention owing to their potential applications in optics, photonics, catalysis, biosensing, filtration, and drug delivery [31,39, 48–52]. Inspired by diatom silicon biomineralization, many different synthesis strategies have been applied for the biomimetic construction of ordered silica and other materials using different types of polyelectrolytes such as poly(amino acids), polyamines, and related compounds [7,42,44, 53–56]. Owing to the diversity and delicacy of the diatom biosilica structures, substantial amount of efforts have been made to fabricate ordered hierarchical structures and functional compositions using natural diatoms as the templates [49]. A typical example for the fabrication of micro- and nanoscale polymer patterns by replica molding using diatom is shown in Fig. 23.2.

The procedure can be summarized in three steps [57]: (i) immobilization of diatom frustule on surface which is used as the biological nanostructured master for replication (Fig. 23.2a); (ii) the transferring of diatom's porous pattern onto elastomeric poly(dimethylsiloxane) (PDMS) polymer (negative replica) by replica molding (Fig. 23.2b and c); and (iii) converting the PDMS replica into positive replica by second replica molding step using desired polymer such as UV curable polymer (Fig. 23.2d and e).

The unique hierarchical structures and high specific surface areas of diatom biosilica and biomimetic analogues (such as metal nanoparticles [58], metal oxides [59–61], boron nitride [62], and other materials [63]) exhibit a wide domain of promising applications in many fields: optics, catalysis, sensors, and biological applications [31,49]. For example, diatom biosilica containing metabolically inserted germanium can be used to fabricate electroluminescent and photoluminescent thin film devices [64]. Antibody-functionalized diatom biosilica frustules have been served as a microscale biosensor platform for selective and label-free photoluminescence-based detection of immunocomplex formation [65]. Recently, a three-dimensional silicon structure was produced from diatom biosilica through

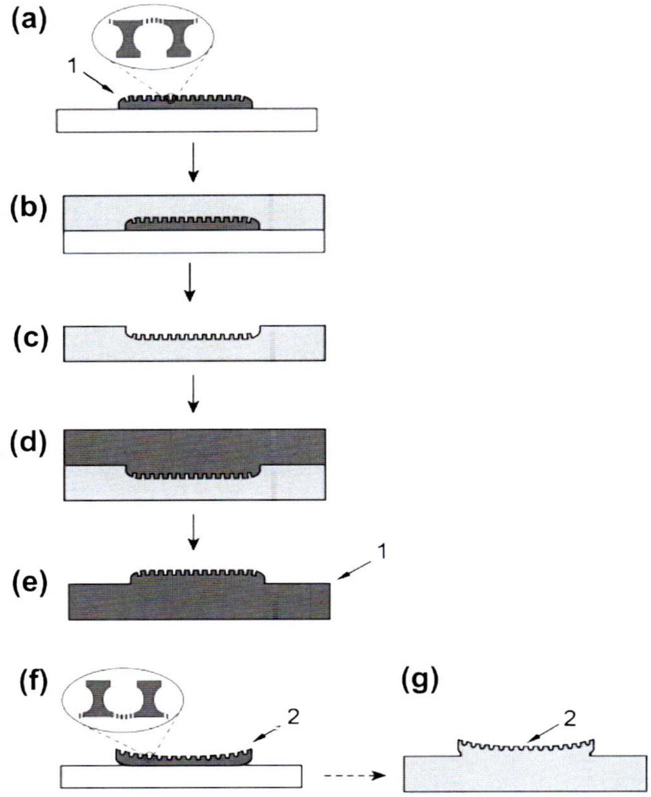

FIGURE 23.2 Schematic diagram of the replica molding process using diatom frustules [57]. (a) Frustule on surface (concave orientations); (b, c) replication of frustule into PDMS (negative replica); (d, e) replication of PDMS replica into mercapto ester type UV-curable polymer (NOA 60); and (f, g) frustule in convex orientations and their corresponding positive replica. *Dusan Losic, James G. Mitchell, Ratnesh Lal, and Nicolas H. Voelcker, Rapid Fabrication of Micro- and Nanoscale Patterns by Replica Molding from Diatom Biosilica, Adv. Funct. Mater. 2007, 17, pages 2439–2446, Copyright Wiley-VCH Verlag GmbH & Co. KGaA. Reproduced with permission.*

a low-temperature magnesiothermic reduction process [66]. The resultant silicon replicas possess a high specific surface area and contain a significant population of micropores, which exhibit rapid changes in impedance upon exposure to gaseous nitric oxide. Furthermore, changes of only 1 ppm in NO(g) concentration could be detected. Some applications of diatom biosilica and biomimetic analogues are being further extended and new properties and uses are being discovered.

23.1.4. Nacre and Its Mimetic Mineralization Materials

23.1.4.1. Background

The main inorganic mineral on Earth produced in biomineralization processes of natural organisms is calcium carbonate ($CaCO_3$). The biomineral structure and the morphology of $CaCO_3$ have been found to be remarkably diverse [67]. Calcium carbonate exists in four crystalline polymorphs: aragonite, calcite, vaterite, and amorphous. Nacre, also known as mother-of-pearl, is an organic–inorganic composite material possessing remarkable mechanical properties, which can be found in the inner layer of many shells (Fig. 23.3a and b). The microstructure of nacre is well organized in a layered "brick and mortar" architecture over several length scales, where the inorganic bricks are organized polygonal aragonite platelets (about 5–10 μm in diameter, 0.2–0.5 μm in thickness) and the mortar is about 5–20 nm thick organic biopolymer (e.g., hydrophobic proteins, chitin, etc.) interlayer (Fig. 23.3c) [68–70]. Nacre is composed of about 95% inorganic aragonite, with only a few percent of organic matrix [69]. The interdigitating brickwork array of crystalline aragonite platelets and biopolymer layers results in a fracture toughness that is about 3000 times greater than that of pure aragonite [71].

23.1.4.2. Mechanism

The nacreous layer has attracted interest of researchers in a broad range of chemistry disciplines owing to its structure–function harmony. In the past several decades, much effort has been devoted to exploring the formation mechanisms of nacreous layer. Generally, biomineralized materials contain only a few weight percent of proteins or other biopolymers, whereas these macromolecules play a critical role in the assembly of elaborately structured composites with remarkable precision over scales ranging from the atomic to macroscopic.

The initial research indicated that the protein matrix plays an important role in determining the crystal type in mollusks and the organic matrices extracted from the nacreous layer induced aragonite crystal formation [72]. At that time, it was still unknown whether the molecular spacing of protein matrix permits profitable speculation as to the formation mechanisms of nacre. Later, it was found that a significant proportion of the soluble protein of the organic matrix of mollusk shells is composed of a repeating sequence of aspartic acid separated by either glycine or serine. The regularly spaced, negatively charged aspartic acid serves as a template when the mineralization occurs [73]. On the basis of the experimental results, a model of interlamellar matrix structure in nacre was proposed [74–76]. The organic matrix comprises thin layers of β-chitin sandwiched between layers of β-sheet silk fibroin-like proteins, which were oriented orthogonal to the chitin fibrils and which supported acidic macromolecules. Crystals are considered to nucleate at specific sites on the preformed insoluble matrix due to binding of the Ca^{2+} ions to the matrix. The individual aragonite tablets are separated by interlamellar organic sheets constructed from thin layers of β-chitin,

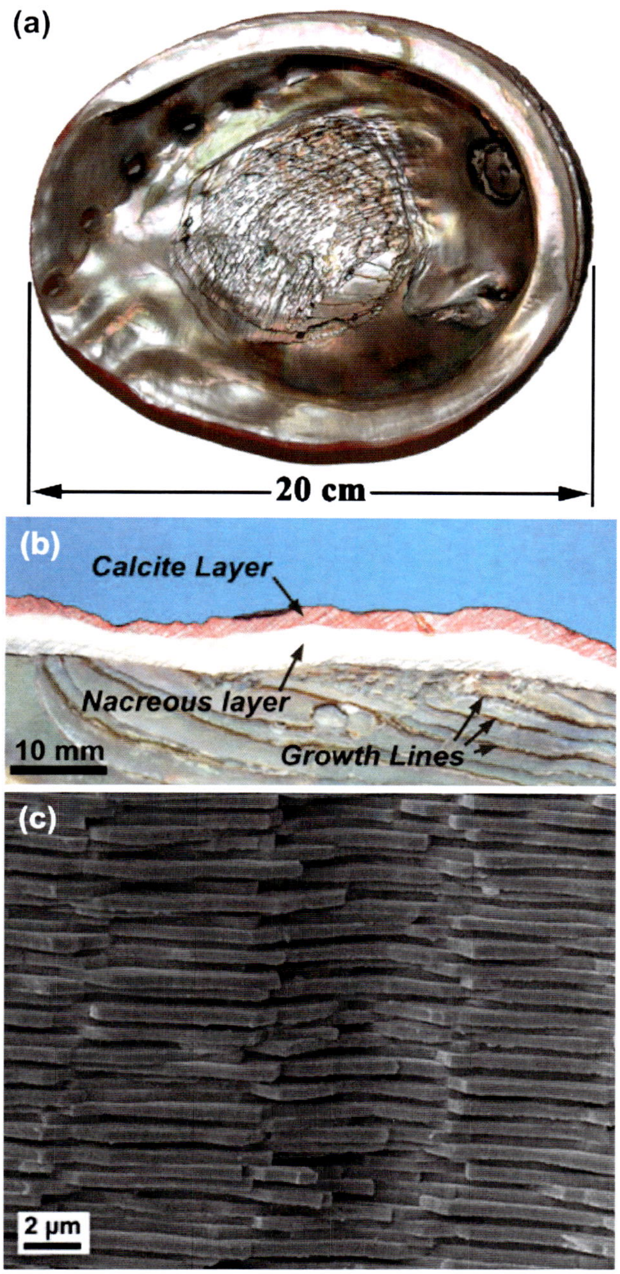

FIGURE 23.3 (a) View of the inner nacreous layer of a red abalone shell; (b) typical arrangement of a red abalone shell; and (c) the scanning electronic microscopy (SEM) image of a fracture surface in nacre [68]. *Reprinted from Journal of the Mechanics and Physics of Solids 55 (2007), F. Barthelat, H. Tang, P.D. Zavattieri, C.-M. Li, H.D. Espinosa, On the mechanics of mother-of-pearl: A key feature in the material hierarchical structure, pages 306–337, Copyright (2007), with permission from Elsevier.*

associated with silk fibroin-like proteins and acidic macromolecules [77]. Insoluble fibers bind to the aragonite tablets at the optimized inorganic interface, acting as a natural adhesive between the layers [1]. After shell formation is complete, the organic matrix probably contributes to the mechanical properties of the shell itself. Some other macromolecules in the nacreous layer were also proposed to play important roles in the formation of aragonite crystals [78–80].

Although a series of matrix proteins have been identified from various mollusk shells [81], no proteins that induce aragonite crystal formation characteristic of nacre have been characterized, nor has any Asp-rich acidic macromolecule been identified. Recently, an acidic matrix protein, Pif, in the pearl oyster *Pinctada fucata* was proved to be a key macromolecule for nacre formation. The Pif complementary deoxyribonucleic acid (cDNA) encoded a precursor protein, which was posttranslationally cleaved to produce Pif 97 and Pif 80. The results from immunolocalization, a knockdown experiment that used ribonucleic acid (RNA) interference, and in vitro calcium carbonate crystallization studies strongly indicate that Pif regulates nacre formation [82]. In the past several decades, some important progresses have been made in the clarification of the nacre formation mechanism, but the formation mechanism of crystal growth and orientational control in nacre is under hot debate. Therefore, further fundamental investigations are still necessary to reveal the biomineralization mechanism of nacre. Deeper understandings on the biomineralization processes would be useful to provide new models for the biomimetic syntheses of new high-performance composite materials and give new insights into the genetic control of biological structure.

23.1.4.3. Mimetic Mineralization Materials

In nature, living organisms build inorganic minerals under mild conditions by using organic macromolecules as templates during the biomineralization process. This process results in the biominerals with special morphology, hierarchical organization, and superior materials properties. Inorganic or inorganic–organic hybrid materials with specific size, shape, orientation, organization, and hierarchy have attracted much interest owing to their great fundamental and practical applications. Inspired by the biomineralization process of nacreous layer, many different synthesis strategies have been developed to mimic biomineralization principles and transfer them to the general control of crystallization processes. A large amount of polymers (such as double hydrophilic block copolymers (DHBCs), polyelectrolytes, biopolymers, and dendrimers) were applied to control the crystallization process of inorganic materials, including crystal sizes, crystallographic orientation, morphology, polymorph, stability, and architecture [15–17,20,67,83–85].

Low-molecular mass polyelectrolytes can not only electrosterically stabilize inorganic colloids, but also control the growth of diverse inorganic materials.

Research indicated that the addition of polyelectrolytes, such as polyacrylic acid or polyaspartic acid, can stabilize the amorphous nanobuilding blocks in the early stage, and then stimulate a mesoscale transformation or act as a material depot in a dissolution–recrystallization process.

Utilizing the polyelectrolytes as crystal growth modifiers or structure-directing agents, a wide variety of inorganic materials ($CaCO_3$, $BaCO_3$, $Ca_3(PO_4)_2$, $Ca_8H_2(PO_4)_6$, $BaSO_4$, $BaCrO_4$, ZnO, etc.) with complex and hierarchical superstructures have been fabricated [14–16]. The mechanism for the formation of the complex superstructures under control of a polyelectrolyte has not yet been well known. A polymer-induced liquid precursor process was proposed to illustrate the possible mechanism for the formation of the complex morphologies [86]. A typical experiment for the synthesis of inorganic materials with hierarchical structures is shown below.

In a typical procedure [87]: 10 mM $CaCl_2$ and Na_2CO_3 solutions were used in the preparation of amorphous $CaCO_3$ thin films. An anionic polymer polyacrylic acid is added to the solution that serves as a process-directing agent which induces liquid–liquid phase separation and ultimately deposition of an amorphous precursor film. The amorphous $CaCO_3$ precursor film is first deposited on the surface of the glass slides that are at the top and the base of the perfusion cell. After the deposition for 45 min, glass slides were dipped into water to remove residuals of $CaCO_3$ solution and left to dry in the air. After heating at 400°C for 2 h, the prepared amorphous thin film was transferred into a crystalline thin film with a patchwork structure. The preformed polycrystalline calcite thin film can serve as a template for epitaxial overgrowth of highly oriented calcite coating with uniform nacre-type structures (Fig. 23.4).

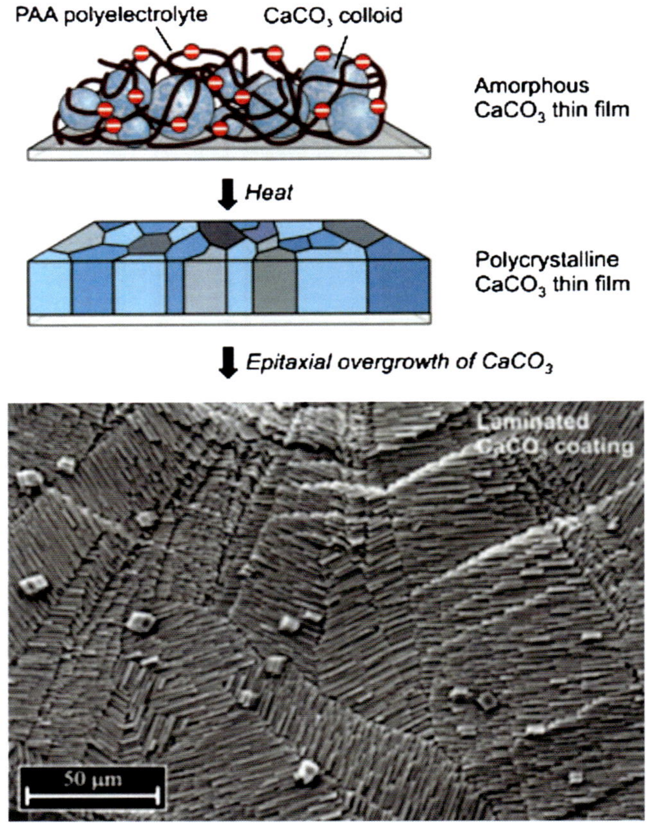

FIGURE 23.4 Scheme of a three-step procedure for the morphosynthesis of nacre-type laminated $CaCO_3$ coatings [87]. In the first step an amorphous highly hydrated $CaCO_3$ thin film is deposited on a glass substrate. Upon heating this precursor film is transformed into a polycrystalline thin film consisting of a mosaic of flat single-crystalline calcite domains. In the last step highly oriented single and multiple layers of calcite crystals are grown epitaxially on the underlying polycrystalline thin film.

Recently, DHBCs have attracted rapidly increasing interest, which are analogous to the structures of proteins involved in biomineralization and are used for the control of mineral crystallization [88,89]. The new class of functional polymer is a bio-inspired DHBC, which consists of two water-soluble blocks of different chemical nature. One hydrophilic block can interact strongly with the appropriate inorganic minerals and surfaces, and the other hydrophilic block does not interact (or only weakly interacts) and mainly promotes solubilization in water. Recent advances have demonstrated that DHBCs are highly effective for the synthesis, stabilization, and crystallization for a variety of inorganic, inorganic–organic hybrids, as well as organic crystals with special shapes and complex forms. A great number of DHBCs with different functional blocks, such as –OH, –COOH, $-SO_3H$, $-SO_4$, $-PO_3H_2$, $-PO_4H_2$, –SCN, $-NR_3$, $-HNR_2$, and H_2NR, have been designed and used as crystal modifiers for the biomimetic synthesis of $CaCO_3$, $BaCO_3$, $CdCO_3$, $MnCO_3$, $PbCO_3$, CaC_2O_4, $Ca_3(PO_4)_2$, $BaSO_4$, $BaCrO_4$, $CdWO_4$, $La(OH)_3$, Ag, ZnO, CdS, etc. Typical inorganic materials synthesized using DHBCs as crystal modifiers are shown in Table 23.2. A typical inorganic mineralization by using DHBC as a crystal growth modifier for self-assembly of complex and unusual architectures is shown below.

In a typical experiment [90]: 500 μL of a $CaCl_2$ solution (0.1 M) was injected into 5 mL PEG-*b*-PHEI solution ($1\ g\ L^{-1}$) under mild stirring. The pH was adjusted to a fixed value using dilute HCl or NaOH. Then the glass substrates were placed at the bottom of bottle to collect the crystals. The bottles were covered with parafilm, which was punched with three needle holes. $CaCO_3$ crystals were grown by slow diffusion of ammonium carbonate in a closed desiccator. After two weeks, calcite pancakes with multiple stacked and porous layers can be obtained (Fig. 23.5).

TABLE 23.2 Typical Inorganic Materials Synthesized Using DHBCs as Crystal Modifiers

Materials	DHBCs	Inorganic precursors	Structures	Ref.
$CaCO_3$	PEG-*b*-PEIPA	$CaCl_2/(NH_4)_2CO_3$	Snow-ball-like calcite	[91]
	PEG-*b*-PMAA	$CaCl_2/(NH_4)_2CO_3$	Spherule calcite	[91]
	PEG-*b*-PMAA	$CaCl_2/Na_2CO_3$	Dumbbell calcite	[92]
	PEO-*b*-PAA	$CaCl_2/C_3H_6O_3$	Amorphous $CaCO_3$	[93]
	PEG-*b*-PMAA	$CaCl_2/(NH_4)_2CO_3$	Hollow calcite	[94]
	CH_3O-PEG-*b*-PEIPA	$CaCl_2/Na_2CO_3$	Vaterite hollow sphere	[95]
	PEO-*b*-PMAA	$CaCl_2/NaHCO_3$	Rhombohedral calcite	[96]
	PEG-*b*-PHEE	$Ca(HCO_3)_2$	Vaterite superstructure	[97]
	PEG-*b*-PGL	$Ca(HCO_3)_2$	Vaterite superstructure	[97]
	PEG-*b*-PHEI	$CaCl_2/(NH_4)_2CO_3$	Pancake-like calcite	[90]
	PEG-*b*-PEIPA-C_{17}	$CaCl_2/(NH_4)_2CO_3$	Vaterite microring	[98]
	PEG(110)-*b*-pGlu(6)	$CaCl_2/(NH_4)_2CO_3$	Vaterite microsphere	[99]
	PEG-*b*-PEI	$CaCl_2/(NH_4)_2CO_3$	Multilayered vaterite	[100]
	PEO-*b*-PMAA	$CaCl_2/(NH_4)_2CO_3$	Peanut-like vaterite	[101]
	PEO-*b*-PMAA + CTAB	$CaCl_2/Na_2CO_3$	Peanut-shaped calcite	[102]
	PEO-*b*-PMAA + SDS	$CaCl_2/Na_2CO_3$	Calcite hollow sphere	[102]
$BaCO_3$	PEG-*b*-PMAA-PO_3H_2	$BaCl_2/(NH_4)_2CO_3$	Spherical aggregate	[94]
	PEG-*b*-PHEE-PO_4H_2	$BaCl_2/(NH_4)_2CO_3$	Oval crystal	[94]
	PEG-*b*- DOAEE	$BaCl_2/(NH_4)_2CO_3$	Helical fiber	[103]
	PEG-*b*-DOAEE	$BaCl_2/(NH_4)_2CO_3$	Ultralong nanofiber	[104]
$CdCO_3$	PEG-*b*-PMAA	$CdCl_2/(NH_4)_2CO_3$	Dumbbell structures	[94]
$MnCO_3$	PEG-*b*- DOAEE	$MnCl_2/(NH_4)_2CO_3$	Dumbbell/sphere structure	[94]
$PbCO_3$	PEG-*b*-PMAA	$Pb(NO_3)_2/(NH_4)_2CO_3$	Scute structure	[94]
	PEG-*b*- DOAEE	$Pb(NO_3)_2/(NH_4)_2CO_3$	Quasi-hexagonal nanoplate	[94]
CaC_2O_4	PEG-*b*-PMAA	$CaCl_2/Na_2C_2O_4$	Tetragonal prism	[105]
$Ca_3(PO_4)_2$	PEO-*b*-PMAA	$CaCl_2$/phosphate	Spherical particle	[106]
	PEO-*b*-PMAA-C_{12}	$CaCl_2$/phosphate	Needlelike filament	[106]
$BaSO_4$	PEG-*b*-PEIPSA	$Ba(OAc)_2/(NH_4)_2SO_4$	Flower-like particle	[107]
	PEG-*b*-PEIPA	$Ba(OAc)_2/(NH_4)_2SO_4$	Peanut-like particle	[107]
	PEG-*b*-PMAA	$Ba(OAc)_2/(NH_4)_2SO_4$	Peanut-like particle	[107]
	PEG-*b*-PMAA-PO_3H_2	$Ba(OAc)_2/(NH_4)_2SO_4$	Nanofilament	[108]
	PEO-*b*-PEI-SO_3H	$Ba(OAc)_2/(NH_4)_2SO_4$	Flower-like structure	[109]
$BaCrO_4$	PEG-*b*-PMAA-PO_3H_2	$BaCl_2/Na_2CrO_4$	Nanofiber bundle	[110]
	PEG-*b*-PMAA-PO_3H_2 + PSS-PAH	$BaCl_2/Na_2CrO_4$	Nanofiber	[111]
$CdWO_4$	PEG-*b*-PMAA	$CdCl_2/Na_2WO_4$	Nanofiber	[112]
	PEG-*b*-PMAA-PO_3H_2	$CdCl_2/Na_2WO_4$	Plate-like particle	[112]

(*Continued*)

TABLE 23.2 Typical Inorganic Materials Synthesized Using DHBCs as Crystal Modifiers—Cont'd

Materials	DHBCs	Inorganic precursors	Structures	Ref.
$La(OH)_3$	PAA-*b*-PAM	$LaCl_3$, $La(NO_3)_3$	Rice grain-like particle	[113]
Ag	PEO-*b*-PMAA + SDS	$AgNO_3$ + ascorbic acid	Hollow sphere	[114]
ZnO	PEO-*b*-PMAA	$Zn(NO_3)_2$/HMT	Hexagonal prismatic crystal	[115]
	PEO-*b*-PSSH	$Zn(NO_3)_2$/HMT	Stack of pancake	[116]
CdS	PEG-*b*-PEI	$CdCl_2/Na_2S$	Monodisperse nanoparticle	[117]

CTAB = Cetyltrimethylammonium bromide, DOAEE = 2-[4-dihydroxyphosphoryl]-2-oxabutyl)acrylate ethyl ester], HMT = hexamethylenetetramine, PAA = poly (acrylic acid), PAH = poly(allylamine hydrochloride), PAM = poly(acrylamide), PEG = poly(ethylene glycol), PEI = poly(ethylene imine), PEIPA = poly(ethylene imine)-poly(acetic acid), PEIPSA = poly(ethylene imine)-poly(sulfonic acid), PEO = Poly(ethylene oxide), PGL = Poly(glycidol), pGlu(6) = poly(L-glutamic acid), PHEE = poly(hydroxyethylethylene), PHEI = poly(1,4,7,10,13,16-hexaazacyclooctadecane ethylene imine), PMAA = poly(methacrylic acid), PSS = poly(sodium 4-styrenesulfonate), PSSH = poly(styrene sulfonic acid), and SDS = sodium dodecyl sulfate.

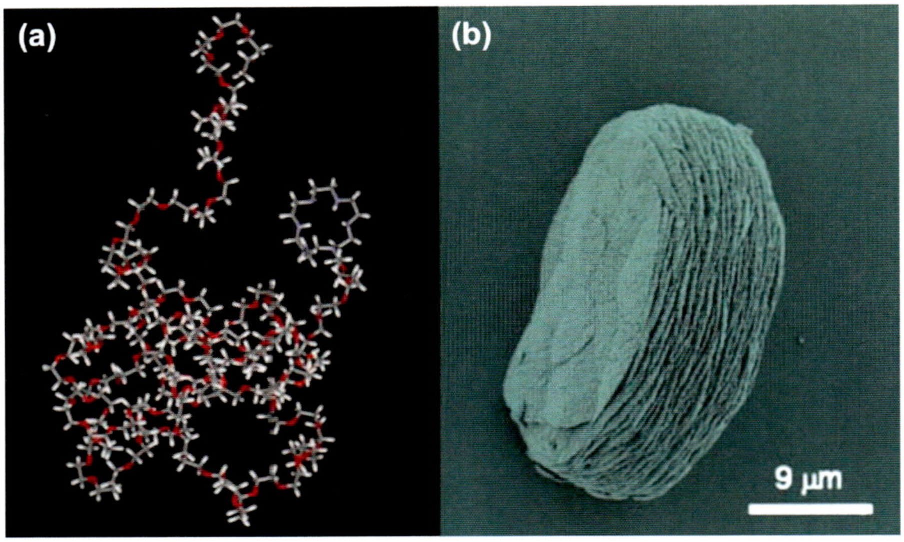

FIGURE 23.5 (a) The structure of PEG-*b*-PHEI (gray: carbon, white: hydrogen, blue: nitrogen, red: oxygen). (b) An SEM image of a pancake-like self-stacked $CaCO_3$ obtained after two weeks gas diffusion reaction in the presence of 1 g L^{-1} PEG-*b*-PHEI, starting pH 4, $[Ca^{2+}]$ = 10 mM [90].

23.2. BIOTEMPLATED INORGANIC MATERIALS

23.2.1. Introduction

After millions of years of evolution, nature goes far beyond humans in producing novel materials. Biological materials exhibit an astonishing variety of sophisticated hierarchical structures from the molecular to the nano-, micro-, and macroscales, which is difficult to achieve by the advanced synthetic strategies. Templating is a simple and fast technique for the replication of surface structures. Although a wide range of templates are used for the synthesis of inorganic materials, in this section we focus on biotemplates. Utilizing the natural materials (such as bacteria, viruses, diatoms, insect wings, plant leaves, shell membranes, wood, DNA, and proteins) as templates, many different synthesis strategies have been developed to design and fabricate inorganic materials with intricate architectures [5,118–123]. Some typical biotemplates used in the inorganic materials synthesis are listed below.

23.2.2. DNA

DNA is a polynucleotide, whose monomer units are composed of a monophosphorylated deoxyribose sugar linked to one of four nitrogenated nucleobases: two purines, called adenine (A) and guanine (G); and two pyrimidines, called thymine (T) and cytosine (C) [124]. DNA is the basic information storage molecule in nature. DNA is one of the most interesting template systems in the programmed nanoparticle assembly, because of its diameter of only 2 nm and the micrometer-long distribution of well-defined sequences of DNA bases. It possesses the ability to form well-defined secondary and tertiary structures. For example, beyond strictly periodic lattices, DNA origami is a technique which can fold a long single-stranded DNA (ssDNA) molecules into arbitrary shapes by hybridization with multiple ssDNA oligonucleotides that anchor the final geometry in place [125]. The resulting nanoscale patterns are roughly 100 nm in diameter and have a spatial resolution of about 6 nm, exhibiting six different shapes, such as square, rectangle, star, smiley face, triangle with rectangular domains, and sharp triangle (Fig. 23.6). Therefore, DNA has been used extensively as a building block to assembly complex inorganic nanostructures, which has attracted much attention [124,126–128]. Present study about DNA-templated nanofabrication can be classified roughly into three complementary areas [124]: (i) fabrication of nanomaterials directly on DNA [129]; (ii) assembly of nano/mesoscale materials with DNA [130]; and (iii) use of DNA templates in top-down nanofabrication schemes [131].

In 1992, the biopolymer calf thymus DNA was first employed to stabilize the formation of quantum-confined crystallites of cadmium sulfide nanocrystallites [132]. The resultant semiconductor material is extremely stable under reasonable storage conditions (5°C). In a typical experiment [132], freshly prepared $Cd(ClO_4)_2 \cdot 6H_2O$ (5.0 μL, 1 M) was diluted to 2 mL and purged with nitrogen (final Cd^{2+} concentration 4×10^{-4} M). In a separate flask, fresh Na_2S (5.0 μL, 1 M) was diluted to 5 mL and also purged thoroughly with nitrogen. In a third flask, approximately 15 mg calf thymus DNA was dissolved in 5.0 mL deionized water in a 50 mL round-bottomed flask. The nucleotide is slow to dissolve, and should be allowed to stand for about 30 min, whereupon it can then be thoroughly mixed. The nucleotide flask was fitted with a septum and purged very slowly with nitrogen for about 20 min. To form quantum-confined CdS clusters, the 2 mL Cd^{2+} solution was added to the nucleotide, and the mixture was purged with nitrogen for another 5 min. The 5 mL S^{2-} solution was then transferred to the reaction flask containing the nucleotide and Cd^{2+} via syringe; the near-instantaneous appearance of a yellow color was observed.

Three-dimensional nanoparticle arrays are likely to be the foundation of future optical and electronic materials.

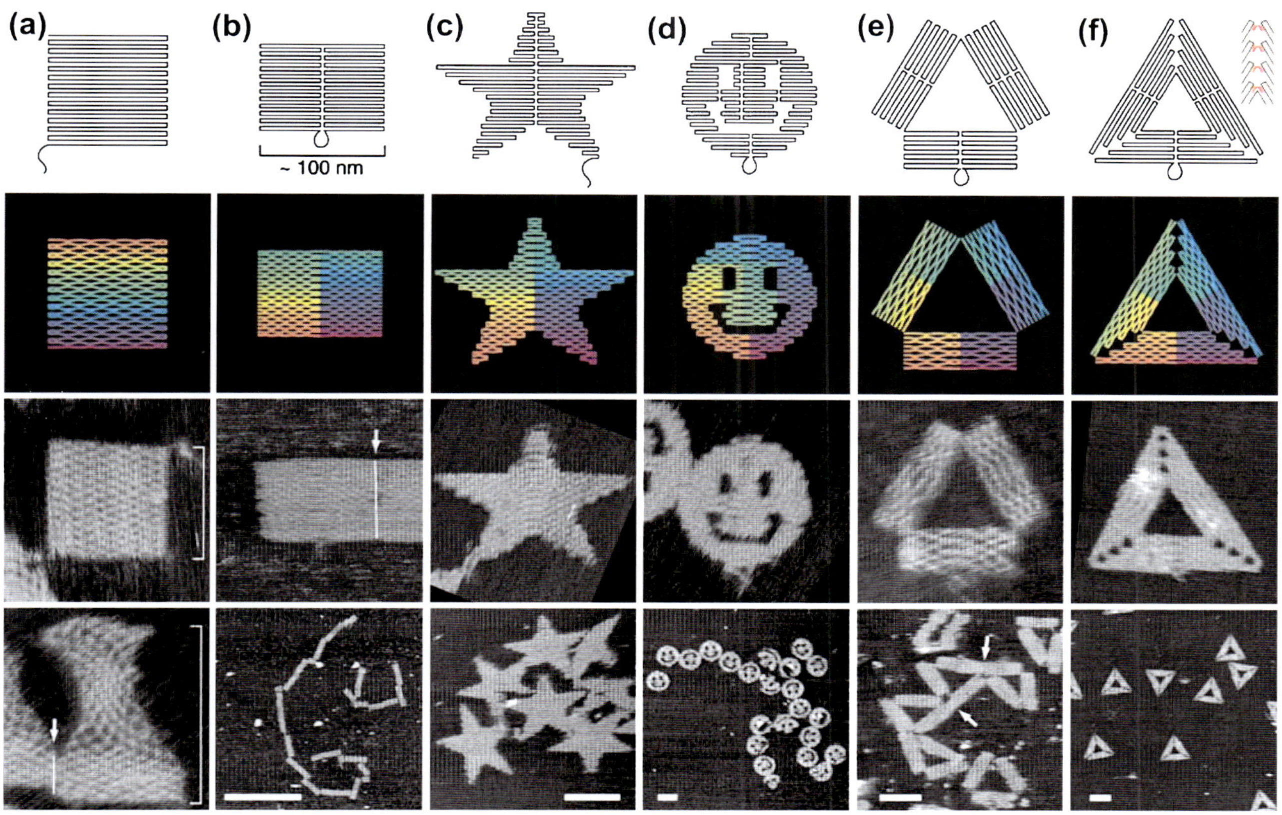

FIGURE 23.6 DNA origami shapes. Top row, folding paths: (a) square, (b) rectangle, (c) star, (d) smiley face, (e) triangle with rectangular domains, and (f) sharp triangle. Second row, computer-generated diagrams showing helix crossovers. Bottom two rows, AFM images of folded DNA specimens. White lines and arrows indicate blunt-end stacking between two independent molecular objects. White brackets in column (a) mark the height of a square and that of a square stretched vertically into an hourglass shape. Images and panels without scale bars are 165 nm × 165 nm. Scale bars for lower AFM images: (b) 1 μm; and (c–f), 100 nm. *Reprinted with permission from Macmillan Publishers Ltd: Nature, [125], Copyright (2006).*

A promising way to assemble them is through the transient pairings of complementary DNA strands [133]. ssDNA oligonucleotides have been used to modify the surface of nanoparticles and endow them with selective interaction and recognition capabilities. Such oligonucleotide-functionalized nanoparticles were proved to be valid in the assembly of nanoparticles possessing designed structures. Recently, three-dimensional crystalline assemblies of gold nanoparticles have been constructed (Fig. 23.7), which were mediated by interactions between complementary DNA molecules attached to the nanoparticles' surface [134]. The results showed that the nanoparticle crystals form reversibly during heating and cooling cycles and the body-centered-cubic lattice structure is temperature-tunable and structurally open, with particles occupying only 4% of the unit cell volume. Particle assembly was carried out at 25°C by combining equimolar amounts of type-A and type-B DNA-capped gold particles in 200 μL ([A] = [B] = ~30 nM) solution of 10 mM phosphate buffer, 0.2 M NaCl, pH = 7.1. These particles were allowed to assemble into aggregates overnight, and the resulting precipitate was collected and transferred in buffer to a quartz capillary (1.0 mm diameter), and sealed with wax. At the same time, the other

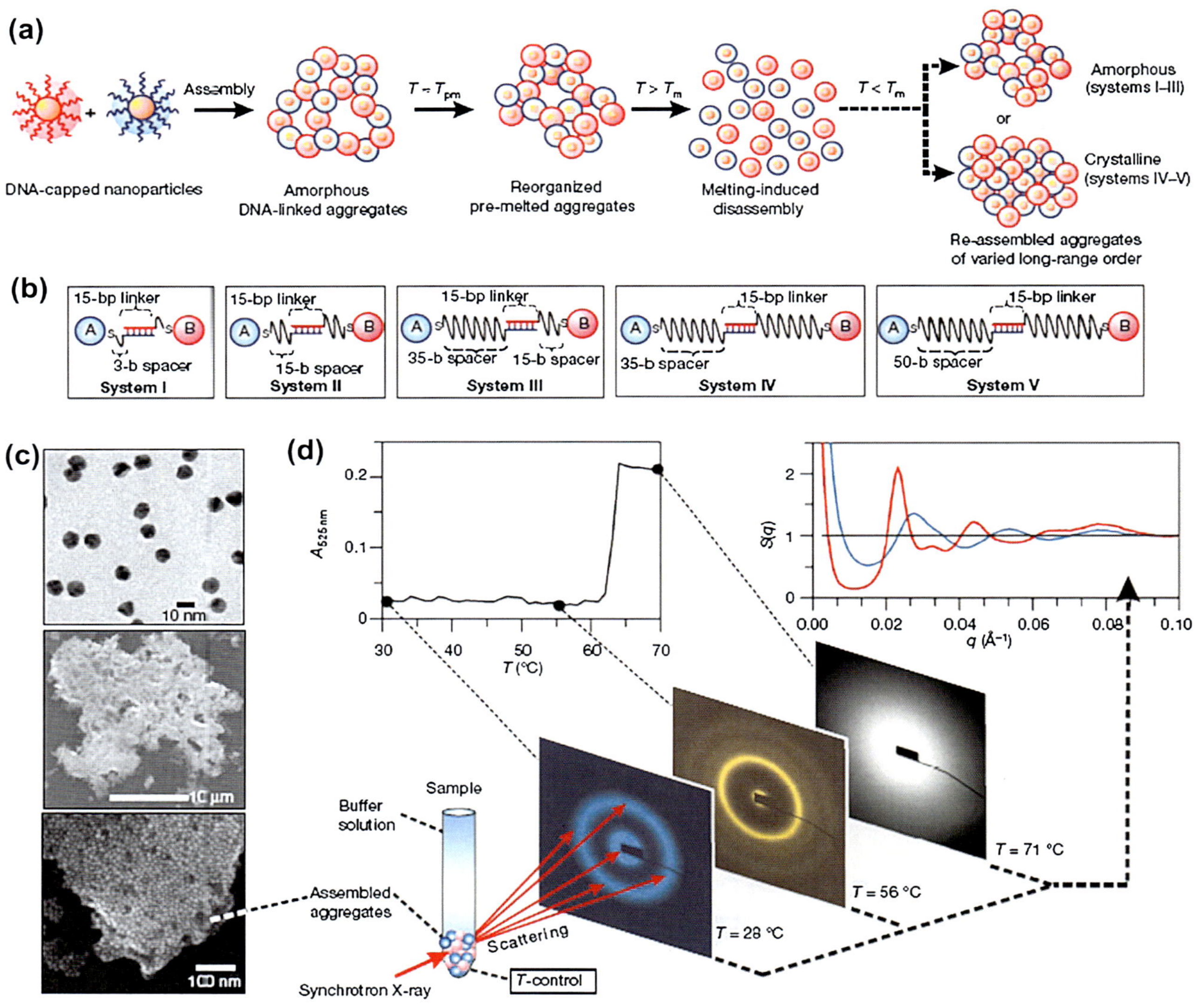

FIGURE 23.7 Schematic of experimental design. (a) The assembly system of DNA-capped nanoparticles, the aggregates of which show a series of structural changes under a variety of thermal conditions. (b) DNA linkages between nanoparticles with recognition sequences for the A (blue) and B (red) sets of DNA capping (bp = base pairs, b = bases, S = thiol termination of DNA). (c) Representative transmission (top) and scanning (middle, bottom) electron microscopy images of nanoparticles before (top) and after (middle, bottom) assembly at room temperature. (d) Typical example of experimental measurements that reveal a correlation between the ultraviolet–visible melting profile of the aggregate and its internal structure as probed by in situ small angle X-ray scattering measurements at room temperature, premelting temperature, and above the disassembly/melting temperature. *Reprinted with permission from Macmillan Publishers Ltd: Nature [134], Copyright 2008.*

study demonstrated that DNA can be used to control the crystallization of nanoparticle–oligonucleotide conjugates to the extent that different DNA sequences guide the assembly of the same type of inorganic nanoparticle into different crystalline states [135]. The choice of DNA sequences attached to the nanoparticle building blocks, the DNA linking molecules and the absence or presence of a nonbonding single-base flexor can be adjusted so that gold nanoparticles assemble into micrometer-sized face-centered-cubic or body-centered-cubic crystal structures. Both experimental results independently demonstrated that the attractive and repulsive potentials can be manipulated by varying the length of the DNA linker attached to the particles, modifying the stiffness by having a double-stranded DNA (dsDNA) region in the linker, and adjusting the strength of the attractive interactions through carefully selecting the length of the ssDNA portion that is complementary to other oligonucleotide-functionalized nanoparticles [124]. This is essential to form ordered structures.

Utilizing DNA as the biotemplate, a wide variety of inorganic materials with different structures have been fabricated. Some typical DNA-templated inorganic materials are shown in Table 23.3.

23.2.3. Bacteria

Bacteria are single-celled microorganisms, typically of a few micrometers in length, which exist either as independent organisms or as parasites. Bacteria possess the following paramount advantages: commercial availability, ease of preparation/handling, potential for genetic manipulation, and well-defined stunning morphologies (i.e., cocci, bacilli, spirilla, etc.). Owing to these attractive characteristics, bacteria can be used as the templates to the fabrication of corresponding three-dimensional nanostructures. For example, *Lactobacillus* can be served as sacrificial templates for the synthesis of inorganic materials, which maintain the size and morphology of the initial bacterium. When *Streptococcus thermophilus* (*S. thermophilus*) and *Lactobacillus bulgaricus* (*L. bulgaricus*) are used as templates, ZnS hollow spheres and nanotubes can be synthesized by the sonochemical method [149]. Figure 23.8 schematically illustrates the formation process of ZnS hollow spheres using *Str. theromophilus* as the template. The experimental procedure is shown below.

In a typical procedure, zinc acetate (110 mg), thioacetamide (37.5 mg) and *Lactobacillus* powder (1 g) were dispersed in deionized water (50 mL), then the mixture was sonicated with an ultrasonic cleaner at room temperature for 6 h under ambient air. The resulting white powder was recovered by centrifugation, washed repeatedly with distilled water and ethanol, and then dispersed in ethanol.

Besides ZnS materials, other functional inorganic materials like zeolite fibers [150], SiO_2 [151], Au nanoparticles [152], and magnetic nickel particles [153] have been fabricated by using the corresponding bacteria as the templates.

TABLE 23.3 Some Typical DNA-templated Inorganic Materials

Products	DNA template	Methods	Structures	Ref.
Au	DNA origami	Lithography and self-assembly	Two-dimensional array	[136]
	DNA-modified diatom	Template method	Three-dimensional structure	[137]
	λ-DNA	Photochemical method	Nanoparticle nanowire	[138]
	λ-DNA	Self-assembly	Necklace-like structure	[139]
	λ-DNA	Solution phase assembly	Lines, ribbons, and branches	[140]
Ag	λ-DNA	Two-step method	Silver wire	[141]
	4 × 4 DNA fragments	Templated self-assembly	Nanowire	[142]
Pd	λ-DNA	Two-step approach	Nanowire	[143]
Pt	Calf thymus DNA	Chemical reduction	Nanochain	[144]
CdS	DNA scaffolds	Photochemical synthesis	Nanowire	[145]
	λ-DNA	Template method	Nanochains, nanowire	[146]
PdS	Calf thymus DNA	Template method	Quantum dot	[147]
$CoFe_2O_4$	λ-DNA	Electrostatic assembly	One-dimensional chain	[148]

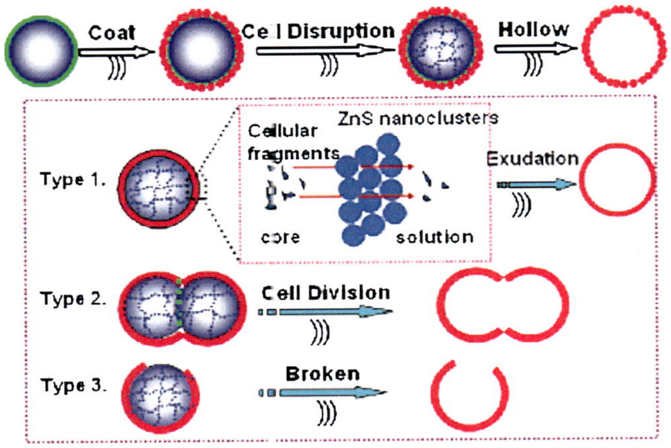

FIGURE 23.8 Schematic illustration of the in situ one-step formation of ZnS hollow spheres using *Str. thermophilus* as templates. *Reprinted with permission from [149]. Copyright 2007 American Chemical Society.*

23.2.4. Insect Wings

In nature, some insect wings exhibit particular photonic and other properties due to their special micro- to nanostructures [154–156]. Utilizing natural insect wings as templates, the replicated inorganic structures may combine the merits offered by both the material and biological structures [157].

The structural blue color of a *Morpho* butterfly originates from the diffraction of light and interference effects due to the presence of the microstructures on the wing of the butterfly [155]. A variety of inorganic materials have been fabricated with the morphology and features inherited from biotemplates of butterfly wings. For example, ordered lead lanthanum zirconate titanate with micrometer and submicrometer structures was fabricated by an impressing sol–gel process, where the butterfly wings were used as biological templates [158]. ZnO microtubes with adjustable arrayed nanopores on the walls were prepared by using butterfly wing scales as natural biotemplates [159]. The unique structures of butterfly wings can be replicated by chemical vapor deposition of silica [160]. Recently, the butterfly wing scales were used to design and synthesize hierarchically periodic microstructure titania photoanode [161]. By using a layer-by-layer sol–gel-based deposition technique and a template of chitin-based butterfly wing scales, three-dimensional rutile titania-based structures with *Morpho* butterfly wing scale morphologies were fabricated [162].

Besides the biotemplates of butterfly wings, the cicada wing is also an important natural template for the fabrication of functional inorganic materials [163]. Arrays of nanosized pillars (about 100 nm in diameter and 300 nm in height) were found on cicada wings, which can greatly minimize the reflectivity on their surfaces over broad angles or frequency ranges [164]. Many efforts have been made to replicate or directly utilize the cicada wing's nanostructures for special applications [165]. For example, the nanostructure of the cicada wing was duplicated by using hot embossing lithography and UV nanoimprint lithography. The diameter and pitch of the replica were the same as those of the natural cicada wing [166]. The nano-nipple arrays on the surface of cicada wings were replicated to a poly (methyl methacrylate) (PMMA) film by a replica molding technique [167]. Figure 23.9 is a schematic diagram showing nano-imprint lithography with cicada-wing stamps, fabrication of gold pillar arrays, and transfer of the pattern to a silicon substrate [168]. A typical example for the fabrication of inorganic materials with an antireflective property using the cicada wings as templates was presented.

PMMA was dissolved in toluene to form a 5 wt% solution. The solution was spun onto a $2 \times 2\ cm^2$ silicon wafer. Imprinting was carried out on a Nanoimprinter at about 40 bar pressure and 190°C. The pressure was released after 180 s, and the wing stamp was carefully removed with tweezers. To transfer the pattern to the substrates, reactive ion etching was carried out for about 3 min. Patterned PMMA layer served as a mold to fabricate gold pillar arrays similar to the structure on the surface of cicada wings.

23.2.5. Shell Membranes

Compared with artificial templates, biological templates are inherently complex and hierarchical. Eggshell membrane is a naturally semipermeable biomembrane consisting of calcium carbonate biominerals embedded in an organic matrix. The eggshell membrane is a macroporous network composed of interwoven and coalescing shell membrane fibers ranging from 0.5 to 1.5 μm in diameter [169]. Furthermore, the eggshell membrane is cheap, abundant, and environmentally benign. Due to these attractive characteristics, using natural eggshell membrane as a biotemplate, a great number of inorganic materials with hierarchical structures have been prepared, including TiO_2 [169–171], ZrO_2 [172], SnO_2 [173–175], $CaCO_3$ [176], $BaSO_4$ [177], $SrCrO_4$ [178], SiC [179], PbS [180], PbSe [181], etc. A typical experimental procedure is shown below.

In a typical synthesis [172], the eggshell membrane was dipped into a closed vessel containing a 7% (w/w) solution of zirconium propoxide in 2-propanol for 4 h, where an equimolar amount of acetyl acetone was added to reduce the hydrolysis rate of zirconium propoxide. After removal from this solution, the membrane coated with the precursor was dried in air at room temperature for 48 h. The resulting yellow hybrid material was then heated to 600°C or 700°C in an oven under

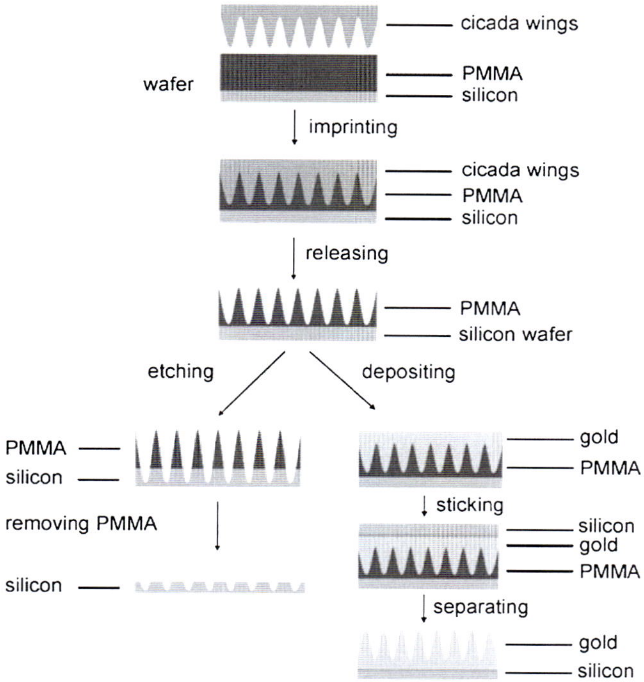

FIGURE 23.9 Schematic diagram of nanoimprint lithography, showing cicada-wing stamps, fabrication of gold pillar arrays, and transfer of the pattern onto a silicon substrate [168]. *Guoming Zhang, Jin Zhang, Guoyong Xie, Zhongfan Liu, and Huibo Shao, Cicada Wings: A Stamp from Nature for Nanoimprint Lithography, Small 2006, 2, No. 12, pages 1440–1443, Copyright Wiley-VCH Verlag GmbH & Co. KGaA. Reproduced with permission.*

air atmosphere to burn off the organics and crystallize zirconia.

23.2.6. Other Typical Biotemplates

Besides the above-mentioned biotemplates, other natural materials, such as tobacco mosaic virus (TMV), chilo iridescent virus (CIV), bacterial surface layer (S-layer), egg-white protein ovalbumin (EWPO), silk, tubulin, agarose gels, wood cell, and pond-weed leaf, have also been used as templates for the synthesis of inorganic materials (Table 23.4).

23.3. BIOMIMETIC SYNTHESIS OF INORGANIC CHIRAL MATERIALS

Chirality, or handedness, means that an object or molecule cannot be superimposed on its mirror image by any translations or rotations [189]. Achiral (not chiral) objects are those objects which are identical to their mirror image. Chirality is commonly observed in nature and life, which exists at all length scales, from the molecular level (the single left- or right-handed amino acid) to the supramolecular level (the right-handed double helix of DNA), and up to macro biomineralized materials (marine and snail shells) [190–193]. Chirality is a central theme in nature, which has fascinated scientists ever since the phenomenon was first observed. Recently, chiral materials have attracted much attention owing to their potential applications in various fields such as biosensing, optics, electronics, photonics, catalysis, nanotechnology, and drug or DNA delivery [194–196]. Inspired by the chirality in nature and life, a wide variety of different synthesis strategies have been developed to extend the chiral materials to inorganic materials [197–200]. In this section, some typical inorganic chiral materials, such as zeolites, molecular sieves, SiO_2, and metals, are described on the basis of the corresponding synthesis methods.

23.3.1. Zeolites and Molecular Sieves

The term zeolites is derived from the Greek words "zeo" (to boil) and "lithos" (stone). Zeolites, one of the most important families of microporous solids, are

TABLE 23.4 Biomimetic Inorganic Materials Templated by Natural Materials

Biotemplates	Products	Methods	Properties	Ref.
CIV	Gold nanoshell	Metallodielectric assembly	Plasmon resonance	[182]
TMV	Aligned nanoparticle	Self-assembly	Magneticity	[183]
EWPO	Hierarchical porous zeolite	Seeded growth strategy	—	[184]
S-layer	Metal nanoparticle array	Electron-beam induction	—	[185]
Silk	Porous metal oxide	Sol–gel process	—	[186]
Tubulin	Ring-like Ag nanoparticle	Template deposition	—	[187]
Agarose gels	Star-shaped calcite	Mineralization	—	[188]
Wood cell	Hierarchical structured silica	Chemical vapor deposition	Self-cleaning	[160]
Pond-weed leaf	Hierarchical structured silica	Chemical vapor deposition	Optical properties	[160]

hydrated, crystalline tectoaluminosilicates that are constructed from TO_4 tetrahedra (T = tetrahedral atom, e.g., Si, Al); each oxygen atom bridging two tetrahedra [201]. Zeolites and molecular sieves have found widespread applications in catalysis, ion exchange, and selective sorption/separation owing to their unique properties arising from their uniformity in pore size (in small molecules scale) [197,202–205]. Although chirality is commonly observed in nature and life systems, its occurrence in zeolites and molecular sieves is particularly rare. Chiral zeolites and molecular sieves are of prime importance with respect to their potential applications in enantioselective sorption, separation and catalysis, chromatography, and fundamental aspects in chirality [197,206–209]. Here, we illustrated some typical chiral zeolites and molecular sieves materials prepared by different synthesis strategies.

Zeolite beta family is one of the most important zeolitic catalysts in industry, which consists of three polymorphs: A, B, and C, all built from the same layer containing 12 rings [210]. Thermally stable silicogermanates SU-32 (polymorph A) possessing an intrinsically chiral zeolite structure can be synthesized under hydrothermal conditions using a simple organic amine, diisopropylamine, as the template [203]. The pore size and shape in zeolites play important roles in the practical applications. The fabrication of zeolitic materials with chiral frameworks, extralarge pores, and high adsorption capacities would extend the practical applications of zeolites. A germanosilicate zeolite (ITQ-37) mesoporous chiral zeolite with extralarge 30-ring windows was fabricated through a two-step method [211], including the synthesis of silicogermanate zeolite ITQ-33 (a stable zeolite with 18 × 10 × 10-ring windows) [212], and the synthesis of chiral zeolite SU-32 (each crystal exhibiting only one handedness) [203]. Three large dicationic organic structure-directing agents were applied to discover the new zeolite ITQ-3 by high-throughput techniques.

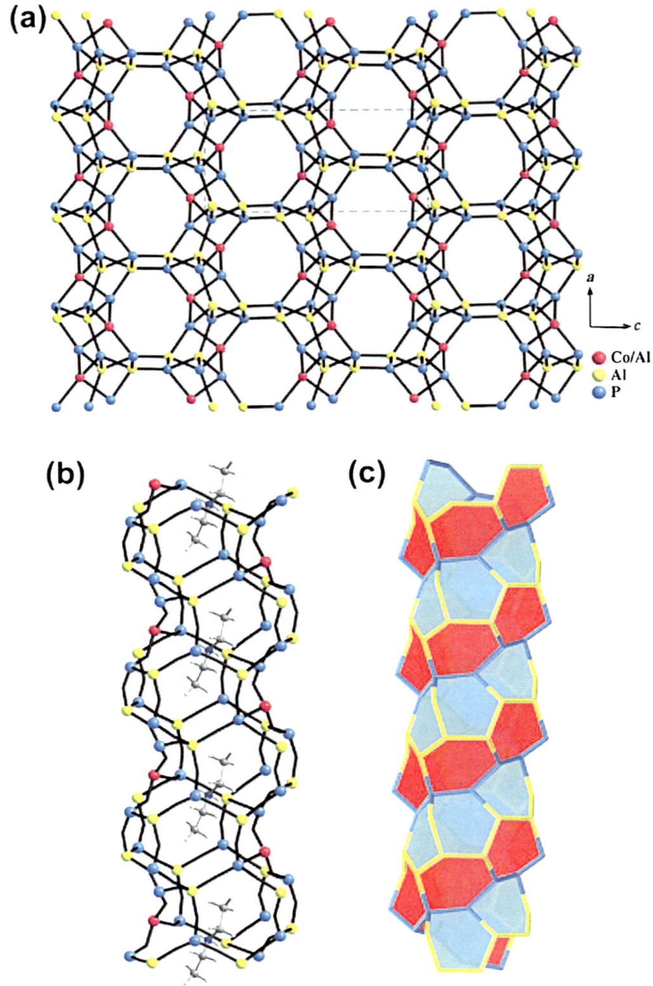

FIGURE 23.10 Framework structure of CoAPO-CJ40 [217]. (a) Viewed along the [010] direction. (b) The helical 10-ring channel and helical arrangement of cobalt atoms (shown in pink). (c) The 10-ring channel enclosed by double-helical ribbons made of the edge-sharing of 6 rings.

Aluminophosphate molecular sieves ($AlPO_4$-n) constitute an important family of zeolite materials [213–215]. The incorporation of transitional-metal ions into the framework sites of $AlPO_4$-n molecular sieves results in the formation of MAPO materials (M = metal ion), which have potential applications in clean technology and green chemistry [216]. Chiral heteroatom-containing $AlPO_4$-n is highly suitable for enantioselective conversions by taking advantage of chiral single-site solid catalysts. Heteroatom-containing chiral aluminophosphate molecular sieves $(C_4H_{12}N)_2[M_2Al_{10}P_{12}O_{48}]$ (denoted MAPO-CJ40, M = Co, Zn) with one-dimensional helical 10-ring channels can be synthesized through the introduction of transition-metal ions into the aluminophosphate framework (Fig. 23.10). The framework is intrinsically chiral, exhibiting a new zeotype structure, which may open up new applications of molecular sieves in the fine-chemical and pharmaceutical industries [217].

In a typical synthesis of CoAPO-CJ40 and ZnAPO-CJ40 [217], H_3PO_4 (2.29 mL, 85 wt% in water) was dispersed in a mixture of tetraethylene glycol (8 mL) and H_2O (2 mL), followed by the addition of Co $(NO_3)_2 \cdot 6H_2O$ (0.227 g) or $Zn(NO_3)_2 \cdot 9H_2O$ (0.365 g) and $Al(O\textit{i}Pr)_3$ (0.840 g) with stirring. Next, diethylamine (2.92 mL, 99 wt% aqueous solution) was added. A homogeneous gel was formed after stirring for 1 h. The gel was sealed in a Teflon-lined stainless steel autoclave and heated at 180°C for 12 days under static conditions. Blue rectangular single crystals of CoAPO-CJ40 and colorless rectangular single crystals of ZnAPO-CJ40 were separated from the remainder of the product by sonication, washed with distilled water, and then dried in air.

23.3.2. SiO_2

According to the classification made by IUPAC [218], porous solids can be arranged in three main categories, depending on their pore size (diameter, *d*), in microporous ($d < 2$ nm), mesoporous (2 nm $< d <$ 50 nm), and macroporous materials ($d > 50$ nm). The above-mentioned zeolites and molecular sieves materials belong to the microporous materials [219]. Mesoporous materials, particularly mesoporous silica, have a wide domain of applications in catalysis, sensing, adsorption, drug delivery, and controlled release, associated with their special nanostructures. Recently, chiral mesoporous silica materials have attracted much attention, owing to their potential applications, particularly in chiral selectivity, chiral recognition, and chiral catalysis. In this section, some representative examples of chiral mesoporous silica are presented.

The first synthesis of silica crystals with chiral mesopores was performed by using tetraethoxylsilane (TEOS) as the inorganic precursor, chiral surfactant *N*-miristoyl-L-alanine sodium salt (C_{14}-L-AlaS) as template, *N*-trimethoxysilylpropyl-*N,N,N*-trimethylammonium chloride (the cation denoted here as TMAPS) and 3-aminopropyltrimethoxysilane (APS) as the co-structure-directing agents (CSDAs) [220]. The resultant materials possess a twisted hexagonal rod-like morphology, with diameter of 130−180 nm and length of 1−6 μm. The presence of hexagonally ordered chiral channels winding around the central axis of the rods is confirmed by transmission electron microscopy (TEM) observations in combination with computer simulations [221]. Such chiral materials extend the application fields of mesoporous silica. The typical synthesis procedure is shown below.

C_{14}-L-AlaS (0.32 g, 1 mmol) was dissolved in deionized water (32 g) with stirring at room temperature. 0.1 M HCl (1.4 g, 0.14 mmol) was added to the surfactant solution under vigorous stirring at room temperature. After stirring for 1 h, a mixture of 1.40 g TEOS and 0.20 g TMAPS (50% in methanol) was added to the mixture with stirring at 22°C. Then, the mixture was allowed to react at 22°C under static conditions for 2 h. The chiral mesostructured product thus formed was cured at 80°C for an additional 15 h. The products were recovered by centrifugal separation and dried at 60°C. Both the anionic surfactant and the organics of the CSDA used were removed by calcination at 650°C for 6 h.

For the synthesis of chiral mesoporous silica, in a TEOS/CSDA/3-aminopropyltriethoxysilane/C_{14}-L-AlaS system, stirring was proved to be a convenient tool for controlling not only the size and shape of chiral particles but also the chiral structure and helical pitch length. Diverse morphologies, such as twisted-ribbon, twisted-rod, and nonhelical (spherical and irregular) structures, can be obtained by tuning the stirring rate [222].

As discussed in the preceding parts, chiral surfactants can be used for the synthesis of chiral mesoporous silica materials. Chiral mesoporous silica structures were also realized by using an achiral surfactant as template. For example, two types (single-axis and dual-axis) of chiral mesostructured silica nanofibers can be fabricated in a sodium silicate (SS)/cetyltrimethylammonium bromide (CTAB)/ethyl acetate (EA)/H_2O media [223]. The SEM and TEM images of the dual-axis nanofiber and the corresponding three-dimensional models are presented in Fig. 23.11. The reaction compositions for the synthesis of single-axis and dual-axis nanofibers were SS/CTAB/EA/H_2O 0.15−0.50/0.12−0.33/1.4−3.0/1000, and 0.15−0.30/0.16−0.50/0.3−0.5/1000 (in molar ratio), respectively [223]. Recently, functional chiral mesoporous silica nanorods were synthesized by using only an achiral surfactant via a one-step process. In this process, self-assembly of achiral CTAB and cocondensation of TEOS and 3-mercaptopropyltrimethoxysilane took place simultaneously in basic solution under static conditions. The obtained chiral mesoporous silica nanorods demonstrated excellent adsorption selectivity for Hg^{2+} ions [224]. By using achiral surfactant sodium dodecyl sulfate (SDS) as the template, TMAPS as the CSDA, and TEOS as the silica source, helical ordered mesoporous silica was also synthesized, where the chemical composition (mole ratios) of the reaction mixture was 1/0.46/5.5/1611 SDS/TMAPS/TEOS/H_2O [225].

Chirality is found mainly in biological molecules such as DNA and proteins. These chiral structures of biomolecules can be used to direct the expression of form of an inorganic material, with morphological control being achieved through the transcription of internal information. Recently, DNA and DNA superstructures were transcribed into pore-structure-tunable mesoporous silicas via the electrostatic interaction between the negatively charged phosphate groups of DNA backbones and the positively charged TMAPS [226]. After removal of DNA by calcination or extraction, silica mesoporous fibers with an inner helical pore can be obtained (Fig. 23.12). This DNA transcription method provides a possible route to create more complicated inorganic materials by using rationally designed natural DNA and artificial DNA structures. A typical synthesis method is presented here [226].

The molar composition of the reaction gel was DNA/HCl or NaOH/TMAPS/TEOS/$H_2O = 1/x/1.27/15/18333$, where x was 0.03−0.26 for HCl and 0.08−2.97 for NaOH, respectively. The pH value of the DNA

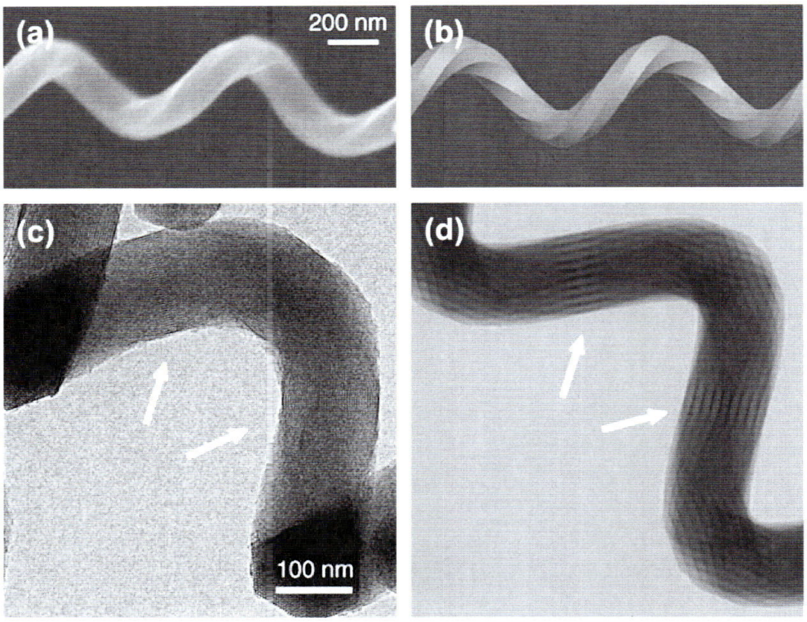

FIGURE 23.11 (a) SEM image, (b) simulated three-dimensinal model, (c) TEM image, and (d) simulated TEM image of a dual-axis nanofiber [223]. *Bo Wang, Cheng Chi, Wei Shan, Yahong Zhang, Nan Ren, Wuli Yang, and Yi Tang, Chiral Mesostructured Silica Nanofibers of MCM-41, Angew. Chem. Int. Ed. 2006, 45, 2088 −2090, Copyright Wiley-VCH Verlag GmbH & Co. KGaA. Reproduced with permission.*

solution before adding TMAPS and TEOS was controlled in the range of 4.3−11.9. The mixture was allowed to react at 15−30°C under static conditions for 4−7 d after homogenization. The resultant white powders were collected by centrifugation, washed and dried. Organic-free mesoporous silicas were obtained by calcination at 650°C for 6 h.

23.3.3. Metals

In recent years, chiral metal materials attracted much attention owing to their importance in both fundamental science and technological applications in catalysis, medicine, optics, sensing, fuel cells, and nanotechnology [198]. Density functional theory indicated that the single-walled helical Au (5,3) nanotubes exhibit high catalytic activity for the CO oxidation [227]. Chiral gold nanoparticles have been extensively reviewed in reference [198]. In order to induce chiral properties in the metal, a number of efforts have been devoted to doping metals with chiral molecules [228]. Doping with L-glutathione or L-quinine or either D- or L-tryptophan chiral molecules, it can reach the chirality induction in bulk gold and silver [229]. In an ultrahigh vacuum system, suspended gold nanowires (0.6 nm in diameter and 6 nm in length) possessing a multishell structure composed of coaxial tubes have been synthesized, where each tube consists of helical atom rows coiled round the wire axis [230]. Gold

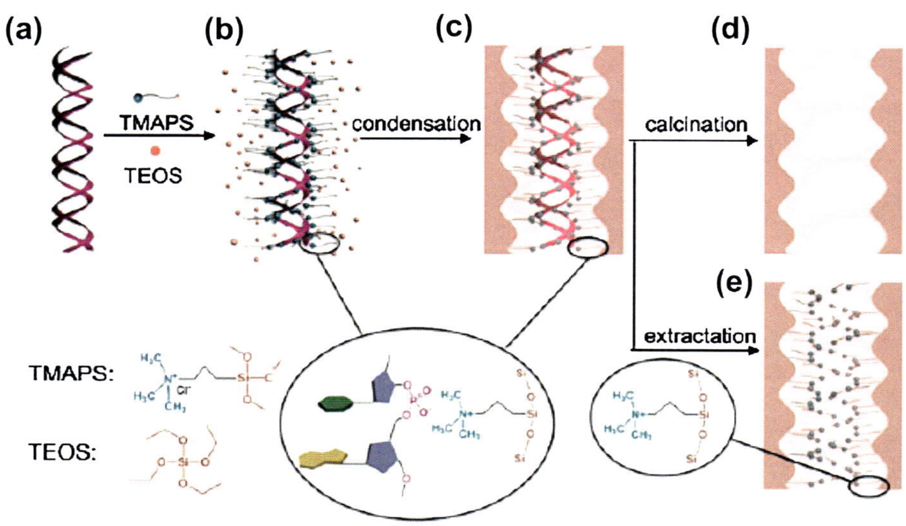

FIGURE 23.12 Schematic representation of DNA transcription into chiral porous silica based on a co-structure-directing route. (a) DNA molecule; (b) electrostatic interaction between phosphate groups of DNA and quaternary ammonium groups of TMAPS; (c) cocondensation of TMAPS and TEOS to produce porous silica; (d) silica with an inner helical pore after DNA removed by calcination; and (e) silica with an inner helical pore with a probable chiral arrangement of quaternary ammonium groups after DNA removed by extraction [226]. *Reproduced with permission from The Royal Society of Chemistry.*

TABLE 23.5 Inorganic Chiral Materials Synthesized Using Different Methods

Products	Formulations	Methods	Structures	Ref.
Borophosphate	$(NH_4)_4[H_2B_2P_4O_{16}]$	Ionothermal synthesis	Three-dimensional open-framework	[233]
Phosphonate	$Zn_2[(S)\text{-}O_3PCH_2NHC_4H_7CO_2]_2$	Hydrothermal synthesis	Helical channel	[234]
Phosphite	$(C_5H_6N_2)Zn(HPO_3)$	Hydrothermal synthesis	Helical structure	[235]
Zinc borate	$Zn_2(OH)BO_3$	Hydrothermal synthesis	Single crystal	[236]
Lead borate	$(Pb_6B_{11}O_{18}(OH)_9)$	Hydrothermal synthesis	Infinite helical chain	[237]
Barium carbonate	$BaCO_3$	Template method	Helix	[103]
Calcite	$CaCO_3$	Template method	helical structure	[238]
Zinc oxide	ZnO	Vapor–solid growth process	nanohelix	[239]
Cadmium sulfide	CdS	Microwave process	Quantum dot	[240]
Titania	TiO_2	Template method	Helical ribbon	[241]
Derivative	SU-MB	Hydrothermal synthesis	Mesoporous material	[242]

SU-MB: germanium oxide derivative, $|(H_2MPMD)_{5:5}(H_2O)_x|([Ge_{10}O_{21}(OH)_2]_2\cdot[Ge_7O_{14}F_3])$.

nanowires with a helical multishell structure can be synthesized at 150 K by electron-beam thinning of a thin gold foil [231]. In the absence of catalysts or templates, single-crystal tellurium nanobelts and nanotubes could be prepared by the hydrothermal disproportionation of Na_2TeO_3 in aqueous ammonia solution. Furthermore, the tellurium nanobelts with a helical pitch can be coaxed into chiral nanotubes [232].

23.3.4. Other Inorganic Materials

As discussed in the above parts, some inorganic chiral materials, such as zeolites, molecular sieves, silica, Au, Ag, and Te, have been synthesized by using templating method, DNA transcription, and other methods. In the past few years, some other inorganic chiral materials were also fabricated through the corresponding synthesis strategies, which are listed in Table 23.5.

23.4. BIO-INSPIRED MULTISCALE INORGANIC MATERIALS

Over millions of years of evolution, nature has been producing inorganic materials and hybrid composites with a remarkable efficiency (such as special wettability, high strength fibers, and toughened organic–inorganic composites) by making use of highly selective multiscale structures. The construction of materials presenting complex hierarchical structures is a particularly interesting challenge for the materials chemists. Learning from nature will give us important inspiration to develop new methods to construct artificial advanced materials [1–4]. Recently, much attention has been paid to bio-inspired multiscale inorganic materials. In this section, some typical bio-inspired inorganic materials, such as special wettability materials, hollow structured materials, ultra-strong and super-tough layered materials, gecko-inspired high adhesive inorganic materials, and other bio-inspired multiscale inorganic materials, will be presented.

23.4.1. Bio-inspired Surfaces With Special Wettability

Wettability is a fundamental property of a solid surface, which plays important roles in daily life, industry, and agriculture. The wettability of solid substrates is governed by their surface free energy and surface geometrical structure. In nature, many biological materials exhibit excellent surface wettability. Beautiful examples are the self-cleaning ability of lotus leaves [243], anisotropic wetting properties of rice leaves and butterfly wings [243,244], nonwetting leg of water strider [245], antifogging functionality of mosquito eyes [246], superhydrophobic antireflection behavior of cicada wings [247–249], superhydrophobic high adhesive ability of gecko feet [250–252], "petal effect" of red roses surfaces [253], water collecting property of the Namib Desert beetle and spider silk [254,255], and self-cleaning ability of fish scales [256]. Biomimetic research indicates that multiscale micro/nanostructures on these natural surfaces play an important role in generating the desired wettability, which gives us much inspiration to realize special wettability on functional surfaces through the cooperation between the

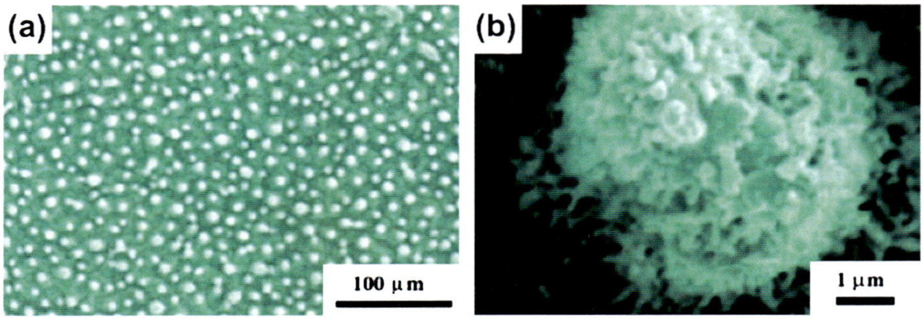

FIGURE 23.13 A superhydrophobic lotus leaf [243]. (a) Low- and (b) high-magnification SEM images of the surface structures on the lotus leaf. Every epidermal cell forms a micrometer-scale papilla and has a dense layer of epicuticular waxes superimposed on it. Each of the papillae consists of branch-like nanostructures.

TABLE 23.6 Representative Examples of Superhydrophobic Inorganic Materials

Products	Methods	Structures	Ref.
Au	Electrochemical deposition + LBL	Coral-like structure	[259]
Ag	Electrochemical deposition + LBL	Branch-like nanostructure	[260]
Cu	Chemical-based deposition method	Ribbed nanoneedle	[261]
Al	Chemisorption	Lamination pattern	[262]
CdS	Chemical bath deposition	Porous binary structure	[263]
ZnO	Electrochemical deposition	Porous structure	[264]
TiO_2	Spin coating	Nanoparticles aggregate	[265]
SnO_2	Thermal pyrolysis	Nanoflowers structure	[266]
Co_2O_3	Sol–gel	Spiral nanorod array	[267]
Sb_2O_3	Hydrothermal synthesis	Hierarchical structure	[268]
Y_2O_3	Template synthesis	Flower-like structure	[269]
LDHs	Ion exchange	Hemispherical protrusion	[270]
Carbon fibers	Pyrolysis	Nanofiber	[271]
Carbon nanotube	Pyrolysis	Honeycomb-like	[272]
Alloy	Two-step method	Flower-like structure	[273]
Stainless steel mesh	Spray-and-dry method	Block-like structure	[274]

LBL: Layer-by-layer and LDHs: layered double hydroxides.

chemical composition and the surface hierarchical micro/nanostructures.

In nature, the lotus leaf is one of the most well known and studied examples owing to the well-documented "lotus effect", where raindrops roll easily across the lotus leaf surface, carrying away dirt and debris [257]. The unique superhydrophobic self-cleaning property of lotus leaves can be attributed to the epicuticular wax and micro–nanoscale hierarchical architectures in the form of cilium-like nanostructures superimposed on top of the micrometer-scale papillae on their surfaces (Fig. 23.13) [243]. Inspired by the self-cleaning effect of lotus leaves, a great variety of artificial superhydrophobic surfaces have been fabricated by creating appropriate surface chemical composition and hierarchical surface geometrical structure. For example, a stable bionic superhydrophobic surface can be fabricated by immersing a copper plate into a solution of fatty acids at ambient temperature, which exhibit a high water contact angle of about 162° and a low sliding angle of about 2°. The typical experimental procedure was presented [258]: a copper plate was immersed in an ethanol solution of *n*-tetradecanoic acid (0.01 M) at room temperature for 3–5 days, the immersed copper plate was rinsed with deionized water and ethanol thoroughly, and then dried in air. Table 23.6

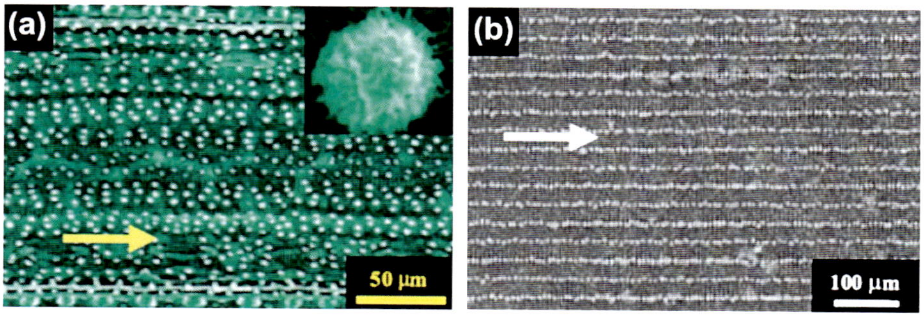

FIGURE 23.14 (a) Large-scale SEM images of the surfaces of a rice leaf (*Oryza sativa*) with different magnification. (b) SEM image of the top view of a rice-like ACNT film [243].

listed some of the superhydrophobic inorganic materials inspired by the lotus leaf.

Anisotropic wettability has been observed on the surfaces of some plants and insects. For instance, natural rice leaves possess a hierarchical structure similar to lotus leaves, exhibiting superhydrophobicity. However, differently, the papillae are arranged in one-dimensional order parallel to the edge of rice leaves (Fig. 23.14a). The water drop can roll off freely along this direction but moves much harder along the perpendicular one [243]. Inspired by the anisotropic arrangement of the papillae on the rice leaf, rice leaf-like aligned carbon nanotube (ACNT) films were fabricated by controlling the surface distribution of catalyst (Fig. 23.14b), where the microscale ACNT arrays were patterned with different spacings in the mutually orthogonal directions. The anisotropic dewetting phenomenon was also observed on such a film [243]. Recently, a two-step phase separation micromolding process was applied to replicate the structures of rice leaves. The replicated artificial rice leaves exhibited not only the very similar structures of the natural rice leaves but also surface anisotropic wetting properties [275].

In nature, mosquitoes possess excellent vision even in a watery and dim habitat (Fig. 23.15). The mosquito eye is a compound structure composed of hundreds of microscale hemispheres. The surface of each microhemisphere is covered with numerous, fine, nanoscale nipples. These nipples are very uniform, with average diameters of 101.1 ± 7.6 nm and interparticle spacings of 47.6 ± 8.5 nm, and organize in an approximately hexagonal nonclose-packed (ncp) array [246]. The elaborate micro- and nanostructures of mosquito eyes result in the striking superhydrophobic antifogging properties. Inspired by the special hierarchical microstructures of mosquito eyes, artificial compound-eye analogues possessing superhydrophobic antifogging properties can be obtained by the soft lithography method (Fig. 23.15). A typical synthesis is shown below [246].

Ordered arrays of circular photoresist posts were fabricated by using photolithography. A layer (ca. 10 μm) of positive photoresist was spin-coated onto the slides at 2000 rpm and placed in an oven at 88°C for 18 min. Subsequently, a second layer (ca. 10 μm) of photoresist was spin-coated onto the solidified samples at 2000 rpm and baked in an oven at 90°C for 30 min. This photoresist layer was exposed to a 1000 W UV light

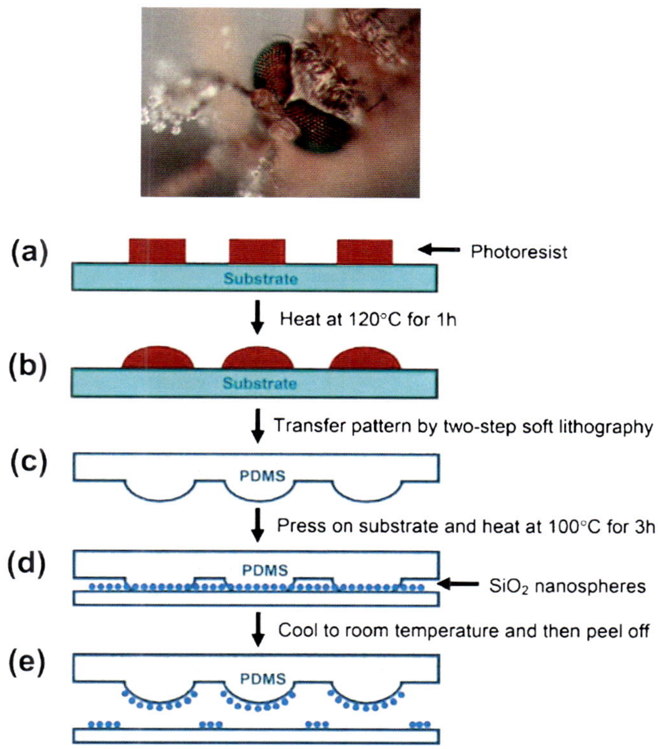

FIGURE 23.15 (Top) A photograph of antifogging mosquito eyes. Even though they are exposed to moisture, the surface of the eyes remains dry and clear while the surrounding hairs nucleate many drops. (Down) A schematic illustrating the procedure used to make the artificial compound-eye micro- and nanostructures [246]. *Xuefeng Gao, Xin Yan, Xi Yao, Liang Xu, Kai Zhang, Junhu Zhang, Bai Yang, and Lei Jiang, The Dry-Style Antifogging Properties of Mosquito Compound Eyes and Artificial Analogues Prepared by Soft Lithography, Adv. Mater. 2007, 19, pages 2213–2217, Copyright Wiley-VCH Verlag GmbH & Co. KGaA. Reproduced with permission.*

source through a transparent mask patterned with 20 μm circles spaced 5 μm apart. This approach was used to generate circular photoresist posts after developing (0.5 wt% NaOH solution), rinsing with deionized water, and drying under nitrogen. After the samples were heated at 160°C for 5–15 min, the circular photoresist posts on the slides melted and formed hemispheres because of the minimization of interfacial energy. The resulting hemisphere arrays of photoresists were replicated on a PDMS stamp, which in turn was used as the master to replicate the PDMS stamp with features of hexagonally close-packed (hcp) microscale hemisphere arrays. The silica nanospheres were centrifuged in ethanol, and the precipitate was re-dispersed in water. A 10–20 μL drop of the colloidal suspension was applied to a slightly tilted silicon substrate. Evaporation of the suspension took place in a closed chamber with control over temperature and ambient humidity. In a typical lift-up process, a PDMS stamp with patterned features was brought into conformal contact with an obtained crystal film on a silicon substrate, and the sample was hot-pressed. The PDMS stamp was carefully peeled away after the sample was cooled to room temperature. The obtained artificial compound-eye microstructures were modified by low-surface-energy fluoroalkylsilane.

Inspired by the special wettability of biological surface, many different synthetic strategies have been developed to fabricate artificial special wettability surfaces, which have been reviewed in a number of publications [3,257,276–286]. Recently, much effort has been devoted to superamphiphobic (both superhydrophobic and superoleophobic) surfaces owing to their many practical applications. Inspired by the natural design principle, a variety of biomimetic artificial superamphiphobic surfaces have been produced via the combination of surface chemical composition, hierarchical micro/nanostructures, and other factors [287–289]. For example, an electrodeposition process was developed for the fabrication of superamphiphobic surfaces on some metal substrates [290]. By using other synthetic strategies, engineering metals, alloys, and bulk metallic glass surfaces with superamphiphobicity have been constructed [291–294].

23.4.2. Biomimetic Hollow Micro–nanomaterials

In nature, hollow structures have been adopted by a number of animals. For example, in order to reduce weight by increasing friction with air and serve as heat shields from intense solar radiation, feathers of many birds possess multichannel inner structure. In the formidable polar environment, pelts of polar bears present excellent thermo-insulation properties owing to their hair with multichamber structures. Inspired by special biological structures, a variety of synthetic strategies have been developed to fabricate hollow materials [295].

Electrospinning is a highly versatile and simple method to generate ultrathin fibers from a rich variety of materials that include inorganic, organic, and inorganic–organic materials [295–298]. Recently, bio-mimic multichannel microtubes were fabricated by a novel multifluidic compound-jet electrospinning technique [299]. Figure 23.16a showed the three-channel tube fabrication system as an example, where three metallic capillaries embedded in a plastic syringe were arranged at three vertexes of an equilateral triangle. These conductive metallic inner capillaries serve as inner fluid vessels and electrode at the same time. Two immiscible viscous liquids were fed separately to the three inner capillaries and an outer syringe in an appropriate flow rate. An ethanol solution of $Ti(O\textit{i}Pr)_4$ and poly(vinyl pyrrolidone) served as outer liquid, while a commercially available innocuous paraffin oil was chosen for inner liquid. After a compound fluidic electrospinning process, a fibrous film was collected on the counter electrode. By removing the organics of as-prepared products through calcination, TiO_2 three-channel tube was obtained (Fig. 23.16b–d). Furthermore, the channel number (from two to five), diameter, and inner morphology of tubes are all controllable [299]. Using the multifluidic compound-jet electrospinning technique, multicomponent microcapsules with a novel multicompartment structure can be obtained [300]. By the electrospinning process, a wide variety of inorganic materials (such as Au [301], Fe [302], Co [302,303], Ni [302], Cu [304], TiO_2 [305–307], ZnO [308–311], SnO_2 [312], Fe_2O_3 [313], ZnS [314], and molecular sieves [315]) possessing different multiscale structures can be realized.

23.4.3. High-performance Organic–inorganic Composites Inspired by Nacre

Among the variety of natural materials, nacre (mother-of-pearl) is one of the most promising biological composite material that exhibits remarkable combinations of high stiffness, strength, toughness, and low weight. As discussed in the preceding section (Section 23.2), nacre is a typical organic–inorganic nanocomposite material with a hierarchical structure, resembling a three-dimensional brick and mortar wall [69]. For abalone nacre, the work of fracture induced brick-and-mortar laminated structure is about 3000 times higher than that of pure mineral [70]. Inspired by the microstructure and mechanism of nacre, a variety of synthetic strategies have been developed to construct biomimetic, synthetic organic–inorganic composites that attempt to reproduce nature's achievements.

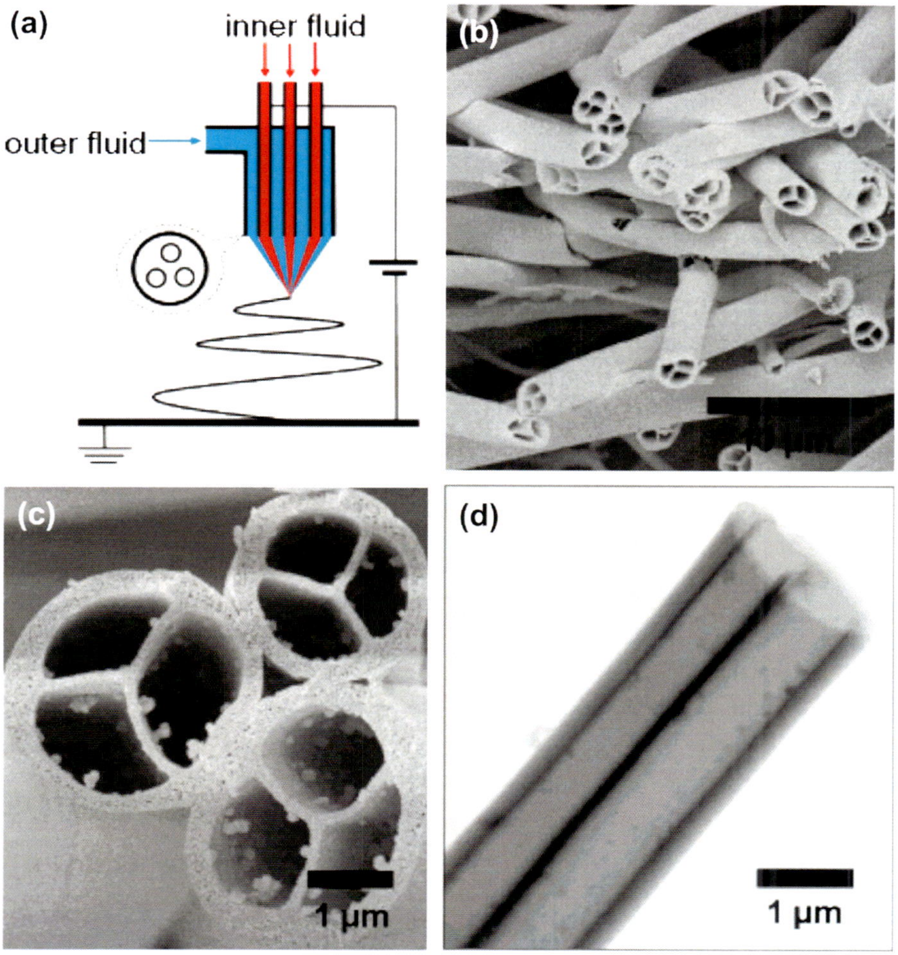

FIGURE 23.16 (a) Schematic illustration of the three-channel tube fabrication system. The immiscible inner and outer fluids (red for paraffin oil and blue for $Ti(OiPr)_4$ solution) were issued out separately from individual capillaries. When an appropriate high electric potential was applied, a liquid thread jetted out from the vertex of the compound drop and then formed a fibrous film on the counter electrode. The inset shows the outlet section of the spinneret. (b) Side-view SEM image of sample after the organics have been removed. (c) Magnified SEM image of tubes in which the channels were divided into three independent flabellate parts by a Y-shape inner ridge. (d) TEM image of a three-channel tube; the individual channels of tube are straight and continuous. *Reprinted with permission from [299]. Copyright 2007 American Chemical Society.*

Through the bottom-up colloidal assembly of strong submicrometer-thick ceramic platelets within a ductile polymer matrix, layered hybrid films combining high tensile strength and ductile behavior can be obtained [316]. Figure 23.17 is a schematic representation of the bottom-up colloidal assembly of multilayered hybrid films. The assembly of colloidal alumina platelets at the air–water interface was carried out by slowly adding approximately 2 mL of the platelet stock suspension onto a water surface in a cylindrical beaker. After spreading the platelet suspension on the water surface, the beaker was sonicated for 15–30 min to form a dense and homogeneous Langmuir film at the air–water interface. Platelet layers assembled with this colloidal approach were transferred to a glass substrate by manual dipping and afterward dried in an oven at 50°C. To form the organic layers of the hybrid films, 0.25–3 wt% of chitosan was dissolved in an aqueous solution of 2 wt% acetic acid (0.34 M) for at least 24 h before use. The chitosan layers were deposited on the platelet-covered glass substrates by spin coating. Spin coating was conducted by statically adding 1 mL of the chitosan solution onto the substrate and subsequently spinning the substrate at 2000 rpm for 40 s. The wet chitosan layers obtained after the spinning process were dried at 50°C. Multilayered hybrid films were fabricated via the sequential deposition of organic and inorganic layers onto the glass substrate. The first and last layers were always made of chitosan. Typical films consisted of 8–10 layers of inorganic platelets.

By using layer-by-layer assembly, ice-templated, biomimetic crystallization, and other methods, a number of mimetic nacre organic–inorganic composites have been synthesized, including clay platelets/polyelectrolytes [317,318], montmorillonite clay/poly(vinyl alcohol) [319], alumina–Al–Si [320], K_2SO_4/poly(acrylic acid) [321], Al_2O_3/polymethyl methacrylate [322], polyurethanes/poly(acrylic acid) [323], and double hydroxides/chitosan composites [324].

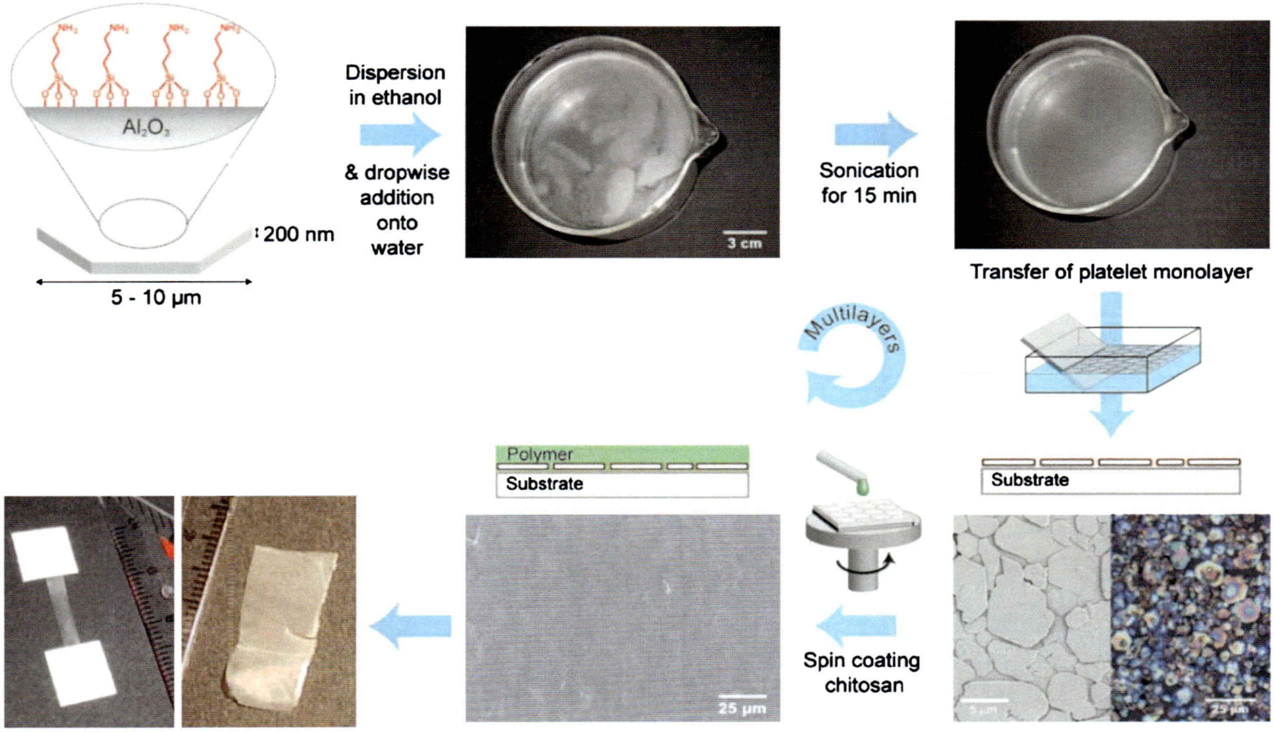

FIGURE 23.17 Bottom-up colloidal assembly of multilayered hybrid films. Surface-modified platelets are assembled at the air–water interface to produce a highly oriented layer of platelets after ultrasonication. Two-dimensional assembled platelets are transferred to a flat substrate and afterward covered with a polymer layer by conventional spin coating. *From Ref. [316]. Reprinted with permission from AAAS.*

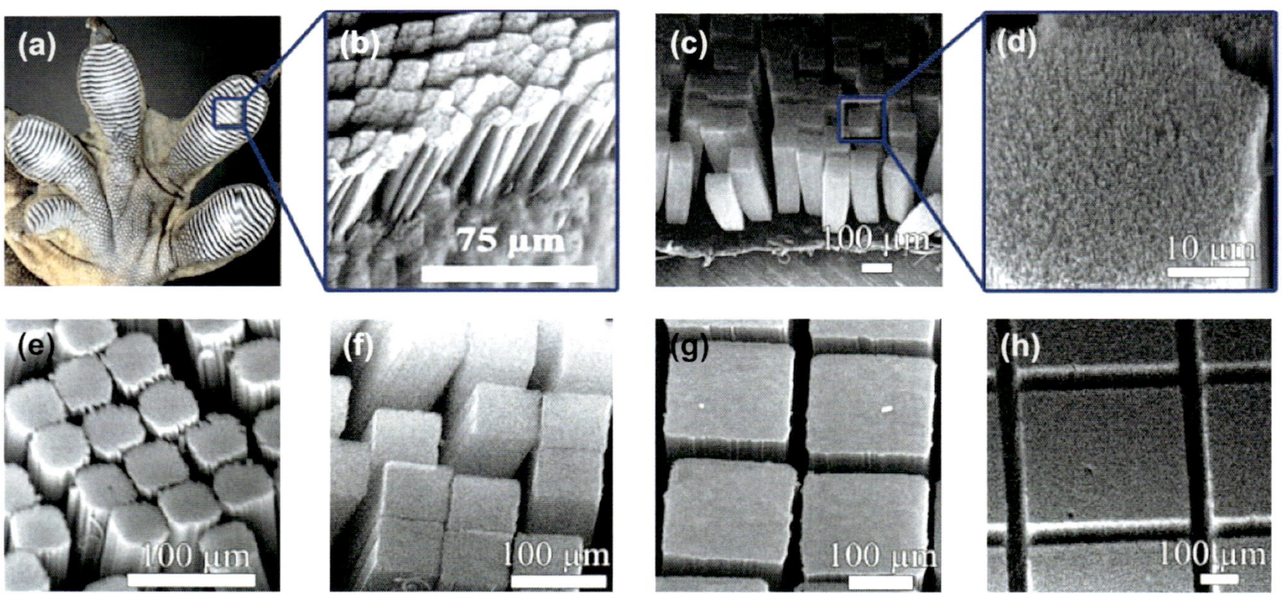

FIGURE 23.18 Microfabricated aligned multiwalled carbon nanotube setae and spatulas. (a) Optical picture of gecko foot showing that the setae are arranged in many lobes along the foot. (b) SEM image of natural gecko setae terminating into thousands of smaller spatulas. (e–h) SEM images of synthetic setae of width 50 (e), 100 (f), 250 (g), and 500 (h) μm. (c and d) Side views (c) and higher-magnification SEM image (d) of the 100 μm setae. *From Ref. [329]. Copyright (2007) National Academy of Sciences, U.S.A.*

23.4.4. Other Multiscale Materials

In nature, some creatures, including insects, spiders, and lizards, have a unique ability to cling to and detach from walls using their attachment systems. Gecko is an amazing animal and has evolved as one of the most versatile adhesives known in nature. The extraordinary ability of geckos was attributed to the hierarchical structure of their feet. The gecko foot is made up of well-aligned microscopic keratinous hairs called setae (30–130 μm in length and 5 μm in diameter), which are further split into hundreds of smaller nanoscale ends (0.2–0.5 μm in diameter) called spatulae (Fig. 23.18) [250,325]. Contact between the gecko spatulae and an opposing solid surface generates the van der Waals forces that are sufficient to allow the gecko to climb vertical walls or across ceilings [250]. In the past few years, many different synthetic strategies have been applied to construct gecko feet-mimetic high adhesive materials [326–328]. For instance, aligned carbon nanotube-based synthetic gecko tapes with hierarchical structures of setae and spatulas have been fabricated by transferring micro-patterned carbon nanotube arrays onto flexible polymer tape (Fig. 23.18). The multiscale structures with length scales of micrometers (setae) and nanometers (spatulas) are necessary to achieve high shear and peeling forces [329].

In order to enhance mechanical properties, tiny amounts of metals (Zn, Mn, Ca, or Cu, etc.) were found in the protein structures of some biological organisms' body parts (such as jaws, mandibles, stingers, cuticles, and claws). Inspired by this mechanism, metals were intentionally infiltrated into inner protein structures of spider dragline silks through multiple pulsed vapor-phase infiltration, resulting in the formation of a metal-incorporated protein matrix and a large enhancement of the mechanical properties of spider dragline silks [330].

References

[1] Y. Bar-Cohen, Biomimetics: Biologically Inspired Technologies, Taylor & Francis CRC Press, Boca Raton, FL, 2006.
[2] R.L. Reis, S. Weiner, Learning from Nature How to Design New Implantable Biomaterials, Kluwer Academic Publishers, New York, 2005.
[3] F. Xia, L. Jiang, Adv. Mater. 20 (2008) 2842.
[4] M.A. Meyers, P.Y. Chen, A.Y.M. Lin, Y. Seki, Prog. Mater. Sci. 53 (2008) 1.
[5] T.X. Fan, S.K. Chow, Z. Di, Prog. Mater. Sci. 54 (2009) 542.
[6] H.D. Espinosa, J.E. Rim, F. Barthelat, M.J. Buehler, Prog. Mater. Sci. 54 (2009) 1059.
[7] P. Behrens, E. Bäuerlein, Handbook of Biomineralization: Biomimetic and Bioinspired Chemistry, Wiley-VCH Verlag GmbH & Co. KGaA, Weinheim, 2007.
[8] H. Lowenstam, S. Weiner, On Biomineralization, Oxford University Press, New York, 1989.
[9] S. Mann, Biomineralization: Principles and Concepts in Bioinorganic Materials Chemistry, Oxford University Press, Oxford, New York, 2001.
[10] E. Bäuerlein, Biomineralization: Progress in Biology, Molecular Biology and Application, Wiley-VCH Verlag GmbH & Co. KGaA, Weinheim, 2004.
[11] E. Bäuerlein, Handbook of Biomineralization: Biological Aspects and Structure Formation, Wiley-VCH Verlag GmbH & Co. KGaA, , Weinheim, 2007.
[12] F.E. Round, R.M. Crawford, D.G. Mann, The Diatoms: Biology and Morphology of the Genera, Cambridge University Press, Cambridge, 1990.
[13] S. Mann, J. Mater, Chem. 5 (1995) 935.
[14] H. Cölfen, S.H. Yu, MRS Bull. 30 (2005) 727.
[15] H. Cölfen, Top. Curr. Chem. 271 (2007) 1.
[16] S.H. Yu, Top. Curr. Chem. 271 (2007) 79.
[17] A.W. Xu, Y.R. Ma, H. Cölfen, J. Mater, Chem. 17 (2007) 415.
[18] C. Sanchez, H. Arribart, M.M.G. Guille, Nat. Mater. 4 (2005) 277.
[19] T.X. Fan, S.K. Chow, D. Zhang, Prog. Mater. Sci. 54 (2009) 542.
[20] J.L. Arias, M.S. Fernandez, Chem. Rev. 108 (2008) 4475.
[21] J.H. Harding, D.M. Duffy, M.L. Sushko, P.M. Rodger, D. Quigley, J.A. Elliott, Chem. Rev. 108 (2008) 4823.
[22] S. Mann, Nature 332 (1988) 119.
[23] B.R. Heywood, S. Mann, Adv. Mater. 6 (1994) 9.
[24] S. Mann, Nature 365 (1993) 499.
[25] S. Mann, Biomimetic Materials Chemistry, John Wiley & Sons, Inc., Weinheim, 1996.
[26] U. Schubert, N. Hüsing, Synthesis of Inorganic Materials, Wiley-VCH Verlag GmbH & Co. KGaA, Weinheim, 2005.
[27] S. Mann, D.D. Archibald, J.M. Didymus, T. Douglas, B.R. Heywood, F.C. Meldrum, et al., Science 261 (1993) 1286.
[28] I. Weissbuch, L. Addadi, M. Lahav, L. Leiserowitz, Science 253 (1991) 637.
[29] S. Mann, N.H.C. Sparks, R.G. Board, Adv. Microb. Phys. 31 (1990) 125.
[30] G. Bevelander, H. Nakahara, Calcif. Tissue Res. 3 (1969) 84.
[31] M. Hildebrand, Chem. Rev. 108 (2008) 4855.
[32] C.B. Field, M.J. Behrenfeld, J.T. Randerson, P. Falkowski, Science 281 (1998) 237.
[33] Z.V. Finkel, M.E. Katz, J.D. Wright, O.M.E. Schofield, P.G. Falkowski, Proc. Natl. Acad. Sci. U.S.A. 102 (2005) 8927.
[34] P.G. Falkowski, M.J. Oliver, Nat. Rev. Microbiol. 5 (2007) 813.
[35] M.E. Katz, J.D. Wright, K.G. Miller, B.S. Cramer, K. Fennel, P.G. Falkowski, Mar. Geol. 217 (2005) 323.
[36] C.E. Hamm, R. Merkel, O. Springer, P. Jurkojc, C. Maier, K. Prechtel, et al., Nature 421 (2003) 841.
[37] M. Sumper, E. Brunner, Adv. Funct. Mater. 16 (2006) 17.
[38] M. Sumper, N. Kröger, J. Mater, Chem. 14 (2004) 2059.
[39] N. Kröger, N. Poulsen, Annu. Rev. Genet. 42 (2008) 83.
[40] T.L. Simpson, B.E. Volcani, Silicon and Siliceous Structures in Biological Systems, Springer, New York, 1981.
[41] N. Kröger, R. Deutzmann, M. Sumper, Science 286 (1999) 1129.
[42] N. Kröger, S. Lorenz, E. Brunner, M. Sumper, Science 298 (2002) 584.
[43] M. Sumper, R. Hett, G. Lehmann, S. Wenzl, Angew, Chem. Int. Ed. 46 (2007) 8405.
[44] N. Kröger, R. Deutzmann, C. Bergsdorf, M. Sumper, Proc. Natl. Acad. Sci. U.S.A. 97 (2000) 14133.
[45] S. Wenzl, R. Hett, P. Richthammer, M. Sumper, Angew, Chem. Int. Ed. 47 (2008) 1729.
[46] E. Brunner, P. Richthammer, H. Ehrlich, S. Paasch, P. Simon, S. Ueberlein, et al., Angew, Chem. Int. Ed. 48 (2009) 9724.
[47] M. Sumper, Science 295 (2002) 2430.

[48] W.J. Crookes-Goodson, J.M. Slocik, R.R. Naik, Chem. Soc. Rev. 37 (2008) 2403.
[49] D. Losic, J.G. Mitchell, N.H. Voelcker, Adv. Mater. 21 (2009) 2947.
[50] M.B. Dickerson, K.H. Sandhage, R.R. Naik, Chem. Rev. 108 (2008) 4935.
[51] P.J. Lopez, J. Descles, A.E. Allen, C. Bowler, Curr. Opin. Biotechnol. 16 (2005) 180.
[52] N. Kroger, Curr. Opin. Chem. Bio. 11 (2007) 662.
[53] E.A. Coffman, A.V. Melechko, D.P. Allison, M.L. Simpson, M.J. Doktycz, Langmuir 20 (2004) 8431.
[54] M. Fujiwara, K. Shiokawa, I. Sakakura, Y. Nakahara, Nano. Lett. 6 (2006) 2925.
[55] M.S. Kent, J.K. Murton, F.J. Zendejas, H. Tran, B.A. Simmons, S. Satija, et al., Langmuir 25 (2009) 305.
[56] K.D. Demadis, E. Neofotistou, Chem. Mater. 19 (2007) 581.
[57] D. Losic, J.G. Mitchell, R. Lal, N.H. Voelcker, Adv. Funct. Mater. 17 (2007) 2439.
[58] Z.H. Bao, E.M. Ernst, S. Yoo, K.H. Sandhoge, Adv. Mater. 21 (2009) 474.
[59] S.J. Lee, C.H. Huang, S. Shian, K.H. Sandhage, J. Am. Ceram. Soc. 90 (2007) 1632.
[60] K.H. Sandhage, M.B. Dickerson, P.M. Huseman, M.A. Caranna, J.D. Clifton, T.A. Bull, et al., Adv. Mater. 14 (2002) 429.
[61] S. Shian, Y. Cai, M.R. Weatherspoon, S.M. Allan, K.H. Sandhage, J. Am, Ceram. Soc. 89 (2006) 694.
[62] U. Kusari, Z. Bao, Y. Cai, G. Ahmad, K.H. Sandhage, L.G. Sneddon, Chem. Commun. (2007) 1177.
[63] M.R. Weatherspoon, S.M. Allan, E. Hunt, Y. Cai, K.H. Sandhage, Chem. Commun. (2005) 651.
[64] C. Jeffryes, R. Solanki, Y. Rangineni, W. Wang, C.H. Chang, G.L. Rorrer, Adv. Mater. 20 (2008) 2633.
[65] D.K. Gale, T. Gutu, J. Jiao, C.H. Chang, G.L. Rorrer, Adv. Funct. Mater. 19 (2009) 926.
[66] Z.H. Bao, M.R. Weatherspoon, S. Shian, Y. Cai, P.D. Graham, S.M. Allan, et al., Nature 446 (2007) 172.
[67] N. Ueyama, K. Takahashi, A. Onoda, T.A. Okamura, H. Yamamoto, Top. Curr. Chem. 271 (2007) 155.
[68] F. Barthelat, H. Tang, P.D. Zavattieri, C.M. Li, H.D. Espinosa, J. Mech, Phys. Solids 55 (2007) 306.
[69] G. Mayer, Science 310 (2005) 1144.
[70] X.D. Li, W.C. Chang, Y.J. Chao, R.Z. Wang, M. Chang, Nano. Lett. 4 (2004) 613.
[71] B.L. Smith, T.E. Schaffer, M. Viani, J.B. Thompson, N.A. Frederick, J. Kindt, et al., Nature 399 (1999) 761.
[72] N. Watabe, K.M. Wilbur, Nature 188 (1960) 334.
[73] S. Weiner, L. Hood, Science 190 (1975) 987.
[74] S. Weiner, Y. Talmon, W. Traub, J. Biol, Macromol. 5 (1983) 325.
[75] S. Weiner, W. Traub, S B. Parker, Phil. Trans. R. Soc. Lond. B 304 (1984) 425.
[76] S. Weiner, W. Traub, FEBS Lett. 111 (1980) 311.
[77] F.C. Meldrum, H. Cölfen, Chem. Rev. 108 (2008) 4332.
[78] L. Addadi, D. Joester, F. Nudelman, S. Weiner, Chem. Eur. J. 12 (2006) 981.
[79] A.M. Belcher, X.H. Wu, R.J. Christensen, P.K. Hansma, G.D. Stucky, D.E. Morse, Nature 381 (1996) 56.
[80] G. Falini, S. Albeck, S. Weiner, L. Addadi, Science 271 (1996) 67.
[81] F. Marin, G. Luquet, B. Marie, D. Medakovic, Curr. Top. Dev. Biol. 80 (2008) 209.
[82] M. Suzuki, K. Saruwatari, T. Kogure, Y. Yamamoto, T. Nishimura, T. Kato, et al., Science 325 (2009) 1388.
[83] K. Naka, Top. Curr. Chem. 271 (2007) 119.
[84] N.A.J.M. Sommerdijk, G. de With, Chem. Rev. 108 (2008) 4499.
[85] S.F. Chen, J.H. Zhu, J. Jiang, G.B. Cai, S.H. Yu, Adv. Mater. 22 (2010) 540.
[86] L.B. Gower, D.J. Odom, J. Cryst, Growth 210 (2000) 719.
[87] D. Volkmer, M. Harms, L. Gower, A. Ziegler, Angew, Chem. Int. Ed. 44 (2005) 639.
[88] S.H. Yu, H. Cölfen, J. Mater, Chem. 14 (2004) 2124.
[89] H. Cölfen, Macromol. Rapid Commun. 22 (2001) 219.
[90] S.F. Chen, S.H. Yu, T.X. Wang, J. Jiang, H. Colfen, B. Hu, et al., Adv. Mater. 17 (2005) 1461.
[91] S.H. Yu, H. Colfen, J. Hartmann, M. Antonietti, Adv. Funct. Mater. 12 (2002) 541.
[92] H. Colfen, L.M. Qi, Chem. Eur. J 7 (2001) 106.
[93] B. Guillemet, M. Faatz, F. Grohn, G. Wegner, Y. Gnanou, Langmuir 22 (2006) 1875.
[94] S.H. Yu, H. Colfen, M. Antonietti, J. Phys, Chem. B 107 (2003) 7396.
[95] H. Colfen, M. Antonietti, Langmuir 14 (1998) 582.
[96] J.M. Marentette, J. Norwig, E. Stockelmann, W.H. Meyer, G. Wegner, Adv. Mater. 9 (1997) 647.
[97] J. Rudloff, H. Colfen, Langmuir 20 (2004) 991.
[98] Y.X. Gao, S.H. Yu, H.P. Cong, J. Jiang, A.W. Xu, W.F. Dong, et al., Chem. B 110 (2006) 6432.
[99] X.H. Guo, S.H. Yu, G.B. Cai, Angew, Chem. Int. Ed. 45 (2006) 3977.
[100] Y.X. Gao, S.H. Yu, X.H. Guo, Langmuir 22 (2006) 6125.
[101] Z.P. Zhang, D.M. Gao, H. Zhao, C.G. Xie, G.J. Guan, D.P. Wang, et al., Chem. B 110 (2006) 8613.
[102] L.M. Qi, J. Li, J.M. Ma, Adv. Mater. 14 (2002) 300.
[103] S.H. Yu, H. Colfen, K. Tauer, M. Antonietti, Nat. Mater. 4 (2005) 51.
[104] T.X. Wang, A.W. Xu, H. Colfen, Angew. Chem. Int. Ed. 45 (2006) 4451.
[105] D.B. Zhang, L.M. Qi, J.M. Ma, H.M. Cheng, Chem. Mater. 14 (2002) 2450.
[106] M. Antonietti, M. Breulmann, C.G. Goltner, H. Colfen, K.K.W. Wong, D. Walsh, et al., Chem. Eur. J. 4 (1998) 2493.
[107] L.M. Qi, H. Colfen, M. Antonietti, Angew. Chem. Int. Ed. 39 (2000) 604.
[108] L.M. Qi, H. Colfen, M. Antonietti, Chem. Mater. 12 (2000) 2392.
[109] H. Colfen, L.M. Qi, Y. Mastai, L. Borger, Cryst. Growth Des. 2 (2002) 191.
[110] S.H. Yu, H. Colfen, M. Antonietti, Chem. Eur. J. 8 (2002) 2937.
[111] S.H. Yu, H. Colfen, M. Antonietti, Adv. Mater. 15 (2003) 133.
[112] S.H. Yu, M. Antonietti, H. Colfen, M. Giersig, Angew. Chem. Int. Ed. 41 (2002) 2356.
[113] F. Bouyer, N. Sanson, M. Destarac, C. Gerardin, New J. Chem. 30 (2006) 399.
[114] D.B. Zhang, L.M. Qi, J.M. Ma, H.M. Cheng, Adv. Mater. 14 (2002) 1499.
[115] M. Oner, J. Norwig, W.H. Meyer, G. Wegner, Chem. Mater. 10 (1998) 460.
[116] A. Taubert, C. Kubel, D.C. Martin, J. Phys, Chem. B 107 (2003) 2660.
[117] L.M. Qi, H. Colfen, M. Antonietti, Nano Lett. 1 (2001) 61.
[118] R.A. Caruso, Angew, Chem. Int. Ed. 43 (2004) 2746.
[119] S. Sotiropoulou, Y. Sierra-Sastre, S.S. Mark, C.A. Batt, Chem. Mater. 20 (2008) 821.
[120] K.J.C. van Bommel, A. Friggeri, S. Shinkai, Angew. Chem. Int. Ed. 42 (2003) 980.
[121] Y. Wang, A.S. Angelatos, F. Caruso, Chem. Mater. 20 (2008) 848.
[122] S.S. Behrens, J. Mater. Chem. 18 (2008) 3788.
[123] S.R. Hall, Proc. Roy. Soc. A 465 (2009) 335.
[124] H.A. Becerril, A.T. Woolley, Chem. Soc. Rev. 38 (2009) 329.
[125] P.W.K. Rothemund, Nature 440 (2006) 297.

[126] L.J. Xu, D.S. Liu, Chem. Soc. Rev. 39 (2010) 150.
[127] S.I. Lim, C.J. Zhong, Acc. Chem. Res. 42 (2009) 798.
[128] S.G. Zhang, Nat. Biotechnol. 21 (2003) 1171.
[129] J.L. Coffer, S.R. Bigham, X. Li, R.F. Pinizzotto, Y.G. Rho, R.M. Pirtle, et al., Appl. Phys. Lett. 69 (1996) 3851.
[130] Y. He, T. Ye, M. Su, C. Zhang, A.E. Ribbe, W. Jiang, et al., Nature 452 (2008) 198.
[131] D.S. Hopkins, D. Pekker, P.M. Goldbart, A. Bezryadin, Science 308 (2005) 1762.
[132] J.L. Coffer, S.R. Bigham, R.F. Pinizzotto, H. Yang, Nanotechnology 3 (1992) 69.
[133] J.C. Crocker, Nature 451 (2008) 528.
[134] D. Nykypanchuk, M.M. Maye, D. van der Lelie, O. Gang, Nature 451 (2008) 549.
[135] S.Y. Park, A.K.R. Lytton-Jean, B. Lee, S. Weigand, G.C. Schatz, C.A. Mirkin, Nature 451 (2008) 553.
[136] A.M. Hung, C.M. Micheel, L.D. Bozano, L.W. Osterbur, G.M. Wallraff, J.N. Cha, Nat. Nanotechnol. 5 (2010) 121.
[137] N.L. Rosi, C.S. Thaxton, C.A. Mirkin, Angew. Chem. Int. Ed. 43 (2004) 5500.
[138] F. Patolsky, Y. Weizmann, O. Lioubashevski, I. Willner, Angew. Chem. Int. Ed. 41 (2002) 2323.
[139] H. Nakao, H. Shiigi, Y. Yamamoto, S. Tokonami, T. Nagaoka, S. Sugiyama, et al., Nano Lett. 3 (2003) 1391.
[140] M.G. Warner, J.E. Hutchison, Nat. Mater. 2 (2003) 272.
[141] E. Braun, Y. Eichen, U. Sivan, G. Ben-Yoseph, Nature 391 (1998) 775.
[142] H. Yan, S.H. Park, G. Finkelstein, J.H. Reif, T.H. LaBean, Science 301 (2003) 1882.
[143] K. Nguyen, M. Monteverde, A. Filoramo, L. Goux-Capes, S. Lyonnais, P. Jegou, et al., Adv. Mater. 20 (2008) 1099.
[144] W.E. Ford, O. Harnack, A. Yasuda, J.M. Wessels, Adv. Mater. 13 (2001) 1793.
[145] S. Kundu, H. Liang, Adv. Mater. 20 (2008) 826.
[146] L.Q. Dong, T. Hollis, B.A. Connolly, N.G. Wright, B.R. Horrocks, A. Houlton, Adv. Mater. 19 (2007) 1748.
[147] L. Levina, W. Sukhovatkin, S. Musikhin, S. Cauchi, R. Nisman, D.P. Bazett-Jones, et al., Adv. Mater. 17 (2005) 1854.
[148] J.M. Kinsella, A. Ivanisevic, J. Phys. Chem. C 112 (2008) 3191.
[149] H. Zhou, T. Fan, D. Zhang, Q. Guo, H. Ogawa, Chem. Mater. 19 (2007) 2144.
[150] B.J. Zhang, S.A. Davis, N.H. Mendelson, S. Mann, Chem. Commun. (2000) 781.
[151] S.A. Davis, S.L. Burkett, N.H. Mendelson, S. Mann, Nature 385 (1997) 420.
[152] V. Berry, S. Rangaswamy, R.F. Saraf, Nano. Lett. 4 (2004) 939.
[153] R. Mogul, J.J.G. Kelly, M.L. Cable, A.F. Hebard, Mater. Lett. 60 (2006) 19.
[154] P. Vukusic, J.R. Sambles, C.R. Lawrence, Nature 404 (2000) 457.
[155] O. Sato, S. Kubo, Z.Z. Gu, Acc. Chem. Res. 42 (2009) 1.
[156] R. Won, Nat. Photonics 1 (2007) 130.
[157] T.E. Mengesha, R.R. Vallance, M. Barraja, R. Mittal, Bioinspir. Biomim. 4 (2009) 36004.
[158] B. Li, J. Zhou, R.L. Zong, M. Fu, Y. Bai, L.T. Li, et al., Ceram. Soc. 89 (2006) 2298.
[159] W. Zhang, D. Zhang, T.X. Fan, J. Ding, Q.X. Gu, H. Ogawa, Nanotechnology 17 (2006) 840.
[160] G. Cook, P.L. Timms, C.G. Spickermann, Angew. Chem. Int. Ed. 42 (2003) 557.
[161] W. Zhang, D. Zhang, T.X. Fan, J.J. Gu, R. Ding, H. Wang, et al., Chem. Mater. 21 (2009) 33.
[162] M.R. Weatherspoon, Y. Cai, M. Crne, M. Srinivasarao, K.H. Sandhage, Angew. Chem. Int. Ed. 47 (2008) 7921.
[163] R.H.A. Ras, E. Sahramo, J. Malm, J. Raula, M. Karppinen, J. Am. Chem. Soc. 130 (2008) 11252.
[164] W.L. Min, B. Jiang, P. Jiang, Adv. Mater. 20 (2008) 3914.
[165] W. Lee, M.K. Jin, W.C. Yoo, J.K. Lee, Langmuir 20 (2004) 7665.
[166] S.H. Hong, J. Hwang, H. Lee, Nanotechnology 20 (2009) 385303.
[167] G.Y. Xie, G.M. Zhang, F. Lin, J. Zhang, Z.F. Liu, S.C. Mu, Nanotechnology 19 (2008) 095605.
[168] G.M. Zhang, J. Zhang, G.Y. Xie, Z.F. Liu, H.B. Shao, Small 2 (2006) 1440.
[169] D. Yang, L.M. Qi, J.M. Ma, Adv. Mater. 14 (2002) 1543.
[170] Q. Dong, H.L. Su, W. Cao, D. Zhang, Q.X. Guo, Y.J. Lai, J. Solid State Chem. 180 (2007) 949.
[171] Q. Dong, H.L. Su, D. Zhang, Z.T. Liu, Y.J. Lai, Microp. Mesop. Mater. 98 (2007) 344.
[172] D. Yang, L.M. Qi, J.M. Ma, J. Mater. Chem. 13 (2003) 1119.
[173] Q. Dong, H.L. Su, W. Cao, D. Zhang, Q.X. Guo, F.Y. Zhang, Eur. J. Inorg. Chem. (2007) 2265.
[174] Q. Dong, H.L. Su, D. Zhang, W. Cao, N. Wang, Langmuir 23 (2007) 8108.
[175] Q. Dong, H.L. Su, D. Zhang, F.Y. Zhang, Nanotechnology 17 (2006) 3968.
[176] P.K. Ajikumar, R. Lakshminarayanan, S. Valiyaveettil, Cryst. Growth Des. 4 (2004) 331.
[177] J.K. Liu, Q.S. Wu, Y.P. Ding, Chem. Res. Chin. Univ. 21 (2005) 243.
[178] X.H. Yang, Q.S. Wu, J.K. Lju, Cryst. Res. Technol. 42 (2007) 211.
[179] A.R. Maddocks, A.T. Harris, Mater. Lett. 63 (2009) 748.
[180] H.L. Su, J. Han, N. Wang, Q. Dong, D. Zhang, C.F. Zhang, Smart Mater. Struct. 17 (2008) 015045.
[181] H.L. Su, N. Wang, Q. Dong, D. Zhang, J. Membr. Sci. 283 (2006) 7.
[182] C. Radloff, R.A. Vaia, J. Brunton, G.T. Bouwer, V.K. Ward, Nano Lett. 5 (2005) 1187.
[183] M. Kobayashi, M. Seki, H. Tabata, Y. Watanabe, I. Yamashita, Nano Lett. 10 (2010) 773.
[184] A.G. Dong, Y.J. Wang, Y. Tang, N. Ren, Y.H. Zhang, J.H. Yue, et al., Adv. Mater. 14 (2002) 926.
[185] R. Wahl, M. Mertig, J. Raff, S. Selenska-Pobell, W. Pompe, Adv. Mater. 13 (2001) 736.
[186] J.H. He, T. Kunitake, Chem. Mater. 16 (2004) 2656.
[187] S. Behrens, W. Habicht, K. Wagner, E. Unger, Adv. Mater. 18 (2006) 284.
[188] D. Yang, L.M. Qi, J.M. Ma, Chem. Commun. (2003) 1180.
[189] A. Guijarro, M. Yus, The Origin of Chirality in the Molecules of Life, The Royal Society of Chemistry, Cambridge, UK, 2009.
[190] R. Purrello, Nat. Mater. 2 (2003) 216.
[191] A.J. Gellman, ACS Nano 4 (2010) 5.
[192] R. Kuroda, B. Endo, M. Abe, M. Shimizu, Nature 462 (2009) 790.
[193] O. Lukin, F. Vogtle, Angew. Chem. Int. Ed. 44 (2005) 1456.
[194] A.B. Buda, T.A. Derheyde, K. Mislow, Angew. Chem. Int. Ed. 31 (1992) 989.
[195] G.A. Hembury, V.V. Borovkov, Y. Inoue, Chem. Rev. 108 (2008) 1.
[196] D. Cahard, X.H. Xu, S. Couve-Bonnaire, X. Pannecoucke, Chem. Soc. Rev. 39 (2010) 558.
[197] J.H. Yu, R.R. Xu, J. Mater. Chem. 18 (2008) 4021.
[198] C. Gautier, T. Burgi, ChemPhysChem 10 (2009) 483.
[199] C.E. Song, S.G. Lee, Chem. Rev. 102 (2002) 3495.
[200] S. Marx, D. Avnir, Acc. Chem. Res. 40 (2007) 768.
[201] M.E. Davis, R.F. Lobo, Chem. Mater. 4 (1992) 756.
[202] D.E. De Vos, M. Dams, B.F. Sels, P.A. Jacobs, Chem. Rev. 102 (2002) 3615.
[203] L.Q. Tang, L. Shi, C. Bonneau, J.L. Sun, H.J. Yue, A. Ojuva, et al., Nat. Mater. 7 (2008) 381.
[204] A. Joy, V. Ramamurthy, Chem. Eur. J 6 (2000) 1287.

[205] M.E. Davis, Acc. Chem. Res. 26 (1993) 111.
[206] C. Dryzun, Y. Mastai, A. Shvalb, D. Avnir, J. Mater. Chem. 19 (2009) 2062.
[207] J. Coronas, Chem. Eng. J 156 (2010) 236.
[208] Y. Li, J.H. Yu, Z.P. Wang, J.N. Zhang, M. Guo, R.R. Xu, Chem. Mater. 17 (2005) 4399.
[209] T.E. Gier, X.H. Bu, P.Y. Feng, G.D. Stucky, Nature 395 (1998) 154.
[210] P. Liu, X.G. Zhang, Y. Yao, J. Wang, Appl. Catal. A Gen. 371 (2009) 142.
[211] J.L. Sun, C. Bonneau, A. Cantin, A. Corma, M.J. Diaz-Cabanas, M. Moliner, et al., Nature 458 (2009) 1154.
[212] A. Corma, M.J. Diaz-Cabanas, J.L. Jorda, C. Martinez, M. Moliner, Nature 443 (2006) 842.
[213] J.H. Yu, R.R. Xu, Acc. Chem. Res. 36 (2003) 481.
[214] Y. Song, J.H. Yu, Y. Li, G.H. Li, R.R. Xu, Angew. Chem. Int. Ed. 43 (2004) 2399.
[215] J.H. Yu, R.R. Xu, Chem. Soc. Rev. 35 (2006) 593.
[216] J.M. Thomas, R. Raja, G. Sankar, R.G. Bell, Acc. Chem. Res. 34 (2001) 191.
[217] X.W. Song, Y. Li, L. Gan, Z.P. Wang, J.H. Yu, R.R. Xu, Angew. Chem. Int. Ed. 48 (2009) 314.
[218] K.S.W. Sing, D.H. Everett, R.A.W. Haul, L. Moscou, R.A. Pierotti, J. Rouquerol, et al., Pure Appl. Chem. 57 (1985) 603.
[219] G.J.D. Soler-illia, C. Sanchez, B. Lebeau, J. Patarin, Chem. Rev. 102 (2002) 4093.
[220] S. Che, Z. Liu, T. Ohsuna, K. Sakamoto, O. Terasaki, T. Tatsumi, Nature 429 (2004) 281.
[221] T. Ohsuna, Z. Liu, S.N. Che, O. Terasaki, Small 1 (2005) 233.
[222] H.Y. Jin, Z. Liu, T. Ohsuna, O. Terasaki, Y. Inoue, K. Sakamoto, et al., Adv. Mater. 18 (2006) 593.
[223] B. Wang, C. Chi, W. Shan, Y.H. Zhang, N. Ren, W.L. Yang, et al., Angew. Chem. Int. Ed. 45 (2006) 2088.
[224] H.M. Chen, J.H. He, Dalton Trans. (2009) 6651.
[225] X.W. Wu, H.Y. Jin, Z. Liu, T. Ohsuna, O. Terasaki, K. Sakamoto, et al., Chem. Mater. 18 (2006) 241.
[226] C.Y. Jin, H.B. Qiu, L. Han, M.H. Shu, S.A. Che, Chem. Commun. (2009) 3407.
[227] A. Mahdavi, L. Ferreira, C. Sundback, J.W. Nichol, E.P. Chan, D.J.D. Carter, et al., Proc. Natl. Acad. Sci. U.S.A. 105 (2008) 2307.
[228] X. Lopez-Lozano, L.A. Perez, I.L. Garzon, Phys. Rev. Lett. 97 (2006) 233401.
[229] H. Behar-Levy, O. Neumann, R. Naaman, D. Avnir, Adv. Mater. 19 (2007) 1207.
[230] Y. Kondo, K. Takayanagi, Science 289 (2000) 606.
[231] Y. Oshima, A. Onga, K. Takayanagi, Phys. Rev. Lett. 91 (2003) 205503.
[232] M.S. Mo, J.H. Zeng, X.M. Liu, W.C. Yu, S.Y. Zhang, Y.T. Qian, Adv. Mater. 14 (2002) 1658.
[233] H.Z. Xing, Y. Li, T. Su, J. Xu, W.T. Yang, E.B. Zhu, et al., Dalton Trans. 39 (2010) 1713.
[234] X. Shi, G.S. Zhu, S.L. Qiu, K.L. Huang, J.H. Yu, R.R. Xu, Angew. Chem. Int. Ed. 43 (2004) 6482.
[235] J. Liang, Y. Wang, J.H. Yu, Y. Li, R.R. Xu, Chem. Commun. (2003) 882.
[236] Z.T. Yu, J.J. Xu, Y.S. Jiang, Z. Shi, Y. Guo, D.J. Wang, et al., Chem. 13 (2003) 2227.
[237] Z.T. Yu, Z. Shi, Y.S. Jiang, H.M. Yuan, J.S. Chen, Chem. Mater. 14 (2002) 1314.
[238] T. Sugawara, Y. Suwa, K. Ohkawa, H. Yamamoto, Macromol. Rapid Commun. 24 (2003) 847.
[239] P.X. Gao, Y. Ding, W.J. Mai, W.L. Hughes, C.S. Lao, Z.L. Wang, Science 309 (2005) 1700.
[240] M.P. Moloney, Y.K. Gun'ko, J.M. Kelly, Chem. Commun. (2007) 3900.
[241] J.H. Jung, H. Kobayashi, K.J.C. van Bommel, S. Shinkai, T. Shimizu, Chem. Mater. 14 (2002) 1445.
[242] X.D. Zou, T. Conradsson, M. Klingstedt, M.S. Dadachov, M. O'Keeffe, Nature 437 (2005) 716.
[243] L. Feng, S.H. Li, Y.S. Li, H.J. Li, L.J. Zhang, J. Zhai, et al., Adv. Mater. 14 (2002) 1857.
[244] Y.M. Zheng, X.F. Gao, L. Jiang, Soft Matter 3 (2007) 178.
[245] X.F. Gao, L. Jiang, Nature 432 (2004) 36.
[246] X.F. Gao, X. Yan, X. Yao, L. Xu, K. Zhang, J.H. Zhang, et al., Adv. Mater. 19 (2007) 2213.
[247] A.R. Parker, H.E. Townley, Nat. Nanotechnol. 2 (2007) 347.
[248] P. Vukusic, J.R. Sambles, Nature 424 (2003) 852.
[249] D.G. Stavenga, S. Foletti, G. Palasantzas, K. Arikawa, Proc. Roy. Soc. B Biol. Sci. 273 (2006) 661.
[250] K. Autumn, M. Sitti, Y.C.A. Liang, A.M. Peattie, W.R. Hansen, S. Sponberg, et al., Proc. Natl. Acad. Sci. U.S.A. 99 (2002) 12252.
[251] W.R. Hansen, K. Autumn, Proc. Natl. Acad. Sci. U.S.A. 102 (2005) 385.
[252] S. Sethi, L. Ge, L. Ci, P.M. Ajayan, A. Dhinojwala, Nano Lett. 8 (2008) 822.
[253] L. Feng, Y.A. Zhang, J.M. Xi, Y. Zhu, N. Wang, F. Xia, et al., Langmuir 24 (2008) 4114.
[254] A.R. Parker, C.R. Lawrence, Nature 414 (2001) 33.
[255] Y.M. Zheng, H. Bai, Z.B. Huang, X.L. Tian, F.Q. Nie, Y. Zhao, et al., Nature 463 (2010) 640.
[256] M.J. Liu, S.T. Wang, Z.X. Wei, Y.L. Song, L. Jiang, Adv. Mater. 21 (2009) 665.
[257] T.L. Sun, L. Feng, X.F. Gao, L. Jiang, Acc. Chem. Res. 38 (2005) 644.
[258] S.T. Wang, L. Feng, L. Jiang, Adv. Mater. 18 (2006) 767.
[259] F. Shi, Z.Q. Wang, X. Zhang, Adv. Mater. 17 (2005) 1005.
[260] N. Zhao, F. Shi, Z.Q. Wang, X. Zhang, Langmuir 21 (2005) 4713.
[261] X. Yao, Q. Chen, L. Xu, Q. Li, Y. Song, X. Gao, et al., Adv. Funct. Mater. 20 (2010) 656.
[262] S. Desbief, B. Grignard, C. Detrembleur, R. Rioboo, A. Vaillant, D. Seveno, et al., Langmuir 26 (2010) 2057.
[263] Y. Liu, T. Tan, B. Wang, R. Zhai, X. Song, E. Li, et al., J. Colloid Interf. Sci. 320 (2008) 540.
[264] M. Li, J. Zhai, H. Liu, Y.L. Song, L. Jiang, D.B. Zhu, J. Phys. Chem. B 107 (2003) 9954.
[265] X. Zhang, H. Kono, Z. Liu, S. Nishimoto, D.A. Tryk, T. Murakami, et al., Chem. Commun. (2007) 4949.
[266] A.C. Chen, X.S. Peng, K. Koczkur, B. Miller, Chem. Commun. (2004) 1964.
[267] Z.G. Guo, W.M. Liu, Appl. Phys. Lett. 90 (2007) 193108.
[268] K.S. Liu, J. Zhai, L. Jiang, Nanotechnology 19 (2008) 165604.
[269] R.K. Jia, Y.B. Bai, T.J. Li, Mater. Lett. 59 (2005) 4010.
[270] F.Z. Zhang, L.L. Zhao, H.Y. Chen, S.L. Xu, D.G. Evans, X. Duan, Angew. Chem. Int. Ed. 47 (2008) 2466.
[271] L. Feng, Z.L. Yang, J. Zhai, Y.L. Song, B.Q. Lin, Y.M. Ma, et al., Angew. Chem. Int. Ed. 42 (2003) 4217.
[272] S.H. Li, H.J. Li, X.B. Wang, Y.L. Song, Y.Q. Liu, L. Jiang, et al., Chem. B 106 (2002) 9274.
[273] K.S. Liu, M.L. Zhang, J. Zhai, J. Wang, L. Jiang, Appl. Phys. Lett. 92 (2008) 183103.
[274] L. Feng, Z.Y. Zhang, Z.H. Mai, Y.M. Ma, B.Q. Liu, L. Jiang, et al., Angew. Chem. Int. Ed. 43 (2004) 2012.
[275] J. Gao, Y.L. Liu, H.P. Xu, Z.Q. Wang, X. Zhang, Langmuir 25 (2009) 4365.
[276] X. Zhang, F. Shi, J. Niu, Y.G. Jiang, Z.Q. Wang, J. Mater. Chem. 18 (2008) 621.
[277] D. Quéré, Annu. Rev. Mater. Res. 38 (2008) 71.

[278] X.M. Li, D. Reinhoudt, M. Crego-Calama, Chem. Soc. Rev. 36 (2007) 1350.
[279] S.T. Wang, Y.L. Song, L. Jiang, J. Photochem. Photobiol. C 8 (2007) 18.
[280] P. Roach, N.J. Shirtcliffe, M.I. Newton, Soft Matter 4 (2008) 224.
[281] X.J. Feng, L. Jiang, Adv. Mater. 18 (2006) 3063.
[282] H. Liu, J. Zhai, L. Jiang, Soft Matter 2 (2006) 811.
[283] C. Dorrer, J. Ruhe, Soft Matter 5 (2009) 51.
[284] M. Callies, D. Quéré, Soft Matter 1 (2005) 55.
[285] M. Liu, Y. Zheng, J. Zhai, L. Jiang, Acc. Chem. Res. 43 (2009) 368.
[286] M. Nosonovsky, B. Bhushan, Curr. Opin. Colloid Interf. Sci. 14 (2009) 270.
[287] A. Tuteja, W. Choi, M.L. Ma, J.M. Mabry, S.A. Mazzella, G.C. Rutledge, et al., Science 318 (2007) 1618.
[288] A. Tuteja, W. Choi, J.M. Mabry, G.H. McKinley, R.E. Cohen, Proc. Natl. Acad. Sci. U.S.A. 105 (2008) 18200.
[289] W. Choi, A. Tuteja, S. Chhatre, J.M. Mabry, R.E. Cohen, G.H. McKinley, Adv. Mater. 21 (2009) 2190.
[290] J.M. Xi, L. Feng, L. Jiang, Appl. Phys. Lett. 92 (2008) 053102.
[291] H.F. Meng, S.T. Wang, J.M. Xi, Z.Y. Tang, L. Jiang, J. Phys. Chem. C 112 (2008) 11454.
[292] K. Zhao, K.S. Liu, J.F. Li, W.H. Wang, L. Jiang, Scripta Mater. 60 (2009) 225.
[293] W.C. Wu, X.L. Wang, D.A. Wang, M. Chen, F. Zhou, W.M. Liu, et al., Chem. Commun. (2009) 1043.
[294] X.J. Liu, W.C. Wu, X.L. Wang, Z.Z. Luo, Y.M. Liang, F. Zhou, Soft Matter 5 (2009) 3097.
[295] Y. Zhao, L. Jiang, Adv. Mater. 21 (2009) 3621.
[296] H.J. Liang, T.E. Angelini, P.V. Braun, G.C.L. Wong, J. Am, Chem. Soc. 126 (2004) 14157.
[297] A. Greiner, J.H. Wendorff, Angew, Chem. Int. Ed. 46 (2007) 5670.
[298] X.F. Lu, C. Wang, Y. Wei, Small 5 (2009) 2349.
[299] Y. Zhao, X.Y. Cao, L. Jiang, J. Am. Chem. Soc. 129 (2007) 764.
[300] H.Y. Chen, Y. Zhao, Y.L. Song, L. Jiang, J. Am. Chem. Soc. 130 (2008) 7800.
[301] V.G. Pol, E. Koren, A. Zaban, Chem. Mater. 20 (2008) 3055.
[302] H. Wu, R. Zhang, X.X. Liu, D.D. Lin, W. Pan, Chem. Mater. 19 (2007) 3506.
[303] M. Graeser, M. Bognitzki, W. Massa, C. Pietzonka, A. Greiner, J.H. Wendorff, Adv. Mater. 19 (2007) 4244.
[304] M. Bognitzki, M. Becker, M. Graeser, W. Massa, J.H. Wendorff, A. Schaper, et al., Adv. Mater. 18 (2006) 2384.
[305] D. Li, Y.N. Xia, Nano Lett. 3 (2003) 555.
[306] K. Onozuka, B. Ding, Y. Tsuge, T. Naka, M. Yamazaki, S. Sugi, et al., Nanotechnology 17 (2006) 1026.
[307] W.K. Son, D. Cho, W.H. Park, Nanotechnology 17 (2006) 439.
[308] H.Q. Liu, J.X. Yang, J.H. Liang, Y.X. Huang, C.Y. Tangz, J. Am. Ceram. Soc. 91 (2008) 1287.
[309] H. Wu, D. Lin, R. Zhang, W. Pan, J. Am. Ceram. Soc. 91 (2008) 656.
[310] H. Wu, W. Pan, J. Am. Ceram. Soc. 89 (2006) 699.
[311] P. Viswanathamurthi, N. Bhattarai, H.Y. Kim, D.R. Lee, Nanotechnology 15 (2004) 320.
[312] A. Yang, X.M. Tao, G.K.H. Pang, K.G.G. Siu, J. Am. Ceram. Soc. 91 (2008) 257.
[313] Y. Zhu, J.C. Zhang, J. Zhai, L. Jiang, Thin Solid Films 510 (2006) 271.
[314] D. Lin, H. Wu, R. Zhang, W. Pan, J. Am. Ceram. Soc. 90 (2007) 3664.
[315] J. Di, H. Chen, X. Wang, Y. Zhao, L. Jiang, J. Yu, et al., Chem. Mater. 20 (2008) 3543.
[316] L.J. Bonderer, A.R. Studart, L.J. Gauckler, Science 319 (2008) 1069.
[317] Z.Y. Tang, N.A. Kotov, S. Magonov, B. Ozturk, Nat. Mater. 2 (2003) 413.
[318] M. Rubner, Nature 423 (2003) 925.
[319] P. Podsiadlo, A.K. Kaushik, E.M. Arruda, A.M. Waas, B.S. Shim, J.D. Xu, et al., Science 318 (2007) 80.
[320] S. Deville, E. Saiz, R.K. Nalla, A.P. Tomsia, Science 311 (2006) 515.
[321] K. Oaki, H. Imai, Angew, Chem. Int. Ed. 44 (2005) 6571.
[322] E. Munch, M.E. Launey, D.H. Alsem, E. Saiz, A.P. Tomsia, R.O. Ritchie, Science 322 (2008) 1516.
[323] P. Podsiadlo, E.M. Arruda, E. Kheng, A.M. Waas, J. Lee, K. Critchley, et al., ACS Nano 3 (2009) 1564.
[324] H.-B. Yao, H.-Y. Fang, Z.-H. Tan, L.-H. Wu, S.-H. Yu, Angew, Chem. Int. Ed. 49 (2010) 2140.
[325] K. Autumn, Y.A. Liang, S.T. Hsieh, W. Zesch, W.P. Chan, T.W. Kenny, R. Fearing, et al., Nature 405 (2000) 681.
[326] H. Lee, B.P. Lee, P.B. Messersmith, Nature 448 (2007) 338.
[327] L.T. Qu, L.M. Dai, M. Stone, Z.H. Xia, Z.L. Wang, Science 322 (2008) 238.
[328] Z.L. Wang, Nat. Nanotechnol. 4 (2009) 407.
[329] L. Ge, S. Sethi, L. Ci, P.M. Ajayan, A. Dhinojwala, Proc. Natl. Acad. Sci. U.S.A. 104 (2007) 10792.
[330] S.M. Lee, E. Pippel, U. Gosele, C. Dresbach, Y. Qin, C.V. Chandran, et al., Science 324 (2009) 488.

CHAPTER

24

Frontier of Inorganic Synthesis and Preparative Chemistry (II)-Designed Synthesis—Inorganic Crystalline Porous Materials

Jihong Yu
Jilin University, China

In "*Synthesizing our future*" [1], when Ryoji Noyori, the 2001 Nobel Laureate for Chemistry said "*Chemistry attempts both to understand the structures and characteristics of substances in minute details at the atomic and molecular levels and to create new compounds with desirable properties and functions*", it pointed out where the synthetic chemists should focus their efforts in the future. Chemical research should sustain our civilized society in twenty-first century and further into the future, therefore the synthetic processes should be highly efficient, atom economical, environmentally benign and rational. Up to now, more than 50 millions of chemical substances have been created by human beings. However, their syntheses are mainly on the basis of the use of empirical, trial-and-error methods. How to rationalize the synthesis of new materials with desired structures and functionalities has become one of the frontier research subjects in modern synthetic chemistry.

Inorganic synthesis and preparative chemistry play a pivotal role in chemistry, which concerns the development of new synthesis and preparative strategies, routes and methods, elucidation of the chemical processes and mechanism of synthesis and preparation, and the establishment of relevant synthesis theories for the discovery of new inorganic materials, etc. One of the major objectives of inorganic synthesis and preparative chemistry is to synthesize the inorganic materials with desired structures and properties. This requires a deep understanding of the relationship and principle of synthesis—structure—property of inorganic materials. Toward the designed synthesis of new inorganic materials with desired structures and properties, the following three key issues have to be considered: (i) the understanding of the chemical basis of properties; (ii) the development of structural design methods; and (iii) the discovery of the rational synthetic approaches for the tailored materials. Even though considerable progress has been achieved in the understanding of the correlations among structure, bonding, and physical properties of some inorganic compounds, many difficult challenges remain in targeting the synthesis with desired structures and properties of inorganic materials, especially inorganic crystalline solids. Unlike the highly developed field of organic synthesis, a main limitation towards the designed synthesis concerning inorganic materials lies in the uncertain formation mechanism and the unclear structure—synthesis relationship. The designed synthesis of new inorganic materials in a rational way, while targeting novel structures and properties is a formidable task in inorganic chemistry.

Engineering the synthesis of new materials with desired structures and functions has aroused considerable attention in the area of chemistry and materials science. To pursue this objective, it needs to start from the specific functional system, to investigate the rules and principles of synthesis—structure—property in order to develop the rational approaches toward the designed synthesis. In this chapter the designed synthesis of inorganic crystalline porous materials will be discussed as an example. Inorganic crystalline porous materials, typically known as zeolites, have regular nanoporous space, large surface area, and tunable surface properties (e.g., acidic center, redox center, other functional groups, etc.) [2—4]. These materials have attracted considerable

Modern Inorganic Synthetic Chemistry, DOI: 10.1016/B978-0-444-53599-3.10024-1

interest since the first discovery of the synthetic zeolites in 1940s because they are very important in the fields of catalysis, adsorption, and ion exchange. Besides, they are finding new applications as advanced functional materials in the areas of medicine, sensor, optics, electronics, magnetism, etc. Their excellent performance is essentially determined by their unique structural character, such as the size of the pore window, accessible void space, dimensionality of the channel system, numbers and sites of cations, and special functional groups. Synthesis of such materials is typically carried out in a gel medium under hydrothermal/solvothermal conditions by using alkali metal ions or organic amines/ammoniums as the templates or structure-directing agents (SDAs) [5]. However, the kinetics of crystallization for these materials is very complex, which is determined by many variables such as source materials, gel composition, pH value, SDAs, solvent, crystallization temperature, and time. The difficulty towards the rational synthesis of zeolitic materials lies in our limited knowledge of their crystallization mechanism. Nevertheless, the knowledge and understanding of the relationship of the synthesis—structure—functionality of zeolitic inorganic porous crystalline materials have increased over the past six decades. Particularly, computational simulations have significantly enhanced the ability of researchers in targeting zeolites with specific structures and properties [6]. Although there is still not a clear picture of formation mechanism at the molecular level of zeolitic inorganic crystalline porous materials, much effort has been made toward the rational design and synthesis of such materials [7,8]. In this chapter, the state of art and future perspective in this area will be shown, which I hope will stimulate the further advance in the designed synthesis of new materials, which is one of the challenging frontiers in the modern inorganic synthetic chemistry.

24.1. STRUCTURE DESIGN OF INORGANIC CRYSTALLINE POROUS MATERIALS

A prerequisite towards the designed synthesis is that we should be able to design the desired structures that can perform specific functions. In the aspects of zeolitic inorganic porous materials, achieving the designed synthesis of such materials calls for our capability to design the porous architectures such as pore system, pore shape, and pore dimension that are responsible for their performances, especially for shape-selective catalysis. Nowadays, computational methods in the elucidation of inorganic crystal structures have been well established [9]. Much work has been devoted to the generation of potential new zeolite structures [10—14]. For example, Treacy and coworkers used a symmetry constrained interstice bond searching method to generate over 2 millions of structures. The Website http://www.hypotheticalzeolites.net/ contains a collection of hypothetical zeolite structures in a searchable database. On the other hand, predicting the structures in a rational way is of particular interest with the increasing demand for zeolitic materials that fulfill the specific applications. For example, extra-large pore structures show advantages for diffusing and processing large molecules, chiral zeolite structures may allow the asymmetric catalysis and separation, and the catalytic performance is generally improved if structures with multidirectional channel systems are achieved.

Some methods have been developed for the design of zeolitic inorganic porous structures with predefined pore geometries and structural building units (SBUs). It should be noted that the thermodynamic stability of the theoretical structures is considered when designing structures.

24.1.1. Design of Zeolite Structures with Predefined Pore Geometries

24.1.1.1. Generation of the Pore Structure Through Constrained Assembly of Atoms Outside of the "Forbidden Zones"

Several computational simulation approaches have been developed for the generation of four-connected zeolite structures. However, without the constraint of the pore information of a zeolite in the simulation, it cannot prevent the generation of nonporous frameworks, i.e., the undesirable structures. In 2003, Li et al. [15] developed a computational methodology for the design of zeolite structures with predefined pore geometries. In their method, the concept of "forbidden zone" is introduced in the simulation. The forbidden zone corresponds to a specific porous pattern inside which no T atom (Tetrahedral atom) can be placed. This method is straightforward and efficient for the design of zeolite frameworks with desired pore geometry.

As illustrated in Fig. 24.1, the pore structures represented by cylinders, called the "forbidden zones" are introduced into the unit cell, within which no atoms are placed during the generation process of a zeolite framework. Such a forbidden zone can be specifically designed to reflect a particular desired porous system in the zeolite such as one-dimensional channel (Fig. 24.1a) [15], three-dimensional intersecting channel (Fig. 24.1b) [16], and chiral channel (Fig. 24.1c) [17]. The atoms are assembled outside the forbidden zones on the basis of specified symmetry and distance constraints. Two constraint conditions must be satisfied when placing the atoms: (i) no T atom is allowed inside a forbidden zone and (ii) the distance between any two

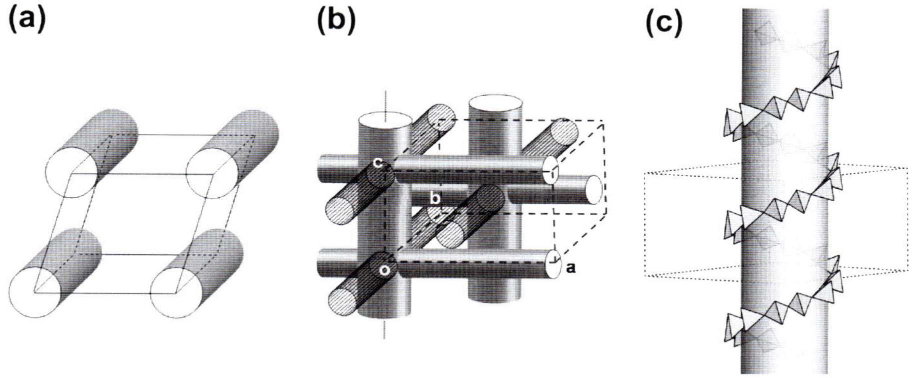

FIGURE 24.1 Forbidden zones defined as cylinders in one unit cell. (a) one-dimensional channel system; (b) three-dimensional channel system; and (c) chiral channel system. *Reprinted with permission from [15]. Copyright 2003 American Chemical Society.*

T atoms should not be less than 3.0 Å keeping the Si−Si distance.

The structure generation procedure mainly involves the following steps:

(1) Unit cell, space group, and forbidden zones are defined.
(2) Unique atoms that are selected on the basis of the zeolite framework density are randomly placed outside the forbidden zones in the unit cell followed by automatic generation of their equivalent atoms based on symmetry operations under specific space group. Atoms violating the above constraint conditions (1) and (2) will be removed from further consideration; otherwise, a data set containing unique T atom coordinates will be saved. For distance calculation, the crystal coordinates are converted to Cartesian coordinates.
(3) Multiple data sets will be generated and saved by repeating step (2) until certain criteria are met, e.g., a specified number of different data sets have been generated. Coordination sequences (CS) [18] are used to distinguish different framework types, as well as to judge whether the generated configuration is a viable one.
(4) A structural model is built based on every data set by using Cerius[2] software package [19]. Bridging atoms are added using Structure Solve_Bridging atom.
(5) The structure is refined using a distance least squares (DLS) refinement assuming an SiO_2 composition [20]. The final determination of symmetry of each generated structure is performed through a "Find-Symmetry" analysis.
(6) Potential energy of each structural model is calculated using the Burchart 1.01 force field [21].

It can be illustrated by an example for the design of a hypothetical zeolite structure H1 with predefined one-dimensional pore system under the hexagonal space group $P6_3/mmc$ (No. 194), $a = b = c = 15$ Å for the unit cell (Fig. 24.2). The pore radius is defined as 6.0 Å, approximately corresponding to a 12-ring channel. Two unique atoms are selected, one is defined at the general position *l* and the other at the special position *j*. The first unique atom is constrained to the pore wall whose position is randomly selected. Then its 23 equivalent atoms are generated using the symmetry operation. Twenty-four atoms must satisfy conditions (1) and (2), i.e., all atoms outside of the forbidden zones and T−T distances should not be less than 3.0 Å (Fig. 24.2a). The second unique atom T2 is placed at the special position *j* followed by generation of its 11 equivalent atoms (Fig. 24.2b). Now 36 atoms must satisfy conditions (1) and (2), then the position of the second unique atom T2 is retained; otherwise, the atoms generated by the second unique atom will be replaced until all the atoms satisfy the conditions. A structural model is built on the basis of each set of unique *T* atom coordinates using Cerius.[2] Bridging atoms are added (Fig. 24.2c). The optimized structure has a unit cell $a = 12.5$ Å and $c = 15.7$ Å with a space group $P6_3/mmc$ (No. 194) and an empirical formula $Si_{36}O_{72}$. As shown in Fig. 24.2d, the structure of H1 contains predefined one-dimensional 12-ring channels along the [001] direction. It contains columns of *can* cages along the *c*-axis that enclose a 12-ring channel. Its simulated X-ray powder diffraction pattern is shown in Fig. 24.2d.

This method, having constrained assembly of atoms outside of the forbidden zones, allows the design of zeolite frameworks with different space groups, cell parameters, pore dimensions, as well as channel systems. Cross-linked channels can be obtained by defining a forbidden zone whose channel direction is perpendicular to the principal axis under a specified space group. For example, assuming a tetragonal space group $I4/mmm$ (No. 139), and defining a forbidden zone (channel) perpendicular to the fourfold rotation axis in the unit cell, as shown in Fig. 24.1b, a channel perpendicular to this defined channel will be naturally

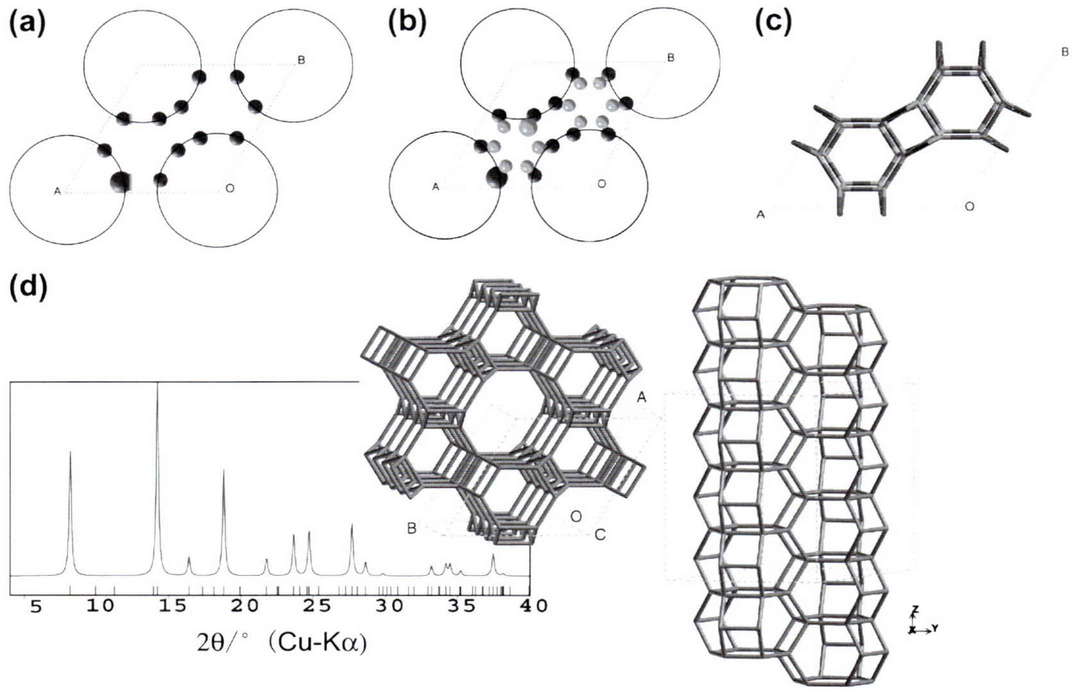

FIGURE 24.2 Illustration of the generation process for a hypothetical framework H1, assuming the space group of $P6_3/mmc$ ($a = b = c = 15$ Å). The pore radius is set at 6.0 Å. The first unique atom represented by a larger ball is confined to the pore wall followed by the generation of 23 equivalent atoms using symmetry operation l (a). The second unique atom represented as a larger ball is randomly placed outside of the forbidden zones followed by the generation of 11 equivalent atoms by symmetry operation j (b). Bridging atoms are added (c). Framework structure containing 12-ring channels. The simulated XRD pattern is shown (d). *Reprinted with permission from [15]. Copyright 2003 American Chemical Society.*

produced by the rotation operation. By using this method, hypothetical structures with cross-linked channels have been predicted. Figure 24.3 shows a hypothetical structure H15 generated by this approach assuming the space group of $I4/mmm$ (No. 139). Its space group changes to $P4/mmm$ (No. 123) after the structure is refined. The structure of H15 has two kinds of parallel 12-ring channels along the [001] direction and 8-ring channels along [110] direction. Notably, H15 is constructed by D4Rs only (Fig. 24.3c).

Strikingly, chiral zeolite structures that are highly desirable in asymmetric catalysis and separation can

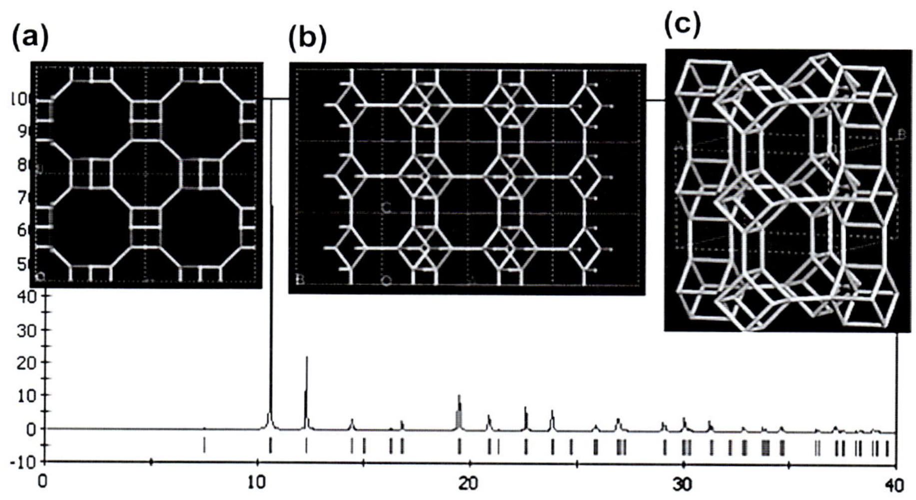

FIGURE 24.3 Structure of H15 (a) view along [001] direction; (b) view along [110] direction; (c) the cages contained in H15 (the simulated X-ray diffraction pattern is shown). $Si_{16}O_{32}$: $P4/mmm$, $a = 11.7886$ Å, $b = 7.1905$ Å.

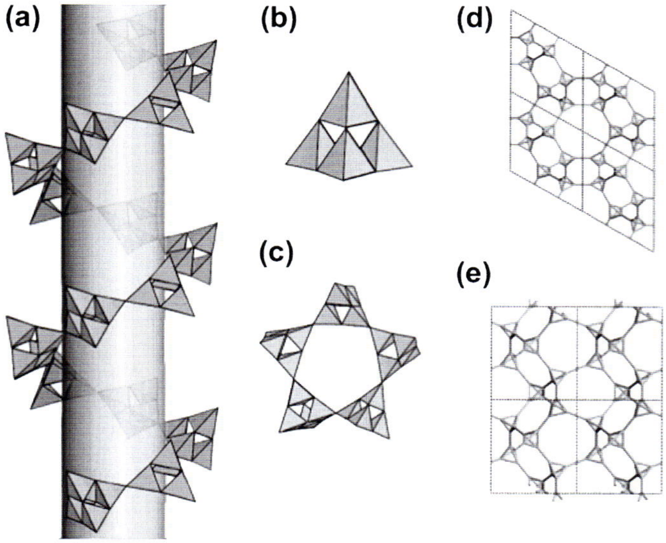

FIGURE 24.4 (a) Chiral channel of the hypothetical framework H178-1 (cylinder diameter of 9.5 Å), (b) supertetrahedron T2 building block, (c) super-5MR constructed by T2 building blocks, and the framework viewed along, (d) the [001] direction, and (e) the [010] direction. *Reprinted with permission from [17]. Copyright 2005 American Chemical Society.*

be generated by using this approach under the chiral space groups [17]. So far, only a few zeolite structures possess chiral frameworks like BEA, CZP, GOO, OSO, and JRY, etc. Theoretically, numerous hypothetical chiral zeolite frameworks can be generated by using this approach. For example, assuming a hexagonal chiral space group $P6_122$ (No. 178), and defining a forbidden zone (channel) along the 6_1 screw axis in the unit cell, as shown in Fig. 24.1c, the TO_4 tetrahedra forms a helical chain by the 6_1 operation and further convolve around the forbidden zone to form a chiral channel.

H178-1 is generated in the space group $P6_122$. Figure 24.4a shows its chiral channel, and parts of the T atoms are omitted for clarity. Its basic building block is a T2 supertetrahedron (Fig. 24.4b). Five such supertetrahedra form a super-5-rings (Fig. 24.4c), and the linkages of super-5-rings further form a chiral channel through edge sharing. Figure 24.4d and e shows its framework viewed along the [001] and [010] directions, respectively.

A variety of hypothetical structures with chiral channels have been generated under the space group $P6_122$ with the following constraint conditions. The forbidden zones are defined as a cylinder running along the 6_1 screw axis (i.e., the [001] direction) with the radii of 3.0, 4.0, 5.0, and 6.0 Å, respectively, and the cell parameters vary from 5 to 25 Å. The numbers of unique T atoms range from 1 to 5. Figure 24.5 shows the plot of the framework energy (assuming the SiO_2 composition) versus the framework density. The black triangles represent the data for the 161 (up to 2005) known structures, while the black squares represent the data for the hypothetical structures. A linear correlation between the framework energy and the framework density can be observed for most of the known frameworks as illustrated by the dashed line in Fig. 24.5. It is noted that the energies of some of the known structures such as

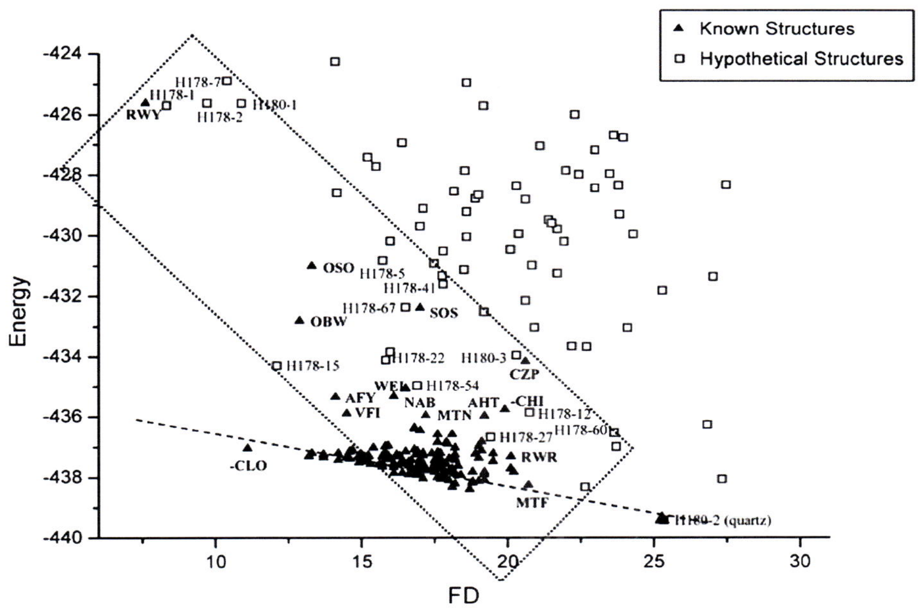

FIGURE 24.5 Framework energy versus framework density. Black triangles denote the data for the 161 known frameworks, and black squares denote the data for hypothetical structures obtained under space group $P6_122$. *Reprinted with permission from [17]. Copyright 2005 American Chemical Society.*

RWY, OSO, OBW, SOS, OBW, CZP, WEI, −CHI, and AFY, are relatively higher than those of the normal silicate frameworks. This is because these frameworks are in fact with non-SiO_2 compositions. According to the calculation results, most of the four-connected frameworks with chiral channels are not energetically favorable for SiO_2 compositions because of the special geometric torsions in such structures. This suggests that introducing other T elements such as beryllium, boron, germanium, and arsenic instead of silicon, and other bridging atoms like sulfur, instead of oxygen, might stabilize such frameworks more effectively. This is because they could offer more reasonable bond distances and bond angles than an SiO_2 composition does.

To sum up, the method of constrained assembly of atoms outside of the forbidden zones is very simple and straightforward. A hypothetical zeolite structure database has been built up by Li et al. which contains the not-yet-discovered zeolite structures predicted by them. The database is available at the Website http://www.hypotheticalzeolites.net/. One can easily search the structures by putting the defined structural characters such as the pore system, pore dimension, and SBU.

24.1.1.2. *Framework Generation in Density Map*

In 2008, Li et al. [22] further developed a method of framework generation in density map (FGDM). In this method, they integrate the pore information extracted from the density map of high-resolution electron microscopy (HRTEM) into the structure modeling. A term that evaluates the agreement between the structure models generated and the HRTEM image has been added to the cost function along with the bonding and connectivity terms for structure simulation. This work shows that by using just one HRTEM image as constraint, the number of degrees of freedom for solving the structure-modeling problem can be reduced to practically manageable level, even for the most complex zeolite structures. The power of this approach is demonstrated with the solutions of the most complex structures of IM-5 ($[Si_{288}O_{576}]$) [23] and TNU-9 ($[Si_{192}O_{384}]$) [24], as well as with the design of desired porous structures.

The generation of structure models includes the following major steps:

(1) A black and white density map is constructed from a symmetry-averaged HRTEM image where the channels along certain directions are easily recognized. The gray scale gives an approximation of the potential projected along this direction.

(2) The two-dimensional projection is then expanded over the whole unit cell. In this way, each point in the unit cell is given a fixed density value determined from its projection on the HRTEM image. Notably, this three-dimensional map is not physically real, but it divides the unit cell into regions where atoms may be located and those where they may not. A threshold value for the potential is then defined, and atoms in regions with values above this threshold will not contribute to the E_{map} term of the cost function, while those in the other regions contribute substantially.

(3) An estimated number of T atoms on the basis of framework density are introduced into the asymmetric unit of the unit cell in a random fashion, and their equivalent atoms are generated by symmetry operators.

(4) During the simulation, a reasonable bonding geometry is sought by moving the T atoms around, and atoms are "pushed" from the low-density area into the high-density area to fit the density map. The cost function (and thereby the structure) is optimized by using a simulated annealing [25] and parallel tempering algorithm [26]. At each temperature stage, configuration-biased Monte Carlo [26] is used for sampling.

(5) The framework with the lowest cost value is saved. The bridging O atoms are added between the nearest T−T pairs, and the bonding geometry of the entire structure is further optimized by using a simple distance least squares (DLS) algorithm. Coordination sequences are calculated for structure identification.

(6) A new simulation cycle is started by reassigning random coordinates to all atoms.

Figure 24.6 illustrates the generation of the zeolite IM-5 framework in a three-dimensional density map derived from one HRTEM image along the [100] direction. First, the density map is generated from the HRTEM image of IM-5 along the [100] direction (Fig. 24.6a and b). Then T atoms are placed randomly in the density map at the beginning of the simulation (Fig. 24.6c) and rearranged in the density map during the simulation (Fig. 24.6d). Finally, structure models that are in agreement with the HRTEM image along the [100] direction is generated (Fig. 24.6e). The cost values of the generated structure models are plotted (Fig. 24.6f). All generated models are unique, and only 10 models with the lowest cost values are shown. Model 1 has the lowest cost value and the best R_{DLS} value, and it corresponds to the correct structure solution. Its framework projections along the [100] and [001] directions are displayed together with those of two other models with low cost values.

The advantage of this method lies not only for the solutions of the highly complex framework structures, but also for the design of new structures with desired pore structures. For example, one can define a density

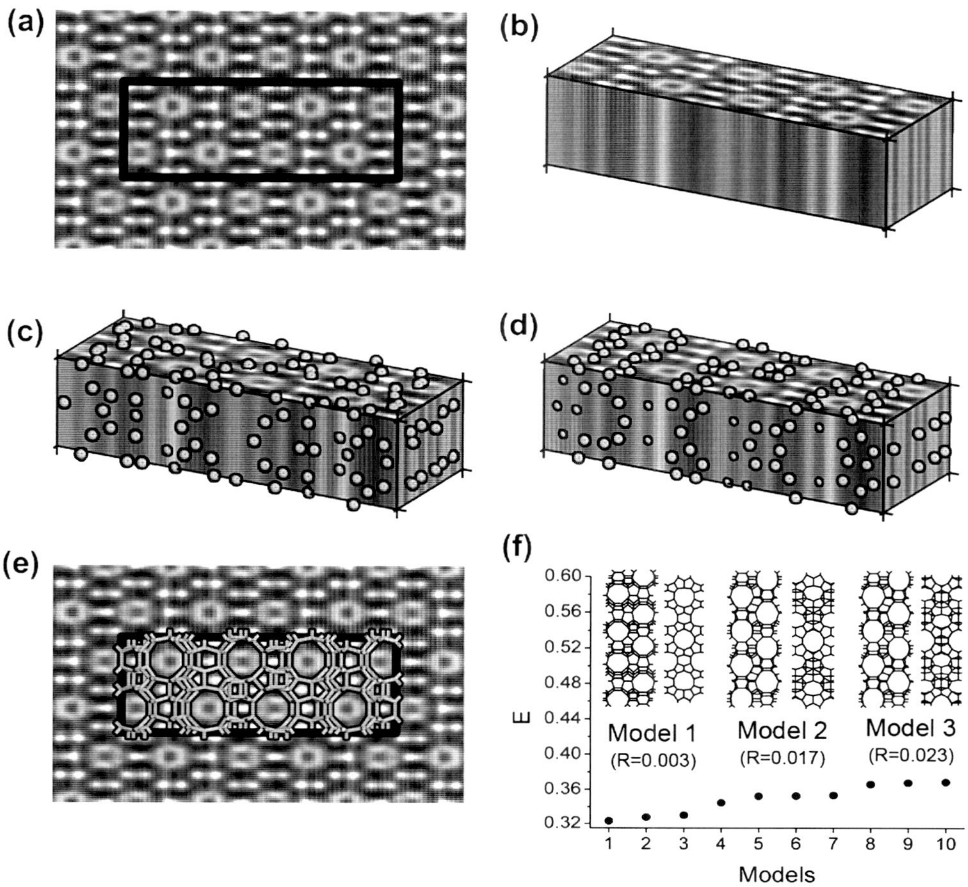

FIGURE 24.6 Generation of the IM-5 framework in a three-dimensional density map derived from one HRTEM image along the [100] direction. (a) The HRTEM image of IM-5 along the [100] direction; (b) the density map generated from the image shown in (a); (c) T atoms placed randomly in the density map at the beginning of the simulation; (d) T atoms rearranged in the density map during the simulation; (e) agreement between the generated structure model and HRTEM image along the [100] direction; and (f) plot of the cost values of the generated structure models. All generated models are unique, and only 10 models with the lowest cost values are shown. Model 1 has the lowest cost value and the best R_{DLS} value, and it corresponds to the correct structure solution. Its framework projections along the [100] and [001] directions are displayed together with those of two other models with low cost values. *Y. Li, J. Yu, R. Xu, Ch. Baerlocher, L.B. McCusker: Combining Structure Modeling and Electron Microscopy to Determine Complex Zeolite Framework Structures. Angewandte Chemie International Edition. 2008. Volume 47, Page 4405. Copyright Wiley-VCH Verlag GmbH & Co. KGaA. Reproduced with permission from [22].*

map that reflects the desired porous pattern, and then generate structures with such pore geometry. Furthermore, the desired channel structures can be designed by tuning the known zeolite pores in terms of simple evolution concepts like reproduction (generation of the same pore topology as the original one), mutation (variation of the pore topology from the original one), and crossover (creation of new pore topology by combining two or more known pore systems).

This method is a powerful tool for zeolite structure prediction according to functional requirement.

24.1.1.3. Framework Generation Using Genetic Algorithm and Exclusion Zones

In 2004, Woodley et al. [27] developed a genetic algorithm (GA) approach to generate zeolite frameworks from the knowledge of unit cell dimensions, constituent atoms, and exclusion zones (EZs) which represent the pore regions in zeolite structures. This approach contains two main stages. In the first stage, candidate structures are generated using genetic algorithm approach. A cost function is employed to evaluate the quality of all generated candidate structures. In the second stage, structures with reasonable bond lengths and coordination numbers are further relaxed through lattice energy minimization.

In stage 1, the unit cell is divided into evenly spaced grids. Si and O atoms are randomly assigned into these grids whereby the initial structures are generated. The initial structures can be represented by sequences of binary numbers, which is known as the "genotype representation". Through crossover and mutation, new candidate structures can be generated. The detailed steps of this genetic algorithm are shown in the following chart:

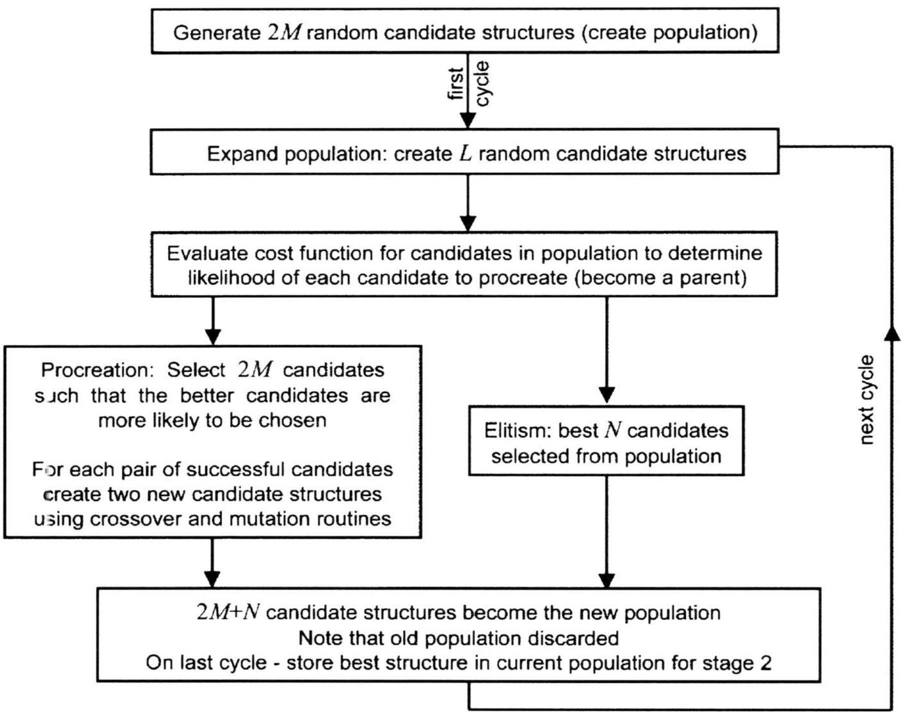

Woodley et al. showed that porous structures are difficult to generate, even for zeolites JBW and BIK which only possess small pores. To facilitate the generation of pore apertures, exclusion zones, i.e., regions within the unit cell where atoms are forbidden, are introduced in this genetic algorithm approach (Fig. 24.7). The exclusion zones can be ellipsoidal, elliptical, cylindrical, or planar. Exclusion zones are incorporated in the first stage of the GA approach either indirectly, by defining additional cost function, or directly, by removing grids in the exclusion regions. Both indirect and direct EZs can greatly enhance the generation of target structures with predefined pore architecture.

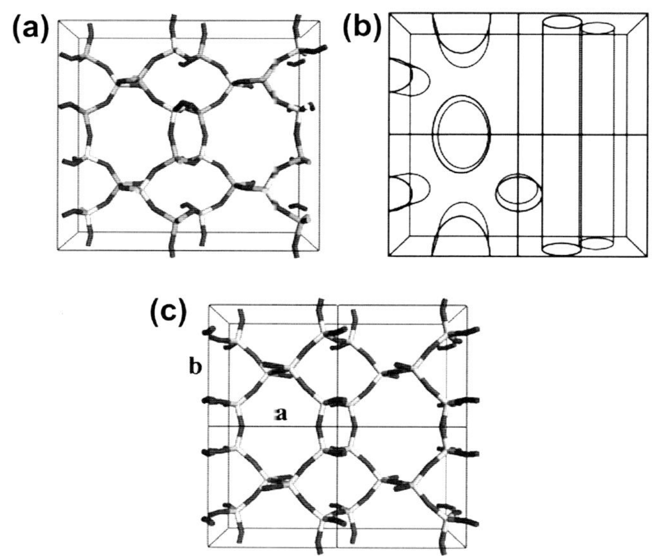

FIGURE 24.7 Perspective view of the framework structure for (a) zeolite JBW and (c) the SiO_2 target structure shown within $2 \times 2 \times 1$ supercell. (b) The EZs (with arbitrary radii) used to aid the generation of (c), where only half of each type of EZ channel is shown for clarity. *Images reproduced with permission from [27].*

24.1.2. Structure Design with Predefined Structural Building Units (SBUs)

24.1.2.1. Building Units Design and Scale Chemistry

Building unit (BU) can be defined as the minimum assembly of atoms, ions, or molecules which by condensation of the group with others (identical or different) gives rise to the final solid [28]. The concept of a BU can be described in two ways: the first is just a topological tool for the description of structures. For example, various BUs, such as D4Rs, D6Rs, sodalite cages, are used to describe the structures of zeolites; the second refers to the real existing one in solution/gel during the formation of a solid such as the Keggin's structure.

Recently, Férey proposed a concept *"Scale Chemistry"* [28] which is powerful in the design of solid structures. As is noticed, many complex structures adopt a simple topology when every BU is considered as a single entity. For example, the structures of ReO_3 and zeolite LTA can both be described by a simple cubic P lattice like polonium metal by replacing the octahedra in ReO_3 and the sodalite cage in LTA by an atom, respectively. This means that, whatever the size of the BUs, their spatial arrangement is the same. This is the so-called *"Scale Chemistry"*. Obviously, increasing the size of the BUs starting from a simple topology is a feasible approach to the design of open-framework solids with a larger pore size. Here two examples based on germanates will illustrate the powerful concept of scale chemistry in the design of extra-large pore open-framework structures.

Recently, much attention has been paid to synthesis of open-framework germanates. Germaniun can be four-, five-, and six-coordinated with oxygen to form large clusters such as Ge_7X_{19} (Ge_7), Ge_8X_{20} (Ge_8), Ge_9X_{26-m} (Ge_9, $m = 0{-}1$), and $Ge_{10}X_{28}$ Ge_{10}, (X = O, OH, F) [29]. It is interesting to identify large structure motifs as possible building units for the prediction of extra-large pore structures. Pan et al. [30] reported a tubular germanate, $[(C_5N_2H_{14})_4(C_5N_2H_{13})(H_2O)_4][Ge_7O_{12}O_{4/2}(OH)F_2][Ge_7O_{12}O_{5/2}(OH)F]_2[GeO_{2/2}(OH)_2]$ (denoted JLG-5), with 12-ring channels built from $6^8 12^6$ cavities. Figure 24.8 shows the structure of JLG-5. The highly symmetric $6^8 12^6$ cavity, built from Ge_7 clusters, is stabilized by an $(H_2O)_{16}$ cluster.

The $6^8 12^6$ cavity can be identified as a cuboctahedron (Fig. 24.9a and b) with each of its 12 vertices decorated by a four-coordinated Ge_7 cluster (Fig. 24.9c). The $6^8 12^6$ cavity can be further used as a BU to decorate the nodes of a simple net to construct three-dimensional framework structures with extra-large pores. For example, **reo-e** net corresponds to the primitive cubic packing of six-coordinated cuboctahedra (Fig. 24.9d). If each vertex of the **reo-e** net is decorated by a six-coordinated Ge_7 cluster, then a new hypothetical three-dimensional germanate framework ($Pm3\text{-}m$, $a = 22.9078$ Å) is formed, which contains two types of cavities, i.e., the $6^8 12^6$ cavity and a $6^8 10^6 12^{12}$ supercavity (Fig. 24.9e). The $6^8 10^6 12^{12}$ supercavity corresponds to a rhombicuboctahedron (**rco**; Fig. 24.9f and g) decorated by Ge_7 clusters. It has a free diameter of 17.8 Å and (Fig. 24.9h). The $6^8 12^6$ cavities themselves adopt a primitive cubic arrangement. Each $6^8 12^6$ cavity is connected to six other cavities by linking their six 12-rings through additional germanium-centered tetrahedra. The arrangement of

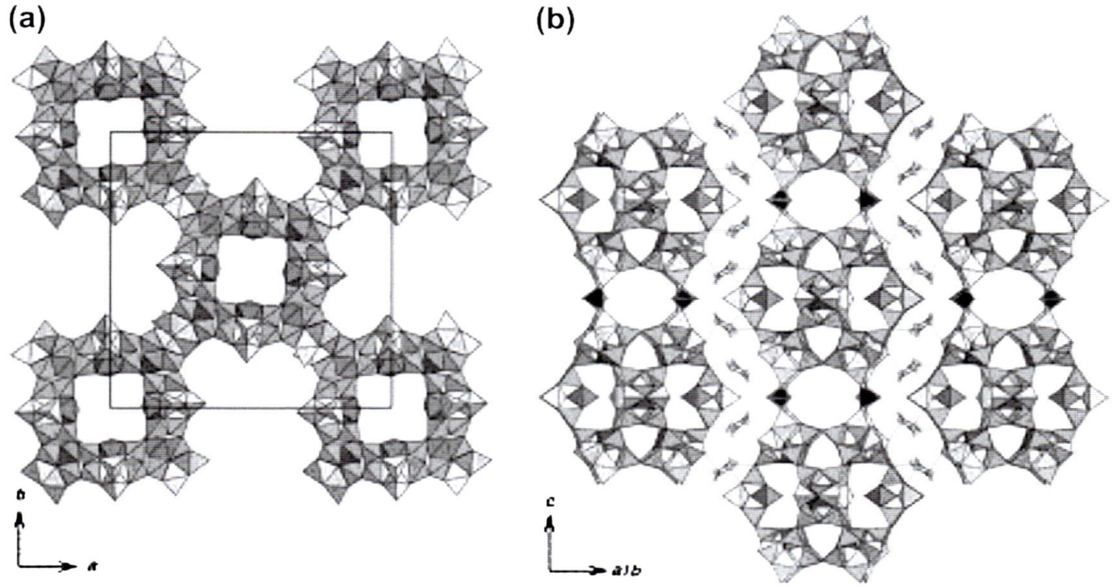

FIGURE 24.8 The structure of JLG-5 viewed along (a) the [001] and (b) the [1 −1 0] directions. The $6^8 12^6$ cavities are packed in a body-centered manner, with the 12-ring windows aligned along the three main axes. The $6^8 12^6$ cavities are connected through additional germanium-centered tetrahedra (black) along the *c*-axis to form12-ring tubes. The tubes are held together by 2-methylpiperazine cations (C gray ball, N black ball) via hydrogen bonding. Within the Ge_7 clusters, the germanium-centered octahedra are dark grey, the germanium-centered trigonal bipyramids are light grey, and the germanium-centered tetrahedra are medium grey. For clarity, only one unique 2-methylpiperazine cation is shown in (b), and the water cluster is omitted. *Q. Pan, J. Li, K. Christensen, C. Bonneau, X. Ren, L. Shi, J. Sun, X. Zou, G. Li, J. Yu, R. Xu: A germanate built from $6^8 12^6$ cavity cotemplated by a $(H_2O)_{16}$ cluster and 2-methylpiperazine. Angewandte Chemie International Edition. 2008. Volume 47, Page 7868. Copyright Wiley-VCH Verlag GmbH & Co. KGaA. Reproduced with permission from [30].*

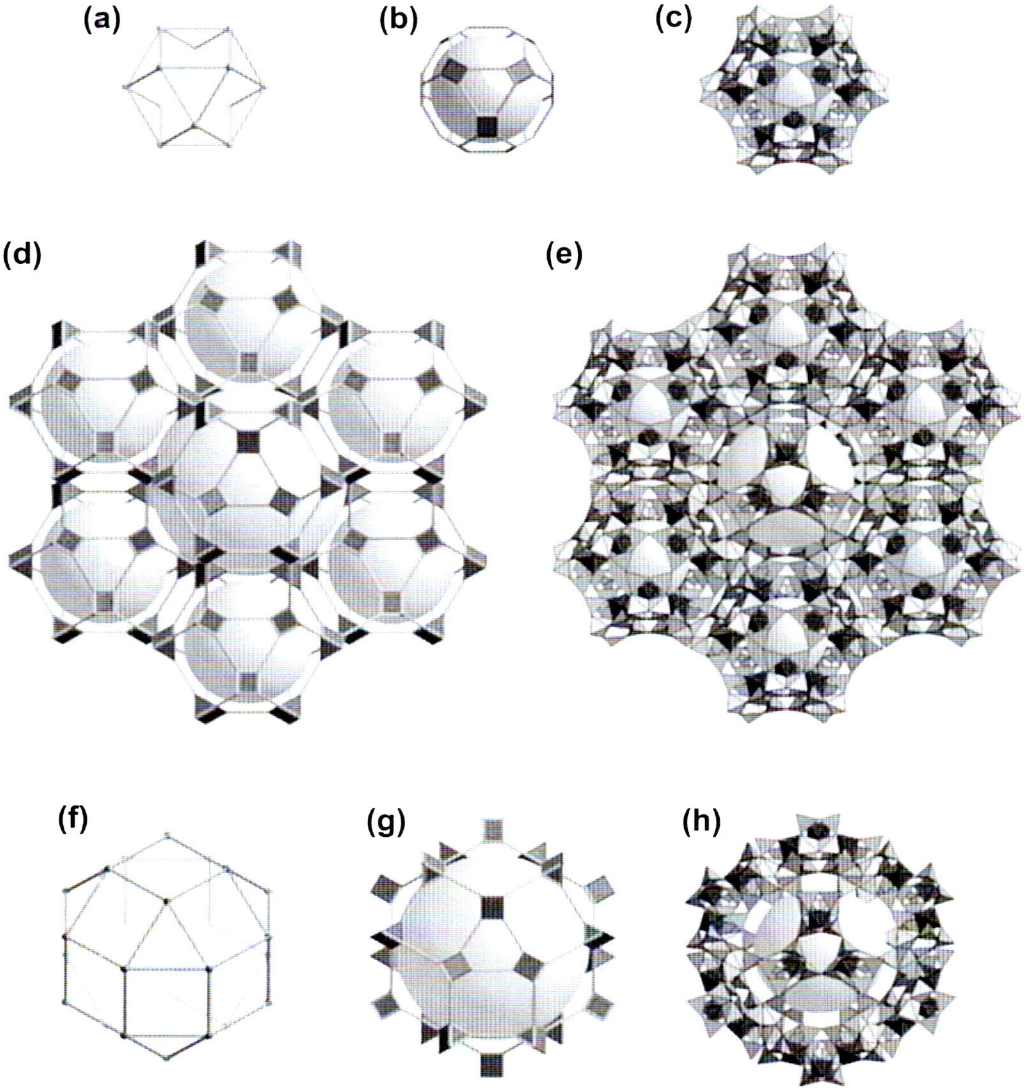

FIGURE 24.9 Abstraction of the $6^8 12^6$ cavity and a hypothetical three-dimensional framework built from the cavity. (a) A cuboctahedron. (b) Augmented cuboctahedron with four-coordinated nodes (gray). (c) The $6^8 12^6$ cavity resulting from the decoration of each node of the cuboctahedron by a four-coordinated Ge7 cluster. (d) Augmented **reo-e** net with six-coordinated nodes (black). (e) A hypothetical three-dimensional framework resulting from the decoration of each node of the **reo-e** net by a six-coordinated Ge7 cluster. (f) The rhombicuboctahedron found in (e). (g) Augmented rhombicuboctahedron with six-coordinated nodes (black). (h) A $6^8 10^{12} 12^6$ cavity resulting from the decoration of each node of the rhombicuboctahedron with a six-coordinated Ge_7 cluster. The light grey and dark grey spheres represent the $6^8 12^6$ and $6^8 10^{12} 12^6$ cavities with free diameters of 7.1 × 17.8 Å, respectively. The germanium-centered octahedra are dark grey, the germanium-centered trigonal bipyramids are light grey, and the germanium-centered tetrahedra are medium grey. The additional germanium-centered tetrahedra are dark grey. *Q. Pan, J. Li, K. Christensen, C. Bonneau, X. Ren, L. Shi, J. Sun, X. Zou, G. Li, J. Yu, R. Xu: A germanate built from $6^8 12^6$ cavity cotemplated by a $(H_2O)_{16}$ cluster and 2-methylpiperazine. Angewandte Chemie International Edition. 2008. Volume 47, Page 7868. Copyright Wiley-VCH Verlag GmbH & Co. KGaA. Reproduced with permission from [30].*

the $6^8 12^6$ cavities in the hypothetical structure is similar to that of the sodalite cages ($4^6 6^8$) in zeolite A, which results in α-cavity ($4^{12} 6^8 8^6$). Furthermore, many extra-large pore germanate structures can be formed by using the $6^8 12^6$ cavity and even the $6^8 10^6 12^{12}$ supercavity as the BUs.

JLG-12 is a crystalline germanate with mesoporous 30-ring channels [31]. The framework is made of strict alternation of Ge_7 and Ge_9 clusters. In the three-dimensional structure of JLG-12, each Ge_7 cluster connects with four neighboring Ge_9 clusters through its four GeO_4 tetrahedra, and each Ge_9 cluster connects with eight neighboring Ge_7 clusters. Such a connection gives rise to a three-dimensional open framework with parallel 30- and 12-ring channels running along the [001] direction (Fig. 24.10a). The free diameters are 13.0 × 21.4 Å for the 30-ring channel, which is

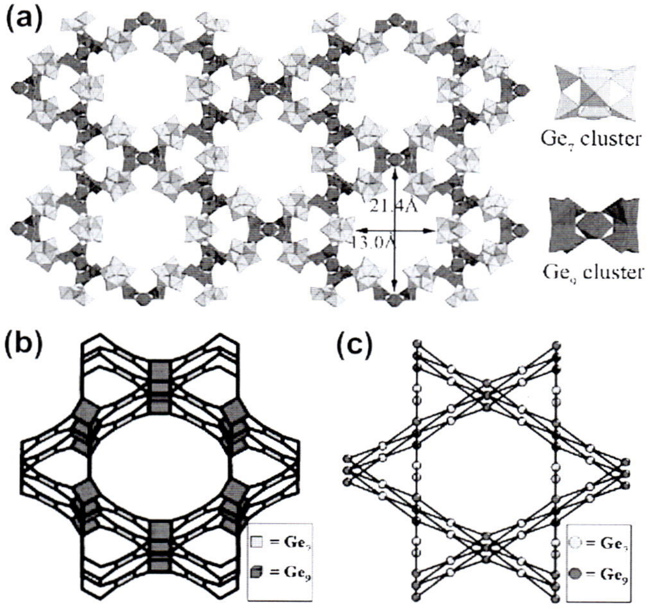

FIGURE 24.10 (a) Framework structure of JLG-12 viewed along the [001] direction. (b, c) The three-dimensional framework of JLG-12 interpreted as a three-dimensional net structure. (b) The **csq-a** net. (c) The **csq** net. The Ge_7 and Ge_9 clusters are presented in light gray and gray, respectively. The partially occupied GeX_2 moieties are not shown for clarity. *Reprinted with permission from [31]. Copyright 2009 American Chemical Society.*

comparable to those of 30-rings (10.0 × 22.4 Å) in SU-M. The framework can be described as a three-dimensional net structure. If all the GeO_4 tetrahedra in the Ge_7 and Ge_9 clusters and the GeO_4X trigonal bipyramids in the Ge_9 clusters are identified as vertices, the framework of JLG-12 can be reduced into a 3,4-heterocoordinated **csq-a** net [14] (Fig. 24.10b). If the whole Ge_7 and Ge_9 clusters are taken as vertices, the framework of JLG-12 can be further reduced into a 4,8-heterocoordinated **csq** net (Fig. 24.10c). Therefore, it is expected that by introducing Ge_7 and Ge_9 clusters into other nets, many more open-framework germanate structures can be predicted. Figure 24.11 shows two hypothetical germanate frameworks generated by introducing Ge_7 and Ge_9 clusters into a 4,8-heterocoordinated **scu** net. The hypothetical framework **scu-1** has 22-ring channels running along the [001] direction, 10-ring channels along the [010] and [110] directions, and 8-ring channels running along the [100] direction (Fig. 24.11a). The 22-ring channel has free diameters of 13.2 × 15.8 Å. Similar to **scu-1**, the framework of **scu-2** has intersecting 22-, 10-, and 8-ring channels (Fig. 24.11b). The 22-ring channel in **scu-2** has free diameters of 11.3 × 16.6 Å. These two hypothetical structures are generated on the basis of the same **scu** net, but the orientations of Ge_7 clusters are different, which leads to different pore geometries and symmetries.

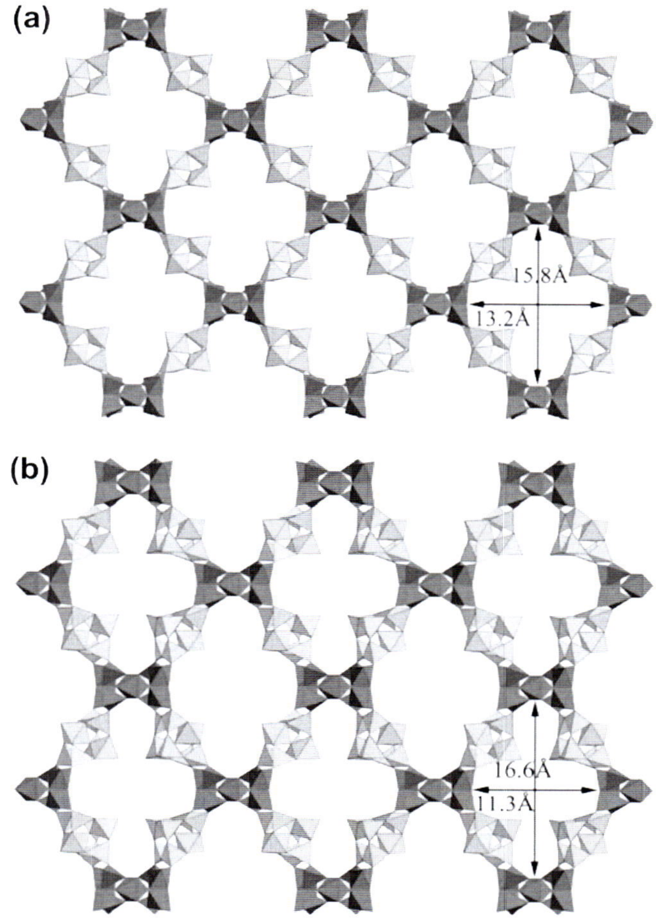

FIGURE 24.11 Hypothetical germanate frameworks. (a) The **scu-1**; (b) the **scu-2** viewed along the [001] direction. *Reprinted with permission from [31]. Copyright 2009 American Chemical Society.*

Scale chemistry is a very general concept which applies all sorts of solids. The structures of metal-organic framework (MOF) compounds are also in agreement with this concept [32,33]. Many complex structures can be described as a simple net through replacing various BUs by the nodes. On the other hand, the length of organic linkers can be scaled up to augment the frameworks without changing the size of BUs.

It has been proved that the concept of BUs as well as scale chemistry is very useful for the design of new topologies. Furthermore, computer simulations for the design of new structures by predefining the BUs have proven to be possible (see the next section), which allows the exploration of the possible ways of assembling predefined BUs. Furthermore, it is also possible to design structures through introducing specific physical properties, e.g., optical, electronic, and magnetic properties within the BUs, therefore to create functional inorganic solid materials.

24.1.2.2. *Automated Assembly of Secondary Building Units (AASBU method)*

Mellot-Draznieks and coworkers developed a computational method for the prediction of inorganic crystal structures through automated assembly of secondary building units (AASBU) [34,35]. It can produce virtual libraries of viable inorganic structures based on predefined secondary building units (SBUs).

The SBUs can be extracted from known inorganic crystal structures. Examples of several inorganic SBUs are shown in Fig. 24.12. Figure 24.12a shows examples of simple SBUs M_xL_y (M: central metal atoms; L: coordinating or "ligand" atoms), including tetrahedra, octahedra, and corner-, edge-, and face-sharing tetrahedra and/or octahedra. Figure 24.12b shows the D4R unit, $M_8B_{12}L_8$ (M: central metal atoms; L: ligand atoms; B: bridging atoms), which is commonly found in a series of inorganic structures. During the simulations, the inorganic SBUs are treated as rigid bodies. The rule that controls the possible assembly of the building units is a force field with some Lennard–Jones expressions which essentially favor the attraction between ligand atoms. This force field favors the formation of L···L "sticky-atom" pairs and allows two building units to assemble through L···L linkages. The Lennard–Jones expression for the energy of interaction between pairs of atoms i and j is defined in Eq. (24.1):

$$E_{ij} = \varepsilon_{ij}[(r_{ij}^{*}r_{ij}^{-1})^{12} - 2(r_{ij}^{*}r_{ij}^{-1})^{6}] \qquad (24.1)$$

There exist several linkages of the SBUs which should be avoided, for example, the $M_i \cdots M_j$ pairs in the assembly of M_xL_y, the $M_i \cdots M_j$ pairs and the $L_i \cdots B_j$ pairs in the assembly of M_xL_yB. To prevent these undesirable linkages, the repulsive potential between some pairs has to be considered to prevent SBUs from overlapping with each other.

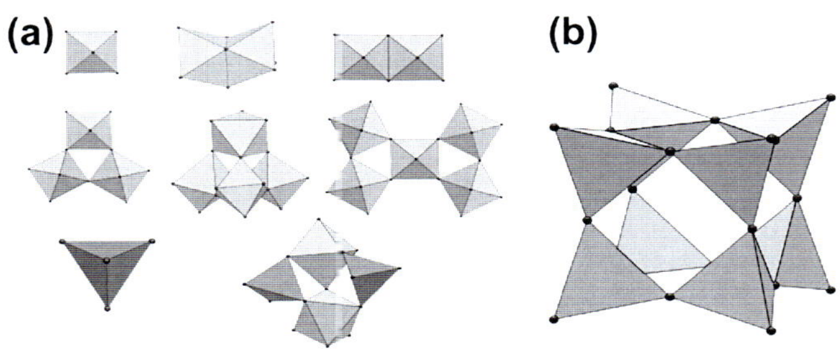

FIGURE 24.12 Examples of building units used in the AASBU method. (a) Simple building units and (b) D4R unit. *C. Mellot-Draznieks, J.M. Newsam, A.M. Gorman, C.M. Freeman, G. Férey: De novo prediction of inorganic structures developed through automated assembly of secondary building units (AASBU method). Angewandte Chemie International Edition. 2000. Volume 39, Page 2270. Copyright Wiley-VCH Verlag GmbH & Co. KGaA. Reproduced with permission from [34].*

The total cost or "energy" of a given arrangement of the building units in a unit cell, E_{total}, is calculated as the sum over the set of building units of the Lennard–Jones terms that involve dissimilar pairs of atoms.

In the case of the assembly of the M_xL_y building units, E_{total} is defined in Eq. (24.2):

$$E_{total} = \sum (E_{L\cdots L} + E_{M\cdots L} + E_{M\cdots M}) \qquad (24.2)$$

In the case of the assembly of the $M_8B_{12}L_8$ (D4R), E_{total} is defined in Eq. (24.3):

$$E_{total} = \sum (E_{L\cdots L} + E_{M\cdots L} + E_{M\cdots M} + E_{L\cdots B}) \qquad (24.3)$$

The magnitude of this cost function provides an estimate of the degree of connectivity of a given arrangement of SBUs. The weight of each term is given by the depth of Lennard–Jones potential well (ε_{ij}).

SIMULATION STEPS FOR THE GENERATION OF CANDIDATE STRUCTURES

The candidate structures are generated by a cascade of simulations that includes the following five major steps:

(1) Periodic trial arrangements of SBUs are randomly generated by a simulated annealing procedure. The angular degrees of freedom for each SBU are sampled by a Metropolis Monte Carlo algorithm. For each step, both the cell parameters and the distances between the SBUs are adjusted to relieve close interatomic contacts, leading to successive phases of cell contractions and expansions. As the simulation proceeds, configurations of lower cost are stored.

(2) The trials arrangements are reduced to a set of unique arrangements corresponding to unique local minima. The redundant arrangements of SBUs are removed.

(3) The resulting set of configurations is minimized with respect to the "cost" function (energy).

(4) After the structure optimization, the redundant arrangements are eliminated. A list of possible SBUs configuration is ranked by cost or degree of connectivity.

(5) For each configuration of the SBUs, the pairs of sticky atoms at very short separation distances are merged into single atoms to form realistic crystal structures. The symmetry of each simulated structure is then re-determined automatically by the Find_Symmetry algorithm.

By using this method, candidate structures can be successfully generated during the simulations.

Taking ML_4 as the building unit, one or two SBUs are used per asymmetric unit with a selected set of space

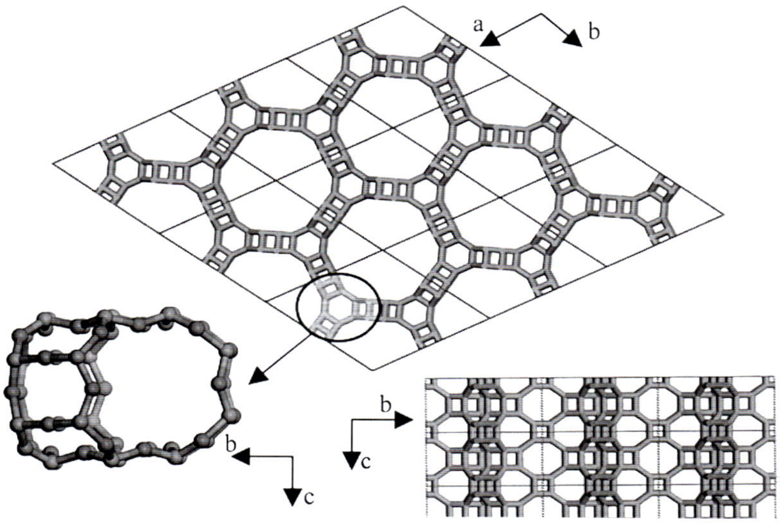

FIGURE 24.13 A hypothetical structure with 24-ring channels predicted in space group *P6/mmm*. *C. Mellot-Draznieks, J.M. Newsam, A.M. Gorman, C.M. Freeman, G. Férey: De novo prediction of inorganic structures developed through automated assembly of secondary building units (AASBU method). Angewandte Chemie International Edition. 2000. Volume 39, Page 2270. Copyright Wiley-VCH Verlag GmbH & Co. KGaA. Reproduced with permission [34].*

groups, the known structure types are generated, including GME, FAU, RHO, and LTL [34]. New hypothetical frameworks are also predicted. Figure 24.13 shows a hypothetical structure found in space group *P6/mmm*. It contains gmelinite cages and has channels running along the [001] direction surrounded by 24 tetrahedra with free aperture dimensions of 17.2 × 19.4 Å. Each 24-ring channel connects with its six neighbors through gmelinite cages and six double 8-rings. The unit cell content is $Si_{48}O_{96}$, and the cell parameters are $a = 24.73$ Å, $c = 10.58$ Å.

Taking D4R as the building units, their simulated assemblies include the known zeotypes ACO, AFY, and LTA. The ACO topology can be obtained in various space groups: *P*1, *P*-1, *C*2, *C*2/*c*, *P*2$_1$, *P*2$_1$/c, *C*222$_1$, and *Pna*2$_1$ [35]. Its tilted derivative can also be generated in other space groups: *P*2, *P*2$_1$, *C*2, *C*c, *P*2$_1$/*c*, *P*222, *P*2$_1$2$_1$2$_1$, and *Pna*2$_1$. The final symmetry after lattice energy minimizations is *Im*-3*m* for the ACO topology in agreement with the experimentally known structure and *P*4/*mnc* for the tilted topology. The AFY and LTA topologies can also be generated in *P*-1 and *R*3 space group, respectively, with a final symmetry of *P*-31*m* for AFY and *Pm*-3*m* for LTA zeotype in its SiO_2 form.

Except for the known zeolite frameworks mentioned above, a number of hypothetical structures can be also generated during the simulations (structures T1–T10), which are shown in Fig. 24.14. Each structure possesses a three-dimensional framework containing cages or channels. Their relative stabilities are estimated in terms of the lattice energy at constant pressure and expressed in kJ mol^{-1}/TO_4 and are compared with known zeolites in the SiO_2, $AlPO_4$, and $GaPO_4$ forms, respectively. The results are given and classified as decreasing relative stabilities:

For SiO_2 phases:

SiO_2 > ACO > LTA > VFI > T3 > T1 > T2* > T2 > T6 > T8 > AFY > T4 > T9 > T10 > T5;

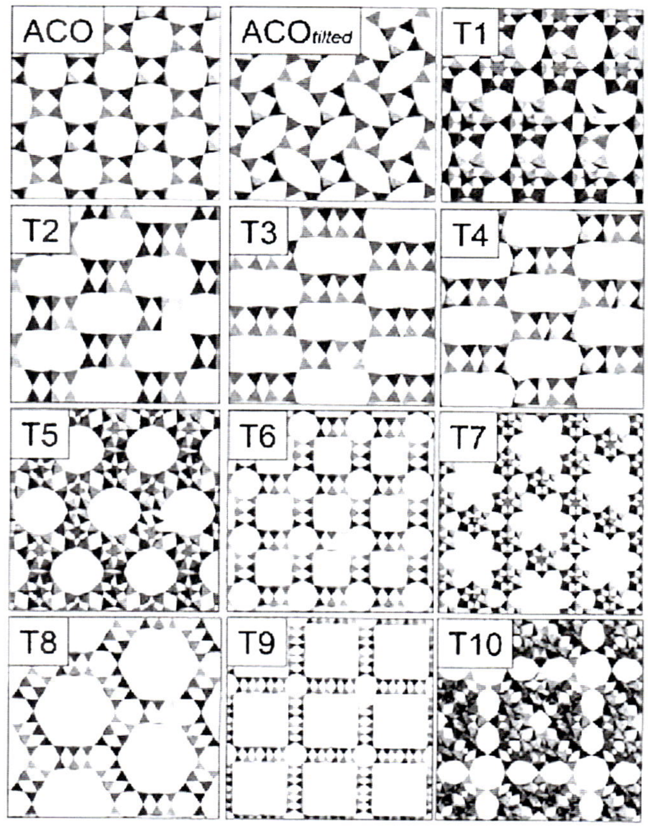

FIGURE 24.14 Hypothetical structures (T1–T10) generated using the D4R as SBU. The known zeolite ACO structure and its titled structure are also predicted. *Reprinted with permission from [35]. Copyright 2002 American Chemical Society.*

For $AlPO_4$ phases:

$AlPO_4$ > VFI > ACO > LTA > T1 > T3 > T2* > T2 > T6 > T8 > AFY > T4 > T9 > T10 > T5;

For $GaPO_4$ phases:

$GaPO_4$ > T3 > T4 > VFI > ACO > T2 > LTA > T1 > T5 > T6 > T8 > T2* > AFY > T9 > T10.

It is believed that the structures whose relative energy is intermediate between LTA and AFY can be considered as "*not-yet-synthesized structures*", while those that are less stable than AFY remain "*virtual*".

AASBU method provides an effective way for the structural design on the basis of predetermined SBUs.

24.1.2.3. *Prediction of Open-framework Aluminophosphate Structures by Using the AASBU Method With Lowenstein's Constraints*

On the basis of the above-described AASBU approach, Li et al. [36] developed an approach to generate hypothetical open-framework aluminophosphate structures with specified Al/P stoichiometry. Open-framework aluminophosphates are constructed from strict alternation of AlO_n ($n = 4, 5, 6$) polyhedra and $P(O{-}Al)_nO_{4-n}$ ($n = 1, 2, 3, 4$) tetrahedra via bridging oxygen atoms, which show rich structural architectures with diverse stoichiometries. In this approach, the additional constraints based on the Lowenstein's rule are introduced, i.e., no Al—O—Al or P—O—P linkages are allowed. The AlO_n ($n = 4, 5, 6$) polyhedra and $P(O{-}Al)_nO_{4-n}$ ($n = 1, 2, 3, 4$) tetrahedra, together with the clusters constructed by them, can be selected as the BUs for the structure generation. To satisfy the Lowenstein's rule, the force field that controls the assembly of the building units is parameterized to favor the formation of Al—O—P linkages while avoiding the formation of Al—O—Al or P—O—P linkages.

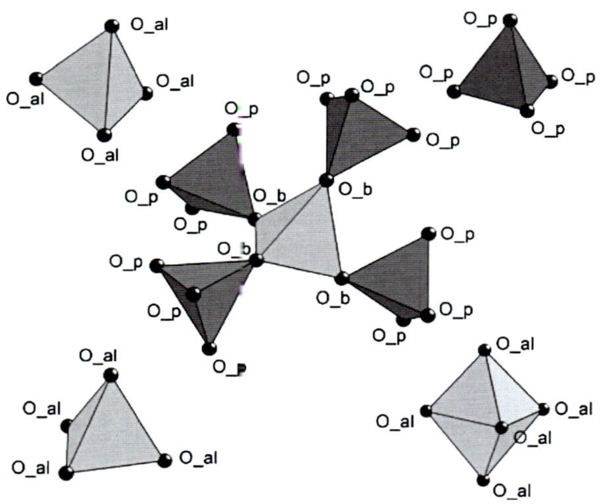

FIGURE 24.15 BUs for the generation of aluminophosphate structures with six $Al(O_b)_4$, two $Al(O_b)_5$, four $P(O_b)_4$, and six $P(O_b)_3(O_t)$ atoms in a unit cell. Only half of the required numbers of building units are needed because of the symmetric operations under space group P1. *Reprinted with permission from [36]. Copyright 2005 American Chemical Society.*

Through alternation of different Al and P coordinations in the lattice, various stoichiometries can be generated. For each specified Al/P stoichiometry, all the possible combinations of Al and P atoms with different coordination states can be calculated according to the Lowenstein's rule. Two criteria must be satisfied during the calculation of the possible combinations for a specified stoichiometry:

(1) The sum of the numbers of each kind of Al atoms must be equal to the number of Al atoms in the unit cell, similarly, the sum of the numbers of each kind of P atoms must be equal to the number of P atoms in the unit cell;
(2) The number of $Al{-}O_b$ bonds must be equal to the number of $P{-}O_b$ bonds.

These two rules can also be summarized in Eq. (24.4):

$$\sum_{i=1}^{I} m_i = \sum_{j=1}^{J} n_j \tag{24.4}$$

where I (J) is the number of Al (P) atoms in the unit cell; m_i (n_j) is the number of O_b atoms connected with the i-th (j-th) Al (P) atom.

All the possible combinations of different kinds of Al and P atoms can be calculated for a specified stoichiometry. Here the prediction of open-framework aluminophosphates with $Al_8P_{10}O_{40}$ stoichiometry is taken as an example. For instance, 6 $Al(O_b)_4$, 2 $Al(O_b)_5$, 4 $P(O_b)_4$, and 6 $P(O_b)_3(O_t)$ atoms are one of the combinations for aluminophosphates with $Al_8P_{10}O_{40}$ stoichiometry. Assuming a P-1 space group, only half of the atoms, i.e., 3 $Al(O_b)_4$, 1 $Al(O_b)_5$, 2 $P(O_b)_4$, and 3 $P(O_b)_3(O_t)$ atoms, are needed to be introduced in the asymmetric unit. All the BUs used are shown in Fig. 24.15 together with the labels of the force field atom types for the oxygen atoms. The simulation is carried out in a unit cell with $a = b = c = 10$ Å and $\alpha = \beta = \gamma = 90°$.

The generation of hypothetical structure H-1 is taken as an example to illustrate the general assembly procedure. First, the BUs are randomly introduced in the unit cell (Fig. 24.16a); then, they are assembled through the attraction between the O_al and O_p atoms under the control of the force field including a Lenard—Jones expression and a repulsive exponential expression (Fig. 24.16b). As the simulation proceeds, the BUs are assembled together and the "sticky-atom" pairs are formed (Fig. 24.16c). At the end of the simulation, "sticky-atom" pairs are glued

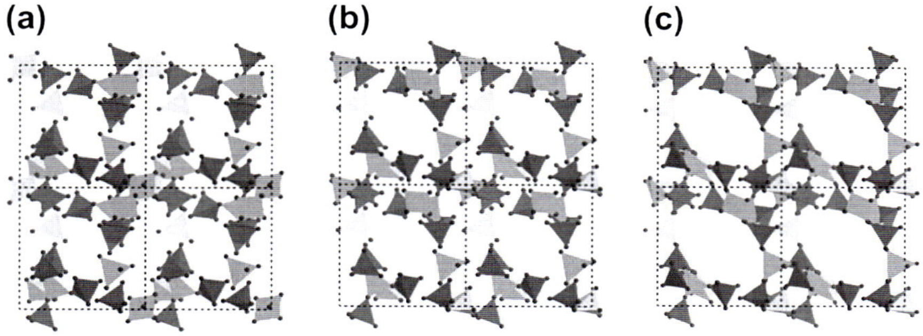

FIGURE 24.16 Generation process of H-1 using the AASBU simulation. (a) The initial BUs are randomly generated in the unit cell; (b) the BUs are assembled together through the attraction of O_al and O_p atoms; (c) the "sticky-atom" pairs are "glued" together to form the structure. The grey polyhedra denote the Al-centered polyhedra, the light grey tetrahedra denote the PO4 tetrahedra, and the black balls denote the oxygen atoms. *Reprinted with permission from [36]. Copyright 2005 American Chemical Society.*

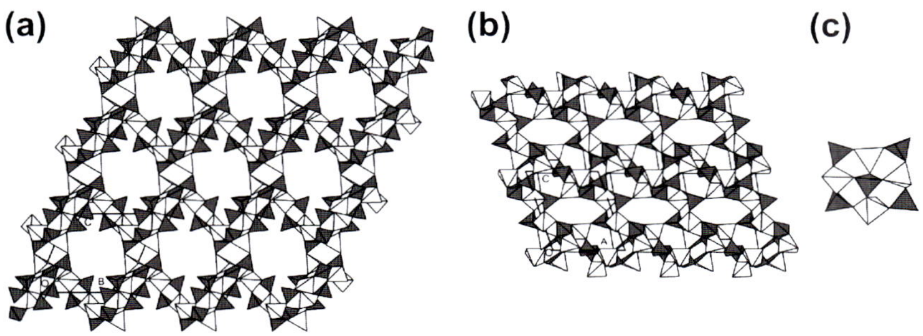

FIGURE 24.17 Hypothetical aluminophosphate structure H-1. It has a distorted 16-ring channel along the [100] direction (a) and an 8-MR channel along the [010] direction (b). This structure could also be seen as constructed by the Al_4P_5 BUs (c). The light gray polyhedra denote the Al-centered polyhedra; the gray tetrahedra denote the P-centered tetrahedra. *Reprinted with permission from [36]. Copyright 2005 American Chemical Society.*

into single O_b atoms, and the hypothetical structure H-1 is formed. The structure of H-1 is further refined using Burchart force field. As shown in Fig. 24.16, H1 has a three-dimensional open framework. Along the [100] direction, there are distorted 16-ring channels inside which pairs of terminal O atoms protrude (Fig. 24.17a). There also exist 8-ring channels running along the [010] direction (Fig. 24.17b). This structure can also be viewed as exclusively constructed by the Al_4P_5 BUs (Fig. 24.17c).

This method not only aids the prediction of hypothetical aluminophosphate frameworks, but also will serve as a useful tool to set up the initial structural models for the solution of unknown aluminophosphate structures.

As demonstrated above, computational simulation has proven to be a powerful tool for designing inorganic porous structures with defined pore geometries, building units, and compositions. Further accessing these novel and not-yet-discovered structures might be achieved by innovative synthetic strategies with the aid of the rational template design.

24.2. CHEMICAL AND PHYSICAL PROBLEMS ON WAY TO THE RATIONAL SYNTHESIS OF INORGANIC CRYSTALLINE POROUS MATERIALS

A typical crystallization procedure of zeolitic inorganic porous crystalline materials involves the use of inorganic cations or organic templates, also called structure-directing agents (SDAs) under hydrothermal or solvothermal conditions. Although zeolite synthesis has been extensively studied during the last six decades, the rational synthesis is still far behind the empirical trial-and-error synthesis. This identifies the challenges and problems that remain in this area. Zeolite crystallization represents one of the most complex chemical problems in crystallization phenomena such as polymerization−depolymerization, solution−gelation, and nucleation−crystallization. It involves complex reactions and equilibria between solid and solution components during the crystallization process. Elucidating the formation pathway at the molecular level has been one of the major challenges in porous materials science [37−41]. However, until now, there is no clear understanding of the formation mechanism of zeolites by which the starting amorphous gels are converted to the porous crystalline solids in the presence of the templates or SDAs. Especially, there is no understanding on how to control the self-assembly of reaction species that determines the formation of a specific zeolitic porous structure under a given set of synthetic conditions. This significantly limits the synthesis of zeolite and related inorganic porous crystalline materials in a predictable way.

The self-assembly of the crystalline open-framework structures with the template or the SDA molecules is a complex chemical process, involving changes over the small (primary units), to medium (nuclei), to large

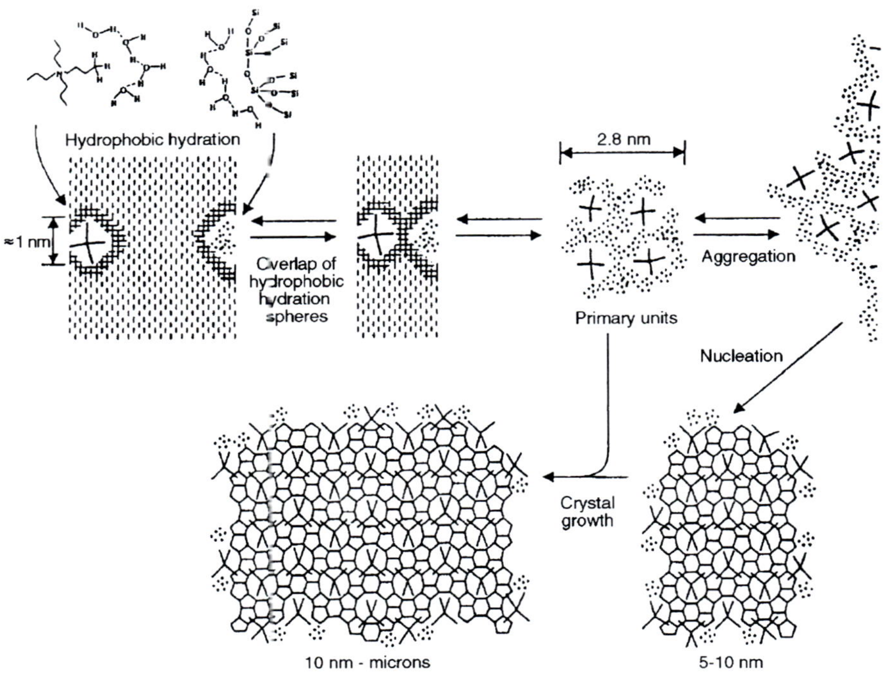

FIGURE 24.18 Schematic illustration of the crystallization mechanism of Si-TPA-MFI. *P.P.E. A. de Moor, T.P.M. Beelen, B.U. Komanschek, L.W. Beck, P. Wagner, M.E. Davis, R.A. van Santen: Imaging the Assembly Process of the Organic-Mediated Synthesis of a Zeolite. Chemistry — A European Journal. 1999. Volume 5, Page 2083. Copyright Wiley-VCH Verlag GmbH & Co. KGaA. Reproduced with permission [42].*

scale (crystallites). Figure 24.18 is a schematic illustration of the assembly process of Si-TPA-MFI (TPA: Tetrapropylammonium) proposed by de Moor et al. [42]. It shows that 2.8 nm entities comprising TPA and silicate as primary building units aggregate to form structures up to 10 nm in size which in turn form viable nuclei that initiate the growth of zeolite crystals. The nucleation mechanism of TPA-MFI thus involves ordering initially on the nanometer scale followed by order at larger scale derived from the primary building units that are specific for the crystal structure formed.

However, unlike the organic synthesis, there are no defined reactions in such complex inorganic—inorganic and organic—inorganic system mediated by water or nonaqueous solvent (e.g., alcohol). In addition, even though some characteristic BUs have been commonly used to describe the zeolite structures, they may not exist as the real ones during their formation process. This is significantly different from the chemistry of metal-organic framework (MOF) materials where the concept of "Reticular Synthesis" concerning the molecular building blocks can be well applied to the designed synthesis. Currently, there is lack of advanced characterization techniques that can detect the reaction species during the crystallization process. Time-resolved in situ experimentation following the changes that occur with time at the molecular level represents a significant physical problem towards the rational synthesis of zeolitic inorganic porous crystalline materials as well.

So far, a variety of analytical techniques such as advanced nuclear magnetic resonance (NMR) spectroscopy, high-resolution electron microscopy (HRTEM), atomic force microscopy (AFM), small-angle X-ray and neutron scattering (SAXS and SANS), and electrospray ionization mass spectrometry (ESI-MS), as well as combinations of them with computer simulations, have been used to study the formation of zeolitic materials. However, none of these techniques could provide a clear picture of detailed chemical steps occurring during the crystallization of a zeolite. Despite considerable research efforts towards understanding the complex synthetic system, the proposed mechanistic models are still unclear at the molecular level and are even debated. To finely control the chemical steps towards the formation of a given zeolitic inorganic porous structure is far beyond our current capabilities. A better understanding of the formation mechanism is needed in order to allow the synthesis of zeolitic inorganic porous crystalline materials in a rational way. Knowledge gained from the studies will undoubtedly guide the designed synthesis of such materials and further contribute to the success of this field.

24.3. ATTEMPTS TO THE RATIONAL SYNTHESIS OF INORGANIC POROUS CRYSTALLINE MATERIALS

One of the great challenges in the synthetic chemistry of the inorganic porous crystalline materials today is the design and synthesis of novel inorganic porous crystalline solids possessing desired structures and properties. Although there exist difficult chemical and physical problems associated with the understanding of the crystallization mechanism of zeolitic inorganic porous crystalline materials, many research efforts have been made to the tailor-made of such materials with targeted porous structures. The database banks of zeolitic materials will help to increase the knowledge and understanding of the relationship of synthesis and structure of such materials. Especially, computational methods have greatly assisted in targeting zeolites with specific structures and properties. The current attempts toward the designed synthesis of zeolitic inorganic porous crystalline materials shall be presented which include: (i) synthesis guided by the structure-directing effect of SDAs, (ii) synthesis guided by substituent element effects, and (iii) synthesis guided by data mining. By means of the new synthetic approaches, zeolitic inorganic porous crystalline materials with specific structures are able to be accessed in a more rational way.

24.3.1. Synthesis Guided by the Structure-directing Effect of Template

In the synthesis of zeolites and related inorganic porous crystalline materials, the organic templates or structure-directing agents (SDAs) play an important role in the assembly of inorganic species to highly ordered open frameworks [43]. Although a direct analogy between the SDA and the resultant channel space cannot be clearly established, there is ample evidence that the SDA does have an influence on defining the channel geometry. The host–guest chemistry has given some direction to the template synthesis of zeolite materials. The synthesis of new porous framework structures by rational design and selection of template molecules has become highly feasible.

A remarkable example is the synthesis of pure silica zeolite LTA (ITQ-29) by Corma et al. [44] by using a supramolecular organic SDA obtained via the self-assembly of two identical organic cationic moieties through $\pi-\pi$ stacking interactions. The selection of the SDAs to synthesize high-silica LTA is based on the following criteria:

(i) The bulky organic SDA should fit the inner surface of the α-cage of the LTA structure with as many van der Waals contacts as possible but with the least deformation; and

(ii) The SDA should have a weak tendency to form complex with the solvent, besides maintaining adequate hydrophobicity.

However, it is difficult to choose a bulky SDA to fulfill the above criteria. Therefore, Corma et al. started from the idea that a very large SDA could be achieved by supramolecular self-assembly of two moieties through $\pi-\pi$ stacking interactions, each with proper rigidity and polarity. The 4-methyl-2,3,6,7-tetrahydro-1*H*,5*H*-pyrido[3.2.1-ij]quinolinium iodide was then selected that could form self-assembled dimers in aqueous solution. Figure 24.19 shows the formation of the LTA structure from the supramolecular self-assembly of the organic SDA molecules.

Significantly, computer modeling has proven to be useful for examining the structure-directing effects of known molecules as well as predicting promising SDAs for a given structure in terms of the host–guest interaction.

24.3.1.1. De Novo Template Design by Computational "Growth" of the Template Molecules

The empirical evidence is that a successful template must be a "good fit" with the host framework. Lewis et al. [45] developed a method for de novo design of

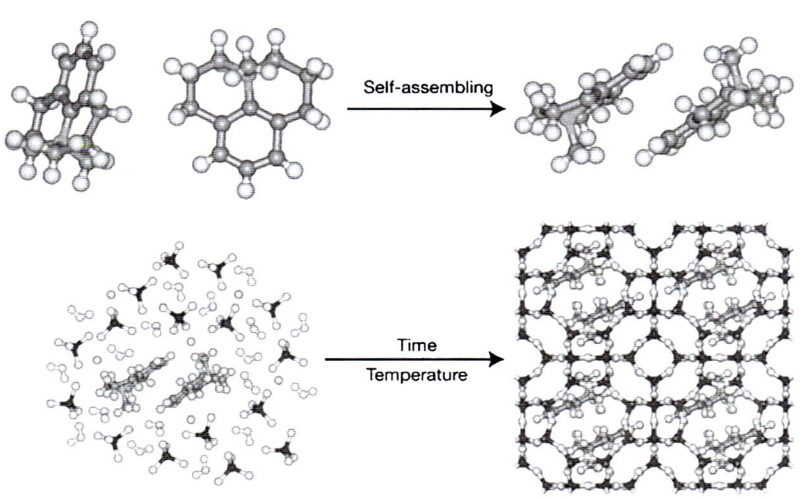

FIGURE 24.19 Formation of the LTA structure from the supramolecular self-assembling of the organic SDA molecules. *Reprinted with permission from Macmillan Publishers Ltd: Nature [44], Copyright (2004).*

SDA molecules, which can be computationally "grown" within the confines of the pore system for a specific microporous structure. This method demonstrates its potential in developing new SDAs for different known types of zeolite topologies, as well as for new zeolitic materials with desired framework structures.

In this method, the growth of the template is initiated from a seed molecule placed in the void space of the host. The molecule then grows by a number of random actions using the fragment library as the source of new atoms including (i) build, (ii) rotate, (iii) shake, (iv) rock, (v) bond twist, (vi) ring formation, and (vii) energy minimization. A successful template, on the whole, must effectively fill the void space of the framework. A cost function based on overlap of van der Waals spheres is used to control the development of new templates:

$$f_c = \sum_t C(tz)/n\,, \qquad (24.5)$$

where $C(tz)$ signifies the closest contact between a template atom t and any host atom z and n is the number of atoms in the template. During actions (ii)–(vi), movements of the template are made as a series of small steps, the action only continues if f_c has increased. Such a strategy allows the template to locate rapidly the largest available space in the host, and to achieve optimum geometric complementarity with the host.

This method demonstrates its potential in developing new templates on different types of zeolite topologies, such as LEV, MFI, EU-1, and CHA. Figure 24.20 illustrates the growth process involved in the formation of a candidate template, 1,2-dimethylcyclohexane in the LEV structure starting from a methane seed. It shows a high binding energy with the host framework. Experimentally, using the amino analogue of this template, 2-methylcyclohexylamine, as a structure-directing agent, a cobalt aluminophosphate DAF-4 with zeolite LEV structure type has been synthesized [46].

24.3.1.2. Predicting the Templates in Terms of the Energies of the Host–template Interactions

Suitable organic templates for the formation of a specific porous structure can be predicted in terms of the host–guest interaction energies. Lewis et al. [47] investigated the ability of organic molecules in templating the microporous materials. They showed that the efficacy of a template can be rationalized in terms of the energies of the host–template interactions. Now some examples to demonstrate how the template design can be useful in guiding the rational synthesis shall be presented.

24.3.1.2.1. SYNTHESIS OF ALUMINOPHOSPHATES WITH AN OPEN FRAMEWORK ANALOGOUS TO $[C_6H_{18}N_2]$ $[AL_4P_5O_{19}(OH)]$ (ALPO-HDA)

Li et al. [48] developed a computational method for the prediction of the templating ability of organic amines in the formation of a family of two-dimensional-layered compounds. Their two-dimensional sheets are stabilized by protonated organic amines through H-bonding interactions. The templating ability of various organic amines for the known inorganic layers has been investigated and well explained in terms of the energies of the host–template interactions ($E_{inter} = E - E_f - E_t$, where E is the total energy of the whole structure, E_f is the energy of the framework, and E_t is the energy of the encapsulated organic template itself). The Burchart 1.01–Dreiding 2.21 force field in the Cerius^2 package was chosen for energy optimization and calculation. Furthermore, some organic templates that can potentially direct the formation of a given host can be successfully predicted. This method was further applied to three-dimensional open-framework aluminophosphate system to investigate the templating ability of organic amines [49]. The addition of template to the channel is attained by using Monte Carlo method.

$[C_6H_{18}N_2][Al_4P_5O_{19}(OH)]$ (AlPO-HDA) has an interrupted open-framework structure with interconnected 12- and 8-ring channels along [010] and [100] directions, respectively [50]. The framework is made up of

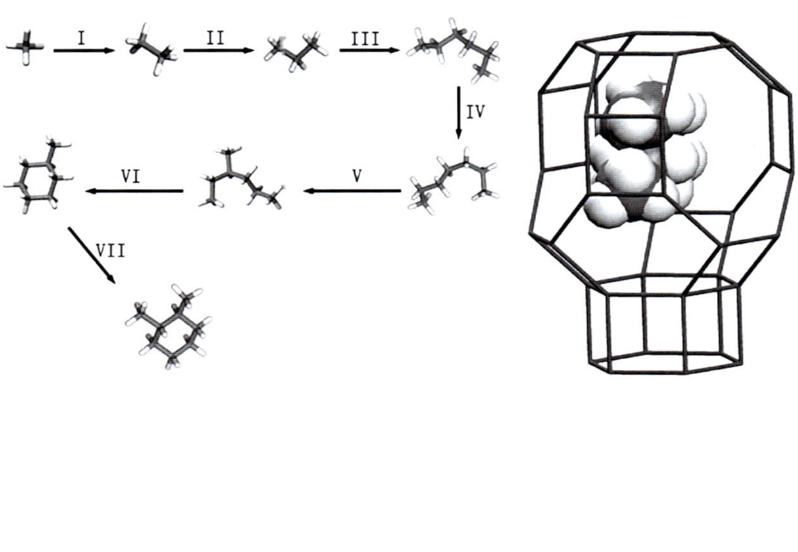

FIGURE 24.20 (a) The growth processes involved in the formation of the substituted cyclohexane derivative in the LEV structure. Starting from a methane seed, the alkyl chain grows inside the restricted void of the LEV structure (I–V). High bias of the ring formation action results in the formation of methylcyclohexane in step VI, when an atom with a fifth-order neighbor within a distance of 3 Å is selected. Further substitution of this ring then occurs; and (b) the calculated position of the template in the *lev* cage [45].

alternating Al units (AlO_4 and AlO_5) and P units (PO_4 and $PO_2(=O)(OH)$). The template molecule, diprotonated 1,6-hexanediamine (HDA), is located in the main 12-ring channels and interacts with the host framework through H-bonds as shown in Fig. 24.21a. Figure 24.21b shows an isolated HDA molecule at its lowest energy state. It can be seen clearly that the conformations of the isolated HDA molecule and the encapsulated HDA molecule in the channel are quite different. The energy of the encapsulated HDA (14.03 kcal/mol) is much higher than that of the isolated HDA (−1.09 kcal/mol). However, the configuration of the encapsulated HDA ensures the lowest host−guest interaction energy.

Theoretical calculations show that the experimental and optimized interaction energies E_{inter} between the framework and template molecules in AlPO-HDA are −17.83 and −18.34 kcal/mol (per unit of $[Al_4P_5O_{19}(OH)]^{2-}$, respectively. In order to find suitable templates that can potentially direct the formation of framework of AlPO-HDA, some typical organic diamines and polyamines, such as 1,5-pentanediamine, 1,4-butanediamine, 1,3-propanediamine, diethylenetriamine, and triethylenetetramine, were placed in the 12-ring channel followed by energy optimization calculations. The optimized interaction energies of the host framework and the theoretical templates are presented in Table 24.1. Some suitable templates that have lower interaction energies with the host framework could be predicted, for example, 1,5-pentanediamine and diethylenetriamine exhibit the lowest host−guest interaction energies of −21.40 and −21.30 kcal/mol, respectively. Experimentally, the aluminophosphates with open-framework structures analogous to AlPO-HDA were successfully synthesized by using 1,5-pentanediamine and diethylenetriamine as the templates. The positions of the templates in the crystal structures are in good agreement with the theoretical prediction, as shown in Fig. 24.22. On the other hand, when organic amines that have higher interaction energies, for example, $H_2N(CH_2)_8NH_2$ (No. 11) and $H_2N(CH_2)_2NH_2$ (No. 8) with host−guest interaction energies of 0.88 and −12.05 kcal/mol, respectively, were used, no open frameworks analogous to AlPO-HDA could be synthesized.

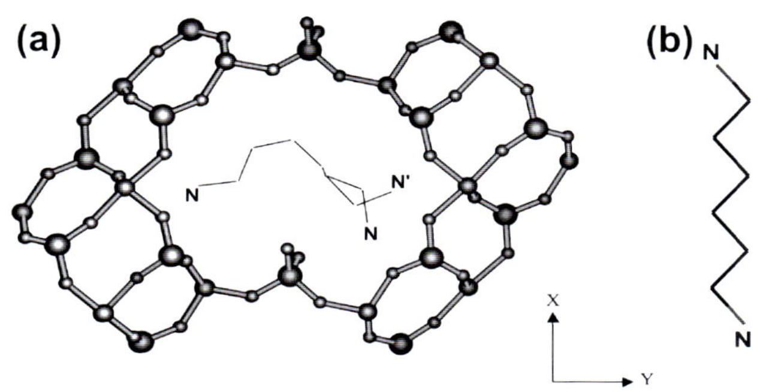

FIGURE 24.21 Open-framework of AlPO-HDA with diprotonated 1,6-hexanediammine (1,6-HDA) molecules residing in the 12-ring channels, and the isolated 1,6-HDA molecule at its lowest energy state. *Reprinted with permission from [49]. Copyright 2000 American Chemical Society.*

TABLE 24.1 Interaction Energies (E_{inter}) of Template−host Per Unit of $[Al_4P_5O_{20}H]^{2-}$ (kcal/mol)

no.	templates[a]	E_{VDW}	$E_{H\text{-}bond}$	E_{inter}
1	$H_2N(CH_2)_6NH_2$	−18.34	0.00	−18.34
2	$HNC_6H_{10}NH$	−16.21	−2.12	−18.33
3	$H_2NCH\ C_6H_{10}CHNH_2$	−12.94	0.00	−12.94
4	$H_2N(CH_2)_5NH_2$	−19.03	−2.37	−21.40
5	$H_2N(CH_2)_4NH_2$	−15.67	−3.48	−19.15
6	$H_2N(CH_2)_7NH_2$	−15.21	−0.17	−15.39
7	$H_2N(CH_2)_3NH_2$	−11.94	−2.31	−14.26
8	$H_2N(CH_2)_2NH_2$	−8.72	−3.33	−12.05
9	TETA	−15.99	−2.61	−18.60
10	DETA	−19.61	−1.69	−21.30
11	$H_2N(CH_2)_8NH_2$	1.16	−0.28	0.88
12	$H_2NCH(CH_3)CH_2NH_2$	−10.78	−3.74	−14.52
13	$H_2NCH_2CH(CH_3)CH_2NH_2$	−11.82	−1.83	−13.36

[a]*TETA: $H_2N(CH_2)_2NH(CH_2)_2NH(CH_2)_2NH_2$; DETA: $H_2N(CH_2)_2NH(CH_2)_2NH_2$.*
Note: The Coulombic interactions between the framework and the template are ignored in the calculation. Only the low-energy nonbonding host−guest interactions including van der Waals (VDW) and H-bonding interactions are considered here.
Reprinted with permission from [49]. Copyright 2000 American Chemical Society.

24.3.1.2.2. SYNTHESIS OF MICROPOROUS ALUMINOPHOSPHATE ALPO$_4$-21 BY COMBINATORIAL APPROACH

The suitable template molecules can be predicted by computer simulations. However, it must also be noted that gel chemistry is also crucial, and that the templating action will take effect only when the right gel environment is achieved. The methodology of combinatorial chemistry plays a significant role in the discovery of new materials [51]. Combinatorial approach has been successfully applied to the hydrothermal or solvothermal synthesis of zeolite and related inorganic porous crystalline materials [52−56]. By this approach an extensive study of complex multi-component system can be easily achieved. Song et al. [57] demonstrated a powerful strategy for the rational synthesis of microporous materials through combination of computational simulation approach and combinatorial approach.

This strategy has been demonstrated by using the synthesis of aluminophosphate molecular sieve $AlPO_4$-21

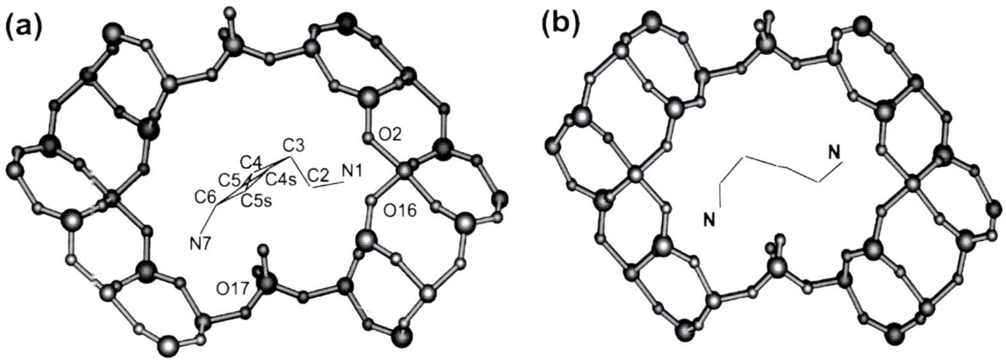

FIGURE 24.22 (a) Experimental 1,5-pentanediamine located in the 12-membered ring of AlPO-HDA and (b) predicted atomic positions of 1,5-pentanediamine inside the 12-membered ring (H-bonds are indicated by dashed lines). *Reprinted with permission from [49]. Copyright 2000 American Chemical Society.*

as an example. The framework of $AlPO_4$-21 (AWO zeotype) consists of 8-ring channels along the [001] direction (Fig. 24.23). It is made up of the alternation of Al units (including AlO_4 and AlO_5) and PO_4 tetrahedra. The templating abilities of various organic amines in the formation of $AlPO_4$-21 were evaluated in terms of nonbonding interaction energies of the host−guest calculated by molecular simulations. Table 24.2 shows the calculated results. It shows that organic amines such as dimethylamine, ethylenediamine, *n*-propylamine, ethylamine, and pyrrolidine have similar interaction energies (−12.55 to −13.66 kcal/mol per unit of $Al_3P_3O_{13}H$)) with $AlPO_4$-21 framework, indicating their similar templating ability in the formation of $AlPO_4$-21. In order to find other suitable templates to direct the formation of $AlPO_4$-21 framework, the templating abilities of several organic amines that have not been used for the synthesis of $AlPO_4$-21 earlier were investigated. Among them, trimethylamine and ethanolamine show lower interaction energies of −13.55 and −13.48 kcal/mol, respectively, which are comparable to those of experimental templates discussed above. Therefore they were predicted as the potential templates for the synthesis of $AlPO_4$-21.

Through selection of the predicted suitable templates, e.g., ethanolamine, ethylamine, dimethylamine, and *n*-propylamine, etc., $AlPO_4$-21 were successfully synthesized by hydrothermal combinatorial approach in the reaction system with molar composition $1.0Al(^iOPr)_3$:xH_3PO_4:yR:$255.0H_2O$ (R: amines) at 180 °C for 5 days. Table 24.3 gives the detailed synthetic conditions. It was noticed that the use of the predicted unsuitable organic amine *N,N,N′,N′*-tetramethylethylenediamine that has

FIGURE 24.23 The framework of $AlPO_4$-21 (AWO zeotype) consisting of 8-ring channels along the [001] direction (gray for AlO_4, light gray for AlO_5, and black for PO_4).

TABLE 24.2 Interaction Energies (E_{inter}) of Template−host Per Unit of $[Al_3P_3O_{13}H]^-$ (kcal/mol)

No.	Templates (R)	E_{VDW}	$E_{H\text{-}bond}$	E_{inter}
1	**Dimethylamine**	**−13.16**	**0.00**	**−13.16**
2	**Ethylamine**	**−13.50**	**−0.16**	**−13.66**
3	***n*-Propylamine**	**−12.87**	**−0.09**	**−12.96**
4	**Ethylenediamine**	**−9.39**	**−3.16**	**−12.55**
5	**Pyrrolidine**	**−13.61**	**0.00**	**−13.61**
6	***N,N,N″,N′*-tetramethyl-1,3-propanediamine**	**−8.00**	**0.00**	**−8.00**
7	*Ethanolamine*	*−13.48*	*0.00*	*−13.48*
8	*Trimethylamine*	*−13.55*	*0.00*	*−13.55*
9	Methylamine	−9.55	−0.25	−9.80
10	*N,N,N′,N′*-Tetramethylamine	−8.19	0.00	−8.19
11	1,3-propanediamine	−5.43	−0.07	−5.50
12	1,6-hexanediamine	−6.23	0.00	−6.23

*Bold: experimental templates; Italic: predicted templates.
With kind permission from Springer Science+Business Media: Topics in Catalysis, Toward Rational Synthesis of Microporous Aluminophosphate $AlPO_4$-21 by Hydrothermal Combinatorial Approach, volume 35, 2005, Yu Song, Figure 2 [57].

TABLE 24.3 The detailed crystallization conditions of $AlPO_4$-21

Samples	Templates (R)	Crystallization ranges	Phases
1	Pyrrolidine	$1.0Al(^iOPr)_3$-(3-4)H_3PO_4-(4-7)R	$AlPO_4$-21
2	Ethylamine	$1.0Al(^iOPr)_3$-(1-4)H_3PO_4-(1-7)R	$AlPO_4$-21
3	Dimethylamine	$1.0Al(^iOPr)_3$-(1-5)H_3PO_4-(3-7)R	$AlPO_4$-21
4	*n*-Propylamine	$1.0Al(^iOPr)_3$-(3-5)H_3PO_4-(5-7)R	$AlPO_4$-21
5	Ethanolamine	$1.0Al(^iOPr)_3$-(1-5)H_3PO_4-(1-7)R	$AlPO_4$-21
6	Trimethylamine	$1.0Al(^iOPr)_3$-(1-2)H_3PO_4-(>7)R	$AlPO_4$-21
7	*N,N,N′,N′*-Tetramethylamine	$1.0Al(^iOPr)_3$-(1-2)H_3PO_4-(1-7)R	$AlPO_4$-21
8	1,3-Propanediamine	$1.0Al(^iOPr)_3$-(1-2)H_3PO_4-(1-3)R	$[NH_3(CH_2)_3NH_3][Al_2P_2O_9]$
9	1,6-Hexanediamine	$1.0Al(^iOPr)_3$-(1-2)H_3PO_4-(1-3)R	AlPO-HDA

With kind permission from Springer Science+Business Media: Topics in Catalysis, Toward Rational Synthesis of Microporous Aluminophosphate $AlPO_4$-21 by Hydrothermal Combinatorial Approach, volume 35, 2005, Yu Song, Figure 2 [57].

higher interaction energy (−8.19 kcal/mol) as the template also produced $AlPO_4$-21 (sample 7). However, single-crystal structure analysis of sample 7 showed that the real template residing in the 8-ring channels is dimethylamine molecule which is believed to be resulted from the decomposition of large *N,N,N′,N′*-tetramethylethylenediamine in the reaction system.

24.3.1.2.3. SYNTHESIS OF EXTRA-LARGE PORE ZINC PHOSPHITES WITH 24-RING CHANNELS

The synthesis of extra-large pore inorganic crystalline materials is of considerable interest [58]. The extra-large pore can be either templated by bulky rigid organic molecules or by multiple small organic amines. Liang et al. [59] reported a zinc phosphite $[(C_4H_{12}N)_2][Zn_3(HPO_3)_4]$ (denoted ZnHPO-CJ1) with extra-large 24-ring channels, which is made up of strictly alternating ZnO_4 tetrahedra and HPO_3 pseudo-pyramids. The approximate size of the 24-ring window is 11.0 × 11.0 Å (Fig. 24.24). Multiple protonated *n*-butylamine molecules reside in each 24-ring window. The alkyl groups of these cations extend into the hollow space of the 24-ring channel and their NH_3^+ groups form H-bonds with the framework oxygen atoms. As with ZnHPO-CJ1, many other extra-large micropore materials with high framework charge density are also found whose extra-large pores are assembled around multiple small amines. This exemplifies the cooperative templating mechanism of multiple small organic cations in the formation of extra-large micropore open-framework materials. This work demonstrated a useful strategy toward the template-guided synthesis of porous materials with extra-large pores.

The templating abilities of several organic amines for the formation of extra-large pore ZnHPO-CJ1 were studied in terms of the space matching, charge−density matching, and the nonbonding interactions of the host−guest [60]. The calculated results are listed in Table 24.4.

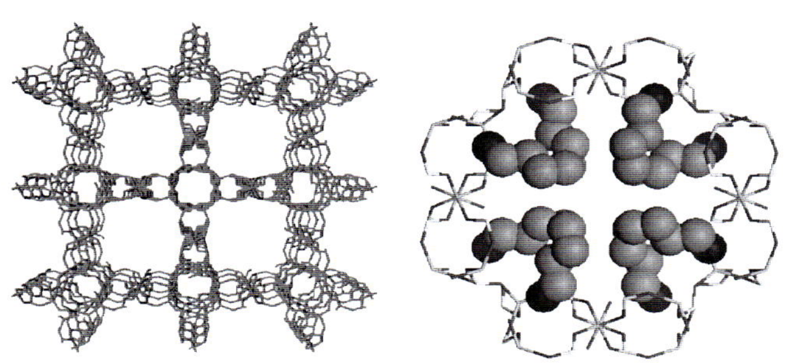

FIGURE 24.24 (a) The open-framework structure of ZnHPO-CJ1 viewed along the [001] direction showing the extra-large 24-ring and the 8-ring channels; and (b) multiple $CH_3(CH_2)_3NH_3^+$ ions reside in one 24-ring pore window. *J. Liang, J. Li, J. Yu, P. Chen, Q. Fang, F. Sun, R. Xu: $[(C_4H_{12}N)_2][Zn_3(HPO_3)_4]$: An open-framework zinc phosphite containing extra-large 24-ring channels. Angew. Chem. Int. Ed. 2006. Volume 37, Page 2546. Copyright Wiley-VCH Verlag GmbH & Co. KGaA. Reproduced with permission [59].*

TABLE 24.4 Calculated charge density (CD_{guest}) and occupied volume (V_{occ}) of several organic amines, and the interaction energy (E_{inter}, kcal/mol) and H-bonding interaction energy ($E_{H\text{-}bonds}$, kcal/mol) of the host–guest

Code	Organic amine[a]	V_{occ} (Å^3)[c]	CD_{guest}	E_{inter}	$E_{H\text{-}bonds}$
1	***n*-Butylamine**	773.44	0.2	−77.92	39.87
2	*n-Propylamine*	665.68	0.25	−82.19	−41.83
3	*Cyclohexanamine*	852.16	0.14	−71.14	−41.80
4	*Cyclopentylamine*	813.44	0.17	−70.75	−44.69
5	1,6-Diaminohexane	561.8	0.25	−42.08	−17.81
6	1,8-Diaminooctane	726.76	0.2	−37.03	−19.09
7	1,9-Diaminononane	854.72	0.18	−43.42	−27.26
8	1,10-Diaminodecane	860.04	0.17	−17.07	−8.26
9	Diethylenetriamine[b]	501.08	0.29	−43.46	−23.52
10	Diaminobicyclooctane	486.52	0.25	−26.81	−3.51

[a]*Bold: experimental template; Italic: predicted templates.*
[b]*Diethylenetriamine is diprotonated.*
[c]*Occupied volumes of all templates in one unit cell are considered, and number of template molecules is determined based on the charge balance.*
Reprinted with permission from [61]. Copyright 2008 American Chemical Society.

The ratios of framework charge/framework T number and template charge/template non-H number were calculated as the charge densities of host framework and organic template, respectively. Materials Studio software was used to calculate the "free volume" of the ZnHPO-CJ1 and the "occupied volume" of guest organic molecules in free status. Cerius^2 package was used to calculate the nonbonding interactions between the host framework and the guest molecules based on the Burchart 1.01–Dreiding 2.21 force field. More detailed information about the calculation could be found in Ref. [60].

The suitable template molecules should first meet the space matching and charge matching with the host framework. Furthermore, they should have lower interaction energies with the host framework. On the basis of the above criteria, some suitable templates that could potentially direct the formation of the extra-large pore structure of ZnHPO-CJ1 were predicted. For example, cyclohexylamine (CHA), cyclopentylamine (CPA), and *n*-propylamine were predicted as favorable templates. Experimentally, utilizing these amines as the templates, $(C_6H_{14}N)_2[Zn_3(HPO_3)_4]$(ZnHPO-CJ2), $(C_5H_{12}N)_2[Zn_3(HPO_3)_4]$ (ZnHPO-CJ3), and $(C_3H_{10}N)_2[Zn_3(HPO_3)_4]$ (ZnHPO-CJ4) were successfully hydrothermally prepared with target extra-large 24-ring channels analogous to $(C_4NH_{12})_2[Zn_3(HPO_3)_4]$ (ZnHPO-CJ1) [61]. Their structures were characterized by powder X-ray diffraction analysis as well as single-crystal crystal structure analysis. Figure 24.25 shows the frameworks of ZnHPO-CJ2 and ZnHPO-CJ3. Each 24-ring channel accommodates multiple protonated cyclohexylamine $(CH_2)_5CHNH_3^+$ cations and cyclopentylamine $(CH_2)_4CHNH_3^+$ cations, respectively.

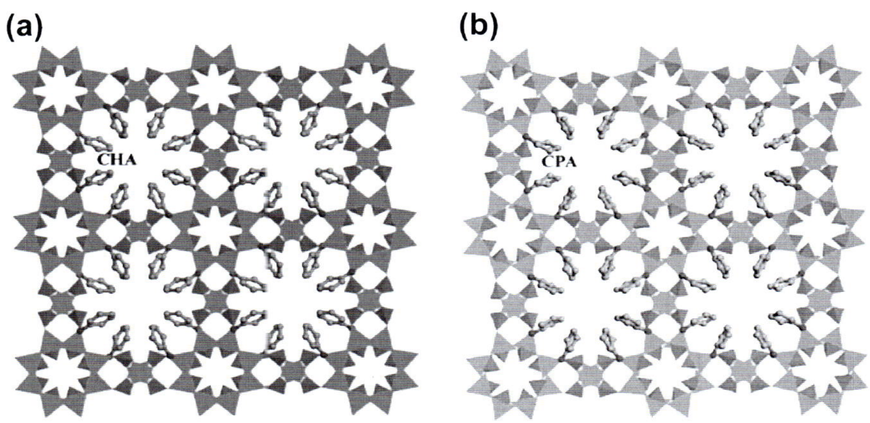

FIGURE 24.25 Crystal structures of (a) ZnHPO-CJ2 and (b) ZnHPO-CJ3 viewed along the [001] direction. *Reprinted with permission from [61]. Copyright 2008 American Chemical Society.*

24.3.1.2.4. CO-TEMPLATING SYNTHESIS OF SILICOALUMINOPHOSPHATES WITH THE SAV AND KFI FRAMEWORK TOPOLOGIES

Castro et al. [62] investigated a co-templating strategy assisted by molecular modeling for the synthesis of silicoaluminophosphates with SAV and KFI zeotype frameworks. STA-7 (SAV) has a three-dimensional 8-ring channel system. Its structure is composed of D6Rs and possesses two types of cages (Fig. 24.26). The cyclam and tetraethylammonium cations (TEA^+) were used for the synthesis of SAPO STA-7. Single-crystal diffraction analysis showed that the TEA^+ cations fit nicely in the smaller cages, and indicated that the cyclam molecules reside in the larger cage disorderedly. If *n*-proplyamine (DPA) and diisopropylamine (DIPA) were used instead of TEA^+ in the synthesis, the product was mainly STA-6 (SAS) with minor admixed STA-7. Modeling of their lowest energy configurations in the smaller cage of STA-7 provided some idea of the role of TEA^+. The binding energies of DPA, DIPA, and TEA^+ in the smaller cage are −83.6, −124.9, and −125.6 kJ mol^{-1}, respectively. The modeled position of TEA^+ is close to the position measured experimentally but the DPA protrudes from the smaller cage. The higher selectivity of TEA^+ over DIPA is attributed to its better fit to the geometry of the cage of STA-7.

KFI was previously reported as an aluminosilicate zeolite ZK-5 [63]. Like the SAV framework, it is built up from D6Rs only, but with a different stacking arrangement (Fig. 24.27). The (001) surface of SAV is topologically identical to the <100> surface of KFI. As with SAV, KFI also has two types of cages, i.e., the α-cage found in $AlPO_4$-42 (LTA) and a smaller cage found in zeolite merlinoite (MER). A co-templating strategy as the synthesis of SAPO SAV was therefore employed. It was believed that the azaoxacryptand 4,7,13,16,21,24-hexaoxa-1,10-diazabicyclo[8.8.8]hexacosane (K222) which was a good template for $AlPO_4$-42 [64] would fit the α-cage, and some smaller organic amine or ammonium cations might fit the smaller *mer* cage of KFI.

A few readily available amines and alkylammonium cations were screened computationally for their fit within the *mer* cages of KFI with an $AlPO_4$ composition, including tetramethylammonium (TMA^+), TEA and tetrapropylammonium (TPA^+) cations, methylamine (MA), ethylamine (EA), propylamine (PA), dimethylamine (DMA), diethylamine (DEA), DPA, DIPA, triethylamine (TREA), and diisopropylethylamine (DIPEA). The nonbonding energies are shown in

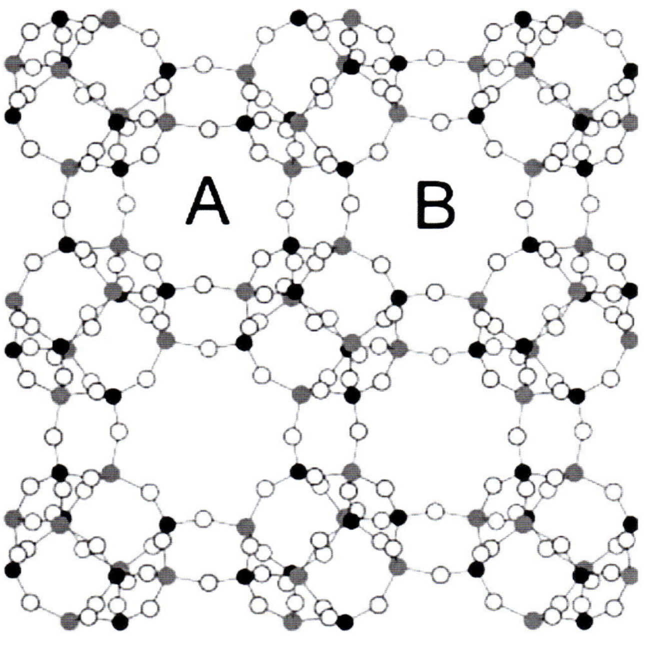

FIGURE 24.26 The structure of STA-7 with two different types of cages, A and B (aluminum atoms are gray, phosphorus black, and oxygen white). *Images reproduced with permission from [62].*

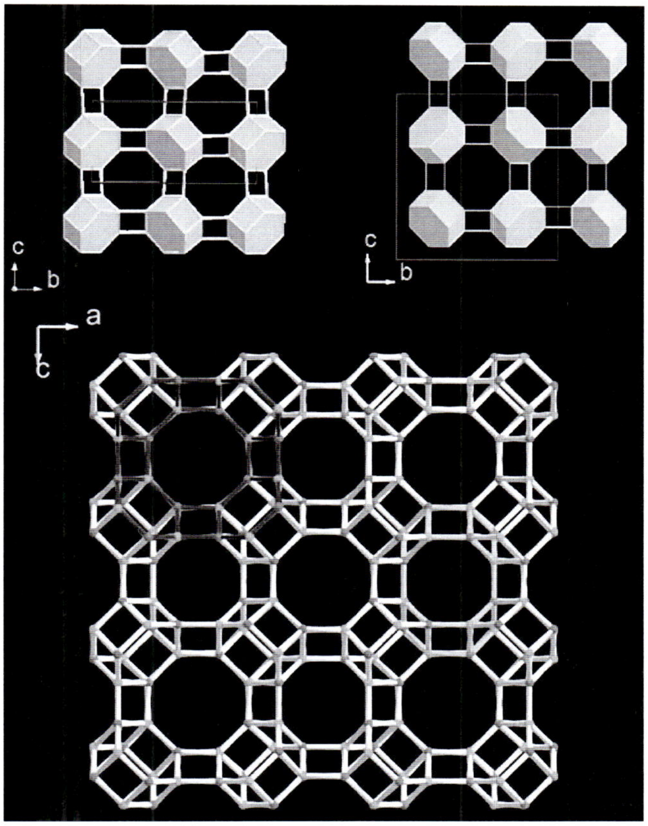

FIGURE 24.27 Above: The structures of SAV (left) and KFI (right). Both are made up entirely of D6Rs (in blue), which can be distinguished when viewed along the *a*-axis. In SAV, layers are stacked along the *c*-axis by a simple translation, whereas in KFI adjacent layers are related by a mirror plane perpendicular to the *c*-axis. Below: The KFI structure and with α- and *mer* cages outlined in grey, respectively. *Images reproduced with permission from [62].*

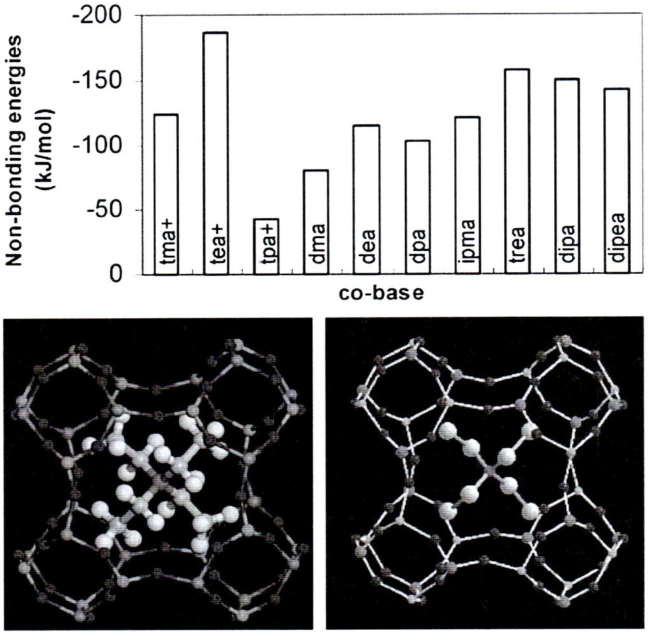

FIGURE 24.28 Above: A histogram of the nonbonding energies of potential co-templates for the *mer* cages of the $AlPO_4$-KFI structure, indicating that TEA^+ ions have most favorable binding energies. Below: The modeled position of the TEA^+ cations within the *mer* cages of the $AlPO_4$-KFI structure (left) and that observed experimentally (right). In the latter, one of two symmetry-related positions is shown. *Images reproduced with permission from [62].*

Fig. 24.28. As can be seen, the most favorable binding energy is -177.8 kJ mol^{-1} for the TEA^+ cation. Experimentally, TEA^+ was used a co-template with K222 in the synthesis from a magnesioaluminophosphate gel system, and gave the desired $AlPO_4$-based KFI product. Single-crystal structure analysis of MgAPO KFI revealed that the TEA^+ cations were located within the MER cages in the configuration predicted by modeling. Using the K222/TEA^+ combination, CoAPO and SAPO compositional variants of KFI were also successfully synthesized.

24.3.2. Synthesis Guided by Substituent Element Effects

24.3.2.1. *Synthesis of D4R-containing Zeolites by using Ge as a Silica Substituent*

Some elements have been found to have a potential ability to direct the formation of a particular building unit. For example, Be and Zn atoms tend to direct the formation of 3-rings in zeolite structures. A series of germanates with D4R structures have been reported [65]. Recently, employing Ge as a silica substituent combined with the use of novel template molecules, Corma's group has successfully synthesized a number of novel D4R-containing germanosilicate zeolites. Ge preferentially occupies positions at the D4R units [66]. The structure-directing effect of Ge towards the formation of D4R-containing zeolites is due to the fact that the smaller Ge–O–Ge angles than the Si–O–Si angles can relax the geometric constraints in the D4R units and thus stabilize the resulting structures.

For example, zeolite ITQ-21 was synthesized in the gel with molar composition $0.33GeO_2$:$0.67SiO_2$:0.50 MSPTOH:0.5HF:$20H_2O$ at 175 °C for 5 days using *N*-methylsparteinium hydroxide (MSPTOH) as SDA [67]. In the absence of Ge in the reaction mixture CIT-5 (CFI) was formed which did not have D4Rs in the structure. The structure-directing role of Ge in the D4R-containing zeolites has been also clearly shown by the synthesis of polymorph C of zeolite beta (BEC) with a variety of organic template molecules [68]. The syntheses were carried out in the gels with molar compositions $(1-x)SiO_2$:$xGeO_2$:(0.5–0.25)SDAOH:0.5HF:wH_2O at 135–175 °C for 15–120 h. Notably, when Ge was introduced into the reaction system, only the D4R-containing BEC structure was obtained irrespective of the type of the template used. In the absence of Ge, different zeolite phases including ITQ-4, beta, and ZSM-12 were obtained depending on the template used.

Remarkably, novel zeolites with extra-large pore structures have been prepared by using Ge as a silica substituent. Notable examples, as shown in Figs 24.29–24.31 (framework structures and three templates used), are ITQ-15 [69] with two-dimensional 14 × 12-ring channels (FD: 15.6), ITQ-33 [70] with three-dimensional 18 × 10 × 10-ring channels (FD: 12.3), and ITQ-37 [71] with three-dimensional 30-ring channels (FD: 10.3). Interestingly, ITQ-37 represents an interrupted zeotype structure that contains extra-large mesoporous 30-ring gyroidal channel with the pore dimension of 7.05 × 22.04 Å and the lowest framework density among all the known four-connected zeolite structures. It was synthesized using the bulky diammonium ion, which contains chiral centers, as the organic SDA. More recently, Jiang et al. [72] reported a new silicogermanate ITQ-44 with three-dimensional 18 × 12 × 12-ring channels by using (2′*R*, 6′*S*)-2′, 6′-dimethylspiro[isoindole-2,1′-piperidin-1′-ium] as an SDA. Interestingly, ITQ-44 is featured by novel D3Rs along with D4Rs. There is also a preferential Ge occupancy of the D3Rs. The structure of ITQ-44 is closely related to that of ITQ-33. The 12-ring channels in ITQ-44 can be viewed as expansion of the 10-ring channels through D3Rs (Fig. 24.32).

24.3.2.2. *Heteroatom-stabilized Chiral Framework of Aluminophosphate Molecular Sieves*

Heteroatoms have also demonstrated their effect on stabilization the chiral zeolite frameworks as demonstrated by the work of Song et al. [73]. Li et al. have developed a method for the design of chiral zeolite frameworks with specified pore geometries through

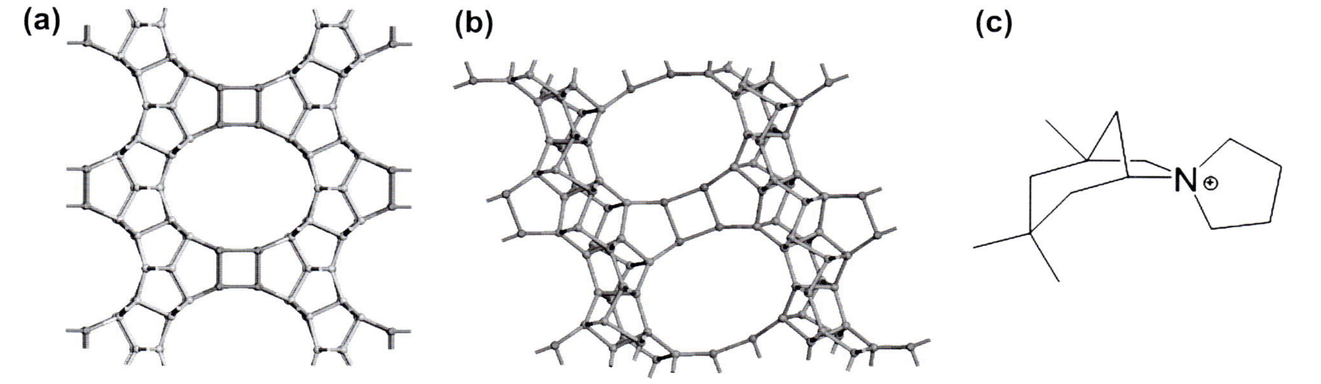

FIGURE 24.29 (a) The 14-ring view along *c*-axis. (b) The 12-ring view along the *b*-axis. (c) The SDA cation used in ITQ-15 synthesis Only the T−T connections are shown.

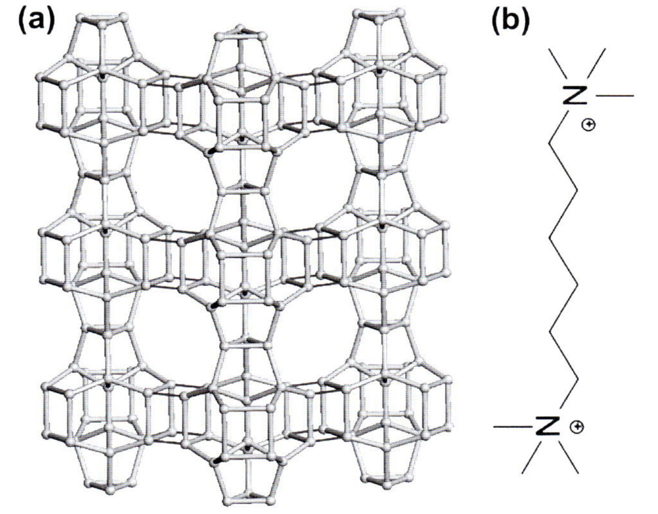

FIGURE 24.30 (a) Lateral view of the ITQ-33 structure, showing the 10-ring and (b) The SDA cation used in ITQ-33.

constrained assembly of atoms (see Section 2.1.1). According to the theoretical results, most four-connected frameworks generated with chiral channels are energetically unfavorable for the SiO_2 composition because of the special geometric strains in such structures. This suggests that introducing other elements such as Be, B, Ge, and transition metals, instead of Si, might be a promising strategy for stabilizing such frameworks, as they could offer a more reasonable bonding geometry, like bond distances and bond angles than an SiO_2 composition does. By the incorporation of metal ions (Co^{2+}, Mg^{2+}, Fe^{2+}, etc.) into the aluminophosphate framework, novel heteroatom-containing chiral aluminophosphate MAPO-CJ40 (M = Co^{2+}, Mg^{2+}, Fe^{2+}, etc.) has been synthesized with one-dimensional helical 10-ring channels. Figure 24.33 shows the structure of CoAPO-CJ40. Its structure is based on the strict alternation of MO_4 (M = Al, Co) and PO_4 tetrahedra forming an anionic $[Co_2Al_{10}P_{12}O_{48}]^{2-}$ framework. Charge neutrality is achieved by protonated diethylamine cations. The helical 10-ring channels are enclosed by double helical ribbons of the same handedness made of the edge sharing of 6-rings along the 2_1 screw axis. The Co atoms, which substitute one of the three unique Al sites, adopt a helical arrangement along the channel. The framework of CoAPO-CJ40 is intrinsically chiral with underlying symmetry of $I2_12_12_1$ and exhibits a new zeotype structure which has been assigned as JRY.

It was noted that pure aluminophosphate AlPO-CJ40 could not be successfully prepared in the absence of Co or Zn ions. In CoAPO-CJ40, the Co atoms exclusively occupy the Al(1) position among the three unique Al sites, with a probability of 50%. This could be explained by a molecular mechanics computation and geometric calculation. A structure model with a pure $AlPO_4$ composition was built by replacing all the Co atoms by Al atoms. The bond angle variance for each AlO_4/PO_4 tetrahedron was calculated after the geometry of this structure model was fully optimized. The results showed that there is a large bond angle variance on the Al(1) site (Table 24.5) which indicates that the Al(1)-centered tetrahedron may suffer from high distortion from the ideal tetrahedron. The incorporation of Co atoms might be necessary for the framework to relax the high distortion, thus stabilizing the whole structure.

Three other structure models were built by replacing Al(1), Al(2), and Al(3) with Co atoms, respectively. The calculated framework energies of the three optimized models are −9401.35, −9374.83, and −9372.36 kcal mol^{-1} per unit cell, respectively. Therefore the first structure model in which cobalt atoms occupy the Al (1) position should be the most reasonable one. This result agrees with the single-crystal structure analysis. In summary, molecular simulations and geometric

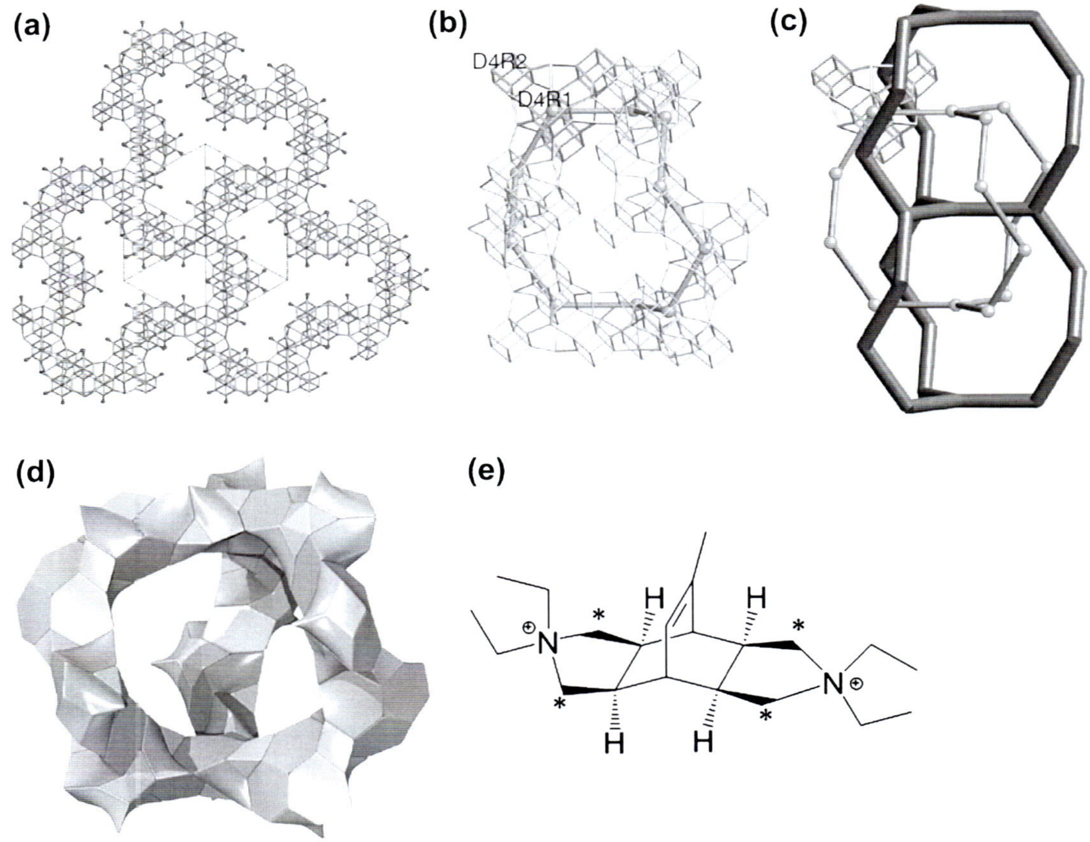

FIGURE 24.31 (a) A slice (15.3 Å thick) viewed down the [111] direction. All double 4-rings have the same orientation. (b) The 30-ring built from 10 tertiary building units. The centers of the tertiary building units fall on the nodes of one srs net. (c, d) The large cavity defined by three 30-rings in the topological and tiling methods. (e) Structure of SDA cation used for synthesizing the ITQ-37 zeolite. SDA contains four chiral centers (marked with asterisks) in a mesoconformation, making the overall molecule achiral. *Reprinted with permission from Macmillan Publishers Ltd: Nature [71], Copyright (2009).*

calculations demonstrated that Co^{2+} ions play an important role in stabilizing the chiral framework.

24.3.3. Synthesis Guided by Data Mining

The difficulty towards the rational synthesis of zeolitic inorganic porous crystalline materials lies in the unclear relationship of the synthetic factors and resulting structural characters. In order to establish such a relationship, Yan et al. [74] established a Zeobank, which contains a database of AlPO synthesis, and a database of AlPO structures. The databases are available at the Website http://izasc.ethz.ch/fmi/xsl/IZA-SC/ol.htm [75]. The Zeobank will allow to apply computational techniques to study complex relationships between the synthetic parameters and the corresponding zeolite structures in a systematic manner. An SVM-based computational study on microporous aluminophosphates has demonstrated the general feasibility in establishing a relationship between the synthetic parameters and the structural features of the zeolite materials. This provides a useful strategy to the rational synthesis of zeolitic inorganic crystalline materials [76].

The AlPO synthesis database contains over 1600 reaction data for ca. 230 AlPO structures. The data were mostly collected from the literature. Each entry in the database consists of four pieces of synthesis information: the source materials, the template, the synthesis conditions, and the structural characteristics of the product. The channel systems of microporous AlPOs can be divided into 17 types according to the sizes of their pore rings. Among them, AlPOs with (6,12)-rings represent a major class, and 396 synthetic records in the database are associated with this channel system. An SVM-based computational study has been performed for predicting the formation of (6,12)-ring-containing microporous aluminophosphates.

For a specified classification problem, an SVM is trained (or supervised) to learn to optimally distinguish the data elements in a *positive* data set from the data elements in a *negative* data set, together forming the *training data set*, based on their respective feature values

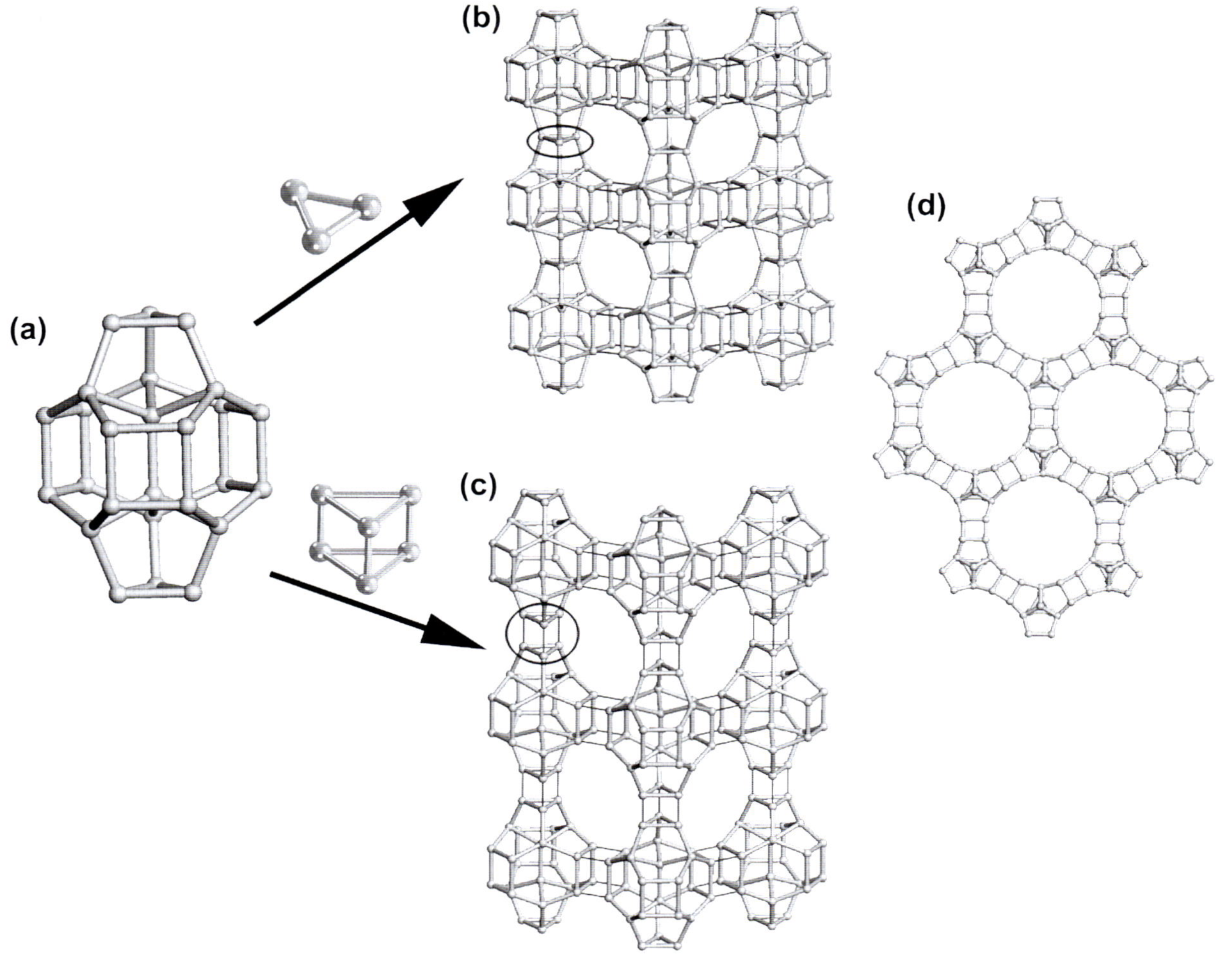

FIGURE 24.32 (a) The ternary building unit of both ITQ-33 and ITQ-44. (b) Condensation of the tertiary units in columns through the 3-ring for ITQ-33 (indicated by the cycle). (c) Condensation of the tertiary building unit in columns through the double 3-ring (D3R) for ITQ-44 (indicated by the cycle). (d) The 18-ring pore view down along the *c*-axis for both ITQ-33 and ITQ-44.

[77]. During the training, an SVM continues to adjust two parallel (separating) hyperplanes in the feature space, attempting to keep as many positive data elements on one side of a hyperplane and the negative points on the other side of the parallel hyperplane as possible, and the two hyperplanes have the largest distance possible. Distinguishing between the (6,12)-ring-containing AlPOs and the other AlPOs in the AlPO synthesis database is essentially a binary classification problem.

Feature selection is crucial for the successful classification by an SVM-based classifier. Through analyses of a database of AlPO synthesis with ca. 1600 reaction data, a number of synthetic parameters are identified as the features (F) such as three gel molar ratios of Al_2O_3 (F1), P_2O_5 (F2), and the organic amine template (F4), as well as a number of parameters associated with the geometric and electronic characteristics of the templates as the input (Table 24.6). Eleven template parameters are considered that are believed to cover the most important features of a template, including the longest distance (F11), the second longest distance (F12), and the shortest distance (F13) associated with the template geometry, the van der Waals (VDW) volume (F14), the dipole moment (F15), the ratio of C/N (F16), the ratio of $N/(C+N)$ (F17), and the ratio of N/VDW volume (F18) in the template, the Sanderson electronegativity (F19), the number of freely rotated single bond (F20), and the max H^+ number (F21). Using these parameters, an SVM-based classifier has been trained on a training data set containing 363 (6,12)-ring-containing AlPOs and 1069 AlPOs without such rings. Three important features related to the molar concentrations of Al_2O_3, P_2O_5, and the organic template in the starting gel and the 11 template parameters are used as the input parameters for training the classifier. The output of the trained classifier is either 0 or 1,

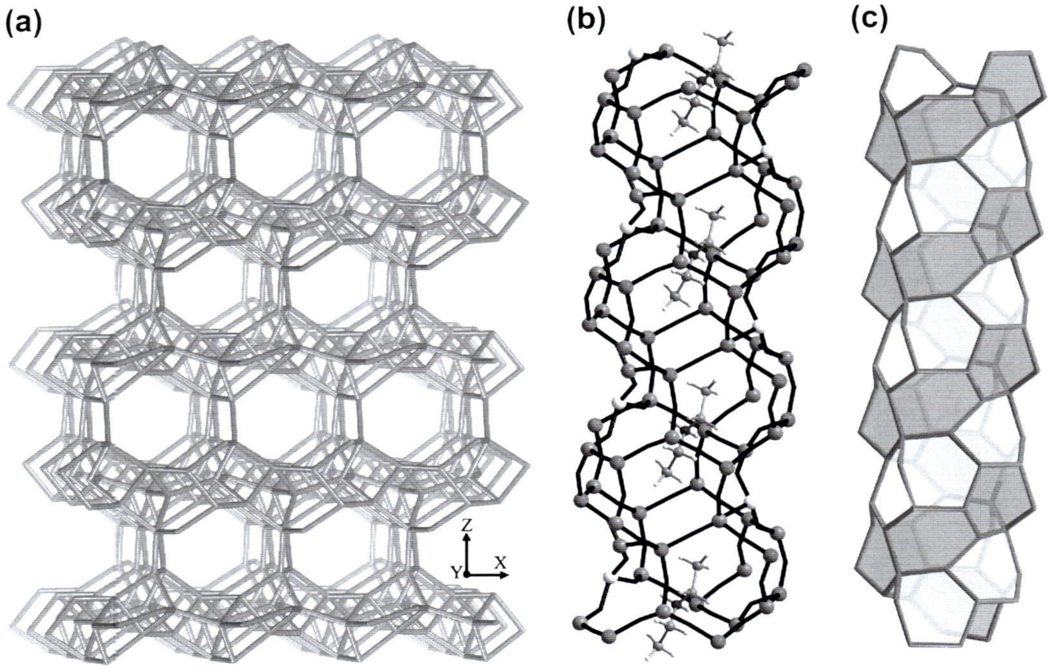

FIGURE 24.33 Framework structure of CoAPO-CJ40. (a) Viewed along the [010] direction; (b) the helical 10-ring channel and helical arrangement of cobalt atoms (white color for Co); (c) the 10-ring channel enclosed by double helical ribbons made up of the edge sharing of 6-rings.

representing a resulting structure containing a (6,12)-ring or not, respectively.

Each synthetic parameter has been tested individually as well as some of them in combinations in order to know which ones may have the predictive power in distinguishing the two classes of AlPOs. For example, it is found that a classifier using only three synthetic parameters involving the molar concentrations of Al_2O_3 (F1), P_2O_5 (F2), and template (F4) cannot predict

TABLE 24.5 The bond angle variance for each AlO_4/PO_4 tetrahedron for CoAPO-CJ40

	Al0	Co1	Co2	Co3	Si
Al1	4.62290	0.48070	6.48750	8.80290	3.07350
Al2	1.76090	6.11130	19.32990	17.14490	1.85990
Al3	3.18480	2.82550	8.67860	20.04860	0.77040
P1	1.04380	0.08400	15.06540	2.93800	1.22310
P2	1.54060	0.70720	3.60340	12.34910	2.45710
P3	0.71630	0.11130	1.28080	2.57930	2.79310
SUM	12.86930	10.32000	54.44560	63.86280	12.17710
AVR	2.14488	1.72000	9.07427	10.64380	2.02952

X. Song, Y. Li, L. Gan, Z. Wang, J. Yu, R. Xu: Heteroatom-stabilized chiral framework of the aluminophosphate molecular sieves. Angewandte Chemie International Edition. 2009. Volume 48, Page 314. Copyright Wiley-VCH Verlag GmbH & Co. KGaA. Reproduced with permission [73].

if an AlPO contains (6,12)-rings or not. Further introducing the 11 aforementioned template parameters along with the three synthetic parameters, individually or grouping combinations, as the classification features, gave substantially different levels of classification performance. This suggested that the choice of suitable template parameters was important to the classification performance. The highest and the lowest predicting accuracies among different combinations are shown in Fig. 24.34. In view of the highest accuracy as a function of the number (N) of the used template parameters, it was found that adding two or more template parameters generally gave better performance in the classification than individual parameter-based classifiers.

By analyzing the classification performances based on different combinations of the template parameters with Al_2O_3 (F1), P_2O_5 (F2), and template (F4), the most relevant template parameters could be identified to the classification problem. Notably, it was found that the second longest distance within each template (F12) has the best prediction performance among all individual parameters, giving a training accuracy of 80.54% and a testing accuracy of 78.08% by itself (Fig. 24.35). Further combination of F12 with another electronic parameter of the template such as F16 (the ratio of C/N) or F18 (the ratio of N/VDW volume), gave the training accuracy 83.66–84.31%, and the testing accuracy 81.61–81.75% (Table 24.7).

TABLE 24.6 Description of the input synthetic parameters

	Code[a]	Description of parameters
Gel composition	F1	The molar amount of Al_2O_3 in the gel composition
	F2	The molar amount of P_2O_5 in the gel composition
	F4	The molar amount of template in the gel composition
Organic template	F11	The longest distance of organic template
	F12	The second longest distance of organic template
	F13	The shortest distance of organic template
	F14	The Van der Waals volume
	F15	The dipole moment
	F16	The ratio of C/N
	F17	The ratio of $N/(C+N)$
	F18	The ratio of N/Van der Waals volume
	F19	The Sanderson electronegativity
	F20	The number of free rotated single bond
	F21	The maximal number of protonated H atoms

[a]*[F3], [F5]–[F10] refer to other synthetic parameters such as the crystallization temperature and time, and the feature of the solvent which are ignored in this study. Molecular dimensions of the organic template are calculated based on the optimized model by using PM3 method in MOPAC 2000. The longest, the second longest, and the shortest distances are defined as the three largest lengths of an organic template in three perpendicular directions of space.*

Reprinted from Microporous and Mesoporous Materials, Jiyang Li, Miao Qi, Jun Kong, Jianzhong Wang, Yan Yan, Weifeng Huo, Jihong Yu, Ruren Xu, Ying Xu, Computational prediction of the formation of microporous aluminophosphates with desired structural features, Copyright (2010), with permission from Elsevier [76].

Computational data mining techniques showed a great promise in guiding the rational synthesis of zeolitic materials. Currently various computation techniques have been used to retrieve information from the data analysis [78] such as neural networks, support vector machines, classification trees, clustering analysis, principal component analysis, and control theory. However, it is worth mentioning that many factors influence the prediction results such as the number and the quality of the data set in the Zeobank database, the feature selection of the input synthetic parameters and the output structural parameters in the classification, and the data mining techniques. Moreover, the successful prediction of the formation of microporous materials with specified structural features will heavily rely on the deep understanding of zeolite synthetic chemistry.

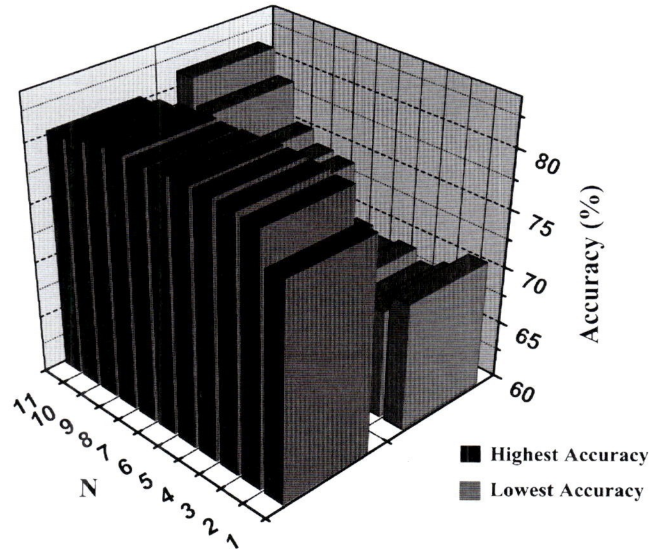

FIGURE 24.34 Highest and lowest prediction accuracies of N-parameter combinations along with the molar concentrations of F1, F2, and F4 by using a Gaussian RBF kernel function. *Reprinted from Microporous and Mesoporous Materials, Jiyang Li, Miao Qi, Jun Kong, Jianzhong Wang, Yan Yan, Weifeng Huo, Jihong Yu, Ruren Xu, Ying Xu, Computational prediction of the formation of microporous aluminophosphates with desired structural features, Copyright (2010), with permission from Elsevier [76].*

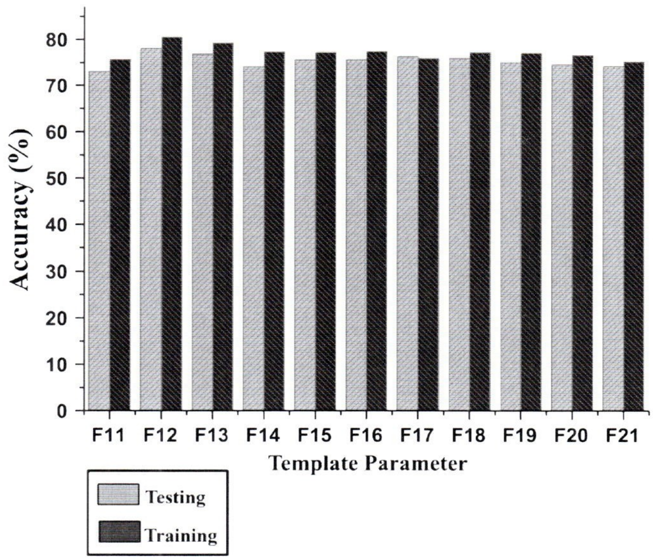

FIGURE 24.35 Effect of single template parameter along with F1, F2, and F4 on the classification result. *Reprinted from Microporous and Mesoporous Materials, Jiyang Li, Miao Qi, Jun Kong, Jianzhong Wang, Yan Yan, Weifeng Huo, Jihong Yu, Ruren Xu, Ying Xu, Computational prediction of the formation of microporous aluminophosphates with desired structural features, Copyright (2010), with permission from Elsevier [76].*

TABLE 24.7 Sets of input template parameters exhibiting highest accuracy

N[a]	Template parameters	Training accuracy (%)	Testing accuracy (%)
1	F12	80.54	78.08
2	F12 F16	84.31	81.75
3	F12 F16 F21	85.74	82.06
4	F12 F15 F16 F17	86.94	82.31
5	F12 F15 F16 F17 F18	87.51	82.44
6	F12 F13 F14 F15 F18 F19	87.86	82.36
7	F12 F13 F14 F15 F17 F19 F21	87.94	82.36
8	F12 F13 F14 F15 F17 F18 F19 F21	87.97	82.17
9	F12 F13 F14 F15 F16 F17 F18 F19 F21	87.97	82.08
10	F12 F13 F14 F15 F16 F17 F18 F19 F20 F21	87.99	81.97
11	F11 F12 F13 F14 F15 F16 F17 F18 F19 F20 F21	87.98	81.64

[a]*N is the number of input template parameters.*

Reprinted from Microporous and Mesoporous Materials, Jiyang Li, Miao Qi, Jun Kong, Jianzhong Wang, Yan Yan, Weifeng Huo, Jihong Yu, Ruren Xu, Ying Xu, Computational prediction of the formation of microporous aluminophosphates with desired structural features, Copyright (2010), with permission from Elsevier [76].

24.4. FUTURE PERSPECTIVE ON THE TAILOR-MADE SYNTHESIS OF DESIRED INORGANIC POROUS CRYSTALLINE MATERIALS

Zeolitic inorganic porous crystalline materials are of importance in many technological processes such as catalysis, adsorption, and separation because of their unique pore architectures coupled with the active reaction sites. Especially, zeolites are very successful catalysts in shape-selective catalysis that have found many industrial applications. Fine-tuning zeolites with controlled pore architectures, such as the pore system, pore dimension, and pore shape, will improve their performance for applications. Clearly, this represents a formidable challenge and will call for a capability to tailor-made synthesis of inorganic porous crystalline materials with desired structures and functionalities [79].

Currently, computational approaches, combined with increasing knowledge and understanding of the property–structure–synthesis have greatly facilitated the designed synthesis of zeolitic inorganic crystalline porous materials. Figure 24.36 depicts the future blueprint for the tailor-made synthesis of desired zeolitic materials. The engineering to access the target functional inorganic porous crystalline materials can be described as following:

- **(i)** A practical application, e.g., a specific catalytic reaction, raises detailed requirements for the structures with defined pore dimension, pore system, pore shape, active sites, etc.
- **(ii)** The desired porous structures are then designed by computational methods.
- **(iii)** By using computational modeling, the candidate SDA molecules are predicted for the given structures. Further data-mining techniques will predict the synthesis conditions for the target structures.
- **(iv)** The synthesis is achieved by using various synthetic techniques under hydrothermal or solvothermal conditions. Especially, the combinatorial techniques will allow exploring in a large experimental space by means of the appropriate experimental design.
- **(v)** The structures of as-made materials are identified by comparing the experimental X-ray diffraction patterns with the simulated ones derived from the theoretical structures.
- **(vi)** The application is eventually accessible by such a rational design and synthesis approach.

It is worth underscoring here that toward the rational synthesis of inorganic porous crystalline materials, there still remain a number of challenges ahead to achieve such a goal. Although great strides have been made in

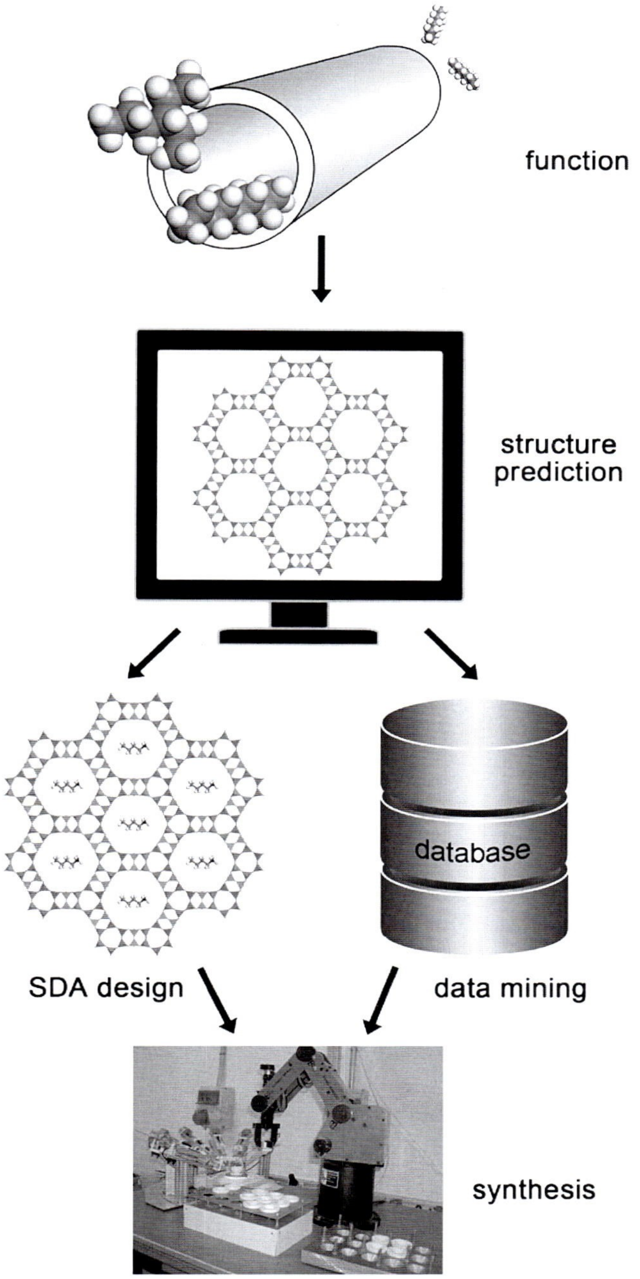

FIGURE 24.36 Future blueprint for engineering the rational synthesis of zeolitic materials with desired functionalities and structures. Copyright 2010 American Chemical Society. *Reprinted with permission from [79]. Copyright 2003 American Chemical Society.*

this area, future advances in understanding the formation mechanism at the molecular level are needed before the promise is fulfilled.

24.5. CONCLUDING REMARKS

Nowadays, the search for new inorganic materials has been changing from the past empirical, trial-and-error methods to the rational design philosophy. Our dream is to control chemically the self-assembly process of inorganic materials with predictable compositions, structures and functionalities, and eventually to replace the classical trial-and-error strategy. This is, in fact, a long journey from possibility to reality. Despite the difficulties, this is the way where the synthetic chemists should concentrate. More research as well as untiring efforts should be driven along these directions.

References

[1] R. Noyori, Nat. Chem. 1 (2009) 5–6.
[2] Ch. Baerlocher, L.B. McCusker, <http://www.iza-structure.org/databases/>
[3] R. Xu, W. Pang, J. Yu, Q. Huo, J. Chen, Chemistry of Zeolites and Related Porous Materials: Synthesis and Structure, Wiley, 2007.
[4] S. Kulprathipanja, Zeolites in Industrial Separation and Catalysis, Wiley, 2010.
[5] J. Yu, Introduction to zeolite science and practice, 3rd revised ed. in: J. Čejka, H. van Bekkum, A. Corma, F. Schüth (Eds.), Stud. Surf. Sci. Catal., vol. 168 Elsevier, Amsterdam, 2007, pp. 39–103.
[6] R. Catlow, R. Bell, F. Cora, B. Slater, Introduction to zeolite science and practice, 3rd revised ed. in: J. Čejka, H. van Bekkum, A. Corma, F. Schüth (Eds.), Stud. Surf. Sci. Catal., vol. 168 Elsevier, Amsterdam, 2007, pp. 659–700.
[7] J. Yu, R. Xu, Recent advances in the science and technology of zeolites and related materials, in: E. van Steen, L.H. Callanan, M. Claeys (Eds.), Stud. Surf. Sci. Catal., vol. 154, Elsevier, Amsterdam, 2004, pp. 1–13.
[8] A. Corma, Recent advances in the science and technology of zeolites and related materials, in: E. van Steen, L.H. Callanan, M. Claeys (Eds.), Stud. Surf. Sci. Catal., vol. 154, Elsevier, Amsterdam, 2004, pp. 25–40.
[9] S.M. Woodley, R. Catlow, Nat. Mater 7 (2008) 937–946.
[10] J.V. Smith, Chem. Rev. 88 (1988) 149–182.
[11] M.W. Deem, J. Newsam, Nature 342 (1989) 260–262.
[12] D.E. Akporiaye, G.D. Price, Zeolites 9 (1989) 23–32.
[13] O.D. Friedrichs, A.W.M. Dress, D.H. Huson, J. Klinowski, A.L. Mackay, Nature 400 (1999) 644–647.
[14] M.M.J. Treacy, I. Rivin, E. Balkovsky, K.H. Randall, M.D. Foster, Microporous Mesoporous Mater. 74 (2004) 121–132.
[15] Y. Li, J. Yu, D. Liu, W. Yan, R. Xu, Chem. Mater 15 (2003) 2780–2785.
[16] Y. Li, M. Guo, J. Yu, J. Li, R. Xu, Stud. Surf. Sci. Catal 154 (2004) 308–316.
[17] Y. Li, J. Yu, Z. Wang, J. Zhang, M. Guo, R. Xu, Chem. Mater 17 (2005) 4399–4405.
[18] W.M. Meier, H.J. Moeck, J. Solid State Chem. 27 (1979) 349–355.
[19] Cerius2, Molecular simulations/Biosym Corporation, San Diego, CA, 1995.
[20] C. Baerlocher, A. Hepp, W.M. Meier, DLS-76, F. Kristallographie, Zürich. 1978.
[21] E. de vos Burchart, J. Chem. Soc. Faraday Trans. 88 (1992) 2761–2769.
[22] Y. Li, J. Yu, R. Xu, Ch. Baerlocher, L.B. McCusker, Angew. Chem. Int. Ed. 47 (2008) 4112–4405.
[23] C. Baerlocher, F. Gramm, L. Massüger, L.B. McCusker, Z. He, S. Hovmöller, et al., Science 315 (2007) 1113–1116.
[24] F. Gramm, Ch. Baerlocher, L.B. McCusker, S.J. Warrender, P.A. Wright, B. Han, et al., Nature 444 (2006) 79–81.
[25] S. Kirkpatrick, C.D. Gelatt, M.P. Vecchi, Science 220 (1983) 671–680.

[26] M. Falcioni, M.W. Deem, J. Chem. Phys.(110) (1999) 1754–1766.
[27] S.M. Woodley, P.D. Battle, J.D. Gale, C.R.A. Catlow, Phys. Chem. Chem. Phys. 6 (2004) 1815–1822.
[28] G. Férey, J. Solid State Chem. 152 (2000) 37–48.
[29] X. Zou, T. Conradsson, M. Klingstedt, M.S. Dadachov, M.O'Keeffe, Nature 437 (2005) 716–719.
[30] Q. Pan, J. Li, K. Christensen, C. Bonneau, X. Ren, L. Shi, et al., Angew. Chem. Int. Ed. 47 (2008) 7868–7871.
[31] X.Y. Ren, Y. Li, Q.H. Pan, J.H. Yu, R.R. Xu, Y. Xu, et al., J. Am. Chem. Soc. 131 (2009) 14128–14129.
[32] O.M. Yaghi, M. O'Keeffe, N.W. Ockwig, K. Hee, M. Eddaoudi, J. Kim, Nature 423 (2003) 705–714.
[33] M. Eddaoudi, J. Kim, N. Rosi, D. Vodak, J. Wachter, M. O'Keeffe, et al., Science 295 (2002) 469–472.
[34] C. Mellot-Draznieks, J.M. Newsam, A.M. Gorman, C.M. Freeman, G. Férey, Angew. Chem. Int. Ed. 39 (2000) 2270–2275.
[35] C. Mellot-Draznieks, S. Girard, G. Férey, J. Am. Chem. Soc. 124 (2002) 15326–15335.
[36] Y. Li, J. Yu, J. Jiang, Z. Wang, J. Zhang, R. Xu, Chem. Mater 17 (2005) 6086–6093.
[37] D.W. Breck, J. Chem. Educ. 41 (1964) 678–689.
[38] G.T. Kerr, J. Phys. Chem. 70 (1966) 1047–1050.
[39] C.S. Cundy, P.A. Cox, Microporous Mesoporous Mater. 82 (2005) 1–78.
[40] C.E.A. Kirschhock, R. Ravishankar, L. Van Looveren, P.A. Jacobs, J.A. Martens, J. Phys. Chem. B 103 (1999) 4972–4978.
[41] T.M. Davis, T.O. Drews, H. Ramanan, C. He, J. Dong, H. Schnablegger, et al., Nat. Mater 5 (2006) 400–408.
[42] P.P.E.A. de Moor, T.P.M. Beelen, B.U. Komanschek, L.W. Beck, P. Wagner, M.E. Davis, et al., Chem. Eur. J 5 (1999) 2083–2088.
[43] M.E. Davis, R.F. Lobo, Chem. Mater 4 (1992) 756–768.
[44] A. Corma, F. Rey, J. Rius, M.J. Sabater, S. Valencia, Nature 431 (2004) 287–290.
[45] D.W. Lewis, D.J. Willock, C.R.A. Catlow, Nature 382 (1996) 604–606.
[46] D.W. Lewis, G. Sankar, J.K. Wyles, Angew. Chem. Int. Ed. 36 (1997) 2675–2677.
[47] D.W. Lewis, C.M. Freeman, C.R.A. Catlow, J. Phys. Chem. 99 (1995) 11194–11202.
[48] J. Li, J. Yu, W. Yan, R. Xu, Chem. Mater 11 (1999) 2600–2606.
[49] J. Yu, J. Li, K. Wang, R. Xu, Chem. Mater 12 (2000) 3783–3787.
[50] J. Yu, K. Sugiyama, S. Zheng, S. Qiu, J. Chen, R. Xu, et al., Chem. Mater 10 (1998) 1208–1211.
[51] B. Jandeleit, D.J. Schaefer, T.S. Powers, H.W. Turner, W.H. Weinberg, Angew. Chem. Int. Ed. 38 (1999) 2494–2532.
[52] D.E. Akporiaye, I.M. Dahl, A. Karlsson, R. Wendelbo, Angew. Chem. Int. Ed. 37 (1998) 609–611.
[53] J.M. Newsam, T. Bein, J. Klein, W.F. Maier, W. Stichert, Microporous Mesoporous Mater. 48 (2001) 355–365.
[54] J. Klein, C.W. Lehmann, H.W. Schmidt, W.F. Maier, Angew. Chem. Int. Ed. 37 (1998) 3369–3372.
[55] Y. Song, J. Yu, G. Li, Y. Li, Y. Wang, R. Xu, Chem. Commun. (2002) 1720–1721.
[56] A. Cantin, A. Corma, M.J. Diaz-Cabanas, J.L. Jorda, M. Moliner, J. Am. Chem. Soc. 128 (2006) 4216–4217.
[57] Y. Song, J. Li, J. Yu, K. Wang, R. Xu, Topics Catal 35 (2005) 3–8.
[58] J. Jiang, J. Yu, A. Corma, Angew. Chem. Int. Ed. 49 (2010) 3120–3145.
[59] J. Liang, J. Li, J. Yu, P. Chen, Q. Fang, F. Sun, et al., Angew. Chem. Int. Ed. 45 (2006) 2546–2548.
[60] J. Li, J. Yu, R. Xu, Microporous Mesoporous Mater. 101 (2007) 406–412.
[61] J. Li, L. Li, J. Liang, P. Chen, J. Yu, Y. Xu, et al., Cryst. Growth Des. 8 (2008) 2318–2323.
[62] M. Castro, R. Garcia, S.J. Warrender, M.Z. Slawin, P.A. Wright, P.A. Cox, et al., Chem. Commun. (2007) 3470–3472.
[63] W.M. Meier, G.T. Kokotailo, Z. Kristallogr. 121(1965) 211.
[64] L. Schreyeck, F. D'Agosto, J. Stumbe, P. Caullet, J.C. Mougenel, Chem. Commun. (1997) 1241–1242.
[65] H. Li, O.M. Yaghi, J. Am. Chem. Soc. 120 (1998) 10569–10570.
[66] L.A. Villaescusa, P.A. Barrett, M.A. Camblor, Angew, Chem. 111 (1999). 2164; Angew. Chem. Int. Ed. 38(1999) 1997–2000.
[67] A. Corma, M.J. Diaz-Cabanas, J. Martinez-Triguero, F. Rey, J. Rius, Nature 418 (2002) 514–517.
[68] A. Corma, M.T. Navarro, F. Rey, J. Rius, S. Valencia, Angew. Chem. Int. Ed. 40 (2001) 2277–2280.
[69] A. Corma, M.J. Díaz-Cabanas, F. Rey, S. Nicolopoulus, K. Boulahya, Chem. Commun. (2004) 1356–1357.
[70] A. Corma, M.J. Díaz-Cabanas, J.L. Jord, C. Martínez, M. Moliner, Nature 443 (2006) 842–845.
[71] J. Sun, C. Bonneau, A. Cantin, A. Corma, M.J. Díaz-Cabanas, M. Moliner, et al., Nature 458 (2009) 1154–1159.
[72] J. Jiang, I. Minuesa, J.L. Jorda, M.J. Diaz-Cabanas, J. Yu, A. Corma, Angew. Chem. Int. Ed. 49 (2010) 4986–4988.
[73] X. Song, Y. Li, L. Gan, Z. Wang, J. Yu, R. Xu, Angew. Chem. Int. Ed. 48 (2009) 314–317.
[74] Y. Yan, J. Li, M. Qi, X. Zhang, J. Yu, R. Xu, Sci. China. Ser. B Chem. 52 (2009) 1734–1738.
[75] J. Li, J. Yu, R. Xu, Database of AlPO Synthesis; Y. Li, J. Yu, R. Xu, Database of AlPO Structure. <http://izasc.ethz.ch/fmi/xsl/IZA-SC/ol.htm.>
[76] J. Li, M. Qi, J. Kong, J. Wang, Y. Yan, W. Huo, et al., Microporous Mesoporous Mater. 129 (2010) 251–255.
[77] J. Han, M. Kamber, Data Mining: Concepts and Techniques, Higher Education Press, 2001, p. 68.
[78] J.M. Serra, L.A. Baumes, M. Moliner, P. Serna, A. Corma, Comb. Chem. High Throughput Screen 10 (2007) 13–24.
[79] J. Yu, R. Xu, Acc. Chem. Res. 2010, 10.1021/ar900293m

Subject Index

N

O

P

Q

R